Area-Wide Integrated Pest Management

Development and Field Application

Area-Wide Integrated Pest Management

Development and Field Application

Edited by

Jorge Hendrichs, Rui Pereira and Marc J.B. Vreysen

CRC Press
Taylor & Francis Group
Boca Raton London New York

CRC Press is an imprint of the
Taylor & Francis Group, an **informa** business

First edition published 2021
by CRC Press
6000 Broken Sound Parkway NW, Suite 300, Boca Raton, FL 33487-2742, USA

and by CRC Press
2 Park Square, Milton Park, Abingdon, Oxon, OX14 4RN, UK

CRC Press is an imprint of Taylor & Francis Group, LLC

ISBN: 978-0-367-76986-4 (hbk)
ISBN: 978-1-003-16923-9 (ebk)

FSC www.fsc.org MIX Paper from responsible sources FSC® C013056

Printed and bound in Great Britain by
TJ Books Limited, Padstow, Cornwall

PREFACE

The concept of area-wide integrated pest management (AW-IPM), in which the total population of a pest in an area is targeted, is central to the effective control of such populations through the integration of genetic, biological and other pest suppression technologies. Insect movement, occurring sometimes over long distances, is generally underestimated. As a consequence, most conventional pest management is implemented as a localized or field-by-field, un-coordinated action against segments of a pest population, not taking in consideration insect movement, resulting very often in an unsustainable spiral of insecticide application and eventual resistance of the pest against the used insecticides. On the other hand, an AW-IPM approach adopts a preventive rather than a reactive strategy, whereby all individuals of the pest population are targeted in time and space and selecting a time when the pest populations are more vulnerable (e.g. during certain times of the year when the population densities are naturally low), requiring in the longer term fewer inputs and resulting in more cost-effective and sustainable pest management. It involves a coordinated effort over often larger areas, including not only agricultural, but also natural and other areas with pest presence. By addressing these sources of re-infestation in the surroundings of the agricultural areas, satisfactory pest control is achieved in the whole area and fewer control actions are required.

This new textbook on AW-IPM assembles a series of selected papers that attempts to address various fundamental components of AW-IPM, e.g. the importance of relevant problem-solving research, the need for essential baseline data, the significance of integrating adequate tools for appropriate control strategies, and the value of pilot trials, etc. Of special interest are the numerous papers on pilot and operational programmes that pay special attention to practical problems encountered during the implementation of insect pest control programmes. A significant number of contributions to this book resulted from oral and poster presentations at the Third FAO/IAEA International Conference on "Area-wide Management of Insect Pests: Integrating the Sterile Insect and Related Nuclear and Other Techniques", which was successfully held from 22-26 May 2017 at the Vienna International Centre, Vienna, Austria. The conference was attended by 360 delegates from 81 countries and six international organization. However, the book contributions were selected beyond the work presented at the conference and a number of experts dealing mainly with action programmes were invited to present their work in this publication.

The book is a compilation of 48 papers that are authored by experts from 30 countries. Each paper was peer-reviewed by two or more independent, outside experts and edited for the English language. In addition, the editors subjected each paper to an in-depth technical quality control process. As a result, we trust that the information provided is accurate, up-to date and of a high international standard. This process of peer-review, editing and formatting has taken considerable time and we appreciate the patience of the authors.

The Editors
August 2020

FOREWORD

The latest report of the UN Intergovernmental Panel on Climate Change concludes that the world faces serious threats to its food supply. Reaching the Food and Agriculture Organization (FAO)'s goal to eradicate hunger and ensuring food security is only possible if we work together in partnership. Only through effective collaboration with governments, civil society, private sector, academia, research centres and cooperatives, and making use of each other's knowledge and comparative advantages, can our goal of "nourishing people while nurturing the planet" be achieved.

There could hardly be a less efficient use of resources than to invest in land, water, fertilizer, seeds, labour, and energy to produce agricultural commodities, only to have these investments partially or totally destroyed by insects and other pests. Pre-harvest losses in developing countries are currently estimated at more than one third of attainable crop production, while post-harvest losses add at least another 10–20%. Insects, followed by pathogens and weeds, cause the largest portion of these losses.

The availability of effective and persistent synthetic organic insecticides immediately after the Second World War marked the onset of chemically-based insect pest control. The availability and easy accessibility of these "off-the-shelf", relatively cheap and often subsidized chemicals offered farmers the freedom and flexibility to implement insect pest control measures on their property at any time. But, as we know today, chemically-based insect pest control also came at a heavy cost. Over 98% of sprayed insecticides and 95% of herbicides reach a destination other than their target species, including non-target species, air, water and soil. In addition, insecticide use reduces biodiversity, contributes to pollinator decline, destroys habitat, and threatens endangered species. And insect pests develop quickly resistance to insecticides, necessitating new formulations or application of higher doses to counteract the resistance, which exacerbates the pollution problem. The World Health Organization estimates that each year, three million workers in agriculture in the developing world experience severe poisoning from insecticides, about 18 000 of whom die.

Integrated pest management (IPM) has been endorsed by the FAO since 1966 and has remained the dominant paradigm of pest control for the past 50 years. IPM offers a strategic approach to solving pest problems in an ecosystem context, while reducing insecticide use and guarding human health and the environment. The integration of a number of different insect pest control methods into the IPM approach, to facilitate the achievement of these goals, is still primarily done at the local field-by-field, orchard-by-orchard or herd-by-herd level, which is often very ineffective as many insects travel freely between commercial, back-yard and abandoned properties, and between wild hosts and cultivated areas.

A much more effective application of the IPM approach is on an area-wide (AW) or population-wide basis, which aims at the management of the total population of a pest within an often larger but delimited area. This requires close coordination among the numerous stakeholders, a centrally managed approach and strong community involvement.

AW-IPM is now increasingly practiced, especially for mobile insect pests, where management at a larger scale is more effective and preferable to the uncoordinated field-by-field approach. For major livestock pests, vectors of human diseases and pests of high-value crops with low pest tolerance, there are compelling economic reasons for participating in AW-IPM.

Nevertheless, issues around public participation, financing of public goods and free riders, all play a significant role in AW-IPM implementation. These social and managerial issues have, in some cases, severely hampered the positive outcome of AW-IPM programmes and emphasise the need for contemplation not only of ecological, environmental and economic aspects, but also of the social and management dimensions.

For more than five decades the FAO and the International Atomic Energy Agency (IAEA), through their Joint FAO/IAEA Division of Nuclear Techniques in Food and Agriculture, have expanded the use of the sterile insect technique (SIT). This has involved both applied research to improve the technique and to develop it for new pest insects, as well as the transfer of the SIT package to member countries so that these can benefit from improved plant, animal and human health, cleaner environments, increased crop and animal production and accelerated economic development.

Today, the Joint FAO/IAEA Programme supports some 35 field projects that integrate the SIT to manage populations of major insect pests, including several species of tsetse flies and fruit flies, screwworm flies and moths. And endeavours are currently underway to develop the SIT for the control of mosquitoes, important vectors of major diseases such as malaria, dengue, chikungunya and Zika, and a main hindrance to economic development in endemic areas and a serious threat to as yet non-endemic areas. The development and validation of effective and sustainable approaches to managing pests and diseases will be essential to meet major future challenges.

Ren Wang

Assistant Director-General
Agriculture and Consumer Protection Department,
Food and Agriculture Organization of the United Nations (FAO)

TABLE OF CONTENTS

SECTION 1: OPERATIONAL AREA-WIDE PROGRAMMES

SECTION 3: CLIMATE CHANGE, GLOBAL TRADE AND INVASIVE SPECIES

SECTION 5: NEW DEVELOPMENTS AND TOOLS FOR AREA-WIDE INTEGRATED PEST MANAGEMENT

SECTION 1

OPERATIONAL AREA-WIDE PROGRAMMES

DEVELOPMENT AND AREA-WIDE APPLICATION OF BIOLOGICAL CONTROL USING THE PARASITOID *Aphidius gifuensis* AGAINST *Myzus persicae* IN CHINA

Y. B. YU[1], H. L. YANG[2], Z. LIN[1], S. Y. YANG[2], L. M. ZHANG[2], X. H. GU[2], C. M. LI[3] AND X. WANG[4]

[1]Yunnan Tobacco Company of China National Tobacco Corporation (CNTC), Kunming, Yunnan 650011, China; yuyanbi@sina.com
[2]Yuxi Branch of Yunnan Tobacco Company of CNTC, Yuxi, Yunnan, 653100, China
[3]Honghe Branch of Yunnan Tobacco Company of CNTC, Honghe, Yunnnan, 652300, China; 3641775@qq.com
[4]Dali Branch of Yunnan Tobacco Company of CNTC, Dali, Yunnan, 671000, China

SUMMARY

Ecologically safe and environment-friendly pest control strategies and technologies are important to ensure the quality of Chinese agricultural products and sustainable agricultural development. Aphids are among the world's major agricultural and forest pests, and *Myzus persicae* Sulzer (Hemiptera: Aphididae) is one of the main agricultural pests in China, transmitting various viral diseases and causing reductions in crop yield and quality that regularly triggered applications of synthetic insecticides. *Aphidius gifuensis* Ashmead (Hymenoptera: Braconidae) is an important endoparasitoid of many aphids. Starting in 2000, the Yunnan Tobacco Company has developed methods for large-scale rearing of this parasitoid on this aphid, and technological systems for augmentative releases of *A. gifuensis*. The augmentative use of this parasitoid has achieved area-wide suppression of *M. persicae* in tobacco and other crops in China. This approach is being applied on large areas, covering more than 3 million ha between 2010 and 2015. This programme is currently the largest biological control programme in China. Over 500 mass-rearing facilities were constructed in 16 provinces with a total surface area of 420 000 m^2 and a breeding capacity of 24 000 million parasitoids per breeding period. This technology has effectively controlled the aphid on tobacco, while other beneficial insects have increased in the absence of insecticide applications, further protecting biodiversity in the fields and providing long-term ecological benefits. The use of this technology has also been expanded to other crops, solving problems of insecticide resistance in the targeted aphids, reducing pesticide residues and environmental pollution, and yielding benefits for society, the economy, and the environment.

J. Hendrichs, R. Pereira and M. J. B. Vreysen (eds.), Area-Wide Integrated Pest Management: Development and Field Application, pp. 3–16. CRC Press, Boca Raton, Florida, USA.

Key Words: Aphididae, aphids, augmentative biological control, tobacco, technology research, technology transfer, large-scale parasitoid breeding, Braconidae, release technique, training, extension, environment-friendly pest control, Yunnan

1. INTRODUCTION

Yunnan Province, China is one of the most important tobacco growing regions in the world with 469 000 ha under cultivation, representing 35% and 20% of tobacco production in China and the world, respectively. Yunnan province is also famous for producing tobacco of high quality, and its tobacco industry has provided a sustainable livelihood and alleviated poverty for more than 800 000 farming families. Tobacco in Yunnan province is usually planted in mountainous areas, of high scenic value, that have fragile ecosystems. The Yunnan tobacco-planting region has five characteristics: (1) a wide distribution of planting sites, (2) planting in a variety of ecological regions, (3) dominance by smallholder farmers, (4) farmers of widely different backgrounds, and (5) a mosaic of factors influencing tobacco leaf yields and quality.

Aphids are among the most destructive pests, are often highly polyphagous and impact a broad range of agricultural crops worldwide. Plant sap-sucking and honeydew production by the tobacco aphid *Myzus persicae* Sulzer (Hemiptera: Aphididae) directly injures the host plant, causing significant yield reduction (Kulash 1949; Starý 1970). In addition, damage from *M. persicae* is exacerbated by its ability to transmit over 100 viral diseases to more than 400 host plants (Mackauer and Way 1976). Most of these viral diseases cause a decline in tobacco yield and quality.

In Yunnan, control of this aphid pest has largely been dependent on insecticides (Zhao et al. 1980), leading to problems of resistance, difficulty of control, destruction of natural enemies, decrease of biodiversity, excessive pesticide residues, reduction in product quality due to repeated applications, inappropriate application methods and incorrect application rates (Gao et al. 1992; Han et al. 1989; He 2013).

In view of these problems, there was a strong need to develop and apply biological control tactics to replace the chemical control systems applied against *M. persicae*, taking into account the population dynamics of the pest, herbivore-natural enemy interactions, and the economic relationship between pest infestation and crop yield loss (Guan et al. 2016).

Aphidius gifuensis Ashmead (Hymenoptera: Braconidae) is one of the most important natural enemies of aphids and is found in many habitats (Chen 1979). This natural enemy is widely distributed all over the world, for example Canada, China, India, Japan, and USA, where a good basis of ecological knowledge has been accumulated for wider application. Also, many biological and ecological studies of this insect have been conducted in China (Bi and Ji 1993; Lu et al. 1993; Lu et al. 1994).

After mating, females of *A. gifuensis* search for hosts and lay their eggs inside aphid bodies. The eggs develop by absorbing the nutrients from the aphid, and the development of the parasitoid results in the death of the aphid (Ohta et al. 2001) (Fig. 1).

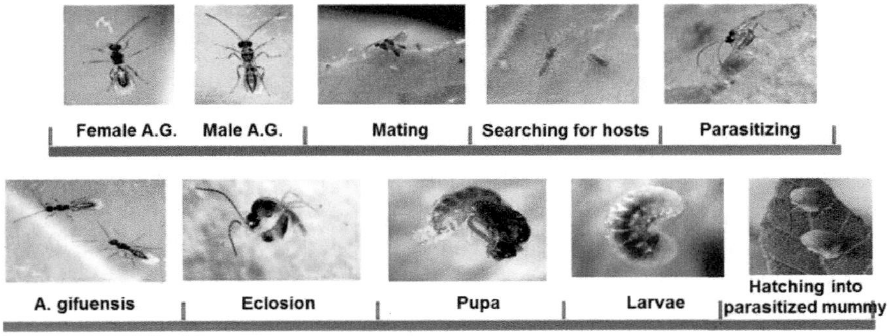

Figure 1. Life cycle and biological stages of Aphidius gifuensis (*A.G.*) (*starting clockwise from top left*).

There are two main challenges with respect to the biological control of aphids by using *A. gifuensis* on a large scale. The first is to develop methods for the large-scale production of the parasitoid and its release. The second is to find mechanisms to transfer this technology to technicians and farmers so that these smallholders can benefit from the technology.

Therefore, after carrying out the research, we have developed and established two systems for effective, economic, and convenient high-density mass-rearing technologies for different application areas, named *adult-plant breeding* and *seedling breeding*, published an industry-level standard, constructed a "one plus two" model of technology extension, and achieved area-wide application of the technology for sustainable aphid control by *A. gifuensis*.

2. MASS-REARING SYSTEMS

The large-scale mass-rearing of *A. gifuensis* is mainly divided into three parts: (1) cultivating of host plants for *M. persicae*, (2) breeding large populations of *M. persicae*, and (3) high-density mass-rearing of *A. gifuensis* on *M. persicae* (Gu et al. 2015). The following summarizes the optimal conditions and procedures for these technologies (Deng et al. 2010, 2011).

2.1. Adult-Plant Breeding System

2.1.1. Cultivating Host Plants

Tobacco plants that are highly resistant to tobacco mosaic virus (TMV) were selected for mass-rearing of the aphids, i.e. tobacco variety *Yunyan 203*, as well as white radish and Chinese cabbage variety *Chinese 82*. The host plants are seeded and transplanted after 70 to 80 days and grown for 25-30 days till the 6-8 leaves stage for further use.

2.1.2. Mass-rearing of Myzus persicae

Each tobacco plant, at the 6-8 leaves stage, is inoculated with 20 healthy aphids (nymphs and adults), followed by rearing of the aphids for 15-20 days at 17-27°C and 50-80% RH in a greenhouse (50 m x 12 m x 4.6 m) (Wu et al. 2000; Deng et al. 2006; Yang and Zhao 2009; Yang et al. 2009) (Fig. 2).

2.1.3. High-density Mass-rearing of A. gifuensis *on Aphids*

When the tobacco leaves are incubated with aphids for 15-20 days, *A. gifuensis* are released at a parasitoid to aphid ratio of between 1:50 and 1:100. After *A. gifuensis* females have laid their eggs, the parasitized aphids form mummies, from which a new generation of *A. gifuensis* emerges. After 10 to 15 days, parasitism rates of > 90% are obtained (Wei et al. 2003, 2005; Yang et al. 2009).

Each tobacco plant can produce 6000 – 10 000 *A. gifuensis* adults, and each small greenhouse (3 m x 3 m x 2 m) containing 28 plants, can produce 160 800 *A. gifuensis* individuals (Fig. 2).

Figure 2. Breeding process of Aphidus gifuensis *in greenhouses (starting clockwise from top left).*

2.2. Seedling Breeding System

2.2.1. Cultivating Host Plants

Tobacco variety *Yunyan 203* with high resistance to TMV is bred according to the China National Standard GB / 25241 (Liu et al. 2010). Tobacco seedlings with 5 leaves and 1 heart can be used to breed aphids by the "Method of breeding aphid and *Aphidus gifuensis* separately".

Alternatively, tobacco seedlings in the 3rd - 4th leaf stage can be used to breed aphids by the "Method of breeding aphid and *Aphidus gifuensis* at the same time" as described below.

2.2.2. Rearing of Large Populations of Myzus persicae

Mass-rearing of aphids is done using two alternative methods:

Breeding aphid and A. gifuensis *separately*: On tobacco plants with 5 leaves and 1 heart, leaves are inoculated at a rate of 10 aphids per plant. After 10 to 12 days, at 20 to 30°C and 60 - 80% RH, when the aphid density reaches 200 per plant on average, the population of aphids is ready for use for rearing natural enemies (Fig. 2).

Breeding aphid and A. gifuensis *at the same time*: On tobacco plants in the 3rd - 4th leaf stage, leaves are inoculated with 2.5 aphids per plant (aphids with a parasitism rate from 40% to 60% are used for the inoculation, or a parasitoid-aphid ratio of between 1:20 and 1:10), at 20 - 30°C and 60 - 80% RH.

2.2.3. High-density Breeding of A. gifuensis *on Aphids*

Parasitoid mass-rearing on aphids is done following two alternative methods:

Breeding aphid and A. gifuensis *separately*: According to the population of aphids per single plant, *A. gifuensis* or parasitised aphids are inoculated onto leaves in the greenhouse. After 17 days, the population of parasitised aphids will reach one 100 000 per square meter (Fig. 2).

Breeding aphid and A. gifuensis *at the same time*: *A. gifuensis* parasitoids are inoculated onto leaves at the same time as aphids. After the *A. gifuensis* parasitoids emerge from the parasitised aphids, they will parasitize other aphids. If the parasitism rate is too high, more aphids need to be added; alternatively, if the parasitism rate is too low, more *A. gifuensis* will need to be added. After 23 days, the population of parasitised aphids will reach 49 000 per square meter (Fig. 3) (Chen et al. 2009).

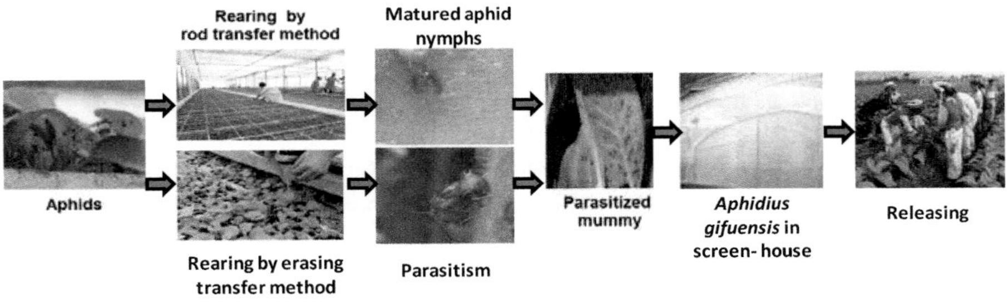

Figure 3. Breeding process of aphids and Aphidius gifuensis.

2.3. Collection and Storage of A. gifuensis

There are two methods for collection of *A. gifuensis*:

Manual collection: Home-made, simple collection devices or automatic collection systems (aspirators) are used to collect *A. gifuensis* in the tents, and *A. gifuensis* are stored in containers (Fig. 4).

Automatic collection: Collection bags are placed in the breeding screen-houses (nylon-net covered cages or tents) in tobacco fields, and *A. gifuensis* adults will fly into the bags as a result of their phototaxis (Fig. 4).

2.4. Release of A. gifuensis

Different methods are used to release *A. gifuensis:*

1. Release of parasitised aphids
2. Release of *A. gifuensis*, and
3. Self-dispersal in the field.

When releasing parasitised aphids, leaves or seedling with parasitised aphids are brought to the field and hung onto plants.

When releasing *A. gifuensis*, parasitoids are taken to the field in collection bags or bottles and released before 12 o'clock in the absence of any rain; total transport time should be less than three hours.

For self-dispersal of parasitoids in the field, they are bred in the screen-houses or breeding tents in the field. The tents are opened when parasitism reaches 90%, and the *A. gifuensis* will disperse naturally to find aphids (Fig. 4).

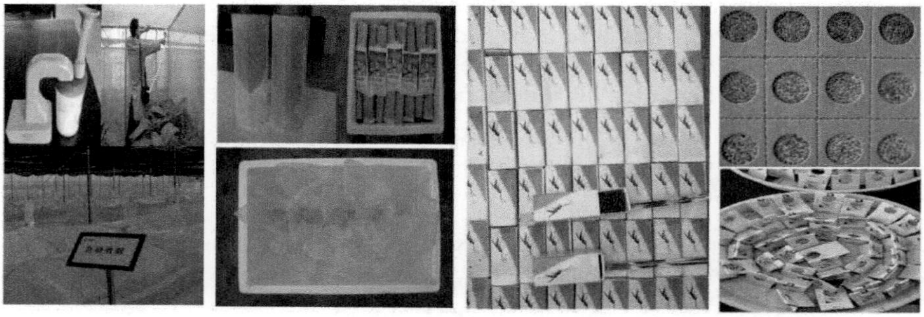

Figure 4. Collection and release processes for Aphidius gifuensis *parasitoids.*

The density of aphids needs to be assessed to determine the parasitoid release time and numbers to be released. When the population of aphids per plant reaches an average of 1 to 5 individuals, *A. gifuensis* parasitoids are released at a rate of 3 000 to 7 500 per ha. The subsequent second and third release are adjusted according to aphid densities.

When the population of aphids per plant reaches an average of 6 to 20, *A. gifuensis* parasitoids are released at a rate of 7 500 to 15 000 per ha. When the population of aphids per plant is more than 20, *A. gifuensis* parasitoids are released at a rate of 15 000-18 000 per ha.

2.5. Conservation of Aphid and A. gifuensis *Colonies in Winter*

2.5.1. Conservation of Aphids in Winter

The main method to maintain a colony of aphids in the winter is by conserving the host plants in greenhouses. Aphids are collected from the wild and inoculated onto healthy tobacco seedlings, cabbage, radish or other host plants. The breeding conditions are held at temperatures between 17 and 27°C and 50 and 80% RH. The status of aphids and host plants is observed, and old and diseased aphids and host plants infected with virus are removed at three different times. Healthy aphids are obtained and used to reinvigorate the colony.

2.5.2. Conservation of A. gifuensis *in Winter*

Holding-over host plants with aphids in the greenhouse or cold storage of parasitised aphids are the two methods used for storing *A. gifuensis* in winter. *A. gifuensis* are collected from the wild and used to parasitize aphids several times. To obtain healthy *A. gifuensis*, the colony needs to be maintained at a temperature of 17-27°C and 50-80% RH.

When in cold storage, parasitised aphids are collected using a brush or other tools to directly collect them from host plants, placing aphids into tubes, which are then held at 4-5°C. The seedlings or larger plants with parasitised aphids can also directly be placed into 4-5°C. Parasitoid emergence rates of up to 90% are obtained after cold storage for 20 days.

3. TECHNOLOGY TRANSFER AND APPLICATION SYSTEMS

3.1. Technical Standard

An industrial standard named "Technical specification for *Myzus persicae* biological control with *Aphidius gifuensis* (YC/T 437-2012)" was published (Yun et al. 2012), including conservation, colony rejuvenation, large-scale breeding, collection and release.

A total of 506 breeding facilities were built in 16 provinces (regions, cities) in China, with a total surface area of 420 000 m^2 and a breeding capacity of 24 000 million parasitoids per breeding period.

3.2. Training

To spread and transfer this technology, a training system was developed and implemented on four levels, i.e. industrial, provincial, municipal and county level. The training programme was developed based on research conducted by the tobacco industry, and also based on field experience by extension services of the government (Fig. 5).

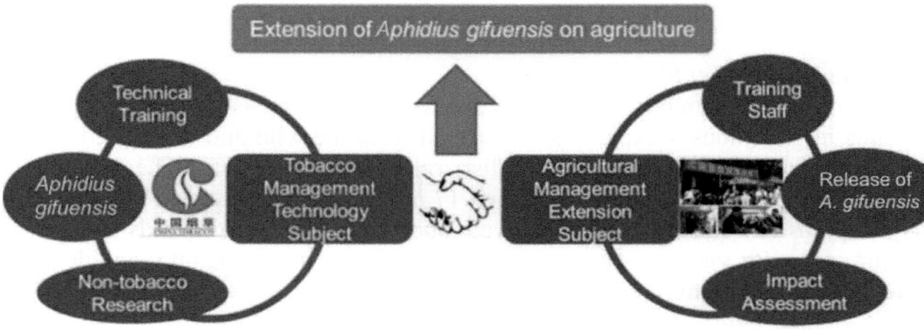

Figure 5. Extension system for biological control of tobacco aphid by parasitoid augmentative releases.

We developed the training platform, carried out theoretical and practical training for the technical experts, technicians, extension workers and farmers, covering step-by-step the key points and difficulties for technicians during programme operation (Fig. 6).

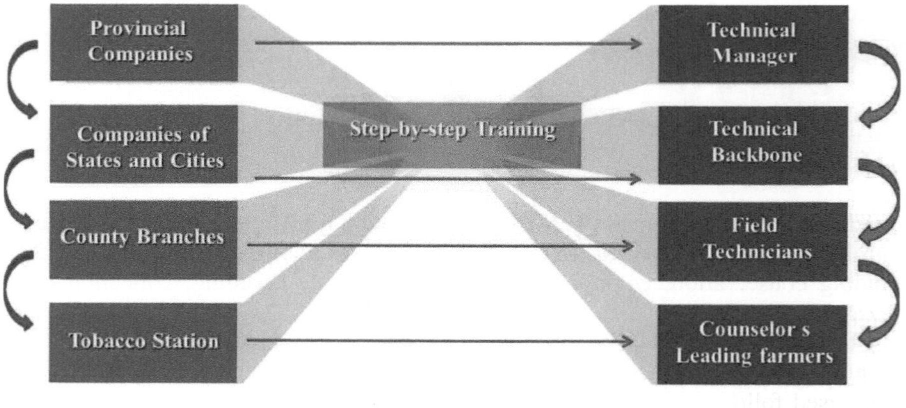

Figure 6. Training system for biological control of tobacco aphid Aphidius gifuensis *by parasitoids.*

Overall, we have trained more than 3000 core experts and more than 20 000 technicians, and also provided more than 1200 thousand copies of technical training books and booklets (Fig. 7). More than 120 000 farms benefited from this training and extension work at all levels (Fig. 8).

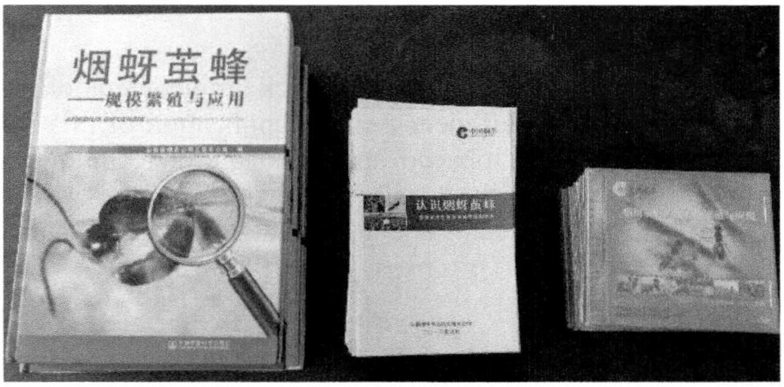

Figure 7. Technical materials on the use of Aphidius gifuensis *parasitoids.*

Figure 8. Four-level training for biological control of tobacco aphid by mass-release of Aphidius gifuensis *parasitoids.*

3.3. Goal Setting

Depending on the densities of the aphids in the farms with tobacco fields, *A. gifuensis* are released following five release densities, i.e. 7500, 12 000, 15 000, 18 000 and 22 500 parasitoids per ha. Also, depending on the aphid infestation pressure in different areas, application rates are 30%, 50%, 80% and 95% of the total tobacco-planting area.

3.4. Matching of Funding

The funding for the technical application is contained within technical project funds. Each provincial tobacco company matches the obtained funding of USD 11.3-22.6 per ha according to set application areas.

3.5. Evaluation

The programme evaluation is carried out to check implementation, supply, and application scale, and to control efficiency of the application of this technology at different levels. To promote the tobacco aphid biocontrol technique, a series of rewards and penalties were established, with the outcomes of the evaluation results directly tied to the salary of technicians and also to the funding for application in the next year.

4. IMPACT ASSESSMENT

Each small greenhouse containing 28 plants can produce ca. 17 000 *A. gifuensis* individuals for release that can protect six ha of tobacco plants. Compared to chemical control of aphids, the cost of aphid control by *A. gifuensis* mass-releases is much lower. The cost of biological control is estimated at about USD 13.3 per ha, compared to USD 244.8 per ha for the chemical control applying insecticides. Table 1 compares the sub-item costs for both treatments.

Table 1. Comparison of costs of biological and chemical control of aphids

Treatments	Total Cost (USD/ha)	Sub-items	Sub-items costs (USD/ha)
Insecticide	244.8	Cost of insecticides	13.6×3 times = 40.8
		Cost of labour	68.0×3 times = 204.0
Biocontrol	13.3	Cost of facilities	3.8
		Cost of mass-rearing	5.0
		Cost of releasing	4.5

In addition, aphids were well controlled because of the sustained, long-term release of *A. gifuensis* (Yang et al. 2010) (Fig. 9 and 10).

Meanwhile, populations of other beneficial insects, such as the predators *Coccinella septempunctata* (L.), *Harmonia axyridis* (Pallas), *Episyrphus balteatus* (De Geer), *Chrysopa sinica* (Tjeder), and *Lycosa pseudoamulata* Boes. et Str. obviously increased in the absence of insecticide applications, further protecting biodiversity in the fields and providing long-term ecological benefits resulting from the biological control (Shen et al. 2018).

Application of this technology was started in 2000 in the tobacco-planting area of Yuxi, Yunnan Province (Yunnan Tobacco Yuxi City Company 2010), and by 2010 the entire tobacco planting area of Yuxi had been covered (AERET 2011; Yang et al. 2011).

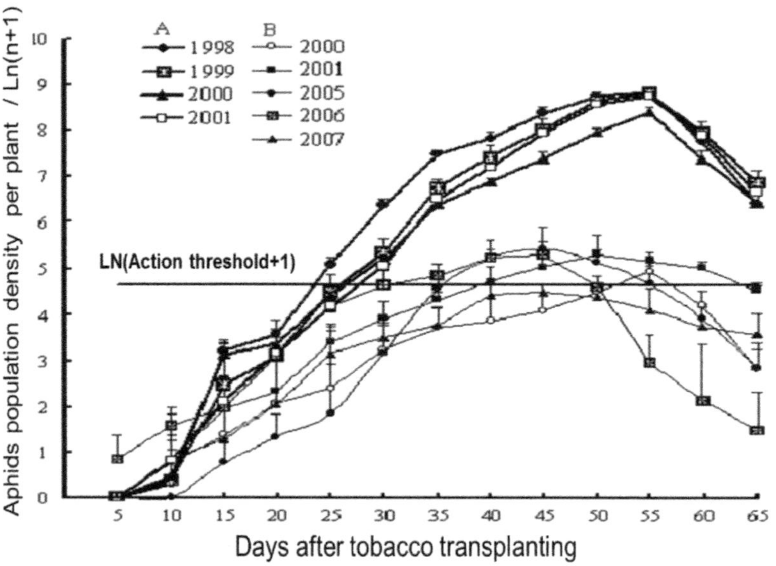

Figure 9. Aphid population control results without A. gifuensis *between 1998-2001 (A) and with* A. gifuensis *between 2000-2007 (B).*

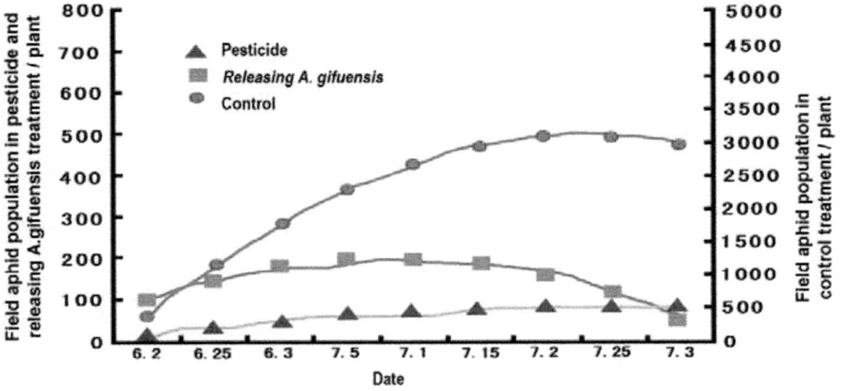

Figure 10. The population dynamics of aphids in tobacco fields treated by insecticides or the release of parasitoids (left y-axis), or not treated (control) (right y-axis), in tobacco fields in Yuxi, China in 2005.

By 2013, 90% of all the Yunnan tobacco fields had been covered by the augmentative biological control programme. Starting in 2014, this technology has been applied throughout China. Step by step, this technology has expanded to cover 100% of the Yunnan tobacco-planting area and 90% of the Chinese tobacco-planting areas.

Effectiveness of aphid control has reached 80%, better than in pesticide-treated fields (Li et al. 2006). The total application area in China has reached 3 017 547 ha over six years (2010 to 2015), total decrease of pesticide use reached 1 966.05 tons, and total decrease of control costs reached USD 230 million. A financial loss of USD 1 326 million was avoided and 1 million farm households have benefited from this technology.

The use of *A. gifuensis* against *M. persicae* is the biological control technology that is most widely adopted in China. Aphid population control has been transformed from mainly insecticide- to largely biological control-based, promoting a pest control strategy that changed from passive, reactive and chemical insecticide-based to active, preventive and ecologically-founded, with significant social and environmental benefits.

The biological control of aphids by the mass-release of *A. gifuensis* parasitoids in some other crops (Fig. 11) has also reached high efficacies on a large scale (Shen et al. 2018).

Figure 11. Examples of biological control of aphids by Aphidius gifuensis *augmentative releases on other crops in China.*

5. ACKNOWLEDGEMENTS

This project has been financially supported by the China Tobacco Company. We thank Professor Zeng-Rong Zhu of Zhejiang University for his critical comments and suggestions for modifications of earlier versions of this chapter.

6. REFERENCES

(AERET) (Association Européenne pour la Recherche et l'Expérimentation Tabacole / European Association for Tobacco Research and Experimentation. 2011. *Aphidius gifuensis* (Ashmead) safeguards ecological leaf tobacco in Yunnan, China. AERET website, published 18 April 2011.

Bi, Z. B., and Z. D. Ji. 1993. Bionomics of *Aphidius gifuensis* Ashmead. I. Development stages and morphology of larval stage. Journal of Hebei Agricultural University 16(2): 1–8 [in Chinese].

Chen, J. H. 1979. The fundamental knowledge of taxonomy of Aphidiidae in China. Entomological Knowledge 16: 265–268 [in Chinese].

Chen, S., Y. Liu, and W. Li. 2009. High-density rearing method for *Aphidius gifuensis* involves inoculating tobacco aphid for twenty days after seeding seedling when seed is small and generating *Aphidius gifuensis* in fifteen days after inoculation process. Chinese patent: CN101548658–A, 7 October 2009 [in Chinese].

Deng, J. H., X. F. Wu, C. M. Song, J. M. Huang, G. H. Liu, and S. Y. Yang. 2006. Rearing effect of *Aphidius gifuensis* with nylon-net covered cages in tobacco fields. Journal of Southwest Agricultural University (Natural Science) 28: 66–73 [in Chinese].

Deng, X. G., W. Wu, and S. Yang. 2010. *Aphidius gifuensis*: Mass rearing and application. First Edition. Environmental Science Press, Beijing, China [in Chinese].

Deng, X., G. Fu, and C. Li. 2011. Large-scale *Aphidius gifuensis* Ashmead propagating technique for preventing agricultural pest, involves forming mummified aphids after certain number of days, and growing aphids to *Aphidius gifuensis* in one week. Chinese patent: CN101455190–A; CN101455190–B, 2 November 2011 [in Chinese].

Gao, X. W., B. Z. Zheng, and B. J. Cao. 1992. Resistance in *Myzus persicae* to organophosphorus and carbamate insecticides in China. Acta Phytophylacica Sinica 19: 365–371 [in Chinese].

Gu, X. H., S. Y. Yang, Y. B. Yu, S. G. Ji, H. L. Yang, J. L. Zhao, L. M. Zhang, and H. G. Zhang. 2015. Application of biological control technology of *Myzus persicae* with *Aphidius gifuensis* in Yunnan Province. Chinese Journal of Biological Control 31: 1–7 [in Chinese].

Guan, Q. L., X. C. Shen, and L. F. Yuan. 2016. Agricultural pest resistance and comprehensive control. Agricultural Technology and Equipment 1: 24–25 [in Chinese].

Han, Q. F., P. J. Zhuang, Z. H. Tang, X. Z. Xu, and X. Q. Deng. 1989. The preliminary study on resistance of *Myzus persicae* to insecticides. Contributions of Shanghai Institute of Entomology 9: 19–27 [in Chinese].

He, X. L. 2013. Pest resistance research and summary of governance. World Pesticides 35(5): 34–38 [in Chinese].

Kulash, W. M. 1949. The green peach aphid as a pest of tobacco. Journal of Economic Entomology 42: 677–680.

Li, M. F., Y. P. Zhang, and X. Z. Wang. 2006. Investigation of the effect to prevent and control aphids by rearing *Aphidius gifuensis*. Chinese Agricultural Science Bulletin 22: 343–346 [in Chinese].

Liu, J. L., X. D. Shi, G. S. Liu, G. Wang, and Y. H. Yang. 2010. Rules for tobacco intensive seedling production. GB/T 25241. China National Standard Press. 24 pp. [in Chinese].

Lu, H., B. C. Shi, and Z. L. Zhang. 1993. Studies on fecundity of *Aphidius gifuensis* Ashmead. Acta Agriculturae Boreali-Sinicae 8: 76–79 [in Chinese].

Lu, H., B. C. Shi, Y. Z. Niu, and Z. L. Zhang. 1994. Development thresholds and thermal constants of *Aphidius gifuensis* and *Diaeretiella rapae*. Acta Agriculturae Boreali-Sinica 9: 72–75 [in Chinese].

Mackauer, M., and M. J. Way. 1976. *Myzus persicae* Sulzer, an aphid of world importance, pp. 51–119. *In* V. L. Delucchi (ed.), Studies in biological control. Cambridge University Press, Cambridge, UK.

Ohta, I., K. Miura, and M. Kobayashi. 2001. Life history parameters during immature stage of *Aphidius gifuensis* Ashmead (Hymenoptera: Braconidae) on green peach aphid, *Myzus persicae* (Sulzer) (Homoptera: Aphididae). Applied Entomology and Zoology 36 (1): 103–109.

Shen, S., G. Xu, F. Chen, D. R. Clements, X. Gu, S. Ji, L. Zhang, H. Yang, F. Zhang, K. Yin, H. Zhang, J. Li, and D. Wu. 2018. Effects of *Aphidius gifuensis* release on insect communities and diversity in tobacco fields of Yunnan Province, China. Pakistan Journal of Biological Sciences 21: 284–291.

Starý, P. 1970. Biology of aphid parasites (Hymenoptera: Aphidiidae) with respect to integrated control. Series Entomologica 6: 1–643.

Wei, J. N., T. F. Li, R. P. Kuang, Z. Wang, T. Yin, X. Wu, L. Zou, W. Z. Zhao, J. Cao, and J. Deng. 2003. Mass rearing of *Aphidius gifuensis* (Hymenoptera: Aphidiidae) for biological control of *Myzus persicae* (Homoptera: Aphididae). Biocontrol Science and Technology 13: 87–97.

Wei, J. N., B. B. Bai, T. S. Yin, Y. Wang, Y. Yang, L. H. Zhao, R. P. Kuang, and R. J. Xiang. 2005. Development and use of parasitoids (Hymenoptera: Aphidiidae & Aphelinidae) for biological control of aphids in China. Biocontrol Science and Technology 15: 533–551.

Wu, X. F., T. F. Li, and J. N. Wei. 2000. Temperature effects on development and fecundity of *Aphidius gifuensis* Ashmead. Zoological Research 21(3): 192–198 [in Chinese].

Yang, S., S. Y. Yang, C. P. Zhang, J. N. Wei, and R. P. Kuang. 2009. Population dynamics of *Myzus persicae* on tobacco in Yunnan Province, China, before and after augmentative releases of *Aphidius gifuensis*. Biocontrol Science and Technology 19(2): 219–228.

Yang, S., S. Y. Yang, C. P. Zhang, and R.P. Kuang. 2010. Changes of population dynamics of *Myzus persicae* and effects of its natural enemies in tobacco fields. Journal of Southwest China Normal University. Natural Science 35: 68–72 [in Chinese].

Yang, S., J. N. Wei, S. Y. Yang, and R. P. Kuang. 2011. Current status and future trends of augmentative release of *Aphidius gifuensis* for control of *Myzus persicae* in China's Yunnan Province. Journal Entomological Research Society 13: 87–99.

Yang, S. Y., and J. L. Zhao. 2009. Mass rearing and release technique of *Aphidius gifuensis*, pp. 186–191. *In* Entomology Annual 2009 Conference in Yunnan Province, China [in Chinese].

Yunnan Tobacco Yuxi City Company. 2010. *Aphidius gifuensis* mass rearing and application. China Environmental Science Press. ISBN: 7511102441, 9787511102447, China [in Chinese].

Yun, D., T. L. Gao, B. J. Yu, S. Y. Yang, J. Zhu, and Z. H. Tian. 2012. Technical specification for *Myzus persicae* biological control with *Aphidius gifuensis*. YC/T 437-2012. China National Standard Press. 12 pp. [in Chinese].

Zhao, W. Y., C. P. Din, and W. L. Zhang. 1980. The bionomics of *Aphidius gifuensis* Ashmead and its utilization for the control of tobacco aphid *Myzus persicae* Sulzer. Zoological Research 1: 405–416 [in Chinese].

BIOLOGICAL CONTROL: CORNERSTONE OF AREA-WIDE-INTEGRATED PEST MANAGEMENT FOR THE CASSAVA MEALYBUG IN TROPICAL ASIA

K. A. G. WYCKHUYS[1], W. ORANKANOK[2], J. W. KETELAAR[3], A. RAUF[4], G. GOERGEN[5] AND P. NEUENSCHWANDER[5]

[1]*CGIAR Research Program on Roots, Tubers and Banana (CRP-RTB), International Center for Tropical Agriculture CIAT, Hanoi, Viet Nam; kagwyckhuys@gmail.com*
[2]*Department of Agricultural Extension, Bangkok, Thailand*
[3]*Food and Agriculture Organization (FAO), Bangkok, Thailand*
[4]*Bogor Agricultural University, Bogor, Indonesia*
[5]*International Institute of Tropical Agriculture (IITA), Cotonou, Benin*

SUMMARY

The cassava mealybug *Phenacoccus manihoti* Mat.-Ferr. (Hemiptera: Pseudococcidae) is a globally important pest of cassava (*Manihot esculenta* Crantz), a crop that is cultivated on nearly 25 million ha across the tropics. Following its continent-wide invasion of Africa during the 1970s and early 1980s, *P. manihoti* was inadvertently introduced to Southeast Asia in late 2008, where it caused important yield drops in local crops. Guided by the widely-acclaimed biological control successes against this mealybug in Africa, the endophagous parasitoid *Anagyrus lopezi* De Santis (Hymenoptera: Encyrtidae) was introduced to Thailand in 2009. Subsequent introductions of *A. lopezi* were made into neighbouring countries, and an integrated campaign was launched to scale-up mealybug biological control. Multi-country field surveys were carried out to map *P. manihoti* geographic distribution, field-level abundance and extent of parasitoid-mediated suppression, and innovative extension programmes were deployed to raise farmer awareness of mealybug pests and associated natural enemies. Survey work from nearly 600 fields throughout mainland Southeast Asia revealed that *P. manihoti* occurred at abundance levels of 14.3 ± 30.8 individuals per tip in the dry-season, and *A. lopezi* parasitism averaged at 38.9%. An applied research programme yielded critical insights into various determinants of *A. lopezi* establishment, spread and biological control efficacy. In close collaboration with national partners, research was carried out on the eventual effects of soil fertility and plant nutrition, landscape composition, and a plant's phytopathogen infection status, amongst others. Our work shows how the host-specific *A. lopezi* effectively suppresses the cassava mealybug across a range of agro-climatic, biophysical and socio-economic contexts in tropical Asia, and constitutes a central component of area-wide integrated pest management (AW-IPM) for this global pest invader. This study also underlines the need for holistic, transdisciplinary approaches to (invasive) pest management, and the tangible yet (largely) untapped potential of coupling social and biological sciences to address crop protection problems in the developing-world tropics.

J. Hendrichs, R. Pereira and M. J. B. Vreysen (eds.), Area-Wide Integrated Pest Management: Development and Field Application, pp. 17–32. CRC Press, Boca Raton, Florida, USA.

Key Words: Classical biological control, ecosystem services, invasive species management, food security, landscape diversity, natural enemy, trophic ecology, tropical agriculture, *Manihot esculenta, Phenacoccus manihoti, Anagyrus lopezi*

1. INTRODUCTION

Cassava mealybug *Phenacoccus manihoti* Mat.-Ferr. (Hemiptera: Pseudococcidae) is a prominent herbivore on cassava (*Manihot esculenta* Crantz) and one of the world's most notorious invasive species. Endemic to the Paraguay River basin in South America, *P. manihoti* was inadvertently introduced into Africa during the early 1970s and rapidly spread across the continent's extensive cassava belt (Herren and Neuenschwander 1991; Bellotti et al. 2012). Capable of inflicting yield losses up to 58-84% (Nwanze 1982; Schulthess et al. 1991), *P. manihoti* devastated local cassava production and caused widespread hunger for farming families across sub-Saharan Africa.

In late 2008, this same pest was detected in Thailand's eastern seaboard, where it caused an 18% drop in aggregate crop yield of cassava, a yearly loss of over 8 million ton fresh root in Thailand alone, and more than 2-fold surges in prices of cassava starch (Muniappan et al. 2009; Wyckhuys et al. unpublished). By 2011, *P. manihoti* had spread extensively in Thailand and had inflicted economic losses on the country's cassava sector of over USD 30 million nationally (TTTA 2011).

In 2014, *P. manihoti* had also entered prime cassava growing areas in neighbouring Cambodia, Indonesia, Lao PDR, Malaysia, and Viet Nam (Sartiami et al. 2015; Graziosi et al. 2016). Climate-based niche modelling revealed that other key cassava production areas in eastern Indonesia and the Philippines are also at risk to *P. manihoti* (Yonow et al. 2017). As Southeast Asia accounts for nearly 95% of the world's cassava exports and is home to a multi-billion-dollar cassava starch industry (Cramb et al. 2017), *P. manihoti* was expected to inflict major socio-economic impacts at a regional level.

1.1. Control of Cassava Mealybug in Africa

Though this mealybug invader evidently posed an immediate threat to the rural economy of several Asian countries, a nearly tailor-made management solution had been successfully developed in Africa more than thirty years ago. In fact, after the 1980 discovery of *P. manihoti* in Paraguay by A. Bellotti (International Center for Tropical Agriculture, CIAT), one of the world's best-known and successful insect classical biological control programmes was initiated (Bellotti et al. 1999; Neuenschwander 2001). In 1981, the Centre for Agriculture and Bioscience International (CABI) and the International Institute of Tropical Agriculture (IITA) teamed up to carry out foreign exploration in the presumed region of origin of *P. manihoti*, ultimately resulting in the collection and subsequent shipment of the *Anagyrus lopezi* De Santis (Hymenoptera: Encyrtidae) (Löhr et al. 1990).

Following its 1981 release in western Nigeria, *A. lopezi* promptly established and suppressed *P. manihoti* population levels from more than 100 to fewer than 10–20 individuals per cassava tip (Hammond et al. 1987). In less than three years following its release, *A. lopezi* had effectively dispersed over 200 000 km^2 in south-western

Nigeria. It had also been mass-reared and distributed across multiple release points in several African countries (Herren et al. 1987). Though multiple endemic primary parasitoids and hyperparasitoids were recorded in mealybug-invaded areas in Africa (Neuenschwander et al. 1987; Neuenschwander and Hammond 1988), these largely did not impede the success of *A. lopezi* as a biological control agent.

Overall, the parasitoid wasp successfully established in 26 African countries, prevented wide-spread famine and generated economic benefits of USD 9400-20 200 million (Zeddies et al. 2001). Moreover, across the highly diverse and vast African continent, no agro-ecological conditions were found under which *A. lopezi* was unable to establish and attack its mealybug host (Neuenschwander 2001).

1.2. Cassava Mealybug in Southeast Asia

Soon after its detection in Asia, Thailand's late Amporn Winotai of the Department of Agriculture (DoA) solicited assistance from CIAT's Anthony Bellotti to tackle the fast-spreading mealybug pests in her country. Well aware of the accomplishments in Africa, A. Bellotti rightly pointed Thai colleagues to G. Goergen and P. Neuenschwander at the IITA station in Cotonou, Benin. In late 2009, *A. lopezi* was then effectively introduced from West Africa into Thailand, and rearing labs were established in different parts of the country, through a joint endeavour between the Food and Agriculture Organization of the United Nations (FAO), centres of the Consultative Group on International Agricultural Research (CGIAR), and Thailand's Royal Government (Winotai et al. 2010).

The *A. lopezi* releases received ample public attention and quickly culminated in an unprecedented, nation-wide campaign to mass-rear and distribute wasps, in which government institutions, grower associations and private sector actors, such as the Thai Tapioca Development Institute (TTDI), all joined forces. Within the context of a regional technical cooperation project, mass-releases of *A. lopezi* were carried out across Thailand, some of which by airplane, and were followed by FAO-led introductions into Cambodia, Lao PDR and Viet Nam. The 2014 release of several hundred *A. lopezi* pairs into Indonesia were enabled by A. Rauf at Bogor Agricultural University (Wyckhuys et al. 2015). All these parasitoid introductions and capacity building interventions were implemented with the Royal Thai Government Ministry of Agriculture and Cooperatives' technical assistance, most notably from key IPM experts in its Department of Agriculture (DoA) and Department of Agricultural Extension (DoAE).

Other methods promoted for *P. manihoti* control include neonicotinoid stake dips (e.g. Parsa et al. 2012) and mass-releases of laboratory-reared predators and entomopathogens (Saengyot and Burikam 2012; Sattayawong et al. 2016). Despite the above efforts to promote a wide range of chemical and biologically-based management tactics, it is widely thought that it is *A. lopezi* that suppressed cassava mealybug populations across mainland Southeast Asia.

In this paper, we provide an in-depth assessment of *P. manihoti* population pressure and Asia-wide distribution, report on the establishment and spatial spread of the introduced *A. lopezi*, and examine biophysical, agro-climatic, and social factors that might enhance or impede mealybug biological control.

2. MAPPING MEALYBUG DISTRIBUTION

Until the appearance of *P. manihoti* in Asia's cassava fields, there was only scant and scattered information about the nature, distribution, etiology, epidemiology and ecology of the primary phytosanitary constraints of this crop in Southeast Asia (Bellotti et al. 2012; Graziosi and Wyckhuys 2017). In late 2013 though, an ambitious surveillance programme was set up together with international and national partners in Cambodia, Lao PDR, Myanmar, southern China, Thailand, and Viet Nam. With backstopping through the late Prabat Kumar at the Asian Institute of Technology, this programme intended to map the geographic distribution of *P. manihoti*, assess its pest pressure in local cassava fields, chart its potential invasion pathways, and understand its relative importance in relation to other arthropod pests and plant diseases. Ultimately, the programme sampled more than 572 cassava fields over 2 years, covering areas as diverse as Viet Nam's Central Highlands, the Ayeyawaddy delta of Myanmar or the remote uplands of southern Lao PDR.

Survey protocols are described in detail in Graziosi et al. (2016). In brief, we randomly selected older fields (>5-6 months old) in the main cassava-growing provinces within each country, with separate plots located at least 1 km apart. Surveys were carried out in January-May 2014 (dry season), October-November 2014 (late rainy season) and January-March 2015 (dry season). Location and elevation of each field were recorded using a handheld GPS unit (Garmin Ltd, Olathe, Kansas, USA). Per field, five linear transects were randomly chosen, with each transect covering 10 plants. By doing so, a total of 50 plants per field were assessed for *P. manihoti* infection status and associated 'bunchy top' symptoms (Neuenschwander et al. 1987), and per-plant mealybug abundance. In-field identification of mealybugs was based on morphological characters such as colour and length of abdominal waxy filaments. Following transect walks, we computed average *P. manihoti* infestation pressure (number of individuals per infected tip) and estimated field-level incidence of this pest (proportion of *P. manihoti*-affected tips, or 'bunchy tops') for each field.

Mealybugs are the most widespread group of arthropods on cassava crops in Southeast Asia, occurring in 70% of cassava fields (Graziosi et al. 2016). In some countries, such as Myanmar and Thailand, mealybugs were found in 95 and 100% of the fields, respectively. In infested fields, mealybugs were found on 27 ± 2% of plants, this representing the highest incidence among cassava-associated arthropods. The resident mealybug community on cassava was composed of 4 non-native species: (1) *P. manihoti*; (2) the papaya mealybug *Paracoccus marginatus* Williams & Granara de Willink; (3) *Pseudococcus jackbeardsleyi* Gimpel & Miller; and (4) the striped mealybug *Ferrisia virgata* Cockerell.

Within this mealybug community, *P. manihoti* represented 19.8% of the species complex (n= 572 fields, across dry and rainy season), and was recorded from 37% of fields during the 2014 dry season. The cassava mealybug was recorded from fields across Cambodia and Thailand, and it was also recorded in southern parts of Lao PDR and Viet Nam (Fig. 1). Across sites and sampling events, *P. manihoti* was recorded at average incidence rates of 7.4 ± 15.8% and dry-season abundance of 14.3 ± 30.8 individuals per infected tip. Maximum incidence rates were 100%, and maximum field-level abundance was 366.6 mealybugs per tip. Field-level abundance and incidence rates were highly variable between settings and countries (Table 1).

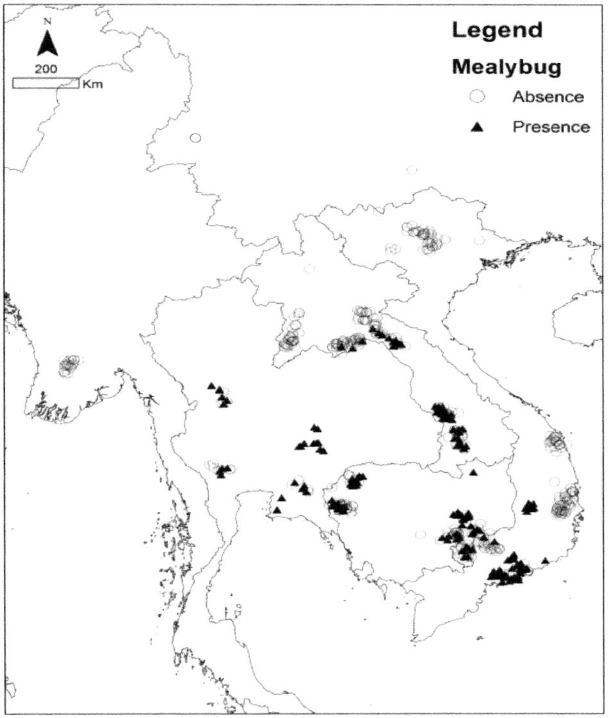

Figure 1. Geographic distribution of Phenacoccus manihoti, *as recorded during 2014-15 surveillance across the Greater Mekong subregion.*

Overall, in Asia's primary cassava cropping areas, current mealybug pest pressure was nearly identical to that in Africa during the mid-1980s. At the time, *P. manihoti* populations collapsed soon after parasitoid introduction and stabilized at incidence rates of 23% and abundance levels below 10 individuals per tip (Hammond and Neuenschwander 1990).

3. PARASITOID ESTABLISHMENT, SPREAD AND INCIDENCE

To assess *A. lopezi* establishment patterns and parasitism rates, and to delineate the parasitoid community associated with *P. manihoti*, we conducted two observational studies. First, over the course of two consecutive growing seasons, bi-monthly sampling was carried out in Tay Ninh, Viet Nam to characterize mealybug-parasitoid population dynamics (Le et al., unpublished). Second, dry-season sampling was done during 2014-2016 at a number of mealybug-invaded sites in eastern Cambodia (n = 15), eastern Thailand (n = 20), and southern Viet Nam (n = 19, 40). In each experiment, sampling consisted of collecting a total of 20 mealybug-infested tips or 'bunchy tops' from local cassava fields that were transferred to the laboratory for subsequent assessment of parasitoid emergence.

Table 1. Average incidence (percentage mealybug-infested plants per field; mean ± SD) and abundance (number of individuals per infected tip) of Phenacoccus manihoti *as recorded during multi-country surveillance in the 2014-15 growing seasons*

Country	Province	Date	Season	Sample size (*n*)	Incidence (%)	Abundance (# / tip)
Thailand	Nakhon Ratchasima	Mar 2014	Dry	10	22.6 ± 16.1	16.3 ± 18.5
	Kampheang Phet	Mar 2014	Dry	9	3.8 ± 5.2	9.2 ± 8.8
	Chachoengsao	Mar 2014	Dry	9	2.9 ± 4.7	6.5 ± 5.4
	Kanchanaburi	Mar 2014	Dry	11	6.6 ± 13.2	37.0 ± 58.9
Lao PDR	Borikhamxay	Feb 2014	Dry	27	0.5 ± 1.9	1.7 ± 1.5
	Vientiane capital	Feb 2014	Dry	22	0.1 ± 0.4	1.0 ± 0.0
	Xiengkhuong	Jan 2014	Dry	6	0.0	0.0
	Xayabuli	Feb 2014	Dry	20	0.0	0.0
	Champasak	Feb 2014	Dry	25	4.2 ± 11.1	3.4 ± 4.2
	Salavan	Feb 2014	Dry	25	13.6 ± 18.3	6.3 ± 9.6
Cambodia	Banteay Meanchey	Feb 2014	Dry	20	13.8 ± 11.9	7.7 ± 4.4
	Kampong Cham	Feb 2014	Dry	20	6.7 ± 18.2	14.4 ± 14.4
	Pailin	Feb 2014	Dry	19	9.5 ± 17.9	5.5 ± 3.9
	Battambang	Mar 2014	Dry	20	0.3 ± 0.9	1.2 ± 0.3
	Kratie	Feb 2014	Dry	20	16.3 ± 14.2	24.7 ± 26.5
Myanmar	Ayeyawaddy	Apr 2014	Dry	20	0.0	0.0
Viet Nam	Dong Nai	Feb 2014	Dry	20	43.7 ± 19.7	32.5 ± 14.6
	Binh Phuoc	Apr 2014	Dry	21	3.0 ± 13.4	7.7 ± 0.0
	Ba Ria-Vung Tau	May 2014	Rainy	20	35.9 ± 29.4	11.1 ± 17.0
	Tay Ninh	May 2014	Rainy	21	5.0 ± 8.2	5.0 ± 2.3
	Phu Yen	Apr 2014	Dry	19	0.0	0.0
	Dak Lak	Mar 2015	Dry	10	11.6 ± 6.4	3.7 ± 2.4
	Quang Ngai	Apr 2014	Rainy	20	0.0	0.0
	Binh Thuan	Mar 2015	Dry	10	9.2 ± 14.5	7.1 ± 12.8
	Yen Bai	Oct 2014	Rainy	20	0.0	0.0
	Phu Tho	Oct 2014	Rainy	19	0.0	0.0
China	Yunnan	Nov 2014	Rainy	25	0.0	0.0

Sampling procedures were adapted from Neuenschwander and Hammond (1988) and consisted of breaking off 20-cm 'tips' of infested plants and placing these in sealed paper bags. Next, bags with plant material were transferred to the laboratory, where each tip was carefully examined and the total number of *P. manihoti* was counted. Next, cassava tips were individualized within transparent polyvinyl chloride (PVC) containers and covered with fine cotton fabric mesh. Over the course of 3 weeks, containers were stored at ambient conditions and inspected on a daily basis for emergence of parasitoid wasps. Next, parasitoids and potential hyperparasitoids were collected by aspirator and stored for subsequent identification.

In the first study, *P. manihoti* occurred at an average incidence of 24.8 ± 17.7% and abundance level of 5.6 ± 5.3 individuals per tip across both growing seasons. In general, mealybug populations built up during the second half of the dry season and remained at low levels during the rainy season. High *A. lopezi* parasitism levels were recorded during each year, at average levels of 50.3% and 43.9% in rainy and dry season, respectively. Though rainfall does indeed cause high mortality of *P. manihoti*, it is believed that *A. lopezi* accounts for the sustained low mealybug population levels across seasons, through direct *P. manihoti* parasitism and host-feeding. The primary parasitoid community was entirely composed of *A. lopezi*, yet three potential hymenopterous hyperparasitoid species were also found from sites in Tay Ninh (Viet Nam): *Chartocerus* sp. near *walkeri* (Signiphoridae), *Promuscidea unfasciativentris* Girault (Eriaporidae) and *Prochiloneurus* sp. (Encyrtidae). Hyperparasitism levels were on average 2.8 ± 5.4%, with maximum rates of 26.4%. In smallholder cassava fields in eastern Cambodia, the hyperparasitoid community was found to be more diverse and species-rich, though locally-recorded species remain to be identified (Wyckhuys et al. 2017c).

In the second study, *A. lopezi* was found in *P. manihoti*-affected fields in Cambodia, Thailand and Viet Nam at parasitism levels of 10-57%, with an overall average of 38.9% (Wyckhuys et al. 2017b). Both studies exemplify how the introduced parasitoid has effectively colonized cassava fields in at least three Asian countries, attaining medium to high parasitism rates and contributing to *P. manihoti* control under a variety of agro-ecological conditions.

4. MULTI-FACETED DETERMINANTS OF BIOLOGICAL CONTROL SUCCESS

Field surveys and observational studies across the tropical Asia region have shown relatively low *P. manihoti* infestation levels, yet highly variable *A. lopezi* parasitism rates. For example, while *A. lopezi* attains dry-season parasitism of 16.3 ± 3.4% in coastal Viet Nam, it attains rates of 52.9 ± 4.3% in intensified cropping systems in the Tay Ninh province (Fig. 2).

To gain a better appreciation of potential constraints to *A. lopezi* success, we examined *P. manihoti* biological control through a number of different lenses, drawing on disciplines such as landscape ecology, plant pathology and soil science. Other factors, such as access to floral nectar and interference through tending ants are being investigated by A. Rauf and students in Indonesia, but they are not reported in this paper.

4.1. Soil-Mediated Effects on Mealybug Biological Control

Soil fertility and structure can determine plant health and shape overall resistance to pests (Amtmann et al. 2008), however, the impact of below-ground processes on above-ground interactions varies and is particularly difficult to predict. Also, alterations in plant nutrients are readily transmitted through trophic chains and affect the relative role of resource ("bottom-up") versus consumer ("top-down") forces in the structuring of ecological communities (Hunter and Price 1992). The success of both native and invasive herbivores has been explained through a range of theories and hypotheses, some of which simultaneously account for the role of the above plant resource availability and natural enemies (Blumenthal 2005; Center et al. 2014). Hence, understanding how certain herbivores (such as *P. manihoti*) and their associated parasitoids such as *A. lopezi* interact and respond to soil fertility and plant nutrient status is extremely valuable.

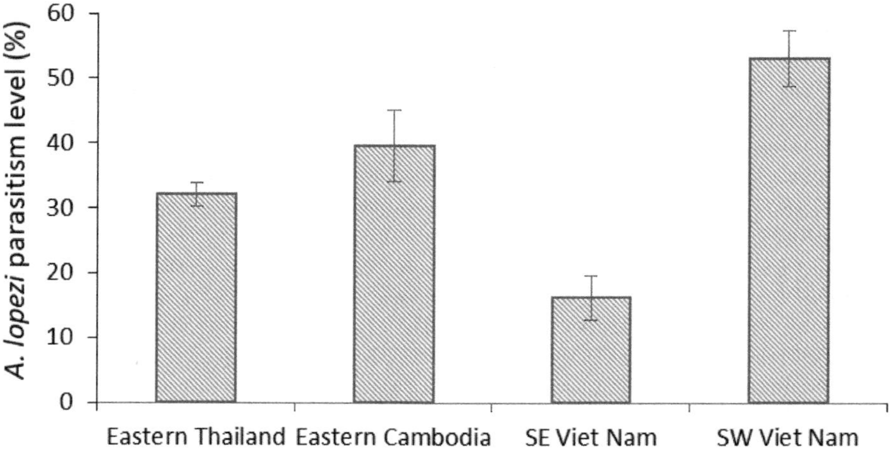

Figure 2. Parasitism levels (mean ± SD) by Anagyrus lopezi, *as recorded from selected fields in the invaded range of* Phenacoccus manihoti *(respective sample size: 20, 15, 19, 40).*

During 2015-2016, a set of manipulative trials and observational studies were carried out to illuminate how soil fertility affects *P. manihoti* x *A. lopezi* interactions (Wyckhuys et al. 2017b). More specifically, potted-plant fertilizer trials were combined with a regional survey of 65 cassava fields with varying soil fertility. Pot trials revealed strong bottom-up effects for *P. manihoti*, with nitrogen and potassium addition equally boosting development and fitness of *A. lopezi.*

Field surveys indicate that mealybug performance is highly species-specific and context-dependent. For *P. manihoti*, in-field abundance is associated with soil texture, i.e. silt content, and mealybug populations are disproportionately favoured in low-fertility conditions. Parasitism by *A. lopezi* varied greatly with field and soil fertility conditions and was highest in soils with intermediate fertility levels and where management practices include the addition of fertilizer supplements.

These findings on the field context show how deficient soil management can further exacerbate mealybug pest problems and ultimately push farmers into 'poverty traps'. On the other hand, our work can help target additional biological control measures and inform management practices, such as mulching, organic matter addition, or corrective nutrient supplementation, to enhance or restore mealybug biological control.

Similar impact across several trophic levels had been documented also in Africa. In Malawi, for instance, biological control despite the presence of *A. lopezi* only failed in the few (10%) fields on sandy, un-mulched soils that did not sustain leafy plants (Neuenschwander et al. 1991).

4.2. Plant-Microbe-Insect Interactions

In recent years, scientific interest in plant-insect-microbe or 'cross-kingdom' interactions has boomed, steadily revealing the multiple, intricate ways in which micro-organisms mediate plant-herbivore interplay (Ponzio et al. 2013; Tack and Dicke 2013). Phytopathogens regularly alter whole repertoires of plant phenotypic traits, and they bring about shifts in key chemical or morphological characteristics of plant hosts (Tack et al. 2012; Biere and Tack 2013). Though largely overlooked, pathogens can also cause cascading effects on higher trophic levels and eventually shape entire plant-associated arthropod communities.

In an observational study in early 2016, I. Graziosi and Cambodian colleagues investigated whether a *Candidatus* Phytoplasma causing cassava witches' broom (CWB) is altering relative abundance and species composition of different invasive mealybugs and determines success of their associated parasitoids, including *A. lopezi*. The CWB is an emerging phytopathogen that occurs at near-pandemic levels in several parts of Southeast Asia, and which causes leaf discoloration, extensive proliferation of leaves and stems, and stunted growth. In their study, samples were taken from multiple sites of CWB-symptomatic and asymptomatic plants (Wyckhuys et al. 2007c). From each plant, the apical part or 'tip' was collected and transferred to the laboratory for further processing. After counting and identifying all mealybug individuals, each cassava tip was transferred individually to transparent PVC containers, closed with fabric mesh. Over a period of 14 days, containers were checked for emergence of parasitoids or hyperparasitoids. Parasitoids were identified to morpho-type and stored in ethanol for subsequent species-level identification. CWB infection was found to positively affect overall mealybug abundance and species richness, and to disproportionately favour the generalist *Paracoccus marginatus* over *P. manihoti* (Fig. 3).

Moreover, CWB phytoplasma infection was positively correlated with an increased parasitoid richness and diversity. Though overall parasitism rate did not differ among CWB-infected and uninfected plants, lower numbers of *A. lopezi* were obtained from infected plants. Also, CWB-infection status affected *A. lopezi* sex ratio, with more male-biased sex ratios on CWB-infected plants (Wyckhuys et al. 2017c). This possibly could be explained by smaller 'undernourished' mealybugs which are more often selected by females for male eggs.

This work underlines how systemic plant pathogens such as CWB do impact parasitoid establishment and efficacy, and how they could influence *P. manihoti* biological control. Hence, entomologists need to work across disciplines and take into consideration plant pathology aspects when assessing field-level parasitism rates and biological control success.

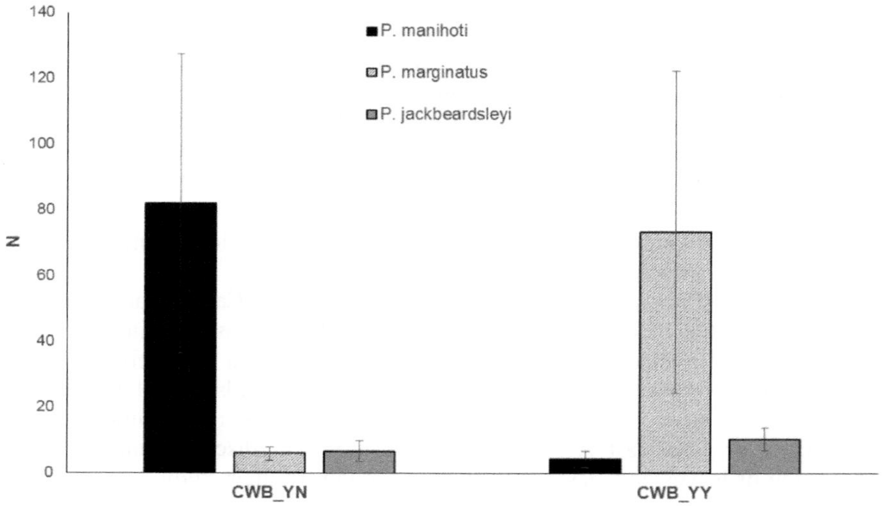

Figure 3. Plant-level abundance of different mealybug species on plants with characteristic symptoms of asymptomatic plants (CWB_YN) and cassava witches' broom (CWB_YY).

4.3. Landscape-Level Drivers

As exemplified in the above Sections, patch-level characteristics, such as soil fertility or plant disease pressure, greatly impact *A. lopezi* abundance and performance. On the other hand, landscape-level variables may equally affect biological control of invasive pests such as *P. manihoti*, though have rarely been taken into consideration.

The impact of landscape structure on natural enemy abundance, diversity, and activity in temperate cropping systems such as grains, canola or cabbage crops is fairly well documented (Bianchi et al. 2006; Chaplin-Kramer et al. 2011; Schellhorn et al. 2015), though much less is known about its overall impact on pest pressure or natural biological control. However, for specialist parasitoids such as *A. lopezi*, landscape simplification could be particularly disruptive (Cagnolo et al. 2009). Also, landscape complexity differentially benefits hyperparasitoids, which potentially could derail biological control of *P. manihoti* (Rand et al. 2012).

In mid-2013, T. T. N. Le and collaborators from Viet Nam's Plant Protection Department (VPPD) embarked upon a two-year study to assess mealybug x parasitoid population dynamics under varying landscape context (Le et al. 2018). Over the course of two consecutive cropping seasons, insect populations were surveyed under small-field and high-diversity or large-field and low-diversity landscape settings. In certain areas, cassava fields are small (1–2 ha in size), embedded within relatively complex and diverse landscape settings (here termed 'high-diversity' sites). Other landscape sectors are primarily made up of larger fields, ranging between 4 and 8 ha (here termed 'low-diversity' sites). Overall, *P. manihoti* colonized fields earlier and attained higher incidence in small plots within high-diversity landscapes as compared to large fields in simplified landscapes (Fig. 4).

Landscape type, however, significantly affected hyperparasitism rate at certain crop ages, but did not impact *P. manihoti* abundance or *A. lopezi* parasitism rate. Also, a slightly more pronounced density-dependent response of *A. lopezi* was found within low-diversity settings, at a scale of both individual cassava tips and entire fields. These landscape-dependent impacts likely directly relate to dispersal modalities and other ecological traits of *A. lopezi*, including its supreme ecological plasticity, exceptional dispersal capacity and ability to equally host-feed and consume cassava extra-floral nectar (Neuenschwander 2001).

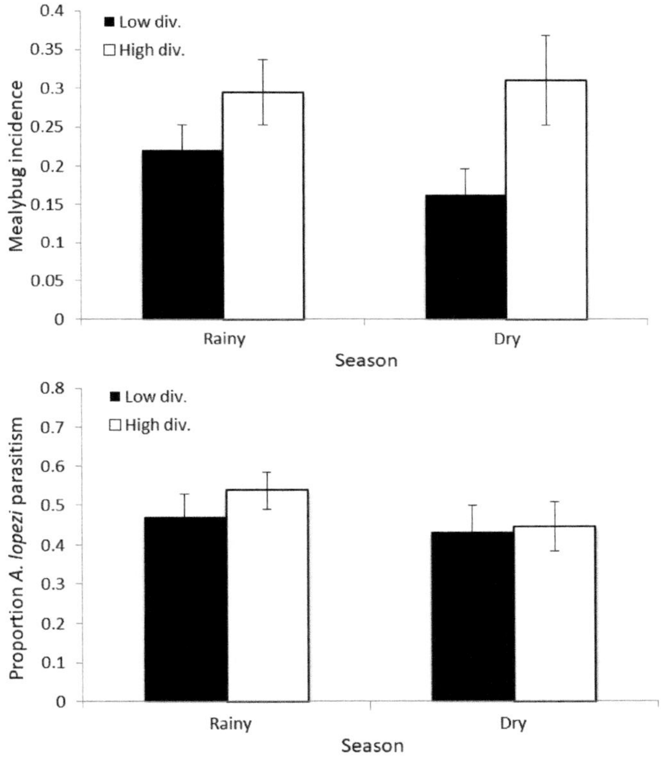

Figure 4. Phenacoccus manihoti *incidence (proportion infected plants) and* Anagyrus lopezi *parasitism rate during dry and rainy season, for fields within high- and low-diversity landscape settings (see Le et al. 2018).*

5. SOCIAL SCIENCE: COMPASS AND PUBLIC AWARENESS FOR A REGIONAL BIOLOGICAL CONTROL CAMPAIGN

Though regularly overlooked or roundly disregarded, social science is of paramount importance for the successful promotion of IPM, and particularly for effective biological insect control. More so, only a fraction of biological control studies over the past 25 years explicitly address social science or technology transfer aspects (Wyckhuys et al. 2017a).

A retrospective analysis of social science studies from the developing-world tropics showed that most farmers have highly-deficient knowledge of natural enemies, and routinely express doubts about the exact value of biological control services on their farm. More so, farmers' knowledge of (fast-spreading) invasive pests and their control is considered to be exceptionally weak. These trends are worrying and could stifle farm-level adoption and subsequent diffusion of knowledge-intensive technologies such as biological control (Catalini and Tucker 2017).

Though classical biological control - as in the case of *A. lopezi* - is largely considered to be self-propelling and requires little or no intervention by farmers (Andrews et al. 1992), it is critical to build and strengthen farmers' agro-ecological knowledge in light of the active promotion of systemic insecticides for mealybug control. At the time of writing this chapter, hundreds of thousands of smallholder growers in one particular Asian country had embraced the use of prophylactic dips with neonicotinoids and considered those as a central component in their cassava crop management. These products cause long-term negative impacts on *A. lopezi* survival and fitness (Lankaew, Tan, Nguyen and Wyckhuys, unpublished), and as such they could hamper biological control.

In late 2014, a two-country survey was started to characterize agro-ecological knowledge, attitudes and pest management practices of local cassava growers (Uphadyay et al. 2018). A parallel study was done by A. Rauf and N. Wardani in Indonesia. In Lao PDR and Viet Nam, farmers had limited awareness of recent invasive pests, such as *P. manihoti*, and their knowledge was highly context- and locality-specific, i.e. shaped by the invasion history of the mealybug. Only the occasional farmer was aware of the existence of natural enemies on his/her farm, and even charismatic and important guilds such as lady beetles, spiders or lacewings were recognized by <10% local growers. Overall, a minority of farmers used preventive tactics and in recently invaded fields frequently resorted to drastic measures such as overhead sprays of insecticides and burning of entire fields. In southern Viet Nam, women guide household-level IPM decision-making (Uphadyay et al. 2018), yet their weak agro-ecological knowledge base could further enable local diffusion of insecticide use.

To counteract some of the above trends and ease obstacles in farmer learning about biological control, a video-mediated extension campaign was launched. Farmer-to-farmer video is particularly suited to transfer complex concepts such as parasitism and insect predation, and it can help secure *P. manihoti* biological control in Asia's cassava systems. A farmer to farmer video was developed by carefully selecting farmers from the FAO-deployed Farmer Field Schools in eastern Thailand and documenting some of their successes with *P. manihoti* control. This allowed production of a multi-lingual video entitled '*Managing Mealybugs in Cassava*' by the Belgium-based company AgroInsight. This video is available for streaming in English and multiple Asian languages through Access Agriculture (2019) or YouTube channels. This farmer-to-farmer video was subsequently distributed through multiple means, including private sector actors, national television and rural extension bureaus, reaching >200 000 growers in a matter of months in Viet Nam alone.

6. CONCLUSIONS

This chapter aims at providing a comprehensive overview of the multi-institutional biological control programme against the invasive mealybug *P. manihoti* in cassava crops across mainland Southeast Asia. Two-year population surveys and area-wide pest surveillance reveal how *P. manihoti* has effectively spread to at least six Asian countries, yet occurs at low to intermediate incidence rates and at abundance levels of 10-20 individuals per tip. Though rainfall and humidity are responsible for important levels of mortality, *P. manihoti* is deemed to be under effective biological control by the introduced *A. lopezi*. Five years after its introduction in eastern Thailand, *A. lopezi* is presently recorded at high though variable population levels in most mealybug-invaded fields in the region. Although not covered in this chapter, chemical and physical exclusion assays have shown how the resulting *P. manihoti* infestation rates only have minor impact on cassava crop yield or harvest indices. Hence, the minute wasp that was originally sourced in southern Brazil and Paraguay in 1981, and released across Africa during the 1980s, now also brings relief to cassava farmers in tropical Asia. The impact of this biological control programme on farmer livelihoods, national economies and rural agro-industries still needs to be assessed, but the economic benefits are expected to equal and probably surpass the multi-billion dollar benefits that were estimated in Africa (Zeddies et al. 2001; Wyckhuys et al. 2018).

Some of the factors that underpinned the outright success of this tropical Asia-wide biological control campaign are the following:

1. Globe-spanning collaboration between FAO, CABI and CGIAR institutions, plus ready access to insect biodiversity in countries such as Brazil and Paraguay, allowed the necessary foreign exploration and effective identification of suitable natural enemies. Next, a swift mobilization of government institutions in Thailand and the strong support from private sector actors such as TTDI proved to be of crucial importance in ensuring establishment and country-wide spread of *A. lopezi*.

2. Extension campaigns that were built upon a sound appreciation of farmers' knowledge, attitudes and practices across farming contexts and sites. Though conventional extension initiatives were effective in Thailand, a farmer-to-farmer educational video proved to be key to transferring growers' experiences, perceptions and innovations from early-adopters and Farmer Field School groups in invaded areas. This undoubtedly boosted preparedness and prevented certain detrimental practices — such as unguided use of insecticides — from gaining a foothold.

3. A near-exclusive focus on herbivore-natural enemy interactions at the level of a single plant is often adopted in today's biological control or IPM studies. Though this yields valuable insights into those particular trophic interactions, it regularly obscures other equally important mechanisms at different trophic, spatial or organizational scales. Hence, we advocate holistic, systems-level approaches that draw upon disciplines beyond conventional entomology or insect ecology.

4. Gaining a thorough understanding of the various factors that shape variability in parasitism not only is valuable from an ecological perspective, but it can equally guide efforts to improve biological control efficacy (Rosenheim 1998). Our assessment of determinants of *A. lopezi* parasitism pointed to options to enhance mealybug pest control through interventions targeting soil nutrients, landscape composition, plant diseases, or crop management scenarios (for the latter, see Delaquis et al. 2018).

5. In September 2015, Nature Magazine (Volume 525) boldly stated that:

> "*to solve the grand challenges facing society and to save the world, scientists and social scientists must work together*" (Nature 2015).

Interdisciplinary science matters (see also Brondizio et al. 2016), and our work underscores that fully collaborative, integrative research is central to effectively solve invasive pest problems and to advance AW-IPM and biological control in developing-world agriculture.

7. ACKNOWLEDGEMENTS

This work presents original data-sets that were generated through collaborative research by counterparts across tropical Asia, CIAT Asia personnel and international co-operators. The bulk of research was conducted as part of an EC-funded, IFAD-managed and CIAT-executed programme (CIAT-EGC-60-1000004285), while additional funding was provided through the CGIAR-wide Research Program on Roots, Tubers and Banana (CRP-RTB). Exploratory research on plant-insect-microbe interactions was carried out as part of a STINT project funded by the Swedish University of Agricultural Sciences (SLU).

8. REFERENCES

Access Agriculture. 2019. Managing mealybugs in cassava (video).

Andrews, K. L., J. W. Bentley, and R. D. Cave. 1992. Enhancing biological control's contributions to Integrated Pest Management through appropriate levels of farmer participation. Florida Entomologist 75: 429–439.

Amtmann, A., S. Troufflard, and P. Armengaud. 2008. The effect of potassium nutrition on pest and disease resistance in plants. Physiologia Plantarum 133: 682–691.

Bellotti, A. C., L. Smith, and S. L. Lapointe. 1999. Recent advances in cassava pest management. Annual Review of Entomology 44: 343–370.

Bellotti, A. C., B. V. Herrera, and G. Hyman. 2012. Cassava production and pest management: Present and potential threats in a changing environment. Tropical Plant Biology 5: 39–72.

Bianchi, F. J. J. A., C. J. H. Booij, and T. Tscharntke. 2006. Sustainable pest regulation in agricultural landscapes: A review on landscape composition, biodiversity and natural pest control. Proceedings of the Royal Society B – Biological Sciences 273: 1715–1727.

Biere, A., and A. J. M. Tack. 2013. Evolutionary adaptation in three-way interactions between plants, microbes and arthropods. Functional Ecology 27: 646–660.

Blumenthal, D., 2005. Interrelated causes of plant invasion. Science 310: 243–244.

Brondizio, E. S., K. O'Brien, X. Bai, F. Biermann, W. Steffen, F. Berkhout, C. Cudennec, M. C. Lemos, A. Wolfe, J. Palma-Oliveira, and C. T. A. Chen. 2016. Re-conceptualizing the Anthropocene: A call for collaboration. Global Environmental Change 39: 318–327.

Cagnolo, L., G. Valladares, A. Salvo, M. Cabido, and M. Zak. 2009. Habitat fragmentation and species loss across three interacting trophic levels: Effects of life-history and food-web traits. Conservation Biology 23: 1167–1175.

Catalini, C., and C. Tucker. 2017. When early adopters don't adopt. Science 357: 135–136.

Center, T. D., F. A. Dray, E. D. Mattison, P. W. Tipping, and M. B. Rayamajhi. 2014. Bottom-up effects on top-down regulation of a floating aquatic plant by two weevil species: The context-specific nature of biological control. Journal of Applied Ecology 51: 814–824.

Chaplin-Kramer, R., M. E. O'Rourke, E. J. Blitzer, and C. Kremen. 2011. A meta-analysis of crop pest and natural enemy response to landscape complexity. Ecology Letters 14: 922–932.

Cramb, R., V. Manivong, C. J. Newby, K. Sothorn, and P. S. Sibat. 2017. Alternatives to land grabbing: Exploring conditions for smallholder inclusion in agricultural commodity chains in Southeast Asia. The Journal of Peasant Studies 44: 939–967.

Delaquis, E., S. de Haan, and K. A. G. Wyckhuys. 2018. On-farm diversity offsets environmental pressures in tropical agro-ecosystems: A synthetic review for cassava-based systems. Agriculture, Ecosystems and Environment 251: 226–235.

Graziosi, I., and K. A. G. Wyckhuys. 2017. Integrated management of arthropod pests of cassava: The case of Southeast Asia, pp. 1-26. *In* Hershey, C. (ed.), Achieving sustainable cultivation of cassava. Volume II. Burleigh Dodds, Swaston, Cambridge, UK. 300 pp.

Graziosi, I., N. Minato, E. Alvarez, D. T. Ngo, T. X. Hoat, T. M. Aye, J. M. Pardo, P. Wongtiem, and K. A. G. Wyckhuys. 2016. Emerging pests and diseases of South-east Asian cassava: A comprehensive evaluation of geographic priorities, management options and research needs. Pest Management Science 72: 1071–1089.

Hammond, W. N. O., and P. Neuenschwander. 1990. Sustained biological control of the cassava mealybug *Phenacoccus manihoti* (Homoptera: Pseudococcidae) by *Epidinocarsis lopezi* (Hymenoptera: Encyrtidae) in Nigeria. Entomophaga 35: 515–526.

Hammond, W. N. O., P. Neuenschwander, and H. R. Herren. 1987. Impact of the exotic parasitoid *Epidinocarsis lopezi* on the cassava mealybug (*Phenacoccus manihoti*) populations. *In* P. Neuenschwander, J. S. Yaninek, and H. R. Herren (eds.), Africa-wide biological control project of cassava pests. Insect Science and its Application 8: 887–891.

Herren, H. R., and P. Neuenschwander. 1991. Biological control of cassava pests in Africa. Annual Review of Entomology 36: 257–283.

Herren, H. R., P. Neuenschwander, R. D. Hennessey, and W. N. O. Hammond. 1987. Introduction and dispersal of *Epidinocarsis lopezi* (Hym., Encyrtidae), an exotic parasitoid of the cassava mealybug, *Phenacoccus manihoti* (Hom., Pseudococcidae), in Africa. Agriculture, Ecosystems and Environment 19: 131–144.

Hunter, M. D., and P. W. Price. 1992. Playing chutes and ladders: Heterogeneity and the relative roles of bottom-up and top-down forces in natural communities. Ecology 73: 724–732.

Le, T. T. N., I. Graziosi, T. M. Cira, M. W. Gates, L. Parker, and K. A. G. Wyckhuys. 2018. Landscape context does not constrain biological control of *Phenacoccus manihoti* in intensified cassava systems of southern Vietnam. Biological Control 121: 129–139.

Löhr, B., A. M. Varela, and B. Santos. 1990. Exploration for natural enemies of the cassava mealybug, *Phenacoccus manihoti* (Homoptera: Pseudococcidae), in South America for the biological control of this introduced pest in Africa. Bulletin of Entomological Research 80(4): 417–425.

Muniappan, R., B. M. Shepard, G. W. Watson, G. R. Carner, A. Rauf, D. Sartiami, P. Hidayat, J. V. K. Afun, G. Goergen, and A. K. M. Ziaur Rahman. 2009. New records of invasive insects (Hemiptera: Sternorrhyncha) in Southeast Asia and West Africa. Journal of Agricultural and Urban Entomology 26: 167–174.

Nature 2015. Why interdisciplinary research matters. Scientists must work together to save the world. A special issue asks how they can scale disciplinary walls. 17 September 2015. Nature 525: 305.

Neuenschwander, P., 2001. Biological control of the cassava mealybug in Africa: A review. Biological Control 21: 214–229.

Neuenschwander, P., and W. N. O. Hammond. 1988. Natural enemy activity following the introduction of *Epidinocarsis lopezi* (Hymenoptera: Encyrtidae) against the cassava mealybug, *Phenacoccus manihoti* (Homoptera: Pseudococcidae) in southwestern Nigeria. Environmental Entomology 17: 894–902.

Neuenschwander, P., W. N. O. Hammond, and R. D. Hennessey. 1987. Changes in the composition of the fauna associated with the cassava mealybug, *Phenacoccus manihoti*, following the introduction of the parasitoid *Epidinocarsis lopezi*. Insect Science and its Application 8: 893–898.

Neuenschwander, P., R. Borowka, G. Phiri, H. Hammans, S. Nyirenda, E. H. Kapeya, and A. Gadabu. 1991. Biological control of the cassava mealybug *Phenacoccus manihoti* (Hom., Pseudococcidae) by *Epidinocarsis lopezi* (Hym., Encyrtidae) in Malawi. Biocontrol Science and Technololgy 1: 297–310.

Nwanze, K. F. 1982. Relationships between cassava root yields and crop infestations by the mealybug, *Phenacoccus manihoti*. International Journal of Pest Management 28: 27–32.

Parsa, S., T. Kondo, and A. Winotai. 2012. The cassava mealybug (*Phenacoccus manihoti*) in Asia: First records, potential distribution, and an identification key. PLoS One 7(10): e47675.

Ponzio, C., R. Gols, C. M. J. Pieterse, and M. Dicke. 2013. Ecological and phytohormonal aspects of plant volatile emission in response to single and dual infestations with herbivores and phytopathogens. Functional Ecology 27: 587–598.

Rand, T. A., F. J. Van Veen, and T. Tscharntke. 2012. Landscape complexity differentially benefits generalized fourth, over specialized third, trophic level natural enemies. Ecography 35: 97–104.

Rosenheim, J. A. 1998. Higher-order predators and the regulation of insect herbivore populations. Annual Review of Entomology 43: 421–447.

Saengyot, S., and I. Burikam. 2012. Bionomics of the apefly, *Spalgis epius* (Lepidoptera: Lycaenidae), predatory on the papaya mealybug, *Paracoccus marginatus* (Hemiptera: Pseudococcidae) in Thailand. Songklanakarin Journal of Science and Technology 34: 1–7.

Sartiami, D., G. W. Watson, M. N. M. Roff, M. Y. Hanifah, and A. B. Idris. 2015. First record of cassava mealybug, *Phenacoccus manihoti* (Hemiptera: Pseudococcidae), in Malaysia. Zootaxa 3957: 235–238.

Sattayawong, C., S. Uraichuen, and W. Suasa-ard. 2016. Larval preference and performance of the green lacewing, *Plesiochrysa ramburi* (Schneider) (Neuroptera: Chrysopidae), on three species of cassava mealybugs (Hemiptera: Pseudococcidae). Agriculture and Natural Resources 50: 460–464.

Schellhorn, N. A., H. R. Parry, S. Macfadyen, Y. Wang, and M. P. Zalucki. 2015. Connecting scales: Achieving in-field pest control from areawide and landscape ecology studies. Insect Scien. 22: 35–51.

Schulthess, F., J. U. Baumgärtner, V. Delucchi, and A. P. Gutierrez. 1991. The influence of the cassava mealybug, *Phenacoccus manihoti* Mat.-Ferr. (Homoptera: Pseudococcidae) on yield formation of cassava, *Manihot esculenta* Crantz. Journal of Applied Entomology 111: 155–165.

Tack, A. J. M., and M. Dicke. 2013. Plant pathogens structure arthropod communities across multiple spatial and temporal scales. Functional Ecology 27: 633–645.

Tack, A. J. M., S. Gripenberg, and T. Roslin. 2012. Cross-kingdom interactions matter: Fungal-mediated interactions structure an insect community on oak. Ecology Letters 15: 177–185.

(TTTA) Thai Tapioca Trade Association. 2011. Annual report. Bangkok, Thailand.

Uphadyay, B., D. D. Burra, T. T. Nguyen, and K. A. G. Wyckhuys. 2018. Caught off guard: Folk knowledge proves deficient when addressing invasive pests in Asian cassava systems. Environment, Development and Sustainability 15(5): 1–21.

Winotai, A., G. Goergen, M. Tamo, and P. Neuenschwander. 2010. Cassava mealybug has reached Asia. Biocontrol News and Information 31: 10N–11N.

Wyckhuys, K. A. G., A. Rauf, and J. Ketelaar. 2015. Parasitoids introduced into Indonesia: Part of a region-wide campaign to tackle emerging cassava pests and diseases. Biocontrol News and Information 35: 29N–38N.

Wyckhuys, K. A. G., J. W. Bentley, R. Lie, T. T. N. Le, and M. Fredrix. 2017a. Maximizing farm-level uptake and diffusion of biological control innovations in today's digital era. BioControl 63: 133–148.

Wyckhuys, K. A. G., D. D. Burra, D. H. Tran, I. Graziosi, A. J. Walter, T. G. Nguyen, H. N. Trong, B. V. Le, T. T. N. Le, and S. J. Fonte. 2017b. Soil fertility regulates invasive herbivore performance and top-down control in tropical agro-ecosystems of Southeast Asia. Agriculture, Ecosystems and Environment 249: 38–49.

Wyckhuys, K. A. G., I. Graziosi, I., D. D. Burra, and A. J. Walter. 2017c. Phytoplasma infection of a tropical root crop triggers bottom-up cascades by favoring generalist over specialist herbivores. PLoS One 12(8): e0182766.

Wyckhuys, K., W. Zhang, S. Prager, D. Kramer, E. Delaquis, C. Gonzalez, and W. van der Werf. 2018. Biological control of an invasive pest eases pressures on global commodity markets. Environmental Research Letters 13(9): 094005.

Yonow, T., D. J. Kriticos, and N. Ota. 2017. The potential distribution of cassava mealybug (*Phenacoccus manihoti*), a threat to food security for the poor. PLoS One 12(3): e0173265.

Zeddies, J., R. P. Schaab, P. Neuenschwander, and H. R. Herren. 2001. Economics of biological control of cassava mealybug in Africa. Agricultural Economics 24: 209–219.

HOLISTIC AREA-WIDE APPROACH FOR SUCCESSFULLY MANAGING CITRUS GREENING (HUANGLONGBING) IN MEXICO

C. J. GARCÍA-ÁVILA, F. J. TRUJILLO-ARRIAGA, A. QUEZADA-SALINAS, I. RUIZ-GALVÁN, D. BRAVO-PÉREZ, J. M. PINEDA-RÍOS, J. G. FLORENCIO-ANASTASIO AND P. L. ROBLES-GARCÍA

General Directorate of Plant Health (DGSV) of Servicio Nacional de Sanidad, Inocuidad y Calidad Agroalimentaria (SENASICA), Guillermo Pérez Valenzuela 127, Coyoacán 04100, Mexico City, Mexico; clemente.garcia@senasica.gob.mx

SUMMARY

The General Directorate of Plant Health (DGSV in Spanish) is recognized as the National Plant Protection Organization of the Federal Government of Mexico that acts under the Plant Health Federal Law. Some relevant plant protection programmes that Mexico is implementing include: The Huanglongbing (HLB) - Asian Citrus Psyllid (ACP) Programme, Mediterranean Fruit Fly Programme, National Fruit Fly Campaign, Pink Hibiscus Mealybug Programme and a permanent Phytosanitary Epidemiological Surveillance Programme to prevent the introduction and spread of regulated non-native pests. HLB or citrus greening is caused by the bacterium, *Candidatus* Liberibacter spp., and considered the most devastating citrus disease in the world. Once infected, it causes the death of orange, mandarin, grapefruit and lemon trees within 3 to 8 years. HLB is transmitted by the ACP, (*Diaphorina citri* Kuwayama), an insect vector widely distributed in most citrus producing regions of the world, including the citrus areas of Mexico. Until 2004, the disease only existed in Asia and Africa. It was first reported to occur in the Americas in 2004 (São Paulo, Brazil) and 2005 (Florida USA). In 2009, it was detected for the first time in Yucatán, Mexico. During that year, the national Mexican citrus production was 6.82 million tons (SIAP 2017). The economic impact evaluation by Salcedo-Baca et al. (2010) indicated that without the intervention of the Federal Government, HLB would be responsible for a reduction of the Mexican citrus production by 2.7 million tons in five years (39.6%). In spite of the spread of HLB, the citrus production in Mexico increased 11% to 7.56 million tons in 2015 (SIAP 2017). Today Mexico has 573 406 hectares (ha) of citrus compared to 545 947 in 2009, an increase of 5%. The first phytosanitary actions implemented on an area-wide basis were: (1) timely detection of HLB in agricultural and urban areas; (2) systematic elimination of infected trees in areas under surveillance; (3) control of the *D. citri* vector and (4) protection of propagative material in nurseries to avoid its infection. As a result of the successful HLB Programme implemented since 2008, adverse effects of the disease have largely been avoided. Management of HLB is organized through Regional Areas of Control (*Areas Regionales de Control* or ARCOs), which implement the following area-wide measures: epidemiological surveillance and monitoring of psyllids based on criteria associated to climate and host presence in urban

J. Hendrichs, R. Pereira and M. J. B. Vreysen (eds.), Area-Wide Integrated Pest Management: Development and Field Application, pp. 33–49. CRC Press, Boca Raton, Florida, USA.

and cultivated areas, and chemical and biological controls. From 2010 to 2015, 31 million parasitoid wasps *Tamarixia radiata* (Waterston) were produced and released in commercial citrus and backyard host areas of Yucatán, Quintana Roo, Campeche, Tabasco, Chiapas, Oaxaca, Hidalgo and Guerrero. The ARCOs are public-private organizations jointly operated by federal and state governments together with citrus grower associations. In 2016, the Mexican government allocated almost USD 8.5 million to the HLB Programme. With these actions, Mexico has largely mitigated the adverse effects of the disease while at the same time slightly increased citrus production. In addition, research programmes have been established together with scientific institutions to generate vegetative material with tolerance or resistance to the disease. Although the government has successfully implemented area-wide strategies for regional control, it is necessary to develop new and improved technologies to eliminate the vector, following the example of the Mediterranean fruit fly Programme in Mexico.

Key Words: Asian citrus psyllid, *Diaphorina citri*, *Candidatus* Liberibacter, economic impact, Mexican states, citrus production, *Tamarixia radiata*

1. INTRODUCTION

Huanglongbing (HLB) or citrus greening is a disease native to China and is considered one of the most destructive citrus diseases in Asian countries where its occurrence was originally reported more than a century ago (Bové 2006). More recently this bacterium reached the American continent together with its insect vector, the Asian citrus psyllid (ACP) (*Diaphorina citri* Kuwayama), capable of infecting and causing considerable damage to plants in the family Rutaceae (Alemán et al. 2007; Bové 2012).

The current geographic distribution of HLB extends to 12 countries in Asia, several islands in the Indian Ocean, Iran, portions of Africa, the Arabian Peninsula, Argentina, Brazil, Paraguay, Central America (except Panama), Barbados, Belize, Cuba, Dominican Republic, Guadeloupe, Jamaica, Martinique, USA (including Puerto Rico and the U.S. Virgin Islands) and Mexico (da Graça and Korsten 2004; Halbert and Manjunath 2004; NAPPO 2005; Bové 2006; Manjunath et al. 2008; Collazo et al. 2009; Trujillo-Arriaga 2011). In 2009, HLB was detected in Mexico for the first time in the state of Yucatán, and now is known to occur in 24 Mexican states (SENASICA 2017a).

The HLB-causing bacterium mainly attacks sweet orange and mandarins (da Graça and Korsten 2004) although all citrus varieties have shown varying degrees of susceptibility to infection as well as other members of the Rutaceae family. Currently, there is no successful control method to cure this disease, and as a result, infected trees die in the course of a few years.

This disease is caused by Gram-negative bacteria of the genus *Candidatus* Liberibacter (Bové 2006). There are three species of HLB-associated bacteria: *Ca.* L. asiaticus, *Ca.* L. africanus and *Ca.* L. americanus. All three have been described as the cause of HLB in different countries and climates world-wide (da Graça and Korsten 2004; Halbert and Manjunath 2004; Bové 2006; Wang et al. 2009), along with its insect vectors: ACP and the African psyllid, *Trioza erytreae* (Del Guercio). The first is the vector of *Ca.* L. asiaticus and *Ca.* L. americanus; while *T. erytreae* is the vector of *Ca.* L. africanus. A fourth species, *Ca.* L. caribbeanus, recently was identified from samples of ACP and *Citrus sinensis* (L.) Osbeck in Córdoba, Colombia. Efforts are underway to determine its pathogenicity and if it causes HLB symptoms (Keremane et al 2015).

The main symptoms of HLB are asymmetrical blotchy mottle yellowing in leaves, chlorosis, fruit drop and foliar loss and tree death. In addition to reducing the size and quality of fruit, the disease causes malformations and bad taste of fruit (Schwarz et al. 1973; Bové 2006). Although infected trees can remain productive for 5 to 8 years, the fruit are of poor quality (Halbert and Manjunath 2004).

To date, no control strategy is available that allows the immediate elimination of the pathogen, so its management in commercial citrus production areas is limited to the control of psyllid vectors through the application of insecticides, elimination of symptomatic trees with the aim of reducing levels of inoculum, isolation of affected areas by quarantine enforcement, and certification of pathogen-free propagative material (da Graça and Korsten 2004; Bové 2006; Manjunath et al. 2008; Gottwald et al. 2012). Because infected orchards become economically non-viable after 7 to 10 years, efforts focus on elimination of infected trees that have caused losses worth billions of dollars on a global basis (Gottwald et al. 1991; Vojnov et al. 2010).

In Florida, introduction of HLB resulted in major changes to pest management practices and corresponding costs. According to Singerman and Burani-Arouca (2017), average annual pest control consisted of two sprays for processed juice fruit and six sprays for fresh market grapefruit mostly to control several minor diseases, mites and weevils. Following a series of hurricanes in 2004 and 2005, that resulted in the catastrophic spread of citrus canker and forced abandonment of eradication actions, the number of sprays increased to 3–4 for processed juice oranges and 10 for fresh market grapefruit. After the discovery of HLB in August 2005 and citrus black spot in 2010, the number of treatments rose to 8–9 sprays for processed juice fruit and 14 sprays for fresh market grapefruit aimed at both disease and ACP control combined with additional fertilizer treatments.

Costs per acre of foliar sprays for producing processed oranges in south-western Florida rose from USD 185.63 in 2003/2004 to USD 666.00 (+240%) in 2014/2015 while fertilizer treatments went from USD 207.69 to USD 486.96 (+134%) during the same timeframes. By comparison, costs per acre of foliar sprays for fresh market grapefruit in the Indian River area increased from USD 493.08 in 2003/2004 to USD 1300.40 (+164%) in 2014/2015 while fertilizer treatments rose from USD 190.56 per acre in 2003/2004 to USD 452.55 (+137%) per acre in 2014/2015. These totals do not include other cultural control costs nor cost for tree replacement. It was found that area-wide control of ACP through Citrus Health Management Areas provides an estimated differential gross economic benefit of USD 714 (USD 1218) for 2012/2013 (2013/2014) (Singerman and Page 2016).

Over the period from 2012/2013 to 2015/2016, HLB caused a cumulative loss of USD 1672 million in grower revenues (average of USD 418 million annually) resulting in average annual economic impacts to the Florida economy of -7945 jobs, -USD 658 million in value added, and USD 1098 million in industry output (Court et al. 2017). Citrus bearing acreage in Florida diminished from 576 400 acres in 2005/2006 to 410 700 (-29%) in 2016/2017 while production value dropped from USD 1 491 136 in 2005/2006 to USD 1 032 227 (-30%) in 2016/2017 (USDA-NASS 2006, 2017).

2. THE HLB IN MEXICO

Mexico is the fourth largest producer of citrus in the world, with a total of 572 000 ha in production, yielding 7.8 million tons annually (Table 1).

Table 1. Areas under citrus production for Mexican states - 2015 (SIAP 2017)

State	Area (ha)	%
Veracruz	246 750	43.13
Michoacán	50 276	8.79
Tamaulipas	44 432	7.77
San Luis Potosí	37 505	6.56
Puebla	32 067	5.61
Nuevo León	31 789	5.56
Oaxaca	25 469	4.45
Colima	19 748	3.45
Yucatán	18 189	3.18
Tabasco	15 532	2.72
Sonora	8 523	1.49
Guerrero	7 135	1.25
Jalisco	6 841	1.20
Hidalgo	5 680	0.99
Campeche	4 731	0.83
Chiapas	4 725	0.83
Quintana Roo	3 089	0.54
Baja California Sur	2 870	0.50
Sinaloa	2 729	0.48
Nayarit	2 469	0.43
Morelos	610	0.11
Baja California	383	0.07
Querétaro	253	0.04
Zacatecas	246	0.04
Total	**572 051**	**100.00**

The major commercial citrus varieties grown in Mexico are orange (*Citrus sinensis*), Key lime (*Citrus aurantifolia*), Persian lime (*Citrus latifolia*), mandarin (*Citrus reticulata*) and grapefruit (*Citrus paradisi*) (Table 2). The value of this crop was estimate at USD 862 million (SIAP 2017).

HLB was first reported in samples of psyllids in July 2009 in Yucatán. In subsequent years more detections occurred in other states of the country (Table 3) (Trujillo-Arriaga 2014; SENASICA 2017a; SENASICA 2017b; SENASICA 2017c). In addition, infected psyllids were detected in Coahuila and Tamaulipas, where mechanisms of control and eradication of the disease vector have been implemented.

Table 2. Production and value of principal citrus varieties in Mexico - 2015 (SIAP 2017)

Citrus crop	Production (million tons)	Value (USD million)
Limes (Key & Persian)	2.33	461
Orange	4.52	345
Grapefruit	0.42	33
Mandarin	0.29	23
Total	**7.56**	**862**

Table 3. HLB detections in Mexico, after initial detection in the state of Yucatán in the month of July 2009 (Trujillo-Arriaga 2014; SENASICA 2017a; SENASICA 2017b; SENASICA 2017c)

State	Detections of HLB	State	Detections of HLB
Quintana Roo	August, 2009	Tabasco	December, 2012
Jalisco	December, 2009	Guerrero	March, 2013
Nayarit	December, 2009	Puebla	September, 2013
Campeche	March, 2010	Zacatecas	September, 2013
Colima	April, 2010	Coahuila	December 2013*
Veracruz	June, 2010	Oaxaca	April, 2014
Sinaloa	July 2010*	Tamaulipas	June 2014*
Michoacán	December, 2010	Querétaro	October, 2015
Morelos	December 2010*	San Luis Potosí	October, 2015
Chiapas	March, 2011	Nuevo León	December, 2015
Sonora	April 2011*	Veracruz	December, 2015
Hidalgo	April, 2011	Tamaulipas	December, 2015
Baja California Sur	April, 2011	Baja California	January, 2016
Nuevo León	August 2011*	Morelos	August, 2016
San Luis Potosí	September 2011*	Sonora	March, 2017

* *Infective psyllids*

As of September 2017, HLB is recorded to infect plant material in all 24 citrus-producing states of Mexico (Fig. 1), where it has been detected in a total of 450 municipalities (SENASICA 2017a).

In Persian lime production in Yucatán, the presence of *Ca.* L. asiaticus caused a reduction in weight of the fruit (17.3%) and a decrease in the volume of juice (18.6%) (Flores-Sánchez et al. 2015). In Key lime, experts estimated a reduction of 183 168 tons should HLB become established throughout Mexico (Salcedo-Baca et al. 2010).

3. RESPONSE TO HLB IN MEXICO

The Mexican citrus industry, federal and state plant protection agencies responded to this situation with the following area-wide phytosanitary actions to manage the HLB disease: epidemiological surveillance in commercial orchards, urban areas, and sentinel gardens; and, chemical and biological controls of the insect vector in both backyards and commercial orchards in those regions where the weather conditions favour infection.

Management of HLB is organized through Regional Control Areas (ARCOs), which implement the area-wide measures, and allow for coordination of monitoring, biological and chemical control actions across hundreds of ha; thereby preventing outbreaks from expanding to thousands of ha (SENASICA 2012).

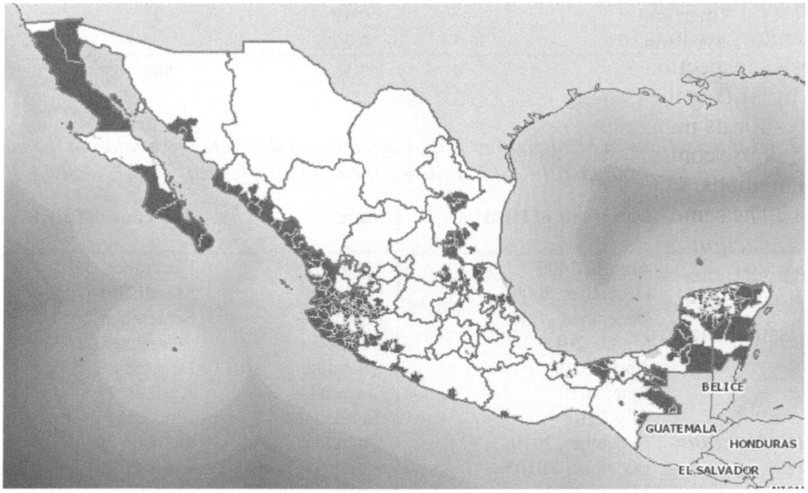

Figure 1. Areas in 24 states of Mexico with presence of HLB as of September 2017 (source SIRVEF 2019).

Since 2002, the vector of HLB has been present in Mexico, posing a significant risk for spreading HLB disease. Following the detection of HLB in Belize during 2008, the National Health, Food Safety and Food Quality Service (SENASICA in Spanish), of the General Directorate of Plant Health (DGSV in Spanish), initiated in 2008 a national priority campaign in 24 citrus states of the country. The aim of the programme was to detect the disease in a timely fashion and provide protection to more than 570 000 ha of citrus. A series of protocols were developed and revised by DGSV with input from State Plant Health Committees and the affected industry (SENASICA 2008a, 2009).

In 2009, the programme received an allocation of USD 2.56 million that allowed for monitoring the disease in 77 192 ha of commercial citrus and backyard trees (Sanchez 2013). As a result of these actions, the first detection of HLB found in infected ACP occurred in July 2009 in the town of Cuyo in the municipality of Tizimín, Yucatán (Trujillo-Arriaga 2011). In that same year and the following years, there were more detections in other states of the country (Table 3).

Due to the importance of the HLB to the Mexican citrus industry, efforts also were made at the federal and scientific levels to involve national universities and research centres to develop better management strategies, as well as better methods for the timely phytosanitary surveillance of the disease and the insect vector. To this end, SENASICA-DGSV established the legal bases for action as follows:

3.1. *Phytosanitary Regulation*

With the detection in 2009 of HLB in Tizimín, Yucatán, the federal government issued as a matter of emergency, the Mexican Official Emergency Norm NOM-EM-047-FITO-2009, which outlined and established phytosanitary actions to mitigate the risks of introduction and dispersal of HLB in Mexico (DOF 2009). It outlined the phytosanitary actions to implement a monitoring programme that included sampling, diagnosis, inspection and surveillance to assess any new introduction and further spread of HLB in the country and, where appropriate, the application of phytosanitary measures for its management. These included the delimitation of infected areas under phytosanitary control, the removal of infected material, the application of vector control methods, sampling, inspections, and restriction of the movement of vegetative material. The same actions are established in the "*Protocol of Action to the Emergency by the Detection of HLB*" and are supported under the agreement that discloses the phytosanitary measures which should be applied for the control of HLB and its vector (DOF 2010). In addition, the DGSV and the State or Regional Technical Working Group issued the Protocol to establish the ARCOs (SENASICA 2012) for HLB and ACP.

3.2. *Sampling and Monitoring to Assess HLB Distribution*

With the presence of HLB in Mexico, one of the main activities carried out in all citrus-producing states of the country was an assessment of the presence of the disease through the collection of plant material and adult *D. citri*. Priority was given to the commercial areas of Key lime, Persian lime and orange. When symptomatic plant material was found, photographs were taken and sent through the Digital Diagnostic System (SIDIADI), so the trained technical staff, based on the visual symptoms, determined whether the material was suspect of HLB infection and further sampling was required. In commercial orchards, sampling of *D. citri* was done mainly in trees located along the periphery of the orchards, where 1 to 100 adults were collected and analysed by molecular techniques for presence of HLB. Urban areas (parks, tree-lined boulevards, harbours, etc.) were also sampled.

3.3. *Phytosanitary Diagnosis*

Given the need for timely diagnoses, four official laboratories were used where plant material and psyllids could be examined: 1) the National Phytosanitary Reference Center (CNRF) in Tecámac, Mexico state, 2) the National Quarantine Station of Epidemiology and Plant Sanitation in Querétaro; 3) State Committee of Plant Protection of Yucatán, and 4) State Committee of Plant Health of Colima. In addition, eight private laboratories were approved to assist with these diagnoses. Furthermore, the DGSV has a mobile phytosanitary diagnostic unit with adequate equipment required for HLB *in situ* detections such as bioclimatic chambers, real-time PCR and End-Point PCR. This excellent infrastructure reduced the response time to a minimum and allowed speedy decision-making to apply local or regional phytosanitary strategies as described in 3.4 through 3.6 below.

3.4. Cultural Control

After an initial find of HLB positive vegetal material in a new area, the cultural control consisted of removing infected plants to avoid further spread from or possible resurgence of infection from new outbreaks. The following phytosanitary measures were implemented:

- All trees with symptoms of HLB should be eliminated within 5 days. For detections in commercial orchards, the owners were responsible for surveying all trees within their groves for the disease. Plants with symptoms were marked with a plastic tape indicating the branch(es) exhibiting symptoms. Plants located on the outer rows of orchards where the symptomatic plants were detected, were also marked with tape. Technical personnel checked each suspect orchard and confirmed whether or not the symptoms were caused by HLB. All positive cases were georeferenced. It was not considered necessary to sample these plants and technical personnel immediately removed the plants based upon a visual diagnosis. After tree removal, herbicides were applied to the stump. It is important to note that pruning cannot be used to manage HLB control and that replanting is not recommended as new plants are more susceptible to the disease (SENASICA 2010a; DOF 2010).
- In 2010 the Mexican government performed these activities in the following states with presence of HLB: Campeche, Colima, Jalisco, Michoacán, Nayarit, Quintana Roo, Sinaloa, and Yucatán. In these entities the exploration was carried out in 1553 localities, in which 1 127 275 citrus plants and 306 138 plants of lakeview jasmine (*Murraya paniculata* (L.) Jack) were inspected in search of suspicious symptoms and psyllids carrying the bacteria. Likewise, 17 539 citrus trees in orchards and 77 522 of lakeview jasmine in backyards were eliminated, as well as 5037 trees of Mexican lemon, Persian lemon and orange, in commercial orchards and 1 360 626 nursery plants (SAGARPA 2011).
- The presence of HLB in all the plants was determined by searching for symptoms and sending samples to the laboratory for diagnosis by molecular techniques, including those in the proximity of citrus nurseries. All hosts present in nurseries without anti-aphid mesh protection were removed. When a detection occurred in a backyard, it was the responsibility of trained Auxiliary Plant Protection Organization staff to identify HLB symptoms (SENASICA 2010a; DOF 2010).
- In orchards where at least 28% of plants showed HLB symptoms and where the clinical diagnosis was positive for the disease, all trees in the orchard were removed within a period of no more than five calendar days. The review of orchards in an outbreak area is being implemented permanently, with a review of such orchards done every 3 weeks to know the status of the disease (SENASICA 2010a; DOF 2010).

Currently HLB control is still based on eradicating sick trees in the states with new detections. In areas with high incidence of HLB disease, the growers have opted to implement intensive nutrition programmes to extend the productive life of the affected plantations. The strategies of the federal government campaign are directed towards the control of the ACP, through the establishment of the ARCOs, which implement the biological and chemical control activities in their respective regional areas (SENASICA 2017d).

3.5. Biological Control of the Vector

Augmentative biological control is a strategy that plays an important role in reducing the population density of *D. citri*, and its area-wide use in ARCOs has significantly contributed to reducing adverse environmental effects and minimizing interference with natural control of agricultural pests as a result of using agrochemicals (DGSV 2016).

In response to the detection of HLB and its vector *D. citri* in Mexico and to mitigate its threat for citriculture, SENASICA-DGSV and the CNRF established a *Biological Control Programme for Asian Citrus Psyllid*, as a complementary strategy to the integrated management of the HLB vector. The main activities in this programme have been:

- Search and selection of biological control agents of *D. citri*
- Mass-production of the species-specific ectoparasitoid *Tamarixia radiata* Waterston (Hymenoptera: Eulophidae)
- The release of adult parasitoids in specific areas not subject to insecticide application (such as urban areas, inaccessible areas, organic orchards, abandoned vegetable gardens, backyard host trees, orchards adjacent to urban areas, areas under integrated pest management, and protected natural areas or reserves)
- Assessment of the effectiveness of biological control agents in the laboratory and in the field
- Training and public education on the recognition and use of biological control agents of ACP
- Advice on the design of rearing facilities for *T. radiata*
- Optimization of the mass-production process of the parasitoid
- Research on different strains of *T. radiata* present in Mexico and their regional impact
- Research on the use of strains of entomopathogenic fungi, and finally
- Obtaining national and international support on parasitoid rearing (FAO-SENASICA 2013).

Therefore, the DGSV through the CNRF, established collaborative agreements with the State Committees of Plant Protection in the states of Colima and Yucatán in 2009. The first agreement with the Entomophagous Insects Department of the National Reference Center for Biological Control (CNRCB), based in Tecomán, Colima, aims to generate technology (basic and applied) for the use of biological control agents of *D. citri*, as well as the production of *T. radiata* in Tecomán, Colima; while the second agreement with the Regional Mass-Rearing Laboratory of *Tamarixia radiata* in the Southeast has as its sole objective the mass-production of the parasitoid in Merida, Yucatán; both agreements are coordinated by the CNRCB (FAO-SENASICA 2013).

From 2010 to 2015, 31 million *T. radiata* were produced and released in citrus orchards and abandoned, urban and backyard areas of Yucatán, Quintana Roo, Campeche, Tabasco, Chiapas, Oaxaca, Hidalgo, and Guerrero. These release activities were supported by the State Plant Health Committees, who are responsible for transporting and release of parasitoids to the infested areas (CNRCB-CNRF-DGSV 2016). Many of these areas began with a parasitism rate ranging between 3-26% that increased to 70-85% after augmentative releases (SENASICA 2016).

The recommended release rate is 100 parasitoids every 50-100 linear meters depending on the level of ACP infestation and density of host plants. If on average more than 20 *D. citri* nymphs were observed per tree shoot, 100 parasitoids were released every 50 meters (SENASICA 2015). The releases are carried out with a minimum interval of one month and a maximum of 3 months (DGSV 2016). These releases directly and indirectly benefit hundreds of growers.

With respect to the use of entomopathogenic fungi as a complement to the control of *D. citri* populations, the following research activities were established in support of the Biological Control Programme of the ACP:

- Exploration of entomopathogenic fungi
- Selection of isolates of entomopathogenic fungi candidates for the control of immature and adult stages of *D. citri*
- Evaluation of conidia production
- Evaluation of types of entomopathogenic fungi formulation
- Evaluation of fungal formulations in the field
- Evaluation of application equipment; and
- Biosafety tests (FAO-SENASICA 2013).

By 2016 three strains of *Isaria javanica* (formerly *fumosorosea*), CHE-CNRCB 303, 305 and 307, formerly Pf15, Pf17 and Pf21, respectively, as well as one of *Metarhizium anisopliae* (CHE-CNRCB 224, formerly Ma59), had been identified. Laboratory tests achieved 93-100% mortality in nymphs and up to 95% in adults of *D. citri*, respectively. Applications of entomopathogenic fungi in preliminary field trials reduced psyllid populations from 48 to 90% (FAO-SENASICA 2013). In 2012 and 2013, the application of two strains of entomopathogenic fungi (Ma59 and Pf21) was carried out on 15 932 ha in the states of Colima, Hidalgo, Jalisco, Nayarit, San Luis Potosí and Veracruz (CNRCB-CNRF-DGSV 2016).

3.6. Chemical Control

In other countries, HLB disease has been managed mainly through the suppression of vector populations using synthetic insecticides. In Mexico, only the use of insecticides authorized by COFEPRIS (DOF 2010) are recommended in accordance with the use of products that have been approved and shown to be efficacious in other countries. Cortéz-Mondaca et al. (2010) conducted tests of the effectiveness of conventional synthetic and organic insecticides with different modes of action, including botanical extracts, mineral oils, soaps, entomopathogenic detergents and growth regulators. Based on these results, the National Campaign against HLB has been rotating the use of the following active ingredients according to the specific local situations: thiamethoxam, imidacloprid + beta cyfluthrin, mineral oil, bifenthrin, tricarboxyls, chlorpyrifos, imidacloprid, dimethoate, thiamethoxam + lambda cyhalothrin, bifenthrin + zeta cypermethrin, azadirachtin, bifenthrin + abamectin, detergent and lime oxide (SENASICA 2012).

In Mexico, chemical control is performed by the ARCOs. The application time is determined by the population dynamics of *D. citri* and the phenology of the citrus in each region. The spraying is done 2-3 times in all the orchards that are part of an ARCO within a two-week period of time. The State Plant Health Committees responsible for HLB disease management oversee the actions taken by ARCOs and inform growers about the timing of annual applications, and the overall pesticide management programme. ARCOs are responsible for any pesticide applications (DGSV 2016). In 2015 and 2016, 216 566 and 273 318 commercial ha of citrus were sprayed, respectively (SENASICA 2017b).

3.7. Vegetative Material

The Mexican Official Standard NOM-079-FITO-2002, "Phytosanitary Specifications for the Production and Mobilization of Propagation Material Free of Citrus Tristeza Virus and Other Pathogens Associated with Citrus", and the agreement that discloses the phytosanitary measures to be applied for the control of HLB and its vector (DOF 2002, 2010; SENASICA 2010a) establishes requirements for propagation and certification of citrus nursery stock through Certified Production Units as a means of providing disease-free trees for commercial sale and planting purposes.

3.8. Training and Outreach

Two international workshops on ACP and Citrus Huanglongbing were held in Mexico during 2008 and 2010, respectively, with the objective of providing training on disease diagnostics to technical staff in charge of field monitoring and sampling. Experts were invited from infected countries such as Belize, Brazil, China, Cuba and the USA, to share their experiences on management and disease prevention practices (SENASICA 2008b, 2010b). In addition, three events on quarantine pests of citrus were organized in 2009, 2011, and 2013 (SENASICA 2013).

In order to create greater awareness among the general public about HLB and the risks posed, an additional 34 667 training events were organized between 2008 and 2011 in different regions of the country. Attendees were encouraged to participate under the motto "All against HLB of citrus and its vector".

In addition, billboards along roadways and avenues in rural villages and information disseminated using printed triptychs, posters, flyers, technical files, postcards, radio spots and videos were used to educate all involved. These materials invited the public to be on the alert for symptoms of HLB in their commercial and backyard orchards, and to immediately report HLB symptoms to the local plant health boards in their region (Trujillo-Arriaga 2011).

4. HLB IMPACT ON CITRUS PRODUCTION FROM 2009 TO 2015

In 2009, the year when HLB was first detected in Mexico, the area planted with lime, orange, mandarin and grapefruit comprised 523 321 ha yielding a total of 6.82 million tons of fruit (SIAP 2017). That same year, SENASICA and the Inter-American Institute for Cooperation on Agriculture (IICA) commissioned a study to estimate the expected impact of HLB in Mexico (Salcedo-Baca et al. 2010).

Various scenarios were developed to assess potential impacts to the industry and economy with and without any governmental intervention. Under the low impact scenario, the study estimated that within five years following the establishment of the disease a loss of 2.7 million tons of citrus fruit would occur nationwide with an overall reduction of 39.6% in orange, 33% in grapefruit, 17% in mandarin, and 10% in lime production respectively. Under the high impact scenario, losses would increase to 3 million tons of fruit equivalent to 41% of orange, 53% of grapefruit, 26% of mandarin, and 18% of lime production.

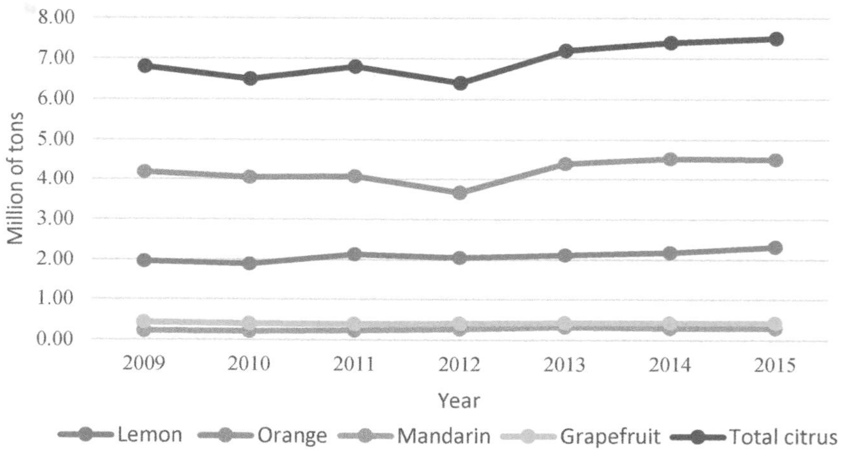

Figure 2. National citrus production in Mexico after HLB arrival in 2009 (SIAP 2017).

Six years after HLB was first detected in Mexico, however, with governmental intervention, the estimated losses in citrus production have yet to occur. Although citrus production in 2010 and 2012 decreased slightly, annual yields have remained fairly stable and have slightly increased by 0.7 million tons above the 2009 level (Fig. 2 and Table 4).

In 2016, the Mexican government allocated approximately USD 8.5 million to the HLB Programme. As a result of the successful application of the area-wide actions outlined above, Mexico largely mitigated the adverse effects of the disease, while at the same time slightly increasing citrus production. In addition, research programmes have been promoted among scientific institutions to generate propagative material tolerant or resistant to the disease.

Table 4. State-by-state comparison of citrus production in Mexico - 2009 and 2015 (SIAP 2017)

State	Production (tons) 2009				Production (tons) 2015			
	Orange	Lime	Mandarin	Grapefruit	Orange	Lime	Mandarin	Grapefruit
Aguascalientes	23	0	0	0	14	46	0	0
Baja California	3, 499	2,978	192	96	3,260	1,171	0	183
Baja California Sur	29, 303	144	0	66	49,995	309	22	495
Campeche	37, 018	5,953	215	15,014	25,122	10,325	332	18,955
Chiapas	15,983	4,185	452	0	16,027	7416	404	0
Colima	3,977	423,040	0	0	4,496	191890	0	158
Durango	804	856	0	612	785	702	0	577
Guanajuato	0	3	0	0	0	12	0	0
Guerrero	4,189	78,404	230	66	4,435	71,867	138	9
Hidalgo	45,481	2,589	262	0	59,041	2,231	239	0
Jalisco	6,779	30,351	6	1,310	6,386	81,198	0	740
México	327	1,227	0	0	234	1,190	39	0
Michoacán	4,374	414,562	75	59,559	3,489	670,613	0	49,566
Morelos	4,430	3,660	82	160	4,857	4,438	190	144
Nayarit	1,576	14,093	41	5	712	18,423	19	0
Nuevo Leon	296,973	0	35,892	17,734	313,439	0	45,751	26,201
Oaxaca	60,626	176,182	0	5,440	56,290	245,137	0	1,161
Puebla	254,841	48,352	680	11,160	214,175	28,211	51,746	5,067
Querétaro	1,345	47	24	0	2,446	26	5	0
Quintana Roo	32,289	1,107	32	0	16,841	26,222	112	0
San Luis Potosí	431,567	8,599	20,358	55	337,717	21,986	22,212	62
Sinaloa	16,970	2,173	1,420	5,195	19,320	1,648	446	2,887
Sonora	167,371	988	581	17,500	142,445	1,564	3,059	16,689
Tabasco	81,519	80,939	380	850	81,451	83,141	387	845
Tamaulipas	539,526	46,411	54,235	29,083	668,935	121,200	11,618	46,544
Veracruz	2,058,040	514,728	102,046	256,064	2,336,427	659,034	147,345	248,927
Yucatán	94,534	104,777	6,441	11,704	147,107	74,463	6,885	5,106
Zacatecas	128	0	72	0	74	1,605	130	0
Total	**4,193,484**	**1,966,345**	**223,718**	**431,671**	**4,515,520**	**2,326,068**	**291,078**	**424,315**

Although the government has successfully implemented area-wide strategies for regional control, it is necessary to further develop new and improved technologies. Mexico, Belize and USA have formed a Tri-National Working Group for purposes of technical exchange, information sharing, planning, coordination, and identification of research priorities.

Mexico currently has 573 406 ha of citrus corresponding to 335 019 ha of oranges, 180 209 ha of limes (Key and Persian), 21 297 ha of mandarin, 17 590 ha of grapefruit, 12 736 ha of tangerine, 5238 has of tangelo and 1317 ha of sweet limes, with an estimated annual production of 8 million tons per year, and whose production value is approximately USD 20 424 million pesos (approximately USD 1.12 billion dollars) (SIAP 2017).

The first phytosanitary actions implemented on an area-wide basis were: timely detection of HLB in citrus and urban areas, elimination of infected plants in areas under control, suppression of the *D. citri* vector, and protection of propagative material within enclosed or screened nurseries.

5. CONCLUSION

In conclusion, as a result of the successful HLB Programme, citrus losses have been largely avoided in Mexico despite the fact that the disease has now been detected in all states of the country where commercial citrus is produced. Six years after the disease first appeared, the surface for the four principal citrus varieties actually increased by 5%, from 523 321 to 553 671 ha (Fig. 3) (SIAP 2017).

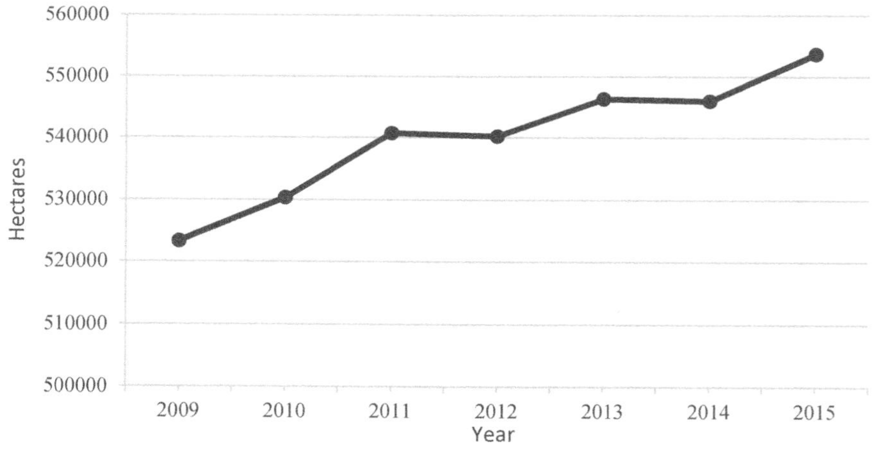

Figure 3. Expanding national areas of the four principal Citrus *spp. produced in Mexico after the HLB arrival in 2009 (SIAP 2017).*

6. REFERENCES

Alemán, J., H. Baños, and J. Ravelo. 2007. *Diaphorina citri* y la enfermedad Huanglongbing: Una combinación destructiva para la producción citrícola. Revista de Protección Vegetal 22: 154–165.

Bové, J. M. 2006. Huanglongbing: A destructive, newly-emerging, century-old disease of citrus. Journal of Plant Pathology 88: 7–37.

Bové, J. M. 2012. Huanglongbing and the future of citrus in Sao Paulo state, Brazil. Journal of Plant Pathology 94: 465–467.

(CNRCB-CNRF-DGSV) Centro Nacional de Referencia de Control Biológico-Centro Nacional de Referencia Fitosanitaria-Dirección General de Sanidad Vegetal. 2016. El control biológico como parte de las campañas fitosanitarias y áreas regionales de control (ARCOs).

Collazo, M. L. C., R. Llauger, E. Blanco., I. Peña, D. López, C. González, J. C. Casín, L. Batista, E. Kitajima, F. A. O. Tanaka, R. B. Salaroli, D. C. Teixeira, E. C. Martins, and J. M. Bové. 2009. Occurrence of citrus huanglongbing in Cuba and association of the disease with *Candidatus* Liberibacter asiaticus. Journal of Plant Pathology 91: 709–712.

Cortéz-Mondaca, E., J. I. López-Arroyo, L. M. Hernández-Fuentes, A. Fú-Castillo, and J. Loera-Gallardo. 2010. Control químico de *Diaphorina citri* Kuwayama en cítricos dulces en México: Selección de insecticidas y épocas de aplicación. Campo Experimental Valle del Fuerte, Los Mochis, Sinaloa. Folleto Técnico Núm. 35, INIFAP, Mexico. 22 pp.

Court, C. D., A. W. Hodges, M. Rahmani, and T. H. Spreen. 2017. Economic contributions of the Florida citrus industry in 2015-16. University of Florida. Institute for Food and Agricultural Sciences. Gainesville, Florida, USA. 35 pp.

da Graça, J. V., and L. Korsten. 2004. Citrus huanglongbing: Review, present status and future strategies. Diseases of Fruits and Vegetables 1: 229–245.

(DGSV) Dirección General de Sanidad Vegetal. 2016. Manual operativo de la campaña contra el Huanglongbing de los cítricos. Versión 1. México, D. F., México.

(DOF) Diario Oficial de la Federación. 2002. NORMA Oficial Mexicana NOM-079-FITO-2002. Requisitos fitosanitarios para la producción y movilización de material propagativo libre de virus tristeza y otros patógenos asociados a cítricos. México, D. F., México.

(DOF) Diario Oficial de la Federación. 2009. NORMA Oficial Mexicana de Emergencia NOM-EM-047-FITO-2009, Por la que se establecen las acciones fitosanitarias para mitigar el riesgo de introducción y dispersión del Huanglongbing (HLB) de los cítricos (*Candidatus* Liberibacter spp.) en el territorio nacional. México, D. F., México.

(DOF) Diario Oficial de la Federación. 2010. ACUERDO por el que se dan a conocer las medidas fitosanitarias que deberán aplicarse para el control del Huanglongbing (*Candidatus* Liberibacter spp.) y su vector. México, D. F., México.

(FAO-SENASICA) Organización de las Naciones Unidas para la Alimentación y la Agricultura – Servicio Nacional de Sanidad, Inocuidad y Calidad Agroalimentaria. 2013. Taller Subregional de Control Biológico de *Diaphorina citri*, Vector del HLB. Panamá, Panamá. 65 pp.

Flores-Sánchez, J. L., G. Mora-Aguilera, E. Loeza-Kuk, J. I. López-Arroyo, S. Domínguez-Monge, G. Acevedo-Sánchez, and P. Robles-García. 2015. Pérdidas en producción inducidas por *Candidatus* Liberibacter asiaticus en limón persa en Yucatán, México. Revista Mexicana de Fitopatología 33: 195–210.

Gottwald, T. R., C. I. Gonzales, and B. G. Mercado. 1991. Analysis of the distribution of citrus greening in groves in the Philippines, pp. 414-420. *In* R. H. Brlansky, R. F. Lee, and L. W. Timmer (eds.), Proceedings 11th Conference of the International Organization of Citrus Virologists (IOCV), Riverside, California, USA.

Gottwald, T. R., J. H. Graham, M. D. S. Irey, T. G. McCollum, and B. W. Wood. 2012. Inconsequential effect of nutritional treatments on huanglongbing control, fruit quality, bacterial titer and disease progress. Crop Protection 36: 73–82.

Halbert, S. E., and K. L. Manjunath. 2004. Asian citrus psyllids (Sternorrhyncha: Psyllidae) and greening disease of citrus: A literature review and assessment of risk in Florida. Florida Entomologist 87: 330–352.

Keremane, M. L., C. Ramadugu, A. Castaneda, J. E. Diaz, E. A. Peñaranda, J. Chen, Y. P. Duan, S. E. Halbert, and R. F. Lee. 2015. Report of *Candidatus* Liberibacter caribbeanus, a new citrus- and psyllid-associated Liberibacter from Colombia, South America. Oral Technical Session 101-O.

Manjunath, K. L., S. E. Halbert, C. Ramadugu, S. Webb, and R. F. Lee. 2008. Detection of *Candidatus* Liberibacter asiaticus in *Diaphorina citri* and its importance in the management of citrus huanglongbing in Florida. Phytopathology 98: 387–396.

(NAPPO) North American Plant Protection Organization. 2005. New Federal Restrictions to Prevent Movement of Citrus Greening. Official Pest Reports. Ottawa, Canada.

Salcedo-Baca, D., R. A. Hinojosa, G. Mora-Aguilera, I. Covarrubias-Gutiérrez, F. J. R DePaolis, C. L Cíntora-González, and J. S. Mora-Flores. 2010. Evaluación del impacto económico de Huanglongbing (HLB) en la cadena citrícola Mexicana. Instituto Interamericano de Cooperación para la Agricultura (IICA). México, D. F., México. 141 pp.

Sanchez, H. M. 2013. Vigilancia estratégica para la detección de HLB. 3er Taller Internacional sobre Plagas Cuarentenarias de los Cítricos. Del 27 al 30 de agosto de 2013. Santiago, Manzanillo, México.

Schwarz, R. E., M. L. C. Knorr, and M. Prommintara. 1973. Presence of citrus greening and its psylla vector in Thailand. FAO Plant Protection Bulletin 21: 132–138.

(SAGARPA) Secretaría de Agricultura, Ganadería, Desarrollo Rural, Pesca y Alimentación. 2011. Boletín Informativo: Invierte SAGARPA más de 268 millones de pesos en la lucha contra el HLB. Delegación en el Estado de Guanjuato. Comunicación Social. 15 de febrero de 2011.

(SENASICA) Servicio Nacional de Sanidad, Inocuidad y Calidad Agroalimentaria. 2008a. Huanglongbing de los cítricos. Protocolos de actuación. México, D. F., México.

(SENASICA) Servicio Nacional de Sanidad, Inocuidad y Calidad Agroalimentaria. 2008b. 1° Taller internacional sobre el Huanglongbing de los cítricos (*Candidatus* Liberibacter spp.) y el psílido asiático de los cítricos (*Diaphorina citri*). Del 7 al 9 de mayo del 2008. Hermosillo, Sonora. México.

(SENASICA) Servicio Nacional de Sanidad, Inocuidad y Calidad Agroalimentaria. 2009. "Protocolo de Actuación para la Detección de Huanglongbing". México, D. F., México.

(SENASICA) Servicio Nacional de Sanidad, Inocuidad y Calidad Agroalimentaria. 2010a. Protocolo de actuación ante la emergencia por la detección del Huanglongbing. México, D. F., México.

(SENASICA) Servicio Nacional de Sanidad, Inocuidad y Calidad Agroalimentaria. 2010b. 2° Taller Internacional sobre el Huanglongbing y el Psílido Asiático de los Cítricos. Del 19 al 23 de julio del 2010. Mérida, Yucatán, México.

(SENASICA) Servicio Nacional de Sanidad, Inocuidad y Calidad Agroalimentaria. 2012. Protocolo para establecer áreas regionales de control (ARCOs) del Huanglongbing y el psílido asiático de los cítricos. Dirección de Protección Fitosanitaria. México, D. F., México.

(SENASICA) Servicio Nacional de Sanidad, Inocuidad y Calidad Agroalimentaria. 2013. 3er Taller Internacional sobre Plagas Cuarentenarias de los Cítricos. Del 27 al 30 de agosto de 2013. Santiago, Manzanillo, México.

(SENASICA) Servicio Nacional de Sanidad, Inocuidad y Calidad Agroalimentaria. 2015. Manual de reproducción masiva de *Tamarixia radiata*. Principal parasitoide del psílido asiático de los cítricos, vector del HLB. Centro Nacional de Referencia de Control Biológico. México, D. F., México.

(SENASICA) Servicio Nacional de Sanidad, Inocuidad y Calidad Agroalimentaria. 2016. Laboratorio de Reproducción Masiva de *Tamarixia radiata* del Sureste. Mérida, Yucatán. 13 pp.

(SENASICA) Servicio Nacional de Sanidad, Inocuidad y Calidad Agroalimentaria. 2017a. Huanglongbing de los cítricos. Informes y Evaluaciones 2017. México, D. F., México.

(SENASICA) Servicio Nacional de Sanidad, Inocuidad y Calidad Agroalimentaria. 2017b. Campañas y programas fitosanitarios - Huanglongbing de los cítricos. México, D. F., México.

(SENASICA) Servicio Nacional de Sanidad, Inocuidad y Calidad Agroalimentaria. 2017c. Huanglongbing de los cítricos. Informes y Evaluaciones 2016. México, D. F., México.

(SENASICA) Servicio Nacional de Sanidad, Inocuidad y Calidad Agroalimentaria. 2017d. Manual operativo de la campaña contra el Huanglongbing de los cítricos. Dirección de Protección Fitosanitaria-DGSV. México, D. F., México.39 pp.

(SIAP) Servicio de Información Agroalimentaria y Pesquera. 2017. Anuario Estadístico de la Producción Agrícola. México, D. F., México.

Singerman, A., and B. Page. 2016. What is the economic benefit of a Citrus Health Management Area (CHMA)? A case study. Document FE982. February 2016. University of Florida. Institute for Food and Agricultural Sciences. Gainesville, Florida, USA. 3 pp.

Singerman, A., and M. Burani-Arouca. 2017. Evolution of citrus disease management programs and their economic implications: The case of Florida's citrus industry. Document FE915. Revised January 2017. University of Florida. Institute for Food and Agricultural Sciences. Gainesville, Florida, USA. 5 pp.

(SIRVEF) Sistema Integral de Referencia para Vigilancia Epidemiológica Fitosanitaria. 2019. Dirección General de Sanidad Vegetal, México.

Trujillo-Arriaga, F. J. 2011. HLB en México. Situación actual, regulación y perspectivas de áreas regionales para el manejo del HLB. 3er Encuentro Internacional de Investigación en Cítricos. Septiembre de 2011. Martínez de la Torre, Veracruz, México.

Trujillo-Arriaga, F. J. 2014. Reporte de México. 38ª Reunión Anual de la NAPPO. Del 20 al 24 de octubre. Huatulco, Oaxaca, México. En línea

(USDA-NASS) United States Department of Agriculture-National Agricultural Statistical Service. 2006. Citrus Fruits 2006 Summary. Washington, DC, USA. 49 pp.

(USDA-NASS) United States Department of Agriculture-National Agricultural Statistical Service. 2017. Citrus Fruits 2017 Summary. Washington, DC, USA. 34 pp.

Vojnov, A. A., A. M. do Amaral, J. M. Dow, A. P. Castagnaro, and M. R. Marano. 2010. Bacteria causing important diseases of citrus utilize distinct modes of pathogenesis to attack a common host. Applied Microbiology and Biotechnology 87: 467–477.

Wang, N., L., Wenbin, M. Irey, G. Aibrigo, K. Bo, and J. Kim. 2009. Citrus huanglongbing. Tree and Forestry Science and Biotechnology 3(Special Issue 2): 66–72.

TECHNOLOGY USED BY FIELD MANAGERS FOR PINK BOLLWORM ERADICATION WITH ITS SUCCESSFUL OUTCOME IN THE UNITED STATES AND MEXICO

R. T. STATEN[1] AND M. L. WALTERS[2]

[1]Center for Plant Health Science and Technology (CPHST), Animal and Plant Health Inspection Service (APHIS), United States Department of Agriculture (USDA), Phoenix, Arizona, USA
Retired, Programme Consultant; azbugdoc@cox.net
[2]CPHST-USDA-APHIS, Deceased

SUMMARY

Pink bollworm, *Pectinophora gossypiella* (Saunders), has been eradicated over a 7-state area in northern Mexico and the southern USA. Over this region, pink bollworm has been a key pest of cotton for 50+ years. The bi-national eradication programme grew out of a long-standing Sterile Insect Technique (SIT) containment/exclusion programme to protect cotton in the San Joaquin Valley of California, as well as numerous area-wide research and demonstration projects in southern California, Baja California, and Arizona. It included all contiguous infested production areas of the states of Chihuahua, Sonora, and Baja California in Mexico. It also included all contiguous generally infested areas of the states of Texas, New Mexico, Arizona, and California in the USA. In this chapter we provide descriptions and key references for the technologies that were integrated in this multi-tactic, area-wide programme over its extensive geographic range. Technology described and used includes state programme-based central data management. The programme covered all activities including extensive GPS mapping, pheromone trap monitoring for adult populations, and the integration of all control operations. Operational information and data were shared among all participants as needed. Control tools included *Bt*-cotton, the release of sterile moths, pheromone mating disruption, cultural control, and on a very limited basis conventional insecticide application. Critical area-wide resistance management using sterile moth release, rather than planting susceptible cotton in refugia, was pioneered in this programme. Success as documented was possible over an enormous and diverse cotton production area because the technologies used were heavily researched, broad-based, and could be tailored to fit each major area. Uniform management within each state was coordinated bi-nationally. This programme was conducted sequentially over time. Summaries for each state provide measurements of progress, success, and experiences gained through time of operation.

Key Words: Pectinophora gossypiella, pink bollworm trap, area-wide management, integrated pest management, *Bt*-cotton, resistance management, gossyplure, mating disruption, Sterile Insect Technique, SIT, pest detection survey, okra

J. Hendrichs, R. Pereira and M. J. B. Vreysen (eds.), Area-Wide Integrated Pest Management: Development and Field Application, pp. 51–92. CRC Press, Boca Raton, Florida, USA.

1. INTRODUCTION

Introduction of the pink bollworm (PBW), *Pectinophora gossypiella* (Saunders), to the south-western USA and north-western Mexico irrigated cotton growing regions was first reported in 1916 near Torreón, in Coahuila, Mexico (Noble 1969). This infestation was presumed to originate from seed shipments from Egypt into Mexico in 1911. It quickly became the key pest of cotton.

With the exception of the San Joaquin Valley of California, no other pest species was as dominant or detrimental to the fortunes and survivability of the cotton farmers of these regions. This was particularly true in the cotton growing areas of the Colorado River Basin. Insecticide use for this pest was extensive for a period of more than 45 years.

Early attempts at management or eradication with conventional insecticides were expensive and difficult. All control efforts resulted in some short-term success and frequent frustration. This was true both when control was on a field-by-field basis and also in the case of a coordinated, insecticide-driven state-wide programme in Arizona (Anonymous 1961; Schmitt Jr. 1967).

The emphasis of this chapter will be to record the technology used and the success of the bi-national Pink Bollworm Eradication Programme to assist others in the development of bio-rational approaches to the pest. We expect such knowledge will be critical in mitigating any seed-borne movement of PBW back to the USA and northern Mexico. The programme herein described evolved only after in-depth research (over 3000 references in a CPHST-APHIS-USDA data base and Naranjo et al. 2002) and numerous large-scale field trials. All this investment in R&D is frequently oversimplified and overlooked when the positive results of the programme are considered.

The programme operations started on a sequential basis in the generally infested areas of the south-western USA and northern Mexico (Fig. 1). This "rolling carpet" approach (Hendrichs et al. 2021), followed as programme phases, was necessary due to the physical limitations of the production output of sterile PBW moths by a single mass-rearing facility located in Phoenix, Arizona.

In this summation we identify the most important tools commonly used by the programme and provide relevant references and accounts of experience used in the design of the programme. Important summary data utilized to measure programme progress are also provided. Units of measurement were those used in the respective countries (metric system in Mexico, "U.S. customary unit system" in the USA). The data came from the records used in day-to-day management and are provided in Sections on a state-by-state basis in both countries.

Each of these Sections starts with a listing of the state managers who were responsible for day-to-day operations and ultimately were the backbone of the programme's achievement. The success of this programme grew out of a long-standing containment/exclusion programme in the San Joaquin Valley of California (Staten et al. 1993) and numerous area-wide research and demonstration projects in geographically-defined locations particularly in southern California, Baja California, and Arizona (Walters et al. 2000). Many of these trials were reported through proceedings of the National Cotton Council Beltwide Cotton Production Conferences.

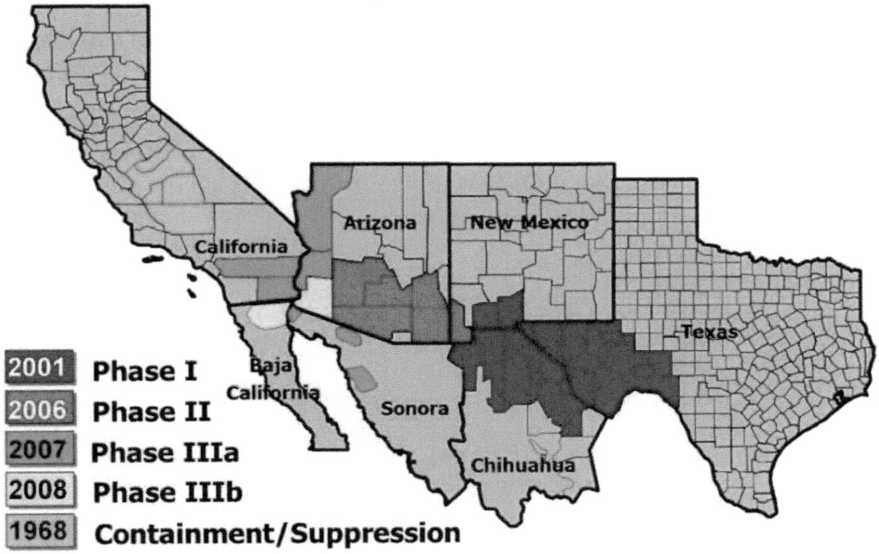

Figure 1. Pink bollworm eradication phases, dates, and areas in south-western USA and north-western Mexico (updated from Grefenstette et al. 2009).

2. MANAGEMENT

Standardized management and organization are critical in integrated area-wide insect pest eradication programmes (Vreysen et al. 2007; Suckling et al. 2014). As this was not a "voluntary" programme, all in-season applications of suppression treatments (insecticides, sterile insect release, and pheromones) for PBW were under a central management and coordinating authority in each state. This area-wide concept of the bi-national programme had to be agreed upon by the majority of all cotton growers in referenda held before the programme could be initiated in each of the states. In Arizona, for example, the programme could only be started after a second grower referendum passed in 2005 with more than a 66 % grower approval (Grefenstette et al. 2009).

A brief outline of management entities in the USA and Mexico involved in the PBW eradication programme are shown in Tables 1 and 2.

Table 1. Brief outline of management entities involved in the USA and their contributions to the pink bollworm eradication programme

ENTITIES IN USA	CONTRIBUTIONS
USDA-APHIS	All sterile insect production, USA release cost, and USA regulatory enforcement
The producer communities:	Within-state cost of all non-SIT[2] in-field treatments and operations (includes *Bt*-cotton, pheromone mating disruption, and insecticides)
1. Texas Boll Weevil Foundation (TBWF)[1]	All field management of treatments, monitoring, evaluation and reporting
2. New Mexico PBW and BW Foundation[1]	All field management of treatments, monitoring, evaluation and reporting
3. Arizona Cotton Research and Protection Council (ACRPC)[1]	All field management of treatments, monitoring, evaluation and reporting
4. California Cotton Pest Control Board (CCPCB), funds managed by CDFA[1]	All field management of treatments, monitoring, evaluation and reporting

[1] All funds were raised via local assessments through organizations 1-4, and from legislative support to USDA via the National Cotton Council
[2] SIT = Sterile Insect Technique

3. TECHNOLOGY USED

The successful PBW eradication was dependent on a multi-tactic approach in which the authors will not designate one control technology as most critical for its success. All technologies, integrated in different ways in the different areas, were essential and born from in-depth research and development efforts over a 100-year time frame (.

This programme was fortunate in following an ongoing Boll Weevil Eradication Programme (Knipling 1971; Allen 2008), which established the benefits of a standardized area-wide approach to programme success. Lessons learned and organizational basics were of extreme importance. El-Lissy et al. (2002) used a simple classification for all programme activities. This paper was the foundation for all field operations used by state organizations for PBW.

All activities were sub-divided into three activities: 1. mapping and data management, 2. surveying (trapping and larval sampling), and 3. control. The authors will generally follow these three critical components, adding detail to each to fully elucidate their scope and interaction. It should be noted that a separate Section is included dedicated to transgenic cotton. Despite being an essential part of the programme's control components, unique issues related to the incorporation of transgenic cotton require additional discussion.

Table 2. Brief outline of management entities involved in Mexico and their contributions to the pink bollworm eradication programme

ENTITIES IN MEXICO	CONTRIBUTIONS
SAGARPA (Ministry of Agriculture, Livestock, Rural Development, Fisheries and Food), SENASICA (National Service of Health, Food Safety, and Agriculture Quality)	Leadership, Technical and managerial support, critical funds (varied year to year dependent on needs and availability at national level)
USDA-APHIS-International Services and Plant Protection and Quarantine (PPQ)	Technical and information technology support, logistical support, bi-national coordination, coordination with USA embassy for security, procurement of some supplies, and some field personnel and SIT[2] coordination
1. Comité Estatal de Sanidad Vegetal (state plant protection committee) de Chihuahua[1]	State level management of operations (treatment, survey, and control), funding via grower assessments and direct contributions
2. Comité Estatal de Sanidad Vegetal (state plant protection committee) de Sonora[1]	State level management of operations (treatment, survey, and control), funding via grower assessments and direct contributions
3. Comité Estatal de Sanidad Vegetal (state plant protection committee) de Baja California[1]	State level management of operations (treatment, survey, and control), funding via grower assessments and direct contributions

[1] *Authors were not involved in funding decisions, but they understand that sources of revenue varied according to available resources from the various entities in the different states*

3.1. Mapping and Data Management

Eradication requires complete control of needed technology over broad or well-defined geographic areas, over which uniform management is of paramount importance. This starts at the beginning of each growing season with the process of finding and mapping of all fields with cotton. Each state's programme had to be able to monitor all these fields. Managers required rapid access to all mapping, survey and treatment data. PBW populations are clustered and non-uniform in distribution within a field and within subsites and definable geographic areas. Management must be able to operate within a spatial context allocating resources where they are most critically needed without regard to ownership or political constraints.

When Texas initiated the first surveys of PBW populations, it modified its boll weevil data management system to include all PBW management needs. This was then made available to all participating states. Details of how this data management system is used today can be found under TBWEF (2019). It was adopted by all state programmes except California, which already had a long-standing data management system in place.

Within the above context each state management had complete access to its mapping and data including the following:

1. Precise GPS locations of all fields with unique identification numbers for every field and its trap or traps

2. Barcoded identification of all traps with GPS location within the programme

3. Storage and access to all trap and capture data for sterile and non-sterile specimens

4. Precise location of all *Bt* and non-*Bt* cotton (*Gossypium hirsutum* L.) fields, including a distinction for Pima cotton, *Gossypium barbadense* L.

5. Access to detailed information on all programme-applied pheromone mating disruption treatments, conventional insecticides, and sterile moth releases – this included access to needed regulatory notifications within each state and flight recordings for all spray and sterile release aircraft, and

6. Reports generated from complete data by servicing date or any other needed time interval and geographically-defined parameter.

The use of this harmonized system expedited communication within and between state programmes.

3.2. Survey Technology

3.2.1. Trap Selection and Use

The eradication of PBW has long relied on the use of the delta trap for surveys. This trap and the modified Frick trap it replaced are fully illustrated by Foster et al. (1977). The trap is deltoid in shape and is 7 inches (17.8 cm) long and 3.5 inches (8.9 cm) on each of its three sides. The inverted triangular opening found on each end is one inch (2.54 cm) on each of the three sides that form the opening.

The delta trap was first used on an area-wide basis in the San Joaquin valley of California in 1976. Staten et al. (1993) reported on multiple years of data from very large numbers of this trap in over one million acres (> 400 000 ha) of cotton each year. The delta trap will overload in high PBW populations when traps quickly exceed 50 moths per service interval. It is however the most sensitive trap known to the authors in detection and monitoring of lower density PBW populations. The most important need in this programme was its ability to successfully find low level populations before they reached levels only allowed in pre-eradication pest management scenarios.

The following are four key requirements of traps needed for operational success:

1. Superiority as a detection tool with the best capture rate in low population densities

2. The trap must facilitate accurate identification, preserving the specimen intact enough for dissection and/or, with special servicing, DNA analysis

3. The trap must be durable enough to withstand "normal" extremes such as wind, rain, handling and routine field and operational hazards, and

4. The trap must be cost- and operationally-effective to use. This includes unit cost, storage, installation, recovery, and replacement.

Throughout the PBW programme, trap density standards were set at one trap per 80 acres (32.4 ha) in the USA and one trap per 20 ha in Mexico for all *Bt*-cotton (cotton genetically modified to express the endotoxins of *Bacillus thuringiensis* Berliner) (*Bt*). All cotton fields which did not express these resistant traits were trapped at one trap per 10 acres (4.05 ha) in the USA or one trap per 4 ha in Mexico. Trap placement at these numbers required 2 considerations. The trap must be in a position from which its emitted pheromone attractant would have a high probability of intersecting the casting flights of the male moths. It must also be reachable by servicing personnel quickly and efficiently.

Studies by Leggett et al. (1994) resolved many questions involved in management of trapping. Traps were placed on field margins preferably where they will not be destroyed in normal field cultivation activities. Trappers are trained to look at such factors as prevailing air movement patterns as fields cool down after sunset. Male moths typically become active in search of females as temperatures decline with an 80°F (26.7°C) threshold (Lingren1989). Traps were serviced or changed at least weekly.

3.2.2. Trap Lure Formulation

The discovery and development of the female sex attractant of the PBW was the single most important entomological breakthrough of the mid 1970's with respect to PBW control. The name "gossyplure" and its characteristics were first published by Hummel et al. in 1973. Bierl et al. (1974) published detailed data illustrating its importance and the role of its specific components. Gossyplure is a near 50/50 ratio mixture of (ZZ) and (ZE)-7,11 hexadecadien-1-ol acetate isomers. This paper also noted the detrimental effect of the EE and EZ isomers of this molecule in reducing attraction. The introduction of a controlled release formulation of gossyplure had a profound impact on the San Joaquin Valley exclusion programme being implemented at that time for PBW in California. Previous adult surveys had relied on the para-pheromone hexalure. Hexalure required 20+ times the attractant for much lower capture rates and detection efficacy.

A number of controlled release formulations have been used in PBW traps. Flint et al. (1974) provided the first published paper showing the advantages of a controlled release trap lure formulation for PBW. Flint used the red rubber septa and a version of this lure was used throughout the programme at 4 mg per lure. There are other formulations which have a flatter emission release rate over longer periods of time, but they are more expensive.

Throughout the eradication programme covered in this publication, lures were replaced with every trap service or at least every two weeks, even though the septa have excellent properties for a longer period. This lure is currently produced for all state programmes at cost by the Arizona programme.

The quality of the gossyplure is as or more important than the substrate used. Staten et al. (1988) illustrated the importance of using trap bioassays in the procurement of gossyplure for surveys. Important differences of PBW attraction still cannot be explained with known chemical analysis alone; traces of an alcohol are suspected. The programme in Arizona maintains a supply of "technical" grade gossyplure for all post-eradication survey in this programme.

3.2.3. Moth Identification

The programme had to face two critical issues, namely species identification (taxonomic) and separation of sterile from native insect specimens. From the first sterile moth releases in the San Joaquin Valley in 1968, moth taxonomic identification used labial palp bands, and genital clasper characteristics to separate *P. gossypiella* males from other species. The survey traps and lures used, with rare exceptions, only attracts the male moth. The trap is not absolutely species-specific and will sometimes capture a few accidental "contaminant" specimens. Most of these do not resemble PBW. There are a few similar-sized moths which may have an attractant similar to gossyplure. If these captures were confused with PBW, they could affect treatment decisions adversely. Good dissection techniques and microscopic examination were used for specimen identification. This was considered sufficient throughout the years of the programme. Late in the programme, DNA signatures were under development at the University of Arizona.

Separation of mass-produced release moths from native moths in the cotton fields was accomplished using a dye incorporated in the sterile moth larval production diet. Calco Red oil food dye was used as a diet induced marker (Graham and Mangum 1971). Marking was very accurate as non-marked moths could not be found among laboratory-produced moths even when extreme searches were conducted periodically throughout the Phoenix rearing facility's history. Searches involved thousands of aged moths crushed on white filter paper. In these searches, moths were routinely taken from discarded egg production cages. This, however, cannot completely represent moths under field conditions.

A simple paper chromatograph technique was in use in support of sterile moth releases as early as 1970 in the Coachella Valley of California, when the first author worked with that trial programme. It has endured through all sterile moth releases. The technique involves the use of a small straight sided vial at or near 25 x 10 mm; exact size is not critical. A moth is crushed in the bottom with an uncontaminated rod, preferably glass. The vial plus moth then receives 1 ml of hexane and a strip of chromatographic paper cut to 9 x 30 mm. The strip is cut to a point at its terminal end so that as the solvent moves upward it concentrates the dye in the tip as it dries, facilitating identification.

As the eradication programme reached completion, the potential for misidentification became extremely critical. As each state programme reached this point, the importance of absolute accuracy increased. If a 1:1 000 000 rate of error was possible, a second independent marker or analysis with high levels of confidence would, in terms of probability, make a missed detection of a non-marked sterile virtually impossible.

Burns et al. (1983a) used an inductively coupled plasma-atomic emission spectrometer to test possible use of 13 elemental markers. Out of those, Strontium (Sr) proved to be the most viable candidate as a second marker, independent from the Calco Red dye. Burns also illustrated its potential in sterile Mediterranean fruit fly *Ceratitis capitata* (Wiedemann) analysis (Burns et al. 1983b).

PBW larval diet preparation had undergone major changes in the mass-rearing facility (Miller et al. 1996), so some re-testing with Sr was required. In this effort in 2011, a 540-ppm level of Sr was found to have excellent retention in moths as old as

45 days (Walters, unpublished reports). This technology was used starting in 2012 for all the sterile moth production.

3.2.4. Larval Sampling

Pre-eradication pest management scouting procedures usually used larval PBW populations at 2-5% as the action threshold to trigger insecticide ground and aerial applications of conventional insecticides. When populations reached this level, non-selective insecticides were considered the only workable solution. All treatment decisions were on a field-by-field basis. In many areas that resulted in frequent (5-15) treatments for PBW alone or in combination with insecticides for other pests in a growing season.

In the case of the PBW eradication programme, protocols used for suppression were designed to prevent development of populations high enough for normal detection of larvae. Boll larval sampling was used most extensively in the first 2-3 years of the programme. Some states used a random selection of non-*Bt* fields for boll collection to assess larval populations. Conversely, during the first years of the Arizona programme all non-*Bt* fields were sampled. This was logistically possible due to a high ratio of *Bt* to non-*Bt* fields. In instances where pockets of higher native moths were detected in cotton fields, targeted searches were also used as the programme progressed.

Two different sampling methods were used. Bolls collected from the field could be processed within boll holding boxes (Fye 1976) or by direct examination of bolls cut open immediately after field collection. When the data are needed for immediate operational decisions, boll cutting is critical. In this case data could trigger an immediate conventional insecticide treatment. This became extremely rare as the programme progressed. For resistance monitoring, or when some assessment of reproduction was desired, boll boxes were used. The detection threshold of a trap is always better than any larval assessment.

3.3. Control Technologies

Throughout the eradication programme, PBW control was the responsibility of state-wide programme management (Tables 1 and 2). These organizations controlled all treatment activity except for the type of cotton to be grown. Producers chose not to plant or to plant *Bt*-cotton, although the latter option was encouraged. Individual growers were responsible for adhering to Environmental Protection Agency (US-EPA) regulations in terms of respecting seed contracts and label compliance. In Texas, USA and Chihuahua, Mexico (Phase I), the PBW programme ran simultaneously with active boll weevil eradication, where boll weevil treatments were concurrent. In the USA, the grower who chose to use *Bt*-cotton contracted with the seed provider to pay the technology fee for that resistant cotton. Where *Bt*-cotton was not in use, producers paid a higher assessment for programme-applied pheromone mating disruption and other control actions needed for suppression. Base costs covered other programme aspects.

3.3.1. Transgenic Cotton

Cotton genetically modified with genes from *B. thuringiensis* (*Bt*) provided the single most important change in PBW control in the late 20th century. In 1990, Wilson et al. (1992) conducted the first tests of experimental lines "that carry an altered version of the insect-controlling protein gene from *B. thuringiensis kurstaki"*. These lines were not commercially available at that time. The technology subsequently developed to a commercial state quickly.

The PBW mortality levels which commercial *Bt*-cotton varieties produced were unprecedented at >99% (Flint et al. 1995; Watson 1995). Staten et al. (1995) noted its potential importance as an eradication tool, proposing its integration with other "soft" technologies because it targeted only the larval stage as it fed within or on the plant. Because *Bt*-cotton varieties did not affect adult PBW it would, in effect, provide an excellent synergistic tool when combined with the inverse density-dependent action mode of the SIT and mating disruption.

The immediate concern for the *Bt* technology in all control contexts, however, was that its extreme efficacy would lead to overuse and thus ultimately to resistance development. This has proven to be a realistic concern. Tabashnik et al. (2013) offered an excellent review "after the 1st billion acres of use". Resistance to both common endotoxins (*Cry 1Ab* and *Cry 2Ac*) found in commercial cotton resulted in major losses or shifts in strategies in India and China (Tabashnik et al. 2013; Wan et al. 2017).

Currently there is discussion of major losses in *Bt*-cotton from PBW in Pakistan (Shahid 2014). Losses from resistant PBW have been reversed by using hybrid cotton in China (Wan et al. 2017). In this case ±25 % of all cotton plants would not express the *Bt* traits. This provides "in the bag refugia" as described by Head and Greenplate (2012).

Within the eradication zones, the commercial use of *Bt*-cotton always had label restrictions requiring resistance management by providing plantings of susceptible cotton (refugia). These enforced EPA label restrictions required that an individual grower entity use one of two choices. The grower could plant at least 20% of his cotton with a non-*Bt* variety, which could be treated with any labelled conventional insecticide. Alternatively, the 2nd choice was that at least 5% of the surface would be of a susceptible variety of cotton, but the grower could not use a long list of conventional insecticides on those refugia.

Additional restrictions published in 2005 for the 2006 growing season added mating disruption and sterile moth release to this list. These restrictions in the use of *Bt*-cotton precluded eradication. Under this scenario, simple calculations could place production of a diapausing PBW population in a 5-acre field at > 500 000 insects in one late-season generation. This is a very conservative number. Late-season cotton produces susceptible bolls for more than one generation, thereby laying the groundwork for a large overwintering population.

Staten et al. (1999) had noted the potential use of sterile moth release over *Bt*-cotton in lieu of structured untreated refugia as part of a resistance management strategy. Sterile insect distribution is more reliable than non-directed capricious movement found in nature from native populations which are non-randomly distributed.

Arizona immediately sought a section 24C special local needs label to utilize a grower choice of up to 100% *Bt*-cotton as long as this acreage would receive an average of 10 or more sterile insects/acre/day (24.7/ha/day). This application ultimately required an extensive formal review before an EPA Science Advisory Panel in 2006 (Antilla and Liesner 2008). This strategy with some variation is in use today. Tabashnik et al. (2010) published an excellent review and assessment of this strategy in Arizona.

The choice of what cotton is to be grown has always been left to the individual producer in this programme. As part of the field mapping procedures all growers are canvassed in early spring for an inventory of expected fields to be planted. The inventory includes a separation of *Bt* and non-*Bt* types. All fields are then checked and tested with an ELISA test after germination (AGDIA Inc. Elkhart, Indiana, USA), as a safety check to ensure accuracy of *Bt*-cotton distribution maps.

3.3.2. Mating Disruption

Within this eradication programme, mating disruption was used on all non-*Bt* cotton during at least the first four years of each state's operations. The hand-applied PBW Rope (Shin-Etsu Chemical Company) was preferred. Aerially applied NoMate Fiber, NoMate Mec (Scentry Biologicals) and Check Mate (Suterra), were also used when circumstances required. These latter formulations had an effective disruption time of 8 to 14 days. A review with product details is found in Staten et al. (1997). The use of gossyplure for mating disruption for PBW represents the most successful early application of this technology (Cardé et al. 1997; Cardé, this volume). A hollow fibre-controlled release formulation was the first EPA registered disruptant (Brooks et al. 1979).

The use of pheromones for mating disruption is fundamentally different than any conventionally applied insecticide (Cardé, this volume). A treatment of a controlled release pheromone does not kill the intended target. In the case of PBW, treating an already reproducing or mated population in even moderate levels is therefore futile. There was an "attract and kill" system (Staten and Conley patent 4671010 now expired), which involved adding very small traces of a pyrethroid insecticide to the adhesive in the NoMate fibre system. It appeared to be of assistance, but its value was not great.

Staten et al. (1997) characterized both low and high-rate systems. Low-rate systems (applicable by air) require frequent reapplication with escalated error potential as each treatment timing decision is made to achieve constant disruption for a 30 to 60-day time frame. The PBW eradication programme used in the aerial spraying the three low-rate formulations described above.

The first high-rate system known as PBW Rope (Flint et al. 1985) was field tested in the Imperial, USA and Mexicali, Mexico valleys in 1986 (Staten et al. 1987). This PBW Rope provided continuous disruption and efficacy over a much longer time frame than 4-8 applications of low-rate systems applied by air. This was true even on a field-by-field basis. From that time to the present, the formulation has only had one major change. Its application, when correctly done, maintains complete trap shut-down for a 50-70-day time frame in low to moderate populations.

PBW Rope was designed to be tied on an individual plant. The programme began to pre-wrap it on a bamboo stake in 2006 in Arizona (Antilla and Liesner 2008). The current formulation was applied at 200/acre or 500/ha. For maximum efficacy in upland cotton, it must be in the field at or before cotton reaches the 6-leaf stage. This is before a female can mate and live long enough to successfully oviposit. In rare cases a second application has been justified.

Area-wide (mandated) use of pheromone has a long history. Baker et al. 1990 reported on a one-year project in the Imperial Valley of California using low-rate systems. It covered > 40 000 acres and targeted the first two generations. Low rate pheromone systems were required before first square (first flower bud) formation. Its goal was to suppress PBW long enough to reduce conventional insecticide use and secondary whitefly problems.

The aforenoted review by Staten (1997) covered two separate, later area-wide trials. These trials were in the Coachella Valley of southern California and the Parker Valley of Arizona. The most important trial in Parker, Arizona is covered in detail over a 5-year period by Antilla et al. (1996). Both trials strongly illustrated the importance of an area-wide approach with pheromone disruption. For the first time, after the first or second year, season-long control without major reliance on conventional insecticides became possible. Results were obtained in an area where the pest was normally severe. These trials both depended partially or completely on the high-rate systems. Additional trials in the Imperial and Palo Verde Valleys were used to develop a better understanding of the SIT when combining the technologies (Staten et al. 1999; Walters et al. 1998, 1999, 2000).

3.3.3. Sterile Insect Release

Releases of sterile moths in this programme had two purposes: a suppression tactic in and of itself, and as a resistance prevention strategy (see discussion in Section 3.3.1.). The release of sterile PBW was started in 1968 in the San Joaquin Valley of California as part of a containment/exclusion strategy to prevent establishment of the pest (Staten et al. 1993, 1999). Releases were continuous in areas of detection from 1970 through 2011.

Over time three sterile moth rearing facilities were established to produce PBW for field release, namely a temporary facility in Harlingen, Texas (1968-1970) and two separate facilities in Phoenix, Arizona, from 1969-1995, and 1995 to the present. The current existing facility is unique in that it was designed to produce twice the known maximum needs of the then-existing San Joaquin Valley programme.

The purchase of a 66 000 square feet (6 131 m^2) building in Phoenix, its extensive renovation and most of its equipment were paid for by the CCPCB (through the California Department of Food and Agriculture (CDFA) and managed by Mr. Wally Shropshire as chairman from 1969 through 2010, and Mr. Ted Sheely, the current chairman).

The Phoenix mass-rearing facility has been managed by USDA with California funding and minimal federal (appropriated) funds until 2005. From 2005 onward it phased into a fully USDA federally funded operation, less a few expenses such as property taxes. The production and funding of sterile moths for release then became a USA government obligation. All other field costs remained with the grower community (Tables 1 and 2).

All moth handling, packaging, chilled shipment, handling pre-release, and release procedures were developed for the San Joaquin Valley containment/exclusion programme. This occurred long before eradication started in the fully infested areas. The reporting of the successful San Joaquin Valley programme is not complete, but segments and procedures are partially covered by Rudig and Keaveny (2008) and Staten et al. (1993). Many of the sub-areas of this valley have more favourable growing conditions for PBW population development than did areas which had moderate to heavy infestations. Examples include Safford, Arizona and the El Paso Valley of Texas. Wind-borne PBW movement into the San Joaquin Valley was documented and monitored by the CDFA programme. In addition, Stern and Sevacherian (1978) and Stern (1979) established its potential through monitoring of the desert between the San Joaquin Valley and with consideration of plant growth analysis and studies of the overwintering potential.

By 2005 the Phoenix mass-rearing facility had enough space, all major equipment needed, and the technology developed to produce an expected 20-28 million moths per day. Previously it had been producing an average of 5 million moths per day. Production for the years in which the "expanded" eradication programme was in effect is shown in Table 3. Range of production resulted from varied season lengths of the multiple treatment areas.

The sterile moths were directly collected after adult emergence from pupae with immediate chilling at ±35°F (1.66°C). They were maintained at as near that temperature as possible through irradiation and packaging. Shipment occurred in specialized pre-chilled shipping containers to the release destinations. Any holding time of the sterile moths at destination was in cold rooms or cold boxes. They were then loaded into specialized aerial release machines installed in small aircraft (Pierce et al. 1995).

The release aircraft must be capable of working speeds of 120 miles/h (193 km/h). A Cessna 206 aircraft served this purpose in this programme. Release height above the cotton fields was maintained at an average of 500 feet (152 m) above ground. Sterile moths were normally released within 24 hours post-collection. When insects were held an additional day, loss in quality was observed. Individual non-*Bt* fields were specifically targeted. Release grids were used for a lower release rate over *Bt*-fields.

Table 3. Daily sterile moths produced at the Phoenix Arizona mass-rearing facility (in millions) used for release in the bi-national PBW eradication programme areas

Year	2004	2005	2006	2007	2008	2009	2010	2011	2012	2013
Production goal/day	5.0	10.0	22.0	22.0	22.0	26.0	20.0	20.0	14.5	5.0
Mean/day	5.4	12.5	22.9	25.1	24.1	27.2	22.2	22.9	16.3	5.8
Range low	6.8	1.3	16.2	19.6	10.2	19.6	20.2	20.3	14.7	4.1
Range high	15.5	6.7	30.1	33.2	30.7	32.9	27.2	32	22.8	7.3

3.3.4. Conventional Insecticides

Insecticides were those applied to kill the pest as a direct result of their application. This form of treatment was and is the only method available to manage a rapidly expanding population when larvae are first found in the cotton bolls. In many areas, before initiation of PBW eradication, this occurred early in the cotton fruiting cycle and lasted for 2-4 months.

In the case of the eradication programme, all other control activities started either at planting (*Bt*-cotton) or before a moth can be mated and lay its first fertile egg on a plant with a susceptible fruiting form. Optimum application of mating disruption was before the 6-leaf crop stage; sterile release started even before.

An indicator of success in the first few years of programme initiation was the reduction in insecticide sprays from those previously used by the producers individually. Choice of insecticide was made based on local recommendations and knowledge. The most commonly used insecticides were chlorpyrifos as Lockon or Lorsban (Dow Chemical), or a pyrethroid labelled for PBW control. As shown in the following outcome Section, very low percentage of the total area required any traditional insecticide treatment. That predominantly occurred in the first two years of programme operations. This approach was particularly applicable in Texas and New Mexico, before sterile moth release was available.

3.3.5. Cultural Control

Among the earliest research conducted on PBW involved cultural practices for its control (Noble 1969). As PBW exploded through Arizona, invading southern California in the 1960's and 1970's, cultural control became a major area of research (Naranjo et al. 2002). Over a 40-year period, each growing area within the generally infested cotton areas has developed a balance between profitable production and essential regulated cultural practices.

PBW eradication used all existing regulations as a standard. Management encouraged or was involved in any needed regulatory enforcement of these best practices. Of greatest importance were the programme actions that were needed for post-harvest crop destruction and host-free periods to minimize carry-over between cotton crops.

4. OUTCOMES

The successful eradication programme outcome is presented on a state by state basis in the general order that the seven states initiated their activities as part of the programme phases (Fig. 1).

Management credits herein provided for each state are for those managers who were responsible for day-to-day decisions, operation, all data acquisition and evaluation. Managers listed were accountable for day-to-day success as well as setbacks. These individuals are considered as the most important contact points in each state. The first listed are those in the operational offices where daily decisions were made.

The programme was feasible because only non-*Bt* cotton areas required centralised control cost and intensive day-to-day attention. We provide total cotton area planted and percent *Bt*-cotton to partially illustrate the magnitude and intensity of operations in each of the states. States and cotton areas within the states with the highest ratios of non-*Bt* required the most intense management per acre or ha. The data presented are in units of measurement found in the respective field records (acres or ha).

First programmatic treatments occurred in Texas in 2001. Overall, the last PBW detected as adult unmarked moths were captured in 2012. The last sterile release treatments occurred in 2013. Population collapse is illustrated by annual adult native moth capture.

An understanding of positive economic impact is found in the reduction and eventually elimination of any detectable larvae. Direct cost-benefit is best illustrated by the decreasing use of annual inputs in high-rate pheromone (PBW Rope), sprayable pheromone systems, and conventionally applied insecticides. Lessons learned from the first state to start were extremely beneficial as the subsequent states entered the programme.

4.1. Texas, USA (Phase I)

Edward Herrera, Supervisor, El Paso /Trans-Pecos District, 1999-2013
Osama El-Lissy, TBWF Programme Director, 1999- 2000
Charles Allen, TBWF Programme Director, 2001-2009
Larry Smith, TBWF Programme Director, 2009-present

The El Paso / Trans-Pecos growers of western Texas were the first group to initiate the PBW suppression and eradication efforts reported here. The Texas Boll Weevil Foundation (TBWF) was, in 1999, fully functional and successfully involved in its part of a USA cotton belt-wide eradication effort for boll weevil, *Anthonomus grandis*

Boheman (Allen 2008). Recently detected weevil populations had become a major concern in western Texas, and treatment was restricted to these populations in 1999 through 2005 in this area.

In 1999, the grower community of western Texas passed a referendum to join the TBWF for boll weevil eradication and to initiate PBW population evaluation. In 2001, the PBW area-wide suppression activities were added to eliminate economic loss from PBW for the growers in the area. Details and complete results for the first years are covered by Allen et al. (2005). Control activities began in 2001 without sterile releases and in the absence of isolation. The programme was modelled after the Arizona Parker project (Antilla et al. 1996).

The Texas PBW programme encompassed two distinct agronomic areas (Fig. 1). The El Paso "Valley" in the USA is separated only by the Rio Grande (Bravo) River from the Juárez "Valley" of Mexico. Mexico did not start programme activities the same year (2001) despite the fact that, in many cases, fields from the two countries were less than 200 meters apart. More than half of the Texas programme was in this valley.

Cotton in the El Paso work unit was more than 50% Pima *(G. barbadense*). This species of cotton is considered to be the most PBW susceptible commercial cotton grown as it was all non-*Bt*. Initial PBW populations were very high as depicted in trap counts in the year 2000 (Table 4). Due to lack of isolation, population suppression and economic loss prevention were the only achievable goals for the first two years of the Texas programme (2001-2002).

In contrast, the area east of El Paso comprised a distinctly different production system. This included the general vicinity of the Pecos River and the town of Fort Stockton, Texas. Land was characterized by a shorter season and high usage of *Bt*-cotton. In this distinctively different area, Pima production was minimal, being grown principally in one isolated organic cotton block. Fields in this zone were dispersed over a very large geographic area and occurred in isolated clusters. Separation of such cotton blocks was frequently more than 50 miles.

From 2001-2004 (Table 4), programme treatment options on non-*Bt* cotton were limited. Required treatments were heavily concentrated in the El Paso Valley area. Aerially applied NoMate fibre, with 0.000586 lb (0.265 g) pyrethroid per acre mixed in the adhesive as an "attract and kill" formulation, represented one important control approach. As the seasons progressed chlorpyrifos at 24 fluid ounce per acre (1.68 kg/ha) was used as an overspray.

Deployment of the high-rate PBW Rope during the first two years was limited to sensitive areas near schools, etc. Its use expanded over time reaching a peak in 2003 and then gradually declining to zero over time. The initiation in 2002 of eradication programme activities in part of New Mexico and the Juárez sector of Chihuahua state in Mexico was a major improvement.

Table 4. Summary data pink bollworm programme in Texas 2000 through 2004

Year	2000	2001	2002	2003	2004
Total acres cotton	48 281	48 222	39 538	36 100	40 826
% *Bt*-cotton		49	44	35	38
PBW Rope (acres)		5399	9056	23 551	19 815
Cumulative acres aerial low-rate pheromone		142 842	123 129	58 017	26 224
Acres pheromone + insecticide*		47 897	43 386	34 945	25 162
Pyrethroid (acres)		0	0	0	2039
Larvae/100 bolls		4.53	0.81	0.13	0.76
Native moths trapped (million)	1.40	0.75	0.27	0.18	0.09

* *Chlorpyriphos at full labelled rate applied at the same time as a low-rate pheromone Acres treated are also included in low-rate pheromone treated acres*

The PBW populations on Pima cotton in the El Paso zone were critical in this state's programme. They accounted for most of the non-*Bt* cotton in the Texas programme. In 2001 Pima made up 51% of the cotton in the El Paso valley. As a result of the suppression, total trap counts of native moths were reduced from 1.4 million in 2000 (pre-control) to 0.74 million in 2001.

By the end of the last year (2004) without state-wide sterile moth releases, the season-long capture of all moths totalled 0.09 million moths. These moths were captured in traps serviced weekly at one trap per 5-10 acres of cotton. This was the highest trap density used in the programme. The annual pre-control number of native moths per trap per service averaged 17.77 in 2000 vs. less than one in 2004 (0.94).

Larval populations in bolls further illustrate progress during this 4-year suppression period. Each year 60 randomly selected fields were sampled season-long. Larval counts decreased in a steady progressive fashion; details are found in Allen (2005). From an economic perspective it is important to note the general decrease in inputs needed for control including intensively managed aerial application of low-rate pheromones and conventional insecticide.

Limited sterile moth releases on isolated fields was initiated in the Trans-Pecos area in 2004. In 2005, funding for a targeted 10 million sterile moths for release per day was obtained for the core programme area of Texas, New Mexico, and northern Chihuahua. Sterile moths were released season-long in 2005 in Texas and New Mexico.

In Juárez, Mexico, the start of sterile releases was administratively delayed for some time. During this delay moths earmarked for Mexico were heavily released in Texas along the river border between the two countries. Sterile moth movement into Mexico and therefore coverage, was far better than expected. Eventually moth distribution in 2005 of the targeted 70 million per week were carried out over the entire three state area. At this time non-*Bt* fields were directly targeted by release aircraft. The ability to release these moths in the cotton fields on already suppressed populations before any mated female moth could deposit eggs on a susceptible plant was a game changer. As populations of native PBW declined, sterile moth release numbers were diverted to expanding treatment areas.

Table 5 provides the data for the period 2005-2012, when general sterile moth releases began, until after the last native or unmarked moth was trapped. In 2005, a total of 1336 million sterile moths were released in Texas. Over 1.4 million sterile moths were trapped together with 11 917 native moths. With the exception of the year 2009, this downward trend in native moth capture was very positive as shown in Table 5.

Table 5. Summary data pink bollworm eradication programme Texas programme 2005-2012

Year	2005	2006	2007	2008	2009	2010	2011
Total acres cotton	43 358	42 304	39 533	33 029	34 299	38 268	48 447
% *Bt*-cotton	35	22	30	48	40	49	55
PBW Rope (acres)	9226	9686	2843	3198	1682	8050	0
Acres aerial low-rate pheromones	6228	7597	4964	0	0	0	0
Chlorpyrifos (acres)	2923	3653	2804	0	0	0	0
Pyrethroid (acres)	0	0	3613	0	0	0	0
Sterile moths recovered (million)	1.45	0.86	3.19	2.23	2.93	1.05	2.09
Native moths trapped	11 917	3302	1363	14	3291	16	60

No native moths were trapped in or after 2012. The last pre-season randomly selected monitoring fields with larvae were found in 2005 and 2006. Season-long larvae/100 boll recovery levels in these fields were 0.01 and 0.20 respectively.

Thereafter no further larvae were found even when fields with positive traps were specifically targeted. The last year conventional insecticides and dual insecticide-pheromone treatments were applied was in 2007. This was the result of one hotspot population along the Texas-Mexican border. The two fields with the greatest difficulty were an upland non-*Bt* cotton field on the Mexican side of the border and a 7.9-acre Pima cotton field in Texas, separated by less than 150 m distance. Larvae had been found late in the season in the Mexican field in 2006. After extensive treatment in 2007 it was not possible to detect larvae. All of these fields were treated with PBW Rope in 2008. In both 2008 and 2009 the only non-sterile inputs were PBW Rope applications.

In 2009, two unexpected events occurred shortly before cotton was harvested. These events were widely separated geographically and with different, but logically, highly probable origins.

The first occurred in and around the Pecos River Valley, where field traps started capturing large numbers of native moths on September 29. Captures were distributed over 6 023 acres, with only 30 acres of non-*Bt* cotton. No native moths had been captured before this time in this area. Historically, wind-borne movement of PBW was common at this time of the year and has been documented frequently in monitoring traps in non-production areas along desert highway trap lines. A storm front movement from an isolated heavily infested organic farm near Midland, Texas, matched this hypothesis. Pierce et al. (2013) strongly verified this conclusion in a 3-year study.

The second unexpected event started in the El Paso Valley, October 19, in an area with intensive Pima cotton. On August 22, two irradiation canisters with ±180 000 moths per container left the Phoenix facility without having proper radiation tags filed. They would have been transported in shipping boxes holding 2.1-2.2 million insects each. All moths captured were in an area documented with GPS flight recorders from this shipment. No native moth captured was more than half a mile (0.8 km) from a release swath from this flight. The vast majority of the 2626 non-marked moths in the El Paso area were directly within the expected swath of this aircraft. No larvae could be found, although exit holes and some characteristic damage was observed. The capture curves fitted with expected late-season life cycle length. The programme had examined a total of 67 246 blooms and bolls prior to the outbreak without any detection. Native moth capture had not occurred in the entire area before October 12. It was only logical to conclude that the unanticipated captures were due to a release of non-irradiated moths. To mitigate this situation, procedures for irradiation safeguarding were reviewed and significantly stiffened at the Phoenix PBW mass-rearing facility.

In 2010, this entire area (8050 acres) was treated with PBW Rope (Table 5) and received enhanced sterile moth releases. The native moths captured in 2010 were scattered throughout the El Paso zone, but not in the PBW Rope treated area. The last native moths captured were in 2011. The Texas PBW programme, as described, has not had a further detected moth or larvae.

4.2. Chihuahua, Mexico (Phase I)

Ing. Alfonso Soto Martinez, Gerente del Comité Estatal de Sanidad Vegetal, 2001-2008
Ing. Antonio Medina Arroyo, Gerente del Comité Estatal de Sanidad Vegetal, 2009
Ing. Juan Carlos Ramirez Sagahon, Gerente del Comité Estatal de Sanidad Vegetal, 2010-2011
Ing. Antonio Medina Arroyo, Gerente del Comité Estatal de Sanidad Vegetal 2011-2013
Ing. Jesús Escárcega Terán, Gerente del Comité Estatal de Sanidad Vegetal, 2013 to declaration.

Chihuahua has the largest cotton growing area of any state in Mexico. It is also a state in which production areas grew rapidly during this programme. Eradication of PBW and boll weevil were and are within the same management programme under the Comité Estatal de Sanidad Vegetal. Maps and details for 2002 through 2007 can be found in Staten and Ramirez-Sagahon (2008).

The state programme was ultimately organized in four work units, Ascensión, Meoqui, Ojinaga and Juárez, all in northern Chihuahua (Fig. 1). Trap placement in Chihuahua was standardized at 1 trap/4 ha and 1trap/20 ha in non-*Bt* and *Bt*-cotton respectively.

4.2.1. Ascensión, Chihuahua

This work unit included all cotton grown in the north-western portion of the state. This cotton, with only a few exceptions, was grown with a shorter season and colder winter. This is typical of all cotton found in the highland Chihuahua deserts, where centre pivot irrigation predominates. Historically boll damage was sporadic and usually occurred late in the season. With this in mind, the programme was designed to be heavily dependent on the planting of *Bt*-cotton and PBW Rope treatment. By 2004, the PBW population had been reduced to a level in which sterile insects would not be required.

During the critical initial three years of the programme (2002-2004), the percentage of *Bt*-cotton was at 67-69% (Table 6). During this time all non-*Bt* cotton was treated with PBW Rope. This cotton was well dispersed throughout the cotton production area. Native moth captures decreased by 3.6 and 3.7-fold each year. As the programme progressed, the application rate of PBW Rope was reduced in areas with moderate risk (based on previous year's moth trap capture) from 500 to 250 per ha.

In 2005 some fields did not receive pheromone treatment. Numbers of non-*Bt* fields treated also generally declined with the exception of the 2006 year. This was preceded by the only season to season escalation in native moth captures in traps. This occurred principally late in 2005, when minimum numbers of larvae were found in non-*Bt* monitoring field searches (less than 10).

By 2006 only 38% of the cotton was *Bt*-cotton (Table 6). 2006 had the only appreciable escalation of PBW Rope use. Conventional insecticides were used in only two years directly for PBW suppression. The last native moth capture in the Ascensión area of Chihuahua was in 2007 (12 specimen). Staten and Ramirez-Sagahon (2008) noted that this represented two native moths per every 10 000 traps serviced.

Table 6. Summary data pink bollworm eradication programme for the Ascensión area of the state of Chihuahua

Year	2002	2003	2004	2005	2006	2007	2008	2009
Total hectares (ha)	11 268	16 499	25 637	23 088	28 430	29 228	26 632	10 209
% *Bt*-cotton	69	68	67	47	38	29	30	31
PBW Rope (ha)	3507	5177	8566	8936	13 110	1118	461	0
Insecticides (ha)*	581	0	0	86.9	0	0	0	0
Native moths	20 256	5489	1467	4204	63	12	0	0

* *Conventional aerially applied insecticides expected locally to be the most effective*

4.2.2. Meoqui, Chihuahua

This unit is Chihuahua's southern-most cotton growing area. This area of Chihuahua is bordered by the states of Coahuila and Durango. It contained all the cotton cultivation found around the cities of Delicias and Jimenez. It is an older, diverse production unit mostly using impounded water from the Rio Conchos. It has enough overwintering habitat to make boll weevil its key pest.

Area-wide treatment for both PBW and boll weevil started in 2002. The weevil required ULV malathion with emphasis on pin square treatments and treatments in mid- and late-season based on weevil trap captures. This was followed by treatments targeting weevils going into diapause. A high percentage of *Bt*-cotton was present in all years except 2006 (Table 7). The PBW Rope was ideally suited for use in these circumstances. In the initial three years (2002-2004), all non-*Bt* cotton was treated with PBW Rope; application was targeted for 6 leaf cotton in the spring.

In 2005, PBW Rope was applied in the spring on non-*Bt* fields that had positive trap captures the previous year or at detection in early and mid-season. Only in the first year were 44.3 ha treated for PBW with a conventional insecticide and a sprayable pheromone, in addition to PBW Rope. Suppression in these fields was triggered where even a single larva was found by targeted scouting. No aerial treatments were applied thereafter.

Table 7. Summary data pink bollworm eradication programme for the Meoqui area of Chihuahua

Year	2002	2003	2004	2005	2006	2007	2008	2009	2010
Total hectares (ha)	930	5151	9332	6754	4332	933	1588	45	1704
Non-*Bt* (ha)	164	380	278	321	2195	114	266	2	254
% *Bt*-cotton	82	93	97	95	49	88	83	99	85
PBW Rope (ha)	164	380	278	180	139	25	12	0	0
Insecticides (ha)	44.3	0	0	0	0	0	0	0	0
Native moths trapped	N/A*	377	2970	1203	13 410	2632	953	265	0

* *N/A= not available; 5046 moths were captured in the combined Meoqui and Ojinaga work units*

Unfortunately, native moth capture summary data for 2002 are not available as they were combined for the areas of the Ojinaga and Meoqui work units. Throughout the programme in Chihuahua, trap grids were not completely established until July 2002. During this programme's 2002 establishment period, boll weevil treatment and trapping, as well as PBW Rope application, required availability of early-season resources.

Weekly PBW trap data as reported in Staten and Ramirez-Sagahon (2008) provide an understanding of important occurrences in the first two years, with graphics showing separate weekly trap captures for each area. There were appreciable trap captures in 2002. In 2003, captures were reduced season-long, however, in Meoqui during September 2004, trap capture of native PBW escalated. From September to the last service date in October 2914 moths were captured with no correlation to non-*Bt* fields. Migration was suspected but not verified.

In 2006 no moth captures occurred anywhere in Meoqui before the week of September 11. From that point on, there was a massive escalation in captures throughout the region with no correlation to non-*Bt* cotton. This included a new growing area with no PBW history and major captures in new *Bt*-fields with no cotton history. Highway trap lines were quickly established to the cotton growing area south of the state of Chihuahua in the La Laguna area, in the states of Coahuila and Durango, and to the Texas-Mexico border in Ojinaga. The conclusion, that a major weather-driven migration was responsible, was inescapable.

In 2007, many of the PBW positive fields were shifted to *Bt*-cotton or other crops. PBW Rope use continued but declined. By 2009 no treatment for PBW was needed. There have been no further PBW captures since 2009. By 2010, the La Laguna area was moving forward for a combined PBW and boll weevil programme.

4.2.3. Ojinaga, Chihuahua

This work unit, as stated above, was managed together with the Meoqui unit for three years. It had two distinct habitats for boll weevil and PBW. A well-established area was in a surface irrigated system at the confluence of the Rio Concho and the Rio Bravo (Grande, USA) rivers. This area is a local commerce centre at Presidio, Texas and Ojinaga, Chihuahua. Initially Texas had some cotton on its side of the border. On the Mexican side, removed from this river valley, where rapidly developing areas of centre pivot irrigation. The growing season in these areas is shorter than in the cotton areas in the river valley. Its rapid expansion is reflected in the increase in cotton areas (Table 8).

During the initial three years all the non-*Bt* cotton was treated with PBW Rope. Need and use of pheromone was reduced thereafter; the last treatment for PBW occurred in 2010 on 78 ha. No conventional insecticides were used for PBW in the Ojinaga unit. Boll weevil was treated with ULV malathion.

Table 8. Summary data pink bollworm eradication programme for the Ojinaga area of Chihuahua

Year	2002	2003	2004	2005	2006	2007	2008	2009	2010
Total hectares (ha)	1382	3987	9541	10 936	20 102	19 885	19 611	12 585	27 220
% *Bt*-cotton	42	52	82	85	81	60	42	69	86
Pheromone (ha)	803	1907	1715	130	245	769	248	315	78
Insecticides (ha)	0	0	0	0	0	0	0	0	0
Native moths trapped	N/A*	390	79	4	5415	96	225	89	0

* *N/A = not available, 5046 native moths were captured in the combined Meoqui and Ojinaga work units*

As noted in Staten and Ramirez-Sagahon (2008), native moth captures were common when traps were deployed in 2002. With area-wide suppression, numbers dropped drastically in the following years through 2005, when only four PBW moths were captured. However, as detailed in the discussion of the Meoqui unit, Ojinaga's 2006 trap captures exploded. A continued declining south to north gradient was apparent, with the largest portion of total captures occurring in the southern part of this area near Camargo, Chihuahua. The lowest captures occurred in the river valley on the Texas-Chihuahua border.

A season total of 5415 PBW moths were recovered in 2006. This escalation started during the reporting week of September 24, as was true of the Meoqui area. Native PBW had not been detected before this time in 2006. Similar patterns of capture occurred through 2009. Nevertheless, the last moth detections were in 2009 at 89 moths for that season. In 2010 no moths were detected with trap services exceeding 4000 traps per week.

4.2.4. Juárez, Chihuahua

This unit contained the state of Chihuahua's most destructive PBW populations. Its situation was inseparable from the El Paso, Texas valley unit as described in the Texas Section. The major differences were in a latter programme starting date (2002) and the absence of Pima cotton. Cotton was largely non-*Bt* in the Juárez unit and non-*Bt* cotton fields on the Mexican side of the border were all *G. hirsutum* varieties (upland cotton).

The first year's start was with a short time frame as described in the previous areas of Chihuahua. This unit started with only the cotton areas in the Rio Bravo (Rio Grande, USA) River Valley. As time progressed, new centre pivot irrigated cotton areas near Villa Ahumada, south of Juárez, were added to the unit. This centre pivot cotton did not create problems. *Bt*-cotton was primarily grown in the outskirts of the city of Juárez's urban interfaces. This scattered cotton production was found in the north-western end of the valley. *Bt*-cotton ranged from 33% in 2002 to 13% in 2008 (Table 9).

Table 9. Summary data pink bollworm eradication programme for the Juárez area of Chihuahua

Year	2002	2003	2004	2005	2006	2007	2008	2009
Total hectares (ha)	5251	7579	8689	7915	8898	7625	7371	5052
% *Bt*-cotton	33	24	29	29	23	20	13	18
PBW Rope (ha)	3528	5736	6177	1070	1063	1159	194	155
Insecticides (ha)	697	60	70	30	0	114	0	0
Sterile moths released (millions)	0	0	0	69.4	205.7	346.5	371.9	344.5
Native moths trapped	9886	5573	1247	1108	1957	447	4	0

Conventional insecticide was most heavily applied at the start of the year 2002; in total 697 aggregate ha were treated over that season. The second highest aggregate treatment occurred in 2007 at 114 ha. Use of aerially applied insecticides was limited to 30-70 ha in 2003-2005. All non-*Bt* fields were treated with PBW Rope during the initial three years of operations (2002-2004). As in other areas of Chihuahua, detection thresholds were used after 2004 and PBW Rope treatments declined after that year.

In 2002 native moth captures were the most intense of any area in Chihuahua. From August 25 to November 10, a total of 9486 native moths were captured in 11 586 total traps serviced. As in other areas of Chihuahua, native moth captures declined during subsequent years (Staten and Ramirez-Sagahon 2008).

By 2006 and 2007, the majority of all captures were recovered in one general area of the state of Chihuahua, which was contiguous with fields on the Texas side of the border. In 2008, the Chihuahua programme successfully encouraged producers to shift most of these fields to *Bt*-cotton. The last native PBW in the Juárez unit were trapped in 2008.

In 2010 and 2012, PBW Rope was applied to 129 and 153 ha respectively, in coordination with Texas treatments targeting finds in adjacent Texas fields still with native moth captures. These treatments were triggered based on late-season captures in 2009 and 2011 in Texas (Table 5). Sterile moth releases, started in 2005, therefore continued through 2012 at 180.2, 119.8, and 118.3 million per year. This was done in concert with Texas captures in adjacent fields. There is no ecological separation between the two areas.

4.3. New Mexico, USA (Phases I and II)

Joe Friesen, Executive Director PBW and BW Foundation, 1999-2013
Patrick Sullivan, Executive Director PBW and BW Foundation, 2013 to present

The New Mexico Phase I programme, managed by the New Mexico PBW and BW Foundation, covered all the cotton along the Rio Grande River in the Mesilla Valley and the Hatch Valley (Fig. 1). These two valleys follow the river from El Paso, Texas, north to the Caballo Lake Dam ($\pm$100 miles), where cotton is no longer a dominant crop. This area, established before 1930, has a diversified agriculture with a high percentage of its area dedicated to pecans, peppers, vegetables, alfalfa, and grain. Urban interface is significant around its biggest city of Las Cruces. The majority of cotton acreage is in these two valleys, where it is irrigated with lake-stored water supplemented with groundwater in close proximity to the river.

A smaller area of cotton production is found further west near Deming, New Mexico. It is pump-irrigated and similar to the Ascensión area of Chihuahua, Mexico south of this part of New Mexico. Details for the PBW programme for the years 2000 through 2007 can be found in Friesen and Staten (2008). In four of the first five years of the programme, non-*Bt* exceeded *Bt*-cotton (Table 10). Non-*Bt* cotton was made up of Pima and upland cotton in order of importance. Of this cotton, 2-5% was certified organic.

From 2002 through 2004 all non-*Bt* cotton was treated with PBW Rope. As outlined in Friesen and Staten (2008) this was complicated by local cultivation practices applied for weed control. Cotton was planted in nearly flat low beds, but then cultivation raised the soil in the plant row while forming a deep furrow. As a result, PBW Rope applied at the 6th leaf node on a young plant was quickly covered in dirt. Pheromone emission was therefore blocked rendering the treatment ineffective.

To overcome this cultivation problem, in 2003 sprayable pheromones and dual insecticide-pheromone treatments were applied from ground spray rigs and by air at a targeted 10-day interval until the PBW Rope could be applied. This was logistically difficult and drastically increased programme input costs. It also contributed to the higher than expected usage of sprayable pheromones and insecticide shown in Table 10.

Table 10. Summary data pink bollworm eradication programme in New Mexico (Phase I)

Year	2002	2003	2004	2005	2006	2007	2008	2010
Total acres	17 061	21 061	21 701	21 722	21 627	16 957	14 664	13 246
% *Bt*-cotton	37	51	46	45	36	80	74	76
PBW Rope (acres)	10 690	9300	9493	1991	627	1325	0	0
Acres aerial pheromone	0	17 025	9843	3445	0	63	0	0
Acres pheromone + insecticide	0	13 115	4806	255	0	0	0	0
Mean traps serviced/week	1782	1906	2371	2231	1652	910	412	633
Sterile moths (millions)				N/A	322.7	365.9	307	
Sterile moths recovered					330 308	394 842	279 385	15 907
Larvae detected		227	2	0	0	0	0	0
Native moths trapped	51 764	126 033	18 126	2978	203	15	0	1

Treatments were concentrated in pockets. By 2005 this level of treatment could be reduced. Even so, with over 11 900 acres of non-*Bt* cotton, only a combined cumulative 3703 acres of sprayable pheromone and conventional insecticides were applied. PBW Rope applications in 2007 were based largely on native moth captures in 2006. The last year of pheromone treatment in the New Mexico programme was 2007.

Sterile moth releases started on a partial, "experimental" basis in Deming, Hatch, and north of Las Cruces in 2004. The entire area received its first full complement of sterile moths in 2005. Although *Bt*-fields were not specifically targeted, sterile moths were present in all fields. In 2007, all non-*Bt* fields received direct sterile moth releases. Less than 50% of these non-*Bt* fields required any additional pheromone treatment.

Native moth captures were again used to document the decline of detectable populations. In 2003, total counts were much higher than in the first year (2002). Nevertheless, in 2001 pre-programme captures of moths per trap per week still peaked at $\pm$7 times greater when compared to similar data for 2003. Post-2003 captures of native moths declined each year through 2007, when only 15 native moths were trapped for the entire year.

In 2010 one moth was captured in a field just north of El Paso, Texas, which abutted a Texas field with concurrent late-season captures. These fields were literally separated by a line on a map representing no more than 20 feet in distance. This small area was included in the Texas sterile moth release operations for the remainder of the 2010 season and in 2011.

By 2006, cotton in Chavis, Eddy, and Lee counties of New Mexico was managed by contract as part of the Texas boll weevil eradication programme. During this time the area was under extensive ULV malathion treatment for boll weevil. The cotton produced was predominantly *Bt*-cotton. Pierce et. al. (2013) discusses the last 9 moth captures in 2009. All evidence indicates these captures were remnants of the same population movement that affected the Pecos valley area of Texas in that same time frame. In that study no New Mexico captures occurred in 2010 and 2011. APHIS records show negative surveys in 2011 to 2015.

In Hidalgo county, a small area of cotton production on the Gila river at the Arizona-New Mexico border was managed in Phase II as part of Arizona's Safford district (Section 4.4.1.) without separation of data.

4.4. Arizona, USA (Phases II, IIIa and IIIb)

Larry Antilla, ACRPC Director, 1991-2011
Leighton R. Liesner, ACRPC Director, 2011 to present

The Arizona Cotton Research and Protection Council (ACRPC) was established in 1984 to deal with an outbreak of boll weevil in Arizona. The outbreak was rapidly expanding in scope. The resulting programme culminated in declared boll weevil eradication in 1991 (Neal and Antilla 2001). The organization remained intact for continued monitoring of boll weevil and PBW populations. Its numerous projects included the successful Parker area-wide mating disruption trial, as noted previously.

It became a model for other areas of what could be accomplished. It was the key example used to convince Texas growers to make a first commitment to eradication (Allen et al. 2005).

The organization became deeply involved in a coordinated programme to address area-wide resistance management of PBW in *Bt*-cotton with the University of Arizona (Tabashnik et al. 2000; Antilla et al. 2001; Dennehy et al. 2004). The understanding of the presence of resistance and its risk potential was intense. As an issue and threat, it was completely controlled as exemplified by declines of resistance expression (Tabashnik et al. 2010, 2013).

When sterile moths became available, the organization was in place for an area-wide eradication programme. It had extensive pre-programme population monitoring data and knowledge of the relevant areas. The ACRPC and the University of Arizona were critical in clearing and implementing a special state 24C label allowing the utilization of sterile insects in lieu of structured non-*Bt* cotton refugia for PBW resistance management. A 2005 revised EPA primary *Bt*-cotton label issued for 2006 could have ended eradication efforts as they would have rendered some cotton untreatable (see transgenic cotton Section 3.3.1. of this chapter).

4.4.1. Eastern Three Fourths of Southern Arizona (Phase II)

With the passage of a grower-approved referendum on PBW eradication funding in 2005, Arizona's ACRPC started its first year with all tools at their disposal in 2006. It started in approximately 85% of cotton state-wide. It constituted all cotton in surface and groundwater irrigated areas within the eastern three fourths of southern Arizona (Fig. 1, Phase II).

As shown in Table 11, of the 165 683 acres, 93.1% were *Bt*-cotton in 2006. The Safford Valley exceeded this overall average, providing the majority of the Phase II area's non-*Bt* cotton. During this time, much of the cotton planted was Pima (*G. barbadense*).

An unpublished cooperative effort with ACRPC, USDA-CPHST, and Pacific BioControl led the development of the application technology of PBW Rope pre-wound on a bamboo stick in lieu of the ropes being directly "tied" on small cotton plants. This allowed ACRPC to mechanize treatments for many of their non-*Bt* fields. In all non-*Bt* fields, treatment could thus be initiated earlier. Arizona was successful in applying PBW Rope on all non-*Bt* cotton in 2006-2009. Late 2011 native moth captures led Arizona to again treat all non-*Bt* cotton with PBW Rope in 2012. As shown in Table 11, aerially applied pheromone treatments declined from 2006 to 2010 (from 6409 to 0).

In 2006, Arizona treated a cumulative 6409 acres with conventional insecticides predominantly on a small cluster of fields near Eloy. This event did not reoccur in subsequent years. All conventional insecticide use was eliminated in 2009. Arizona conducted boll surveys in late-season as part of an ongoing *Bt* resistance monitoring programme. Larval PBW populations ceased to be detectable by 2009 in this portion of the state. Adult PBW capture decline from 2006, the first year of treatment, through 2011, was equally impressive.

Table 11. Summary data pink bollworm eradication programme in Arizona (Phase II - Arizona Zone 1)

Year	2006	2007	2008	2009	2010	2011	2012	2013
Total acres	165 683	145 947	118 435	124 191	168 255	213 413	159 244	136 244
Non-*Bt* acres (%)	11 465 (6.9)	6389 (4.4)	2029 (1.7)	2144 (1.7)	7387 (4.4)	11 946 (5.6)	6607 (4.1)	4427 (3.2)
PBW Rope (acres)	11 465	6389	2029	2144	0	0	6607	0
Acres aerial pheromone	6409	1458	31	64	0	0	0	0
Insecticides (acres)	2907	1197	62	0	0	0	0	0
Sterile moth releases (millions)*	1682	1443	1410	1990	1123	1593	1300	382
Sterile moths recovered (millions)	1.447	0.918	0.551	0.307	1.137	1.047	0.489	0.012
Larvae detected (%)	1126 (2.0)	31 (0.08)	2 (0.02)	0	0	0	0	
Native moths trapped	657 752	199 726	2306	866	453	566	0	0

* *In 2006, California Phase IIIa's release numbers are included in this table*

4.4.2. North of Yuma Arizona along the Colorado River (Phase IIIa)

Eradication treatments of cotton areas north of the Yuma, Arizona area along the Colorado River (Fig. 1) began in 2007. It included the Parker and Mojave valleys along the Colorado River. Southern California started its eradication treatments at the same time. The area had high ratios of *Bt*- to non-*Bt* cotton (Table 12). Prior to 2007 many of the growers in this area used the un-treated non-*Bt* cotton option (5%) for resistance management. At least some of this was not harvested (sacrificed) due to PBW damage.

With the initiation of sterile moth releases for resistance management, all non-*Bt* cotton could be and was treated at or just before 6 leaf with PBW Rope on a bamboo stake. This occurred from 2007 through 2010, and in 2012. The decision to treat in 2012 was state-wide after unexpected increases in late 2011 (see above). With a preponderance of *Bt*-cotton, PBW Rope treatment, and sterile moth releases, the need for sprayable pheromone and conventional insecticides ended quickly (Table 12).

Throughout the pheromone-treatment years, the majority of all native moth captures were actually in *Bt*-cotton, either near a non-*Bt* cotton field or in fields which were rotated out of a non-*Bt* refuge field the previous year. Captures were reduced by 97.9% between 2007 and 2008. Larval populations were no longer detectable after 2007.

Table 12. Summary data pink bollworm eradication programme in Arizona (Phase IIIa - Arizona Zone 2)

Year	2006*	2007	2008	2009	2010	2011	2012	2013
Total acres	N/A	16 546	12 835	12 108	18 654	28 940	24 204	19 485
Non-*Bt* acres (%)	N/A	509 (3.1)	993 (7.7)	1812 (15)	1116 (6)	1278 (4.4)	299 (1.2)	285 (1.5)
PBW Rope (acres)	N/A	509	993	1 812	1 116	0	299	0
Areal pheromone (acres)	N/A	290	0	111.5	0	0	0	0
Conventional (acres)	N/A	117	0	0	0	0	0	0
Sterile moths recovered	N/A	56 689	47 432	32 546	18 125	17 150	21 904	10
Larvae detected	784	31	0	0	0	0	0	0
Native moths trapped	N/A	155 104	3259	1183	36	17	0	0

* *Pre-programme treatment*

4.4.3. Yuma and Lower Colorado Basin (Phase IIIb)

The profile of the Yuma area, Phase IIIb was similar to Phase IIIa and the rest of the Lower Colorado Basin (Fig. 1). In all these areas, when non-*Bt* cotton is produced full season, the PBW population development potential is extensive, as most of the winter is frost-free. Fortunately, in Yuma most of the cotton is rotated with high value winter vegetables. That portion of the area's cotton crop is terminated in August and September. Consequently, the PBW population growth potential from September through November is severely curtailed in these fields. This, and high percentages of *Bt*-cotton, was extremely important in potential PBW population reduction (Table 13).

All non-*Bt* fields were treated with PBW Rope for the first five years of programme operations. A portion of the non-*Bt* fields were treated with aerially applied pheromone in 2008 and 2009 after PBW Rope efficacy ended. Conventional insecticides were only needed in 2008 and were minimal. Larval populations were not detected after 2009. After 2007, pre-programme survey moth captures declined sharply with no native moth captures in 2011 and one in 2012.The one capture in 2012 was in early April before hostable fruit set. It was the last native moth detection in the programme in the USA.

Table 13. Summary data pink bollworm eradication programme in Arizona (Phase IIIb - Arizona Zone 3)

Year	2007*	2008	2009	2010	2011	2012	2013
Total acres	N/A	10 358	12 957	15 571	24 676	19 116	11 502
Non-*Bt* acres (%)	N/A	236 (2.3)	717 (5.5)	344 (2.2)	2010 (8.1)	458 (2.4)	178 (1.5)
PBW Rope	N/A	236	714	344	2010	458	0
Acres aerial pheromone	N/A	248	930	0	0	0	0
Conventional (acres)	N/A	386	0	0	0	0	0
Sterile moths recovered	N/A	11 861	75 027	39 718	75 822	75 961	130 978
Larvae detected	44	7	2	0	0	0	0
Native moths trapped	61 166	21 032	4175	85	0	1	0

* *Pre-programme operations*

4.5. Southern California, USA (Containment Programme, Phases IIIa and IIIb)

Jodie Brigman District, Supervisor CDFA, 2001-2017
Jim Rudig, Programme Manager CDFA, 2006-2011
Victoria Hornbaker, Programme Manager CDFA, 2011, interrupted, current

The California organization (California Cotton Pest Control Board, CCPCB) was in place and operating in southern California before 2007, with all needed CDFA staff working on PBW activities, including the long-standing San Joaquin Valley containment/exclusion programme. They were deeply involved in population monitoring particularly in long standing monitoring of seasonal PBW movement throughout the area. Important efforts included monitoring for *Bt* resistance management in cooperation with the other states, led by the University of Arizona (Dennehy et al. 2004).

When PBW invaded southern California in 1963-64, the Imperial Valley, the Coachella Valley, and the Blythe-Palo Verde Valley had an extensive and prosperous cotton industry, frequently comprising more than 100 000 acres. Before PBW establishment, southern California produced very high yields with limited, targeted insecticide. The tenets of integrated pest management pioneered here, as described by Stern et al. (1959), had become a world standard.

In this southern California area, cotton was produced using a February-March planting window and harvested well into December. As part of the Lower Colorado River Basin (Fig. 1), with its neighbours of Yuma in Arizona, San Luis Río Colorado in Sonora, and Mexicali in Baja California, cotton's long-season growth regimen allowed the introduction and establishment of PBW to produce more generations per season than anywhere else in its North American range. Results of the invasion were significant yield losses and extensive insecticide use. Secondary pests, exemplified by whitefly, became common.

By the time southern California entered the PBW eradication programme in 2007, neither the Coachella Valley, nor the Imperial Valley had any commercial cotton. Only the Blythe area, the Palo Verde Valley, Bard/Winterhaven area (adjacent to Yuma), and Needles north of Parker, still produced some cotton in southern California. Most importantly, southern California had the highest ratio of *Bt*- to non-*Bt* cotton of any area in the programme (Table 14).

Southern California was unique in that it also had a small but important okra production. Okra is a weak host for PBW in the south-western USA and PBW has a decided preference for cotton. In addition, okra pods are harvested before larvae can develop, serving as a mechanical control. In the above agronomic environment, no conventional insecticides or sprayable pheromone systems were used. All non-*Bt* cotton was treated with PBW Rope from 2007-2011. Okra in the Imperial Valley was treated with PBW Rope for the initial three years. PBW Rope was also applied in the Coachella valley in 2008.

The small amount of non-*Bt* cotton was targeted at the onset of the programme at the standard mean of 250 sterile moths/acre/day (618/ha/day). All okra was targeted at a mean of 200 sterile moths/acre/day. In their most important role, sterile moth releases were used for PBW resistance prevention and management. *Bt*-cotton was targeted at a standard of 10 sterile moths/acre/day (24.7/ha/day). Adult moth monitoring showed a consistently decreasing number of native PBW moths captured each year that the programme progressed (Table 14).

The last native moth was captured in southern California in 2011. It was captured in a highway trap line, not in a field trap.

Table 14. Summary data pink bollworm eradication programme in southern California

Year	2007	2008	2009	2010	2011	2012	2013
Total acres cotton	16 555	9 635	6 132	10 445	19 175	18 293	16 171
Non-*Bt* acres (%)	130 (0.8)	30 (0.3)	71 (1.1)	0 (0.0)	85 (0.4)	105 (0.6)	30 (0.2)
Okra (acres)	205	460	597	560	513	555	515
PBW Rope cotton (acres)	130	30	71	0	85	0	0
PBW Rope okra (acres)	205	460	327	0	0	0	0
Sterile moths released (millions)	76.46	72.15	36.71	124.74	5.27	1.25	0
Native moths trapped	447 067	16 395	6142	147	1	0	0

4.6. Northern Sonora, Mexico (Phase IIIa)

Ing. Javier Valenzuela Lagarda, Gerente de Comité de Sanidad Vegetal, 2006 to declaration.

The cotton growing area of northern Sonora is found predominantly along the lower Colorado River in San Luis Río Colorado. In addition, it includes cotton in Sonoyta on the Arizona/Sonora border more than 200 km to the east. The Sonoyta data include scattered fields near the city of Caborca. In the latter, like much of Sonora's southern coastal production areas, agriculture has shifted to vegetable production. At this time no cotton production remains in Caborca. The only other cotton in Sonora during this programme was in the state's southern coastal areas. The Sonoyta area was limited and of a shorter season than San Luis. The San Luis growing area is separated from the Yuma, Arizona growing area by the small twin cities of San Luis Río Colorado and contiguous San Luis, Arizona. It is separated from the Mexicali Valley of Baja California only by the normally dry Colorado River and riparian area. The cotton is planted in early March through mid-April. In San Luis irrigation for cotton production is limited after August. This limits reproduction in later generations of PBW.

During 2007 and 2008, only non-*Bt*-cotton was produced in the Sonoyta / Caborca area. By 2013, this trend was completely reversed in favour of *Bt*-cotton in Sonoyta, while after 2010 the scattered fields around Caborca were no longer in cotton. The increased ratio of *Bt*- to non-*Bt* cotton was more pronounced in San Luis proper as well. By 2013, 99% of all cotton was *Bt* as illustrated in summary Table 15.

Table 15. Northern Sonora, Mexico pink bollworm eradication programme summary data
(Phase IIIb, Figure 1)

Year	2007	2008	2009	2010	2011	2012	2013
Total hectares (ha)	3126	3885	2974	3657	6881	6751	3515
% *Bt*-cotton	54	72	75	87	91	92	99
Pheromones (ha)	1425	1093	751	478	550	237	10
Insecticide + (ha)	454	197	55	33	0	0	0
Sterile moth releases (millions)	0	0	152.5	198.4	208.8	210.3	117.2
Larvae detected	22 fields	9 fields	7 fields	1 field			
Native moths trapped	1 139 586	159 421	35 771	1139	163	7	0

In 2007 and 2008, sterile insect release was not available until 2009. The programme objectives were to drive populations down below conventional pest management field treatment thresholds. Northern Sonora's objectives were to treat all non-*Bt* cotton with PBW Rope on bamboo stakes. Limited conventional insecticide plus sprayable pheromone was required, but progressively reduced through 2010 (Table 15). Programme management treated fields with the highest risk with a second application of PBW Rope at 50-65 days, when triggered by native moth captures. Insecticide plus sprayable pheromone was used only when a larva was detected in boll samples.

In the first year (2007), with a full trap grid (1 trap per 4 ha non-*Bt* and 1 trap per 20 ha *Bt*-cotton), captures totalled 1 139 586 native moths season-long. The San Luis Rio Colorado area contributed most of these captures. It was influenced by its adjacent neighbour, the Mexicali valley, which was not yet in the eradication programme. In 2008 captures declined significantly to 159 421 (Table 15).

Perspective pre-programme data on file show season-long mean native moths per trap per week from 2006 through 2008 to be 106, 46, and 16. This downward progression continued to zero native moths captured in 2013. Solid evidence of reproduction is found when any larvae can be detected in a field. During the first year, 22 fields were positive, all in San Luis Río Colorado. By 2010, even with extensive "directed" sampling of fields with native moth captures, only in one field a larva was detected.

No PBW has been detected in this programmatic area after May of 2012. The last seven native moth captures coincide in time and space with the last adults captured in neighbouring Mexicali and Arizona.

4.7. *Mexicali, Baja California, Mexico (Phase IIIb)*

Ing. Enrique Montano, Gerente del Comité Estatal de Sanidad Vegetal, 2006-2007
Ing. Roberto Roche Uribe, Gerente del Comité Estatal de Sanidad Vegetal, 2008 to declaration.

The Mexicali Valley, during the course of this eradication programme, contained all the cotton in Baja California (Fig. 1). It was established as a production area in 1912. Its water source, like San Luis Río Colorado, is the Colorado River, its eastern border. The northern limit of this valley is the California, USA border with Mexico. As with San Luis Rio Colorado, if cotton were to be grown for its longest potential season, it would generate extremely high populations of PBW. Irrigation for cotton has long been terminated in late August. Yield potential is still high, with planting and harvest windows consistent with its Sonoran neighbour. This area has no PBW population separation from northern Sonora.

When PBW entered this system, as in San Luis Río Colorado, it drastically affected production practices. Insecticide use escalated for secondary pests as well as PBW. Shorter growing cycles became the reality. As in all areas, the introduction of *Bt*-cotton was profound. Its use escalated even as growers ceased to find PBW resulting from eradication activities in non-*Bt* cotton (Table 16).

Pre-programme monitoring data for 2007 has been provided, indicating extremely high native moth captures. Programmatic control activities in non-*Bt* cotton started in 2008 with pheromone treatment of all fields. Two different high-rate systems were used in 2008. The PBW Rope was used on 73% of non-*Bt* fields. A second high-rate system was used on the remaining non-*Bt* fields.

In 2009-2012 all non-*Bt* fields were treated with the PBW Rope, targeting a pre-6 node cotton development window. After 2010, all PBW Rope applications at or before 6 leaf were applied on the bamboo stake. Dual insecticide-pheromone applications were used on fields in which trap captures exceeded one moth per trap per night, or in which larvae could be found. This occurred in 2008, from the week of 4 August through 15 September. Field re-treatment varied depending on trap captures. By 2011, only one field required treatment (36 ha, Table 16). Sterile moth releases started in 2009 with the majority directed over non-*Bt* fields. This continued through 2013 in areas where native moths were captured in 2012.

In population assessment, no larvae were detected after 2009. These data were from a programme evaluation survey of randomly selected non-*Bt* fields. Fields were selected when mapping of cotton fields was complete early in the season. Field selection occurred before boll set.

Table 16. Mexicali valley, Baja California, Mexico pink bollworm eradication programme summary data

Year	2007	2008	2009	2010	2011	2012	2013
Total (ha)	20 643	19 984	17 385	20 153	33 671	32 829	22 814
% *Bt*-cotton	62	68	78	84	96	96	96
Pheromone (ha)	0	6505	3771	3170	1455	1511	0
Insecticide+ (ha)	0	2750	1652	264	36	0	0
Sterile moth releases (millions)	0	0	822.6	766.3	825.5	755.4	258
Larvae detected (%)	1450 (29)	181 (3.2)	4 (0.007)	0	0	0	0
Native moths trapped	2 705 400	709 203	162 226	15 258	401	18	0

All native moth capture data in Table 16 were totals from the programmatic standard season-long trap grids. Pre-programme valley monitoring in 2007 produced a season-long moths/trap/week average of 103.9 native moths. Many of these delta traps were past trap capacity and no longer capturing all moths which entered the trap. At the end of the first year of programmatic treatments, season-long average capture/trap/week dropped to 9.2.

As was true in Sonora and Arizona, 2012 was the last year in which native adult moths were captured. In 2012, a total of 18 native moths were all captured on or before the week of 26 May. These were the last native moths captured and the last detection of any life form of PBW anywhere in this bi-national programme.

5. CONCLUSIONS

On November 22, 2012 ten municipalities in north-western Chihuahua were declared free of PBW (as officially eradicated). This area was the Ascensión work area which had not had a detected population for 5 years. Subsequently, on December 8, 2014, eradication was declared for the remainder of the state of Chihuahua. On February 3, 2016, PBW was declared eradicated from Sonora and Baja California (SENASICA 2018).

Though not covered in this report, PBW has in the meantime also recently been eradicated from the Mexican states of Coahuila and Durango (Diario Oficial 2018). This latter cotton area (La Laguna) centres around the city of Torreón in the states of Coahuila and Durango, which was the first reported area infested with PBW in continental North America (Noble 1969). The state of Tamaulipas, which is contiguous to Texas and has likewise been involved in PBW eradication activities, also had no PBW captures in 2018, but has not yet been declared PBW-free (SADER 2018).

In the USA, eradication could only be declared after *Bt*-cotton labelling issues for refugia (grower variety selection) were resolved. This occurred after 6 years of continuous negative surveys. The United States Secretary of Agriculture signed the eradication proclamation for all USA cotton production areas on October 19, 2018 (USDA 2018). Eradication has been successfully achieved over a very diverse geographic and ecological range because many years of research and development had provided multiple surveillance and control tools (Noble 1969; Naranjo et al. 2002). These were tools which could be used synergistically. The area-wide integration of tools was then successfully tailored to varied habitats over the pest's broad range.

Resistance to *Bt*-cotton is, in the view of the first author, the most important entomological issue concerning cotton worldwide. Movement of a multi-gene resistant PBW population back into this bi-national programme area would be of the gravest concern.

6. ACKNOWLEDGEMENTS

Dr. Michelle Walters passed away due to cancer before this manuscript could be completed. She worked extensively for PBW eradication throughout her career. Larry Antilla, Theodore N. Boratynski, Edwardo Gutierrez, Tish Bond, and Eoin B. Davis were invaluable in data verification and manuscript preparation.

It is not tenable within the scope of this paper to list all the people and organizations that were important for achieving success. In the USA, leaders from the producer community and "field" were the principal forces for initiation and execution. The project in Mexico followed the presidencies found in many of its state phytosanitary committees. Direct field managers for each state have already been listed in the text.

Important acknowledgments for Mexico agricultural authorities are: Dr. Jorge Hernandez Baeza, Director Sanidad Vegetal; Dr. Javier Trujillo Director, SENASICA; Ing. Hector Sanchez Anguiano, DGSV; and Ing. Juan Carlos Ramirez-Sagahon, SENASICA.

In Chihuahua we worked with: Ing. Rubin Ortega Rodrigues, Ing. Lionel Gutierrez Estrada, Ing. Arnulfo Nunez Carbajal, Ing. Carlos Garcia Duran, Ing. Jesus Antonio Escarcega Terin, Ing. Izabel Roman Medina, Ing. Alfredo De La Torre Rivera (USDA FSN), Ing. Juan Angel Guzman M., Ing. Francisco Cardenas, Ing. Epifanio Hernandez G. Ing. Ricardo Alvarado Garcia, Ing. Luis Omar Jimenez Quintana, and Ing. Luis Carlos Ortega Duran.

In Sonora programme colleges include: Ing. Gilberto Valdez, Ing. Urgujo, Ing. Erick Cortes Onofre, Ing. Mauricio Chavarria Onofre, Ing. Rene Yescas Dominguez, and Ing. Ricardo Vazquez.

Acknowledgements of importance for Baja California include: Ing. Enrique Montano, Ing. Roberto Roche Uribe, Ing. Hector Aguirre Romero, and Ing. Ricardo Mora Armento.

The USDA-APHIS-IS network collaborated with all the above and included: Elba Quintero, Nicholas Gutiérrez, Theodore Boratynski, Ing. Francisco Corrales Dorame (FSN), and Ing. Edwardo Gutiérrez (FSN).

In the USA the National Cotton Council provided the framework for all legislative, technical, managerial and budgetary decisions through its producer member Pink Bollworm Action Committee (PBWAC). Chairmen included Ted Pierce (Arizona), Bill Lovelady (Texas), Denis Palmer (Arizona), Clyde Sharp (Arizona), and Ted Sheely (California).

A Technical Advisory Subcommittee offered its recommendations. Vice President John Maguire and staff members Frank Carter and Don Parker are noted.

The most important contributions of the USDA-APHIS-PPD were in sterile insect production, regulatory issues and programmatic support. Fred D. Stewart, Ernie Miller, and Eoin Davis managed all rearing and shipping in sequence. The USDA-APHIS staff officers William J. Grefenstette, Osama A. El-Lissy, and James A. Schoenholz are noted.

El Paso, Texas producers that fostered the eradication programme were led by Bill Lovelady and Jim-Ed Miller. They worked with TBWF and its board of directors, the Foundation Director Lindy Patton and the Boll Weevil Programme Manager Osama El-Lissy. Woody Anderson provided primary grower leadership as Foundation Chairman.

The New Mexico cotton producers who were a constant throughout are represented by Keith Deputy and Robert Sloan. Work unit supervisors were Leighton Liesner and Allen Van Tassel.

Arizona's strength came from the Arizona Cotton Growers Association, Rick C. Lavis (Vice-President), and its operations arm, the producer-led Arizona Cotton Research and Protection Council. Council Chairmen include: Paul Ollerton, Clyde Sharp, Denis Palmer, Adam Hatley, and Jerry Rovey. Arizona directors previously listed and Mike Whitlow, Donna Fairchild, Mike Woodward, Bob Ellington, Penny Malone, Jerry Kerr, and Bobby Soto.

The California Cotton Pest Control Board (CCPCB) were primary funders of the 60-year SIT containment programme and the PBW rearing facility. Wally Shropshire, Chairman, and Jack Stone were 50+ year influencers on this board. Funds were managed through CDFA with partnership with USDA.

Important early managers from CDFA included Dr. Isi Siddique and Robert Roberson. They were critical in administration of CCPCB funding and in construction of the rearing facility. Agricultural Commissioners in all California cotton production counties were the local regulatory arm.

7. REFERENCES

Allen, C. T., L. E. Smith, S. E. Herrera, and L. W. Patton. 2005. Pink bollworm eradication in Texas – a progress report, pp. 1219–1224. Proceedings Beltwide Cotton Conferences, New Orleans, Louisiana. National Cotton Council of America. Nashville, Tennessee, USA.

Allen, C. T. 2008. Boll weevil eradication: An areawide pest management effort, pp. 467–559. *In* O. Koul, G. Cuperus, and N. Elliott (eds.), Areawide pest management: Theory and implementation. CAB International. Wallingford, UK.

Anonymous. 1961. The pink bollworm in Arizona 1958–1960. USDA, Agricultural Research Service, Plant Protection Pest Control Division.

Antilla, L., and L. Liesner. 2008. Program advances in the eradication of the pink bollworm *Pectinophora gossypiella* in Arizona cotton, pp. 1162–1168. Proceedings Beltwide Cotton Conferences, Nashville, Tennessee. National Cotton Council of America. Nashville, Tennessee, USA.

Antilla, L., M. Whitlow, R.T. Staten, O. El-Lissy, and F. Meyers. 1996. An integrated approach in areawide pink bollworm management in Arizona, pp. 1083–1086. *In* P. Dugger, and D. A. Richter (eds.), Proceedings Beltwide Cotton Conferences, Memphis, Tennessee. National Cotton Council of America. Nashville, Tennessee, USA.

Antilla, L., M. Whitlow, B. Tabashnik, T. Dennehy, and Y. Carrière. 2001. Benefits of multi-level monitoring activities for a pink bollworm resistance management program in transgenic (*Bt*) cotton in Arizona, pp. 1173–1175. Proceedings Beltwide Cotton Conferences. National Cotton Council of America. Nashville, Tennessee, USA.

Baker, T. G., R. T. Staten, and H. M. Flint. 1990. Use of pink bollworm pheromone in the southwestern United States, pp 417–436. *In* R. L. Ridgway, R. M. Silverstein, and M. N. Inscoe (eds.), Behavior modifying chemicals for insect management: Application of pheromones and other attractants. Marcel Dekker, New York, USA.

Bierl, B. A., M. Beroza, R. T. Staten, P. E. Sonnet, and V. E. Adler. 1974. The pink bollworm sex attractant Journal of Economic Entomology 67: 211–216.

Brooks, T. W., C. C. Doane, and R. T. Staten. 1979. Experience with the first commercial pheromone communication disruptive for suppression of an agricultural pest, pp 375–388. *In* F. J. Ritter (ed.), Chemical ecology: Odour communication in animals. Elsevier/North-Holland Biomedical Press, Amsterdam, The Netherlands.

Burns, D. W., M. P. Murphy, M. L. Parsons, L. A. Hickle, and R. T. Staten. 1983a. The evaluation of internal elemental discriminators for pink bollworm by inductively coupled plasma-atomic emission spectrometry. Applied Spectroscopy 37: 120–123.

Burns, D. W., M. P. Murphy, K. L. Jones, M. L. Parsons, P. Farnsworth, E. T. Ozaki, and R. T. Staten. 1983b. The evaluation of internal elemental markers for Mediterranean fruit fly (Diptera: Tephritidae) reared in tagged artificial diets. Journal of Economic Entomology 76: 1397–1420.

Cardé, R. T., A. Mafra-Neto, R. T. Staten, and L. P. S. Kuenen. 1997. Understanding mating disruption in the pink bollworm. *In* P. Witzgall, and H. Arn (eds.), Technology transfer in mating disruption. Pheromones and other semiochemicals in integrated production. Proceedings IOBC-WPRS Symposium, 1996, Montpellier, France.

Dennehy, T. J., G. C. Unnithan, S. A. Brink, B. D. Wood, Y. Carrière, B. E. Tabashnik, L. Antilla, and M. Whitlow. 2004. Update on pink bollworm resistance to *Bt* cotton in the southwest, pp. 1569–1577. Proceedings Beltwide Cotton Conferences. National Cotton Council of America. Nashville, Tennessee, USA.

Diario Oficial. 2018. Acuerdo por el que se declara como zona libre del gusano rosado (*Pectinophora gossypiella*) a los estados de Coahuila de Zaragoza y Durango. Diario Oficial de la Federación, Primera Sección. 20 de diciembre de 2018. Secretaría de Agricultura y Desarrollo Rural (SADER). Estados Unidos Mexicanos. Ciudad de México, México.

El-Lissy, O., R. T. Staten, and B. Grefenstette. 2002. Pink bollworm eradication plan in the U.S, pp. 973–971. Proceedings Beltwide Cotton Conferences, Atlanta, Georgia. National Cotton Council of America. Nashville, Tennessee, USA.

Flint, H. M., S. Kuhn, B. Horn, and H. A. Sallam. 1974. Early season trapping of pink bollworm with gossyplure. Journal of Economic Entomology 67: 738–740.

Flint, H. M., J. R. Merkle, and A. Yamamoto. 1985. Pink bollworm (Lepidoptera Gelechiidae): Field testing a new polyethylene tube dispenser for gossyplure. Journal of Economic Entomology 78: 1431–1436.

Flint, H. M., T. J. Henneberry, F. D. Wilson, E. Holguin, N. Parks, and R. E. Buehler. 1995. The effects of transgenic cotton *Gossypium hirsutum* lines containing *Bacillus thuringiensis* toxin genes for the control of the pink bollworm, *Pectinophora gossypiella* (Saunders) Lepidoptera Gelechiidae and other arthropods. The Southwestern Entomologist 20: 281–292.

Foster, R. N., R. T. Staten, and E. Miller. 1977. Evaluation of traps for pink bollworm. Journal of Economic Entomology 70: 289–291.

Friesen, J., and R. T. Staten. 2008. Progress report for the New Mexico pink bollworm, *Pectinophora gossypiella* (Saunders), eradication program, pp. 1179–1185. Proceedings Beltwide Cotton Conferences, Nashville, Tennessee. National Cotton Council of America. Nashville, Tennessee, USA.

Fye, R. E. 1976. Improved method for holding cotton bolls for detecting pink bollworm. U. S. Department of Agriculture, Agricultural Research Service Publication 1, ARS W 37, Western Region, 3 pp.

Graham, H. M., and C. L. Mangum. 1971. Larval diets containing dyes for tagging pink bollworm moths internally. Journal of Economic Entomology 64: 376–379.

Grefenstette, B., O. El-Lissy, and R. T. Staten. 2009. Pink bollworm eradication plan in the United States. U.S. Department of Agriculture, Animal and Plant Health Protection Service. Washington, DC, USA.

Head, G. P., and J. Greenplate. 2012. The design and implementation of insect resistance management programs for *Bt* crops. GM Crops & Food: Biotechnology in Agriculture and the Food Chain 3: 144–153.

Hendrichs, J., M. J. B. Vreysen, W. R. Enkerlin, and J. P. Cayol. 2021. Strategic options in using sterile insects for area-wide integrated pest management, pp. 841–884. *In* V. A. Dyck, J. Hendrichs, and A. S. Robinson (eds.), Sterile Insect Technique – Principles and practice in Area-Wide Integrated Pest Management. Second Edition. CRC Press, Boca Raton, Florida, USA.

Hummel, H. E., L. K. Gaston, H. H. Shorey, R. S. Kaae, K. J. Byrne, and R. M. Silverstein. 1973. Clarification of the chemical status of the pink bollworm sex pheromone. Science 181: 873–875.

Knipling, E. F. 1971. Boll weevil and pink bollworm eradication: Progress and plans. Cotton Ginners Journal Yearbook 39: 23–30.

Leggett, J. E., O. El-Lissy, and L. Antilla. 1994. Pink bollworm moth catches with perimeter and in field gossyplure baited delta traps. The Southwestern Entomologist 19: 140–155.

Lingren, P. D., T. J. Henneberry, and T. W. Popham. 1989. Pink bollworm (Lepidoptera Gelechiidae): Nightly and seasonal activity patterns of male moths as measured in gossyplure-baited traps. Journal of Economic Entomology 82: 782–787.

Miller, E., F. Stewart, A. Lowe, and J. Bomberg. 1996. New method of processing diet for mass rearing pink bollworm, *Pectinophora gossypiella* (Saunders) (Lepidoptera: Gelechiidae). Journal of Agricultural Entomology 13: 129–137.

Naranjo, S. E., G. D. Butler, and T. J. Henneberry. 2002. A bibliography of the pink bollworm, *Pectinophora gossypiella* (Saunders). Bibliographies and Literature of Agriculture No. 136. USDA, Agricultural Research Service, USA. 156 pp.

Neal, C. R., and L. Antilla. 2001. Boll weevil establishment and eradication in Arizona and northwest Mexico pp. 213–224. *In* W. A. Dickerson, A. L. Brashear, J. T. Brumley, F. L. Carter, W. J. Grefenstette, and F. A. Harris (eds.), Boll weevil eradication in the United States through 1999. Reference Book Series No. 6. The Cotton Foundation Publisher, Memphis, Tennessee, USA.

Noble, L. W. 1969. Fifty years of research on the pink bollworm in the United States. U.S. Department of Agriculture, Agricultural Research Service, Agricultural Handbook No. 357.

Pierce, D. L., M. L. Walters, A. J. Patel, and S. P. Swanson. 1995. Flight path analysis of sterile pink bollworm release using GPS and GIS, pp. 1059–1060. *In* D. A. Richter, and J. Armour (eds.), Proceedings Beltwide Cotton Conferences, Memphis, Tennessee. National Cotton Council of America. Nashville, Tennessee, USA.

Pierce, J. B., C. Allen, W. Multer, T. Doederlein, M. Anderson, S. Russell, J. Pope, R. Zink, M. Walters, D. Kerns, J. Westbrook, and I. Smith. 2013. Pink bollworm (Lepidoptera: Gelechiidae) in the southern plains of Texas and New Mexico: Distribution and eradication of a remnant population. Southwestern Entomologist 38(3): 369–378.

Rudig, J. F., and D. F. Keaveny. 2008. Progress report and overview of pink bollworm, *Pectinophora gossypiella* (Saunders) eradication in California, pp. 1156–1192. Proceedings Beltwide Cotton Conferences. National Cotton Council of America. Nashville, Tennessee, USA.

(SADER) Secretaría de Agricultura y Desarrollo Rural. 2018. Sanidad Vegetal: Décimo segundo informe mensual campaña contra plagas reglamentadas del algodonero. Dirección General de Sanidad Vegetal, Dirección de Protección Fitosanitaria. Servicio Nacional de Sanidad, Inocuidada y Calidad Agroalimentaria (SENASICA), SADER, Gobierno de Mexico.

Schmitt Jr., T. J. 1967. The pink bollworm in Arizona, second addendum. Agricultural Research Service, Plant Protection Pest Control Division, USDA, USA.

(SENASICA) Servicio Nacional de Sanidad, Inocuidad y Calidad Agroalimentaria. 2018. Declara SAGARPA Zona Libre del gusano rosado del algodonero a tres estados y un municipio de Coahuila. Chihuahua, Baja California y Sonora, así como el municipio de Sierra Mojada en Coahuila obtuvieron el estatus libre de plaga. 7 de noviembre de 2018. Secretaría de Agricultura, Ganadería, Desarrollo Rural, Pesca y Alimentación. Ciudad de Mexico, Mexico.

Shahid, J. 2014. Bollworms develop resistance against *Bt* cotton crop. Islamabad: Farmers and agriculture scientists are alarmed by the destructive attack of bollworms this year that seem to have developed resistance against the genetically-modified (GM) cotton crop. Dawn, July 14, 2014. Islamabad, Pakistan.

Staten, R. T., and J. C. Ramirez-Sagahon. 2008. The bi-national pink bollworm eradication program – an overview, pp. 1169–1178. Proceedings Beltwide Cotton Conference, Nashville, Tennessee. National Cotton Council of America. Nashville, Tennessee, USA.

Staten, R. T., R. W. Rosander, and D. F. Keaveny. 1993. Genetic control of cotton insects: The pink bollworm as a working program, pp. 269–283. *In* Management of insect pests: Nuclear and related molecular and genetic techniques. International Atomic Energy Agency, Vienna, Austria.

Staten, R. T., L. Antilla, and M. L. Walters. 1995. Pink bollworm management: Prospects for the future, pp. 153–156. *In* D. A. Richter, and J. Armour (eds.), Proceedings Beltwide Cotton Conferences, Memphis, Tennessee. National Cotton Council of America. Nashville, Tennessee, USA.

Staten, R. T., O. El-Lissy, and L. Antilla. 1997. Successful area-wide program to control pink bollworm by mating disruption, pp. 383–396. *In* R. Cardé, and A. K. Minks (eds.), Insect pheromone research: New directions. Chapman Hall, New York, USA.

Staten R. T., M. Walters, R. Roberson, and S. Birdsall. 1999. Area-wide management / maximum suppression of pink bollworm in Southern California, pp. 985–988. Proceedings Beltwide Cotton Conferences, Memphis, Tennessee. National Cotton Council of America. Nashville, Tennessee, USA.

Staten, R. T., E. Miller, M. Grunnet, P. Gardner, and E. Andress. 1988. The use of pheromones for pink bollworm management in western cotton, pp. 206–209. *In* J. M. Brown, and D. A. Richter (eds.), Proceeding Beltwide Cotton Conferences, Memphis, Tennessee. National Cotton Council of America. New Orleans, Louisiana, USA.

Staten, R. T., H. M. Flint, R. C. Weddle, E. Quintero, R. E. Zarate, C. M. Finnell, M. Hernández, and A. Yamamoto 1987. Pink bollworm (Lepidoptera Gelechiidae): Large scale field trials with a high rate gossyplure formulation. Journal of Economic Entomology 80: 1207–1271.

Stern, V. M. 1979. Long- and short-range dispersal of the pink bollworm *Pectinophora gossypiella* over southern California. Environmental Entomology 8: 524–527.

Stern, V. M., and V. Sevacherian. 1978. Long-range dispersal of pink bollworm into the San Joaquin Valley. California Agriculture 32(7): 4–5.

Stern, V. M., R. F. Smith, R. van der Bosch, and K. S. Hagen. 1959. The integration of chemical and biological control of the spotted alfalfa aphid: The integrated control concept. Hilgardia 29(2): 81–154.

Suckling, D. M., L. Stringer, A. Stephens, B. Woods, D. Williams, G. Baker, and A. El-Sayed. 2014. From integrated pest management to integrated pest eradication: Technologies and future needs. Pest Management Science 70: 179–189.

Tabashnik, B. E., T. Brevault, and Y. Carriere. 2013. Insect resistance to *Bt* crops: Lessons from the first billion acres. Nature Biotechnology 31: 510–521.

Tabashnik, B. E., A. L. Patin, T. J. Dennehy, Y. B. Liu, Y. Carrière, M. A. Sims, and L. Antilla. 2000. Frequency of resistance to *Bacillus thuringiensis* in field populations of pink bollworm. Proceedings National Academy of Sciences USA 97(24): 12980–12984.

Tabashnik, B. E., M. S. Sisterson, P. C. Ellsworth, T. J. Dennehy, L. Antilla, L. Liesner, M. Whitlow, R. T. Staten, J. A. Fabrick, G. C. Unnithan, A. J. Yelich, C. Ellers-Kirk, V. S. Harpold, X. Li, and Y. Carriere. 2010. Suppressing resistance to *Bt* cotton with sterile insect release. Nature Biotechnology 28: 1304–1307.

(TBWEF) Texas Boll Weevil Eradication Foundation. 2019.

(USDA) United States Department of Agriculture. 2018. USDA announces pink bollworm eradication significantly saving cotton farmers in yearly control costs. Press Release No. 0222.18. October 19, 2018. Washington, DC, USA.

Vreysen, M. J. B., A. S. Robinson, and J. Hendrichs (eds.). 2007. Area-wide control of insect pests: From research to field implementation. Springer. Dordrecht, The Netherlands. 789 pp.

Walters, M., R. T. Staten, and R. C. Roberson. 1988. Pink bollworm integrated management technology under field trial conditions in the Imperial Valley, California, pp. 1282–1285. *In* P. Dugger, and D. A. Richter (eds.), Proceedings Beltwide Cotton Conferences, Memphis, Tennessee. National Cotton Council of America. Nashville, Tennessee, USA.

Walters, M., Staten, R. Sequeira, and T. Dennehy. 1999. Preliminary analysis of pink bollworm population distributions in a large acreage of genetically engineered cotton with regard to resistance management, pp. 989–991. *In* P. Dugger, and D. A. Richter (eds.), Proceedings Beltwide Cotton Conferences, Memphis, Tennessee. National Cotton Council of America. Nashville, Tennessee, USA.

Walters, M. L., R. T. Staten, and R. C. Roberson. 2000. Pink boll worm integrated management using sterile insects under field trial conditions, Imperial Valley, California, pp. 201–206. *In* K.-H. Tan (ed.), Area-wide control of fruit flies and other insect pests. Penerbit Universiti Sains Malaysia, Pulau Pinang, Malaysia.

Wan, P., D. Xu, S. Cong, Y. Jiang, Y. Huang, J. Wang, H. Wu, L. Wang, K. Wu, Y. Carriere, A. Mathias, X. Li, and B. E. Tabashnik. 2017. Hybridizing transgenic *Bt* cotton with non-*Bt* cotton counters resistance to pink bollworm. Proceedings National Academy of Sciences USA 114(21): 5413–5418.

Watson, T. F. 1995. Impact of transgenic cotton on pink bollworm and other Lepidopteran insects, pp. 784–796. *In* D. A. Richter, and J. Armour (eds.), Proceedings Beltwide Cotton Conferences, Memphis, Tennessee. National Cotton Council of America. Nashville, Tennessee, USA.

Wilson, R. D., H. M. Flint, W. R. Deaton, D. A. Fischhoff, F. J. Perlak, T. A. Armstrong, R. L. Fuchs, S. A. Berberich, N. J. Parks, and B. R. Stapp 1992. Resistance of cotton lines containing a *Bacillus thuringiensis* toxin to pink bollworm (Lepidoptera: Gelechiidae) and other insects. Journal of Economic Entomology 85: 1516–1521.

THE SUPPRESSION OF THE FALSE CODLING MOTH IN SOUTH AFRICA USING AN AW-IPM APPROACH WITH A SIT COMPONENT

N. BOERSMA

Xsit (Pty) Ltd., 2 Schalk Patience Street, Citrusdal 7340, South Africa; nb@xsit.co.za

SUMMARY

The false codling moth, *Thaumatotibia leucotreta* (Meyrick) (Lepidoptera: Tortricidae), is native to sub-Saharan Africa, where it infests various commercial, and wild, fruit-bearing plants. This major pest is not present in the Americas, Europe, and Asia, and therefore has phytosanitary implications, which impose severe limitations on potential South African exports. Consequently, this pest represents a severe threat to the fruit industry of South Africa, in terms of socio-economic impacts on both fruit production and job security. Although the pest can be managed to some extent with insecticides, mating disruption, and orchard sanitation, a long-term environment-friendly solution was needed. This became more evident as *T. leucotreta* developed resistance to available insecticides, while stricter quarantine measures were enforced by importers of African citrus. In 2002, research commenced on an area-wide integrated pest management (AW-IPM) programme in conjunction with the development of the Sterile Insect Technique (SIT) for the false codling moth. Commercial sterile insect releases started in the 2007-2008 season over 1500 ha of citrus orchards in Citrusdal, Western Cape Province, but by 2017-2018 had gradually expanded to almost 19 000 ha in three different citrus producing regions of South Africa. The programme is currently owned by the Citrus Growers Association (CGA) that have contributed to the steady growth of the SIT programme in the citrus industry. Over the past ten years the status of *T. leucotreta* as a pest threat was systematically reduced in areas where the SIT was practiced on an area-wide basis, compared to non-release areas.

Key Words: Citrus, navel orange, area-wide, Sterile Insect Technique, pest management, *Thaumatotibia leucotreta,* Tortricidae, resistance, quarantine, South African exports, Western Cape

1. THE PROBLEM

The false codling moth *Thaumatotibia leucotreta* (Meyrick) (Lepidoptera: Tortricidae) is a polyphagous indigenous pest of both cultivated crops and wild plants in sub-Saharan Africa. False codling moth was first noted in the Paarl region of the

J. Hendrichs, R. Pereira and M. J. B. Vreysen (eds.), Area-Wide Integrated Pest Management: Development and Field Application, pp. 93–109. CRC Press, Boca Raton, Florida, USA.

Western Cape Province (South Africa) around 1969 (Hofmeyr et al. 2015). Although it attacks many different deciduous, subtropical, and tropical plants, it prefers citrus as one of its primary hosts.

By the mid-1970s *T. leucotreta* was detected at a holiday resort, 170 km north of the Paarl region near Citrusdal, an important citrus exporting region in the Western Cape Province (Hofmeyr et al. 2015). By the end of the 1970s it had spread through some parts of the valley, with heavy infestations in navel orange orchards (Hofmeyr et al. 2015).

The presence of this insect represents a high phytosanitary risk for South African fruit exports to the USA, Asia, and Europe. The economic threats imposed by the pest to the fruit growers and the industry of South Africa may also have severe socio-economic consequences for food and job security. The situation was exacerbated after *T. leucotreta* developed resistance against available registered insecticides and stricter regulations were imposed on exporters (Hofmeyr and Pringle 1998). This included a zero tolerance for *T. leucotreta* and the requirement of a post-harvest cold treatment (Hofmeyr et al. 2016a, 2016b).

2. PRE-OPERATIONAL ACTIVITIES

Although the pest has been managed to some extent, by integrating control tactics such as insecticides, mating disruption, and orchard sanitation, a longer-term solution was needed. In 2002, research was conducted to develop an area-wide integrated pest management system (AW-IPM programme) with an SIT component (Hendrichs et al. 2007; Klassen and Vreysen 2021). Citrus Research International (CRI) (Pty) Ltd, the Citrus Growers Association of South Africa (CGA), the Joint Food and Agriculture Organization of the United Nations/International Atomic Energy Agency Division (FAO/IAEA), the United States Department of Agriculture (USDA) through its Agricultural Research Service (ARS) and Centre for Plant Health Science and Technology (CPHST), joined resources and efforts to develop and test the efficacy of a SIT programme for *T. leucotreta.*

During the first phase of research, the radiation biology and inherited sterility of *T. leucotreta* (Bloem et al. 2003) was investigated, which was followed by field cage trials to evaluate mating compatibility and competitiveness (Hofmeyr et al. 2005). The results of these biological studies on the effect of gender and irradiation dose on *T. leucotreta* are shown in Fig. 1. As expected, fertility of both male and female moths declined with increasing irradiation dose (Bloem et al. 2003). This dose effect was greater for crosses involving irradiated female moths, which were almost completely sterile when treated with a dose of 200 Gy, while irradiated males still had a residual fertility of 5.2% when treated with a dose of 350 Gy (Fig. 1) (Bloem et al. 2003). Similar to other Lepidoptera, *T. leucotreta* exhibited inherited sterility when partially sterile male moths copulated with wild female counterparts (Carpenter et al. 2004). The resulting F_1 progeny was shown to be fully sterile, mostly male, and took longer to develop (Bloem et al. 2003). Furthermore, the F_1 generation would either fail to hatch or would develop into sterile, but fully competitive F_1 adults, that would provide additional pest population suppression in the subsequent generation.

The promising results led to the next research phase in the Olifants River Valley, which involved a SIT pilot study in citrus orchards during the 2005-2006 season (Hofmeyr et al. 2015). This involved the release of sterile *T. leucotreta* adults in a 35·ha navel orange orchard, surrounded by natural vegetation, while another navel orchard was used as a control.

Releases were performed over a 29-week period, with a total of 2000 sterile moths released per week. Thirteen delta traps evenly spaced over the orchards, equipped with synthetic pheromone (Cardiff Chemicals, Cardiff, UK) in Lorelei dispensers, were used. The goal was to create an overflooding ratio within the orchards and this was maintained at no less than 1 wild: 10 sterile moths per week (Hofmeyr et al. 2005; Hofmeyr et al. 2015). The encouraging results from this pilot study indicated a 77% decrease of wild *T. leucotreta* trap catches and an approximate 95% reduction of *T. leucotreta* related fruit infestations, compared to the non-SIT area.

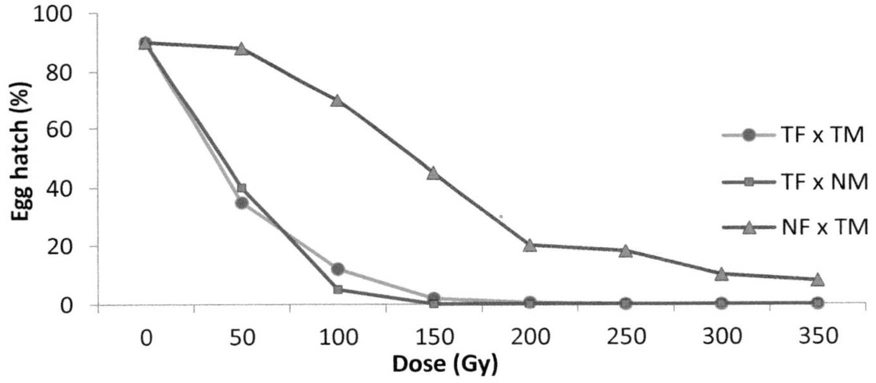

Figure 1. Effect of irradiation dose administered to T. leucotreta *adults on the mean egg hatch (%) per mated female. Males and females were treated (T) with 50, 100, 150, 200, 250, 300, and 350 Gy and inbred (TF x TM) or out-crossed (TF x NM, NF x TM) to untreated adults (N) (adapted from Bloem et al. 2003).*

The results of this 2005-2006 season pilot project in the Citrusdal region are shown in both Figs. 2 and 3. From the results obtained in this initial trial, the South African citrus industry was convinced to fast-track the commercial introduction of the SIT programme for *T. leucotreta*.

As a result, in 2006, the private company *Xsit* (Pty) Ltd. was established to manage the production, sterilisation, and the release of sterile *T. leucotreta* in the Citrusdal region. New equipment was designed to upscale moth production and to replace the insufficient infrastructure being used for the small-scale rearing of *T. leucotreta* (Hofmeyr et al. 2015). The new mass-rearing facility became operational in early 2007, and the release of irradiated moths commenced in November of that year.

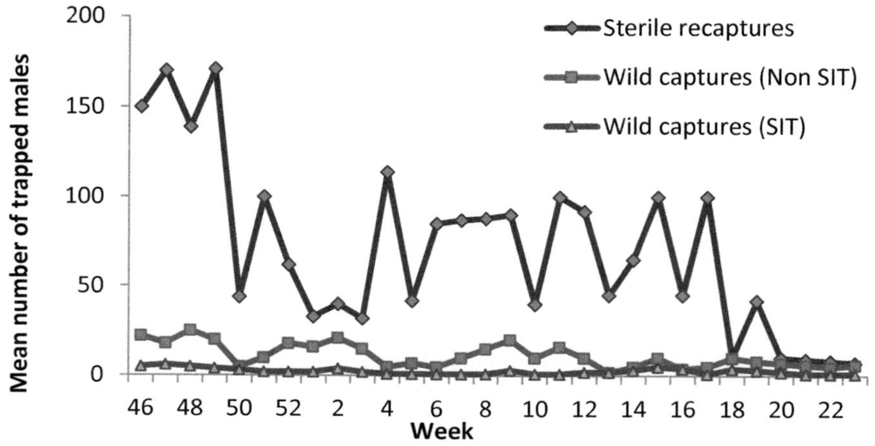

Figure 2. Capture of released (irradiated and topically marked) and wild T. leucotreta *males in SIT-treated and non-SIT-treated navel orange orchards as part of a SIT pilot project carried out in the Citrusdal region during the 2005-2006 season. A minimum ratio of 1:10 wild:sterile moths were maintained throughout the pilot trial in the SIT-treated orchard (adapted from Hofmeyr et al. 2015).*

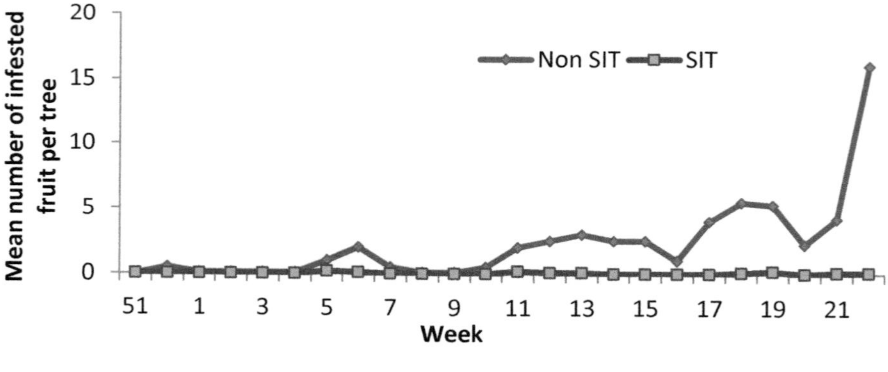

Figure 3. Fruit drop due to T. leucotreta *infestation in non-SIT and SIT-treated citrus orchards (35 ha) as part of a SIT pilot project carried out in the Citrusdal region, Western Cape Province, during the 2005-2006 season.*

3. THE FIRST DAYS OF THE SIT PROGRAMME, EXPANSIONS AND IMPROVEMENTS

Since the commercialization of the AW-IPM programme with a SIT component in Citrusdal in 2007, many new systems and equipment have been designed, developed, and manufactured in a relatively short time. Production monitoring systems for traceability, cold chain management, and quality management of reared moths were developed and constantly improved.

From 2007 to 2018 the initial 1500 ha SIT-treated area in the Citrusdal region was gradually increased to 4800 ha by incorporating the rest of the Olifants River Valley. Sterile insect releases are also being carried out over 6500 and 2200 ha of the Sundays River and Gamtoos River valleys in the Eastern Cape Province, respectively. During the 2016-2017 season, the release of sterile *T. leucotreta* was also expanded to the lower Orange River area in the Northern Cape Province and the Hex River Valley of the Western Cape Province, the latter, an important table grape export region, treating 1500 and 4000 ha respectively. At the time of writing (2018), this privately-owned programme was providing sterile insects on a weekly basis to cover more than 18 000 ha.

There was a progressive seasonal improvement in wild *T. leucotreta* suppression following routine releases of sterile moths in all treated areas. The results showed a reduction in crop losses and fewer rejections of fruit consignments destined for exports due to *T. leucotreta* presence. While this rapid growth of the programme was very exciting, it was accompanied by many challenges and hardships that sometimes threatened its existence.

3.1. Rearing Equipment

During the initial days of programme implementation, some of the old equipment and processes developed for the pilot trial were utilized in the commercial rearing programme. However, with the expansion of the programme, it became apparent that most of these systems were inadequate, and only suitable for small-scale rearing of *T. leucotreta* (Hofmeyr et al. 2015).

3.1.1. Larval Diet Preparation Equipment

A new artificial diet (Moore 2002; Moore et al. 2014) was introduced for rearing purposes in 2007. The diet was prepared using large-scale equipment from the baking industry, and then pulsed into 500 ml glass jars fitted with breathable replaceable paper membranes, allowing for gas exchange, in the metallic screw-lids (Hofmeyr et al. 2015). The individual jars were placed by hand into stainless steel baskets, containing 25 rearing jars each. Baskets were then stacked on a steel trolley, holding up to 16 baskets before they were pulled into an oven. Although the oven was an innovative piece of equipment, it was not ideally suited, as evidenced by the uneven cooking of the diet, and lack of sterilisation. After baking two trolleys per oven cycle with 400 glass jars each, they were placed in a room for cooling. Each jar of diet was inoculated with approximately one thousand 24-h old eggs via placement of an egg sheet, sterilised by an 8% formaldehyde solution, on top of the diet (Hofmeyr et al. 2015).

As the programme expanded and diet preparation increased from 6000 to 20 000 bottles a day, the handling and diet preparation processes had to be re-considered. Disadvantages of the jars included relatively high costs, susceptibility to breakage, and requirement for individual handling and cleaning after larvae emerged. In 2013, new technologies to prepare the diet were investigated. These included radiowave, microwave, infrared, steam, and extrusion. After completing the initial trials, only radiowaves, microwaves, and extrusion seemed potentially viable.

Different cooking times and temperatures were tested to develop a cost-effective method for delivering a diet producing high larval yields without a negative impact on larval development. Although several parameters were tested, the following criteria were critical in validating the optimum cooking/sterilisation process:

1. Proportion of large 5th instar larvae (0.04 g ± 1 and >10 mm) produced.
2. Feed conversion ratio, denoted by the amount of diet required to produce one large 5th instar larva.
3. Absence of viral and bacterial infected larvae - denoted in the amount (g) of diet required to produce one healthy larva.

To validate the correct process and cooking protocol (time vs. temperature), the number of large larvae (5th instar) produced was counted. It was evident that both the microwave and the extrusion processes resulted in the highest number of large larvae with the least feed (best feed conversion). The original Xsit diet required 0.35 g to rear one large larva, while both the microwave and the extrusion required only between 0.21 g and 0.22 g, respectively to rear one large larva. It was clear that the best results were obtained with the laboratory microwave and the extruder, where approximately 90% of the reared larvae reached 5th instar on day 12, while only 40% of the control diet reached 5th instar at the same time (Fig. 4).

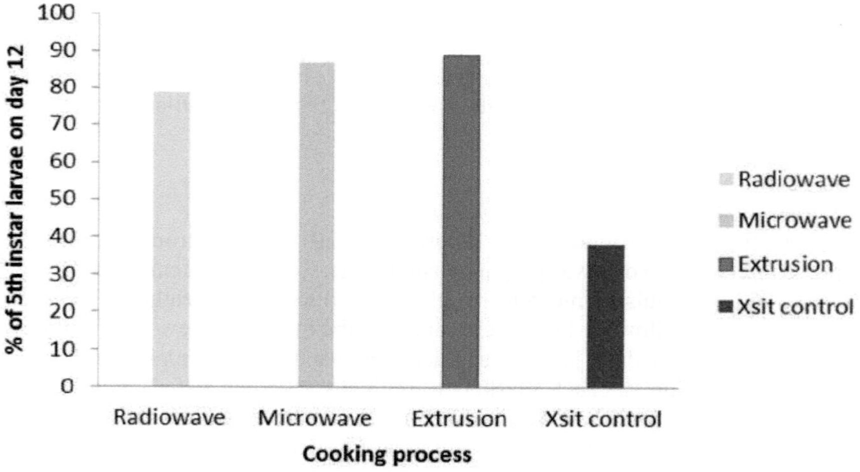

Figure 4. Percentage of large larvae per diet preparation treatment. Results of the preliminary laboratory studies were confirmed in this laboratory trial. Microwave and extrusion samples resulted in the highest amount of 5th instar larvae on day 12.

The next step was to test both processes in a commercial trial. In 2015, several experiments were conducted in a commercial microwave. These trials were repeated three times under different settings, but the positive results obtained with the laboratory microwave (1000 W) could not be replicated in the commercial microwave. After further investigation in obtaining a commercial microwave, the lack of support of the industry was evident, and it was decided to continue with the extrusion process. Commercial trials with different extruder processes and settings were tested until the end of 2016, when the correct settings and consistency of the diet were obtained, with proven results of growth and yield of larvae replicated on a semi-commercial scale; nearly 90% of larvae had grown on the extruded diet, and reached 5th instar on day 12, compared to only 40% on the Xsit diet.

Xsit purchased its own extruder plant in July 2017. However, after the commissioning of the extruder, it was clear that the larval diet was not similar or even comparable to the product developed over the past four years. The extruded larval diet was sticky and could hardly be packed or handled, while larval growth was retarded, and yields were low. The larval diet produced was therefore, not fit for use. For six months, extensive trials were carried out, testing all variables, including raw product variability and extruder conditions. It was concluded that the main cause of the problem was (a) the dextrinisation and gelatinisation of the starch in the diet, and (b) the inconsistent granule size of the maize meal, which comprises 80% of the diet. During the extrusion process the starch granules swell when pressurised under high temperatures during cooking and drying, and then shrink as soon as the product cools down. In the event, when the diet is cooked at a too high temperature, the starch molecules lose the ability to swell, leading them to shear and burst. This will lead to the loss of the semi-crystalline structure of the starch, while the smaller amylose molecules leach from the granule to form new chemical structures which cannot be digested by the larvae. During cooling, the semi-crystalline structure recovers and, provided that the granules did not burst, will re-align to a similar position or structure prior to cooking. This phenomenon is known as retrogradation (Oates 1997; Wang et al. 2015). As the initial research was conducted in different stages, the diet was cooled prior to drying, while at the new commercial plant the extruded diet was immediately transferred to the oven for drying, leading to dextrinisation and gelatinisation of the starch. This assumption was confirmed during trials where the diet was allowed to cool before drying. As a final outcome, excellent results were obtained which were similar to those obtained in the research done over the previous four years. An additional cooling unit was introduced after extrusion, before drying.

3.1.2. Rearing Containers

During the original rearing of the larvae in glass jars, the screw lids and membranes were removed when larvae reached 5th instar. The jars were placed on their sides to assist larval exit. An integrated aperture below each basket held the pupation substrate, which is a square of polycarbonate honeycomb material, 570 mm x 530 mm x 11 mm with 6 mm diameter aperture, placed on a 570 mm x 530 mm x 3 mm fibre sheet.

In 2012, the whole false codling moth colony suffered from a severe bacterial infection in the facility. After an intense investigation it appeared that miniscule holes between the honeycomb cells became breeding sites for *Bacillus cereus*, Gram-negative opportunistic bacteria that undermine the immune system of the larvae, killing them in a matter of days. In response, the polycarbonate sheets were replaced with disposable, pre-manufactured sheets of corrugated, single-face cardboard.

In parallel with the development of the new diet preparation system described above, a replacement of the glass jars was pursued. Several types of containers were investigated, including paper bags, starch bags, polyethylene cups, and disposable polyethylene bags. The most significant challenge was to find a similar membrane to the one used in the jars. The new extruded diet was more prone to drying out, but one had to keep in mind that the regulation and exchange of gasses were critical. A material with these specific properties, while preserving moisture, had to be found.

In 2016, disposable polyethylene bags (280 mm x 160 mm), with a breathable polyethylene-based microporous membrane, were introduced. These were automatically filled by a diet dispensing machine. A volume of water equal to 47% of the total volume of the dry mix was added to form a fluffy diet, while 250 g of larval diet was required to produce at least 550 larvae per unit. The bags were consequently phased in to replace the glass jars. Egg sheets containing 800 ± 100 *T. leucotreta* eggs, dipped in an 8% formaldehyde solution to prevent contamination by any bacteria and/or virus, were then placed on the diet and sealed. The bags were then assembled on a rearing cart containing 480 bags. When ready for pupation, usually on day 12 of the rearing cycle, the larvae chew their way out of the bags and descend on silk threads to the cardboard pupation substrate, placed 30 mm below the bags.

After the implementation of this new system, significant production losses were experienced due to larvae dying in the bags before reaching the 5th instar. Upon investigation it was determined that the HVAC system was not capable of handling the large volume of CO_2 generated by the large number of larvae reared per m^2. Consequently, a new HVAC system with increased capacity and higher air change rate was designed and installed by the end of the 2016-2017 season to ensure sustainable production of sterile insects on a continuous basis.

3.1.3. Moth Emergence Cabinets

The pupation boards were placed into custom designed steel emergence cabinets to permit moth emergence and collection. Each cabinet, 1550 mm x 630 mm x 940 mm, was welded on a 900 mm supporting framework (Fig. 5). The cabinets were divided longitudinally with a perforated stainless-steel sheet separating two compartments: a back compartment, 740 mm deep with an access door, containing 50 horizontally placed pupation sheets, and a front compartment, 200 mm deep with a glass door to the outside, allowing moths to move phototactically from the back into the front compartment (Boersma and Carpenter 2016). The front compartment was lightly dusted with talcum powder and was fitted with a collection cone at the bottom, attached to a plenum-based air-braking moth collection system.

Establishing the correct speed of the airstream that transferred the moths from the moth cabinet to the collection room was challenging, as too high airflow resulted in damaged moths, while too low airflow caused clogging of moths. This was resolved by adjusting the airspeed to 12 m/s for transferring moths to the collection pans, while reducing the airstream to 3 m/s as they enter the collection room, allowing for a soft landing (Hofmeyr and Pretorius 2010).

Scaling up the release area from 3300 ha to more than 18 000 ha induced a lot of pressure on these emergence cabinets, resulting in problems with temperature consistency, clogging of moths due to overcrowding, and production losses due to moths escaping from these old cabinets and equipment. In 2015 cabinets were re-designed, allowing for better airflow by making the following improvements:

1) the steel sheets of the cabinets were folded rather than welded for better durability;

2) extra space was provided at the back to allow increased airflow between pupation boards, resulting in fewer temperature spikes;

3) use of rubber sealed Perspex doors for better sealing.

Figure 5. Moth emergence cabinets.

In the moth collection pans adult moths need to be kept immobile. This is a critical procedure to prevent mating in the collection pans as well as the prevention of damage to adult moth's wings. Mated adults or those with damaged wings may have a negative effect on their field performance. However, no set or established temperature range were used during the moth collection, handling, and transport. This led to moths being exposed to temperatures below their critical thermal limits (below 6°C), resulting in poor field recaptures in the warmer months of the season (Boersma and Carpenter 2016; Boersma et al. 2017).

In 2015, new cooling and handling protocols were introduced with a cold chain with a set temperature range from the moth cabinets to the orchard to ensure moths were kept between 6-10°C, which resulted in better quality and recapture of sterile moths in the field.

3.2. Sterilisation Dose

As soon as the moths reached a required temperature of 10°C in the collection pans, they were placed into cardboard boxes (140 mm x 140 mm x 50 mm) and irradiated with a dose of 150 Gy (Bloem et al. 2003) in a 20 kCi ^{60}Co source panoramic irradiator. During the 2017-2018 season, the irradiation dose was increased to 200 Gy to compensate for an apparent reduced sterilisation effect of the 150 Gy dose with reduced dose-rate of the cobalt source. Although the strength of the cobalt source has weakened, and moth exposure time adjusted accordingly, the reason for this decrease in radiation impact is not known and is currently being investigated.

3.3. Release Methods and Devices

The irradiated moths are stored in a holding room between 6-8°C for approximately 12-24 h, and then transported to orchards or an airfield in a refrigerated vehicle at the same, regulated temperature range.

From 2007 to 2010, sterile moths were released with all-terrain vehicles (ATVs) or "quad bikes", manned by a driver and an assistant responsible for releasing the moths by hand into the trees (Hofmeyr et al. 2015). Later the ATVs were equipped with a release box with a release auger. Although this release method was relatively inexpensive, it had a few disadvantages:

1. Human factor: releasing an accurate and constant number of moths in orchards was not possible.
2. The terrain where some of the orchards are located is rough, making driving while releasing an equal number of sterile moths difficult, leading to inadequate moth distribution.
3. Access to farms was sometimes difficult.
4. Logistical constraints: covering releases twice a week in a valley which stretches more than 100 km in its length and 60 km in its width became a logistical constraint; this led to an increasing cost of maintaining the ATV's and preventing breakdowns with a constant challenge of completing the releases in time, versus maintaining the quality of the product.

Since 2010, releases of the moths have progressed from ground releases with ATVs to aerial releases using gyrocopters, and later to fixed-wing aircrafts. Moth releases with fixed-wing Piper Pawnee aircrafts commenced at the end of 2015 but were gradually replaced by helicopters in 2017.

In 2010, Xsit outsourced the releases to a company that used gyrocopters. The release system and holding boxes (hopper) of the sterile moths were slightly modified, and fitted to the gyrocopter, making aerial releases possible. The results obtained were excellent. The recapture rate of the sterile moths increased, while the wild false codling moth population decreased to the lowest levels since the start of the Xsit programme. Unfortunately, after the tragic loss of two pilots, in two separate incidents, the gyrocopters were grounded by the South African Civil Aviation Authority, and releases had to be continued using fixed-wing aircraft. Although the results obtained from the fixed-wing aircraft were comparable to the gyrocopter in certain areas, it had a few disadvantages:

1. The minimum speed the aircraft flew were significantly higher than the gyrocopters (160 km/h vs. 100 km/h), resulting in poorer recaptures.

2. The minimum height the aircraft flew were higher than the gyrocopter (160 feet vs. 100 feet), leading to poorer recaptures.

3. Flying in small valleys and mountainous areas were impossible, adding to logistical constrains by filling gaps with ground releases.

4. Quality degradation due to the prop wash of the aircraft (the force of wind generated behind a propeller) causing moths to be blown into a swirl.

Starting the 2017-2018 season, fixed-wing releases were gradually phased out and replaced with small helicopters (R22) to simulate the conditions of gyrocopter releases. This increased efficiency and resulted in sustainable results, contributing to an even greater suppression of wild false codling moth over the past two seasons. Currently the possibility of releasing the sterile insects by unmanned aerial vehicles is being investigated, with the first experimental releases occurring in early 2018. Current aviation legislation in South Africa, in conjunction with costs, still make this venture impossible on a commercial scale.

4. THE RESULTS AND IMPACT

The SIT programme for *T. leucotreta* is governed by phytosanitary requirements, demanding a zero-tolerance level of pest incidence in fruit. The SIT programme is one component of an area-wide approach integrating multiple tactics to mitigate the threat posed by the *T. leucotreta*. This includes obligatory orchard sanitation by growers, with various alternative control measures.

The frequency of sterile male releases varies for each insect species and depends mainly on the survival of the sterile insects in the target area. The ability of sterile insects to survive and remain sexually active as long as possible in the field is essential, and if their longevity declines, the frequency of releases needs to be increased to ensure optimal overflooding ratios at all times (Dowell et al. 2021). The success of the sterile *T. leucotreta* release programme is very much determined by the ability to ensure the release of pre-determined numbers of sterile moths into the orchards that will guarantee a minimum sterile:wild male overflooding ratio of 10:1 (as assessed by trap catches).

In the warmer months, sterile *T. leucotreta* has a shortened life span, therefore requiring two releases of 1000 sterile adults per week. These double releases take place between November to April, while only one release of 2000 sterile adults per week takes place in the cooler months (September to October, and May to June), when the longevity of the moths increases. If the desired overflooding ratio is not achieved, supplementary releases of sterile moths are conducted. Maintaining a continuous optimal overflooding ratio maximises the probability that a wild moth will mate with a sterile moth in the field, thereby resulting in no viable or fertile offspring and an eventual population decline (Carpenter et al. 2004; Hofmeyr et al. 2015).

Once the wild population has been reduced to such a level that no wild moths are captured on a consistent basis, the release of sterile moths can become the main or sole control method, as is currently the case in numerous citrus orchards in various valleys. Furthermore, the pre- and post-harvest absence of *T. leucotreta* infested fruit, are additional indicators of programme success.

In recent years, the success of the SIT programme is evident by a marked reduction of the wild *T. leucotreta* populations, in some areas below economic thresholds. The released and wild populations are monitored at weekly intervals, using delta traps baited with the female sex pheromone to attract male moths. Sterile males are differentiated from their wild counterparts by their pink intestines caused by a food dye that is mixed with the larval diet (Hofmeyr et al. 2015).

Commercial results for the three main growing areas currently serviced by the programme, one in the Western Cape, and two in the Eastern Cape, are encouraging. Despite many challenges, the SIT has proven to be a sustainable approach to reduce the occurrence of wild *T. leucotreta* to well below economic thresholds (results from the recently incorporated Hex River Valley in the Western Cape, and the lower Orange River Valley in the Northern Cape will not be reported here).

4.1. Olifants River Valley, Western Cape

The positive achievements of the area-wide programme were evidenced by the progressive increase in the numbers of sterile male *T. leucotreta* moths trapped from 2007 to 2010 (Hofmeyr et al. 2015). In addition, trap catches of wild male adults declined from 13.0 moths per trap per week prior to the sterile moth releases in 2006, to 2.0, 0.4, and 0.1 moths per trap per week in 2012, 2013, and 2014, respectively. In addition, infestation of *T. leucotreta* in citrus fruit was reduced from 2.6% in the 2010-2011 season to 0.1% in 2013 (Barnes et al. 2015; Hofmeyr et al. 2015).

During the 2006-2007 season, i.e. before the start of the area-wide SIT programme, each tree had on average 30 fruit damaged by larvae of the false codling moth. As the SIT programme advanced, and the ratio of sterile to wild adult *T. leucotreta* increased, the average infestation rate declined to only 0.2 damaged citrus fruits per tree per season. The Perishable Products Export Control Board of South Africa reported a substantial reduction of pre-harvest crop losses (Hofmeyr et al. 2015).

Despite this initial success, the number of trapped wild *T. leucotreta* males increased during the 2010-2011 season, as did crop damage, to an average of 1.56 infested fruits per tree (Fig. 6). However, this was still 95% lower than the damage level before the release programme started. Higher than normal ambient temperatures experienced during the spring and summer of 2010-2011 were attributed as the cause of the increased wild male captures. This resulted in at least one additional generation, leading to more pressure on the SIT programme resulting in the increased fruit damage.

Despite this temporary upsurge in wild male catches, the suppression of the wild *T. leucotreta* population in the SIT-treated areas was restored and progressively improved from 1.5 moths per trap per week in the 2010-2011 season to 0.1 moths per trap per week in the 2015-2016 season (Fig. 6). The gyrocopter accidents coupled with production issues during the 2016-17 season resulted in a slight increase of average wild male catches from 0.3 and 0.4 males per trap per week. This, however, did not result in an increase in fruit infestation. This could be explained by the increased rearing efficiency, resulting in the release of better-quality moths, and the fact that the wild population of *T. leucotreta* was so low that sterile released adults were more effective, outcompeting the low numbers of the wild population (Hofmeyr et al. 2015).

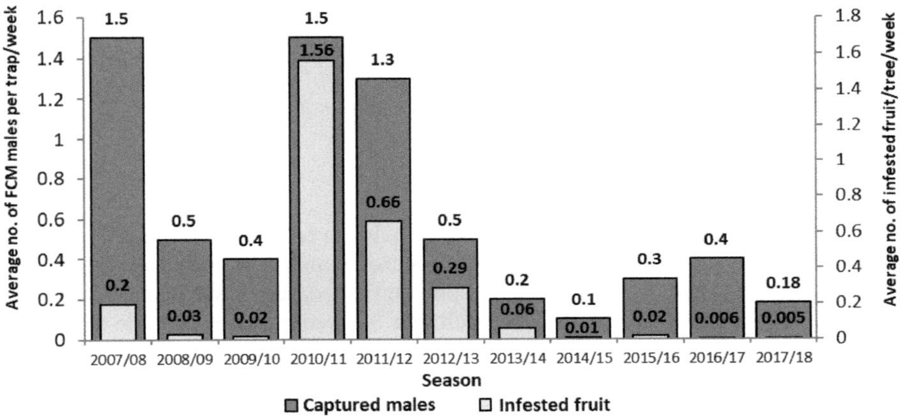

Figure 6. Reduction in numbers of wild T. leucotreta *males and infested fruit in sterile insect release areas in the Olifants River Valley, Western Cape Province from 2007-2008 to 2017-2018 seasons (data obtained from Xsit).*

4.2. Sundays River Valley, Eastern Cape

Sterile moth releases were initiated in the Sundays River Valley in 2011-2012, and since then the density of the wild *T. leucotreta* population has progressively declined with successive seasons, resulting in less infested fruit (Fig. 7). There was, however, an increase in the average number of trapped wild males in 2013-2014, but this was due to areas with historically high population densities being added to the SIT programme.

A similar trend of higher wild trap catches was seen in the Sundays River Valley in 2016-2017 due to challenges experienced in the mass-rearing facility. During the 2017-2018 season wild false codling moth catches decreased to only 1 wild male per trap and 0.02 infested fruit per tree respectively, the lowest since the start of the SIT programme.

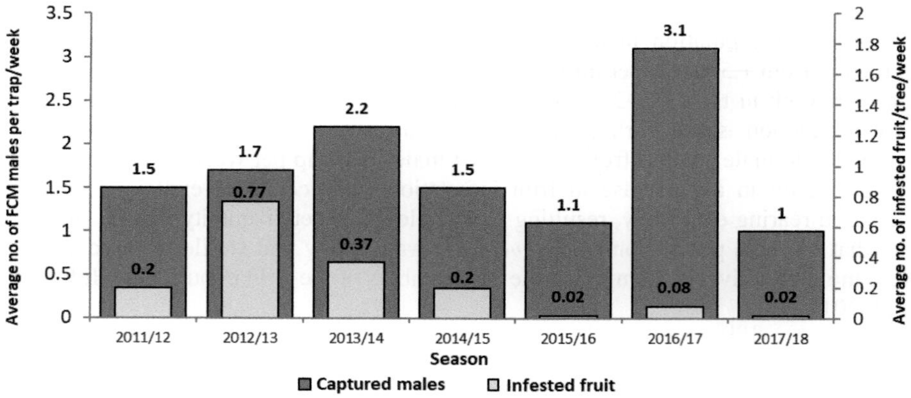

Figure 7. Reduction in numbers of wild T. leucotreta *males and infested fruit in the Sundays River Valley, Eastern Cape Province from 2011-2012 to 2017-2018, as a result of sterile insect releases (data obtained from Xsit).*

4.3. Gamtoos River Valley, Eastern Cape

The natural population of *T. leucotreta* was much lower in the Gamtoos Valley as compared to the other areas, causing much less crop damage. This was evidenced by much lower trap catches of wild males during the first season of sterile moth releases in comparison to the other areas. As a result, the SIT programme was able to reduce the wild moth population density and the number of infested fruits within the first release season (2014-2015).

In the following season, wild males were suppressed to such a low level that basically no infested fruit were recorded for the entire season, while a slight increase was recorded in the 2016-2017 season for the same reasons mentioned above (challenges at the rearing facility) (Fig. 8).

5. THE REASONS FOR SUCCESS

This AW-IPM programme with a sterile male release component against the *T. leucotreta* has had a significant impact on the citrus industry in South Africa; securing exports to the rest of the world in a sustainable manner. The success of the programme can be attributed to the following factors:

1. *Single crop industry*: The citrus trade, unlike other fruit sectors such as the deciduous fruit industry, is a single crop industry led by the Citrus Growers Association. This results in easier decision-making, management, and funding, since all stakeholders have the same vision. This played a significant role in the success of the programme as it was an industry-driven project to secure sustainable citrus exports for the growers.

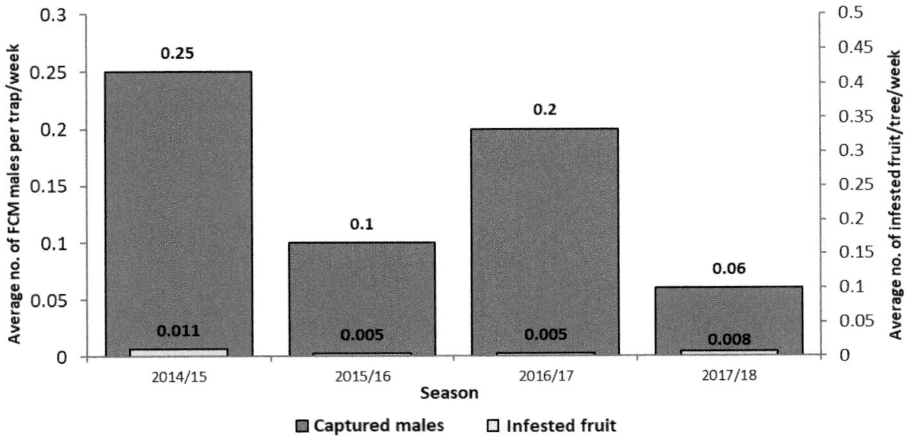

Figure 8. Reduction in numbers of wild T. leucotreta *males and infested fruit in the Gamtoos River Valley, Eastern Cape Province from 2014-2015 to 2017-2018 seasons, as a result of sterile insect releases (data obtained from Xsit).*

2. *Area-wide integration of suppression methods:* The SIT programme was managed as part of an area-wide programme. Xsit did not only take responsibility of both the monitoring of wild *T. leucotreta* and infestation, but also played a significant role in the monitoring of sanitation practises and the treatment of hot spots (an area with a high wild *T. leucotreta* population), in conjunction with other integrated pest control practises.

3. *Management to ensure sustainable sterile moth production*: Well-experienced and capable management was in place which ensured sustainable production of sterile insects, while the shareholders of the programme were also industry-related individuals, which ensured that the interests of the programme were always well-managed.

4. *Support of farmers and industry*: Most farmers were in favour of the programme, understanding the advantages of the AW-IPM approach. However, education and training were provided throughout the programme to ensure farmers were kept informed about industry-related matters.

5. *Phytosanitary regulations*: Since *T. leucotreta* is a regulated quarantine pest, governed by phytosanitary regulations in line with a systems approach (FAO 2017) for the controlling of the pest for export purposes, a zero-tolerance policy is enforced. Although there are several choices of control measures for *T. leucotreta* under the systems approach, it encouraged farmers to take part in the SIT programme if they wanted to export their fruit.

6. *In-house research*: Xsit employed its own researchers, constantly exploring better means of rearing, processing, and releasing insects, staying informed of the newest technology.

7. *Set protocols and procedures*: The use of standard protocols and procedures combined with continuous training of employees are essential for the efficient rearing of insects. The correct handling and distribution of insects are also essential to ensure good quality of insects in the field. Maintaining a cold chain proved essential to prevent damage to the sterile insects during transport, while the proper handling temperature has to be selected as this affects the competitiveness of the adults in the field (Boersma and Carpenter 2016).

6. FUTURE PLANS

The need for AW-IPM programmes with a SIT component to manage *T. leucotreta* in other countries where this pest is present, has become more apparent. Export crops such as avocados in Angola, and chilies in Kenya may also require the use of a SIT-based AW-IPM in the future to deal with this polyphagous pest.

Meanwhile, many export crops in South Africa, such as table grapes, stone fruits, and citrus grown in other regions are anticipating the introduction of the SIT. Xsit currently services 14 000 ha of the 70 000 ha of citrus in South Africa, while 4000 ha of table grapes are already enrolled in the programme. This indicates that there is scope for integration and expansion of this valuable technology to production areas not yet under this area-wide pest control method.

The potential of *T. leucotreta* becoming a major invasive pest in different countries globally is a reality, representing a threat to agriculture and food security. With the SIT now developed for this pest, if invasive false codling moth outbreaks are detected early, efficient integration of the SIT on an area-wide basis will allow eliminating them in an effective and environment-friendly way.

7. REFERENCES

Barnes, B. N., J. H. Hofmeyr, S. Groenewald, D. F. Conlong, and M. Wohlfarter. 2015. The Sterile Insect Technique in agricultural crops in South Africa: A metamorphosis .... but will it fly? African Entomology 23: 1–18.

Bloem, S., J. E. Carpenter, and J. H. Hofmeyr. 2003. Radiation biology and inherited sterility in false codling moth (Lepidoptera: Tortricidae). Journal of Economic Entomology 96: 1724–1731.

Boersma, N., and J. E. Carpenter. 2016. Influence of holding temperature and irradiation on field performance of mass-reared *Thaumatotibia leucotreta* (Lepidoptera: Tortricidae). Florida Entomologist 99 (Special Issue 1): 215–221.

Boersma, N., L. Boardman, M. Gilbert, and J. S. Terblanche. 2017. Sex-dependent thermal history influences cold tolerance, longevity and fecundity in false codling moth *Thaumatotibia leucotreta* (Lepidoptera: Tortricidae). Agriculture and Forest Entomology 20: 41-50.

Carpenter, J. E., S. Bloem, and J. H. Hofmeyr. 2004. Acceptability and suitability of eggs of false codling moth (Lepidoptera: Tortricidae) from irradiated parents to parasitism by *Trichogrammatoidea cryptophlebiae* (Hymenoptera: Trichogrammatidae). Biological Control 30: 351–359.

Dowell, R. V., J. Worley, P. J. Gomes, P. Rendón and R. Argilés Herrero. 2021. Supply, emergence, and release of sterile insects, pp. 441–484. *In* V. A. Dyck, J. Hendrichs, and A. S. Robinson (eds.), Sterile Insect Technique – Principles and practice in Area-Wide Integrated Pest Management. Second Edition. CRC Press, Boca Raton, Florida, USA.

(FAO) Food and Agriculture Organization of the United Nations. 2017. The use of integrated measures in a systems approach for pest risk management. International Standard for Phytosanitary Measures (ISPM) No. 14. International Plant Protection Convention. FAO, Rome, Italy.

Hendrichs, J., P. Kenmore, A. S. Robinson, and M. J. B. Vreysen. 2007. Area-Wide Integrated Pest Management (AW-IPM): Principles, practice and prospects, pp. 3–33. *In* M. J. B. Vreysen, A. S. Robinson, and J. Hendrichs (eds.), Area-wide control of insect pests. From research to field implementation. Springer, Dordrecht, The Netherlands.

Hofmeyr, J. H., and K. L. Pringle. 1998. Resistance of false codling moth, *Cryptophlebia leucotreta* (Meyrick) (Lepidoptera: Tortricidae), to the chitin synthesis inhibitor, triflumuron. African Entomology 6: 373–375.

Hofmeyr, J. H., and J. Pretorius. 2010. Development of a device to collect mass-reared false codling moth, *Thaumatotibia leucotreta* (Meyrick) (Lepidoptera: Tortricidae), in a commercial insectary. African Entomology 18: 374–378.

Hofmeyr, J. H., J. E. Carpenter, and S. Bloem. 2005. Developing the Sterile Insect Technique for *Cryptophlebia leucotreta* (Lepidoptera: Tortricidae): Influence of radiation dose and release ratio on fruit damage and population growth in field cages. Journal of Economic Entomology 98: 1924–1929.

Hofmeyr, J. H., J. E. Carpenter, S. Bloem, J. P. Slabbert, M. Hofmeyr, and S. S. Groenewald. 2015. Development of the Sterile Insect Technique to suppress false codling moth *Thaumatotibia leucotreta* (Lepidoptera: Tortricidae) in citrus fruit: Research to implementation (Part 1). African Entomology 23: 180–186.

Hofmeyr, J. H., M. Hofmeyr, V. Hattingh, and J. P. Slabbert. 2016a. Postharvest phytosanitary disinfestation of *Thaumatotibia leucotreta* (Lepidoptera: Tortricidae) in citrus fruit: Determination of ionising radiation and cold treatment conditions for inclusion in a combination treatment. African Entomology 24: 208–216.

Hofmeyr, J. H., V. Hattingh, M. Hofmeyr, and J. P. Slabbert. 2016b. Postharvest phytosanitary disinfestation of *Thaumatotibia leucotreta* (Lepidoptera: Tortricidae) in citrus fruit: Validation of an ionising radiation and cold combination treatment. African Entomology 24: 217–224.

Klassen, W., and M. J. B. Vreysen. 2021. Area-Wide Integrated Pest Management and the Sterile Insect Technique, pp. 75–112. *In* V. A. Dyck, J. Hendrichs, and A. S. Robinson (eds.), Sterile Insect Technique – Principles and practice in Area-Wide Integrated Pest Management. Second Edition. CRC Press, Boca Raton, Florida, USA.

Moore, S. D. 2002. The development and evaluation of *Cryptophlebia leucotreta* granulovirus (ClGV) as a biological control agent for the management of false codling moth, *Cryptophlebia leucotreta,* on citrus. Ph.D. thesis, Rhodes University, Grahamstown, South Africa. 308 pp.

Moore, S. D., G. I. Richards, C. Chambers, and D. Hendry. 2014. An improved larval diet for commercial mass rearing of the false codling moth, *Thaumatotibia leucotreta* (Meyrick) (Lepidoptera: Tortricidae). African Entomology 22: 216–219.

Oates, C. G. 1997. Towards an understanding of starch granule structure and hydrolysis. Trends in Food Science and Technology 8: 375–382.

Wang, S., C. Li, L. Copeland, Q. Niu, and S. Wang. 2015. Starch retrogradation: A comprehensive review. Comprehensive Reviews in Food Science and Food 14: 568–585.

PUTTING THE STERILE INSECT TECHNIQUE INTO THE MODERN INTEGRATED PEST MANAGEMENT TOOLBOX TO CONTROL THE CODLING MOTH IN CANADA

C. NELSON[1], E. ESCH[1], S. KIMMIE[1], M. TESCHE[1], H. PHILIP[2] AND S. ARTHUR[3]

[1]Okanagan-Kootenay Sterile Insect Release Programme, 1450 KLO Road, Kelowna, BC V1W 3Z4 Canada; CNelson@oksir.org
[2]IPM 2 GO Consulting Service, 465 Knowles Road, Kelowna, BC V1W 1H2 Canada
[3]Okanagan-Kootenay Sterile Insect Release Programme, 11401-115 Street, Osoyoos BC V0H 1V5 Canada

SUMMARY

The Okanagan-Kootenay Sterile Insect Release (OKSIR) programme, in southern British Columbia, Canada, has been successfully applying the Sterile Insect Technique (SIT) as part of a sustainable area-wide integrated pest management (AW-IPM) programme to control the codling moth *Cydia pomonella* L. in pome fruits in the region for over 20 years. Chemical, cultural and biological techniques that complement the SIT are also integrated into orchard and regional pest management plans by the programme and/or individual growers. The AW-IPM programme is supported by close monitoring of codling moth populations in orchards and adjacent urban properties; enforcing suppression of codling moth infestations in orchards and urban areas; removing derelict orchards, wild host trees and poorly managed host trees; and increasing public awareness and education. Successful collaboration between the OKSIR programme, the pome fruit industry, area residents and various government organizations has reduced codling moth populations by 94%, relative to pre-programme levels, and codling moth damage to less than 0.2% of fruit, in more than 90% of the orchards in the programme area. Local pesticide sales indicate a 96% reduction in the amount of active ingredient used against the codling moth since 1991. Implementing the SIT through an innovative social approach to local community-centred area-wide pest management has posed many challenges and created many learning opportunities. The codling moth mass-rearing facility in Osoyoos, British Columbia, has the capacity to produce 780 million sterile codling moths annually, but only a portion of that is used seasonally to treat 3400 hectares (ha) of pome fruit made up of small orchards intermixed with residential areas in the Okanagan Valley. As a result of climate change and increasing global trade, destructive insect pests are migrating to new habitats throughout the world. These new threats must be managed in ways that protect both the agrifood industries and the natural environments in which the industries operate. The OKSIR programme is an effective and easily transferred model to meet these challenges, especially as a supplement to other biological control methods. The programme is also a

J. Hendrichs, R. Pereira and M. J. B. Vreysen (eds.), Area-Wide Integrated Pest Management: Development and Field Application, pp. 111–127. CRC Press, Boca Raton, Florida, USA.

compelling model of success that can encourage other regions to use the SIT in their pest management toolbox to combat codling moth infestations across multiple local community jurisdictions using environment-friendly, cost-effective methods based on proven technology. The OKSIR programme is exploring the sale of surplus sterile moths, egg sheets or possible virus production as an opportunity to offset costs of incorporating additional area-wide approaches to combat other invasive pests.

Key Words: beneficial insects, biological control, virus, pheromone-mediated, mating disruption, pesticide resistance, area-wide integrated pest management, IPM, SIT, *Cydia pomonella,* British Columbia

1. INTRODUCTION

The codling moth *Cydia pomonella* L., the proverbial "worm in the apple", damages pome fruit directly, and is a key pest of this crop in most of the areas where it is cultivated (Beers et al. 2003). If left uncontrolled, the codling moth can damage 50 to 90% of an apple crop (Ontario Ministry of Agriculture, Food and Rural Affairs 2011). Though the codling moth originated in Asia Minor, it arrived in the Okanagan region in the early 1900s, making it a pest for nearly as long as apples have been produced commercially in the area (Bloem et al. 2007). The Okanagan region in southern British Columbia, the western-most province of Canada (Fig. 1), is unique in Canada because of its dry, sunny climate (hot, dry summers and mild winters). Tree fruit production has been a hallmark of the region for over 100 years.

In this region, and most other areas where pome fruits are produced, broad-spectrum insecticides were previously used heavily to control codling moth populations (Madsen and Morgan 1970). Codling moth populations in many areas had started developing resistance to some classes of insecticides, and concerns over development of cross-resistance were mounting (Dunley and Welter 2000). The use of broad-spectrum insecticides also had indirect costs related to the loss of natural enemies and pollinators; it was a source of environmental contamination, and consumers were concerned over insecticide residues on food (Vreysen et al. 2010). For these reasons, negative public attitudes towards the use of pesticides stimulated support for codling moth control strategies that did not rely on synthetic pesticides (Madsen and Morgan 1970).

Twenty years of research and planning culminated in the early 1990s with the implementation of the Okanagan-Kootenay Sterile Insect Release (OKSIR) programme (Dyck et al. 1993). Though the programme was initially more expensive than a conventional insecticide programme, a number of factors contributed to its adoption in the region. Restrictions on pesticide use, particularly those most effective at controlling the codling moth, were increasing. Concern for the impacts of insecticides on beneficial insects, the environment, and surrounding communities was mounting. Finally, reducing pesticides and codling moth populations created marketing advantages for local pome-fruit growers. In weighing these costs and benefits, and taking a long-term view of their implications, the programme was deemed to have a net benefit to the industry and community (Holm 1985, 1986; Jeck and Hansen 1987) and was therefore initiated in 1991.

2. THE STERILE INSECT TECHNIQUE

The concept of the Sterile Insect Technique (SIT) for insect control was conceived by E. F. Knipling in the 1930s, and first developed and applied in the 1950s to successfully control the New World screwworm *Cochliomyia hominivorax* Coquerel (Klassen et al. 2021). The SIT is a biological insect control method in which insects are mass-reared, irradiated to make them sterile, and then released into the environment at regular intervals to mate with wild insects. Wild female insects that mate with the sterilized male insects produce no offspring, thereby reducing the number of insects in the next generation. Continued use of the SIT at appropriate overflooding ratios thus leads to successively smaller generations, and can, if applied correctly on an area-wide basis, result in an area of low pest prevalence or even eradication of the population in that area (Dyck et al. 2021).

Since the 1950s, the SIT has been successfully used around the world to suppress, prevent, contain or eradicate many dipteran insect populations such as the New World screwworm, several tsetse flies *Glossina* spp., and fruit flies such as the melon fly *Zeugodacus cucurbitae* Coquillet, Mediterranean fruit fly *Ceratitis capitata* Wiedemann and Mexican fruit fly *Anastrepha ludens* Loew (Dyck et al. 2021).

The SIT has also been applied with success against various lepidopteran pests such as the cactus moth *Cactoblastis cactorum* Berg, the painted apple moth *Orgyia anartoides* Walker, the false codling moth *Thaumatotibia leucotreta* Meyrick, and the pink bollworm *Pectinophora gossypiella* Saunders (Carpenter et al. 2007; Suckling et al. 2007; Dyck et al. 2021; Bello et al., this volume; Boersma, this volume; Staten and Walters, this volume).

3. OKANAGAN-KOOTENAY STERILE INSECT RELEASE PROGRAMME

At the inception of the OKSIR programme, other non–insecticide-based insect control methods, such as pheromone-mediated mating disruption, were considered in order to reduce codling moth populations in southern British Columbia (Judd et al. 1996). Due to the heterogeneous landscape of orchard and residential areas in most of the region (creating a non-contiguous orchard area (Cardé 2007; Witzgall et al. 2008)), the SIT was considered a more suitable solution to manage the codling moth. The OKSIR programme was initially conceived in 1991 as an eradication programme, and was based on research conducted by Proverbs and colleagues in the 1970s and 1980s (Proverbs et al. 1978, 1982).

The mandate and objectives changed in 1997 when the goal became suppression of the codling moth below economic levels through delivery of an efficient, effective and sustainable AW-IPM programme using the SIT as the main control tool. It was concluded that permanent area-wide suppression rather than eradication was a more realistic goal because of the large programme area, the limited human and financial resources available, and because an expensive quarantine programme that would be necessary during and after eradication for which the Federal Government of Canada, which regulates quarantines, had no plans (Bloem et al. 2007).

A community-based, AW-IPM approach was essential to reduce costs and increase the effectiveness of the programme. Area-wide measures applied over a geographically- or politically-defined area enable control of an entire pest population, a prerequisite to sustainable pest management. Insects do not respect property boundaries, and a field-by-field approach that focusses narrowly on the value loss to individual crops cannot successfully control pest populations because unmanaged cultivated and wild host trees on neighbouring public, private or abandoned land are recurrent sources of reinfestation (Hendrichs et al. 2007).

An AW approach requires a strong partnership among all stakeholders to succeed. For OKSIR, a partnership was therefore developed between the region's local governments and the tree-fruit industry. The Province of British Columbia enacted legislation establishing a mandatory community-based, AW-IPM programme (Municipalities Enabling and Validating Act [RSBC 1960] 1989) that required all property owners, including both residential and commercial host-tree owners, to control codling moth infestations and participate in local funding of the programme.

Researchers at Agriculture Canada, local governments and growers selected the SIT to reduce the use of insecticides because of concerns about excessive pesticides in the environment, an attempt to delay insecticide resistance, and to support agritourism opportunities in the area. This collective action, i.e. action taken jointly by the stakeholders in pursuit of their perceived shared interests, made it possible to deliver results in a much larger geographic scale than could be provided or protected by a single farmer (Lefebvre et al. 2015b).

3.1. Programme Area

The programme area covers approximately 600 km^2, and at its onset serviced approximately 8 900 ha (22 000 acres) of pome fruit; the surface of pome fruit within this area was gradually reduced to 3 440 ha (at the time of writing). The large area to be serviced, and the need for pre-release sanitation (discussed below), required the programme to be implemented sequentially across three zones (Fig. 1). Pre-release sanitation and construction of the rearing facility began in zone 1 in 1992 followed by moth release in 1994. Pre-release sanitation started in 1998 and 2000 in zones 2 and 3, respectively, with moth release occurring after two years of sanitation efforts.

3.2. Rearing Facility

At the heart of the OKSIR programme is a state-of-the art insect mass-rearing facility. Construction of the facility located in Osoyoos, British Columbia was completed in 1993. The facility is capable of producing more than 2 million sterilized moths per day or 780 million sterile codling moths annually. Maintenance and operational costs, as well as capital upgrades, are funded by local property tax requisitions.

3.3. *Programme Services*

The OKSIR programme services include pre-release sanitation, mandatory SIT application, surveillance, enforcement and education:

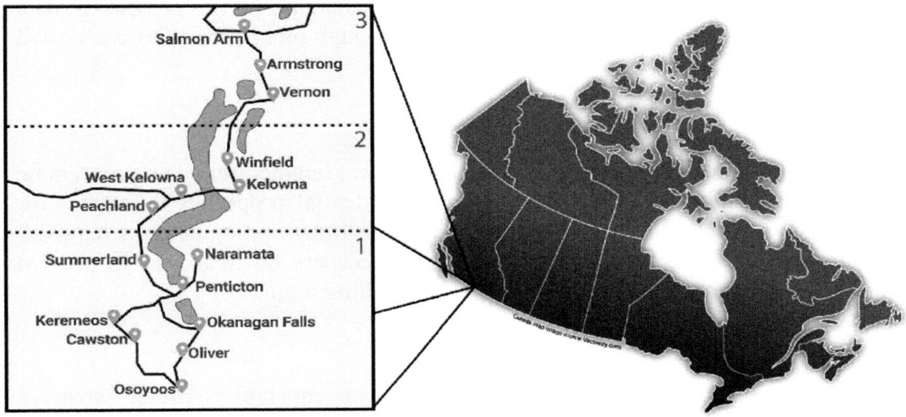

Figure 1. Location of the OKSIR programme. The map of Canada (right) indicates where the OKSIR Programme is located in British Columbia, and the inset (left) illustrates how the programme area, covering a linear distance of ca. 175 km, was divided into three zones.

3.3.1. *Pre-Release Sanitation*

The first phase of the programme was pre-release sanitation. This entailed the removal of thousands of unmanaged/abandoned host trees to reduce refugia for the codling moth. The programme also coordinated and supported the suppression of codling moth populations in orchards through the use of conventional insecticides, cultural practices and pheromone-mediated mating disruption. Wild codling moth populations had to be reduced as much as possible throughout all communities to increase the efficiency of subsequent SIT application. The programme continues to remove unmanaged/abandoned host trees and derelict orchards as needed.

3.3.2. *Mandatory SIT Application*

The programme delivers a mandatory area-wide control application of sterile codling moths to every orchard property, i.e. 2000 sterile codling moths of mixed sex (1:1)/ha/week for approximately 20 weeks per season. As necessary and practical, additional releases are made to address high pest pressure "hot spots."

3.3.3. *Surveillance*

Every orchard property is monitored with pheromone-baited traps (1 trap/ha) that are checked once a week. These spatially explicit trapping data are disseminated in real-time through the programme's website to allow growers to respond rapidly with supplementary controls if needed (OKSIR 2017).

Other monitoring techniques include: in-season fruit inspections, end-of-season assessment of fruit damage, and banding of host trees (corrugated cardboard strips wrapped around trees to trap mature larvae; later the strips are removed and destroyed). Codling moth host trees are also monitored on non-orchard properties (properties with less than 20 host trees) by visual fruit inspections and banding. It is important to recognize that the monitoring of non-orchard properties is focussed on a 200-m buffer zone around commercial orchards, though other properties are visited as needed.

3.3.4. Enforcement

Codling moth control is enforced throughout the programme area. The programme has the legal authority to enter orchards and residential properties to inspect for codling moth infestation and issue control orders for fruit stripping and tree removal. Dedicated enforcement ensures that all host-tree owners do their part to prevent outbreaks and ensure proper management of the codling moth.

3.3.5. Education

Education is essential to reduce enforcement actions as much as possible. Extensive education and outreach about the programme and codling moth control is done each season with growers and residential tree owners in the region. This is critical, especially during the first stages of programme implementation. Outreach occurs via media advertising, publications, newsletter articles, the programme website, field visits and public meetings.

An information technology specialist maintains the website that provides real-time trapping and phenological data to growers. The growers themselves become sources of positive promotion once they are convinced of the benefits that the OKSIR programme is bringing to their economies and the environment (Bloem et al. 2007).

3.4. Governance, Funding and Budget

The OKSIR programme is governed by a Board of Directors, comprising five elected community representatives from each of the four regional municipal governments within the programme's service area, and three grower representatives nominated by the pome fruit industry.

The programme has a central administration and is funded through local taxation. All properties in the region pay a tax based on assessed land value. Currently, the average residential property pays ca. CAD 11/year (USD 9/year), and growers pay a parcel tax of ca. CAD 340/ha. Overall, the programme obtains 60% of its funding from local property owners in the community and 40% from commercial growers. The percentage share of the funding allocation has been arbitrarily determined based on political will. The annual programme budget is ca. CAD 3.2 million. In 2016, CAD 1.7 million was collected from general taxpayers via land value tax and CAD 1.16 million from growers (approximately 3440 ha) via parcel tax (cost/ha). The programme received an additional ca. CAD 350 000 from the sale of excess products, interest income and grants.

Of the total operating budget in 2016, CAD 1.1 million was used for field operations, CAD 1.1 million to operate the mass-rearing facility, and CAD 681 000 to cover administrative costs. The programme services are delivered by 16 full-time staff and up to 75 seasonal staff working in the field and mass-rearing facility. From 2010 to 2017, the programme operated without increase in either land value or parcel taxes.

4. SUCCESS OF THE OKSIR PROGRAMME OVER 20 YEARS

Over the more than 20 years that the OKSIR programme has been operating, it has achieved sustained codling moth suppression (Fig. 2). Overall, there was a 94% reduction in codling moth population, relative to pre-programme levels in the OKSIR programme area as measured by pheromone traps (Fig. 2).

It must be noted that in zones 2 and 3, on recommendation of an operational advisory committee, the OKSIR programme piloted zone-wide use of pheromone-mediated mating disruption in commercial orchards as an alternative to the SIT from 2011-2014. Pheromone trap captures during these years are not directly comparable to other years or zone 1 during these years because a different trapping system was used. For economic and other reasons, in 2015 the OKSIR programme returned to using the SIT in commercial orchards across the entire programme area (Cartier 2014). Since 2000, and in all zones, mean codling moth trap captures remain well below the recommended treatment threshold of two codling moths per trap per week for two consecutive weeks (Vakenti 1972) (Fig. 2).

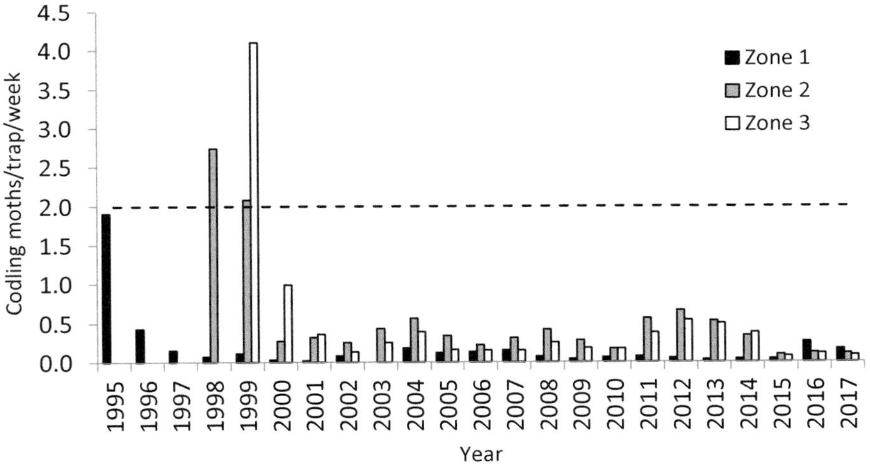

Figure 2. Mean wild codling moth captures per trap per week from 1995 to 2017 for each zone managed by the OKSIR programme in zone 1 (from 1995), in zone 2 (from 1998), and in zone 3 (from 1999), averaged over each fruit-growing season. The dashed line indicates the recommended threshold (two codling moths per trap/week for two consecutive weeks) at which insecticide controls supplementary to the SIT would be required.

In zone 1, codling moth fruit damage was effectively suppressed after approximately five years, and has remained that way in more than 90% of orchards (Fig. 3).

Codling moth suppression followed similar trends in zones 2 and 3, except that population suppression occurred at a slower rate (Figs. 2 and 3). It is difficult to pinpoint exactly why this was the case, and no single factor alone was likely responsible for this result. Reasons for slower population decline include:

- Larger urban centres in zones 2 and 3 with more infested backyard trees acting as refugia for codling moths.
- Pheromone-mediated mating disruption that was used extensively in zones 2 and 3 during the pre-release sanitation phase, may have been less effective than the extensive initial application of organophosphate insecticides for suppression in zone 1.
- The increased service area created greater demand on the programme's limited expert knowledge and human resources.
- The organic fruit producers reside predominantly in zone 1, meaning there may have been less grower buy-in or more scepticism in zones 2 and 3.

Ultimately, by 2015 more than 90% of orchards had less than 0.2% fruit damage due to codling moth (Fig. 3) in all zones.

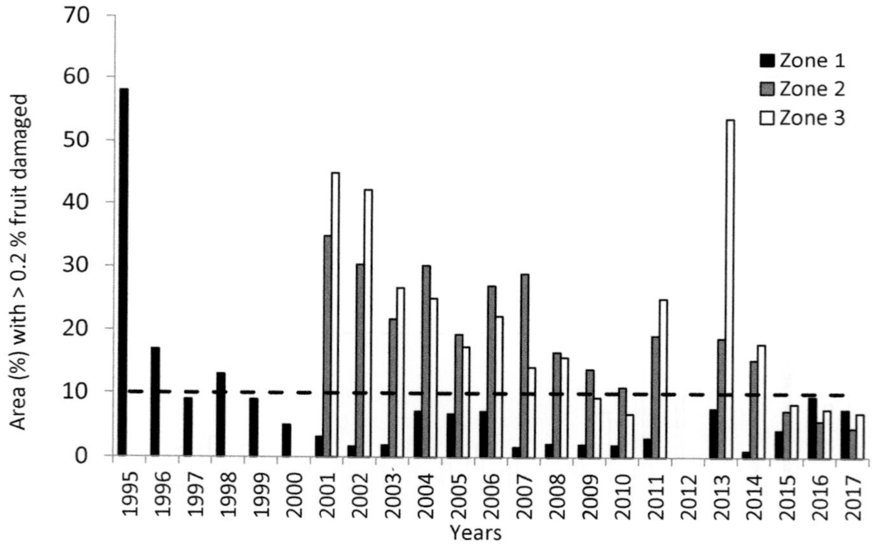

Figure 3. Percent of programme area with >0.2% of fruit damaged by the codling moth. Bars show data from 1995 (zone 1) and from 2001 (zones 2 and 3) to 2017 for each zone managed by the SIR programme. The dashed line indicates 10% of the programme area, an economic target set by the Programme's Board; 2012 data not available.

On average, 85% of the planted area was sampled every year for fruit infestation just prior to harvest (though this varied significantly across years due to limited human resources) ranging from 15 to 100%. Orchards that were not sampled were those with no evidence of codling moth populations, as evidenced by trap captures, early-season inspections, and previous sample history, and thus were placed in the ≤0.2% damage category.

The programme has contributed to a dramatic decrease in the amount of insecticides applied per ha of pome fruit. From 1991 to 2016, there was an estimated 96% reduction in insecticides used against the codling moth (Fig. 4). Other factors, such as changes in spray application rates in spindle versus traditional planting systems, new product formulations, etc., contributed in part to this reduction. Personal testimonies from growers indicated that many have not needed to spray insecticides to control the codling moth in more than 15 years.

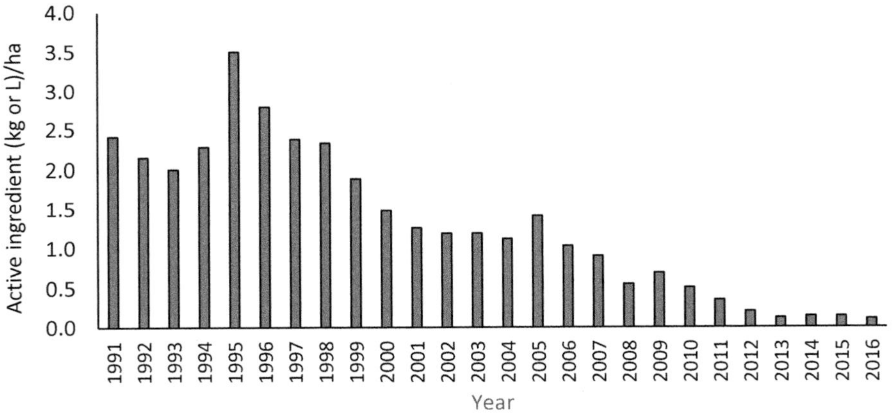

Figure 4. Estimated pesticide active ingredient (kg or L) applied per ha per year for all zones managed by the SIR programme from 1991 to 2016 based on the estimated proportion of sales for the 15 products registered for use against the codling moth (note: a number of these insecticides are also applied for other pests and/or crops). The estimates of active ingredients are divided by the area (ha) of planted pome fruit in the programme area to account for changes in sales due to amount of pome fruit under cultivation.

Anecdotal evidence collected during an external review of the OKSIR programme also suggests that there are collateral pest management benefits occurring (Carpenter et al. 2014). Leafrollers *Spilonata ocellana* Denis and Schiffermüller, *Choristoneura rosaceana* Harris and *Archips* spp., were reported to be on the decline similar to what organic growers have experienced, i.e. when insecticide use is reduced the natural enemy populations are allowed to increase and reduce some pest populations (Leach and Mumford 2008).

5. BENEFITS AND OPPORTUNITIES

The OKSIR programme is an AW-IPM programme that is able to combat insect infestations across multiple local community jurisdictions. It is a programme that requires close collaboration between four different regional governments, growers, urban host-tree owners, fruit packers, industry advisors, tree-fruit retailers and researchers. Therefore, there are benefits and responsibilities at all levels, and the continued commitment of all partners is vital.

Prior to the AW-IPM programme and the release of sterile insects, spraying organochlorine and later organophosphate insecticides was for many years the main method to control the codling moth. Growers had to apply multiple cover sprays each season, and concerns that the codling moth might become resistant to insecticides mounted. In addition, there were growing concerns regarding chemical residues on the crop and the insecticide load in the environment. Residents of the Okanagan Valley have a strong preference for reduced insecticide use on their properties as well as on neighbouring farms, and for reduced insecticide residues on food (Cartier 2014). Insecticide poisoning of farmworkers and groundwater contamination are also serious concerns in many fruit-growing areas (Witzgall et al. 2008), and run-off from farmers' fields may have direct toxic effects on the water supply and aquatic organisms.

Following the introduction of the OKSIR programme, the reduced reliance on insecticides used for codling moth control in the region has significantly minimized the potential risks to the environment, including the local biodiversity and water for both people and crops, and has significantly minimized the risks to workers' safety while maintaining the economic viability of the crop (Tracy 2014). Residents who live near the orchards now have lower exposure to insecticides and access to sustainably produced fruit, and they enjoy the benefits of a thriving local agriculture and the green spaces it provides. Also, the local economy is improved because the agritourism and agricultural industries are supported by the programme.

Currently, a trend towards "social economy" is becoming apparent in Canada, in which individuals and groups take into account the social consequences of economic activity, including the increased attention to social issues. Local non-insecticide control of the codling moth can be seen as a "shared value," which is defined as "policies and operating practices that enhance the competitiveness of a company while simultaneously advancing the economic and social conditions in the communities in which it operates" (European Commission 2013).

This trend is evidenced by the willingness of the vast majority (90.4%) of non-agricultural residents in the area to pay for the programme. Under the current funding structure of the area-wide approach, growers pay 65% less than they would for the same service without the programme (Cartier 2014).

Both the provincial and federal governments contributed toward the cost of constructing the codling moth mass-rearing facility (CAD 7.4 million), which is a major seasonal employer in the region. Total employment associated with the programme contributes CAD 2.2 million, or CAD 620 per ha, to the regional gross domestic product (Cartier 2014). Cartier's (2014) benefit-cost analysis revealed that, for every CAD 1 in cost there was CAD 2.5 in benefit, for both producers and society as a whole.

As the mass-rearing facility has the capacity to produce 780 million sterile codling moths annually, of which only a portion is used seasonally to treat the areas in the Okanagan Valley, this has opened opportunities to serve other regions or countries through sales of sterile moths or egg sheets, and offers options for codling moth virus production. As an example, the OKSIR programme currently sells excess production of sterile moths to researchers in the USA working on pheromone research. The New Zealand Institute for Plant and Food Research is conducting a pilot project for codling moth control using sterile codling moth shipped from Canada (Horner et al. 2016). The OKSIR programme also sells egg sheets for commercial virus production.

There are examples of for-profit operations selling sterile insects and related services commercially, such as the onion fly *Delia antiqua* (Meigen) in the Netherlands (Vreysen et al. 2006), Mediterranean fruit fly in Israel and South Africa (Bassi et al. 2007; Barnes et al. 2015), and the false codling moth in South Africa (Boersma, this volume).

As costs of controlling the codling moth decrease because of the success of area-wide application of the SIT, other resources can be redirected to address threats of other/new invasive insect pests. Due to climate change and increasing global trade, invasive insect pests are migrating to new habitats throughout the world.

Subsequent to the successful suppression of the codling moth to below economic levels, the local fruit industry has requested the programme to use its surveillance infrastructure to monitor other existing pests, e.g. leafrollers, and new invasive pests, such as apple clearwing moth *Synanthedon myopaeformis* Borkhausen, apple maggot *Rhagoletis pomonella* Walsh, and brown marmorated stink bug *Halyomorpha halys* Stål. Since programme staff visit the pome orchard properties weekly for codling moth management purposes, there is an opportunity to incrementally add monitoring services for other pests and take advantage of economies of scale.

Insecticide resistance is also increasing. The SIT reduces insecticide use and, therefore, reduces the likelihood of insecticide resistance. There is an opportunity to use codling moth SIT to supplement the use of other biocontrol techniques, such as pheromone-mediated mating disruption or biological insecticides, e.g. *Cydia pomonella* granulosis virus (CpGV) (Judd and Gardiner 2005; Eberle and Jehle 2006; Cardé 2007; Witzgall et al. 2008). The SIT is very efficacious and sustainable, and hence is the ideal tool to supplement other methods as needed (Dyck et al. 2021).

A further benefit to a region of an AW-IPM programme with an SIT component is the provision of employment opportunities for area residents. A major challenge can be the availability and recruitment of sufficient staff for the planned facility and field operations when choosing the physical location of the programme. Central administration of the programme is important because it ensures continuity between areas and allows for the consistent provision of full IPM support services, including monitoring, education and enforcement, at a lower cost. The commitment of staff to the outcome is also critically important; experienced staff is the programme's most valuable resource.

Looking globally, apples and cherries from the area can now be exported into high-value restricted markets, such as Taiwan, China and Japan (the presence of the codling moth restricts the movement of cherries and apples in some high-value Asian markets). The SIT promotes areas of low pest prevalence, which can allow for the biosecurity of the product to be almost guaranteed, while the cost for a rejected shipment of fruit in such restricted markets and the closure of such markets can be very costly to the exporting nation.

The OKSIR programme is also working with leaders worldwide in addressing invasive pests in a changing climate by developing projects that transfer knowledge using AW-IPM approaches and the SIT. In addition, the programme has successfully exported quality sterile moths to New Zealand, South Africa and the USA for use in IPM pilot projects and research (Blomefield et al. 2011; Carpenter et al. 2012; Horner et al. 2016; Adams et al. 2017).

In recent years, the visibility of the OKSIR programme on the international scene has increased; in 2015 at the 8th International IPM Symposium the programme received the International IPM Award of Excellence for a regional integrated pest management programme (IPM 2015). Also, a panel of international experts recommended the OKSIR programme as a role model for AW-IPM using the SIT (Carpenter et al. 2014).

The Food and Agriculture Organization of the United Nations, the World Bank and other international organizations promote the deployment of biological control agents to suppress pests and replace chemical controls (IAASTD 2009). The Stockholm Convention on Persistent Organic Pollutants, a global environmental treaty effective from 2004, aims to eliminate or restrict persistent organic pollutants (Stockholm Convention 2009). These actions epitomize the global need for environment-friendly pest management programmes like the OKSIR programme. Following improved technology and research, the SIT could play an important role in this area, particularly as a supplement to other biocontrol techniques applied on an area-wide basis (Cardé and Minks 1995; Hendrichs et al. 2007).

6. CHALLENGES

According to Tracy (2014), common obstacles to adopting a programme like the OKSIR are:

1. Low levels of investment in research and development.
2. Lack of coordination among growers in adopting the different methods.
3. Weak or conflicting regulatory framework.
4. Absence of market incentives and consumer awareness.

The OKSIR programme's experience of more than 20 years has revealed vital lessons. The goal in sharing these lessons is to assist other regions and pome fruit growers who want to incorporate the SIT for codling moth control into their pest management system and start their programmes as efficiently and smoothly as possible (Carpenter et al. 2014).

Support from political partners (policy/regulatory) and the public is critically important for establishing and sustaining an AW-IPM programme like OKSIR. The impact of the pest and the techniques used to manage it can be influenced by both politics and emotions. It is vital that all stakeholders share common values, and that their expectations are identified from the beginning and managed along the way. This is crucial for building trust in the programme, especially in its initial phases before the programme results can provide convincing proof.

Champions for the programme outside the administration who represent the interests of each of the stakeholder groups are extremely useful to move the initiative forward and maintain ongoing support for an AW-IPM programme.

Committed partners in the programme are critical to its success. It may be easy to identify and join forces with potential partners, but what happens if this collaboration fails or is unproductive? It is crucial to define how the partnership is constituted. The responsibilities of each partner must be mutually agreeable, such that the integrity and efficacy of the programme can be maintained. Not doing so can jeopardize the success of the entire programme. Defining procedures to allow exit and entry of partners is invaluable.

In the initial phase of the OKSIR programme, many growers resented having the programme "imposed" on them. In addition, it was difficult to convince growers, whose livelihoods depend on their crops, to trust the SIT component of the AW-IPM approach and to apply supplemental chemical control only when needed. However, trust grew over time, in large part due to the real-time trapping and monitoring data made available to growers on the website, and especially with the obvious decline in codling moth populations and crop damage. Growers are now very supportive of the AW-IPM approach because it protects them from potential infestation coming from their neighbours' poor management practices (both commercial and residential). In addition, the high number of hits on the website's real-time data pages shows that growers are checking their monitoring data and scrutinizing moth thresholds before spraying, rather than simply applying "comfort sprays" (Lefebvre et al. 2015a).

This was the first area-wide use of the SIT for controlling the codling moth in commercial orchard plantings. Challenges in the early years of the programme included overcoming public resistance to the mandatory requirement to control the pest on all properties, and for the growers to receive a mandatory applied control measure. This was caused in part by ineffective communication with the public and philosophical disagreements between the programme and individual property owners. However, over time, due to improved communication, the roles and responsibilities of all stakeholder groups were better understood, resulting in better cooperative actions and more effective codling moth control. Furthermore, the legal authority for programme staff to enter properties to conduct activities was initially met with resistance from growers and the public. This authority is vitally important for the success of the programme.

Convincing consumers to embrace AW-IPM practices has also been a challenge. Compared with the sale price of fruit produced by conventional production systems, it has been shown that consumers are willing to pay significantly more for certified organic fruit, and moderately more for fruit produced through IPM practices (Lefebvre et al. 2015a). The OKSIR programme supports a relatively small organic pome fruit industry, i.e. less than 10% of British Columbia's pome fruit production (Macey 2013). These growers enjoy the economic benefits of marketing their fruit as certified organic. However, the remaining conventional growers have not yet taken steps to capitalize on market opportunities afforded by agroecological and socially responsible practices in the same way that growers in the USA have benefited from the "Responsible Choice" label (Tracy 2014).

During the planning stages of an AW-IPM programme, it is important to understand what the benefits and costs of the programme are, and to understand the long-term expectations of the stakeholders once the programme achieves its goals. A major challenge to maintain a sustainable programme is to obtain continued funding when the pest is no longer deemed a problem. The success of an AW-IPM approach means that the reason for its existence appears to diminish. However, removing the AW-IPM measures will create a resurgence of the pest to levels existing prior to the programme. For example, once success has been achieved, keeping the population levels below treatment thresholds requires a risk management approach.

The local governments in the community-based OKSIR programme are also under pressure to allocate tax revenue to other uses, and they must justify the continued area-wide service to their constituents (who are also benefactors of the programme). In addition, there has been an overall decrease in pome-fruit area in the region because of low market returns. Because of their northern location, growers in British Columbia are faced with higher production costs than growers in the nearby USA. Also, higher profit margins can be made growing alternative crops such as sweet cherries and wine grapes. Even though a decreasing pome-fruit area equates to reduced parcel-tax revenue, the OKSIR programme has remained financially strong, and in the past seven years there has been no tax increase.

Expansion of the programme also presents challenges. There are challenges for a public-sector regulatory service when considering diversifying the revenue model to include expansion of scale and scope, such as producing a commercial supply for sales of sterile moths or other by-products, and in assisting the setting up of programmes in other regions/countries. There can be complications due to the governance/business model, as well as different languages and cultures when providing technical support to those areas purchasing sterile insects. There are regulatory, shipping and logistics timing issues involved in selling a product internationally (Blomefield et al. 2011; Adams et al. 2017; Suckling et al. 2017). Biological control agents and beneficial insects are still a developing sector. Unfamiliarity with import regulations of these new products by the many customs officials and private companies along the transport chain can impede movement over international borders (FAO 2017).

A small, locally funded programme cannot afford large-scale investment in research to improve the technology, and generally it can afford to focus only on critical operational improvements as situations arise. The OKSIR programme is currently working with collaborators in New Zealand and the USA on advancing the use of unmanned aerial systems for more efficient release of the sterile codling moth in field implementation trials using the SIT to control the codling moth.

The OKSIR programme has relied on the senior government and international research community for technical assistance, and for advancing AW-IPM and the SIT for the codling moth. Due to the increasing public pressure to reduce the use of chemical insecticides around the world, a global investment in SIT research is needed to advance the technology. A few areas worthy of investigation include if or how released sterile female codling moths could create an effect similar to commercially applied pheromone-mediated mating disruption, how SIT affects parasitoid populations, what are effective sterile: wild ratios for extremely low wild populations, and methods to monitor extremely low wild densities (Judd and Gardiner 2005; Witzgall et al. 2008; Bau and Cardé 2016).

7. CONCLUSIONS

The OKSIR programme is a highly effective and easily transferrable model to control orchard pest infestations. There is greater demand for biological control methods to meet mounting pressures of climate change, increasing insect pest populations, impacts of global trade, and concerns over affordability and availability of methods. Concerns are mounting that the codling moth is becoming resistant to the insecticides used commonly for control, with few replacement chemical options being developed. Biological control options also face pressure because they are not stand-alone technologies, often relying on compatible, supplemental control products and crop management practices.

The OKSIR programme clearly illustrates that area-wide integration of the SIT can successfully manage codling moth populations in an environmentally sound way. In addition, it can easily be integrated with other biological control methods such as pheromone-mediated mating disruption and CpGV. Most importantly, the SIT can replace control products that are no longer environmentally or economically viable, and hence provide an excellent biologically sustainable solution for controlling insect pests.

Despite its proven success, the SIT is often considered a curiosity rather than an effective, environment-friendly technology that dovetails into many modern IPM programmes. When considering the success of the OKSIR programme, other regions and pome fruit growers around the world will be compelled positively to consider AW-IPM with a SIT component as a viable management strategy for the codling moth.

8. REFERENCES

Adams, C. G., J. H. Schenker, P. S. McGhee, L. J. Gut, J. F. Brunner, and J. R. Miller. 2017. Maximizing information yield from pheromone-baited monitoring traps: Estimating plume reach, trapping radius, and absolute density of *Cydia pomonella* (Lepidoptera: Tortricidae) in Michigan apple. Journal of Economic Entomology 110: 305–318.

Barnes, B. N., J. H. Hofmeyr, S. Groenewald, D. E. Conlong, and M. Wohlfarter. 2015. The Sterile Insect Technique in agricultural crops in South Africa: A metamorphosis.... but will it fly? African Entomology 23: 1–18.

Bassi, Y., S. Steinberg, and J. P. Cayol. 2007. Private sector investment in Mediterranean fruit fly mass-production and SIT operations – The "sheep" of the private sector among the "wolves" of the public good?, pp. 457–472. *In* M. J. B. Vreysen, A. S. Robinson, and J. Hendrichs (eds.), Area-wide control of insect pests: From research to field implementation. Springer, Dordrecht, The Netherlands.

Bau, J., and R. T. Cardé. 2016. Simulation modeling to interpret the captures of moths in pheromone-baited traps used for surveillance of invasive species: The gypsy moth as a model case. Journal of Chemical Ecology 42: 877–887.

Beers, E. H., D. M. Suckling, R. J. Prokopy, and J. Avilla. 2003. Ecology and management of apple arthropod pests, pp. 489–519. *In* D. C. Ferree, and I. N. Warrington (eds.), Apples: Botany, production and uses. CAB International, Wallingford, United Kingdom.

Bloem, S., A. McCluskey, R. Fugger, S. Arthur, S. Wood, and J. Carpenter. 2007. Suppression of the codling moth *Cydia pomonella* in British Columbia, Canada using an area-wide integrated approach with an SIT component, pp. 591–601. *In* M. J. B. Vreysen, A. S. Robinson, and J. Hendrichs (eds.), Area-wide control of insect pests: From research to field implementation. Springer, Dordrecht, The Netherlands.

Blomefield, T., J. E. Carpenter, and M. J. B. Vreysen. 2011. Quality of mass-reared codling moth (Lepidoptera: Tortricidae) after long-distance transportation: 1. Logistics of shipping procedures and quality parameters as measured in the laboratory. Journal Economic Entomology 104: 814–822.

Cardé, R. T., 2007. Using pheromones to disrupt mating of moth pests, pp.122–169. *In* M. Kogan, and P. C. Jepson (eds.), Perspectives in ecological theory and Integrated Pest Management (No. 632.9 P46). Cambridge University Press, Cambridge, UK.

Cardé, R. T., and A. K. Minks. 1995. Control of moth pests by mating disruption: Successes and constraints. Annual Review of Entomology 40: 559–585.

Carpenter, J. E., S. Bloem, and H. Hofmeyr. 2007. Area-wide control tactics for the false codling moth Thaumatotibia leucotreta in South Africa: A potential invasive species, pp. 351–359. *In* M. J. B. Vreysen, A. S. Robinson, and J. Hendrichs (eds.), Area-wide control of insect pests: From research to field implementation. Springer, Dordrecht, The Netherlands.

Carpenter, J. E., T. Blomefield, and M. J. B. Vreysen. 2012. A flight cylinder bioassay as a simple, effective quality control test for *Cydia pomonella*. Journal Applied Entomology 136: 711–720.

Carpenter, J. E., D. Conlong, P. McGhee, G. Simmons, G. Taret, and M. J. B. Vreysen. 2014. Report of an external review of the Okanagan-Kootenay sterile insect release program, June 9–13 2014. 35 pp.

Cartier, L. 2014. A benefit-cost analysis of the Okanagan Kootenay sterile insect release program. Okanagan School of Business, Okanagan College, Kelowna, BC, Canada.

Dunley, J. E., and S. C. Welter. 2000. Correlated insecticide cross-resistance in azinphosmethyl resistant codling moth (Lepidoptera: Tortricidae). Journal of Economic Entomology 93: 955–962.

Dyck, V. A., S. H. Graham, and K. A. Bloem. 1993. Implementation of the sterile insect release programme to eradicate the codling moth, *Cydia pomonella* (L.) (Lepidoptera: Olethreutidae), in British Columbia, Canada, pp. 285–297. *In* Management of insect pests: Nuclear and related molecular and genetic techniques. Proceedings of the FAO/IAEA Symposium, October 1992, Vienna, IAEA.

Dyck, V. A., J. Hendrichs, and A. S. Robinson (eds.). 2021. Sterile Insect Technique – Principles and practice in Area-Wide Integrated Pest Management. Second Edition. CRC Press, Boca Raton, Florida, USA. 1200 pp.

Eberle, K. E., and J. A. Jehle. 2006. Field resistance of codling moth against *Cydia pomonella* granulovirus (CpGV) is autosomal and incompletely dominant inherited. Journal of Invertebrate Pathology 93: 201–206.

European Commission. 2013. Social economy and social entrepreneurship: Social Europe guide. Volume 4. Luxembourg: Publications Office of the European Union. ISBN: 978-92-79-26866-3.

(FAO) Food and Agriculture Organization of the United Nations. 2017. ISPM No. 3: Guidelines for the export, shipment, import and release of biological control agents and other beneficial organisms. International Standards for Phytosanitary Measures. International Plant Protection Convention (IPPC), Rome, Italy.

Hendrichs, J., P. Kenmore, A. S. Robinson, and M. J. B. Vreysen. 2007. Area-wide integrated pest management (AW-IPM): Principles, practice and prospects, pp. 3–33. *In* M. J. B. Vreysen, A. S. Robinson, and J. Hendrichs (eds.), Area-wide control of insect pests: From research to field implementation. Springer, Dordrecht, The Netherlands.

Holm, W. R. 1985. An evaluation of the commercial cost of a sterile insect release program for the codling moth in British Columbia. Agriculture Canada, New Westminster, BC, Canada.

Holm, W. R. 1986. Analysis of risks and costs of a sterile insect release program for control of the codling moth in the Okanagan region of British Columbia. Agriculture Canada, New Westminster, BC, Canada.

Horner, R. M., J. T. S. Walker, D. J. Rogers, P. L. Lo, and D. M. Suckling. 2016. Use of the Sterile Insect Technique in New Zealand: Benefits and constraints. New Zealand Plant Protection 68: 296–304.

(IAASTD) International Assessment of Agricultural Knowledge, Science and Technology for Development. 2009. Agriculture at a crossroads: International assessment of agricultural knowledge, science and technology for development. North America and Europe report. Washington, DC, USA.

(IPM) International IPM Symposium. 2015. 8th International IPM Symposium. IPM: Solutions for a Changing World, March 23–26, 2015, Salt Lake City, Utah, USA.

Jeck, S., and J. Hansen. 1987. Economics of codling moth control by sterile insect release: Benefit-cost approach. Agriculture Canada, New Westminster, British Columbia, Canada.

Judd, G. J. R., and M. G. T. Gardiner. 2005. Towards eradication of codling moth in British Columbia by complimentary actions of mating disruption, tree banding and Sterile Insect Technique: Five-year study in organic orchards. Crop Protection 24: 718–733.

Judd, G. J. R., M. G. T. Gardiner, and D. R. Thomson. 1996. Commercial trials of pheromone-mediated mating disruption with Isomate-C® to control codling moth in British Columbia apple and pear orchards. Journal of the Entomological Society of British Columbia 93: 23–34.

Klassen, W., C. F. Curtis, and J. Hendrichs. 2021. History of the Sterile Insect Technique, pp. 1–44. *In* V. A. Dyck, J. Hendrichs, and A. S. Robinson (eds.), Sterile Insect Technique – Principles and practice in Area-Wide Integrated Pest Management. Second Edition. CRC Press, Boca Raton, Florida, USA.

Leach, A. W., and J. D. Mumford. 2008. Pesticide environmental accounting: A method for assessing the external costs of individual pesticide applications. Environmental Pollution 151: 139–147.

Lefebvre, M., S. R. H. Langrell, and S. G. Paloma. 2015a. Incentives and policies for Integrated Pest Management in Europe: A review. Agronomy for Sustainable Development 35: 27–45.

Lefebvre, M., M. Espinosa, S. G. Paloma, J. L. Paracchini, A. Piorr, and I. Zasada. 2015b. Agricultural landscapes as multi-scale public good and the role of the Common Agricultural Policy. Journal of Environmental Planning and Management 58: 2088–2112.

Macey, A. 2013. Organic agriculture in British Columbia, organic statistics 2012. Canadian Organic Growers, Ottawa, Canada. 19 pp.

Madsen, H. F., and C. V. G. Morgan. 1970. Pome fruit pests and their control. Annual Review of Entomology 15: 295–320.

Municipalities Enabling and Validating Act [RSBC 1960]. 1989. Section 283: Okanagan-Kootenay Sterile Insect Release Board. British Columbia, Canada.

(OKSIR) Okanagan-Kootenay Sterile Insect Release Program. 2017. Trap viewer mapping application.

Ontario Ministry of Agriculture, Food and Rural Affairs. 2016. Integrated Pest Management for Apples- Publication 310. Queen's Printer for Ontario. Guelph, Ontario, Canada.

Proverbs, M. D., J. R. Newton, and D. M. Logan. 1978. Suppression of codling moth, *Laspeyresia pomonella* (Lepidoptera: Olethreutidae), by release of sterile and partially sterile moths. Canadian Entomologist 110: 1095–1102.

Proverbs, M. D., J. R. Newton, and C. J. Campbell. 1982. Codling moth: A pilot program of control by sterile insect release in British Columbia. Canadian Entomologist 114: 363–376.

Stockholm Convention. 2009. Stockholm Convention on Persistent Organic Pollutants (POPs). Text and annexes as amended in 2009, Secretariat of the Stockholm Convention on Persistent Organic Pollutants, United Nations Environment Programme (UNEP). Geneva, Switzerland

Suckling, D. M., A. M. Barrington, A. Chhagan, A. E. A. Stephens, G. M. Burnip, J. G. Charles, and S. L. Wee. 2007. Eradication of the Australian painted apple moth *Teia anartoides* in New Zealand: Trapping, inherited sterility, and male competitiveness, pp. 603–615. *In* M. J. B. Vreysen, A. S. Robinson, and J. Hendrichs (eds.), Area-wide control of insect pests: From research to field implementation. Springer, Dordrecht, The Netherlands.

Suckling, D. M., D. E. Conlong, J. E. Carpenter, K. A. Bloem, P. Rendon, and M. J. B. Vreysen. 2017. Global range expansion of pest Lepidoptera requires socially acceptable solutions. Biological Invasions 19: 1107–1119.

Tracy, E. F. 2014. The promise of biological control for sustainable agriculture: A stakeholder-based analysis. Journal of Science Policy and Governance 5: 1–13.

Vakenti, J. 1972. Utilization of synthetic codling moth pheromone in apple pest management systems. Master of Science thesis, Simon Fraser University. Vancouver, Canada.

Vreysen, M. J. B., J. Hendrichs, and W. R. Enkerlin. 2006. The sterile insect technique as a component of sustainable Area-Wide Integrated Pest Management of selected horticultural insect pests. Journal of Fruit and Ornamental Plant Research 14: 107–132.

Vreysen, M. J. B., J. E. Carpenter, and F. Marec. 2010. Improvement of the Sterile Insect Technique for codling moth *Cydia pomonella* (Linnaeus)(Lepidoptera Tortricidae) to facilitate expansion of field application. Journal of Applied Entomology 134: 165–181.

Witzgall, P., L. Stelinski, L. Gut, and D. Thomson. 2008. Codling moth management and chemical ecology. Annual Review of Entomology 53: 503–522.

AREA-WIDE MANAGEMENT OF MEDITERRANEAN FRUIT FLY WITH THE STERILE INSECT TECHNIQUE IN SOUTH AFRICA: NEW PRODUCTION AND MANAGEMENT TECHNIQUES PAY DIVIDENDS

J-H. VENTER[1], C. W. L. BAARD[2] AND B. N. BARNES[3]

[1]*Department of Agriculture, Forestry and Fisheries, Private Bag X14, Gezina 0031, South Africa; byjhventer@gmail.com*
[2]*FruitFly Africa (Pty) Ltd., P. O. Box 1231, Stellenbosch 7600, South Africa*
[3]*Technical Advisor: FruitFly Africa (Pty) Ltd., P. O. Box 5092, Helderberg 7135, South Africa*

SUMMARY

A mass-rearing facility to produce sterile male Mediterranean fruit flies, *Ceratitis capitata* (Wiedemann), for a Sterile Insect Technique (SIT) programme in the Hex River Valley in the Western Cape Province started in the late 1990s. The programme was initially underfunded and could only produce about 5 million sterile male flies per week. The resultant aerial release rate of 500 sterile males/ha/week reduced wild Mediterranean fruit fly populations substantially, but not to sufficiently low levels. Due to financial considerations, in 2003 aerial releases were replaced with ground releases targeting all gardens, other hotspots and neglected host plants. It was clear that with more funding, fruit fly mass-rearing facility and field operations could be improved, better quality control could be implemented, and more and better quality male sterile flies could be produced and released. Increased government support in 2001 resulted in a larger mass-rearing facility, and further improvements included the implementation of a quality control management system and the introduction of a new genetic sexing strain (VIENNA 8). The resultant increase in the production of sterile Mediterranean fruit flies of better quality enabled the SIT programme to be systematically introduced to additional fruit production areas. The Mediterranean fruit fly SIT programme was privatised in 2003 and is now operated by *FruitFly Africa (Pty) Ltd.* In 2009 a new approach to funding was adopted with a renewable Memorandum of Understanding (MoU) between the Department of Agriculture, Forestry and Fisheries (DAFF) and the deciduous fruit and table grape industry. Under the MoU, the DAFF provides 50% of the necessary funding, while 50% is collected from growers through statutory levies. In 2010 a new state of the art mass-rearing facility became operational and subsequent improvements in production processes and facility maintenance resulted in improved fruit fly production and quality. By 2016 sterile male production had increased to 56 million flies per week. After 12 years of ground releases of sterile Mediterranean fruit flies, aerial releases were resumed in three main production areas, and, at the time of writing, include approximately 15 000 ha of commercial deciduous fruit and table grapes. As a result of this well-funded area-wide integrated pest management (AW-IPM) programme, average wild Mediterranean fruit fly populations in the SIT areas have decreased by as much as 73%. The

J. Hendrichs, R. Pereira and M. J. B. Vreysen (eds.), Area-Wide Integrated Pest Management: Development and Field Application, pp. 129–141. CRC Press, Boca Raton, Florida, USA.

South African Mediterranean fruit fly SIT programme now aims to manage some of the fruit production areas as areas of low pest prevalence. Increased funding and a stable income stream also enabled *FruitFly Africa* to apply early detection and rapid response programmes for invasive pests such as *Bactrocera dorsalis* in relevant areas.

Key Words: Ceratitis capitata, Ceratitis rosa, Ceratitis quilicii, Tephritidae, SIT, public/private partnership, sterile male releases, suppression, Western Cape, Northern Cape, Eastern Cape

1. INTRODUCTION

The Western Cape Province is the centre of the South African deciduous fruit industry, followed by the Northern and Eastern Cape Provinces and to a lesser extent by smaller production areas in other provinces. Two fruit fly species of economic importance were previously recorded as occurring in the Western Cape, i.e. the Mediterranean fruit fly (*Ceratitis capitata* (Wiedemann)) and the Natal fruit fly (*C. rosa* Karsch) (Blomefield et al. 2015). However recent studies have revealed distinct morphological and molecular differences within populations of *C. rosa,* differentiated further by environmental requirements such as by temperature and altitude (Virgilio et al. 2013; Karsten et al. 2016). This resulted in the description of a new species viz. *Ceratitis quilicii* (De Meyer et al. 2016; FAO/IAEA 2019). Comprehensive surveys have not yet been conducted to determine the prevalence of either species. It can be deducted, however, from the studies of Karsten et al. (2016) that *C. quilicii* is more prevalent in the Western Cape than *C. rosa*.

Mediterranean fruit fly is the predominant species in most areas in the Western Cape (De Villiers et al. 2013; Manrakhan and Addison 2014) and is categorised as a quarantine pest for most of South Africa's export markets for deciduous fruit. Globally, more than 260 different fruit species, including citrus, are hosts of Mediterranean fruit fly, and it can cause enormous crop losses to commercially-produced fruit and also some vegetables if not controlled (USDA 2019). Small-scale farmers, as well as communities with backyard fruit trees, are also seriously affected by this species (White and Elson-Harris 1994).

Deciduous fruit (pome and stone fruit) and table grapes are mostly grown in mountain valleys in the Western and Eastern Cape, and in a semi-desert area alongside the Lower Orange River in the Northern Cape. The valleys are fairly isolated by surrounding mountains, and the Lower Orange River production area by the surrounding semi-desert area, making possible area-wide integrated pest management (AW-IPM) programmes incorporating the Sterile Insect Technique (SIT) covering a number of separate and relatively isolated areas.

South Africa is a net exporter of fruit and for many decades has had an established, well organised and integrated deciduous fruit industry, which in 2016 exported approximately 880 000 metric tons of deciduous fruit with an estimated value of USD 1200 million (DAFF 2017). The deciduous fruit industry in South Africa is one of the largest employers in horticulture, representing a significant investment both in terms of human resources and foreign exchange earnings (DAFF 2017). The country cannot afford to jeopardise future exports by allowing fruit flies to hinder international trade.

It is therefore essential for the South African export fruit industry to reduce fruit fly interceptions to a minimum to ensure market access and to reduce production losses. The European Union is the destination of more than 40% of fruit exports, and other markets include Japan, Taiwan and the USA (DAFF 2017). The European Union will intercept and detain fruit consignments for any non-European fruit fly larvae detected in fruit, which often includes unidentified larvae of the Mediterranean fruit fly, but also other *Ceratitis* spp., if the consignment does not originate from the European Union.

The use of chemical insecticides has become increasingly complex due to pest resistance, environmental concerns, and restrictions on residue levels by importing countries. In the interests of reducing insecticide use, as well as pre- and post-harvest crop losses, while maintaining sustainable agricultural systems, AW-IPM programmes integrating the SIT have proved effective in supporting safe and environment-friendly international trade.

South Africa is one of the largest deciduous fruit exporting countries in the southern hemisphere, but with a relatively small SIT programme to suppress or eradicate fruit flies. South Africa's major competitors on the international fruit export market, such as Chile, are either fruit fly-free, well advanced in achieving this or have at least low pest prevalence status. Use of the SIT, which has successfully contributed to eradicating the Mediterranean fruit fly in Chile and North America, as well as parts of Argentina, Australia, Peru, and Central America, has resulted in substantial savings to these countries (Enkerlin 2021).

A key factor for the success of any SIT programme is availability of adequate funding and long-term commitment of stakeholders. Funds generated from growers by making use of, e.g. a statutory grower levy, need to be supplemented with government funding in view of the public benefits of such programmes. Political will to support AW-IPM programmes which include the SIT is therefore needed to ensure sustainable funding (Dyck et al. 2021a).

2. HISTORIC OVERVIEW OF THE SOUTH AFRICAN SIT PROGRAMME

The Mediterranean fruit fly SIT programme in South Africa originated in 1996 when the Agricultural Research Council's (ARC) Infruitec-Nietvoorbij Institute for Fruit, Vine and Wine in Stellenbosch approached the Joint Food and Agriculture Organization of the United Nations (FAO)/International Atomic Energy Agency (IAEA) Programme of Nuclear Techniques in Food and Agriculture for technical support for a project to investigate the feasibility of integrating the SIT to suppress or eradicate Mediterranean fruit fly in the Hex River Valley. The pilot area was chosen mainly because of its relative geographic isolation, its large production area of 5000 ha of table grapes, a major export crop, and the fact that Mediterranean fruit fly was the dominant fruit fly pest (Barnes 2016).

Aerial releases of sterile Mediterranean fruit flies using a fixed-wing aircraft started in 1999, and flies were dispersed at a density of 500 sterile male Mediterranean fruit flies per ha per week. Wild Mediterranean fruit fly populations were subsequently reduced by 80% (Barnes et al. 2015), but still inadequately.

Nevertheless, encouraged by these early results, the governing body of the deciduous fruit industry, the then Deciduous Fruit Producers' Trust (DFPT, later HORTGRO) assisted the ARC with limited funding for the implementation of the project. Overall management of project funding at local level was through a formal 'SIT Partnership' agreement between the ARC and the DFPT. Additional funding in 2001 from the Western Cape Department of Agriculture allowed for improved infrastructure.

In 2002, a quality management system was incorporated into the mass-rearing process, and in 2003 a new genetic sexing strain of Mediterranean fruit fly based on a temperature sensitive lethal (*tsl*) mutation, i.e. the VIENNA 8 strain, was introduced and reared as the main colony (Franz et al. 2021). All these factors significantly improved mass-production levels and quality of the sterile males (Barnes et al. 2015; Barnes 2016).

In 2003 the programme was privatised with the establishment of *SIT Africa (Pty) Ltd.* (Barnes 2007), and in 2004 the sterile male release programme was extended to two additional production areas (Barnes 2016). Financial considerations resulted in the replacement of aerial releases with ground releases in 2003. These were focussed on farm and town gardens and other hotspots where wild Mediterranean fruit fly populations remained high (Barnes 2016). The rationale behind this strategy was to achieve high sterile to wild fly ratios in these localities where wild flies overwinter in low numbers (Barnes 2008), thus minimizing the number of wild flies which are able to migrate back to commercial fruit plantings in summer. However, still insufficient sterile to wild fly ratios during summer often occurred (Manrakhan and Addison 2014).

Subsequently, substantial funds were made available by the fruit industry to introduce extensive fruit fly monitoring programmes in production areas in order to identify and further suppress fruit fly hotspots prior to sterile male releases to ensure that momentum in the SIT programme was maintained. Greater detail on the development, progression and results of the Mediterranean fruit fly SIT programme is given in Barnes (2007, 2016).

When it became clear that a new approach was required to ensure sustainable funding and industry-wide roll out of the Mediterranean fruit fly SIT programme, a 50:50 contractual funding partnership was formed in 2008 between the DFPT and the then National Department of Agriculture (NDA) (now Department of Agriculture, Forestry and Fisheries, DAFF) in the form of a Memorandum of Understanding (MoU).

The broad AW-IPM programme with a SIT component includes, at the time of writing, eight distinct fruit production areas at different levels of SIT implementation. The sterile flies are released in the Elgin, Grabouw and Vyeboom area (9600 ha), the Hex River area (Hex River Valley, De Wet and Brandwacht, 5700 ha), and the Warm Bokkeveld, Wolseley and Tulbagh area (7000 ha), all in the Western Cape Province. Pre-SIT baseline data collection is being carried out in the Hemel and Aarde Valley in the Western Cape (300 ha), the Langkloof Valley in the Eastern Cape (4700 ha), and in the Lower Orange River area in the Northern Cape (Kakamas and Keimoes, 4200 ha).

Fruit fly densities in commercial orchards are monitored with Chempac® bucket traps baited with a three-component lure (Biolure) that are deployed at a density of 1 trap per 20 ha (Barnes 2016). Baiting no longer includes organophosphate insecticides, but the organically certified spinosad-based product GF-120 NF NATURALYTE ™ bait.

The programme contributes towards various government priorities, such as export competitiveness, economic growth and development, job creation, food security, and reduced insecticide use. Monitoring for non-native fruit flies, including *Bactrocera dorsalis* Hendel, which is already established in northern parts of South Africa (Manrakhan et al. 2015), forms part of the national exotic fruit fly surveillance programme, which is augmented through the SIT monitoring programme. For these species, Chempac® bucket traps baited with methyl eugenol lures are used at a density of 1 trap per 100 ha.

3. FINANCIAL MODEL

By 2008, several factors had influenced the economic viability of the programme. These included the absence of sustained investment from government, an inability to raise venture capital from private institutions, a fruit industry which was under economic stress and the resultant difficulties in getting grower buy-in, and a too-small and aging Mediterranean fruit fly mass-rearing facility which could not produce the numbers of sterile flies required (Barnes 2007). It was in this context that the DAFF/DFPT MoU was formalised in 2008.

The DAFF/DFPT MoU is a 3-year renewable contract; in 2009 DAFF's contribution was approximately USD 460 000 (current value) and included an annual consumer price increase. Under the MoU, approximately the same amount was contributed by the fruit industry towards the monitoring and sterile Mediterranean fruit fly production components of the programme. The aerial baiting component of the programme is funded solely by producers.

The objectives of the MoU focussed on the concept that reduced fruit fly population levels can lead to areas of low pest prevalence, pest free areas and, possibly, eradication of invasive fruit flies. This would ensure maintenance of market access and increase South Africa's ability to export high quality, residue safe fruit. Furthermore, all producers within the relevant area would be able to participate.

Subsequently, the MoU has been renewed twice. The current MoU for the business years 2015/16 to 2017/18 has ensured a financial contribution to the SIT programme by DAFF of USD 770 000, USD 930 000 and USD 1.1 million, respectively.

In 2013 *SIT Africa* evolved into *FruitFly Africa (Pty) Ltd.* (FruitFly Africa 2019), which was recognised by DAFF as the implementation structure for the MoU. DAFF funding is allocated for major projects within the SIT programme, namely awareness and educational programmes, optimisation of fruit fly monitoring, preparation of new areas for sterile Mediterranean fruit fly releases, supplementary fruit fly bait applications, remedial action in hotspot areas, effective radiation sterilisation of pupae, increased production and quality of sterile flies, area-wide release techniques, and continuation of releases in existing areas.

4. ORGANIZATIONAL STRUCTURE

The current (2017) organizational structure of *FruitFly Africa* is given in Fig. 1. During the initial stages of sterile fly production at the Stellenbosch mass-rearing facility, 5 million sterile flies per week were produced by 15 rearing technicians. *FruitFly Africa* currently (2017) has eight rearing technicians; their production output per week has risen from 15 million in 2014 to the 56 million VIENNA 8 sterile males per week in 2017. Production efficiency has been improved not by additional automation, but by the efficient use of labour and the adoption of improved production and quality practices.

A quality control officer monitors the quality of the sterile male flies produced, ensures adherence to the protocol of aerial applications of GF-120 baits, and evaluates all other fruit fly management measures applied in the field.

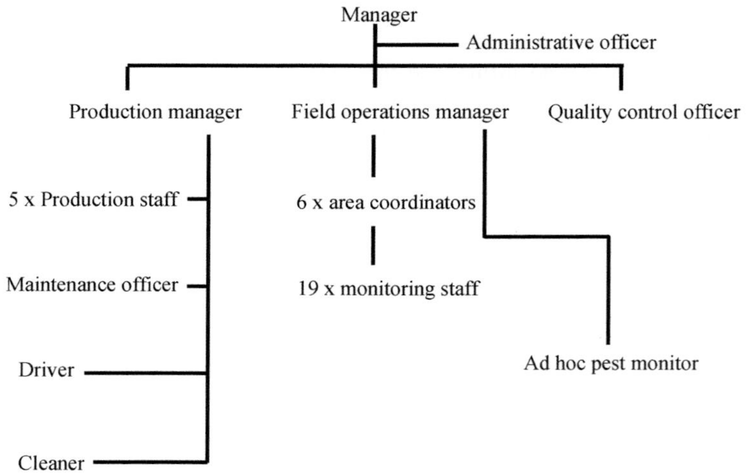

Figure 1: The current organizational structure of FruitFly Africa (Pty) Ltd.

Field surveillance staff carry out the day-to-day tasks in the areas that participate in the programme. Each SIT area has a coordinator who is responsible for planning and public relations in that area. Field monitoring staff report to the area coordinator in that area, who in turn reports to the field operations manager who supervises the implementation of monitoring and management strategies within that area.

5. MANAGEMENT

In a SIT programme, sterile insects must be of the best quality, strategies and decisions need to be technically correct, and customer service has to be excellent. This, coupled with the fact that the broader community which is impacted also needs to contribute to fruit fly suppression, has a unique impact on programme management practices

(Dyck et al. 2021b). As the current programme evolved, it became evident that programme management needs to be able to rely on staff that are technically competent, and that are able to build good working relationships with all fruit industry stakeholders, e.g. community organizations and individual producers. People possessing both these skills are relatively rare, and in this programme much emphasis is put on recruiting suitably-skilled people for these tasks.

Since 2010, area-wide aerial baiting with GF-120 has also formed a crucial part of the fruit fly suppression strategy; in most areas four to six applications per season are applied shortly before harvest as an additional crop protection exercise. Ground-released sterile flies are used as the main intervention in urban areas, farm gardens, on alternate hosts and in Mediterranean fruit fly hotspots.

In view of the encouraging results of aerial sterile fly releases during the 1999-2003 pilot phase in the Hex River Valley (Barnes 2016), and the improved funding base, aerial releases of sterile males were reconsidered and a pilot trial over 2200 ha was implemented during the 2014/15 season using a gyrocopter. Reasonable success was obtained (Barnes 2016), and in 2016/17 season-long, area-wide helicopter releases were carried out at the standard release rate of a 1000 sterile males/ha/week over ±39 000 ha in three areas where the SIT forms part of the fruit fly management strategy. "Attract and kill" bait stations as well as mass-trapping were also used as part of the management strategy in programme areas, although their use has been limited to backyards and hotspots on farms.

Historically, fruit fly monitoring has been at a density of 1 trap per 20 hectares. These traps are not used in orchard-level decision making, but they are used to determine the area-wide distribution of both wild and sterile fruit flies, as well as to determine the ratios between the two. Because all farmers contribute financially in equal amounts to the programme, each of them feels entitled to the same AW-IPM service from *FruitFly Africa* in equal quantities. Once a strategy has been decided on for a season, it needs to be implemented equitably across all participating areas. Fruit fly trap catches have thus been not so much a tool for weekly management decisions, but more as an indication of whether a particular strategy for the season has been successful in that area. They are also a useful tool for timing other control interventions (e.g. host plant management), using historical trends. Trap catches have thus been used to compare fruit fly populations for a whole area across weeks and between seasons. Increasingly, farmers are opting for higher trap densities to enable them to make their own management decisions. For this, more detailed and timely information is necessary, and *FruitFly Africa* is thus developing an electronic database system that will be available to the farmers and provide detailed and real-time reporting on trap catches, and population levels and distribution.

6. PRODUCTION PRACTICES AND QUALITY CONTROL

During the period 2011 to 2015, systematic changes were made to the sterile Mediterranean fruit fly mass-rearing process based on experience gained by the *FruitFly Africa* quality control officer during a visit to the Moscamed El Pino Mediterranean fruit fly facility in Guatemala. These included a 20% reduction in adult fly density in the oviposition cages (from 4400 to 3600 adults – this equates to 0.0027 to 0.0034 cm^3 per fly), twice-daily brushing of eggs off the adult cage screen walls through which the females oviposit into water troughs, and the use of 1 kg 'starter packs' of diet for rearing first instar larvae.

The reduction in the number of adults per oviposition cage reduced the amount of stress on the flies that need to feed, mate and lay their eggs. Brushing the eggs more frequently from the screens through which they have been laid reduced the number of eggs that stuck to the screens and become desiccated, thereby increasing the percentage of viable eggs. The use of the 1 kg starter packs concentrated the recently-hatched larvae in a smaller volume of diet, thereby retaining essential metabolic heat.

Through a cascade effect, these measures resulted in better quality larvae, pupae and adults in the main colony, which in turn equated to better quality eggs, larvae, pupae and adults in the release stream. The end result is the release of better-quality sterile males (Barnes, 2016). The outcome of these changes is given below.

Additionally, a more stable flow of increased funding enabled the production team to make improvements to the facility infrastructure, as well as to the production equipment. This included improved illumination in the adult room, replacing egg-bubbling aeration pumps with a single air supply line, better climate control equipment, and emergency standby services for equipment that is essential to production.

The stable flow of funding also enabled the mass-production facility to procure raw materials of a higher and more uniform quality from reliable sources. Examples of this are bran that is free of pesticide residues, vermiculite with low moisture levels, and yeast with a high and stable protein content.

The quality control parameters and production targets, calculated weekly, include daily egg production (volume), daily pupal production (volume), egg hatch (%), egg to pupa recovery (%), pupal weight (mg), adult flight ability (%), sterility in the release stream (%), and fertility of the main colony (%). All tests are carried out in accordance with the standards set out in the standard operating procedures and in the international product quality control manual (FAO/IAEA/USDA 2019).

7. RESULTS AND DISCUSSION

Following the inception of changes to the infrastructure and mass-rearing procedures described above, there was a marked improvement in the following production and quality control parameters in the release stream (Barnes 2016):

- Daily egg production per cage increased by 45.3%, with a decrease in standard deviation (SD) of 18.2%.
- Mean egg hatch improved from 39.6% to 42.6%, an increase of 7.6% (SD decreased by 50.0%). In 2011 the target of 40% hatch was often not met; this rarely happened in 2015.
- Egg to pupa recovery improved from 16.9% to 20.6%, an increase of 21.9%; there was no change in the SD.
- Mean flight ability increased from 82.2% to 87.5% (SD decreased by 64.8%). In 2011, the target of 80% was not achieved on a number of occasions; in 2015 it never dropped below 81%.

Egg to pupa recovery is a good indicator of the cost-effectiveness of production, since a large percentage of the variable costs of production is spent on rearing the fly from the egg stage to the pupal stage. With an increase in egg to pupa recovery it was not unexpected that the unit cost of production from 2011 to 2015 was reduced by 37% (nominal). In addition to the increased production efficiency, the increase in numbers of sterile males produced, coupled with a minimal increase in fixed costs, translated into lower unit costs, since the increase in total costs were not proportionate to that of total volumes. While the quality of sterile flies together with the cost-effectiveness of production are important to the success of an SIT programme, the effect of the programme on the degree of wild Mediterranean fruit fly population reduction has to be taken into account when considering the cost-effectiveness of such a programme.

The trend in wild Mediterranean fruit fly populations in the sterile male release areas during the period 2007 to 2017 is shown in Figure 2 (note the difference in the flies/trap/day (FTD) scale between the three areas).

The average wild fly population levels during the harvest season (first 20 weeks of the calendar year when ripe fruit is most abundant), in the large areas where the SIT forms part of the management strategy, decreased as follows:

- When comparing the average FTD for 2007-2008 (period before the MoU) with that of 2015-2017, the FTDs in the Hex River Valley decreased by 73% from an average of 4.32 to 1.14. This average includes hotspots that are focally supressed.
- The same comparison for the Elgin/Grabouw area indicates a population reduction of 19% (although the reduction from the 3-year period immediately following the MoU is 32%), from a FTD of 0.50 to 0.41.
- No reliable data prior to 2010 (implementation of the full programme) are available for the Warm Bokkeveld area. When comparing the average FTD for 2010-2011 with that of the period 2015-17, the FTDs in the Warm Bokkeveld decreased by 78% from an average of 1.46 to 0.32.

Over the period 2007-2017 the FTD values for the Hex River Valley were much higher than the FTD values of Elgin/Grabouw. This is mainly due to differences in wild and commercial host plants present and varieties cultivated, as well as harvesting processes and sanitation between the two areas (Barnes 2016), which made fruit fly management throughout the Hex River Valley more difficult. A further factor is climate; long-term data show that the Hex River Valley has higher average maximum temperatures than the Elgin/Grabouw area (Barnes et al. 2015), conditions which favour development of Mediterranean fruit fly (Nyamukondiwa et al. 2013).

Increased funding from DAFF from 2008 to 2017 allowed for essential improvements at the Mediterranean fruit fly mass-rearing facility. The timely integration of a combination of fruit fly management techniques made a positive difference to the outcomes of the programme. Production increased from 15 million sterile male flies per week in 2014 to 56 million per week by 2016, which enabled better sterile male to wild male overflooding ratios. Improved quality and quantity of sterile fruit flies produced and released, better release techniques, and overall, better on-farm management of fruit fly populations, have resulted in a generally steady decrease in average wild Mediterranean fruit fly populations over time, as illustrated in Fig. 2.

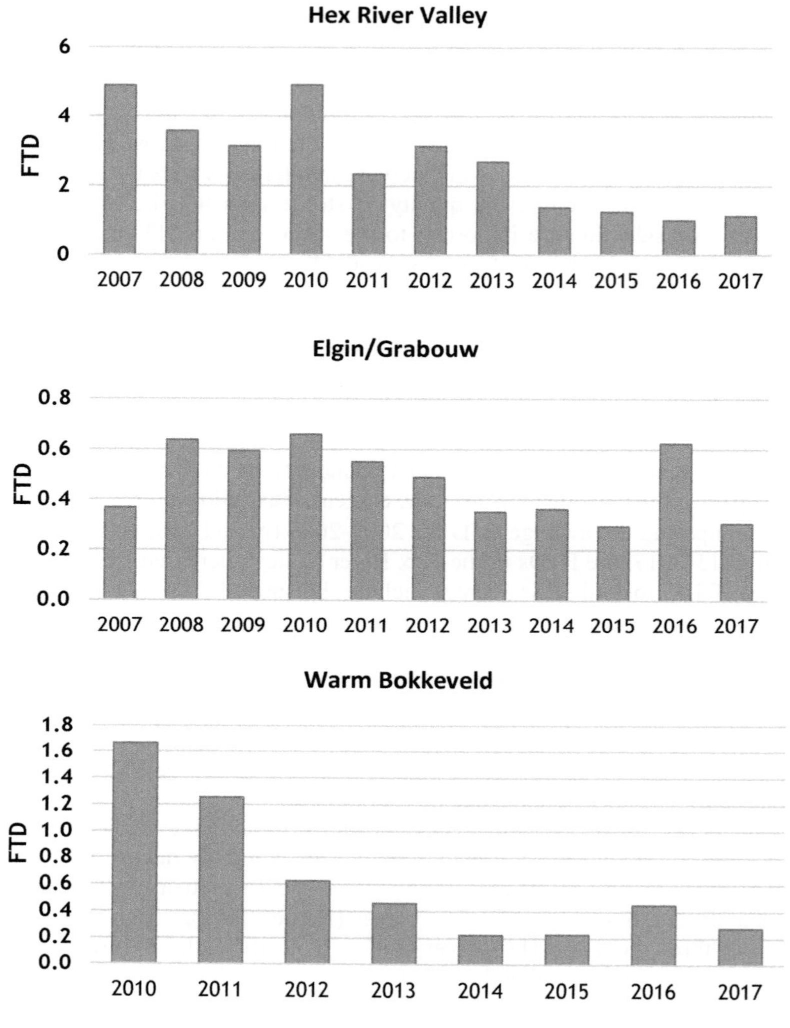

Figure 2. Average numbers of wild Mediterranean fruit flies/trap/day (FTD) trapped in three fruit production areas under SIT application during the first 20 weeks of the year (= harvest period) from 2007/2010 to 2017 (no data available for Warm Bokkeveld before 2010).

During the last 18 years, the South Africa Mediterranean fruit fly SIT programme has faced many challenges such as outdated infrastructure and inadequate equipment due to a poor local funding base, and initially a hesitant grower community, but, with excellent and sustained support from the FAO/IAEA, and later, better co-funding from government, determination by the SIT team, improved facilities and equipment, and a grower community steadily starting to believe in the programme, it has grown to a programme serving a total of 30 000 ha, and is destined for further growth.

Many lessons have been learnt along the way, the most important being:

- Broad-based, multi-organizational and sustainable funding must be available from the start of the AW-IPM programme.
- No single control measure is a stand-alone technology. All available tools for the management of the pest must be used in combination with each other. Efforts must be made to educate stakeholders accordingly.
- The area for the AW-IPM programme, especially at its initiation, must be carefully selected. Besides geographic or topographic isolation, the target pest should already be well managed by conventional methods and sanitation, with growers who are progressive in their pest management outlook.
- Effective management of alternative fruit fly host plants and active orchard/vineyard sanitation at farm level is crucial to the success of an AW-IPM programme.
- There must be buy-in and long-term commitment to the programme by all growers in the selected area(s). Ideally, there should be a 'push-pull' approach by the stakeholders: SIT technologists should 'push' (advocate) AW-IPM (including SIT) where it is appropriate, with a simultaneous 'pull' (a willingness/receptiveness) for the SIT on the part of the growers.
- Good relations and communication between AW-IPM service providers, growers, and the broader public is crucial, and should be based on transparent real-time reporting on trap catches, and population levels and distribution.
- High specification infrastructure, equipment and human capital must be available to produce good quality insects. A good quality management system must be in place in the rearing and release facilities and must include regular internal and external audits of procedures, processes and performance.
- Ground releases of sterile *C. capitata* are not a long-term solution to area-wide population suppression. Above a certain scale, releases should be by air if at all possible.
- Programme managers must keep abreast of the latest international developments in the field of AW-IPM and make good use of knowledge and input from international experts.
- AW-IPM programmes are not quick fixes to a problem. Population reduction exercises can take a couple of seasons to show results. Stakeholder expectations should be managed in this regard.
- Applied research and development should be on-going, and all cost-effective improvements in procedures and processes should be implemented.

South Africa is now aiming to identify some of the existing deciduous fruit and table grape areas in the AW-IPM programme and manage them as areas of low pest prevalence.

The invasion of *B. dorsalis* is officially controlled by DAFF in South Africa and preparedness plans are in place to immediately initiate eradication programmes in case of outbreaks. Official control actions include quarantine, delimiting surveys and eradication measures with the application of the male annihilation technique and bait application (Manrakhan et al. 2012).

Although SIT application for *B. dorsalis* is not envisaged at this stage, future expansion of the fruit production areas to be covered by an AW-IPM approach, which will include the SIT for Mediterranean fruit fly, is planned (Manrakhan 2020). A new MoU with DAFF is planned, which will provide the necessary support for such expansion. This expansion of SIT activities will mainly be within areas where other area-wide control measures (e.g. monitoring and aerial baiting) are already being implemented to effectively suppress populations. Such areas have already been identified in the Northern and Eastern Cape Provinces.

8. REFERENCES

Barnes, B. N. 2007. Privatizing an SIT programme: A conflict between business and technology? pp. 449–456. *In* M. J. B. Vreysen, A. S. Robinson and J. Hendrichs (eds.), Area-wide control of insect pests. From research to field implementation. Springer, Dordrecht, The Netherlands.

Barnes, B. N. 2008. Why have fruit flies become so troublesome? South African Fruit Journal 7: 56–57.

Barnes, B. N. 2016. Sterile Insect Technique (SIT) for fruit fly control – the South African experience, pp. 435–464. *In* S. Ekesi, S. A. Mohamed and M. De Meyer (eds.), Fruit fly research and development in Africa – Towards a sustainable management strategy to improve horticulture. Springer, Switzerland.

Barnes, B. N., J. H. Hofmeyr, S. Groenewald, D. E. Conlong, and M. Wohlfarter. 2015. The Sterile Insect Technique in agricultural crops in South Africa: A metamorphosis .... But will it fly? African Entomology 23 (1): 1–18.

Blomefield, T. L, B. N. Barnes, and J. H. Giliomee. 2015. Peach and nectarine, pp. 320–339. *In* G. L. Prinsloo, and V. M. Uys (eds.), Insects of cultivated plants and natural pastures in southern Africa. Entomological Society of Southern Africa, Hatfield, South Africa.

(DAFF) Department of Agriculture, Forestry and Fisheries. 2017. Economic review of the South African agriculture 2016/17. Directorate of Statistics and Economic Analysis. Department of Agriculture, Forestry and Fisheries. Pretoria, South Africa. 10 pp.

De Meyer, M., M. Mwatawala, R. S. Copeland, and M. Virgilio. 2016. Description of new *Ceratitis* species (Diptera: Tephritidae) from Africa, or how morphological and DNA data are complementary in discovering unknown species and matching sexes. European Journal of Taxonomy 233: 1–23.

De Villiers, M., A. Manrakhan, P. Addison, and V. Hattingh. 2013. The distribution, relative abundance, and seasonal phenology of *Ceratitis capitata, Ceratitis rosa*, and *Ceratitis cosyra* (Diptera: Tephritidae) in South Africa. Environmental Entomology 42: 831–840.

Dyck, V. A., E. E. Regidor Fernández, B. N. Barnes, P. Gómez Riera, T. Teruya, J. Reyes Flores, G. Iriarte, R. Reuben, and D. Lindquist. 2021a. Communication and Stakeholder Engagement in Area-Wide Integrated Pest Management programmes that integrate the Sterile Insect Technique, pp. 813–838. *In* V. A. Dyck, J. Hendrichs and A. S. Robinson (eds.), Sterile Insect Technique – Principles and practice in Area-Wide Integrated Pest Management. Second Edition. CRC Press, Boca Raton, Florida, USA.

Dyck, V. A., J. Reyes Flores, M. J. B. Vreysen, E. E. Regidor Fernández, B. N. Barnes, M. Loosjes, P. Gómez Riera, T. Teruya and D. Lindquist. 2021b. Management of Area-Wide Integrated Pest Management programmes that integrate the Sterile Insect Technique, pp. 781–814. *In* V. A. Dyck, J. Hendrichs, and A. S. Robinson (eds.), Sterile Insect Technique – Principles and practice in Area-Wide Integrated Pest Management. Second Edition. CRC Press, Boca Raton, Florida, USA.

Enkerlin, W. R. 2021. Impact of fruit fly control programmes using the Sterile Insect Technique, pp. 979–1006. *In* V. A. Dyck, J. Hendrichs, and A. S. Robinson (eds.), Sterile Insect Technique – Principles and practice in Area-Wide Integrated Pest Management. Second Edition. CRC Press, Boca Raton, Florida, USA.

(FAO/IAEA) Food and Agriculture Organization of the United Nations/International Atomic Energy Agency. 2019. A guide to the major pest fruit flies of the world. R. Piper, R. Pereira, J. Hendrichs, W. Enkerlin, and M. De Meyer (eds.). Scientific Advisory Services Pty Ltd. Mourilyan, Queensland, Australia. 80 pp.

(FAO/IAEA/USDA) Food and Agriculture Organization of the United Nations/International Atomic Energy Agency/United States Department of Agriculture. 2019. Product quality control for sterile mass-reared and released tephritid fruit flies, Version 7.0. International Atomic Energy Agency, Vienna, Austria. 148 pp.

Franz, G., K. Bourtzis, and C. Cáceres. 2021. Practical and operational genetic sexing systems based on classical genetic approaches in fruit flies, an example for other species amenable to large-scale rearing for the Sterile Insect Technique, pp. 575–604. *In* V. A. Dyck, J. Hendrichs, and A. S. Robinson (eds.), Sterile Insect Technique – Principles and practice in Area-Wide Integrated Pest Management. Second Edition. CRC Press, Boca Raton, Florida, USA.

FruitFly Africa. 2019. http://www.fruitfly.co.za/

Karsten, M., P. Addison, B. J. van Vuuren, and J. S. Terblanche. 2016. Investigating population differentiation in a major African agricultural pest: Evidence from geometric morphometrics and connectivity suggests high invasion potential. Molecular Ecology 25: 3029–3032.

Manrakhan, A. 2020. Pre-harvest management of the Oriental fruit fly. CAB Reviews 15: 003.

Manrakhan, A., and P. Addison. 2014. Assessment of fruit fly (Diptera: Tephritidae) management practices in deciduous fruit growing areas in South Africa. Pest Management Science 70: 651–660.

Manrakhan, A., J. H. Venter, and V. Hattingh. 2012. Action plan for the control of the African invader fruit fly, *Bactrocera invadens* Drew, Tsuruta and White. Department of Agriculture, Forestry and Fisheries, Republic of South Africa, Pretoria. 28 pp.

Manrakhan, A., J. H. Venter, and V. Hattingh, 2015. The progressive invasion of *Bactrocera dorsalis* (Diptera: Tephritidae) in South Africa. Biological Invasions 17: 2803–2809.

Nyamukondiwa, C., C. W. Weldon, S. L. Chown, P. C. Le Roux, and J. S. Terblanche. 2013. Thermal biology, population fluctuations and implications of temperature extremes for the management of two globally significant insect pests. Journal of Insect Physiology 59: 1199–1211.

(USDA) United States Department of Agriculture. 2019. Mediterranean fruit fly, *Ceratitis capitata*, host list. Animal and Plant Health Inspection Service, Riverdale, Maryland, USA.

Virgilio, M., H. Delatte, S. Quilici, T. Backeljau, and M. De Meyer. 2013. Cryptic diversity and gene flow among three African agricultural pests: *Ceratitis rosa, Ceratitis fasciventris* and *Ceratitis anonae* (Diptera, Tephritidae). Molecular Ecology 22: 2526–2539.

White, I. M., and M. M. Elson-Harris. 1994. Fruit flies of economic significance: Their identification and bionomics. Second Edition. International Institute of Entomology, London, UK. 600 pp.

THE CHINESE CITRUS FLY, *Bactrocera minax* (DIPTERA: TEPHRITIDAE): A REVIEW OF ITS BIOLOGY, BEHAVIOUR AND AREA-WIDE MANAGEMENT

M. A. RASHID[1], Y. DONG[1], A. A. ANDONGMA[1], Z. CHEN[1], Y. WANG[1], P. XU[1], P. LI[2], P. YANG[2], A. R. CLARKE[3] AND C. NIU[1]

[1]Hubei Key Laboratory of Insect Resource Application and Sustainable Pest Control, College of Plant Science & Technology, Huazhong Agricultural University, Wuhan 430070, China; niuchangying88@163.com
[2]Pest Control Division, National Agricultural Technology Extension and Service Center, Ministry of Agriculture, Beijing 100125, China
[3]School of Earth, Environmental and Biological Sciences, Faculty of Science and Technology, Queensland University of Technology (QUT), P.O. Box 2434, Brisbane QLD 4001, Australia

SUMMARY

The Chinese citrus fly *Bactrocera minax* (Enderlein) is a major pest of citrus in some Asian countries. It is a univoltine, oligophagous pest, which strictly infests *Citrus* species and varieties, and has an exceptionally long pupal diapause. *B. minax* has great socio-economic importance in China and its neighbouring countries because citrus production is a key fruit industry in these countries. We review the biology and management of this pest with a focus on its distribution, life cycle, diapause, behavioural ecology, and host preferences. We further review potential area-wide integrated pest management (AW-IPM) strategies, including chemical control, but also various eco-friendly, locally developed and adopted techniques applied mainly in China and Bhutan. After years of continuous efforts in AW-IPM of *B. minax*, significant progress has been achieved in suppressing *B. minax* populations to a level of less than 5% infested fruit and a 60-80% reduction in the use of synthetic insecticides against this pest in China.

Key Words: temperate fruit fly, life cycle, behavioural ecology, diapause, *Tetradacus*, Dacinae, oranges, *Citrus*, AW-IPM, China, Bhutan, IPM

J. Hendrichs, R. Pereira and M. J. B. Vreysen (eds.), Area-Wide Integrated Pest Management: Development and Field Application, pp. 143–159. CRC Press, Boca Raton, Florida, USA.

1. INTRODUCTION

Citrus fruits rank first across the world in the international fruit trade in terms of value (Liu et al. 2012; Srivastava 2012). The Chinese citrus fly *Bactrocera minax* (Enderlein) (Diptera: Tephritidae) is a major pest of *Citrus* spp. in China, Bhutan, India, Viet Nam and other neighbouring countries (White and Wang 1992; Dong et al. 2014a). It is a univoltine insect with a long pupal diapause that feeds solely on different *Citrus* species and varieties (Chen et al. 2016). The Chinese citrus fly is thought to be endemic to China as its presence could have been recorded in a poem written about 1000 years ago during the Song dynasty:

> *"The yellow oranges drop to the ground of the garden due to the wind in autumn. When the oranges opened, there were maggots inside the oranges instead of dragon"* (Yang et al. 2013).

In the 1940s *B. minax* was only recorded in Guizhou and Sichuan, China (Chen and Wong 1943). Currently, it is reported to occur in the major citrus growing provinces of China (Chongqing, Guangxi, Guizhou, Hubei, Hunan, Shanxi, and Sichuan, see Fig. 1), climatically ranging from temperate to subtropical (Wang and Zhang 2009; Gao et al. 2013).

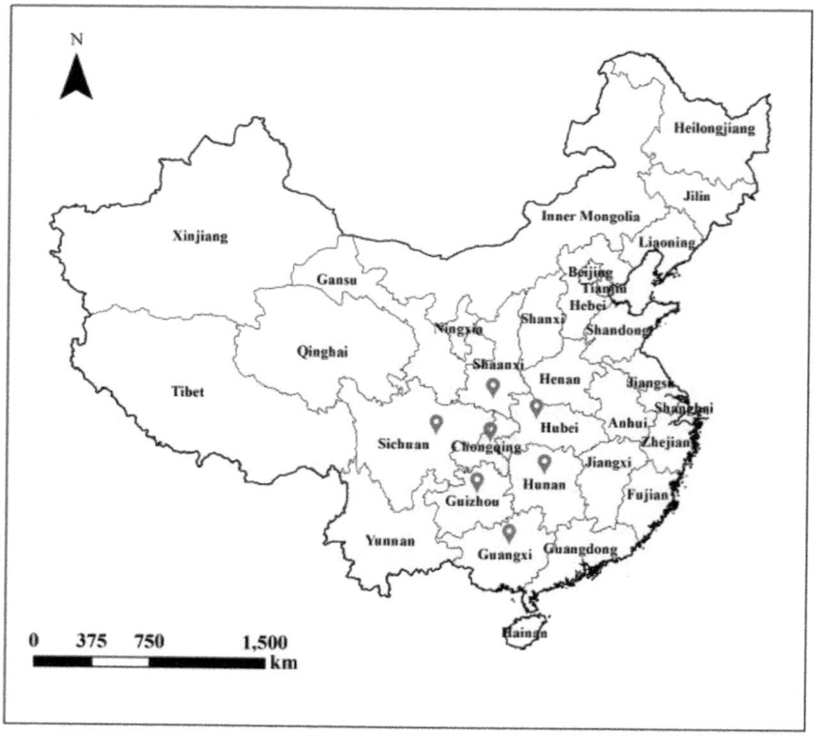

Figure 1. The occurrence and distribution of Bactrocera minax *in China (indicated with dark marks showing location within different provinces).*

Following a population outbreak in 2008 in south-western China, and the resulting heavy losses incurred by farmers, *B. minax* was removed from the national quarantine plant pest list in 2009 (Announcement No. 1216 of the Ministry of Agriculture of the People's Republic of China). Thus, the current management against *B. minax* in China has shifted from eradication to suppression.

1.1. Nature of Damage and Economic Importance

The adult female fly oviposits its eggs under the peel of green, immature citrus fruits with the aid of its elongated ovipositor (Fig. 2). The eggs hatch when fruit reach mid-level development and the larvae feed on the fruit flesh leading to premature ripening and fruit drop, ultimately resulting in economic damage and yield loss (Allwood et al. 1999; Liu et al. 2015). The larval stage is therefore considered as the most destructive life stage (Dorji et al. 2006). Heavy economic losses of USD 200 million were reported in 2008 due to the above-mentioned outbreak of *B. minax* in Guangyuan, Sichuan Province, China. Almost 1 million tons of oranges were destroyed during this flare-up. This outbreak and heavy infestation resulted in a ban on international trade of citrus commodities from China. The outbreak was reported by "The China Daily" in its headlines to highlight the serious damage to the citrus industry (Liu et al. 2015).

Similarly, *B. minax* is a major pest in the eastern Himalayan kingdom of Bhutan, where mandarin *Citrus reticulata* Blanco is one of the major fruit crops. Crop losses caused by *B. minax* infestation ranging from 35 to 75% are common in mid- and high-altitude orchards (>1100 m), and the fly is considered as one of the major barriers to citrus production (van Schoubroeck 1999; Dorji et al. 2006; Xia et al. 2018).

Figure 2. Female Bactrocera minax *using its elongated ovipositor to lay its eggs into small green citrus fruit (20~30 mm diameter).*

1.2. Taxonomy, Distribution and Host Plants

The Chinese citrus fly was for the first time described in 1920 from specimens collected from Sikkim, India by Enderlein and named *Polistomimetes minax* (Thompson 1998). In 1940, the species was also collected in Sichuan Province, China. Drew (1979) provided a detailed description of the *B. minax* based on the specimens collected in 1920 and placing the species in the subgenus *Bactrocera* (Polistomimetes).

Subsequently, a lectotype of *B. minax* was ascribed to the subgenus *Bactrocera* (*Tetradacus*) by White and Wang (1992), who also recorded that *Bactrocera citri* Chen should be regarded as a junior synonym of *B. minax*. The fly is currently placed in the subgenus *Bactrocera* (*Tetradacus*), a small monophyletic clade evolutionary basal to all other *Bactrocera* species (Krosch et al. 2012).

Currently, *B. minax* is regarded as present in Bhutan, China, India (West Bengal and Sikkim), Nepal, and Viet Nam (Dorji et al. 2006; Drew et al. 2006). The host range of Chinese citrus fly is almost exclusively restricted to *Citrus* species and varieties. It has been recorded in citron *Citrus medica* L., lemon *Citrus limon* (L.) Burman f., meiwa kumquat *Fortunella crassifolia* Swingle, pummelo *Citrus maxima* (Burm.) Merr., sour orange *Citrus aurantium* L., sweet orange *Citrus sinensis* (L.) Osbeck, tangerine *Citrus reticulata* Blanco, navel orange *Citrus sinensis* Osb. var. *brasiliensis Tanaka*, pyunkyul *Citrus tangerina* Hort. ex Tanaka, grapefruit *Citrus paradisi* Macfad., and trifoliate orange *Poncirus trifoliata* L. (Nath 1972; Chao and Ming 1986; Liu et al. 2014). Among these, the preferred citrus host plant is sweet orange (Liu et al. 2014).

2. BIOLOGY AND LIFE CYCLE

2.1. Seasonal Phenology

The phenology of *B. minax*'s life cycle may vary subtly depending on local climate conditions. However, the general pattern appears quite fixed. Based on the population of *B. minax* in Guizhou, China and in Bhutan, we summarize the life cycle of *B. minax* as follows:

1. The female oviposits eggs in small unripe fruit from mid-June to mid-July. Usually it takes two months for the eggs to hatch, much longer than other species in the *Bactrocera* genus; the eggs hatch in late August, after which the larvae go through three larval instars.
2. The larval stage lasts until the end of October, which is then followed by a pupal stage.
3. To survive the cold winter temperatures, the pupae enter a six-month overwintering diapause.
4. Adult emergence usually begins in May and mating starts about 25 to 30 days after adult emergence (Wang and Luo 1995; Dorji et al. 2006).

2.2. *Eggs of* B. minax

The egg of *B. minax* is milky white, oblong and curved in shape, slightly pointed at one end and round at the other. It has a length of 1.1 to 1.5 mm and a maximum width of 0.2 to 0.4 mm (Sun 1961). Female *B. minax* lay eggs in clutches. Usually, each female oviposits about 14 to 17 eggs per oviposition event, with a maximum of 35 eggs per clutch. During its lifetime, a female can produce more than 100 eggs (Zhang 1989). After depositing eggs, the fruit surface is covered with juice around the oviposition wound. In the beginning, this juice is transparent to translucent for one to three days, then it gradually becomes yellow. The oviposition site bulges out, cracks and fruit skin around the oviposition site turns yellow to crimson-purple (Wang and Zhang 1993). Egg hatch starts in July and reaches its peak in late August (Xiong et al. 2016).

2.3. *Larvae of* B. minax

After eclosion, the larva feeds internally on the citrus flesh. The mature larva is milky white or pale yellow, 15 to 18 mm long, conical in shape and nearly transparent at one end. The mouth is equipped with sclerotized mouth hooks, and the body has 11 segments. The larval stage lasts 52 to 72 days depending on temperature, with an average of 63 days (Lu et al. 1997). Young larvae usually feed in a small group on a single fruit segment, later spreading to other segments: the average number of larvae in an infected fruit is 9.5 (Zhang 1989). Larval infestation leads to premature fruit-fall from October to November (Liu et al. 2015).

The larvae stay within the fruit for about 18-52 days after fruit drop. Such a long pre-pupal period is very unusual among the Dacinae and it suggests that long larval development occurs after fruit drop (Dorji et al. 2006). A mature, third instar larva usually leaves the fruit in the early morning and pupates within a day. Pupation starts in late October and reaches a peak in early to mid-November (Dong et al. 2013; Chen et al. 2016).

2.4. *Pupae of* B. minax

The mature third instar larva pupates in the soil at 3 to 5 cm depth (Zhang 1989; Dorji et al. 2006). The long overwintering pupal diapause in *B. minax* is highly unusual within the genus *Bactrocera* and is considered as an adaptive strategy to survive the cold winter periods which occur in its native range (Fan et al. 1994). In Yichang city, where *B. minax* is present, the mean winter temperature usually ranges from 5 to 15 °C (Dong et al. 2013). The overwintering pupal phase lasts for 160-170 days, with the emergence of adults synchronised with the early fruiting season of citrus (Wang and Luo 1995). The pupa of *B. minax* is 9 to 10 mm long with a diameter of 4 mm, weight average of 77.5 mg. It is oval in shape and yellow-brown in colour. Prior to adult emergence, the pupal case becomes slightly dark brown (Zhang 1989; Wang and Luo 1995).

2.5. *Diapause Termination in* B. minax

Research on the development of efficient and sustainable *B. minax* management techniques is still very difficult because the fly has only one generation per year, accompanied by six months of pupal diapause (White and Wang 1992; van Schoubroeck 1999). This bottleneck made the mass-production of this fly very difficult, thereby limiting research capability and its potential use in area-wide integrated pest management (AW-IPM) programmes that have a Sterile Insect Technique (SIT) component (Lü et al. 2014). Therefore, options to break the diapause are considered essential for research and the development of the SIT package for this fly (see Section 4.2.3).

Pupal diapause is usually an evolved response in univoltine temperate tephritids in order to survive harsh environmental conditions and seasonal periods of host scarcity (Teixeira and Polavarapu 2001, 2005; Ragland et al. 2009; Papanastasiou et al. 2011; Moraiti et al. 2012). In *B. minax* pupal diapause is a vital strategy to tolerate cold stress and face the seasonal adversity. Research into the underlying mechanisms in terms of diapause termination, such as major cellular shifts, protein processing, differentially expressed genes and pathways, are still ongoing (Lü et al. 2014; Dong et al. 2014a; Wang et al. 2016, 2017).

Pupal diapause in *B. minax* is influenced by both chilling temperature and duration. A higher chilling temperature, coupled with longer chilling duration, results in a shorter pupal developmental time and improves the synchronisation of adult emergence (Dong et al. 2013). Apart from providing this chilling exposure, hormonal application of 20-hydroxyecdysone (20E) is considered a faster and more efficient method to break pupal diapause in *B. minax* (Dong et al. 2014a; Wang et al. 2014; Chen et al. 2016). Either injection or topical application of 20E can trigger a rapid termination of the pupal diapause in *B. minax* and the morphological changes are observed within 1 week at 22°C. On the tenth day after 20E treatment, the head, thorax and abdomen of the insect can clearly be distinguished, and the colour of body and eyes are milky (Chen et al. 2016). The 20E early-response genes, including *ecr*, *broad* and *foxo*, are up-regulated within 72 h of 20E exposure, indicating these genes are involved in diapause termination processes and pupal metamorphosis (Chen et al. 2016).

The gene sets involved in protein and energy metabolisms vary throughout early- late- and post-diapause insects in response to cold stress. When diapause is terminated by 20E, many genes involved in ribosome and metabolic pathways are differentially expressed, which may mediate diapause transition (Dong et al. 2014a). The variation of transcriptomic and metabolomic profiles of pupae at five stages (pre-, early-, middle-, late-, and post-diapause) suggests major shifts in metabolism and signal transduction, as well as changes in the endocrine and digestive systems. Nine metabolites significantly contribute to the variation in the metabolomic profiles, especially proline and trehalose, which are well-known cryoprotective agents (Wang et al. 2017).

2.6. Adults of B. minax

The adult fly is 10 to 13.2 mm long, not including the female ovipositor, with a wingspan of ~10.8 mm (Drew 1979). The female possesses a long ovipositor of approximately 6.5 mm. The flies are of a brownish colour with yellow markings, the wings have a dark band along the outer margin, and the general appearance is wasp-like (Chen and Xie 1955). A morphological description of the adult is provided in Drew et al. (2007), who also note that the fly is probably the largest of all *Bactrocera* species (Fig. 2).

3. BEHAVIOUR AND ECOLOGY

Since the Chinese citrus fly is not attracted to methyl eugenol or cue-lure as are many species of the genus *Bactrocera*, thorough behavioural and ecological studies have been carried out over the past years in order to develop effective monitoring and control strategies in the long run.

3.1. Feeding Behaviour

It is well known that for most studied tephritid species, both males and females are anautogenous and forage for sugar and protein to fuel metabolic activities and to meet reproductive requirements (Aluja and Norrbom 1999; Drew and Yuval 2000; Taylor et al. 2013). Females need a protein diet for vitellogenesis and ovarian development (Harwood et al. 2015), while males feed on protein to reach sexual maturity leading to copulation which is crucial in achieving reproduction (Lushchak et al. 2013). Adult *B. minax* forage on non-host plants for honeydew, nectar, sooty mould and fruit juices to meet their dietary requirements during sexual maturation. The flies then shift to licking sooty moulds, bird faeces and, to a lesser extent, an unknown substance (probably leaf phylloplane bacteria and plant leachates) on citrus leaves and fruits during the mating and oviposition period (Hendrichs et al. 1993; Dong et al. 2014b).

3.2. Mating Behaviour

Male aggression and territoriality have been reported as a typical behaviour in some *Bactrocera* species (Shelly 1999; Weldon 2005; Benelli et al. 2014, 2015). For most tropical polyphagous *Bactrocera* species, males aggregate at a common place on foliage to attract and court females for mating, i.e. a non-resource-based lek mating system (Emlen and Oring 1977; Maan and Seehausen 2011). However, in *B. minax*, as in many temperate and oligophagous tephritids, male courtship behaviour is absent and the mating system is a resource-based defence polygyny, consisting of two phases: 1) males defend a resource (host fruit) (intrasexual selection), 2) where copulation takes place (intersexual selection) (Opp et al. 1996). In the wild, all mating events take place on citrus fruit (Dong et al. 2014b). Territory formation and copulation usually occurs on immature green fruits. It has been suggested that male flies that try to copulate in the vicinity of the ovipositional site have more chances to encounter and court receptive females (Prokopy 1976; Smith and Prokopy 1980).

Mating behaviour in *B. minax* is closely synchronised with the host fruiting season and has been described in the field as follows: (i) the male establishes its territory close to a potential oviposition substrate (citrus fruit); (ii) the female lands on the fruit and begins inspection and ovipositor probing on the fruit surface; and (iii), the male mounts and copulates with the female (Dong et al. 2014b). In this mating system females face trade-offs associated with the cost of additional, apparently unneeded matings on each fruit in return for access to resources. Like in the case of *Rhagoletis* species the resource is assumed to be the oviposition site (Opp et al. 1996; Opp and Prokopy 2000; Prokopy and Papaj 2000).

3.3. Oviposition and Host Preference

In tephritids, fruit flies use different cues for host finding behaviour and egg-laying behaviour. Usually, long-distance volatile chemicals are important before landing on host trees, visual stimuli act as short range once on the tree and contact chemicals on and inside host fruit influence female egg-laying decision.

Bactrocera minax oviposits solely into citrus fruits (Family *Rutaceae*) (Wang and Luo 1995; Dong et al. 2013). It is a large, powerful insect with a long ovipositor adapted for piercing through the thick skin of young, green citrus fruit (Liu and Zhou 2016). Visual cues including fruit shape, colour, and size are important for host finding (Prokopy and Owens 1983; Piñero et al. 2017). On the other hand, the egg-laying behaviour is greatly influenced by chemical stimuli, for example, semiochemicals, sugar content, levels of secondary plant compounds and physical properties of fruit (Bush 1969). The preference of *B. minax* oviposition on different citrus varieties is as follows, *Citrus sinensis* cv. Navel and *C. aurantium* > *C. sinensis* cv. Bintang, Amakusa and *C. reticulata* cv. Satsuma > *C. maxima* cv. Shatian > *C. reticulata* cv. Ponka. This ovipositional preference is positively correlated with larval survival and development; while in the field greater egg-laying occurs on those citrus fruits which are close to the surrounding vegetation and trees (Liu et al. 2014).

3.3.1. Visual Cues (Colour, Shape, and Size) for Oviposition

The hardness of a citrus fruit peel has an impact on the female insect's decision to oviposit (Lin et al. 2011). On the basis of egg oviposition marks on citrus fruit, it appears that *B. minax* significantly prefers to oviposit on the distal hemisphere rather than the basal hemisphere (Liu and Zhou 2016). Apart from peel hardness, the ovipositional behaviour of tephritid female flies is also influenced by fruit colour and shape (Alyokhin et al. 2000). It has been shown that tephritid flies respond to fruit-mimics of the same colour or reflecting similar levels of light than host fruit of a particular fly species (Aluja and Norrbom 1999). For example, the Queensland fruit fly *Bactrocera tryoni* (Froggatt) showed attraction to spheres painted with cobalt blue pigments, which reflected the same UV spectrum as favoured blue-coloured host fruits occuring in its native rainforest environment (Drew et al. 2003). The oriental fruit fly *Bactrocera dorsalis* (Hendel) is attracted to white-yellow colour (Vargas et al. 1991) and the apple maggot *Rhagoletis pomonella* (Walsh), a pest of apples, is attracted to fruit-mimicking traps such as a red sphere (Duan and Prokopy 1995).

Both sexes of *B. minax* are attracted to orange or yellow/green spheres of 50 mm diameter (Drew et al. 2006). However, we found that *B. minax* adults prefer green over other colours, and this preference is significantly increased in sexually mature flies over immature flies (author submitted results).

3.3.2. Chemical Cues (Semiochemicals) for Oviposition

Chemical cues play an important role in foraging and oviposition of fruit flies (Sarles et al. 2015), and these chemicals are widely exploited in integrated pest management (Shrivastava et al. 2010). For example, the application of oviposition marking pheromone reduced *Rhagoletis cerasi* L. infestation up to 100% in cherry orchards (Katsoyannos and Boller 1976; Boller and Hurther 1998). Together with visual cues, semiochemical cues may also influence host finding and egg-laying of *B. minax*. The peel odours of different varieties of orange preferred by *B. minax* produce different volatile blends, including acids, aldehydes, alcohols, and oils. It is presumed that these volatile compounds directly influence the olfactory orientation of *B. minax* females. However, there is as yet no proof if these volatiles released by host plants have a direct impact on the oviposition preference of *B. minax* (Liu and Zhou 2016).

From the perspective of the biology, ecology and behaviour, *B. minax* is more reminiscent of flies in the temperate genus *Rhagoletis* than other pest species in the *Bactrocera* genus. Therefore, some components of pest management can be drawn from the extensive scientific literature on apple maggot fly *R. pomonella,* European cherry fruit fly *Rhagoletis cerasi*, and other fruit flies (Vargas et al. 2016). For example, the application of fruit volatiles in conjunction with visual traps may yield good results for the control of *B. minax*.

4. TOWARDS THE AREA-WIDE MANAGEMENT OF *B. MINAX*

Several different control tactics have been used to manage populations of the Chinese citrus fly. These include chemical control and "attract and kill" techniques using protein/food baits and fruit-mimicking traps. In addition, pilot trials of the Sterile Insect Technique (SIT) have been assessed for *B. minax* control (Wang et al. 1990; Wang and Luo 1995), and farmers in China and Bhutan have adopted locally developed suppression techniques.

4.1. Use of Chemicals

Pesticide applications (cover sprays) are the most commonly used conventional control practices against insects pests, especially in the case of outbreaks. Though they are effective in reducing the losses caused by fruit fly infestation, the negative impacts of pesticides on humans, the environment and non-target organisms have raised much concern. Different insecticides have been used to suppress *B. minax* populations, including phoxim, dichlorvos, chlorpyrifos, abamectin, botanically derived pesticides, and pyrethroids. Amongst these, abamectin and dichlorvos proved to have the highest and lowest toxicity, respectively. However, chlorpyrifos had the strongest effect on pupae, and phoxim had the strongest influence on emergence.

These chemicals are not recommended due to toxicity to non-target organisms and long residual effects, but they are effective against *B. minax* (Liu et al. 2015).

4.2. *Eco-friendly Management*

4.2.1. *Field Sanitation*

Field sanitation is an effective and important strategy to reduce *B. minax* populations for the next fruiting season. The collection of infested fruits from the ground every week from mid-September to late-November is essential to remove the breeding population from orchards. The protocol demands that these infested fruits are transferred into thick plastic bags (20-25 kg per bag), that can be supplemented with aluminium phosphide to facilitate the killing of larvae. However, this is not critical if the bags are kept in the field under the sun for 7~10 days. Finally, the rotting fruits serve as fertilizer. The plastic bags can be recycled and used again (Liu et al. 2011; Li et al. 2013).

In Bhutan, cultural practices such as the application of soil tillage, along with natural predation (pupae picking by birds), seem to have a role in reducing the number of pupae. However, this reduction is not significant. Thus, it is not recommended as the only control measure for reducing the overall *B. minax* population in the wild (Dorji et al. 2010).

4.2.2. *Protein and Food Baits*

Spraying a mixture of protein bait, with a small quantity of insecticide added, has proven to be an effective strategy for large-scale control of fruit fly populations (Conway and Forrester 2011). Bait sprays are effective for fruit fly population control as newly emerged females require protein to become sexually mature (Perez-Staples et al. 2007; McQuate 2009). In China, it is a widely accepted approach for the farmers to use vinegar, sugar and wine mixtures, plus detergent, as baits station/spots spray for the control of *B. minax*, which is simple and cheap (Zhou et al. 2012). Attractants such as GF-120, and other locally available commercial products, are also used for the suppression of the Chinese citrus fly. Fresh enzymatically-hydrolysed beer yeast (H-protein) liquid protein bait effectively attracted and killed more *B. minax* flies than GF-120 sprayed in the field (Zhou et al. 2012).

4.2.3. *Use of Fruit-Mimicking Traps*

Usually semiochemicals and plant derived volatiles are used to trap fruit flies (Díaz-Fleischer et al. 2014), but the males of *B. minax* are not attracted to either of the standard *Bactrocera* male lures (i.e. methyl eugenol and cue-lure) (Drew et al. 2006). Visual traps have been used as an alternative, and in Bhutan, both sexes of the fly were most attracted to green-yellow or orange fruit-mimicking spheres in the field (Drew et al. 2006). In recent years, a specific fruit-mimicking trap (spherical green sticky trap) has been developed and widely applied to monitor and control *B. minax* in China. After field deployment of spherical traps in sweet orange orchards in Zigui, China, the infested fruit rate dropped to 2.7% compared with 28.6% in untreated control orchards (Yi et al. 2015). Efficiency of the control effort was closely associated with appropriate trap deployment density and time (Chen et al. 2017). Considering the cost of commercial traps, as well as efficacy, spherical green sticky traps with a diameter of 7 cm were recommended at a deployment rate of 20~30 traps per 1000 m^2 in citrus orchards (Chen et al. 2017; Gong et al. 2017).

4.2.4. Sterile Insect Technique (SIT)

The SIT has been successfully used to manage fruit flies including *Ceratitis capitata* (Wiederman), *B. dorsalis,* the melon fly *Zeugodacus cucurbitae* (Coquillett) and the Mexican fruit fly *Anastrepha ludens* (Loew) (Calkins et al. 1994; Koyama et al. 2004; Dhillo et al. 2005; Enkerlin et al. 2017). In AW-IPM programmes that have an SIT component, the production of sterile insects in large numbers is of paramount importance (Enkerlin 2021). Sterilisation can be achieved by irradiation or genetic manipulation. Irradiating *B. minax* pupa two days before emergence with a dose of 90 Gy is recommended to ensure adult sterility (Zhang and Li 1990).

Due to the serious damage in major citrus production regions in China, a mass-release of sterilized *B. minax* flies was carried out in Guizhou, China in the late 1980s and early 1990s. In these pilot projects, mature 3rd instar larvae were collected in fruit from the field and then allowed to pupate in the lab (Wang et al. 1990; Wang and Luo 1995). Although the SIT trials in Guizhou Province resulted in a significant reduction of the pest population (Wang et al. 1990; Wang and Luo 1995), the technology is so far not integrated into *B. minax* control programmes in China. This is mainly related to problems with the development of mass-rearing methods due to technical barriers such as the extreme long pupal diapause period. This is the reason that there is currently no foreseeable plan for an SIT approach against *B. minax*, although the selection of non-diapausing strains is being explored.

4.2.5. Natural Enemies

Knowledge on the parasitoids associated with the Chinese citrus fly remains very scarce. Only one parasitoid, *Diachasmimorpha feijeni* van Achterberg, has been associated with this fly in Bhutan (van Achterberg 1999) and in China (authors' unpublished data). However, there is still no detailed information available on the parasitoid's interactions with Chinese citrus fly or the potential for parasitoid manipulation.

4.2.6. From Pilot Trials to Area-Wide Management in China

Since 2009, the Ministry of Agriculture and National Agro-Tech Extension and Service Center of China has promoted and organized annual nation-wide conferences and training courses on tephritid control for local technicians and farmers, with the aim to educate and transfer new developments and technologies. Over the years, pilot trials of AW-IPM approaches against *B. minax* have been implemented in the provinces of Guizhou, Hubei, Hunan, Shanxi, and Sichuan where *B. minax* is a serious pest. The validated demonstration practices in the main citrus producing regions led to the establishment of a *B. minax* AW-IPM programme in China.

In the beginning of every year, the Ministry of Agriculture announces and issues online management programmes for *B. minax* control, along with that for other major agricultural pests, to facilitate the sustainable management of the pests. In the case of *B. minax*, the integrated environmental-friendly measures can be summarized as follows:

- First, population monitoring in April using field traps and pupal emergence cages, aimed at accurately identifying the timing of pupal developmental state and adult emergence, is carried out by local plant protection stations annually: this is critical to guide the implementation of control practices.
- Second, "attract and kill" strategies are applied from May to July including the systematic use of spherical green sticky traps, protein bait sprays and sugar-vinegar-wine liquid in bait stations or spot sprays.
- Third, field sanitation of habitats/orchards by removing the fallen and infested fruits and weeds from September to November has proved to be an important and effective population control measure. Recyclable plastic bags are widely adopted to keep-and-kill the mature larvae by a combination of hypoxia and heat.

Thus, through years of continuous efforts in AW-IPM of *B. minax*, significant progress has been achieved in suppressing *B. minax* populations to a level resulting in less than 5% of infested commercial fruit and a reduction in the use of synthetic pesticides by 60-80%. The *B. minax* AW-IPM programme will now be complying with national standards for 'green control', which results in notable direct and indirect economic, ecological and social benefits for China.

5. FUTURE PERSPECTIVES

With increased understanding of the biology and behaviour of *B. minax*, effective operational AW-IPM strategies against this pest have been established in China and Bhutan. However, SIT and natural enemies have not yet been exploited due to various biological and physiological obstacles. Great efforts are required to overcome these gaps for the future sustainable management of *B. minax*:

1. Mass-rearing of *B. minax* still remains a big challenge considering its univoltine and oligophagous traits. Future work should be focused on improving (a) understanding of egg hatch and (b) artificial diet formulation for the newly hatched young larvae.
2. Little information has been published on the insect-plant interactions of *B. minax*. It is widely known that the adults have a close relationship with *Citrus* spp., but how these adults utilize visual, olfactory and tactile cues to orientate to host plants for mating and oviposition has received very little attention.
3. Symbiotic organisms, including *Wolbachia* (Stouthamer et al. 1999), that affect biology and reproduction of *B. minax* should be characterized. A thorough screening of microorganisms by culture-dependent and high-throughput technology, in combination with related functional studies, will help to better understand the complex relationship between symbionts and *B. minax*. Such knowledge may lead to potential development and application of the Incompatible Insect Technique (IIT) (Zabalou et al. 2004).

4. Natural enemies have not been used as a component of AW-IPM against *B. minax* most likely because very little knowledge is available about them and their effectiveness. The possible synchronised diapause of parasitoids and *B. minax* pupae needs to be investigated. In addition, other agents such as predators or fungi causing pupal mortality deserve to be further investigated.

5. There are currently no effective semiochemical or plant derived volatile lures to attract *B. minax* males or females available. The volatile chemicals from the host fruits, as well as a sex attractant for monitoring and mass-trapping the Chinese citrus fly, urgently needs to be identified and exploited.

6. The growing published online resources on transcriptome, proteome, and genome of *B. minax*, RNA interference and CRISPR-Cas9 technologies targeting specific gene functions, will facilitate further investigations of molecular mechanisms responsible for the biology, behaviour, physiology and evolution of the Chinese citrus fly. The comprehensive understanding of *B. minax* is the most promising way to develop sustainable management of this economically important citrus pest in the long run.

6. ACKNOWLEDGEMENTS

The authors thank all the reviewers who enthusiastically made comments and suggestions to improve this paper. This study was funded by National Natural Science Foundation of China (31661143045, 31371945), International Atomic Energy Agency (CRP No. 17153 and No. 18269), Crop Disease and Insect Pest Monitoring and Control Program supported by the Ministry of Agriculture of People's Republic of China (10162130108235049) and the Fundamental Research Funds for the Central Universities (2662015PY148).

7. REFERENCES

Allwood, A., A. Chinajariyawong, S. Kritsaneepaiboon, R. Drew, E. Hamacek, D. Hancock, C. Hengsawad, J. Jipanin, M. Jirasurat, and C. K. Krong. 1999. Host plant records for fruit flies (Diptera: Tephritidae) in Southeast Asia. Raffles Bulletin of Zoology 47: 1–92.

Aluja, M., and A. Norrbom. 1999. Fruit flies (Tephritidae): Phylogeny and evolution of behavior. CRC Press, Boca Raton, Florida, USA. 984 pp.

Alyokhin, A. V., R. H. Messing, and J. J. Duan. 2000. Visual and olfactory stimuli and fruit maturity affect trap captures of Oriental fruit flies (Diptera: Tephritidae). Journal of Economic Entomology 93: 644–649.

Benelli, G., K. M. Daane, A. Canale, C. Y. Niu, R. H. Messing, and R. I. Vargas. 2014. Sexual communication and related behaviours in Tephritidae: Current knowledge and potential applications for Integrated Pest Management. Journal of Pest Science 87: 385–405.

Benelli, G., N. Desneux, D. Romano, G. Conte, R. H. Messing, and A. Canale. 2015. Contest experience enhances aggressive behaviour in a fly: When losers learn to win. Scientific Reports 5: 9347.

Boller E.F., and J. Hurther. 1998. The marking pheromone of the cherry fruit fly: A novel non-toxic and ecologically safe technique to protect cherries against cherry fruit fly infestation, pp 99–101. *In* Proceedings International Symposium on Insect Pheromones. March 1998. Wageningen, The Netherlands.

Bush, G. L. 1969. Sympatric host race formation and speciation in frugivorous flies of genus *Rhagoletis* (Diptera: Tephritidae). Evolution 23: 237–251.

Calkins, C. O., K. Bloem, S. Bloem, and D. L. Chambers. 1994. Advances in measuring quality and assuring good field performance in mass reared fruit flies, pp. 85–96. *In* C. O. Calkins, W. Klassen, and P. Liedo (eds.), Fruit flies and the Sterile Insect Technique. CRC Press, Boca Raton, Florida, USA.

Chao, Y., and Y. Ming. 1986. The investigation on fruit-flies (Trypetidae-Diptera) injurious to fruits and vegetables in south China. Technical Bulletin of Plant Quarantine Research 10: 1–61.

Chen, F., and F. Wong. 1943. Study on citrus maggot in Jiangjin County. Agricultural Science 1: 46.

Chen, S. X., and Y. Z. Xie. 1955. The classification and characterization of *Bactrocera (Tetradacus) minax* Enderlein. Acta Entomologica Sinica 5: 123.

Chen, Z. Z., Y. C. Dong, Y. H. Wang, A. A. Andongma, M. A. Rashid, P. Krutmuang, and C. Y. Niu. 2016. Pupal diapause termination in *Bactrocera minax*: An insight on 20-hydroxyecdysone induced phenotypic and genotypic expressions. Scientific Reports 6: 27440.

Chen, Z. Z., A. S. Deng, Q. Zhu, W. B. Chen, X. M. Xu, H. B. Zheng, and C. Y. Niu. 2017. Assessing attracting and killing effects of the spherical traps against Chinese citrus fly, *Bactrocera minax*. Journal of Huazhong Agricultural University 36: 33–37.

Conway, H. E., and O. T. Forrester. 2011. Efficacy of ground spray application of bait sprays with malathion or spinosad on Mexican fruit fly (Diptera: Tephritidae) in Texas citrus. Journal of Economic Entomology 104: 452–458.

Dhillo, M. K., R. Singh, J. S. Naresh, and H. C. Sharma. 2005. The melon fruit fly, *Bactrocera cucurbitae*: A review of its biology and management. Journal of Insect Science 5: 1–16.

Díaz-Fleischer, F., J. C. Piñero, and T. E. Shelly. 2014. Interactions between tephritid fruit fly physiological state and stimuli from baits and traps: Looking for the pied piper of Hamelin to lure pestiferous fruit flies, pp. 145–172. *In* T. E. Shelly, N. Epsky, E. B. Jang, J. Reyes-Flores, and R. Vargas (eds.), Trapping and the detection, control, and regulation of tephritid fruit flies. Springer, Dordrecht, The Netherlands.

Dong, Y. C., N. Desneux, C. L. Lei, and C. Y. Niu. 2014a. Transcriptome characterization analysis of *Bactrocera minax* and new insights into its pupal diapause development with gene expression analysis. International Journal of Biological Sciences 10: 1051–1063.

Dong, Y. C., L. Wan, R. Pereira, N. Desneux, and C. Y. Niu. 2014b. Feeding and mating behaviour of Chinese citrus fly *Bactrocera minax* (Diptera, Tephritidae) in the field. Journal of Pest Science 87: 647–657.

Dong, Y. C., Z. J. Wang, A. R. Clarke, R. Pereira, N. Desneux, and C. Y. Niu. 2013. Pupal diapause development and termination is driven by low temperature chilling in *Bactrocera minax*. Journal of Pest Science 86: 429–436.

Dorji, C., A. R. Clarke, R. A. Drew, B. S. Fletcher, P. Loday, K. Mahat, S. Raghu, and M. C. Romig. 2006. Seasonal phenology of *Bactrocera minax* (Diptera: Tephritidae) in western Bhutan. Bulletin of Entomological Research 96: 531–538.

Dorji, C., K. Mahat, and P. Loday. 2010. Effect of tillage on pupal mortality of the Chinese citrus fruit fly, *Bactrocera minax* (Enderlein) (Diptera:Tephritidae). Journal of Renewable Natural Resources 6: 24–30.

Drew, R. 1979. The genus *Dacus* Fabricius (Diptera: Tephritidae)–two new species from northern Australia and a discussion of some subgenera. Australian Journal of Entomology 18: 71–80.

Drew, R. A. I., and B. Yuval. 2000. The evolution of fruit fly feeding behavior, pp. 731–749. *In* M. Aluja and A. Norrbom (eds.), Fruit flies (Tephritidae): Phylogeny and evolution of behavior. CRC Press, Boca Raton, Florida, USA.

Drew, R. A. I., R. J. Prokopy, and M. C. Romig. 2003. Attraction of fruit flies of the genus *Bactrocera* to colored mimics of host fruit. Entomologia Experimentalis et Applicata 107: 39–45.

Drew R. A. I., M. C. Romig, and C. Dorji. 2007. Records of dacine fruit flies and new species of *Dacus* (Diptera: Tephritidae) in Bhutan. The Raffles Bulletin of Zoology 55: 1–21.

Drew, R. A. I., C. Dorji, M. C. Romig, and P. Loday. 2006. Attractiveness of various combinations of colors and shapes to females and males of *Bactrocera minax* (Diptera: Tephritidae) in a commercial mandarin grove in Bhutan. Journal of Economic Entomology 99: 1651–1656.

Duan, J. J., and R. J. Prokopy. 1995. Control of apple maggot flies (Diptera: Tephritidae) with pesticide-treated red spheres. Journal of Economic Entomology 88: 700–707.

Emlen, S. T., and L. W. Oring. 1977. Ecology, sexual selection, and the evolution of mating systems. Science 197: 215–223.

Enkerlin, W. R. 2021. Impact of fruit fly control programmes using the Sterile Insect Technique, pp. 977–1004. *In* V. A. Dyck, J. Hendrichs and A. Robinson (eds.), Sterile Insect Technique – Principles and practice in Area-Wide Integrated Pest Management. Second Edition. CRC Press, Boca Raton, Florida, USA.

Enkerlin W. R., J. M. Gutiérrez Ruelas, R. Pantaleón, C. Soto-Litera, A. Villaseñor-Cortés, J.L. Zavala-López, D. Orozco-Dávila, P. Montoya-Gerardo, L. Silva-Villarreal, E. Cotoc-Roldán, F. Hernández-López, A. Arenas-Castillo, D. Castellanos-Domínguez, A. Valle-Mora, P. Rendón-Arana, C. Cáceres-Barrios, D. Midgarden, C. Villatoro-Villatoro, E. Lira-Prera, O. Zelaya-Estradé, R. Castañeda-Aldana, J. López-Culajay, P. Liedo-Fernández, G. Ortíz-Moreno, J. Reyes-Flores, F. Ramírez y Ramírez, J. Trujillo-Arriaga, J. Hendrichs. 2017. The Moscamed regional programme: A success story of area-wide Sterile Insect Technique application. Entomologia Experimentalis et Applicata 264(3): 188–203.

Fan, J. A., X. Q. Zhao, and J. Zhu. 1994. A study on the cold-resistance and diapause in *Tetradacus citri* Chen. Journal of Southwest Agricultural University 6: 532–534.

Gao, L. Z., Y. H. Liu, X. W. Wan, J. Wang, and F. Hong. 2013. Screening of microsatellite markers in *Bactrocera minax* (Diptera: Tephritidae). Scientia Agricultura Sinica 46: 3285–3292.

Gong, B. Y., F. L. Xiao, B. C. Mo, W. G. Dong, Z. Q. Wen, S. Z. Yang, and X. X. Li. 2017. Research on the attracting effects of the spherical traps against Chinese citrus fly, *Bactrocera minax*. Plant Protection 43: 218–221.

Harwood, J. F., K. Chen, P. Liedo, H. G. Müller, J. L. Wang, A. E. Morice, and J. R. Carey. 2015. Female access and diet affect insemination success, senescence and the cost of reproduction in the male Mexican fruit fly *Anastrepha ludens*. Physiological Entomology 40(1): 65–71.

Hendrichs, J., S. S.Cooley, and R. J. Prokopy. 1993. Uptake of plant surface leachates by apple maggot flies, pp. 173–175. *In* M. Aluja, and P. Liedo (eds.), Fruit flies, biology and management. Springer, New York, NY, USA.

Katsoyannos B. I., and E. F. Boller. 1976. First field application of oviposition-deterring marking pheromone of European cherry fruit fly. Environmental Entomology 5:151–152.

Koyama, J., H. Kakinohana, and T. Miyatake. 2004. Eradication of the melon fly, *Bactrocera cucurbitae*, in Japan: Importance of behavior, ecology, genetics, and evolution. Annual Review of Entomology 49: 331–349.

Krosch, M. N., M. K. Schutze, K. F. Armstrong, G. C. Graham, D. K. Yeates, and A. R. Clarke. 2012. A molecular phylogeny for the tribe Dacini (Diptera: Tephritidae): Systematic and biogeographic implications. Molecular Phylogenetics & Evolution 64: 513–523.

Li, P., C. Y. Niu, Q. He, G. Z. Zhou, X. M. Xu, P. Y. Yang, H. B. Peng, Z. C. Jiang, M. Zhang, F. Ren, Y. H. Wang, W. Z. Hu, and J. X. Yang. 2013. A poster on green control strategies for *Bactrocera minax*. Agriculture Press, Beijing, China.

Lin, W. L., S. Z. Yang, M. S. Pan, H. L. Chen, Z. P. Huang, J. G. Long, and F. L. Xiao. 2011. The damage characteristics of *Tetradacus citri* Chen on different *Citrus* varieties in Hunan Province. Hunan Agricultural Sciences 23: 95–97.

Liu, C., X. H. Xiao, H. H. Wu, C. H. Yang, and C. J. Song. 2011. Experimental trial on plastic bag use to manage the Chinese citrus fruit fly larvae infested fruits. Plant Doctor 24: 32–34.

Liu, H. Q., G. F. Jiang, Y. F. Zhang, F. Chen, X. J. Li, J. S. Yue, C. Ran, and Z. M. Zhao. 2015. Effect of six insecticides on three populations of *Bactrocera (Tetradacus) minax* (Diptera: Tephritidae). Current Pharmaceutical Biotechnology 16: 77–83.

Liu, L., and Q. Zhou. 2016. Olfactory response of female *Bactrocera minax* to chemical components of the preference host citrus volatile oils. Journal of Asia-Pacific Entomology 19: 637–642.

Liu, L., Q. Zhou, A. Q. Song, and K. X. You. 2014. Adult oviposition and larval feeding preference for different *Citrus* varieties in *Bactrocera minax* (Diptera: Tephritidae). Acta Entomologica Sinica 57: 1037–1044.

Liu, Y. Q., E. Heying, and S. A. Tanumihardjo. 2012. History, global distribution, and nutritional importance of citrus fruits. Comprehensive Reviews in Food Science and Food Safety 11: 530–545.

Lu, H. X., K. P. He, H. F. Ruan, and B. Z. Mou. 1997. The biological features of Chinese citrus fly *Dacus citri* (Chen). Journal of Hubei Agricultural College 17: 169–170.

Lü, Z., L. Wang, G. Zhang, F. Wan, J. Guo, H. Yu, and J. Wang. 2014. Three heat shock protein genes from *Bactrocera (Tetradacus) minax* Enderlein: Gene cloning, characterization, and association with diapause. Neotropical Entomology 43: 362–372.

Lushchak, V., D. V. Gospodaryov, B. M. Rovenko, I. S. Yurkevych, N. V. Perkhulyn, and V. I. Lushchak. 2013. Specific dietary carbohydrates differentially influence the life span and fecundity of *Drosophila melanogaster*. The Journals of Gerontology Series A: Biological Sciences and Medical Sciences 69: 3–12.

Maan, M. E., and O. Seehausen. 2011. Ecology, sexual selection and speciation. Ecology Letters 14: 591–602.

McQuate, G. 2009. Effectiveness of GF-120NF fruit fly bait as a suppression tool for *Bactrocera latifrons* (Diptera: Tephritidae). Journal of Applied Entomology 133: 444–448.

Moraiti, C. A., C. T. Nakas, and N. T. Papadopoulos. 2012. Prolonged pupal dormancy is associated with significant fitness cost for adults of *Rhagoletis cerasi* (Diptera: Tephritidae). Journal of Insect Physiology 58: 1128–1135.

Nath, D. 1972. *Callantra minax* (Enderlein) (Tephritidae: Diptera), a new record of a ceratitinid fruit fly on orange fruits (*Citrus reticulata* Blanco) in India. Indian Journal of Entomology 34: 246.

Opp, S. B., S. A. Spisak, A. Telang, and S. S. Hammond. 1996. Comparative mating systems of two *Rhagoletis* species: The adaptive significance of mate guarding, pp. 43–50. *In* B. A. McPheron, and G. J. Steck (eds.), Fruit fly pests: A world assessment of their biology and management. St. Lucie Press, Delray Beach, Florida, USA.

Opp, S. B., and R. J. Prokopy. 2000. Multiple mating and reproductive success of male and female apple maggot flies, *Rhagoletis pomonella* (Diptera: Tephritidae). Journal of Insect Behavior 13: 901–914.

Papanastasiou, S. A., D. Nestel, A. D. Diamantidis, C. T. Nakas, and N. T. Papadopoulos. 2011. Physiological and biological patterns of a highland and a coastal population of the European cherry fruit fly during diapause. Journal of Insect Physiology 57: 83–93.

Perez-Staples, D., V. Prabhu, and P. W. Taylor. 2007. Post-teneral protein feeding enhances sexual performance of Queensland fruit flies. Physiological Entomology 32: 225–232.

Piñero, J.C., S. K. Souder, R. I. Vargas. 2017. Vision-mediated exploitation of a novel host plant by a tephritid fruit fly. PLoS One 12(4): e0174636.

Prokopy, R. J. 1976. Feeding, mating, and oviposition activities of *Rhagoletis fausta* flies in nature. Annals of the Entomological Society of America 69: 899–904.

Prokopy, R. J., and D. R. Papaj. 2000. Behavior of flies of the genera *Rhagoletis, Zonosemata,* and *Carpomya* (Trypetinae: Carpomyina), pp. 219–252. *In* M. Aluja, and A. L. Norrbom (eds.), Fruit flies (Tephritidae): Phylogeny and evolution of behavior. CRC Press, Boca Raton, Florida, USA.

Prokopy R. J., and E. D. Owens. 1983. Visual detection of plants by herbivorous insects. Annual Review of Entomology 28: 337–364.

Ragland, G. J., J. Fuller, J. L. Feder, and D. A. Hahn. 2009. Biphasic metabolic rate trajectory of pupal diapause termination and post-diapause development in a tephritid fly. Journal of Insect Physiology 55(4): 344–350.

Sarles, L., A. Verhaeghe, F. Francis, and F. J. Verheggen. 2015. Semiochemicals of *Rhagoletis* fruit flies: Potential for Integrated Pest Management. Crop Protection 78: 114–118.

Shelly, T. E. 1999. Defense of oviposition sites by female Oriental fruit flies (Diptera: Tephritidae). Florida Entomologist 82: 339–346.

Shrivastava, G., M. Rogers, A. Wszelaki, D. R. Panthee, and F. Chen. 2010. Plant volatiles-based insect pest management in organic farming. Critical Reviews in Plant Sciences 29: 123–133.

Smith, D. C., and R. J. Prokopy. 1980. Mating behavior of *Rhagoletis pomonella* (Diptera: Tephritidae) VI. Site of early-season encounters. The Canadian Entomologist 112(6): 585–590.

Srivastava, A. K. 2012. Advances in citrus nutrition. Springer, Dordrecht, The Netherlands. 477 pp.

Stouthamer, R., J. Breeuwer, and G. Hurst. 1999. *Wolbachia pipientis*: Microbial manipulator of arthropod reproduction. Annual Reviews in Microbiology 53: 71–102.

Sun, Z. Y. 1961. The study and control of *Dacus citri*. Chinese Plant Protection Science, Scientific Publishing House, Beijing, China.

Taylor, P., D. Pérez-Staples, C. Weldon, S. Collins, B. Fanson, S. Yap, and C. Smallridge. 2013. Post-teneral nutrition as an influence on reproductive development, sexual performance and longevity of Queensland fruit flies. Journal of Applied Entomology 137: 113–125.

Teixeira, L. A., and S. Polavarapu. 2001. Postdiapause development and prediction of emergence of female blueberry maggot (Diptera: Tephritidae). Environmental Entomology 30: 925–931.

Teixeira, L. A., and S. Polavarapu. 2005. Diapause development in the blueberry maggot *Rhagoletis mendax* (Diptera: Tephritidae). Environmental Entomology 34: 47–53.

Thompson, F. C. 1998. Fruit fly expert identification system and systematic information database: A resource for identification and information on fruit flies and maggots, with information on their classification, distribution and documentation. Backhuys Publishers for the North American Dipterists' Society, Leiden, Netherlands.

van Achterberg, C. 1999. The Palaearctic species of the genus *Diachasmimorpha* Viereck (Hymenoptera: Braconidae: Opiinae). Zoologische Mededeelingen 73: 1–10.

van Schoubroeck, F. 1999. Learning to fight a fly: Developing citrus IPM in Bhutan. Unpublished PhD thesis, Wageningen Universiteit, The Netherlands.

Vargas, R., J. D. Stark, R. J. Prokopy, and T. A. Green. 1991. Response of Oriental fruit fly (Diptera: Tephritidae) and associated parasitoids (Hymenoptera: Braconidae) to different-color spheres. Journal of Economic Entomology 84: 1503–1507.

Vargas, R. I., J. C. Piñero, L. Leblanc, N. C. Manoukis, and R. F. L. Mau. 2016. Area-wide management of fruit flies (Diptera: Tephritidae) in Hawaii, pp. 673–693. *In* S. Ekesi, S. Mohamed, and M. Meyer (eds.), Fruit fly research and development in Africa: Towards a sustainable management strategy to improve horticulture. Springer International Publishing, Switzerland.

Wang, H. S., and H. Q. Zhang. 1993. Control of the Chinese citrus fly, *Dacus citri* (Chen), using the Sterile Insect Technique, pp. 505–512. *In* Proceeding: Management of insect pests: Nuclear and related molecular and genetic techniques. FAO/IAEA International Symposium, 19–23 October 1992, Vienna, Austria. STI/PUB/909. IAEA, Vienna, Austria.

Wang, H. S., C. D. Zhao, H. X. Li, H. Z. Lou, Q. R. Liu, W. Tang, J. G. Hu, and H. Q. Zhang. 1990. Control of Chinese citrus fly *Dacus citri* by male sterile technique. Acta Agriculturae Nucleatae Sinica 4(3): 135–138.

Wang, J., H. Fan, K. C. Xiong, and Y. H. Liu. 2017. Transcriptomic and metabolomic profiles of Chinese citrus fly, *Bactrocera minax* (Diptera: Tephritidae), along with pupal development provide insight into diapause program. PLoS One 12: e0181033.

Wang, J., H. Y. Zhou, Z. M. Zhao, and Y. H. Liu. 2014. Effects of juvenile hormone analogue and ecdysteroid on adult eclosion of the fruit fly *Bactrocera minax* (Diptera: Tephritidae). Journal of Economic Entomology 107: 1519–1525.

Wang, J., K. C. Xiong, and Y. H. Liu. 2016. De novo transcriptome analysis of Chinese citrus fly, *Bactrocera minax* (Diptera: Tephritidae), by high-throughput illumina sequencing. PLoS One 11: e0157656.

Wang, X. J., and L. Y. Luo. 1995. Research progress in the Chinese citrus fruit fly. Entomological Knowledge 32: 310-315.

Wang , X. L., and R. J. Zhang. 2009. Review on biology, ecology and control of *Bactrocera (Tetradacus) minax* Enderlein. Journal of Environmental Entomology 31: 73–79.

Weldon, C. W. 2005. Mass-rearing and sterilisation alter mating behaviour of male Queensland fruit fly, *Bactrocera tryoni* (Froggatt) (Diptera: Tephritidae). Australian Entomology 44: 158–163.

White, I. M., and X. Wang. 1992. Taxonomic notes on some dacine (Diptera: Tephritidae) fruit flies associated with citrus, olives and cucurbits. Bulletin of Entomological Research 82: 275–279.

Xia, Y., X. L. Ma, B. H. Hou, and G. C. Ouyang. 2018. A review of *Bactrocera minax* (Diptera: Tephritidae) in China for the purpose of safeguarding. Advances in Entomology 6: 35–61.

Xiong, K. C., J. Wang, J. H. Li, Y. Q. Deng, P. Pu, H. Fan, and Y. H. Liu. 2016. RNA interference of a trehalose-6-phosphate synthase gene reveals its roles during larval-pupal metamorphosis in *Bactrocera minax* (Diptera: Tephritidae). Journal of Insect Physiology 91: 84–92.

Yang, W. S., C. R. Li, J. Lan, H. L. An. 2013. The spread mode and dispersal history of Chinese citrus fly. Journal of Yangtze University 10: 8–11.

Yi, J. P., S. H. Li, G. G. Zhang, F. Xiang, G. Z. Zhou, and H. G. Luo. 2015. Attracting and killing effects of the green-yellow spherical traps against Chinese citrus fly, *Bactrocera minax.* China Plant Protection 35: 34–37.

Zabalou, S., A. Apostolaki, I. Livadaras, G. Franz, A. Robinson, C. Savakis, and K. Bourtzis. 2009. Incompatible Insect Technique: Incompatible males from a *Ceratitis capitata* genetic sexing strain. Entomologia Experimentalis et Applicata 132: 232–240.

Zhang, W., and Y. Y. Li. 1990. Effect of ^{60}Co γ irradiation on germ cell of Chinese citrus fly *Dacus citri*. Acta Agriculturae Nucleatae Sinica 4: 115–119.

Zhang, Y. A. 1989. Citrus fruit flies of Sichuan province (China). EPPO Bulletin 19: 649–654.

Zhou, X. W., C. Y. Niu, P. Han, and N. Desneux. 2012. Field evaluation of attractive lures for the fruit fly *Bactrocera minax* (Diptera: Tephritidae) and their potential use in spot sprays in Hubei province (China). Journal of Economic Entomology 105: 1277–1284.

AREA-WIDE FRUIT FLY PROGRAMMES IN LATIN AMERICA

P. RENDÓN[1] AND W. ENKERLIN[2]

[1]*International Atomic Energy Agency - Technical Cooperation TCLAC, Programa Moscamed/USDA, Guatemala City, Guatemala; pedro.rendon@usda.gov*
[2]*Joint FAO/IAEA Division of Nuclear Techniques in Food and Agriculture, International Atomic Energy Agency, Vienna, Austria; W.R.Enkerlin@iaea.org*

SUMMARY

The chapter presents an overview of fruit fly (Tephritidae) pests and their economic impact in the Latin America and Caribbean (LAC) region, with a focus on the damage they inflict to horticultural production, as well as national and international commercialization. It reviews global trends that have favoured the establishment of several invasive fruit fly species in the region and the need to avoid further transboundary movement of invasive species. It also discusses the opportunities to increase fruit and vegetable production in the region despite the fruit fly problem and how integrated fruit fly management approaches within the framework of the International Plant Protection Convention (IPPC) can be applied for effective fruit fly control and to facilitate the international commercialization of horticultural commodities. The need for increased consumption of fruit and vegetables worldwide and in the Latin America and Caribbean region to mitigate the growing incidence of non-communicable diseases is discussed, as well as the trends in human population growth that will require increased provision of adequate diets. It also discusses the opportunities for Latin America and Caribbean countries to commercialize produce taking advantage of the global trend towards healthier food and less animal protein consumption. It presents available mechanisms for technical cooperation that facilitate technology transfer for more sustainable area-wide fruit fly management. It also provides case examples in the Latin America and Caribbean region of successful area-wide fruit fly programmes that have increased production, opened markets and generated significant return on investment, as well as job opportunities. Future perspectives and challenges to address the fruit fly problem in the Latin America and Caribbean region are described.

Key Words: Tephritidae, *Anastrepha*, *Bactrocera*, *Ceratitis*, *Rhagoletis*, Caribbean, Central America, South America, invasive, horticultural exports, losses, pest free areas, fruit and vegetable demand, economic impact

J. Hendrichs, R. Pereira and M. J. B. Vreysen (eds.), Area-Wide Integrated Pest Management: Development and Field Application, pp. 161–195. CRC Press, Boca Raton, Florida, USA.

1. INTRODUCTION

Among the most important Sustainable Development Goals (SDGs), established by the United Nations in 2015, are: No Poverty, Zero Hunger, Good Health and Well-Being, and Life on Land (UN 2015).

Food is a common thread linking all 17 SDGs, given the interconnected economic, social and environmental dimensions of food systems (EIU 2018). Policy makers, politicians, health officials, and entrepreneurs are currently faced with the need to end hunger, achieve food security and improve nutrition, which are key steps toward sustainable development (UN 2016).

Food insecurity is one of the main challenges the world is facing to achieve the 2030 set milestones for at least these SDGs. Food insecurity is defined by the Food and Agriculture Organization of the United Nations (FAO) as the:

> *"situation when people lack secure access to sufficient amounts of safe and nutritious food for normal growth and development and an active and healthy life"* (FAO/IFAD/UNICEF/WFP/WHO 2017).

Globally, the total number of people defined as moderately or severely food insecure was nearly 1.8 billion in 2015. On the other hand, over the next four decades, the world's population is forecast to increase by 2 billion people to exceed 9 billion by 2050 (Worldometers 2019).

Recent estimates indicate that to meet the projected food demand, global agricultural production will have to increase by 25-70% from its 2005-2007 levels, while nutrient losses and greenhouse gas emissions from agriculture must drop dramatically to restore and maintain the functioning of ecosystems (Hunter et al. 2017).

No one can question that pests, in particular insects, are significantly contributing to food insecurity worldwide. On average, insect pests are responsible for substantial pre- and post-harvest losses of horticultural products estimated according to some sources at least 18 to 20% at an annual estimated value of USD 470 billion (Sharma et al. 2017).

Losses are considerably higher in the developing world, especially, in the tropics of Africa, Asia and Latin America where pest control practices are much less effective and where most of the human population increase is expected. Moreover, the human population is facing an epidemic increase of non-communicable diseases due to insufficient consumption of fruit and vegetables among others (Hall et al. 2009). Treating these diseases between now and 2030 will cost USD 30 trillion globally (Nierenberg 2018).

Due to the climatic and soil conditions, the countries of Latin America could produce fruit and vegetables in larger quantities and at improved quality, taking advantage of the increasing demand from an increasing human population in the region and the world. The increased production of fruit and vegetables would further develop the horticultural industry creating diversification of the region's income. However, to take full advantage of the opportunity, a number of insect pest problems need to be addressed and overcome.

From the list of so-called key pests of agriculture, some fruit fly species are considered among the most devastating insects pests. They cause direct damage to horticultural production by reducing yields and indirect damage by disrupting national and international trade (White and Elson-Harris 1992). For example, the devastating impact of fruit flies in Africa, including direct damage to horticultural production and bans on trade, result in estimated losses of at least USD 2 billion annually on this continent alone (Ekesi et al. 2016). By preventing and significantly reducing fruit fly damage by means of effective and environment-friendly pest control measures, additional food supply would be made available, contributing to alleviating the large global deficit projected by 2050 (Hunter et al. 2017).

To reduce the transboundary risks and global burdens caused by invasive pests and at the same time continue facilitating international trade of horticultural products, the World Trade Organization (WTO) provides an instrument through its Agreement on Sanitary and Phytosanitary Measures (SPS). The SPS agreement is enforced by the International Plant Protection Convention (IPPC), based at FAO, and its contracting parties. The IPPC drafts and adopts *International Standards for Phytosanitary Measures* (ISPMs), aimed at providing a framework to the contracting parties for best practices in pest control and for mitigating pest risk in international trade. This includes a number of ISPMs specifically for fruit flies that have been recently harmonized in one suit of ISPMs to facilitate interpretation and use by IPPC contracting parties, as will be presented and discussed later.

This chapter presents a summary of the tephritid fruit fly pests and their economic impact in the Latin America and Caribbean (LAC) region. The main focus is on the damage they inflict to horticultural production and commercialization in the region, as well as on area-wide integrated pest management (AW-IPM) strategies for fruit flies that are applied, based on the international phytosanitary framework, for effective fruit fly control and to facilitate the commercialization of horticultural commodities.

2. FRUIT FLIES AND THEIR ECONOMIC IMPACT WORLDWIDE

2.1. *Numbers and Current Distribution*

To date, worldwide, approximately 4223 fruit fly species have been identified belonging to around 500 genera (Norrbom et al. 1998). From the total recorded so far,

250 species (less than 6%) are of some economic importance and from these, only some 78 species (1.8%) in 11 genera can be classified of major economic and quarantine significance (FAO/IAEA 2019). Nevertheless, the damage inflicted by these species to the horticultural industry worldwide is devastating, amounting to billions of US dollars every year.

These pests are found in almost all fruit and vegetable growing areas in all continents with the exception of the Antarctic (White and Elson-Harris 1992). The most important genera include: *Anastrepha* (Schiner), *Bactrocera* (Macquart), *Ceratitis* (MacLey), *Dacus* (Fabricius), *Rhagoletis* (Loew), *Toxotrypana* (Gerstaecker) (recently synonymised with *Anastrepha*) and *Zeugodacus* (Hendel). The genus *Anastrepha* and corresponding species are indigenous to the Americas, the *Bactrocera* to Asia and Oceania/Australia, the *Dacus* and *Ceratitis* to Africa, *Rhagoletis* to more temperate areas of Europe, North and South America, and the *Zeugodacus* to Asia.

Some fruit fly species belonging to these genera inflict serious economic damage to the horticultural industry in the countries and regions of endemism. With some exceptions (some temperate species), fruit fly pests are polyphagous, infesting a wide range of host species including some of the fruits and vegetables with the highest commercial value. Fruit flies have a high reproductive rate; thus, can produce several generations per year. These characteristics place fruit flies among the most important group of insect pests affecting horticultural production and trade worldwide. Fruit flies cause damage by laying eggs inside horticultural crops after which larvae hatch and feed on the mesocarp. As a result, infested fruits will prematurely drop from the tree or remain on the tree until harvest.

Depending on the fruit fly species and host, damage can range from 10 to nearly 100% of the crop when effective control practices are not implemented. In addition to the direct damage on fruits and vegetables, and significant yield reduction, fruit fly presence may seriously disrupt trade by quarantine restrictions imposed by importing countries which are free of the pests. This is illustrated by the recent case in the Dominican Republic where the Mediterranean fruit fly *Ceratitis capitata* (Wiedemann) was detected in 2015 and ten months later USD 40 million had been lost due to an immediate ban to Dominican exports imposed by importing countries (Zavala-López et al., this volume).

Fruit flies are effective invaders capable of spreading and establishing in regions outside their natural distribution range. For example, the oriental fruit fly *Bactrocera dorsalis* (Hendel) is an Asian species of great economic importance, infesting some 200 species of fruits and vegetables and causing direct losses estimated in millions of US dollars/year. Over the last decade, this species has invaded sub-Saharan Africa, causing losses ranging from 30 to 100% in some fruit crops such as mango, closing trade routes and triggering the loss of export opportunities that are vital for both the smallholder farmers and the more commercial fruit industry in the continent (Ekesi et al. 2016).

Other classic examples of fruit fly tephritid pest species spreading outside their natural distribution range and causing significant damage include the Mediterranean fruit fly, spreading from Central and North Africa to Europe, the Americas, Indian Ocean and Oceania (Gutiérrez-Samperio 1976); the melon fly *Zeugodacus cucurbitae* (Coquillet) spreading from Southeast Asia into Central Africa, and the Indian and Pacific Oceans (FAO/IAEA 2019); the peach fruit fly *Bactrocera zonata* (Saunders) spreading from Central Asia into North Africa and the Indian Ocean (FAO/IAEA 2019); the olive fruit fly *Bactrocera oleae* (Rossi) invading California and northern Mexico (Yokohama 2015; and the carambola fruit fly *Bactrocera carambolae* (Drew & Hancock) spreading from Asia to Suriname (Malavasi et al. 2000).

2.2. *Factors that Contribute to Pest Movement and Establishment*

Global trends including increased travel and trade, human movement and climate change are positively correlated with the significant increase in transboundary movement of invasive insect pests. Hulmes (2009) estimates that in the past 200 years the rate of non-native species introductions has increased 76-fold. The way these factors affect the spread of invasive pests is briefly described below.

2.2.1. *Global Trade and Transport*

With open market economies developing further through the integration of countries into new economic blocks and the expansion of current economic regions, transboundary trade and transportation of goods, including agricultural commodities, has been increasing significantly and is expected to increase further. The commercial movement of goods results in effective pathways for invasive pest species such as fruit flies, which are moving along with the commodities that they infest. For example, an analysis to assess the risk of Mediterranean fruit fly incursions into California, USA, indicated that the pathways with the highest probability for pest introduction were air passengers and crew baggage from foreign countries, express mail carriers from Asia and Hawaii to California containing packages with small amounts of fruits sent to relatives living in communities around Los Angeles Basin and cargo ships from Central America and other foreign countries (USDA/APHIS 1992).

2.2.2. *Human Movement and Travel*

Humans crossing borders to escape from violence, hunger and lack of opportunities in their countries, is undeniably increasing. Unintentional movement of insect pests by migrating humans is also an effective pathway for disseminating and introducing invasive pests. This is particularly true for fruit fly pests, that are often moved in fruits and vegetables that migrants carry as they travel for long distances across borders. Therefore, a rise in human movement and travel also results in an increased movement of invasive insect pest species that are not present in the country of transit or destination.

Humans are not only moving more frequently and in larger numbers because of social pressure or economic reasons, but also, as standards of living increase, travel has been increasing for segments of the population who travel long distance for leisure. For example, the number of tourists arriving to the Americas from all over the world has increased from 109 million in 1995 to 192.6 million in 2015, i.e. an increase of more than 75% in 20 years. Tourists often carry small amounts of fruits and vegetables to and from the site of destination, becoming effective pathways for invasive pests as well (Statista 2019).

2.2.3. Climate Change

Climate change plays an increasingly important role in the survival and establishment of invasive pests (Pimentel 2002). Some pest species which are of tropical and subtropical origin, are now able to survive in regions of the world where climate has been gradually changing from cold winters with freezing temperatures to milder winters. This has in many cases, allowed pest binvasion and expansion of their distributions into new territories. For example, in the past 25 years, the Mediterranean fruit fly has been expanding its geographic distribution from North Africa and South Europe, to Central and East Europe as average winter temperatures are raising (Bjelis et al. 2016).

In addition, other factors related to climate change such as the increasing frequency of the El Niño Southern Oscillation (ENSO) in the tropics of Central and South America. Average temperatures increase, affecting biological cycles of pests by reducing development time and increasing the yearly numbers of generations and population density, thereby exacerbating pest problems. For example, the effects of the El Niño on Mediterranean fruit fly populations was assessed in Guatemala (Lira and Midgarden, this volume). This was done through modelling the increase by fractions in the average temperatures and observing the effect on population growth rates. A rapid and steep population growth exerts greater pressure over the containment barrier in southern Mexico that protects the Mediterranean fruit fly-free areas from infested areas in Central America. A positive correlation was observed during periods characterized as El Niño years, with increasing numbers of outbreaks in the Mediterranean fruit fly-free areas in northern Guatemala and the free areas in the state of Chiapas bordering Guatemala (Enkerlin et al. 2015).

Another relevant factor contributing to invasive pest introductions is the increase in the frequency and intensity of tropical storms, with high speed winds easily disseminating invasive pests over very long distances (Bhattarai and Cronin 2014). Calamitous events, such as hurricanes, tidal waves/tsunamis, droughts, floods, civil strife, often result in mass-displacement of people that can promote pest movement. Aid and assistance brought in as relief can pose risks if perishable provisions are brought in from countries infested with non-native fruit fly species.

3. FRUIT FLIES AND THEIR ECONOMIC IMPORTANCE IN THE LAC REGION

3.1. *Fruit Flies in the Americas*

Of the total number of tephritid fruit fly recorded, roughly, 23% (ca. 977 species) occur on the American continent. Most of them are present in the Neotropical region, extending from Mexico to Argentina (Norrbom et al. 1998). From this number of endemic species, around 15 (1.5%) are of economic significance.

Endemic to the more subtropical and tropical regions of the Americas is the genus *Anastrepha* with nine species known to be of economic significance, as follows: Mexican fruit fly *A. ludens* (Loew), West Indian fruit fly *A. obliqua* (Macquart), South American fruit fly (*A. fraterculus* Wiedemann) (a complex of about seven species with different hosts and behaviours), guava fruit fly (*A. striata* Schiner), sapote fruit fly *A. serpentina* (Wiedemann), inga fruit fly *A. distincta* Greene, Caribbean fruit fly *A. suspensa* (Loew), papaya fruit fly *A. curvicauda* (Gerstaecker) (formerly *Toxotrypana curvicauda*), and South American cucurbit fruit fly *Anastrepha grandis* (Macquart) (Weems Jr. et al. 2017; FAO/IAEA 2019).

In most of the countries in the Latin American region, these fruit fly species are still largely responsible for reduced horticultural production yields and for commercial production being mostly marketed domestically (Enkerlin et al. 1989). Exports of fruit and vegetable commodities are limited as many markets, including lucrative foreign ones, maintain restrictions from countries where major fruit fly pests are known to occur. Where horticultural trade does occur, it is largely among countries within the region that share similar pest problems or must undergo post-harvest commodity treatments.

For example, Mexico has more than one million hectares (ha) planted to fruit crops which are affected by fruit fly pests with an estimated annual production value of more than USD 4850 million (Gobierno de México 2018). The fruit industry is significantly hindered by fruit fly species, in particular by *A. ludens*, *A. obliqua*, *A. striata* and *A. serpentina*. In 1991, the annual direct damage caused by these indigenous fruit flies was estimated at more than USD 230 million despite control activities (Reyes et al. 1991). To address this serious constraint in a more systematic and coordinated way, the National Plant Protection Organization (NPPO) of Mexico established in the early 1990s the National Fruit Fly Campaign, which has effectively controlled fruit fly pests on an area-wide basis in large regions of the country, generating a very significant return on investment as discussed in Section 6.1.2 of this chapter (Reyes et al. 2000; IICA 2010; Gutiérrez-Ruelas et al. 2013).

The genus *Rhagoletis* is also endemic to the Americas but is generally present in more temperate regions. The species of economic concern include: the eastern cherry fruit fly *R. cingulata* (Loew), the walnut husk fly *R. completa* Cresson, the black-bodied cherry fruit fly *R. fausta* (Osten Sacken), the western cherry fruit fly *R. indifferens* Curran, the blueberry maggot *R. mendax* Curran, and the apple maggot *R. pomonella* (Walsh).

All these species inflict serious economic damage to fruit production and commercialization. For example, a report has indicated that the overall domestic and export cost of *R. pomonella* could be USD 392.5 million annually in the state of Washington, USA (DEFRA 2018). The total reduction in net returns to producers from *R. completa* injured walnuts amounts from 50 to 75 % (Boyce 1934).

In addition to the endemic fruit fly species, a number of non-native invasive species have invaded and become established in some parts of the Latin America and Caribbean region and are causing severe direct and indirect losses to the horticultural industries in the region (Fig. 1).

From these species, the Mediterranean fruit fly is the most devastating as it is capable of infesting more than 300 species of fruits and vegetables (USDA/APHIS 2019) and it is subject to rigorous quarantine restrictions by importing countries free from the pest (Fig. 2). This species is responsible for economic losses estimated at USD 242 million/year in Brazil alone (Oliveira et al. 2013). Moreover, citrus in Central America covers an area of approximately 84 000 ha; the damage without control has been estimated at 28% in orange, 50% in tangerine and 24% in grapefruit. The combined damage of the endemic *A. ludens* and *A. obliqua* and the non-native *C. capitata* in mango, *Mangifera indica* L. amounts from 15 to 20% when left without control actions (Daxl 1978; Rhode et al. 1971; Vo et al. 2003).

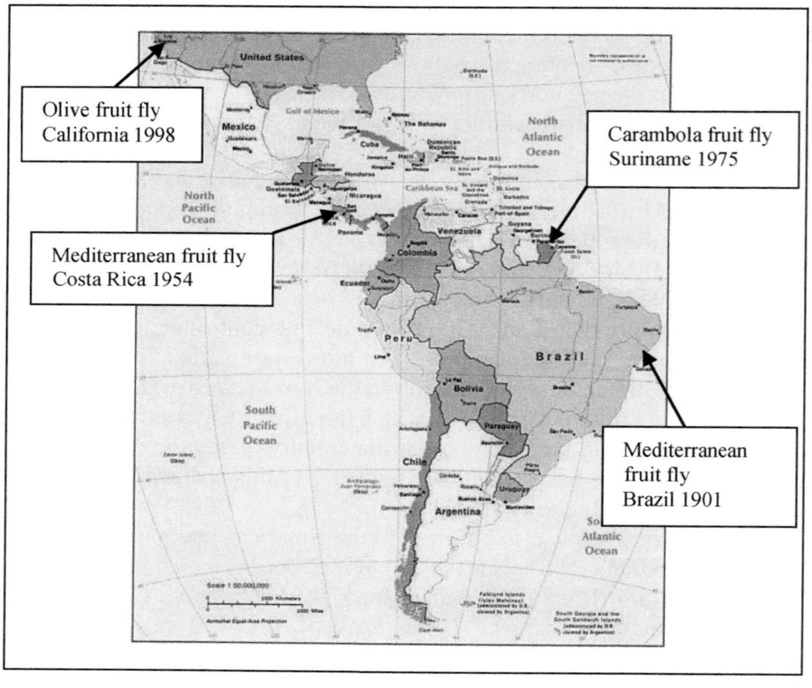

Figure 1. Introductions, establishment and spread of non-native tephritid fruit fly species in the Americas.

Figure 2. Coffee berries are a host of the Mediterranean fruit fly. The coffee belt extends from southern Mexico to Brazil (photo from W. Enkerlin).

Despite this serious fruit fly problem, as will be presented in Section 7, some countries in the Latin America and Caribbean region have been able to effectively control fruit fly pests and overcome trade barriers by establishing and maintaining fruit fly free and low prevalence areas through the systematic implementation of phytosanitary strategies, including AW-IPM strategies (Hendrichs et al. 2007), that in certain situations also integrate the Sterile Insect Technique (SIT).

3.2. Current Situation of Horticultural Production in the LAC Region

The Latin America and Caribbean region has a surface area of 19.2 million km^2 that represents nearly 13% of the Earth's surface and is currently home to 640 million inhabitants. The region has a wide range of climatic conditions, ecological zones, soil types and an overall positive balance in water supply (Peel et al. 2007).

The favourable subtropical and tropical conditions of Latin America and Caribbean countries allow them to produce fruits and vegetables for their inhabitants as well as for export markets. The Latin America and Caribbean region is a net exporter of agricultural commodities to the world, with ca. 16% of global food and agriculture exports between 2012 and 2014. The Latin America and Caribbean region has always maintained a strong comparative advantage in agricultural and in particular in horticultural production (World Bank 2013).

The sustainability of food production is a desirable path that is being promoted, and the production of seasonal vegetables, and in particular fruits in orchards, not only allows regenerative agricultural processes (soil recovery, protection from erosion, CO_2 capture) to take place, but also results in a low ranking with respect to ecological footprints among foods (Nierenberg 2018).

Many Latin America and Caribbean countries have steadily increased their fruit and vegetable exports during the past years and have enjoyed increasing investments in the production of fruit and vegetables (Prensa Libre 2019). Experts have especially increased from fruit fly free and low prevalence areas, where investments of the horticultural industry are protected from the presence of native and non-native fruit fly species (SARH/DGSV-USDA/APHIS 1990; SAG 1996; SAGAR 1999; Braga Sobrinho et al. 2004; Noe-Pino 2016).

3.3. LAC Region – The Need to Increase Horticultural Production

Despite the progress made in the past decades in fruit and vegetable production in the Latin America and Caribbean region, it is certainly insufficient to face the challenges ahead, including satisfying the demand of an increasing population. The projected increase in human population in Latin America alone will require additional production of fruit and vegetables in order to provide it with adequate diet options (Fig. 3). The countries of the Latin America and Caribbean region are suffering in different degrees the 'triple burden' of malnutrition, which consists of:

- Undernourishment, affecting 5.5% of the population in the region is (ca. 35 million)
- Micronutrient deficiencies, and
- Overweight and obesity.

Malnutrition results in an increase of non-communicable diseases, the incidence of which needs to be reduced (FAO 2017a). Fruits and vegetables are loaded with vitamins, minerals, antioxidants and fibre, which are considered to reduce many health problems, including cancer. However, fruit and vegetable consumption, even though less acute than in some other subtropical/tropical regions of the world, remains generally low (Hall et al. 2009).

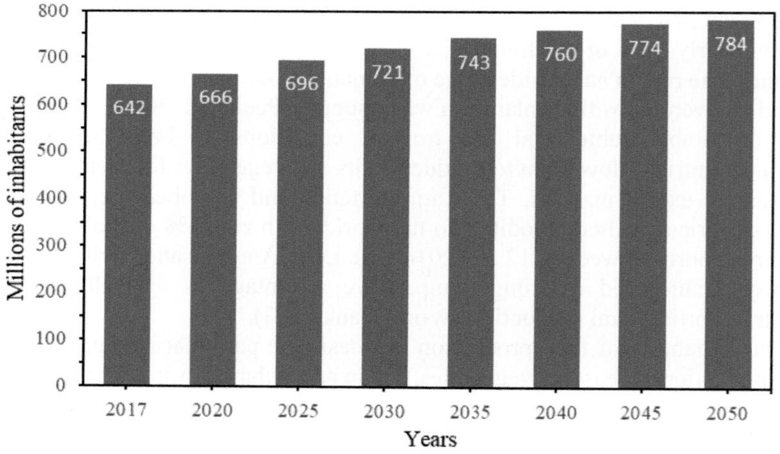

Figure 3. Projected human population increase in Latin-America 2017-2050 (based on data available at Worldometers 2019).

Fresh fruits and vegetables consumption could be increased by:

1. Subsidies to reduce the fruit and vegetable prices
2. Expanding access to healthy diets and income generation strategies (FAO 2017a), and
3. Introducing new educational programmes regarding nutrition, and modification to policies and best practices, which range from the development of eating guidelines to imposing taxes that discourage unhealthy consumption patterns (EIU 2018).

Combining all or parts of these strategic actions might be needed to address the projected increase in human non-communicable diseases, which stresses the need for visionary and timely decision-making, including policies to reduce food losses and waste.

3.4. Opportunities for Horticultural Exports

As mentioned in previous Sections, the Latin America and Caribbean region produces and exports large amounts of produce for international markets. Currently, due to the awareness of the effects of food choices on human health, more customers are changing their preferences towards more healthy food. This is also favoured by rising incomes in some countries (i.e. per capita vegetable consumption has significantly risen in China).

World-wide, there are a number of countries with a population that enjoys increased life expectancy and improved health due to the increased consumption of healthy foods. This mega-trend provides the opportunity for producers, entrepreneurs and existing companies to offer their services for these expanding export markets. Also, countries need to review their internal policies and nutrition guidelines, which combined with education and school lunches, could open additional opportunities for sustainable horticultural production.

Food safety is a key concern for production and handling of fruit and vegetable produce, as unsafe food remains a major cause of disease and death (WHO 2015). There is also an increased need for safe non-residual pest control tactics which include the area-wide application of environment-friendly and therefore sustainable tactics (e.g. the EU food safety public standards are established in the General Food Law or Regulation (EC 2002).

The Latin America and Caribbean region needs to strengthen its business sector, while embracing sustainable farming. Such actions will build up agricultural diversification, increase fruit and vegetable production and exports, and further job creation, i.e. factors that contribute towards a prosperous and stable economy.

For the past 100 years or more, the international trade of horticultural products has been subjected to phytosanitary regulations. Given the presence of fruit fly pests, post-harvest disinfestation treatments were, since the 1920's, the only alternative to overcome export barriers. Trade was prohibited for fruit and vegetable commodities for which no post-harvest treatment was available, or it was too costly, or reduced the quality of the product, or for which there was no feasibility for the establishment of a pest free area, the first of which was only recognized in 1988 in Sonora, Mexico (SARH/DGSV-USDA/APHIS 1990; SAGAR 1999; USDA 2018).

Given this situation, and with the aim of facilitating agricultural international trade where the phytosanitary risk is mitigated to an acceptable level, an alternative is the SPS agreement of the WTO that includes the possibility to combine areas of low pest prevalence with other pest mitigating measures in a "systems approach" that provide a negligible risk to the importing countries. Section 5 presents a brief description of the phytosanitary approaches available to reduce pest risk and increase the potential for exporting horticultural products that are fruit fly hosts, following ISPMs of the IPPC.

4. CHALLENGES TO PREVENT THE INTRODUCTION AND ESTABLISHMENT OF PESTS AND DISEASES AND THEIR CONTROL

In order to maintain and increase productivity and food security, in particular for middle- and low-income countries, national and regional plant protection organizations (RPPOs) need to actively participate in the prevention of transboundary movement of pests and diseases. Prevention has proven to be the most cost-effective strategy, minimizing the use of insecticides, negative environmental impacts and reducing high costs associated with remedial control practices (FAO 2017a). Prevention and rapid response might be the only way to protect crops, while other options may be limited or more expensive (i.e. desert locust prevention in western and north-western Africa is estimated at USD 3.3 million per year, while a control campaign during 2003-05 was close to USD 600 million, or equivalent to 170 years of prevention (FAO 2014).

Prevention, suppression and eradication of pests and diseases will require greater coordination at the international and regional levels to understand the risks and strategize how to deal with them. One example of such coordination and cooperation is the Practical Arrangement subscribed between the Joint FAO/IAEA Division of Nuclear Techniques in Food and Agriculture and the regional plant protection organization for Central America, namely the Organismo Internacional Regional de Sanidad Agropecuaria (OIRSA). This collaboration allows training of plant protection staff from the Central American region, through the implementation of surveillance practices and emergency response drills against incursions of invasive species of quarantine significance such as the oriental fruit fly and other major fruit fly pests of economic importance.

Other challenges include the timely detection and reporting of pest presence that can prevent the national or regional spread of pests. This is feasible through the development and adoption of regional surveillance databases that include pest alert systems and discussion platforms.

Late reporting has had unfortunate consequences, such as the carambola fruit fly introduction in Suriname in 1975 which spread from there to other areas in French Guiana and other countries in the region (Marchioro 2016).

5. APPROACHES TO OVERCOME TRADE BARRIERS

Countries need to aim at effective fruit fly surveillance and control to increase quality and production of horticultural products, and to foster opportunities for trade in international markets. To achieve these goals, the initial step is to apply basic IPPC standards; these include:

- *ISPM No. 6 "Surveillance"* that includes general surveillance aimed at providing NPPOs with elements such as phytosanitary import/export requirements, documentation on pest free areas, pest reporting and eradication strategies, and specific surveillance aimed at providing technical information such as pest detection and population dynamics in an area (FAO 2018).
- *ISPM No. 8 "Determination of a Pest Status in an Area"* aimed at providing information on the presence or absence of a pest (FAO 2017b), and
- *ISPM No. 11 "Pest Risk Analysis for Quarantine Pests"* aimed at providing details to conduct pest risk analysis (PRA) to determine if pests are of quarantine importance. It describes the integrated processes to be used for risk assessment as well as the selection of risk management options (FAO 2019a).

If it has been determined that the pest is absent from the target area (ISPM No. 8) or if the commodity of interest is not a fruit fly host, then the commodity should not be subjected to quarantine regulations for trade. Host status is assessed through applying *ISPM No. 37 "Determination of host status of fruit to fruit flies (Tephritidae)"* (FAO 2019b). For example, by applying a research protocol mutually agreed upon by the NPPOs of Mexico and the USA, findings showed that Hass avocado produced in Mexico could be recognized as a non-fruit fly host of *Anastrepha* species of concern. As a result, a quarantine that had been imposed for 82 years by USDA on Hass avocado exports was revoked in 1997, resulting in the opening of the USA market (Enkerlin et al. 1993; Aluja et al. 2004; Gutiérrez-Ruelas et al. 2013). Over one million tonnes of Hass avocado are exported annually to the USA under a bilateral workplan subscribed by the NPPOs of Mexico and the USA, generating over two billion USD per year, creating thousands of jobs and a high demand for materials and services.

If, on the contrary, the regulated pest of concern is present in the area and the commodity is a host (even if only a conditional host), then pest management strategies need to be applied against the pest to mitigate the risk posed by it to the importing country.

In the case of fruit fly pests, the IPPC has adopted a suite of fruit fly-specific ISPMs (IPPC 2017). Depending on the objective of the fruit fly control programme and the situation of the pests and hosts in the area, the following ISPMs may be applied:

1. *ISPM No. 37* on determination of host status as described above,
2. *ISPM No. 26 "Establishment of Pest Free Areas (PFA) for Fruit Flies (Tephritidae)"* (FAO 2015), and

3. *ISPM No. 35 "Systems Approach for Pest Risk Management of Fruit Flies (Tephritidae)"* (FAO 2019c). This international standard combines the application of monitoring and control practices to mitigate pest risk that may or may not include a post-harvest treatment.

These fruit fly ISPM's have a number of technical annexes which are considered to be part of the standards, such as *Annex 1 of ISPM No. 35 "Establishment of Areas of Low Pest Prevalence (ALPP) for Fruit Flies (Tephritidae)"* and *Annex 3 of ISPM No. 26 "Phytosanitary Procedures for Fruit Fly Management"*.

In addition, the fruit fly ISPMs refer to a series of (non-binding) appendices which provide additional information such as the *Appendices of ISPM No. 26 "Fruit Fly Trapping"* and *"Fruit Sampling"*.

Through its standard implementation facility, the IPPC promotes the provision of technical assistance to contracting parties with the objective of facilitating the implementation of the international standards. This can be done by the NPPOs together with stakeholders through a number of technical cooperation mechanisms, as will be presented in the following Section.

6. TECHNICAL COOPERATION MECHANISMS

Effective implementation of SPS measures requires capabilities and competencies in the public and private sectors of each country, as well as good communication and collaboration between the various public sector organizations involved and with the private sector. Typically, governments are responsible for the establishment and oversight of an enabling regulatory framework for food safety, animal health, veterinary services, plant health and/or trade, and for ensuring the compliance of agri-food exports with SPS requirements of trading partners. Ultimately, it is the private sector that plays the leading role in food and agricultural production and trade, and that is responsible for meeting SPS requirements in export markets.

The IAEA and FAO support Member States in creating capacities for implementation of SPS measures, including transferring technologies for fruit fly prevention and control through technical cooperation projects. Technology transfer regarding the area-wide application of the SIT is carried out with the support of professional staff of the Joint FAO/IAEA Division of Nuclear Techniques in Food and Agriculture.

In the Latin America and Caribbean region, this mechanism for technology transfer has been used in support of SIT technology transfer since the late 1970's when it was transferred to southern Mexico for the eradication of the invading Mediterranean fruit fly. This represented the first large-scale use of the SIT technology for fruit flies and resulted in the eradication of the pest from 800 000 ha in the state of Chiapas, Mexico (Hendrichs et al. 1983). Previously, in 1975, a small infestation of this pest in Los Angeles, California, USA, was eradicated using sterile flies reared in the USDA-ARS Hawaiian Fruit Fly Laboratory and shipped to California (Harris 1977).

Other more recent mechanisms that support capacity building and technology transfer to Member States are the Standards and Trade Development Facility (STDF 2019), as well as the Collaborating Centres (CC) scheme, through which the IAEA officially recognizes the technical capacity of specific institutions in Member States. One relevant CC is the Programa Nacional de Moscas de la Fruta in México. Through this CC, international training courses on fruit fly AW-IPM are regularly offered, expert advice is provided and technology for fruit fly surveillance and control is advanced through research and development.

Regarding the STDF, it provides a platform for organizations to come together to discuss SPS capacity building needs, share experiences and good practice, leverage additional funding, and work on coordinated and coherent solutions, including solutions to fruit fly problems. The goal of the STDF is to increase the capacity of developing countries to implement international SPS standards, guidelines and recommendations, and hence the ability to gain and maintain market access.

7. CASE STUDIES OF FRUIT FLY AW-IPM PROGRAMMES IN THE LAC REGION

Some cases of successful fruit fly area-wide programmes, several of which have been supported through the above mechanisms and international standards, are presented and described in this Section.

7.1. Guatemala-Mexico-USA Moscamed Programme for the Containment and Eradication of the Mediterranean Fruit Fly

After invading Costa Rica in 1955 and gradually spreading through Central America, most likely in infested fruits carried in small amounts by migrants moving north looking for better living conditions and through commercial trade of horticultural products throughout the region, the Mediterranean fruit fly was first detected in Guatemala in 1975. The establishment of the pest in Guatemala posed a significant threat to high value fruit and vegetable industries, as well as neighbouring countries of Belize, Mexico, and the Caribbean. It also posed a serious threat to producers in the USA.

To address this threat, the NPPOs of Guatemala, Mexico and the USA, established bi-lateral cooperative agreements and, in 1977, created the Mediterranean fruit fly eradication programme (Moscamed Programme). The Moscamed programme based its control strategy on the area-wide integration of the SIT with other methods to contain and eradicate the pest. The SIT technology was transferred to the Moscamed Programme through technical cooperation projects with the IAEA and FAO, and the technical guidance of the Joint FAO/IAEA Division as well as with the support of USDA-ARS in Honolulu, Hawaii.

A large Mediterranean fruit fly mass-rearing and sterilisation facility (with a production capacity of 500 million per week) was constructed (1977-78) in southern Mexico at Metapa, Chiapas (a state on Mexico's southern Pacific coast bordering Guatemala) with the first sterile flies released in 1979 (Fig. 4).

Figure 4. Fruit fly mass-rearing and sterilisation facilities at Metapa, Chiapas, Mexico SENASICA-SAGARPA (photo from Moscamed Programme; reproduced with permission).

Four years later (in 1982), after releasing billions of sterilized flies, the Mediterranean fruit fly was declared eradicated from approximately 800 000 ha in Chiapas (Hendrichs et al. 1983; Enkerlin et al. 2015).

A second rearing facility funded by USDA producing Mediterranean fruit fly standard strain (San Miguel Petapa) was inaugurated in 1983 in Guatemala. In 1985, a modular section was added at this location (Tween 1986).

The Petapa facility maintained its production of standard (non-genetic sexing) strains until it was superseded years later by a second and much larger mass-rearing facility in El Pino, Guatemala (Fig. 5), which was designed and constructed (1994-95) using modular design for the production of a genetic sexing strain, i.e. based on a temperature sensitive lethal (*tsl*) mutation. The mass-production capacity of this facility is up to 2000 million sterile males per week.

Figure 5. El Pino Mediterranean fruit fly mass-rearing and sterilisation facility in Guatemala (photo from Moscamed Programme; reproduced with permission).

This regional programme has greatly contributed to maintaining the biological containment barrier and the goal of protecting the Mediterranean fruit fly-free areas in Petén, Guatemala and preventing northern spread of the pest (Fig. 6) (Enkerlin et al. 2017).

To continue to maintain the barrier today, the Mediterranean fruit fly mass-rearing facilities in Mexico and Guatemala currently rear over 1.5 billion insects per week (Figs. 4 and 5). The El Pino and Metapa facilities also shipped sterile pupae to support SIT programmes in a number of countries, including preventive release programmes in California and Florida, USA; eradication programmes in Chile, Dominican Republic, Mexico (Tijuana and Manzanillo), and the USA; and, for some periods suppression programmes in Argentina (Patagonia), Israel, and more recently in Ecuador (DIR-SIT 2018). This has been done through bilateral arrangements between the NPPOs of interested countries, and some also received technical support within the framework of IAEA technical cooperation projects.

Keeping the USA and Mexico Mediterranean fruit fly-free has created favourable conditions for the development of multi-billion dollar horticulture industries in these countries and paved the way to increase production and export of fruits and vegetables from Guatemala and Belize (IICA 2013). The return on investment measured in macroeconomic terms through a benefit-cost analysis, gives an extremely favourable 150 to 1 benefit-cost ratio (BCR) in spite of the programme's annual operational cost of ca. USD 35 million (Enkerlin et al. 2015, 2017; Enkerlin 2021).

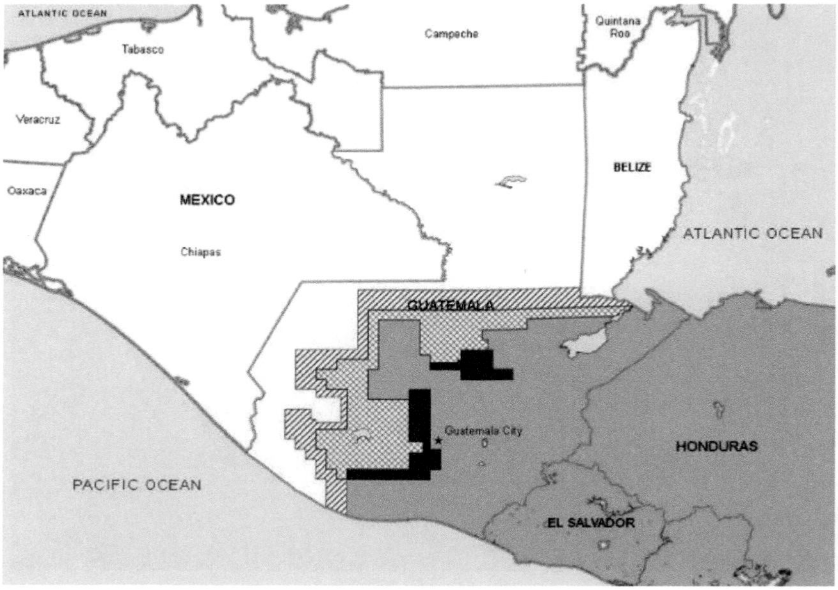

Figure 6. Location of the Mediterranean fruit fly containment barrier in Guatemala in 2015 (reproduced from Enkerlin et al. 2017).

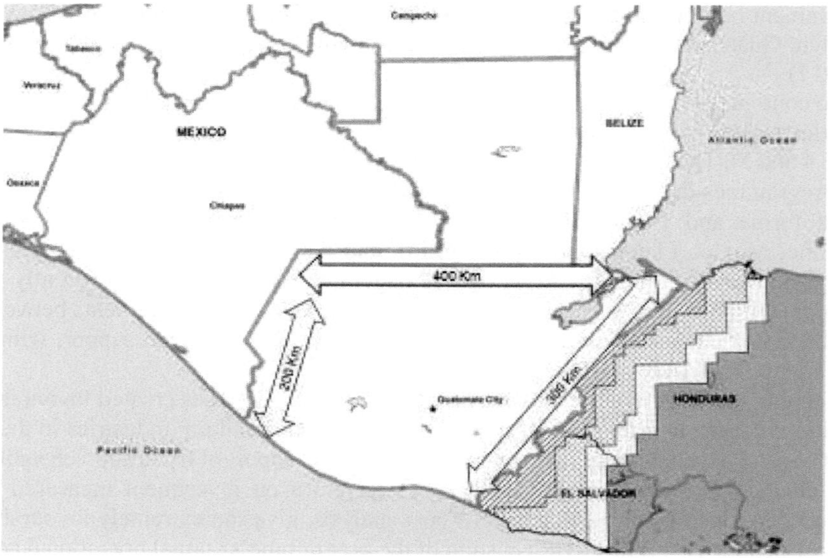

Figure 7. Potential, more sustainable, containment barrier at the El Salvador and Honduras border with Guatemala (reproduced from Enkerlin et al. 2017).

A prospective benefit-cost analysis (IICA 2013) projected for the period 2012 to 2021, with the corresponding investment, presented a scenario where using current improved technology the Mediterranean fruit fly could be eradicated from Guatemala and the containment barrier moved to the border with El Salvador and Honduras, where the Central American Isthmus begins and the length of the containment barrier would be reduced by half (Fig. 7).

Moving the containment barrier would make it more sustainable in economic terms, but also in technical terms, in view of the easier topography for programme activities, significantly reduced host areas, and quarantine measures in place at land border crossings, seaports, and airports (IICA 2013).

7.2. The Mexican and West Indian Fruit Fly Suppression and Eradication Programme – The National Fruit Fly Programme in Mexico

In 1992, the Mexican federal government (SENASICA-SAGARPA) approved the National Fruit Fly Programme for the control of indigenous fruit fly species, primarily the Mexican fruit fly and the West Indian fruit fly. The programme applies an AW-IPM approach including the SIT (Reyes et al. 2000; Gutiérrez-Ruelas et al. 2013). Strategic alliances between federal and state governments, and the horticultural industry, proved to be an effective way to operate a national programme aimed at suppressing and eradicating populations of fruit flies of economic significance for the establishment of ALPP and PFA.

For the area-wide application of the SIT against these two major *Anastrepha* species, a multi-species mass-rearing and sterilisation facility was built in Metapa, Chiapas, Mexico (Fig. 4). The federal government supplies the sterile flies and provides the infrastructure for their processing and release, the state governments contribute financial resources for operations, and the horticultural industry implements activities in the commercial orchards, including trapping and fruit sanitation in orchards. In 1997, these two fruit fly species were eradicated from more than 35 000 ha of commercial plantations of citrus, mango, apple, and peach in north-western Mexico, completely freeing the states of Chihuahua, Sonora, Baja California, and Baja California Sur of these pest insects (SAGAR/IICA 2001).

In 2001, after fruit fly eradication in north-western Mexico was officially declared and PFAs established, the direct benefits (reduced fruit fly damage and increased yield) amounted to USD 25 million per year. In addition, during the same period, the benefits obtained from the price differential paid by export markets and savings in post-harvest treatments, totalled approximately USD 35 million. Thus, the total benefits from these fruit fly PFAs over four years amounted to USD 60 million, with a total programme implementation cost of USD 4 million over this period, resulting in a benefit-cost ratio (BCR) of 7.5 to 1 (SAGAR/IICA 2001).

An economic study covering the period from 1994 to 2008, shows that the return on investment of the National Fruit Fly Programme for the mango industry resulted in a BCR of twenty-two dollars for each dollar invested (22 to 1) and a net revenue at present value (NPV) of USD 1.1 billion, and for the citrus industry a BCR of 19 to 1 and a NPV of USD 2.0 billion (IICA et al. 2010).

7.3. Chile's National Fruit Fly Programme

Chile's fruit fly-free status has allowed the development of one of the most important export-oriented horticultural industries in the world. To obtain this status and protect this valuable asset, the Government of Chile, through the Agricultural and Livestock Service (SAG) of the Ministry of Agriculture, created Chile's National Fruit Fly Programme in 1980. Its objectives have been to free Chile of the Mediterranean fruit fly, established in the north, and to prevent the introduction and establishment of any fruit fly species of economic importance, including the Mediterranean fruit fly and other pest species of the genera *Anastrepha* and *Bactrocera* anywhere else in Chile (Olalquiaga and Lobos 1993; Rodríguez et al. 2016).

The National Fruit Fly Programme in Chile operates through a centralised organizational structure under the Ministry of Agriculture, that includes a mass-rearing facility in Arica near the border with Peru (Fig. 8).

Figure 8. Mediterranean fruit fly mass-rearing and sterilisation facility in Arica, Chile (photo from R. Rodríguez SAG Chile; reproduced with permission).

In addition, as part of a regional approach to the fruit fly problem, the Government of Chile has also subscribed bi-national agreements with neighbouring Argentina and Peru (Wedekind 2007). Chile has achieved and maintained its fly-free status by implementing two major strategic activities:

- An effective national and international quarantine system (including interprovincial quarantine road stations when there is an outbreak and international quarantine at ports of entry), and an extensive and highly sensitive fruit fly-trapping network to detect fruit fly introductions at an early stage. Numerous outbreaks of non-native fruit flies, mainly the Mediterranean fruit fly, have been eradicated through the effective execution of an emergency eradication plan based on detecting and eradicating infestations (McInnis et al. 2017; Shelly et al. 2017). A *B. dorsalis* outbreak on Easter Island was eradicated in 2011 at a cost of USD 100 000 (AGROMEAT 2011; FAO/IAEA 2011).
- In Arica province, at the border with Peru, a containment barrier that integrates the release of sterile males (Fig. 8) to avoid the natural or artificial spread of Mediterranean fruit fly populations into northern Chile, in order to protect the main fruit and vegetable production areas in the central and southern parts of the country.

After six years of an intensive integrated area-wide programme based on the SIT, the Mediterranean fruit fly was eradicated from Arica province in 1995, and all of Chile was declared a fruit fly-free country (SAG 1995, 1996). USDA-APHIS recognized Chile as a pest free area for Mediterranean fruit fly in 2010 (Federal Register 2010).

As a result of the fruit fly-free status, exports have grown to 320 million boxes of fruits per year, mainly table grapes, apples, stone fruits, kiwis, and avocados, valued in 2016 at USD 4000 million (ASOEX 2018).

7.4. Patagonia, Argentina – Mediterranean Fruit Fly PFA

A programme to eradicate Mediterranean fruit fly from fruit production areas in northern Patagonia was launched in late 1996 by the Programa de Control y Erradicación de Mosca del Mediterráneo (PROCEM-SENASA) and the Fundación Barrera Zoofitosanitaria Patagónica (FUNBAPA) (Borges et al. 2016). Mediterranean fruit fly eradication activities started in 2001 and concluded in 2004 with the official declaration of Patagonia as a Mediterranean fruit fly-free area (Guillen and Sanchez 2007).

Trading partners, including Chile, Mexico and the USA, recognized Patagonia as a Mediterranean fruit fly-free area (Borges et al. 2016). Eradication was achieved through an intensive area-wide programme integrating the SIT. Sterile flies were shipped from the mass-rearing and sterilisation facility located in the Province of Mendoza, Argentina (De Longo et al. 2000). As a result of the eradication of Mediterranean fruit fly, costly quarantine treatments could be eliminated for most of the three million boxes of quality pears and apples that this region exports yearly. Other mayor benefits included gaining access to previously closed markets (Villareal et al. 2018). Of fundamental importance to protect this pest free area is the extensive quarantine barrier that is effectively operated by FUMBAPA (Fig. 9) (Wedekind 2007).

Figure 9. Inspection at a FUNBAPA quarantine road station in Patagonia, Argentina (photo from E. Rial, PROCEM Patagonia, reproduced with permission).

7.5. Mediterranean Fruit Fly-Free Places and Sites of Production, Honduras, Central America

Through a careful review of international phytosanitary standards, the National Health and Agrifood Safety of Honduras (SENASA) determined that the pest risk mitigation scheme that could apply to the melon (*Cucumis melo* L.) production sites of Montelíbano (400 ha) and Santa Rosa (800 ha) was ISPM 10 "*Pest free places of production and pest free production sites*". Following international fruit fly trapping guidelines (Appendix 1 of ISPM 26; IAEA 2013), SENASA established a fruit fly surveillance network for these sites in July 2011 (Noe-Pino 2016). Trapping results clearly confirmed the absence of fruit fly pests in the areas of interest. These results, and the fact that melon is recognized as a conditional host of the target fruit fly species, were the critical technical factors used in the bilateral negotiations between the national phytosanitary authorities of Honduras and Taiwan that resulted in an agreement to export melons using a pest risk mitigation scheme based on pest free production sites (Fig. 10).

A major advantage of this pest risk mitigation scheme is that no internal quarantine checkpoints are required, and places and sites of production need to be fruit fly free only during the entire fruit production and harvest period (FAO/IAEA 2017a).

Figure 10. Melons from fruit fly free places of production in Honduras (photo from SENASA Honduras, reproduced with permission).

7.6. Establishment of Fruit Fly ALPP and PFA in Central America

For the past decades, countries in Central America have been affected by low international prices of the traditional export crops coffee, banana and sugarcane. The governments of these countries and Panama have therefore been seeking new alternatives for international trade through production and export of non-traditional fruit and vegetables. To assists them in this task, from 2001 to 2006, IAEA and FAO provided support through a regional technical cooperation project to strengthen the countries' phytosanitary framework to allow them to establish fruit fly ALPP and fruit fly PFA using an AW-IPM approach that included, in some cases, the SIT (Reyes et al. 2007).

To achieve the goals, an approach was proposed to overcome existing constraints by integrating three main elements: 1) the development of a multi-institutional strategic alliance, 2) the use of pilot areas as a territorial strategy for suppression and eradication of fruit flies, and 3) a focus on promoting the export of fruits and vegetables.

The project outcome included: 1) the establishment of a number of fruit fly ALPP and PFAs in each of the participating countries, 2) investment by the fruit and vegetable industries in Costa Rica, El Salvador, Guatemala, and Nicaragua and of around USD 150 million in support of establishing and maintaining areas of fruit fly low prevalence through a systems approach for exports of tomatoes and bell peppers to the USA, and 3) exports of papaya from Mediterranean fruit fly-free areas in the Department of Petén, in northern Guatemala, without the need for quarantine treatments (Fig. 11).

This project demonstrates that exports of non-traditional fruits and vegetables are a viable economic alternative to the traditional crops in the region by establishing fruit fly PFAs and ALPP integrated with a systems approach and creating more rural jobs than traditional crops (Reyes et al. 2007). Attempts to establish more of these areas throughout the region would be successful if:

1. The Ministries of Agriculture are the driving force of any such area-wide initiatives
2. The horticultural industry is convinced of the potential benefits that these areas can bring and is an active partner in the activities, and
3. Alliances are established between technical and financing organizations present in the region and they commit to working together sharing a common vision.

Strategic approaches for fruit fly control, which focus on specifically selected horticultural production areas, are in some instances easier to implement and more realistic than approaches which aim to initiate extensive and costly suppression and eradication programmes.

Figure 11. Papayas from Mediterranean fruit fly-free area in Petén, Guatemala (photo from Moscamed Programme; reproduced with permission).

7.7. Mediterranean Fruit Fly Eradication from the Dominican Republic

The presence of the Mediterranean fruit fly in the Dominican Republic was officially reported in March 2015. The pest had already spread to 2053 km^2 in the eastern part of the country, constituting a major outbreak in the Caribbean that up to then had been free of the Mediterranean fruit fly. An immediate ban on most exports of fruit and vegetables was imposed by trading partners, causing a loss of over USD 40 million for the remaining nine months of 2015.

Given the emergency situation, the Ministry of Agriculture established the Moscamed-DR Programme, with adequate financial resources and an effective organizational structure for its coordination and operations. The Guatemala-Mexico-USA Moscamed Programme and international organizations, including the FAO, IAEA, IICA (Inter-American Institute for Cooperation on Agriculture) and OIRSA, joined efforts with the Ministry of Agriculture to address the Mediterranean fruit fly outbreak. A technical advisory committee of experts provided oversight throughout the eradication campaign. An AW-IPM approach, including the application of the SIT as a final component, was used to eradicate the pest.

Official eradication was announced in July 2017 after six Mediterranean fruit fly generations of zero catches (Zavala-López et al., this volume). The Dominican Republic is now on the list of countries that have successfully eradicated the Mediterranean fruit fly, thereby avoiding the establishment of a major pest in the Caribbean, and substantially strengthening its fruit fly surveillance system and emergency response capacity (Fig. 12).

Figure 12. Plaque presented in July 2017 by the IAEA to the Minister of Agriculture of the Dominican Republic as a recognition for his leadership in eradicating the Mediterranean fruit fly (photo from W. Enkerlin).

As a spin-off of the successful eradication of the Mediterranean fruit fly, the Ministry of Agriculture of the Dominican Republic established a national fruit fly programme to manage native fruit flies, and to maintain the surveillance and response capacities for invasive fruit flies and other pests.

7.8. Carambola Fruit Fly Containment Programme in Guyana, Surinam, French Guiana and Brazil

The carambola fruit fly *B. carambolae* is native from Southeast Asia. It is known to infest a wide range of fruits and vegetables including carambola, guava, mango and others. The pest was first detected in the Americas in Suriname in 1975. Given its invasiveness and the risk it represented to the horticultural industry in Suriname and neighbouring countries, including Brazil, an eradication programme based on the male annihilation technique (MAT) was launched (Malavasi et al. 2000). The programme was funded by the International Fund for Agricultural Development (IFAD), France, the Netherlands, and the USA, and officially started in 1998 (Midgarden et al. 2016). In addition, during 1994-95, FAO provided capacity building assistance to the Brazilian Ministry of Agriculture (MAPA) to strengthen exclusion, detection and emergency response to new pest introductions. This enabled MAPA to train personnel, install detection traps within the state of Amapa, and take emergency response measures whenever detections of CFF occurred.

Considering the transboundary nature of this pest, programme operations were implemented on a regional level, including activities in Suriname, French Guiana, Guyana, and states in north-eastern Brazil. As a result, by 2001, the distribution of *B. carambolae* was reduced to limited areas of Suriname and French Guiana (Midgarden et al. 2016). Containing the spread of the pest resulted in 1) important economic benefits in reduced direct damage to fruits and vegetables, 2) social benefits by protecting important jobs associated to the horticultural industry, and 3) environmental benefits by preventing the massive use of insecticides that would be needed to control the pest.

Despite these results, in 2002 some of the donors significantly reduced their funding and the programme had to close in 2003. As a result, the pest reinvaded areas that had already been freed. Since then, only Brazil has continued implementing and financing the programme to contain its advance into Brazil. Nevertheless, in the following years *B. carambolae* expanded its distribution with detections as far southeast as Curralinho in the state of Pará in Brazil (Fig. 13). The presence of the pest in Guyana means that continuous incursions into both northern Brazil and Venezuela are likely, if not inevitable (Godoy et al. 2020).

Closing this programme before its completion has resulted in increasing costs to South American agriculture, and increased risk to Central and North America and the Caribbean. A specimen of *B. carambolae* was identified in Puerto Rico in 2015, and two putative *B. carambolae* specimens were trapped in Orlando, Florida in 2008 (Midgarden et al. 2016).

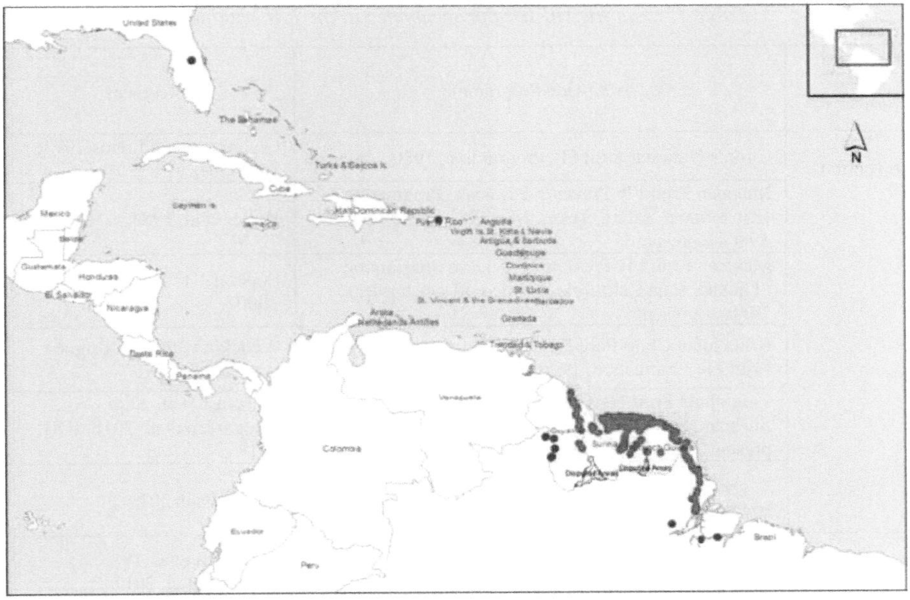

Figure 13. Presence of carambola fruit fly in South America up to 2016 (red dots present, blue dots detections eradicated) (reproduced from Midgarden et al. 2016).

A recent economic assessment shows a BCR of up to 37 to 1 for eradicating *B. carambolae* from the currently infested areas and from preventing further spread and invasion of the free areas in Brazil (IDB 2018). Based on this, in 2018, the Government of Brazil jointly with the Interamerican Development Bank (IDB) commissioned the preparation of a regional project for the control and eradication of the carambola fruit fly (IDB 2018). A coordinated programme amongst infested countries could still mitigate the risk of the spread of *B. carambolae* in the region (Midgarden et al. 2016).

Area-wide fruit fly control programmes in Latin America that apply an AW-IPM approach are listed in Table 1.

8. FUTURE PERSPECTIVES

- Given its devastating effects in the region, the fruit fly problem should receive high priority on the agendas of the Ministerial Agricultural Organizations, namely the Comité Internacional Regional de Sanidad Agropecuaria (CIRSA), the Consejo Agropecuario del Sur (CAS), and the Caribbean Agricultural Health and Food Safety Agency (CAHFSA).

Table 1. Fruit fly AW-IPM programmes in the LAC region

Strategic Objective	Area-wide Programme	References
Prevention	Chile's National Fruit Fly Programme, 1980 – present	Olalquiaga and Lobos 1993; Rodríguez et al. 2016
	Mexican Fruit Fly Preventive Release Programme (Rio Grande Valley, Texas, Mexico – USA border), 1980s – present	Holler et al. 1984
	Mexican Fruit Fly Preventive Release Programme (Tijuana, Baja California, Mexico – USA border), 1960s – present	Lopez 1970; Dowell et al. 2000
	Bi-national Chile-Peru Programme for Mediterranean Fruit Fly Eradication, 1996 – present	Wedekind 2007; Rodríguez et al. 2016
	Carambola Fruit Fly Containment Programme in Surinam, French Guiana, Guyana, and Brazil, 1998 – present	Malavasi et al. 2000; Midgarden et al. 2016; IDB 2018
	A. grandis prevention programme in north-eastern Brazil to protect PFA, 1990 – present	Razera Papa 2019
Eradication	Guatemala-Mexico-USA Moscamed Programme for the Containment and Eradication of the Mediterranean Fruit Fly, 1975 – present	Hendrichs et al. 1983; Enkerlin et al. 2017
	Mediterranean Fruit Fly Eradication Programme "PROCEM" (Patagonia – Mendoza – San Juan, Argentina), 1992 – present	De Longo et al. 2000; Guillen and Sanchez 2007; Wedekind 2007; Borges et al. 2016; Quiroga et al. 2016
	Mediterranean Fruit Fly-Free Places and Sites of Production in Honduras, 2017 – present	Noe Pino 2016
	Mediterranean Fruit Fly Eradication Programme (Altagracia, Dominican Republic), 2015 – 2017	Zavala et al., this volume
Suppression	Mediterranean Fruit Fly Control Programme (Peru), 1970s – present	Guillen and Quintanilla 2008; Rivera-Tejada 2011; Manrique and Rivera 2016
	Fruit fly control programme in Ecuador in localized production sites, 2018 – present	Vilatuña 2018
	Plan Nacional Moscas de la Fruta (PNMF) in Colombia, 2008 – present	Arevalo 2016
	Programa Nacional de Control y Erradicación de Mosca de los Frutos in north-eastern Argentina – PROCEM NEA, 2015 – present	Morilla et al. 2016
	Moscasul programme pilot project to suppress the South American fruit fly in southern Brazil, 2014 – present	Kovaleski and Mastrangelo, this volume
	Mediterranean fruit fly suppression on table grape export areas along Rio San Francisco irrigation zone (Bahia/Pernambuco), 2015 – present	Baronio et al. 2018
Suppression or Eradication	The Mexican and West Indian Fruit Flies Suppression and Eradication Programme – The National Fruit Fly Programme in Mexico, 1991 – present	Reyes et al. 2000; Gutierrez et al. 2013; Liedo et al., this volume
	Establishment of Fruit Fly ALPP and PFA in Central America, 2007 – present	Reyes et al. 2007

- These regional government organizations and their institutions should show strong political will by mobilizing and committing national resources and establishing regional phytosanitary policies that enable the enforcement of phytosanitary strategies such as the establishment of fruit fly PFAs and ALPP.
- The policies and actions should be enforced through instruments such as "Regional Fruit Fly Strategic Plans" against endemic fruit fly species, as well as against invasive species of quarantine significance. This would provide a systematic framework and guidance on the necessary actions required to achieve the objectives in controlling fruit fly pests in the Latin America and Caribbean region.
- Given the transboundary nature of the fruit fly problem in the Latin America and Caribbean region, RPPOs should have a more proactive role in preventing the introduction and establishment of invasive fruit fly pests and providing guidance on the phytosanitary approaches available for effective control of endemic fruit flies. They should promote the implementation of the ISPMs and propose/coordinate specific actions together with the NPPOs and main beneficiaries and stakeholders, including the growers and exporters.
- To address climate change and globalization that induce new pest problems, current legislation and policies need to be amended. Among these is the need to support the development of innovative control approaches to mitigate and manage these biological incursions, and to contain the geographic expansion of non-indigenous pest populations.
- The surveillance systems and emergency response capacities need to be strengthened on a regional basis to detect early introductions of non-native species still not present and that are of quarantine significance.
- The strategic approach against endemic fruit flies of economic significance should be, in most cases, the establishment of carefully selected areas free of pests or at low pest prevalence levels from which horticultural products can be produced and sold, rather than major eradication or suppression programmes that extend over very large areas, sometimes entire countries.
- Countries from the Latin America and Caribbean region should consider the possibility of promoting increased production and consumption of horticultural products in order to reduce the incidence of non-communicable diseases, plan for expected population growth, and take advantage of the trends in consumer preference to generate commercial opportunities to strengthen their economic performance.
- Countries interested in applying area-wide the SIT should take advantage of the existing sterile fly production capacity in the region (over 2.5 billion sterile flies per week) rather than each country aiming at building their own rearing and sterilisation facility. The investment should be focused on building sterile fly emergence and release facilities (FAO/IAEA 2017b), to hold and feed adult flies emerged from sterile pupae purchased from reliable external sources. The capital investment of such an approach would be much lower, as well as the risks of successfully integrating the SIT.

9. REFERENCES

AGROMEAT. 2011. SAG erradicó mosca oriental de la fruta en Isla de Pascua. Con éxito terminó la campaña de erradicación de *Bactrocera dorsalis* que se desarrollaba en el territorio insular desde diciembre de 2010. 4 de mayo de 2011. Buenos Aires, Argentina.

Aluja, M., F. Diaz-Fleisher, and J. Arredondo. 2004. Non-host status of commercial *Persea americana* cultivar "Hass" to *Anastrepha ludens*, *Anastrepha obliqua*, *Anastrepha serpentina*, and *Anastrepha striata* (Diptera: Tephritidae) in Mexico. Journal of Economic Entomology 97: 293–309.

Arevalo, P. E. 2016. Plan nacional moscas de la fruta (PNMF) en Colombia, pp. 111. *In* Book of Abstracts, 9a Reunión del Grupo de Trabajo en Moscas de la Fruta del Hemisferio Occidental, 16-22 October 2016, Buenos Aires, Argentina.

(ASOEX) Asociación de Exportadores de Frutas de Chile, A. G. 2018. Frutas de Chile: Plan 2018-2019. Cruz del Sur N°133, Piso 2, Las Condes, Santiago de Chile.

Baronio, C. A., D. Bernardi, B. A. J. Paranhos, F. R. M. Garcia, and M. Botton. 2018. Population suppression of *Ceratitis capitata* (Wiedemann) on table grapes using toxic baits. Anais da Academia Brasileira de Ciências 90 (4) October/December 2018, Rio de Janeiro, Brasil.

Bhattarai, G. P., and J. T. Cronin. 2014. Hurricane activity and the large-scale pattern of spread of an invasive plant species. PLoS One 9(5): e98478.

Bjelis, M., L. Popovic, M. Kiridzija, G. Ortiz, and R. Pereira. 2016. Suppression of Mediterranean fruit fly using the Sterile Insect Technique in Neretva river valley of Croatia, pp. 29–45. *In* B. Sabater-Muñoz, T. Vera, R. Pereira, and W. Orankanok (eds.), Proceedings of the 9th International Symposium on Fruit Flies of Economic Importance, Bangkok, Thailand.

Borges, C. A., A. P. Mongabure, C. Stazionati, and E. Rial. 2016. Patagonia Argentina, 11 años de exportaciones como área libre de mosca de los frutos, pp. 110. *In* Book of Abstracts, 9a Reunión del Grupo de Trabajo en Moscas de la Fruta del Hemisferio Occidental, 16–22 October 2016, Buenos Aires, Argentina.

Boyce, A. M. 1934. Bionomics of the walnut husk fly, *Rhagoletis completa*. Hilgardia 8(11): 363–579. University of California, Berkeley.

Braga Sobrinho, R., R. N. Lima, M. A. Peixoto, and A. L. M. Mesquita. 2004. South American cucurbit fruit fly-free area in Brazil, pp. 173–177. *In* B. N. Barnes (ed.), Proceedings of the 6th International Symposium on Fruit Flies of Economic Importance, 6-10 May 2002, Stellenbosch, South Africa. Isteg Scientific Publications, Irene, South Africa.

Daxl, R. 1978. Mediterranean fruit fly ecology in Nicaragua and a proposal for integrated control. FAO Plant Protection Bulletin 26(4): 150–157.

(DEFRA) Department for Environment, Food & Rural Affairs. 2018. Pest specific plant health response plan: *Rhagoletis pomonella* (apple maggot fly). Sands Hutton, York, UK.

De Longo, O., A. Colombo, P. Gomez-Riera, and A. Bertolucci. 2000. The use of massive SIT for the control of the medfly, *Ceratitis capitata* (Wied.), strain SEIB 6-96, in Mendoza, Argentina, pp. 351–359. *In* K. H. Tan (ed.), Proceedings: Area-wide control of fruit flies and other insect pests. International Conference on Area-wide Control of Insect Pests, and the 5th International Symposium on Fruit Flies of Economic Importance, 28 May–5 June 1998, Penang, Malaysia. Penerbit Universiti Sains Malaysia, Pulau Pinang, Malaysia.

DIR-SIT. 2018. World-Wide Directory of SIT Facilities (DIR-SIT). Joint FAO/IAEA Division of Nuclear Techniques in Food and Agriculture. Vienna, Austria.

Dowell, R. V., I. A. Siddiqui, F. Meyer, and E. L. Spaugy. 2000. Mediterranean fruit fly preventative release programme in southern California, pp. 369–375. *In* K. H. Tan (ed.), Proceedings: Area-wide control of fruit flies and other insect pests. International Conference on Area-wide Control of Insect Pests, and the 5th International Symposium on Fruit Flies of Economic Importance, 28 May–5 June 1998, Penang, Malaysia. Penerbit Universiti Sains Malaysia, Pulau Pinang, Malaysia.

(EC) European Communities. 2002. EC Regulation No 178/2002 of the European Parliament and of the Council of 28 January 2002, laying down the general principles and requirements of food law, establishing the European Food Safety Authority and laying down procedures in matters of food safety. Official Journal of the European Communities L 31/1.

Ekesi, S., S. A. Mohamed, and M. de Meyer (eds). 2016. Fruit fly research and development in Africa: Towards a sustainable management strategy to improve horticulture. Springer International Publishing, Switzerland. 778 pp.

Enkerlin, W. 2021. Impact of fruit fly control programmes using the Sterile Insect Technique, pp. 979–1006. In V. A. Dyck, J. Hendrichs, and A. S. Robinson (eds.), Sterile Insect Technique – Principles and practice in Area-Wide Integrated Pest Management. Second Edition. CRC Press, Boca Raton, Florida, USA.

Enkerlin, D., L. García R., and F. López M. 1989. Mexico, Central and South America, pp. 83–90. *In* A. S. Robinson and G. Hooper (eds.), World crop pests, Volume 3A. Fruit flies. Their biology, natural enemies and control. Elsevier, Amsterdam, The Netherlands.

Enkerlin, W., J. Reyes, A. Bernabe, J. L. Sánchez, J. Toledo, and M. Aluja. 1993. El aguacate "Hass" como hospedante de tres especies de *Anastrepha* (Diptera: Tephritidae), en condiciones forzadas y naturales. Agrociencia, Serie Protección Vegetal 4: 329–348.

Enkerlin, W., J. M. Gutiérrez-Ruelas, A. Villaseñor Cortes, E. Cotoc Roldan, D. Midgarden, E. Lira, J. L. Zavala López, J. Hendrichs, P. Liedo, and F. J. Trujillo Arriaga. 2015. Area freedom in Mexico from Mediterranean fruit fly (Diptera: Tephritidae): A review of over 30 years of a successful containment program using an integrated area-wide SIT approach. Florida Entomologist 98: 665–681.

Enkerlin, W. R., J. M. Gutiérrez Ruelas, R. Pantaleon, C. Soto Litera, A. Villaseñor Cortes, J. L. Zavala Lopez, D. Orozco Davila, P. Montoya Gerardo, L. Silva Villarreal, E. Cotoc Roldan, F. Hernandez Lopez, A. Arenas Castillo, D. Castellanos Dominguez, A. Valle Mora, P. Rendon Arana, C. Caceres Barrios, D. Midgarden, C. Villatoro Villatoro, E. Lira Prera, O. Zelaya Estrade, R. Castaneda Aldana, J. Lopez Culajay, F. Ramírez y Ramírez, P. Liedo Fernández, G. Ortiz Moreno, J. Reyes Flores, and J. Hendrichs. 2017. The Moscamed Regional Programme: A success story of area-wide Sterile Insect Technique application. Entomologia Experimentalis et Applicata 164(Special Issue): 188–203.

(EIU) Economist Intelligence Unit. 2018. Fixing Food 2018. Best practices towards the sustainable development goals. Report by Sarah Murray, edited by Martin Koehring of The EIU. November 2018. 44 pp.

(FAO) Food and Agriculture Organization of the United Nations. 2014. Transboundary plant pests and diseases: Management and challenges. Committee on Agriculture (COAG), 24th session, 29 September–3 October 2014. Rome, Italy.

(FAO) Food and Agriculture Organization of the United Nations. 2015. Fruit fly pest free areas. International Standard for Phytosanitary Measures (ISPM) No. 26, International Plant Protection Convention (IPPC). FAO, Rome, Italy.

(FAO) Food and Agriculture Organization of the United Nations. 2017a. The future of food and agriculture. Trends and challenges. Rome, Italy. 180 pp.

(FAO) Food and Agriculture Organization of the United Nations. 2017b. Determination of pest status in an area. International Standard for Phytosanitary Measures (ISPM) No. 8, International Plant Protection Convention (IPPC). FAO, Rome, Italy.

(FAO) Food and Agriculture Organization of the United Nations. 2018. Surveillance. International Standard for Phytosanitary Measures (ISPM) No. 6, International Plant Protection Convention (IPPC). FAO, Rome, Italy.

(FAO) Food and Agriculture Organization of the United Nations. 2019a. Pest risk analysis for quarantine pests. International Standard for Phytosanitary Measures (ISPM) No. 11. International Plant Protection Convention (IPPC). FAO, Rome, Italy.

(FAO) Food and Agriculture Organization of the United Nations. 2019b. Determination of host status of fruit to fruit flies (Tephritidae). International Standard for Phytosanitary Measures (ISPM) No. 37, International Plant Protection Convention (IPPC). FAO, Rome, Italy.

(FAO) Food and Agriculture Organization of the United Nations. 2019c. Systems approach for pest risk management of fruit flies (Tephritidae). International Standard for Phytosanitary Measures (ISPM) No. 35, International Plant Protection Convention (IPPC). FAO, Rome, Italy.

(FAO/IAEA) Food and Agriculture Organization of the United Nations/International Atomic Energy Agency. 2011. Oriental fruit fly eradicated from Easter Island, Chile, pp. 28–29. *In* Insect Pest Control Newsletter No. 77. Vienna, Austria.

(FAO/IAEA) Food and Agriculture Organization of the United Nations/International Atomic Energy Agency. 2017a. Montelíbano and Santa Rosa Mediterranean fruit fly free places and sites of production, Honduras, Central America, pp 32–33. *In* Insect Pest Control Newsletter No. 89. Vienna, Austria.

(FAO/IAEA) Food and Agriculture Organization of the United Nations/International Atomic Energy Agency. 2017b. Guideline for packing, shipping, holding and release of sterile flies in area-wide fruit fly control programmes. J. L. Zavala-López, and W. R. Enkerlin (eds.), Food and Agriculture Organization of the United Nations. Rome, Italy. 155 pp.

(FAO/IAEA) Food and Agriculture Organization of the United Nations/International Atomic Energy Agency. 2019. A guide to the major pest fruit flies of the world. R. Piper, R. Pereira, J. Hendrichs, W. Enkerlin and M. De Meyer (eds.). Copyright Scientific Advisory Services Pty Ltd, Mourilyan, Australia. 44 pp.

(FAO/IFAD/UNICEF/WFP/WHO) Food and Agriculture Organization of the United Nations/International Fund for Agricultural Development/United Nations Children's Fund/World Food Programme/World Health Organization. 2017. The state of food security and nutrition in the world 2017. Building resilience for peace and food security. Rome, Italy.

Federal Register. 2010. Notice of determination of pest-free areas in the Republic of Chile. Volume 75, No. 111, June 10, 2010. Animal and Plant Health Inspection Service, United States Department of Agriculture, Washington, DC, USA.

Gobierno de México. 2018. Anuario estadístico de la producción agrícola. Servicio de Información Agroalimentaria y Pesquera. Secretaria de Agricultura y Desarrollo Rural. México, DF, México.

Godoy, M. J. S., W. S. Pinto, C. A. Brandão, C. V. Vasconcelos, and J. M. Pires. 2020. Eradication of an outbreak of *Bactrocera carambolae* (carambola fruit fly) in the Marajo Archipelago, State of Para, Brazil, pp. 315–323. *In* D. Perez-Staples, F. Diaz-Fleischer, P. Montoya, and M. T. Vera (eds.), Area-Wide Management of Fruit Fly Pests. CRC Press, Boca Raton, Florida, USA.

Guillen, D., and R. Sanchez. 2007. Expansion of the national fruit fly control programme in Argentina, pp. 653–660. *In* M. J. B. Vreysen, A. S. Robinson, and J. Hendrichs (eds.), Area-wide control of insect pests: From research to field implementation. Springer, Dordrecht, The Netherlands.

Guillen E. R. E., and M. G. F. S. Quintanilla. 2008. Sistema integrado de información de moscas de la fruta del Perú, pp. 15. *In* 7a Reunión del Grupo de Trabajo en Moscas de la Fruta del Hemisferio Occidental. Sinaloa, México.

Gutiérrez-Ruelas, J. M., G. S Martínez, A. Villaseñor Cortes, W. R. Enkerlin, and F. Hernández López. 2013. Los programas de moscas de la fruta en México. Su historia reciente. Talleres de S y G Editores. México, DF, México. 89 pp.

Gutiérrez-Samperio, J. 1976. La mosca del Mediterráneo *Ceratitis capitata*, y los factores ecológicos que favorecerían su establecimiento y propagación en México. Secretaría de Agricultura y Recursos Hidraúlicos, Dirección General de Sanidad Vegetal. México, DF, México.

Hall, J. N., S. Moore, S. B. Harper, and J. W. Lynch. 2009. Global variability in fruit and vegetable consumption. American Journal of Preventive Medicine 36: 402–409.

Harris, E. J. 1977. The threat of the Mediterranean fruit fly to American agriculture and efforts being made to counter this threat. Proceedings Hawaiian Entomological Society XXII (3): 475–480.

Hendrichs, J., Ortíz G., Liedo P., Schwarz A. 1983. Six years of successful medfly program in Mexico and Guatemala, pp. 353–365. *In* R. Cavalloro (ed.), Fruit flies of economic importance. A. A. Balkema, Rotterdam, The Netherlands.

Hendrichs, J., A. S. Robinson, P. Kenmore, and M. J. B. Vreysen. 2007. Area-wide Integrated Pest Management (AW-IPM): Principles, practice and prospects, pp. 3–33. *In* M. J. B. Vreysen, A. S. Robinson, and J. Hendrichs (eds.), Area-wide control of insect pests. From research to field implementation. Springer, Dordrecht, The Netherlands.

Holler, T., J. Davidson, A. Suárez, and R. García. 1984. Release of sterile Mexican fruit flies for control of feral populations in the Rio Grande Valley of Texas and Mexico. Journal of the Rio Grande Horticultural Society 37: 113–121.

Hulme, P. E. 2009. Trade, transport and trouble: Managing invasive species pathways in an era of globalization. Journal of Applied Ecology 46: 10–18.

Hunter, M. C., R. G. Smith, M. E. Schipanski, L. W. Atwood, and D. A. Mortensen. 2017. Agriculture in 2050: Recalibrating targets for sustainable intensification. BioScience 67: 386–391.

(IDB) Inter-American Development Bank. 2018. Diagnóstico de la situación actual y propuestas para el control y erradicación de la mosca de la carambola en Brasil. Prepared by a Technical Team MAPA-BID. March–April 2018. Ministerio de Agricultura, Pecuaria y Abastecimiento Secretaría de Defensa Agropecuaria - Dirección de Protección Vegetal (MAPA-SDA-DSV) Brasil, Banco Interamericano de Desarrollo (BID). pp. 109.

(IICA) Instituto Interamericano de Cooperacion para la Agricultura. 2010. Evaluación económica de la campaña nacional contra moscas de la fruta en los Estados de Baja California, Guerrero, Nuevo León, Sinaloa, Sonora, y Tamaulipas (1994–2008). D. Salcedo, J. R. Lomelí, G. H. Terrazas, and E. Rodríguez-Leyva (eds.). Septiembre 2010. IICA, México. 204 pp.

(IICA) Instituto Interamericano de Cooperación para la Agricultura. 2013. Evaluación económica del Programa Moscamed en Guatemala y sus impactos en ese país, México, Estados Unidos y Belice. D. Salcedo, J. R. Lomelí, G. H. Terrazas, J. Suarez, and E. Muniz (eds.). Junio 2013. IICA, México. 188 pp.

(IPPC) International Plant Protection Convention. 2017. Teamwork award from FAO-AG Department to IPPC-IAEA joint team on the fruit fly standards. International Plant Protection Convention (IPPC). FAO, Rome, Italy.

Lopez, F. 1970. Sterile Insect Technique for the eradication of the Mexican and Caribbean fruit flies: Review of current status, pp. 111–117. *In* Sterile-male technique for control of fruit flies. Proceedings Panel IAEA 1969. Vienna, Austria.

Malavasi, A., A. Van-Sauers-Muller, D. Midgarden, V. Kellman, V. Didelot, P. Caolong, and O. Ribeiro. 2000. Regional programme for the eradication of the carambola fruit fly in South America, pp. 395–399. *In* K. H. Tan (ed.), Proceedings: Area-wide control of fruit flies and other insect pests. International Conference on Area-wide Control of Insect Pests, and the 5th International Symposium on Fruit Flies of Economic Importance, 28 May–5 June 1998, Penang, Malaysia. Penerbit Universiti Sains Malaysia, Pulau Pinang, Malaysia.

Manrique, L. J., and C. Rivera. 2016. Control de las moscas de la fruta en Perú, pp. 23. *In* Book of Abstracts, 9a Reunión del Grupo de Trabajo en Moscas de la Fruta del Hemisferio Occidental, 16–22 October 2016, Buenos Aires, Argentina.

Marchioro, C. A. 2016. Global potential distribution of *Bactrocera carambolae* and the risks for fruit production in Brazil. PLoS One 11(11): e0166142.

McInnis, D. O., J. Hendrichs, T. Shelly, N. Barr, K. Hoffman, R. Rodríguez, D. R. Lance, K. Bloem, D. M. Suckling, W. Enkerlin, P. Gomes, and K. H. Tan. 2017. Can polyphagous invasive tephritid pest populations escape detection for years under favorable climatic and host conditions? American Entomologist 63: 89–99.

Midgarden, D., A. van Sauers-Muller, M. J. Signoretti Godoy, and J. F. Vayssières. 2016. Overview of the programme to eradicate *Bactrocera carambolae* in South America, pp. 705–736. *In* Fruit fly research and development in Africa: Towards a sustainable management strategy to improve horticulture. S. Ekesi, S. A. Mohamed, and M. De Meyer (eds.). Springer International Publishing, Switzerland.

Morilla, R. C., L. López, A. Gaiga, R. Chiovetta, G. Reniero, B. Repice, S. Velo, and R. Giménez. 2016. Programa nacional de control y erradicación de mosca de los frutos en el Noreste Argentino – PROCEM NEA, pp. 112. *In* Book of Abstracts, 9a Reunión del Grupo de Trabajo en Moscas de la Fruta del Hemisferio Occidental, 16–22 October 2016, Buenos Aires, Argentina.

Nierenberg, D. (ed.) 2018. Nourished planet: Sustainability in the global food system. Barilla Center for Food and Nutrition. Island Press, Washington, DC, USA. 250 pp.

Noe-Pino, C. A. 2016. Fincas Montelíbano y Santa Rosa, lugares y sitios de producción libres de mosca del Mediterráneo (*Ceratitis capitata*), Honduras, Centro América. 9a Reunión del Grupo de Trabajo en Moscas de la Fruta del Hemisferio Occidental. Poster presentation, 16–22 October 2016, Buenos Aires, Argentina.

Norrbom, A. L., L. E. Carroll, F. C. Thompson, I. M. White, and A. Freidberg. 1998. Systematic database of names, pp. 65–299. *In* F.C. Thompson (ed.), Fruit Fly Expert Identification System and Systematic Information Database. Backhuys Publishers, Leiden, The Netherlands.

Olalquiaga, G., and C. Lobos. 1993. La mosca del Mediterraneo en Chile: Introducción y erradicación. Ministerio de Agricultura, Servicio Agrícola y Ganadero. Santiago, Chile.

Oliveira, C. M., A. M. Auad, S. M. Mendes, and M. R. Frizzas. 2013. Economic impact of exotic insect pests in Brazilian agriculture. Journal of Applied Entomology 137: 1–15.

Peel, M. C., B. L. Finlayson, and T. A. McMahon. 2007. Updated world map of the Köppen-Geiger climate classification. Hydrology Earth System Sciences 11: 1633–1644.

Pimentel, D. 2002. Biological invasions. Economic and environmental costs of alien plant, animal and microbe species. CRC Press LLC, Boca Raton, Florida, USA. 369 pp.

Prensa Libre. 2019. Empresa ecuatoriana invierte US$6 millones en Petén y Retalhuleu. Natiana Gándara, 2 de julio de 2019, Guatemala City, Guatemala.

Quiroga, D., W. Ramírez, P. Fedyszak, E. Garavelli, M. F. Vazquez, and C. Ruiz. 2016. Programa nacional de control y erradicación de moscas de los frutos de Argentina (PROCEM): Lecciones aprendidas y desafíos a futuro, pp. 21. *In* Book of Abstracts, 9a Reunión del Grupo de Trabajo en Moscas de la Fruta del Hemisferio Occidental, 16–22 October 2016, Buenos Aires, Argentina.

Razera Papa, R. C. 2019. Establishment and maintenance of the South American cucurbit fruit fly (*Anastrepha grandis*) PFA in the states of Rio Grande do Norte and Ceará, Brazil, pp 65–66. *In* IPPC Guide for establishing and maintaining pest free areas. Published by FAO on behalf of the Secretariat of the International Plant Protection Convention. Rome, Italy.

Reyes, O. P., J. Reyes, W. Enkerlin, J. Galvez, M. Jimeno, and G. Ortíz. 1991. Estudio beneficio-costo de la Campaña Nacional Contra Moscas de la Fruta. Secretaria de Agricultura y Recursos Hidráulicos (SARH), México, DF, México.

Reyes, J., G. Santiago M., and P. Hernández M. 2000. The Mexican fruit fly eradication programme, pp. 377–380. *In* K. H. Tan (ed.), Proceedings: Area-wide control of fruit flies and other insect pests. International conference on area-wide control of insect pests, and the 5th international symposium on fruit flies of economic importance, 28 May–5 June 1998, Penang, Malaysia. Penerbit Universiti Sains Malaysia, Pulau Pinang, Malaysia.

Reyes, J., X. Carro, J. Hernandez, W. Mendez, C. Campo, H. Esquivel, E. Salgado, and W. Enkerlin. 2007. A multi-institutional approach to create fruit fly low prevalence and fly-free areas in Central America, pp. 627–640. *In* M. J. B. Vreysen, A. S. Robinson, and J. Hendrichs (eds). Area-wide control of insect pests: From research to field implementation. Springer, Dordrecht, The Netherlands.

Rivera-Tejada, C. J. 2011. Control y erradicación de moscas de la fruta en Perú. Ministerio de Agricultura. Lima, Perú.

Rhode, L. H., J. Simon, A. Perdomo, J. Gutiérrez, C. F. Dowling, Jr., and D. A. Lindquist. 1971. Application of the sterile insect release technique in Mediterranean fruit fly suppression. Journal of Economic Entomology 64(3): 708–713.

Rodríguez, P. R., C. Lobos A., R. Castro F., J. Yevenes F., A. Barra P. and P. Jara R. 2016. Sistema nacional de detección de moscas de la fruta (SNDMF) – Chile. pp. 22. *In* Book of Abstracts, 9a Reunión del Grupo de Trabajo en Moscas de la Fruta del Hemisferio Occidental. 16–22 October 2016, Buenos Aires, Argentina.

(SAG) Servicio Agrícola y Ganadero. 1995. Chile: A medfly-free country. Pamphlet. Government of Chile, Santiago, Chile.

(SAG) Servicio Agrícola y Ganadero. 1996. Chile: País libre de mosca de la fruta. Departamento de Protección Agrícola, Proyecto 335, moscas de la fruta. Segunda edición, Julio 1996. Ministerio de Agricultura, Servicio Agrícola y Ganadero, Santiago, Chile.

(SAGAR) Secretaría de Agricultura y Ganadería. 1999. La presencia de sanidad vegetal en la agricultura mexicana del siglo XX, J. Reyes (ed.). Fitófilo (edición especial) No. 89, April 1999. México, DF, México.

(SAGAR/IICA) Secretaria de Agricultura Ganadería y Desarrollo Rural/Instituto Interamericano de Cooperación para la Agricultura. 2001. Campaña Nacional Contra Moscas de la Fruta. Government of México, México, DF, México.

(SARH/DGSV-USDA/APHIS) Secretaria de Agricultura y Recursos Hidráulicos/Direccion de Sanidad Vegetal- United States Department of Agriculture/Animal and Plant Health Inspection Service. 1990. Work plan for the Sonora fruit fly free zone program for the 1990 export season. Bilingual English-Spanish. 21 pp.

Sharma, S., R. Kooner, and R. Arora. 2017. Insect pests and crop losses, pp. 45–66. *In* R. Arora, and S. Sandhu (eds.), Breeding insect resistant crops for sustainable agriculture. Springer, Dordrecht, The Netherlands.

Shelly, T., D. R. Lance, K. H. Tan, D. M. Suckling, K. Bloem, W. Enkerlin, K. M. Hoffman, K. Barr, R. Rodríguez, P. J. Gomes, and J. Hendrichs. 2017. To repeat: Can polyphagous invasive tephritid pest populations remain undetected for years under favorable climatic and host conditions? American Entomologist 63: 224–231.

Statista. 2019. International tourist arrivals worldwide from 1995 to 2017 by region (in millions). S. Lock (ed.).

(STDF) Standards and Trade Development Facility. 2019. Geneva, Switzerland.

Tween, G. 1986 A modular approach to fruit fly production facilities for the Mediterranean fruit fly Central American program, pp. 283–291. *In* A.P. Economopoulos (ed.), Fruit flies. Proceedings of Second International Symposium of Fruit Flies of Economic Importance, Crete, Greece.

(UN) United Nations. 2015. Sustainable development goals. Knowledge platform. 2030 Agenda for Sustainable Development. New York, NY, USA.

(UN) United Nations. 2016. The sustainable development goals report. New York, NY, USA.

(USDA) United States Department of Agriculture. 2018. Pest-free areas. 09/2018.

(USDA/APHIS) United States Department of Agriculture-Animal and Plant Health Inspection Service. 1992. Risk assessment: Mediterranean fruit fly, pp. 113. *In* C. E. Miller (ed.), Planning and risk analysis systems policy and program development. Riverdale, Maryland, USA.

(USDA/APHIS) United States Department of Agriculture-Animal and Plant Health Inspection Service. 2019. Fruit fly host lists and host assessments. Riverdale, Maryland, USA.

Vilatuña, J. 2018. AGROCALIDAD, Ecuador. Pest Control Newsletter 90: 34–35.

Villareal, P., A. Mongabure, C. A. Borges, and C. Gómez Segade. 2018. Evaluación del impacto económico del programa nacional de control y erradicación de mosca de los frutos Procem Patagonia. INTA/SENASA, Secretaría de Agroindustria. Imprenta Minigraf, Carmen de Patagones, Argentina. 95 pp.

Vo, T., W. Enkerlin, C. E. Miller, G. Ortíz, and J. Pérez. 2003. Economic analysis of the suppression/eradication of the Mediterranean fruit fly and other fruit flies in Central America and Panama. Report prepared by an ad hoc expert group. Presented to the Organismo Internacional Regional de Sanidad Agropecuaria (OIRSA), San Salvador, El Salvador.

Wedekind, L. 2007. Science, sex, superflies, pp. 32–40. *In* International Atomic Energy Agency (IAEA) Bulletin 48/2. Vienna, Austria.

Weems Jr., H. V., J. B. Heppner, T. R. Fasulo, and J. L. Nation. 2017. Caribbean fruit fly *Anstrepha suspensa* (Loew) (Insecta: Diptera: Tephritidae). University of Florida/Institute of Food and Agricultural Sciences. Gainesville, Florida, USA.

White, I. M., and M. M. Elson-Harris. 1992. Fruit flies of economic significance: Their identification and bionomics. CAB International, Wallingford, Oxon UK. 601 pp.

World Bank. 2013. Agricultural exports from Latin America and the Caribbean: Harnessing trade to feed the world and promote development. World Bank Group. Washington, DC, USA.

Worldometers. 2019. Current world population.

Yokohama, V. Y. 2015. Olive fruit fly (Diptera: Tephritidae) in California table olives, USA: Invasion, distribution, and management implications. Journal of Integrated Pest Management 6(1): 1–18.

AREA-WIDE MANAGEMENT OF FRUIT FLIES IN A TROPICAL MANGO GROWING AREA INTEGRATING THE STERILE INSECT TECHNIQUE AND BIOLOGICAL CONTROL: FROM A RESEARCH TO AN OPERATIONAL PROGRAMME

P. LIEDO[1], P. MONTOYA[2] AND J. TOLEDO[1]

[1]*El Colegio de la Frontera Sur, Tapachula, Chiapas, México; pliedo@ecosur.mx*
[2]*Programa Moscafrut, SAGARPA-IICA, Metapa, Chiapas, México*

SUMMARY

The Sterile Insect Technique (SIT) has been successfully used for the control of fruit flies in a number of places in the world. One requirement for its successful application is that wild populations should be at low densities to achieve effective sterile to wild fly overflooding ratios. This has been an important reason that has limited its integration in fruit fly management in tropical fruit growing areas, where climate conditions and the availability of hosts all year-round results in high population densities. Here we report the results of a project where SIT integration into fruit fly management was evaluated under the tropical conditions of the mango growing area in the Soconusco region of Chiapas, Mexico. The basis for the area-wide integrated pest management (AW-IPM) approach was the knowledge of the population dynamics of the pest fruit flies in the region and of the fruit phenology. The main commercial mango growing areas are in the lowlands, where fruit fly populations are very low outside of the mango production season. Population densities are higher in the midlands and highlands, where alternate hosts are common in backyards and as part of the natural vegetation. We call these refuge areas, and the AW-IPM approach aimed at establishing a biological barrier with releases of parasitoids and sterile male fruit flies to suppress the fruit fly populations and prevent or minimize the dispersal of wild flies from the refuge areas to the mango orchards. In 2014, after two years of releases, fruit fly population densities were suppressed more than 70% in the release area and 65% in the entire area, including the lowlands with the mango orchards. With the support of fruit growers, state and federal governments, this project was continued and established as an operational AW-IPM programme. In 2016, after 4 years of programme implementation, the detection of wild flies was significantly reduced, and the number of batches of fruit that were rejected at the packing houses due to the detection of infested fruits was the lowest in the past 12 years, since the recording of these data was initiated. These indicators declined even further in 2017. The results obtained demonstrate that AW-IPM integrating the SIT can be applied successfully against fruit flies under tropical conditions with naturally high pest densities, providing there is adequate knowledge on the population dynamics of the fruit fly species present in the region.

J. Hendrichs, R. Pereira and M. J. B. Vreysen (eds.), Area-Wide Integrated Pest Management: Development and Field Application, pp. 197–214. CRC Press, Boca Raton, Florida, USA.

Key Words: Integrated pest management, autocidal control, SIT, augmentative biological control, *Anastrepha* fruit flies, Tephritidae, *Anastrepha ludens*, *Anastrepha obliqua*, *Diachasmimorpha longicaudata*, Soconusco, Chiapas, Mexico

1. INTRODUCTION

In Mexico, mango represents one of the most important fruit production and export value chains, with more than 180 000 ha of cultivation, giving an annual production of approximately 1.8 million tons. Mexico ranks 6th by area and 4th by production in the world, and on the international market Mexico and India are among the most important exporting countries by volume (SIAP 2015).

Among the factors that limit or affect mango production and marketing are insect pests, and within these, fruit flies are among the most devastating. In view of their importance, these are considered of public interest and for this reason the National Campaign against Fruit Flies was established in 1992. The Campaign has succeeded in achieving fruit fly pest free areas (FF-PFA, FAO 2016) in 52.8% of the national territory, and fruit fly areas of low pest prevalence (FF-ALPP, FAO 2008) in another 10.4%, while the remaining 36.8% is considered a zone under phytosanitary management (Liedo 2016; Ramírez y Ramírez et al. 2019). Due to their agro-ecological requirements, mango producing areas are located mostly in the subtropical and tropical zones of the country, most of which are in the area under phytosanitary management. Under these favourable ecological conditions that promote fruit fly abundance, the development of technologies and the design of pest management strategies are required to allow the production of fruits free of fruit fly damage.

There are modern and appropriate and more sustainable technologies to deal with these pests, such as the Augmentative Biological Control (ABC) and the Sterile Insect Technique (SIT), among others (Montoya and Toledo 2010; Enkerlin et al. 2021). These technologies, integrated with other control methods have been applied successfully in Mexico to prevent for over 35 years the invasion of the Mediterranean fruit fly, *Ceratitis capitata* (Wiedemann), into southern Mexico along the border with Guatemala (Enkerlin et al. 2015) and for the implementation of FF-PFA for native *Anastrepha* fruit flies in the north of the country (Reyes et al. 2000; Liedo 2016). However, under the tropical conditions in which commercial mango is extensively grown, native *Anastrepha* fruit flies have high rates of population growth and therefore the effective application of these technologies is much more challenging (Montoya et al. 2000).

The characteristics of the SIT and the ABC, as well as the high mobility of fruit flies, make it necessary to adopt an area-wide integrated pest management (AW-IPM) approach, which considers the management of the total population of the pest and its spatial distribution (Hendrichs et al. 2007; Montoya et al. 2007).

In the mango producing region of Soconusco, Chiapas in southern Mexico, there is ample knowledge about the population dynamics of *Anastrepha* fruit flies (Aluja et al. 1996; Celedonio-Hurtado et al. 1995; Montoya et al. 2000). In the case of *Anastrepha ludens* (Loew) and *Anastrepha obliqua* (Macquart), which are the species that infest mango, it is known that their populations are high in the midlands and in the highlands (120-600 m elevation) with little temporal fluctuations.

In the commercial mango production areas of the lowlands (0-120 m elevation), the *Anastrepha* populations are low most of the year and only increase during the mango fruiting season, when growers have to apply repeated ground bait sprays to minimize fruit fly infestation.

Based on the available background information, and with the objective to reduce the bait sprays and fruit fly infestation in mangoes, a project was submitted to validate the use of the ABC and the SIT as elements of an AW-IPM approach, for the management of these native fruit flies in mango. We proposed to carry out this project in the Soconusco region in Chiapas State, considering the available knowledge on fruit fly populations, their hosts and their seasonality, and taking advantage of the infrastructure of the National Fruit Fly Campaign (Mexican Plant Protection Organization SENASICA, and IICA) in the region with respect to mass-rearing and release of sterile fruit flies and parasitoids. This 4-year project was funded by the National Council of Science and Technology (CONACYT) and the Ministry of Agriculture (SAGARPA) sectorial fund. In view of the increasing support and interest of the mango growers, the project was converted into an operational programme during the fourth and last year of the project.

2. MATERIALS AND METHODS

2.1. *Strategy*

It was assumed that the *Anastrepha* populations are maintained by year-round host availability in the refuge sites of the midlands and highlands outside the mango season. Thus, the designed strategy consisted of implementing a "biological barrier" based on the release of sterile flies and parasitoids in the intermediate zone between the high and the lowlands (ca. 100-200 m elevation), seeking to suppress the populations there and avoiding or minimizing their movement to the lowlands, where most of the commercial mango orchards are located. To facilitate achieving favourable sterile to fertile male ratios, releases of sterile insects were initiated at the end of November or early December 2012 and continued to 2015, i.e. the period when historically the lowest population levels have been observed (Aluja et al. 1996).

In the first year of the project (2011-2012), two trapping transects were established from the highlands to the lowlands, to monitor the populations of fruit flies along an altitudinal gradient (Fig. 1). Before initiating the releases in the second year, a third trapping route was established with the objective of monitoring the populations of sterile flies in the release polygon and adjacent areas.

The initial experimental design for the 4-year project (2011-2015) was: (a) to monitor populations during the first year without the application of the ABC and the SIT, (b) to apply the ABC and the SIT in the intermediate zone during the second and third years, and (c) depending on results, to transfer the technology to mango growers during the fourth year. Based on the results obtained during the second and third years, when the control methods were applied on an area-wide basis, it was decided to continue with the releases for another year. During this fourth year, the technology was transferred, and from week 10 of 2015, the fruit growers' union was in charge of funding and coordinating the parasitoid and sterile fruit fly releases through the Local Board of Plant Health and the State Committee of Plant Health of Chiapas. During this fourth year, the research project was converted into an operational action programme, recognized by the National Fruit Fly Campaign.

Based on the monitoring of the first year and the availability of biological material, a 15 000 ha polygon (21.0 x 7.14 km) was established for the release of parasitoids and sterile fruit flies. This polygon was located between the cities of Mazatán and Tapachula, in the transition zone between the high and the lowlands. The location of the release polygon and the traps deployed are shown in Fig. 1.

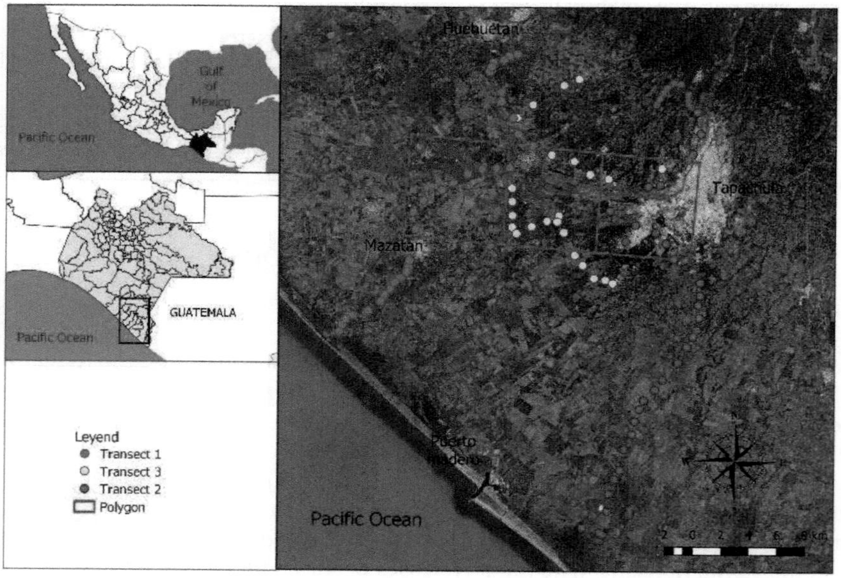

Figure 1. Location of the sterile fly release polygon in the Soconusco region of the state of Chiapas, covering 15 000 ha, and the three subquadrants for the release and assessment of parasitoids of 5000 ha each, as well as the location of the traps for population monitoring. The blue dots indicate the location of the traps deployed along transect 1, the red dots correspond to transect 2, and the yellow dots to transect 3.

For the evaluation of the ABC, the polygon was subdivided into 3 quadrants of 5000 ha each in which to make and assess the release of parasitoids. Initially, it was planned to release the parasitoids only in the sub-quadrant to the East -the one with higher trap captures- and leave the central one as the buffer area and the one on the West as a control, with the intention of alternating the quadrants in the third year and thus be able to make comparisons in time and space. However, based on the results obtained in the second year, it was decided to repeat the release area in the third year (East quadrant) in order to evaluate the effect of the modifications in the release method (see below) and also to contribute to the suppression of pest populations in the area of highest infestation.

2.2. *Monitoring of Populations*

For the monitoring of the *Anastrepha* populations, Multilure® traps baited with Biolure® (ammonium acetate + putrescine) were used. Propylene glycol was used for the retention and conservation of the trapped specimen. The traps were checked weekly and the trapped flies identified by species and sex. In the case of *A. ludens* and *A. obliqua*, the dye marking used for the sterile flies was used to discriminate between the released sterile and wild flies. Along transect 1, 35 traps were deployed from the town of Huehuetán in the highlands to Barra de San Simón in the lowlands. Forty-four traps were deployed along transect 2 from Canton Pumpuapa to the northwest of the city of Tapachula, in the highlands, to the Ejido Conquista Campesina, in the lowlands. Transect 3 contained 30 traps of which 15 were deployed within the release polygon, 10 were located in the area adjacent to the north of the polygon and another 5 to the south of it (Fig. 1).

2.3. *Biological Material*

Every week, the project received 7.5-8.5 million sterile males of *A. ludens* Tapachula-7 genetic sexing strain, 5-10 million sterile males and females of *A. obliqua*, as well as 5 million of the parasitoid *Diachasmimorpha longicaudata* (Ashmead). With these quantities and based on the experience of the National Campaign, the target release densities were 533 male *A. ludens* / ha, 333 male *A. obliqua* / ha, and 1000 parasitoids / ha. The weekly amounts varied slightly depending on the weekly production of the Moscafrut facility in Metapa, as well as the needs of the National Campaign. For some time during the second year of releases, batches of the bisexual *A. ludens* strain (males and females) were also received, increasing the number of sterile insects released. The opposite was true for *A. obliqua* as the needs of the National Campaign resulted in smaller quantities being received sometimes (see Table 1 and Fig. 4), reducing the release densities.

Table 1. Number of sterile flies received during three release seasons

	2012-13	2013-14	2014-15*
A. ludens Tap-7 males			
Total	499 771 956	436 226 000	269 320 000
Average / week	9 610 999	8 388 961	9 974 815
A. ludens bisexual			
Total	85 729 000	517 932 000	241 365 000
Average / week	1 617 528	9 772 302	7 785 968
A. ludens sum of males			
Total	542 636 456	695 192 000	390 002 500
Average /week	10 419 763	13 275 112	13 867 799
A. obliqua			
Total	470 633 000	505 576 006	244 436 000
Average / week	8 879 868	9 520 302	8 147 867

* *In the 2014-15 season, data are included only until week 22 (June 6, 2015)*

During the three release seasons, a total of 1205 million sterile *A. ludens* Tap-7 strain (only males), 845 million sterile *A. ludens* bisexual strain (males and females), 1219 million sterile *A. obliqua* (males and females) and 385 million *D. longicaudata* parasitoids were released. The number of sterile flies received in each of the three release seasons, as well as weekly averages, are shown in Table 1.

The quality of these sterile insects is shown in Table 2. From the series of quality control tests that were applied, we selected the percentage of flying flies after chilling as a parameter representing quality of this biological material. In the quality control manual this test is known as "absolute-post-chill flyers" (FAO/IAEA/USDA 2019).

Table 2. Quality of the sterile flies received during three release seasons, as percentage of flying flies after chilling

Strain	2012-13	2013-14	2014-15*
A. ludens Tap-7	81.05	86.14	85.42
A. ludens bisexual	87.47	89.05	85.23
A. obliqua	88.46	89.39	88.26

**For the 2014-15 season, data are included only until week 22 (June 6, 2015)*

The mass-production facility in Metapa delivered the biological material to the Mediterranean Fruit Fly Emergence and Release Facility (CEMM by its initials in Spanish), located near the Tapachula airport. All biological material was received as irradiated pupae under hypoxic conditions. The CEMM staff placed the pupae in "Mexico" type emergence towers that were provided with water and food. The food was a mixture of sugar and enzymatic hydrolysed yeast in a 24:1 ratio. The towers were kept in the emergence rooms at 23 ± 1 °C and 65 ± 5% relative humidity. Adults emerged 2-3 days later and were released once they were sexually mature, i.e. 5 days later in the case of *A. ludens* and 4 days in the case of *A. obliqua*. On the release day, the towers were placed in a cold room at 2-4 °C for one hour to immobilize the adults, and then they were placed in specially designed refrigeration boxes and transported to the airport for chilled aerial release (Hernández et al. 2010). Images of the emergence towers and the release box are shown in Fig. 2.

The services of the Mubarqui® company were contracted to implement the aerial releases of the sterile flies. This company has adapted aircrafts, appropriate release machines, and support infrastructure to implement these releases. For each release flight, a report was generated indicating the time of the flight, the route followed, the number of sterile insects released and the corresponding density. Samples were taken from each batch of sterile insects to assess standardized quality parameters (FAO/IAEA/USDA 2019).

Parasitoids were released from the ground with the support of the staff and vehicles of the State Committee for Plant Health. During the first year of releases, the parasitized pupae were placed in PARC® boxes (Plastic Aerial Release Container) with a mixture of honey and paper as a food source (Fig. 3A).

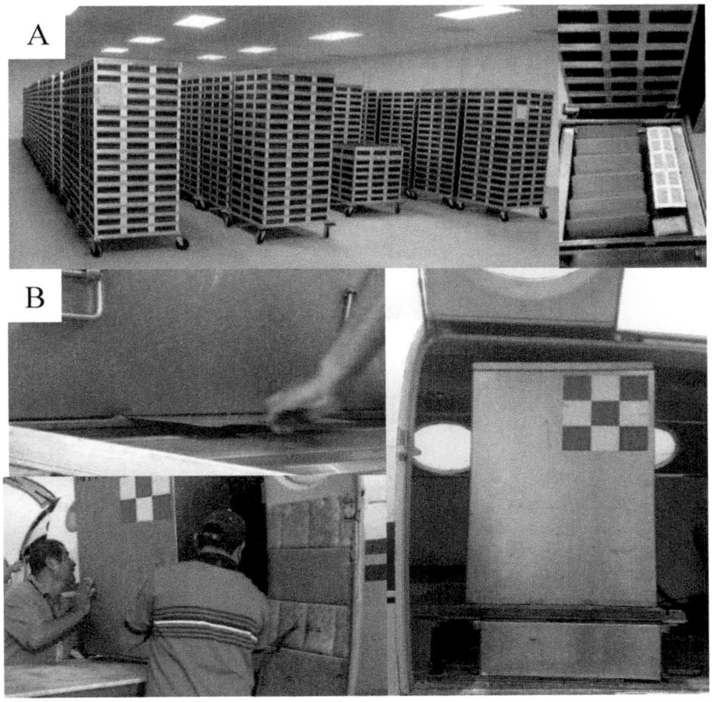

Figure 2. Towers type "México" used for the emergence of sterile adults (A) and chilled release box being loaded into the aircraft (B).

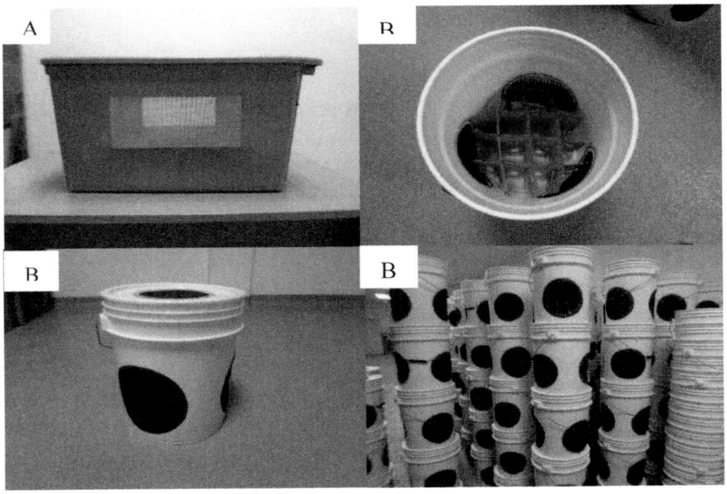

Figure 3. PARC box (A) and "R2D2" devices (B) used for the emergence and feeding of parasitoids and their subsequent ground release in the field.

After eight days, the emerged parasitoid adults were released. With the aim of improving the performance of these parasitoids, in the following year the release was implemented using “R2D2” devices, which are 20 litres plastic containers with mesh windows on the wall and cover. To increase the surface area, a corrugated plastic honeycomb was placed inside the containers. Each device contained approximately 2000 parasitized pupae (Montoya et al. 2012). The “R2D2” devices are shown in Fig. 3B. These “R2D2" devices replaced the PARC boxes in the second and third release years.

3. RESULTS

3.1. Biological Material

Weekly variations in the number of sterile males released during the three release seasons are shown in Fig. 4.

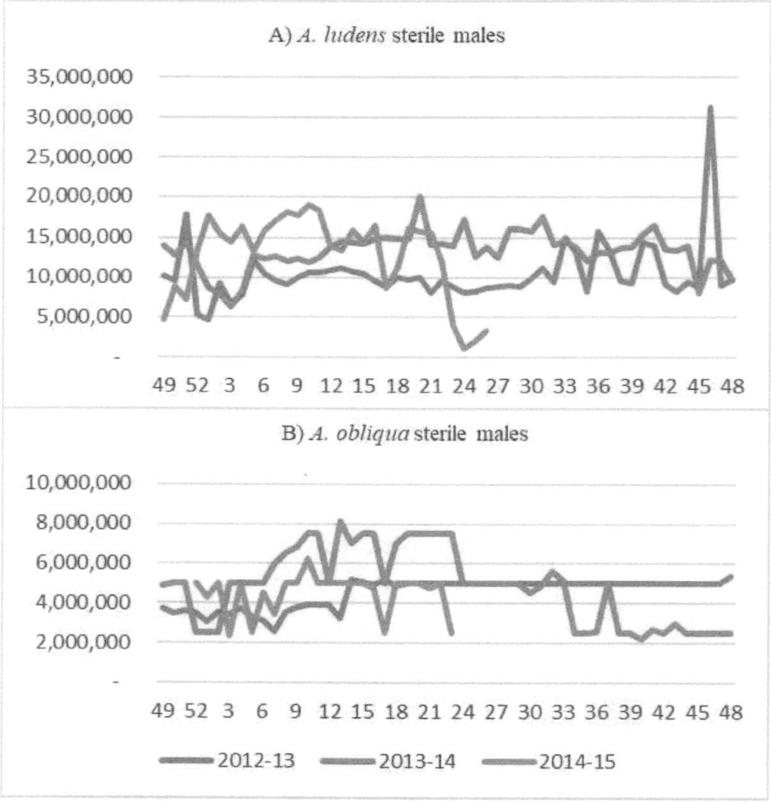

Figure 4. Number of sterile males released per week in each of the three release seasons. A) A. ludens *males of the Tapachula-7 and bisexual strains. B)* A. obliqua *males. Season 2012-13 is shown in blue, 2013-14 in orange, and 2014-15 in grey.*

For the third season (2014-15), data are shown only up to week 22 when, based on the number and locations of wild fly detections, it was decided to modify the release polygon (see Section 4). It should be noted that since week 10 of this last season, the association of fruit growers took over the funding and coordinating of field activities, fulfilling the goal of technology transfer, and thus transforming a research project into an operational programme.

The number of *A. ludens* sterile males released ranged between 10 and 20 million per week for most of the time for the three release seasons. Numbers for the 2013-14 and 2014-15 seasons were consistently higher. With these quantities, the average densities released were 696, 891 and 849 males per ha for the seasons 2012-13, 2013-14, and 2014-15, respectively. The highest density was 2084 males / ha in week 46 of 2013 and the lowest was 314 males / ha in week 1 of 2013. Considering the natural population dynamics of this species, in the first weeks of the calendar year it is key to have high sterile fly densities in the field, which was achieved in the 2014-15 season.

For *A. obliqua* the quantities of sterile flies were smaller and the variation greater. The average densities released were 302, 323 and 299 males / ha for the seasons 2012-13, 2013-14, and 2014-15, respectively. The highest density was 543 males / ha in week 13 of 2014 and the lowest was 148 males / ha in week 40 of the same year.

As of week 34 of 2014, there was a significant reduction in *A. obliqua* sterile fly availability and from week 49 to 52 of 2014 no biological material was received. This was attributed to production problems at the Moscafrut facility and due to the demand for this species by the National Campaign.

3.2. Sterile Fly Densities

The releases of sterile flies were monitored with Multilure traps deployed inside the area of the release polygon. Out of the total of 109 traps were deployed as part of the three trapping transects, 30 were located inside this polygon. Average fly per trap per day (FTD) values for trapped sterile males in these 30 traps for each season and species are shown in Table 3.

Table 3. Average of sterile males captured per trap per day (FTD) in the release polygon in each of the three release seasons for A. ludens *and* A. obliqua

Season	*A. ludens*	*A. obliqua*
2012-13	0.766	0.643
2013-14	1.363	0.531
2014-15*	0.472	0.135

**Only until week 22, since the release polygon was modified afterwards*

Another parameter used to monitor sterile fly releases was the percentage of traps that trapped sterile flies, regardless of the amount captured. This parameter informs about the uniformity of the releases. Fig. 5 shows how this percentage varied throughout the year for the two species in the three seasons. The average percentage of traps that trapped sterile flies in each season ranged from 53 to 66% for *A. ludens* and 27 to 60% for *A. obliqua*. The lower value in *A. obliqua* corresponded to the 2014-15 season and was related to the suspension of releases from week 49 to 52 due to lack of biological material.

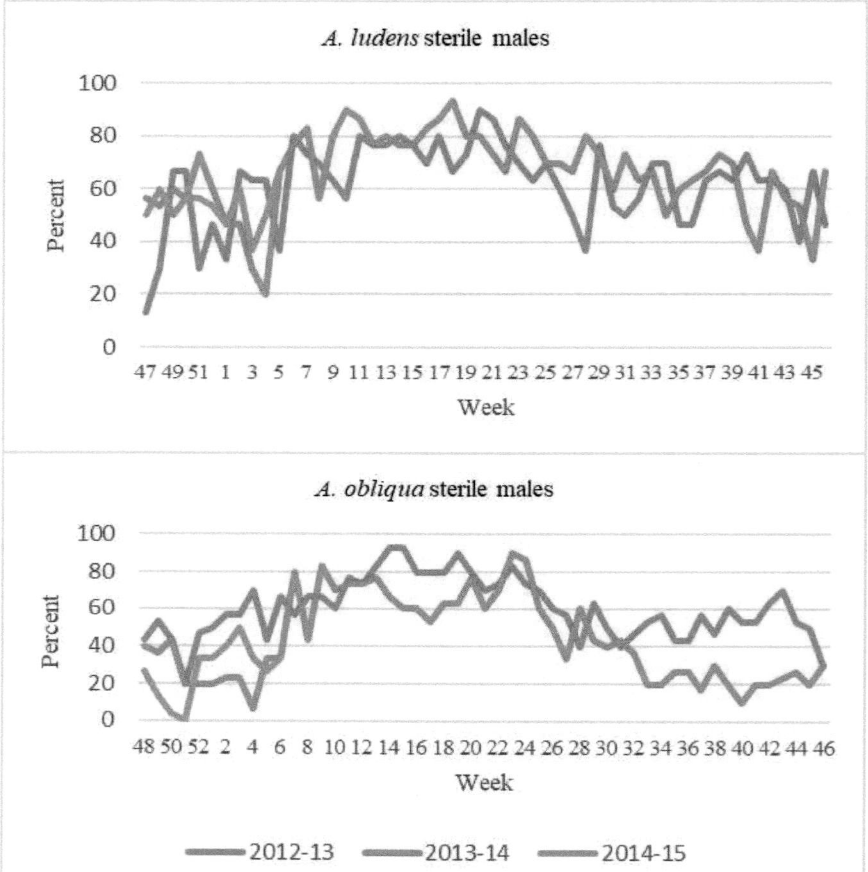

Figure 5. Percentage of traps with capture of sterile flies in each of the three release seasons for A. ludens *and* A. obliqua.

The sterile: fertile ratio refers to the number of sterile males compared to the number of fertile or wild males in the monitoring traps. This relationship is the basis of the Knipling model (Knipling 1955), which in turn is the foundation of SIT application. The ratios achieved for each species and in each season are shown in Table 4. In the case of *A. ludens*, considering the total catches throughout the complete season, the sterile: fertile ratios were always higher than 200:1 and in the 2013-14 season the ratio was 519:1. In *A. obliqua* these ratios were much lower, even in those cases where the number of sterile males trapped was similar to *A. ludens* (i.e. season 2012-13).

Table 4. Number of sterile and fertile or wild males captured in traps and the corresponding sterile: fertile ratio for A. ludens *and* A. obliqua *in the three seasons*

Season	*A. ludens*			*A. obliqua*		
	Sterile	Fertile	S:F	Sterile	Fertile	S:F
2012-13	7 936	31	256	6 956	618	11
2013-14	15 058	29	519	5 929	145	41
2014-15*	2 131	10	213	575	25	25

* *Only includes the first 22 weeks, since the release polygon was modified afterwards*

Empirical evidence indicates that to obtain good suppression, the sterile: fertile ratio must be greater than 30:1, but ideally greater than 100:1 (Flores et al. 2014, 2017). The achieved sterile: fertile ratios were satisfactory in the case of *A. ludens*, particularly in the 2013-14 season, when the average release density was 891 males / ha. While in the case of *A. obliqua*, the lower release densities (averages <350 males / ha) together with the larger wild populations resulted in sterile: fertile ratios much lower than the target of 100:1.

3.3. *Wild Population Densities*

To make an estimate of wild population levels we used wild female catches. During the four monitoring seasons, considering all the traps, a total of 3792 *A. ludens* and 37 445 *A. obliqua* females were trapped. This represents a 9.8 times greater catch for *A. obliqua*. To assess the effect of the sterile fly releases, only the traps located in the release polygon were considered (Fig. 6).

The wild populations in the first 2012-13 season were higher in the release polygon than in the preceding year without them. In the subsequent 2013-14 and 2014-15 seasons, the density of sterile flies was increased, and this situation was reversed. The populations were significantly suppressed. The suppression effect ranged from 76 - 81% for *A. ludens*, while for *A. obliqua*, despite the low sterile: fertile ratio, suppression ranged from 89 to 91%. The suppression is confirmed by a comparison of the mean numbers of females per trap per two seasons between the release polygon and the traps located at the north of the polygon (Fig. 7), where no control measures were applied, and south of the polygon, where most commercial mango orchards are located and which was the protected area.

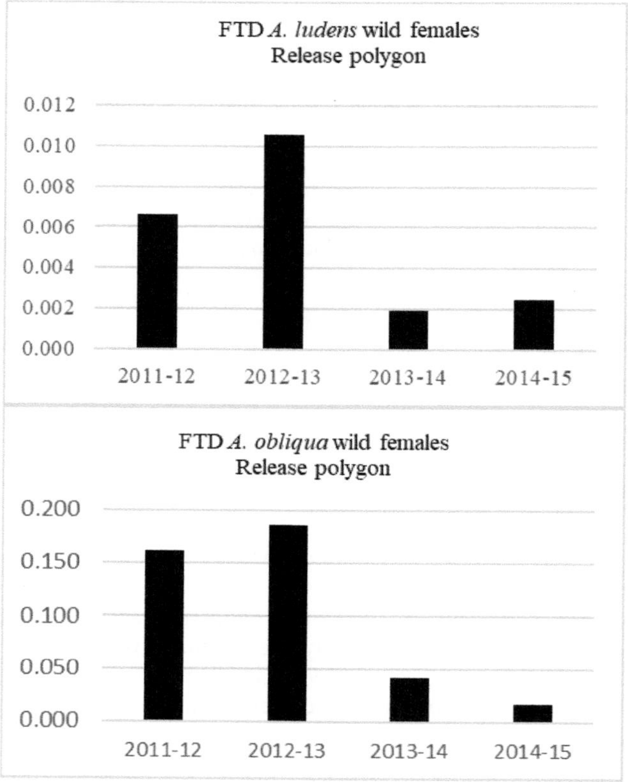

Figure 6. Capture of wild females, expressed in FTD (flies per trap per day), for A. ludens *and* A. obliqua *during four seasons in the release polygon. The 2011-12 season without and the other three seasons with ABC and SIT releases.*

Releases of *D. longicaudata* parasitoids resulted in a significant increase in the parasitisation rate. Table 5 shows these rates in the area of the polygon where parasitoids were released and in the control area where no releases were made. Parasitism by *D. longicaudata* in the area where releases were made was 15.12%; the other 1.38% was by native parasitoids. Montoya et al. (2017) provide more detailed information on the effects of the *D. longicaudata* releases.

Table 5. Number of flies, parasitoids and parasitisation rate in the zone with augmentative releases of D. longicaudata *and the control zone without releases*

Zone	Flies	Parasitoids	Parasitism (%)	Fruit samples with parasitoids (%)
Without releases	468	5	1.05	3.8
With releases	5 271	947	16.50	27.7

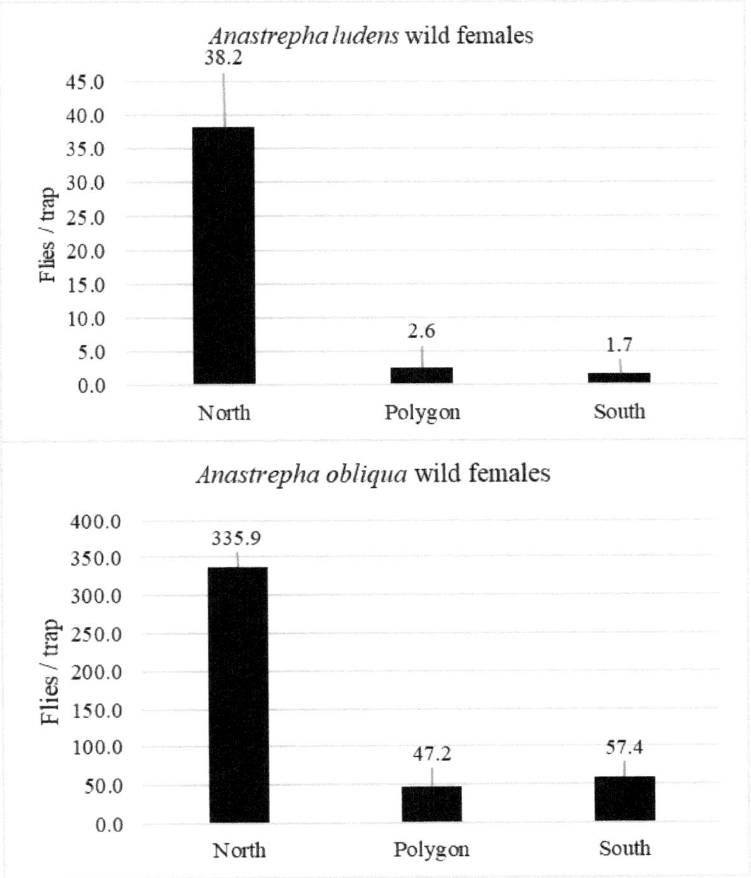

Figure 7. Mean (SE) capture of wild females expressed in flies per trap for A. ludens *and* A. obliqua *in the 2012-13 and 2013-14 seasons within the release polygon and in the traps located north and south of the release polygon.*

3.4. Infested Batches and the Species of Concern

The differences in abundance observed between the two species - *A. ludens* and *A. obliqua* - were consistent and in agreement with what was previously reported (Aluja et al. 1996). However, this difference did not match the detection of infested batches of mango in the packing houses: more than 94% of the infested lots were by *A. ludens* (information provided by the State Committee for Plant Health of Chiapas) (Table 6).

In 2014, Aluja et al. (2014) reported that mango cultivar Ataulfo was not infested by *A. obliqua* when fruits were still on the trees. Only the fruits known as "mango niño" (baby mango) were infested. These are small fruits of the same cultivar that do not grow as the normal fruits. The Ataulfo cultivar is the one that occupies the largest surface area in Chiapas and represents more than 90% of mango exports. This observation made us conclude that *A. ludens* was the important species to suppress to minimize or prevent fruit infestation. This new knowledge was fortunate as the programme could receive sufficient sterile male *A. ludens* to achieve densities of 800 to 1000 males per ha, reaching sterile: fertile ratios of more than 100:1.

Table 6. Number of infested batches at the packing houses per species and the percent of those infested by Anastrepha ludens

Season	*A. obliqua*	*A. ludens*	% *A. ludens*
2012	21	390	94.9
2013	9	401	97.8
2014	6	236	97.5
2015	3	241	98.8

4. FROM A RESEARCH PROJECT TO AN ACTION PROGRAMME

Considering the four seasons of the research project, we were able to verify and validate the effectiveness of the SIT and the ABC for the suppression of populations of these two species of fruit flies. The strategy of establishing a "biological barrier" between the midlands and highlands with high fruit fly populations, and the lowlands, where the largest area planted with commercial mango is located, seemed to be appropriate to reduce fruit fly infestation in mango from 76% to 91%. Nevertheless, for an action programme, monitoring should be expanded so that the location of this "biological barrier" is dynamic and can be adapted to the situation of the pest. Also, the importance of achieving high sterile to fertile ratios before the start of the mango season should always be kept in mind.

At this stage of the research project, growers were very interested in continuing the releases of sterile insects and parasitoids. They were willing to contribute the needed funds, and to fulfil the requirements to be considered for participation in the National Campaign. It was decided to focus only on *A. ludens*, and the National Campaign offered to provide the programme with 15 million sterile males (Tapachula-7 genetic sexing strain) and 5 million *D. longicaudata* for the ABC.

A technical group to follow up the programme was established with participants from the mango growers' union, the State Plant Health Committee, the National Fruit Fly Campaign and ECOSUR (El Colegio de la Frontera Sur, public research centre). This technical group meets every two weeks or weekly, depending on the time of the year. The technical group agreed on the design of a new release polygon of 15 000 ha based on the pest situation and the availability of biological material. Releases of sterile flies and parasitoids have not been stopped since then. As an example of this follow-up, the situation for week 25 in 2016 is shown in Fig. 8, with the location of the modified release polygon (40 x 3.75 km), the aerial release lines, and the location of the monitoring traps indicating flies captured.

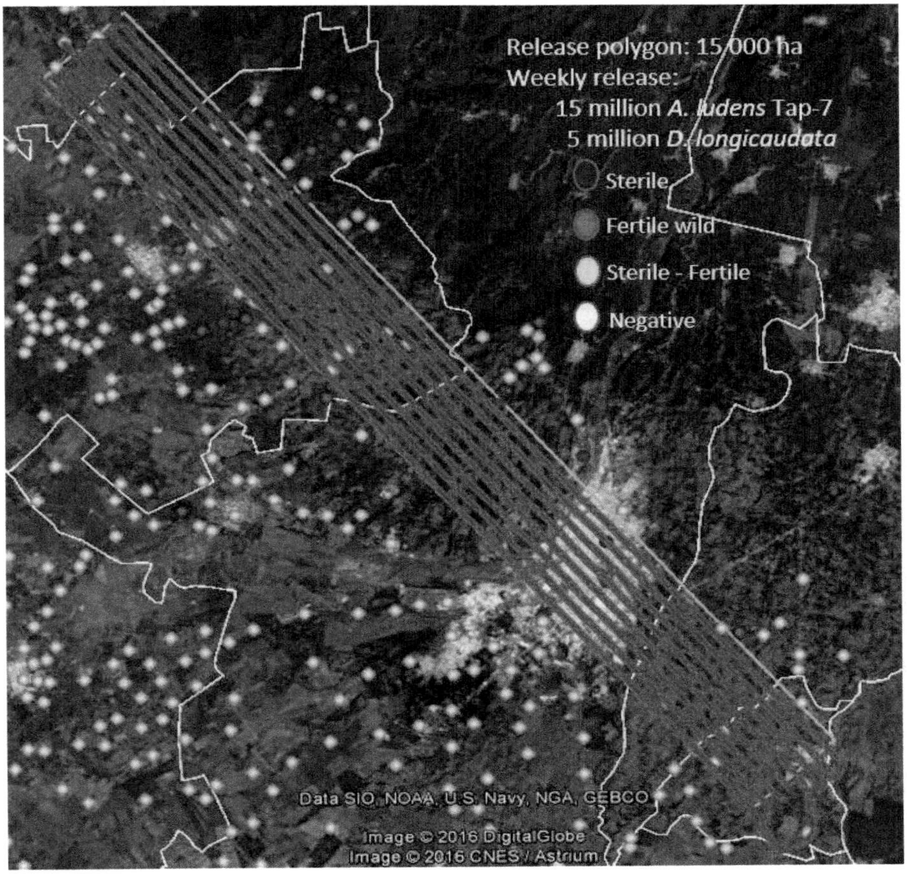

Figure 8. Modified release polygon of the action programme showing flight lines and the location of traps. Colour codes indicate type of trapped flies; sterile - fertile means that both types of flies were captured in the trap; these data correspond to week 25, 2016.

Although there was no immediate suppression effect of sterile flies and parasitoids releases, the continued releases from 2012 to 2017 resulted in a gradual suppression of wild *A. ludens* populations and as a result the number of infested mango batches detected at the packing houses has been greatly reduced. The number of infested batches per season, and the corresponding index of infested batches per ton of exported mangoes are shown in Fig. 9. These results are encouraging and demonstrate the long-term effects of area-wide integration of the SIT and the ABC. They also show that with good knowledge of the dynamics of fruit fly populations it is feasible to design AW-IPM programmes, integrating the SIT and the ABC in the management of fruit flies under tropical conditions where pest populations are normally high.

The challenge now is to maintain and refine the releases of sterile flies and parasitoids and assess whether their synergistic effect will further decrease the pest populations, and in the long term minimize or avoid bait insecticide sprays in the region.

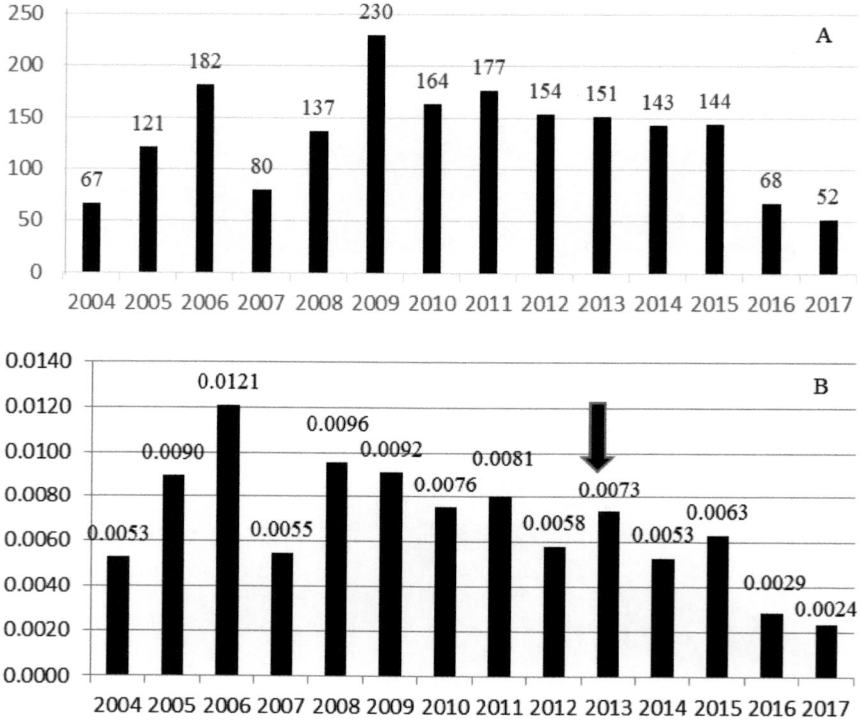

Figure 9. Number of infested batches detected at packing houses per season (A) and index of infested batches per ton of exported mangoes (B). The arrow shows when releases of sterile flies and parasitoids were initiated.

5. ACKNOWLEDGMENTS

We thank Ezequiel de León, Reyna Bustamante, Azucena Oropeza, Lucy Tirado, Fredy Gálvez, Rolando Cabrera, and Pedro Leal for technical assistance. The Mexican National Campaign of Fruit Flies (SENASICA, SAGARPA) provided biological material, infrastructure and technical and logistical support. We especially thank the Local Board of Plant Health, the Soconusco Fruit Growers Association, and the State Plant Health Committee of Chiapas for having made possible the transformation of a research project into an action programme. The research project was funded by the SAGARPA-CONACYT sectorial fund to project 163431.

6. REFERENCES

Aluja, M., H. Celedonio-Hurtado, P. Liedo, M. Cabrera, F. Castillo, J. Guillen, and E. Rios. 1996. Seasonal population fluctuations and ecological implications for management of *Anastrepha* fruit flies (Diptera: Tephritidae) in commercial mango orchards in southern México. Journal of Economic Entomology 89: 654–667.

Aluja, M., J. Arredondo, F. Díaz-Fleischer, A. Birke, J. Rull, J. Niogret, and N. Epsky. 2014. Susceptibility of 15 mango (Sapindales: Anacardiaceae) cultivars to the attack by *Anastrepha ludens* and *Anastrepha obliqua* (Dipetera: Tephritidae) and the role of underdeveloped fruit as pest reservoirs: Management implications. Journal of Economic Entomoogy 107: 375–388.

Celedonio-Hurtado, H., M. Aluja, and P. Liedo. 1995. Adult population fluctuations of *Anastrepha* species (Diptera: Tephritidae) in tropical orchard habitats of Chiapas, Mexico. Environmental Entomology 24: 861–869.

Enkerlin, W. R. 2021. Impact of fruit fly control programmmes using the Sterile Insect Technique, pp. 977–1004. *In* V.A. Dyck, J. Hendrichs, and A. S. Robinson (eds.), Sterile Insect Technique – Principles and practice in Area-Wide Integrated Pest Management. Second Edition. CRC Press, Boca Raton, Florida, USA.

Enkerlin, W., J. M. Gutiérrez-Ruelas, A. Villaseñor-Cortes, E. Cotoc-Roldan, D. Midgarden, E. Lira, J. L. Zavala-López, J. Hendrichs, P. Liedo, and F. J. Trujillo-Arriaga. 2015. Area freedom in Mexico from Mediterranean fruit fly (Diptera: Tephritidae): A review of over 30 years of a successful containment program using an integrated area-wide SIT approach. Florida Entomologist 98: 665–681.

(FAO) Food and Agriculture Organization of the United Nations. 2008. Establishment of areas of low pest prevalence for fruit flies (Tephritidae). International Standard for Phytosanitary Measures (ISPM) No. 30. International Plant Protection Convention. Rome, Italy.

(FAO) Food and Agriculture Organization of the United Nations. 2016. Establishment of pest free areas for fruit flies (Tephritidae). International Standard for Phytosanitary Measures (ISPM) No. 26. International Plant Protection Convention. Rome, Italy.

FAO/IAEA/USDA (Food and Agriculture Organization of the United Nations/International Atomic Energy Agency/United States Department of Agriculture). 2019. Product quality control for sterile mass-reared and released tephritid fruit flies. Version 7.0. International Atomic Energy Agency, Vienna, Austria. 148 pp.

Flores, S., P. Montoya, J. Toledo, W. Enkerlin, and P. Liedo. 2014. Estimation of populations and sterility induced in *Anatrepha ludens* (Diptera: Tephiritidae) fruit flies. Journal of Economic Entomology 107: 1502–1507.

Flores, S., E. Gómez-Escobar, P. Liedo, J. Toledo, and P. Montoya. 2017. Density estimation and optimal sterile-to-wild ratio to induce sterility in *Anastrepha obliqua* populations. Entomologia Experimentalis et Applicata 164(3): 284–290.

Hendrichs, J., P. Kenmore, A. S. Robinson, and M. J. B. Vreysen. 2007. Area-wide integrated pest management (AW-IPM): Principles, practice and prospects, pp. 1-31. *In* M. J. B. Vreysen, A. S. Robinson, and J. Hendrichs (eds.), Area-wide control of insect pests: From research to field implementation. Springer, Dordrecht, The Netherlands.

Hernández, E., A. Escobar, B. Bravo, and P. Montoya. 2010. Chilled packing systems for fruit flies (Diptera: Tephritidae) in the Sterile Insect Technique. Neotropical Entomology 39 (4): 601–607.

Knipling, E. F. 1955. Possibilities of insect control or eradication through the use of sexually sterile males. Journal of Economic Entomology 48: 459–462.

Liedo, P. 2016. Management of fruit flies in Mexico, pp. 695–704. *In* S. Ekesi, S. A. Mohamed, M. DeMeyer (eds.), Fruit fly research and development in Africa – Towards a sustainable management strategy to improve horticuluture. Springer International Publishing, Switzerland.

Montoya, P. and J. Toledo. 2010. Estrategias de control biológico, pp. 169–182. *In* P. Montoya, J. Toledo, and E. Hernández (eds.), Moscas de la fruta: Fundamentos y procedimientos para su manejo. SyG Editores, México, DF, México.

Montoya, P., J. Cancino, and L. Ruiz. 2012. Packing of fruit fly parasitoids for augmentative releases. Insects 3: 889–899.

Montoya, P., P. Liedo, B. Benrey, J. Cancino, J. F. Barrera, J. Sivinski, and M. Aluja. 2000. Biological control of *Anastrepha* spp. (Diptera: Tephritidae) in mango orchards through augmentative releases of *Diachasmimorpha longicaudata* (Ashmead) (Hymenoptera: Braconidae). Biological Control 18: 216–224.

Montoya, P., J. Cancino, M. Zenil, G. Santiago, and J. M. Gutiérrez. 2007. The augmentative biological control component in the Mexican campaign against *Anastrepha* spp. fruit flies, pp. 661–670. *In* M. J. B. Vreysen, A. S. Robinson, and J. Hendrichs (eds.), Area-wide control of insect pests: From research to field implementation. Springer. Dordrecht, The Netherlands. ISBN 978-4020-6058-8.

Montoya, P., P. López, J. Cruz, F. López, C. Cadena, J. Cancino, P. Liedo. 2017. Effect of *Diachasmimorpha longicaudata* releases on the native parasitoid guild attacking *Anastrepha* spp. larvae in disturbed zones of Chiapas, Mexico. BioControl 62: 581–593.

Ramírez y Ramírez, F., R. A. Hernández-Livera, and A. Bello-Rivera. 2019. El Programa nacional de moscas de la fruta en México, pp. 3–20. *In* P. Montoya, J. Toledo, and E. Hernández (eds.), Moscas de la fruta: Fundamentos y procedimientos para su manejo, 2nd edición. SyG Editores, México, DF, México.

Reyes, J., G. Santiago, and P. Hernández. 2000. The Mexican fruit fly eradication program, pp. 377–380. *In* K. H. Tan (ed.), Area-wide control of fruit flies and other insect pests. Penerbit Universiti Sains Malaysia, Pulau, Pinang, Malaysia.

(SIAP) Servicio de Información Agroalimentaria y Pesquera. 2015. México, D. F., México.

MOSCASUL PROGRAMME: FIRST STEPS OF A PILOT PROJECT TO SUPPRESS THE SOUTH AMERICAN FRUIT FLY IN SOUTHERN BRAZIL

A. KOVALESKI[1] AND T. MASTRANGELO[2]

[1]EMBRAPA Uva e Vinho, Estação Experimental de Vacaria, BR 285–Km 115, 95200-000, Vacaria, Rio Grande do Sul, Brazil
[2]Center for Nuclear Energy in Agriculture / Universidade de São Paulo, Av. Centenario 303, 13400-970, Piracicaba, São Paulo, SP, Brazil; piaui@cena.usp.br

SUMMARY

In Brazil, 99% of the apple growing areas are concentrated in the southern region, with an annual harvest of more than 1.2 million tons of fruits and a supply chain amounting to USD 1900 million. Despite the occurrence of several species of tephritid fruit flies in the region, the South American fruit fly, *Anastrepha fraterculus* (Wiedemann) (Tephritidae) represents 98.5% of the flies captured in commercial apple orchards. The gross value of yield losses and the cost of associated chemical control of this pest were estimated at close to USD 8 million per year. Moreover, the infestation rate by *A. fraterculus* has increased during the past four years, as the most commonly used insecticides have been banned. Brazilian researchers, along with state institutes and the Brazilian Association of Apple Producers (ABPM) have been promoting environment-friendly alternatives to insecticide application, such as sterile insects and parasitoids, to suppress the pest, and created the *Moscasul Biological Control and Integrated Fruit Fly Management Center*. After receiving the support of the Ministry of Agriculture at the end of 2014, almost USD 600 000 have been invested in constructing a rearing facility for *A. fraterculus* at the Agricultural Experiment Station of EMBRAPA (Brazilian Agricultural Research Corporation), Vacaria, Rio Grande do Sul, Brazil. The first containerized laboratory modules were installed in May 2016. During the pre-operational phase of the project, pilot trials were planned in apple orchards at Vacaria. As the Center for Nuclear Energy in Agriculture (CENA) has a 250 m^2 facility to produce fruit flies, including radiation sources, the sterile flies for the pilot trials will be initially provided by CENA. Both sterile flies and *Diachasmimorpha longicaudata* (Ashmead) parasitoids will be released in the surrounding non-commercial vegetation located within a 50-100 m buffer zone in the periphery of the target orchards, as these areas are the native breeding sites and repositories of the wild flies. Based on the wild population densities of the in the target orchards (114 ha) and the surrounding forested areas (111 ha), about 150 000 sterile flies per week will be required for the first pilot phase. After the fine-tuning of all rearing and sterilisation procedures in the beginning of 2017, CENA will ship more than 200 000 irradiated pupae weekly by air to Vacaria for 6 months, starting in September when the level of the wild fly population is lower. The results may influence the direction of future control tests and benefit the area-wide management of *A. fraterculus* involving hundreds of apple producers in the region and other temperate fruit growing farmers from southern Brazil.

J. Hendrichs, R. Pereira and M. J. B. Vreysen (eds.), Area-Wide Integrated Pest Management: Development and Field Application, pp. 215–230. CRC Press, Boca Raton, Florida, USA.

Key Words: Anastrepha fraterculus, Diachasmimorpha longicaudata, parasitoids, SIT, sterile insects, area-wide, apple orchards, wild hosts, Rio Grande do Sul, Santa Catarina, Paraná

1. INTRODUCTION

Refuting USA technical reports from the 1960s claiming that producing apples on a commercial scale would be impossible in Brazil, the apple industry has become one of the most recent success stories of fruit producers in the country (Klanovicz 2010). The joint efforts of industry, research institutions and extension services has resulted in an increase in apple yield from 2-4 to 28-30 tons per ha (Klanovicz 2010). In 1986, the First Santa Catarina Apple Festival was celebrated at Fraiburgo, where the Brazilian Association of Apple Producers (*Associação Brasileira dos Produtores de Maça* – ABPM) launched the campaign "The Brazilian Apple: the sin that worked out right". By 1989, commercial apple production had become firmly established in Brazil (Brazilian Apple Yearbook 2017).

Recently, in only 30 years, Brazil almost reached self-sufficiency in apple production, with a total planted area of 34 399 ha, a yield of 38.9 tons/ha and a total of 1 247 088 tons produced in 2016/2017; the supply chain of the sector amounts to USD 1900 million per year (Brazilian Apple Yearbook 2017). More than 4300 growers are distributed throughout the three southern states of Brazil (Rio Grande do Sul, Santa Catarina and Paraná), where 99% of Brazil's apple production is located (GAIN Report 2016).

The municipalities of São Joaquim (Santa Catarina state) and Vacaria (Rio Grande do Sul state) have taken turns leading the production volume (ca. 400 000 tons in each one, depending of the year) (Brazilian Apple Yearbook 2017). These localities present the most favourable weather conditions, with colder winter temperatures (i.e. more than 900 hours below 7.2°C), mean annual temperature of 15.2°C, altitudes higher than 800 m, as well as mild summer and autumn days, but with cold nights (10-15°C), ideal for the physiological processes of temperate fruit trees (Petri et al. 2011).

Despite all the technological advances that allowed the successful establishment of the apple production in the highlands of the states of southern Brazil, the orchards are constantly under threat of important pests like the European red spider mite (*Panonychus ulmi* Koch), woolly apple aphids (*Eriosoma lanigerum* Hausmann), Brazilian apple leafroller (*Bonagota salubricola* Meyrick), oriental fruit moth (*Grapholita molesta* Busk) and the South American fruit fly (*Anastrepha fraterculus* Wiedemann).

In October 1991, the codling moth, *Cydia pomonella* L., was detected in urban areas of four municipalities, but never invaded the commercial farms. Brazil was declared free of this pest in May 2014 after a successful eradication programme that was mainly based on host-tree removal in household backyards of urban and suburban areas (Kovaleski and Mumford 2007).

In 2015, the South American fruit fly became the target of another integrated pest management (IPM) programme.

2. IMPORTANCE OF THE SOUTH AMERICAN FRUIT FLY IN SOUTHERN BRAZIL

The South American fruit fly is a complex of cryptic species that comprises at least eight different morphotypes under the same species designation (Hernández-Ortiz et al. 2012). This complex is distributed from Texas to Argentina and can attack more than 80 species of host fruit trees (Steck 1998; Norrbom 2004). The flies develop within the range of 15-30°C, and the main biological characteristics at 25°C are: pre-oviposition period of 7-14 days; oviposition until 46-62 days (fecundity can reach 40 eggs/female/day, with an average of 25.2 eggs/female/day, and one female can lay up to 979 eggs during her lifetime); embryogenesis lasts 1-3 days; larval development of 10-14 days and a pupal period of 11-21 days (Machado et al. 1995; Salles 1993, 2000; Vera et al. 2007; Cladera et al. 2014).

The damage caused by *A. fraterculus* on apples occurs soon after the beginning of fruit development (ca. 2 cm diameter) and is caused by the piercing of the fruit skin by the female ovipositor, causing lesions in the fruit that result in fruit deformations (Magnabosco 1994; Sugayama et al. 1997). In case fully developed fruits get infested, the external symptoms usually do not appear, but the pulp may be destroyed by larvae (Kovaleski et al. 2000).

The pomiculture in southern Brazil has suffered heavy losses due to the attacks of *A. fraterculus* over several years, and the annual yield losses and associated annual costs of chemical control of this pest alone have amounted to more than USD 7.9 million (Salles 1998; Nora and Hickel 2002; Kovaleski and Ribeiro 2003).

Despite the presence of the Mediterranean fruit fly *Ceratitis capitata* (Wiedemann), *Rhagoletis* spp., and more than 10 other fruit fly species of the *Anastrepha* genus in southern Brazil (Kovaleski et al. 2000), *A. fraterculus* represents 98.5% of the flies captured in commercial apple orchards (Canal Daza et al. 1994; Nora et al. 2000; Santos et al. 2017).

Pre-harvest management of fruit flies has been implemented almost exclusively with chemical methods, and the control effort has been guided by population monitoring with McPhail traps with a solution of 25% grape juice or hydrolysed proteins (McPhail 1937; FAO/IAEA 2018) (Fig. 1).

Figure 1. Weekly field monitoring of Anastrepha fraterculus *populations in commercial apple orchards with McPhail traps baited with fruit juice or hydrolysed proteins.*

The number of traps deployed depends on the size of the orchards, i.e. 4 traps/ha for areas up to 2 ha, and 2 traps/ha for areas between 2 and 5 ha (Kovaleski et al. 2000). Growers mostly have used a solution of 25% grape juice as the common attractant with the traps, but there are concerns with respect to their efficiency for fruit fly monitoring in apple orchards (Bortoli et al. 2016). Recent studies have demonstrated that protein-based lures of plant or animal origin (e.g. BioAnastrepha™ and Ceratrap™, respectively) are better attractants for *A. fraterculus* (Scoz et al. 2006; Rosa et al. 2017).

As the *A. fraterculus* populations usually invade apple orchards from the surrounding areas, where they develop on non-commercial preferred hosts (Kovaleski et al. 1996; Sugayama et al. 1997; Santos et al. 2017), growers have traditionally applied weekly toxic bait sprays at the periphery of the apple orchards. The use of these baits (composed of water + insecticide + attractant such as hydrolysed proteins or 5-7% sugarcane molasses) is intensified in the first months of fruit growth, when the introduction of ovipositor may lead to external fruit deformations (Kovaleski et al. 2000).

When the number of adult flies captured inside the orchard exceeds a threshold level of 0.5 flies per trap per day (FTD) (a threshold level adopted by Brazilian growers since the 1980s), insecticides with systemic action and long residual effects are applied as cover-sprays, usually requiring 8 to 10 applications per season at a cost of about USD 240/ha/year.

Due to their low cost and residual properties (Harter et al. 2015), organophosphate insecticides have been heavily used by Brazilian growers for more than 20 years (Puzzi and Orlando 1957; Salles and Kovaleski 1990; Kovaleski and Ribeiro 2003). Many organophosphate insecticides, however, are being gradually banned or the maximum tolerable residue levels have been drastically reduced for exported fruits and derivatives like juice, concentrates and purees (Rawn et al. 2006; Eddleston et al. 2012). For example, fenthion was the most commercialized insecticide to control *A. fraterculus* in Brazilian apple orchards until 1997, when the growers stopped using it due to the risk of rejection of the fruits on international markets (Kovaleski et al. 2000). In addition, there is the risk of insecticide resistance developing in *A. fraterculus* as reported for *C. capitata* (Couso-Ferrer et al. 2011).

In order to meet the requirements of international markets for low residues on fruit, growers are increasingly being pushed to avoid insecticide applications against *A. fraterculus* during longer periods before harvest. Consequently, many apple orchards are being left unprotected from this pest during periods when the fruits are most susceptible.

In the absence of chemical control, the yields of apple orchards can be reduced with up to 30% due to *A. fraterculus* damage (Kovaleski et al. 2000). The shrinking choice of insecticides available to control *A. fraterculus* in Brazil, coupled with public demand for sustainable alternatives, have created a significant opportunity for promoting the area-wide augmentative biological control using sterile insects and parasitoids.

3. POPULATION DYNAMICS OF THE SOUTH AMERICAN FRUIT FLY IN APPLE ORCHARDS OF SOUTHERN BRAZIL

Tephritid pest population dynamics are largely affected by climatic features and availability of hosts (Aluja et al. 2012). This was confirmed by the general dynamics of the *A. fraterculus* populations of the highlands of the midwestern plateau of Santa Catarina and the mountainous region of Rio Grande do Sul, where dynamics of this fly have remained relatively constant for the last 20 years (Salles 1995; Garcia et al. 2003; Calore et al. 2013; Santos et al. 2015, 2017).

Trap captures over a period of 10 years showed that *A. fraterculus* population densities were higher between November and February (spring-summer), but between May and September (autumn-winter) they practically disappear from the commercial apple orchards when mean daily temperatures drop below 15°C and host availability is very low (Fig. 2) (Salles and Kovaleski 1990). This pattern of population fluctuation has remained unchanged in most apple orchards to date (Santos et al. 2017).

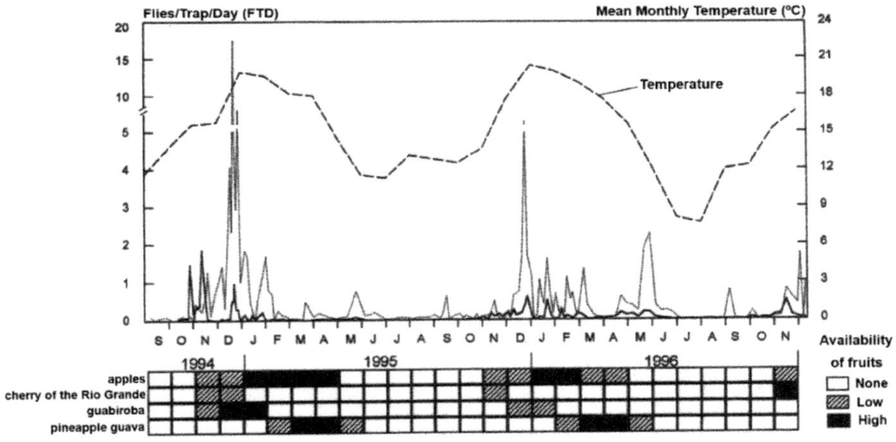

Figure 2. Mean monthly temperatures and population fluctuation of Anastrepha fraterculus *(FTD) from commercial apple orchards (represented by a dark solid line) and from adjacent native forest (by a lighter broken line), together with the availability of apples and wild hosts at Vacaria, Rio Grande do Sul from September 1994 to November 1996 (from Kovaleski et al. 2000).*

The presence of natural hosts from the family Myrtaceae (Malavasi et al. 1980) in the forested areas, bordering the apple orchards, provide an opportunity for *A. fraterculus* populations to be maintained throughout the year in the region. The natural hosts are mainly the cherry of the Rio Grande (*Eugenia involucrata* DC), guabiroba (*Campomanesia xanthocarpa* Berg) and pineapple guava (*Feijoa sellowiana* Berg) (Fig. 2), which bear fruits in November, December-January, and February-May, respectively (Kovaleski et al. 2000).

In the beginning of November, the cherries of the Rio Grande can be infested by *A. fraterculus* females who are residuals from the previous autumn season. When the adults that emerged from the infested cherries become sexually mature, they infest the guabirobas, but available mature apples can also be attacked. After January, pineapple guavas are preferably infested. As the temperatures start to drop in April, the larval development in these guavas can be prolonged until 80 days, and the pupal period can last up to 120 days (Kovaleski 1997; Kovaleski et al. 2000). The flies that emerge in August and beyond May survive until the appearance of new cherries, guabirobas and apples in November (adult overwintering). Thus, autumns with high availability of pineapple guavas and mild winters can be followed by high *A. fraterculus* populations in the spring. The late infestation of apples during March-April may also produce a certain amount of pupae, whose adults can emerge at the end of winter (or overwintering as immatures) (Kovaleski et al. 2000). Therefore, the 4-5 months (June - October), when mean temperatures are below 15°C and no host fruits are available, cause a natural decline in *A. fraterculus* populations each year, creating a perfect window of opportunities for the initiation of the release of parasitoids and sterile flies in the forested areas surrounding apple orchards.

The area-wide management of *A. fraterculus* in the commercial apple orchards in southern Brazil is facilitated by the absence of resident populations in the apple orchards, as the natural breeding sites of this pest are the native forests (Kovaleski et al. 1996, 1999; Santos et al. 2017). In the sierra region, where apple orchards and forested areas intermingle, traps located closer to the forested areas generally catch much more flies (e.g. values can even reach 20 FTD) than those deployed inside the orchards (Fig. 3).

Almost 80% of the damage to fruits occurs in the periphery of the first lines of the apple orchards, because of the prevalence of the foraging behaviour of *A. fraterculus* for oviposition sites and food resources rather than migratory movements (Sugayama 1995; Kovaleski et al. 1996, 1999).

Sugayama et al. (1997) described the diel pattern of *A. fraterculus* in Brazilian apple orchards. When an apple orchard is located closely to forests, most of the flies do not remain in the orchard during the night. The females mostly oviposit their eggs between 16 and 17 h and at nightfall they return to the forested areas. The fact that apple trees in Brazilian orchards do not form dense canopies that could serve as shelter can contribute to this behaviour. Furthermore, apples can be considered alternative hosts for *A. fraterculus* (Salles 1995) as most varieties behave as poor hosts. Immature apples are unsuitable hosts, with less than 1% of survival to pupal stage, and the reproductive rates of *A. fraterculus* in mature apples is usually low (e.g. under field conditions, infestation levels reach 600-800 pupae/kg of apple) (Sugayama 1995; Sugayama et al. 1998; Sugayama and Malavasi 2000).

A decade-long study (Kovaleski et al. 2000) that included fly monitoring, release-recapture trials and host surveys has demonstrated that *A. fraterculus* populations are primarily present in the forested areas that contain native hosts surrounding the apple orchards (Fig. 3). Consequently, these areas should be the primary targets for suppression and the release of sterile flies and parasitoids starting in September. Maintaining sufficient overflooding ratios throughout the final winter months and spring-summer should theoretically suppress the wild populations.

4. THE STERILE INSECT TECHNIQUE AGAINST THE SOUTH AMERICAN FRUIT FLY

The Sterile Insect Technique (SIT) is an effective and environment-friendly control technology that relies on inundative releases of mass-reared insects, sterilized by ionizing radiation (Dyck et al. 2021). This technique has been applied as a component of many area-wide integrated pest management (AW-IPM) programmes against fruit flies, moths, screwworms and tsetse flies (Vreysen et al. 2007).

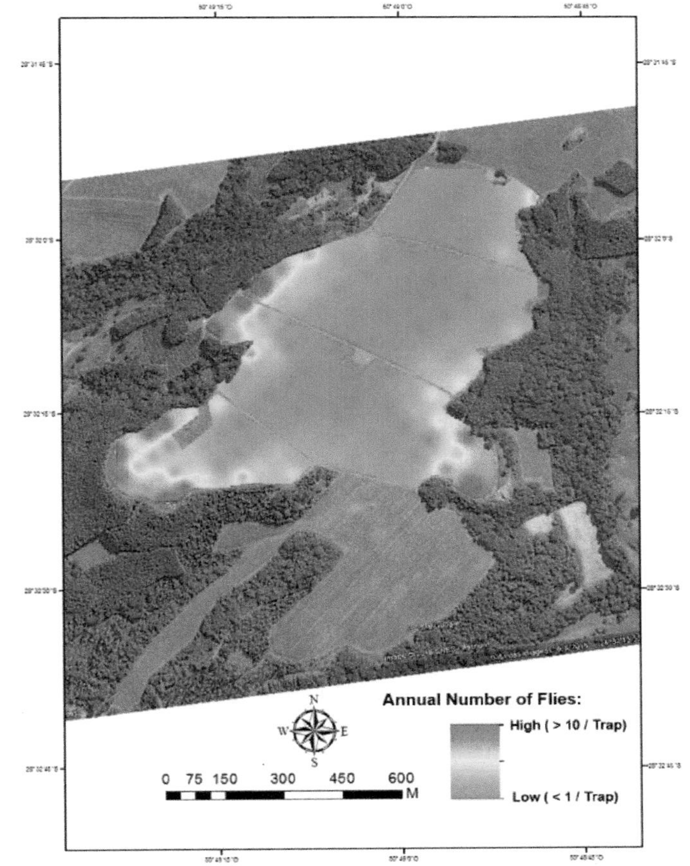

Figure 3. Annual trap catches of wild Anastrepha fraterculus *in McPhail traps baited with grape juice in a commercial apple orchard largely surrounded by forest in Vacaria, Rio Grande do Sul, during the 2016-2017 harvest.*

For example, almost 300 million sterile flies are produced per week at the Moscafrut facility for the control of several *Anastrepha* species of economic importance in southern Mexico (Orozco-Davila et al. 2017). In British Columbia, Canada, the wild populations of *C. pomonella* are being kept at minimum levels since

1997 through the release of sterile moths from the Okanagan-Kootenay Sterile Insect Release (OKSIR) Programme, with populations reduced by 94% and damage reduced to less than 0.2% of fruits in more than 90% of the orchards in the programme area (Judd and Thompson 2012; Simmons et al. 2021; Nelson et al., this volume).

Despite all the knowledge gathered on the biology and genetics of *A. fraterculus* (Cladera et al. 2014), AW-IPM programmes integrating the SIT against this pest are still not being implemented. The SIT has several technical components and major requirements that need to be met: (1) availability of accurate baseline data on the target wild population (e.g. population density in space and time, dispersal patterns etc.); (2) methods available to mass-rear the insect at reasonable cost; (3) irradiation procedures for proper sterilisation of large batches of the mass-produced insect; (4) a reliable quality control management system of the sterile insects that is applied routinely; (5) transport, fly emergence, handling and release technologies available; (6) adequate sterile to wild male overflooding ratios in the field in order to guarantee a significant induction of sterility in the wild population (Dyck et al. 2021). Most of these issues have been addressed by studies conducted with *A. fraterculus*, especially by researchers from Argentina and Brazil (Ortiz 1999).

Morphological and genetic studies have revealed that *A. fraterculus* is actually a complex of cryptic species (Morgante et al. 1980; Steck 1991), and as a consequence, each morphotype should be treated separately for the successful implementation of the SIT (Whitten and Mahon 2021). At least eight morphotypes have been recognized so far based on integrative taxonomy and their geographic distribution in Latin America has been defined (Hernández-Ortiz et al. 2012, 2015; Devescovi et al. 2014; Hendrichs et al. 2015; Prezotto et al. 2017). The results obtained by Dias et al. (2016) showed the existence of full mating compatibility among *A. fraterculus* populations from southern Brazil (populations from Vacaria, Pelotas, Bento Gonçalves and São Joaquim). Thus, southern Brazilian populations and Argentinean morphotypes are likely to belong to the same species within the *A. fraterculus* complex (Rull et al. 2012, 2013).

Significant progress has been made with the domestication and artificial rearing of *A. fraterculus* since the FAO/IAEA Workshop held in Viña del Mar, Chile, in November 1996 (Ortiz 1999), and large colonies have been successfully established in Argentina and Brazil (Salles 1992, 1999; Jaldo et al. 2001; Walder et al. 2006; Vera et al. 2007; Oviedo et al. 2011; Nunes et al. 2013). Walder et al. (2014) developed an artificial rearing system that allows rapid colony built-up and production of enough sterile insects for use in pilot-programmes. The available rearing system can still be optimized to increase insect yields. For example, the rearing costs at the Center for Nuclear Energy in Agriculture from the University of São Paulo (CENA/USP) were reduced by half in 2016 when agar for the larval diet of Salles (1992) was replaced with carrageenan. Ninety litres of pupae (ca. 3 million pupae) of a strain from Vacaria were produced by the F_3 to the F_{12} generation with mean values of 77.4%, 77.0% and 0.49 for egg hatch, adult emergence and sex ratio (♀/♂+♀), respectively (Mastrangelo, unpublished data).

Radiation experiments with gamma and X-rays have shown that treating pupae of *A. fraterculus* with a dose of 40-60 Gy can induce 99% sterility in adult male flies (Bartolucci et al. 2006; Allinghi et al. 2007; Mastrangelo et al. 2010). Although the recommended sterilisation dose for treating pupae 48 h before emergence has been 70 Gy in Argentina (Cladera et al. 2014; Alba et al. 2016), radiation studies with both gamma and X-rays and field cage tests carried out at CENA/USP with the Vacaria strain in 2016 demonstrated that treating pupae 72 h before emergence with 40 Gy is sufficient to produce 99% sterile flies (that are competitive against wild flies). This dose is sufficient as doses higher than 15 Gy induce complete atrophy of the females' ovaries, and a sterile:wild male overflooding ratio of 45:1 induced more than 95% sterility in wild populations (Mastrangelo et al. 2018).

The sterile flies can be released by static ground-based devices (such as cardboard, plastic or PVC boxes), mobile ground-based systems (such as bags or cardboard containers being released from a mechanical device) (Dominiak et al. 2010; Bjeliš et al. 2013), or as chilled adult flies delivered from small aircrafts or even drones (Tan and Tan 2011; Mubarqui et al. 2014; FAO/IAEA 2017).

Despite all the advances made, some issues must be addressed yet, like automation of rearing processes, strain management and sex separation. Progress has been made to study the cuticular hydrocarbons and the chemical composition of the volatiles emitted by *A. fraterculus* males (Vanickova et al. 2012; Milet-Pinheiro et al. 2015), but no specific lure is so far commercially available that would increase the accuracy of field-monitoring. Overall, however, nearly all technical problems have been solved by scientists, and the implementation of the SIT against *A. fraterculus* can already be performed at pilot scales.

Most of the AW-IPM campaigns that include the SIT are composed of three phases of implementation: a preparatory pre-operational phase, a population reduction phase applying suppression measures and, then, the sterile insect release phase (Hendrichs et al. 2021). The pre-operational phase includes obtaining the commitment of all stakeholders, the development of funding mechanisms, of physical infrastructure (the establishment of mass-rearing, sterilisation, packing, fly emergence and release facilities) and securing appropriate management and human resources (i.e. strong leadership, dedicated full-time staff, development of institutional capacity, flexible and independent management structure), collection of baseline data on the distribution and population dynamics of the target species, public awareness, pilot trials in the field, continuity of the implementation of the critical components of the project, and independent reviews of it. Although Cladera et al. (2014) stated that most of these human and managerial components were still missing in Argentina, the case in southern Brazil is different. The Brazilian apple industry is strongly organized and competent to support an endeavour against another pest, as demonstrated by the successful eradication of *C. pomonella* achieved in 2014 after a 17-year campaign (Kovaleski and Mumford 2007; Capra 2014).

In north-eastern Brazil, some progress has also been made in the past few years with the management of fruit flies and mosquitoes after the establishment of Moscamed Brasil in the San Francisco River Valley. This programme has focused on the suppression of populations of *C. capitata* in the valley with the integration of sterile males with other suppression methods (Malavasi et al. 2007). More than 20 million sterile males *C. capitata* were produced for a pilot project in 2006-2007, that successfully suppressed wild populations in more than 2000 ha of mostly mango orchards (Moscamed Brasil 2007). However, most *C. capitata* populations from the San Francisco River Valley are still very high (FTD > 2) due to the presence of alternative hosts almost all year round and an excessive number of neglected orchards (França 2016). In southern states, on the other hand, temperate fruit growers can count on a unique climate advantage (natural suppression during the winter) that makes the management of fruit fly populations less costly.

5. HISTORY OF THE MOSCASUL PILOT PROJECT

The *Moscasul Biological Control and Integrated Fruit Fly Management Center* project, including a mass-rearing facility for *A. fraterculus* and its parasitoids, to be established by EMBRAPA Grape & Wine and the ABPM, was first presented to the Federal Government of Brazil on September 18th 2013, when it obtained the support of the Ministry of Science, Technology and Innovation. In December 2014, the Ministry of Agriculture, Livestock and Food Supply (MAPA) signed a cooperative agreement that included an investment of ca. USD 630 000 for the establishment of the Moscasul Center, and that was followed by parliamentary amendments from two senators and one congressman that increased that budget by USD 329 000. The total amount of resources obtained from authorities through these efforts by EMBRAPA and ABPM amounted to almost USD 959 000 for the initial phases of the project.

After this initial support, almost USD 600 000 were invested in constructing the mass-rearing facility at the Agricultural Experiment Station of EMBRAPA Grape & Wine at Vacaria, and the first containerized rearing modules were installed in May 2016 (Fig. 4). Twenty-one containers (fifteen larger ones: 12.9 m length x 2.9 m height x 2.4 width, and six smaller ones: 5.9 m x 2.9 m x 2.4 m) were installed in front of the agricultural station.

Figure 4. Containerized rearing modules installed for the establishment of the Moscasul Biological Control and Integrated Fruit Fly Management Center at Vacaria, Rio Grande do Sul, Brazil.

Each stage of rearing (i.e. adult colonies, larval rearing, pupal maturation and holding) will take place in separate containers. The selection of a modular system has several advantages: (1) less costly than building an entire single-unit brick facility,

(2) more species can be reared separately, (3) less susceptible to perturbations of daily operations (Tween 1987), and (4) insect production can be increased depending on the demand and availability of funds.

After the political and economic turmoil in Brazil in 2015, the new Minister of Agriculture visited Vacaria on August 9th 2016, when partnership protocols were signed with representatives of research institutes (Universidade Federal do Rio Grande do Sul and CENA/USP) and growers' associations (ABPM and the state farmer's associations AGAPOMI – Associação Gaúcha dos Produtores de Maçã, SINDOCOPEL – Sindicato das Indústrias de Doces e Conservas Alimentícias de Pelotas, and APPRP – Associação dos Produtores de Pêssego da Região de Pelotas), and an additional USD 158 000 were raised from federal funds. In August 2017, EMBRAPA approved an internal project of USD 185 000 to support and implement *A. fraterculus* monitoring and SIT activities.

A gamma or X-ray irradiator still needs to be acquired for the Moscasul Center for the sterilisation of the mass-reared flies. In the meantime, to avoid this constraint and more delays in field tests of the pre-operational phase, the CENA/USP at Piracicaba, São Paulo state, accepted to sterilize the mass-reared flies during the first years of pilot projects. Since the 1970s, research on fruit flies and the SIT has been carried out at this institute, also with the support of the Joint FAO/IAEA Division, and a pilot facility (250 m^2) is present since 1998 dedicated to the production of sterile insects of several species (e.g. up to 2 million sterile *C. capitata* can be reared per week) and to training (Walder 2002). This centre is equipped with two gamma irradiators, i.e. a GammaCell-220™ and a panoramic Gammabeam-650™ (MDS Nordion International Inc., Canada), and an X-ray irradiator RS 2400V (RadSource Technologies Inc., Buford, Georgia, USA).

A pilot trial has been planned in three large apple orchards in Vacaria to demonstrate to growers the feasibility of using the SIT against *A. fraterculus*. Sterile flies will be released between the months of September and March only in surrounding zones of forests, covering a 50-100 m periphery of the target orchards, since these areas serve as reservoir of wild flies. Based on the monitoring data from 15 consecutive years of the selected orchards, approximately 150 000 sterile flies per week would be required for the first pilot release phase. The *A. fraterculus* colonies of the Vacaria strain have been well established at both the EMBRAPA and CENA laboratories. After the fine-tuning of all rearing and sterilisation procedures during the first half of 2017, CENA is planning to ship more than 200 000 irradiated pupae weekly by air to Vacaria for 6 months, starting in September when the level of the wild fly population is extremely low in the three pilot areas following the winter (average FTD < 0.5).

Most of the studies on marking and shipment procedures have been completed and the teams have received training in terms of surveillance, distribution of the sterile flies and identification of the caught insects. Depending on the level of production of *Diachasmimorpha longicaudata* (Ashmead) at EMBRAPA and CENA/USP at the time of the trials, this parasitoid is also intended to be released in some of the areas. The feasibility of shipping irradiated *A. fraterculus* eggs for the mass-rearing *of D. longicaudata* and *Doryctobracon areolatus* (Szépligeti) at Vacaria is also being assessed (Nunes et al. 2011; Costa et al. 2016).

Arrangements to establish a pilot trial releasing sterile flies and parasitoids to suppress *A. fraterculus* in peach orchards from the region of Pelotas are also being made. The results of these pilot trails will have the potential to influence the direction of future control tests and to lead towards the sustainable management of *A. fraterculus* by hundreds of apple farmers and other temperate fruit-growing farmers from the southern states of Brazil.

6. REFERENCES

Alba, M. G., D. Segura, M. M. Terrada, and S. Lopez. 2016. Estudio comparativo sobre el efecto de la radiación X y gamma sobre pupas de *Anastrepha fraterculus* (Wied). *In* D. Quiroga (ed.), Libro de Resumenes: 9th Meeting of Tephritid Workers of the Western Hemisphere, 17-21 October 2016, Buenos Aires, Argentina.

Allinghi, A., C. Gramajo, E. Willink, and J. C. Vilardi. 2007. Induction of sterility in *Anastrepha fraterculus* (Diptera: Tephritidae) by gamma radiation. Florida Entomologist 90: 96–102.

Aluja, M., M. Ordano, L. Guillen, and J. Rull. 2012. Understanding long-term fruit fly (Diptera: Tephritidae) population dynamics: Implications for areawide management. Journal of Economic Entomology 105: 823–836.

Bartolucci, A., M. T. Vera, V. Yusef, and A. Oviedo. 2006. Morphological characterization of the reproductive system of irradiated *Anastrepha fraterculus*, pp. 45–52. *In* R. L. Sugayama, R. A. Zucchi, S. M. Ovruski, and J. Sivinski (eds.), Proceedings of the 7th International Symposium on Fruit Flies of Economic Importance, 10-15 September 2006, Salvador, Brazil.

Bortoli, L. C., R. Machota Jr., F. R. M. Garcia, and M. Botton. 2016. Evaluation of food lures for fruit flies (Diptera: Tephritidae) captured in a citrus orchard of the Serra Gaucha. Florida Entomologist 99: 381–384.

Bjeliš, M., D. Radunić, and P. Bulić. 2011. Pre- and post-release quality of sterile *Ceratitis capitata* males released by an improved automated ground release machine. Journal of Applied Entomology 137: 154–162.

Brazilian Apple Yearbook. 2017. Anuário brasileiro da maçã de 2017. Editora Gazeta Santa Cruz, Santa Cruz do Sul, Rio Grande do Sul, Brasil. 56 pp.

Calore, R. A., J. C. Galli, W. C. Pazini, R. T. Duarte, and J. A. Galli. 2013. Fatores climáticos na dinâmica populacional de *Anastrepha* spp. (Diptera: Tephritidae) e de *Scymnus* spp. (Coleoptera: Coccinellidae) em um pomar experimental de goiaba (*Psidium guajava* L.). Revista Brasileira de Fruticultura 35: 67–74.

Canal Daza, N. A., R. A. Zucchi, N. M. Silva, and F. L. Leonel Jr. 1994. Reconocimiento de las especies de parasitóides (Hym., Braconidae) de moscas de las frutas (Dip., Tephritidae) en dos municipios del Estado de Amazonas, Brasil. Boletín del Museo de Entomología de la Universidad del Valle 2: 1–17.

Capra, G. A. 2014. Brazil is free from codling moth. Embrapa Grape & Wine, Ministry of Agriculture, Livestock and Food Supply. Brasilia, Brasil.

Costa, M. L. Z., M. G. Pacheco, L. A. Lopes, V. W. Botteon, and T. Mastrangelo. 2016. Irradiation of *Anastrepha fraterculus* (Diptera: Tephritidae) eggs to inhibit fly emergence in the mass-rearing of *Diachasmimorpha longicaudata* (Hymenoptera: Braconidae). Journal of Insect Science 16: 98–107.

Couso-Ferrer, F., R. Arouri, B. Beroiz, N. Perera, A. Cervera, V. Navarro-Llopis, P. Castañera, P. Hernández-Crespo, and F. Ortego. 2011. Cross-resistance to insecticides in a malathion-resistant strain of *Ceratitis capitata* (Diptera: Tephritidae). Journal of Economic Entomology 104: 1349–1356.

Cladera, J. L., J. C. Vilardi, M. Juri, L. E. Paulin, M. C. Giardini, P. V. Gomez Cendra, D. F. Segura, and S. B. Lanzavecchia. 2014. Genetics and biology of *Anastrepha fraterculus*: Research supporting the use of the Sterile Insect Technique (SIT) to control this pest in Argentina. BMC Genetics 15 (Suppl. 2): S12.

Devescovi, F., S. Abraham, A. K. P. Roriz, N. Nolazco, R. Castañeda, E. Tadeo, C. Cáceres, D. F. Segura, M. T. Vera, I. Joachim-Bravo, N. Canal, and J. Rull. 2014. Ongoing speciation within the *Anastrepha fraterculus* cryptic species complex: The case of the Andean morphotype. Entomologia Experimentalis et Applicata 152: 238–247.

Dias, V. S., J. C. Silva, K. M. Lima, C. S. C. D. Petitinga, V. Hernández-Ortiz, R. A. Laumann, B. J. Paranhos, K. Uramoto, R. A. Zucchi, and I. S. Joachim-Bravo. 2016. An integrative multidisciplinary approach to understanding cryptic divergence in Brazilian species of the *Anastrepha fraterculus* complex (Diptera: Tephritidae). Biological Journal of the Linnean Society 117: 725–746.

Dominiak, B. C., S. Sundaralingam, L. Jiang, A. J. Jessup, and I. M. Barchia. 2010. Impact of marker dye on adult eclosion and flight ability of mass-produced Queensland fruit fly *Bactrocera tryoni* (Froggatt) (Diptera: Tephritidae). Australian Journal of Entomology 49: 166–169.

Dyck, V. A., J. Hendrichs, A. S. Robinson (eds.). 2021. Sterile Insect Technique – Principles and practice in Area-Wide Integrated Pest Management. Second Edition. CRC Press, Boca Raton, Florida, USA. 1200 pp.

Eddleston, M., S. Adhikari, S. Egodage, H. Ranganath, F. Mohamed, G. Manuweera, S. Azher, S. Jayamanne, E. Juzczak, M. R. Sheriff, A. H. Dawson, and N. A Buckley. 2012. Effects of a provincial ban of two toxic organophosphorus insecticides on pesticide poisoning hospital admissions. Clinical Toxicology 50: 202–209.

(FAO/IAEA) Food and Agriculture Organization of the United Nations/International Atomic Energy Agency. 2017. Guideline for packing, shipping, holding and release of sterile flies in area-wide fruit fly control programmes. J. L. Zavala-López and W. R. Enkerlin (eds.). Second Edition, Rome, Italy. 140 pp.

(FAO/IAEA) Food and Agriculture Organization of the United Nations/International Atomic Energy Agency. 2018. Trapping guidelines for area-wide fruit fly programmes. W. R. Enkerlin, and J. Reyes-Flores, J. (eds.). Second edition. Rome, Italy. 65 pp.

França, P. R. P. 2016. Flutuação populacional de moscas-das-frutas (Diptera: Tephritidae) em pomares comerciais de mangueira e viabilidade de implantação de área de baixa prevalência em Petrolina, Pernambuco. Dissertation, Universidade Federal de Viçosa, Minas Gerais, Brasil. 50 pp.

(GAIN) Global Agricultural Information Network Report. 2016. USDA Foreign Agricultural Service. Brazil Annual Fresh Deciduous Fruit Report. GAIN Report Number: BR 1617.

Garcia, F. R. M., J. V. Campos, and E. Corseiul. 2003. Flutuação populacional de *Anastrepha fraterculus* (Wiedemann, 1830) (Diptera, Tephritidae) na região oeste de Santa Catarina, Brasil. Revista Brasileira de Entomologia 47: 415–420.

Hendrichs, J., T. Vera, M. de Meyer, and A. Clarke. 2015. Resolving cryptic species complexes of major tephritid pests. ZooKeys 540: 5–39.

Hendrichs, J., M. J. B. Vreysen, W. R. Enkerlin, and J. P. Cayol. 2021. Strategic options in using sterile insects for area-wide integrated pest management, pp. 839–882. *In* V.A. Dyck, J. Hendrichs, and A.S. Robinson (eds.), Sterile Insect Technique – Principles and practice in Area-Wide Integrated Pest Management. Second Edition. CRC Press, Boca Raton, Florida, USA.

Harter, W. R., M. Botton, D. E. Nava, A. D. Grutzmacher, R. S. Gonçalves, R. M. Junior, D. Bernardi, and O. Z. Zanardi. 2015. Toxicities and residual effects of toxic baits containing spinosad or malathion to control the adult *Anastrepha fraterculus* (Diptera: Tephritidae). Florida Entomologist 98: 202–208.

Hernández-Ortiz, V., A. F. Bartolucci, P. Morales-Valles, D. Frias, and D. Selivon. 2012. Cryptic species of the *Anastrepha fraterculus* complex (Diptera: Tephritidae): A multivariate approach for the recognition of South American morphotypes. Annals of the Entomological Society of America 105: 305–318.

Hernández-Ortiz, V., N. A. Canal, J. O. T. Salas, F. M. Ruíz-Hurtado, and J. F. Dzul-Cauich. 2015. Taxonomy and phenotypic relationships of the *Anastrepha fraterculus* complex in the Mesoamerican and Pacific Neotropical dominions (Diptera, Tephritidae). ZooKeys 540: 95–124.

Jaldo, H. E., M. C. Gramajo, and E. Willink. 2001. Mass rearing of *Anastrepha fraterculus* (Diptera: Tephritidae): A preliminary strategy. Florida Entomologist 84: 716–718.

Judd, G., and D. Thompson. 2012. Taking a flexible approach to mating disruption in British Columbia. *In* Abstracts from the 86th Orchard Pest Management Conference, January 11–13 2012, Portland, Oregon, USA.

Klanovicz, J. 2010. Toxicity and apple production in southern Brazil. História, Ciências, Saúde – Manguinhos, vol. 17, no. 1 (Jan.-Mar 2010), Rio de Janeiro, Brazil. 18 pp.

Kovaleski, A. 1997. Processos adaptativos na colonização da maçã (*Malus domestica*) por *Anastrepha fraterculus* (Wied.) (Diptera: Tephritidae) na região de Vacaria, Rio Grande do Sul. Thesis Doutorado em Ciências. Instituto de Biociências, Universidade São Paulo, São Paulo Brasil. 122 pp.

Kovaleski, A., and L. G. Ribeiro. 2003. Manejo de pragas na produção integrada de maçãs. Circular Técnica, 34. Embrapa Uva e Vinho, Bento Gonçalves, Rio Grande do Sul, Brasil. 7 pp.

Kovaleski, A., and J. Mumford. 2007. Pulling out the evil by the root: The codling moth *Cydia pomonella* eradication programme in Brazil, pp. 581–590. *In* M. J. B Vreysen, A. S. Robinson, and J. Hendrichs (eds.), Area-wide control of insect pests: From research to field implementation. Springer, Dordrecht, The Netherlands.

Kovaleski, A., K. Uramoto, R. L. Sugayama, N. A. Canal, and A. Malavasi. 1996. A survey of *Anastrepha* species present in apple growing area from Rio Grande do Sul, Brazil, pp. 32. *In* Proceedings 2nd Meeting of the Working Group on Fruit Flies of the Western Hemisphere, Viña del Mar, Chile.

Kovaleski, A., R. L. Sugayama, and A. Malavasi. 1999. Movement of *Anastrepha fraterculus* from native breeding sites into apple orchards in southern Brazil. Entomologia Experimentalis et Applicata 91: 457–463.

Kovaleski, A., R. L. Sugayama, K. Uramoto and A. Malavasi. 2000. Rio Grande do Sul, pp. 285–290. *In* A. Malavasi, and R. A. Zucchi (eds.), Moscas-das-frutas de importância econômica no Brasil: Conhecimento básico e aplicado. FAPESP-Holos, Ribeirão Preto, São Paulo, Brasil.

Machado, A. E., L. A. B. Salles, and A. E Loeck. 1995. Exigências térmicas de *Anastrepha fraterculus* (Wied.) e estimativa do número de gerações em Pelotas, RS. Anais da Sociedade Entomológica do Brasil 24: 573–579.

Magnabosco, A. L. 1994. Influência de fatores físicos e químicos de maçãs, cv. *Gala*, no ataque e desenvolvimento larval de *Anastrepha fraterculus* (Wied., 1830) (Diptera: Tephritidae). Dissertation Mestrado em Fitossanidade, Curso de Pós-graduação em Agronomia, Universidade Federal de Pelotas, Rio Grande do Sul, Brasil. 95 pp.

Malavasi, A., J. S. Morgante, and R. A. Zucchi. 1980. Biologia de 'moscas-das-frutas' (Diptera: Tephritidae). I. Lista de hospedeiros e ocorrência. Revista Brasileira de Biologia 40: 9–16.

Malavasi, A., A. Nascimento, B. A. J. Paranhos, M. L. C. Costa, and J. M. M. Walder. 2007. Implementation of a medfly, fruit fly parasitoids and codling moth rearing facility in northeastern Brazil, pp. 527–534. *In* M. J. B. Vreysen, A. S. Robinson, and J. Hendrichs (eds.), Area-wide control of insect pests: From research to field implementation. Springer, Dordrecht, The Netherlands.

Mastrangelo, T., A. G. Parker, A. Jessup, R. Pereira, D. Orozco-Dávila, A. Islam, T. Dammalage, and J. M. Walder. 2010. A new generation of X ray irradiators for insect sterilisation. Journal of Economic Entomology 103: 85–94.

Mastrangelo, T., A. Kovaleski, V. Botteon, W. Scopel, and M. L. Z. Costa. 2018. Optimization of the sterilizing doses and overflooding ratios for the South American fruit fly. PLoS One 13:e0201026.

Milet-Pinheiro, P., D. M. A. Navarro, N. C. De Aquino, L. L. Ferreira, R. F. Tavares, R. C. C. Da Silva, A. Lima Mendonça, L. Vaníčková, A. L. Mendonça, and R. R. do Nascimento. 2015. Identification of male-borne attractants in *Anastrepha fraterculus* (Diptera: Tephritidae). Chemoecology 25: 115–122.

Morgante, J. S., A. Malavasi, and G. L. Bush. 1980. Biochemical systematics and evolutionary relationships of neotropical *Anastrepha*. Annals of Entomological Society of America 73: 622–630.

Moscamed Brasil. 2007. Relatório anual de atividades da biofábrica Moscamed Brasil de 2007.

Mubarqui, R. L., R. C. Perez, R. A. Kladt, J. L. Z. Lopez, A. Parker, M. T. Seck, B. Sall, and J. Bouyer. 2014. The smart aerial release machine, a universal system for applying the Sterile Insect Technique. PLoS One 9(7): e103077.

McPhail, M. 1937. Relation of time of day, temperature, and evaporation to attractiveness of fermenting sugar solution to Mexican fruit fly. Journal of Economic Entomology 30: 793–799.

Nora, I., and E. R. Hickel. 2002. Pragas da macieira, pp. 463–498. *In* A Cultura da macieira. Epagri-Itajaí, Florianópolis, Santa Catarina, Brasil.

Nora, I., E. R. Hickel, and H. F. Prado. 2000. Moscas-das-frutas nos estados brasileiros: Santa Catarina, pp. 271–275. *In* A. Malavasi and R.A. Zucchi (eds.), Moscas-das-frutas de importância econômica no Brasil: Conhecimento básico e aplicado. FAPESP-Holos, Ribeirão Preto, São Paulo, Brasil.

Norrbom, A. L. 2004. Host plant database for *Anastrepha* and *Toxotrypana* (Diptera: Tephritidae); Diptera data dissemination disk. North American Dipterist's Society, Washington, DC, USA.

Nunes, A. M., D. E. Nava, F. A. Muller, R. S. Gonçalves, and M. S. Garcia. 2011. Biology and parasitic potential of *Doryctobracon areolatus* on *Anastrepha fraterculus* larvae. Pesquisa Agropecuária Brasileira 46: 669–671.

Nunes A. M., K. Z. Costa, K. M. Faggioni, M. L. Z. Costa, R. S. Gonçalves, J. M. M. Walder, M. S. Garcia, and D. E. Nava. 2013. Dietas artificiais para a criação de larvas e adultos da mosca-das-frutas sul-americana. Pesquisa Agropecuária Brasileira 48: 1309–1314.

Orozco-Davila, D., L. Quintero, E. Hernández, E. Solis, T. Artiaga, R. Hernández, C. Ortega, and P. Montoya. 2017. Mass rearing and sterile insect releases for the control of *Anastrepha* spp. pests in Mexico–A review. Entomologia Experimentalis et Applicata 164: 176–187.

Ortiz, G. 1999. Introduction, pp. 1–2. *In* The South American fruit fly, *Anastrepha fraterculus* (Wied.): Advances in artificial rearing, taxonomic status and biological studies. Proceedings of FAO/IAEA *Anastrepha fraterculus* Workshop, 1-2 November 1996, Viña del Mar, Chile. IAEA-TECDOC-1064, IAEA, Vienna, Austria.

Oviedo, A., D. Nestel, N. T. Papadopoulos, M. J. Ruiz, S. C. Prieto, E. Willink, and M. T. Vera. 2011. Management of protein intake in the fruit fly *Anastrepha fraterculus*. Journal of Insect Physiology 57: 1622–1630.

Petri, J. L., G. B. Leite, M. Couto, and P. Francescatto. 2011. Advances of the apple crop in Brazil. Revista Brasileira de Fruticultura 33: 48–56.

Puzzi, D., and A. Orlando. 1957. Ensaios de combate às "Mosca-das-frutas" *Ceratitis capitata* (Wied.) e *Anastrepha* spp. por meio de pulverizações de iscas envenenadas. O Biológico 23: 21–25.

Prezotto, L. F., A. L. P. Perondini, V. Hernández-Ortiz, C. L. Marino, and D. Selivon. 2017. *Wolbachia* strains in cryptic species of the *Anastrepha fraterculus* complex (Diptera, Tephritidae) along the Neotropical region. Systematic and Applied Microbiology 40: 59–67.

Rawn, D. F. K., S. C. Quade, J. Shields, G. Conca, W. F. Sun, G. M. A. Lacroix, M. Smith, A. Fouquet, and A. Belanger. 2006. Organophosphate levels in apple composites and individual apples from a treated Canadian orchard. Journal of Agricultural and Food Chemistry 54: 1943−1948.

Rosa, J. M., C. J. Arioli, J. P. dos Santos, A. C. Menezes-Netto, and M. Botton. 2017. Evaluation of food lures for capture and monitoring of *Anastrepha fraterculus* (Diptera: Tephritidae) on temperate fruit trees. Journal of Economic Entomology 110: 995−1110.

Rull, J., S. Abraham, A. Kovaleski, D. F. Segura, A. Islam, V. Wornoayporn, T. Dammalage, U. S. Tomas, and M. T. Vera. 2012. Random mating and reproductive compatibility among Argentinean and southern Brazilian populations of *Anastrepha fraterculus* (Diptera: Tephritidae). Bulletin of Entomological Research 102: 435–443.

Rull, J., S. Abraham, A. Kovaleski, D. F. Segura, M. Mendoza, M. C. Liendo, and M. T. Vera. 2013. Evolution of pre-zygotic and post-zygotic barriers to gene flow among three cryptic species within the *Anastrepha fraterculus* complex. Entomologia Experimentalis et Applicata 148: 213–222.

Santos, J. P., L. R. Redaelli, J. Sant'Ana, and E. R. Hickel. 2015. Suscetibilidade de genótipos de macieira a *Anastrepha fraterculus* (Diptera: Tephritidae) em diferentes condições de infestação. Revista Brasileira de Fruticultura 37: 77–83.

Santos, J. P., L. R. Redaelli, J. Sant'Ana, and E. R. Hickel. 2017. Population fluctuation and estimate of generations number of *Anastrepha fraterculus* (Diptera: Tephritidae) in apple orchard in Caçador, Santa Catarina, Brazil. Arquivos do Instituto Biológico 84: 1–7, e0482015.

Salles, L. A. B. 1992. Metodologia de criação de *Anastrepha fraterculus* (Wied.) (Diptera: Tephritidae) em dieta artificial em laboratório. Anais da Sociedade Entomológica do Brasil 21: 479–486.

Salles, L. A. B. 1993. Efeito da temperatura constante na oviposição e no ciclo de vida de *Anastrepha fraterculus*. Anais da Sociedade Entomológica do Brasil 22: 57–62.

Salles, L. A. B. 1995. Bioecologia e controle da mosca-das-frutas sul-americana. Embrapa-CPACT, Pelotas, Rio Grande do Sul, Brasil. 58 pp.

Salles, L. A. B. 1998. Principais pragas e seu controle, pp. 205–242. *In* M. C. B. R. Raseira (ed.), A cultura do pessegueiro. Brasília: Embrapa-SPI; Embrapa-CPACT, Pelotas, Rio Grande do Sul, Brasil.

Salles, L. A. B. 1999. Rearing of *Anastrepha fraterculus* (Wiedemann), pp. 95-100. *In* The South American fruit fly, *Anastrepha fraterculus* (Wied.): Advances in artificial rearing, taxonomic status and biological studies. Proceedings of FAO/IAEA *Anastrepha fraterculus* Workshop, 1-2 November 1996, Viña del Mar, Chile. IAEA-TECDOC-1064. IAEA, Vienna, Austria.

Salles, L. A. B. 2000. Biologia e ciclo de vida de *Anastrepha fraterculus* (Wied.), pp. 81–86. *In* A. Malavasi and R. A. Zucchi (eds.), Moscas-das-frutas de importância econômica no Brasil: Conhecimento básico e aplicado. Holos Editora, Ribeirão Preto, São Paulo, SP, Brasil.

Salles, L. A. B., and A. Kovaleski. 1990. Inseticidas para controle da mosca-das-frutas. Horti Sul 1(3): 10–11.

Simmons, G. S., K. A. Bloem, S. Bloem, J. E. Carpenter, and D. M. Suckling. 2021. Impact of moth suppression/eradication programmes using the Sterile Insect Technique or inherited sterility, pp.

1005–1048. *In* V.A. Dyck, J. Hendrichs, and A.S. Robinson (eds.), Sterile Insect Technique – Principles and practice in Area-Wide Integrated Pest Management. Second Edition. CRC Press, Boca Raton, Florida, USA.

Sugayama, R. L. 1995. Comportamento, demografia, e ciclo de vida de *Anastrepha fraterculus* Wied. (Diptera: Tephritidae) associada a três cultivares de maçã no sul do Brasil. Dissertation Mestrado, Instituto de Biosciências, Universidade de São Paulo, Piracicaba, SP, Brasil. 97 pp.

Sugayama, R. L. 2000. *Anastrepha fraterculus* (Wiedeman) (Diptera: Tephritidae) na região produtora de maçãs do Rio Grande do Sul: Relação com os inimigos naturais e potencial para o controle biológico. PhD thesis, Instituto de Biociências, Universidade de São Paulo, São Paulo, SP, Brasil.

Sugayama, R. L., and A. Malavasi. 2000. Ecologia comportamental, pp. 99–108. *In* A. Malavasi and R. A. Zucchi (eds.), Moscas das frutas de importância econômica no Brasil: Conhecimento básico e aplicado. FAPESP-Holos Editora, Ribeirão Preto, São Paulo, SP, Brasil.

Sugayama, R. L., E. S. Branco, A. Malavasi, A. Kovaleski, and I. Nora. 1997. Oviposition behavior and preference of *Anastrepha fraterculus* in apple and dial pattern of activity in an apple orchard in Brazil. Entomologia Experimentalis et Applicata 83: 239–245.

Sugayama, R. L., A. Kovaleski, P. Liedo, and A. Malavasi. 1998. Colonization of a new fruit crop by *Anastrepha fraterculus* (Diptera: Tephritidae) in Brazil: A demographic analysis. Environmental Entomology 27: 642–648.

Scoz, P. L., M. Botton, M. S. Garcia, and P. L. Pastori. 2006. Avaliação de atrativos alimentares e armadilhas para o monitoramento de *Anastrepha fraterculus* (Wiedemann, 1830) (Diptera: Tephritidae) na cultura do pessegueiro (*Prunus pérsica*). Idesia (Arica) 24: 7–13.

Steck, G. J. 1998. Taxonomic status of *Anastrepha fraterculus*, pp. 13-20. *In* The South American fruit fly, *Anastrepha fraterculus* (Wied.): Advances in artificial rearing, taxonomic status and biological studies. Proceedings of FAO/IAEA *Anastrepha fraterculus* Workshop, 1-2 November 1996, Viña del Mar, Chile. IAEA-TECDOC 1064. IAEA, Vienna, Austria.

Steck, G. J. 1991. Biochemical systematics and population genetic structure of *Anastrepha fraterculus* and related species (Diptera: Tephritidae). Annals of Entomological Society of America 84:10–28.

Tan, L. T., and K. H. Tan. 2011. Alternative air vehicles for Sterile Insect Technique aerial release. Journal of Applied Entomology 137: 126–141.

Tween, G. 1987. A modular approach to fruit fly production facilities for the Mediterranean fruit fly Central American program, pp. 283–291. *In* A.P. Economopoulos (ed.), Proceedings of Second International Symposium on Fruit Flies of Economic Importance, 16–21 September 1986, Crete, Greece. Elsevier Science Publishers, Amsterdam, The Netherlands.

Vanickova, L., A. Svatos, J. Kroiss, M. Kaltenpoth, R. R. Nascimento, M. Hoskovec, R. Břízová, and B. Kalinova. 2012. Cuticular hydrocarbons of the South American fruit fly *Anastrepha fraterculus*: variability with sex and age. Journal of Chemical Ecology 38: 1133–1142.

Vera, M. T., S. Abraham, A. Oviedo, and E. Willink. 2007. Demographic and quality control parameters of *Anastrepha fraterculus* (Diptera: Tephritidae) maintained under artificial rearing. Florida Entomologist 90: 53–57.

Vreysen, M. J. B., J. Gerardo-Abaya, and J. P. Cayol. 2007. Lessons from Area-Wide Integrated Pest Management (AW-IPM) programmes with an SIT component: An FAO/IAEA perspective, pp. 723–744. *In* M. J. B. Vreysen, A. S. Robinson, and J. Hendrichs (eds.), Area-wide control of insect pests: From research to field implementation. Springer, Dordrecht, The Netherlands.

Walder, J. M. M. 2002. Produção de moscas-das-frutas e seus inimigos naturais: Associação de moscas estéreis e controle biológico, pp. 181–190. *In* J. R. P. Parra, P. S. M. Botelho, B. S. Correa-Ferreira, and J. M. S. Bento (eds.), Controle biológico no Brasil: Parasitóides e predadores. Editora Manole, São Paulo, SP, Brasil.

Walder, J. M. M., M. L. Z. Costa, and T. A. Mastrangelo. 2006. Developing mass-rearing system for *Anastrepha fraterculus* and *Anastrepha obliqua* for future SIT-AWIPM procedures in Brazil. *In* 1st. Progress Report: FAO/IAEA 2nd Research Coordination Meeting, 5-9 September, Salvador, Bahia, Brazil.

Walder, J. M. M., R. Morelli, K. Z. Costa, K. M. Faggioni, P. A. Sanches, B. A. J. Paranhos, J. M. S. Bento, and M. L. Z Costa. 2014. Large scale artificial rearing of *Anastrepha* sp.1 *aff. fraterculus* (Diptera: Tephritidae) in Brazil. Scientia Agricola 71: 281–286.

Whitten, M., and R. Mahon. 2021. Misconceptions and constraints driving opportunities, pp. 45–74. *In* V. A. Dyck, J. Hendrichs, and A. S. Robinson (eds.), Sterile Insect Technique – Principles and practice in Area-Wide Integrated Pest Management. Second Edition. CRC Press, Boca Raton, Florida, USA.

SECTION 2

ANIMAL AND HUMAN HEALTH

AREA-WIDE MANAGEMENT OF STABLE FLIES

D. B. TAYLOR

Agroecosystem Management Research Unit, USDA-ARS, Lincoln, Nebraska, USA; Dave.Taylor@ARS.USDA.GOV

SUMMARY

Stable flies are highly vagile and their dispersal ability appears to be limited only by the availability of hosts. In addition, stable fly larval developmental substrates are diverse, dispersed and often difficult to locate. This life history necessitates the use of area-wide integrated pest management (AW-IPM) strategies if effective control of stable flies is to be achieved, but complicates the use of the Sterile Insect Technique (SIT) and mating disruption technologies often employed in such programmes against other insect pests. Area-wide management of stable flies will require nationally or regionally coordinated implementation of traditional control methods, including sanitation/cultural, biological, and chemical technologies. An administrative structure will need to be implemented to coordinate, monitor, inspect and enforce compliance, especially if agronomic crop residues are integral to stable fly infestations. Research on stable fly developmental substrates and their management, larval and adult population dynamics, efficient and economical adult suppression systems, including traps and targets, is needed to improve the efficiency and economy of area-wide management of stable flies.

Key Words: Stomoxys calcitrans, control, IPM, livestock, dispersal, crop residues, regional coordination, filth flies

1. INTRODUCTION

Stable flies, *Stomoxys calcitrans* (L.), are important, pests of livestock throughout much of the world. Their painful bites disrupt feeding and other behaviours of livestock (Dougherty et al. 1993, 1994, 1995; Mullens et al. 2006), reducing productivity (Campbell et al. 1993, 2001) and, in extreme infestations, resulting in mortality (Bishopp 1913). In addition to their effects upon cattle, stable flies disrupt human recreational activities (Newson 1977) and molest companion animals (Yeruham and Braverman 1995) and wildlife (Elkan et al. 2009) throughout their range. Landing counts of 80-100 flies per minute on humans have been observed on the beaches of north-western Florida (Hogsette et al. 1987).

Adult stable flies are obligate hematophages, both males and females require blood prior to mating (Anderson 1978). Females need 3-4 blood meals to develop

J. Hendrichs, R. Pereira and M. J. B. Vreysen (eds.), Area-Wide Integrated Pest Management: Development and Field Application, pp. 233–250. CRC Press, Boca Raton, Florida, USA.

their first batch of eggs and 2 more for each additional batch (Bishopp 1913; Anderson and Tempelis 1970). After feeding, stable flies retire to a nearby surface, frequently warmed by the sun, to digest their blood meal. Immature stable flies develop in decaying or fermenting vegetative materials frequently contaminated with animal dung or urine (Simmons and Dove 1941, 1942; Silverly and Schoof 1955; Hafez and Gamal-Eddin 1959; Campbell and McNeal 1979; Hall et al. 1982) where larval densities can exceed 20 000 per square-meter of substrate (Patterson and Morgan 1986; Broce et al. 2005).

Four species of flies, face fly (*Musca autumnalis* De Geer), house fly (*Musca domestica* L.), horn fly (*Haematobia irritans* (L.)), and stable fly are frequently found in association with livestock. Often, these flies are referred to collectively as "filth flies." Although morphologically similar, the behaviour and biology of these flies are distinct (Moon 2002; Zumpt 1973). Stable fly and horn fly are obligate parasites, primarily of livestock, with biting mouth parts. Face fly and house fly are non-biting flies with sponging mouthparts. Face fly and horn fly larvae develop in fresh, undisturbed bovine dung. Stable fly and house fly larvae develop in older or aged manure, frequently mixed with decomposing vegetative material as well as decomposing non-manure substrates. Horn flies are semi-permanent parasites, spending the majority of their adult life on a host, whereas stable flies are temporary parasites, visiting the host only to blood-feed. Face flies are obligate parasites as well, but rather than feeding on blood, they feed on mucus and other fluids around the eyes and mouth of the host. Because of these biological differences, many of the technologies and methods used for their management are species-specific. Proper identification of the offending fly species is essential before initiating a management programme.

2. DEVELOPMENT IN CROP RESIDUES

Many types of decomposing and fermenting organic materials support stable fly larval development (Hogsette et al. 1987), although in North America most practitioners consider residues from livestock production systems and barnyards to be the primary sources. This, however, has not always been the case. In the first half of the 20th century, straw of oats, rice, barley, and wheat were reportedly the most common developmental substrates. Population levels were correlated with grain production (Bishopp 1913) and severe stable fly outbreaks were attributed to development in peanut straw, celery and bay grass (Dove and Simmons 1941, 1942). Recently, agronomic crop residues have re-emerged as important sources of stable flies. Serious outbreaks associated with pineapple production have been reported in Costa Rica (Herrero et al. 1989, 1991), sugarcane in Brazil and Mauritius (Kunz and Monty 1976; Koller et al. 2009; Souza Dominghetti et al. 2015), and vegetable crop residues in Western Australia (Cook et al. 1999). Counts of greater than 2000 stable flies per animal are being associated with development in crop residues (Fig. 1).

This gives rise to one of the primary differences between the stable fly situation in the USA and that in Australia, Brazil, Costa Rica, and potentially other countries. In the USA, stable fly larval developmental sites are typically associated with livestock production. In a sense, livestock producers are responsible for the problem they perceive. Where stable fly developmental sites are being attributed to agronomic production and crop residues (e.g. Australia, Brazil and Costa Rica), another industry, or someone else, is responsible for the problem. Calls for regulation and government action are louder when someone else is to blame.

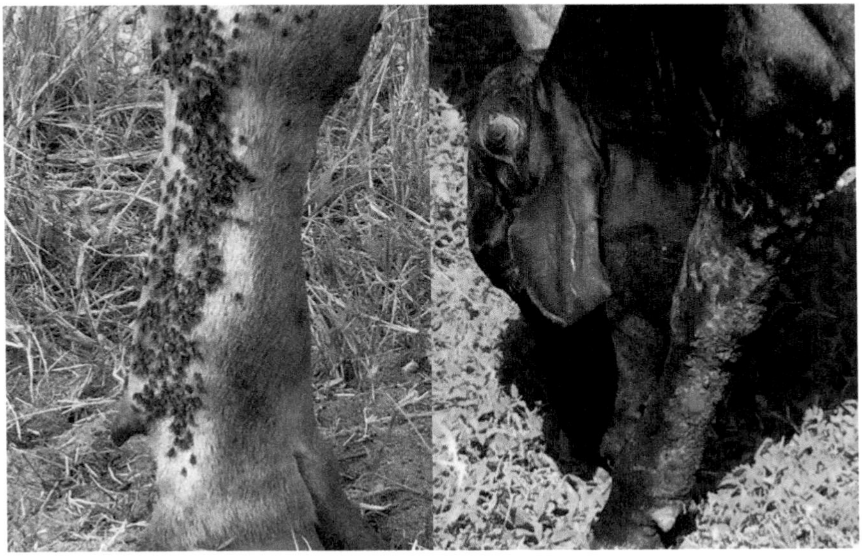

Figure 1. Stable flies on leg of steer (left, photographer David Cook) and damage from stable fly bites in Costa Rica (right, photographer Jose Solórzano).

3. AREA-WIDE MANAGEMENT

3.1. Need for Area-wide Management

The biology of stable flies necessitates the incorporation of area-wide concepts for their management. Adults are highly vagile, capable of flying up to 30 km in 24 hours on a flight mill (Bailey et al. 1973) and 8 km in less than 2 hours in the field. Mean dispersal distance from a natural larval developmental site was 1.5 km (Taylor et al. 2010), however Hogsette and Ruff (1985) reported individual flies dispersing over 225 km. The dispersal ability of stable flies appears to be limited only by the availability of hosts. They disperse until suitable hosts are located. Because both male and female stable flies require a blood meal prior to mating, most of the dispersing flies appear to be physiologically young. The efficacy of managing stable flies on individual premises, or focusing control efforts to locations where populations exceed the economic threshold, is limited by the ability of flies to disperse from premises and locations with no stable fly control to those attempting to control this pest.

Insects are best targeted for control when they are concentrated, immobile and accessible (Horsfall 1985). For stable flies, as for most pestiferous Diptera, this would be during the immature or larval stage. However, stable fly larval developmental substrates are diverse, dispersed and often difficult to locate. Stable fly larvae have been observed in a broad variety of substrates including flotsam containing decomposing mayfly exuvia (Pickard 1968), aquatic plants (King and Lenert 1936; Simmons and Dove 1941), livestock wastes (Meyer and Petersen 1983; Broce et al. 2005), agronomic wastes (Bishopp 1913; Dove and Simmons 1941; Solórzano et al. 2015; Cook et al. 2011, 2017), grass clippings (Silverly and Schoof 1955; Todd 1964) and sewage sludge (Doud et al. 2012).

Beyond "fermenting organic material", little is known about the biological, chemical and physical factors defining developmental substrates (Gilles et al. 2008; Wienhold and Taylor 2012; Friesen et al. 2016). An active microbial community is necessary for larval development (Lysyk et al. 1999; Romero et al. 2006). As the number of coliform bacteria declines with microbial succession, so does the suitability of the substrate for stable fly development (Talley et al. 2009). Because substrates are suitable for stable fly development only during specific phases of decomposition, developmental sites are most often ephemeral, supporting only one generation of flies (Talley et al. 2009; Taylor and Berkebile 2011). These sites are broadly dispersed throughout rural and urban landscapes. Even relatively small sites can produce large numbers of flies (Todd 1964; Patterson and Morgan 1986). These characteristics complicate our ability to locate larval developmental sites for management prior to adult emergence.

Because of their painful bites and persistent feeding behaviour, just a few stable flies can reduce the productivity of livestock, harass companion animals and disrupt human recreational activities. The economic threshold for stable flies on feeder cattle in feedlots has been established at five flies per front leg (≈15 per animal as stable flies preferentially bite the front legs (Campbell and Berry 1989; Berry et al. 1983)), although cattle often exhibit defensive behaviours with fewer flies (Mullens et al. 2006). When infestations reach very high levels, cattle may no longer resist, and mortality may follow (Bishopp 1913).

Classic integrated pest management (IPM) programmes are based upon the concept of initiating control measures only after the pest population reaches an economic threshold (Metcalf and Luckman 1975). Because stable fly larval developmental sites are difficult to locate prior to adult emergence and larvae are intrinsically innocuous, economic thresholds are based upon counts of adult flies biting animals (Campbell and Berry 1989). By the time adult counts exceed the economic threshold, most have already emerged, and it is too late to initiate larval control procedures. Because of the ability of stable flies to move from property to property, the broad range of development sites and substrates, and the relatively low numbers of flies needed to inflict economic damage, effective management must be approached from a preventive, area-wide perspective.

3.2. Challenges for Area-wide Management

Area-wide management programmes often involve the use of mating disruption with pheromones or mass-production and release of insects with reduced reproductive potential. Unfortunately, aspects of stable fly biology are not conducive to the use of mating disruption. Unlike many Lepidoptera, muscoid Diptera such as stable flies lack volatile pheromones (Blomquist et al. 1987) suitable for mating disruption. Rather, their mate recognition pheromones are non-volatile cuticular hydrocarbons which act on contact, or at very close range (Muhammed et al. 1975; Uebel et al. 1975; Carlson and Mackley 1985).

Release of flies with genetic changes, whether induced by irradiation or transgenesis (Box 1), is complicated by three aspects of stable fly biology:

First, both male and female stable flies are obligate hematophages, blood-feeding 1-3 times per day for their entire life (Harris et al. 1974). Releasing large numbers of biting flies will increase the burden on livestock significantly.

Secondly, stable fly populations can be very large. Huge numbers of flies with non-persistent or threshold dependent genetic modifications must be released to attain the ratios necessary for control. Releasing such numbers of painful biters will meet with public protests. Added to this, their ability to disperse requires that even greater numbers be released over wider areas.

Thirdly, because of their high reproductive rate, short of eradication, small populations can recover to outbreak proportions quickly, precluding the concept of releasing flies with genetic changes for a limited period of time when natural populations are low with the hope of retarding later population growth.

Box 1. Non-Persistent and Persistent Genetic Changes

Genetic changes caused for example by irradiation, or the insertion of external genetic constructs (*transgenesis*) through modern biotechnology, can reduce the reproductive fitness of an insect. These genetic changes can be non-persistent or persistent in the target pest population (Carter and Friedman 2016). For example, the genetic changes of released sterile insects are non-persistent because they are not expected to persist in the environment. The released insects mate with wild insects reducing their fitness, but their genetic changes are not passed to their progeny. Therefore, programmes releasing insects with non-persistent genetic changes such as the Sterile Insect Technique (SIT) must release them continuously in numbers that greatly exceed the target population, often in the range of 10:1 to 200:1 (Knipling 1955). Because the random dominant mutations induced by irradiation render flies sterile, genetic changes do not persist in the environment, thus these SIT programmes have met broad public acceptance, and several are currently active (Klassen et al. 2021).

Persistent genetic modifications are designed to be, at least temporarily, incorporated into the gene pool of the target reducing either fitness or pathogenicity (Champer et al. 2016). Persistent modifications are often linked with a genetic drive mechanism to allow them to increase their frequency in the pest population. Genetic drive constructs can be subdivided into threshold dependent and threshold independent (Carter and Friedman 2016). The frequency of threshold dependent constructs must exceed a given level, the threshold, before increasing in frequency. Threshold independent constructs can theoretically be introduced into a population at a very low level and they will increase their frequency to fixation, replacing the original or natural population. Because these constructs can persist in nature and even replace the natural population, they are receiving a great deal more regulatory and ethical scrutiny than non-persistent technologies.

Given these constraints, the most viable option for genetic control of stable flies would be to release small numbers of persistent genetically modified flies with a threshold independent gene drive construct. Pending the development of such constructs and public acceptance of the release of such genetically modified organisms that are expected to become established and spread in the pest population (Box 1), our area-wide options for stable fly control are limited to the integrated implementation of traditional management technologies including cultural, biological and chemical methods.

3.3. Prerequisites for Area-Wide Management

3.3.1. Public Support / Consensus / Demand

Area-wide management programmes are administratively complex and require longer-term commitment (Hendrichs et al. 2007; Vreysen et al. 2007). A primary prerequisite for establishing such a programme is stakeholder collaboration and public recognition of the costs and benefits. This requires effective outreach to ensure that the public is aware of the damages and knowledgeable of the etiological agent.

Outreach is especially important for a pest such as stable flies. Producers often fail to differentiate among the species of muscoid flies associated with livestock. These flies are morphologically similar to the untrained eye. When querying producers about stable fly problems, one frequently hears "no, I don't have a stable fly problem, I have a fly problem." Similarly, when fielding calls from producers seeking assistance with flies, they are rarely able to identify the species of fly with which they are dealing. Smaller species, such as horn flies, are frequently mischaracterized as young flies that will "grow up" into larger flies (flies do not grow after metamorphosis to the adult stage). The biology and management methods for these species differ significantly, making proper identification essential prior to developing management strategies.

In Costa Rica, livestock producers refer to stable flies developing in pineapple fields as "mosca de la piña" and are insistent that they are a different species from the stable fly, "mosca del establo" that they observed prior to the large-scale pineapple production in the country. The importance of education and outreach to gain public support for an area-wide management programme cannot be over-emphasized. Economic assessments of the damage are also essential. Annual production losses to the cattle industry from stable flies are estimated to be USD 2.2 billion in the USA (Taylor et al. 2012a), USD 340 million in Brazil (Grisi et al. 2014), and USD 6.8 million in Mexico (Rodríguez-Vivas et al. 2017).

3.3.2. Regulatory Authority

Common concerns for area-wide management programmes are "free-riders", individuals who take advantage of the programme, but fail to contribute. This problem is exacerbated when the "problem", in this case stable flies, does not affect source producers, for example crop producers. Stable flies have no negative effects on crop production. Without regulatory authority, it will be very difficult to convince those producers to control the flies developing on their farms.

In Australia, Brazil, and Costa Rica, there have been public calls and demands for regulatory actions by the governments to address stable flies. The governments of those countries have enacted policies requiring agronomic producers to manage stable flies developing on their properties. In the USA, no such public demands have been made and regulatory policies have not been enacted. Public demand and pressure may ultimately lead to the development of regulation of stable fly source industries.

3.3.3. Funding

Regional differences in the sources and nature of stable fly infestations make detailed discussion of funding for area-wide management programmes beyond the scope of this discussion. In most cases, some degree of public funding will be needed to support the administrative and regulatory framework. Where an industry or agronomic system is deemed responsible for economically significant outbreaks, stable fly management should be considered a production expense. Sources of funding for management of non-commercial sources and research will need to be identified by the regionally interested parties.

3.4. Management Options

3.4.1. Cultural / Sanitation Methods

Elimination of larval development substrates has always been the primary recommendation for stable fly control (Greene 1993). In the USA, where substrates associated with livestock production are considered primary developmental sites, this largely involves manure management. Piling manure reduces the surface area suitable for stable fly development and allows metabolic heat to raise the substrate temperature to a level where stable fly larvae cannot survive. Covering manure and silage excludes ovipositing females. Spreading manure thinly on fields permits it to dry before stable fly larvae can complete development. Avoiding and removing spilled feed reduces the amount of substrate available for larval development.

Cultural methods can be applied to stable flies developing in agronomic wastes as well. Burying post-harvest vegetable residues with several different types of agricultural machinery and then compacting the soil with a landroller has proven effective for reducing stable fly development in Western Australia (David Cook, personal communication). Burial of waste is less effective for pineapple because of the quantity, 230 tons per hectare (Solórzano et al. 2015). Removal of pineapple waste would rapidly deplete soil fertility and is technically not feasible due to the quantity.

Some cultural methods for reducing stable flies have negative environmental ramifications. For example, burning sugarcane prior to harvest reduces the amount of substrate available for stable fly larval development, but also has serious consequences for air quality. Likewise, disposal of vinasse (a byproduct of ethanol distillation) in bodies of water renders it unsuitable for stable fly development, but it pollutes aquatic ecosystems (Souza Dominghetti et al. 2015).

3.4.2. Biological Methods

Biological control agents for stable flies can be divided into three categories, parasitoids, predators, and pathogens. Under natural conditions, egg to adult mortality of stable flies is estimated to exceed 95%, about half of which can be attributed to parasitoids and predators (Smith et al. 1985). The remainder is the result of pathogens and environmental stressors.

Pupal parasitoids are the most commonly used biological control agent for filth flies (Rueda and Axtell 1985; Machtinger et al. 2015). Two genera of pteromalid wasps, *Muscidifurax* and *Spalangia,* are frequently observed parasitizing stable flies in North America with 2 and 4 relatively common species, respectively. Several species, including both genera, can be seen in individual collections. How these parasitoids partition their resources is not clear.

The efficacy of augmentative releases of parasitoids is equivocal. Several studies indicated released parasitoids decreased fly populations (Weinzier and Jones 1998; Skovgård 2004; Geden and Hogsette 2006), while others failed to show a significant effect (Meyer et al. 1990; Andress, and Campbell 1994; Skovgård and Nachman 2004). In Costa Rica, two species of parasitoids have been collected from stable fly pupae in pineapple residues, *Muscidifurax raptoroides* Kogan & Legner and *Spalangia gemina* Boucek (unpublished observations), and a pilot programme using inundative releases of *Spalangia endius* Walker is showing promising results (Solórzano et al. 2017).

Several predators have been observed feeding on immature stable flies including macrochelid mites and staphylinid beetles (Smith et al. 1987; Seymour and Campbell 1993). Augmenting predator populations has not been evaluated for stable fly control.

Pathogens of stable fly were reviewed by Greenberg (1977). Entomopathogenic fungi have been evaluated for control of immature (Moraes et al. 2008, 2010; Alves et al. 2012; Machtinger et al. 2016) and adult (López-Sánchez et al. 2012; Cruz-Vázquez et al. 2015; Weeks et al. 2017) stable flies. Various formulations are commercially available. Several studies have evaluated entomopathogenic nematodes in the genera *Heterorhabditis* and *Steinernema* for filth fly control. In laboratory assays, results have been very promising (Taylor et al. 1998; Mahmoud et al. 2007). However, field trials have been disappointing (unpublished data). Although we observed slightly reduced numbers of flies emerging from sites treated with nematodes, we were unable to find infected fly larvae or detect infective juvenile nematodes more than 24 hours after treatment using sentinel greater wax moth larvae *Galleria mellonella* L.

3.4.3. Traps and Targets

The majority of stable fly traps are based upon visual attractants with a sticky surface to catch the flies. Williams (1973) recognized that Alsynite® fiberglass panels selectively attracted stable flies and Broce (1988) modified the trap making it more efficient and resistant to windy conditions. The next generation of traps was derived from blue and black fabric traps designed for tsetse fly (*Glossina* spp.) control (Mihok et al. 1995). The blue and black fabric traps are of limited utility in temperate parts of North America where sticky traps outperform them and they are susceptible to damage from gnawing insects such as grasshoppers (Orthoptera: Acrididae)

(Taylor and Berkebile 2006). However, in tropical regions such as La Réunion Island, they have proven to be very effective, especially the Vavoua trap (Laveissière and Grébaut 1990; Gilles et al. 2007).

In Costa Rica, improvised traps constructed from white plastic bags coated with an adhesive (Fig. 2) have been employed by the thousands for control of stable flies around pineapple plantations (Solórzano et al. 2015). These traps must be replaced every 1-2 days because they become saturated with insects and lose their effectiveness (Beresford and Sutcliffe 2017). Because of the environmental impact of disposing of such large numbers of plastic bags, research is currently underway to replace the white traps with insecticide-treated Vavoua traps.

Figure 2. Sticky traps used for stable fly control in Costa Rica.

Targets are like traps, but they intoxicate the attracted insects rather than catch them. Therefore, they do not need to be emptied or replaced routinely. Meifert et al. (1978) developed an early target system for stable flies by applying permethrin to the fiberglass panels of the William's trap. They indicated that the system was able to reduce the stable fly population by 30% per day when employed at a density of one target for every five animals. Blue and black targets are a modification of the blue and black traps (Foil and Younger 2006). When impregnated with 0.1% λ-cyhalothrin or 0.1% ζ-cypermethrin targets remain effective for ≈4 months (Hogsette et al. 2008). In a study in Louisiana, an average of 220 stable flies landed per hour on targets long enough to be intoxicated (Hogsette and Foil 2018).

A disadvantage of the targets relative to traps is that they cannot be used to quantify the number of flies in the population nor the number of flies killed. In addition, targets provide less psychological satisfaction because dead or trapped flies are not apparent. However, both of these concerns can be mitigated by placement of sticky traps adjacent to selected targets (Foil and Younger 2006).

3.4.4. Chemical Control

- *Immatures.* Because substrates for stable fly larval development tend to be microbially very active (Romero et al. 2006; Talley et al. 2009; Scully et al. 2017), most insecticides applied to substrates tend to degrade quickly and have little residual activity. Two classes of insect growth regulators (IGRs), cyromazine and benzoylureas have proven to be the most effective for controlling stable fly larvae (Taylor et al. 2012b, 2014; Solórzano et al. 2015). A single application of these compounds can provide 12 or more weeks of control and they have relatively low vertebrate toxicity (Tunaz and Uygun 2004). Cyromazine and benzoylureas belong to different insecticide mode of action classes with distinct resistance mechanisms (Keiding et al. 1991; IRAC 2017). Therefore, they are suitable for rotation to reduce the development of insecticide resistance. In addition, cyromazine and benzoylureas are compatible with biological control using parasitoids (Ables et al. 1975; Morgan and Patterson 1990).
- *Adults.* Chemical options for controlling adult stable flies associated with food animals such as cattle are limited. Premise or area sprays should be reserved as a last resort for outbreaks where other control measures have failed. Pyrethroids remain effective, although resistance has been detected (Cilek and Greene 1994; Olafson et al. 2011). Their continued effectiveness is probably a reflection of the low efficiency of treatments (Greene 1993). Insecticide-impregnated netting provided as resting sites near livestock are showing promise, especially in the dairy environment. In a study in Nebraska, ≈1000 meters of netting were installed on the periphery of two dairy barns. Up to 60 stable flies per linear meter per day were collected dead beneath the netting. Based upon observations, we estimate that the collections represented less than 10% of the flies that were lethally intoxicated (unpublished data). As methods for targeting stable flies with insecticides improve, resistance will become a greater problem.

3.4.5. On-Animal

On-animal strategies include physical protection or barriers such as boots, masks, sheets, etc., and chemical agents such as repellents and insecticides. Physical protection is frequently used for high value animals such as horses, but it is not practical for livestock such as cattle.

On-animal chemical technologies such as ear tags and pour-on insecticides are commonly used to protect livestock from horn flies. However, because stable flies spend little time on the host and bite primarily on the lower legs, these technologies are less effective against them (Foil and Hogsette 1994; Broce et al. 2005). The primary disadvantage of on-animal chemical treatments is that they have short residual activity against stable flies, less than 3-4 days for most and less than 6-8 hours for many (Foil and Hogsette 1994; Mullens et al. 2009; Benelli and Pavela 2018). A combination of fipronil and permethrin provided 5 weeks of repellence when applied to dogs in the laboratory (Fankhauser et al. 2015); however, this formulation has not been tested on livestock in pastures.

3.5. Area-Wide Strategy for Stable Flies

With the current state of technology, management strategies incorporating the release of large numbers of biting, sterile or genetically modified, stable flies are unlikely to be accepted by livestock producers or the public. Pending the development, and public acceptance, of threshold independent genetic drive mechanisms for stable flies, management options are limited to the area-wide application and integration of traditional methods such as cultural, biological and chemical.

Cultural management of animal and vegetative wastes should be the first priority. In an area-wide programme, especially if agronomic systems are contributing significant numbers of flies, such control will need to be mandated along with inspection and enforcement systems. Most of the currently recognized larval developmental substrates originate from human activities, and therefore are more manageable. Those developmental substrates that cannot be rendered unsuitable for stable fly development by cultural methods will need to be treated with biological and/or chemical control agents. Although biological control programmes on stable flies have had inconsistent results, pteromalid parasitoids are the most developed option. IGRs are the most effective and environmentally sound chemical alternatives available. Insecticide resistance management including rotation of insecticides with distinct modes of action must be included for a sustainable management plan.

A concerted effort must be made to identify and remediate all larval developmental sites within the control region. Management of larval developmental sites must be the primary emphasis of an area-wide stable fly management programme. However, outbreaks of adult flies due to control failure or unanticipated developmental sites are still likely to occur. Adult stable flies need to be managed in the vicinity of the developmental sites and susceptible hosts including humans and livestock. Traps, targets and insecticide impregnated artificial resting sites are the best options for managing adult stable flies. On-animal insecticides and repellents may be necessary for short-term remediation in cases where other control measures failed, but these are best applied on a premise by premise basis and in pest hot spots, rather than an area-wide basis.

Depending upon the situation, one cultural method such as burying vegetable residues may be adequate to control a stable fly problem. Alternatively, multiple strategies including both larval and adult control may be required if no single technology is adequately effective. Reliance upon chemical control alone is short-sighted and will lead to insecticide resistance and eventual loss of control. Cultural, and often biological, control efforts should accompany chemical control.

3.6. Research Needs

Because area-wide management of stable flies is dependent upon reducing and eliminating larval developmental sites, it is imperative that we develop a better understanding of the biological, environmental and physical characteristics of developmental substrates. In addition to the developmental substrates discussed in Section 2, developmental sites which do not fit into the current paradigm appear to be contributing to the adult stable fly populations (unpublished data). Recognized larval developmental sites tend to have high densities of larvae restricted to small areas. Is it possible that we are overlooking a second type of developmental sites, those with low densities of larvae, possibly one or two per square meter, but distributed over many hectares of land? Possibilities include crop residues in agronomic fields and grass and other plant residues (thatch) in grasslands. If such "low-density, large-area" developmental sites are widespread, then a very different approach to stable fly management will be needed. A better understanding of developmental substrates will help with the development of cultural and mechanical methods to render substrates unsuitable for stable flies as well.

A second research priority is a better understanding of the population dynamics of both larval and adult stable flies. How are females locating oviposition sites and how are larvae utilizing the substrates? What environmental factors are driving dispersal and population fluctuations? How far are adults dispersing? Incorporation of this information into area-wide management projects will improve their efficiency greatly.

Lastly, improved adult suppression systems are needed; more efficient traps and targets requiring less maintenance and novel adult suppression methods will add greatly to management programmes. It is unlikely we will ever be able to locate and remediate all larval developmental sites within the potential dispersal distance of stable flies. Therefore, adult suppression will remain an important component of any management programme.

3.7. Education and Outreach

An area-wide management programme for stable flies must include an educational component. Primary to this effort is information on the types of flies associated with livestock, their biology and effects on the productivity and comfort of the animals. Education will improve public support from both political and applied perspectives. Without such education, a successful area-wide programme may be perceived by the public as a failure if infestations of other species of muscoid flies continue and cannot be differentiated from stable flies. Livestock producers and landowners should also be aware of the natural enemies of flies and methods to preserve and augment their populations.

All levels of the distribution chain for chemical control agents from producers and suppliers to cattlemen must know their proper use for the species of flies affecting livestock production systems.

4. REFERENCES

Ables, J. R., R. P. West, and M. Shepard. 1975. Response of the house fly and its parasitoids to Dimilin (TH-6040)12. Journal of Economic Entomology 68: 622–624.

Alves, P. S. A., A. P. R. Moraes, C. M. C. de Salles, V. R. E. P. Bittencourt, and A. J. Bittencourt. 2012. *Lecanicillium lecanii* for control of the immature stage of *Stomoxys calcitrans*. Revista Brasileira de Medicina Veterinaria 34 (Suppl. 1): 66–72.

Anderson, J. R. 1978. Mating behavior of *Stomoxys calcitrans*: Effects of a blood meal on the mating drive of males and its necessity as a prerequisite for proper insemination of females. Journal of Economic Entomology 71: 379–386.

Anderson, J. R., and C. H. Tempelis. 1970. Precipitin test identification of blood meals of *Stomoxys calcitrans* (L.) caught on California poultry ranches, and observations of digestion rates of bovine and citrated human blood. Journal of Medical Entomology 7: 223–229.

Andress, E. R., and J. B. Campbell. 1994. Inundative releases of pteromalid parasitoids (Hymenoptera: Pteromalidae) for the control of stable flies, *Stomoxys calcitrans* (L.) (Diptera: Muscidae) at confined cattle installations in west central Nebraska. Journal of Economic Entomology 87: 714–722.

Bailey, D. L., T. L. Whitfield, and B. J. Smittle. 1973. Flight and dispersal of the stable fly. Journal of Economic Entomology 66: 410–411.

Benelli, G., and R. Pavela. 2018. Beyond mosquitoes—Essential oil toxicity and repellency against bloodsucking insects. Industrial Crops and Products 117: 382–392.

Beresford, D. V., and J. F. Sutcliffe. 2017. Evidence for sticky-trap avoidance by stable fly, *Stomoxys calcitrans* (Diptera: Muscidae), in response to trapped flies. Journal of the American Mosquito Control Association 33: 250–252.

Berry, I. L., D. A. Stage, and J. B. Campbell. 1983. Populations and economic impacts of stable flies on cattle *Stomoxys calcitrans*, Nebraska, production losses. Transactions of the American Society of Agricultural Engineers 26: 873–877.

Bishopp, F. C. 1913. The stable fly (*Stomoxys calcitrans* L.), an important livestock pest. Journal of Economic Entomology 6: 112–126.

Blomquist, G. J., J. W. Dillwith, and T. S. Adams. 1987. Biosynthesis and endocrine regulation of sex pheromone production in Diptera, pp. 217–250. *In* G. D. Prestwich and G. J. Blomquist (eds.), Pheromone biochemistry. Academic Press, New York, NY, USA.

Broce, A. B. 1988. An improved alsynite trap for stable flies, *Stomoxys calcitrans* (Diptera: Muscidae). Journal of Medical Entomology 25: 406–409.

Broce, A. B., J. Hogsette, and S. Paisley. 2005. Winter feeding sites of hay in round bales as major developmental sites of *Stomoxys calcitrans* (Diptera: Muscidae) in pastures in spring and summer. Journal of Economic Entomology 98: 2307–2312.

Campbell, J. B., and C. D. McNeal. 1979. A guide to Integrated Pest Management at feedlots and dairies. Nebraska University College of Agriculture and Home Economics Extension Circular EC 80-1536. Lincoln, Nebraska, USA.

Campbell, J. B., and I. L. Berry 1989. Economic threshold for stable flies on confined livestock, pp. 18–22. *In* J. J. Petersen and G. L. Greene (eds.), Current status of stable fly (Diptera: Muscidae) research. Miscellaneous Publications of the Entomological Society of America 74.

Campbell, J. B., M. A. Catangui, G. D. Thomas, D. J. Boxler, and R. Davis. 1993. Effects of stable flies (Diptera, Muscidae) and heat stress on weight gain and feed conversion of feeder cattle. Journal of Agricultural Entomology 10: 155–161.

Campbell, J. B., S. R. Skoda, D. R. Berkebile, D. J. Boxler, G. D. Thomas, D. C. Adams, and R. Davis. 2001. Effects of stable flies (Diptera: Muscidae) on weight gains of grazing yearling cattle. Journal of Economic Entomology 94: 780–783.

Carlson, D. A., and J. W. Mackley. 1985. Polyunsaturated hydrocarbons in the stable fly. Journal of Chemical Ecology 11: 1485–1496.

Carter, S. R., and R. M. Friedman. 2016. Policy and regulatory issues for gene drives in insects. Workshop report. J. Craig Venter Institute's Policy Center and University of California at San Diego, USA. 21 pp.

Champer, J., A. Buchman, and O. S. Akbari. 2016. Cheating evolution: Engineering gene drives to manipulate the fate of wild populations. Nature Reviews Genetics 17: 146–159.

Cilek, J. E., and G. L. Greene. 1994. Stable fly (Diptera: Muscidae) insecticide resistance in Kansas cattle feedlots. Journal of Economic Entomology 87: 275–279.

Cook, D. F., I. R. Dadour, and N. J. Keals. 1999. Stable fly, house fly (Diptera: Muscidae), and other nuisance fly development in poultry litter associated with horticultural crop production. Journal of Economic Entomology 92: 1352–1357.

Cook, D. F., I. R. Dadour, and S. C. Voss. 2011. Management of stable fly and other nuisance flies breeding in rotting vegetable matter associated with horticultural crop production. International Journal of Pest Management 57: 315–320.

Cook, D. F., D. V. Telfer, J. B. Lindsey, and R. A. Deyl. 2017. Substrates across horticultural and livestock industries that support the development of stable fly, *Stomoxys calcitrans* (Diptera: Muscidae). Austral Entomology 57: 344–348.

Cruz-Vázquez, C., J. Carvajal Márquez, R. Lezama-Gutiérrez, I. Vitela-Mendoza, and M. Ramos-Parra. 2015. Efficacy of the entomopathogenic fungi *Metarhizium anisopliae* in the control of infestation by stable flies, *Stomoxys calcitrans* (L.), under natural infestation conditions. Veterinary Parasitology 212: 350–355.

Dougherty, C. T., F. W. Knapp, P. B. Burrus, D. C. Willis, P. L. Cornelius, and N. W. Bradley. 1993. Multiple releases of stable flies (*Stomoxys calcitrans* L.) and behavior of grazing beef cattle. Applied Animal Behaviour Science 38: 191–212.

Dougherty, C. T., F. W. Knapp, P. B. Burrus, D. C. Willis, and P. L. Cornelius. 1994. Moderation of grazing behavior of beef cattle by stable flies (*Stomoxys calcitrans* L.). Applied Animal Behaviour Science 40: 113–127.

Dougherty, C. T., F. W. Knapp, P. B. Burrus, D. C. Willis, and P. L. Cornelius. 1995. Behavior of grazing cattle exposed to small populations of stable flies (*Stomoxys calcitrans* L.). Applied Animal Behaviour Science 42: 231–248.

Doud, C. W., D. B. Taylor, and L. Zurek. 2012. Dewatered sewage biosolids provide a productive larval habitat for stable flies and house flies (Diptera: Muscidae). Journal of Medical Entomology 49: 286–292.

Dove, W. E., and S. W. Simmons. 1941. Control of dog fly breeding in peanut litter. USDA, Bureau of Entomology and Plant Quarantine, E–542.

Dove, W. E., and S. W. Simmons. 1942. Creosote oil with water for control of the stable fly, or "dog fly," in drifts of marine grasses. Journal of Economic Entomology 35: 589–592.

Elkan, P. W., R. Parnell, and J. L. David Smith. 2009. A die-off of large ungulates following a *Stomoxys* biting fly out-break in lowland forest, northern Republic of Congo. African Journal of Ecology 47: 528–536.

Fankhauser, B., J. P. Irwin, M. L. Stone, S. T. Chester, and M. D. Soll. 2015. Repellent and insecticidal efficacy of a new combination of fipronil and permethrin against stable flies (*Stomoxys calcitrans*). Parasites & Vectors 8: 61.

Foil, L. D., and J. A. Hogsette. 1994. Biology and control of tabanids, stable flies and horn flies. Scientific and technical review of the Office International des Epizooties (Paris) 13: 1125–1158.

Foil, L. D., and C. D. Younger. 2006. Development of treated targets for controlling stable flies (Diptera: Muscidae). Veterinary Parasitology 137: 311–315.

Friesen, K. M., D. R. Berkebile, B. J. Wienhold, L. M. Durso, J. J. Zhu, and D. B Taylor. 2016. Environmental parameters associated with stable fly (Diptera: Muscidae) development at hay feeding sites. Environmental Entomology 45: 570–576.

Geden, C. J., and J. A. Hogsette. 2006. Suppression of house flies (Diptera: Muscidae) in Florida poultry houses by sustained releases of *Muscidifurax raptorellus* and *Spalangia cameroni* (Hymenoptera: Pteromalidae). Environmental Entomology 35: 75–82.

Gilles, J., J.-F. David, G. Duvallet, S. de la Rocque, and E. Tillard. 2007. Efficiency of traps for *Stomoxys calcitrans* and *Stomoxys niger niger* on Réunion Island. Medical and Veterinary Entomology 21: 65–69.

Gilles, J., J.-F. David, P. Lecomte, and E. Tillard. 2008. Relationships between chemical properties of larval media and development of two *Stomoxys* species (Diptera: Muscidae) from Réunion Island. Environmental Entomology 37: 45–50.

Greenberg, B. 1977. Pathogens of *Stomoxys calcitrans* (stable flies). Bulletin of the World Health Organization 55 (Suppl. 1): 259–261.

Greene, G. L. 1993. Chemical, cultural, and mechanical control of stable flies and house flies, pp. 83–90. *In* G. D. Thomas and S. R. Skoda (eds.), Rural flies in the urban environment. North Central Regional Research Publication No. 335, Institute of Agriculture and Natural Resources, University of Nebraska, Lincoln, Nebraska, USA.

Grisi, L., R. Cerqueira-Leite, J. R. de Souza Martins, A. T. Medeiros de Barros, R. Andreotti, P. H. D. Cançado, A. A. Pérez de León, J. Barros Pereira, and H. Silva Villela. 2014. Reassessment of the potential economic impact of cattle parasites in Brazil. Brazilian Journal of Veterinary Parasitology 23: 150–156.

Hafez, M., and F. M. Gamal-Eddin. 1959. Ecological studies on *Stomoxys calcitrans* L. and *sitiens* Rond. in Egypt, with suggestions on their control (Diptera: Muscidae). Bulletin de la Société Entomologique d'Égypte 43: 245–283.

Hall, R. D., G. D. Thomas, and C. E. Morgan. 1982. Stable flies, *Stomoxys calcitrans* (L.), breeding in large round hay bales: Initial associations (Diptera: Muscidae). Journal of the Kansas Entomological Society 55: 617–620.

Harris, R. L., J. A. Miller, and E. D. Frazer. 1974. Horn flies and stable flies feeding activity. Annals Entomological Society of America 67: 891–894.

Hendrichs, J., P. Kenmore, A. S. Robinson, and M. J. B. Vreysen. 2007. Area-Wide Integrated Pest Management (AW-IPM): Principles, practice and prospects, pp. 3-33. *In* M. J. B. Vreysen, A. S. Robinson, and J. Hendrichs (eds.), Area-wide control of insect pests: From research to field implementation. Springer, Dordrecht, The Netherlands.

Herrero, M. V., L. Montes., C. Sanabria, A. Sánchez, and R. Hernández. 1989. Estudio inicial sobre la mosca de los establos, *Stomoxys calcitrans* (Diptera: Muscidae), en la región del Pacífico Sur de Costa Rica. Ciencias Veterinarias (Heredia, Costa Rica) 11: 11–14.

Herrero, M. V., L. Montes-Pico, and R. Hernández. 1991. Abundancia relativa de *Stomoxys calcitrans* (L.) (Diptera: Muscidae) en seis localidades del Pacífico Sur de Costa Rica. Revista de Biología Tropical 39: 309–310.

Hogsette, J. A., and J. P. Ruff. 1985. Stable fly (Diptera: Muscidae) migration in northwest Florida. Environmental Entomology 14: 170–175.

Hogsette, J. A., and L. D. Foil. 2018. Blue and black cloth targets: Effects of size, shape and color on stable fly (L.) (Diptera: Muscidae) attraction. Journal of Economic Entomology 111: 974–979.

Hogsette, J. A., J. P. Ruff, and C. J. Jones. 1987. Stable fly biology and control in northwest Florida. Journal of Agricultural Entomology 4: 1–11.

Hogsette, J. A., A. Nalli, and L. D. Foil. 2008. Evaluation of different insecticides and fabric types for development of treated targets for stable fly (Diptera: Muscidae) control. Journal of Economic Entomology 101: 1034–1038.

Horsfall, W. R. 1985. Mosquito abatement in a changing world. Journal of the American Mosquito Control Association 1: 135–138.

(IRAC) Insecticide Resistance Action Committee. 2017. IRAC mode of action classification scheme. Version 9.1. CropLife International.

Keiding, J., J. B. Jespersen, and A. S. El-Khodary. 1991. Resistance risk assessment of two insect development inhibitors, diflubenzuron and cyromazine, for control of the housefly *Musca domestica*. Part I: Larvicidal tests with insecticide-resistant laboratory and Danish field populations. Pesticide Science 32: 187–206.

King, W. V., and L. G. Lenert. 1936. Outbreaks of *Stomoxys calcitrans* L. ("dog flies") along Florida's northwest coast. Florida Entomologist 19: 33–39.

Klassen, W., C. F. Curtis, and J. Hendrichs. 2021. History of the Sterile Insect Technique, pp. 1–44. *In* V. A. Dyck, J. Hendrichs, and A. S. Robinson (eds.), Sterile Insect Technique – Principles and practice in Area-Wide Integrated Pest Management. Second Edition. CRC Press, Boca Raton, Florida, USA.

Knipling, E. F. 1955. Possibilities of insect control or eradication through the use of sexually sterile males. Journal of Economic Entomology 48: 459–462.

Koller, W. W., J. B. Catto, I. Bianchin, C. O. Soares, F. Paiva, L. E. R. Tavares, and G. Graciolli. 2009. Surtos da mosca-dos-estábulos, *Stomoxys calcitrans*, em Mato Grosso do Sul: novo problema para as cadeias produtivas da carne e sucroalcooleira? Documentos 175, Embrapa Gado de Corte, Campo Grande, Mato Grosso do Sul, Brazil. 31 pp.

Kunz, S. E., and J. Monty. 1976. Biology and ecology of *Stomoxys nigra* Marquart and *Stomoxys calcitrans* (L.) (Diptera: Muscidae) in Mauritius. Bulletin of Entomological Research 66: 745–755.

Laveissière, C., and P. Grébaut. 1990. Recherches sur les pièges à glossines (Diptera: Glossinidae). Mise au point d'un modèle économique: le piège "Vavoua". Tropical Medicine and Parasitology 41: 185–192.

López-Sánchez, J., C. Cruz-Vázquez, R. Lezama-Gutiérrez, and M. Ramos-Parra. 2012. Effect of entomopathogenic fungi upon adults of *Stomoxys calcitrans* and *Musca domestica* (Diptera: Muscidae). Biocontrol Science and Technology 22: 969–973.

Lysyk, T., L. Kalischuk-Tymensen, L. Selinger, R. Lancaster, L. Wever, and K. Cheng. 1999. Rearing stable fly larvae (Diptera: Muscidae) on an egg yolk medium. Journal of Medical Entomology 38: 382–388.

Machtinger, E. T., C. J. Geden, P. E. Kaufman, and A. M. House. 2015. Use of pupal parasitoids as biological control agents of filth flies on equine facilities. Journal of Integrated Pest Management 6: 16.

Machtinger, E. T., E. N. I. Weeks, and C. J. Geden. 2016. Oviposition deterrence and immature survival of filth flies (Diptera: Muscidae) when exposed to commercial fungal products. Journal of Insect Science 16: 33253.

Mahmoud, M. F., N. S. Mandour, and Y. I. Pomazkov. 2007. Efficacy of the entomopathogenic nematode *Steinernema feltiae* cross N 33 against larvae and pupae of four fly species in the laboratory. Nematologia Mediterranea 35: 221–226.

Meifert, D. W., R. S. Patterson, T. Whitfield, G. C. LaBrecque, and D. E. Weidhaas. 1978. Unique attractant-toxicant system to control stable fly populations. Journal of Economic Entomology 71: 290–292.

Metcalf, R. L., and W. H. Luckman. 1975. Introduction to insect pest management. John Wiley & Sons, New York, USA.

Meyer, J. A., and J. J. Petersen. 1983. Characterization and seasonal distribution of breeding sites of stable flies and house flies (Diptera: Muscidae) on eastern Nebraska feedlots and dairies. Journal of Economic Entomology 76: 103–108.

Meyer, J. A., B. A. Mullens, T. L. Cyr, and C. Stokes. 1990. Commercial and naturally occurring fly parasitoids (Hymenoptera: Pteromalidae) as biological control agents of stable flies and house flies (Diptera: Muscidae) on California dairies. Journal of Economic Entomology 83: 799–806.

Mihok, S., E. K. Kang'ethe, and G. K. Kamau. 1995. Trials of traps and attractants for *Stomoxys* spp. (Diptera: Muscidae). Journal of Medical Entomology 32: 283–289.

Moon, R. D. 2002. Muscid flies (Muscidae), pp. 279–316. *In* G. Mullen, and L. Durden (eds.), Medical and veterinary entomology. Academic Press, San Diego, California, USA.

Moraes, A. P. R., I. D. C. Angelo, E. K. K. Fernandes, V. R. E. P. Bittencourt, and A. J. Bittencourt. 2008. Virulence of *Metarhizium anisopliae* to eggs and immature stages of *Stomoxys calcitrans*. Annals of the New York Academy of Sciences 1149: 384–387.

Moraes, A. P. R., V. R. E. P. Bittencourt, and A. J. Bittencourt. 2010. Pathogenicity of *Beauveria bassiana* on immature stages of *Stomoxys calcitrans*. Ciencia Rural 40: 1802–1807.

Morgan, P. B., and R. S. Patterson. 1990. Efficiency of target formulations of pesticides plus augmentative releases of *Spalangia endius* Walker (Hymenoptera: Pteromalidae) to suppress populations of *Musca domestica* L. (Diptera: Muscidae) at poultry installations in the southeastern United States, pp. 69-78. *In* D. A. Rutz, and R. S. Patterson (eds.), Biocontrol of arthropods affecting livestock and poultry. Westview Press, Boulder, Colorado, USA.

Muhammed, S., Butler, J. F, and Carlson, D. A. 1975. Stable fly sex attractant and mating pheromones found in female body hydrocarbons. Journal of Chemical Ecology 1: 387–398.

Mullens, B. A., W. G. Reifenrath, and S. M. Butler. 2009. Laboratory trials of fatty acids as repellents or antifeedants against house flies, horn flies and stable flies (Diptera: Muscidae). Pest Management Science 65: 1360–1366.

Mullens, B. A., K.-S. Lii, Y. Mao, J. A. Meyer, N. G. Peterson, and C. E. Szijj. 2006. Behavioural responses of dairy cattle to the stable fly, *Stomoxys calcitrans*, in an open field environment. Medical and Veterinary Entomology 20: 122–137.

Newson, H. D. 1977. Arthropod problems in recreation areas. Annual Review of Entomology 22: 333–353.

Olafson, P. U., J. B. Pitzer, and P. E. Kaufman. 2011. Identification of a mutation associated with permethrin resistance in the para-type sodium channel of the stable fly (Diptera: Muscidae). Journal of Economic Entomology 104: 250–257.

Patterson, R. S., and P. B. Morgan. 1986. Factors affecting the use of an IPM scheme at poultry installations in a semitropical climate, pp. 101–107. *In* R. S. Patterson, and D. A. Rutz (eds.), Biological control of muscoid flies. Miscellaneous Publications of the Entomological Society of America No. 61.

Pickard, E. 1968. *Stomoxys calcitrans* (L.) breeding along TVA reservoir shorelines. Mosquito News 28: 644–646.

Rodríguez-Vivas, R. I., L. Grisi, A. A. Pérez de León, H. Silva Villela, J. F. de Jesús Torres-Acosta, H. Fragoso Sánchez, D. Romero Salas, R. Rosario Cruz, F. Saldierna, and D. García Carrasco. 2017. Potential economic impact assessment for cattle parasites in Mexico. Revista Mexicana de Ciencias Pecuarias 8: 61–74.

Romero, A., A. Broce, and L. Zurek. 2006. Role of bacteria in the oviposition behaviour and larval development of stable flies. Medical and Veterinary Entomology 20: 115–121.

Rueda, L. M., and R. C. Axtell. 1985. Guide to common species of pupal parasites (Hymenoptera: Pteromalidae) of the house fly and other muscoid flies associated with poultry and livestock manure. North Carolina Agricultural Research Service Technical Bulletin 278.

Scully, E., K. Friesen, B. Wienhold, and L. M. Durso. 2017. Microbial communities associated with stable fly (Diptera: Muscidae) larvae and their developmental substrates. Annals of the Entomological Society of America 110: 61–72.

Seymour, R. C., and J. B. Campbell. 1993. Predators and parasitoids of house flies and stable flies (Diptera: Muscidae) in cattle confinements in west central Nebraska. Environmental Entomology 22: 212–219.

Silverly, R. E., and H. F. Schoof. 1955. Utilization of various production media by muscoid flies in a metropolitan area. I. Adaptability of different flies for infestation of prevalent media. Annals Entomological Society of America 48: 258–262.

Simmons, S. W., and W. E. Dove. 1941. Breeding places of the stable fly or "dog fly" *Stomoxys calcitrans* (L.) in northwestern Florida. Journal Economic Entomology 34: 457–462.

Simmons, S. W., and W. E. Dove. 1942. Waste celery as a breeding medium for the stable fly or "dog fly" with suggestions for control. Journal Economic Entomology 35: 709–715.

Skovgård, H. 2004. Sustained releases of the pupal parasitoid *Spalangia cameroni* (Hymenoptera: Pteromalidae) for control of house flies, *Musca domestica* and stable flies *Stomoxys calcitrans* (Diptera: Muscidae) on dairy farms in Denmark. Biological Control 30: 288–297.

Skovgård, H., and G. Nachman. 2004. Biological control of house flies *Musca domestica* and stable flies *Stomoxys calcitrans* (Diptera: Muscidae) by means of inundative releases of *Spalangia cameroni* (Hymenoptera: Pteromalidae). Bulletin of Entomological Research 94: 555–567.

Smith, J. P., R. D. Hall, and G. D. Thomas. 1985. Field studies on mortality of the immature stages of the stable fly (Diptera: Muscidae). Environmental Entomology14: 881–890.

Smith, J. P., R. D. Hall, and G. D. Thomas. 1987. Arthropod predators and competitors of the stable fly, *Stomoxys calcitrans* (L.) in central Missouri. Journal Kansas Entomological Society 60: 562–567.

Solórzano, J.-A., J. Gilles, O. Bravo, C. Vargas, Y. Gomez-Bonilla, G. Bingham, and D. B. Taylor. 2015. Biology and trapping of stable flies (Diptera: Muscidae) developing in pineapple residues (*Ananas comosus*) in Costa Rica. Journal Insect Science 15: 145.

Solórzano, J.-A., H. Mena, R. Romero, J. Treviño, J. Gilles, C. Geden, D. Taylor, and H. Skovgård. 2017. Biological control of livestock pest biting fly *Stomoxys calcitrans* at agriculture pineapple residues using the parasitoid *Spalangia endius* reared on irradiated Mediterranean fruit fly: Assessment of parasitism in field and laboratory in Costa Rica, pp. 242–243. *In* Third FAO/IAEA International Conference on Area-wide Management of Insect Pests, Book of Abstracts. Vienna, Austria.

Souza Dominghetti, T. F. de, A. T. Medeiros de Barros, C. Oliveira Soares, and P. H. Duarte Cançado. 2015. *Stomoxys calcitrans* (Diptera: Muscidae) outbreaks: Current situation and future outlook with emphasis on Brazil. Brazilian Journal of Veterinary Parasitology 24: 387–395.

Talley, J., A. Broce, and L. Zurek. 2009. Characterization of stable fly (Diptera: Muscidae) larval development habitat at round hay bale feeding sites. Journal of Medical Entomology 46: 1310–1319.

Taylor, D. B., and D. R. Berkebile. 2006. Comparative efficiency of six stable fly traps. Journal Economic Entomology 99: 1415–1419.

Taylor, D. B., and D. R. Berkebile. 2011. Phenology of stable fly (Diptera: Muscidae) larvae in round bale hay feeding sites in eastern Nebraska. Environmental Entomology 40: 184–193.

Taylor, D. B., R. D. Moon, and D. R. Mark. 2012a. Economic impact of stable flies (Diptera: Muscidae) on cattle production. Journal of Medical Entomology 49: 198–209.

Taylor, D. B., K. Friesen, and J. Zhu. 2014. Stable fly control in cattle winter feeding sites with Novaluron. Arthropod Management Tests 39: 1–2.

Taylor, D. B., A. L. Szalanski, B. J. Adams, and R. D. Peterson II. 1998. Susceptibility of house fly, *Musca domestica* (Diptera: Muscidae) larvae to entomopathogenic nematodes (Rhabditida: Heterorhabditidae, Steinernematidae). Environmental Entomology 27: 1514–1519.

Taylor, D. B., K. Friesen, J. J. Zhu, and K. Sievert. 2012b. Efficacy of cyromazine to control immature stable flies (Diptera: Muscidae) developing in winter hay feeding sites. Journal of Economic Entomology 105: 726–731.

Taylor, D. B., R. D. Moon, J. B. Campbell, D. R. Berkebile, P. J. Scholl, A. B. Broce, and J. A. Hogsette. 2010. Dispersal of stable flies (Diptera: Muscidae) from larval developmental sites. Environmental Entomology 39: 1101–1110.

Todd, D. H. 1964. The biting fly *Stomoxys calcitrans* (L.) in dairy herds in New Zealand. New Zealand Journal of Agricultural Research 7: 60–79.

Tunaz, H., and N. Uygun. 2004. Insect growth regulators for insect pest control. Turkish Journal of Agriculture and Forestry 28: 377–387.

Uebel, E. C., P. E. Sonnet, and R. W. Miller. 1975. Sex pheromone of the stable fly: Isolation and preliminary identification of compounds that induce mating strike behavior. Journal of Chemical Ecology 1: 377–385.

Vreysen, M. J. B., J. Gerardo-Abaya, and J. P. Cayol. 2007. Lessons from Area-Wide Integrated Pest Management (AW-IPM) programmes with an SIT component: An FAO/IAEA perspective, pp. 723–744. *In* M. J. B. Vreysen, A. S. Robinson, and J. Hendrichs (eds.), Area-wide control of insect pests. From research to field implementation. Springer, Dordrecht, The Netherlands.

Weeks, E. N. I., E. T. Machtinger, S. A. Gezan, P. E. Kaufman, and C. J. Geden. 2017. Effects of four commercial fungal formulations on mortality and sporulation in house flies (*Musca domestica*) and stable flies (*Stomoxys calcitrans*). Medical and Veterinary Entomology 31: 15–22.

Weinzier, R. A., and C. J. Jones. 1998. Releases of *Spalangia nigroaenea* and *Muscidifurax zaraptor* (Hymenoptera: Pteromalidae) increase rates of parasitism and total mortality of stable fly and house fly (Diptera: Muscidae) pupae in Illinois cattle feedlots. Journal Economic Entomology 91: 1114–1121.

Wienhold, B. J., and D. B. Taylor. 2012. Substrate properties of stable fly developmental sites associated with round bale hay feeding sites in eastern Nebraska. Environmental Entomology 41: 213–221.

Williams, D. F. 1973. Sticky traps for sampling populations of *Stomoxys calcitrans*. Journal of Economic Entomology 66: 1279–1280.

Yeruham, I., and Y. Braverman. 1995. Skin lesions in dogs, horses and calves caused by the stable fly *Stomoxys calcitrans* (L.) (Diptera: Muscidae). Revue d'Élevage et de Médecine Vétérinaire des Pays Tropicaux 48: 347–349.

Zumpt, F. 1973. The stomoxyine biting flies of the world. Gustav Fischer Verlag, Stuttgart, Germany. 175 pp.

ADVANCES IN INTEGRATED TICK MANAGEMENT RESEARCH FOR AREA-WIDE MITIGATION OF TICK-BORNE DISEASE BURDEN

A. A. PÉREZ DE LEÓN[1], R. D. MITCHELL III[1], R. J. MILLER[2] AND K. H. LOHMEYER[1]

[1]USDA-ARS, Knipling-Bushland U.S. Livestock Insects Research Laboratory and Veterinary Pest Genomics Center, 2700 Fredericksburg Road, Kerrville, Texas 78028, USA; Beto.PerezdeLeon@ARS.USDA.GOV
[2]USDA-ARS, Cattle Fever Tick Research Laboratory and Veterinary Pest Genomics Center, 22675 N. Moorefield Road, Edinburg, Texas 78541, USA

SUMMARY

In some parts of the world, ticks are the most dangerous animals followed by mosquitoes as ectoparasites and vectors of infectious agents, causing morbidity and mortality in domestic animals including wildlife and humans. The majority of tick-borne diseases are zoonotic. The global importance of ticks and tick-borne diseases in veterinary medicine and public health keeps growing. Some ticks are invasive and transmit pathogens causing transboundary diseases of high consequence for populations of domestic animals and humans. Integrated management pursues the optimized use of compatible methods to manage pests in a way that is safe, economically viable, and environmentally sustainable. The area-wide approach augments and expands the benefits of integrated pest management strategies. Issues challenging the implementation, adoption, and viability of area-wide tick management programmes include funding and socio-political aspects, the availability of support systems related to extension and veterinary services, and stakeholder involvement. Management strategies need to adapt and integrate novel technologies to decrease significantly the use of pesticide and address the complex problem of ticks and tick-borne diseases effectively. Applying the *One Health* concept, the strategy to optimize health outcomes for humans, animals, and the environment, facilitates research on the interplay between climate, habitat, and hosts driving tick population dynamics. It enhances our understanding of the epidemiology of tick-borne diseases and advances their management. This overview of research for adaptive area-wide integrated management concentrates on ticks affecting livestock. Examples focus on *Rhipicephalus microplus* (Canestrini) as one of the tick disease vectors most studied worldwide. Highlights of integrated management research for ticks

J. Hendrichs, R. Pereira and M. J. B. Vreysen (eds.), Area-Wide Integrated Pest Management: Development and Field Application, pp. 251–274. CRC Press, Boca Raton, Florida, USA.

of public health importance transmitting zoonotic diseases are reviewed to document opportunities for integrated control that mitigate the health burden of tick-borne diseases on humans, domestic animals, and wildlife. Implementation of the research conducted so far is needed to accelerate advancements in area-wide management of tick populations that can be applied to improve prevention across tick-borne diseases, while decreasing pesticide application and contributing to vector control globally.

Key Words: Acari, *Rhipicephalus annulatus*, *Rhipicephalus microplus,* babesiosis, ectoparasites, disease vectors, tick-borne pathogens, acaricides, resistance, cattle fever tick reservoirs, livestock vaccination, area-wide tick management, integrated tick-borne disease prevention, One Health, global change, invasive

1. INTRODUCTION

In some parts of the world, ticks (Acari) are the most dangerous animals followed by mosquitoes as ectoparasites and vectors of infectious agents causing morbidity and mortality in domestic animals, wildlife and humans (Ahmed et al. 2007; Socolovschi et al. 2008; Heyman et al. 2010; Barker et al. 2014; Paddock et al. 2016). Approximately 80% of the cattle in tropical and subtropical regions of the world are affected by economically important ticks and tick-borne pathogens (McCosker 1979; de Castro 1997). In addition, estimates indicate that Lyme disease and other diseases caused by tick-borne pathogens could burden over 30% of the global human population by 2050 (Davidsson 2018; Sakamoto 2018). Most tick-borne diseases affecting people are zoonotic because they can be transmitted from wild and domestic animals to humans through the bite of an infected tick (Lorusso et al. 2016; Ojeda-Chi et al. 2019).

Life history traits afford ticks considerable importance as pests and vectors of pathogens. Ticks are ancient arthropods that parasitize vertebrate hosts by feeding on blood to be able to complete their life cycle (Mans et al. 2011; Peñalver et al. 2018). Tick-borne pathogens include protozoa, bacteria, and viruses that co-infect their vectors and hosts (Brites-Neto et al. 2015; Talactac et al. 2018; Wikel 2018). Being local specialists and global generalists in their host associations underlie the global distribution of ticks and their ability to adapt to diverse environmental niches (McCoy et al. 2013; de la Fuente et al. 2015b; Beati and Klompen 2019).

There are ca. 920 described tick species in the world, but the diversity of ticks remains to be fully established (Dantas-Torres 2018; Mans et al. 2019). The so-called hard ticks belong to the Ixodidae family that have a sclerotized scutal plate in their dorsum (Sonenshine and Roe 2014). By comparison, soft ticks in the family Argasidae lack the scutum and have a flexible leathery cuticle (Uspensky 2008). Depending on the tick species, the parasitic larva, nymph, and adult stages are completed in one, two, or three hosts (Estrada-Peña 2015). After blood-engorged, females that mated on the host, then drop off and lay their eggs in the environment (Needham and Teel 1991).

Some ticks are invasive and transmit pathogens causing transboundary diseases of high consequence for populations of domestic animals and humans (Minjauw and McLeod 2003; Burridge 2011; Fernández and White 2016; Higgs 2018; Robles et al. 2018; Spengler et al. 2018).

Non-anthropogenic and anthropogenic factors associated with global change, including environmental disturbance and climate variability (Benavides Ortiz et al. 2016; Ogden and Lindsay 2016; Singer and Bulled 2016), increased international trade and travel (Abdullah et al. 2018; Hansford et al. 2018), and the wildlife-livestock-human interface (Gortazar et al. 2015), have increased tick densities resulting in a greater prevalence of tick-borne disease cases (Gasmi et al. 2018; Rasi et al. 2018; Sonenshine 2018). Furthermore, several of the newly discovered tick-borne microbes are pathogenic to humans and domestic animals (Mansfield et al. 2017; Harvey et al. 2019).

Discoveries by Smith and Kilborne (1893), documenting that *Rhipicephalus annulatus* Say was a vector of *Babesia bigemina* (Smith et Kilborne 1893), were important in the history of science by showing for the first time that arthropods can transmit pathogens to their hosts (Smith and Kilborne 1893; McCosker 1993; Egerton 2013).

Smith and Kilborne (1893) suggested the destruction of all *R. annulatus* infesting cattle to treat the disease after noting that outbreaks of bovine babesiosis, caused by *B. bigemina*, also known as redwater or cattle tick fever, and considered to be the most economically important arthropod-borne disease of cattle worldwide (Bock et al. 2008), could not happen without tick parasitism. In retrospect, this research association is an example of the One Health concept described below because T. Smith was a physician and F. L. Kilborne a veterinarian (Schultz 2008).

By 1893, cattle in the USA, Australia, and parts of Africa were already immersed in dipping vats containing various chemical pesticides active against ticks commonly referred to through time as tickicides, ixodicides, or acaricides, to manage infestations associated with what we now know are tick-borne diseases (Angus 1996; George 2000; Alonso-Díaz et al. 2006). The term acaricide used here refers to pesticides used to kill ticks of veterinary and public health importance following the conventions of most literature published on the topic. Vaccination against the pathogen is another approach to prevent and control tick-borne diseases. Attempts by Connaway and Francis (1899) to protect cattle from bovine babesiosis were among the first ones to vaccinate against a tick-borne disease. Several vaccines are commercially available in Europe to prevent tick-borne encephalitis (Riccardi et al. 2019). Nevertheless, the need remains for improved and cost-effective vaccines to prevent tick-borne diseases affecting humans (Šmit and Postma 2016; Reece et al. 2018), as well as domestic animals (Perry 2016; Pruneau et al. 2018; Suarez et al. 2019).

Effective and safe tick and tick-borne disease management requires integration of rational tactics involving multiple biological, chemical, physical and vaccine technologies on and off hosts. They can include the judicious application of safer acaricides to address the concerns with chemical treatments (de Meneghi et al. 2016; Pfister and Armstrong 2016; Ginsberg et al. 2017).

Here we review highlights of integrated management research for ticks of public health importance transmitting zoonotic diseases to document opportunities for combined interventions that mitigate the health burden of tick-borne diseases, benefitting humans, domestic animals, and wildlife (Drexler et al. 2014; Khamesipour et al. 2018; Wang et al. 2018).

2. AREA-WIDE TICK MANAGEMENT AND RESEARCH

2.1. Research Needs for Integrated Area-wide Tick Management

This overview concentrates on research to enable the area-wide integrated management of livestock ticks. Examples focus on *Rhipicephalus microplus* (Canestrini), a one-host tick commonly known as the Asian blue tick or southern cattle fever tick, originally described as *Haemaphysalis micropla* by Canestrini (1887). It is one of the ticks most studied worldwide as it is a vector of *B. bigemina* and *B. bovis* Babes causing bovine babesiosis (Pérez de León et al. 2014b; Gray et al. 2019), and *Anaplasma marginale* Theiler causing anaplasmosis (Atif 2015). *R. microplus* is an invasive species considered the most economically important ectoparasite of livestock globally (Rodríguez-Vivas et al. 2017a; Betancur-Hurtado and Giraldo-Ríos 2018; Sungirai et al. 2018).

The synonym concepts of area-wide integrated pest management, system-, or area-wide pest management, convey the need for research that can be applied to address the complex problem with ticks and tick-borne diseases (Brévault and Bouyer 2014; Pérez de León et al. 2014a; Bourtzis et al. 2016). Efficiency and cost-effectiveness are fundamental to area-wide approaches dealing with societal problems for centuries, including those related to tick disease vectors (Hendrichs et al. 2007; Koul et al. 2008; Shepard et al. 2014).

The goal of integrated pest management is to optimize the use of compatible methods in a way that is safe, economically viable, and ecologically sustainable (Jørs et al. 2017; Mullens et al. 2018). The area-wide approach augments and expands to the population level the benefits of integrated pest management strategies. Tick suppression and eradication can be considered as a continuum in the spectrum of area-wide strategies to manage tick-borne diseases. Approaches for sustainable area-wide control of tick populations recognize the need for translational research to develop new and improved technologies before eradication can be contemplated (Bram and Gray 1979; Pegram et al. 2007; Pluess et al. 2012; Suckling et al. 2014). A common theme for these strategies is the continued need to re-evaluate our understanding of tick biology and ecology (Tatchell 1992; Schmidtmann 1994; Esteve-Gassent et al. 2016; Canevari et al. 2017).

2.2. Unifying Area-wide Tick-borne Disease Mitigation and One Health through Integrated Tick Management Research

Applying the One Health concept, i.e. a strategy to optimize health outcomes for humans, animals, and the environment, facilitates research on the interplay between climate, habitat, and hosts driving tick population dynamics. It enhances our understanding of the epidemiology of tick-borne diseases and advances their management (Dantas-Torres et al. 2012; Vayssier-Taussat et al. 2015; Laing et al. 2018; World Bank 2018) (Fig. 1).

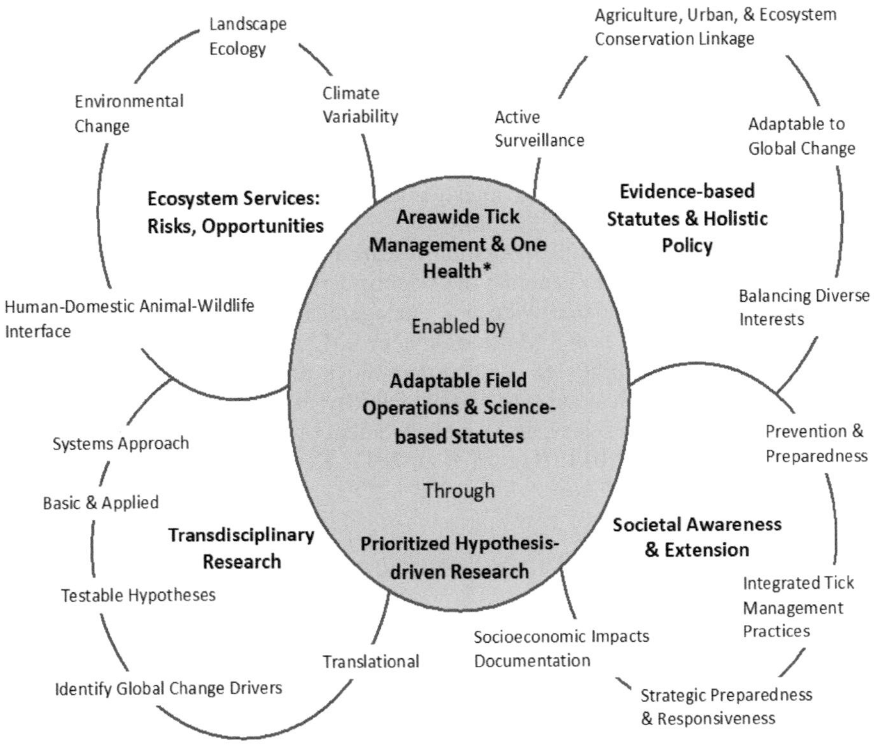

*Figure 1. Suggested research and implementation framework toward sustainable area-wide integrated tick management to prevent tick-borne diseases in the context of global change and the One Health approach (*adapted from Pérez de León et al. 2012).*

Previous efforts indicate socio-economic and cultural aspects must be considered in the planning and evaluation of area-wide tick management programmes (Pegram et al. 2000; Hendrichs et al. 2007; Rushton 2009; Mutavi et al. 2018). This can be done using algorithms to select area-wide tick management interventions where the evidence thus generated is used to enhance model predictions that improve area-wide tick management practices (Sutherst et al. 2007; Wang et al. 2017).

Ideal characteristics of technologies for broad acceptability and integrated use against ticks include low cost, minimal effort required for their application, spectrum of efficacy covering as many tick species as possible, and residual activity (Graf et al. 2004; Playford et al. 2005; Eisen and Eisen 2018). Control technologies can target ticks on or off the host.

In addition to acaricides, parasitoids and predators, alteration of the environment, and physical methods were identified for off-host tick control (FAO 1961). The Sterile Insect Technique, botanical repellents and acaricides, host resistance, pheromone-based approaches, and anti-tick vaccines are potential approaches to be integrated for the control of ticks infesting hosts (IAEA 1968; Ghosh and Nagar 2014; de Oliveira Filho et al. 2017). However, reducing to practice the integration of acaricides with other technologies in area-wide tick management remains to be fully accomplished (Jonsson 2004; de la Fuente et al. 2015a).

Adequate protocols and implementation research to evaluate technologies under field conditions are needed to generate the scientific evidence required to justify the investment of resources for area-wide tick management programmes (Piesman and Eisen 2008; Bautista-Garfias and Martínez-Ibañez 2012).

The adoption and viability of these programmes require attention to resource allocation and socio-political aspects, the availability of support systems related to extension and veterinary services, and the engagement of stakeholders (Walker 2011; Estrada-Peña and Salman 2013; Bugeza et al. 2017; Kerario et al. 2018; Mihajlović et al. 2019).

2.3. *Alternatives to Acaricide Use and Strategies to Solve Resistance to Chemical Treatments*

Chemical treatment practices in livestock production systems are under scrutiny because of the impact acaricides and endectocides like ivermectin have on public health, the environment, and the international trade of livestock and animal products (González and Hernández 2012; Arisseto-Bragotto et al. 2017; Miraballes and Riet-Correa 2018).

Intense chemical treatment of infested hosts exerts strong selection pressure for acaricide resistance among tick populations (Guerrero et al. 2014b; de Miranda Santos et al. 2018; Rodríguez-Vivas et al. 2018). Resistance to multiple classes of acaricides keeps spreading among tick populations due to intensive application (Miller et al. 2013; Cuore et al. 2015; Klafke et al. 2017b; Vudriko et al. 2018).

Acaricide resistance resulting from chemical treatment intended to control other parasites can exacerbate the problem with ticks and tick-borne diseases (Foil et al. 2004). Ivermectin used to treat gastrointestinal parasitic infections in cattle simultaneously infested with *R. microplus* selected for resistance due to exposure of the ticks to sublethal doses of that endectocidal drug (Alegría-López et al. 2015). *R. microplus* ranks sixth among the arthropods most resistant to pesticides in the world (Whalon et al. 2008).

Resistance to organophosphates, pyrethroids, amitraz, and ivermectin was reported in the brown dog tick, *Rhipicephalus sanguineus* sensu lato (Latreille) (Klafke et al. 2017a, Rodríguez-Vivas et al. 2017b).

Amblyomma cajennense s.l. (Fabricius) is another three-host tick that was found to be resistant to organophosphates and amitraz in Mexico (Alonso-Díaz et al. 2013). Different acaricide resistance profiles have been reported for two- and three-host tick species parasitizing cattle in South Africa (Ntondini et al. 2008). Widespread resistance to commonly used acaricides has not been reported for other important vectors of zoonotic tick-borne diseases parasitizing humans in the USA and Europe (Coles and Dryden 2014; EMA 2018).

Strategies to diminish acaricide use in domestic animals need to consider the concept of integrated parasite/vector management to maximize the contributions of veterinary public health towards sustainable development (Henrioud 2011; Scasta 2015; Narladkar 2018).

The commercial availability of a technology based on the recombinant protein Bm86 in the 1990s to vaccinate cattle against *R. microplus* represented a significant research achievement towards sustainable area-wide tick management (de la Fuente et al. 2007; Willadsen 2008). Integrating the use of a Bm86-based anti-*R. microplus* vaccine in an area-wide management programme confirmed that this approach decreases the frequency of acaricide treatments and diminishes the amount of chemicals used to control infestations, while reducing tick-borne cases in a cost-effective manner (de la Fuente et al. 1998; Redondo et al. 1999; Valle et al. 2004; Suarez et al. 2016). This is a rational and environment-friendly approach to manage *R. microplus* populations that are resistant to multiple classes of acaricides. Various research efforts to develop vaccines against other hard and soft ticks and the application of an anti-tick vaccine to protect humans, domestic animals, and wildlife from tick-borne diseases are ongoing (Évora et al. 2017; Almazán et al. 2018; de la Fuente et al. 2018).

Collaborative partnerships established to improve efficiencies in the research and development process of those anti-tick vaccines are examples of how global efforts could fully realize the benefits of international cooperation to enable breakthroughs allowing the adaptation of area-wide tick management practices to protect livestock and humans from tick-borne diseases (Sprong et al. 2014; Schetters et al. 2016; Rodríguez-Mallon et al. 2018; Ybañez et al. 2018). These joint international efforts have also resulted in the sequencing the genome of *R. microplus* to mine the information therein for the innovation of management technologies (Barrero et al. 2017).

Integrative taxonomy studies reinstated *Rhipicephalus australis* as a species and revealed that *R. microplus* consists of 3 clades (Estrada-Peña et al. 2012; Roy et al. 2018). Furthermore, some of the new microbes found to be associated with *R. microplus* are known livestock pathogens, while the pathogenicity of others remains unknown (Andreotti et al. 2011; Biguezoton et al. 2016; de Souza et al. 2018; de Oliveira Pascoal et al. 2019).

3. RESEARCH FOR ADAPTIVE AREA-WIDE TICK MANAGEMENT

3.1 Eradication Efforts Exemplify Challenges with Tick Management in the Context of Global Change

In 1906 the USA established the Cattle Fever Tick Eradication Program (CFTEP) to eliminate bovine babesiosis by exterminating the tick vectors of the disease based on the scientific evidence generated by the research of Curtice (1891) and Smith and Kilborne (1893) (Curtice 1910). In the context of complex socio-economic dynamics (Strom 2010), the CFTEP efforts has involved cooperation between federal, state governments, and the livestock industry. In 1943, with the exception of a Permanent Quarantine Zone along the Rio Grande in south Texas at the border with Mexico, the USA was declared free of the tick vectors (Graham and Hourrigan 1977).

The cattle fever ticks *R. annulatus* and *R. microplus* remain widespread in Mexico (Bautista-Garfias and Martínez-Ibañez 2012), and incursions from Mexico into the free zone, which comprises the rest of the continental USA, are buffered by the Permanent Quarantine Zone (George 1989; Klassen 1989). After 112 years of operations, events related to global change, such as land use changes, livestock-wildlife interface intricacies, and climate variability, complicate efforts by the CFTEP to keep the USA cattle fever tick-free (George 2008; Esteve-Gassent et al. 2014; Rutherford 2019).

A surge of cattle fever tick outbreaks in the free zone during the first decade of this century prompted a re-evaluation of the research agenda in support of the CFTEP (Pérez de León et al. 2010; Lohmeyer et al. 2011). Action was taken based on research needs in consultation with stakeholders to address the main concerns with acaricide resistance (Pérez de León et al. 2013), the role of native and exotic ungulates as cattle fever tick reservoirs (Pound et al. 2010), climate variability as a driver for the reintroduction of cattle fever ticks into the free zone (Giles et al. 2014), and a re-evaluation of financial losses associated with these concerns and events (Anderson et al. 2010).

Research in support of integrated cattle fever tick eradication provides a pathway to generate scientific evidence that could be used to adapt CFTEP operations that minimize the impact of global change (Pérez de León et al. 2012). Aspects related to the mechanism of action of organophosphate acaricides allowed charging the dipping vats at 0.3% coumaphos to mitigate concerns by the CFTEP with organophosphate-resistant cattle fever tick outbreak populations (Miller et al. 2005). The detection of pyrethroid resistance in cattle fever ticks infesting cattle and white-tailed deer, *Odocoileus virginianus* (Zimmermann), a wild ungulate species native to the Americas and a host of cattle fever ticks that is abundant in south Texas, limits treatments with permethrin by the CFTEP (Busch et al. 2014). Use of an injectable formulation of 1% doramectin was adapted by the CFTEP as an alternative eradication procedure (Davey et al. 2012). Macrocyclic lactone resistance among cattle fever tick outbreak populations has so far not been reported.

Further studies are required to determine if research can be translated into protocols involving the use by the CFTEP of safer acaricides to treat cattle and wild ungulates (Costa-Júnior et al. 2016; Gross et al. 2017).

A device using corn as bait to attract white-tailed deer for self-treatment was developed for use by the CFTEP because white-tailed deer cannot be gathered for treatment as it is done with cattle (Pound et al. 2012). The white-tailed deer consuming corn rub against acaricide-impregnated rollers placed on the device during the hunting season and are thus treated topically, whereas corn medicated with ivermectin acting systemically to control cattle fever ticks is used to bait white-tailed deer during the off-hunting season (Lohmeyer et al. 2013). However, complex white-tailed deer behaviours and social interactions to access the bait stations and other logistical aspects limit the use of this technology in the Permanent Quarantine Zone (Currie 2013).

Impediments for cattle fever tick eradication associated with the presence of nilgai (*Boselaphus tragocamelus* (Pallas) in parts of south Texas, where they can coexist with cattle and white-tailed deer, further exemplify the challenges presented by the livestock-wildlife interface for area-wide tick management (Wang et al. 2016; Singh et al. 2017; Lohmeyer et al. 2018). Nilgai are introduced bovid hosts of cattle fever ticks and suspected reservoirs of *B. bovis* and *B. bigemina* with home ranges larger than white-tailed deer (Foley et al. 2017; Olafson et al. 2018). Research is underway to determine if nilgai can be attracted to sites where they would be treated against cattle fever tick infestation (Goolsby et al. 2017).

The high efficacy of the Bm86 antigen against *R. annulatus* prompted efforts to research the use of an anti-tick vaccine as part of integrated cattle fever tick eradication procedures (Miller et al. 2012). Research involved reverse vaccinology to pursue the discovery of antigens that could be formulated for use by the CFTEP with efficacy against *R. microplus* equivalent to that of Bm86-based vaccines against *R. annulatus* (Guerrero et al. 2014a). In the interim, a public-private partnership enabled the use of a Bm86-based vaccine by the CFTEP (Pérez de León et al. 2018). This was a significant event in the history of cattle fever tick eradication in the USA because federal and state statutes, more than a century old governing the CFTEP, were adapted to use the anti-tick vaccine technology. This Bm86-based vaccine was used in a research project for integrated *R. microplus* management in Puerto Rico (Wang et al. 2019).

Integrating vaccination of white-tailed deer against cattle fever ticks would complement the effects of the self-treatment bait stations described above (Carreón et al. 2012; Estrada-Peña et al. 2014). However, delivery systems remain to be refined to vaccinate free-ranging white-tailed deer against cattle fever ticks in the Permanent Quarantine Zone.

3.2. *Research Perspectives to Mitigate Tick-borne Disease Burden Focused on Integrated Tick Management*

Applying the concept of precision agriculture and making use of newly available technologies, provides the opportunity to establish exact and targeted interventions to realize substantial savings in inputs for area-wide tick management (Urdaz-Rodríguez et al. 2015; Pérez de León 2017).

Experiments with unmanned aerial vehicles or drones showed this technology can support surveillance by the CFTEP (Goolsby et al. 2016), but it could also be integrated with remote sensing using ground-truth data for strategic cattle fever tick suppression (Phillips et al. 2014; Leal et al. 2018). Robotic technology is being adapted for tick control as well, showing potential in reducing tick densities (Gaff et al. 2015).

Precision tick management could facilitate the adoption of safer control technologies for effective area-wide campaigns. These include commercially available alternatives to the conventional use of acaricides such as acaropathogenic fungi or nematodes, and botanical acaricides, although they require further testing for adoption by the CFTEP (Thomas et al. 2017; Goolsby et al. 2018; Singh et al. 2018). Additionally, big data strategies facilitate the translation of genomic information into knowledge that can be applied to develop technologies which specifically target cattle fever ticks (Munoz et al. 2017; Brock et al. 2019).

Tick-borne diseases threaten public health in the USA. Around twenty human diseases or clinical conditions are associated with tick bites (USHHS 2018). Current trends indicate that >75% of the vector-borne disease cases reported are tick-borne (Rosenberg et al. 2018). Among the ticks commonly found biting humans (Eisen et al. 2017), the black-legged tick *Ixodes scapularis* Say is known to transmit seven pathogens of human diseases (Eisen and Eisen 2018). Controlling tick populations, together with personal protection measures, reduce exposure of the public to infected ticks, which prevents tick-borne diseases (Stafford III et al. 2017; White and Gaff 2018).

A higher level of public acceptability is associated with area-wide interventions employing technologies that are safe for people, pets, and the environment (Aenishaenslin et al. 2016; Keesing and Ostfeld 2018). These include the integrated use of host-targeted devices delivering minimal acaricide quantities with broadcast application of acaropathogenic fungus, as well as white-tailed deer reduction to decrease the risk of human exposure to *I. scapularis* infected with *Borrelia burgdorferi* Johnson et al. (Telford 2017; Williams et al. 2018). Additionally, internet-based surveillance tools and citizen science participation may enhance area-wide integrated tick management practices (Pollett et al. 2017; Nieto et al. 2018; Jongejan et al. 2019).

The detection in 2017 of *Haemaphysalis longicornis* Neumann, commonly known as the Asian longhorned tick, and subsequent reports of infestations in humans, domestic animals, and wildlife in the USA is a reminder of the threat posed by invasive ticks to the health of humans and other animals (Rainey et al. 2018). *H. longicornis* is a known vector of pathogens affecting humans, domestic animals, and wildlife in its native range and previously invaded areas, but it remains to be determined if it is transmitting pathogens in the USA (Beard et al. 2018).

Habitat suitability analyses indicate that *H. longicornis* could become established also in other parts of North America (Magori 2018; Hutcheson et al. 2019; Rochlin 2019). Challenges managing the spread of this Asian longhorned tick in the USA present an opportunity to apply the One Health concept where governmental agencies, academic institutions, public organizations, and private industry representing the agricultural, public health, medical, and veterinary sectors operate under a national strategy to prevent cases of *H. longicornis*-borne diseases in humans and other animal species.

Implementation research is needed to accelerate advancements in area-wide tick management. Achieving this goal will facilitate the adaptation and adoption of those advancements to improve prevention across tick-borne diseases while contributing to vector control globally (WHO 2017; Theobald et al. 2018; Fouet and Kamdem 2019; Petersen et al. 2019).

4. CONCLUSIONS

Ticks and tick-borne diseases continue to present new and emerging threats to humans, domestic animals, and wildlife. Constraints faced by the CFTEP to continue maintaining the USA cattle fever tick-free, a successful area-wide programme that has been operating in the USA since its establishment in 1906, illustrate how global change impacts area-wide tick management efforts.

Current issues are complex and need to be addressed by veterinary and public health programmes dealing with ticks and tick-borne diseases. This grand challenge requires a reassessment of strategies to manage tick populations. The One Health approach provides a framework to mitigate the health burden of tick-borne diseases on humans, domestic animals, and wildlife.

Advances in transdisciplinary scientific research present opportunities to adapt the strategy for area-wide tick management. The integration of novel technologies can decrease the use of acaricides significantly. Pilot field studies help determine the utility of integrated tick management strategies under real-life conditions. Outcomes from those pilot field studies inform decisions on the extent of interventions to prevent tick-borne diseases through improved tick population management. Progressive tick control affords flexibility to fine-tune the integration of technologies through the exchange of scientific information between stakeholders engaged in the adaptation process and provides feedback to revise the research agenda.

Implementation research can accelerate the translation of earlier research efforts to area-wide tick management practice. It is important for scientists to also understand the socio-economic context of research. Grasping the expectations of end-users of technology is paramount to realize the common vision of improving the outcomes of tick control interventions. This process will enhance the quality of evidence delivered by scientific research. Such scientific evidence can be used to generate the support for resources to establish the capacities required for the effective management of ticks to mitigate the burden of tick-borne diseases.

5. REFERENCES

Abdullah, S., C. Helps, S. Tasker, H. Newbury, and R. Wall. 2018. Prevalence and distribution of *Borrelia* and *Babesia* species in ticks feeding on dogs in the UK. Medical and Veterinary Entomology 321: 14–22.

Aenishaenslin, C., P. Michel, A. Ravel, L. Gern, J. P. Waaub, F. Milord, and D. Bélanger. 2016. Acceptability of tick control interventions to prevent Lyme disease in Switzerland and Canada: A mixed-method study. BMC Public Health 16: 12–21.

Ahmed, J., H. Alp, M. Aksin, and U. Seitzer. 2007. Current status of ticks in Asia. Parasitology Research 1012: 159–162.

Alegría-López, M., R. Rodríguez-Vivas, J. Torres-Acosta, M. Ojeda-Chi, and J. Rosado-Aguilar. 2015. Use of ivermectin as endoparasiticide in tropical cattle herds generates resistance in gastrointestinal nematodes and the tick *Rhipicephalus microplus* (Acari: Ixodidae). Journal of Medical Entomology 522: 214–221.

Almazán, C., G. A. Tipacamu, S. Rodríguez, J. Mosqueda, and A. Pérez de León. 2018. Immunological control of ticks and tick-borne diseases that impact cattle health and production. Frontiers in Bioscience (Landmark edition) 23: 1535–1551.

Alonso-Díaz, M., R. Rodríguez-Vivas, H. Fragoso-Sánchez, and R. Rosario-Cruz. 2006. Ixodicide resistance of the *Boophilus microplus* tick to ixodicides. Archivos de Medicina Veterinaria 382: 105–113.

Alonso-Díaz, M., A. Fernández-Salas, F. Martínez-Ibáñez, and J. Osorio-Miranda. 2013. *Amblyomma cajennense* (Acari: Ixodidae) tick populations susceptible or resistant to acaricides in the Mexican tropics. Veterinary Parasitology 197: 326–331.

Anderson, D., A. Hagerman, P. Teel, G. Wagner, J. Outlaw, and B. Herbst. 2010. Economic impact of expanded fever tick range. Agricultural & Food Policy Center, Texas A&M University, College Station, Texas, USA.

Andreotti, R., A. A. Pérez de León, S. E. Dowd, F. D. Guerrero, K. G. Bendele, and G. A. Scoles. 2011. Assessment of bacterial diversity in the cattle tick *Rhipicephalus* (*Boophilus*) *microplus* through tag-encoded pyrosequencing. BMC Microbiology 11: 6.

Angus, B. M. 1996. The history of the cattle tick *Boophilus microplus* in Australia and achievements in its control. International Journal for Parasitology 2612: 1341–1355.

Arisseto-Bragotto, A. P., M. M. C. Feltes, and J. M. Block. 2017. Food quality and safety progress in the Brazilian food and beverage industry: Chemical hazards. Food Quality and Safety 12: 117–129.

Atif, F. A. 2015. *Anaplasma marginale* and *Anaplasma phagocytophilum*: Rickettsiales pathogens of veterinary and public health significance. Parasitology Research 114: 3941–3957.

Barker, S. C., A. R. Walker, and D. Campelo. 2014. A list of the 70 species of Australian ticks; diagnostic guides to and species accounts of *Ixodes holocyclus* (paralysis tick), *Ixodes cornuatus* (southern paralysis tick) and *Rhipicephalus australis* (Australian cattle tick); and consideration of the place of Australia in the evolution of ticks with comments on four controversial ideas. International Journal for Parasitology 4412: 941–953.

Bautista-Garfias, C., and F. Martínez-Ibañez. 2012. Experiences on the control of cattle tick *Rhipicephalus (Boophilus) microplus* in Mexico, pp. 205-216. *In* M. Woldemeskel (ed.), Ticks: Disease, management, and control. Nova Science Publishers, Inc. New York, NY, USA.

Beard, C. B., J. Occi, D. L. Bonilla, A. M. Egizi, D. M. Fonseca, J. W. Mertins, B. P. Backenson, W. I. Bajwa, A. M. Barbarin, M. A. Bertone, J. Brown, N. P. Connally, N. D. Connell, R. J. Eisen, R. C. Falco, A. M. James, R. K. Krell, K. Lahmers, N. Lewis, S. E. Little, M. Neault, A. A. Pérez de León, A. R. Randall, M. G. Ruder, M. N. Saleh, B. L. Schappach, B. A. Schroeder, L. L. Seraphin, M. Wehtje, G. P. Wormser, M. J. Yabsley, and W. Halperin. 2018. Multistate infestation with the exotic disease-vector tick *Haemaphysalis longicornis* - United States, August 2017-September 2018. Morbidity and Mortality Weekly Report 6747: 1310–1313.

Beati, L., and H. Klompen. 2019. Phylogeography of ticks (Acari: Ixodida). Annual Review of Entomology 64: 379–397.

Benavides Ortiz, E., J. Romero Prada, and L. C. Villamil Jiménez. 2016. Las garrapatas del ganado bovino y los agentes de enfermedad que transmiten en escenarios epidemiológicos de cambio climático: Guía para el manejo de garrapatas y adaptación al cambio climático. Instituto Interamericano de Cooperación para la Agricultura (IICA), San José, Costa Rica.

Betancur-Hurtado, O. J., and C. Giraldo-Ríos. 2018. Economic and health impact of the ticks in production animals, pp. 1–19. *In* M. Abubakar (ed.), Ticks and tick-borne pathogens. IntechOpen, London, UK.

Biguezoton, A., V. Noel, S. Adehan, H. Adakal, G. K. Dayo, S. Zoungrana, S. Farougou, and C. Chevillon. 2016. *Ehrlichia ruminantium* infects *Rhipicephalus microplus* in West Africa. Parasites & Vectors 9: 354.

Bock, R. E., L. A. Jackson, A. J. de Vos, and W. K. Jorgensen. 2008. Babesiosis of cattle, pp. 281–307. *In* A. S. Bowman, and P. Nuttal (eds.), Ticks: Biology, disease, and control. Cambridge University Press, New York, NY, USA.

Bourtzis, K., R. S. Lees, J. Hendrichs, and M. J. Vreysen. 2016. More than one rabbit out of the hat: Radiation, transgenic and symbiont-based approaches for sustainable management of mosquito and tsetse fly populations. Acta Tropica 157: 115–130.

Bram, R. A., and J. H. Gray. 1979. Eradication - An alternative to tick and tick-borne disease control. World Animal Review 30: 30–35.

Brévault, T., and J. Bouyer. 2014. From integrated to system-wide pest management: Challenges for sustainable agriculture. Outlooks on Pest Management 253: 212–213.

Brites-Neto, J., K. M. R. Duarte, and T. F. Martins. 2015. Tick-borne infections in human and animal population worldwide. Veterinary World 83: 301–315.

Brock, C. M., K. B. Temeyer, J. Tidwell, Y. Yang, M. A. Blandon, D. Carreón-Camacho, M. T. Longnecker, C. Almazán, A. A. Pérez de León, and P. V. Pietrantonio. 2019. The leucokinin-like peptide receptor from the cattle fever tick, *Rhipicephalus microplus*, is localized in the midgut periphery and receptor silencing with validated double-stranded RNAs causes a reproductive fitness cost. International Journal for Parasitology 49: 287–299.

Bugeza, J., C. Kankya, J. Muleme, A. Akandinda, J. Sserugga, N. Nantima, E. Okori, and T. Odoch. 2017. Participatory evaluation of delivery of animal health care services by community animal health workers in Karamoja region of Uganda. PLoS One 126: e0179110.

Burridge, M. J. 2011. Non-native and invasive ticks: Threats to human and animal health in the United States. University Press of Florida, Gainesville, Florida, USA.

Busch, J. D., N. E. Stone, R. Nottingham, A. Araya-Anchetta, J. Lewis, C. Hochhalter, J. R. Giles, J. Gruendike, J. Freeman, and G. Buckmeier. 2014. Widespread movement of invasive cattle fever ticks (*Rhipicephalus microplus*) in southern Texas leads to shared local infestations on cattle and deer. Parasites & Vectors 7: 188.

Canestrini, G. 1887. Intorno ad alcuni Acari ed Opilionidi dell' America. Atti della Società Veneto-Trentina di Scienze Naturali 11: 100–111.

Canevari, J. T., A. J. Mangold, A. A. Guglielmone, and S. Nava. 2017. Population dynamics of the cattle tick *Rhipicephalus (Boophilus) microplus* in a subtropical subhumid region of Argentina for use in the design of control strategies. Medical and Veterinary Entomology 311: 6–14.

Carreón, D., J. M. P. de la Lastra, C. Almazán, M. Canales, F. Ruiz-Fons, M. Boadella, J. A. Moreno-Cid, M. Villar, C. Gortázar, and M. Reglero. 2012. Vaccination with BM86, subolesin and akirin protective antigens for the control of tick infestations in white-tailed deer and red deer. Vaccine 302: 273–279.

Coles, T. B., and M. W. Dryden. 2014. Insecticide/acaricide resistance in fleas and ticks infesting dogs and cats. Parasites & Vectors 71: 8.

Connaway, J. W., and M. C. Francis. 1899. Texas fever. Experiments made by the Missouri Experiment Station and the Missouri State Board of Agriculture, in cooperation with the Texas Experiment Station in immunizing northern breeding cattle against Texas fever for the southern trade. Missouri Agricultural Experiment Station Bulletin 48: 1–66.

Costa-Júnior, L. M., R. J. Miller, P. B. Alves, A. F. Blank, A. Y. Li, and A. A. Pérez de León. 2016. Acaricidal efficacies of *Lippia gracilis* essential oil and its phytochemicals against organophosphate-resistant and susceptible strains of *Rhipicephalus (Boophilus) microplus*. Veterinary Parasitology 228: 60–64.

Cuore, U., W. Acosta, F. Bermúdez, O. Da Silva, I. García, R. Pérez Rama, L. Luengo, A. Trelles, and M. A. Solari. 2015. Tick generational treatment: Implementation of a methodology to eradicate *Rhipicephalus* (*Boophilus*) *microplus* tick resistant to macrocyclic lactones in a population management. Veterinaria (Montevideo) 51: 14–25.

Currie, C. 2013. Influence of white-tailed deer on cattle fever tick eradication efforts in southern Texas. PhD dissertation. Texas A&M University-Kingsville, Kingsville, Texas, USA.

Curtice, C. 1891. The biology of the cattle tick. Journal of Comparative Medical and Veterinary Archives 12: 313–319.

Curtice, C. 1910. Progress and prospects of tick eradication. US Department of Agriculture, Bureau of Animal Industry. Twenty-seventh Annual Report: 255–265.

Dantas-Torres, F. 2018. Species concepts: What about ticks? Trends in Parasitology 34(12): 1017–1026.

Dantas-Torres, F., B. B. Chomel, and D. Otranto. 2012. Ticks and tick-borne diseases: A One Health perspective. Trends in Parasitology 28: 437–446.

Davey, R. B., J. M. Pound, J. A. Klavons, K. H. Lohmeyer, J. M. Freeman, and P. U. Olafson. 2012. Analysis of doramectin in the serum of repeatedly treated pastured cattle used to predict the probability of cattle fever ticks (Acari: Ixodidae) feeding to repletion. Experimental and Applied Acarology 56: 365–374.

Davidsson, M. 2018. The financial implications of a well-hidden and ignored chronic Lyme disease pandemic. Healthcare (Basel) 6: 16.

de Castro, J. J. 1997. Sustainable tick and tickborne disease control in livestock improvement in developing countries. Veterinary Parasitology 71: 77–97.

de la Fuente, J., K. Kocan, and M. Contreras. 2015a. Prevention and control strategies for ticks and pathogen transmission. Scientific and Technical Review 34: 249–264.

de la Fuente, J., A. Estrada-Peña, A. Cabezas-Cruz, and R. Brey. 2015b. Flying ticks: Anciently evolved associations that constitute a risk of infectious disease spread. Parasites & Vectors 8: 538.

de la Fuente, J., M. Villar, A. Estrada-Peña, and J. A. Olivas. 2018. High throughput discovery and characterization of tick and pathogen vaccine protective antigens using vaccinomics with intelligent Big Data analytic techniques. Expert Review of Vaccines 17: 569–576.

de la Fuente, J., M. Rodríguez, M. Redondo, C. Montero, J. García-García, L. Méndez, E. Serrano, M. Valdés, A. Enriquez, and M. Canales. 1998. Field studies and cost-effectiveness analysis of vaccination with Gavac™ against the cattle tick *Boophilus microplus*. Vaccine 16: 366–373.

de la Fuente, J., C. Almazán, M. Canales, J. M. P. de la Lastra, K. M. Kocan, and P. Willadsen. 2007. A ten-year review of commercial vaccine performance for control of tick infestations on cattle. Animal Health Research Reviews 8: 23–28.

de Meneghi, D., F. Stachurski, and H. Adakal. 2016. Experiences in tick control by acaricide in the traditional cattle sector in Zambia and Burkina Faso: Possible environmental and public health implications. Frontiers in Public Health 4: 239.

de Miranda Santos, I. K., G. R. Garcia, P. S. Oliveira, C. J. Veríssimo, L. M. Katiki, L. Rodrigues, M. P. Szabó, and C. Maritz-Olivier. 2018. Acaricides: Current status and sustainable alternatives for controlling the cattle tick, *Rhipicephalus microplus*, based on its ecology, pp. 91–134. *In* C. Garros, J. Bouyer, W. Takken, and R. C. Smallegange (eds.), Pests and vector-borne diseases in the livestock industry. Wageningen Academic Publishers, Wageningen, The Netherlands.

de Oliveira Filho, J. G., L. L. Ferreira, A. L. F. Sarria, J. A. Pickett, M. A. Birkett, G. M. Mascarin, A. A. Pérez de León, and L. M. F. Borges. 2017. Brown dog tick, *Rhipicephalus sanguineus* sensu lato, infestation of susceptible dog hosts is reduced by slow release of semiochemicals from a less susceptible host. Ticks and Tick-borne Diseases 8: 139–145.

de Oliveira Pascoal, J., S. M. de Siqueira, R. da Costa Maia, M. P. J. Szabó and J. Yokosawa. 2019. Detection and molecular characterization of Mogiana tick virus (MGTV) in *Rhipicephalus microplus* collected from cattle in a savannah area, Uberlândia, Brazil. Ticks and Tick-borne Diseases 101: 162–165.

Drexler, N., M. Miller, J. Gerding, S. Todd, L. Adams, F. S. Dahlgren, N. Bryant, E. Weis, K. Herrick, J. Francies, K. Komatsu, S. Piontkowski, J. Velascosoltero, T. Shelhamer, B. Hamilton, C. Eribes, A. Brock, P. Sneezy, C. Goseyun, H. Bendle, R. Hovet, V. Williams, R. Massung, and J. H. McQuiston. 2014. Community-based control of the brown dog tick in a region with high rates of Rocky Mountain spotted fever, 2012–2013. PLoS One 912: e112368.

Egerton, F. N. 2013. History of ecological sciences, part 46: From parasitology to germ theory. The Bulletin of the Ecological Society of America 942: 136–164.

Eisen, R. J., and L. Eisen. 2018. The blacklegged tick, *Ixodes scapularis*: An increasing public health concern. Trends in Parasitology 344: 295–309.

Eisen, R. J., K. J. Kugeler, L. Eisen, C. B. Beard, and C. D. Paddock. 2017. Tick-borne zoonoses in the United States: Persistent and emerging threats to human health. ILAR Journal 58: 319–335.

(EMA) European Medicines Agency. 2018. Reflection paper on resistance in ectoparasites, Draft. EMA Committee for Medicinal Products for Veterinary Use, London, UK. 30 pp.

Esteve-Gassent, M. D., A. A. Pérez de León, D. Romero-Salas, T. P. Feria-Arroyo, R. Patino, I. Castro-Arellano, G. Gordillo-Pérez, A. Auclair, J. Goolsby, and R. I. Rodríguez-Vivas. 2014. Pathogenic landscape of transboundary zoonotic diseases in the Mexico–US border along the Rio Grande. Frontiers in Public Health 2: 177.

Esteve-Gassent, M. D., I. Castro-Arellano, T. P. Feria-Arroyo, R. Patino, A. Y. Li, R. F. Medina, A. A. Pérez de León, and R. I. Rodríguez-Vivas. 2016. Translating ecology, physiology, biochemistry, and population genetics research to meet the challenge of tick and tick-borne diseases in North America. Archives of Insect Biochemistry and Physiology 921: 38–64.

Estrada-Peña, A. 2015. Ticks as vectors: Taxonomy, biology and ecology. Scientific and Technical Review 34: 53–65.

Estrada-Peña, A., and M. Salman. 2013. Current limitations in the control and spread of ticks that affect livestock: A review. Agriculture 3: 221–235.

Estrada-Peña, A., D. Carreón, C. Almazán, and J. de la Fuente. 2014. Modeling the impact of climate and landscape on the efficacy of white-tailed deer vaccination for cattle tick control in northeastern Mexico. PLoS One 97: e102905.

Estrada-Peña, A., J. M. Venzal, S. Nava, A. Mangold, A. A. Guglielmone, M. B. Labruna, and J. de La Fuente. 2012. Reinstatement of *Rhipicephalus (Boophilus) australis* (Acari: Ixodidae) with redescription of the adult and larval stages. Journal of Medical Entomology 494: 794–802.

Évora, P. M., G. S. Sanches, F. D. Guerrero, A. Pérez de León, and G. H. Bechara. 2017. Immunogenic potential of *Rhipicephalus (Boophilus) microplus* aquaporin 1 against *Rhipicephalus sanguineus* in domestic dogs. Revista Brasileira de Parasitologia Veterinária 26: 60–66.

(FAO) Food and Agriculture Organization of the United Nations. 1961. The control of ticks on livestock. FAO Agricultural Studies No. 54. FAO, Rome, Italy.

Fernández, P. J., and W. R. White. 2016. Atlas of transboundary animal diseases, Second Edition. World Organisation for Animal Health (OIE), Paris, France.

Foil, L., P. Coleman, M. Eisler, H. Fragoso-Sanchez, Z. Garcia-Vazquez, F. Guerrero, N. Jonsson, I. Langstaff, A. Li, N. Machila, R. J. Miller, J. Morton, J. H. Pruett, and S. Torr. 2004. Factors that influence the prevalence of acaricide resistance and tick-borne diseases. Veterinary Parasitology 125: 163–181.

Foley, A. M., J. A. Goolsby, A. Ortega-S Jr, J. A. Ortega-S, A. Pérez de León, N. K. Singh, A. Schwartz, D. Ellis, D. G. Hewitt, and T. A. Campbell. 2017. Movement patterns of nilgai antelope in South Texas: Implications for cattle fever tick management. Preventive Veterinary Medicine 146: 166–172.

Fouet, C., and C. Kamdem. 2019. Integrated mosquito management: Is precision control a luxury or necessity? Trends in Parasitology 35: 85–95.

Gaff, H. D., A. White, K. Leas, P. Kelman, J. C. Squire, D. L. Livingston, G. A. Sullivan, E. W. Baker, and D. E. Sonenshine. 2015. TickBot: A novel robotic device for controlling tick populations in the natural environment. Ticks and Tick-borne Diseases 6: 146–151.

Gasmi, S., C. Bouchard, N. H. Ogden, A. Adam-Poupart, Y. Pelcat, E. E. Rees, F. Milord, P. A. Leighton, R. L. Lindsay, and J. K. Koffi. 2018. Evidence for increasing densities and geographic ranges of tick species of public health significance other than *Ixodes scapularis* in Québec, Canada. PLoS One 138: e0201924.

George, J. E. 1989. Cattle fever tick eradication programme in the USA: History, achievements, problems and implications for other countries, pp. 1–7. *In* Proceedings Expert Consultation on the Eradication of Ticks with Special Reference to Latin America. FAO International Symposium, 22-26 June 1987, Mexico City, Mexico. FAO Animal Production and Health Paper 75, Rome, Italy.

George, J. E. 2000. Present and future technologies for tick control. Annals of the New York Academy of Sciences 916: 583–588.

George, J. E. 2008. The effects of global change on the threat of exotic arthropods and arthropod-borne pathogens to livestock in the United States. Annals of the New York Academy of Sciences 1149: 249–254.

Ghosh, S. and G. Nagar. 2014. Problem of ticks and tick-borne diseases in India with special emphasis on progress in tick control research: A review. Journal of Vector Borne Diseases 51: 259–270.

Giles, J. R., A. T. Peterson, J. D. Busch, P. U. Olafson, G. A. Scoles, R. B. Davey, J. M. Pound, D. M. Kammlah, K. H. Lohmeyer, and D. M. Wagner. 2014. Invasive potential of cattle fever ticks in the southern United States. Parasites & Vectors 71: 189.

Ginsberg, H. S., T. A. Bargar, M. L. Hladik, and C. Lubelczyk. 2017. Management of arthropod pathogen vectors in North America: Minimizing adverse effects on pollinators. Journal of Medical Entomology 54: 1463–1475.

González Sáenz Pardo, J., and R. Hernández Ortiz. 2012. *Boophilus microplus*: Current status of acaricide resistance on the Mexican American border and its impact on commerce. Revista Mexicana de Ciencias Pecuarias 3 (Supplement 1): 1–8.

Goolsby, J., J. Jung, J. Landivar, W. Mccutcheon, R. Lacewell, R. Duhaime, and A. Schwartz. 2016. Evaluation of Unmanned Aerial Vehicles (UAVs) for detection of cattle in the Cattle Fever Tick Permanent Quarantine Zone. Subtropical Agriculture and Environments 67: 24–27.

Goolsby, J. A., N. K. Singh, A. Ortega-S Jr, D. G. Hewitt, T. A. Campbell, D. Wester, and A. A. Pérez de León. 2017. Comparison of natural and artificial odor lures for nilgai (*Boselaphus tragocamelus*) and white-tailed deer (*Odocoileus virginianus*) in south Texas: Developing treatment for cattle fever tick eradication. International Journal for Parasitology: Parasites and Wildlife 62: 100–107.

Goolsby, J., N. Singh, D. Shapiro-Ilan, R. Miller, P. Moran, and A. Pérez de León. 2018. Treatment of cattle with *Steinernema riobrave* and *Heterorhabditis floridensis* for control of the southern cattle fever tick, *Rhipicephalus (= Boophilus) microplus*. Southwestern Entomologist 432: 295–301.

Gortazar, C., I. Diez-Delgado, J. A. Barasona, J. Vicente, J. de La Fuente, and M. Boadella. 2015. The wild side of disease control at the wildlife-livestock-human interface: A review. Frontiers in Veterinary Science 1: 27.

Graf, J.-F., R. Gogolewski, N. Leach-Bing, G. Sabatini, M. Molento, E. Bordin, and G. Arantes. 2004. Tick control: An industry point of view. Parasitology 129 (Supplement): S427–S442.

Graham, O., and J. Hourrigan. 1977. Eradication programs for the arthropod parasites of livestock. Journal of Medical Entomology 13: 629–658.

Gray, J. S., A. Estrada-Peña, and A. Zintl. 2019. Vectors of babesiosis. Annual Review of Entomology 64: 149–165.

Gross, A. D., K. B. Temeyer, T. A. Day, A. A. Pérez de León, M. J. Kimber, and J. R. Coats. 2017. Interaction of plant essential oil terpenoids with the southern cattle tick tyramine receptor: A potential biopesticide target. Chemico-Biological Interactions 263: 1–6.

Guerrero, F. D., R. Andreotti, K. G. Bendele, R. C. Cunha, R. J. Miller, K. Yeater, and A. A. Pérez de León. 2014a. *Rhipicephalus (Boophilus) microplus* aquaporin as an effective vaccine antigen to protect against cattle tick infestations. Parasites & Vectors 71: 475.

Guerrero, F. D., A. A. Pérez de León, R. I. Rodríguez-Vivas, N. Jonsson, R. J. Miller, and R. Andreotti. 2014b. Acaricide research and development, resistance and resistance monitoring, pp. 353–381. *In* D. E. Sonenshine, and R. M. Roe (eds.), Biology of ticks, Volume 2. Oxford Unversity Press, New York, NY, USA.

Hansford, K. M., M. E. Pietzsch, B. Cull, E. L. Gillingham, and J. M. Medlock. 2018. Potential risk posed by the importation of ticks into the UK on animals: Records from the tick surveillance scheme. Veterinary Record 182: 107.

Harvey, E., K. Rose, J.-S. Eden, N. Lo, T. Abeyasuriya, M. Shi, S. L. Doggett, and E. C. Holmes. 2019. Extensive diversity of RNA viruses in Australian ticks. Journal of Virology 93: e01358-01318.

Hendrichs, J., M. Vreysen, A. Robinson, and P. Kenmore. 2007. Area-Wide Integrated Pest Management (AW-IPM): Principles, practice and prospects, pp. 3–33. *In* M. J. B. Vreysen, A. S. Robinson, and J. Hendrichs (eds.), Area-wide control of insect pests: From research to field implementation. Springer, Dordrecht, The Netherlands.

Henrioud, A. N. 2011. Towards sustainable parasite control practices in livestock production with emphasis in Latin America. Veterinary Parasitology 180: 2–11.

Heyman, P., C. Cochez, A. Hofhuis, J. Van Der Giessen, H. Sprong, S. R. Porter, B. Losson, C. Saegerman, O. Donoso-Mantke, and M. Niedrig. 2010. A clear and present danger: Tick-borne diseases in Europe. Expert Review of Anti-infective Therapy 8: 33–50.

Higgs, S. 2018. African swine fever – A call to action. Vector-Borne and Zoonotic Diseases 18: 509–510.

Hutcheson, H. J., L. R. Lindsay, and S. J. Dergousoff. 2019. *Haemaphysalis longicornis*: A tick of considerable importance, now established in North America. Canadian Journal of Public Health 110: 118–119.

(IAEA) International Atomic Energy Agency. 1968. Control of livestock insect pests by the Sterile-Male Technique. Proceedings of a panel, 23-27 January 1967, Vienna, Austria.

Jongejan, F., S. de Jong, T. Voskuilen, L. van den Heuvel, R. Bouman, H. Heesen, C. Ijzermans, and L. Berger. 2019. "Tekenscanner": A novel smartphone application for companion animal owners and veterinarians to engage in tick and tick-borne pathogen surveillance in the Netherlands. Parasites & Vectors 12: 116.

Jonsson, N. N. 2004. Integrated control programs for ticks on cattle: An examination of some possible components. Food and Agriculture Organization of the United Nations Animal Production and Health Paper: 1–78.

Jørs, E., A. Aramayo, O. Huici, F. Konradsen, and G. Gulis. 2017. Obstacles and opportunities for diffusion of Integrated Pest Management strategies reported by Bolivian small-scale farmers and agronomists. Environmental Health Insights 11: 1178630217703390.

Keesing, F., and R. S. Ostfeld. 2018. The tick project: Testing environmental methods of preventing tick-borne diseases. Trends in Parasitology 34: 447–450.

Kerario, I. I., M. Simuunza, E. L. Laisser, and S. Chenyambuga. 2018. Exploring knowledge and management practices on ticks and tick-borne diseases among agro-pastoral communities in Southern Highlands, Tanzania. Veterinary World 11: 48–57.

Khamesipour, F., G. O. Dida, D. N. Anyona, S. M. Razavi, and E. Rakhshandehroo. 2018. Tick-borne zoonoses in the Order Rickettsiales and Legionellales in Iran: A systematic review. PLoS Neglected Tropical Diseases 12: e0006722.

Klafke, G., R. Miller, J. Tidwell, R. Barreto, F. Guerrero, P. Kaufman, and A. Pérez de León. 2017a. Mutation in the sodium channel gene corresponds with phenotypic resistance of *Rhipicephalus sanguineus* sensu lato (Acari: Ixodidae) to pyrethroids. Journal of Medical Entomology 54: 1639–1642.

Klafke, G., A. Webster, B. D. Agnol, E. Pradel, J. Silva, L. H. de La Canal, M. Becker, M. F. Osório, M. Mansson, and R. Barreto. 2017b. Multiple resistance to acaricides in field populations of *Rhipicephalus microplus* from Rio Grande do Sul state, Southern Brazil. Ticks and Tick-borne Diseases 8: 73–80.

Klassen, W. 1989. Eradication of introduced arthropod pests: Theory and historical practice. Miscellaneous Publications of the Entomological Society of America 73: 1–29.

Koul, O., G. W. Cuperus, and N. Elliott. 2008. Areawide pest management: Theory and implementation. CABI, Cambridge, Massachusetts, USA. 590 pp.

Laing, G., M. Aragrande, M. Canali, S. Savic, and D. de Meneghi. 2018. Control of cattle ticks and tick-borne diseases by acaricide in Southern Province of Zambia: A retrospective evaluation of animal health measures according to current One Health concepts. Frontiers in Public Health 6: 45.

Leal, B., D. B. Thomas, and R. K. Dearth. 2018. Population dynamics of off-host *Rhipicephalus (Boophilus) microplus* (Acari: Ixodidae) larvae in response to habitat and seasonality in south Texas. Veterinary Sciences 5: 33.

Lohmeyer, K. H., J. Pound, M. May, D. Kammlah, and R. Davey. 2011. Distribution of *Rhipicephalus* (*Boophilus*) *microplus* and *Rhipicephalus* (*Boophilus*) *annulatus* (Acari: Ixodidae) infestations detected in the United States along the Texas/Mexico border. Journal of Medical Entomology 48: 770–774.

Lohmeyer, K. H., J. M. Pound, J. A. Klavons, and R. Davey. 2013. Liquid chromatographic detection of permethrin from filter paper wipes of white-tailed deer. Journal of Entomological Science 48: 258–260.

Lohmeyer, K. H., M. A. May, D. B. Thomas, and A. A. Pérez de León. 2018. Implication of nilgai antelope (Artiodactyla: Bovidae) in reinfestations of *Rhipicephalus (Boophilus) microplus* (Acari: Ixodidae) in south Texas: A review and update. Journal of Medical Entomology 55: 515–522.

Lorusso, V., M. Wijnveld, A. O. Majekodunmi, C. Dongkum, A. Fajinmi, A. G. Dogo, M. Thrusfield, A. Mugenyi, E. Vaumourin, and A. C. Igweh. 2016. Tick-borne pathogens of zoonotic and veterinary importance in Nigerian cattle. Parasites & Vectors 9: 217.

Magori, K. 2018. Preliminary prediction of the potential distribution and consequences of *Haemaphysalis longicornis* (Ixodida: Ixodidae) in the United States and North America, using a simple rule-based climate envelope model. bioRxiv: 389940.

Mans, B. J., D. De Klerk, R. Pienaar, and A. A. Latif. 2011. *Nuttalliella namaqua*: A living fossil and closest relative to the ancestral tick lineage: Implications for the evolution of blood-feeding in ticks. PLoS One 6: e23675.

Mans, B. J., J. Featherston, M. Kvas, K. A. Pillay, D. G. de Klerk, R. Pienaar, M. H. de Castro, T. G. Schwan, J. E. Lopez, P. Teel, A. A. Pérez de León, D. E. Sonenshine, N. I. Egekwu, D. K. Bakkes, H. Heyne, E. G. Kanduma, N. Nyangiwe, A. Bouattour, and A. A. Latif. 2019. Argasid and ixodid systematics: Implications for soft tick evolution and systematics, with a new argasid species list. Ticks and Tick-borne Diseases 10: 219–240.

Mansfield, K. L., L. Jizhou, L. P. Phipps, and N. Johnson. 2017. Emerging tick-borne viruses in the twenty-first century. Frontiers in Cellular and Infection Microbiology 7: 298.

McCosker, P. J. 1979. Global aspects of the management and control of ticks of veterinary importance, pp. 45–53. *In* J. G. Rodriguez (ed.), Recent advances in acarology, Volume II. Academic Press, New York, NY, USA.

McCosker, P. J. 1993. Ticks in a changing world. World Animal Review 74-75: 1–3.

McCoy, K. D., E. Léger, and M. Dietrich. 2013. Host specialization in ticks and transmission of tick-borne diseases: A review. Frontiers in Cellular and Infection Microbiology 3: 57.

Mihajlović, J., J. Hovius, H. Sprong, P. Bogovič, M. Postma, and F. Strle. 2019. Cost-effectiveness of a potential anti-tick vaccine with combined protection against Lyme borreliosis and tick-borne encephalitis in Slovenia. Ticks and Tick-borne Diseases 10: 63–71.

Miller, R. J., R. B. Davey, and J. E. George. 2005. First report of organophosphate-resistant *Boophilus microplus* (Acari: Ixodidae) within the United States. Journal of Medical Entomology 42: 912–917.

Miller, R., A. Estrada-Peña, C. Almazán, A. Allen, L. Jory, K. Yeater, M. Messenger, D. Ellis, and A. A. Pérez de León. 2012. Exploring the use of an anti-tick vaccine as a tool for the integrated eradication of the cattle fever tick, *Rhipicephalus* (*Boophilus*) *annulatus*. Vaccine 30: 5682–5687.

Miller, R. J., C. Almazán, M. Ortíz-Estrada, R. B. Davey, J. E. George, and A. Pérez de León. 2013. First report of fipronil resistance in *Rhipicephalus* (*Boophilus*) *microplus* of Mexico. Veterinary Parasitology 191: 97–101.

Minjauw, B., and A. McLeod. 2003. Tick-borne diseases and poverty: The impact of ticks and tick-borne diseases on the livelihoods of small-scale and marginal livestock owners in India and eastern and southern Africa. Department for International Development, Animal Health Programme, Centre for Tropical Veterinary Medicine, University of Edinburgh, UK.

Miraballes, C., and F. Riet-Correa. 2018. A review of the history of research and control of *Rhipicephalus* (*Boophilus*) *microplus*, babesiosis and anaplasmosis in Uruguay. Experimental and Applied Acarology 75: 383–398.

Mullens, B. A., N. C. Hinkle, R. Trout Fryxell, and K. Rochon. 2018. Past, present, and future contributions and needs for veterinary entomology in the United States and Canada. American Entomologist 64: 20–31.

Munoz, S., F. D. Guerrero, A. Kellogg, A. M. Heekin, and M.-Y. Leung. 2017. Bioinformatic prediction of G protein-coupled receptor encoding sequences from the transcriptome of the foreleg, including the Haller's organ, of the cattle tick, *Rhipicephalus australis*. PLoS One 12: e0172326.

Mutavi, F., N. Aarts, A. Van Paassen, I. Heitkönig, and B. Wieland. 2018. Techne meets metis: Knowledge and practices for tick control in Laikipia County, Kenya. NJAS - Wageningen Journal of Life Sciences 86-87: 136–145.

Narladkar, B. 2018. Projected economic losses due to vector and vector-borne parasitic diseases in livestock of India and its significance in implementing the concept of integrated practices for vector management. Veterinary World 11: 151–160.

Needham, G. R., and P. D. Teel. 1991. Off-host physiological ecology of ixodid ticks. Annual Review of Entomology 36: 659–681.

Nieto, N. C., W. T. Porter, J. C. Wachara, T. J. Lowrey, L. Martin, P. J. Motyka, and D. J. Salkeld. 2018. Using citizen science to describe the prevalence and distribution of tick bite and exposure to tick-borne diseases in the United States. PLoS One 13: e0199644.

Ntondini, Z., E. Van Dalen, and I. G. Horak. 2008. The extent of acaricide resistance in 1-, 2-and 3-host ticks on communally grazed cattle in the eastern region of the Eastern Cape Province, South Africa. Journal of the South African Veterinary Association 79: 130–135.

Ogden, N. H., and L. R. Lindsay. 2016. Effects of climate and climate change on vectors and vector-borne diseases: Ticks are different. Trends in Parasitology 32: 646–656.

Ojeda-Chi, M. M., R. I. Rodríguez-Vivas, M. D. Esteve-Gassent, A. A. Pérez de León, J. J. Modarelli, and S. L. Villegas-Pérez. 2019. Ticks infesting dogs in rural communities of Yucatan, Mexico and molecular diagnosis of rickettsial infection. Transboundary and Emerging Diseases 66: 102–110.

Olafson, P. U., D. B. Thomas, M. A. May, B. G. Buckmeier, and R. A. Duhaime. 2018. Tick vector and disease pathogen surveillance of nilgai antelope, *Boselaphus tragocamelus*, in southeastern Texas, USA. Journal of Wildlife Diseases 54: 734–744.

Paddock, C. D., R. S. Lane, J. E. Staples, and M. B. Labruna. 2016. Appendix 8: Changing paradigms for tick-borne diseases in the Americas, pp. 221–258. *In* Global health impacts of vector-borne diseases: Workshop summary. Forum on Microbial Threats, National Academies of Sciences, Engineering, and Medicine, 16-17 September 2014, Washington, DC, USA. National Academies Press, Washington, DC, USA.

Pegram, R. G., D. D. Wilson, and J. W. Hansen. 2000. Past and present national tick control programs: Why they succeed or fail. Annals of the New York Academy of Sciences 916: 546–554.

Pegram, R., A. Wilsmore, C. Lockhart, R. Pacer, and C. Eddi. 2007. The Carribean *Amblyomma variegatum* eradication programme: Success or failure? pp. 709–720. *In* M. J. B. Vreysen, A. S. Robinson, and J. Hendrichs (eds.), Area-wide control of insect pests: From research to field implementation. Springer, Dordrecht, The Netherlands.

Peñalver, E., A. Arillo, X. Delclòs, D. Peris, D. A. Grimaldi, S. R. Anderson, P. C. Nascimbene, and R. Pérez-de la Fuente. 2018. Ticks parasitised feathered dinosaurs as revealed by Cretaceous amber assemblages. Nature Communications 9: 472.

Pérez de León, A. A. 2017. Integrated Tick Management: Challenges and opportunities to mitigate tick-borne disease burden. Revista Colombiana de Ciencias Pecuarias 30 (Supplement): 280–285.

Pérez de León, A. A., D. A. Strickman, D. P. Knowles, D. Fish, E. Thacker, J. de la Fuente, P. J. Krause, S. K. Wikel, R. S. Miller and G. G. Wagner, C. Almazán, R. Hillman, M. T. Messenger, P. O. Ugstad, R. A. Duhaime, P.D. Teel, A. Ortega-Santos, D. G. Hewitt, E. J. Bowers, S. J. Bent, M. H. Cochran, T. F. McElwain, G. A. Scoles, C. E. Suarez, R. Davey, J. M. Howell Freeman, K. Lohmeyer K, A. Y. Li, F. D. Guerrero, D. M. Kammlah, P. Phillips, J. M. Pound, and the Group for Emerging Babesioses and One Health Research and Development in the U.S. 2010. One Health approach to identify research needs in bovine and human babesioses: Workshop report. Parasites & Vectors 3: 36.

Pérez de León, A. A., P. D. Teel, A. N. Auclair, M. T. Messenger, F. D. Guerrero, G. Schuster, and R. J. Miller. 2012. Integrated strategy for sustainable cattle fever tick eradication in USA is required to mitigate the impact of global change. Frontiers in Physiology 3: 195.

Pérez de León, A. A., R. I. Rodríguez-Vivas, F. D. Guerrero, Z. García-Vázquez, K. B. Temeyer, D. I. Domínguez-García, A. Li, N. Cespedes, R. J. Miller, and R. Rosario Cruz. 2013. Acaricide resistance in *Rhipicephalus* (*Boophilus*) *microplus:* Impact on agro-biosecurity and cattle trade between Mexico and the United States of America, pp. 18–35. *In* D. I. Domínguez-García, R. Rosario Cruz, and M. Ortiz Estrada (eds.), Proceeedings 3th International Symposium on Pesticide Resistance in Arthropods: Integrated Cattle Tick and Fly Control and Mitigation of Pesticide Resistance, 24 June 2013, Ixtapa, Zihuatanejo, Mexico. Universidad Autónoma de Guerrero Press, Chilpancingo, Guerrero, Mexico.

Pérez de León, A. A., P. D. Teel, A. Li, L. Ponnusamy, and R. M. Roe. 2014a. Advancing Integrated Tick Management to mitigate burden of tick-borne diseases. Outlooks on Pest Management 256: 382–389.

Pérez de León, A. A., E. Vannier, C. Almazán, and P. J. Krause. 2014b. Tick-borne protozoa, pp. 147–179. *In* D. E. Sonenshine and R. M. Roe (eds.), Biology of ticks, Volume 2. Oxford Unversity Press, New York, NY, USA.

Pérez de León, A. A., S. Mahan, M. Messenger, D. Ellis, K. Varner, A. Schwartz, D. Baca, R. Andreotti, M. R. Valle, R. R. Cruz, D. I. Domínguez García, M. Comas Pagan, C. Oliver Canabal, J. Urdaz, F. Collazo Mattei, F. Soltero, F. Guerrero, and R. J. Miller. 2018. Public-private partnership enabled use of anti-tick vaccine for integrated cattle fever tick eradication in the USA, pp. 275–298. *In* C. Garros, J. Bouyer, W. Takken, and R. C. Smallegange (eds.), Pests and vector-borne diseases in the livestock industry. Wageningen Academic Publishers, Wageningen, The Netherlands.

Perry, B. 2016. The control of East Coast fever of cattle by live parasite vaccination: A science-to-impact narrative. One Health 2: 103–114.

Petersen, L. R., C. B. Beard, and S. N. Visser. 2019. Combatting the increasing threat of vector-borne disease in the United States with a national vector-borne disease prevention and control system. American Journal of Tropical Medicine and Hygiene 100: 242–245.

Pfister, K., and R. Armstrong. 2016. Systemically and cutaneously distributed ectoparasiticides: A review of the efficacy against ticks and fleas on dogs. Parasites & Vectors 9: 436.

Phillips, P. L., J. B. Welch, and M. Kramer. 2014. Development of a spatially targeted field sampling technique for the southern cattle tick, *Rhipicephalus microplus*, by mapping white-tailed deer, *Odocoileus virginianus*, habitat in south Texas. Journal of Insect Science 14 (88): 1–21.

Piesman, J., and L. Eisen. 2008. Prevention of tick-borne diseases. Annual Review of Entomology 53: 323–343.

Playford, M., A. R. Rabiee, I. J. Lean, and M. Ritchie. 2005. Review of research needs for cattle tick control, Phases I and II. Meat & Livestock Australia Ltd., Sydney, Australia.

Pluess, T., R. Cannon, V. Jarošík, J. Pergl, P. Pyšek, and S. Bacher. 2012. When are eradication campaigns successful? A test of common assumptions. Biological Invasions 14: 1365–1378.

Pollett, S., B. M. Althouse, B. Forshey, G. W. Rutherford, and R. G. Jarman. 2017. Internet-based biosurveillance methods for vector-borne diseases: Are they novel public health tools or just novelties? PLoS Neglected Tropical Diseases 11: e0005871.

Pound, J., J. George, D. Kammlah, K. Lohmeyer, and R. Davey. 2010. Evidence for role of white-tailed deer (Artiodactyla: Cervidae) in epizootiology of cattle ticks and southern cattle ticks (Acari: Ixodidae) in reinfestations along the Texas/Mexico border in south Texas: A review and update. Journal of Economic Entomology 103: 211–218.

Pound, J. M., K. H. Lohmeyer, R. B. Davey, L. A. Soliz, and P. U. Olafson. 2012. Excluding feral swine, javelinas, and raccoons from deer bait stations. Human - Wildlife Interactions 6: 169–177.

Pruneau, L., K. Lebrigand, B. Mari, T. Lefrancois, D. F. Meyer, and N. Vachiery. 2018. Comparative transcriptome profiling of virulent and attenuated *Ehrlichia ruminantium* strains highlighted strong regulation of *map1*- and metabolism related genes. Frontiers in Cellular and Infection Microbiology 8: 153.

Rainey, T., J. L. Occi, R. G. Robbins, and A. Egizi. 2018. Discovery of *Haemaphysalis longicornis* (Ixodida: Ixodidae) parasitizing a sheep in New Jersey, United States. Journal of Medical Entomology 55: 757–759.

Rasi, T., I. Majlath, M. Bogdziewicz, K. Dudek, V. Majlathova, J. Wlodarek, M. Almasi, B. Vargova, and P. Tryjanowski. 2018. Tick distribution along animal tracks: Implication for preventative medicine. Annals of Agricultural and Environmental Medicine 25: 360–363.

Redondo, M., H. Fragoso, C. Montero, J. Lona, J. A. Medellín, R. Fría, V. Hernández, R. Franco, H. Machado, and M. Rodríguez. 1999. Integrated control of acaricide-resistant *Boophilus microplus* populations on grazing cattle in Mexico using vaccination with Gavac™ and amidine treatments. Experimental and Applied Acarology 23: 841–849.

Reece, L. M., D. W. Beasley, G. N. Milligan, V. V. Sarathy, and A. D. Barrett. 2018. Current status of Severe Fever with Thrombocytopenia Syndrome vaccine development. Current Opinion in Virology 29: 72–78.

Riccardi, N., R. M. Antonello, R. Luzzati, J. Zajkowska, S. Di Bella, and D. R. Giacobbe. 2019. Tick-borne encephalitis in Europe: A brief update on epidemiology, diagnosis, prevention, and treatment. European Journal of Internal Medicine 62: 1–6.

Robles, N. J. C., H. J. Han, S.-J. Park, and Y. K. Choi. 2018. Epidemiology of severe fever and thrombocytopenia syndrome virus infection and the need for therapeutics for the prevention. Clinical and Experimental Vaccine Research 7: 43–50.

Rochlin, I. 2019. Modeling the Asian longhorned tick (Acari: Ixodidae) suitable habitat in North America. Journal of Medical Entomology 56: 384–391.

Rodríguez-Mallon, A., J. L. C. Anadón, A. A. L. Pérez, G. H. Bechara, R. Z. Machado, R. L. Cruz, A. Domingos, and A. R. T. Sosa. 2018. CYTED Network to develop an immunogen compatible with integrated management strategies for tick control in cattle. Vaccine 36: 6581–6586.

Rodríguez-Vivas, R. I., N. N. Jonsson, and C. Bhushan. 2018. Strategies for the control of *Rhipicephalus microplus* ticks in a world of conventional acaricide and macrocyclic lactone resistance. Parasitology Research 117: 3–29.

Rodríguez-Vivas, R. I., L. Grisi, A. A. Pérez de León, H. Silva Villela, J. F. d. J. Torres-Acosta, H. Fragoso Sánchez, D. Romero Salas, R. Rosario Cruz, F. Saldierna, and D. García Carrasco. 2017a. Potential economic impact assessment for cattle parasites in Mexico. Review. Revista Mexicana de Ciencias Pecuarias 8: 61–74.

Rodríguez-Vivas, R. I., M. M. Ojeda-Chi, I. Trinidad-Martinez, and A. A. Pérez de León. 2017b. First documentation of ivermectin resistance in *Rhipicephalus sanguineus* sensu lato (Acari: Ixodidae). Veterinary Parasitology 233: 9–13.

Rosenberg, R., N. P. Lindsey, M. Fischer, C. J. Gregory, A. F. Hinckley, P. S. Mead, G. Paz-Bailey, S. H. Waterman, N. A. Drexler, G. J. Kersh, H. Hooks, S. K. Partridge, S. N. Visser, C. B. Beard, and L. R. Petersen. 2018. Vital signs: Trends in reported vectorborne disease cases—United States and territories, 2004–2016. Morbidity and Mortality Weekly Report 67: 496–501.

Roy, B. C., A. Estrada-Peña, J. Krücken, A. Rehman, and A. M. Nijhof. 2018. Morphological and phylogenetic analyses of *Rhipicephalus microplus* ticks from Bangladesh, Pakistan and Myanmar. Ticks and Tick-borne Diseases 9: 1069–1079.

Rushton, J. 2009. The economics of animal health and production. CABI, Cambridge, Massachusetts, USA. 384 pp.

Rutherford, B. 2019. A long, thin line. Beef Magazine. December 4, 20-19.

Sakamoto, J. M. 2018. Progress, challenges, and the role of public engagement to improve tick-borne disease literacy. Current Opinion in Insect Science 28: 81–89.

Scasta, J. D. 2015. Livestock parasite management on high-elevation rangelands: Ecological interactions of climate, habitat, and wildlife. Journal of Integrated Pest Management 6: 8.

Schetters, T., R. Bishop, M. Crampton, P. Kopáček, A. Lew-Tabor, C. Maritz-Olivier, R. Miller, J. Mosqueda, J. Patarroyo, and M. Rodriguez-Valle. 2016. Cattle tick vaccine researchers join forces in CATVAC. Parasites & Vectors 9: 105.

Schmidtmann, E. T. 1994. Ecologically based strategies for controlling ticks, pp. 240–280. *In* D. E. Sonenshine, and T. N. Mather (eds.), Ecological dynamics of tick-borne zoonoses. Oxford University Press, New York, NY, USA.

Schultz, M. 2008. Theobald Smith. Emerging Infectious Diseases 14: 1940–1942.

Shepard, D. S., Y. A. Halasa, D. M. Fonseca, A. Farajollahi, S. P. Healy, R. Gaugler, K. Bartlett-Healy, D. A. Strickman, and G. G. Clark. 2014. Economic evaluation of an Area-Wide Integrated Pest Management program to control the Asian tiger mosquito in New Jersey. PLoS One 9: e111014.

Singer, M., and N. Bulled. 2016. Ectoparasitic syndemics: Polymicrobial tick-borne disease interactions in a changing anthropogenic landscape. Medical Anthropology Quarterly 30: 442–461.

Singh, N. K., J. A. Goolsby, A. Ortega-S Jr, D. G. Hewitt, T. A. Campbell, and A. Pérez de León. 2017. Comparative daily activity patterns of Nilgai, *Boselaphus tragocamelus* and white-tailed deer, *Odocoileus virginianus* in South Texas. Subtropical Agriculture and Environments 68: 7–12.

Singh, N. K., R. J. Miller, G. M. Klafke, J. A. Goolsby, D. B. Thomas, and A. A. Pérez de León. 2018. In-vitro efficacy of a botanical acaricide and its active ingredients against larvae of susceptible and acaricide-resistant strains *of Rhipicephalus* (*Boophilus*) *microplus* Canestrini (Acari: Ixodidae). Ticks and Tick-borne Diseases 9: 201–206.

Šmit, R., and M. J. Postma. 2016. Vaccines for tick-borne diseases and cost-effectiveness of vaccination: A public health challenge to reduce the diseases' burden. Expert Review of Vaccines 15: 5–7.

Smith, T., and F. L. Kilborne. 1893. Investigations into the nature, causation, and prevention of Texas or southern cattle fever. US Department of Agriculture, Bureau of Animal Industry Bulletin 1: 1–301.

Socolovschi, C., B. Doudier, F. Pages, and P. Parola. 2008. Tiques et maladies transmises a l'homme en Afrique. Médecine Tropicale 68: 119–133.

Sonenshine, D. E. 2018. Range expansion of tick disease vectors in North America: Implications for spread of tick-borne disease. International Journal of Environmental Research and Public Health 15: 478.

Sonenshine, D. E., and R. M. Roe (eds.). 2014. Biology of ticks. Second Edition, Volumes 1 and 2. Oxford University Press, New York, NY, USA. 560 pp. and 496 pp.

Souza, W. M., M. J. Fumagalli, A. O. Torres Carrasco, M. F. Romeiro, S. Modha, M. C. Seki, J. M. Gheller, S. Daffre, M. R. T. Nunes, P. R. Murcia, G. O. Acrani, and L. T. M. Figueiredo. 2018. Viral diversity of *Rhipicephalus microplus* parasitizing cattle in southern Brazil. Scientific Reports 8: 16315.

Spengler, J. R., D. A. Bente, M. Bray, F. Burt, R. Hewson, G. Korukluoglu, A. Mirazimi, F. Weber, and A. Papa. 2018. Meeting report: Second International Conference on Crimean-Congo Hemorrhagic Fever. Antiviral Research 150: 137–147.

Sprong, H., J. Trentelman, I. Seemann, L. Grubhoffer, R. O. Rego, O. Hajdušek, P. Kopáček, R. Šíma, A. M. Nijhof, and J. Anguita. 2014. ANTIDotE: Anti-tick vaccines to prevent tick-borne diseases in Europe. Parasites & Vectors 7: 77.

Stafford III, K. C., S. C. Williams, and G. Molaei. 2017. Integrated Pest Management in controlling ticks and tick-associated diseases. Journal of Integrated Pest Management 8: 28.

Strom, C. 2010. Making catfish bait out of government boys: The fight against cattle ticks and the transformation of the yeoman South. University of Georgia Press, Athens, Georgia, USA. 197 pp.

Suarez, M., J. Rubi, D. Pérez, V. Cordova, Y. Salazar, A. Vielma, F. Barrios, C. A. Gil, N. Segura, Y. Carrillo, R. Cartaya, M. Palacios, E. Rubio, C. Escalona, C. Ramirez, R. Basulto Baker, H. Machado, Y. Sordo, J. Bermudes, M. Vargas, C. Montero, A. Cruz, P. Puente, J. L. Rodriguez, E. Mantilla, O. Oliva, E. Smith, A. Castillo, B. Ramos, Y. Ramirez, Z. Abad, A. Morales, E. M. Gonzalez, A. Hernandez, Y. Ceballo, D. Callard, A. Cardoso, M. Navarro, J. L. Gonzalez, R. Pina, M. Cueto, C. Borroto, E. Pimentel, Y. Carpio, and M. P. Estrada. 2016. High impact and effectiveness of Gavac™ vaccine in the national program for control of bovine ticks *Rhipicephalus microplus* in Venezuela. Livestock Science 187: 48–52.

Suarez, C. E., H. F. Alzan, M. G. Silva, V. Rathinasamy, W. A. Poole, and B. M. Cooke. 2019. Unravelling the cellular and molecular pathogenesis of bovine babesiosis: Is the sky the limit? International Journal for Parasitology 49: 183–197.

Suckling, D. M., L. D. Stringer, A. E. Stephens, B. Woods, D. G. Williams, G. Baker, and A. M. El-Sayed. 2014. From Integrated Pest Management to integrated pest eradication: Technologies and future needs. Pest Management Science 70: 179–189.

Sungirai, M., S. Baron, N. A. Van der Merwe, D. Z. Moyo, P. De Clercq, C. Maritz-Olivier, and M. Madder. 2018. Population structure and genetic diversity of *Rhipicephalus microplus* in Zimbabwe. Acta Tropica 180: 42–46.

Sutherst, R., G. F. Maywald, and A. S. Bourne. 2007. Including species interactions in risk assessments for global change. Global Change Biology 13: 1843–1859.

Talactac, M. R., E. P. Hernandez, K. Fujisaki, and T. Tanaka. 2018. A continuing exploration of tick–virus interactions using various experimental viral infections of hard ticks. Frontiers in Physiology 9: 1728.

Tatchell, R. 1992. Ecology in relation to Integrated Tick Management. International Journal of Tropical Insect Science 13: 551–561.

Telford, S. R. 2017. Deer reduction is a cornerstone of integrated deer tick management. Journal of Integrated Pest Management 8: 25.

Theobald, S., N. Brandes, M. Gyapong, S. El-Saharty, E. Proctor, T. Diaz, S. Wanji, S. Elloker, J. Raven, and H. Elsey. 2018. Implementation research: New imperatives and opportunities in global health. The Lancet 392: 2214–2228.

Thomas, D., J. Tidwell, and A. Pérez de León. 2017. In vitro efficacy testing of a commercial formulation of the acaropathogenic fungus *Metarhizium brunneum* Petch (Hypocreales: Clavicipitaceae) strain F52 against the southern cattle fever tick *Boophilus microplus* Canestrini (Acari: Ixodidae). Subtropical Agriculture and Environments 68: 1–6.

Urdaz-Rodríguez, J., R. Miller, P. Teel, I. Castro-Arellano, F. Guerrero, M. T. Messenger, F. Soltero, W. E. Grant, H.-H. Wang, C. Oliver-Canabal, M. Comas-Pagan, and A. Pérez de León. 2015. Integrated Tick Management to mitigate the impact of *Rhipicephalus microplus*, bovine anaplasmosis, and bovine babesiosis in livestock farming systems in Puerto Rico, pp. 619. *In* Proceedings 14th Symposium of the International Society for Veterinary Epidemiology and Economics. International Symposium on Veterinary Epidemiology and Economics, 3-7 November 2015, Mérida, Mexico.

(USHHS) United States Health and Human Services. 2018. Tick-Borne Disease Working Group 2018 Report to Congress. Washington, DC, USA.

Uspensky, I. 2008. Argasid (soft) ticks (Acari: Ixodida: Argasidae), pp. 283–288. *In* J. L. Capinera (ed.), Encyclopedia of entomology. Springer. Dordrecht, The Netherlands.

Valle, M. R., L. Mèndez, M. Valdez, M. Redondo, C. M. Espinosa, M. Vargas, R. L. Cruz, H. P. Barrios, G. Seoane, and E. S. Ramirez. 2004. Integrated control of *Boophilus microplus* ticks in Cuba based on vaccination with the anti-tick vaccine Gavac. Experimental and Applied Acarology 34: 375–382.

Vayssier-Taussat, M., J. F. Cosson, B. Degeilh, M. Eloit, A. Fontanet, S. Moutailler, D. Raoult, E. Sellal, M.-N. Ungeheuer, and P. Zylbermann. 2015. How a multidisciplinary 'One Health' approach can combat the tick-borne pathogen threat in Europe. Future Microbiology 10: 809–818.

Vudriko, P., J. Okwee-Acai, J. Byaruhanga, D. S. Tayebwa, R. Omara, J. B. Muhindo, C. Lagu, R. Umemiya-Shirafuji, X. Xuan, and H. Suzuki. 2018. Evidence-based tick acaricide resistance intervention strategy in Uganda: Concept and feedback of farmers and stakeholders. Ticks and Tick-borne Diseases 9: 254–265.

Walker, A. R. 2011. Eradication and control of livestock ticks: Biological, economic and social perspectives. Parasitology 138: 945–959.

Wang, H.-H., M. S. Corson, W. E. Grant, and P. D. Teel. 2017. Quantitative models of *Rhipicephalus* (*Boophilus*) ticks: Historical review and synthesis. Ecosphere 8: e01942.

Wang, H.-H., R. J. Miller, A. Pérez de León, and P. D. Teel. 2019. Simulation tools for assessment of tick suppression treatments of the southern cattle fever tick, *Rhipicephalus (Boophilus) microplus*, on non-lactating dairy cattle in Puerto Rico. Parasites & Vectors 12: 185.

Wang, H.-H., P. D. Teel, W. E. Grant, G. Schuster, and A. Pérez de León. 2016. Simulated interactions of white-tailed deer (*Odocoileus virginianus*), climate variation and habitat heterogeneity on southern cattle tick *(Rhipicephalus (Boophilus) microplus)* eradication methods in south Texas, USA. Ecological Modelling 342: 82–96.

Wang, Y., K. Li, P. Li, J. Sun, L. Ye, Y. Dai, A. Tang, J. Jiang, C. Chen, Z. Tong, and J. Yan. 2018. Community-based comprehensive measures to prevent severe fever with thrombocytopenia syndrome, China. International Journal of Infectious Diseases 73: 63–66.

Whalon, M. E., D. Mota-Sanchez, and R. M. Hollingworth. 2008. Global pesticide resistance in arthropods. CABI, Cambridge, Massachusetts, USA.

White, A., and H. Gaff. 2018. Application of tick control technologies for blacklegged, lone star, and American dog ticks. Journal of Integrated Pest Management 9: 12.

(WHO) World Health Organization of the United Nations. 2017. Global vector control response 2017-2030. Geneva, Switzerland. Licence: CC BY-NC-SA 3.0 IGO.

Wikel, S. 2018. Ticks and tick-borne infections: Complex ecology, agents, and host interactions. Veterinary Sciences 5: 60.

Willadsen, P. 2008. Anti-tick vaccines, pp. 424–446. *In* A. S. Bowman, and P. Nuttal (eds.), Ticks: Biology, disease, and control. Cambridge University Press, New York, NY, USA.

Williams, S. C., K. C. Stafford III, G. Molaei, and M. A. Linske. 2018. Integrated control of nymphal *Ixodes scapularis*: Effectiveness of white-tailed deer reduction, the entomopathogenic fungus *Metarhizium anisopliae*, and fipronil-based rodent bait boxes. Vector-Borne and Zoonotic Diseases 18: 55–64.

World Bank. 2018. One Health: Operational framework for strengthening human, animal, and environmental public health systems at their interface. Working paper, report number 122980. Washington, DC, USA.

Ybañez, A. P., C. N. Mingala, and R. H. D. Ybañez. 2018. Historical review and insights on the livestock tick-borne disease research of a developing country: The Philippine scenario. Parasitology International 67: 262–266.

AREA-WIDE INTEGRATED MANAGEMENT OF A *Glossina palpalis gambiensis* POPULATION FROM THE NIAYES AREA OF SENEGAL: A REVIEW OF OPERATIONAL RESEARCH IN SUPPORT OF A PHASED CONDITIONAL APPROACH

M. J. B. VREYSEN[1], M. T. SECK[2], B. SALL[3], A. G. MBAYE[4], M. BASSENE[2], A. G. FALL[2], M. LO[3] AND J. BOUYER[1,2,5,6]

[1]*Insect Pest Control Laboratory, Joint FAO/IAEA Programme of Nuclear Techniques in Food and Agriculture, A-1400, Vienna, Austria; M.Vreysen@iaea.org*

[2]*Institut Sénégalais de Recherches Agricoles, Laboratoire National d'Elevage et de Recherches Vétérinaires, BP 2057, Dakar – Hann, Sénégal*

[3]*Direction des Services Vétérinaires, BP 45 677, Dakar, Sénégal*

[4]*Services Régionales de l'Elevage de Dakar*

[5]*Unité Mixte de Recherche INTERTRYP, Centre de Coopération Internationale en Recherche Agronomique pour le Développement (CIRAD), 34398, Montpellier, France*

[6]*Unité Mixte de Recherche CMAEE, CIRAD, 34398, Montpellier, France*

SUMMARY

In 2005, the Government of Senegal initiated a project entitled "Projet de lutte contre les glossines dans les Niayes" (Tsetse control project in the Niayes) with the aim of creating a zone free of *Glossina palpalis gambiensis* in that area. The project received technical and financial support from the International Atomic Energy Agency (IAEA), the Food and Agriculture Organization of the United Nations (FAO), the Centre de Coopération Internationale en Recherche Agronomique pour le Développement (CIRAD) and the US Department of State through the Peaceful Uses Initiative (PUI). It was implemented in the context of the Pan African Tsetse and Trypanosomosis Eradication Campaign (PATTEC) following a phased conditional approach (PCA) that entails implementation in distinct phases, in which support to the next phase is conditional upon completion of all (or at least the majority of) activities in the previous phase. In the case of the tsetse project in Senegal, the PCA consisted of 4 phases: (1) commitment of all stakeholders and training, (2) baseline data collection, feasibility studies and strategy development, (3) preparatory pre-operational activities and (4) operational activities. This paper provides an overview of the main activities that were carried out within each phase, with emphasis on the operational research carried out in phases 2

J. Hendrichs, R. Pereira and M. J. B. Vreysen (eds.), Area-Wide Integrated Pest Management: Development and Field Application, pp. 275–303. CRC Press, Boca Raton, Florida, USA.

and 3, that was instrumental in guiding the project's decision-making. Activities of phase 2 focused on the collection of entomological, veterinary, socio-economic and environmental baseline data, and a population genetics study that proved the isolated character of the *G. p. gambiensis* population of the Niayes. These data enabled the tsetse-infested area to be delimited to 1000 km^2, the impact of animal trypanosomosis on the farmers' welfare to be quantified (annual benefits of 2 million Euro in the tsetse-infested zone), and the formulation of an area-wide integrated pest management (AW-IPM) strategy that included a sterile insect (SIT) component to eradicate the isolated tsetse populations from the Niayes. In view of the extreme fragmentation of the remaining favourable habitat of the Niayes and the high human population density (peri-urban area), which excluded the possibility of using the Sequential Aerosol Technique, the IPM strategy that was selected comprised the suppression of the tsetse population with insecticide-impregnated traps/targets and the use of "pour-on" for cattle, followed by the release of sterile males to eliminate the remaining relic pockets. During phase 3, the pre-operational phase, a series of activities were carried out that were needed to implement the operational phase. These included the establishment of a colony of tsetse originating from the target area in Senegal, competitiveness studies between the sterile flies and those from the target area, development of transport methods for long-distance shipments of sterile male pupae, competitiveness of the sterile male flies after release in the target area, development of aerial release methods (including a new chilled adult release system) and development of a Maxent-based distribution model to guide the suppression, sterile male releases and monitoring of the eradication campaign. To be able to properly manage the eradication campaign in different phases, the entire target area was divided into 3 operational blocks. This paper demonstrates how, during the operational phase, scientific principles continued to guide the implementation process. The results to date are encouraging, i.e. the deployment of 269 insecticide-impregnated Vavoua traps in favourable habitat of Block 1 reduced the apparent density of the *G. p. gambiensis* population significantly (from 0.42 (SD 0.39) to 0.04 (SD 0.11) flies/trap/day). This was followed by the aerial release of sterile males that reduced the apparent density to zero after six months of releases. The last wild fly was trapped on August 9, 2012 in Block 1. In Block 2, during the suppression, the apparent fly density dropped from 1.24 (SD 1.23) to 0.005 (SD 0.017) flies/trap/day. Sterile male releases were initiated in February 2014 and expanded to cover the entire Block 2 in January 2015. The apparent fly density has so far been reduced to < 0.001 fly per trap per day until the end of 2018 and releases are still ongoing. The results of the campaign are discussed with respect to the "adaptive management approach" used, which was deemed critical for the success of the campaign.

Key Words: African animal trypanosomosis, *Trypanosoma vivax*, *Trypanosoma congolense*, *Trypanosoma brucei*, nagana, livestock, Sterile Insect Technique, SIT, vector control, elimination, tsetse flies, integrated vector management, adaptive management

1. THE TSETSE AND TRYPANOSOMOSIS PROBLEM IN THE NIAYES AND THE POLITICAL WILL TO FIND A SUSTAINABLE SOLUTION

In the sub-humid savannah of West Africa, riverine tsetse species such as *Glossina palpalis gambiensis* (Vanderplank 1949) inhabit riparian forests where they are major vectors of African animal trypanosomosis (AAT) or nagana (Bouyer et al. 2006; Guerrini et al. 2008) and human African trypanosomosis (HAT) or sleeping sickness (Camara et al. 2006). In Senegal, as in other parts of West Africa, AAT is a major obstacle to the development of more efficient and sustainable livestock production (Itard et al. 2003) and the presence of tsetse flies is considered a major cause of hunger and poverty (Feldmann et al. 2021).*Glossina p. gambiensis* normally thrives in areas that receive a minimum of 600 mm annual rainfall (Brunhes et al. 1998), but in western Senegal, annual precipitation is limited to 400-500 mm. Here, *G. p. gambiensis* populations are mainly confined to a specific ecosystem called the "Niayes" (Morel and Touré 1967; Touré 1971, 1973, 1974) that are situated around Dakar. These habitats are characterized by remnants of Guinean forests that are located in low-lying inter-dune depressions that are periodically or permanently flooded. However, in the last decades, these habitats have been drastically changed due to human intrusion. The second similar but drier ecosystem, "La Petite Côte", is situated south of Dakar and extends along the Atlantic coast towards Joal and the Sine Saloum River (Fig. 1).

In the Niayes, temperature is lower and rainfall higher as compared with the interior of the country, and these conditions facilitate intensive cropping and cattle production even during the dry season. Horses are present in high numbers and are mainly used for the transport of food crops. The bites from tsetse flies pose a continuous nuisance for human populations, especially in Sebikotane and Pout. In addition, the flies seem to have adapted to peri-urban, densely populated areas such as the "Parc de Hann", located in the city centre of Dakar.

The *G. p. gambiensis* populations that inhabited the Niayes and La Petite Côte belonged to one of the most north-western distributions of the tsetse belt in West Africa (Fig. 1). In 2007, a parasitological and serological survey of resident cattle revealed the seriousness of the tsetse and trypanosomosis problems in the area with AAT herd prevalence rates of 10–90% (Baba Sall, unpublished data). This survey showed that *Trypanosoma vivax* Ziemann was the most prevalent species, followed by *T. congolense* Broden. However, the parasitological prevalence may be grossly underestimated, due to the poor sensitivity of the buffy coat technique that was used (Pinchbeck et al. 2008).

In the 1970s, the first attempt was made to eliminate *G. p. gambiensis* populations from more than 150 km of linear habitat in the Niayes, using selective bush clearing and residual ground spraying with dieldrin. Although no tsetse flies were detected after the campaign (Touré 1973), they reappeared in the 1980s, necessitating a second campaign combining insecticide spraying with the deployment of traps and insecticide-impregnated screens. The tsetse problem seemed to have disappeared until in 1998 flies were again detected (Baba Sall, unpublished data).

Staff of the Direction de l'Elevage (DIREL) (now called Direction des Services Vétérinaires (DSV)) and the Institut Sénégalais de Recherches Agricoles (ISRA), in collaboration with the Joint FAO/IAEA Division of Nuclear Techniques in Food and Agriculture (the Food and Agriculture Organization of the United Nations (FAO), the International Atomic Energy Agency (IAEA)) and the IAEA's Department of Technical Cooperation, carried out more extensive surveys in 2002–2003. These surveys confirmed the presence of *G. p. gambiensis* and in view of the isolated nature of the Niayes population (Solano et al. 2010) it is highly likely that the resurgence of the tsetse fly population can be attributed to a population build-up from small residual pockets inside the Niayes, rather than to reinvasion from the main tsetse belt of the Sine Saloum region that is located more than 100 km southeast of Dakar (S. Leak, unpublished reports to the IAEA; Baba Sall, unpublished data).

Following confirmation of *G. p. gambiensis* presence in the Niayes, the DSV and FAO/IAEA initiated a tsetse control campaign that officially started in 2005. Entitled "Projet de lutte contre les glossines dans les Niayes" (Tsetse control project in the Niayes), it was mainly funded and implemented by the DSV of the Ministry of Livestock and Animal Production and ISRA of the Ministry of Agriculture and Rural Equipment.

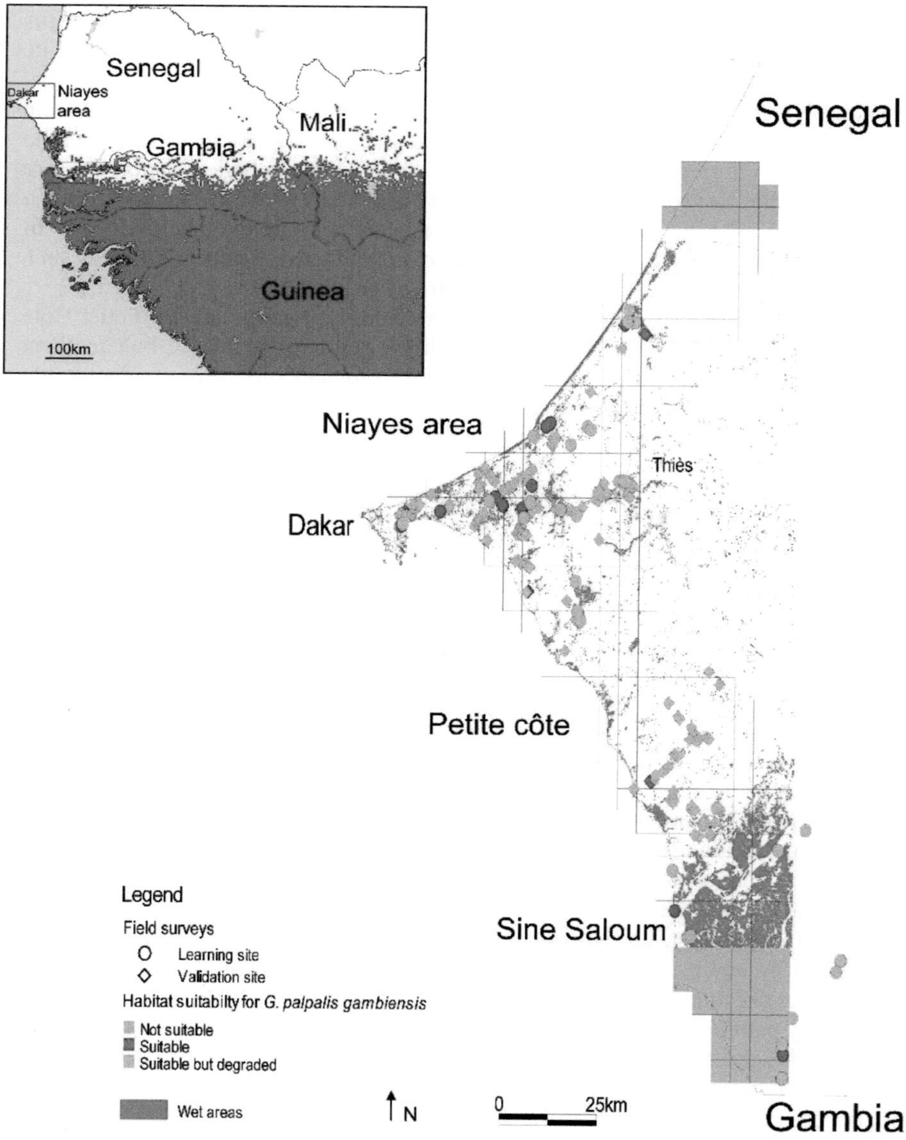

Figure 1. Map top left: Distribution of Glossina palpalis gambiensis *in West Africa and location of the Niayes area around Dakar in Senegal. The red area represents suitable habitats predicted from a Maxent model. Map right: The project area indicating the suitability of the vegetation for harbouring* G. p. gambiensis *after a phytosociological study, and the "wet areas" as obtained from a supervised classification (modified after Bouyer et al. 2010, 2015b).*

The project received technical and financial support from the IAEA, the FAO, the Centre de Coopération Internationale en Recherche Agronomique pour le Développement (CIRAD), and the US Department of State through the Peaceful Uses Initiative (PUI). The Centre International de Recherche-Développement sur l'Elevage en zone Sub-humide (CIRDES), Burkina Faso, the Slovak Academy of Sciences (SAS), Slovakia, and l'Institut de Recherche pour le Développement (IRD), France, were other full- or part-time partners in the project. The project was implemented in the context of the Pan African Tsetse and Trypanosomosis Eradication Campaign (PATTEC), a political initiative of the African Heads of State that called for increased efforts to better manage the tsetse and trypanosomosis problem on the African continent (PATTEC 2019).

2. PHASED CONDITIONAL APPROACH

From the onset, it was decided that the project would be implemented following a phased conditional approach (PCA), whereby project implementation follows distinct phases and in which support to the next phase is conditional upon completion of all (or at least the majority of) activities in the previous phase (Feldmann et al. 2018). Whereas the diverse phases of the PCA might differ with the target pest species, or if a suppression rather than an eradication strategy is selected (Hendrichs et al. 2021), the PCA consisted of 4 phases for the tsetse project in Senegal, i.e. (1) commitment of all stakeholders and training, (2) baseline data collection and feasibility studies, (3) pre-operational activities and (4) operational activities.

2.1. Phase 1: Stakeholder Commitment and Training

After the FAO/IAEA-supported surveys of 2002–2003, discussions within and between the Government of Senegal and the FAO/IAEA culminated in the submission of an official request by the Government of Senegal to the IAEA for technical and financial support. A technical cooperation project entitled "Feasibility Study to Create a Tsetse-free Zone Using the Sterile Insect Technique" was approved in 2005, which provided substantial support to phase 1 of the PCA.

The commitment of the Government was evidenced by the involvement of various Ministries in the project, i.e. the DSV of the Ministry of Livestock and Animal Production took responsibility for coordinating and implementing the project, the ISRA of the Ministry of Agriculture and Rural Equipment was given responsibility for operating the insectary/sterile male emergence and dispersal centre in Dakar and to guide the operational research that accompanied the project, and the Ministry of Environment and Sustainable Development provided the license to operate the project as it was considered environment-friendly.

Initially, training of essential project staff was emphasized, and a total of 16 veterinary field staff received training in tsetse biology, baseline data collection and control. This was a crucial step for the smooth implementation of the project in view of the limited experience of the field and insectary staff with tsetse flies, due to its absence from the Niayes for almost 20 years.

2.2. *Phase 2: Collection of Baseline Data, Feasibility Studies and Strategy Development (2007-2010)*

The importance of the availability of relevant baseline data (phase 2) cannot be overemphasized, as an appropriate control strategy cannot be developed without such detailed and accurate data. Data were required on the geographic distribution of the target tsetse population, their spatial and temporal dynamics, their spatial occupation of the habitat, their genetic profile, the correlation between tsetse presence/density and the parasitological and serological disease prevalence, the socio-economic impact of AAT on the farming community and the potential impact of the selected strategy on the environment (Fig. 2).

At the onset of the project, only limited data was available; therefore, during the first four years, all efforts were focused on collecting these data as part of a feasibility study. The data collected greatly assisted the decision-making process for selecting an appropriate strategy to sustainably manage the tsetse and trypanosomosis problem in the Niayes (Vreysen et al. 2007). The baseline data also enabled accurate monitoring of the operational eradication phase and continuous assessment of the progress made (Leak et al. 2008; Vreysen 2021).

The feasibility study was initiated with the development of a specific entomological sampling protocol aimed at accurately defining the distribution of the *G. p. gambiensis* populations in the Niayes and La Petite Côte. To enable the practical implementation of the protocol, a 5 x 5 km grid (286 cells) was superimposed over the entire initially defined project area of 7150 km^2 to facilitate the field sampling procedures (Leak et al. 2008) (Figs. 1 and 2). Spatial analytical tools were used to facilitate a preliminary phytosociological census that identified eight different types of habitat suitable to harbour *G. p. gambiensis*, which were denominated "wet areas" (Fig. 1).

In early 2009, 683 unbaited Vavoua traps Laveissière and Grébaut 1990) were strategically deployed in the area and the trapping data indicated that tsetse flies were present in 21 grid cells representing an area of 525 km^2. In the area of zero catches adjacent to the infested area (84 grid cells or 2100 km^2), a mathematical model was used to assess the risk that flies were present despite a sequence of zero catches (Barclay and Hargrove 2005; Bouyer et al. 2010).

The analysis showed a risk of tsetse presence > 0.05 in 16 grid cells or 400 km^2 which represented 19% of the area, which was therefore considered potentially infested and included in the target area. The remote sensing analysis identified 285 km^2 as wet areas, which comprised only 4% of the total project area of 7150 km^2, whereas the mathematical model provided an efficient method to improve the accuracy and the robustness of the sampling protocol (Bouyer et al. 2010). Thus, the total area that could be considered as potentially infested with tsetse flies and that could be subjected to the control effort was estimated at approximately 1000 km^2.

The entomological baseline data survey already indicated a high probability that the *G. p. gambiensis* populations of the Niayes were isolated from the remainder of the tsetse belt in the south-eastern part of Senegal. This assumption was mainly based on the absence of tsetse fly captures in La Petite Côte and the lack of any suitable tsetse habitat between the Niayes and the Sine Saloum, the nearest tsetse-infested area in the southeast.

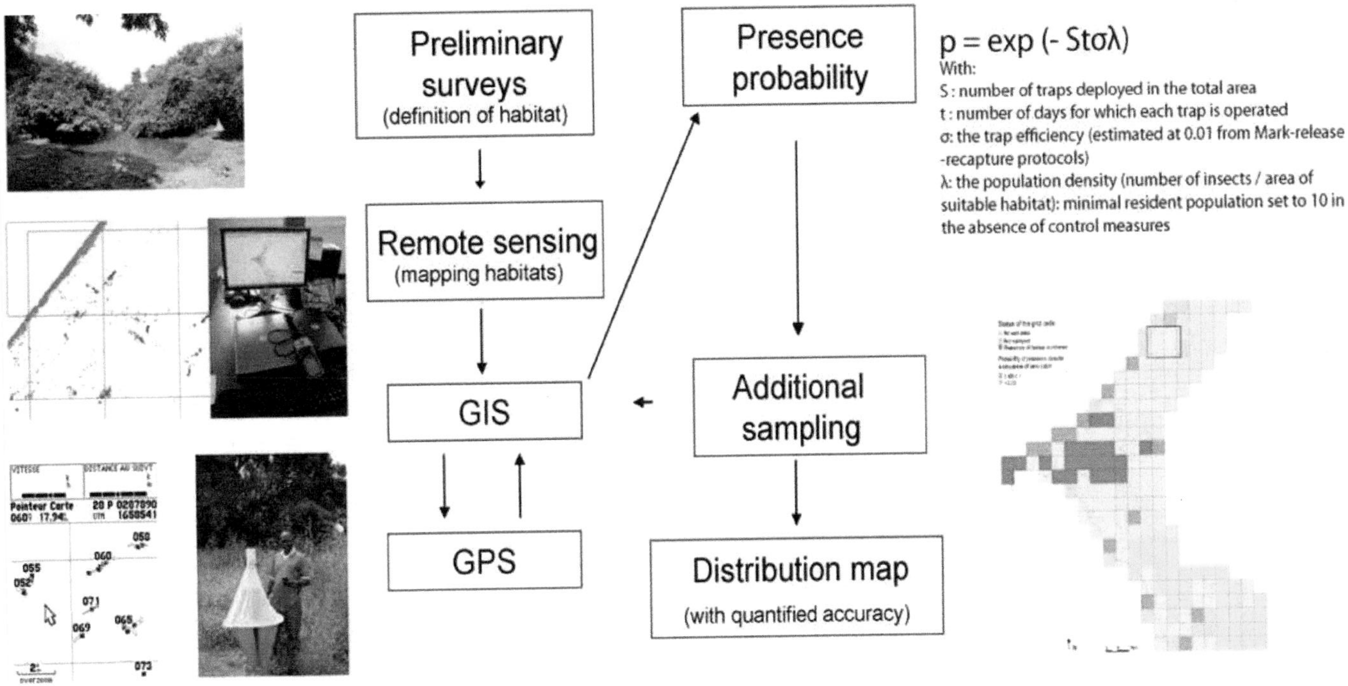

Figure 2. Sampling strategy used in the Niayes of Senegal to delimit the target population of G. p. gambiensis *(modified from Bouyer et al. 2010).*

To confirm this assumption, the genetic differentiation between the *G. p. gambiensis* populations from the Niayes and those from the south-eastern tsetse belt (Missira) was assessed. Using microsatellite DNA, mitochondrial COI DNA and geometric morphometrics of the wings of 153 individuals, complete genetic isolation of the *G. p. gambiensis* populations of the Niayes was confirmed. In addition, the *G. p. gambiensis* tsetse population from the Parc de Hann in Dakar proved to be isolated from other populations in the Niayes (Solano et al. 2010).

A third study focused on the parasitological and serological prevalence of AAT in cattle residing inside and outside the tsetse-infested areas of the Niayes. Before any control efforts were implemented, a mean parasitological prevalence of 2.4% was detected at the herd level in the tsetse-infested area, whereas serological prevalences of 28.7, 4.4, and 0.3% were obtained for *T. vivax*, *T. congolense* and *T. brucei brucei* Plimmer and Bradford, respectively (Seck et al. 2010). Moreover, the observed risk of cattle becoming infected with *T. congolense* and *T. vivax* was 3 times higher in the tsetse-infested as compared with the assumed tsetse-free areas. Furthermore, AAT prevalence decreased significantly with distance from the nearest tsetse sampled, indicating that cyclical transmission of trypanosomes by tsetse flies predominated over any potential mechanical transmission by other biting flies present in the area (Seck et al. 2010).

In addition to these studies, a socio-economic study was carried out to assess potential benefits from the sustainable removal of *G. p. gambiensis* from the Niayes. The study identified three main cattle farming systems, i.e. (1) a traditional system using trypano-tolerant cattle, and (2) two "improved" systems using more productive cattle breeds for milk and meat production. Herd size in improved farming systems was 45% lower and annual cattle sales amounted to €250 per head as compared with €74 per head in the traditional farming system. Tsetse distribution significantly impacted the frequency of occurrence of these farming systems with 34% and 6% of farmers owning improved breeds in the tsetse-free and tsetse-infested areas, respectively.

Two scenarios were considered with respect to potential increases of cattle sales as a result of the sustainable removal of the *G. p. gambiensis* population from the Niayes, i.e. a conservative scenario with a 2% annual replacement rate of the traditional system with improved ones, which was the rate observed just after tsetse eradication in Zanzibar (Vreysen et al. 2014), and a scenario with an increased replacement rate of 10% five years after the removal of the tsetse fly population. The final increase of cattle sales was estimated at ~€2800/km^2/year as compared with the total cost of the eradication campaign of ~€6400/km^2. The benefit-cost analysis indicated that the project was highly cost-effective, with internal rates of return of 9.8% and 19.1% and payback periods of 18 and 13 years for the two scenarios, respectively. In addition to an increase in farmer's income, the benefits of the eradication project included a reduction of grazing pressure on the already fragile ecosystem (Bouyer et al. 2014) (Fig. 3).

The project was considered an ecologically sound approach to achieving intensified cattle production without having a significant negative impact on the environment. Although the strategy included an initial insecticide (deltamethrin) component to suppress the tsetse fly population, the insecticide use was limited to impregnation of cloth traps, targets (Laveissière et al. 1985) and nets (around pig pens), and direct application to cattle (Bauer et al. 1995). The Sterile Insect Technique (SIT) (Knipling 1955) used as the final eradication component in the operational phase is a non-polluting control tactic that is very environment-friendly.

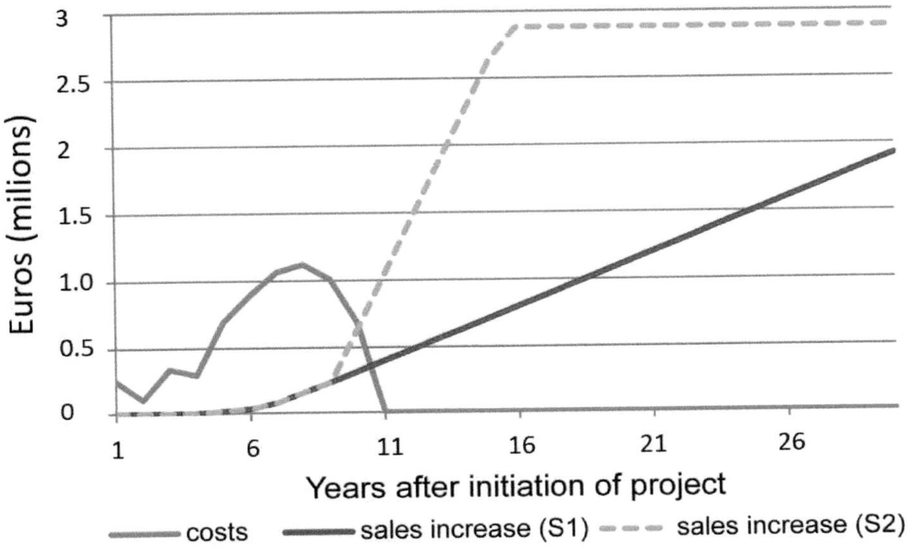

Figure 3. Comparison of the total costs of the eradication project and increase in global cattle sales per year (year 1 = 2007) taking into account two scenarios: a 2% annual replacement rate (S1) of local cattle with improved breeds, and an increased replacement rate of 10% five years after tsetse eradication (S2) (from Bouyer et al. 2014).

The SIT requires the production of large numbers of the target insect in mass-rearing centres, the sterilisation of the male insects using ionizing radiation (gamma rays or X-rays) and the sustained and sequential dispersal of the sterile insects over the target area in numbers large enough to outcompete the wild males for mating with wild females (Vreysen et al. 2013). The transfer of sterile sperm to wild virgin females results in embryonic arrest and hence the absence of offspring (Dyck et al. 2021). With each generation, the ratio of sterile to wild males will increase and as a result, the SIT becomes more efficient as population densities decline (inverse density dependent action of the SIT) (Vreysen and Robinson 2011).

In order to assess the potential impact of the eradication project and of the control tactics on the non-target fauna, an environmental monitoring project was implemented in five sites, one outside the tsetse-infested area (Mbour-centre IRD, a private protected area) and four within the targeted tsetse-infested areas (Dakar-Hann, Kayar, Thiès and Pout). Two fruit-feeding insect families (Coleoptera: Scarabaeidae (Cetoniinae) and Lepidoptera: Nymphalidae) were selected as indicator species as they have been shown to be highly appropriate for measuring the impact of various management practices on general ecosystem health in similar savannah areas in West Africa (Bouyer et al. 2007).

Monitoring with banana-baited traps indicated that of the ten most abundant Cetoniinae species, only one (*Pachnoda interrupta* Olivier) showed a significant reduction in apparent density in Block 1 (Kayar) during the operational phase (when insecticide-impregnated traps were deployed and cattle were treated with "pour-on"), but the population recovered to pre-suppression levels one year later. Similar observations were made with another Cetoniinae species, *Pachnoda marginata* spp. (predominantly *Pachnoda marginata aurantia* (Herbst) in Block 2. No significant impact was observed with the four most abundant Nymphalidae species (*Charaxes* butterflies). These data indicate that the overall impact of the project, as assessed using these sensitive non-target indicator species, was transient and very limited (Bouyer, unpublished data).

In addition to the above-mentioned studies, data were collected on population dynamics of the wild *G. p. gambiensis* populations in four different ecological sites. Apparent densities were shown to fluctuate both in space and time. Natural abortion rates were also highly variable in space and time and were modelled using MODIS satellite data, allowing the correction of apparent abortion rates during the sterile male releases (Bouyer, unpublished data). This in turn allowed the Fried competitiveness index to be estimated (Fried 1971), considering the observed abortion rate under a given sterile to wild ratio during the pre-operational phase. These data were crucial to make correct interpretations of the monitoring data during the control operations.

In conclusion, the data that emanated from these studies contributed to the strategic decision-making and the development of a control strategy. It prompted the Government of Senegal to adopt once more an eradication strategy (Hendrichs et al. 2021), as the isolated character of the *G. p. gambiensis* population of the Niayes and the integration of the SIT in this third attempt offered an opportunity to create a sustainable zone free of tsetse flies and trypanosomosis. In addition, it was decided to implement a project following area-wide integrated pest management (AW-IPM) principles that aimed to integrate the various control tactics (Vreysen et al. 2007) against the entire tsetse population within the circumscribed area to ensure that no population remnants would be left after the campaign.

Moreover, there were several aspects that made the inclusion of the SIT as a component of the AW-IPM strategy a prerequisite, these were: the fragmented nature of the preferred tsetse habitat, the two earlier failures to eradicate the target population in the 1970s and 1980s (Touré 1973) and the low impact/efficiency of insecticide-based bait methods on low-density populations of the targeted species (Bouyer, unpublished data).

It needs to be emphasised that most AW-IPM projects, especially those that incorporate a SIT component, are management-intensive and technically complex. In addition to a complete set of relevant baseline data, AW-IPM projects need to be implemented following sound scientific principles (Vreysen et al. 2007), and embarking on such a project without sound baseline data and a resulting comprehensive control strategy will have a high probability of failure. The probability of success will increase significantly when the project is accompanied by an all-inclusive operational research component to solve emerging problems during its implementation.

2.3. Phase 3: Pre-operational Phase (2009-2011)

2.3.1. Successful Suppression Trial in a Suitable Area of Kayar

A pre-release suppression trial, using insecticide-impregnated Vavoua traps, was carried out in the most northern part of the target area (Kayar) between November 2009 and December 2010, to assess the efficiency of this suppression tactic. Geographic information systems (GIS) and remote sensing were used to select favourable habitat sites at which to deploy the traps at a density of 40 traps per km^2 of suitable habitat, corresponding to 3.2 traps per km^2 in the test area (8% of suitable habitat) (Bouyer, unpublished data). Monitoring data indicated that the *G. p. gambiensis* fly populations were reduced to very low numbers, which confirmed the suitability of the suppression tactic selected for this ecological zone.

2.3.2. Establishment of an Insectary/Dispersal Centre at Dakar

In preparation for the development of colonies (see next Section), a building at the ISRA was refurbished and modified into an insectary/dispersal centre. Essential rearing and release equipment was provided through the IAEA's Department of Technical Cooperation to enable the rearing and maintenance of the tsetse flies.

2.3.3. G. p. gambiensis *Strains and Colony Establishment*

Since the 1970s, a colony of the target species, *G. p. gambiensis,* has been maintained at the CIRDES, Burkina Faso (denoted BKF strain), and was used for the successful eradication of a target population from 1500 km^2 of agro-pastoral land in Sidéradougou (Cuisance et al. 1984; Politzar and Cuisance 1984). From the onset of the project in Senegal, the Government decided not to develop its own mass-rearing facility to produce and sterilize the insects required for the SIT component, as the project area was judged too small to justify the expense of constructing and operating a tsetse mass-rearing facility.

Instead, it was proposed to procure the sterile male flies from the CIRDES. Although a recent study indicated that sterile males from this BKF strain were still competitive in riparian forests in Burkina Faso (Sow et al. 2012), relatively poor survival rates were obtained when released in the Parc de Hann of Dakar (B. Sall and M. Seck, unpublished data). It was speculated that this poor performance could be related to the extreme environmental conditions of this special micro-habitat in an urban setting.

To mitigate the risk that sterile males from the BKF strain would not perform in certain ecosystems in the Niayes, a decision was taken early on in the project to establish a *G. p. gambiensis* colony with pupae originating from Senegal (denoted SEN strain). Between October 2009 and September 2010, a total of 2185 pupae produced by wild-collected females were received at the FAO/IAEA Insect Pest Control Laboratory (IPCL), Seibersdorf, Austria, to develop a SEN colony. By the end of December 2010, the SEN colony had increased to about 450 producing female flies, and by mid-2012 the colony reached a maximum size of 4500 females. Thereafter the colony was maintained with around 1500 females (M. Vreysen, unpublished data).

In addition, a colony was established at the IPCL with pupae derived from the BKF colony in CIRDES to develop a back-up colony for the eradication project and to provide material for experimental work, such as mating compatibility studies between the target strain (SEN) and the strain used for release (BKF), development of transport protocols of the sterile male pupae under low temperatures and the development of an introgressed strain (BKF-SEN).

In view of the fact that colonization of a wild tsetse strain is a labour-intensive and lengthy process, an introgressed strain with a genetic background of 99% from SEN, that would also retain the adaptation to an artificial rearing environment (BKF strain) was developed. However, the strain proved to have a very low fecundity and the idea was abandoned.

2.3.4. Mating Compatibility and Competitiveness of the BKF and the SEN Strains

In view of the marked differences between the ecosystems of Burkina Faso and Senegal, and the large genetic differences between the two populations (BKF and SEN) (Solano et al. 2010), it was important to assess under semi-natural conditions the presence or absence of any potential mating barriers between the BKF and SEN strains that could jeopardize the release component and hence the outcome of the eradication campaign.

The mating performance of the BKF strain was compared with that of the 'wildish' SEN strain (that was a few generations from the wild) in walk-in field cages. The laboratory-adapted BKF strain showed close to equal competitiveness and mating compatibility with the SEN strain, which indicated the potential of using BKF strain males for the SIT component against the *G. p. gambiensis* populations in the Niayes (Mutika et al. 2013). These data were later confirmed during pilot trials in the target area (Bouyer, unpublished data).

2.3.5. Development of Protocols to Irradiate and Transport Male Tsetse Pupae

After the decision to procure the sterile males from the CIRDES, the Government of Senegal requested the IPCL to develop irradiation and transport protocols that would allow the shipment of (only) male *G. p. gambiensis* pupae over long distances, whilst retaining the female flies in the colony at the CIRDES. As female tsetse flies emerge two days before male flies, a scheme was proposed that would expose the male pupae to low temperatures after most of the female flies had emerged. The low temperatures would arrest male emergence from the pupae, making transport of irradiated male pupae to Senegal possible, whilst maintaining the required low temperature.

In the first series of experiments, exposing male pupae of *G. p. gambiensis* to low temperatures (10 and 12.5°C) for 3, 5, or 7 days immediately prior to emergence had no effect on emergence of male flies, whereas emergence of flies held at 15°C started before the simulated transport period was over. Survival of the experimental males and fecundity of females inseminated by males that emerged from pupae held at low temperature for different periods varied within the experimental groups, but mating performance of the experimental males was not impaired (Mutika et al. 2014).

A second series of experiments assessed the combined effect of irradiation and low-temperature period. Emergence and survival of adult male flies which were irradiated as pupae with 70, 90, 110 and 130 Gy on days 25, 27, and 29 post-larviposition was similar to that of un-irradiated pupae. Males that were irradiated with 110 Gy 24 h after initial exposure to the low temperatures and chilled for 5 days at 10°C were as competitive as un-irradiated males of the same age when competing with them in walk-in field cages for virgin untreated females (Mutika et al., unpublished data).

In addition to pupal irradiation and low temperature during their transport, the release protocol required a chilling period for adult males to allow immobilization and collection immediately prior to the aerial release (Mubarqui et al. 2014). A significantly lower proportion of males that had been irradiated (110 Gy) and held at low temperature as pupae (10°C for 5 days) and adults (5.1 ± 0.02°C for 6 or 30 hours six days after emergence) succeeded in mating compared to untreated colony males. Female insemination levels were slightly lower for males held at low temperature for 30 h compared to 6 h or not exposed to low temperature (standard colony conditions). The data confirmed the feasibility of transporting irradiated pupae at low temperatures for long distances followed by releases of chilled males using an adult release system, but it was found necessary to minimize the time that the adults remain chilled (Mutika et al., unpublished data).

2.3.6. Validation of Protocol for Long-distance Shipment of Irradiated Male Pupae

The use of isothermal boxes that contained phase change material (Phase Change Material Products Limited, Cambridgeshire, UK) packs to transport the male pupae was validated during weekly shipments from 2011 to 2013. More than 900 000 *G. p. gambiensis* pupae were transported in 132 shipments from the CIRDES in Burkina Faso, the SAS in Slovakia, and the IPCL in Austria to the ISRA in Dakar, Senegal, using a commercial courier service. The average temperature and humidity inside the insulated transport boxes were 10.1 ± 2.3°C and 81.4 ± 14.3% relative humidity, respectively. Pupae were collected on different days at the source insectary and depending on the date of collection, they were kept for different periods at low temperatures (4°C).

At the emergence and dispersal centre in Senegal (ISRA), the emergence rate from pupae that had been chilled at 4°C for one day in the source insectary before transport (batch 2) was significantly higher than that of pupae that had been chilled at 4°C for two days in the source insectary before transport (batch 1), i.e. an average emergence rate (± SD) of 76.1 ± 13.2% and 72.2 ± 14.3% respectively, with a small proportion emerging during transport (0.7 ± 1.7% and 0.9 ± 2.9% respectively). Among the emerged flies at the dispersal centre, the percentage with deformed (not fully expanded) wings was significantly higher for flies from batch 1 (12.0 ± 6.3%) than from batch 2 (10.7 ± 7.5%). The quantity of sterile males available for release as a percentage of the total pupae shipped was 65.8 ± 13.3% and 61.7 ± 14.7% for batch 1 and 2 pupae, respectively. The results showed that the temperature inside the boxes, during shipment, must be controlled around 10°C with a maximal deviation of 3°C to maximize the male yield (Pagabeleguem et al. 2015).

2.3.7. Quality Control Procedures to Assess Sterile Male Quality after Long-distance Shipment

Routine quality control procedures were required to regularly monitor the biological quality of the shipped and received biological material. This was important to ensure that the flies that were released, especially those released by air, were adequately competitive. A quality control test derived from the one used in fruit flies in Central America (Enkerlin et al. 2015) was developed to monitor the quality of *G. p. gambiensis* males that emerged from pupae produced and irradiated in Burkina Faso (irradiation done at CIRDES) and Slovakia (irradiation done at the IPCL) and transported weekly under low temperature conditions to Dakar.

For each consignment, a subsample of 50 pupae was taken before shipment and at destination to assess emergence, flight ability of the adult flies from a cylinder and survival of the flyers without access to blood meals. The quality protocol proved a good proxy of fly quality, explaining a large part of the variances of emergence rates, percentage of flies with deformed wings and flight ability in the field. Initially only 35.8 ± 18.4% of the transported pupae produced sterile males that showed a propensity to fly, thereafter named "operational flies" (Seck et al. 2015). However, these operational males were very competitive after release, which has already resulted in eradication of some of the target populations (Bouyer et al. 2012). Over time, the handling procedures and transport protocols were fine-tuned, resulting in a significant improvement in the percentage of operational flies from an initial 36% (SD 18%) in 2012 to 59% (SD 15%) in 2016 (Fig. 4). Unfortunately, this percentage dropped again in 2017 and 2018, mainly due to problems with environmental control and blood-feeding in the mass-rearing facilities producing the flies. Improving the quality of the flies will be crucial to ensure the success of the operational phase, as a significant positive correlation was observed between the recapture rate of sterile males in the field and this quality indicator (Bouyer and Seck, unpublished data).

2.3.8. Environmental Suitability of Available Strains for Release in the Niayes

At the CIRAD in Montpellier, a study was carried out to determine the critical environmental thresholds for survival of *G. p. gambiensis* flies from the three strains (BKF, SEN and the introgressed SEN-BKF strain). The study provided information on which strain would be best adapted to a particular environment or ecosystem. The optimal temperatures for maintaining flies of the BKF, SEN-BKF and SEN strains were 25 ± 1, 24.6 ± 1 and 23.9 ± 1°C, respectively. The survival of this tsetse species was governed by temperature alone and unaffected by changing humidity within the tested range. The BKF strain better survived temperatures above these optima than the SEN and SEN-BKF strains, but a temperature of about 32°C was the limit for survival for all strains. The relative humidity ranging from 40 to 75% had no effect on productivity at 25-26°C (Pagabeleguem et al. 2016b).

2.3.9. Field Competitiveness of the BKF Strain after Release in the Niayes

The competitiveness, mortality and dispersal of BKF flies was measured in the field in 2010-2011 (Bouyer et al. 2012) using mark-release-recapture studies in four different ecosystems (Hann, Diaksao Peuhl, Pout and Kayar). Data were collected on recapture rate, trap efficiency, daily mortality of the sterile males, dispersal capacity and mating competitiveness in both space and time. Female abortion rates (i.e. rate of induced sterility) were assessed through dissection of all captured wild females (Van der Vloedt and Barnor 1984; Vreysen et al. 1996) and corrected for natural abortion rates using developed models.

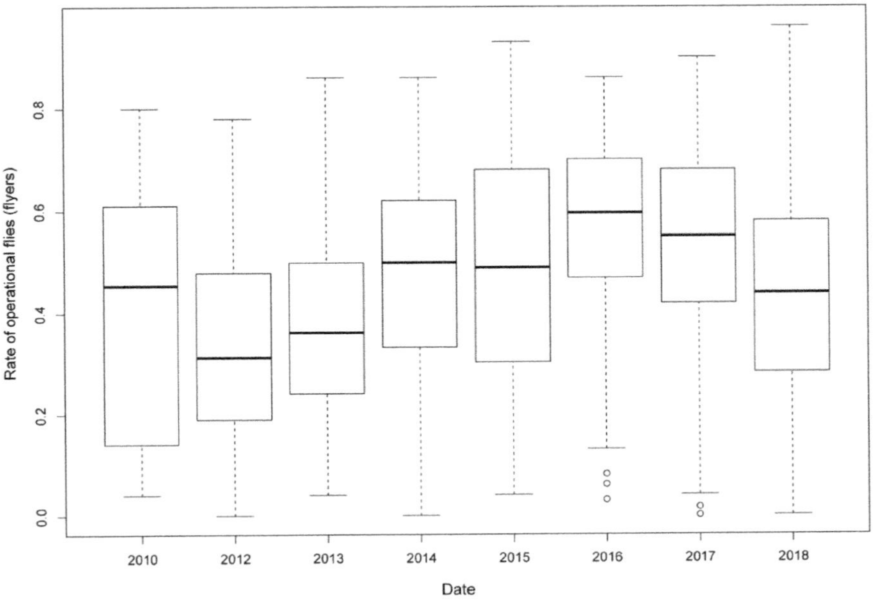

Figure 4. Rate of operational flies measured as the proportion of flies capable of flying out of a flight cylinder as part of routine quality tests at the ISRA (Institut Sénégalais de Recherches Agricoles) dispersal center (2012-2018). Boxplots present the median and quartiles and bars the 95% confidence intervals (updated from data published in Seck et al. 2015).

Trap efficiency (measured as the probability that a trap catches a fly present within 1 km^2 within 1 day (Barclay and Hargrove 2005)) was estimated at 0.03 (SD 0.04) and its variability in space and time was low. The daily mortality rate was quite homogeneous, but higher in the urban ecosystem (Parc de Hann) as compared with the more natural habitats. Although the dispersal rates were lower as compared with values obtained in riparian forests in Burkina Faso (Cuisance et al. 1984; Bouyer et al. 2007) they were, nonetheless, considered sufficient to obtain a homogeneous dispersal of sterile males using swaths of 500 m between aerial release lines. Finally, Fried indices obtained (Fried 1971) were high (> 0.35) but varied with the ecosystem. These data were instrumental in the development of an efficient release strategy for the sterile males.

2.3.10. Molecular Tools to Discriminate Sterile and Wild G. p. gambiensis *Flies*

In any AW-IPM project with a SIT component, the impact of the released sterile males needs to be assessed at regular intervals to monitor project progress and allow quick mitigation of emerging problems. Monitoring usually relies on an adult trapping system that captures both sterile and wild insects in a similar way (Vreysen 2021). This requires procedures that allow discriminating between the trapped wild and sterile male insects.

In the tsetse eradication project in Senegal, sterile adult male *G. p. gambiensis* were marked with a fluorescent dye powder (DayGlo®, 1% dye by weight mixed with sand) during emergence from the pupae (Parker 2005). A similar procedure was used in the *Glossina austeni* Newstead eradication project on Unguja Island of Zanzibar (Vreysen 1995). This type of marking is effective, although not infallible and in some cases, sterile male flies were only slightly marked; conversely, some wild flies could become contaminated with a few dye particles in the cages of the monitoring traps (which leads to incorrect interpretation of the trapping results).

In some cases, predatory ants also damaged the trapped flies, making discrimination between wild and sterile males using a fluorescence camera and / or a fluorescence microscope difficult.

A molecular technique, based on the determination of cytochrome oxidase haplotypes of *G. p. gambiensis*, was therefore developed to discriminate wild from sterile males with a high level of accuracy. DNA was isolated from the fly heads and a portion of the 5' end of the mitochondrial gene cytochrome oxidase I was amplified for sequencing. All sterile males from the BKF strain displayed the same haplotype and differed from that of wild male flies trapped in Senegal (and in Burkina Faso). The method allowed complete and fail-proof discrimination between sterile and wild male *G. p. gambiensis* and might be used in other tsetse control campaigns with a SIT component (Pagabeleguem et al. 2016a).

2.3.11. Aerial Release Trials

Sterile male tsetse flies were released by air for the first time in the *G. austeni* eradication campaign on the Island of Unguja, Zanzibar (Vreysen et al. 2000), using biodegradable carton boxes that contained un-chilled sterile adult insects. The fixed-wing aircraft were equipped with an appropriate chute that allowed the cartons to be released through the fuselage of the aircraft (Vreysen et al. 2000).

In the Niayes project, the area that needed to be covered with sterile males was large enough to opt for aerial releases to efficiently disperse the sterile insects, rather than ground releases which were considered too costly, inefficient and not conducive to an area-wide coverage. The release vehicle of choice was the gyrocopter (Fig. 5), which was initially adapted to release sterile males in carton release containers. A gyrocopter is an autogyro that is characterized by a free-spinning rotor that turns because of the passage of air through the rotor from below which sustains the autogyro in the air, and a separate engine driven propeller that provides forward thrust (Wikipedia 2019).

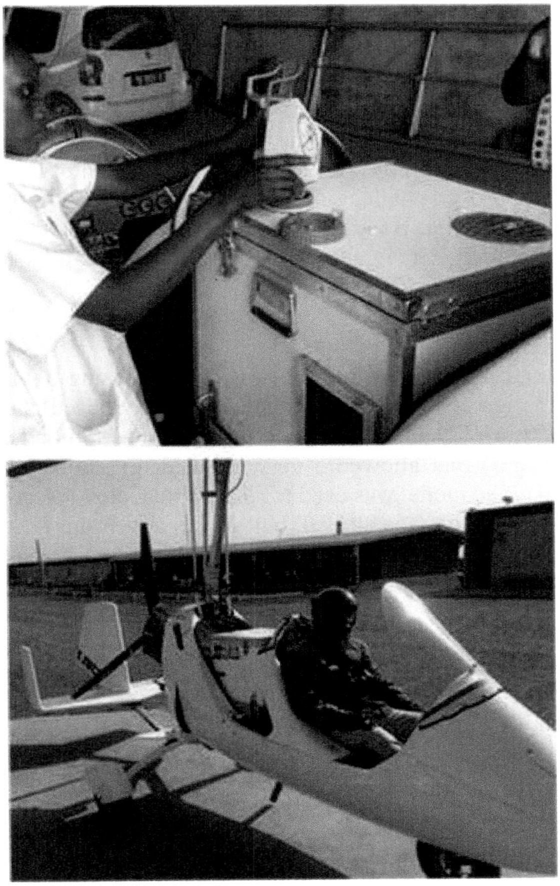

Figure 5. Loading of the chilled adult aerial release device with immobilized adult tsetse males (top) and the gyrocopter ready for take-off to release the sterile insects (bottom).

Gyrocopters have been used for the SIT component in other AW-IPM projects, such as the release of sterile false codling moth, *Thaumatotibia leucotreta* Meyrick, in South Africa (Boersma, this volume).

The aerial release of sterile males using carton boxes was tested in a sub-unit of the first block (Kayar) along 4 release lines that were separated with a swath of 500 m. For 11 weeks (from March 2013 to June 2013), 32 boxes were released each week separated by a distance of 500 m over each release line. A total of 65 000 sterile males were released of which 316 flies (0.5%) were recaptured, giving an estimated daily mortality rate of 28% (SD 12%) and a mean daily displacement of 917 m (SD 477 m).

Although the release with carton boxes was very successful, a new approach for the aerial release of sterile tsetse flies was developed in collaboration with the Mubarqui group of Mexico (Mubarqui et al. 2014). This innovative system (Fig. 5) allowed the release of small numbers of tsetse flies per surface area (between 10–100 per km^2) and was based on the use of a vibrating mechanism. The device is guided by a GIS that can adjust flexibly the density of sterile males to be released depending on the requirements of the different target areas being treated. The GIS is installed on an android tablet which enables the pilot to concentrate on navigating the predefined release lines; the machine will automatically start releasing the required number of sterile insects for each target zone. However, the calibration of the release rate using the release machine proved challenging because of significant (unwanted) secondary vibrations of the gyrocopter. As a consequence, a new release device was designed, based on a rotating cylinder, which provided improved results (patent deposition number 1653994 by CIRAD and ISRA).

2.3.12. Use of a Maxent Distribution Model

All suppression and release activities were optimized using a Maxent distribution model that mixed high spatial resolution data (four supervised classifications of the vegetation Landsat 7ETM+ images from four seasons) with high temporal resolution data (MODIS images) that allowed a very good identification of suitable habitats (Dicko et al. 2014). The model was used to select and deploy insecticide-impregnated traps in suitable vegetation (see above), but also to adjust the release density of the sterile males in relation to the availability of suitable habitat (the reference was 10 and 100 sterile males per km^2 in unsuitable and suitable habitat, respectively).

2.4. Phase 4: Operational Phase (since 2011)

2.4.1. External Review of the Project

An external team of experts visited the project in May 2012 and reviewed all past activities since the initiation of the project. The evaluation team highlighted the thoroughness of the baseline data collection effort that enabled the project area to be defined. The reviewers likewise emphasized the good collaboration, complementarities and interaction between the persons involved in the project as a key factor for the project's success. The team concluded that the project was ready to enter the full operational eradication phase (unpublished report to the IAEA of an external review team–May 2012).

2.4.2. The "Rolling Carpet" Strategy

Although the *G. p. gambiensis* populations in the target area were genetically isolated from the remainder of the tsetse belt in the south-eastern part of Senegal, the lack of sufficient manpower in the field and insufficient numbers of sterile males available on a weekly basis made it impossible to tackle the entire project area at once.

During the baseline data collection, it became apparent that the project area contained three distinct tsetse populations in areas of suitable habitat that were separated from each other by zones of unsuitable habitat (or very fragmented suitable habitat), limiting the potential for tsetse dispersal. The project area was therefore divided into three main operational blocks, i.e. Kayar in the north (Block 1), Pout/Sebikotane/Diacksao Peulh in the middle (Block 2) and Dakar (Block 3B) and Thiès (Block 3B), west and east of Block 2, respectively (Fig. 6).

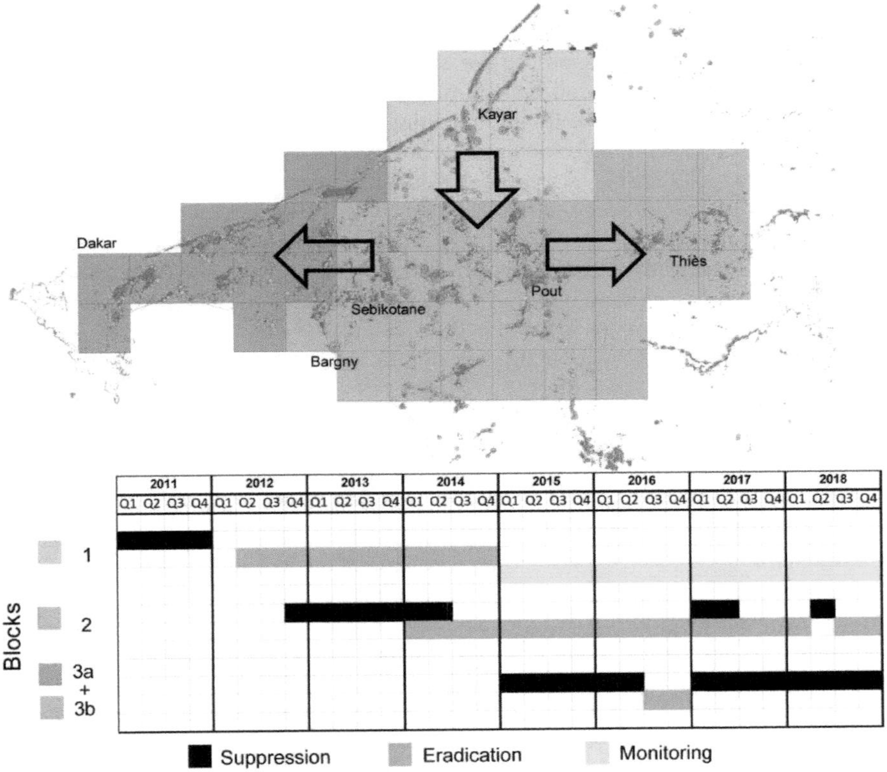

Figure 6. Map of project area with each grid cell corresponding to 5 x 5 km. Diagram of the three main activities of the operational phase in the different blocks of the project area.

An operational "rolling carpet" strategy (Hendrichs et al. 2021) was adopted and implemented whereby the different blocks were treated in sequence (suppression, followed by sterile male releases, and then monitoring of the status of eradication) (Fig. 6). In each block, insecticide-impregnated targets/traps were removed before the start of sterile male releases.

In Block 1, January 2011 marked the start of the operational phase of the project with the deployment of 269 insecticide-impregnated Vavoua traps (Laveissière and Grébaut 1990) in the favourable habitat areas, a density which corresponded to 19.4 traps per km^2 of suitable habitat. The apparent density of the *G. p. gambiensis* population dropped from an average of 0.42 (SD 0.39) flies per trap per day before the start of the suppression to an average of 0.04 (SD 0.11) flies per trap per day at the end of the trap deployment. This was followed by the aerial release of sterile males in March 2012 using biodegradable cardboard boxes over 185 release points, following 23 release lines over a total surface area of 72 km^2. In February 2014, the "boxed release" system was abandoned, and a "chilled adult" release system became operational (Mubarqui et al. 2014).

The apparent density of the *G. p. gambiensis* fly population was reduced to zero catches after six months of sterile male releases. In Block 1, the last wild fly was trapped on August 9, 2012, i.e. an old female (> 40 days), which was in her fourth oviposition cycle and which had an empty uterus. The next follicle in ovulation sequence was still immature and small, indicating an abortion of the larvae or an egg in embryonic arrest. This female showed a copulation scar and a spermathecal fill of 85%, indicating that its sterility was probably induced through a mating with a sterile male (Van der Vloedt and Barnor 1984; Vreysen et al. 1996).

From the beginning of the releases in Block 1 (March 16, 2012) to the date corresponding to the last capture, only three other wild females could be dissected, and all had indications of having mated with a sterile male. The average percentage of sterile males as a proportion of the total catch was then 99.2% (SD 1.6%), corresponding to a sterile-to-wild male ratio of 130:1. The percentage of sterile males remained 100% thereafter (no wild fly has been captured for the subsequent 78 weekly collections with 25 monitoring traps). Sterile male releases were suspended in late 2014 and as of January 2015, all sterile flies were released in Block 2.

The monitoring in Block 1 was continued on a monthly basis and is still ongoing at the time of writing. Since 2012, no wild flies have been trapped in Block 1, corresponding to a very high likelihood of eradication (probability of not detecting potential remaining flies $< 10^{-6}$ at the time of writing, considering that the population would have recovered to at least 10 flies during almost 2 years of monitoring without control) (Fig. 7, upper graph).

In Block 2, remote sensing and land cover maps were used to select 1205 suitable habitat sites for the deployment of insecticide-impregnated traps (corresponding to 16.7 traps per km^2 of suitable habitat and 2.7 traps per km^2 of the total targeted area). Deployment of the suppression traps in Block 2 was initiated in December 2012 and was supplemented with an additional 300 insecticide-impregnated traps in early 2013. In addition, at 6 monthly intervals, 2970 cattle were treated three times with a "pour-on" insecticide as a complementary method to suppress the *G. p. gambiensis* fly population.

In Block 2, the apparent fly density dropped from an average of 1.24 (SD 1.23) flies per trap per day before the suppression to an average of 0.005 (SD 0.017) flies per trap per day at the end of the suppression phase. Sterile male releases were started in Block 2 in February 2014, initially covering a quarter of the block, which was expanded based on sterile male availability to half of the block in April 2014. In January 2015, releases were expanded to cover the entire Block 2.

The apparent fly density was reduced to < 0.001 fly per trap per day by the end of 2018. The releases are scheduled to continue for another 10–12 months after the last wild fly has been trapped (Fig. 7, middle graph). In the beginning of 2017 and 2018 unexpected upsurges in the density of the wild fly population were observed in Block 2 in 3-5 areas. The reasons for these upsurges are not clear, but mitigating action was taken immediately, and suppression traps were deployed in the affected areas (Fig. 6). In addition, emergency insecticide spraying of *Euphorbia* hedges was carried out in selected areas, that brought the fly situation rapidly again under control. Depending on availability of sterile male flies, these areas received higher concentrations of sterile flies as compared to the rest of the area.

Ground releases were carried out in an area of 114 km^2 in Block 2, where recapture rates of sterile males released by air were consistently zero. This was later assumed to be correlated with the opening of a cement factory that apparently had a negative impact of fly survival. As a result, the aerial releases in that area were abandoned and replaced with releases from the ground. Additional ground releases were used in the hot spot areas of Block 2 (Pout and Diacksao Peulh) to supplement the aerial releases.

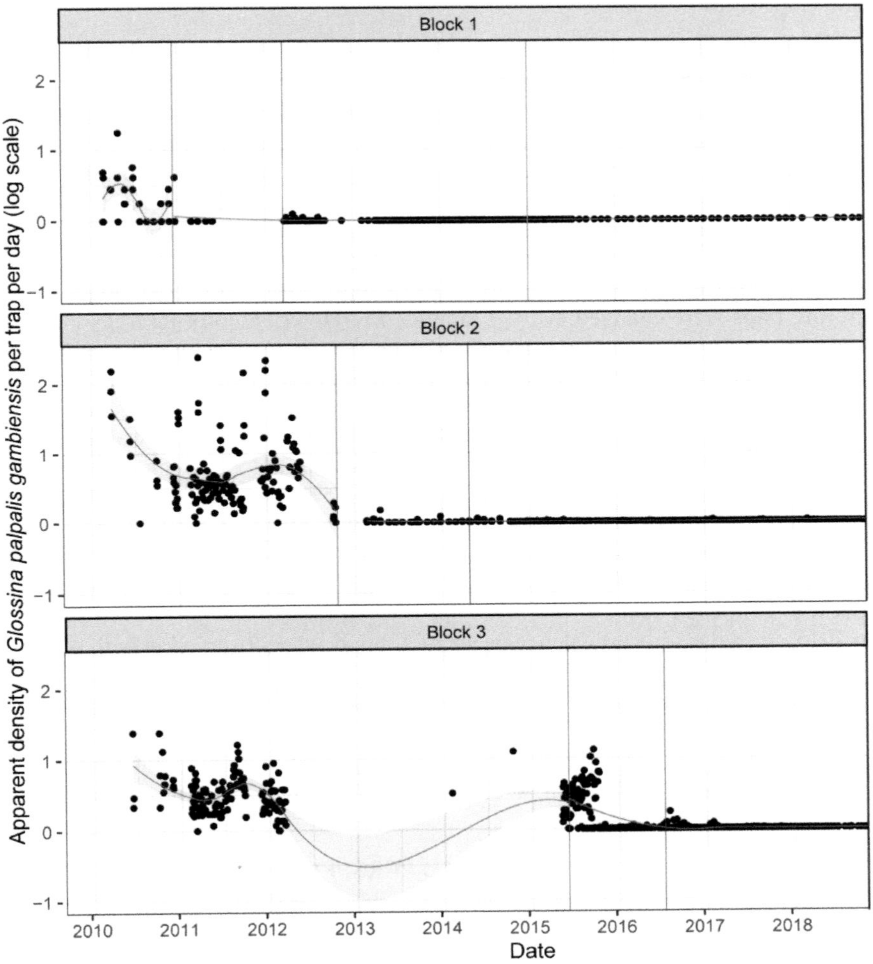

Figure 7. Apparent density (number of flies/trap/day) of the Glossina palpalis gambiensis *populations in Block 1, 2 and 3 of the Niayes during the pre-suppression, suppression and eradication activities. Pre-suppression activities are shown before the blue line, suppression between the blue and red lines, eradication activities with sterile male releases between the red and green lines, and post-SIT monitoring only after the green line.*

In Block 3, the suppression activities started in May 2015 with the deployment of 191 insecticide-impregnated traps in suitable habitat (12.6 traps per km^2 of suitable habitat), thereafter reinforced with an additional 43 traps. Before suppression, the initial apparent density of the fly population was 1.50 (SD 2.12) flies per trap per day; this dropped to 0.008 (SD 0.039) flies per trap per day in June 2016, i.e. a reduction of 99.4%. Sterile male releases were started in July 2016 in 100 km^2 of Block 3, but were suspended in early 2017, to accommodate the releases in the problem areas in Block 2 (Fig. 7, lower graph).

2.4.3. Monitoring the Progress of the Campaign

The Maxent distribution model was also used to guide the monitoring of the eradication campaign by deploying monitoring traps in suitable habitats (Dicko et al. 2014). As eradication was the selected strategy, the suitability threshold was set to provide a high sensitivity (0.96). The model was continuously improved during the project to increase its specificity from an initial 0.43 using the supervised classifications of the vegetation to 0.57 using the Maxent.

The areas around the monitoring traps were regularly cleared of vegetation and the monitoring traps were changed every 3 months. Moreover, monitoring traps in sites with no capture for one year were moved to other sites, but still within the predicted suitable habitats, and were labelled as temporary monitoring sites.

Regular parasitological monitoring of sentinel herds, each composed of ~100 tagged cattle, was carried out every year in three sites, of which one site was in a non-infested area and two were in the target area (in Blocks 1 and 2). In the non-infested area, the overall AAT seroprevalence remained below 5% between 2009 and 2017. In the target area, the AAT prevalence reduced quickly as control operations advanced ($p < 0.001$), i.e. from an initial value of >20% in 2009 to below 1% in 2014 in Block 1, and from 60-85% in 2009–2010 to below 5% in 2016 and 2017 in Block 2 (Bouyer and Seck, unpublished data).

In 2015, irregular sero-prevalence peaks of *T. vivax* were observed in both blocks, i.e. 12% in Block 1 and 16% in Block 2, which might be attributed to mechanical transmission (Desquesnes and Dia 2003; Desquesnes et al. 2009) facilitated by the presence of trypanosomes in tsetse in Block 3 and a small persistence in Block 2.

A blanket treatment of all cattle using trypanocidal drugs will be carried out in the Niayes area after tsetse eradication, to also ensure the eradication of trypanosomes.

3. DISCUSSION AND PERSPECTIVES

Many successful AW-IPM projects with a SIT component were or are implemented by management structures that were/are flexible and independent, with a high degree of financial and political freedom and not affected by strangling government bureaucracies and regulations (Vreysen et al. 2007). The New World Screwworm Commission, initially established between Mexico and the USA and later other countries in Central America, is a good example in this respect (Wyss 2006). The commission had to account for all financial, physical and human resources, could hire and fire staff based on merit and performance, and all staff were employed full-time without any other responsibilities (Vreysen et al. 2007).

Another good example is the Programa Moscamed, a cooperative agreement between Mexico, Guatemala and the USA, that has contained the Mediterranean fruit fly, *Ceratitis capitata* (Wiedemann) for the last 30 years in Guatemala and has prevented its spread into Mexico and the USA, which are free from this pest (Enkerlin et al. 2015).

There are however examples of AW-IPM projects, albeit smaller than the examples given above, that operated successfully outside such an organization, e.g. the tsetse eradication project on Unguja Island of Zanzibar that was implemented within the Ministry of Agriculture, Livestock and Environment of Zanzibar (Vreysen et al. 2000). The success of the project, however, was made possible by the full autonomy and independence that was given to the senior project managers by the Government of Zanzibar to implement the project as required.

The project in the Niayes is likewise operated within the Ministries of Livestock and Agriculture (EXPO Milano 2015) and not implemented by an independent organization. The project adopted an "adaptive management" approach which included monthly project coordination meetings with the different stakeholders (Fig. 8). It is believed that this approach was critical to the project's success. This management approach involved all the stakeholders, including researchers, ensuring transparency and decision-making by consensus. The important decisions in the project were based on scientific principles (never political, personal, or emotional) and were guided by analysed field or other data. Day-to-day operational and financial problems were openly discussed, leading to consolidated solutions being found. Any decision that required follow-up actions was immediately acted upon and was always implemented according to plan, as the DSV and the ISRA had full authority over regional veterinary staff and technicians employed for the SIT component of the project, respectively. It is believed that the collaboration between the internal stakeholders, international partners and the policy of "non-interference" of the respective Ministries have been instrumental for the smooth implementation of the project.

The stability in project staffing, with basically no turn-over experienced in 12 years, both at the management and at the technical (insectary/field staff) level is considered another important factor for the project's success. This created a personnel culture of reliability, transparency and trust, and ensured the necessary institutional memory. The main outputs of the research component of this innovative project were the development of methods that allowed an optimization of the implementation of the SIT to eradicate the tsetse fly using an adaptive management scheme. The involvement of the public sector in the innovation processes guaranteed top-down control of the use of the technology from the central veterinary services to regional veterinary services or dedicated personnel (Devaux-Spatarakis et al. 2016).

All data generated within the project were transferred to and managed within a relational database that was accessible on the web with information displayed in graphs, featuring specific queries that allowed all stakeholders and the general public to make assessments of the progress of the project at any time and at a glance (Projet de Lutte contre la Mouche Tsé-tsé dans le Niayes 2019). This provided transparency on project progress for all stakeholders in the project and also facilitated statistical analyses of the field data to better inform the decision-making process.

Before the start of the operational phase of the project and after a critical review of its components, the Senegalese Ministry of the Environment issued a permit to implement the planned project, provided that it was accompanied by an environmental monitoring scheme for the entire life span of the project. This monitoring revealed a slight and transitory impact of the suppression activities on non-target fauna (Ciss et al. 2019). The removal of the tsetse fly and AAT from the Niayes is expected to result in an improvement of farming systems (i.e. a replacement of traditional, low-productive cattle with more productive cross- and/or exotic breeds–this replacement is already apparent in Block 1 and certain areas of Block 2), but at the same time in an anticipated reduction (up to 45%) of the average size of cattle herds (Bouyer et al. 2014, 2015a). This will actually significantly reduce overgrazing which is a major cause of land degradation in Senegal, and as such, the removal of the tsetse fly will have a positive impact on this already fragile ecosystem and environment (Budde et al. 2004). Despite the experienced upsurges of the wild fly population in the beginning of 2017-2018 period, the apparent density of the wild fly population has been significantly reduced in the entire project area and transmission of AAT has basically stopped in the Niayes at the time of writing. Consequently, milk production, resulting from an increased rate of replacement of local with exotic cattle, has significantly increased and milk import has significantly been reduced. In 2016-2017, Senegal imported more than 1000 exotic cattle into the Niayes area as compared to 100-200 in earlier years.

An important part of the operational funding was provided by international partners, such as the IAEA's Department of Technical Cooperation and the US Department of State's PUI. The socio-economic studies which were carried out documented the processes of innovation that increased the impact of the eradication project (Bouyer et al. 2014, 2015a), and the outcome of these studies were important to convince external partners to continue financing the project, even though it could take some time for the economic impact of the project to become visible.

Like many other AW-IPM projects with a SIT component, the AW-IPM campaign in the Niayes was accompanied by an extensive public relations campaign. The inhabitants of the Niayes were informed from the beginning and during the different phases of the project about the justification, activities, future advantages of the project through meetings organized by the Chiefs of the local veterinary centres in collaboration with administrative (sub-prefects) and local (village chiefs and locally elected politicians) authorities. It is believed that these meetings were instrumental in informing the general public about the project and soliciting their support. Even in the beginning of the project, the period of baseline data collection was taken as an opportunity to inform and intensify contacts with the local farming community regarding the project. In addition, T-shirts and hats were distributed that carried the logo of the project to increase the visibility of the project. Finally, two video films produced in 2012 and 2013 were aired on the national TV.

As was done for Block 1, probability models will also be used to verify eradication over the entire project area (Barclay and Hargrove 2005). These calculations might be complemented by a new innovative diagnostic technique that is based on the prevalence of specific antibodies against tsetse saliva in the host that can persist for 4-6 weeks, which is being developed as an indirect - but very sensitive - measure of tsetse presence (Somda et al. 2013, 2016).

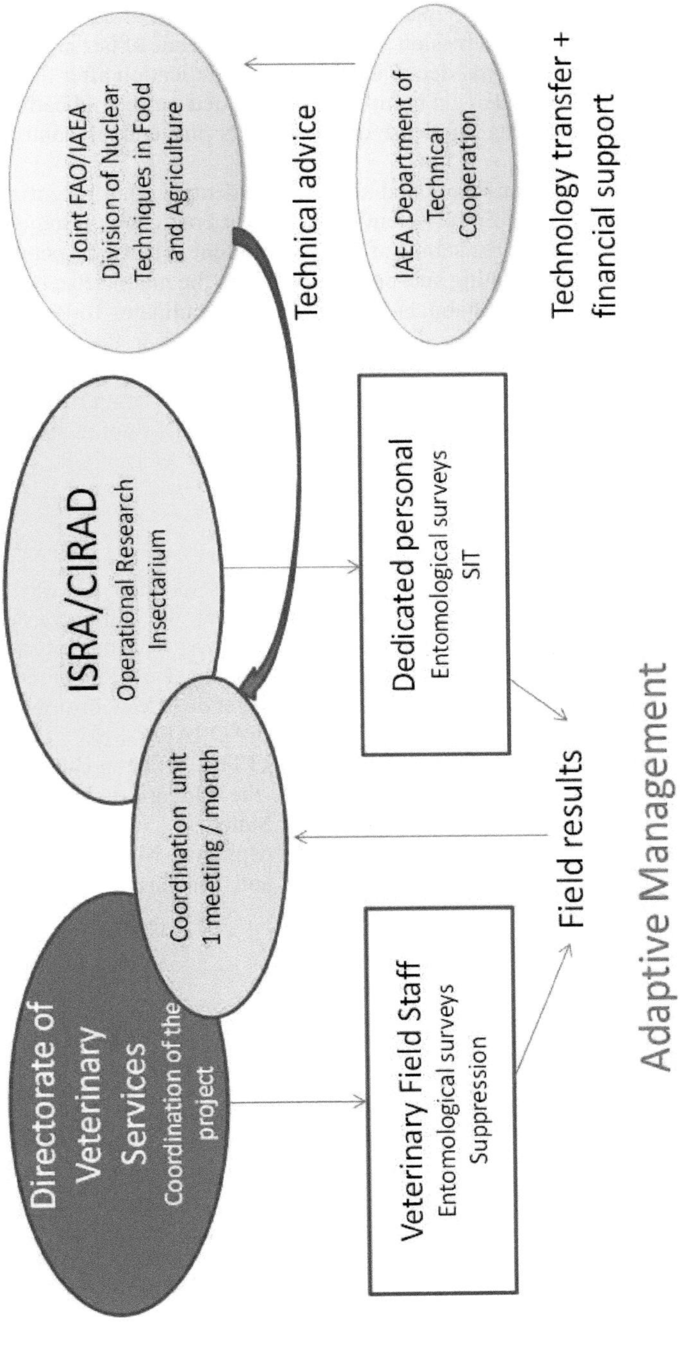

Figure 8. Organigramme of the adaptive management philosophy adopted by the project. (CIRAD = Centre de Coopération Internationale en Recherche Agronomique pour le Développement; ISRA = Institut Sénégalais de Recherches Agricoles; IAEA: International Atomic Energy Agency; FAO: Food and Agriculture Organization of the United Nations

The absence of circulation of the AAT parasites will also be used as an indirect measure of the absence of cyclical transmission and hence, the absence of tsetse flies. All these data will permit the provisional declaration of tsetse eradication after there have been zero fly catches for a period of six months and confirmed tsetse eradication if no wild tsetse are captured during for least one year after the end of control operations (Barclay et al. 2021).

Finally, friction models have been developed and used to identify other potential *G. p. gambiensis* populations that could be potential targets for eradication (Bouyer et al. 2015b). These models allow the resistance of the environment to tsetse dispersal to be mapped, i.e. natural barriers isolating sub-populations from the main tsetse belt. These "ecological islands" of suitable habitats might be good candidates for tsetse eradication projects, but feasibility studies similar to those described in the present paper will be necessary to accurately assess their boundaries and confirm their isolated status with respect to neighbouring populations. The method could be used to prioritize intervention areas elsewhere in Africa within the PATTEC initiative and is applicable to the strategic management of other vector and pest species.

4. ACKNOWLEDGMENTS

We would like to thank the staff of the following organizations for their technical and/or financial support during the implementation of this project: the Ministère de l'Elevage et des Productions Animales, the Ministère de l'Agriculture et de l'Equipement Rural, the Ministère de l'Environnement et du Développement Durable, the ANACIM (Agence Nationale de l'Aviation Civile et de la Météorologie, the military airstrip of Thiès, the CIRAD, the IRD, the FAO/IAEA Insect Pest Control Laboratory, the FAO, the SAS, the CIRDES, the PATTEC office in Burkina Faso, the IAEA's Department of Technical Cooperation, the Mubarqui group in Mexico, l'Ecole Senegal-Japon, and the US Department of State.

This publication is dedicated to the memory of Mr Alphonse Manga of the Ministry of Livestock and Animal Production, Senegal and Mr Idrissa Kaboré, CIRDES, Burkina Faso.

5. REFERENCES

Barclay, H. J., and J. W. Hargrove. 2005. Probability models to facilitate a declaration of pest-free status, with special reference to tsetse (Diptera: Glossinidae). Bulletin Entomological Research 95: 111.

Barclay, H. J., J. W. Hargrove, A. Meats, and A. Clift. 2021. Procedures for declaring pest free status, pp. 923–948. *In* V. A. Dyck, J. Hendrichs, and A. S. Robinson (eds.), Sterile Insect Technique – Principles and practice in Area-Wide Integrated Pest Management. Second Edition. CRC Press, Boca Raton, Florida, USA.

Bauer, B., S. Amsler-Delafosse, P. Clausen, I. Kabore, and J. Petrich-Bauer. 1995. Successful application of deltamethrin pour-on to cattle in a campaign against tsetse flies (*Glossina* spp.) in the pastoral zone of Samorogouan, Burkina Faso. Tropical Medicine and Parasitology 46: 183–189.

Bouyer, F., M. T. Seck, A. Dicko, B. Sall, M. Lo, M. Vreysen, E. Chia, J. Bouyer, and A. Wane. 2014. Ex-ante cost-benefit analysis of tsetse eradication in the Niayes area of Senegal. PLoS Neglected Tropical Diseases 8: e3112.

Bouyer, F., J. Bouyer, M. T. Seck, B. Sall, A. H. Dicko, R. Lancelot, and E. Chia. 2015a. Importance of vector-borne infections in different production systems: Bovine trypanosomosis and the innovation dynamics of livestock producers in Senegal. Revue Scientifique et Technique (International Office of Epizootics) 34: 199–212.

Bouyer, J., L. Guerrini, M. Desquesnes, S. de la Rocque, and D. Cuisance. 2006. Mapping African animal trypanosomosis risk from the sky. Veterinary Research 37: 633–645.

Bouyer, J., Y. Sana, Y. Samandoulgou, J. César, L. Guerrini, C. Kabore-Zoungrana, and D. Dulieu. 2007. Identification of ecological indicators for monitoring ecosystem health in the trans-boundary W Regional park: A pilot study. Biological Conservation 138: 73–88.

Bouyer, J., M. T. Seck, B. Sall, L. Guerrini, and M. J. B. Vreysen. 2010. Stratified entomological sampling in preparation of an Area-Wide Integrated Pest Management project: The example of *Glossina palpalis gambiensis* in the Niayes of Senegal. Journal of Medical Entomology 47(4): 543–552.

Bouyer, J., M. T. Seck, S. Pagabeleguem, B. Sall, M. Lo, M. J. B. Vreysen, T. Balenghien, and R. Lancelot. 2012. Study of the competitiveness of allochtonous sterile males during the tsetse eradication campaign in Senegal. *In* European Society of Vector Ecology (ed.), 18th Conférence E-SOVE 2012. EID, CIRAD, IRD, Montpellier, France.

Bouyer, J., A. H. Dicko, G. Cecchi, S. Ravel, L. Guerrini, P. Solano, M. J. B. Vreysen, T. De Meeûs, and R. Lancelot. 2015b. Mapping landscape friction to locate isolated tsetse populations candidate for elimination. Proceedings National Academy of Sciences of the USA 112: 14575–14580.

Brunhes, J., D. Cuisance, B. Geoffroy, and J.-P. Hervy. 1998. Les glossines ou mouches tsé-tsé. CIRAD/ORSTOM, Montpellier, France.

Budde, M. E., G. Tappan, J. Rowland, J. Lewis, and L. L. Tieszen. 2004. Assessing land cover performance in Senegal, West Africa using 1-km integrated NDVI and local variance analysis. Journal Arid Environments 59: 481–498.

Camara, M., H. Harling Caro-Riaño, S. Ravel, J.-P. Dujardin, J.-P. Hervouet, T. de Meeüs, M. S. Kagbadouno, J. Bouyer, and P. Solano. 2006. Genetic and morphometric evidence for isolation of a tsetse (Diptera: Glossinidae) population (Loos islands, Guinea). Journal of Medical Entomology 43: 853–860.

Ciss, M., M. D. Bassène, M. T. Seck, B. Sall, A. G. Fall, M. J. B. Vreysen, and J. Bouyer. 2019. Environmental impact of tsetse eradication in Senegal. Scientific Reports 9:20313.

Desquesnes, M., and M. L. Dia. 2003. *Trypanosoma vivax:* Mechanical transmission in cattle by one of the most common African tabanids, *Atylotus agrestis.* Experimental Parasitology 103: 35–43.

Desquesnes, M., F. Biteau-Coroller, J. Bouyer, M. L. Dia, and L. D. Foil. 2009. Development of a mathematical model for mechanical transmission of trypanosomes and other pathogens of cattle transmitted by tabanids. International Journal of Parasitology 39: 333–346.

Devaux-Spatarakis, A., D. Barret, J. Bouyer, C. Cerdan, M.-H. Dabat, G. Faure, T. Ferré, E. Hainzelin, I. Medah, L. Temple, and B. Triomphe. 2016. How can international agricultural research better contribute to innovations' impacts: Lessons from outcomes analysis. *In* Social and technological transformation of farming systems: Diverging and converging pathways. 14 pp. European IFSA Symposium, July 2016, Newport, UK.

Dicko, A. H., R. Lancelot, M. T. Seck, L. Guerrini, B. Sall, M. Lo, M. J. B. Vreysen, T. Lefrançois, F. Williams, S. L. Peck, and J. Bouyer. 2014. Using species distribution models to optimize vector control: The tsetse eradication campaign in Senegal. Proceedings of the National Academy of Sciences of the USA 111: 10149–10154.

Dyck, V. A., J. Hendrichs, and A. S. Robinson (eds.). 2021. Sterile Insect Technique – Principles and practice in Area-Wide Integrated Pest Management. Second Edition. CRC Press, Boca Raton, Florida, USA. 1200 pp.

Enkerlin, W., J. M. Gutiérrez-Ruelas, A. V. Cortes, E. C. Roldan, D. Midgarden, E. Lira, J. L. Z. López, J. Hendrichs, P. Liedo, and F. J. T. Arriaga. 2015. Area freedom in Mexico from Mediterranean fruit fly (Diptera: Tephritidae): A review of over 30 years of a successful containment program using an integrated area-wide SIT Approach. Florida Entomologist 98: 665–681.

EXPO Milano 2015. Eradicating the tsetse fly to save farms in Senegal. Five questions for the Directorate of Veterinary Services. Removing the tsetse fly results in tripling milk and meat sales in Senegal.

Feldmann, H. U., S. Leak, and J. Hendrichs. 2018. Assessing the feasibility of creating tsetse and trypanosomosis-free zones. International Journal of Tropical Insect Science 38: 77–92.

Feldmann, U., V. A. Dyck, R. C. Mattioli, J. J. Jannin, and M. J. B. Vreysen. 2021. Impact of tsetse fly eradication programmes using the Sterile Insect Technique, pp. 1051–1080. *In* V. A. Dyck, J. Hendrichs, and A. S. Robinson (eds.), Sterile Insect Technique – Principles and practice in Area-Wide Integrated Pest Management. Second Edition. CRC Press, Boca Raton, Florida, USA.

Fried, M. 1971. Determination of sterile-insect competitiveness. Journal of Economic Entomology 64: 869–872.

Guerrini, L., J. P. Bord, E. Ducheyne, and J. Bouyer. 2008. Fragmentation analysis for prediction of suitable habitat for vectors: The example of riverine tsetse flies in Burkina Faso. Journal of Medical Entomology 45: 1180–1186.

Hendrichs, J., M. J. B. Vreysen, W. R. Enkerlin, and J. P. Cayol. 2021. Strategic options in using sterile insects for Area-Wide Integrated Pest Management, pp. 841–884. *In* V. A. Dyck, J. Hendrichs, and A. S. Robinson (eds.), Sterile Insect Technique – Principles and practice in Area-Wide Integrated Pest Management. Second Edition. CRC Press, Boca Raton, Florida, USA.

Itard, J., D. Cuisance, and G. Tacher. 2003. Trypanosomoses: Historique - répartition géographique, pp. 1607-1615. *In* P.-C. Lefèvre, J. Blancou, and R. Chermette (eds.), Principales maladies infectieuses et parasitaires du bétail. Europe et Régions Chaudes, Vol. 2. Lavoisier, Paris, France.

Knipling, E. F. 1955. Possibilities of insect population control through the use of sexually sterile males. Journal of Economic Entomology 48: 443–448.

Laveissière, C., and P. Grébaut. 1990. Recherches sur les pièges à glossines (Diptera, Glossinidae). Mise au point d'un modèle économique: Le piège "Vavoua". Tropical Medicine and Parasitology 41: 185–192.

Laveissière, C., D. Couret, and T. Traoré. 1985. Tests d'efficacité et de rémanence d'insecticides utilisés en imprégnation sur tissus pour la lutte par piégeage contre les glossines. 1. Protocole experimental, l'effet "knock-down" des pyréthrynoïdes. Cah. ORSTOM Sér. Ent. Méd. et Parasitol. 23: 61–67.

Leak, S. G. A., D. Ejigu, and M. J. B. Vreysen. 2008. Collection of entomological baseline data for tsetse Area-Wide Integrated Pest Management projects. FAO Animal Production and Health Guidelines, Food and Agriculture Organization of the United Nations, Rome, Italy. 215 pp.

Morel, P. C., and S. Touré. 1967. *Glossina palpalis gambiensis* Vanderplank 1949 (Diptera) dans la région des Niayes et sur la Petite Côte (République du Sénégal). Revue d'Elevage et de Médecine Véterinaire des Pays Tropicaux 20: 571–578.

Mubarqui, R. L., R. C. Perez, R. Angulo Kladt, J. L. Zavala Lopez, A. Parker, M. T. Seck, B. Sall, and J. Bouyer. 2014. The smart aerial release machine, a universal system for applying the Sterile Insect Technique. PLoS One 9: e103077.

Mutika, G. N., I. Kabore, A. G. Parker, and M. J. B. Vreysen. 2014. Storage of male *Glossina palpalis gambiensis* pupae at low temperature: Effect on emergence, mating and survival. Parasites & Vectors 7: 465.

Mutika, G. N., I. Kabore, M. T. Seck, B. Sall, J. Bouyer, A. G. Parker, and M. J. B. Vreysen. 2013. Mating performance of *Glossina palpalis gambiensis* strains from Burkina Faso, Mali and Senegal. Entomologia Experimentalis et Applicata 146: 177–185.

Pagabeleguem, S., M. T. Seck, B. Sall, M. J. B. Vreysen, G. Gimonneau, A. G. Fall, M. Bassene, I. Sidibé, J. B. Rayaisse, A. Belem, and J. Bouyer. 2015. Long distance transport of irradiated male *Glossina palpalis gambiensis* pupae and its impact on sterile male yield. Parasites & Vectors 8: 259.

Pagabeleguem, S., G. Gimonneau, M. T. Seck, M. J. B. Vreysen, B. Sall, J.-B. Rayaissé, I. Sidibé, J. Bouyer, and S. Ravel. 2016a. A molecular method to discriminate between sterile and wild tsetse flies during eradication projects that have a Sterile Insect Technique component. PLoS Neglected Tropical Diseases 10: e0004491.

Pagabeleguem, S., S. Ravel, A. H. Dicko, M. J. B. Vreysen, A. Parker, P. Taback, K. Huber, G. Gimonneau, and J. Bouyer. 2016b. The influence of temperature and relative humidity on survival and fecundity of three *Glossina palpalis gambiensis* strains. Parasites & Vectors 9: 520.

(PATTEC) Pan African Tsetse and Trypanosomosis Eradication Campaign. 2019.

Pinchbeck, G. L., L. J. Morrison, A. Tait, J. Langford, L. Meehan, S. Jallow, J. Jallow, A. Jallow, and R. M. Christley. 2008. Trypanosomosis in the Gambia: Prevalence in working horses and donkeys detected by whole genome amplification and PCR, and evidence for interactions between trypanosome species. BMC Veterinary Research 4: 7.

Politzar, H., and D. Cuisance. 1984. An integrated campaign against riverine tsetse *flies Glossina palpalis gambiensis* and *Glossina tachinoides* by trapping and the release of sterile males. Insect Science and its Application 5: 439–442.

Projet de Lutte contre la Mouche Tsé-tsé dans le Niayes. 2019. Statistiques. Dakar, Senegal.

Seck, M. T., J. Bouyer, B. Sall, Z. Bengaly, and M. J. B. Vreysen. 2010. The prevalence of African animal trypanosomoses and tsetse presence in Western Senegal. Parasite 17: 257–265.

Seck, M. T., S. Pagabeleguem, M. D. Bassene, A. G. Fall, T. A. R. Diouf, B. Sall, M. J. B. Vreysen, J.-B. Rayaissé, P. Takac, I. Sidibé, A. G. Parker, G. N. Mutika, J. Bouyer, and G. Gimonneau. 2015. Quality of sterile male tsetse after long distance transport as chilled, irradiated pupae. PLoS Neglected Tropical Diseases 9: e0004229.

Solano, P., D. Kaba, S. Ravel, N. Dyer, B. Sall, M. J. B. Vreysen, M. T. Seck, H. Darbyshir, L. Gardes, M. J. Donnelly, T. de Meeûs, and J. Bouyer. 2010. Tsetse population genetics as a tool to choose between suppression and elimination: The case of the Niayes area in Senegal. PLoS Neglected Tropical Diseases 4: e692.

Somda, M. B., Z. Bengaly, E. Dama, A. Poinsignon, G.-K. Dayo, I. Sidibé, F. Remoué, A. Sanon, and B. Bucheton. 2013. First insights into the cattle serological response to tsetse salivary antigens: A promising direct biomarker of exposure to tsetse bites. Veterinary Parasitology 197: 332–340.

Somda, M. B., S. Cornelie, Z. Bengaly, F. Mathieu-Daudé, A. Poinsignon, E. Dama, J. Bouyer, I. Sidibé, E. Demettre, and M. Seveno. 2016. Identification of a $Tsal1_{52-75}$ salivary synthetic peptide to monitor cattle exposure to tsetse flies. Parasites & Vectors 9: 149.

Sow, A., I. Sidibé, Z. Bengaly, Z. Bancé, G. J. Sawadogo, P. Solano, M. J. B. Vreysen, R. Lancelot, and J. Bouyer. 2012. Irradiated male *Glossina palpalis gambiensis* (Diptera: Glossinidae) from a 40-years old colony are still competitive in a riparian forest in Burkina Faso. PLoS One 7: e37124.

Touré, S. 1971. Les glossines (Diptera, Glossinidae) du Sénégal: Ecologie, répartition géographique et incidence sur les trypanosomoses. Revue d'Elevage et de Médecine Véterinaire des Pays Tropicaux 24: 551–563.

Touré, S. 1973. Lutte contre *Glossina palpalis gambiensis* dans la région des Niayes du Sénégal. Revue d'Elevage et de Médecine Véterinaire des Pays Tropicaux 26: 339–347.

Touré, S. 1974. Note sur quelques particularités dans l'habitat de *Glossina palpalis gambiensis* Vanderplank, 1949 (Diptera, Glossinidae) observées au Sénégal. Revue d'Elevage et de Médecine Véterinaire des Pays Tropicaux 27: 81–94.

Van der Vloedt, A. M. V., and H. Barnor. 1984. Effects of ionizing radiation on tsetse biology. Their relevance to entomological monitoring during integrated control projects using the Sterile Insect Technique. International Journal of Tropical Insect Science 5: 431–437.

Vreysen, M. J. B. 1995. Radiation induced sterility to control tsetse flies: The effect of ionising radiation and hybridisation on tsetse biology and the use of the Sterile Insect Technique in integrated tsetse control. PhD thesis, Landbouwuniversiteit te Wageningen, The Netherlands.

Vreysen, M. J. B. 2021. Monitoring sterile and wild insects in Area-Wide Integrated Pest Management programmes, pp. 485–528. *In* V. A. Dyck, J. Hendrichs, and A. S. Robinson (eds.), Sterile Insect Technique – Principles and practice in Area-Wide Integrated Pest Management. Second Edition. CRC Press, Boca Raton, Florida, USA.

Vreysen, M., and A. S. Robinson. 2011. Ionising radiation and area-wide management of insect pests to promote sustainable agriculture. A review. Agronomy for Sustainable Development 2: 671–692.

Vreysen, M. J. B., A. M. V. Van der Vloedt, and H. Barnor. 1996. Comparative gamma radiation sensitivity of *G. tachinoides* Westw., *G. f. fuscipes* Newst., and *G. brevipalpis* Newst. International Journal of Radiation Biology 69: 67–74.

Vreysen, M., A. S. Robinson, and J. Hendrichs (eds.). 2007. Area-wide control of insect pests. From research to field implementation. Springer, Dordrecht, The Netherlands. 789 pp.

Vreysen, M. J. B., M. T. Seck, B. Sall, and J. Bouyer. 2013. Tsetse flies: Their biology and control using Area-Wide Integrated Pest Management approaches. Journal of Invertebrate Pathology 112: S15–S25.

Vreysen, M. J. B., K. M. Saleh, M. Y. Ali, A. M. Abdulla, Z.-R. Zhu, K. G. Juma, V. A. Dyck, A. R. Msangi, P. A. Mkonyi, and H. U. Feldmann. 2000. *Glossina austeni* (Diptera: Glossinidae) eradicated on the island of Unguja, Zanzibar, using the Sterile Insect Technique. Journal of Economic Entomology 93: 123–135.

Vreysen, M. J. B., K. Saleh, F. Mramba, A. Parker, U. Feldmann, V. A. Dyck, A. Msangi, and J. Bouyer. 2014. Sterile insects to enhance agricultural development: The case of sustainable tsetse eradication on Unguja island, Zanzibar using an Area-Wide Integrated Pest Management approach. PLoS Neglected Tropical Diseases 8: e2857.

Wikipedia. 2019. Autogyro. Principle of operation.

PHYLOGEOGRAPHY AND INSECTICIDE RESISTANCE OF THE NEW WORLD SCREWWORM FLY IN SOUTH AMERICA AND THE CARIBBEAN

L. W. BERGAMO[1,2], P. FRESIA[3] AND A. M. L. AZEREDO-ESPIN[1,2]

[1]Department of Genetics, Evolution, Microbiology and Immunology, Institute of Biology, University of Campinas (UNICAMP), Campinas, SP, Brazil; amlazeredo@gmail.com
[2]Center of Molecular Biology and Genetic Engineering, University of Campinas (UNICAMP), Campinas, SP, Brazil
[3]Pasteur Institute, Montevideo, Uruguay

SUMMARY

Insect pests have a widespread negative impact on livestock production, resulting in large economic losses. Monitoring and surveillance of pest species are fundamental to manage their populations and reduce the damage they inflict on livestock. In addition, resistance to pest control methods, such as the use of insecticides, is becoming an increasingly important issue. Inferring population structure, the phylogeographic pattern of pest species, and the connectivity among populations is key to understanding migration patterns, which can be used to delineate area-wide pest surveillance and management schemes such as the Sterile Insect Technique (SIT). This review provides a summary of phylogeographic patterns of the New World screwworm (NWS) fly, *Cochliomyia hominivorax* Coquerel, a myiasis-causing fly that leads to significant losses in livestock production, based on molecular markers and the monitoring of insecticide resistance to improve its management. The species' current geographic distribution comprises most of the Neotropical region, having been eradicated in North and Central America after area-wide integration of the SIT with other methods. Introducing similar management programmes in South America and the Caribbean could be a strategic alternative to the permanent and exclusive use of insecticides, which has a negative environmental impact and is a growing challenge because of increasing resistance development in NWS. Such an area-wide approach requires NWS population delineation at regional and geographic scales, and the monitoring of mutations that are involved in insecticide resistance in natural populations.

Key Words: Cochliomyia hominivorax, myiasis, livestock, phylogeographic patterns, carboxylesterase, management unit, microsatellites, mitochondrial DNA, molecular markers, population structure, Neotropical region, Caribbean, South America

J. Hendrichs, R. Pereira and M. J. B. Vreysen (eds.), Area-Wide Integrated Pest Management: Development and Field Application, pp. 305–317. CRC Press, Boca Raton, Florida, USA.

1. INTRODUCTION

Successful eradication of the New World screwworm (NWS) fly *Cochliomyia hominivorax* Coquerel from North and Central America, using an area-wide integrated pest management (AW-IPM) (Klassen and Vreysen 2021) approach that included a Sterile Insect Technique (SIT) component, has triggered discussions about its potential eradication in the Caribbean and South America (Vargas-Terán et al. 2021). However, the high livestock density and wildlife distribution in the NWS fly's current habitat area, with the geographical and environmental settings including large rainforests, wetlands, and huge grasslands, make area-wide management and eventual eradication a great challenge.

The efficient area-wide management of a pest requires the control of all its target populations in a delimited geographic region, requiring a minimum area sufficiently large to guarantee that natural dispersion only occurs inside it (Klassen and Vreysen 2021). E. F. Knipling (1972) showed that the survival of a small remnant fraction of the population (i.e. 1% of the original population) is enough for it to recover to a density capable of causing economic damages in a few generations.

In this sense, the delimitation of adequate target regions and geographic scales is extremely important as well the understanding of gene flow pattern among populations (Tabachnick and Black 1995). Several studies, reviewed below, have aimed to characterize the structure of NWS fly populations and infer gene flow patterns at different geographic scales, from local to continental, providing a basis for distinct hypotheses about the distribution of genetic variability and its possible effects on control strategies.

Another important requisite for the effective application of the SIT is a low density of the target field populations (Knipling 1979). Due to the relatively high density of NWS populations in some local situations (Krafsur et al. 1979), complementary actions need to be taken to ensure their reduction prior to the release of sterile insects. Wound and myiasis treatment, which relies on the application of insecticides (e.g. organophosphates and pyrethroids), is the standard method to reduce NWS fly populations in the first step of a management programme (reviewed in Mangan and Bouyer 2021; Vargas-Terán et al. 2021). However, chemical treatment will not succeed if populations are resistant to the used compounds.

Thus, studies that aimed to discover the main genes involved in NWS fly insecticide resistance and to monitor the frequencies of mutations in the genes associated with this resistance in natural populations are also reviewed here.

2. POPULATION GENETICS AND PHYLOGEOGRAPHY

Over the last three decades, technological advances in molecular biology have led to the introduction of many types of molecular markers to assay genetic variation. Accompanying these advances, the genetic variability and structure of NWS fly natural populations in South America and the Caribbean region have been extensively studied and characterized (see Table 1 for a summary).

2.1. NWS Population Genetic Studies from South America and Caribbean Region

Restriction fragment length polymorphism of mitochondrial DNA (mtDNA RFLP) was the first method used and a seasonal analysis of a single population from Brazil (Caraguatatuba, São Paulo) indicated a high genetic heterogeneity for some restriction sites over time, with seven haplotypes exclusively found during summer and fall (Azeredo-Espin 1993). A study of four other populations from the same state, São Paulo, showed 15 haplotypes, with a small number of haplotypes widely distributed and a large number that appeared to be local (Infante-Vargas and Azeredo-Espin 1995). Similarly, Infante-Malachias et al. (1999) explored the nuclear genome with Random Amplified Polymorphic DNA (RAPD) markers, and detected moderate genetic differentiation among 6 populations from south-eastern Brazil and one from northern Argentina.

Table 1. NWS population genetic studies from South America and Caribbean region[*]

Reference	Region	Marker	Var.	F_{ST}
Azeredo-Espin (1993)	South-eastern BR	RFLP	-	-
Infante-Vargas and Azeredo-Espin (1995)	South-eastern BR	RFLP	H	-
Infante-Malachias et al. (1999)	Northern AR, south-eastern BR	RAPD	-	0.122
Taylor et al. (1996)	CB, CR, DR, JM, TT, southern BR	PCR-RFLP	H	-
Lyra et al. (2005)	UY	PCR-RFLP	H	0.145[**]
Torres et al. (2007)	UY	SSR	M	0.031
Torres and Azeredo-Espin (2009)	CB, DR, JM, TT	SSR	M/H	0.157
Griffiths et al. (2009)	BR, JM, TT, UY	SSR	M/H	-
McDonagh et al. (2009)	BR, CB, CO, DR, EC, JM, PE, TT, USA, UY, VE	mtDNA, Nuc	-	-
Lyra et al. (2009)	BR, CB, CO, DR, EC, JM, PY, TT, UY, VE	PCR-RFLP	L/H	0.130
Fresia et al. (2011)	AR and Lyra et al. (2009)	mtDNA	H	0.496
Fresia et al. (2013)	BL, CR, MX, US and Fresia et al. (2011)	mtDNA	-	0.155-0.718
Mastrangelo et al. (2014)	Amazon Basin BR	mtDNA, SSR	H	0.24(mtDNA) 0.099(SSR)
Fresia et al. (2014)	Fresia et al. (2011) and Mastrangelo et al. (2014)	mtDNA	-	-

[*] *Var., variability; H, high; M, moderate; L, low;* F_{ST}, *fixation index; SSR, microsatellites; mtDNA, mitochondrial DNA; Nuc, nuclear marker; AR, Argentina; BL, Belize; BR, Brazil; CO, Colombia; CB, Cuba; CR, Costa Rica; DR, Dominican Republic; EC, Ecuador; JM, Jamaica; MX, Mexico; PY, Paraguay; PE, Peru; TT, Trinidad and Tobago; UY, Uruguay; USA, United States of America; VE, Venezuela.*
[**] *Value not statistically significant*

Subsequently, polymerase chain reaction-restriction fragment length polymorphism (PCR-RFLP) analysis was used to characterize mtDNA variation in Caribbean, Central, and South American NWS fly populations (Taylor et al. 1996). Fourteen mtDNA haplotypes were observed among 18 flies, indicating high variability. These haplotypes, based on phenetic analysis, were divided into three discontinuous assemblages: "North and Central America", "South America", and "Jamaica". Notably, the Cuban sample seemed to be more closely related to Central American populations, while Dominican Republic samples were grouped with those from South America, suggesting a scenario of multiple origins of the NWS fly throughout the Caribbean.

The mtDNA variation was also investigated by PCR-RFLP in seven populations from Uruguay (Lyra et al. 2005). High genetic variability and no evidence of subpopulation differentiation were observed, indicating the existence of a single panmictic population. This lack of differentiation was attributed to the absence of geographic and/or climatic barriers and to the fact that Uruguay is almost at the southern extreme of the species' distribution. These same populations from Uruguay were also investigated by Torres et al. (2007) using nuclear microsatellites. A moderate degree of polymorphism and an excess of observed homozygosity were found, which could have been caused by demographic changes in response to the decrease in temperature and humidity in the Uruguayan winter and/or persistent insecticide treatment. It is likely that the low population differentiation was caused by passive migration of larvae through the movement of infested animals, as well as by recent recolonization events.

Microsatellite markers were also used to investigate ten populations from four Caribbean islands (Torres and Azeredo-Espin 2009) and, contrary to expectations, the level of genetic variability of some Caribbean populations was not lower than that of continental samples. In fact, moderate to high levels of genetic variability and a high level of population differentiation were found, even among populations within the same island.

Despite small sample sizes, an analysis of nine populations from South America and the Caribbean islands found microsatellite differences between Jamaica and Trinidad and Tobago, and in relation to the mainland (Griffiths et al. 2009). Population structure in mainland South America was more difficult to describe, but some weak signals of structure were detected, suggesting that population differentiation may exist between NWS flies from at least some areas.

McDonagh et al. (2009), utilizing the sequences of two mitochondrial (COI and 12S) and one nuclear (EF1α) gene investigated the phylogenetic relationship of NWS fly populations from the Caribbean, South America, and Texas ("historical" North American samples). This study found that NWS fly populations of the Caribbean islands were structured and suggested a period of isolation and/or founder effects following colonization from South America. The data did not support a North American origin of the Cuban NWS population, as previously hypothesized by Taylor et al. (1996). The NWS samples from Texas were in a different lineage as compared with South American and Caribbean samples, indicating a possible north-south division.

Lyra et al. (2009) conducted the first study on a continental-scale that encompassed NWS fly populations covering its entire distribution area. Thirty-four populations from 10 South American and Caribbean countries were analysed using mitochondrial PCR-RFLP. Population structure with significant fixation indices and low variability were found in the Caribbean, indicating that island populations have been evolving independently due to geographic isolation, but are connected by restricted gene flow. In contrast, mainland populations presented high genetic variability and low differentiation, with no correlation of genetic and geographic distances. The moderate and non-homogeneous level of genetic differentiation of the NWS fly in its current distribution area, as well as its high genetic variability, was described as being the product of several historical demographic processes.

In order to highlight and test the results obtained by Lyra et al. (2009), the same NWS fly samples and samples from four other populations were investigated using mtDNA sequences (Fresia et al. 2011). This study found that genetic diversity is distributed in four main groups of populations, corresponding to Cuba (CG), the Dominican Republic (DRG), and North and South Amazon regions (NAG and SAG, respectively). This phylogeographic structure of the NWS populations over its entire range was characterized by distinct historical events:

1. Island colonization from the mainland (a North American and/or Central American colonization was suggested for Cuba, whereas the other Caribbean islands were colonized from South America).
2. Recent separation of NAG and SAG probably associated to a barrier in the Amazon region resulting in separate populations in NAG and SAG.
3. Population expansion that started ca. 20–25 000 years ago and that increased exponentially up to date; it was probably linked with climatic oscillations in the late Pleistocene and resource availability. The population expansion probably caused the low divergence detected within SAG, erasing genetic and geographic correlations even among distant populations (maximum distance of 10 000 km).

In analysing mtDNA sequences from 60 populations (see Fig. 1), a north to south colonization was proposed for the continental Americas (Fresia et al. 2013). According to the best population divergence model chosen by Approximate Bayesian Computation (ABC), a first split occurred between North/Central American and South American populations at the end of the Last Glacial Maximum. A second split occurred between the North and South Amazonian populations in the transition between the Pleistocene and Holocene eras. The NWS fly went through a population expansion during its dispersal toward its current geographic range, with the strongest signals in SAG. This work concluded that climatic oscillations only were not sufficient to explain the phylogeographic patterns observed, and human activity might have played a crucial role in shaping the current distribution of the NWS fly.

The most recent survey of genetic variability was conducted on under-explored NWS populations of the highly important region in Amazonia, in an attempt to better understand the NAG-SAG evolutionary relationships (Mastrangelo et al. 2014). Based on 3 mtDNA genes and 8 microsatellite loci, a high genetic diversity and differentiation was revealed among 9 populations. These Amazonian populations only share mtDNA haplotypes with SAG, suggesting that the NAG-SAG split is the result of a barrier north of the Amazon Basin rather than of the basin environment itself.

Finally, pairwise F_{ST} among South American NWS fly populations were mapped with a geographic information system (GIS) on a friction layer derived from the Maxent niche modelling in order to identify connection corridors between NAG and SAG (Fresia et al. 2014). Despite methodological limitations, it was possible to identify two strong connections between the populations of the NAG and SAG: one along the Atlantic Ocean passing through the northwest of Brazil and the other passing through Peru. The main limitations for this approach are the sampling strategy based mainly on larvae, because it does not capture with precision the adults′ habitat, and the genetic distances estimation based only on mitochondrial DNA sequences.

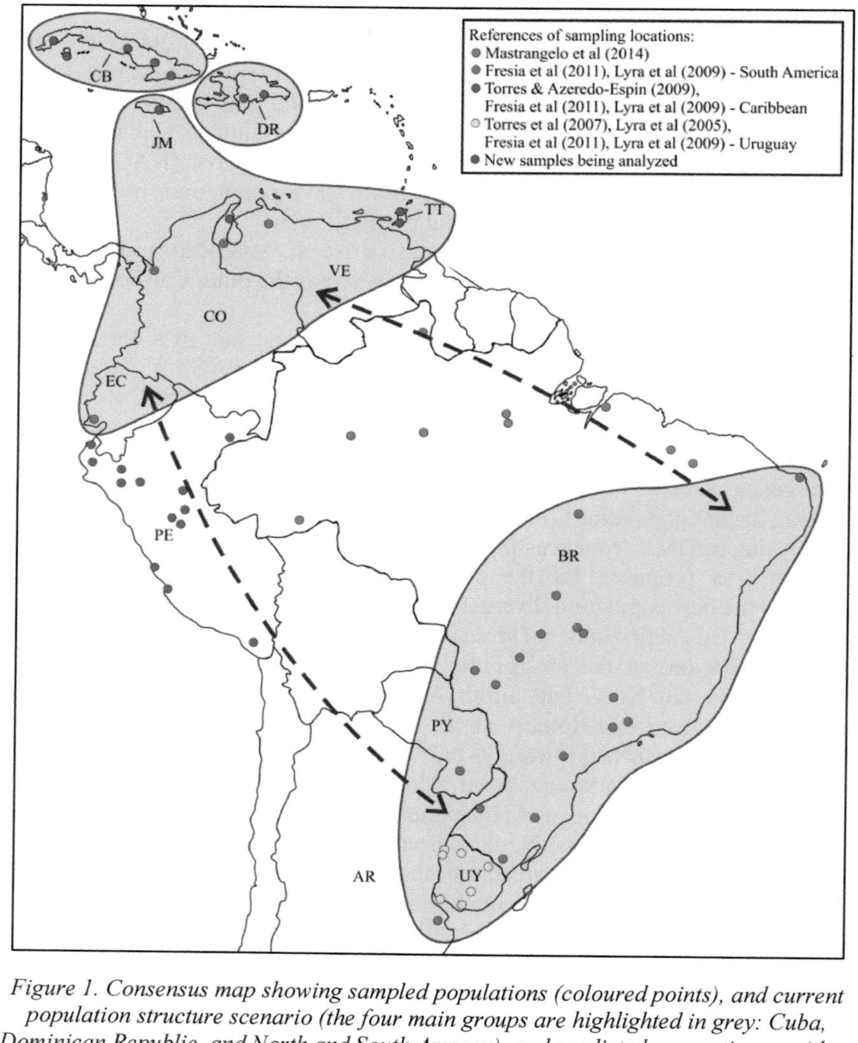

Figure 1. Consensus map showing sampled populations (coloured points), and current population structure scenario (the four main groups are highlighted in grey: Cuba, Dominican Republic, and North and South Amazon), and predicted connection corridors (dashed arrows) for NWS fly populations from the Caribbean and South America.

2.2. Consensus Scenario and Main Conclusions

Synthesizing the results from the previous studies presented above, we established the distribution of genetic diversity and population structure of the NWS populations in the Caribbean and South America (Fig. 1).

Caribbean populations are structured (Taylor et al. 1996; Griffiths et al. 2009; Lyra et al. 2009; McDonagh et al. 2009; Torres and Azeredo-Espin 2009; Fresia et al. 2011) and several events hypothetically resulted in their current distribution, such as Cuba having been originally colonized by North and/or Central American populations and the other Caribbean islands colonized by South American populations (Torres and Azeredo-Espin 2009; Fresia et al. 2011). However, the lack of congruence between nuclear (Torres and Azeredo-Espin 2009) and mtDNA (Lyra et al. 2009) genetic diversity in the Caribbean suggests a complex scenario of population structure.

Unlike the Caribbean populations, South American patterns of genetic variability and structure are not completely clear, but, in general, populations present a high genetic variability and low differentiation with no correlation to geographic distance. There are two distinct genetic groups, NAG and SAG (Fresia et al. 2011), probably separated by a barrier in the north of the Amazon Basin (Mastrangelo et al. 2014) during the transition between the Pleistocene and Holocene eras (Fresia et al. 2013). Populations experienced an expansion during the north-south colonization, mainly SAG, which is probably the cause of its low genetic divergence. All these historical factors and climatic oscillations are important to explain the pattern observed in South America, but current factors may also be influencing it, such as livestock movement and human activity.

Results of NWS phylogeography and population genetics studies can be of relevance to the operation of SIT programmes (Krafsur 1985). However, the significant genetic differences found in these studies do not result in mating incompatibility. Strains from three different locations in Brazil (i.e. Pará state, in the Amazon Basin; Piauí state, in the northeast; and São Paulo state, in the southeast) were crossed and showed no significant differences in all biological parameters assessed and no evidence of hybrid dysgenesis (Mastrangelo et al. 2014). Similarly, the crossing between a Brazilian strain from Goiás state in central Brazil, and the Jamaica-06 wt-strain, which is currently being mass-reared in Panama, did not show any evidence of genetic incompatibility or hybrid dysgenesis (Mastrangelo et al. 2012). The absence of mating incompatibility indicates that sterile males from the Jamaica strain reared in Panama could be used in future SIT-based control programmes throughout Brazil and, possibly, South America. Target management units still need to be determined within the NAG and SAG large geographic distribution area and a better understanding of the distribution of genetic variability in Amazonia is required before considering starting an AW-IPM programme against NWS in these regions. Regional-scale studies in South America were conducted only in Uruguay, a region that coincides with the southern-most distribution of the species, and different degrees of population polymorphism and structure were reported (Lyra et al. 2005; Torres et al. 2007). These differences can be associated with the distinct molecular markers used (Table 1), as they present different modes of inheritance (the effective size of mtDNA populations is one-quarter the size of nuclear DNA ones) and/or mtDNA can present a sex-biased gene flow among the populations.

2.3. Perspectives

In 2009, a study was carried out in a 100 x 60 km area situated at the Brazil-Uruguay border with samples collected during a pilot SIT project against the NWS (Pontes et al. 2009). The high genetic diversity and absence of population structure indicate that the target population limits are certainly larger than the pilot area, and consequently, the management unit should be larger than this pilot project.

Population analyses can be further refined through the use of new and more genetic markers. To reach this goal, we standardized a Genotyping-By-Sequencing (GBS) protocol for the species and the sequencing of the first library, which contains samples from one Uruguayan population, resulting in approximately 1000 filtered single-nucleotide polymorphisms (SNPs).

Another library is being constructed with individuals from the same population that were sampled one year later. After generating these data, we aim to evaluate if the obtained SNPs will give an increased resolution for temporal population genetic analyses in comparison to other molecular markers (mtDNA and microsatellites).

Recently the evolutionary relationships and the phylogeographic structure of populations from the northwest of Brazil and Peru (i.e. the predicted corridor connecting NAG and SAG (Fresia et al. 2014)) were investigated using samples of 13 NWS populations from Peru that were obtained with the assistance of the Servicio Nacional de Sanidad Agraria (SENASA) of Peru (Fig. 1), and three mitochondrial regions (COI, COII, and CR) are being sequenced. Preliminary results suggest the presence of genetically distinct groups with some geographic isolation, high haplotype diversity, low nucleotide diversity, and significant negative values of Tajima's D and Fu's Fs, indicating population expansion.

3. INSECTICIDE RESISTANCE

3.1. Investigation of the Molecular Basis of Resistance Mechanisms

NWS fly management throughout South America is mostly carried out independently on each farm and the farmer decides on the used control strategy. Topical insecticide application on livestock is the most popular and effective suppression method, and two main classes of compounds are used, i.e. organophosphates (OPs) and pyrethroids, which can be applied separately or in combination (Coronado and Kowalski 2009; SINDAN 2010).

A decrease in carboxylesterase activity has been observed in OP resistant strains of some arthropod species (Van Asperen and Oppenoorth 1959; Towsend and Busvine 1969; Hughes and Raftos 1985), that has resulted in the formulation of a mutant ali-esterase hypothesis. This suggests that a structural mutation in a carboxylesterase results in a reduced ability to hydrolyse aliphatic ester substrates, but also in an acquired ability to hydrolyse OP substrates (Claudianos et al. 1999).

In the Australian sheep blowfly, *Lucilia cuprina* (Wiedemann), which belongs to the same family Calliphoridae as the NWS fly, the *LcαE7* gene encodes the ali-esterase E3 isozyme. Biochemical assays with proteins produced by different *LcαE7* alleles showed that an amino acid substitution at position 137 (Gly137Asp) abolished the ali-esterase activity and increased diethyl-OP hydrolase activity, while a second amino acid substitution (Trp251Leu) increased dimethyl-OP hydrolase activity (Campbell et al. 1998). These two amino acid substitutions confer insecticide resistance because they are part of the active site of the enzyme (Newcomb et al. 1997, Campbell et al. 1998).

Based on these previous studies and in view that OPs are commonly used to suppress the NWS fly, the E3 gene in this species (*ChαE7*) was partially characterized (Carvalho et al. 2006, Carvalho et al. 2009). Mutations at the positions responsible for conferring OP resistance in *L. cuprina* (Gly137 and Trp251) were identified, but unlike with *L. cuprina*, NWS fly samples with a mutation in the Trp251 residue showed the substitution of a tryptophan for a serine. It is suggested that this new substitution has the same effect of reducing esterase activity (Taşkin et al. 2004) and may also be involved in pyrethroid resistance and be the molecular basis of cross-resistance between OPs and pyrethroids (Heidari et al. 2005). The strong association between this mutation (Trp251Ser) and dimethyl-OP resistance was later confirmed (Carvalho et al. 2010a).

Population genetic analyses assessed the selective pressures that have shaped carboxylesterase E3 evolution in NWS (Bergamo et al. 2015) and found a negative association between the Gly137Asp and Trp251Ser mutations. Fay & Wu's H value was significantly negative for the exons in which these mutations occur, which suggests that the E3 gene has evolved under positive selection, which is indirect evidence of its role in insecticide resistance.

This association between carboxylesterase E3 mutations and insecticide resistance were not directly proven by bioassays. Only the study involving bioassays by Silva and Azeredo-Espin (2009) indicated a correlation between the Trp251Ser mutation and moderate resistance to the pyrethroid cypermethrin. However, the high conservation of mutations in this gene among dipteran species suggests that the same resistance mechanism could have evolved in the NWS fly. Moreover, mutation-mediated resistance conferred by the E3 gene appears to be the main resistance mechanism selected in this species.

Other mechanisms of insecticide resistance were also investigated for the NWS fly: point mutations in the sodium channel, known as "knockdown resistance" (kdr) (Silva and Azeredo-Espin 2009); point mutations in acetylcholinesterase (AChE) (Carvalho et al. 2010a, Silva et al. 2011); changes in the expression levels of glutathione S-transferases and cytochrome P450 monooxygenases (Carvalho et al. 2010a); and glutamate-gated chloride channels (Lopes et al. 2014). However, no evidence of their association to insecticide resistance was detected.

3.2. Field Monitoring of Mutations in the Carboxylesterase E3 Gene Associated with Organophosphate Insecticide Resistance in South America

In view of the mutations of the carboxylesterase E3 gene that were identified as an important insecticide resistance mechanism in the NWS fly, the characterization of this gene in natural populations of the species throughout its current geographic distribution area can be an important tool for area-wide monitoring of resistance to insecticides. This information can then be used to select and implement more effective pest management programmes.

The Trp251Ser and Gly137Asp mutations were screened in ten NWS fly populations from Brazil, Colombia, Cuba, Paraguay, Uruguay and Venezuela (Silva and Azeredo-Espin 2009; Silva et al. 2011, respectively). Although sample size was small, with only one population from each country (except for Brazil), the Trp251Ser mutation was detected in all populations. In Brazil, allelic frequencies varied from 15.6% to 46.7%. In Cuba, the frequency was 16.7%. In Uruguay, where the use of pyrethroids seems to be common, the frequency was 28.1%, while the highest frequencies were found in Colombia and Venezuela (93.7% and 100%, respectively). The Gly137Asp mutation, however, was not detected in Colombia, Cuba, and Venezuela, although it was present in high frequencies in Brazilian and Uruguayan NWS populations.

The changes in the frequency of both mutations in three different regions of Uruguay in two years (2003 and 2009) were investigated by Carvalho et al. (2010b). The NWS populations of the three regions showed high frequencies of mutated alleles, but whereas the frequency of the Gly137Asp mutation was reduced in 2009 as compared with 2003, the frequency of the Trp251Ser mutation was significantly higher in 2009. This change is probably associated with the current intense use of pyrethroids and dimethyl-OP compounds for NWS fly control in Uruguay.

Analysis of the structure of 21 NWS populations in the SAG area showed three distinct population groups when considering the carboxylesterase E3 gene, with some differences related to both mutation frequencies (Bergamo et al. 2015). Resistant genotypes were observed in high frequencies in all sampled areas, but the frequency of the Trp251Ser and Gly137Asp mutations was higher at lower and higher latitudes.

There is a need for further resistance monitoring studies that would cover the largest possible area of the current distribution of the NWS fly, in addition to studies that would measure changes in temporal frequencies of mutations associated with insecticide resistance. However, the studies presented above clearly indicate that insecticide resistance is widespread throughout studied South American NWS populations.

3.3. Perspectives

Frequencies of both mutations of the E3 gene associated with OP resistance are being monitored in strategic regions of South America that have not been analysed before. The first region of interest is Amazonia, whose NWS populations showed, based on our preliminary results, a considerable frequency of mutant individuals (24% and 16% of the Gly137Asp and Trp251Ser mutations, respectively).

The other important region that is currently being analysed for both E3 mutations is Peru, which is located along a putative connection corridor for the species (Fresia et al. 2014) and consequently can be a key region for the spread of resistance mutations among populations.

4. CONCLUSIONS

Identification of isolated populations or groups of populations is very important to determine target management units for effective AW-IPM programmes of the NWS fly in its current geographic distribution area. Many insights on genetic variability, population structure, and even migration patterns have been obtained, but, except for the Caribbean islands, the identified mainland areas (NAG and SAG regions) are very large and have no identifiable barriers that limit NWS dispersion. The identification of restricted areas and populations within NAG and SAG will be essential for the success of NWS area-wide programmes, both for managing the logistics of implementing the SIT and other suppression methods, and also for the economic implications.

Furthermore, monitoring the spread of insecticide resistance among NWS fly natural populations is equally important, as the effective use of insecticides will be necessary for population suppression activities as part of future area-wide management programmes that integrate the SIT. However, already the current resistance scenario represents a significant challenge.

5. REFERENCES

Azeredo-Espin, A. M. L. 1993. Mitochondrial DNA variability in geographic populations of screwworm fly from Brazil, pp. 161–165. *In* Management of insect pests: Nuclear and related molecular and genetic techniques. International Atomic Energy Agency. Vienna, Austria.

Bergamo, L. W., P. Fresia, and A. M. L. Azeredo-Espin. 2015. Incongruent nuclear and mitochondrial genetic structure of New World screwworm fly populations due to positive selection of mutations associated with dimethyl- and diethyl-organophosphates resistance. PLoS One 10(6): e0128441.

Campbell, P. M., R. D. Newcomb, R. J., Russell, and J. G. Oakeshott. 1998. Two different amino acid substitutions in the ali-esterase, E3, confer alternative types of organophosphorus insecticide resistance in the sheep blowfly, *Lucilia cuprina*. Insect Biochemistry and Molecular Biology 28: 139–150.

Carvalho, R. A., T. T. Torres, and A. M. L. Azeredo-Espin. 2006. A survey of mutations in the *Cochliomyia hominivorax* (Diptera: Calliphoridae) esterase E3 gene associated with organophosphate resistance and the molecular identification of mutant alleles. Veterinary Parasitology 140: 344–351.

Carvalho, R. A., T. T. Torres, M. G. Paniago, and A. M. L. Azeredo-Espin. 2009. Molecular characterization of esterase E3 gene associated with organophosphorus insecticide resistance in the New World screwworm fly, *Cochliomyia hominivorax*. Medical and Veterinary Entomology 23 (Suppl. 1): 86–91.

Carvalho, R. A., A. M. L. Azeredo-Espin, and T. T. Torres. 2010a. Deep sequencing of New World screw-worm transcripts to discover genes involved in insecticide resistance. BMC Genomics 11: 695.

Carvalho, R. A., C. E. G. Limia, C. Bass, and A. M. L. Azeredo-Espin. 2010b. Changes in the frequency of the G137D and W251S mutations in the carboxylesterase E3 gene of *Cochliomyia hominivorax* (Diptera: Calliphoridae) populations from Uruguay. Veterinary Parasitology 170: 297–301.

Claudianos, C., R. J. Russell, and J. G. Oakeshott. 1999. The same amino acid substitution in orthologous esterases confers organophosphate resistance on the house fly and a blowfly. Insect Biochemistry and Molecular Biology 29: 675–686.

Coronado, A., and A. Kowalski. 2009. Current status of the New World screwworm *Cochliomyia hominivorax* in Venezuela. Medical and Veterinary Entomology 23: 106–110.

Fresia, P., M. L. Lyra, A. Coronado, and A. M. L. Azeredo-Espin. 2011. Genetic structure and demographic history of New World screwworm across its current geographic range. Journal of Medical Entomology 48: 280–290.

Fresia, P., A. M. L. Azeredo-Espin, and M. L. Lyra. 2013. The phylogeographic history of the New World screwworm fly, inferred by approximate Bayesian computation analysis. PLoS One 8(10), p.e76168.

Fresia, P., M. Silver, T. Mastrangelo, A. M. L. Azeredo-Espin, and M. L. Lyra. 2014. Applying spatial analysis of genetic and environmental data to predict connection corridors to the New World screwworm populations in South America. Acta Tropica 138 (Suppl.): S34–S41.

Griffiths, A. M., L. M. Evans, and J. R. Stevens. 2009. Characterization and utilization of microsatellite loci in the New World screwworm fly, *Cochliomyia hominivorax*. Medical and Veterinary Entomology 23: 8–13.

Heidari, R., A. L. Devonshire, B. E. Campbell, S. J. Dorrian, J. G. Oakeshott, and R. J. Russell. 2005. Hydrolysis of pyrethroids by carboxylesterases from *Lucilia cuprina* and *Drosophila melanogaster* with active sites modified by in vitro mutagenesis. Insect Biochemistry and Molecular Biology 35: 597–609.

Hughes, P., and D. Raftos. 1985. Genetics of an esterase associated with resistance to organophosphorus insecticides in the sheep blowfly, *Lucilia cuprina* (Wiedemann) (Diptera: Calliphoridae). Bulletin of Entomological Research 75: 535–544.

Infante-Malachias, M. E., K. S. C. Yotoko, and A. M. L. Azeredo-Espin. 1999. Random amplified polymorphic DNA of screwworm fly populations (Diptera: Calliphoridae) from southeastern Brazil and northern Argentina. Genome 42: 772–779.

Infante-Vargas, M. E. I., and A. M. L. Azeredo-Espin. 1995. Genetic variability in mitochondrial DNA of the screwworm, *Cochliomyia hominivorax* (Diptera: Calliphoridae), from Brazil. Biochemical Genetics 33: 237–256.

Klassen, W., and M. J. B. Vreysen. 2021. Area-Wide Integrated Pest Management and the Sterile Insect Technique, pp. 75–112. *In* V. A. Dyck, J. Hendrichs, and A. S. Robinson (eds.), Sterile Insect Technique – Principles and practice in Area-Wide Integrated Pest Management. Second Edition. CRC Press, Boca Raton, Florida, USA.

Knipling, E. F. 1972. Entomology and the management of man's environment. Australian Journal of Entomology 11: 153–167.

Knipling, E. F. 1979. The basic principles of insect population suppression and management, U.S. Department of Agriculture. Washington, DC, USA. 659 pp.

Krafsur, E. S. 1985. Screwworm flies (Diptera: Calliphoridae): Analysis of sterile mating frequencies and covariates.Bulletin of the Entomological Society of America 4: 36–40.

Krafsur, E. S., B. G. Hightower, and L. Leira. 1979. A longitudinal study of screwworm populations, *Cochliomyia hominivorax* (Diptera: Calliphoridae), in northern Veracruz, Mexico. Journal of Medical Entomology 16: 470–481.

Lopes, A. M. M., R. A. Carvalho, and A. M. L. Azeredo-Espin. 2014. Glutamate-gated chloride channel subunit cDNA sequencing of *Cochliomyia hominivorax* (Diptera: Calliphoridae): cDNA variants and polymorphisms. Invertebrate Neuroscience 14: 137–146.

Lyra, M. L., L. B. Klaczko, and A. M. L. Azeredo-Espin. 2009. Complex patterns of genetic variability in populations of the New World screwworm fly revealed by mitochondrial DNA markers. Medical and Veterinary Entomology 23 (Suppl. 1): 32–42.

Lyra, M. L., P. Fresia, S. Gama, J. Cristina, L. B. Klaczko, and A. M. L. Azeredo-Espin. 2005. Analysis of mitochondrial DNA variability and genetic structure in populations of New World screwworm flies (Diptera: Calliphoridae) from Uruguay. Journal of Medical Entomology 42: 589–595.

Mangan, R. L., and J. Bouyer. 2021. Population suppression in support of the Sterile Insect Technique, pp. 549–574. *In* V. A. Dyck, J. Hendrichs, and A. S. Robinson (eds.), Sterile Insect Technique – Principles and practice in Area-Wide Integrated Pest Management. Second Edition. CRC Press, Boca Raton, Florida, USA.

Mastrangelo, T., M. F. Chaudhury, S. R. Skoda, J. B. Welch, A. Sagel, and J. M. M. Walder. 2012. Feasibility of using a Caribbean screwworm for SIT campaigns in Brazil. Journal of Medical Entomology 49: 1495–1501.

Mastrangelo, T., P. Fresia, M. L. Lyra, R. A. Rodrigues, and A. M. L. Azeredo-Espin. 2014. Genetic diversity and population structure of the New World screwworm fly from the Amazon region of Brazil. Acta Tropica 138 (Suppl.): S26–S33.

McDonagh, L., R. García, and J. R. Stevens. 2009. Phylogenetic analysis of New World screwworm fly, *Cochliomyia hominivorax*, suggests genetic isolation of some Caribbean island populations following colonization from South America. Medical and Veterinary Entomology 23: 14–22.

Newcomb, R. D., P. M. Campbell, R. J. Russell, and J. G. Oakeshott. 1997. cDNA cloning, baculovirus-expression and kinetic properties of the esterase, E3, involved in organophosphorus resistance in *Lucilia cuprina*. Insect Biochemistry and Molecular Biology 27: 15–25.

Pontes, J. B., J. E. V. Severo, E. F. C. Garcia, R. Colares, I. Kohek Junior, and M. S. Reverbel. 2009. Projeto demonstrativo de controle e possível erradicação da mosca da bicheira. Hora Veterinária, Porto Alegre 171: 27–30.

Silva, N. M., and A. M. L. Azeredo-Espin. 2009. Investigation of mutations associated with pyrethroid resistance in populations of the New World Screwworm fly, *Cochliomyia hominivorax* (Diptera: Calliphoridae). Genetics and Molecular Research 8: 1067–1078.

Silva, N. M., R. A. Carvalho, and A. M. L. Azeredo-Espin. 2011. Acetylcholinesterase cDNA sequencing and identification of mutations associated with organophosphate resistance in *Cochliomyia hominivorax* (Diptera: Calliphoridae). Veterinary Parasitology 177: 190–195.

(SINDAN) Sindicato Nacional da Indústria de Produtos para Saúde Animal do Brasil. 2010.

Tabachnick, W. J., and W. C. Black. 1995. Making a case for molecular population genetic studies of arthropod vectors. Parasitology Today 11: 27–30.

Taşkin, V., M. Kence, and B. Göçmen. 2004. Determination of malathion and diazinon resistance by sequencing the *MdαE7* gene from Guatemala, Colombia, Manhattan, and Thailand housefly (*Musca domestica* L.) strains. Russian Journal of Genetics 40: 377–380.

Taylor, D. B., A. L. Szalanski, and R. D. Peterson. 1996. Mitochondrial DNA variation in screwworm. Medical and Veterinary Entomology 10: 161–169.

Torres, T. T., and A. M. L. Azeredo-Espin. 2009. Population genetics of New World screwworm from the Caribbean: Insights from microsatellite data. Medical and Veterinary Entomology 23 (Suppl 1): 23–31.

Torres, T. T., M. L. Lyra, P. Fresia, and A. M. L. Azeredo-Espin. 2007. Assessing genetic variation in New World screwworm *Cochliomyia hominivorax* populations from Uruguay, pp. 183–191. *In* M. J. B. Vreysen, A. S. Robinson, and J. Hendrichs (eds.), Area-wide control of insect pests: From research to field implementation. Springer, Dordrecht, The Netherlands.

Towsend, M. G., and J. R. Busvine. 1969. The mechanism of malathion-resistance in the blowfly *Chrysomya putoria*. Entomologia Experimentalis et Applicata 12: 243–267.

Van Asperen, K. and F. J. Oppenoorth. 1959. Organophosphate resistance and esterase activity in house flies. Entomologia Experimentalis et Applicata 2: 48–57.

Vargas-Terán, M., J. P. Spradbery, H. C. Hofmann, and N. E. Tweddle. 2021. Impact of screwworm eradication programmes using the Sterile Insect Technique, pp. 949–978. *In* V. A. Dyck, J. Hendrichs, and A. S. Robinson (eds.), Sterile Insect Technique – Principles and practice in Area-Wide Integrated Pest Management. Second Edition. CRC Press, Boca Raton, Florida, USA.

AREA-WIDE MOSQUITO MANAGEMENT IN LEE COUNTY, FLORIDA, USA

E. W. FOLEY IV, R. L. MORREALE, D. F. HOEL AND A. M. LLOYD

Lee County Mosquito Control District, 15191 Homestead Road, Lehigh Acres, Florida 33971, USA; Hoel@lcmcd.org

SUMMARY

Located in South Florida, the Lee County Mosquito Control District (LCMCD) is the largest single county mosquito abatement programme in the USA based on sheer necessity to combat the extremely high populations of mosquitoes found naturally in the area. South Florida is one of the largest, flattest, wettest, subtropical areas on the planet, making it prime habitat to produce enormous numbers of mosquitoes. LCMCD operates independently an integrated mosquito management (IMM) programme, funded by local taxation, which effectively and responsibly controls mosquitoes minimizing risk to human health, while reducing the environmental footprint. LCMCD incorporates a broad-based approach of control measures ranging from physical or mechanical control, to biological control, larviciding, and adulticiding, as well as mosquito and arbovirus surveillance, public education, and comprehensive evaluation of products and techniques. LCMCD also strives to be at the forefront of advancing technologies, such as the Sterile Insect Technique (SIT) and unmanned aerial systems to assist with the implementation of ongoing suppression efforts. LCMCD continues to be a leader state- and nation-wide with a focus on sound and effective mosquito control for the citizens of Lee County, Florida since 1958.

Key Words: Mosquito abatement programme, control district, integrated mosquito management (IMM), arbovirus surveillance, *Aedes aegypti*, *Aedes albopictus*, *Aedes taeniorhynchus*, *Culex nigripalpus*, *Culex quinquefasciatus*, *Psorophora columbiae*, *Toxorhynchites rutilus rutilus*, *Gambusia holbrooki*

1. LEE COUNTY MOSQUITO CONTROL DISTRICT

Mosquitoes have played a prominent role in Florida's history (Patterson 2004). The discovery that yellow fever, malaria, and dengue fever were mosquito-borne diseases prompted the formation of the Florida State Board of Health in 1889 and the establishment of the Florida Anti-Mosquito Association in 1922, followed shortly by legislation allowing the creation of mosquito control Special Taxing Districts (Connelly and Carlson 2009).

J. Hendrichs, R. Pereira and M. J. B. Vreysen (eds.), Area-Wide Integrated Pest Management: Development and Field Application, pp. 319–338. CRC Press, Boca Raton, Florida, USA.

Lee County Mosquito Control District (LCMCD) was established as an independent taxing district in 1958 by an act of the Florida Legislature, and has been providing mosquito control services to the citizens of Lee County for over sixty years. Additionally, the Lee County Hyacinth Control District was formed by the Florida Legislature in 1961 to serve Lee County in controlling water hyacinth (*Eichhornia crassipes*), water lettuce (*Pistia stratiotes*), both mosquito-breeding plants, and other noxious aquatic weeds impeding navigation in the Caloosahatchee River and within other water bodies located in Lee County.

Both the mosquito and hyacinth control districts are situated at the same physical location and governed by the same seven-member board of commissioners; commissioners are elected to serve a four-year term. Both independent districts collect *ad valorem* taxes needed to perform their respective control activities. LCMCD is governed according to the laws of Florida, Statue Chapter 388 and the rules of the Florida Department of Agricultural and Consumer Service Administration Code 5E/13. Act 98-462, Laws of Florida, is the enabling legislation creating Lee County Hyacinth Control District. The districts are led by a single executive director.

Lee County Mosquito Control District is the largest single county mosquito-abatement district of more than 700 districts and programmes in the USA, of which 66 are in Florida (Challet 1994; McKenna 2016; Kerzee 2019). With an annual budget of ca. USD 24 million, LCMCD has remained at the forefront of mosquito control by helping to develop control technologies that are effective and sensitive to Florida's unique natural habitats and wildlife. Over 97 per cent of Lee County's mosquitoes are controlled by LCMCD, the rest are controlled by the Ft. Myers Beach Mosquito Control District, formed in 1949 by referendum election for the purpose of providing mosquito control for the town of Ft. Myers Beach. The creation of Ft. Myers Beach Mosquito Control District precedes the formation of LCMCD by nine years.

Lee County, Florida is located in the south-eastern USA on the south-western coast of Florida (Fig. 1). Bordered by the Gulf of Mexico on the west, Charlotte County to the north and Collier County to the south. Lee County is known for its popular white sandy beaches and its large estuary habitat at the base of the Caloosahatchee River. With over 56 000 acres (22 662 ha) of salt marsh mangrove habitat and several large, populated barrier islands, Lee County is unique in the scale of mosquito breeding habitats that are in close proximity to urbanized environments.

1.1. Conservation and Land Management Agencies

As concern for conservation increases, a large portion of land in Lee County is protected by various land management agencies, such as the Florida Department of Environmental Protection (FDEP), the Florida Fish and Wildlife Conservation Commission (FFWCC), the U.S. Fish and Wildlife Service (USFWS), and the Environmental Protection Agency (US-EPA) (Connelly and Carlson 2009). LCMCD collaborates with several local, state, and federal land managers to conduct mosquito abatement activities on these lands (Fig. 2). Due to the biodiversity and individual geographic challenges, many of these lands have their own individual management requirements and restrictions pertaining to mosquito abatement (Batzer and Resh 1992).

To better work together towards a common goal, LCMCD holds annual meetings each spring with all land managing agencies to discuss any issues brought forth from the previous year. During this time, future projects and operations are discussed as a way to develop future operations and build working relationships between agencies and LCMCD.

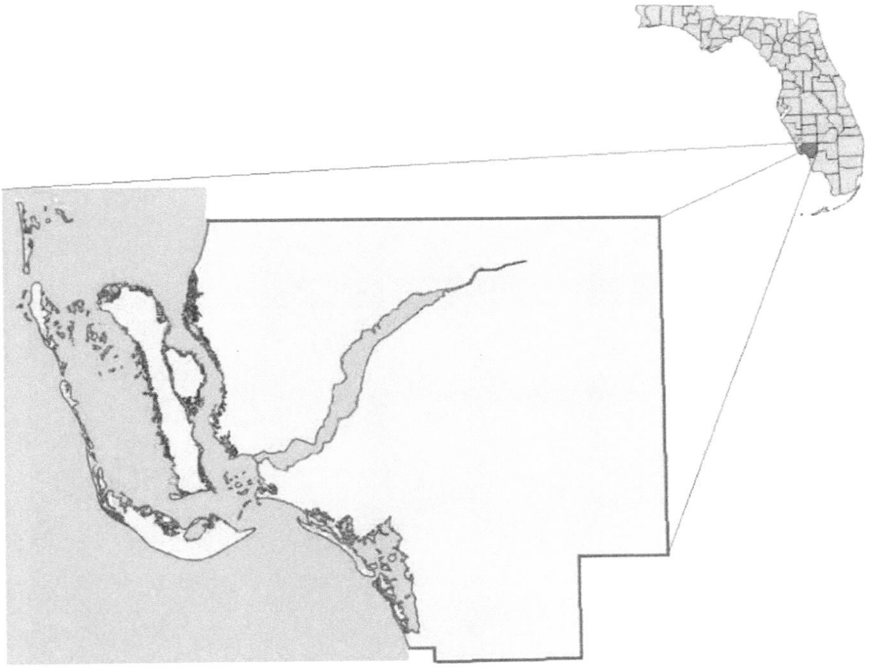

Figure 1. Map showing the location of Lee County in south-western Florida, USA.

1.2. Public Education

As Lee County is one of the counties with the fastest growing human population in the USA, LCMCD dedicates significant resources to educate the public concerning the importance of a strong mosquito abatement programme and why it is needed to live comfortably in south-western Florida. LCMCD believes in strong community engagement and participates in public outreach events throughout the year.

LCMCD also aims to produce a more informed community through a hands-on approach by collaborating with the local Lee County School District and employing a team of licensed educators. LCMCD has developed a unique working relationship with the local school district to fund licensed teachers that offer courses in the school district classrooms across the county teaching mosquito biology and mosquito control essentials to students from kindergarten through high school. All curricula follow the most current standards put forward by the state of Florida and engages students in real world science focused on mosquito control. This instrumental programme gets students excited about science at an early age. LCMCD teachers developed coursework that incorporates biology, ecology, and chemistry with mosquito control, helping the students recognize the science behind what it takes to control mosquitoes. The result is an educated Lee County population, knowledgeable of mosquito control, understanding of why operations occur, the environmental protections in place and judicious use of insecticides.

Figure 2. Map of land managing partners in Lee County, Florida. Red represents FDEP lands, blue represents USFWS lands, and yellow represents FFWCC lands.

1.3. Primary Mosquito Species

The black salt marsh mosquito, *Aedes taeniorhynchus* (Wiedemann) (Agramonte and Connelly 2014), reproduces in extremely high numbers in the 56 000 acres of protected salt marsh within LCMCD (Fig. 3). The aquatic habitat for this species is vast and covers a significant portion of LCMCD's 450 000 acres (1821 km^2). This species is known to oviposit up to 45 000 eggs per square foot (0.1 m^2) (Provost 1969) and is capable of autogeny, the ability to lay an initial batch of eggs without the benefit of a blood meal, believed to be a survival mechanism when hosts are scarce. After developing into adults, *Ae. taeniorhynchus* fly from 20 to 30 miles (32.2 to 48.3 km) in search of a blood meal. It is an aggressive biter and is a major pest along the coastal areas of LCMCD, primary being a nuisance biter and a vector of dog heartworm (*Dirofilaria immitis* Leidy) (Nayar and Connelly 2017), as well as a potential vector for eastern equine encephalitis (EEE) (Agramonte and Connelly 2014).

High tides that flood coastal marsh areas and summer rains cause explosive production of these *Ae. taeniorhynchus* mosquitoes. With a potential of 2 billion eggs per acre (0.4 ha), managing this mosquito on an area-wide basis over the large salt marsh surfaces is of great importance.

Figure 3. Inspector, Sean Christman, searches for Ae. taeniorhynchus *larvae in salt marsh habitat.*

Culex nigripalpus Theobald is a freshwater species with larvae found in roadside ditches, retention/detention ponds, agricultural fields and flooded areas (Day 2017). These sites develop more decomposing organic material later in the rainy season that become increasingly attractive to these mosquitoes. This species is responsible for the transmission of Saint Louis encephalitis and West Nile virus (Day and Curtis 1999). *Cx. nigripalpus* is a major health threat to residents of LCMCD and is a priority for control.

Culex quinquefasciatus Say, the southern house mosquito, is a freshwater species most notable for their association with residential habitats and can be readily found in Lee County (CABI 2019). Similar to *Cx. nigripalpus*, this species can be found in roadside ditches as well as storm drains, containers, and other sites with high organic matter. In contrast to *Cx. nigripalpus*, the larvae of this species are able to survive in waters with higher levels of pollution. *Cx. quinquefasciatus* is the primary vector for St. Louis encephalitis virus throughout the southern USA as well as a potential vector for West Nile virus (Hill and Connelly 2009).

Aedes aegypti (L.), the yellow fever mosquito, is a dusk and dawn biting species, which can also be found biting during the daytime in the shade. This species is closely associated with natural and artificial containers (Zettel and Kaufman 2019). Females have a relatively short flight range (100-500 m) (McDonald 1977; Muir and Kay 1998) and are typically found close to a nearby water source. This species is responsible for the transmission of several disease agents such as yellow fever, dengue, chikungunya and Zika viruses. *Ae. aegypti* is a health threat to residents of LCMCD and is a priority for control.

Psorophora columbiae (Dyar & Knab) is a pestiferous freshwater species (Bibbs et al. 2019). It is found extensively throughout the county in roadside ditches, retention/detention ponds, irrigated agricultural fields, pastures, and low-lying areas that regularly flood, both within and surrounding LCMCD. A floodwater mosquito, it is produced in large numbers as water levels rise and low areas flood during the rainy season and is a major nuisance in the inland areas of LCMCD.

Other nuisance species that sometimes occur in high numbers include *Psorophora ferox* Humboldt, *Ps. ciliata* Fabricius, *Anopheles quadrimaculatus* Say, *An. atropos* Dyar & Knab, *An. cruicians*, *Mansonia titillans* (Walker), *Mn. dyari* Belkin, Heinemann & Page, *Aedes albopictus* (Skuse), and others.

1.4. Weather

Lee County has a subtropical climate, distinguished by warm humid weather year-round, with minimal temperature differences between seasons. Mosquito production in southern Florida is dependent on the presence of standing water throughout the year. Even during the dry winter season, temperatures are rarely sustained low enough to prevent larval development or cause mortality in adults. Seasonal summer rains begin in May or June in south-western Florida and continue through September or October. While the average annual rainfall per year is 53 inches (1.35 m), this amount can be exceptionally variable, especially after hurricanes or tropical storms, and contributes to mosquito production year-round. Rainfall totals and tidal activity are monitored all year with increased monitoring throughout the summer months.

2. SURVEILLANCE

2.1. Population Surveillance

Due to south-western Florida's subtropical climate and mild winters, mosquito surveillance is conducted year-round with increased mosquito collections in the summer months. To address the over 45 mosquito species present in Lee county, an illustration of the nuisance problem being faced, LCMCD deploys multiple trap systems that include eighteen Centers for Disease Control and Prevention (CDC) light traps (Kline 2006; AMCA 2017), seven Biogents BG-Sentinel traps (Regensburg, Germany) (Rose et al. 2006; AMCA 2017), and 6 trap trucks (Fig. 4) to survey 54 pre-determined routes for collecting mosquitoes in flight. CDC and BGS traps are set on a weekly basis, while the trap trucks operate across the county every night from early May through 30 October.

CDC light traps are baited with carbon dioxide in the form of dry ice blocks and set for two trap nights/week. Trap collections are identified the following morning. BGS traps are utilized in urban/suburban areas once per week for *Ae. aegypti* and *Ae. albopictus* surveillance. Along with carbon dioxide, these traps are baited with octenol lures, as well as proprietary BG-lures. BGS traps are very effective in collecting these day-biting mosquitoes.

LCMCD conducts a one-step, Triplex Real-Time PCR (polymerase chain reaction) assay on *Ae. aegypti* and *Ae. albopictus* mosquitoes collected in the weekly BGS trappings. Collected mosquitoes are tested biweekly for the presence of dengue virus, chikungunya virus, and Zika virus.

Figure 4. Trap truck for collecting mosquitoes in flight along pre-determined routes.

Between the months of May and October, LCMCD operates an extensive trap truck programme. An LCMCD trap truck consists of a large conical shaped collection screen (7 feet (2.13 m) wide by 2.5 feet (0.76 m) tall) affixed atop a vehicle (Fig. 4). The trap body measures 11 feet (3.35 m) long and tapers rearward to a 6 by 6-inch (15 by 15 cm) outlet at the rear of the vehicle. At the start of each predetermined route, a collection bag is secured over the collection screen outlets. Routes are driven at a speed of 20 miles (32.2 km) per hour for a three-mile (4.8 km) run and bags collected immediately after finishing. Rainfall gauges are stationed at the beginning and ending locations of each route, providing additional precipitation data important to mosquito production.

2.2. *Arbovirus Surveillance*

LCMCD maintains 16 sentinel chicken coops stationed around the county for the purpose of monitoring arbovirus transmission (Fig. 5). Six birds are kept at each location with blood samples taken once every two weeks and sent to the Florida Department of Health Laboratory in Tampa for analysis.

Figure 5. Sentinel chicken coop on location.

Blood is collected over the course of two days and processed in the LCMCD laboratory prior to shipping to the state laboratory. Half of each processed blood sample is reserved for in-house ELISA testing separate from the state laboratory. Testing samples independently allows for a quicker turnaround time for operational response if a location indicates the presence of an arbovirus. However, the samples sent to the state laboratory are considered the official record of arbovirus detection for the state.

2.3. *Landing Rates*

Landing rates are an effective and quick tool to determine the scale of a mosquito problem in a specific area (Connelly and Carlson 2009; AMCA 2017). Measuring landing rates involves an inspector visiting a citizen complaint location and counting the number of host-seeking adult mosquitoes within a sixty-second period. This relativity simple technique allows for a good understanding of mosquito bite pressure in an area. The landing rate surveillance method allows a single inspector to cover a larger geographic area more efficiently than setting collection traps overnight.

2.4. *Service Request Calls*

Concerned citizens are encouraged to call our office and enter a request for service if they are experiencing a mosquito problem at their residence. LCMCD logs all of the calls into our database and uses them as another form of surveillance. Citizens are also able to enter a request through our website (LCMCD 2019) rather than calling directly if they prefer. By mapping the callers address into our geographic information system-based data management programme, LCMCD is able to use these requests as a way to view problematic areas and dispatch inspectors accordingly.

Inspectors responding to individual calls make every effort to meet with the callers directly and search for mosquito problems on their property. If mosquito breeding is found, the inspectors take the time to educate the homeowner on proactive steps they can take to limit future problems. Inspectors log their findings on laptop computers before moving onto the next site. Service request calls sometimes identify areas needing treatment.

2.5. Field Validation

Field validation at LCMCD is a comprehensive programme designed to evaluate new products and technologies, monitor for the development of insecticide resistance and conduct droplet size characterization on all adulticiding equipment used. The field validation programme also maintains laboratory colonies of four different mosquitoes to include the locally pestiferous or disease vectoring species of *Ae. taeniorhynchus, Cx. quinquefasciatus, Ae. albopictus*, and *Ae. aegypti.*

In addition, the field validation programme maintains colonies of the predatory mosquito *Toxorhynchites rutilus rutilus* (Coquillett) and the predatory mosquito fish *Gambusia holbrooki* Girard that are used for biological control of larval mosquitoes; it also oversees the releases of these predators.

LCMCD conducts bioassays on both larval and adult mosquitoes to evaluate product efficacy in controlling local mosquito populations. Products used to control adult mosquitoes are evaluated using the CDC bottle bioassay protocol (Connelly and Carlson 2009). Products used to control larval mosquitoes are evaluated using a serial dilution larval assay (WHO 2016). Laboratory colony susceptibilities are compared against results of wild mosquitoes to establish a control baseline.

The field validation programme is responsible for conducting droplet characterization on all adulticiding spray systems annually to ensure equipment is in proper working order prior to use. Droplet characterization is conducted using one inch (2.54 cm) and three-millimetre Teflon-coated slides to capture droplets for analysis. Using automated computer software, the slides are analysed under a compound microscope to determine droplet Volume Median Diameter (VMD) (Connelly and Carlson 2009). Droplet sizes must fall within an acceptable range as determined by product label and approved by EPA. Droplet characterization is conducted any time an adulticiding spray system is altered with a minimum of once per year to ensure the equipment is working properly before use.

Additionally, the field validation programme evaluates new products prior to their incorporation into field operations. New product formulations are first evaluated under laboratory conditions to establish the appropriate application rate under ideal conditions. Products with favourable laboratory results are then applied in small-scale field sites for possible operational selection. During these trials, products are monitored for efficacy, duration, and any potential adverse effects to local non-target species. These trials are crucial to determine how a product is going to work under local conditions prior to their implementation as part of the LCMCD treatment programme.

3. BIOLOGICAL CONTROL

Biological control is a vital component of an integrated mosquito management (IMM) programme of any size. Therefore, LCMCD also incorporates biological methods for mosquito control to minimize the use of insecticides. LCMCD accomplishes this by introducing the predatory native mosquito fish (*G. holbrooki*) into mosquito breeding areas (Cassiano et al. 2018) and releasing the predatory mosquito *Tx. rutilus rutilus* (Focks et al. 1980).

In addition, mosquito ditches were installed throughout the 1960s as a form of water management that provides access of natural predators into mosquito breeding habit during times of high tides (Fig. 6). When water levels rise during a high tide, these ditches can introduce juvenile fish species into areas that otherwise would have been inaccessible. As water levels recede these species make their way back to the safety of the ditches.

LCMCD has a mosquito fish programme designed to raise native *G. holbrooki* for release into problematic areas (Fig. 7). In 2019, LCMCD released around one thousand fish into various sites with the goal of natural larval suppression. *Gambusia* is a native freshwater genus of fish that are ferocious predators of mosquito larvae. Often, once this species establishes breeding populations in a body of water, they will suppress mosquito larvae to levels where insecticides are no longer needed.

Figure 6. Network of mosquito ditches on Pine Island, Florida. Darker green foliage shows area where ditches are present.

LCMCD also maintains a colonized population of *Tx. rutilus rutilus* for the purpose of biological control. Native to south-western Florida, these beneficial mosquitoes are predatory on other mosquito larvae and adults do not require a blood meal for reproduction (Focks et al. 1980). By introducing these beneficial mosquitoes into isolated habitats, such as abandoned properties and cemeteries, the goal is to promote the natural suppression of sanguivorous mosquito species.

Figure 7. Biologist, Kara Tyler-Julian, tends to Gambusia *fish rearing tanks.*

4. MOSQUITO BREEDING SOURCE REDUCTION

Source reduction is an important component of an IMM programme. LCMCD inspectors work in the field everyday surveying for larvae. When appropriate, all known larval sources are inspected to determine if the breeding site can be reduced or eliminated before considering chemical and biological treatment methods. This can include, but is not limited to, filling-in tree holes, dumping buckets/containers, removing waste that holds water, drilling holes to drain containers, placing screen covers over rain barrels, etc.

Shortly after LCMCD was established in 1958, crews began work installing mosquito drainage ditches across much of its salt marsh habitat. Digging was accomplished via dragline machinery to depths of five by six feet (1.52 by 1.83 m) wide. By the early 1970s, LCMCD had installed a complex network of canals through much of its problematic coastal areas with the purpose of removing water from the marsh during periods of low tide. By allowing water a place to recede naturally, it limits mosquito breeding habitat and greatly reduces the amount of pesticide needed for control. Although these ditches were installed up to sixty years ago, they continue to function as designed and remain a valuable mosquito control tool in Lee County (Fig. 8).

Figure 8. Mosquito ditch on Pine Island, Florida.

LCMCD continues to implement manual control methods in areas where applicable. Recently, LCMCD collaborated with the National Wildlife Refuge (NWR) of the USFWS to control mosquitoes on a remote island used as an active rookery for several species of shore birds. A depression in the interior of the island would fill with water in the summer months and breed *Ae. taeniorhynchus* mosquitoes. Due to the sensitive nature of nesting birds, getting access to inspect and treat the island was virtually impossible. In 2017, crews from the NWR and LCMCD met at the remote island and hand dug a ditch from the exterior of the island towards the problematic interior. In a couple of hours, a mosquito ditch was installed that drained the stagnant water from the interior of the island. During a high tide, the ditch can introduce natural predators such as fish and other macroinvertebrates into the ecosystem. As the tide retreats, the natural predators leave the island along with the water and mosquito larvae, virtually eliminating mosquitoes naturally.

5. LARVAL MOSQUITO CONTROL

5.1. Ground Larviciding

The ground larviciding programme at LCMCD focuses mainly on inland roadside ditches and residential neighbourhoods. Although some areas are affected by tidal fluctuations, most of ground larviciding is conducted in response to rainfall events for freshwater mosquito species. Ground larviciding crews survey areas of recent rainfall and treat with a variety of methods including a truck-mounted spray system, handheld equipment (backpack unit or squirt bottle) or single use treatment items (water-soluble pouches or briquette-formulated larvicides).

Vehicle-mounted spray systems (Fig. 9) primarily utilize products with the active ingredient temephos for larval control.

Figure 9. Vehicles used for ground larviciding.

Products dispensed with handheld equipment ranges widely depending on the situation, but mostly consists of monomolecular films, larviciding oils, and *Lysinibacillus (Bacillus) sphaericus*. Various formulations of *L. sphaericus*, spinosad, and methoprene products are available for single use treatment items ranging from 30 to 150 days of residual control and are used to treat more permanent bodies of water that will be problematic throughout the season. *Bacillus thuringiensis* is seldom used in ground larviciding operations, but it is commonly used in aerial larviciding operations at LCMCD.

All ground larviciding vehicles are equipped with a Global Positioning System (GPS) monitoring device to record vehicle location and speed. This system also records the activity of the vehicle-mounted spray system. All inspection and treatment information is recorded by the technician onsite with a laptop computer.

In the dry season, ground larviciding crews continue to survey for mosquitoes often found in breeding sites such as containers, tyres, and neighbourhood drainage basins. Without the consistent summer afternoon rains to flush these habitats, *Culex* species become established and cause problems for nearby residences. Consistent surveillance and treatment are critical to control mosquitoes in urbanized ecosystems.

To address more cryptic mosquito habitats with limited inspector access, LCMCD uses a truck-mounted A1 mist sprayer (A1 Mist Sprayers Resources, Inc., Ponca, Nebraska, USA). By driving residential roadways in the evening hours, this unit treats for mosquito larvae by blowing small droplets of liquid larvicide upwards of 50 feet (15.24 m) into the air enabling it to drift into residential areas that would otherwise be difficult to access. LCMCD has integrated this technique as a way to efficiently treat for mosquito larvae that otherwise would require a team of individuals going door-to-door to inspect and treat cryptic breeding sites in areas that may have limited access.

5.2. *Aerial Larviciding*

Impoundments are areas of salt marsh surrounded by a dike to allow control of water levels for mosquito control, thereby negatively affecting wetland function and vegetation (Rey and Rutledge 2006). Even though Lee County is unique in having over 56 000 acres of mangrove salt marsh habitat, little of it is managed through the use of impoundments, as is more common in other parts of the state. This habitat is home to several species of mosquitoes, most notably *Ae. taeniorhynchus*. This species is a ferocious biter and a prolific breeder with an extensive flight range that extends across the county (Provost 1952; Elmore Jr. and Schoof 1963). To best target these mosquitoes, LCMCD operates a robust aerial larviciding programme aimed at controlling these mosquitoes at their source while in their juvenile life stage.

LCMCD owns and operates a fleet of six Airbus (Herndon, Virginia, USA) H125 helicopters for the purpose of accessing and treating remote breeding sites (Fig. 10). LCMCD biologists constantly monitor salt marsh habitat for rainfall and tidal fluctuations throughout the year. Following a high tide or rain event, biologists fly to remote landing sites to inspect the new water for the presence of newly hatched juvenile mosquitoes. They will take various water samples around a geographic area and check for the presence of mosquito larvae. If larval densities exceed individual site thresholds, biologists record inspection data and schedule the area for treatment.

Figure 10. LCMCD Inspector searching for mosquito larvae in salt marsh habitat.

Treatments are conducted once the biologists are able to develop a site-specific treatment plan, often as soon as that same day. LCMCD also owns eight remote heliport locations along the western edge of the county bordering salt marsh habitat. These locations serve as secure outpost facilities to refuel and reload products onto helicopters in areas closest to treatment sites. Computers equipped at these locations give biologists the ability to develop treatment plans on site without the need to return to LCMCD. Depending on site-specific needs, a variety of products are available for use, including temephos, *B. thuringiensis israelensis, L. sphaericus*, spinosad, methoprene, and larviciding oil. Product formulations also range from liquid to granular formulations as well as single use products providing 30-day residual control.

All helicopters are equipped with an on-board computer to control the helicopters' spray system. This system works harmoniously to upload the individualized treatment polygons with spraying turning on when the helicopter flies into the targeted polygon. Once the pilot exits the pre-programmed treatment zone, the spray system turns off. This GIS-based system operates with pinpoint accuracy that increases pilot safety by simplifying inflight procedures and prevents spraying of off-target sites, saving insecticide and money.

Following a treatment, biologists will return to their inspection site to complete a post-treatment inspection. All inspection data are recorded onsite at time of collection with a custom iPhone application. Once synchronised, all data are available for viewing at the office and are recorded in an organized format. All treatment data are captured by the system's on-board computer and are available for viewing post-treatment in a similar fashion.

6. ADULT MOSQUITO CONTROL

6.1. Ground Adulticiding

LCMCD operates 13 vehicles equipped with ultra-low volume (ULV) spray systems used to target adult mosquitoes in and around neighbourhoods (Fig. 11). Ground adulticiding missions are conducted between sunset and 02:00 to target flying mosquitoes when they are most active. Formulated products are applied without dilution or mixing, and equipment is calibrated to treat a 300-feet (91.4 m) swath at a speed of 10 miles (16.1 km) per hour. A variable flow spray system is equipped to keep the targeted application rate even when the vehicle speed increases or decreases from the 10 miles per hour targeted spray rate. As the vehicle changes speed, within a range of 2-20 miles (3.2-32.2 km) per hour, the appropriate amount of product is dispensed according to label directions. When the vehicle speed surpasses 20 miles per hour, the spray system shuts off preventing spray.

Figure 11. Ground adulticiding vehicles with rear-mounted ULV machine.

The vehicle's spray system is operated remotely from inside the cab of the vehicle with a handheld controller. This design prevents the driver from coming into contact with chemicals during spray operations and limits exposure. The spray system also records various parameters throughout the evening such as vehicle speed, vehicle location, spray activity, miles sprayed, acres treated, and total chemical dispensed. Chemical usage information is compared each morning to the amount recorded by the driver at the start and end of their shift to ensure proper calibration.

Small isolated locations that are not large enough to warrant a ULV truck application are easily treated with small handheld ULV sprayers. These units are typically reserved for areas with easy access that can be walked by a technician. Common treatment sites for such handheld applications include used tyre shops or dumps targeting *Ae. aegypti* and small natural areas targeting freshly emerged *Ae. taeniorhynchus* to prevent dispersal.

6.2. Aerial Adulticiding

The aerial adulticiding operations in Lee County is an important programme designed to efficiently control biting mosquitoes across large geographic areas. LCMCD owns and operates a fleet of eight fixed-wing airplanes outfitted with spray equipment designed to target flying adult mosquitoes (Fig. 12). Adulticiding missions are conducted at night between the hours of 21:00 and 02:00 when night-active mosquitoes are typically most active. Applications are made at an altitude of 350 feet (107 m) above ground level and pilots are equipped with night vision goggles for maximum visibility.

Figure 12. Douglas DC3-TP with four 50-gallon chemical tanks used for aerial adulticiding.

Similar to the LCMCD aerial larviciding system, the adulticiding spray system is controlled via an on-board computer with a pre-programmed mission. Once pilots arrive on-site the spray system is automatically turned on and remains spraying until the pilot exits the treatment area. This automation increases precision of the application and enhances pilot safety when flying in such an unconventional manner. Depending on how large the problematic area is, treatments sites can be as large as 23 000 acres per mission per aircraft. Flights typically occur at 130 or 150 knots, depending on aircraft type and chemical flow rate.

LCMCD primarily utilizes naled and malathion for aerial adulticiding. Products are dispensed with a high-pressure nozzle system or with a rotary atomizer at a rate of 0.5 oz/acre, 0.66 oz/acre, or 1.5 oz/acre depending on the pesticide used and the targeted mosquito species. LCMCD does not utilize set treatment frequencies for scheduling treatments of any kind, but rather relies entirely on surveillance data to determine if treatments are warranted. Each surveillance method has an associated treatment threshold that must be met based on inspection type and location baselines.

For an aerial adulticiding treatment to be conducted, surveillance data are evaluated first and considered prior to scheduling. Surveillance methods include landing rate counts, truck trap collections, spray zone thresholds (that were obtained over many years of trap data), arbovirus detection in sentinel chicken flocks, and mosquito trapping results. If criteria mandated by the state of Florida are met and a wide-scale problem is determined, an aerial adulticiding application is scheduled as early as that same night.

7. NOVEL TECHNIQUES

It is essential to keep up with evolving mosquito populations, increasingly sophisticated control technologies, climate change, a constantly increasing human population density, and increases in exotic disease agents and vector invasions. As such, a programme can fall behind and become less efficient than it once was if these changes are not taken into account. To best combat these dynamic circumstances, LCMCD is committed to staying abreast of new technology and the advancement of various control measures.

Applying sterilisation techniques for the control of insect populations is not a new concept, however the application of it on mosquitoes is an emerging field (Lees et al. 2021; Baton et al., this volume). The Sterile Insect Technique (SIT) was first utilized in the late 1950s to successfully control the screwworm fly (*Cochliomyia hominivorax* Coquerel) on the isolated habitat of Sanibel Island in Lee County, Florida (Bushland and Hopkins 1953; Bushland 1960). Since the first trial on Sanibel Island, the SIT has been employed to effectively suppress, contain and eradicate a variety of medically- and agriculturally-important insects (Dyck et al. 2021).

LCMCD is currently in the process of establishing the first SIT programme for *Ae. aegypti* solely operated by a mosquito control district in the state of Florida. This novel programme aims to reduce *Ae. aegypti* in Lee County using X-ray irradiation for sterilisation. To accomplish this, LCMCD will be mass-rearing locally collected populations of *Ae. aegypti*, irradiating the mosquitoes using X-rays, and releasing the sterilized male adults on an area-wide basis into the field to breed with wild female populations. The goal of this programme is to become a valuable complement to traditional mosquito control techniques in the fight to prevent the spread of diseases such as Zika, dengue, yellow fever, and chikungunya, which are transmitted via the bite of the *Ae. aegypti* mosquito.

LCMCD is also interested in using more conventional technology in innovative ways to improve operations. Unmanned aerial systems (UAS) technology has been available to the commercial market for several years now and is utilized primarily for their photographic abilities. LCMCD owns two UAS for the purpose of aiding in inspections of mosquito breeding habitat and have recently purchased one UAS capable of carrying and spraying a payload of insecticide. As the rules and regulations surrounding UAS continue to evolve (Benavente et al., this volume), LCMCD plans on being there along the way to incorporate these new technologies into the mosquito control industry as a part of its commitment to protect the health of the citizens of Lee County.

8. CONCLUSIONS

LCMCD operates a comprehensive IMM programme in an effort to provide the most effective mosquito abatement possible for the citizens of Lee County. As with any IMM programme, the efficient integration of all methods together achieves the most advantageous results.

The larviciding programme, aimed at suppressing mosquitoes in their juvenile life stage, offers the most efficient means of control by targeting mosquitoes when they are at their most concentrated state of development and unable to bite. Source reduction and biological control measures, although varying differently in application, offer a natural and potentially longer duration of control than insecticides.

A strong adulticiding programme plays a vital role in suppressing the biting pressure on the local population and interrupts the disease transmission in the event of an arbovirus outbreak.

The implementation of novel control measures, such as the SIT, complement conventional control methods to aid in the control of disease vectoring agents. All of these methods offer specific advantages, however, if utilized on their own they would prove wildly inadequate. The harmonious integration of all control measures is best supported with a backbone of strong surveillance and a well-educated staff to oversee its implementation.

9. REFERENCES

Agramonte, N. M., and C. R. Connelly. 2014. Black salt marsh mosquito *Aedes taeniorhynchus* (Wiedemann) (Insecta: Diptera: Culicidae). Publication number EENY-591, University of Florida/Institute of Food and Agricultural Sciences Extension.

(AMCA) American Mosquito Control Association. 2017. Best practices for integrated mosquito management: A focused update. Sacramento, California, USA. 58 pp.

Batzer, D. P., and V. H. Resh. 1992. Wetland management strategies that enhance waterfowl habitats can also control mosquitoes. Journal of the American Mosquito Control Association 8: 117–125.

Bibbs, C. S., D. Mathias, and N. Burkett-Cadena. 2019. Dark rice field mosquito *Psorophora columbiae* (Dyar & Knab) (Insecta: Diptera: Culicidae). Publication number EENY-735, University of Florida/Institute of Food and Agricultural Sciences Extension.

Bushland, R. C., and D. E. Hopkins. 1953. Sterilisation of screwworm flies with X-rays and gamma rays. Journal of Economic Entomology 46: 648–656.

Bushland, R. C. 1960. Male sterilisation for the control of insects, pp. 1-25. *In* R. L. Metcalf (ed.), Advances in pest control research, Volume III. Interscience Publishers, New York, NY, USA.

(CABI) Centre for Agriculture and Bioscience International. 2019. *Culex quinquefasciatus* (southern house mosquito) [original text be D. A. Lapointe]. *In* Invasive species compendium. Wallingford, UK.

Cassiano, E. J., J. Hill, Q. Tuckett, and C. Watson. 2018. Eastern mosquitofish, *Gambusia holbrooki*, for control of mosquito larvae. Document FA202, School of Forest Resources, Program in Fisheries and Aquatic Sciences, University of Florida/Institute of Food and Agricultural Sciences Extension.

Challet, G. L. 1994. Mosquito abatement district programs in the United States. The Kaohsiung Journal of Medical Sciences (Gaoxiong Yi Xue Ke Xue Za Zhi) 10 (Supplement): S67–S73.

Connelly, C. R., and D. B. Carlson (eds.). 2009. Florida mosquito control: The state of the mission as defined by mosquito controllers, regulators, and environmental managers. Florida Coordinating Council on Mosquito Control. University of Florida, Institute of Food and Agricultural Sciences, Florida Medical Entomology Laboratory, Vero Beach, Florida, USA. 259 pp.

Day, J. F. 2017. The Florida St. Louis encephalitis mosquito *Culex nigripalpus* Theobald (Insecta: Diptera: Culicidae). Publication number EENY-10, University of Florida/Institute of Food and Agricultural Sciences Extension.

Day, J. F., and G. A. Curtis. 1999. Blood feeding and oviposition by *Culex nigripalpus* (Diptera: Culicidae) blood feeding and oviposition before, during and after a widespread St. Louis encephalitis epidemic in Florida. Journal of Medical Entomology 36: 176–181.

Dyck, V. A., J. Hendrichs, and A. S. Robinson (eds.). 2021. Sterile Insect Technique – Principles and practice in Area-Wide Integrated Pest Management. Second Edition. CRC Press, Boca Raton, Florida, USA. 1200 pp.

Elmore Jr., C. M., and H. E. Schoof. 1963. Dispersal of *Aedes taeniorhynchus* Wiedemann near Savannah, Georgia. Mosquito News 23(1): 1–7.

Focks, D. A., D. A. Dame, A. L. Cameron, and M. D. Boston. 1980. Predator-prey interaction between insular populations of *Toxorhynchites rutilus rutilus* and *Aedes aegypti*. Environmental Entomology 9: 37–42.

Hill, S., and C. R. Connelly. 2009. Southern house mosquito *Culex quinquefasciatus* Say (Insecta: Diptera: Culicidae). Publication Number EENY-457, University of Florida/Institute of Food and Agricultural Sciences Extension.

Kerzee, K. 2019. The whats and whys of mosquito abatement districts. Midwest Pesticide Action Center.

Kline, D. L. 2006. Mosquito population surveillance techniques. Technical Bulletin of the Florida Mosquito Control Association 6: 2–8.

(LCMCD) Lee County Mosquito Control District. 2019. Lehigh Acres, Florida, USA.

Lees, R. S., D. O. Carvalho, and J. Bouyer. 2021. Potential impact of integrating the Sterile Insect Technique into the fight against disease-transmitting mosquitoes, pp. 1081–1118. *In* V. A. Dyck, J. Hendrichs, and A. S. Robinson (eds.), Sterile Insect Technique – Principles and practice in Area-wide Integrated Pest Management. Second Edition. CRC Press, Boca Raton, Florida, USA.

McKenna, M. 2016. Disorganized mosquito control will make US vulnerable to Zika. February 29, 2016. National Geographic, Washington, DC, USA.

McDonald, P. T. 1977. Population characteristics of domestic *Aedes aegypti* (Diptera: Culicidae) in villages on the Kenya coast. II. Dispersal within and between villages. Journal of Medical Entomology 14: 49–53.

Muir, L. E., and B. H. Kay. 1998. *Aedes aegypti* survival and dispersal estimated by mark-release-recapture in northern Australia. American Journal of Tropical Medicine and Hygiene 58: 277–282.

Nayar, J. K., and C. R. Connelly. 2017. Mosquito-borne dog heartworm disease. SP486: Pests in and around the southern home. University of Florida/Institute of Food and Agricultural Sciences Extension.

Patterson, G. 2004. The mosquito wars: A history of mosquito control in Florida. University Press of Florida. Gainesville, Florida. 288 pp.

Provost, M. W. 1952. The dispersal of *Aedes taeniorhynchus*. I. Preliminary studies. Mosquito News 12(3): 174–190.

Provost, M. W. 1969. Man, mosquitoes and birds. The Florida Naturalist 41: 63–67.

Rey, J. R., and C. R. Rutledge. 2006. Mosquito control impoundments. Publication number ENY-648, University of Florida/Institute of Food and Agricultural Sciences Extension.

Rose, A., U. Kröckel, R. Bergbauer, M. Geier, and Á. E. Eiras. 2006. The BG-Sentinel, a novel mosquito trap for research and surveillance. Mitteilungen der Deutschen Gesellschaft für allgemeine und angewandte Entomologie 15: 345–348.

(WHO) World Health Organization. 2016. Monitoring and managing insecticide resistance in *Aedes* mosquito populations: Interim guidance for entomologists. Geneva, Switzerland.

Zettel, C., and P. Kaufman. 2019. Yellow fever mosquito *Aedes aegypti* (Linnaeus) (Insecta: Diptera: Culicidae). Publication number EENY-434, University of Florida/Institute of Food and Agricultural Sciences Extension.

Aedes aegypti CONTROL PROGRAMMES IN BRAZIL

H. R. C. ARAÚJO[1,3], D. O. CARVALHO[2] AND M. L. CAPURRO[1,3]

[1]*Department of Parasitology, Institute of Biomedical Sciences, University of São Paulo, São Paulo, SP, Brazil; mlcapurro@gmail.com*
[2]*Joint FAO/IAEA Division, Insect Pest Control Laboratory, Seibersdorf, Austria; d.carvalho@iaea.org*
[3]*National Institute of Science and Technology in Molecular Entomology, Medical Biochemistry Institute, Federal University of Rio de Janeiro, Brazil*

SUMMARY

Mosquito-borne diseases are among the most significant challenges facing societies around the world. In Brazil, current official epidemiological reports show increasing numbers of cases of mosquito-borne diseases, such as chikungunya, dengue, yellow fever and Zika, which are spreading to new areas of the country. Therefore, it can be stated that current methods used for the management of mosquito vectors in Brazil, established since 2002, have been ineffective. Thus, there is a necessity for readjustment or updating of the *Aedes aegypti* control programmes that are being applied in Brazil. As recommended by the World Health Organization (WHO), the best way to combat these pathogen vectors is an integrated approach where several convenient and compatible control techniques are combined to efficiently reduce or potentially eliminate a targeted insect vector population. In this manuscript, we updated a review published in 2015 by the same authors about *Aedes* control programmes in Brazil showing their basic concept and the principal components of *Aedes* integrated control programmes. Strategies such as public education, community engagement and responsibility; mechanical elimination of mosquito breeding habitats; the use of larvicides and adulticides; massive collection of eggs and adults using traps; and the reduction in the vector population through the promotion of sterility of mosquitoes by ionizing radiation, use of symbiont bacteria such as *Wolbachia,* or genetic modification, are discussed. The Brazilian experience to test and evaluate some of these technologies is described and compared with strategies to prevent and manage mosquito populations in other countries. It is concluded that there are new control methods that can be integrated on an area-wide basis to suppress mosquito populations successfully. Nevertheless, epidemiological studies are also needed to evaluate their impact on disease transmission, in addition to the proof-of-concept that they suppress mosquito populations.

Key Words: Mosquito control methods, community engagement, Projeto Aedes Transgênico, population suppression, integrated vector management (IVM), genetically modified mosquitoes, vector-borne diseases, Plano Nacional de Controle da Dengue (PNCD)

J. Hendrichs, R. Pereira and M. J. B. Vreysen (eds.), Area-Wide Integrated Pest Management: Development and Field Application, pp. 339–366. CRC Press, Boca Raton, Florida, USA.

1. INTRODUCTION

1.1. Aedes aegypti *Primary Vector of Arboviruses*

Aedes aegypti (L.), the yellow fever mosquito, is the primary mosquito vector of various arboviruses such as yellow fever (YFV; genus Flavivirus), dengue (DENV; genus Flavivirus), Zika (ZIKV; genus Flavivirus) and chikungunya (CHIKV, genus *Alphavirus*).

Yellow fever is endemic in tropical areas of Africa, as well as Central and South America. Symptoms include fever, headache, jaundice (origin of the name "yellow" fever), muscle pain, nausea, vomiting, and fatigue. A small proportion of patients develop severe symptoms and approximately half of those die within 7 - 10 days (WHO 2018).

Dengue is endemic in more than 100 countries and is one of the most serious public health problems in the world. Clinical manifestations of dengue virus infection include high fever (40°C) that can be accompanied by severe headache, pain behind the eyes, muscle and joint pains, nausea, vomiting, swollen glands or rash. It is estimated that worldwide, about 40% (2500 million people) of the human population is at risk of contracting dengue fever, and about 390 million people are each year becoming infected with the disease. In 2016, more than 2.38 million cases of dengue were reported in the Americas, of which 1.5 million cases occurred in Brazil, i.e. a threefold increase in cases as compared with 2014 (WHO 2009, 2016a).

A ZIKV infection brings complications and consequences such as microcephaly in babies and the Guillain-Barré syndrome, and their neurological complications are being intensively investigated. Symptoms are generally mild and include fever, rash, conjunctivitis, muscle and joint pain, malaise or headache that last for 2–7 days. However, most people with Zika virus infection do not develop symptoms. Since the 1960s, ZIKV disease has been reported in Africa, Asia, the Pacific islands, and the Americas, but since 2015, its geographic range has expanded rapidly (WHO 2016b). Currently, the ZIKV has been reported in more than 84 countries, territories or subnational areas in the world (WHO 2017a). Between 2015 and 2017, more than 200 000 confirmed autochthonous cases of ZIKV were reported in the countries and territories in the Americas, as well as 3323 confirmed cases of congenital syndrome associated with ZIKV infections. Of these ZIKV cases, a majority (134 057) were reported in Brazil (PAHO/WHO 2016).

CHIKV has been identified in over 60 countries in Asia, Africa, the Americas and Europe. The disease is characterized by fever and is frequently accompanied by joint pain, which is often very debilitating and lasts for a few days or weeks. In 2016, there were more than 150 000 laboratory confirmed cases of chikungunya fever in the Americas. Brazil reported 146 914 confirmed cases, followed by Argentina (322 confirmed cases) and Paraguay (38 confirmed cases) (PAHO/WHO 2014; WHO 2016c, 2017b).

All the above-mentioned viruses are transmitted by *Aedes* spp. when the female mosquitoes take a blood meal from a viremic human host and bites another non-viremic human host. These mosquitoes are distributed throughout tropical and subtropical territories, where they largely overlap, explaining their current scenario of co-infection (Furuya-Kanamori et al. 2016; Rückert et al. 2017).

1.2. Why Vector Control?

The co-distribution and/or co-transmission of vector-borne diseases pose a challenge for public health in endemic and epidemic regions of the world, in particular also in Latin America (Furuya-Kanamori et al. 2016; Rodriguez-Morales et al. 2016; Carrillo-Hernández et al. 2018; O Silva et al. 2018; Suwanmanee et al. 2018). More than 80 % of the global human population lives in areas where they are at risk of contracting at least one vector-borne disease and more than half lives in areas where they are at risk of contracting two or more of these diseases (PAHO 2016). Vector-borne diseases mainly affect poorer populations and impede economic development through direct and indirect medical and other costs such as loss of productivity and tourism (WHO 2017c).

Despite the emergence of new viruses transmitted by *Ae. aegypti*, dengue continues to be one of the most important public health problems in Brazil, considering the burden of disease and the great potential for evolution to death (Martelli et al. 2015; Araújo et al. 2017). Between 2013 and 2016, the cost of hospitalizations for dengue paid by Brazil's publicly funded health care system (known by the acronym SUS) was BRL 68.1 million (SHS 2017). In addition, dengue contributes to the loss of healthy years of life, affecting a large number of people from all age groups, causing some degree of disability during the infection period and deaths, mainly in children (Araújo et al. 2017). The application of remediation measures during epidemic periods can drastically reduce its cost through a more effective prevention programme using entomological surveillance, integrated with area-wide vector control strategies, resulting in the prevention of several diseases and increasing human population life quality in the target area.

In Singapore, for example, the haemorrhagic fever induced by dengue infection became a significant cause of death in the 1960s, affecting especially children. A vector control programme was implemented from 1968 to 1973, using data from entomological and epidemiological surveys to develop a strategy that was based on entomological surveillance, larval source reduction, public education, and law enforcement. The philosophy of the programme was to carry out vector control before the disease is detected as a means to reduce disease transmission. Singapore successfully controlled *Ae. aegypti* population and as a result, DENV infections were reduced and disease incidence remained low for a 15-year period. However, this success proved to be temporary, and the disease incidence increased again in the country in the 1990s (Ooi et al. 2006). The development of a local entomological index correlates the increase of new areas with breeding sites more supceptible to dengue transmission (Ong et al. 2019). In addition, cases of other arboviruses were reported, such as CHIKV in 2008 (Leo et al. 2009) and ZIKV in 2016 (Maurer-Stroh et al. 2016).

The reduction in the density of the *Ae. aegypti* population, the resurgence of DENV and appearance of other diseases transmitted by this vector seems like a paradox. However, it is speculated that several factors may have contributed to an increase in dengue incidence in Singapore: 1) decreased herd immunity after 30 years of low dengue exposure, 2) an increase in the proportion of adult infections, 3) virus transmission occurring outside houses, 4) the adoption of a reactive rather than a proactive approach to vector control, 5) the presence of asymptomatic persons, and 6) a continued introduction of the virus through increasing numbers of travelers returning from endemic areas (Ooi et al. 2006; Ooi and Gubler 2009). Moreover, peridomestic areas, where other competent vectors were present (*Aedes malayensis* (Colless) and *Aedes albopictus* (Skuse)), were not included in the vector management programme (Mendenhall et al. 2017), and hence the programme was not following area-wide principles.

In the 20th century, classic vector control strategies to reduce populations of mosquitoes that transmitted malaria, yellow fever and dengue temporarily reduced the impact of these diseases in several countries using mainly insecticides (NASEM 2016). However, the current distribution of vector-borne diseases in the world shows that these and other disorders are re-emerging and/or spreading to new areas. This means that the full potential for preventing disease transmission is not applied as it should. There are factors that contribute to the failures such as technical complexity, costs and logistic needs, complacency and environmental concerns about insecticides (Townson et al. 2005).

1.3. Aedes aegypti *Control in Brazil*

Control of *Ae. aegypti* in Brazil has been implemented according to the guidelines outlined in the National Plan for Dengue Control (in Portuguese, *Plano Nacional de Controle da Dengue* - PNCD) (MS/FNS 2002), which is aligned with the Integrated Management Strategy for Dengue Prevention and Control in the Region of the Americas (known by the abbreviation IMS-Dengue) approved in the Resolution CD44.R9 adopted by the 44th Directing Council of the Pan American Health Organization/World Health Organization (PAHO/WHO 2003; San Martín and Brathwaite-Dick 2007).

The main objective has been to promote a model for prevention and control of dengue that incorporates national and international experiences and emphasizes the need for change in previous models, including also the decentralisation of the vector control programme so that each municipality is responsible for the control with the support from the State Department of Health and the Ministry of Health (MS/FNS 2002; Tauil 2002; Brasil 2009; PAHO 2018). The main actions involve epidemiological surveillance, vector control, patient care, integration with primary health care, environmental sanitation actions, integrated health education actions, communication and social mobilization, training of human resources, social and political support and evaluation of the programme (MS/FNS 2002, 2009; Braga and Valle 2007; Araújo et al. 2015).

The strategies involve the participation of 'Community Health Agents' (CHA) that are responsible, together with the local community, to promote the mechanical (removal or elimination of potential breeding sites) and chemical control (insecticides) with the objective of guaranteeing the sustainability of the elimination of breeding sites by real estate owners, in an attempt to break the chain of transmission of dengue. Other actions also recommended by the Brazilian Ministry of Health include installations of screens on the doors and windows to prevent the entry of the adult mosquito, in addition to the use of predators or pathogens with potential to reduce the vector population (biological control). Among the available predators are fish and aquatic invertebrates, which eat the larvae and pupae, and pathogens that release toxins including bacteria, fungi and parasites (Zara et al. 2016).

Since the mid-1980s, temephos (organophosphate) has been the main insecticide used against larvae of *Ae. aegypti* in Brazil. However, since 2002, mosquito populations in half of the country have become resistant to temephos (Chediak et al. 2016). As a result, pyriproxyfen, an insect growth regulator that mimics a natural hormone and interrupts insect development, was introduced in 2014 for the suppression of *Ae. aegypti* larvae (MS 2014a). Since 2009, malathion (organophosphate insecticide) has been used to control adults, replacing the use of pyrethroids after the identification of high levels of knockdown resistance registered (Martins et al. 2009). Concentrations of the all insecticides currently used, as well as the applied bioassay protocols, are those recommended by the WHO (2013) and the Brazilian Ministry of Health (MS 2014b).

Despite the risk of favouring the rise and dispersal of resistant populations, and the consequent lack of alternative insecticides to the currently available malathion against *Ae. aegypti* adults, the Brazilian Ministry of Health intensified insecticide spraying against *Ae. aegypti,* as a response to the Zika and chikungunya epidemics. The reliance on a strategy that was mainly based on chemicals to bring the *Ae. aegypti* population under control gave the human population in Brazil a false conception of security (Augusto et al. 2016). The unprecedented spread of vector-borne diseases clearly highlights the challenges faced by everyone, not just the health agencies. Multiple control tactics will need to be used for the management of vector-borne diseases, and this will only be possible if an integrated vector management (IVM) approach is selected. An IVM approach was adopted in 2004 by WHO for all vector-borne diseases and involves a rational decision-making process for the optimal use of resources, to improve cost-effectiveness, ecological soundness, and sustainability of disease-vector control (WHO 2004, 2008, 2017c). The outcome of IVM is improved human capacity and strengthened infrastructure to increase the well-being, and not only protecting human population against disease. The WHO recommends integrated control of the mosquito vectors, mainly those of dengue. Control activities should target *Ae. aegypti* (or any of the other vectors depending on the evidence of transmission) in all its immature (egg, larva, and pupa) and adult stages (WHO 2017d). The critical components of *Aedes* integrated vector management programme in Brazil are illustrated in Fig. 1.

2. EDUCATION, COMMUNITY ENGAGEMENT AND RESPONSIBILITY

2.1. General Overview

Of primary importance in any IVM strategy is training of health personnel in community-based participation so that the local population can understand and hence participate in several aspects of vector control (Gubler and Clark 1996; Ulibarri et al. 2016). Vector control also requires national level support to provide strategic direction, technical expertise, and training, aside from the development of norms and indicators to monitor the progress of operational activities.

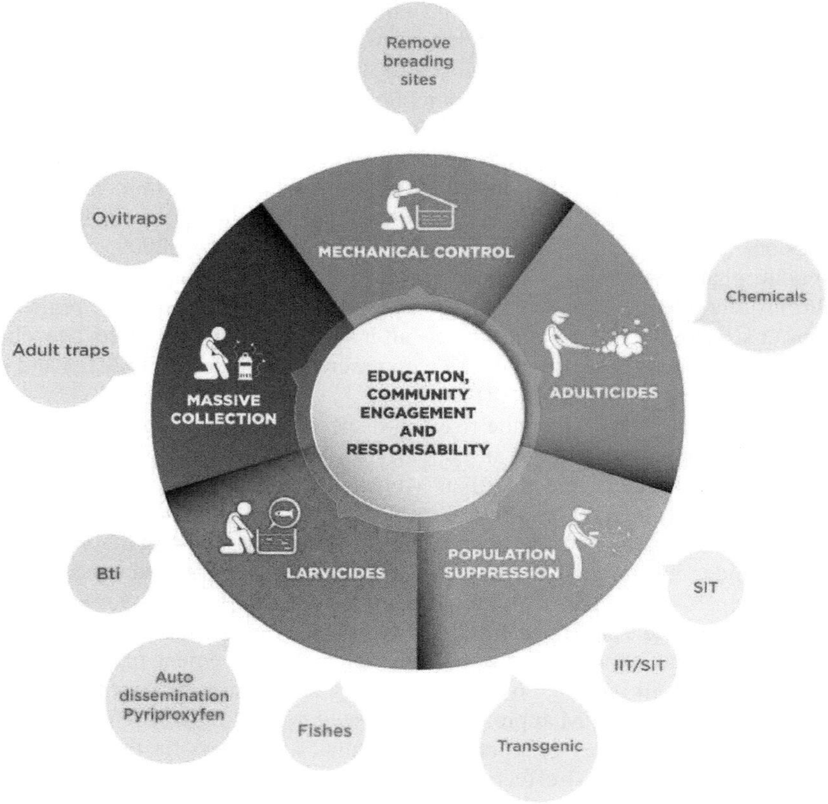

Figure 1. Integrated control programme for Aedes aegypti *populations in Brazil.*

The distribution and incidence of vector-borne diseases are determined by ecological factors, but they are also influenced by the behaviour of humans. Thus, vector control interventions that incorporate human population engagement are more likely to be successful as they offer the opportunity to take into account community problems (Townson et al. 2005).

The WHO has prepared and made available guidelines to assist national programmes with the design and implementation of social mobilization and communication strategies aimed at dengue fever prevention and control. The approaches to social mobilization are known by the initials "COMBI" (Communication-for-Behavioural-Impact) that integrate the participation of different members of the community, from households to political leaders. COMBI represents a set of marketing, education, communication, promotion, advocacy and mobilization approaches with the same goal, i.e. to ensure sustained community participation to combat *Ae. aegypti* and as such, to promote the health of community members (Parks and Lloyd 2004; Tapia-Conyer et al. 2012).

2.2. *The Brazilian Perspective*

In Brazil, the Municipal Health Secretariats have begun to manage and execute PNCD actions with the support of the States and the Ministry of Health, with most funding provided at the federal level. Engagement of the communities and education of the public in the control of *Ae. aegypti* does not mean to bombard people with information about mosquitoes or vector-borne diseases. In Brazil, it has happened that despite growing levels of public knowledge about mosquitoes and their control, many people are not taking the required basic actions such as the elimination of larval sources (Claro et al. 2006). Nevertheless, a study conducted in Ribeirão Petro (south-eastern Brazil) reveals the relevance of educational campaigns and educational health programmes using different types of media to reach different community levels to transmit the necessary information (Alves et al. 2016).

Caprara et al. (2015) developed an eco-health programme, based on community engagement, developing and distributing educational and informative material. They also promoted workshops for the community and developed activities to involve the community directly, such as mobilization of schoolchildren and the elderly, organization of meetings and active participation during campaigns to remove/relocate breeding sites. Although the overall result shows that there was still an increase in the mosquito population after the rainy season (which also corresponded to the end of the experiment), the non-treated site had a significantly higher increase in mosquito density compared to the treated area during the same period (Caprara et al. 2015).

The key to educational campaigns is achieving a long-term modification of the behavioural of the general public that must be conscious of its own actions and be responsible for the surrounding environment. In support, a recent Brazilian sanitary legislation allows the application of fines in the case of impediment or difficulty when implementing sanitary measures that aim at the prevention of the diseases and their dissemination (Brasil 2016).

2.3. *Innovations and Experiences of Other Countries*

Community engagement and information activities were performed in Brazil during the entire mosquito population suppression *Projeto Aedes Transgênico* (2010-2013 described in Section 7) that relied on the release of genetically modified mosquitoes (GMM) (Capurro et al. 2016). These activities, carried out before and during the mosquito release project, showed positive results and provided guidance for the design of similar public engagement plans in other regions or countries. This pioneering study in continental America showed that full transparency was crucial to make the public aware of all aspects of the mosquito release project, particularly in this case involving genetically modified organisms.

The work from Sommerfeld and Kroeger (2015) reviews community-based vector control interventions in different countries in Latin America that are fighting against dengue and using educational campaigns, chemical and non-chemical strategies, including new approaches such as waste management. The authors mention that these strategies involving the community require establishment of a prolonged interaction with control services, municipalities and other public actors, proving to be rewarding during the process and with excellent potentials for sustainability, however, they were time-consuming and costly at the beginning. The results of community participation programmes used in Mexico showed that continuity of these activities in long-term campaigns is a prerequisite to achieve the desired goals (Tapia-Conyer et al. 2012). However, governments are often reluctant to invest and support these initiatives, and consequently, these programmes are often relegated to serve as epidemiological projects during dengue outbreaks.

3. MECHANICAL CONTROL

3.1. *General Overview*

In general, mechanical control consists of the elimination of *Ae. aegypti* larval breeding sites from domestic and peridomestic areas, and the application of measures that prevent the contact between humans and the vector. The interventions include changing the environment through cleaning and removal of possible habitats suitable for any stage of *Ae. aegypti* and *Ae. albopictus* to prevent or minimize vector propagation. This entails covering water storage containers, disposing of non-biodegradable waste, and installing mosquito screens on windows, doors and other entry points, in addition to the use of mosquito bed nets. Local government agencies must take responsibility for the clean-up of public spaces and to eliminate illegal dumps and discarded tyres (Arunachalam et al. 2012; US-EPA 2017; WHO 2017e).

3.2. *The Brazilian Perspective*

Dengue is a disease that has ecological, biological and social factors involved in its transmission. The dynamics of *Ae. aegypti* breeding sites are closely linked to human behaviour; therefore, elimination of larval sites through household interventions is an efficient way to reduce the mosquito population. In Brazil, the removal of breeding sites is the responsibility of households. Periodically the community health agents, and the 'Endemic Disease Control Agents' (EDCA) visit houses looking for possible breeding sites, but they are mainly responsible for non-residential properties, and if necessary, integrate chemical (insecticide) application (Zara et al. 2016).

Chaebo and Medeiros (2017) investigated five conditions for an effective strategy for dengue control policy implementation through co-production, which they defined as the strategy for policy implementation resulting from technological, economic, and institutional influences. Initially the technical, economic, normative, cognitive and structural conditions were analysed and as a result they stated that technical, economic and normative conditions are interdependent, and changing one will change the others. In addition, the authors added two extra conditions to implement policy using co-production that they defined as cognitive and structural conditions. Including these conditions to the main study, the authors state:

> *"We believe, it is impossible to successfully undertake policy implementation via co-production unless users recognize that an important problem exists and are able and willing to undertake the necessary co-production actions"* (Chaebo and Medeiros 2017).

Unfortunately, besides the responsibilities of authorities, the communities often wait for the public vector control services to carry out the task of controlling mosquito breeding sites. In some cases, communities are fully aware of the threats leaving breeding sites and their responsibility to eliminate them, but they are not involved in the programme, and this paradigm needs to change. In integrated vector control, the householders must be stimulated to interact with vector control staff and to ensure appropriate interpersonal communication (Arunachalam et al. 2010).

3.3. Innovations and Experiences of Other Countries

The effect of encouraging community members to eliminate *Ae. aegypti* breeding sites showed a favourable impact in studies carried out in the Caribbean (Rosenbaum et al. 1995), Latin America (Tapia-Conyer et al. 2012), Thailand (Suwannapong et al. 2014), Pakistan (Zahir 2016), USA (Healy et al. 2014), and many other parts of the world (Spiegel et al. 2002; Kay and Nam 2005; Vanlerberghe et al. 2009; Sanchez et al. 2012). There is a consensus among health authorities that this measure is an essential component of environmentally sustainable mosquito control programmes. A recent mathematical model for dengue control developed by Carvalho et al. (2019) confirms that, even though the combination of mechanical and chemical approaches is the most suitable one instead of using them separately, it is still insufficient to eliminate disease transmission completely.

4. MASS-TRAPPING

4.1. General Overview

Different models of traps are available to monitor *Ae. aegypti* and *Ae. albopictus* populations, and they can generate baseline data that are essential to guide control operations. They can also be included in the entomological surveillance to improve mosquito population density prediction prior to epidemic periods (Honório et al. 2009; Degener et al. 2014). An increased number of deployed traps can be used to reduce the target mosquito population, i.e. the gravid females are attracted to the oviposition traps (ovitrap) and are killed when making contact with the oviposition substrate that is impregnated with insecticides, or lethal ovitraps, collecting eggs that are subsequently killed by an insecticide-treated ovistrip (Paz-Soldan et al. 2016).

According to a review on mass-trapping interventions for suppression of urban *Aedes* by Johnson et al. (2017), successful deployment is achieved with a high area coverage (>80%), a pre-intervention and/or additional source reduction, the direct involvement of community members for sustainability, and the use of new-generation traps (such as the Autocidal Gravid Ovitrap – AGO, or Gravid *Aedes* Trap – GAT) to outcompete remaining water-holding containers.

In areas where *Ae. albopictus* co-exists with *Ae. aegypti*, eggs or larvae collected in ovitraps need to be taken to the laboratory for species identification at the larval stage or maintained until adult emergence. In those areas, the AGO is a good alternative to monitor mosquito populations (Caputo et al. 2015). These traps are simple, specific and efficient for gravid females, and their integration with other chemical or biological control methodologies can contribute significantly to decrease mosquito populations. However, their use is laborious, which is a disadvantage for deployment over large areas of action. Combining mass-trapping of adults, with the use of larvicides, can have a more significant impact on *Ae. aegypti* populations than using each of these methods alone (Regis et al. 2008).

4.2. The Brazilian Perspective

Mass-trapping is not currently used for mosquito control in any vector control programme in the country; however, several works have assessed the effect of lethal ovitraps, and the results were promising in several situations (Regis et al. 2008, 2013). A modified ovitrap containing *Bacillus thuringiensis israelensis* (*Bti* – see next Section) that kills any larvae developing inside was evaluated in north-eastern Brazil. The *Bti*-treated trap can safely remain in the field for up to two months and during that time can collect more than 7000 eggs/trap (Regis et al. 2008), of course depending on the initial population density.

Deployment in urban areas of ovitraps treated with the pyrethroid deltamethrin reduced the density of the adult female population by 40% (Perich et al. 2003). The study involved the placement of 10 ovitraps/residence (five inside and five outside) for 12 weeks in two municipalities in Rio de Janeiro, and the sampling of 30 houses per intervention neighbourhood. The authors mentioned that although lethal ovitraps were not designed to be a control method to be used alone, their results show that lethal ovitraps could provide an inexpensive, simple, environmentally benign way to be integrated into vector control strategies (Perich et al. 2003).

Sticky ovitraps with an adhesive strip, rather than an insecticide-treated oviposition surface that traps the ovipositing females when they land, have been used for surveillance in areas with high mosquito insecticide resistance. A study was conducted for 17 months to suppress mosquito populations in the Amazon region through a mass-trapping system using sticky ovitraps. The authors conclude that this intervention alone was not able to show mosquito population suppression, and they indicate as probable reasons a lack of buffer zones, which allowed mosquito migration from other areas, the lack of an area-wide approach due to the small size of the treated area, and insufficient collection efficacy of the trap or inadequate number of traps/household (Degener et al. 2015).

4.3. Innovations and Experiences of Other Countries

Like the Brazilian experience, *Ae. aegypti* populations were significantly reduced in Thailand when lethal permethrin-treated ovitraps were deployed in conjunction with other interventions such as source reduction, use of screen covers, and biological control. In this case, they also evaluated the impact on dengue transmission and the proportion of DENV IgG–IgM positives in the treated areas, which were reduced from 13.46% to 0%, whereas those from untreated areas increased from 9.43 to 19.15% (Kittayapong et al. 2008).

A previous study using lethal ovitraps, also in Thailand, showed a 49-80% reduction in the mosquito population in an experiment over 30 weeks (Sithiprasasna et al. 2003). In Cairns, Australia, the acceptance by households of a mass-trapping scheme allowed the comparison of different types of lethal ovitraps in three separate trials. The results suggest that a high trap density can collapse a mosquito population over time (Ritchie et al. 2009).

The AGO, baited with an attractant and containing an adhesive card placed inside the trap entrance that serves as an autocidal oviposition substrate, was developed by the Centers for Disease Control and Prevention (CDC) to catch gravid *Ae. aegypti* females (Mackay et al. 2013). The AGOs placed in 85% of residences in four communities of two municipalities in Puerto Rico between November 2015 and February 2016 to control *Ae. aegypti* mosquitoes significantly reduced the prevalence of CHIKV IgG antibodies in participating communities without any other control tactic used (Lorenzi et al. 2016).

5. LARVAL CONTROL

5.1. General Overview

Larvicides are biocides used against immature mosquito stages and their use fits well within environment-friendly management strategies (except in emergency situations). Larval control can minimize the need for widespread use of insecticides to kill adult mosquitoes. Larvicides are used by vector control staff to treat water-holding structures and containers in public places, whereas the general citizen is supposed to do the same to treat fountains, septic tanks, pots and pools on private properties. The use of larvicides should be restricted to containers that are not used for drinking, and that cannot be covered, dumped or removed (CDC 2017a). Widely used is *Bti,* a bacterium marketed commercially as a biological larvicide to control insects relevant to public health. It is safe for humans, but when ingested by mosquito larvae, lethal endotoxins proteins are produced during the bacterium sporulation, killing the larvae before reaching adulthood (Federici et al. 2007; Ibrahim et al. 2010). An alternative is the auto-dissemination approach, that can be augmented by the release of males which were tainted with pyriproxyfen, a juvenile hormone analogue, and who will contaminate females during mating or directly the larval habitats (Bouyer and Lefrançois 2014).

5.2. The Brazilian Perspective

In Brazil, *Bti* is used since 2002 when resistance to the organophosphate larvicide, temephos, was observed (Suter et al. 2017). In those cases, *Bti* can be used alone or in association with different chemical larvicides such as pyriproxyfen (MS/FNS 2009; Suter et al. 2017). Recent bioassays with Brazilian populations of *Ae. aegypti* and *Ae. albopictus* that have been exposed for many years to insecticides, in particular *Bti*, showed that both species are equally susceptible to *Bti*, suggesting that the same application rates may be used where the species co-exist (Suter et al. 2017).

A study in Manaus (northern Brazil), using pyriproxyfen, showed not only mosquito mortality, but also that adult emergence was reduced more than 10 times (Abad-Franch et al. 2015). They concluded that this approach is very promising to complement current mosquito control strategies, which heavily rely on the difficult task of detecting vector breeding sites and therefore perform poorly.

In some contexts, however, the application of larvicides by public health services can be complicated. Many *Aedes* breeding sites are small, sheltered and difficult to locate (cryptic habitats). Therefore, depending entirely on breeding site treatment or removal is complex, requiring a combined strategy. Therefore, auto-dissemination methods are an alternative to overcome these limitations, as they rely on the oviposition behaviour of adult mosquitoes and their attraction to breeding sites. The auto-dissemination method to control *Aedes* mosquitoes requires artificial adult resting sites (dissemination stations) to which adult females are attracted and where they are contaminated with pyriproxyfen when entering the station and then contaminate breeding sites with lethal levels of pyriproxyfen (Caputo et al. 2012; Unlu et al. 2017).

5.3. Innovations and Experiences of Other Countries

In a study carried out in Thailand, about 61.8% of water containers were treated with *Bti* and temephos, and the rate of positive containers (with larvae) was reduced from 13.8% in untreated areas to 3.7% in treated areas ($P < 0.001$) showing the combined approach of *Bti* and insecticide were effective in achieving the result in the target area (Arunachalam et al. 2010).

The autodissemination approach was tested in the USA with pyriproxyfen-treated males and showed, in combination with another insecticide, a decline in the *Ae. albopictus* adult population by around 74-78% (Unlu et al. 2018). In a similar approach using only pyriproxyfen, the male mosquitoes were shown to be vehicles of insecticide in areas with low mosquito densities to intoxicate potential breeding sites before the seasonal emergence of the target population (Mains et al. 2015). These males can also contaminate the females, increasing even more the affected breeding sites, interrupting the development of immature offspring. On the other hand, a study only using pyriproxyfen conducted in Florida showed that there was no apparent pupal mortality during the study period (Lloyd et al. 2017).

In Southeast Asia, larvivorous fish, e.g. from the genus *Gambusia*, that feed on mosquito larvae are often used in pots that decorate houses and terraces (Araújo et al. 2015) (Fig. 2). This practice is also employed as a non-insecticidal method to control malaria vectors in India and Africa (Kamareddine 2012; Kant et al. 2013; Walshe et al. 2013). However, the use of fish to control mosquito larvae is feasible and effective only in breeding sites that are easily identified and in those as observed in Asian culture (Chandra et al. 2008).

Studies carried out in Mexico have shown that larvivorous fish can reduce larval and pupal numbers in household water containers, but there was no evidence of a reduction in DENV infection (Morales-Pérez et al. 2017). In villages of Karnataka, South India, the introduction of fish, e.g. the guppy *Poecilia reticulata* (Peters) and *Gambusia affinis* (Baird & Girard), combined with information, education and communication campaigns, had a significant impact on the density of the *Aedes* population and decreased the prevalence of chikungunya (Ghosh et al. 2011). This method is harmless to humans and exhibits minimal risks of mosquito resistance. Besides, the fish are cheap to produce in most cases, saving resources that could serve for other needs. However, in some cases, these invasive fish can be negative effects on biodiversity (El-Sabaawi et al. 2016).

Figure 2. Larvivorous fish in water containers in Southeast Asia (Araújo et al. 2015).

6. ADULT CONTROL

6.1. General Overview

Adulticides are intended to impact a significant number of infected adult mosquitoes in a short time through surface (indoor) and/or spatial (outdoor) treatments with insecticides of residual or low residual activity. The indoor residual spraying (IRS) consists of the application of long-acting chemical insecticides on the walls or others surfaces of houses in a given area using backpack sprayers. A recent review using seven databases evaluated the effectiveness of indoor spraying of insecticides and showed the effect on adult mosquitoes is high immediately after application (Samuel et al. 2017). These spraying activities are usually carried out by staff of the vector control programmes, but the general public can also buy commercial adulticides for use in their homes.

Space spraying is recommended only in emergency situations when people in a large area are at risk of infection, or mosquito densities are very high. The insecticides can be applied by backpack sprayers, trucks or airplanes. When cases of the disease are detected in the early stages of an epidemic, emergency space spraying can reduce disease transmission quickly. However, applying other vector control measures such as larviciding or environmental modification help provide longer-term control as a part of an integrated mosquito management programme (CDC 2017b; MS 2017a; WHO 2017b, 2017f).

6.2. The Brazilian Perspective

In 2016, the Brazilian Government published law No. 13.301 of July 27, 2016 that allows the incorporation of adult vector control mechanisms through aerial spraying upon approval of sanitary authorities and scientific evidence on the efficacy of the measure (MS 2017b). However, in the same year, the Oswaldo Cruz Foundation (FIOCRUZ) issued a technical note stating that there are risks to human health related to the spraying of a neurotoxic product such as malathion in urban areas. They considered that it not only posed a threat to the environment and the population's health, but it is also of little efficacy in the combat of *Ae. aegypti*, which in its adult stage lives mainly within the domiciles (FIOCRUZ 2016).

6.3. Innovations and Experiences of Other Countries

The Florida Keys Mosquito Control District used aerial sprays with insecticide (naled) and bacterial larvicides to reduce *Ae. aegypti* populations in urban areas of Key West, Florida, USA (CDC 2017c; Pruszynski et al. 2017). The aerial applications of *Bti* caused a significant decrease in adult female populations throughout the summer because, in Key West, larvae of this mosquito develop in micro-containers around human habitations. The advantage of aerial spraying of larvicide is the area-wide coverage over and around urban areas achieved in a short period and in the case of Key West, the aerial application of larvicide was effective in controlling the *Ae. aegypti* outbreak (Pruszynski et al. 2017).

7. POPULATION SUPPRESSION INTEGRATING THE STERILE INSECT TECHNIQUE, THE INCOMPATIBLE INSECT TECHNIQUE, AND GENETICALLY MODIFIED MOSQUITOES

To improve the Brazilian dengue vector control programme, it is mandatory to use the principles of IVM to minimise financial and personnel requirements and be able to cover the target geographic area to be treated with the chosen vector control methods. Furthermore, improved monitoring and evaluation tools for vector control should be developed and applied, and relevant training must be performed based on necessity (Horstick et al. 2010).

As described above, suppression of disease-transmitting mosquito populations is still mainly based on insecticides (larvicides and adulticides). A reduction in mosquito densities is the most reliable method to decrease pathogen-host contact, which will reduce the probability of humans becoming infected. Nevertheless, the long-term use of chemical compounds has selected for mosquito populations resistant to them, resulting in the increase of the number of cases in endemic areas and the spread of diseases transmitted by these insects into entirely new areas (Campos et al. 2015; Díaz et al. 2015; Zanluca et al. 2015; Luksic et al. 2017).

Other population suppression approaches are therefore under development and evaluation, and these could be integrated into the currently used IVM approaches. These methods have the benefit that they can reduce vector populations in a target area, without causing the selection of resistance as promoted by insecticides (Bourtzis et al. 2016). They have in common the release of sterilized male insects (because male mosquitoes do not blood-feed and therefore do not transmit diseases), and the monitoring of the sterile and wild male populations in the target area (Lees et al. 2014). These males must be mass-reared to achieve the required numbers to promote suppression of the target population. After release, an efficient monitoring system is needed to be able to follow the vector population fluctuation and if required, to adjust male production and release rates (Hood-Nowotny et al. 2006; Vreysen 2021).

The first of these approaches is the Sterile Insect Technique (SIT), which uses an ionizing radiation source (gamma or X-ray) to sterilize the mass-reared males that will be released into the open field in numbers 10-100 times larger than the wild-type population. The high sterile to wild male overflooding ratios increase the probability of a mating of a wild virgin female with a sterile male (Vreysen et al. 2014; Dyck et al. 2021). For more than 50 years, the SIT has proven to be an effective control tactic to suppress agricultural insect pests such as moths, fruit flies, screwworm and tsetse flies (Hendrichs and Robinson 2009; Klassen et al. 2021). With support from the international scientific community through the International Atomic Energy Agency (IAEA) and the Food and Agricultural Organization of the United Nations (FAO), several countries like Brazil, Cuba, Italy, France (La Réunion), Mauritius, Mexico, Thailand, USA and others have or are initiating pilot trials against mosquitoes on a small to medium scale as a proof-of-concept (Lees et al. 2021).

A similar approach is the release of males that are infected with symbionts that cause sterility without the use of ionizing radiation. The intracellular bacterium, *Wolbachia*, is a symbiont that is sexually transmitted and maternally inherited and can promote cytoplasmic incompatibility in embryos when the father is infected with a particular strain but not the mother (Sinkins 2004). This approach is called the Incompatible Insect Technique (IIT) (Zabalou et al. 2009) and is already under evaluation in several countries like USA (Mains et al. 2016), China, and in French Polynesia for *Aedes polynesiensis* (O'Connor et al. 2012). A related approach under evaluation in Australia, Brazil (Niteroi and Rio de Janeiro), Colombia (Bello and Medellín), and Indonesia (Yogyakarta) through the Eliminate Dengue and other campaigns (De Barro et al. 2011; Maciel-de-Freitas et al. 2012; Flores and O'Neill 2018) involves the release of both *Wolbachia*-infected males and females, resulting in population replacement by substituting the original population with a *Wolbachia*-infected population, this approach takes advantage of *Wolbachia*'s capacity to block pathogen transmission to the human host (Van den Hurk et al. 2012; Frentiu et al. 2014; Dutra et al. 2016).

The release of genetically modified mosquitoes (GMM) is the third population suppression alternative that is under evaluation, and so far, some programmes or trials have demonstrated success in reducing the mosquito population in the target areas Carvalho et al. (2015). This transgenic approach, which requires regulatory approvals and involves other issues, has a broad range of possibilities to interfere and trigger mosquito population suppression or population replacement by blocking disease transmission, due to its potential to manipulate, exclude or include new features at the genomic level of the target mosquito species (Handler 2002; Travanty et al. 2004; Catteruccia et al. 2009).

Brazil is one of the best locations to test and evaluate these new technologies, due to its diverse environments and extensive prevalence of arboviruses. Since 2002, Brazil has been implementing the PNCD to control dengue transmission and related diseases, such as chikungunya, Zika, and yellow fever. However, the efforts and strategies that are combined in the PNCD cannot entirely prevent disease dissemination; on the contrary, the number of reported cases only increases every year (Pessanha et al. 2009; SS-PE 2015; MS 2017c). Therefore, the inclusion of new technologies cannot alone change the vector density and transmission situation if their deployment is not carefully planned according to the specific characteristics and needs of each target area and taking advantage of the best characteristic of each of the technologies. Thus, it is necessary to combine and better apply where appropriate all these techniques as part of effective IVM approaches (Horstick et al. 2010; Bourtzis et al. 2016; Van den Berg et al. 2012).

7.1. Two-step Male Release Strategy – Integration of Techniques

Several models on the use of these inovative technologies, such as GMM, the IIT, the SIT and others, predict that it will take several seasons to suppress a targetted mosquito population, and even when achieving it, some virus transmission can still occur (Andraud et al. 2012; Chen and Hsieh 2012; Okamoto et al. 2013; Ndii et al. 2015). The IVM approaches can be improved by applying the suppression methods more effectively and based on mosquito biology. Models combining several techniques demonstrate the advantages of targeting different developmental stages and integrating different ways to suppress a population.

A two-step strategy was proposed to reduce mosquito populations and then block efficiently disease transmission (Carvalho et al. 2014). A first step involves the integration of any methods which have a significant impact decreasing the target vector population, such as the use of larvicides and adulticides, educational campaigns, breeding site elimination, the release of sterile males, which also can be carrying pyriproxyfen to suppress a population. Once the population has been suppressed, this should be followed by a second step, which could involve releases targeting population substitution (for example *Wolbachia*-infected females or GMM), in order to disrupt disease transmission entirely. The idea is first to reduce the mosquito population to extremely low levels, and then to substitute this residual population by one that is no longer able to transmit viruses, thereby obtaining the advantage of this low-cost combination strategy that can be implemented as part of IVM over larger areas (Fig. 3).

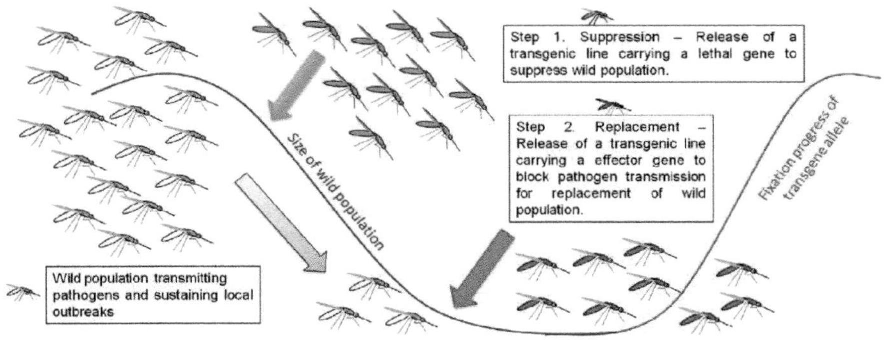

Figure 3. A schematic diagram of two-step male release strategy (Carvalho et al. 2014) using as an example a genetically modified mosquito (GMM) strain for population suppression and replacement.

Recently, some studies proposed transgenic constructions able to block the virus transmission with low impact on the overall mosquito fitness (Jupatanakul et al. 2017; Buchman et al. 2019). Strategies using RNA interference or RNA-based strategies targeting critical virus RNA's were already developed and tested under laboratory conditions and their fitness evaluated (Franz et al. 2006, 2014; Buchman et al. 2019). In addition, other strains targeting malaria parasites have also been developed for population replacement, entirely blocking the parasite transmission (Kokoza et al. 2010).

Nevertheless, it is a common evolutionary fact that without a stable gene drive mechanism such systems alone may not be enough to replace the population successfully, and over time will be displaced (James 2005). However, the use of new gene editing techniques, such as CRISPR-Cas9, provide an easier way not only to create strains for population suppression, but also for population replacement (Häcker and Schetelig, this volume). They may also eventually overcome the issue of gene drive resistance mechanisms that emerge in field populations (Champer et al. 2018). This resistance can originate from the drive itself, when cleavage is repaired and it changes the sequence of the target site, so that it can no longer be recognized, becoming resistant to future conversion.

A model provides that more than 100 generations are needed for the wild-type population to reach 50% of resistant alleles and the use of CRISPR-Cas9 can be an efficient way to provide stable strains for vector control programmes without the accumulation of genetical instability (Unckless et al. 2017). It is a matter of time for the availability and developing state of art of the gene drive technologies to provide further information on their behaviour in the genome and their ecological impact and long-term effects.

Further discussion among the scientific community, stakeholders and population regarding the advantages, risk assessment, and regulatory issues of using them are needed (Carter and Friedman 2016; Häcker and Schetelig, this volume; Nielsen, this volume).

7.2. *Open Field Release Using the OX513A* Ae. aegypti *Transgenic Line and its Evaluation*

Brazil and other countries have initiated the field assessment of the impact of some of these new technologies as part of IVM approaches. The first continental GMM release to suppress an *Ae. aegypti* mosquito population was carried out between 2011 to 2015 in two different cities, Juazeiro and Jacobina in the state of Bahia, in north-eastern Brazil. This *Projeto Aedes Transgênico* aimed at evaluating various aspects of a full IVM programme by using the OX513A transgenic line developed by the commercial company Oxitec Ltd. (Lee et al. 2009). Before the release of the genetically modified male mosquitoes in Brazil, several regulatory steps, as described by Carvalho and Capurro (2015), had to be performed. The most crucial approval was provided by the Brazilian National Committee of Biosafety, which regulates all research projects and products directly and indirectly involving genetically modified organisms, including a public review of the project that had no vote against it.

Due to all apprehensions around genetically modified organisms in plants and other organisms, the *Projeto Aedes Transgênico* initiated a pioneering communication plan to create adequate public awareness regarding the use of this technology and its purpose. Emphasis was likewise placed on community engagement and stakeholder participation during the execution of the project. This experience can serve as a model for other initiatives using the same approach (Carvalho et al. 2015; Capurro et al. 2016).

In the initial phase, some quality parameters of these GMM males were assessed in the target area/environment, such as flight range and longevity under field conditions. Based on this first phase, an assessment was made of the number of sterile male mosquitoes to be released to achieve population suppression, and the data compared with those of the first trial in Grand Cayman Island (Harris et al. 2011, 2012). This range-finding process, consisting of six weeks of releases and the three following weeks for evaluation (around 2800 males/ha/week were released in this first phase), was helpful in optimizing the release number and mass-rearing process. It was also crucial for the next phase involving overflooding the target area in Juazeiro with male mosquitoes for suppression purposes because it provided and confirmed parameters to initiate this suppression phase (Carvalho et al. 2015).

After the 17 months release period, around 95% of population suppression was achieved in Juazeiro, based on an indirect evaluation using a monitoring system with ovitraps. Afterwards, the study kept track of the GMM and wild-type populations after the suppression effect. The outcome was that when the releases were discontinued, the wild mosquito population returned rapidly to pre-control levels within 17 weeks due to immigration and other factors, such the eclosion of eggs that remained unhatched during the release phase (Garziera et al. 2017).

The second part of *Projeto Aedes Transgênico* included procedures for ground shipment of pupae to the city of Jacobina, around 300 km from the mass-rearing facility. This step also included optimizing the monitoring system, increased community engagement and awareness activities, and improved efficiency of mass-rearing procedures and release methods. The *Projeto Aedes Transgênico* was terminated at the end of the contract with the Bahia State Health Department. Again, genetic monitoring of the GMM and wild-type populations continued post-suppression, indicating that portions of the transgenic strain genome became incorporated into the target population (Evans et al. 2019).

In parallel, as an independent initiative, Oxitec started a trial in Piracicaba and Juiz de Fora municipalities (in São Paulo state, south-eastern Brazil), following a similar approach and using the predetermined parameters established during the first two initials trials in the country (Paes de Andrade et al. 2016). So far these trials are service contracts directly performed with the municipalities without any support of the Brazilian Ministry of Health.

8. FINAL CONSIDERATIONS

Numerous activities are currently being integrated in Brazil to suppress *Ae. aegypti,* the vector of various arboviruses. Box 1 presents some important bullets summarizing the strengths, weaknesses, opportunities, and threats (SWOT) for the Brazilian vector control strategy.

Box 1. Strengths, weaknesses, opportunities, and threats of the Brazilian vector control strategy.

STRENGTHS	WEAKNESSES
• Strong research institutions • Decentralisation of the vector control • Existence of the National Plan for Dengue Control (PNCD) • Historical record of successful vector elimination • Reference research laboratories	• Insufficient budget / trained staff • Insufficient public mobilization resulting in low community commitment and household participation and acceptance • Limited time required for data analysis resulting in poor management • Lack of consistent and frequent control strategy application
OPPORTUNITIES	**THREATS**
• Possibility to test new techniques in different biomes and urbanization levels • Use of different combinations and methods for vector control for specific conditions • Learning from different models and previous experiences (including other countries)	• The continuous increase in reported arboviruses cases • Different vector species participating in disease transmission • Entry of new arboviruses promoting illness • Difficulty of treating increasing cases in big cities

PNCD activities and efforts are not enough to interfere with disease transmission by this vector. All have been recommended by WHO (Brasil 2009), however they have to be adapted to different levels of difficulty in different situations (for example, vector control in isolated small areas vs non-isolated, large and densely populated urban areas). Among the main reasons for the insufficient control are the low budget, lack of trained staff, insufficient insecticide application, insufficient public mobilization, and poor management. There is a need to increase vector control efforts all over the country, but at the same time to complement the adopted strategies with promising innovative approaches (Zara et al. 2016; Coelho 2012). There is the potential to exploit and include new methods as part of the IVM package in order to suppress more effectively and sustainably the mosquito populations and control disease transmission.

The current PNCD activities being performed should not be interrupted due to the advent of new technologies, but these can be validated and implemented as part of the IVM package. The range of approaches integrating new technologies is huge, and they have demonstrated that they can successfully contribute to mosquito population suppression and reduce disease transmission. In view of the proof-of-concept of these techniques (most of them carried out under Brazilian conditions), they are ready for the next step, which is their application as part of a long-term programme, not only to demonstrate their effect on mosquito populations, but also their impact on disease transmission.

9. REFERENCES

Abad-Franch, F., E. Zamora-Perea, G. Ferraz, S. D. Padilla-Torres, and S. L. B. Luz. 2015. Mosquito-disseminated pyriproxyfen yields high breeding-site coverage and boosts juvenile mosquito mortality at the neighborhood scale. PLoS Neglected Tropical Diseases 9 (4): e0003702.

Alves, A. C., A. L. dal Fabbro, A. D. Costa Passos, A. F. T. Mendes Carneiro, T. Martins Jorge, and E. Zangiacomi Martinez. 2016. Knowledge and practices related to dengue and its vector: A community-based study from southeast Brazil. Revista da Sociedade Brasileira de Medicina Tropical 49 (2): 222–226.

Andraud, M., N. Hens, C. Marais, and P. Beutels. 2012. Dynamic epidemiological models for dengue transmission: A systematic review of structural approaches. PLoS One 7: e49085.

Araújo, H. R. C., D. Carvalho, R. Ioshino, A. Costa-da-Silva, and M. Capurro. 2015. *Aedes aegypti* control strategies in Brazil: Incorporation of new technologies to overcome the persistence of dengue epidemics. Insects 6: 576–594.

Araújo, V. E. M., J. M. T. Bezerra, F. F. Amâncio, V. M. de A. Passos, and M. Carneiro. 2017. Aumento da carga de dengue no Brasil e unidades federadas, 2000 e 2015: Análise do Global Burden of Disease Study 2015. Revista Brasileira de Epidemiologia 20 (Suppl. 1): 205–216.

Arunachalam, N., S. Tana, F. Espino, P. Kittayapong, W. Abeyewickreme, K. T. Wai, B. K. Tyagi, A. Kroeger, J. Sommerfeld, and M. Petzold. 2010. Eco-bio-social determinants of dengue vector breeding: A multicountry study in urban and periurban Asia. Bulletin of the World Health Organization 88: 173–184.

Arunachalam, N., B. K. Tyagi, M. Samuel, R. Krishnamoorthi, R. Manavalan, S. C. Tewari, V Ashokkumar, A. Kroeger, J. Sommerfeld, and M. Petzold. 2012. Community-based control of *Aedes aegypti* by adoption of eco-health methods in Chennai City, India. Pathogens and Global Health 106: 488–496.

Augusto, L. G. S., A. M. Gurgel, A. M. Costa, F. Diderichsen, F. A. Lacaz, G. Parra-Henao, R. M. Rigotto, R. Nodari, and S. L. Santos. 2016. *Aedes aegypti* control in Brazil. The Lancet 387: 1052–1053.

Bourtzis, K., R. S. Lees, J. Hendrichs, and M. J. B. Vreysen. 2016. More than one rabbit out of the hat: Radiation, transgenic and symbiont-based approaches for sustainable management of mosquito and tsetse fly populations. Acta Tropica 157: 115–130.

Bouyer, J., and T. Lefrançois. 2014. Boosting the Sterile Insect Technique to control mosquitoes. Trends in Parasitology 30(6): 271–273.

Braga, I., and D. Valle. 2007. *Aedes aegypti*: Histórico do controle no Brasil. Epidemiologia e Serviços de Saúde 16: 113–118.

Brasil. 2009. Diretrizes nacionais para a prevenção e controle de epidemias de dengue. Ministerio da Saude, Secretaria de Vigilância em Saúde, and Departamento de Vigilância Epidemiológica. 1st ed. Série A. Normas e Manuais Técnicos. Brasilia, D. F., Brasil.

Brasil. 2016. Lei No 13.301, de 27 de junho de 2016. Dispõe sobre a adoção de medidas de vigilância em saúde quando verificada situação de iminente perigo à saúde pública pela presença do mosquito transmissor do vírus da dengue, do vírus chikungunya e do vírus da Zika; e altera a Lei no 6.437, de 20 de agosto de 1977. Diário Oficial da União, Brasilia, DF, 28 de junho de 2016. Seção 1, p. 1. ISSN 1677–7042

Buchman, A., S. Gamez, M. Li, I. Antoshechkin, H.-H. Li, H.-W. Wang, C.-H. Chen, M. J. Klein, J.-B. Duchemin, P. N. Paradkar, and O. S. Akbari. 2019. Engineered resistance to Zika virus in transgenic *Aedes aegypti* expressing a polycistronic cluster of synthetic small RNAs. Proceedings of the National Academy of Sciences of the United States of America 116 (9): 3656–3661.

Campos, G. S., A. C. O. Pinho, C. J. de Freitas, A. C. Bandeira, and S. I. Sardi. 2015. Dengue virus 4 (DENV-4) re-emerges after 30 years in Brazil: Cocirculation of DENV-2, DENV-3, and DENV-4 in Bahia. Japanese Journal of Infectious Diseases 68: 45–49.

Caprara, A., J. W. De Oliveira Lima, A. C. Rocha Peixoto, C. M. Vasconcelos Motta, J. M. Soares Nobre, J. Sommerfeld, and A. Kroeger. 2015. Entomological impact and social participation in dengue control: A cluster randomized trial in Fortaleza, Brazil. Transactions of the Royal Society of Tropical Medicine and Hygiene 109 (2): 99–105.

Capurro, M. L., D. O. Carvalho, L. Garziera, C. Pedrosa, I. Damasceno, I. Lima, B. Duarte, J. Fernandez-Virginio, R. S. Lees, and A. Malavasi. 2016. Description of social aspects surrounding releases of transgenic mosquitoes in Brazil. International Journal of Recent Scientific Research 7: 10363–10369.

Caputo, B., A. Ienco, D. Cianci, M. Pombi, V. Petrarca, A. Baseggio, G. J. Devine, and A. della Torre. 2012. The "auto-dissemination" approach: A novel concept to fight *Aedes albopictus* in urban areas. PLoS Neglected Tropical Diseases 6(8): e1793.

Caputo, B., A. Ienco, M. Manica, V, Petrarca, R. Rosà, and A. della Torre. 2015. New adhesive traps to monitor urban mosquitoes with a case study to assess the efficacy of insecticide control strategies in temperate areas. Parasites & Vectors 8: 134.

Carrillo-Hernández, M. Y., J. Ruiz-Saenz, L. Jaimes-Villamizar, S. Y. Gómez-Rangel, and M. Martínez-Gutierrez. 2018. Co-circulation and simultaneous co-infection of dengue, chikungunya, and Zika viruses in patients with febrile syndrome at the Colombian-Venezuelan border. BMC Infectious Diseases 18 (1): 61.

Carter, S. R., and R. M. Friedman. 2016. Policy and regulatory issues for gene drives in insects: Workshop report. J. Craig Venter Institute, San Diego, California, USA. 21 pp.

Carvalho, D. O., and M. L. Capurro. 2015. Community engagement, pp. 409–422. *In* J. Leonard (ed.), Genetic control of malaria and dengue. Elsevier, Amsterdam, The Netherlands.

Carvalho, D. O., A. L. Costa-da-Silva, R. S. Lees, and M. L. Capurro. 2014. Two step male release strategy using transgenic mosquito lines to control transmission of vector-borne diseases. Acta Tropica 132: 1–8.

Carvalho, D. O., A. R. McKemey, L. Garziera, R. Lacroix, C. A. Donnelly, L. Alphey, A. Malavasi, and M. L. Capurro. 2015. Suppression of a field population of *Aedes aegypti* in Brazil by sustained release of transgenic male mosquitoes. PLoS Neglected Tropical Diseases 9: e0003864.

Carvalho, S. A., S. O. da Silva, and I. D. C. Charret. 2019. Mathematical modeling of dengue epidemic: Control methods and vaccination strategies. Theory in Biosciences: 1–17.

Catteruccia, F., A. Crisanti, and E. A. Wimmer. 2009. Transgenic technologies to induce sterility. Malaria Journal 8 (Suppl. 2): S7.

(CDC) Centers for Disease Control and Prevention. 2017a. Integrated Mosquito Management - Zika virus. Atlanta, Georgia, USA.

(CDC) Centers for Disease Control and Prevention. 2017b. Integrated Mosquito Management for *Aedes aegypti* and *Aedes albopictus* mosquitoes. Atlanta, Georgia, USA.

(CDC) Centers for Disease Control and Prevention. 2017c. Information on aerial spraying - Zika virus. Atlanta, Georgia, USA.

Chaebo, G., and J. J. Medeiros. 2017. Conditions for policy implementation via co-production: The control of dengue fever in Brazil. Public Management Review 19 (10): 1381–1398.

Champer, J., J. Liu, S. Y. Oh, R. Reeves, A. Luthra, N. Oakes, A. G. Clark, and P. W. Messer. 2018. Reducing resistance allele formation in CRISPR gene drive. Proceedings of the National Academy of Sciences of the United States of America 115 (21): 5522–5527.

Chandra, G., I. Bhattacharjee, S. N. Chatterjee, and A. Ghosh. 2008. Mosquito control by larvivorous fish. The Indian Journal of Medical Research 127: 13–27.

Chediak, M., F. G. Pimenta, G. E. Coelho, I. A. Braga, J. B. P. Lima, K. R. L. Cavalcante, L. C. de Sousa, M. A. V. de Melo-Santos, M. L. G. Macoris, A. P. de Araújo, C. F. J. Ayres, M. T. M. Andrighetti, R. Gonçalves de Gomes, and R. N. C. Guedes. 2016. Spatial and temporal country-wide survey of temephos resistance in Brazilian populations of *Aedes aegypti*. Memorias Do Instituto Oswaldo Cruz 111: 311–321.

Chen, S. C., and M. H. Hsieh. 2012. Modeling the transmission dynamics of dengue fever: Implications of temperature effects. Science of the Total Environment 431: 385–391.

Claro, L. B. L., H. Kawa, L. T. Cavallini, and M. L. Garcia Rosa. 2006. Community participation in dengue control in Brazil. WHO Regional Office for South-East Asia. Dengue Bulletin 30: 214–222.

Coelho, G. E. 2012. Challenges in the control of *Aedes aegypti*. Revista do Instituto de Medicina Tropical de São Paulo 54 (Suppl. 18): S13–14.

De Barro, P. J., B. Murphy, C. C. Jansen, and J. Murray. 2011. The proposed release of the yellow fever mosquito, *Aedes aegypti* containing a naturally occurring strain of *Wolbachia pipientis*, a question of regulatory responsibility. Journal of Consumer Protection and Food Safety 6: 33–40.

Degener, C. M., T. M. F. de Ázara, R. A. Roque, C. T. Codeço, A. Araújo Nobre, J. J. Ohly, M. Geier, and Á. E. Eiras. 2014. Temporal abundance of *Aedes aegypti* in Manaus, Brazil, measured by two trap types for adult mosquitoes. Memórias Do Instituto Oswaldo Cruz 109: 1030–1040.

Degener, C. M., T. M. F. de Ázara, R. A. Roque, S. Rösner, E. S. Oliveira Rocha, E. G. Kroon, C. T. Codeço, A. Araújo Nobre, J. J. Ohly, M. Geier, and A. E. Eiras. 2015. Mass trapping with MosquiTRAPs does not reduce *Aedes aegypti* abundance. Memórias do Instituto Oswaldo Cruz 110 (4): 517–527.

Díaz, Y., J.-P. Carrera, L. Cerezo, D. Arauz, I. Guerra, J. Cisneros, B. Armién, A. M. Botello, A. B. Araúz, V. Gonzalez, Y. López, L. Moreno, S. López-Vergès, and B. A. Moreno. 2015. Chikungunya virus infection: First detection of imported and autochthonous cases in Panama. The American Journal of Tropical Medicine and Hygiene 92: 482–485.

Dutra, H. L. C., M. N. Rocha, F. B. S. Dias, S. B. Mansur, E. P. Caragata, and L. A. Moreira. 2016. *Wolbachia* blocks currently circulating Zika virus isolates in Brazilian *Aedes aegypti* mosquitoes. Cell Host & Microbe 19: 771–774.

Dyck, V. A., J. Hendrichs, and A. S. Robinson (eds.). 2021. Sterile Insect Technique – Principles and practice in Area-Wide Integrated Pest Management. Second Edition. CRC Press, Boca Raton, Florida, USA. 1200 pp.

El-Sabaawi, R. W., T. C. Frauendorf, P. S. Marques, R. A. Mackenzie, L. R. Manna, R. Mazzoni, D. A. Phillip, M. L. Warbanski, and E. Zandonà. 2016. Biodiversity and ecosystem risks arising from using guppies to control mosquitoes. Biology Letters 12: 20160590.

Evans, B. R., P. Kotsakiozi, A. L. Costa-da-Silva, R. S. Ioshino, L. Garziera, M. C. Pedrosa, A. Malavasi, J. F. Virginio, M. L. Capurro, and J. R. Powell. 2019. Transgenic *Aedes aegypti* mosquitoes transfer genes into a natural population. Scientific Reports 9: 13047.

Federici, B. A., H.-W. Park, D. K. Bideshi, M. C. Wirth, J. J. Johnson, Y. Sakano, and M. Tang. 2007. Developing recombinant bacteria for control of mosquito larvae. Journal of the American Mosquito Control Association 23 (Suppl. 2): 164–175.

(FIOCRUZ) Oswaldo Cruz Foundation. 2016. Nota Técnica No. 4/2016/IOC-FIOCRUZ/DIRETORIA. Considerações técnicas sobre a aplicação aérea de inseticidas em área urbana.

Flores, H. A., and S. L. O'Neill. 2018. Controlling vector-borne diseases by releasing modified mosquitoes. Nature Reviews Microbiology 16(8): 508–518.

Franz, A. W. E., I. Sanchez-Vargas, Z. N. Adelman, C. D. Blair, B. J. Beaty, A. A. James, and K. E. Olson. 2006. Engineering RNA interference-based resistance to dengue virus type 2 in genetically modified *Aedes aegypti*. Proceedings of the National Academy of Sciences of the United States of America 103 (11): 4198–4203.

Franz, A. W. E., I. Sanchez-Vargas, R. R. Raban, W. C. Black IV, A. A. James, and K. E. Olson. 2014. Fitness impact and stability of a transgene conferring resistance to dengue-2 virus following introgression into a genetically diverse *Aedes aegypti* strain. PLoS Neglected Tropical Diseases 8 (5): e2833.

Frentiu, F. D., T. Zakir, T. Walker, J. Popovici, A. T. Pyke, A. van den Hurk, E. A. McGraw, and S. L. O'Neill. 2014. Limited dengue virus replication in field-collected *Aedes aegypti* mosquitoes infected with *Wolbachia*. PLoS Neglected Tropical Diseases 8: e2688.

Furuya-Kanamori, L., S. Liang, G. Milinovich, R. J. S. Magalhaes, A. C. A. Clements, W. Hu, P. Brasil, F. D. Frentiu, R. Dunning, and L. Yakob. 2016. Co-distribution and co-infection of chikungunya and dengue viruses. BMC Infectious Diseases 16: 84.

Garziera, L. , M. C. Pedrosa, F. A. Souza, M. Gomez, M. B. Moreira, J. F. Virginio, M. Capurro, and D. O. Carvalho. 2017. Effect of interruption of over-flooding releases of transgenic mosquitoes over wild population of *Aedes aegypti*: Two case studies in Brazil. Entomologia Experimentalis et Applicata 164: 327–339.

Ghosh, S. K., P. Chakaravarthy, S. R. Panch, P. Krishnappa, S. Tiwari, V. P. Ojha, R. Manjushree, and A. P. Dash. 2011. Comparative efficacy of two poeciliid fish in indoor cement tanks against chikungunya vector *Aedes aegypti* in villages in Karnataka, India. BMC Public Health 11: 599.

Gubler, D. J., and G. C. Clark. 1996. Community involvement in the control of *Aedes aegypti*. Acta Tropica 61: 169–179.

Handler, A. M. 2002. Prospects for using genetic transformation for improved SIT and new biocontrol methods. Genetica 116: 137–149.

Harris, A. F., D. Nimmo, A. R. McKemey, N. Kelly, S. Scaife, C. A. Donnelly, C. Beech, W. D. Petrie, and L. Alphey. 2011. Field performance of engineered male mosquitoes. Nature Biotechnology 29: 1034–1037.

Harris, A. F., A. R. McKemey, D. Nimmo, Z. Curtis, I. Black, S. A. Morgan, M. N. Oviedo, R. Lacroix, N. Naish, N. I. Morrison, A. Collado, J. Stevenson, S. Scaife, T. Dafa'alla, G. Fu, C. Phillips, A. Miles, N. Raduan, N. Kelly, C. Beech, C. A. Donnelly, W. D. Petrie, and L. Alphey. 2012. Successful suppression of a field mosquito population by sustained release of engineered male mosquitoes. Nature Biotechnology 30: 828–830.

Healy, K., G. Hamilton, T. Crepeau, S. Healy, I. Unlu, A. Farajollahi, and D. M. Fonseca. 2014. Integrating the public in mosquito management: Active education by community peers can lead to significant reduction in peridomestic container mosquito habitats. PLoS One 9: e108504.

Hendrichs, J., and A. S. Robinson. 2009. Sterile Insect Technique, pp. 953–957. *In* V. H. Resh, and R. T. Cardé (eds.), Encyclopedia of insects. Second Edition. Academic Press, Burlington, Massachussetts, USA.

Honório, N. A., C. T. Codeço, F. C. Alves, M. A. F. M. Magalhães, and R. Lourenço-De-Oliveira. 2009. Temporal distribution of *Aedes aegypti* in different districts of Rio de Janeiro, Brazil, measured by two types of traps. Journal of Medical Entomology 46: 1001–1014.

Hood-Nowotny, R., L. Mayr, and B. Knols. 2006. Use of carbon-13 as a population marker for *Anopheles arabiensis* in a Sterile Insect Technique (SIT) context. Malaria Journal 5(1): 6.

Horstick, O., S. Runge-Ranzinger, M. B. Nathan, and A. Kroeger. 2010. Dengue vector-control services: How do they work? A systematic literature review and country case studies. Transactions of the Royal Society of Tropical Medicine and Hygiene 104(6): 379–386.

Ibrahim, M. A., N. Griko, M. Junker, and L. A. Bulla. 2010. *Bacillus thuringiensis*. Bioengineered Bugs 1: 31–50.

James, A. A. 2005. Gene drive systems in mosquitoes: Rules of the road. Trends in Parasitology 21 (2): 64–67.

Johnson, B. J., S. A. Ritchie, and D. M. Fonseca. 2017. The state of the art of lethal oviposition trap-based mass interventions for arboviral control. Insects 8(1): 5.

Jupatanakul, N., S. Sim, Y. I. Angleró-Rodríguez, J. Souza-Neto, S. Das, K. E. Poti, S. L. Rossi, N. Bergren, N. Vasilakis, and G. Dimopoulos. 2017. Engineered *Aedes aegypti* JAK/STAT pathway-mediated immunity to dengue virus. PLoS Neglected Tropical Diseases 11 (1): e0005187.

Kamareddine, L. 2012. The biological control of the malaria vector. Toxins 4(9): 748–767.

Kant, R., S. Haq, H. C. Srivastava, and V. P. Sharma. 2013. Review of the bioenvironmental methods for malaria control with special reference to the use of larvivorous fishes and composite fish culture in central Gujarat, India. Journal of Vector Borne Diseases 50: 1–12.

Kay, B., and V. S. Nam. 2005. New strategy against *Aedes aegypti* in Vietnam. The Lancet 365 (9459): 613–617.

Kittayapong, P., S. Yoksan, U. Chansang, C. Chansang, and A. Bhumiratana. 2008. Suppression of dengue transmission by application of integrated vector control strategies at sero-positive GIS-based foci. The American Journal of Tropical Medicine and Hygiene 78: 70–76.

Klassen, W., C. F. Curtis, and J. Hendrichs. 2021. History of the Sterile Insect Technique, pp. 1–44. *In* V. A. Dyck, J. Hendrichs, and A. S. Robinson (eds.), Sterile Insect Technique – Principles and practice in Area-Wide Integrated Pest Management. Second Edition. CRC Press, Boca Raton, Florida, USA.

Kokoza, V., A. Ahmed, S. Woon Shin, N. Okafor, Z. Zou, and A. S. Raikhel. 2010. Blocking of *Plasmodium* transmission by cooperative action of Cecropin A and Defensin A in transgenic *Aedes aegypti* mosquitoes. Proceedings of the National Academy of Sciences of the United States of America 107 (18): 8111–8116.

Lee, H. L., H. Jokob, W. A. Naznia, and S. S. Vasanc. 2009. Comparative life history parameters of transgenic and wild strains of *Ae. aegypti* in the laboratory. Dengue Bulletin 33: 103–114.

Lees, R. S., B. Knols, R. Bellini, M. Q. Benedict, A. Bheecarry, H. C. Bossin, D. D. Chadee, J. Charlwood, R. K. Dabiré, L. Djogbenou, A. Egyir-Yawson, R. Gato, L. C. Gouagna, M. M. Hassan, S. A. Khan, L. L. Koekemoer, G. Lemperiere, N. C. Manoukis, and J. R. L. Gilles. 2014. Review: Improving our knowledge of male mosquito biology in relation to genetic control programmes. Acta Tropica 132 (Suppl.): S2–S11.

Lees, R. S., D. Carvalho, and J. Bouyer. 2021. Potential impact of the integration of the Sterile Insect Technique in the fight against disease-transmitting mosquitoes, pp. 1081–1118. *In* V. A. Dyck, J. Hendrichs, and A. S. Robinson (eds.), Sterile Insect Technique – Principles and practice in Area-Wide Integrated Pest Management. Second Edition. CRC Press, Boca Raton, Florida, USA.

Leo, Y. S., A. L. P. Chow, L. K. Tan, D. C. Lye, L. Lin, and L. C. Ng. 2009. Chikungunya outbreak, Singapore, 2008. Emerging Infectious Diseases 15: 836–837.

Lloyd, A. M., M. Farooq, A. S. Estep, R.-D. Xue, and D. L. Kline. 2017. Evaluation of pyriproxyfen dissemination via *Aedes albopictus* from a point-source larvicide application in northeast Florida. Journal of the American Mosquito Control Association 33 (2): 151–155.

Lorenzi, O. D., C. Major, V. Acevedo, J. Perez-Padilla, A. Rivera, B. J. Biggerstaff, J. Munoz-Jordan, S. Waterman, R. Barrera, and T. M. Sharp. 2016. Reduced incidence of chikungunya virus infection in communities with ongoing *Aedes aegypti* mosquito trap intervention studies — Salinas and Guayama, Puerto Rico, November 2015–February 2016. Morbidity and Mortality Weekly Report 65(18): 479–480.

Luksic, B., N. Pandak, E. Drazic-Maras, S. Karabuva, M. Radic, A. Babic-Erceg, L. Barbic, V. Stevanovic, and T. Vilibic-Cavlek. 2017. First case of imported chikungunya infection in Croatia, 2016. International Medical Case Reports Journal 10: 117–121.

Maciel-de-Freitas, R., R. Aguiar, R. V. Bruno, M. C. Guimarães, R. Lourenço-de-Oliveira, M. H. F. Sorgin, C. J. Struchiner, D. Valle, S. L O'Neill, and L. A. Moreira. 2012. Why do we need alternative tools to control mosquito-borne diseases in Latin America? Memorias Do Instituto Oswaldo Cruz 107: 828–829.

Mackay, A. J., M. Amador, and R. Barrera. 2013. An improved autocidal gravid ovitrap for the control and surveillance of *Aedes aegypti*. Parasites & Vectors 6(1): 225.

Mains, J. W., C. L. Brelsfoard, and S. L. Dobson. 2015. Male mosquitoes as vehicles for insecticide. PLoS Neglected Tropical Diseases 9(1): e0003406.

Mains, J. W., C. L. Brelsfoard, R. I. Rose, and S. L. Dobson. 2016. Female adult *Aedes albopictus* suppression by *Wolbachia*-infected male mosquitoes. Scientific Reports 6 (September): 33846.

Martelli, C. M. T., J. B. Siqueira, M. P. P. D. Parente, A. L. S. A. Zara, C. S. Oliveira, C. Braga, F. G. Pimenta Junior, F. Cortes, J. G. Lopez, L. R. Bahia, M. C. Ooteman Mendes, M. Q. Machado da Rosa, N. T. Siqueira Filha, D. Constenla, and W. V. Souza. 2015. Economic impact of dengue: Multicenter study across four Brazilian regions. Edited by M. S. Carvalho. PLoS Neglected Tropical Diseases 9 (9): e0004042.

Martins, A. J., J. B. Lima, A. A. Peixoto, and D. Valle. 2009. Frequency of Val1016Ile mutation in the voltage-gated sodium channel gene of *Aedes aegypti* Brazilian populations. Tropical Medicine & International Health 14: 1351–1355.

Maurer-Stroh, S., T.-M. Mak, Y.-K. Ng, S.-P. Phuah, R. G. Huber, J. K. Marzinek, D. A Holdbrook, R. T. C. Lee, L. Cui, and R. T. Lin. 2016. South-east Asian Zika virus strain linked to cluster of cases in Singapore, August 2016. Euro Surveillance 21(38): 30347.

Mendenhall, I. H., M. Manuel, M. Moorthy, T. T. M., Lee, D. H. W. Low, D. Missé, D. J. Gubler, B. R. Ellis, E. E. Ooi, and J. Pompon. 2017. Peridomestic *Aedes malayensis* and *Aedes albopictus* are capable vectors of arboviruses in cities. PLoS Neglected Tropical Diseases 11(6): e0005667.

Morales-Pérez, A., E. Nava-Aguilera, J. Legorreta-Soberanis, A. J. Cortés-Guzmán, A. Balanzar-Martínez, E. Harris, J. Coloma, V. M. Alvarado-Castro, M. V. Bonilla-Leon, L. Morales-Nava, R. J. Ledogar, A. Cockcroft, and N. Andersson. 2017. Where we put little fish in the water there are no mosquitoes: A cross-sectional study on biological control of the *Aedes aegypti* vector in 90 coastal-region communities of Guerrero, Mexico. BMC Public Health 17 (Suppl. 1): 433.

(MS) Ministério da Saúde. 2014a. Larvicidas. Brasilia, DF, Brasil.

(MS) Ministério da Saúde. 2014b. Aplicação espacial de inseticidas. Brasilia, DF, Brasil.

(MS) Ministério da Saúde. 2017a. Controle de vetores / Inseticidas e larvicidas. Brasilia, DF, Brasil.

(MS) Ministério da Saúde. 2017b. Zika virus in Brazil: The SUS Response. Brasilia, DF, Brasil.

(MS) Ministério da Saúde. 2017c. Monitoramento dos casos de dengue, febre de chikungunya e febre pelo vírus Zika até a Semana Epidemiológica 4, 2017. Brasilia, DF, Brasil.

(MS/FNS) Ministério da Saúde / Fundação Nacional de Saúde. 2002. Programa Nacional de Controle da Dengue (PNCD). Ministério da Saúde. Fundação Nacional de Saúde Brasilia, DF, Brasil.

(MS/FNS) Ministério da Saúde / Fundação Nacional de Saúde. 2009. Diretrizes nacionais para a prevenção e controle de epidemias de dengue. Secretaria de Vigilancia em Saúde. Brasilia, DF, Brasil.

(NASEM) National Academies of Sciences, Engineering, and Medicine. 2016. Global health impacts of vector-borne diseases: Workshop summary. National Academies Press. Washington DC, USA. 396 pp.

Ndii, M. Z., R. I. Hickson, D. Allingham, and G. N. Mercer. 2015. Modelling the transmission dynamics of dengue in the presence of *Wolbachia*. Mathematical Biosciences 262: 157–166.

O'Connor, L., C. Plichart, A. C. Sang, C. L. Brelsfoard, H. C. Bossin, and S. L. Dobson. 2012. Open release of male mosquitoes infected with a *Wolbachia* biopesticide: Field performance and infection containment. Edited by Jason L. Rasgon. PLoS Neglected Tropical Diseases 6 (11): e1797.

Okamoto, K. W., M. A. Robert, A. L. Lloyd, and F. Gould. 2013. A reduce and replace strategy for suppressing vector-borne diseases: Insights from a stochastic, spatial model. PLoS One 8(12): e81860.

Ong, J., X. Liu, J. Rajarethinam, G. Yap, D. Ho, and L. Ching Ng. 2019. A novel entomological index, *Aedes aegypti* breeding percentage, reveals the geographical spread of the dengue vector in Singapore and serves as a spatial risk indicator for dengue. Parasites & Vectors 12 (1): 17.

Ooi, E.-E., and D. J. Gubler. 2009. Dengue in Southeast Asia: Epidemiological characteristics and strategic challenges in disease prevention. Cadernos de Saúde Pública 25 (Suppl. 1): S115–S124.

Ooi, E.-E., K.-T. Goh, and D. J. Gubler. 2006. Dengue prevention and 35 years of vector control in Singapore. Emerging Infectious Diseases 12(6): 887–893.

O Silva, M. M., L. B. Tauro, M. Kikuti, R. O. Anjos, V. C. Santos, T. S. F. Gonçalves, I. A. D. Paploski, P. S. S. Moreira, L. C. J. Nascimento, G. S. Campos, A. I. Ko, S. C. Weaver, M. G. Reis, U. Kitron, and G. S. Ribeiro. 2018. Concomitant transmission of dengue, chikungunya and Zika viruses in Brazil: Clinical and epidemiological findings from surveillance for acute febrile illness. Clinical Infectious Diseases 68 (8): 1353–1359.

Paes de Andrade, P., F. J. L. Aragão, W. Colli, O. A. Dellagostin, F. Finardi-Filho, M. H. Hirata, A. de Castro Lira-Neto, M. Almeida de Melo, A. L. Nepomuceno, F. G. da Nóbrega, G. D. de Sousa, F. H. Valicente, and M. H. B. Zanettini. 2016. Use of transgenic *Aedes aegypti* in Brazil: Risk perception and assessment. Bulletin of the World Health Organization 94(10): 766–771.

(PAHO) Pan American Health Organization. 2016. Vector borne diseases in the region of the Americas (2013–2016). Washington, DC, USA.

(PAHO) Pan American Health Organization. 2018. Integrated management strategy for dengue prevention and control in the region of the Americas. Washington, DC, USA.

(PAHO/WHO) Pan American Health Organization / World Health Organization. 2003. 44th Directing Council. 22-26 September 2003. Washington, DC, USA.

(PAHO/WHO) Pan American Health Organization / World Health Organization. 2014. Number of reported cases of chikungunya fever in the Americas by country or territory 2013–2014. Washington, DC, USA.

(PAHO/WHO) Pan American Health Organization / World Health Organization. 2016. Zika suspected and confirmed cases reported by countries and territories in the Americas, 2015–2017, cumulative cases. Washington, DC, USA.

Parks, W., and L. Lloyd. 2004. Planning social mobilization and communication for dengue fever prevention and control: A step-by-step guide. UNDP/World Bank/WHO Special Programme for Research and Training in Tropical Diseases. World Health Organization. Geneva, Switzerland. 138 pp.

Paz-Soldan, V. A., J. Yukich, A. Soonthorndhada, M. Giron, C. S. Apperson, L. Ponnusamy, C. Schal, A. C. Morrison, J. Keating, and D. M. Wesson. 2016. Design and testing of novel lethal ovitrap to reduce populations of *Aedes* mosquitoes: Community-Based Participatory Research between Industry, Academia and Communities in Peru and Thailand PLoS One 11 (8): e0160386

Perich, M. J., A. Kardec, I. A. Braga, I. F. Portal, R. Burge, B. C. Zeichner, W. A. Brogdon, and R. A. Wirtz. 2003. Field evaluation of a lethal ovitrap against dengue vectors in Brazil. Medical and Veterinary Entomology 17: 205–210.

Pessanha, J. E. M., W. T. Caiaffa, C. C. César, and F. A. Proietti. 2009. Avaliação do plano nacional de controle da dengue. Cadernos de Saúde Pública 25(7): 1637–1641.

Pruszynski, C. A., L. J. Hribar, R. Mickle, and A. L. Leal. 2017. A large scale biorational approach using *Bacillus thuringiensis israelensis* (strain AM65-52) for managing *Aedes aegypti* populations to prevent dengue, chikungunya and Zika transmission. PLoS One 12(2): e0170079.

Regis, L., A. M. Monteiro, M. A. V. de Melo-Santos, J. C. Silveira Jr, A. F. Furtado, R. V. Acioli, G. M. Santos, M. Nakazawa, M. S. Carvalho, P. J. Ribeiro Jr, and W. V. de Souza. 2008. Developing new approaches for detecting and preventing *Aedes aegypti* population outbreaks: Basis for surveillance, alert and control system. Memorias Do Instituto Oswaldo Cruz 103: 50–59.

Regis, L. N., R. V. Acioli, J. C. Silveira Jr, M. A. V. Melo-Santos, W. V. Souza, C. M. N. Ribeiro, J. C. S. da Silva, A. M. Vieira Monteiro, C. M. F. Oliveira, R. M. R. Barbosa, C. Braga, M. A. Benedetti Rodrigues, M. Gomes N. M. Silva, P. J. Ribeiro Jr., W. H. Bonat, L. C. de Castro Medeiros, M. Sa Carvalho, and A. F. Furtado. 2013. Sustained reduction of the dengue vector population resulting from an integrated control strategy applied in two Brazilian cities. PLoS One 8(7): e67682.

Ritchie, S. A., L. P. Rapley, C. Williams, P. H. Johnson, M. Larkman, R. M. Silcock, S. A. Long, and R. C. Russell. 2009. A lethal ovitrap-based mass trapping scheme for dengue control in Australia: I. Public acceptability and performance of lethal ovitraps. Medical and Veterinary Entomology 23 (4): 295–302.

Rodriguez-Morales, A. J., W. E. Villamil-Gómez, and C. Franco-Paredes. 2016. The arboviral burden of disease caused by co-circulation and co-infection of dengue, chikungunya and Zika in the Americas. Travel Medicine and Infectious Disease 14 (3): 177–179.

Rosenbaum, J., M. B. Nathan, R. Ragoonanansingh, S. Rawlins, C. Gayle, D. D. Chadee, and L. S. Lloyd. 1995. Community participation in dengue prevention and control: A survey of knowledge, attitudes, and practice in Trinidad and Tobago. The American Journal of Tropical Medicine and Hygiene 53(2): 111–117.

Rückert, C., J. Weger-Lucarelli, S. M. Garcia-Luna, M. C. Young, A. D. Byas, R. A. Murrieta, J. R. Fauver, and G. D. Ebel. 2017. Impact of simultaneous exposure to arboviruses on infection and transmission by *Aedes aegypti* mosquitoes. Nature Communications 8: 15412.

Samuel, M., D. Maoz, P. Manrique, T. Ward, S. Runge-Ranzinger, J. Toledo, R. Boyce, and O. Horstick. 2017. Community effectiveness of indoor spraying as a dengue vector control method: A systematic review. PLoS Neglected Tropical Diseases 11(8): e0005837.

San Martín, J. L., and O. Brathwaite-Dick. 2007. La estrategia de gestión integrada para la prevención y el control del dengue en la región de las Américas. Revista Panamericana de Salud Pública 21: 55–63.

Sanchez, L., J. Maringwa, Z. Shkedy, M. Castro, N. Carbonell, and P. Van der Stuyft. 2012. Testing the effectiveness of community-based dengue vector control interventions using semiparametric mixed models. Vector-Borne and Zoonotic Diseases 12(7): 609–615.

(SHS) Secretariat of Health Surveillance. 2017. Health Brazil 2015/2016. An analysis of health situation and the epidemic caused by Zika virus and other diseases transmitted by *Aedes aegypti*. Ministry of Health of Brazil. Braslia, DF, Brazil. 132 pp.

Sinkins, S. P. 2004. *Wolbachia* and cytoplasmic incompatibility in mosquitoes. Insect Biochemistry and Molecular Biology 34(3): 723–729.

Sithiprasasna, R., P. Mahapibul, C. Noigamol, M. J. Perich, B. C. Zeichner, B. Burge, S. L. Norris, J. W. Jones, S. S. Schleich, and R. E. Coleman. 2003. Field evaluation of a lethal ovitrap for the control of *Aedes aegypti* (Diptera: Culicidae) in Thailand. Journal of Medical Entomology 40(4): 455–462.

Sommerfeld, J., and A. Kroeger. 2015. Innovative community-based vector control interventions for improved dengue and chagas disease prevention in Latin America: Introduction to the Special Issue. Transactions of the Royal Society of Tropical Medicine and Hygiene 109 (2): 85–88.

Spiegel, J., A. Yassi, and R. Tate. 2002. Dengue in Cuba: Mobilisation against *Aedes aegypti*. The Lancet Infectious Diseases 2(4): 207–208.

(SS-PE) Secretaria de Saúde de Pernambuco (2015). Boletim epidemiológico arboviroses transmitidas pelo *Aedes aegypti*. Pernambuco, Brasil.

Suter, T., M. M. Crespo, M. F. de Oliveira, T. S. A. de Oliveira, M. A. V. de Melo-Santos, C. M. F. de Oliveira, C. F. J. Ayres, R. M. Rodrigues Barbosa, A. P. Araújo, L. N. Regis, E. Flacio, L. Engeler, P. Müller, and M. H. N. L. Silva-Filha. 2017. Insecticide susceptibility of *Aedes albopictus* and *Ae. aegypti* from Brazil and the Swiss-Italian border region. Parasites & Vectors 10(1): 431.

Suwannapong, N., M. Tipayamongkholgul, A. Bhumiratana, C. Boonshuyar, N. Howteerakul, and S. Poolthin. 2014. Effect of community participation on household environment to mitigate dengue transmission in Thailand. Tropical Biomedicine 31: 149–158.

Suwanmanee, S., P. Surasombatpattana, N. Soonthornworasiri, R. Hamel, P. Maneekan, D. Missé, and N. Luplertlop. 2018. Monitoring arbovirus in Thailand: Surveillance of dengue, chikungunya and Zika virus, with a focus on coinfections. Acta Tropica 188: 244–250.

Tapia-Conyer, R., J. Méndez-Galván, and P. Burciaga-Zúñiga. 2012. Community participation in the prevention and control of dengue: The *patio limpio* strategy in Mexico. Pediatrics and International Child Health 32 (Suppl. 1): 10–14.

Tauil, P. L. 2002. Aspectos críticos do controle do dengue no Brasil. Cadernos de Saúde Pública 18 (June): 867–871.

Townson, H., M. B. M. Nathan, M. Zaim, P. Guillet, L. Manga, R. Bos, and M. Kindhauser. 2005. Exploiting the potential of vector control for disease prevention. Bulletin of the World Health Organization 83(12): 942–947.

Travanty, E. A., Z. N. Adelman, A. W. Franz, K. M. Keene, B. J. Beaty, C. D. Blair, A. A. James, and K. E. Olson. 2004. Using RNA interference to develop dengue virus resistance in genetically modified *Aedes aegypti*. Insect Biochemistry and Molecular Biology 34(7): 607–613.

Ulibarri, G., B. Betanzos, M. Betanzos, and J. J. Rojas. 2016. Control of *Aedes aegypti* in a remote Guatemalan community vulnerable to dengue, chikungunya and Zika virus: Prospective evaluation of an integrated intervention of web-based health worker training in vector control, low-cost ecological ovillantas, and community engagement. F1000Research 5: 598.

Unckless, R. L., A. G. Clark, and P. W. Messer. 2017. Evolution of resistance against CRISPR/Cas9 gene drive. Genetics 205(2): 827–841.

Unlu, I., D. S. Suman, Y. Wang, K. Klingler, A. Faraji, and R. Gaugler. 2017. Effectiveness of autodissemination stations containing pyriproxyfen in reducing immature *Aedes albopictus* populations. Parasites & Vectors 10(1): 139.

Unlu, I., G. M. Williams, I. Rochlin, D. Suman, Y. Wang, K. Chandel, and R. Gaugler. 2018. Evaluation of lambda-cyhalothrin and pyriproxyfen barrier treatments for *Aedes albopictus* (Diptera: Culicidae) management in urbanized areas of New Jersey. Journal of Medical Entomology 55 (2): 472–76.

(US-EPA) United States Environmental Protection Agency. 2017. Success in mosquito control: An integrated approach. Washington, DC, USA.

Van den Berg, H., C. M. Mutero, and K. Ichimori. 2012. Guidance on policy-making for Integrated Vector Management. Department of Control of Neglected Tropical Diseases, World Health Organization, Geneva, Switzerland. 11 pp.

Van den Hurk, A. F., S. Hall-Mendelin, A. T. Pyke, F. D. Frentiu, K. McElroy, A. Day, S. Higgs, and S. L. O'Neill. 2012. Impact of *Wolbachia* on infection with chikungunya and yellow fever viruses in the mosquito vector *Aedes aegypti*. PLoS Neglected Tropical Diseases 6(11): e1892.

Vanlerberghe, V., M. E. Toledo, M. Rodríguez, D. Gomez, A. Baly, J. R. Benitez, and P. Van der Stuyft. 2009. Community involvement in dengue vector control: Cluster randomised trial. BMJ 338: b1959.

Vreysen, M. J. B., K. Saleh, F. Mramba, A. Parker, U. Feldmann, V. A. Dyck, A. Msangi, and J. Bouyer. 2014. Sterile insects to enhance agricultural development: The case of sustainable tsetse eradication on Unguja Island, Zanzibar, using an Area-Wide Integrated Pest Management approach. PLoS Neglected Tropical Diseases 8(5): e2857.

Vreysen, M. J. B. 2021. Monitoring sterile and wild insects in Area-Wide Integrated Pest Management programmes, pp. 485–528. *In* V. A. Dyck, J. Hendrichs, and A. S. Robinson (eds.), Sterile Insect Technique – Principles and practice in Area-wide Integrated Pest Management. Second Edition. CRC Press, Boca Raton, Florida, USA.

Walshe, D. P., P. Garner, A. A. Abdel-Hameed Adeel, G. H. Pyke, and T. Burkot. 2013. Larvivorous fish for preventing malaria transmission. Cochrane Database of Systematic Reviews 10 (12): CD008090.

(WHO) World Health Organization. 2004. Global strategic framework for Integrated Vector Management. Geneva, Switzerland.

(WHO) World Health Organization. 2008. Position statement on Integrated Vector Management. Geneva, Switzerland.

(WHO) World Health Organization. 2009. Dengue guidelines for diagnostics, treatment, prevention and control. Geneva, Switzerland.

(WHO) World Health Organization. 2013. WHO recommended insecticides for indoor residual spraying against malaria vectors. Geneva, Switzerland.

(WHO) World Health Organization. 2016a. Dengue and severe dengue. Geneva, Switzerland.

(WHO) World Health Organization. 2016b. Zika Strategic Response Plan Quarterly. Geneva, Switzerland.

(WHO) World Health Organization. 2016c. Number of reported cases of Chikungunya fever in the Americas in 2016. Geneva, Switzerland.

(WHO) World Health Organization. 2017a. Situation report: Zika virus, microcephaly, Guillain-Barré Syndrome, 10 March 2017. Situation Report. Geneva, Switzerland.

(WHO) World Health Organization. 2017b. Chikungunya. Geneva, Switzerland.

(WHO) World Health Organization. 2017c. Global vector control response 2017-2030. Geneva, Switzerland.

(WHO) World Health Organization. 2017d. Dengue control strategies. Geneva, Switzerland.

(WHO) World Health Organization. 2017e. Environmental management. Geneva, Switzerland.

(WHO) World Health Organization. 2017f. Chemical control. Geneva, Switzerland.

(WHO) World Health Organization. 2018. Yellow fever.

Zabalou, S., A. Apostolaki, I. Livadaras, G. Franz, A. S. Robinson, C. Savakis, and K. Bourtzis. 2009. Incompatible Insect Technique: Incompatible males from a *Ceratitis capitata* genetic sexing strain. Entomologia Experimentalis et Applicata 132: 232–240.

Zahir, A. 2016. Community participation, dengue fever prevention and practices for control in Swat, Pakistan. International Journal of Maternal and Child Health and AIDS 5: 39–45.

Zanluca, C., V. C. A. de Melo, A. L. P. Mosimann, G. I. V. dos Santos, C. N. D. dos Santos, and K. Luz. 2015. First report of autochthonous transmission of Zika virus in Brazil. Memórias Do Instituto Oswaldo Cruz 110: 569–572.

Zara, A. L. S. A., S. M. Santos, E. S. Fernandes-Oliveira, R. G. Carvalho, and G. E. Coelho, 2016. *Aedes aegypti* control strategies: A review. Epidemiologia e Serviços de Saúde 25(2): 391–404.

COMBINING THE INCOMPATIBLE AND STERILE INSECT TECHNIQUES FOR PEST AND VECTOR CONTROL

L. A. BATON[1], D. ZHANG[2], Y. LI[2] AND Z. XI[1,2]

[1]Department of Microbiology and Molecular Genetics, Michigan State University, East Lansing, Michigan 48824, USA; xizy@msu.edu
[2]Key Laboratory of Tropical Disease Control of the Ministry of Education, Sun Yat-sen University - Michigan State University Joint Center of Vector Control for Tropical Diseases, Zhongshan School of Medicine, Sun Yat-Sen University, Guangzhou, Guangdong 510080, China

SUMMARY

The Incompatible Insect Technique (IIT) is a Sterile Insect Technique (SIT)-related approach that uses the reproductive parasitism caused by infection with maternally-inherited bacterial endosymbionts to make released males reproductively incompatible with the wild-type females of the target population. The most common and widespread of such endosymbionts is *Wolbachia*, which is found throughout many insect orders, and often causes cytoplasmic incompatibility (CI), a form of conditional sterility where the fertilized eggs of females not infected with the same *Wolbachia* strain as the males with which they are mated undergo embryonic death. An advantage of IIT is that the incompatibility induced by *Wolbachia* often has either no or only minor effects on the quality of infected males. In addition, such endosymbionts can also have other desirable phenotypic effects on their hosts, such as reducing the ability of target species to act as disease vectors, thus allowing the undesirable sex(es) to be tolerated among the sterile insects to be released. However, an inherent problem with IIT, which has so far restricted its operational use, is that, unlike SIT, the accidental release of endosymbiont-infected females may prevent further population suppression by causing unintended population replacement, whereby the original target population is replaced with individuals infected with the same endosymbiont strain as the released males. A solution to this problem, at least for the majority of insects whose females are more sensitive than males to radiation, is to combine IIT with SIT, such that all endosymbiont-infected individuals destined for release are also first subjected to low-dose radiation, which completely sterilizes any contaminant females without affecting the incompatibility or quality of the irradiated males. Here, we discuss the biology and general theoretical principles underlying the use of IIT alone, and the rationale and necessity of combining IIT with SIT, as well as the logistical problems encountered, and technological developments required, for the mass-production and release of irradiated endosymbiont-infected individuals as part of area-wide integrated pest control programmes. We primarily illustrate our discussion with examples involving mosquitoes, for which the majority of the relevant research has been conducted, including the first open-release field trial of combined IIT/SIT application against the important arboviral vector *Aedes albopictus* (Skuse). However,

J. Hendrichs, R. Pereira and M. J. B. Vreysen (eds.), Area-Wide Integrated Pest Management: Development and Field Application, pp. 367–404. CRC Press, Boca Raton, Florida, USA.

the combined IIT/SIT approach should be broadly applicable to a wide range of other insect pests and vectors, and so of interest to entomologists in general.

Key Words: Aedes, Wolbachia, combined IIT/SIT, cytoplasmic incompatibility, symbiosis, arboviruses, mosquitoes, radiation, vector-borne diseases

1. INTRODUCTION

The irradiation-based Sterile Insect Technique (SIT) has been successfully used to suppress the populations of a number of insect pests and vectors of agricultural and veterinary importance (Dyck et al. 2021). However, the application of the SIT against other important groups of insects, especially the mosquito vectors of human pathogens, has, so far, been limited (Benedict and Robinson 2003; Dame et al. 2009; Bourtzis et al. 2016; Scott and Benedict 2016). There are various reasons for this, and they have been debated, but the development and implementation of the SIT for such insects continues, and is still an active and productive area of research, and it is hoped that, with further investigation and optimization, the SIT can be successfully and operationally deployed against mosquitoes in the not so distant future (Alphey et al. 2010; Bourtzis et al. 2016; Lees et al. 2015, 2021).

Concurrently, however, other approaches have also been, and continue to be, explored (Alphey 2014; Scott and Benedict 2016). One alternative approach is the Incompatible Insect Technique (IIT), which uses infection with maternally-inherited prokaryotic endosymbionts, such as the alpha-proteobacterium *Wolbachia,* to suppress host populations. Like other sterile-male-based methods of population suppression, these endosymbionts make the released males reproductively incompatible with wild-type females in the target population.

In this chapter, we outline the biological and theoretical basis of endosymbiont-mediated IIT and argue why it is an attractive alternative to the SIT for some groups of insects. However, in the absence of a perfect sex separation system, IIT application has a fundamental constraint necessitating its combination with the SIT. Our discussion primarily concerns mosquitoes, but there is an increasingly large research literature on endosymbionts causing reproductive parasitism in a wide range of insect hosts, which readers are encouraged to explore for themselves.

2. THE STERILE INSECT TECHNIQUE (SIT)

Historically, there have been a number of laboratory tests and pilot field trials of irradiation-based SIT against mosquitoes (Benedict and Robinson 2003; Dame et al. 2009). Despite this, the SIT against mosquitoes has not yet been deployed operationally on a larger scale. Various reasons have been given for this, such as the failure to effectively optimize the timing and magnitude of the irradiation dose used on mass-reared individuals, the high rate of intrinsic increase of these insects and the failure to release sufficient males, an inability to efficiently and cost-effectively mass-produce and/or release them, a lack of knowledge regarding the basic biology and ecology of the target species, and inadequate methods of sex separation (Dame et al. 2009; Scott and Benedict 2016).

More recently, interest in using the SIT against mosquitoes has been revived, and the positive results of small-scale field trials (Bellini et al. 2007, 2013b) support the notion that the SIT is feasible against mosquitoes. In addition, over the past decade, the Food and Agriculture Organization of the United Nations (FAO) and the International Atomic Energy Agency (IAEA), in response to increasing requests from their member states, have increased their efforts to explore and disseminate the possibilities of integrating the SIT within AW-IPM approaches to manage better mosquito populations (Bourtzis et al. 2016; Lees et al. 2013, 2015; 2021).

2.1. *Reduced Quality of Irradiation-sterilized Insects*

A requirement of the SIT is that radiation-based sterilisation does not have serious adverse effects on male competitiveness or overall quality (Knipling 1955). Such effects may derive either from the direct deleterious effects of radiation itself, or indirectly through the mass-rearing and handling procedures and ambient conditions required for the administration of radiation (Bourtzis and Robinson 2006; Bakri et al. 2021). Insects vary in their radio-sensitivity, with some species being inherently more sensitive to the effects of irradiation, such that irradiation doses inducing high levels of male sterility often also have appreciable negative effects on male quality (e.g. mating competitiveness and survival) for some insect species (Bakri et al. 2005; Helinski et al. 2009). For mosquitoes, the process of irradiation has frequently been reported to reduce male competitiveness and survival (Arunachalam and Curtis 1985; Dame et al. 2009; Helinski et al. 2009; Oliva et al. 2012; Maïga et al. 2014; Yamada et al. 2014a, 2014b; Zhang et al. 2016; Zheng et al. 2019). However, whether such negative effects are due to the radiation itself, or the conditions and procedures under which the radiation is administered is often unclear (Scott and Benedict 2016; Yamada et al. 2019). The latter can impose significant fitness costs independent of the effects of radiation itself, which can reduce their quality for use in the SIT.

Many pest/vector species are fragile, with complex holometabolous life cycles, complicating their handling and irradiation, especially under the conditions of mass-rearing and mass-release required for the SIT. Different life cycle stages may also vary in their radio-sensitivity (e.g. late pupae versus adults), and careful timing of irradiation can help to minimize radiation-induced damage, as well as maximize sex-specific differences in radio-sensitivity, which is important for sterilizing contaminant females without adversely affecting male quality (see Section 4.2) (Andreasen and Curtis 2005; Helinski et al. 2006; Brelsfoard et al. 2009; Balestrino et al. 2010; Ndo et al. 2014; Zhang et al. 2015b).

Regardless of whether the adverse effects of irradiation are direct or indirect – and whether or not it might be possible in future to ameliorate such effects through optimization of irradiation protocols and development of better technology – for some groups of insects there is currently a necessary trade-off between sterility and quality, such that, as higher irradiation doses increase male sterility, they simultaneously decrease male quality (Helinski and Knols 2008; Balestrino et al. 2010; Bellini et al. 2013a). In many instances, intermediate irradiation doses can be identified that provide an optimal balance between male sterility and quality (Parker and Mehta 2007; Helinski et al. 2009). Consequently, the use of the SIT may not be precluded

(Bellini et al. 2007, 2013b; Scott and Benedict 2016), although its overall efficiency may be reduced, and its cost-effectiveness decreased, through necessitating larger numbers of insects to be produced and released during area-wide control programmes. For target insects with very high reproductive potential, like some mosquito species (Alphey et al. 2010), the problem of trading-off residual fertility against male quality could be particularly acute, because population can rebound easily through those survived eggs as seeds. With low residual fertility, the number of emerging adults in the wild may be relatively high as the low number of hatching eggs is compensated by low competition for resources among surviving larvae, and hence higher survival rates during the development. Therefore, minimum levels of sterility are necessarily required to overcome the intrinsic growth of the target population (Barclay 2021).

2.2. *Imperfect Sex Separation*

Another problem for the implementation of the SIT, as well as all other sterile-insect-based methods, against insects like mosquitoes where adult females (and not males) of the target species are the pests/vectors, is the absence of perfect sex separation methods (Gilles et al. 2014). Thus, the release of females, even as relatively small numbers of contaminant individuals, is considered unacceptable for SIT applications to control those pests/vectors due to the risk of increased crop destruction, parasitism or pathogen transmission. In other instances, where either males or both sexes act as pests/vectors, the release of any sterile individuals has to be carefully managed, e.g. feeding tsetse males with trypanocidal drugs before their release or using strains with enhanced vector refractoriness (Kariithi et al. 2018). New methods are, therefore, needed that either completely remove any females from among the insects to be released, or reduce the ability of target species to act as pests/vectors, thus allowing the undesirable sex(es) to be tolerated among the sterile insects to be released.

3. THE INCOMPATIBLE INSECT TECNIQUE (IIT)

3.1. Wolbachia, *Cytoplasmic Incompatibility and Population Suppression*

The IIT is an analogue of the SIT, using infection with naturally-occurring maternally-inherited bacterial endosymbionts that cause reproductive parasitism – instead of radiation – to make released males reproductively incompatible with females of the target field population (Bourtzis et al. 2014; Scott and Benedict 2016; Xi and Joshi 2016). *Wolbachia* is the most common and widespread of such endosymbionts (Werren et al. 2008), being found throughout many insect orders, i.e. it is estimated to infect between approximately 48 to 57% of all terrestrial arthropods (Hilgenboecker et al. 2008; Zug and Hammerstein 2012; Weinert et al. 2015). One of the manipulations of host reproduction caused by *Wolbachia* is cytoplasmic incompatibility (CI), a form of conditional sterility whereby the fertilized eggs of females not infected with the same *Wolbachia* strain as the males with which they are mated, undergo embryonic death (Sinkins 2004; Werren et al. 2008; Hurst and Frost 2015). In contrast, *Wolbachia*-infected females produce off-spring normally, whether mated with uninfected males or with males infected with the same *Wolbachia* strain.

The level of CI induced can vary considerably between *Wolbachia* strains: some strains do not cause CI – or any other reproductive manipulation – while others cause either partial or complete CI that either only kills some or all embryos, respectively. In general, *Wolbachia* and other similar endosymbiotic reproductive parasites are only maternally-inherited, but the level of transmission from mother-to-offspring can vary considerably. In mosquitoes, native *Wolbachia* infections typically exhibit very high levels (~100%) of both CI and maternal transmission (Sinkins 2004; Baton et al. 2013), while these characteristics are often markedly lower (<50%) and more variable in other Diptera, such as well-studied drosophilids.

As CI prevents uninfected females – or those infected with a different incompatible *Wolbachia* strain – from having off-spring, infected females leave more off-spring. The consequence is that, over succeeding generations, uninfected females can be driven to extinction as the number of infected females increases, potentially resulting in complete replacement of the original uninfected host population with *Wolbachia*-infected individuals (Caspari and Watson 1959; Fine 1978). The speed and extent with which population replacement occurs, i.e. replacing the original uninfected host population with *Wolbachia*-infected individuals, primarily depends on the level of CI-induced and the rate of endosymbiont maternal transmission, as well as whether or not the endosymbiont has any fitness costs or benefits for its hosts. When CI is complete, maternal transmission is perfect, and the endosymbiont has no fitness costs, complete population replacement is expected, and, at least theoretically, is predicted to be very rapid: occurring in about 100 generations, from a very low (~1%) initial proportion. If the initial proportion is higher (e.g. >10%), population replacement could occur in <10 generations (Caspari and Watson 1959; Fine 1978).

When CI is partial, and/or maternal transmission imperfect, and/or there are fitness costs, endosymbiont-infected individuals either will go extinct, or will only partially if not completely replace uninfected individuals if they constitute a certain proportion of the host population, known as the invasion threshold of the endosymbiont. During the process of population replacement, the size of the uninfected part of the host population is reduced due to the inhibition of reproduction by uninfected females as a result of mating with incompatible infected males (Dobson et al. 2002a). This creates a positive feedback-loop that increases the relative proportion of infected individuals in the population, as well as creating vacant niche space to be filled, and thereby accelerates and drives the rate of both replacement of uninfected individuals and the degree of population suppression (as uninfected females are increasingly more likely to mate with infected males). It is this naturally-occurring mechanism of host population suppression that is exploited by IIT. However, as only males should be released during IIT application, the subsequent population replacement by endosymbiont-infected individuals that occurs in natural systems does not occur during target population suppression, as there are no infected females to maternally-transmit the endosymbiont to the next generation. The consequence is target population elimination.

An alternative strategy for vector control which is currently being extensively investigated and actively implemented – and which we do not discuss further here – involves intentionally releasing endosymbiont-infected females in order to deliberately trigger population replacement (Sinkins et al. 1997; Iturbe-Ormaetxe et al. 2011; Bourtzis et al. 2014; Xi and Joshi 2016). As described in Section 3.4.2 below, some *Wolbachia* infections can reduce vector competence for vector-borne pathogens, such that population replacement with such endosymbiont variants would reduce or prevent pathogen transmission by a vector population. However, we regard the aim of population suppression as preferable to population replacement, because it can be guaranteed to completely prevent any future pathogen transmission (endosymbiont-mediated reduction in vector competence may not be complete, and/or may be lost over time due to the evolution of resistance by the transmitted pathogen and/or changes in the vector-endosymbiont association), and is likely to have greater public acceptance due to male-only releases, the reduction of nuisance biting, and the possibility of vector eradication (Zheng et al. 2019).

3.2. A Brief History of the IIT

CI was first observed in the mosquito *Culex pipiens* L. (Marshall and Staley 1937; Marshall 1938; Roubaud 1941), shortly after the independent discovery of the endosymbiont *Wolbachia* in the same mosquito species (Hertig and Wolbach 1924; Hertig 1936). However, it was not until more than three decades later that the causal link between *Wolbachia* and CI was hypothesized, and then empirically proven through curing mosquitoes of their bacterially-induced CI by antibiotic treatment (Yen and Barr 1971, 1973).

The notion of using CI for suppression of vector populations was developed during the 1960s – that is, prior to the realization that maternally-inherited endosymbionts cause CI – as part of a World Health Organization (WHO)-sponsored programme instigated and led by the German entomologist Hannes Laven (WHO 1964; Pal 1966; Knipling et al. 1968; Laven 1971; Davidson 1974). In unpublished studies, it was first shown, using cage experiments, that the release of incompatible males at an initial 1:1 ratio with target-compatible males, could eradicate a stable target population in only 3 or 4 generations (Pal 1966; Laven 1967, 1971).

Consequently, a small-scale open-release pilot trial in the field was undertaken in a relatively isolated rural village (Okpo) near Rangoon in Myanmar (Burma) against the local vector of filariasis, *Culex quinquefasciatus* Say (Laven 1967, 1971). This trial was a resounding success, effectively eliminating the local mosquito population by the end of the 12-week intervention period, although there were some reservations about the significance, and general applicability, of this "proof-of-principle" demonstration (Laird 1967; Barr 1970; Weidhaas and Seawright 1976).

Subsequently, a larger-scale joint WHO / Indian Council of Medical Research (ICMR)-backed project to further investigate the feasibility of using the IIT was established in the 1970s in India (Grover and Sharma 1974; Pal 1974), resulting in a number of studies characterizing the incompatibility, mating competitiveness and vector competence of endosymbiont-infected mosquitoes (Subbarao et al. 1974, 1977; Grover et al. 1976; Singh et al. 1976; Curtis 1977; Krishnamurthy 1977; Thomas and Singh 1977; Curtis and Reuben 2007), as well as the first attempts of combining the IIT with genetic modification (Laven and Aslamkhan 1970; Krishnamurthy and Laven 1976; Curtis 1977). The results of the field trials were less convincing than before with only partial population suppression ($<70\%$) achieved, apparently due to unexpected high levels of immigration of previously inseminated females from the areas surrounding the release sites (Brooks et al. 1976; Curtis 1977; Curtis et al. 1982).

During this latter period, with the discovery of CI in tephritid flies (the European cherry fruit fly *Rhagoletis cerasi* L., Boller and Bush 1974), and pyralid moths (the almond moth *Cadra cautella* (Walker), Brower 1976), there was interest for using CI to suppress other pest insects (Russ and Faber 1979; Neuenschwander et al. 1983; Blümel and Russ 1989; Boller 1989), with the term "IIT" being coined (Boller et al. 1976), and several promising laboratory studies and semi-field trials undertaken (Brower 1979, 1980; Ranner 1990).

During the 1980s, interest in the IIT (and the SIT) for mosquitoes waned (Scott and Benedict 2016), partly due to the premature termination of the joint WHO/ICMR project (Anonymous 1975; Curtis and Reuben 2007), but also because of doubts about the practical feasibility and economics of rearing large numbers of mosquitoes, as well as the possibility/sustainability of population suppression/elimination in the presence of immigration from outside control areas (Sinkins et al. 1997; Scott and Benedict 2016).

From the 1990s to the present, a new generation of researchers and their academic descendants have given fresh impetus to investigating the use of *Wolbachia* for pest and vector control (Iturbe-Ormaetxe et al. 2011; Bourtzis et al. 2014; Xi and Joshi 2016), resulting in a renewed interest in the IIT and its operational deployment (O'Connor et al. 2012; Mains et al. 2016, 2019; Zheng et al. 2019). The IIT has been under consideration for controlling the Mediterranean fruit fly *Ceratitits capitata* (Wiedemann) (Zabalou et al. 2004, 2009), the olive fruit *fly Bactrocera oleae* Rossi (Apostolaki et al. 2011), and the spotted wing drosophila *Drosophila suzukii* (Matsumura) (Cattel et al. 2018; Nikolouli et al. 2018), as well as tsetse flies (*Glossina* spp.) (Alam et al. 2011; Bourtzis et al. 2016). However, the development and implementation of the IIT was, and remains, the most advanced for mosquitoes, with open-release field trials planned or already recently undertaken for the arboviral and/or filarial vector species *Aedes aegypti* L. in Australia, Mexico, Singapore, and the USA (Xi and Manrique-Saide 2018; Yeung 2018; Corbel et al. 2019; Mains et al. 2019); *Aedes albopictus* (Skuse) in China and the USA (Mains et al. 2016; Zheng et al. 2019); and *Aedes polynesiensis* Marks in French Polynesia (Brelsfoard et al. 2008; O'Connor et al. 2012); as well as for *Cx. quinquefasciatus* on the four islands in the south-western Indian Ocean (La Réunion, Mauritius, Grande Glorieuse and Mayotte) (Atyame et al. 2011, 2015).

3.3. *Generating and Characterizing Novel Endosymbiont Infections*

In order to control a target pest or vector species using the IIT, it is necessary to have incompatible individuals for mass-rearing and release. The simplest method for obtaining such individuals is to collect them from the field. This is possible, for example, if different geographic populations of a target species naturally possess different incompatible endosymbiont infections (Brower 1976; Chen et al. 2013). This was the origin of the incompatible individuals used for the first IIT trials against *Cx. quinquefasciatus* (Laven 1967, 1971), as well as other pest insects. However, there are several problems that require attention.

3.3.1. Introgressing the Nuclear Genome of the Target Population

The first problem is that individuals from one geographic area may not be well-adapted to another location and may have lower mating competitiveness compared to local males from the target population (Barr 1966). This problem can be solved by using backcrosses to introgress the nuclear genome of the target population into the incompatible colony to be used for releases (Barr 1966; Krishnamurthy 1977), which is achieved by mating males from the target population to females from the incompatible colony. The process is repeated but using the daughters of each cross instead. As *Wolbachia* and other endosymbiotic reproductive parasites are maternally inherited, the outcome is a new line that possesses the cytoplasmic organelles (mitochondria and endosymbionts) of the original incompatible line, but now with the nuclear genome of the target population.

This laborious technique is still used to create *Wolbachia*-infected lines (Atyame et al. 2011), and it remains a fundamental method for matching the genetic background of released individuals to the target population in the field. This technique was also used more recently to transfer through inter-specific introgression a naturally occurring *Wolbachia* infection into a target species (*Ae. polynesiensis*) from a closely related non-target sister species (*Aedes riversi* Bohart & Ingram) (Brelsfoard et al. 2008).

3.3.2. Generating Artificial Wolbachia *Infections through Transinfection*

The second – and more significant – problem with relying on naturally-occurring endosymbiont infections is that it limits the availability and diversity of incompatible individuals, and, therefore, the insect species that are amenable to control using the IIT. Although endosymbionts are widespread among arthropods, many important pest and vector species, such as the mosquito *Ae. aegypti*, as well as the many mosquito species in the *Anopheles* genus of malaria vectors, are thought not to be naturally infected with *Wolbachia* (Bourtzis et al. 2014).

Even in those species that are infected with *Wolbachia*, there is often no intra-specific geographic variation in the endosymbionts and their mating compatibilities (as might be expected given the nature of their reproductive parasitism), and closely-related infected sister taxa capable of inter-specific interbreeding may not exist. The many intra-strain mating types observed in *Cx. pipiens*, which enable the IIT against this target species (Laven 1967; Atyame et al. 2011), are apparently atypical (Bourtzis et al. 2014). Although host species are not infrequently superinfected with two or more endosymbiont strains, these are often found throughout their geographic range (e.g. *Ae. albopictus*) (Bourtzis et al. 2014).

The ability to generate artificial *Wolbachia* infections in the laboratory through transinfection between individuals within the same or different host species was, therefore, a major breakthrough (Boyle et al. 1993; Braig et al. 1994). The application of these techniques has since provided renewed impetus to the use of endosymbionts for pest and vector control (Xi et al. 2005b; Hughes and Rasgon 2014), enabling the first open-release in the field of artificially-transinfected *Wolbachia*-infected mosquitoes for IIT application (Mains et al. 2016).

For mosquitoes, transinfection can be achieved either through embryonic (Xi and Dobson 2005) or intra-thoracic microinjection of adults (Ruang-areerate and Kittayapong 2006). However, in order to establish stable germline infections, the former method is regarded as the most efficient, due to the low likelihood of the somatic infections resulting from the latter colonizing the gonads and being maternally transmitted (Hughes and Rasgon 2014). For other insects, inoculation of larval or pupal stages has also been occasionally reported, as has transfer through co-rearing and predation.

Xi et al. (2005b) successfully established the first artificial *Wolbachia* infection of mosquitoes through embryonic microinjection of the cytoplasm from endosymbiont-infected donor eggs (Xi and Dobson 2005), and since then, a number of different artificial germline *Wolbachia* infections have been established in mosquitoes, including *Ae. aegypti* (Xi et al. 2005b; McMeniman et al. 2009; Walker et al. 2011; Ant and Sinkins 2018), *Ae. albopictus* (Xi et al. 2005a; Xi et al. 2006; Suh et al. 2009; Calvitti et al. 2010; Fu et al. 2010; Blagrove et al. 2012; Ant and Sinkins 2018; Moretti et al. 2018b; Zheng et al. 2019), and *Ae. polynesiensis* (Andrews et al. 2012), as well as the malaria vector *Anopheles stephensi* Liston (Bian et al. 2013).

Artificial transfer of CI-inducing *Wolbachia* has now also been achieved in a number of different insect groups, including other Diptera (Drosophilidae and Tephritidae), as well as Lepidoptera, Hemiptera and Coleoptera (Hughes and Rasgon 2014).

Transinfection allows both naturally uninfected, as well as already infected, target species to be artificially infected with *Wolbachia*. In the latter instance, pre-existing native endosymbiont infections can either be first removed by antibiotics or moderately high temperature (Yen and Barr 1973; Portaro and Barr 1975; Dobson and Rattanadechakul 2001), and then replaced with a different strain of *Wolbachia* (Suh et al. 2009; Calvitti et al. 2010; Andrews et al. 2012).

Alternatively, novel superinfections can be generated by adding new artificial infestations of *Wolbachia* strains to the endosymbiont strains already present in the target populations (Fu et al. 2010; Joubert et al. 2016; Ant and Sinkins 2018; Zheng et al. 2019). Establishing superinfections (especially triple infections) may be trickier than replacing pre-existing endosymbiont infections (due to competitive interactions and/or incompatibilities between different *Wolbachia* strains, e.g. Ant and Sinkins 2018), but come with the added benefit of higher endosymbiont densities and broader somatic tissue distributions, which is thought to be of importance for altering pest/vector status (see Section 3.4.2) (Moretti et al. 2018b; Zheng et al. 2019), but not necessarily for the induction of CI.

3.3.3. Selection of Endosymbionts for Transinfection

The potentially unconstrained ability to transfer any *Wolbachia* strain between any host species raises the issue of selecting which endosymbiont strains to transinfect (Hoffmann et al. 2015). So far, only a relatively limited number of *Wolbachia* strains from well-studied hosts have been tried (Hughes and Rasgon 2014), but *Wolbachia* is an ancient, phenotypically diverse, and vast bacterial clade spread across phylogenetically-distant host taxa, with potentially more strains (i.e. millions) than infected host species (due to the occurrence of superinfection) (Werren et al. 2008).

Although the characteristics of endosymbionts after host transfer can sometimes be unpredictable, and depend upon host background (Hoffmann et al. 2015), the behaviour of a given *Wolbachia* strain in one host generally provides a reasonable "rule-of-thumb" for predicting its behaviour in other hosts, especially if those hosts are phylogenetically-related, enabling some guidance in the selection of endosymbiont strains to be transferred.

For example, the unusually virulent *Wolbachia* strain *w*MelPop (Min and Benzer 1997), generally retains its pathogenicity, whether present in closely or more distantly related dipteran hosts. Similarly, native *Wolbachia* infections in mosquito species generally have the same characteristics when transferred to new mosquito hosts (see below).

An important exception to this pattern seems to be that novel *Wolbachia* infections often have higher endosymbiont strain-specific densities and/or broader somatic tissue distributions, which are associated with host fitness and other phenotypic effects (Hoffmann et al. 2015; Xi and Joshi 2016; Ant and Sinkins 2018).

3.3.4. Characterization of New Host-endosymbiont Associations

Once endosymbiont-infected individuals have been found from the field or generated de novo in the laboratory, their host-endosymbiont association needs to be thoroughly characterized to determine if it is suitable for IIT application. The basic requirements for an endosymbiont to be used for the IIT are to induce CI, have favourable levels of maternal transmission, and, in general, to have low fitness costs. CI is required to generate the male incompatibility that enables sterilisation of wild-type females in the target population. Stable maternal transmission is required to ensure that males can cause CI, and to enable their efficient mass-production.

If males are not infected, they cannot induce CI, and if maternal transmission is low then many uninfected individuals will be produced in each generation. As there is currently no method to separate the uninfected from the infected individuals, their presence during factory rearing requires more individuals to be mass-produced, and more males to be released, for a given level of target population suppression. In addition, if maternal transmission is unstable, it can result in self-incompatibility between superinfected individuals, compromising colony maintenance and preventing mass-production of appropriately infected individuals (Ant and Sinkins 2018).

Low fitness costs of *Wolbachia* infection are required to enable efficient mass-production of large numbers of factory-reared individuals for release, as well as to ensure the mating competitiveness of the released males.

Many, although not all, of the artificially infected mosquito lines have been shown to have these characteristics, inducing high levels (~100%) of CI, when the transinfected males mate with wild-type females, causing high levels (~100%) of stable maternal inheritance, and having no or only low fitness costs (Section 3.4.1) (Xi et al. 2005b; Bian et al. 2010, 2013; Calvitti et al. 2010, 2012; Blagrove et al. 2012, 2013; Joshi et al. 2014; Zheng et al. 2019).

3.4. Advantages of Using Endosymbionts

The fundamental difference between the IIT and the SIT is the sterilizing procedure: infection with CI-inducing endosymbionts in the former, and irradiation in the latter (Bourtzis and Robinson 2006). Other aspects of the IIT and the SIT tend to be common to all sterile-male-based methods (Alphey et al. 2010; Bourtzis et al. 2016; Dyck et al. 2021), although the use of endosymbionts entails some specific considerations (Section 5).

As discussed above (Section 2.1), there are potentially direct and/or indirect harmful effects associated with irradiation-based sterilisation, which can both be circumvented by using CI-inducing endosymbionts. Although the initial introduction of a novel endosymbiont strain into a target species is not trivial, requiring considerably more effort, time and specialist skill than administering a single dose of irradiation (Hughes and Rasgon 2014), it only needs to be done once. As CI-inducing endosymbionts are maternally-inherited, once stably introduced into the germline of a target species, incompatibility is self-perpetuating and maintained across generations, so that there is no need for repeated rounds of sterilisation – with their associated economic and biological costs – within and across generations, as is the case for irradiation-based sterilisation. The use of radiation also entails various logistical and bureaucratic requirements (e.g. infrastructure and regulatory frameworks), which are not necessary when using endosymbiont infection. In addition, use of CI-inducing endosymbionts allows greater flexibility with regard to the life cycle stages of the target species that can be released (Bourtzis and Robinson 2006), while the SIT is often restricted by the life cycle stage at which irradiation is optimally performed.

Another overlooked advantage of the IIT is that the released individuals are conveniently "tagged" by their endosymbiont infections: there is no need to additionally mark released insects using chemical dyes – which may impose fitness costs (Curtis et al. 1982) – in order to track them during control programmes (Bourtzis and Robinson 2006). Identification of infected or sterile males, or their sperm, can be done by PCR (O'Connor et al. 2012; Juan-Blasco et al. 2013; Mains et al. 2016, 2019; Zheng et al. 2019). In addition, *Wolbachia* may have beneficial effects on larval development, such as promoting faster development, and thus lower rearing costs (Zhang et al. 2015a; Puggioli et al. 2016).

Some potential disadvantages of using endosymbionts, other than the major one of accidental female release resulting in unintended population replacement (see Section 3.5), are that incompatibility may decline with increasing adult male age (Tortosa et al. 2010), and with male sperm depletion following multiple mating (Bourtzis and Robinson 2006). So far, the decline in male incompatibility with age, which occurs with some native *Wolbachia* infections (Singh et al. 1976; Krishnamurthy et al. 1977; Calvitti et al. 2015), has not been reported for artificial infections (Moretti and Calvitti 2013), possibly due to the higher endosymbiont densities of the latter (Calvitti et al. 2015). Sperm depletion also affects the SIT, and, again, may occur during native *Wolbachia* infections, but it has been reported to have no effect for artificial endosymbiont infections (Turley et al. 2013).

3.4.1. *Cytoplasmic Incompatibility without Male Fitness Costs*

A widely-perceived advantage of the male incompatibility caused by CI-inducing endosymbionts, such as *Wolbachia*, is that it often has either no or only minor effects on male quality (Pal 1966; Laven 1974; Boller et al. 1976; Brower 1976; Sinkins et al. 1997; Scott and Benedict 2016). Although *Wolbachia* infections can be highly virulent (Min and Benzer 1997; McMeniman et al. 2009; Suh et al. 2009; Rasgon 2012), this is apparently atypical. In general, in their co-evolved native hosts, maintained under field conditions, these endosymbionts are thought to more commonly reside in the commensal to mutualist region of the spectrum of symbiosis (if their parasitic and "spiteful" reproductive manipulations are not considered) (Xi and Joshi 2016). CI-inducing endosymbionts can be expected to have been optimized over many millennia of natural selection to specifically induce sterility, while minimizing any harmful effects on male quality, as this would reduce their capacity to invade host populations (Segoli et al. 2014).

Consistent with this theoretical understanding, native *Wolbachia* infections of mosquitoes have generally been reported to have no effects on male quality (Dobson et al. 2002b; Calvitti et al. 2009; Baton et al. 2013). Although several studies have reported reduced mating competitiveness of field released incompatible males (~30 to 70%) (possibly due to the possession of a sterility-inducing chromosomal translocation in one study, and the use of chemical marker dyes in another) (Grover et al. 1976; O'Connor et al. 2012), the majority of studies have shown both native and artificially infected males to have mating competitiveness equal to that of wild-type males in both laboratory and field settings (Brower 1978; Curtis et al. 1982; Arunachalam and Curtis 1985; Blagrove et al. 2013; Moretti and Calvitti 2013; Joshi et al. 2014; Segoli et al. 2014; Atyame et al. 2015; Axford et al. 2016; Puggioli et al. 2016; Zhang et al. 2016; Zheng et al. 2019), with some even suggesting increased mating competitiveness for incompatible males (Puggioli et al. 2016; Moretti et al. 2018b).

In comparison to other aspects of endosymbiotic reproductive parasites, the effects of CI-inducing endosymbionts on the individual components of male fitness, such as sperm competition, are relatively under-studied, although the highly virulent *w*MelPop strain has been found to have no effect on insemination rates, or sperm quantity and viability (Turley et al. 2013). However, the mating competitiveness studies described above imply that the individual components of male fitness are generally unaffected by *Wolbachia* infection.

Many studies have found that *Wolbachia* infection has no effect on male longevity, while some studies have even found that artificial *Wolbachia* infections significantly increase adult male survival, which might increase the efficiency of incompatible males in IIT programmes (Blagrove et al. 2013; Joshi et al. 2014). A few studies have also compared irradiation-based sterilisation with endosymbiont-induced incompatibility, but these have not always used the most appropriate comparison (i.e. uninfected and irradiated individuals compared to non-irradiated *Wolbachia*-infected individuals with the same genetic background) (Atyame et al. 2016; Puggioli et al. 2016; Zhang et al. 2016; Zheng et al. 2019). In addition, standardized protocols need to be developed to enable robust comparison between these different sterilisation methods (Bourtzis et al. 2016).

3.4.2. Other Useful Phenotypes Enabling Tolerance of the Undesired Sex(es)

In addition to causing incompatibility, endosymbiotic reproductive parasites like *Wolbachia* can also have a range of other phenotypic effects on their invertebrate hosts, including effects on both adult male and female fitness. One of the most important phenotypic effects is that *Wolbachia* can inhibit viral pathogens (Hedges et al. 2008; Teixeira et al. 2008), and artificially-transinfected mosquitoes have been shown to often strongly inhibit or completely block a variety of vector-borne pathogens, especially arboviruses, including dengue, chikungunya, Mayaro, West Nile, yellow fever, and Zika viruses, and to a lesser extent filaria and malaria parasites (Kambris et al. 2009; Moreira et al. 2009; Bian et al. 2010, 2013; Glaser and Meola 2010; Andrews et al. 2012; van den Hurk et al. 2012; Aliota et al. 2016; Dutra et al. 2016; Joshi et al. 2017; Pereira et al. 2018; Zheng et al. 2019). As well as direct effects on pathogen infection, endosymbionts might also indirectly reduce pathogen transmission, for example by reducing the survival of their adult female vectors (Brownstein et al. 2003; Rasgon et al. 2003; Cook et al. 2008).

In contrast, native *Wolbachia* infections tend to have less predictable effects on vector-borne pathogens and have been reported to inhibit, enhance or have no effect upon them (Curtis et al. 1983; Dutton and Sinkins 2005; Bian et al. 2010; Blagrove et al. 2012; Graham et al. 2012; Baton et al. 2013; Bourtzis et al. 2014; Zélé et al. 2014). If possible, any released insects should be lower pests/vectors than the target population (Laven and Aslamkhan 1970; Thomas and Singh 1977). Given the potentially variable effects of endosymbiont infection on pest/vector status, this aspect of target species biology should be thoroughly characterized prior to releasing *Wolbachia*-infected insects (e.g. Zheng et al. 2019).

The ability of *Wolbachia* to reduce the ability of target species to act as pests or vectors, enables the release of the sex(es) which are pests/vectors to be tolerated, compensating for imperfect sex separation (Section 2.2) (Moretti et al. 2018b; Zheng et al. 2019), and could be an additional means to make the release of male pests/vectors more tolerable (e.g. tsetse flies) (Bourtzis et al. 2016; Kariithi et al. 2018). The ability of CI-inducing endosymbionts to reduce the ability of insects to act as pests/vectors also provides an important fail-safe during IIT implementation, given the high probability of accidental female release during operational programmes (Section 3.5), as unintended population replacement would then reduce pathogen transmission, while target population suppression would not be achieved (Zheng et al. 2019).

3.5. The Problem of Unintended Population Replacement

An inherent and significant problem with the IIT, in the absence of perfect sex separation methods, which has been recognized since the idea of IIT use was first conceived, and has so far prevented its operational use, is that, unlike SIT, the accidental release of *Wolbachia*-infected females may prevent further population suppression by causing unintended population replacement (Barr 1966; Pal 1974; Curtis 1977).

When the IIT is deployed, release of incompatible males will prevent reproduction by the wild-type females of the target population, and the target population will become suppressed. However, if any *Wolbachia*-infected females are accidentally released along with the released males, the former may successfully begin to reproduce in the field, because they are compatible with the released males infected with same *Wolbachia* strain, as well as the wild-type males in the original target population. Consequently, after an initial period of population suppression, during which the reproduction of wild-type females is inhibited, the original target population may become replaced by a population with *Wolbachia*-infected individuals (Section 3.1).

Whether population replacement actually occurs will depend upon a number of parameters, such as the number of females accidentally released, and the characteristics of the *Wolbachia* infection used to make released males incompatible with the target population. In principle, a single female could trigger population replacement, if her *Wolbachia* infection causes high levels of CI, has high maternal transmission, and no fitness costs (Section 3.1). In reality, stochastic and population density-dependent processes mean that a single female is unlikely to leave surviving off-spring, even if her *Wolbachia* infection has no invasion threshold. Just how many released females are required to inevitably trigger population replacement is unknown, and difficult to quantify accurately in the absence of relevant empirical data. However, the risk of population replacement will clearly increase as the original uninfected target population is suppressed, because this will inevitably increase the relative proportion of any accidentally-released *Wolbachia*-infected females (and their descendants) in relation to the wild-type individuals of the original target population, increasing the likelihood that the former passes its invasion threshold (Section 3.1).

Furthermore, if the original target population is eliminated, rather than merely suppressed, there will be a vacant niche that will be filled, by default, by any accidentally released females (Curtis 1977), which are now more likely to be able to establish a field population, as any inhibitory stochastic and density-dependent processes related to intra-specific competition with wild-type uninfected individuals will now be relaxed (Berryman et al. 1973; Weidhaas and Seawright 1976; Dobson et al. 2002b). Whether population replacement occurs will then mostly depend on the characteristics of the *Wolbachia* infection used to induce incompatibility (i.e. the *Wolbachia*-infected individuals might go extinct if the *Wolbachia* infection imposes severe fitness costs, but otherwise they would be expected to persist, given the general characteristics of the *Wolbachia* strains so far used for IIT application – see Sections 3.1 and 3.4).

Laboratory cage experiments and mathematical modelling both indicate that inundative releases of incompatible males, contaminated with some females, facilitates *Wolbachia* invasion and population replacement (Hancock et al. 2011; Bian et al. 2013; Moretti et al. 2018a; Zheng et al. 2019). In addition, artificial *Wolbachia* infections, with the requisite characteristics, have been able to invade and persist in field populations (Hoffmann et al. 2011, 2014). These observations reinforce the notion that the risk of unintended population replacement following accidental female release is real, and not merely a "hypothetical" or purely academic concern.

Some researchers have claimed that the risk of unintended population replacement can be minimized by releasing *Wolbachia*-infected insects that are bidirectionally incompatible with their wild-type target field population (Sharma et al. 1979; Calvitti et al. 2012; Bourtzis et al. 2014). Bidirectional CI occurs when the target population is infected with its own native *Wolbachia* strain, which causes incompatibility when wild-type field males mate with released females infected with a different *Wolbachia* strain that is used for IIT. The reproduction of any accidentally released females, therefore, will be prevented if they mate with the wild-type field males. This contrasts with unidirectional CI, which occurs when the target field population is either uninfected or infected with *Wolbachia* strain(s) that are not incompatible with the *Wolbachia* strain used for IIT, enabling any accidentally released females to successfully reproduce when mated wild-type field males.

However, explicit mathematical modelling and laboratory cage experiments indicate that bidirectional CI will only provide protection against population replacement if the frequency of the released *Wolbachia* strain does not exceed its own invasion threshold (Dobson et al. 2002a; Moretti et al. 2018a). Although this is unlikely to happen for bidirectional CI at low or intermediate levels of target population suppression, this is not, in general, the intended endpoint of sterile-male-based methods, which aim for high levels of population suppression, if not population elimination, at which point any accidentally-released bidirectionally incompatible females will exceed their invasion threshold. In practice, therefore, bidirectional CI does not provide appreciably greater protection from population replacement than unidirectional CI, if the aim is either population elimination or merely to reduce target population densities below that which causes a pest/vector problem (i.e. high levels of population suppression).

In addition, any released residual females are much more likely to mate with the males with which they are released, than with males in the field population (as the former are held together overnight in containers before their release, and afterwards released males vastly outnumber those in the field) (Zheng et al. 2019). Thus, the advantage from incompatible matings between released residual females and field males due to bi-directional CI may be negligible from a practical standing point.

Although several small-scale short-term field trials have reported the use of the IIT without the apparent consequence of population replacement (O'Connor et al. 2012; Mains et al. 2016, 2019; Zheng et al. 2019), it should be noted that in none of these instances was the target population sufficiently suppressed to enable the rapid replacement by any released *Wolbachia*-infected individuals which might occur as population elimination is approached (i.e. within the ~6 month time-scale of the reported results of these field trials). These studies also involved the release of relatively small numbers of mosquitoes using manual sex separation, which ensured lower female contamination rates than when using mechanical sex separation. However, manual separation is impractical for medium- to large-scale area-wide applications (Pal 1974; Brelsfoard et al. 2008; Zheng et al. 2019). As such, the above described small-scale field trials were probably unlikely to have released enough *Wolbachia*-infected females to enable establishment of *Wolbachia*-infected field populations.

4. COMBINED IIT/SIT APPLICATION

4.1. The Solution to an "Intractable" Problem

In order to solve the inevitable problem of accidental female release, the second series of IIT field trials undertaken in India during the 1970s (Section 3.2) combined the IIT with genetic modification of the released insects, such that they carried a chromosomal translocation that induced semi-sterility when they mated amongst themselves (Laven and Aslamkhan 1970; Brooks et al. 1976; Krishnamurthy and Laven 1976; Curtis 1977; Curtis et al. 1982). Although this approach seemed to show promise in the laboratory, under field conditions it had little impact.

As an alternative, Curtis (1977), who described the problem of unintended population replacement following accidental female release during IIT implementation as "intractable", proposed, nonetheless, a practical, and, as it turns out, viable solution: combined IIT/SIT use. His solution was to exploit the fact that female insects are often more sensitive than males to radiation (Bakri et al. 2005, 2021), and to combine the IIT with the SIT, such that all incompatible individuals destined for field release were first subjected to low-dose radiation, which would completely sterilize any contaminant females without affecting the incompatibility or quality of the simultaneously irradiated males. The combination of the SIT and the IIT for mosquito control is shown in Fig. 1. Although some preliminary laboratory investigations were undertaken by Curtis and others (Sharma et al. 1979; Arunachalam and Curtis 1985; Shahid and Curtis 1987), the notion of using the combined IIT/SIT was neglected for several decades until its re-assessment in more recent times (Bourtzis and Robinson 2006; Brelsfoard et al. 2009; Zhang et al. 2015b, 2016; Kittayapong et al. 2018, 2019; Zheng et al. 2019; Kittayapong, this volume; Liew et al., this volume).

4.2. Optimum Irradiation Dose for Female Sterilisation

Combined IIT/SIT application requires that females are more sensitive to the effects of irradiation than males, and that this difference is sufficiently large to enable complete sterilisation of females without appreciably or only minimally impacting male quality (Curtis 1977). If this inherent biological condition is met, it is necessary to determine the optimum irradiation dose. At first thought, this might be considered to be the minimum irradiation dose required to completely sterilize females, under the assumption that higher doses of radiation would begin to negatively affect males, and thereby undermine the main rationale for using endosymbiont-induced CI.

Accordingly, a number of preliminary studies have characterized the relative susceptibility of females and males to irradiation, with the aim of identifying the minimum irradiation dose required to completely sterilize females (Sharma et al. 1979; Shahid and Curtis 1987; Arunachalam and Curtis 1985; Brelsfoard et al. 2009; Zhang et al. 2015b, 2016). These irradiation studies showed that, at least for mosquitoes, females are indeed more sensitive than males to irradiation, and that there are levels of irradiation that can completely sterilize females, without appreciably impacting on male quality. Importantly, these and other studies have also shown that the irradiation treatment used to completely sterilize females has no effect on the level of CI induction by the co-irradiated *Wolbachia*-infected *Ae. albopictus* males (Zheng et al. 2019).

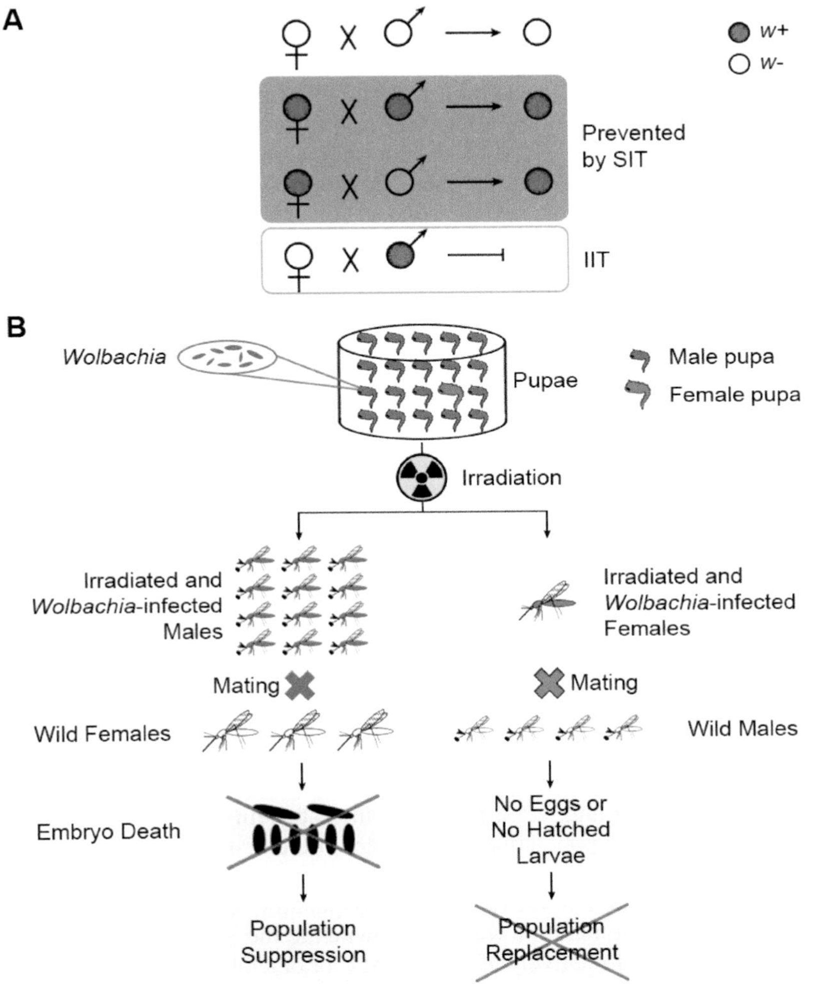

Figure 1. Schematic diagrams illustrating the combined IIT/SIT approach. (A) The four different types of crosses possible between wild-type uninfected and wPip*-infected* Aedes albopictus, *and the role of irradiation in prevention of residual infected females from reproducing in the field. Red indicates* Wolbachia*-infected individuals (W+), while white indicates uninfected individuals (W-). (B) Illustration of production, irradiation, and release of* Wolbachia*-infected males with the residual females, respectively, and their mating with the wild population in the field.*

The optimal irradiation dose for the deployment of the combined IIT/SIT approach, however, is not necessarily the minimum irradiation dose for "complete" female sterilisation observed in laboratory studies, as the latter use small sample sizes, which defines a minimum detectable level of sterility, and different irradiation protocols (in particular, where there is no necessity to overcrowd pupae, as required during mass-irradiation, which may induce radio-protective hypoxia) (Yamada et al. 2019). It is vital for the IIT that any released females have no residual fertility, as this could render any current and future implementations using the same endosymbiont strain ineffectual. Consequently, combined IIT/SIT field releases involving millions of individuals need somewhat higher doses of irradiation than required in small-scale laboratory studies to ensure that all released females are fully sterilized (Yamada et al. 2019).

4.3. Sequential IIT/SIT Application

An alternative strategy related to the combined IIT/SIT releases to prevent population replacement resulting from accidental female release is a "sequential IIT/SIT" approach. This would involve initial IIT only releases, followed by SIT only releases (as opposed to the simultaneous combined IIT/SIT application in the same released individuals) (Atyame et al. 2016). The rationale here is that if the IIT is more efficient than the SIT because of the higher mating competitiveness and higher induced sterility of endosymbiont-infected males, large-scale only IIT releases can be used initially to suppress the target population, followed immediately by smaller-scale only SIT releases to eradicate – "mop-up" – any endosymbiont-infected individuals resulting from females inadvertently released during the initial phase of the IIT.

Whether a sequential IIT/SIT approach is preferable or superior to the combined IIT/SIT approach is not obvious and requires a careful quantitative comparison of the relative costs and benefits of the two strategies. Sequential IIT/SIT releases have the advantage that the males released during IIT application are not irradiated, maximizing their mating competitiveness, and removing the logistical costs/difficulties associated with large-scale irradiation (e.g. reduced male quality because of increased handling, etc.). However, these benefits would come with an increased risk of triggering population replacement in the first place (as many fertile females might now be released), the requirement to more carefully and rigorously monitor the target population to identify when/if population replacement occurs, and the risk of missing the optimal time window to switch to the SIT only releases (such that large-scale releases of relatively inefficient SIT, or other methods, would then be required).

Sequential IIT/SIT releases may be more convenient and effective in highly localized short-term programmes against geographically-restricted and low-abundance target populations, where, overall, relatively few incompatible males need to be initially released, and therefore the risk of accidental female release is inherently lower, while the combined IIT/SIT is likely to be more appropriate under the opposite conditions (i.e. area-wide long-term programmes against geographically-widespread and high-abundance target populations).

When sequential IIT/SIT releases are used, if irradiation-based sterilisation of any females released during the SIT step is not complete, then these individuals could also establish a field population. For this reason, it might be prudent *not* to use for the SIT releases the same endosymbiont-infected insect line used for the initial IIT releases, so as to maintain the effectiveness of the initial insect line originally used for IIT should it be required for this latter purpose again (i.e. in multiple alternate rounds of IIT and SIT application).

In a similar manner to that envisaged for the sequential IIT/SIT, only SIT releases could also be used as a fail-safe after combined IIT/SIT application, should the latter fail to prevent population replacement.

5. THE FIRST OPEN-RELEASE FIELD TRIAL OF COMBINED IIT/SIT

Despite the previous exploratory laboratory studies determining the possibility and optimal dose for differentially sterilizing females and males for use in the combined IIT/SIT approach (Section 4.2.), there had been no previous experimental or field evaluation of this combined strategy. Consequently, a project was initiated, involving collaboration between Sun Yat-sen University, Michigan State University and other partners, to develop and field test combined IIT/SIT releases against the important mosquito arboviral vector *Ae. albopictus.*

This project involved a series of stages (Figs. 2 to 4), as described below, including initial laboratory studies to generate and then characterize an incompatible artificially-*Wolbachia*-infected *Ae. albopictus* line, subsequent "proof-of-concept" semi-field trials of the combined IIT/SIT approach, and then finally an open-field trial to demonstrate the feasibility of area-wide application of combined IIT/SIT releases for the management of an *Ae. albopictus* population.

5.1. Generation and Characterization of Novel Wolbachia *Infection*

The first requirement was to create a novel *Wolbachia* infection in *Ae. albopictus* that would generate incompatibility with wild-type males in our study area. To do this, the *Wolbachia* strain *w*Pip was transferred by embryonic microinjection (Fig. 2) from its native mosquito host *Cx. pipiens* into *Ae. albopictus*, to generate the new mosquito line HC. This line had a similar nuclear genetic background to individuals from the area of our field trial in Guangzhou, China, but in addition a novel triple *Wolbachia* infection (the artificially-transinfected *w*Pip plus its two native *Wolbachia* strains) (Zheng et al. 2019). *w*Pip was chosen because in its native mosquito host it has characteristics appropriate for IIT: it causes complete CI, has perfect maternal transmission, and no appreciable fitness costs. Indeed, upon transfer to its new host, these properties were retained.

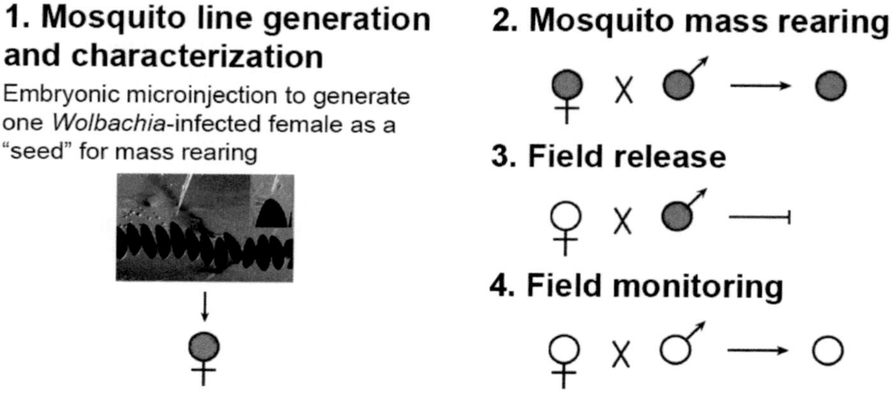

Figure 2. Overview of the IIT or combined IIT/SIT approach. First, a mosquito line with the novel Wolbachia *infection is generated by embryonic microinjection. The hallmark of success in this step is the generation of an infected female with ~100% maternal transmission efficiency to pass* Wolbachia *into their offspring (2.1). Then, the infected individuals are mass-produced in the factory (2.2), or also irradiated. The infected males are subsequently released into the field to induce sterility in the wild population (2.3). The density of the wild-type uninfected population is monitored to measure the effect of population suppression (2.4). Red indicates individuals carrying the novel* Wolbachia *infection, while white represents wild-type individuals.*

In laboratory studies, HC males caused complete CI when mated with wild-type females and had perfect maternal transmission (Zheng et al. 2019). There were also no differences between HC and wild-type *Ae. albopictus* in fecundity (number of eggs laid), fertility (egg hatch), larval/ pupal/ adult male or female survival, sex ratio or body size, although HC had a slightly faster larval development and adult emergence times (Zhang et al. 2015a). In addition, female HC had higher *Wolbachia* densities than wild-type females, and lower susceptibility to dengue and Zika virus infection, with both horizontal and vertical transmission of these arboviruses significantly reduced (Zheng et al. 2019). Although target population suppression/elimination was the aim of our field trial, the reduced vector competence of HC provided an important fail-safe should accidental female release and subsequently population replacement have occurred (Section 3.4.2) (Zheng et al. 2019).

5.2. *Laboratory and Semi-field IIT/SIT Trials*

A series of experimental laboratory studies were undertaken to characterize the HC-line further, and to confirm that it could be used for combined IIT/SIT releases. Laboratory cage experiments indicated that HC male mating competitiveness was equal to that of wild-type males (Zhang et al. 2016; Zheng et al. 2019). They also showed that HC females could cause population replacement, when seeded into cages containing wild-type individuals, with the speed of this population replacement being enhanced when excess HC males (4:1 ratio with wild-type males) were simultaneously released (Zheng et al. 2019). As these experiments demonstrated the potential of accidentally released HC females to trigger population replacement during IIT application, the minimum irradiation dose (28 Gy) necessary to completely sterilize HC females was identified. This dose caused extensive damage to the ovaries, and hence prevented egg-laying as well as the establishment of *w*Pip-infected individuals in a small laboratory cage population (Zhang et al. 2015b). This irradiation dose also did not affect mating competitiveness or survival of HC males, nor did it reduce CI induction (Zhang et al. 2016).

A subsequent semi-field trial was undertaken in field cages, which simulated the accidental release of HC females, in order to provide the first "proof-of-concept" that combined IIT/SIT application could prevent unintended population replacement (Zheng et al. 2019). Replicate control and experimental wild-type populations were established in large cages, into which irradiated excess HC males were repeatedly released (5:1 ratio with wild-type males), each time together with sufficient irradiated HC females to mimic a 2.0% contamination rate of the released HC males. Successful eradication of the wild-type populations occurred in all three of the field cages, without the occurrence of population replacement by the released *w*Pip-infected HC mosquitoes, demonstrating that the combined IIT/SIT strategy works. Having demonstrated experimentally that the combined approach works, an open-release field trial was implemented in Guangzhou, China.

5.3. *Mass-production for Field Release*

In order to produce sufficient numbers of irradiated-incompatible males for open-release during our combined SIT/IIT field trial, it was necessary to optimize rearing protocols and to develop new equipment to enable factory-scale mass-rearing and pupal irradiation (Fig. 3). Artificially-*Wolbachia*-infected mosquitoes do not require special rearing conditions and can be reared using the same protocols as those used for uninfected/wild-type individuals. However, some care should be given to ensure that *Wolbachia*-infected mosquitoes are not exposed to high temperatures or antibiotics (e.g. via their larval food or adult blood meals), as this could potentially remove their endosymbionts. In addition, larval rearing conditions may affect *Wolbachia* density, and, hence, possibly the level of CI expression, as well as maternal transmission of the endosymbiont (Puggioli et al. 2016). Consequently, it should be confirmed during the early stages of mass-production that the rearing conditions used do not adversely affect either the reproductive incompatibility or the quality of the males produced (Zhang et al. 2017, 2018).

For mass-rearing of larval mosquitoes, many rearing trays are required, thus several units for holding and storing large numbers of trays in order to improve space utilization have been developed at the FAO/IAEA Insect Pest Control Laboratory (IPCL) in Seibersdorf, Austria, and at the Wolbaki Institute of Biological Sciences in Guangzhou, China (Balestrino et al. 2012, 2014a; Zhang et al. 2017). The first generation "Wol-unit" holds 40 larval rearing trays, while only occupying 0.68 m^2 of floor space, and enables simultaneous rearing of 264 000 larvae, generating up to 89 000 male pupae per rearing cycle (Zhang et al. 2017).

Figure 3. The different stages of the first combined IIT/SIT field trial against Ae. albopictus *in Guangzhou, China. Photographs illustrating the nine different stages of the combined IIT/SIT field trial. In stage 1-3, artificially-triply-*Wolbachia*-infected adults, eggs and larvae were mass-produced in the mosquito rearing factory. In stage 4, a Fay-Morlan sorter was used for sex separation of pupae, followed by stage 5 with the Wolbaki® X-ray irradiator custom-made for the field trial to enable pupal irradiation. In stage 6, the sex-separated males were packed into buckets for mass-release. After quality control of emerged adult males by manual checking for contaminant adult females (stage 7), those buckets were delivered by vehicle to release sites (stage 8), as shown in the satellite images of the control and release sites (map data: Google, DigitalGlobe). Field populations were monitored through samples collected each week for diagnosis of* wPip *infection using PCR (stage 9).*

The "FAO/IAEA-unit" (Balestrino et al. 2012) holds 50 larval rearing trays, covers 0.94 m^2 ground area, has a capacity to hold 900 000 larvae, and can generate 314 000 male pupae per rearing cycle (Zhang et al. 2017). The second generation "Wol-unit 2.0" is based on the FAO/IAEA-unit and holds 100 larval rearing trays, covers 1.2 m^2 ground area, has a capacity to hold up to 1.5 million larvae, and can generate 550 000 male pupae per rearing cycle. A comparison between these three larval rearing units for *Ae. albopictus* is summarized in Table 1. The Wol-unit 2.0 is recommended for medium to large-scale applications as it requires relatively less space and enables more male pupae to be generated per unit.

Table 1. Comparison between three larval rearing units for production of one million Aedes albopictus *males*

Parameter	Wol-unit	FAO/IAEA-unit	Wol-unit 2.0
Number of trays per unit	40	50	100
Number of larvae reared per unit (10^5)	2.64	9.0	15.0
Number of male pupae acquired per unit (10^5)	0.89	3.14	5.5
Dimensions per unit (m, L * W * H)	0.97 × 0.70 × 1.85	0.78 × 1.2 × 2.10	1.41 x 0.84 x 2.1
Ground area per unit (m^2)	0.68	0.94	1.2
Quantity (unit)	11.2	3.2	1.9
Total space (m^2)	7.6	3.0	2.3
Labour - Adding water	Manual operation	Semi-automatic operation	Semi-automatic operation
Labour - Pupae/Larvae collection	Manual operation	Semi-automatic operation	Semi-automatic operation
Labour - Cleaning	Manual operation	Semi-automatic operation	Semi-automatic operation
Price	Low	High	Medium
Application	Small size factory	Medium size factory	Medium/Large size factory

For holding of adult mosquitoes, a suitable cage structure is important to maximize egg production. A prototype mass-production cage based on a design originally used for Mediterranean fruit flies, had been previously developed at the IPCL that allowed sugar and blood-feeding, as well as a simplified egg collection system that minimized the risk of mosquito escapes (Balestrino et al. 2014b; Mamai et al. 2017). However, we found that the egg production of *Ae. albopictus* was quite low, i.e. an average of ~16 eggs per female per blood meal. As rearing density seems to be the main factor causing low egg production (Balestrino et al. 2014b), the cage height was reduced and, together with the addition of ATP to the blood meal, we were able to increase egg production to an average of ~70 eggs per female given two blood meals (Zhang et al. 2018). The modified mass-production cage and mass-rearing protocol described currently enables the Wolbaki factory to produce 10 million *Ae. albopictus* eggs every 15 days (Zhang et al. 2018).

Male and female pupae of mosquitoes in the genera *Aedes* and *Culex* can be separated on the basis of size differences by using sieves or glass separators, although the traditional equipment is laborious to use (McCray 1961; Focks 1980; Balestrino et al. 2014a). An automated glass separator has been developed at Wolbaki to reduce manual operation and improve sex separation efficiency.

5.4. Irradiation of Pupae for Release

An irradiator specific for mosquito pupae was required for our field trial. Gamma rays have been the most common type of radiation used for insect sterilisation, because of their high energy and penetration (Bakri et al. 2021). However, the use of gamma rays is challenging because of regulatory, logistical and economic issues, related to safety, security, recycling, transportation, storage and initial cost. Consequently, in the past decade the use of X-rays has been suggested as a potential alternative to gamma rays (Mastrangelo et al. 2010; Ndo et al. 2014; Yamada et al. 2014a; FAO/IAEA 2017).

For insect sterilisation, a dose uniformity ratio (DUR: the maximum dose divided by the minimum dose) below 1.2 is required, in order to ensure a uniform dose is given to the irradiated individuals (Yamada et al. 2019). Dose uniformity is required to ensure that males do not receive unnecessarily high doses of radiation, which might needlessly reduce their quality, and is important for the combined IIT/SIT, where it is vital that *all* contaminant females are sufficiently irradiated to ensure complete sterilisation. However, the X-ray irradiators currently available on the market with the recommended DUR are not suitable for larger-scale applications using mosquitoes, because either only a small number of pupae can be simultaneously irradiated (RS 2000, Biological System Irradiator, RadSource, Georgia, USA), or they require relatively frequent replacement of the costly X-ray tube and are inconvenient for pupal irradiation (RS 2400) (Yamada et al. 2014a).

Consequently, Wolbaki in cooperation with the FAO/IAEA, developed a new X-ray irradiator – "the Wolbaki irradiator" – specifically designed for pupal irradiation, which meets the technical requirements and large-scale processing capacity required for our field trial. The irradiator is equipped with a ray tube at a 40-degree angle, and with a maximum power of 4.5 kW. At a horizontal distance of 30 cm from the radiation source, the dose rate is measured at 3.2 Gy/min through a 0.3 mm copper filter. A rotary table for holding canisters is set up for horizontal rotation during exposure. Two separated canisters, with a total loading capacity of one litre male pupae, can be vertically swapped at half target dose. The DUR is reduced to 1.07 by rotating and swapping during exposure.

As described above, the optimum irradiation dose for sterilizing of contaminant females for field release is likely to be appreciably higher than that indicated by a naïve interpretation of laboratory data based on very small sample sizes. Accordingly, we erred on the side of caution, and chose an irradiation dose of 45 Gy to ensure the success of our field trial (i.e. no fertile contaminant females released).

5.5. *Open-Release in the Field and Entomological Surveillance*

The open-release field trial was undertaken over a 2 to 3-year period (2016-2018) on two residential islands in Guangzhou, with each release site having its own control sites (Fig. 3) (Zheng et al. 2019). The field trial started a year earlier in Release Site 1 (2014 compared to 2015 for Release Site 2), with an initial pilot test of the IIT only in Release Site 1 during 2015, and a test of the combined IIT/SIT strategy being performed simultaneously in both sites during 2016 and 2017.

In the years prior to the male HC releases, base-line entomological surveys were carried out in both sets of control and release sites to confirm their suitability for the field trial (2014 for Site 1 and 2015 for Site 2; Fig. 4).

For the pilot test of IIT only, non-irradiated HC mosquitoes, from which the females had been removed by a combination of mechanical and manual sex sorting, were released during the mosquito breeding season (March to October). Initially, males were released throughout the entire area of Release Site 1, and the target field population was suppressed by as much as 55% (March to May; Fig. 5). However, as the mosquito breeding season peaked (late May to early June), the level of population suppression diminished, as it was not possible to release sufficient numbers of *Wolbachia*-infected males throughout the entire release site in order to attain the critical overflooding ratio. This was due to the labour-intensive checks required to manually remove contaminant females from the released males, a rate-limiting step which constrained how many sex-sorted mosquitoes could be produced per week (given the number of staff available for our field trial). Therefore, in an attempt to achieve the critical overflooding ratio for the remainder of the IIT trial in 2015 (mid-June to October), we reduced the treated area within Release Site 1 in which males were released (Fig. 4B), and subsequently expanded it following the "rolling carpet" approach (Dyck et al. 2021), so that the local density of released males would be increased, without the need to release a larger number of males overall.

After reducing the size of the release area, population suppression within the area of continuing releases was striking and significant, whereas very high mosquito densities were found in the immediately neighbouring area of Release Site 1 without continued releases (Fig. 4B), as well as within the control site (Fig. 5). These observations demonstrate the feasibility of using the IIT only for mosquito population suppression, and its potential for population elimination, if technological developments can be made that enable the large-scale mass-production – at a reasonable cost-effectiveness – of sufficient numbers of incompatible males lacking appreciable female contamination.

A trial of the combined IIT/SIT approach was then subsequently undertaken, in which irradiated HC mosquitoes were released. In this instance, females were removed from the released mosquitoes using mechanical separation only, resulting in a higher level of female contamination, but which could be tolerated as the residual females were sterilized by irradiation. As manual checks for contamination were no longer used or required, it was possible to release much larger numbers of male mosquitoes (>10-fold) for the combined IIT/SIT approach than for the IIT alone, and so HC releases could be undertaken throughout the entire area of both release sites for the entire duration of the two-year combined IIT/SIT field trial.

A

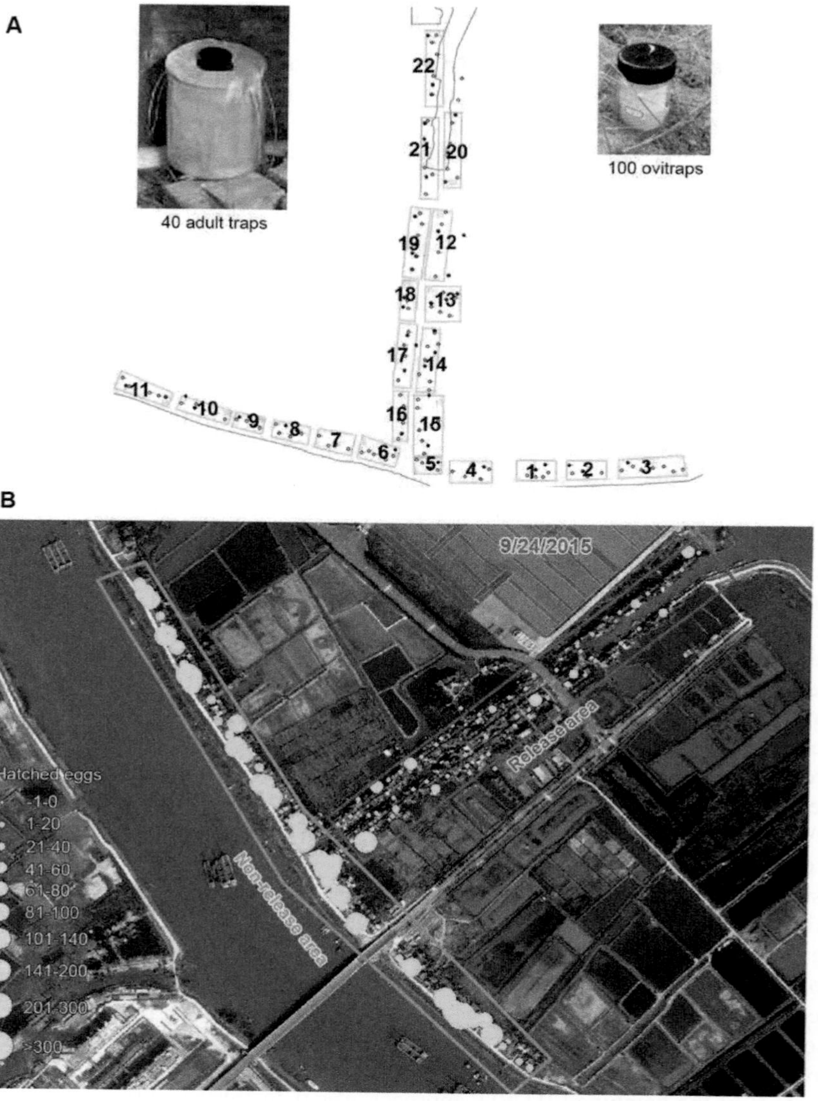

Figure 4. Monitoring and comparison of egg hatch rates during IIT and combined IIT/SIT application against Ae. albopictus *in Release Site 1 in Guangzhou, China. (A) Schematic diagram illustrating the division of Release Site 1 into 22 zones (green boxes), and the location of the ovitraps and adult-collecting Biogents BG-Sentinel traps that were used weekly to monitor* Aedes albopictus *populations during the field trial. (B) Satellite image showing the non-release (blue box; zones 1 to 11) and release (red box; zones 12 to 22) areas within Release Site 1 during the IIT only phase of the field trial in September 2015 (Map data: Google, DigitalGlobe). Yellow circles indicate ovitrap locations, with areas proportional to the number of hatched eggs collected in each for that week in September 2015.*

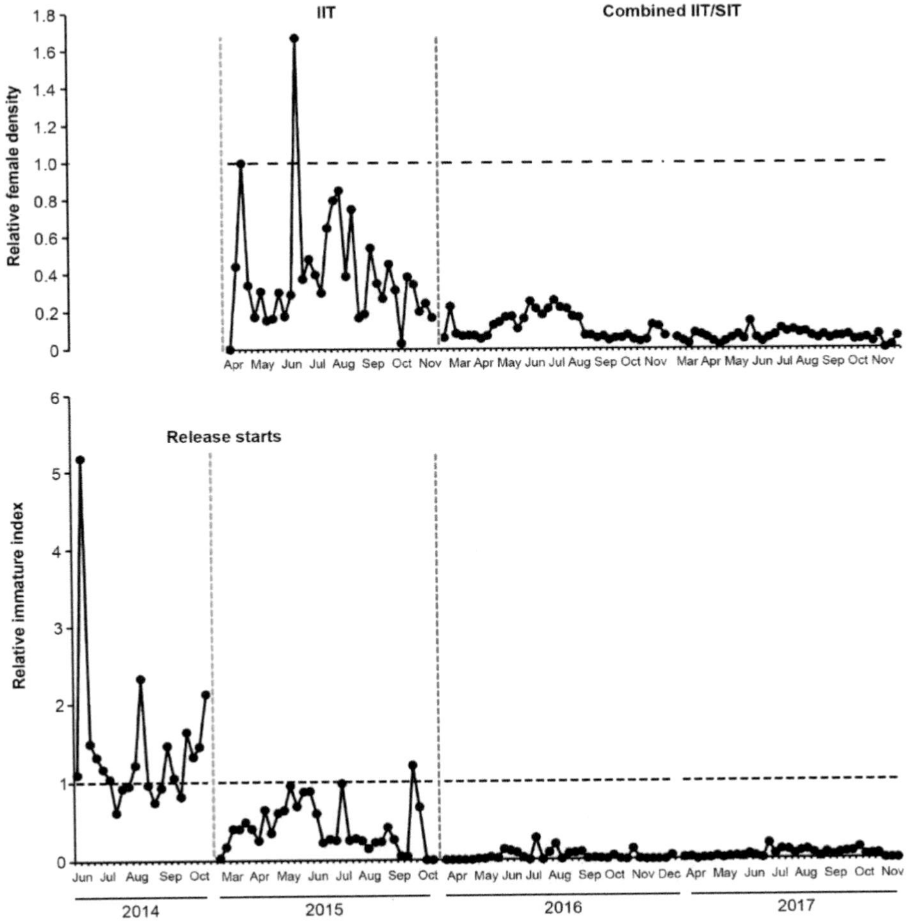

Figure 5. Suppression of Aedes albopictus *in Release Site 1 during the field trial in Guangzhou, China. The solid black lines indicate, respectively, the densities of wild-type adult females (A) and larvae (B) collected in Release Site 1 standardized by dividing by the corresponding number of individuals collected in Control Site 1 (see stage 9 in Figure 3). In 2014, baseline data were collected during the "pre-release" period before any compatible males were released. In 2015, IIT only releases were undertaken. In 2016 and 2017, combined IIT/SIT releases were performed. The horizontal black dashed lines indicate the relative level of larvae/adults in Control Site 1. The vertical green dashed line indicates the onset of IIT-only releases of incompatible males, while the vertical blue dashed line indicates onset of combined IIT/SIT releases of irradiated incompatible males.*

Overall, during the mosquito breeding seasons of 2016 and 2017, over 197 million factory-reared irradiated HC males were released using buckets from which adults emerged (Fig. 3). On average, 0.2-0.3% of the released insects were contaminant females. The sterile to wild male overflooding ratio was estimated at between 8.7:1 to 15.8:1, which resulted in the near-elimination of wild-type adult female *Ae. albopictus* from both release sites, i.e. > 94% reduction in egg hatch and up to 94% reduction in the apparent density of wild adult females (Fig. 5). The failure to eliminate completely the target populations in our release sites appeared to have been due to a low level of immigration.

Importantly, we found no evidence of population replacement during the three-year period of our field trial: throughout the period of male releases, we carefully monitored not only the wild-type target population, but also used PCR to screen collected larvae for the *w*Pip *Wolbachia* strain infecting released HC individuals. Although we did find a very low level of *w*Pip-positive larvae (0.87%, 16/1844 ovitrap samples), confirming the potential risk of population replacement, their collection was spatially and/or temporally-isolated, and they did not seem to constitute a viable breeding population. The field population did not increase in size after its initial suppression, nor was there a delayed rebound increase in egg hatch over time, which would have been expected, as a result of compatible matings becoming more frequent, if *w*Pip-infected mosquitoes had established in the field.

Overall, these observations demonstrate that the combined IIT/SIT approach can (i) suppress and effectively eliminate mosquito vector populations, and (ii) provide protection against the risk of population replacement resulting from the accidental release of fertile compatible endosymbiont-infected females.

6. FUTURE AREA-WIDE COMBINED IIT/IIT RELEASES

Despite our successful field trial, doubts about the area-wide implementation of the combined IIT/SIT persist (Armbruster 2019). As with other sterile-male-based methods, concerns include the affordability and sustainability of large-scale mass-release programmes. We believe that with optimization of the protocols used in our field trial, the combined IIT/SIT approach can be both affordable and sustainable for lower-income countries, and have an important and leading role as part of area-wide integrated pest control programmes (see Supplementary Information to Zheng et al. 2019).

To completely remove the risk of population replacement and to obtain population suppression or elimination, pupae were irradiated with a relatively high dose in the mass-rearing facility (Zheng et al. 2019). In addition, the current design of the X-ray irradiator and canister require a large number (up to 200 000) of pupae to stay in an overcrowded condition for an extended period (up to 15 min). Both result in a negative impact on male mating competitiveness and reduced cost-effectiveness of combined IIT/SIT releases. Thus, efforts will be made to optimize the approach for radiation exposure and to further improve the design of the X-ray irradiator.

As with any sterile-male-based method, large-scale area-wide deployment of the combined IIT/SIT approach would benefit from the development of improved and/or new methods/technologies to facilitate the efficient mass-production and mass-release of sufficient incompatible males to achieve population suppression/elimination. Many of these requirements are not unique to combined IIT/SIT releases (Alphey et al. 2010; Bourtzis et al. 2016), and, as such, we do not review them here in detail, other than to indicate how improvements in sex separation might impact the combined IIT/SIT approach.

Although the development of perfect sex separation methods is highly desirable (Gilles et al. 2014), the existence of such methods would negate the need and necessity for combined IIT/SIT, enabling IIT to be conducted without the risk of accidental female release resulting in unintended population replacement, and enabling the SIT to be conducted without risk of increased pest/vector activity. However, as described by Franz et al. (2021), even the best genetic sexing systems available are not perfect under large-scale operational programmes. In addition, where CI-inducing endosymbionts and irradiation reduce the ability of insects to act as pests/vectors (Sections 3.4.2), the low levels of contaminant females is less problematic.

Improvements in sex separation are likely to have their greatest impact on combined IIT/SIT releases by enabling the application of this method to target species for which there are currently either insufficient sex separation methods to enable mass-releases, or they are not available (e.g. *Anopheles* mosquitoes, which cannot be easily separated on the basis of size).

7. CONCLUSIONS

The combined IIT/SIT strategy integrates the strengths of the IIT with those of the SIT, and in so doing overcomes the current technological limitations of each approach. It can be used as an environment-friendly biopesticide to meet the current need for a novel solution to suppress mosquito populations and their transmitted diseases. Our successful field trial demonstrates the feasibility of area-wide application of combined IIT/SIT releases for *Aedes* mosquitoes.

8. REFERENCES

Alam, U., J. Medlock, C. Brelsfoard, R. Pais, C. Lohs, S. Balmand, J. Carnogursky, A. Heddi, P. Takac, A. Galvani, and S. Aksoy. 2011. *Wolbachia* symbiont infections induce strong cytoplasmic incompatibility in the tsetse fly *Glossina morsitans*. PLoS Pathogens 7: e1002415.

Aliota, M. T., S. A. Peinado, I. D. Velez, and J. E. Osorio. 2016. The *w*Mel strain of *Wolbachia* reduces transmission of Zika virus by *Aedes aegypti*. Scientific Reports 6: 28792.

Alphey, L. 2014. Genetic control of mosquitoes. Annual Review of Entomology 59: 205–224.

Alphey, L., M. Benedict, R. Bellini, G. G. Clark, D. A. Dame, M. W. Service, and S. L. Dobson 2010. Sterile-insect methods for control of mosquito-borne diseases: An analysis. Vector-Borne and Zoonotic Diseases 10: 295–311.

Andreasen, M. H., and C. F. Curtis. 2005. Optimal life stage for radiation sterilisation of *Anopheles* males and their fitness for release. Medical and Veterinary Entomology 19: 238–244.

Andrews, E. S., P. R. Crain, Y. Fu, D. K. Howe, and S. L. Dobson. 2012. Reactive oxygen species production and *Brugia pahangi* survivorship in *Aedes polynesiensis* with artificial *Wolbachia* infection types. PLoS Pathogens 8: e1003075.

Anonymous. 1975. Oh New Delhi, Oh Geneva. Nature 256: 355–357.

Ant, T. H., and S. P. Sinkins. 2018. A *Wolbachia* triple-strain infection generates self-incompatibility in *Aedes albopictus* and transmission instability in *Aedes aegypti*. Parasites & Vectors 11: 295.

Apostolaki, A., I. Livadaras, A. Saridaki, A. Chrysargyris, C. Savakis, and K. Bourtzis. 2011. Transinfection of the olive fruit fly *Bactrocera oleae* with *Wolbachia:* Towards a symbiont-based population control strategy. Journal of Applied Entomology 135: 546–553.

Armbruster, P. A. 2019. Tiger mosquitoes tackled in a trial. Nature 572 (7767): 39–40.

Arunachalam, N., and C. F. Curtis. 1985. Integration of radiation with cytoplasmic incompatibility for genetic control in the *Culex pipiens* complex (Diptera: Culicidae). Journal of Medical Entomology 22: 648–653.

Atyame, C. M., N. Pasteur, E. Dumas, P. Tortosa, M. L. Tantely, N. Pocquet, S. Licciardi, A. Bheecarry, B. Zumbo, M. Weill, and O. Duron. 2011. Cytoplasmic incompatibility as a means of controlling *Culex pipiens quinquefasciatus* mosquito in the islands of the south-western Indian Ocean. PLoS Neglected Tropical Diseases 5: e1440.

Atyame, C. M., J. Cattel, C. Lebon, O. Flores, J. S. Dehecq, M. Weill, L. C. Gouagna, and P. Tortosa. 2015. *Wolbachia*-based population control strategy targeting *Culex quinquefasciatus* mosquitoes proves efficient under semi-field conditions. PLoS One 10: e0119288.

Atyame, C. M., P. Labbé, C. Lebon, M. Weill, R. Moretti, F. Marini, L. C. Gouagna, M. Calvitti, and P. Tortosa. 2016. Comparison of irradiation and *Wolbachia* based approaches for sterile-male strategies targeting *Aedes albopictus.* PLoS One 11: e0146834.

Axford, J. K., P. A. Ross, H. L. Yeap, A. G. Callahan, and A. A. Hoffmann. 2016. Fitness of *w*AlbB *Wolbachia* infection in *Aedes aegypti:* Parameter estimates in an outcrossed background and potential for population invasion. The American Journal of Tropical Medicine and Hygiene 94: 507–516.

Bakri, A., N. Heather, J. Hendrichs, and I. Ferris. 2005. Fifty years of radiation biology in entomology: Lessons learned from IDIDAS. Annals of the Entomological Society of America 98: 1–12.

Bakri, A., K. Mehta, and D. R. Lance. 2021. Sterilizing insects with ionizing radiation, pp. 355–398. *In* V. A. Dyck, J. Hendrichs, and A. S. Robinson (eds.), Sterile Insect Technique – Principles and practice in Area-Wide Integrated Pest Management. Second Edition. CRC Press, Boca Raton, Florida, USA.

Balestrino, F., M. Q. Benedict, and J. R. L. Gilles. 2012. A new larval tray and rack system for improved mosquito mass rearing. Journal of Medical Entomology 49: 595–605.

Balestrino, F., A. Puggioli, J. R. L. Gilles, and R. Bellini. 2014a. Validation of a new larval rearing unit for *Aedes albopictus* (Diptera: Culicidae) mass rearing. PLoS One 9: e91914.

Balestrino, F., I. A. Puggioli, R. Bellini, D. Petric, and J. R. L. Gilles. 2014b. Mass production cage for *Aedes albopictus* (Diptera: Culicidae). Journal of Medical Entomology 51: 155–163.

Balestrino, F., A. Medici, G. Candini, M. Carrieri, B. Maccagnani, M. Calvitti, S. Maini, and R. Bellini. 2010. Gamma ray dosimetry and mating capacity studies in the laboratory on *Aedes albopictus* males. Journal of Medical Entomology 47: 581–591.

Barclay, H. J. 2021. Mathematical models for using sterile insects, pp. 201–244. *In* V. A. Dyck, J. Hendrichs, and A. S. Robinson (eds.), Sterile Insect Technique – Principles and practice in Area-Wide Integrated Pest Management. Second Edition. CRC Press, Boca Raton, Florida, USA.

Barr, A. R. 1966. Cytoplasmic incompatibility as a means of eradication of *Culex pipiens* L. Proceedings and Papers of the Annual Conference of the California Mosquito Control Association 34: 32–35.

Barr, A. R. 1970. Partial compatibility and its effect on eradication by the incompatible male method. Proceedings and Papers of the Annual Conference of the California Mosquito Control Association (Thirty-Seventh Annual Conference, Los Angeles, 27-29th January 1969) 37: 19–24.

Baton, L.A., E. C. Pacidônio, D. da Silva Gonçalves, and L. A. Moreira. 2013. *w*Flu: Characterization and evaluation of a native *Wolbachia* from the mosquito *Aedes fluviatilis* as a potential vector control agent. PLoS One 8: e59619.

Bellini, R., M. Calvitti, A. Medici, M. Carrieri, G. Celli, and S. Maini. 2007. Use of the sterile insect technique against *Aedes albopictus* in Italy: First results of a pilot trial, pp. 505–515. *In* M. J. B. Vreysen, A. S. Robinson, and J. Hendrichs (eds.), Area-wide control of insect pests: From research to field implementation. Springer, Dordrecht, The Netherlands.

Bellini, R., F. Balestrino, A. Medici, G. Gentile, R. Veronesi, and M. Carrieri. 2013a. Mating competitiveness of *Aedes albopictus* radio-sterilized males in large enclosures exposed to natural conditions. Journal of Medical Entomology 50: 94–102.

Bellini, R., A. Medici, A. Puggioli, F. Balestrino, and M. Carrieri. 2013b. Pilot field trials with *Aedes albopictus* irradiated sterile males in Italian urban areas. Journal of Medical Entomology 50: 317–325.

Benedict, M. Q., and A. S. Robinson. 2003. The first releases of transgenic mosquitoes: An argument for the sterile insect technique. Trends in Parasitology 19: 349–355.

Berryman, A. A., T. P. Boo'co, and L. C. Dickmann. 1973. Computer simulation of population reduction by release of sterile insects: II. The effects of dynamic survival and multiple mating, pp. 31–43. *In* Computer models and application of the Sterile-Male Technique: Proceedings of a Panel, Vienna, 13-17 December 1971, organized by the Joint FAO/IAEA Division of Atomic Energy in Food and Agriculture, International Atomic Energy Agency, Vienna, Austria.

Bian, G., Y. Xu, P. Lu, Y. Xie, and Z. Xi. 2010. The endosymbiotic bacterium *Wolbachia* induces resistance to dengue virus in *Aedes aegypti*. PLoS Pathogens 6: e1000833.

Bian, G., D. Joshi, Y. Dong, P. Lu, G. Zhou, X. Pan, Y. Xu, G. Dimopoulos, and Z. Xi. 2013. *Wolbachia* invades *Anopheles stephensi* populations and induces refractoriness to *Plasmodium* infection. Science 340: (6133) 748–751.

Blagrove, M. S. C., C. Arias-Goeta, A.-B. Failloux, and S. P. Sinkins. 2012. *Wolbachia* strain *w*Mel induces cytoplasmic incompatibility and blocks dengue transmission in *Aedes albopictus*. Proceedings of the National Academy of Sciences of the USA 109: 255–260.

Blagrove, M. S. C., C. Arias-Goeta, C. Di Genua, A.-B. Failloux, and S. P. Sinkins. 2013. A *Wolbachia* *w*Mel transinfection in *Aedes albopictus* is not detrimental to host fitness and inhibits Chikungunya virus. PLoS Neglected Tropical Diseases 7: e2152.

Blümel, S., and K. Russ. 1989. Manipulation of races, pp. 387–389. *In* A. S. Robinson and G. Hooper (eds.), Fruit flies: Their biology, natural enemies and control. World Crop Pests, Volume 3B. Elsevier, Amsterdam, The Netherlands.

Boller, E. F. 1989. Cytoplasmic incompatibility in *Rhagoletis cerasi*, pp. 69–74. *In* A. S. Robinson and G. Hooper (eds.), Fruit flies: Their biology, natural enemies and control. World Crop Pests, Volume 3B. Elsevier, Amsterdam, The Netherlands.

Boller, E. F., and G. L. Bush. 1974. Evidence for genetic variation in populations of the European cherry fruit fly, *Rhagoletis cerasi* (Diptera: Tephritidae) based on physiological parameters and hybridization experiments. Entomologia Experimentalis et Applicata 17: 279–293.

Boller, E. F., K. Russs, V. Vallo, and G. L. Bush. 1976. Incompatible races of European cherry fruit fly, *Rhagoletis cerasi* (Diptera: Tephritidae), their origin and potential use in biological control. Entomologia Experimentalis et Applicata 20: 237-247.

Bourtzis, K. and A. S. Robinson. 2006. Insect pest control using *Wolbachia* and/or radiation, pp. 225–246. *In* K. Bourtzis and T. A. Miller (eds.), Insect symbiosis. Volume 2. CRC Press, Boca Raton, USA.

Bourtzis, K., R. S. Lees, J. Hendrichs, and M. J. B. Vreysen. 2016. More than one rabbit out of the hat: Radiation, transgenic and symbiont-based approaches for sustainable management of mosquito and tsetse fly populations. Acta Tropica 157: 115–130.

Bourtzis, K., S. L. Dobson, Z. Xi, J. L. Rasgon, M. Calvitti, L. A. Moreira, H. Bossin, R. Moretti, L. A. Baton, G. L. Hughes, P. Mavingui, and J. R. L. Gilles. 2014. Harnessing mosquito-*Wolbachia* symbiosis for vector and disease control. Acta Tropica 132 (Supplement): S150–S163.

Boyle, L., S. L. O'Neill, H. M. Robertson, and T. L. Karr. 1993. Interspecific and intraspecific horizontal transfer of *Wolbachia* in *Drosophila*. Science 260: (5115) 1796–1799.

Braig, H. R., H. Guzman, R. B. Tesh, and S. L. O'Neill. 1994. Replacement of the natural *Wolbachia* symbiont of *Drosophila simulans* with a mosquito counterpart. Nature 367: (6462) 453–455.

Brelsfoard, C. L., Y. Séchan, and S. L. Dobson. 2008. Interspecific hybridization yields strategy for South Pacific filariasis vector elimination. PLoS Neglected Tropical Diseases 2: e129.

Brelsfoard, C. L., W. St Clair, and S. L. Dobson. 2009. Integration of irradiation with cytoplasmic incompatibility to facilitate a lymphatic filariasis vector elimination approach. Parasites & Vectors 2: (1) 38.

Brooks, G. D., C. F. Curtis, K. K. Grover, B. S. Krishnamurthy, P. L. Rajagopalan, L. S. Sharma, V. P. Sharma, D. Singh, K. R. P. Singh, M. Yasuno, M. A. Ansari, T. Adak, H. V. Aggarwal, C. P. Batra, R. K. Chandrahas, P. R. Malhotra, P. K. B. Menon, R. Menon, S. Das, R. K. Razdan, and V. Vaidyanathan. 1976. A field trial on control of *Culex pipens fatigans* Wied. by release of males of a strain integrating cytoplasmic incompatibility and a translocation. WHO/VBC/76.635. World Health Organization, Geneva, Switzerland.

Brower, J. H. 1976. Cytoplasmic incompatibility: Occurrence in a stored product pest *Ephestia cautella*. Annals of the Entomological Society of America 69: 1011–1015.

Brower, J. H. 1978. Propensity of interstrain mating in cytoplasmically incompatible strains of the almond moth. Journal of Economic Entomology 71: 585–586.

Brower, J. H. 1979. Suppression of laboratory populations of *Ephestia cautella* (Walker) (Lepidoptera: Pyralidae) by release of males with cytoplasmic incompatibility. J. Stored Products Research 15: 1–4.

Brower, J. H. 1980. Reduction of almond moth populations in simulated storages by the release of genetically incompatible males. Journal of Economic Entomology 73: 415–418.

Brownstein, J. S., E. Hett, and S. L. O'Neill. 2003. The potential of virulent *Wolbachia* to modulate disease transmission by insects. Journal of Invertebrate Pathology 84: 24–29.

Calvitti, M., R. Moretti, D. Porretta, R. Bellini, and S. Urbanelli. 2009. Effects on male fitness of removing *Wolbachia* infections from the mosquito *Aedes albopictus*. Medical and Veterinary Entomology 23: 132–140.

Calvitti, M., R. Moretti, E. Lampazzi, R. Bellini, and S. L. Dobson. 2010. Characterization of a new *Aedes albopictus* (Diptera: Culicidae)-*Wolbachia pipientis* (Rickettsiales: Rickettsiaceae) symbiotic association generated by artificial transfer of the *w*Pip strain from *Culex pipiens* (Diptera: Culicidae). Journal of Medical Entomology 47: 179–187.

Calvitti, M., R. Moretti, A. R. Skidmore, and S. L. Dobson. 2012. *Wolbachia* strain *w*Pip yields a pattern of cytoplasmic incompatibility enhancing a *Wolbachia*-based suppression strategy against the disease vector *Aedes albopictus*. Parasites & Vectors 5: 254.

Calvitti, M., F. Marini, A. Desiderio, A. Puggioli, and R. Moretti. 2015. *Wolbachia* density and cytoplasmic incompatibility in *Aedes albopictus*: Concerns with using artificial *Wolbachia* infection as a vector suppression tool. PLoS One 10: e0121813.

Caspari, E., and G. S. Watson. 1959. On the evolutionary importance of cytoplasmic sterility in mosquitoes. Evolution 13: 568–570.

Cattel, J., K. Nikolouli, T. Andrieux, J. Martinez, F. Jiggins, S. Charlat, F. Vavre, D. Lejon, P. Gibert, and L. Mouton. 2018. Back and forth *Wolbachia* transfers reveal efficient strains to control spotted wing drosophila populations. Journal of Applied Ecology 55: 2408–2418.

Chen, L., C. Zhu, and D. Zhang. 2013. Naturally occurring incompatibilities between different *Culex pipiens pallens* populations as the basis of potential mosquito control measures. PLoS Neglected Tropical Diseases 7: e2030.

Cook, P. E., C. J. McMeniman, and S. L. O'Neill. 2008. Modifying insect population age structure to control vector-borne disease. Advances in Experimental Medicine and Biology 627: 126–140.

Corbel, V., C. Durot, N. L. Achee, F. Chandre, M. B. Coulibaly, J. P. David, G. J. Devine, I. Dusfour, D. M. Fonseca, J. Griego, W. Juntarajumnong, A. Lenhart, S. Kasai, A. J. Martins, C. Moyes, L. C. Ng, J. Pinto, J. F. Pompon, P. Muller, K. Raghavendra, D. Roiz, H. Vatandoost, J. Vontas, and D. Weetman. 2019. Second WIN International Conference on "Integrated approaches and innovative tools for combating insecticide resistance in vectors of arboviruses". 1-3 October 2018, Singapore. Parasites & Vectors 12 (1): 331.

Curtis, C. F. 1977. Testing systems for the genetic control of mosquitoes, pp. 106–116. *In* J. S. Packer and D. White (eds.), XV International Congress of Entomology, Entomological Society of America, College Park, Maryland, USA.

Curtis, C. F., and R. Reuben. 2007. Destruction in the 1970s of a research unit in India on genetic control of mosquitoes and a warning for the future management of transgenic research. Antenna 31: 214–216.

Curtis, C. F., G. D. Brooks, M. A. Ansari, K. K. Grover, B. S. Krishnamurthy, P. K. Rajagopalan, L. S. Sharma, V. P. Sharma, D. Singh, K. R. P. Singh, and M. Yasuno. 1982. A field trial on control of *Culex quinquefasciatus* by release of males of a strain integrating cytoplasmic incompatibility and a translocation. Entomologia Experimentalis et Applicata 31: 181–190.

Curtis, C. F., D. S. Ellis, P. E. Doyle, N. Hill, B. D. Ramji, L. W. Irungu, and H. Townson. 1983. Susceptibility of aposymbiotic *Culex quinquefasciatus* to *Wuchereria bancrofti*. Journal of Invertebrate Pathology 41: 214–223.

Dame, D. A., C. F. Curtis, M. Q. Benedict, A. S. Robinson, and B. G. Knols. 2009. Historical applications of induced sterilisation in field populations of mosquitoes. Malaria Journal 8 (Supplement 2): S2.

Davidson, G. 1974. Genetic control of insect pests. Academic Press, London, UK. 168 pp.

Dobson, S. L., and W. Rattanadechakul. 2001. A novel technique for removing *Wolbachia* infections from *Aedes albopictus* (Diptera: Culicidae). Journal of Medical Entomology 38: 844–849.

Dobson, S. L., C. W. Fox, and F. M. Jiggins. 2002a. The effect of *Wolbachia*-induced cytoplasmic incompatibility on host population size in natural and manipulated systems. Proceedings of the Royal Society, Series B: Biological Sciences 269: (1490) 437–445.

Dobson, S. L., E. J. Marsland, and W. Rattanadechakul. 2002b. Mutualistic *Wolbachia* infection in *Aedes albopictus*: Accelerating cytoplasmic drive. Genetics 160: 1087–1094.

Dutra, H. L. C., M. N. Rocha, F. B. S. Dias, S. B. Mansur, E. P. Caragata, and L. A. Moreira. 2016. *Wolbachia* blocks currently circulating Zika virus isolates in Brazilian *Aedes aegypti* mosquitoes. Cell Host and Microbe 19: 771–774.

Dutton, T. J., and S. P. Sinkins. 2005. Filarial susceptibility and effects of *Wolbachia* in *Aedes pseudoscutellaris* mosquitoes. Medical and Veterinary Entomology 19: 60–65.

Dyck, V. A., J. Hendrichs, and A. S. Robinson (eds.). 2021. Sterile Insect Technique – Principles and practice in Area-Wide Integrated Pest Management, Second Edition. CRC Press, Boca Raton, Florida, USA. 1200 pp.

(FAO/IAEA) Food and Agriculture Organization / International Atomic Energy Agency. 2017. Technical specification for an X-ray system for the irradiation of insects for the sterile insect technique and other related technologies. Joint FAO/IAEA Programme of Nuclear Techniques in Food and Agriculture, Vienna, Austria. 11 pp.

Fine, P. E. M. 1978. On the dynamics of symbiote-dependent cytoplasmic incompatibility in culicine mosquitoes. Journal of Invertebrate Pathology 31: 10–18.

Focks, D. A. 1980. An improved separator for the developmental stages, sexes, and species of mosquitoes (Diptera: Culicidae). Journal of Medical Entomology 17: 567–568.

Franz, F., K. Bourtzis, and C. Cáceres. 2021. Practical and operational genetic sexing systems based on classic genetic approaches in fruit flies, an example for other species amenable to large-scale rearing for the Sterile Insect Technique, pp. 575–604. *In* V. A. Dyck, J. Hendrichs, and A. S. Robinson (eds.), Sterile Insect Technique – Principles and practice in Area-Wide Integrated Pest Management. Second Edition. CRC Press, Boca Raton, Florida, USA.

Fu, Y., L. Gavotte, D. R. Mercer, and S. L. Dobson. 2010. Artificial triple *Wolbachia* infection in *Aedes albopictus* yields a new pattern of unidirectional cytoplasmic incompatibility. Applied and Environmental Microbiology 76: 5887–5891.

Gilles, J. R. L., M. F. Schetelig, F. Scolari, F. Marec, M. L. Capurro, G. Franz, and K. Bourtzis. 2014. Towards mosquito sterile insect technique programmes: Exploring genetic, molecular, mechanical and behavioural methods of sex separation in mosquitoes. Acta Tropica 132 (Supplement): S178–S187.

Glaser, R. L., and M. A. Meola. 2010. The native *Wolbachia* endosymbionts of *Drosophila melanogaster* and *Culex quinquefasciatus* increase host resistance to West Nile virus infection. PLoS One 5: e11977.

Graham, R. I., D. Grzywacz, W. L. Mushobozi, K. Wilson, and D. Ebert. 2012. *Wolbachia* in a major African crop pest increases susceptibility to viral disease rather than protects. Ecology Letters 15: 993–1000.

Grover, K. K., and V. P. Sharma. 1974. The present status of the work on induced sterility mechanisms for control of mosquitoes *Culex pipiens fatigans* and *Aedes aegypti* at the WHO/ICMR Research Unit on Genetic Control of Mosquitoes. The Journal of Communicable Diseases 6: 91–97.

Grover, K. K., C. F. Curtis, V. P. Sharma, K. R. P. Singh, K. Dietz, H. V. Aggarwal, R. K. Razdan, and V. Vaidyanathan. 1976. Competitiveness of chemosterilised males and cytoplasmically incompatible translocated males of *Culex pipiens fatigans* Wiedemann (Diptera, Culicidae) in the field. Bulletin of Entomological Research 66: 469–480.

Hancock, P. A., S. P. Sinkins, and H. C. J. Godfray. 2011. Strategies for introducing *Wolbachia* to reduce transmission of mosquito-borne diseases. PLoS Neglected Tropical Diseases 5: e1024.

Hedges, L. M., J. C. Brownlie, S. L. O'Neill, and K. N. Johnson. 2008. *Wolbachia* and virus protection in insects. Science 322: (5902) 702.

Helinski, M. E. H., and B. G. J. Knols. 2008. Mating competitiveness of male *Anopheles arabiensis* mosquitoes irradiated with a partially or fully sterilizing dose in small and large laboratory cages. Journal of Medical Entomology 45: 698–705.

Helinski, M. E. H., A. G. Parker, and B. G. J. Knols. 2006. Radiation-induced sterility for pupal and adult stages of the malaria mosquito *Anopheles arabiensis*. Malaria Journal 5: 41.

Helinski, M. E. H., A. G. Parker, and B. G. J. Knols. 2009. Radiation biology of mosquitoes. Malaria Journal 8 (Supplement 2): S6.

Hertig, M., 1936. The rickettsia, *Wolbachia pipientis* (gen. et sp.n.) and associated inclusions of the mosquito, *Culex pipiens*. Parasitology 28: 453–486.

Hertig, M., and S. B. Wolbach. 1924. Studies on rickettsia-like micro-organisms in insects. The Journal of Medical Research 44: 329–374.

Hilgenboecker, K., P. Hammerstein, P. Schlattmann, A. Telschow, and J. H. Werren. 2008. How many species are infected with *Wolbachia*? A statistical analysis of current data. FEMS Microbiology Letters 281: 215–220.

Hoffmann, A. A., P. A. Ross, and G. Rasic. 2015. *Wolbachia* strains for disease control: Ecological and evolutionary considerations. Evolutionary Applications 8: 751–768.

Hoffmann, A. A., B. L. Montgomery, J. Popovici, I. Iturbe-Ormaetxe, P. H. Johnson, F. Muzzi, M. Greenfield, M. Durkan, Y. S. Leong, Y. Dong, H. Cook, J. Axford, A. G. Callahan, N. Kenny, C. Omodei, E. A. McGraw, P. A. Ryan, S. A. Ritchie, M. Turelli, and S. L. O'Neill. 2011. Successful establishment of *Wolbachia* in *Aedes populations* to suppress dengue transmission. Nature 476: (7361) 454–457.

Hoffmann, A. A., I. Iturbe-Ormaetxe, A. G. Callahan, B. L. Phillips, K. Billington, J. K. Axford, B. Montgomery, A. P. Turley, and S. L. O'Neill. 2014. Stability of the *w*Mel *Wolbachia* infection following invasion into *Aedes aegypti* populations. PLoS Neglected Tropical Diseases 8: e3115.

Hughes, G. L., and J. L. Rasgon. 2014. Transinfection: A method to investigate *Wolbachia*-host interactions and control arthropod-borne disease. Insect Molecular Biology 23: 141–151.

Hurst, G. D. D., and C. L. Frost. 2015. Reproductive parasitism: Maternally inherited symbionts in a biparental world. Cold Spring Harbor Perspectives in Biology 7: a017699.

Iturbe-Ormaetxe, I., T. Walker, and S. L. O'Neill. 2011. *Wolbachia* and the biological control of mosquito-borne disease. EMBO Reports 12: 508–518.

Joshi, D., M. J. McFadden, D. Bevins, F. Zhang, and Z. Xi. 2014. *Wolbachia* strain *w*AlbB confers both fitness costs and benefit on *Anopheles stephensi*. Parasites & Vectors 7: 336.

Joshi, D., X. Pan, M. J. McFadden, D. Bevins, X. Liang, P. Lu, S. Thiem, and Z. Xi. 2017. The maternally inheritable *Wolbachia w*AlbB induces refractoriness to *Plasmodium berghei* in *Anopheles stephensi*. Frontiers in Microbiology 8: 366.

Joubert, D. A., T. Walker, L. B. Carrington, J. T. De Bruyne, D. H. Kien, N. T. Hoang, N. V. Chau, I. Iturbe-Ormaetxe, C. P. Simmons, and S. L. O'Neill. 2016. Establishment of a *Wolbachia* superinfection in *Aedes aegypti* mosquitoes as a potential approach for future resistance management. PLoS Pathogens 12: e1005434.

Juan-Blasco, M., A. Urbaneja, V. San Andrés, P. Castañera, and B. Sabater-Muñoz. 2013. Improving the sterile sperm identification method for its implementation in the area-wide sterile insect technique program against *Ceratitis capitata* (Diptera: Tephritidae) in Spain. Journal of Economic Entomology 106: 2541–2547.

Kariithi, H. M., I. K. Meki, D. I. Schneider, L. De Vooght, F. M. Khamis, A. Geiger, G. Demirbaş-Uzel, J. M. Vlak, I. A. Ince, S. Kelm, F. Njiokou, F. N. Wamwiri, I. I. Malele, B. L. Weiss, and A. M. M. Abd-Alla. 2018. Enhancing vector refractoriness to trypanosome infection: Achievements, challenges and perspectives. BMC Microbiology 18 (Supplement 1): 179.

Kambris, Z., P. E. Cook, H. K. Phuc, and S. P. Sinkins. 2009. Immune activation by life-shortening *Wolbachia* and reduced filarial competence in mosquitoes. Science 326: (5949) 134–136.

Kittayapong, P., N. Kaeothaisong, S. Ninphanomchai, and W. Limohpasmanee. 2018. Combined sterile insect technique and incompatible insect technique: Sex separation and quality of sterile *Aedes aegypti* male mosquitoes released in a pilot population suppression trial in Thailand. Parasites & Vectors 11 (Supplement 2): 657.

Kittayapong, P., S. Ninphanomchai, W. Limohpasmanee, C. Chansang, U. Chansang, and P. Mongkalangoon. 2019. Combined sterile insect technique and incompatible insect technique: The first proof-of-concept to suppress *Aedes aegypti* vector populations in semi-rural settings in Thailand. PLoS Neglected Tropical Diseases 13: e0007771.

Knipling, E. F. 1955. Possibilities of insect control or eradication through the use of sexually sterile males. Journal of Economic Entomology 48: 459–462.

Knipling, E. F., H. Laven, G. B. Craig, R. Pal, J. B. Kitzmiller, C. N. Smith, and A. W. A. Brown. 1968. Genetic control of insects of public health importance. Bulletin of the World Health Organization 38: 421–438.

Krishnamurthy, B. S. 1977. Evaluation of the D3 cytoplasmic incompatible strain of *Culex pipiens fatigans* in laboratory and field cages. The Indian Journal of Medical Research 65 (Supplement): 13–20.

Krishnamurthy, B. S., and H. Laven. 1976. Development of cytoplasmically incompatible and integrated (translocated incompatible) strains of *Culex pipiens fatigans* for use in genetic control. Journal of Genetics 62: 117–129.

Krishnamurthy, B. S., C. F. Curtis, S. K. Subbarao, K. R. P. Singh, R. K. Chandrahas, and T. Adak. 1977. Further studies on the effect of aging and mating history of males on cytoplasmic incompatibility in *Culex pipiens fatigans*. Journal of Genetics 63: 31–37.

Laird, M. 1967. Eradication of *Culex pipiens fatigans* through cytoplasmic incompatibility. Nature 216: (5122) 1358.

Laven, H. 1967. Eradication of *Culex pipiens fatigans* through cytoplasmic incompatibility. Nature 216: (5113) 383–384.

Laven, H. 1971. Une expérience de lutte génétique contre *Culex pipiens fatigans* Wied. 1828. Annales de Parasitologie Humaine et Comparée 46 (3, Supplément): 117–148.

Laven, H. 1974. Genetic control of mosquitoes, pp. 19-26. *In* Proceedings Tall Timbers Conference on Ecological Animal Control by Habitat Management, Number 5. Tall Timbers Research Station, Tallahassee, Florida, USA.

Laven, H., and M. Aslamkhan. 1970. Control of *Culex pipiens pipiens* and *C. p. fatigans* with integrated genetical systems. Pakistan Journal of Science 22: 303–312.

Lees, R. S., D. Carvalho, and J. Bouyer. 2021. Potential impact of the fight against disease-transmitting mosquitoes using the sterile insect technique, pp. 1081–1118. *In* V. A. Dyck, J. Hendrichs, and A. S. Robinson (eds.), Sterile Insect Technique – Principles and practice in Area-Wide Integrated Pest Management. Second Edition. CRC Press, Boca Raton, Florida, USA.

Lees R. S., B. Knols, R. Bellini, M. Q. Benedict, A. Bheecarry, H. C. Bossin, D. D. Chadee, J. Charlwood, R. K. Dabiré, L. Djogbenou, A. Egyir-Yawson, R. Gato, L. C. Gouagna, M. M. Hassan, S. A. Khan, L. L. Koekemoer, G. Lemperiere, N. C. Manoukis, and J. R. L. Gilles. 2013. Review: Improving our knowledge of male mosquito biology in relation to genetic control programmes. Acta Tropica 13 (Supplement): S2–S11.

Lees, R. S., J. R. L Gilles, J. Hendrichs, M. J. B. Vreysen, and K. Bourtzis. 2015. Back to the future: The Sterile Insect Technique against mosquito disease vectors. Current Opinion in Insect Science 10: 156–162.

Maïga, H., D. Damiens, A. Niang, S. P. Sawadogo, O. Fatherhaman, R. S. Lees, O. Roux, R. K. Dabiré, G. A. Ouedraogo, F. Tripet, A. Diabate, and J. R. L. Gilles. 2014. Mating competitiveness of sterile male *Anopheles coluzzii* in large cages. Malaria Journal 13: 460.

Mains, J. W., C. L. Brelsfoard, R. I. Rose, and S. L. Dobson. 2016. Female adult *Aedes albopictus* suppression by *Wolbachia*-infected male mosquitoes. Scientific Reports 6: 33846.

Mains, J. W., P. H. Kelly, K. L. Dobson, W. D. Petrie, and S. L. Dobson. 2019. Localized control of *Aedes aegypti* (Diptera: Culicidae) in Miami, FL, via inundative releases of *Wolbachia*-infected male mosquitoes. Journal of Medical Entomology 56: 1296–1303.

Mamai, W., N. S. Bimbile-Somda, H. Maiga, J. G. Juarez, Z. A. I. Muosa, A. B. Ali, R. S. Lees, and J. R. L. Gilles. 2017. Optimization of mosquito egg production under mass rearing setting: Effects of cage volume, blood meal source and adult population density for the malaria vector, *Anopheles arabiensis*. Malaria Journal 16: 41.

Marshall, J. F. 1938. The British Mosquitoes. William Clowes and Sons, Limited, London and Beccles, UK. 341 pp.

Marshall, J. F., and J. Staley. 1937. Some notes regarding the morphological and biological differentiation of *Culex pipiens* Linnaeus and *Culex molestus* Forskål (Diptera, Culicidae). Proceedings of the Royal Entomological Society of London, Series A: General Entomology 12: 17–26.

Mastrangelo, T., A. G. Parker, A. Jessup, R. Pereira, D. Orozco-Davila, A. Islam, T. Dammalage, and J. Walder. 2010. A new generation of x ray irradiators for insect sterilisation. Journal of Economic Entomology 103: 85–94.

McCray, E. M. 1961. A mechanical device for the rapid sexing of *Aedes aegypti* pupae. Journal of Economic Entomology 54: 819.

McMeniman, C. J., R. V. Lane, B. N. Cass, A. W. C. Fong, M. Sidhu, Y.-F. Wang, and S. L. O'Neill. 2009. Stable introduction of a life-shortening *Wolbachia* infection into the mosquito *Aedes aegypti*. Science 323: (5910) 141–144.

Min, K.-T., and S. Benzer. 1997. *Wolbachia*, normally a symbiont of *Drosophila*, can be virulent, causing degeneration and early death. Proceedings of the National Academy of Sciences of the USA 94: 10792–10796.

Moreira, L.A., I. Iturbe-Ormaetxe, J. A. Jeffery, G. Lu, A. T. Pyke, L. M. Hedges, B. C. Rocha, S. Hall-Mendelin, A. Day, M. Riegler, L. E. Hugo, K. N. Johnson, B. H. Kay, A. E. McGraw, A. F. van den Hurk, P. A. Ryan, and S. L. O'Neill. 2009. A *Wolbachia* symbiont in *Aedes aegypti* limits infection with dengue, Chikungunya, and Plasmodium. Cell 139: 1268–1278.

Moretti, R., and M. Calvitti. 2013. Male mating performance and cytoplasmic incompatibility in a *w*Pip *Wolbachia* trans-infected line of *Aedes albopictus* (*Stegomyia albopicta*). Medical and Veterinary Entomology 27: 377–386.

Moretti, R., G. A. Marzo, E. Lampazzi, and M. Calvitti. 2018a. Cytoplasmic incompatibility management to support Incompatible Insect Technique against *Aedes albopictus*. Parasites & Vectors 11 (Supplement 2): 649.

Moretti, R., P.-S. Yen, V. Houé, E. Lampazzi, A. Desiderio, A.-B. Failloux, and M. Calvitti. 2018b. Combining *Wolbachia*-induced sterility and virus protection to fight *Aedes albopictus*-borne viruses. PLoS Neglected Tropical Diseases 12: e0006626.

Ndo, C., H. Yamada, D. D. Damiens, S. N'do, G. Seballos, and J. R. L. Gilles. 2014. X-ray sterilisation of the *An. arabiensis* genetic sexing strain 'ANO IPCL1' at pupal and adult stages. Acta Tropica 131: 124–128.

Neuenschwander, P., K. Russ, E. Höbaus, and S. Michelakis. 1983. Ecological studies on *Rhagoletis cerasi* L. in Crete for the use of the incompatible insect technique, pp. 41–51. *In* R. Calvalloro (ed.), Fruit flies of economic importance. Proceedings of the CEC/IOBC International Symposium. Athens, Greece, 16-19 November 1982. A. A. Balkema, Rotterdam, The Netherlands.

Nikolouli, K., H. Colinet, D. Renault, T. Enriquez, L. Mouton, P. Gibert, F. Sassu, C. Caceres, C. Stauffer, R. Pereira, and K. Bourtzis. 2018. Sterile insect technique and *Wolbachia* symbiosis as potential tools for the control of the invasive species *Drosophila suzukii*. Journal of Pest Science 91: 489–503.

O'Connor, L., C. Plichart, A. C. Sang, C. L. Brelsfoard, H. C. Bossin, and S. L. Dobson. 2012. Open release of male mosquitoes infected with a *Wolbachia* biopesticide: Field performance and infection containment. PLoS Neglected Tropical Diseases 6: e1797.

Oliva, C. F., M. Jacquet, J. R. L. Gilles, G. Lemperiere, P.-O. Maquart, S. Quilici, F. Schooneman, M. J. B. Vreysen, and S. Boyer. 2012. The sterile insect technique for controlling populations of *Aedes albopictus* (Diptera: Culicidae) on Réunion Island: Mating vigour of sterilized males. PLoS One 7: e49414.

Pal, R. 1966. Genetic control of vectors of disease with special reference to *Culex pipiens fatigans*. WHO/Vector Control/66.198. World Health Organization, Geneva, Switzerland.

Pal, R. 1974. WHO/ICMR programme of genetic control of mosquitoes in India, pp. 73–95. *In* R. Pal and M. J. Whitten (eds.), The use of genetics in insect control. Elsevier / North-Holland Publishing Company, Amsterdam, The Netherlands.

Parker, A., and K. Mehta. 2007. Sterile insect technique: A model for dose optimization for improved sterile insect quality. Florida Entomologist 90: 88–95.

Pereira, T. N., M. N. Rocha, P. H. F. Sucupira, F. D. Carvalho, and L. A. Moreira. 2018. *Wolbachia* significantly impacts the vector competence of *Aedes aegypti* for Mayaro virus. Scientific Reports 8: 6889.

Portaro, J. K., and A. R. Barr. 1975. "Curing" *Wolbachia* infections in *Culex pipiens*. Journal of Medical Entomology 12: 265.

Puggioli, A., M. Calvitti, R. Moretti, and R. Bellini. 2016. *w*Pip *Wolbachia* contribution to *Aedes albopictus* SIT performance: Advantages under intensive rearing. Acta Tropica 164: 473–481.

Ranner, H. 1990. Untersuchungen zur Biologie und Bekämpfung der Kirschfruchtfliege, *Rhagoletis cerasi* L. (Diptera, Trypetidae) - V. Versuche zur Bekämpfung der Kirschfruchtfliege mit Hilfe der Incompatible Insect Technique (IIT). Pflanzenschutzberichte 51: 1–16.

Rasgon, J. L. 2012. *Wolbachia* induces male-specific mortality in the mosquito *Culex pipiens* (LIN strain). PLoS One 7: e30381.

Rasgon, J. L., L. M. Styer, and T. W. Scott. 2003. *Wolbachia*-induced mortality as a mechanism to modulate pathogen transmission by vector arthropods. Journal of Medical Entomology 40: 125–132.

Roubaud, É. 1941. Phénomènes d'amixie dans les intercroisements de Culicides du groupe *pipiens*. Comptes Rendus Hebdomadaires des Séances de l'Académie des Sciences 212: 257–259.

Ruang-areerate, T., and P. Kittayapong. 2006. *Wolbachia* transinfection in *Aedes aegypti*: A potential gene driver of dengue vectors. Proceedings of the National Academy of Sciences of the USA 103: 12534–12539.

Russ, K., and B. Faber. 1979. The possible use of IIT to control *Rhagoletis cerasi* L., the European cherry fruit fly in Austria, pages 38–39. *In* IOBC-WPRS Bulletin, Volume 2, Proceedings of a Joint Meeting of the Working Groups "Biological Control of Olive Pests", "Genetic Control of *Rhagoletis cerasi*", "Genetic Control of *Ceratitis capitata*" and "Genetic Methods of Pest Control". 15-20 May 1978. Sassari, Sardinia, Italy.

Scott, M. J., and M. Q. Benedict. 2016. Concept and history of genetic control, pp. 31–54. *In* Z. N. Adelman (ed.), Genetic control of malaria and dengue. Academic Press, London, UK.

Segoli, M., A. A. Hoffmann, J. Lloyd, G. J. Omodei, and S. A. Ritchie. 2014. The effect of virus-blocking *Wolbachia* on male competitiveness of the dengue vector mosquito, *Aedes aegypti*. PLoS Neglected Tropical Diseases 8: e3294.

Shahid, M. A., and C. F. Curtis. 1987. Radiation sterilisation and cytoplasmic incompatibility in a "tropicalized" strain of the *Culex pipiens* complex (Diptera: Culicidae). Journal of Medical Entomology 24: 273–274.

Sharma, V. P., S. K. Subbarao, T. Adak, and R. K. Razdan. 1979. Integration of gamma irradiation and cytoplasmic incompatibility in *Culex pipiens fatigans* (Diptera: Culicidae). Journal of Medical Entomology 15: 155–156.

Singh, K. R. P., C. F. Curtis, and B. S. Krishnamurthy. 1976. Partial loss of cytoplasmic incompatibility with age in males of *Culex fatigans*. Annals of Tropical Medicine and Parasitology 70: 463–466.

Sinkins, S. P. 2004. *Wolbachia* and cytoplasmic incompatibility in mosquitoes. Insect Biochemistry and Molecular Biology 34: 723–729.

Sinkins, S. P., C. F. Curtis, and S. L. O'Neill. 1997. The potential application of inherited symbiont systems to pest control, pp. 155–175. *In* S. L. O'Neill, A. A. Hoffmann, and J. H. Werren (eds.), Influential passengers: Inherited microorganisms and arthropod reproduction. Oxford University Press, Oxford, UK.

Subbarao, S. K., C. F. Curtis, K. R. P. Singh, and B. S. Krishnamurthy. 1974. Variation in cytoplasmic crossing type in population of *C. p. fatigans* Wied. from Delhi area. The Journal of Communicable Diseases 6: 80–82.

Subbarao, S. K., C. F. Curtis, B. S. Krishnamurthy, T. Adak, and R. K. Chandrahas. 1977. Selection for partial compatibility with aged and previously mated males in *Culex pipiens fatigans* (Diptera: Culicidae). Journal of Medical Entomology 14: 82–85.

Suh, E., D. R. Mercer, Y. Fu, and S. L. Dobson. 2009. Pathogenicity of life-shortening *Wolbachia* in *Aedes albopictus* after transfer from *Drosophila melanogaster*. Applied and Environmental Microbiology 75: 7783–7788.

Teixeira, L., A. Ferreira, and M. Ashburner. 2008. The bacterial symbiont *Wolbachia* induces resistance to RNA viral infections in *Drosophila melanogaster*. PLoS Biology 6: e2.

Thomas, V., and K. R. P. Singh. 1977. Comparative susceptibility to *Wuchereria bancrofti* of *Culex pipiens fatigans* Delhi strain and of strains cytoplasmically incompatible with it. The Indian Journal of Medical Research 65 (Supplement): 102–106.

Tortosa, P., S. Charlat, P. Labbé, J.-S. Dehecq, H. Barré, and M. Weill. 2010. *Wolbachia* age-sex-specific density in *Aedes albopictus*: A host evolutionary response to cytoplasmic incompatibility? PLoS One 5: e9700.

Turley, A. P., M. P. Zalucki, S. L. O'Neill, and E. A. McGraw. 2013. Transinfected *Wolbachia* have minimal effects on male reproductive success in *Aedes aegypti*. Parasites & Vectors 6: 36.

van den Hurk, A. F., S. Hall-Mendelin, A. T. Pyke, F. D. Frentiu, K. McElroy, A. Day, S. Higgs, and S. L. O'Neill. 2012. Impact of *Wolbachia* on infection with Chikungunya and yellow fever viruses in the mosquito vector *Aedes aegypti*. PLoS Neglected Tropical Diseases 6: e1892.

Walker, T., P. H. Johnson, L. A. Moreira, I. Iturbe-Ormaetxe, F. D. Frentiu, C. J. McMeniman, Y. S. Leong, Y. Dong, J. Axford, P. Kriesner, A. L. Lloyd, S. A. Ritchie, S. L. O'Neill, and A. A. Hoffmann. 2011. The *w*Mel *Wolbachia* strain blocks dengue and invades caged *Aedes aegypti* populations. Nature 476: (7361) 450–453.

Weidhaas, D. E., and J. Seawright. 1976. Comments on the article "Genetic control of mosquitoes" (Laven, 1974), pp. 211–220. *In* Proceedings Tall Timbers Conference on Ecological Animal Control by Habitat Management, Number 6. Tall Timbers Research Station, Tallahassee, Florida, USA.

Weinert, L. A., E. V. Araujo-Jnr, M. Z. Ahmed, and J. J. Welch. 2015. The incidence of bacterial endosymbionts in terrestrial arthropods. Proceedings of the Royal Society, Series B: Biological Sciences 282: (1807) 20150249.

Werren, J. H., L. Baldo, and M. E. Clark. 2008. *Wolbachia*: Master manipulators of invertebrate biology. Nature Reviews Microbiology 6: 741–751.

(WHO) World Health Organization. 1964. Genetics of vectors and insecticide resistance. Report of a WHO Scientific Group meeting held in Geneva from 5 to 9 August 1963. Number 268. Geneva, Switzerland. 40 pp.

Xi, Z., and S. L. Dobson. 2005. Characterization of *Wolbachia* transfection efficiency by using microinjection of embryonic cytoplasm and embryo homogenate. Applied and Environmental Microbiology 71: 3199–3204.

Xi, Z., and D. Joshi. 2016. Genetic control of malaria and dengue using *Wolbachia*, pp. 305–333. *In* Z. N. Adelman (ed.), Genetic control of malaria and dengue. Academic Press, London, UK.

Xi, Z., and P. Manrique-Saide. 2018. Scale up from field trial to operation: The combined IIT/SIT approach to eliminate the primary dengue mosquito vectors in China and Mexico. Reunión Nacional del Programa de Enfermedades Transmitidas por Vectores. Los Cabos, Baja California Sur, Mexico.

Xi, Z., J. L. Dean, C. C. H. Khoo, and S. L. Dobson. 2005a. Generation of a novel *Wolbachia* infection in *Aedes albopictus* (Asian tiger mosquito) via embryonic microinjection. Insect Biochemistry and Molecular Biology 35: 903–910.

Xi, Z., C. C. H. Khoo, and S. L. Dobson. 2005b. *Wolbachia* establishment and invasion in an *Aedes aegypti* laboratory population. Science 310: (5746) 326–328.

Xi, Z., C. C. H. Khoo, and S. L. Dobson. 2006. Interspecific transfer of *Wolbachia* into the mosquito disease vector *Aedes albopictus*. Proceedings of the Royal Society, Series B: Biological Sciences 273 (1592): 1317–1322.

Yamada, H., A. G. Parker, C. F. Oliva, F. Balestrino, and J. R. L. Gilles. 2014a. X-ray-induced sterility in *Aedes albopictus* (Diptera: Culicidae) and male longevity following irradiation. Journal of Medical Entomology 51: 811–816.

Yamada, H., M. J. B. Vreysen, J. R. L. Gilles, G. Munhenga, and D. D. Damiens. 2014b. The effects of genetic manipulation, dieldrin treatment and irradiation on the mating competitiveness of male *Anopheles arabiensis* in field cages. Malaria Journal 13: 318.

Yamada, H., H. Maiga, J. Juarez, D. De Oliveira Carvalho, W. Mamai, A. Ali, N. S. Bimbile-Somda, A. G. Parker, D. Zhang, and J. Bouyer. 2019. Identification of critical factors that significantly affect the dose-response in mosquitoes irradiated as pupae. Parasites & Vectors 12: (1) 435.

Yen, J. H., and A. R. Barr. 1971. New hypothesis of the cause of cytoplasmic incompatibility in *Culex pipiens* L. Nature 232: (5313) 657–658.

Yen, J. H., and A. R. Barr. 1973. The etiological agent of cytoplasmic incompatibility in *Culex pipiens*. Journal of Invertebrate Pathology 22: 242–250.

Yeung, J. 2018. Australian experiment wipes out over 80% of disease-carrying mosquitoes. CNN Health. 10th July 2018.

Zabalou, S., M. Riegler, M. Theodorakopoulou, C. Stauffer, C. Savakis, and K. Bourtzis. 2004. *Wolbachia*-induced cytoplasmic incompatibility as a means for insect pest population control. Proceedings of the National Academy of Sciences of the USA 101: 15042–15045.

Zabalou, S., A. Apostolaki, I. Livadaras, G. Franz, A. S. Robinson, C. Savakis, and K. Bourtzis. 2009. Incompatible insect technique: Incompatible males from a *Ceratitis capitata* genetic sexing strain. Entomologia Experimentalis et Applicata 132: 232–240.

Zélé, F., A. Nicot, A. Berthomieu, M. Weill, O. Duron, and A. Rivero. 2014. *Wolbachia* increases susceptibility to *Plasmodium* infection in a natural system. Proceedings of the Royal Society, Series B: Biological Sciences 281: (1779) 20132837.

Zhang, D., X. Zheng, Z. Xi, K. Bourtzis, and J. R. L. Gilles. 2015a. Combining the sterile insect technique with the incompatible insect technique: I - Impact of *Wolbachia* infection on the fitness of triple- and double-infected strains of *Aedes albopictus*. PLoS One 10: e0121126.

Zhang, D., R. S. Lees, Z. Xi, J. R. L. Gilles, and K. Bourtzis. 2015b. Combining the sterile insect technique with *Wolbachia*-based approaches: II - A safer approach to *Aedes albopictus* population suppression programmes, designed to minimize the consequences of inadvertent female release. PLoS One 10: e0135194.

Zhang, D., R. S. Lees, Z. Xi, K. Bourtzis, and J. R. L. Gilles. 2016. Combining the sterile insect technique with the incompatible insect technique: III - Robust mating competitiveness of irradiated triple *Wolbachia*-infected *Aedes albopictus* males under semi-field conditions. PLoS One 11: e0151864.

Zhang, D., M. Zhang, Y. Wu, J. R. L. Gilles, H. Yamada, Z. Wu, Z. Xi, and X. Zheng. 2017. Establishment of a medium-scale mosquito facility: Optimization of the larval mass-rearing unit for *Aedes albopictus* (Diptera: Culicidae). Parasites & Vectors 10: 569.

Zhang, D., Y. Li, Q. Sun, X. Zheng, J. R. L. Gilles, H. Yamada, Z. Wu, Z. Xi, and Y. Wu. 2018. Establishment of a medium-scale mosquito facility: Tests on mass production cages for *Aedes albopictus* (Diptera: Culicidae). Parasites & Vectors 11: 189.

Zheng, X., D. Zhang, Y. Li, C. Yang, Y. Wu, X. Liang, Y. Liang, X. Pan, L. Hu, Q. Sun, X. Wang, Y. Wei, J. Zhu, W. Qian, Z. Yan, A. G. Parker, J. R. L. Gilles, K. Bourtzis, et al. and Z. Xi. 2019. Incompatible and sterile insect techniques combined eliminate mosquitoes. Nature 572: 56–61.

Zug, R., and P. Hammerstein. 2012. Still a host of hosts for *Wolbachia*: Analysis of recent data suggests that 40% of terrestrial arthropod species are infected. PLoS One 7: e38544.

COMBINED STERILE INSECT TECHNIQUE AND INCOMPATIBLE INSECT TECHNIQUE: CONCEPT, STUDY DESIGN, EXPERIENCE AND LESSONS LEARNED FROM A PILOT SUPPRESSION TRIAL IN THAILAND

P. KITTAYAPONG[1,2]

[1]*Center of Excellence for Vectors and Vector-Borne Diseases, Faculty of Science, Mahidol University, 999 Phuttamonthon 4 Road, Nakhon Pathom 73170, Thailand*
[2]*Department of Biology, Faculty of Science, Mahidol University, 272 Rama VI Road, Bangkok 10400, Thailand; pkittayapong@gmail.com*

SUMMARY

Climate change, rapid global transport and land use change leading to urbanization and agricultural intensification have facilitated disease emergence in vulnerable regions like Southeast Asia, and also the global expansion of vectors and vector-borne diseases into other regions like the Americas and Europe. Important vector-borne diseases, i.e. dengue, chikungunya, yellow fever, and Zika are transmitted by the major mosquito vector species, *Aedes aegypti* (L.) and *Aedes albopictus* (Skuse). Management of *Ae. aegypti* populations in countries endemic to these diseases, especially in Southeast Asia, is not sufficiently effective, resulting in high morbidity and mortality in the region. Insecticide resistance has become an important issue, causing failure in insecticide-based vector control. Innovative or alternative tools/approaches are needed to effectively reduce mosquito vector populations and consequently reduce the diseases they transmit. A trial integrating the environment-friendly Sterile Insect Technique (SIT) and the insect incompatible technique (IIT) was successfully carried out on a small-scale in a semi-rural setting in Thailand. In this chapter, we report on the design and methodology, as well as the experience and lessons learned from the baseline preparation and implementation of the pilot trial.

Key Words: *Aedes aegypti*, *Aedes albopictus*, SIT, IIT, *Wolbachia*, mosquito control, integrated vector management, vector-borne diseases

J. Hendrichs, R. Pereira and M. J. B. Vreysen (eds.), Area-Wide Integrated Pest Management: Development and Field Application, pp. 405–432. CRC Press, Boca Raton, Florida, USA.

1. INTRODUCTION

Vector-borne diseases are becoming increasingly a public health problem and globally a significant economic burden. According to the World Health Organization (WHO), about half of the world's people in over 100 countries are at risk of contracting dengue (WHO 2019a). Chikungunya, another viral disease transmitted to humans by mosquito vectors, was originally confined to Africa but has recently been spreading rapidly across the Indian Ocean, Europe, the Americas, Asia, and Oceania. In the last decade, outbreaks of Zika in several parts of the world epitomized the need for new and effective methodologies to manage mosquito populations vectoring these diseases.

With the number of dengue cases and the number of countries affected rising dramatically in recent years, the socio-economic impact of mosquito-transmitted diseases is enormous. The overall estimated annual economic burden of dengue in Southeast Asia was USD 950 million, with the average annual direct costs being USD 451 million and the indirect costs being USD 499 million (Shepard et al. 2013). In Thailand alone, a recent study estimated the mean economic cost of dengue at USD 135 million per annum (Shepard et al. 2013). In the absence of affordable and effective vaccines and drugs to combat dengue, chikungunya, and Zika, population control of mosquito vectors is the most effective way of managing these diseases. Most vector control strategies are insecticide-based, and their widespread use has resulted in increased insecticide resistance among the mosquitoes. Therefore, there is an urgent need for alternative novel approaches for vector control.

Aedes aegypti (L.), the yellow fever mosquito, is considered the main mosquito vector for dengue, chikungunya, and Zika in many parts of the world (Calvez et al. 2017; Kotsakiozi et al. 2017; Trewin et al. 2017). Attempts have been made to control this invasive species, but traditional mosquito control methods, such as insecticide applications and source reduction by eliminating larval breeding sites have been insufficient for suppressing this mosquito vector and reducing disease incidences (Fredericks and Fernandez-Sesma 2014; Trewin et al. 2017).

Several novel *Ae. aegypti* control methods, namely the Sterile Insect Technique (SIT) that is based on the release of irradiated sterile males (Dyck et al. 2021); the Incompatible Insect Technique (IIT), which depends on *Wolbachia*-induced cytoplasmic incompatibility (CI) by releasing *Wolbachia*-infected males (Bourtzis et al. 2014, Mains et al. 2016; Zheng et al. 2019); and the application of genetically modified mosquito strains, such as those carrying RIDL (Release of Insects carrying a Dominant Lethal) constructs (Thomas et al. 2000; Morrison et al. 2010; Carvalho et al. 2015), have recently been endorsed by the WHO to help contain the recent Zika virus outbreak (Zheng et al. 2015; Yakob et al. 2017; WHO 2019b).

Both the SIT and the IIT are based on the repeated inundated release of large numbers of high quality sterile male mosquitoes to compete with their wild male counterparts in mating with wild females in a target area, thus inducing female sterility, which results in a reduction in the target populations (Zheng et al. 2015; Mains et al. 2016; WHO 2017; Zhang et al. 2017; Zheng et al. 2019) and consequently a potential reduction or prevention of the transmission of mosquito-borne diseases.

As a component of area-wide integrated pest management (AW-IPM) programmes, the implementation of the SIT and the IIT depends on several important components, including mass-rearing, sex separation, sterilisation, transportation, release, and monitoring (Zhang et al. 2017; Nikolouli et al. 2018; Dyck et al. 2021). Hence, the number of released sterile males must significantly surpass the number of wild males in the release area to compensate for any negative effect associated with domestication, mass-rearing, storage, and their overall handling, so that they can compete with wild males for matings with wild females, allowing the introduction of sufficient sterility into the wild populations (Vreysen et al. 2007; Barnes et al. 2015; Nikolouli et al. 2018; Dyck et al. 2021).

The combination of *Wolbachia*-induced IIT and the SIT was applied together with initial source reduction to suppress natural populations of *Ae. aegypti* in a semi-rural village in Chachoengsao Province, eastern Thailand. Results of this pilot trial indicated successful reduction of local *Ae. aegypti* populations after 6 months of repeated releases of sterile males (Kittayapong et al. 2019).

In this chapter, we report on the study design and methodology of this pilot trial. In addition, experience and lessons learned from the baseline experiments and from the implementation of this small-scale pilot trial are discussed.

2. COMBINED SIT/IIT APPROACH: CONCEPT AND PROGRESS

2.1. *Sterile Insect Technique (SIT) for Mosquito Control*

The SIT is a method of insect pest control with a strong record of success against a wide range of agricultural pests and which potentially can work against mosquitoes (Dyck et al. 2021). The technique consists of repeated area-wide releases of large numbers of sterile males in the target area, where they will mate with native females. Eggs will be produced but they will not hatch. When adequate sterile to wild male overflooding ratios are maintained, the number of native insects decreases with each generation, potentially driving the native population to very low numbers or, under complete isolation, to local extinction.

The SIT has been successfully implemented in large-scale operations to control agricultural insect pests and to prevent losses in livestock or crops of economic importance. Because it has no environmental impact and its relatively unobtrusive means of deployment, the SIT had been well accepted, even in urban areas. This technique has been successfully proven for over 50 years and is cost-effective for the population control of some major agricultural and livestock pests (Vreysen et al. 2000; Dyck et al. 2021). For public health pests, the SIT has been the subject of extended research since the late-1950s. However, it has never reached an operational level (Dame et al. 2009), even though it is considered to be a highly sustainable and environment-friendly method with, so far, no negative effect on human health (Alphey et al. 2010).

The first experimental sterile mosquito releases were conducted by the United States Department of Agriculture (USDA) in southern Florida. A total of 32 000 sterile *Anopheles quadrimaculatus* Say males that emerged from pupae irradiated with 120 Gy were released for three months in 1959 and in 1960; this amount was increased to 300 000 released over a period of nine months (Weidhaas et al. 1962;

Dame et al. 1964, 2009). However, the project was considered not successful as insufficient sterility was induced in the wild population (Dame et al. 1964, 2009).

The Centers for Disease Control and Prevention (CDC) carried out a release trial in Pensacola, Florida with 110-180 Gy treated *Ae. aegypti*. Although 3.9 million sterile males were released over four months in 1960 and 6.7 million over six months in 1961, the project was considered a failure due to reduced sterile male competitiveness caused by the irradiation of the pupae (Morlan et al. 1962; Dame et al. 2009).

Between 1967 and 1974, the World Health Organization/Indian Council of Medical Research (WHO/ICMR) and the USDA released male *Culex quinquefasciatus* Say irradiated with 60-120 Gy in India and Florida, respectively. The daily release rate ranged between 9000 to 15 000 sterile males. Nevertheless, these studies confirmed previous laboratory findings that the somatic damage was greater when younger pupae were treated as compared with older pupae (Patterson et al. 1975, 1977; Dame et al. 2009).

In 1980, a total of 71 000 sterile *Culex tarsalis* Coquillett males, sterilized by 60 Gy irradiation at the adult stage, were released in California, USA and results showed that these sterile males were fully competitive. However, in 1981, 85 000 sterile males were released, but these sterile males were not capable of seeking out the wild females and transferring the sterile sperm (Reisen et al. 1982; Dame et al. 2009).

Mosquito releases have been carried out for numerous purposes related to SIT application, but most of them were directed at answering a specific research question without any anticipation of population suppression. However, a few suppression and/or elimination projects have been attempted, but only modest effects were observed on sterility of the oviposited eggs and reduction of the wild population density (Benedict and Robinson 2003).

More recently, several SIT pilot projects have been initiated to answer specific questions (Lees et al. 2021). The effect of irradiation on sexual maturation and mating success of males, and the sexual competitiveness of sterile versus wild males in the presence of wild females of *Aedes albopictus* (Skuse) were studied under semi-field conditions in La Réunion Island (Oliva et al. 2012). In Sudan, participation of irradiated *Anopheles arabiensis* Patton males in mating swarms during the evening after their release was demonstrated, but their competitiveness and achieving successful copulation in the field was not proven (Ageep et al. 2014).

In Mauritius, the Ministry of Health and Quality of Life has been developing an operational plan to assess the SIT for population reduction of *Ae. albopictus* to prevent and control chikungunya and dengue, and guidelines for site selection were developed with the beginning of population surveillance (Iyaloo et al. 2014).

The first successful SIT mosquito pilot project was initiated during the summer of 2004 in three small towns in northern Italy. Approximately 900-1600 irradiated *Ae. albopictus* pupae were released per hectare, per week, and this continued for five years. The trial induced up to 68% egg sterility in the target population, demonstrating the potential of sterile males to suppress populations of *Ae. albopictus* (Bellini et al. 2007, 2013; Lees et al. 2015).

To date, there have been no large operational mosquito SIT projects, but operational programmes should eventually become established and more efficient over time (Dame et al. 2009). Experimentation and preparation processes for SIT application tend to be longer-term. Also, it does not have immediate effects on vector numbers, but impacts the size of the wild population in the next generation. In addition, entomological surveillance of the vector population before and during releases is essential to monitor the impact of any releases (Dame et al. 2009; Alphey et al. 2010). Nevertheless, the SIT is robust in term of both efficacy and cost when used in combination with other compatible methods, resulting in successful and sustainable vector control. Apart from being an environmentally-sound biological control approach, the SIT can be easily integrated with other biological control strategies (parasitoids, predators, and pathogens) (Vreysen et al. 2007; Barnes et al. 2015; Nikolouli et al. 2018).

2.2. Wolbachia*-based Approach for Mosquito Control*

Wolbachia are intracellular endosymbionts belonging to Alpha-proteobacteria. They are found in many arthropods and nematodes, and the overall species infection rate is as high as 66% (Hilgenboecker et al. 2008). *Wolbachia* bacteria have attracted the interest of the scientific community because of their potential to block arbovirus infections in mosquitoes (Moreira et al. 2009; Bian et al. 2013), as well as their capacity to replace natural populations of insects through their CI properties (Turelli and Hoffmann 1995). *Wolbachia*-infected male insects are not compatible with their non-infected natural females, leading to a reduction in the egg hatch of *Wolbachia*-uninfected populations and then the replacement by *Wolbachia*-infected populations (Hoffmann et al. 2011). The benefit of CI has been widely recognized for mosquito vector control (Clark et al. 2002; Atyame et al. 2014; Altinli et al. 2018; Baton et al., this volume).

A few years after the development of *Wolbachia*-transinfected *Aedes* mosquitoes (Xi et al. 2005), open field releases of these mosquitoes were carried out to evaluate whether CI induced by *Wolbachia* and their antiviral ability could be used for population suppression in vector control programmes. An open field release of *Aedes polynesiensis* Marks fluorescent-marked males infected with *Wolbachia* was launched in French Polynesia in 2009. The study showed that *Wolbachia*-transinfected *Ae. polynesiensis* males were competitive under field conditions; and after 30 weeks of releases, the egg hatch rate was significantly reduced in the release area, resulting in a reduction of the density of the local mosquito population (O'Connor et al. 2012).

The feasibility of using *Wolbachia* triple infected *Ae. albopictus* as a biopesticide against natural *Wolbachia* double infected *Ae. albopictus* was demonstrated in Lexington, USA and Guangzhou, China. Both the egg hatch rate and the number of adult *Ae. albopictus* were significantly reduced following the release of *Wolbachia*-infected males in these trials (Mains et al. 2016; Zheng et al. 2019). In addition, IIT was demonstrated to successfully suppress natural populations of *Ae. aegypti* in South Miami, USA in order to prevent the Zika disease by releasing *w*AlbB *Wolbachia*-infected *Ae. aegypti* males (Mains et al. 2019).

Strict male release is required for IIT application to obtain vector suppression (O'Connor et al. 2012; Nikolouli et al. 2018; Baton et al., this volume). Indeed, the accidental release of females infected by *Wolbachia* may cause the replacement of the targeted population by a population carrying the *Wolbachia* infection, resulting in field populations being compatible with the released males. Therefore, IIT application requires the development of an efficient method for sex separation at mass-rearing scales, in order to strictly release only *Wolbachia*-infected males (O'Connor et al. 2012; Nikolouli et al. 2018). Different techniques like phenotypic sorting or genetic sexing methods based on classical genetic or molecular methods have been reported for separation or sexing methods (Gilles et al. 2014). However, these methods are not available for all target species, and some techniques involve the release of genetically modified organisms (GMOs), the use of which is of concern in the European Union, as they face public opposition. In addition to public acceptance, GMO releases also face regulation difficulties in some countries, including China and India (O'Connor et al. 2012; Nikolouli et al. 2018).

In Australia, a risk analysis was carried out before the first release of *Wolbachia*-transinfected *Ae. aegypti* male and female mosquitoes into the environment for the purpose of population replacement (Murray et al. 2016). The first release into the field of *Ae. aegypti* males and females infected with *w*Mel was approved and took place in 2011 near Cairns in north-eastern Australia. The study showed that *Wolbachia*-transinfected *Ae. aegypti* successfully invaded and completely replaced uninfected wild *Ae. aegypti* populations (Hoffmann et al. 2011). A follow-up study indicated that field *w*Mel-infected *Ae. aegypti* mosquitoes (F_1), collected one year following the field release, had very low levels of dengue virus replication and dissemination. The frequency of *w*Mel-infected *Ae. aegypti* remained at more than 90% in the mosquito populations for more than 3 years (Frentiu et al. 2014). The success of this first release has led to small- and large-scale releases of *w*Mel-transinfected *Ae. aegypti* in other countries, in order to evaluate the effectiveness of population replacement in controlling dengue disease in human populations (Joubert et al. 2016).

2.3. Development of the Combined SIT/IIT-based Approach

Much progress has been made in recent years towards developing the required technology and methodology to bring mosquito sterility to field application. Hence, pilot releases have begun in a number of sites around the world (Lees et al. 2015). Since the key mosquito disease vectors are all relatively amenable to colonization and rearing, and in many situations the natural population densities are low, the SIT, the IIT, or a combination of the two, are well suited for their management (Lees et al. 2015). The SIT/IIT combination could in principle be applied to any targeted species for which an adequate and highly effective sexing system is not available (Zhang et al. 2015a, 2015b, 2016; Nikolouli et al. 2018; Zheng et al. 2019). However, successful SIT/IIT programmes will also depend on having *Wolbachia* strains with good CI and maternal transmission phenotypes, apart from an effective sexing system.

As the female mosquitoes are more radiation sensitive than males, the minimum dose of radiation that leads to complete sterility in females, whilst not negatively affecting male mating competitiveness, has been identified (Zhang et al. 2015b, 2016, 2017; Nikolouli et al. 2018). As a result, any accidentally released *Wolbachia*-infected females are sterile, and the risk of population replacement is minimised (Lees et al. 2015; Bourtzis et al. 2014; Zhang et al. 2017; Nikolouli et al. 2018; Dyck et al. 2021). Integration of the low irradiation dose with CI when using the virus-resistant strains of *Wolbachia* also minimizes any potential disease transmission by accidentally released sterile females. This has proven to be an efficient strategy in programmes targeting population suppression of *Ae. albopictus* (Zhang et al. 2015a, 2015b, 2016; Nikolouli et al. 2018; Zheng et al. 2019) and *Ae. aegypti* (NEA 2019; Liew et al., this volume). As stated by the WHO, this combined SIT/IIT technology has potential for long-term control of *Ae. aegypti* and *Ae. albopictus* mosquito populations, and this approach is considered an effective and safe strategy for the management of mosquito populations (WHO 2017).

2.4. *Combined SIT/IIT Pilot Trial in Thailand*

A field application of the combined SIT/IIT approach to reduce a local *Ae. aegypti* population was first demonstrated on a small-scale pilot trial with a total study area of 2.19 km^2 in the Plaeng Yao District of Chachoengsao Province, in eastern Thailand. Using the direct microinjection method, two *Wolbachia* strains from *Ae. albopictus* collected from rubber plantations in Thailand were introduced into *Ae. aegypti*. This newly developed *Ae. aegypti* line produced progeny infected with the *Wolbachia* strains *w*AlbA and *w*AlbB, with maternal transmission efficiency as high as 85% after 6 generations (Ruang-areerate and Kittayapong 2006). For the combined SIT/IIT approach, the CI property of *Wolbachia* was used to sterilize natural *Ae .aegypti* mosquito vector populations, while radiation was used to avoid population replacement by assuring that no fertile females were accidentally released.

Once released into nature, *Wolbachia*-transinfected *Ae. aegypti* male mosquitoes not only induced sterility in the females of the natural populations, but any potential virus transmission was also blocked in case a few female mosquitoes were inadvertently present in the releases into nature (Moreira et al. 2009). As the SIT/IT method aims at developing a vector suppression tool that is environment-friendly, no propagation of released mosquitoes should happen in nature. To achieve this goal, sterility of the *Wolbachia*-transinfected male mosquitoes was ensured by exposing the males to an appropriate irradiation dose (Kittayapong et al. 2018).

3. STUDY DESIGN AND METHODOLOGY

3.1. *Study Site Selection*

Selection of the study site to assess SIT, IIT, or combined SIT/IIT application is important for the success of any pilot field trial, and general guidelines for site selection were considered using Mauritius as a case study (Iyaloo et al. 2014). However, due to the differences and uniqueness of any study site, local considerations on specific details are needed. For the ideal study site, the following general criteria should be considered:

a) geographically- or ecologically-isolated
b) targeted mosquito species are dominant
c) manageable size for surveillance and monitoring, and
d) good cooperation of the local government and local communities.

The study site selected for the pilot field trial of the SIT/IIT approach in Thailand was located in the Plaeng Yao District of Chachoengsao Province in the eastern part of the country, which is about 120 km southeast of Bangkok. Three study areas were selected: Nong Satit as the treatment area, Pleang Mai Daeng as the adjacent area, and Nong Sarika as the control area. The distance between the treatment and the control areas was approximately 12 km, whereas the distance between the treatment and the adjacent area was approximately 500-800 meters.

The study areas were located among rice and cassava fields, as well as rubber and other plantations, which formed an ideal partial barrier to the movement of *Ae. aegypti* mosquito vectors. The selected study site was considered a typical semi-rural village similar to most other villages in Thailand (Fig. 1). In addition, it met the general criteria for site selection as previously described.

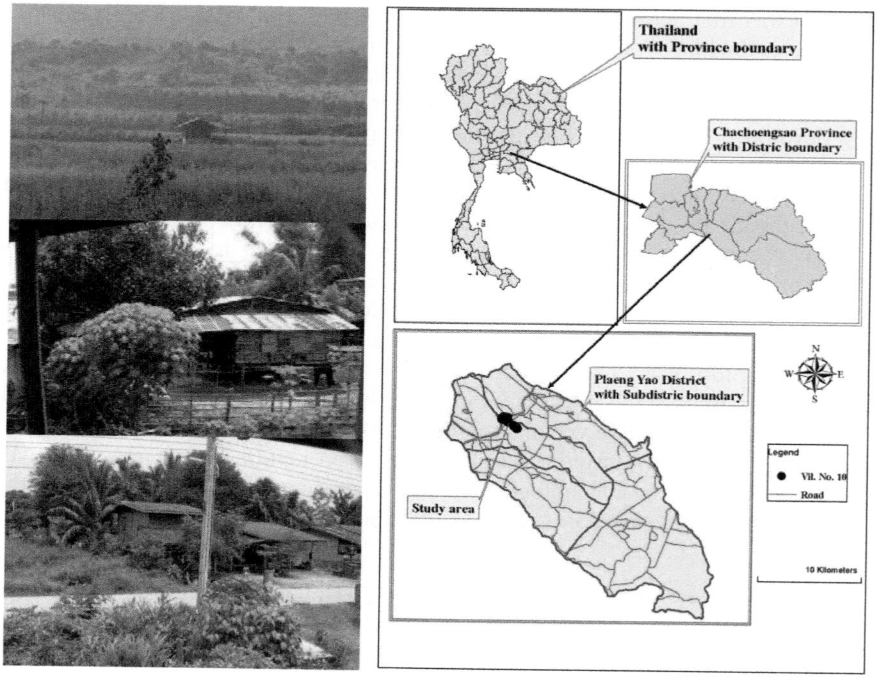

Figure 1. Maps and pictures showing the pilot treatment site located in Nong Satit Village (Village No. 10), Hua Sam Rong Sub-District, Plaeng Yao District, and surrounding areas in Chachoengsao Province, eastern Thailand, where sterile Aedes aegypti *male mosquitoes were released for the first time.*

3.2. Spatial Baseline Data Collection/Mapping of Study Sites

Spatial data obtained from a geographic information system (GIS), supplemented with 'ground truthing,' were used to characterize spatial distribution and patterns of households located at the study areas. Handheld Global Positioning System (GPS) sets were used to record all houses in the study site, and ArcMap software (ESRI, Redlands, California) was used to develop a GIS map (Chansang and Kittayapong 2007; Kittayapong et al. 2008). This GIS map was useful in determining the sampling households (Fig. 2).

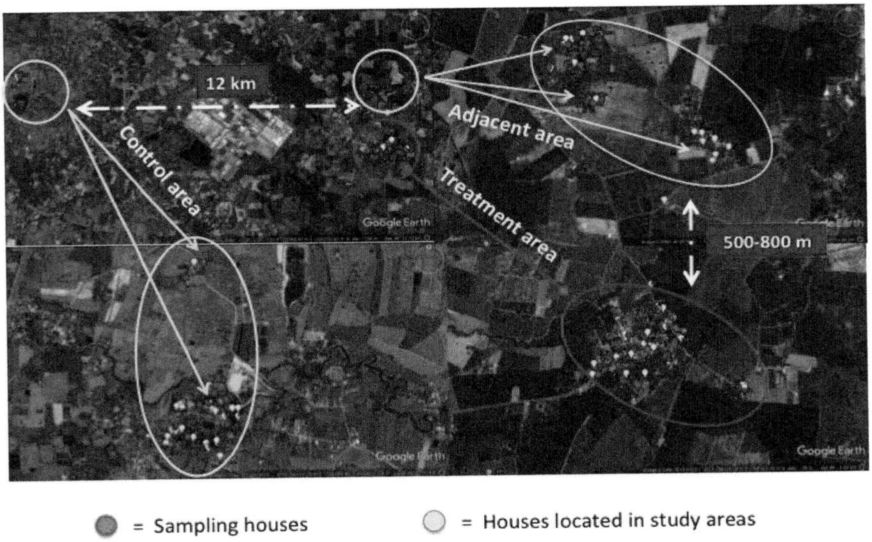

Figure 2. GIS map of the study site (upper left), including the treatment, adjacent, and control areas, and the sampling houses and other houses located in Pleang Yao District, Chachoengsao Province, eastern Thailand.

3.3. Community Engagement Strategy

Government authorities were officially informed about the project objectives and methodologies after the project had obtained institutional ethical approval. A community education campaign was organized for the communities in the study area, before implementation of the project, to raise awareness of vector-borne diseases and alternative vector control using combined SIT/IIT. Furthermore, the local leaders were invited to the meetings, and the key message delivered was that male mosquitoes cannot bite and do not feed on blood. In addition, sterile male mosquitoes in screened cages were brought to the community to demonstrate the key message.

The field release of sterile *Ae. aegypti* male mosquitoes was emphasized as an additional tool, in combination with other conventional control methods, to prevent vector-borne diseases. The message that other vector control measures, such as breeding container removal, could be applied prior to and during the release of sterile male mosquitoes was also emphasized, so that the community was not under the impression that they were fully protected from vector-borne diseases during the implementation of the sterile male releases. Routine classical vector control measures, i.e. fogging and source reduction, were applied to both treatment and control sites prior to the intervention.

3.4. *Entomological Surveillance and Monitoring*

Mosquito abundance and egg hatch rate of the targeted *Ae. aegypti* population were estimated in 60 and 90 households, respectively, in the treatment, adjacent, and control areas within the selected study site during the one-year baseline and six-month intervention. Mosquito abundance was determined using MosHouse sticky traps and MosVac portable vacuum aspirators (Go Green Co., Ltd., Nakhon Pathom, Thailand). MosHouse sticky traps (Fig. 3) were distributed to 60 households to sample adult mosquitoes in the treatment, adjacent, and control areas.

Figure 3. Picture showing the MosHouse sticky trap, the portable mosquito vacuuming aspirator, and the ovitrap that were used for surveillance and monitoring of natural Aedes aegypti *populations in the study site.*

The MosHouse traps were left in houses for one week before the sticky panels were collected and brought back to the laboratory to identify mosquito species and to determine relative mosquito abundance. In addition, the portable MosVac aspirators (Fig. 3) were used to collect resting mosquitoes in the same 60 households.

Sampled mosquitoes were killed by freezing and transported to the field laboratory station for species identification. The total number of *Ae. aegypti* males and females was recorded monthly for a total of two years to determine the dynamics of the different wild *Ae. aegypti* populations.

Ovitraps (Fig. 3) were distributed in 90 households to allow oviposition and then to collect *Ae. aegypti* eggs. The filter papers with eggs were collected on a weekly basis and brought back to the laboratory .The eggs that dried on the filter paper were counted and then hatched in water after 2-3 days. The number of eggs hatched was used to determine the sterility of natural *Ae. aegypti* populations.

4. EXPERIMENTS REQUIRED BEFORE THE PILOT TRIAL

4.1. Rearing of Wolbachia-infected Mosquitoes

Rearing is a crucial step for SIT/IIT implementation. Genomic adaptation to the mass-rearing environment, such as reduction in developmental time, life span, dispersal, and stress resistance, as well as early fertility and increased fecundity, is known to occur. This adaptation could make the individuals in the mass-rearing environment significantly different from the wild populations and affect the quality of the released male mosquitoes and hence, the efficacy of SIT/IIT applications (Nikolouli et al. 2018). Moreover, artificially *Wolbachia*-infected mosquito lines were observed to have increased larval mortality and decreased adult longevity when compared with aposymbiotic ones (Brelsfoard and Dobson 2011). Therefore, a strategy to maintain genetic diversity, biological quality, and competitiveness is required (Nikolouli et al. 2018). A high level of vigilance and consistent standardization of all processes, rearing conditions, and quality control needs to be maintained (Carvalho et al. 2014).

For the SIT/IIT trial in Thailand, mosquitoes were reared at the Center of Excellence for Vectors and Vector-Borne Diseases, Faculty of Science, Mahidol University at Salaya, Nakhon Pathom, Thailand and maintained in aluminium cages (40 x 40 x 40 cm) in a screened insectary at a temperature of 27 ± 2°C, a humidity of 75 ± 2%, and a photoperiod of L12:D12 (Kittayapong et al. 2018). Both male and female mosquitoes had access to a 10% sucrose solution, and females were fed with pig blood obtained from a qualified slaughterhouse.

The females were offered a blood meal for 3-4 consecutive days after mating using a Hemotek blood-feeding system (Hemotek Ltd., UK). Thereafter, plastic containers with the egg papers were placed inside the cages and were collected after 3-4 days. The eggs were dried and transferred to glass containers with screw-top covers filled with deionized water for egg hatching. After the eggs hatched into first-instar larvae, they were counted manually and transferred into plastic trays (32 cm x 42 cm x 5 cm), each containing about 2,000 larvae.

The larval diet had the following ingredients: mixed fish meal (Chanpongcharoen Kankaset Supplier, Thailand), pork liver powder, and yeast (*Saccharomyces cerevisiae*) (Cheese Powder Supplier, Thailand) at a ratio of 5:4:1 respectively. Each tray received 6.5 g of the diet every day. After 6-7 days, the developed pupae were placed inside plastic containers prior to sex separation.

4.2. *Sex Separation before Sterilisation*

Population suppression using the combined SIT/IIT approach requires release of a large number of male mosquitoes; therefore, an efficient system to separate the males from the females is essential to release only sterile males into the environment. Many studies have attempted to develop sex separation methods, based on biological, genetic, and transgenic approaches to support the application of the SIT for mosquito control. Sieving techniques were introduced in view of size differences between male and female pupae (Sharma et al. 1972; Bellini et al. 2007).

The development of genetic sexing strains (GSS), as well as other sex separation strategies, is currently under development and/or refinement, but none of them have so far succeeded in eliminating all females in order to achieve male-only releases for large-scale SIT or other applications (Benedict et al. 2009; Papathanos et al. 2009, 2018; Gilles et al. 2014).

Larval-pupal glass separators (Model 5412, John W. Hock, Co., Ltd., Gainesville, Florida, USA) were used to mechanically separate male and female pupae into different layers. The female pupae are larger in size and are collected in an upper layer between the two adjusted glass plates, while male pupae are drained into a receiving container placed below. Water circulation is supplied all along the process to push and wash the pupae down into the container. The female pupae are eventually flushed into a second receiving container, and the cycle of sex separation is complete.

In the experiments of sex separation, one litre of water that contained about 1500 to 2000 mixed male and female pupae were introduced each cycle into the system. One cycle took on average between 2 and 5 minutes, but it could take longer if the sample was mixed with larvae. After counting, the male pupae were transferred into a plastic cup and transported to the radiation source.

4.3. *Appropriate Irradiation Dose for Male Sterilisation*

Appropriate irradiation doses are different for different species of mosquitoes. Our preliminary studies showed that when *Wolbachia*-infected *Ae. aegypti* mosquitoes were irradiated at the pupal stage, an irradiation dose of 50 Gy was sufficient to obtain complete sterility in females, while males were fully sterilized with a dose of 70 Gy. *Wolbachia*-infected male pupae irradiated with 50 Gy could still produce some viable eggs when mated as adults with non-irradiated *Wolbachia*-infected females, the average percentage of egg hatch being 8%. However, egg hatch was zero when *Wolbachia*-infected males and females were irradiated with 70 Gy and then mated with non-irradiated *Wolbachia*-infected females and males, respectively (Kittayapong et al. 2018).

Since the *Wolbachia*-infected *Ae. aegypti* mosquitoes used in the pilot SIT/IIT trial in Thailand did not express complete CI (Ruang-areerate and Kittayapong 2006; Kittayapong et al. 2019), the complete sterility in these experiments was obtained through appropriate irradiation doses. If *Wolbachia* strains expressing strong CI are used, lower irradiation doses can be applied in order to obtain complete sterility of *Ae. aegypti* mosquitoes.

4.4. Mating Competitiveness and Release Ratio

Mating competitiveness of sterile male mosquitoes needs to be assessed before implementing a pilot SIT/IIT trial. In the past, many SIT trials were not successful in reducing natural mosquito populations due to the low competitiveness of sterile males after they were irradiated with too high doses (Dame et al. 2009). The advantage of the combined SIT/IIT approach is that lower dose radiation can be applied, as the sterility can also be induced in *Ae. aegypti* mosquitoes by the CI property of *Wolbachia* bacteria.

In the cage study, under controlled laboratory conditions, sterile *Ae. aegypti* males were evaluated for their mating competitiveness with wild males and females at different ratios (Fig. 4). Results indicated that a ratio of 10:1:1 or above was effective, as it reduced egg hatch significantly. The hatched eggs/total eggs of the 10:1:1 ratio experimental group was 3/619 ($0.27 \pm 0.65/103.17 \pm 9.09$). Complete sterility was observed with no egg hatch at a ratio of 20:1:1. Therefore, ratios between 10:1:1 and 20:1:1 were determined to be the optimal release ratios for the mass-reared sterile *Ae. aegypti* males, as they could compete with the wild males and induce near complete or complete sterility in the wild females. Our results also indicated that an irradiation dose of 70 Gy did not reduce mating competitiveness of the irradiated *Ae. aegypti* males (Kittayapong et al. 2019). In conclusion, our laboratory experiments demonstrated no significant difference in competitiveness of sterile males when compared to wild ones.

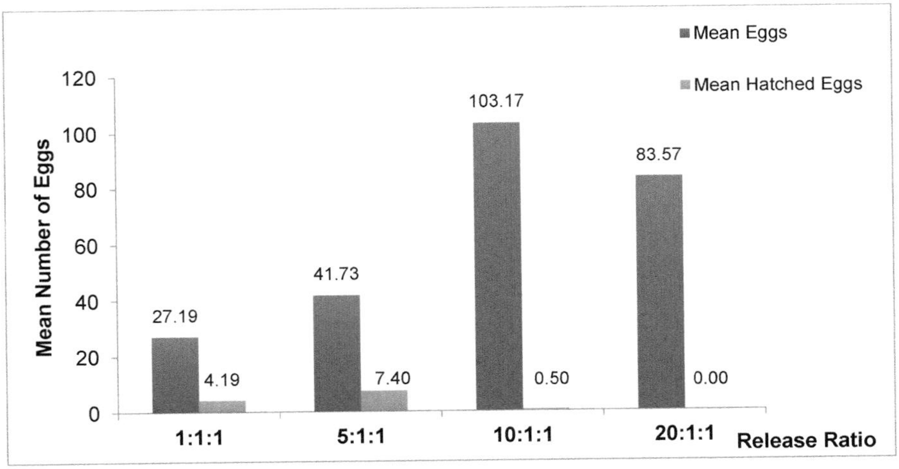

Figure 4. Mating competitiveness of sterile Aedes aegypti *males at different sterile male: wild male: wild female release ratios in cages under controlled laboratory conditions (modified from Kittayapong et al. 2019).*

4.5. *Survival and Longevity of Sterile Males*

Survival and longevity of the released sterile male mosquitoes are other parameters that may have an impact on the success of a SIT/IIT programme. The longer the sterile males can live, the higher the probability of mating with a wild female. In nature, both wild and sterile males should have shorter life spans than those kept under optimal controlled conditions. The same applies to wild and sterile females.

In our baseline experiments, carried out under controlled laboratory conditions, we observed that there was no significant difference in longevity between wild and sterile *Ae. aegypti* males (Fig. 5). On average, the wild males survived for 23.3 ± 0.9 days, while the sterile males survived for 23.8 ± 12.1 days. However, wild females lived significantly longer than sterile females, i.e. an average life span of 29.6 ± 1.0 days and 18.5 ± 9.8 days, respectively ($p = 0.000$) (Kittayapong et al. 2018).

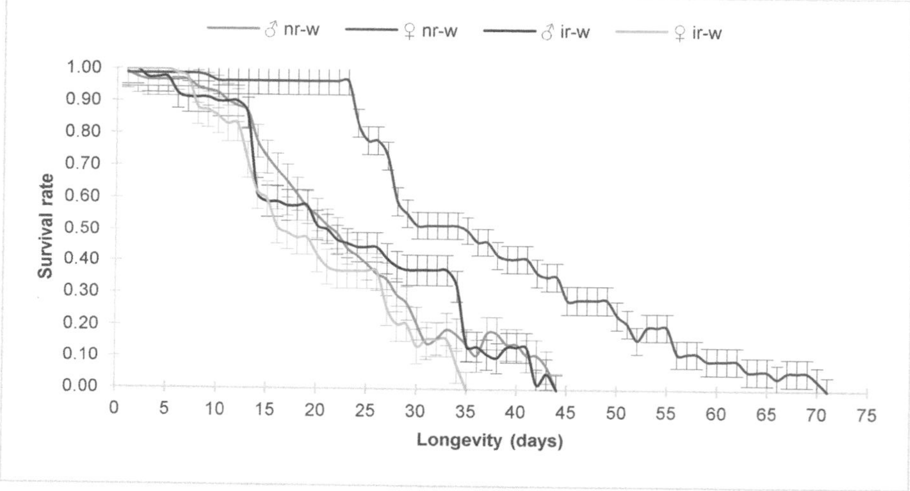

Figure 5. Mean longevity and survival rate of non-irradiated (nr) and irradiated (ir) Wolbachia*-infected* Aedes aegypti *male and female mosquitoes, after being sex-separated by using larval-pupal glass separators (modified from Kittayapong et al. 2018).*

5. EXPERIENCE AND LESSONS LEARNED FROM THE PILOT TRIAL

5.1. *Community/Public/Stakeholder Engagement*

Local stakeholders were identified. Engagement of local government authorities was initiated and followed up. Local government authorities coordinated with community leaders to initiate community engagement and facilitate community participation in implementing further on-site research activities, such as entomological surveillance and sterile male mosquito delivery using local health volunteers. Public education through media was carried out to raise awareness regarding vector-borne diseases and potential risk reduction using an alternative SIT/IIT approach.

From our experience, engagement of the community and public was more effective when it was initiated during times of epidemics. Our public engagement was initiated when both dengue and Zika were epidemic and people were aware of the consequences. Furthermore, we took advantage of related public events to draw the attention of the public. The first open release of sterile male mosquitoes was carried out on ASEAN Dengue Day, when the Ministry of Public Health of Thailand was an official host of regional activities (Fig. 6); hence representatives of many countries were present and witnessed the opening ceremony and first release of sterile males to fight dengue in Thailand.

In this pilot project, the general public was engaged through several national media reports, and TV news and radio programmes. A total of 109 media items, including a documentary, international news, national news, national radio, newspaper and online articles, and TV shows, were produced from January 2016 to February 2018 for public education of the sterile male release in Thailand.

Figure 6. Picture showing the first release of sterile Aedes aegypti *male mosquitoes on ASEAN Dengue Day at the study site in Pleang Yao District, Chachoengsao Province, eastern Thailand.*

A high number of views and sharing on social media, with the online articles published by reliable media publishers, was experienced (Kittayapong et al. 2019). Therefore, social media is an interesting additional channel to be used to communicate key messages to most of the general public. In addition, TV shows and documentaries on the topic of controlling the dengue vector through the SIT/IIT approach gained a lot of public attention in Thailand.

5.2. Handling and Transport of Sterile Mosquitoes for Open Field Release

Ideally, the radiation source used to sterilize the male mosquitoes should be in the same place as the rearing facility, while the subsequent transport to the release sites could be either as pupae or adults. In our case, male pupae had to be transported to a laboratory located 120 km from the rearing site for the irradiation treatment. *Wolbachia*-transinfected *Ae. aegypti* male mosquitoes were produced in the screened insectary at the Center of Excellence for Vectors and Vector-Borne Diseases, Faculty of Science, Mahidol University, Salaya Campus, Nakhon Pathom, and were then transported to the Thailand Institute of Nuclear Technology (TINT) in Nakhon Nayok for sterilisation at the pupal stage of 1-2 days old using a ^{60}Co source.

Sterile male pupae were transported weekly to the study sites using temperature-controlled containers. Our preliminary experiments showed 100% survival of the chilled sterile male pupae at temperatures between 8-12°C for up to 6 hours (Kittayapong et al. unpublished data).

Even though sterilisation at the pupal stage was shown to have no impact on mating competitiveness and the longevity of irradiated mosquitoes, it would be more practical to irradiate adult mosquitoes and then use the same containers for release. This would be less time-consuming, as the adult mosquitoes would not have to be transferred to the different release containers. Further experiments are needed to decide on the best temperature and container for chilling and transporting adults.

5.3. Management and Implementation of Sterile Male Release

A simple field laboratory station was set up in the city of Chachoengsao, 20 km from the release site in Plaeng Yao District. The research team worked with local workers to transfer male pupae to the release containers for adult emergence. Sterile *Ae. aegypti* males were provided with a 10% sucrose solution as food source after emergence for at least one day. They were then transported to the selected study sites, and the public health volunteers released the sterile mosquitoes at a rate of 100-200 per household per week. A total number of 437 980 sterile *Ae. aegypti* males, ranging from 9000 to 25 000 males per week, were released. The weekly releases were carried out only in the treatment area of the study site for a period of 24 weeks.

In this pilot trial, ULV fogging was used by local government staff to reduce the natural populations of *Ae. aegypti* to low densities before the releases of sterile male *Ae. aegypti* were initiated. The vector control activities by local government staff were conducted in both treatment and control areas. Public health volunteers provided assistance with the delivery of the sterile male mosquitoes to their respective households (Fig. 7).

Figure 7. Pictures showing activities related to the release of sterile males in the treatment community in Plaeng Yao District, Chachoengsao Province, eastern Thailand, by the health volunteers and homeowners under the supervision of the research team.

From our experience, frequent entering of private property caused some reluctance of a few homeowners to continue to cooperate. A visit of our staff together with the public health volunteers to these few houses was necessary to keep them cooperating. Therefore, the future planning of the release strategy should consider reducing disturbance of the privacy of homeowners. Open release using drones is recommended, especially in urban and crowded communities where intrusion into households is difficult. However, successful application of drone releases of sterile mosquitoes will need authorized and skilled operation.

5.4. Surveillance and Monitoring of Sterile Mosquitoes after Release

Both MosHouse sticky traps and MosVac portable vacuuming aspirators were used in the households of the study site for collecting both male and female mosquitoes of various species. MosHouse traps seemed to be more efficient in collecting *Ae. aegypti* females, while resting males were collected in higher numbers by using the MosVac portable vacuuming aspirators. When comparing the relative abundance of the *Ae. aegypti* mosquito populations, the average number of *Ae. aegypti* females sampled in the treatment area significantly decreased ($p < 0.05$) when compared to those in the control area, while those of males were not significantly different ($p > 0.05$), even though a large number of sterile males were released during the six-month intervention period in the treated areas (Table 1).

Table 1. Comparison of the mean numbers of Aedes aegypti *males and females collected by using MosHouse sticky traps and MosVac portable mosquito vacuuming aspirators in the treatment, adjacent, and control areas in Plaeng Yao District, Chachoengsao Province, eastern Thailand during the six-month intervention period (modified from Kittayapong et al. 2019)*

Variable	No. House	No. Mos-House traps	No. positive household (Mean ± SD)	Total mosquitoes (Mean ± SD)	Odds Ratio (total mos-quitoes)	95%CI	*P*
Males							
Control	20	20	15.83 ± 1.6	193 (32.17 ± 4.07)	1		
Adjacent	20	20	10.00 ± 3.4	92 (15.33 ± 6.31)	0.263	0.149-0.464	0.000*
Treatment	20	20	16.50 ± 2.9	137 (22.83 ± 6.55)	1.242	0.651-2.373	0.511
Females							
Control	20	20	17.00 ± 3.2	185 (30.83 ± 7.05)	1		
Adjacent	20	20	4.83 ± 1.83	35 (5.83 ± 2.64)	0.056	0.029-0.108	0.000*
Treatment	20	20	2.67 ± 1.75	16 (2.67 ± 1.75)	0.027	0.013-0.056	0.000*

* *Significant difference at* $p < 0.05$

It is possible that *Ae. aegypti* males mostly rested outside households where trapping and vacuuming activities took place (Table 1). The lower numbers of *Ae. aegypti* females in the treatment area compared to the control area indicate the effect of the sterile male releases that produced sterility in the wild *Ae. aegypti* females, resulting in a reduction in the numbers of *Ae. aegypti* female populations in nature by up to 97.30% (Fig. 8).

MosHouse sticky trap and MosVac portable vacuuming aspirators can be employed as tools for monitoring SIT, IIT, or combined SIT/IIT interventions, especially in view of their low cost and uncomplicated deployment. The advantage of the MosHouse traps as compared with the MosVac aspirators is that they can be placed either inside or around households without disturbing the homeowners. In addition, large numbers of the low-cost MosHouse sticky traps can be distributed in different locations in the study areas, resulting in better estimates of natural *Ae. aegypti* populations, compared to using a few high-cost traps placed in only a few locations.

As *Ae. aegypti* mosquitoes are more domestic, placing a few traps in a few locations could lead to a biased estimation of the total natural populations in the study areas. However, additional methods for collecting mosquitoes could be applied in combination to obtain more reliable data sets for entomological evaluation.

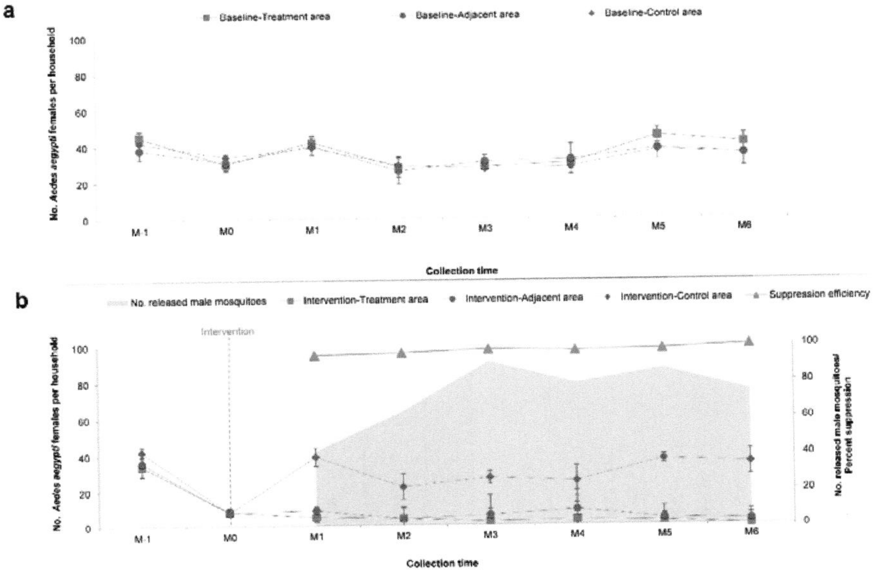

Figure 8. Mean numbers of Aedes aegypti *female mosquitoes collected per households in Pleang Yao District, Chachoengsao Province, Thailand during the baseline (Fig. 8a) and during the intervention (Fig. 8b) periods. Percent suppression efficiency in relation to the number of released sterile males per month is demonstrated in Fig. 8b (modified after Kittayapong et al. 2019).*

Weekly ovitrap data showed that the overall mean egg hatch was lowest in the treatment area, confirming the effectiveness of the sterile male release. The mean egg hatch for the treatment, adjacent, and control areas were 0.20 ± 0.10, 0.24 ± 0.14, and 0.41 ± 0.08 respectively; while those for the second twelve weeks were 0.18 ± 0.09, 0.25 ± 0.16, and 0.54 ± 0.11 respectively (Fig. 9).

There was a significant difference ($p < 0.05$) in mean egg hatch between the first and the second twelve weeks of sterile male releases (Table 2 and Table 3). The released sterile males seemed to show positive effects in reducing hatched eggs in the natural *Ae. aegypti* mosquito populations, in both the treatment area and the adjacent area, when compared to the households monitored in the control area (Fig. 9).

Except for a few outliers, egg hatch decreased to zero or near zero in most of the households monitored in the treatment area and adjacent area (Table 2). Since very low numbers of *Ae. albopictus* were found in this study area, especially in households, we could assume that the unhatched eggs were mostly from *Ae. aegypti* (Table 2 and Table 3).

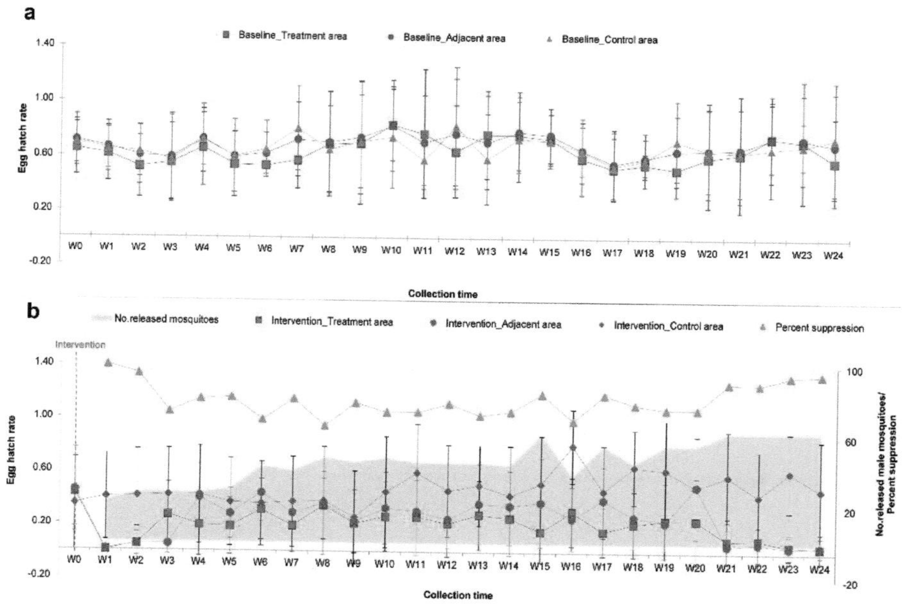

Figure 9. Mean egg hatch rate of natural Aedes aegypti *mosquito populations over time in the treatment, adjacent, and control areas of the study sites during the baseline (a) and during the intervention (b) periods. Percent suppression efficiency in relation to the number of released sterile males per week is demonstrated in Fig. 9b (modified after Kittayapong et al. 2019).*

5.5. Quality Control of Sterile Mosquito Production

Quality control to test for sterility of irradiated *Wolbachia*-infected *Ae. aegypti* males can be done through mating studies between mass-produced sterile males and untreated females from established mosquito colonies with no *Wolbachia* infection, originally collected from the same study site. Zero or near zero egg hatch was expected for each production lot. Mating tests performed during the 24-week open field trial in Plaeng Yao District, Chachoengsao Province, eastern Thailand indicated that the average numbers of hatched eggs in most production lots were quite low, i.e. 1.04 ± 2.18, which demonstrated that the irradiated *Wolbachia*-infected male mosquitoes were highly sterile (Kittayapong et al. 2018).

In addition, *Wolbachia* detection by PCR was conducted in 40 sterile *Ae. aegypti* males sampled from each of the production lots, and the results showed that the mean percentage of *Wolbachia* infection was 50.21 ± 0.49% in released males (Kittayapong et al. 2019).

Table 2. Progressive egg hatch rate of Aedes aegypti *mosquitoes in households in the treatment area, before and after 12 and 24 weeks of releases, when compared to those in households in the adjacent and the control areas*

No.	Before releases			After 12 weeks of releases			After 24 weeks of releases		
	W0			W12			W24		
	Control area	Adjacent area	Treatment area	Control area	Adjacent area	Treatment area	Control area	Adjacent area	Treatment area
1	0.60	0.00	0.00	0.11	0.29	0.10	1.00	0.00	0.00
2	0.00	0.25	0.28	0.37	0.00	0.00	0.43	0.00	0.00
3	0.09	0.81	0.38	0.59	0.00	0.44	0.80	0.00	0.16
4	0.29	0.23	0.64	0.87	0.00	0.00	0.27	0.05	0.38
5	0.39	0.34	0.00	0.26	0.46	0.57	0.68	0.00	0.00
6	0.00	0.52	0.48	0.43	0.00	0.02	0.00	0.09	0.00
7	0.00	0.00	0.03	0.96	0.19	0.25	0.15	0.00	0.00
8	0.74	0.68	0.10	0.36	0.17	0.00	0.60	0.00	0.00
9	0.00	0.94	0.67	1.00	0.34	0.00	0.32	0.04	0.07
10	0.73	0.95	0.87	0.81	0.28	0.00	0.77	0.00	0.00
11	0.22	0.52	0.64	0.00	0.00	0.26	0.44	0.00	0.07
12	0.94	0.31	0.39	1.00	0.16	0.54	0.44	0.18	0.00
13	0.67	0.26	0.18	0.56	0.08	0.00	0.00	0.15	0.00
14	0.94	0.20	0.39	0.47	0.00	0.61	0.20	0.00	0.04
15	0.00	0.35	0.00	0.94	0.84	0.39	0.00	0.01	0.00
16	0.12	0.94	0.00	0.00	0.22	0.00	0.00	0.00	0.00
17	0.00	0.09	0.00	0.31	0.67	0.00	0.50	0.00	0.00
18	0.94	0.53	0.00	0.18	0.18	0.50	0.30	0.14	0.10
19	0.19	0.28	0.74	0.95	0.29	0.00	0.94	0.00	0.00
20	0.09	0.80	0.37	0.94	0.59	0.50	0.60	0.03	0.46
21	0.00	0.20	0.74	0.47	0.42	0.23	0.00	0.13	0.00
22	0.94	0.00	0.21	0.17	0.26	0.50	1.00	0.07	0.00
23	0.00	0.23	0.00	0.45	0.14	0.16	0.00	0.27	0.00
24	0.21	0.62	0.29	0.33	0.26	0.84	0.70	0.00	0.07
25	0.13	0.47	0.00	0.34	0.04	0.00	1.00	0.15	0.00
26	0.00	0.82	0.00	1.00	0.39	0.00	0.00	0.07	0.00
27	0.00	0.80	0.71	1.00	0.00	0.26	0.31	0.00	0.00
28	0.00	0.00	0.55	0.00	0.37	0.00	1.00	0.03	0.00
29	0.00	0.41	0.00	0.31	0.49	0.00	0.10	0.00	0.00
30	0.33	0.94	0.00	0.00	0.25	0.25	1.00	0.13	0.00

High hatch rate (>0.50)

Moderate hatch rate (0.25-0.50)

Low hatch rate (0.01-0.24)

Zero hatch rate

Table 3. Statistical analysis of the egg hatch rate of Aedes aegypti *during the six-month (weeks 1-12 and weeks 13-24) intervention period in the treatment, adjacent, and control areas located in Plaeng Yao District, Chachoengsao Province, eastern Thailand (modified from Kittayapong et al. 2019)*

Variable	No. house	No. ovitrap	No. positive household (Mean ± SD)	Egg hatch rate (Mean ± SD)	Odds Ratio (Egg hatch rate)	95% CI	*P*
W1-W12							
Control	30	60	22.00 ± 0.43	0.41 ± 0.08	1		
Adjacent	30	60	24.50 ± 0.39	0.24 ± 0.14	1.620	0.679 – 3.862	0.277
Treatment	30	60	18.00 ± 0.50	0.20 ± 0.10	0.545	0.252 – 1.179	0.123
W13-W24							
Control	30	60	24.50 ± 0.39	0.54 ± 0.11	1		
Adjacent	30	60	19.00 ± 0.47	0.25 ± 0.16	0.388	0.168 – 0.897	0.027*
Treatment	30	60	12.50 ± 0.48	0.18 ± 0.09	0.160	0.070 – 0.368	0.000*

* *Significant difference at* $p < 0.05$

Our experience confirmed that an irradiation dose of 70 Gy is optimal to induce sterility in *Ae. aegypti* male mosquitoes. In our experiments, we also observed that *Ae. aegypti* females were more radio-sensitive and that a treatment with 50 Gy was sufficient to obtain complete female sterility. Therefore, accidentally released irradiated *Wolbachia*-infected *Ae. aegypti* females in our field trial, if any, were fully sterile after exposure to 70 Gy, and there was no danger of further propagation or *Wolbachia* establishment in the target population.

5.6. Female Contamination during SIT/IIT *Implementation*

Sustainable suppression of *Ae. aegypti* populations by integration of the SIT/IIT depends on the release of only sterile males. Hence, sex separation of mass-produced male and female mosquitoes is an important step, as female contamination could lead to an increase in disease transmission, although it is unlikely due to their *Wolbachia* infection. Inspection of female contamination was carried out weekly during the twenty-four weeks of the pilot field release of sterile *Ae. aegypti* males at the selected study site in Plaeng Yao District, Chachoengsao Province, eastern Thailand. Our results indicate a low percentage of female contamination among sterile males, i.e. 0.06 ± 0.10%, when *Ae. aegypti* pupae were separated through mechanical larval-pupal glass separators. Therefore, at least 99% of sterile males were purely separated from females (Kittayapong et al. 2018).

In this study, we also observed a significant difference in the percentage of female contamination during the first and the second 12-week periods of sterile male releases, i.e. 0.10 ± 0.13% vs 0.02 ± 0.02% ($p < 0.05$). The percentage of female contamination was remarkably reduced in the second twelve weeks of intervention. This was most likely due to the increasing skills of the technicians operating the mechanical sex separation machine. As such, we recommend hands-on training for operating technicians before project implementation to obtain a high efficiency in the manual sex separation process, and hence achieve the lowest possible female contamination during sterile male release.

5.7. Impact of SIT/IIT on the Environment and Ecosystem

The SIT/IIT approach for *Ae. aegypti* mosquito vectors was implemented using a two-step sterilisation process, combining the *Wolbachia*-induced IIT with the SIT using radiation to obtain sterile males. When these sterile males are systematically released into the target area, they can induce sterility in wild females after mating. Mated females lay eggs that cannot hatch, resulting in significant reduction in natural *Ae. aegypti* populations and subsequently, an "assumed" reduction in disease incidence that needs to be verified. In general, the SIT/IIT intervention is assessed to have little or no impact on the environment for the following reasons:

1. Mosquitoes released into the environment are irradiated males to ensure sterility. Also, any accidentally released females do not transmit disease if they are infected with pathogen-resistant *Wolbachia* strains. Thus, using both CI, the property of the *Wolbachia* endosymbiont that induces sterility in wild females, and radiation to sterilise the *Wolbachia*-infected mosquitoes, makes sure that they cannot become established in nature. These sterile mosquitoes have shorter life spans due to either the *Wolbachia* life-shortening effect or irradiation effect, and they will not survive in the natural environment longer than 2-3 weeks after release (Kittayapong et al. 2019). However, this means that the mosquitoes need to be released systematically into the target area to obtain the population reducing effect. Once the native mosquito population is at a low level, fewer sterile males can be released. In view of their short life span, there should be no residual mosquitoes left in the environment a few weeks after termination of the release activities.

2. As this method is species-specific, interfering only with the reproduction of the target population, it has no impact on beneficial insects or any other animals or humans, unlike chemical spraying which impacts the environment, affects non-target organisms, and can leave some residues.

3. The ecosystem will obviously experience a reduction of the *Ae. aegypti* vectors, and hence a reduction of available food for animals that feed on them. However, as there are over hundred species of mosquitoes in the tropical zone, together with the low biomass of the target population, other mosquito species should be able to serve in the food chain for some predators; therefore, the impact on the ecosystem in this regard should be very low or negligible.

Risk assessment on the use of *Wolbachia* for controlling mosquito vectors, both in terms of replacement and suppression approaches, was evaluated in the past, and a very low risk for the environment was reported (Popovici et al. 2010, Murray et al. 2016; NEA 2016).

6. CONCLUSIONS

The successful development and implementation of an operational SIT/IIT programme depends on several factors, and therefore these programmes require extensive and thorough planning based on available knowledge of the genetics, biology, and ecology of the target insect species. These include establishing and maintaining a *Wolbachia*-infected colony of the target species, understanding the field conditions and target population dynamics, assuring community participation, and assessing the potential side effects on humans and the environment.

The SIT is an environment-friendly method. Being species-specific and leaving no toxic residues, it has only minimal or no non-target impact, which has been demonstrated for over 50 years in large scale applications against agricultural pests. Moreover, it can be easily integrated with other biological control strategies. In terms of the IIT, it has already been proven successful in pilot field trials for suppressing *Aedes* mosquito vectors.

The pilot field trial of the combined SIT/IIT technology that was reported in this chapter represents the first clear proof-of-concept for the release of sterile male *Ae. aegypti* mosquitoes in Thailand (Kittayapong et al. 2019), one of the highly arboviral endemic countries in Southeast Asia. Our results show that the combined SIT/IIT approach for controlling mosquito vectors has potential for practical application as part of integrated vector management, working together with traditional control efforts to achieve better and more efficient outcomes (Zheng et al. 2019).

Potential large-scale application of this integrated SIT/IIT approach is possible through a commitment by the relevant vector control organizations, who should be informed of the technology, especially now that it has already been proven to work well in a pilot field trial.

7. REFERENCES

Ageep, T. B., D. Damiens, B. Alsharif, A. Ahmed, E. H. Salih, F. T. Ahmed, A. Diabaté, R. S. Lees, J. R. L. Gilles, and B. B. El Sayed. 2014. Participation of irradiated *Anopheles arabiensis* males in swarms following field release in Sudan. Malaria Journal 13: 484.

Alphey, L., M. Benedict, R. Bellini, G. G. Clark, D. A. Dame, M. W. Service, and S. L. Dobson. 2010. Sterile-insect methods for control of mosquito-borne diseases: An analysis. Vector Borne and Zoonotic Diseases 10(3): 295–311.

Altinli, M., F. Gunay, B. Alten, M. Weill, and M. Sicard. 2018. *Wolbachia* diversity and cytoplasmic incompatibility patterns in *Culex pipiens* populations in Turkey. Parasites & Vectors 11: 198.

Atyame, C. M., P. Labbé, E. Dumas, P. Milesi, S. Charlat, P. Fort, and M. Weill. 2014. *Wolbachia* divergence and the evolution of cytoplasmic incompatibility in *Culex pipiens*. PLoS One 9(1): e87336.

Bian, G., D. Joshi, Y. Dong, P. Lu, G. Zhou, X. Pan, Y. Xu, G. Dimopoulos, and Z. Xi. 2013. *Wolbachia* invades *Anopheles stephensi* populations and induces refractoriness to *Plasmodium* infection. Science 340: 748–751.

Barnes, B. N., J. H. Hofmeyr, S. Groenewald, D. E. Conlong, and M. Wohlfarter. 2015. The Sterile Insect Technique in agricultural crops in South Africa: A metamorphosis .... but will it fly? African Entomology 23(1): 1–18.

Benedict, M. Q., and A. S. Robinson. 2003. The first releases of transgenic mosquitoes: An argument for the sterile insect technique. Trends in Parasitology 19(8): 349–355.

Benedict, M. Q., B. G. J. Knols, H. C. Bossin, P. I. Howell, E. Mialhe, C. Caceres, and A. S. Robinson. 2009. Colonisation and mass rearing: Learning from others. Malaria Journal 8 (Supplement 2): S4.

Bellini, R., M. Calvitti, A. Medici, M. Carrieri, G. Celli, and S. Maini. 2007. Use of the Sterile Insect Technique against *Aedes albopictus* in Italy: First results of a pilot trial, pp. 505–515. *In* M. J. B. Vreysen, A. S. Robinson, and J. Hendrichs (eds.), Area-wide control of insect pests: From research to field implementation. Springer, Dordrecht, The Netherlands.

Bellini, R., A. Medici, A. Puggioli, F. Balestrino, and M. Carrieri. 2013. Pilot field trials with *Aedes albopictus* irradiated sterile males in Italian urban areas. Journal of Medical Entomology 50(2): 317–325.

Bourtzis, K., S. L. Dobson, Z. Xi, J. L. Rasgon, M. Calvitti, L. A. Moreira, H. C. Bossin, R. Moretti, L. A. Baton, G. L. Hughes, P. Mavingui, and J. R. L. Gilles. 2014. Harnessing mosquito–*Wolbachia* symbiosis for vector and disease control. Acta Tropica 132: S150–S163.

Brelsfoard, C. L., and S. L. Dobson. 2011. *Wolbachia* effects on host fitness and the influence of male aging on cytoplasmic incompatibility in *Aedes polynesiensis* (Diptera: Culicidae). Journal of Medical Entomology 48(5): 1008–1015.

Calvez, E., L. Guillaumot, D. Girault, V. Richard, O. O'Connor, T. Paoaafaite, M. Teurlai, N. Pocquet, V. M. Cao-Lormeau, and M. Dupont-Rouzeyrol. 2017. Dengue-1 virus and vector competence of *Aedes aegypti* (Diptera: Culicidae) populations from New Caledonia. Parasites & Vectors 10(1): 381.

Carvalho, D. O., D. Nimmo, N. Naish, A. R. McKemey, P. Gray, A. B. B. Wilke, M. T. Marrelli, J. F. Virginio, L. Alphey, and M. L. Capurro. 2014. Mass production of genetically modified *Aedes aegypti* for field releases in Brazil. Journal of Visualized Experiments 83: e3579.

Carvalho, D. O., A. R. McKemey, L. Garziera, R. Lacroix, C. A. Donnelly, L. Alphey, A. Malavasi, and M. L. Capurro. 2015. Suppression of a field population of *Aedes aegypti* in Brazil by sustained release of transgenic male mosquitoes. PLoS Neglected Tropical Diseases 9(7): e0003864.

Chansang, C., and P. Kittayapong. 2007. Application of mosquito sampling count and geospatial methods to improve dengue vector surveillance. American Journal of Tropical Medicine and Hygiene 77: 897–902.

Clark, M. E., Z. Veneti, K. Bourtzis, and T. L. Karr. 2002. The distribution and proliferation of the intracellular bacteria *Wolbachia* during spermatogenesis in *Drosophila*. Mechanisms of Development 111(1): 3–15.

Dame, D. A., D. B. Woodard, H. R. Ford, and D. E. Weidhaas. 1964. Field behavior of sexually sterile *Anopheles quadrimaculatus* males. Mosquito News 24: 6–14.

Dame, D. A., C. F. Curtis, M. Q. Benedict, A. S. Robinson, and B. G. Knols. 2009. Historical applications of induced sterilisation in field populations of mosquitoes. Malaria Journal 8 (Supplement 2): S2.

Dyck, V. A., J. Hendrichs, and A. S. Robinson (eds.). 2021. Sterile Insect Technique – Principles and practice in Area-Wide Integrated Pest Management. Second Edition. CRC Press, Boca Raton, Florida, USA. 1200 pp.

Fredericks, A. C., and A. Fernandez-Sesma. 2014. The burden of dengue and chikungunya worldwide: Implications for the southern United States and California. Annals of Global Health 80(6): 466–475.

Frentiu, F. D., T. Zakir, T. Walker, J. Popovici, A. T. Pyke, A. van den Hurk, E. A. McGraw, and S. L. O'Neill. 2014. Limited dengue virus replication in field-collected *Aedes aegypti* mosquitoes infected with *Wolbachia*. PLoS Neglected Tropical Diseases 8(2): e2688.

Gilles, J. R. L., M. F. Schetelig, F. Scolari, F. Marec, M. L. Capurro, G. Franz, and K. Bourtzis. 2014. Towards mosquito sterile insect technique programmes: Exploring genetic, molecular, mechanical and behavioural methods of sex separation in mosquitoes. Acta Tropica 132: S178–S187.

Hilgenboecker, K., P. Hammerstein, P. Schlattmann, A. Telschow, and J. H. Werren. 2008. How many species are infected with *Wolbachia*? A statistical analysis of current data. FEMS Microbiology Letters 281(2): 215–220.

Hoffmann, A. A., B. L. Montgomery, J. Popovici, I. Iturbe-Ormaetxe, P. H. Johnson, F. Muzzi, M. Greenfield, M. Durkan, Y. S. Leong, Y. Dong, H. Cook, J. Axford, A. G. Callahan, N. Kenny, C. Omodei, E. A. McGraw, P. A. Ryan, S. A. Ritchie, M. Turelli, and S. L. O'Neill. 2011. Successful establishment of *Wolbachia* in *Aedes* populations to suppress dengue transmission. Nature 476(7361): 454–457.

Iyaloo, D. P., K. B. Elahee, A. Bheecarry, and R. S. Lees. 2014. Guidelines to site selection for population surveillance and mosquito control trials: A case study from Mauritius. Acta Tropica 132: S140–S149.

Joubert, D. A., T. Walker, L. B. Carrington, J. T. De Bruyne, D. H. T. Kien, N. L. T. Hoang, N. V. V. Chau, I. Iturbe-Ormaetxe, C. P. Simmons, and S. L. O'Neill. 2016. Establishment of a *Wolbachia* superinfection in *Aedes aegypti* mosquitoes as a potential approach for future resistance management. PLoS Pathogens 12(2): e1005434.

Kittayapong, P., S. Yoksan, U. Chansang, C. Chansang, and A. Bhumiratana. 2008. Suppression of dengue transmission by application of integrated vector control strategies at sero-positive GIS-based foci. American Journal of Tropical Medicine and Hygiene 78: 70–76.

Kittayapong, P., N. Kaeothaisong, S. Ninphanomchai, and W. Limohpasmanee. 2018. Combined sterile insect technique and incompatible insect technique: Sex separation and quality of sterile *Aedes aegypti* male mosquitoes released in a pilot population suppression trial in Thailand. Parasites & Vectors 11 (Supplement 2): 657.

Kittayapong, P., S. Ninphanomchai, W. Limohpasmanee, C. Chansang, U. Chansang, and P. Mongkalangoon. 2019. Combined sterile insect technique and incompatible insect technique: The first proof-of-concept to suppress *Aedes aegypti* vector populations in semi-rural settings in Thailand. PLoS Neglected Tropical Diseases 13(10): e0007771.

Kotsakiozi, P., A. Gloria-Soria, A. Caccone, B. Evans, R. Schama, A. J. Martins, and J. R. Powell. 2017. Tracking the return of *Aedes aegypti* to Brazil, the major vector of the dengue, chikungunya and Zika viruses. PLoS Neglected Tropical Diseases 11(7): e0005653.

Lees, R. S., D. O. Carvalho, and J. Bouyer. 2021. Potential impact of integrating the Sterile Insect Technique into the fight against disease-transmitting mosquitoes, pp. 1081–1118. *In* V. A. Dyck, J. Hendrichs, and A. S. Robinson (eds.), Sterile Insect Technique – Principles and practice in Area-Wide Integrated Pest Management. Second Edition. CRC Press, Boca Raton, Florida, USA.

Lees, R. S., J. R. L. Gilles, J. Hendrichs, M. J. B. Vreysen, and K. Bourtzis, K. 2015. Back to the future: The sterile insect technique against mosquito disease vectors. Current Opinion in Insect Science 10: 156–162.

Mains, J. W., C. L. Brelsfoard, R. I. Rose, and S. L. Dobson. 2016. Female adult *Aedes albopictus* suppression by *Wolbachia*-infected male mosquitoes. Scientific Reports 6: 33846.

Mains, J. W., P. H. Kelly, K. L. Dobson, W. D. Petrie, and S. L. Dobson. 2019. Localized control of *Aedes aegypti* (Diptera: Culicidae) in Miami, FL, via inundative releases of *Wolbachia*-infected male mosquitoes. Journal of Medical Entomology 56(5): 1296–1303.

Murray, J. V., C. C. Jansen, and P. De Barro. 2016. Risk associated with the release of *Wolbachia*-infected *Aedes aegypti* mosquitoes into the environment in an effort to control dengue. Frontiers in Public Health 4: 43.

Moreira, L. A., I. Iturbe-Ormaetxe, J. A. Jeffery, G. J. Lu, A. T. Pyke, L. M. Hedges, B. C. Rocha, S. Hall-Mendelin, A. Day, M. Riegler, L. E. Hugo, K. N. Johnson, B. H. Kay, E. A. McGraw, A. F. van den Hurk, P. A. Ryan, and S. L. O'Neill. 2009. A *Wolbachia* symbiont in *Aedes aegypti* limits infection with dengue, chikungunya, and *Plasmodium*. Cell 139(7): 1268–1278.

Morlan, H. B., E. M. McCray, and J. W. Kilpatrick. 1962. Field tests with sexually sterile males for control of *Aedes aegypti*. Mosquito News 22: 295–300.

Morrison, N. I., G. Franz, M. Koukidou, T. A. Miller, G. Saccone, and L. S. Alphey. 2010. Genetic improvements to the Sterile Insect Technique for agricultural pests. Asia-Pacific Journal of Molecular Biology and Biotechnology 18(2): 275–295.

(NEA) National Environmental Agency, Singapore. 2016. Risk assessment for the use of male *Wolbachia*-carrying *Aedes aegypti* for suppression of the *Aedes aegypti* mosquito population. Risk Analysis Report of the National Environmental Agency, Singapore, 14 pp.

(NEA) National Environment Agency, Singapore. 2019. New NEA facility to boost production of male *Wolbachia-Aedes aegypti* mosquitoes to benefit more residents. 2 December 2019.

Nikolouli, K., H. Colinet, D. Renault, T. Enriquez, L. Mouton, P. Gibert, F. Sassu, C. Cáceres, C. Stauffer, R. Pereira, and K. Bourtzis. 2018. Sterile insect technique and symbiosis as potential tools for the control of the invasive *Drosophila suzukii*. Journal of Pest Science 91(2): 489–503.

O'Connor, L., C. Plichart, A. C. Sang, C. L. Brelsfoard, H. C. Bossin, and S. L. Dobson. 2012. Open release of male mosquitoes infected with a *Wolbachia* biopesticide: Field performance and infection containment. PLoS Neglected Tropical Diseases 6(11): e1797.

Oliva, C. F., M. Jacquet, J. R. L. Gilles, G. Lemperiere, P. O. Maquart, S. Quilici, F. Schooneman, M. J. B. Vreysen, and S. Boyer. 2012. The sterile insect technique for controlling populations of *Aedes albopictus* (Diptera: Culicidae) on Réunion Island: Mating vigour of sterilized males. PLoS One 7(11): e49414.

Papathanos, P. A., H. C. Bossin, M. Q. Benedict, F. Catteruccia, C. A. Malcolm, L. Alphey, and A. Crisanti. 2009. Sex separation strategies: Past experience and new approaches. Malaria Journal 8(2): S2–S5.

Papathanos, P. A., K. Bourtzis, F. Tripet, H. C. Bossin, J. F. Virginio, M. L. Capurro, M. C. Pedrosa, A. Guindo, L. Sylla, M. B. Coulibaly, F. A. Yao, P. S. Epopa, and A. Diabate. 2018. A perspective on the need and current status of efficient sex separation methods for mosquito genetic control. Parasites & Vectors. 11 (Supplement 2): 654.

Patterson, R. S., V. P. Sharma, K. R. P. Singh, G. C. LaBrecque, P. L. Seetharam, and K. K. Grover. 1975. Use of radio-sterilized males to control indigenous populations of *Culex pipiens quinquefasciatus* Say: Laboratory and field studies. Mosquito News 35: 1–7.

Patterson, R. S., R. E. Lowe, B. J. Smittle, D. A. Dame, M. D. Boston, and A. L. Cameron. 1977. Release of radiosterilized males to control *Culex pipiens quinquefasciatus* (Diptera: Culicidae). Journal of Medical Entomology 14: 299–304.

Popovici, J., L. A. Moreira, A. Poinsignon, I. Iturbe-Ormaetxe, D. McNaughton, and S. L. O'Neill. 2010. Assessing key safety concerns of a *Wolbachia*-based strategy to control dengue transmission by *Aedes* mosquitoes. Memórias do Instituto Oswaldo Cruz 105(8): 957–964.

Reisen, W. K., M. M. Milby, S. M. Asman, M. E. Bock, R. P. Meyer, P. T. McDonald, and W. C. Reeves. 1982. Attempted suppression of a semi-isolated *Culex tarsalis* population by the release of irradiated males: A second experiment using males from a recently colonized strain. Mosquito News 42: 565–575.

Ruang-areerate, T., and P. Kittayapong. 2006. *Wolbachia* transinfection in *Aedes aegypti*: A potential gene driver of dengue vectors. Proceedings of the National Academy of Science of the USA 103: 12534–12539.

Sharma, V. P., R. S. Patterson, and H. R. Ford. 1972. A device for the rapid separation of male and female mosquito pupae. Bulletin WHO 47(3): 429–43.

Shepard, D. S., E. A. Undurraga, and Y. A. Halasa. 2013. Economic and disease burden of dengue in Southeast Asia. PLoS Neglected Tropical Diseases 7(2): e2055.

Thomas, D. D., C. A. Donnelly, R. J. Wood, and L. S. Alphey. 2000. Insect population control using a dominant, repressible, lethal genetic system. Science 287(5462): 2474–2476.

Trewin, B. J., J. M. Darbro, C. C. Jansen, N. A. Schellhorn, M. P. Zalucki, T. P. Hurst, and G. J. Devine. 2017. The elimination of the dengue vector, *Aedes aegypti*, from Brisbane, Australia: The role of surveillance, larval habitat removal and policy. PLoS Neglected Tropical Diseases 11(8): e0005848.

Turelli, M., and A. A. Hoffmann. 1995. Cytoplasmic incompatibility in *Drosophila simulans*: Dynamics and parameter estimates from natural populations. Genetics 140(4): 1319–1338.

Vreysen M. J. B., J. Gerado-Abaya, and J. P. Cayol. 2007. Lessons from area-wide integrated pest management (AW-IPM) programmes with an SIT component: An FAO/IAEA perspective, pp. 723–744. *In* M. J. B. Vreysen, A. S. Robinson, and J. Hendrichs (eds.), Area-wide control of insect pests: From research to field implementation. Springer, Dordrecht, The Netherlands.

Vreysen, M. J. B., K. M. Saleh, M. Y. Ali, A. M. Abdulla, Z. R. Zhu, K. G. Juma, V. A. Dyck, A. R. Msangi, P. A. Mkonyi, and H. U. Feldmann. 2000. *Glossina austeni* (Diptera: Glossinidae) eradicated on the island of Unguja, Zanzibar, using the sterile insect technique. Journal of Economic Entomology 93(1): 123–135.

Weidhaas, D. E., C. H. Schmidt, and E. L. Seabrook. 1962. Field studies on the release of sterile males for the control of *Anopheles quadrimaculatus*. Mosquito News 22: 283–291.

(WHO) World Health Organization. 2017. Fifth Meeting of the Vector Control Advisory Group, Geneva, Switzerland, 2–4 November 2016. WHO/HTM/NTD/VEM/2017.02. Licence: CC BY-NC-SA 3.0 IGO. Geneva, Switzerland.

(WHO) World Health Organization. 2019a. Dengue and severe dengue. Geneva, Switzerland.

(WHO) World Health Organization. 2019b. Mosquito control: Can it stop Zika at source? Geneva, Switzerland.

Xi, Z., C. C. Khoo, and S. L. Dobson. 2005. *Wolbachia* establishment and invasion in an *Aedes aegypti* laboratory population. Science 310(5746): 326–328.

Yakob, L., S. Funk, A. Camacho, O. Brady, and W. J. Edmunds. 2017. *Aedes aegypti* control through modernized, integrated vector management. PLoS Currents Outbreaks 9: Jan 30.

Zhang, D., Zheng, X., Xi, Z., Bourtzis, K., and J. R. L. Gilles. 2015a. Combining the sterile insect technique with the incompatible insect technique: I. Impact of *Wolbachia* infection on the fitness of triple and double-infected strains of *Aedes albopictus*. PLoS One 10: e0121126.

Zhang, D., R. S. Lees, Z. Xi, J. R. L. Gilles, and K. Bourtzis. 2015b. Combining the sterile insect technique with *Wolbachia*-based approaches: II. A safer approach to *Aedes albopictus* population suppression programmes, designed to minimize the consequences of inadvertent female release. PLoS One 10: e0135194.

Zhang, D., R. S. Lees, Z. Xi, K. Bourtzis, and J. R. L. Gilles. 2016. Combining the sterile insect technique with the incompatible insect technique: III. Robust mating competitiveness of irradiated triple *Wolbachia*-infected *Aedes albopictus* males under semi-field conditions. PLoS One 11(3): e0151864.

Zhang, D., M. Zhang, Y. Wu, J. R. L. Gilles, H. Yamada, Z. Wu, Z. Xi, and X. Zheng. 2017. Establishment of a medium-scale mosquito facility: Optimization of the larval mass-rearing unit for *Aedes albopictus* (Diptera: Culicidae). Parasites & Vectors 10: 569.

Zheng, M. L., D. J. Zhang, D. D. Damiens, R. S. Lees, and J. R. L. Gilles. 2015. Standard operating procedures for standardized mass rearing of the dengue and chikungunya vectors *Aedes aegypti* and *Aedes albopictus* (Diptera: Culicidae): II. Egg storage and hatching. Parasites & Vectors 8: 348.

Zheng, X., D. Zhang, Y. Li, Y. Wu, X. Liang, Y. Liang, X. Pan, L. Hu, Q. Sun, X. Wang, Y. Wei, J. Zhu, W. Qian, Z. Yan, A. G. Parker, J. R. L. Gilles, K. Bourtzis, J. Bouyer, M. Tang, B. Zheng, J. Yu, J. Liu, J. Zhuang, Z. Hu, M. Zhang, J. Gong, X. Hong, Z. Zhang, L. Lin, Q. Liu, Z. Hu, Z. Wu, L. A. Baton, A. A. Hoffmann, and Z. Xi. 2019. Incompatible and sterile insect techniques combined eliminate mosquitoes. Nature 572: 56–61.

ECOLOGY, BEHAVIOUR AND AREA-WIDE CONTROL OF THE FLOODWATER MOSQUITO *Aedes sticticus*, WITH POTENTIAL OF FUTURE INTEGRATION OF THE STERILE INSECT TECHNIQUE

J. O. LUNDSTRÖM[1,2], M. L. SCHÄFER[2] AND P. KITTAYAPONG[3,4]

[1]*Zoonosis Research Center, Department of Medical Biochemistry and Microbiology, Uppsala University, Box 582, SE-75123 Uppsala, Sweden; jan.lundstrom@mygg.se*
[2]*Biologisk Myggkontroll, Nedre Dalälven Utvecklings AB, Vårdsätravägen 5, SE-75646 Uppsala, Sweden*
[3]*Center of Excellence for Vectors and Vector-Borne Diseases, Faculty of Science, Mahidol University, 999 Phuttamonthon 4 Road, Nakhon Pathom 73170, Thailand*
[4]*Department of Biology, Faculty of Science, Mahidol University, 272 Rama VI Road, Bangkok 10400, Thailand*

SUMMARY

The strategy of aerial control of the floodwater mosquito *Aedes sticticus* (Meigen) in the floodplains of River Dalälven, central Sweden, was developed to directly address specific larval breeding areas in temporary flooded wet meadows and swamps. Using the *Bti*-based larvicide VectoBac G®, a very strong reduction of larval abundance is achieved, resulting in a massive decrease of blood-seeking females that could otherwise spread from the wetlands to feast on blood from humans and animals within 5 km or more from the larval biotopes. However, there is also a political demand to reduce the usage of the control agent through hypothetical alternatives, such as cattle grazing and mowing of the meadows, as well as hydrological changes of the River Dalälven. An evaluation of these measures showed that they are either insufficient or unrealistic in reducing floodwater mosquito abundance. Thus, we searched for other potential population suppression methods. Using the criteria of efficacy, environmental neutrality and compatibility within an integrated suppression approach, we conclude that Sterile Insect Technique (SIT) and the Incompatibility Insect Technique (IIT) would qualify for a pilot-scale test of their feasibility for the integrated control of the floodwater mosquito *Ae. sticticus*. The SIT and the IIT are similar strategies involving the release of sterile males which mate with local fertile females and result in infertile eggs. Prerequisites for a sterile male strategy to control *Ae. sticticus* include: a laboratory colony of the species, a facility for mass-rearing of mosquitoes, the sterilisation of males, a transport strategy, a dispersal system, assay systems for several life stages, and a method capable of reducing the population of this superabundant

J. Hendrichs, R. Pereira and M. J. B. Vreysen (eds.), Area-Wide Integrated Pest Management: Development and Field Application, pp. 433–459. CRC Press, Boca Raton, Florida, USA.

species before commencing the sterile male release. One factor in favour of implementing the SIT or IIT against *Ae. sticticus* is that mating occurs in or near well-defined larval breeding areas with specific relation to flood events. Another factor in favour of the SIT or the IIT is the availability of existing methods to measure gender, larvae and egg abundance. Also, existing *Bti*-treatments can substantially lower the population size before sterile male release. Other prerequisites, like the successful colonization of *Ae. sticticus* will require more tests and adaptations of existing mosquito rearing protocols. A pilot study is suggested for an isolated study area, protected from reinvasion by *Ae. sticticus*-females and included in routine *Bti*-treatments.

Key Words: Aedes sticticus, Sweden, River Dalälven, floodplains, wetlands, *Bacillus thuringiensis israelensis* (*Bti*), larvicide, VectoBac G, SIT, Incompatibility Insect Technique (IIT)

1. INTRODUCTION

Sweden, located in the north of Europe, is a country where mosquitoes are pervasive. While mosquito abundances were assumed to be highest in the northernmost part of the country, mosquito diversity increases towards the south (Schäfer and Lundström 2001). Generally, snow pool mosquitoes, e.g. *Aedes communis* (De Geer) and *Aedes punctor* (Kirby) are the most common species found throughout the country. Nuisance by these univoltine mosquito species can be severe, but occurs mainly in spring and early summer, followed by rapidly declining numbers.

When people in the River Dalälven floodplains in central Sweden complained about mosquito problems in the 1980's and 1990's, they were not taken seriously and often met with the conventional wisdom that mosquito problems are much more severe in the north. For a long time, the actual nuisance mosquito species was unknown, since knowledge on the mosquito fauna in the River Dalälven region was insufficient. In a study from 1985, the floodwater mosquito *Aedes rossicus* Dolbeskin, Gorickaja and Mitrofanova was reported as the most abundant species (Jaenson 1986). Ten years later, researchers studying Sindbis virus in the area needed to use protective clothing due to the enormous abundance of mosquitoes, but no general identification of nuisance species was performed (Lundström et al. 1996).

In the summer of 2000, we studied mosquito species diversity in the central part of the region at Lake Färnebofjärden, which coincided with one of the worst mosquito nuisance years due to massive floods. Mosquito sampling with CDC miniature light traps baited with dry ice resulted in enormous numbers (up to 61 500 mosquitoes per trap and night) and the predominant species was *Aedes sticticus* (Meigen) (Schäfer et al. 2008). This can be compared to the maximum number of 4500 mosquitoes per trap and night (trap-night) from a wetland in northern Sweden (unpublished information).

The people of the River Dalälven floodplains were desperate, and children had to be transported away from the area by buses to be able to swim and play outside during their summer vacation. Media awakened and the mosquito-infected towns in the region became known in the whole country. The major breakthrough in the people's struggle to continue living in this area was a visit from the Minister of Environment, Mr. Kjell Larsson, who is still the only minister to experience massive floodwater mosquito nuisance. His words, "*You cannot have it like that*" became historic; and resulted in the development of the first professional mosquito control in Sweden.

The identification of the flood-water mosquito *Ae. sticticus* as the main cause of the horrendous nuisance made it the prime target species for control. Larviciding with

Bacillus thuringiensis israelensis (*Bti*) was the method of choice for its low environmental impact, efficacy, and practical application over large areas.

During recent years, political pressure has created a demand for alternative methods to control this superabundant day-active and long-range dispersing mosquito, motivating us to search for new, less intrusive mosquito control methods suitable for area-wide use in natural wetlands.

Below we describe development of our high-tech GIS-based strategy of direct *Bti*-based larval control (Section 2), the way forward for adapting SIT-based birth control for area-wide control of *Ae. sticticus* in natural wetlands (Section 5), and a section on perspectives (Section 6).

2. DEVELOPING AREA-WIDE CONTROL OF A*edes sticticus* IN NATURAL WETLANDS

2.1. The River Dalälven Floodplains

The River Dalälven covers a catchment area of approximately 29 000 km^2, originating in the mountains along the Swedish-Norwegian border and outflowing into the Baltic Sea (Fig. 1).

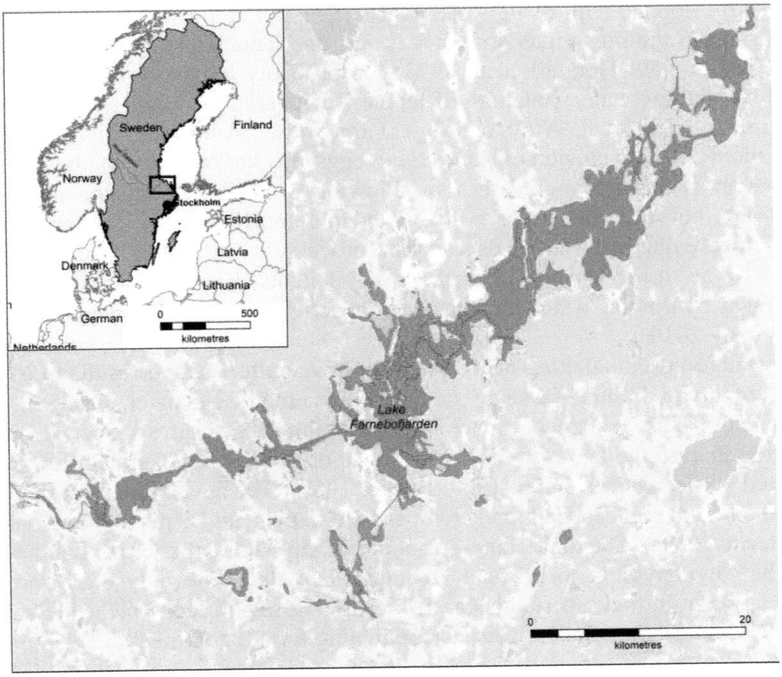

Figure 1. The location of the River Dalälven in central Sweden (inlet) and the floodplains with its many lakes in the lower part of River Dalälven. The areas with permit for mosquito control treatments by Bti *for 2019 are shown in orange.*

The River Dalälven, with its main branches Västerdalälven and Österdalälven, is partly regulated for production of hydro-electric power. In its lower part, the river forms a chain of lakes connected by rapids. In these floodplains, temporary flooded marshes, wet meadows and alder swamps cover several thousand hectares (ha) and most of this area is protected by both national regulations and EU-regulations. Water level fluctuations are most dramatic and frequent in Lake Färnebofjärden, which is protected as a National Park and contains several protected Nature Reserves and Natura 2000 areas. Flooding is induced by melting snow and/or heavy rainfall that causes increased waterflow in River Dalälven and other smaller watercourses in the floodplain area, and the water level can increase by 2.0 m or more in the Lake Färnebofjärden area. This is an area of enormous mosquito abundance and a hotspot for mosquito diversity (Schäfer et al. 2018).

2.2. Ecology and Behaviour of Aedes sticticus

Floodwater mosquitoes, in particular *Ae. sticticus* and *Aedes vexans* (Meigen), are the predominant mosquito species in areas influenced by large rivers or lakes with water level fluctuations in adjacent lowlands (Becker and Ludwig 1981; Merdic and Lovakovic 2001; Minar et al. 2001; Schäfer et al. 2008). These mosquito species oviposit their eggs on moist soil, into small depressions or in moss, which are subsequently flooded with rising water levels (Horsfall et al. 1973). The eggs are in diapause during autumn, winter and early spring, and remain viable for at least 4 years (Gjullin et al. 1950; Horsfall et al. 1973) but probably longer. When the eggs are flooded by shallow water, hatching of larvae is triggered by water temperature and decreasing oxygen level. After melting of the snow, water temperature needs to exceed about 8°C for eclosion of *Ae. sticticus* eggs (Becker et al. 2010), thus avoiding larval hatching during the cold seasons. Flowing water inundating the wetlands is oxygen-rich, but once the water in the inundated areas becomes stagnant, oxygen levels decrease due to bacterial degradation processes. This signals the appropriate time for larvae to hatch from the eggs. Newly hatched larvae no longer risk being carried away by flowing water, and the bacterial activity ensures adequate food supply (Becker et al. 2010).

The synchronised hatching of *Ae. sticticus* larvae after a flood results in massive amounts of larvae at about the same time, although not all eggs hatch during the same flood event. This so-called 'hatching-in-installment' ensures survival of the population in case the larval breeding site dries out before development to adults is completed (Wilson and Horsfall 1970; Becker 1989). The development of the larvae to pupae and emergence of adults is temperature-dependent (Trpis and Shemanchuk 1970; Becker 1989). The males emerge about one day before the females and need to rotate their hypopygium to be ready to mate. Females mate only once and store sufficient sperm in their spermathecae for fertilizing several egg batches (Becker et al. 2010). After mating, the females start searching for a blood meal to develop eggs. The blood-seeking *Ae. sticticus* females are known for their long-distance dispersal behaviour, covering distances of at least 10 km (Brust 1980; Sudarić Bogojević et al. 2011).

Floodwater mosquitoes are multivoltine and each flood during spring and summer can produce a new generation of mosquitoes. Together with their capability for mass-reproduction, this explains the enormous numbers and the lengthy occurrence of floodwater mosquito nuisance over several months in summer and fall. These mosquitoes cause an enormous nuisance affecting every aspect of living, working and visiting the mosquito-infested areas, as well as the health of the human and livestock populations, resulting in reduced property prices.

2.3. Area-wide Control of Ae. sticticus *using* Bti

In the summer of 2000, when people once again were attacked by horrendous numbers of floodwater mosquitoes, the desperate call for help to reduce mosquito nuisance became major and repetitive news in the media at all levels. Officials of one of the seven affected municipalities then made the decision to initiate professional mosquito control operations and the other six municipalities followed the lead.

It was rapidly clear that the only possible and realistic solution was larviciding using a *Bti*-product. In view of the multitude of protected areas in the River Dalälven floodplains and the high environmental awareness in Sweden, chemical control or the application of less specific control agents were excluded. When applied correctly, *Bti*-products are highly selective against target mosquitoes without any known negative effects on non-target organisms or the environment (Lundström et al. 2010a, 2010b; Caquet et al. 2011; Lagadic et al. 2013, 2016). We decided to use the ready-to-use product VectoBac G®, consisting of corn-cob granules coated with *Bti* attached to the granules with corn oil.

Successful application of VectoBac G® requires detailed knowledge on the ecology of the target species to direct the treatments to the correct sites at the appropriate time. Thus, the first step for the programme against *Ae. sticticus* was precise mapping of the larval breeding sites. Mapping in the field started in the autumn of 2000 using a high precision GPS, amounting to a total area of 1170 ha near the two most affected towns Österfärnebo and Tärnsjö. This method was based on vegetation as indicators for temporary flooded areas and was very labour-intensive.

To speed up the mapping process and get more precise information on the geographic extent of inundations, another approach was needed. We decided to develop a high-precision digital elevation model (DEM) based on laser-scanning of the relevant areas. In 2003, the entire lower part of the River Dalälven was covered by air-borne laser-scanning. The multitudinous point measurements of the laser-scanning were used to create a DEM with sufficient resolution to discern height with centimetre precision. Since then, we use modelling with this DEM to discern the shallow flooded areas harbouring *Ae. sticticus* larvae, and to prepare the polygons to precisely direct the VectoBac G® larval treatments.

The first mosquito control operation in Sweden was carried out in 2002 and covered in total 443 ha. In the beginning, only temporary flooded areas outside nature reserves, the national park and close to the towns of Österfärnebo and Tärnsjö were included (Fig. 2).

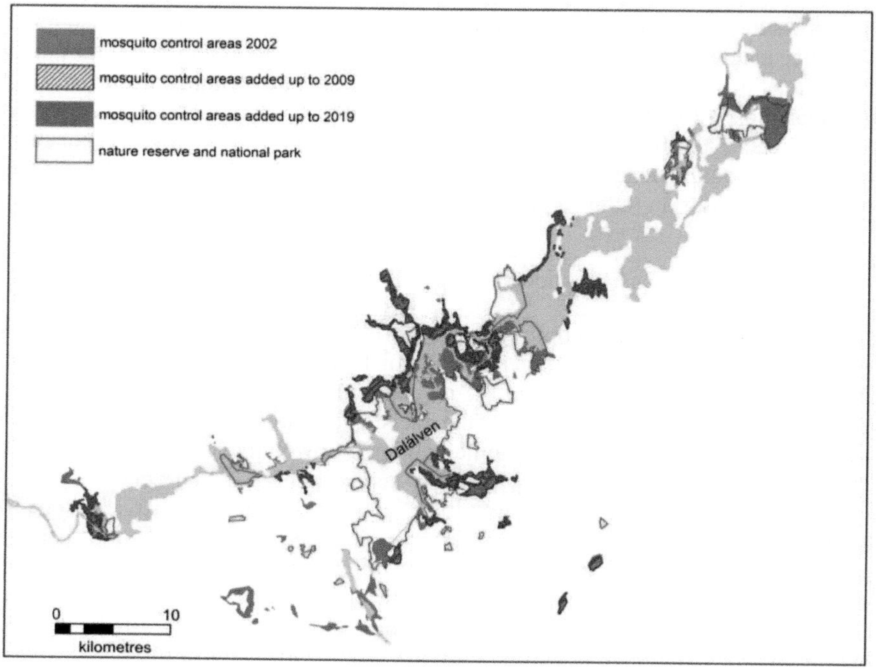

Figure 2. The gradual increase of areas with permissions for mosquito control based on dispersal of VectoBac G® granules by helicopter in natural wetlands, exemplified by the years 2002, 2009 and 2019. In 2009, some protected areas were finally included because of a case won at the Supreme Environmental Court of Sweden.

Over the years, we gradually increased the coverage of treatment areas, and another milestone was reached in 2009 with the first permission for treatments in protected areas (Fig. 2). This achievement however required a court case that was decided in favour of treatments at the Supreme Environmental Court of Sweden.

Since 2016, we have permission to treat more than 10 000 ha of swamps, marshes and meadows and the single largest treatment so far covered 4411 ha in May 2018. The need for VectoBac G® application varied between the years, from no application at all during some years when no floods took place, e.g. 2004 and 2017, to a maximum total of 9345 ha in 2015 (Fig. 3). In addition, the applied dosage of VectoBac G® was gradually reduced from 15-17 kg/ha during the first years to 11-13 kg/ha during recent years. This dose reduction was achieved by technical improvements regarding the helicopter application and navigation system used.

2.4. Routine Control Operations

From middle of April until end of August, the water flow fluctuations of the River Dalälven is followed seven days per week, and through collaboration with water regulation authorities we have access to a professional water flow prognosis. If there is an indication of rising water levels in the lakes of the floodplain, actual inundations are monitored in the field and the presence of newly hatched larvae of the target floodwater mosquitoes assessed. It is crucial to find the first-instar larvae of *Ae. sticticus* as early as possible to maximize the time window for *Bti*-treatments. During these first days, several two person teams visit selected sites to measure the abundance of mosquito larvae with a standard larval dipper and to map the waterline with a handheld high-precision GPS. These field-collected data provide the baseline for all decisions on treatments.

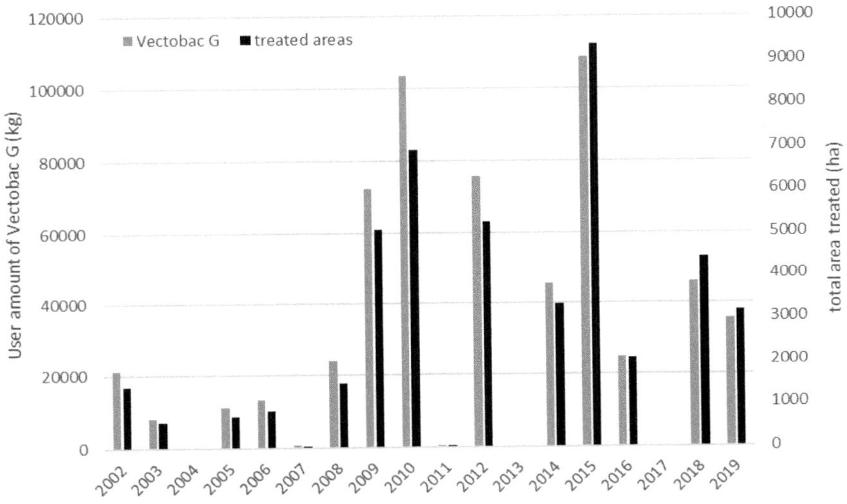

Figure 3. The amount of VectoBac G applied by helicopter in the River Dalälven floodplains for mosquito control and the total area treated per year for the years 2002 to 2019.

The GPS-points are transferred to a GIS-software and plotted on the DEM. The lakes in the floodplain are situated at different elevations, and the inundated areas are therefore modelled for each sub-area. As floodwater mosquito larvae are rare or absent in open deep water, the DEM is used to exclude those areas from treatments. The water depth limit for application depends on vegetation density and height as vegetation provides retreats for mosquito larvae in open waters. During spring floods, when there is little and low vegetation, areas up to 40 cm water depth are included in the applications, while during summer with plenty of vegetation, the limit is set at 60 cm water depth. In the GIS, the relevant areas are defined and prepared as polygons for the helicopter treatments.

Application of VectoBac G by helicopter can start approximately two to three days after detecting the first floodwater mosquito larvae, including the time needed to collect and analyse all necessary technical and biological information. Two sling buckets with rotating discs are calibrated for application of VectoBac G®. Using two buckets makes application very efficient, allowing for simple change of bucket for the helicopter pilot without landing (Fig. 4).

In the helicopter, a navigation system connected to a GPS with a high update-frequency reads the polygons as areas to be treated. The pilot prepares the appropriate flight routes with a defined distance (20-30m, exact distance is based on calibration results) between flight lines, guiding flight routes ensure complete coverage within each treatment area. All the VectoBac G®-applications are logged and transferred to the GIS for assessing the areas covered.

Figure 4. The use of two sling buckets, combined with change of bucket without landing the helicopter, allows for increased speed and reduced cost for aerial application of VectoBac G in the River Dalälven floodplains, central Sweden (credit J. O. Lundström).

All *Bti*-treatments should be completed before the mosquito larvae reach fourth instar. Therefore, especially during warm summer weeks, large floods require very rapid and efficient operations. Fortunately, there is almost 24 hours of daylight during summer in this part of Sweden. If necessary, the helicopter can apply VectoBac G®-from approximately 04:00 in the morning until 24:00 at night. These intensive 20 hrs working days require double crew on duty both on the ground and in the air.

In May 2018, a total area of 4411 ha was treated, including areas in seven municipalities along an approximately 100 km stretch of the River Dalälven, the largest mosquito control operation so far. Treatments were completed in 5 days with successful reduction of floodwater mosquito larvae. Currently, more than 1100 ha of natural wetlands can be treated by helicopter per day.

2.5. *Eighteen Years of* Bti-*based* Ae. sticticus *Control*

The goal for the floodwater mosquito control is to reduce mosquito abundance to less than 500 mosquitoes per trap-night to ensure that people in the River Dalälven floodplains can live normal lives during the short summer in Sweden. As mentioned, mosquito abundance in the area was extremely high before initiation of control and people were plagued by blood-seeking *Ae. sticticus* females, even in the centre of towns in the middle of the day. For example, in the centre of Österfärnebo village we measured 23 000 mosquitoes per CDC trap-night in August 2000 (Fig. 5). This measurement was before the first VectoBac G® application.

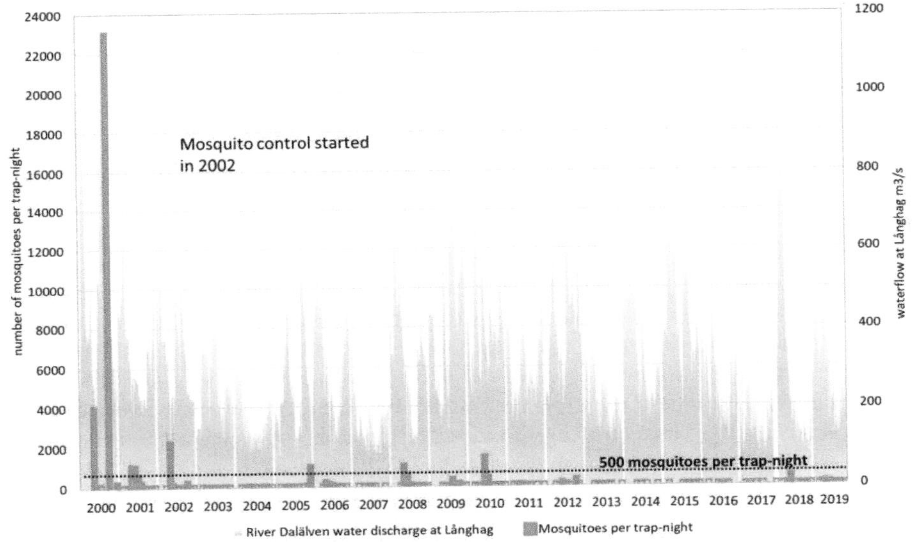

Figure 5. The number of mosquitoes per trap night in central Österfärnebo and River Dalälven water discharge at an upstream station from 2000 to 2019. The goal of mosquito control is to reduce mosquito abundance to less than 500 mosquitoes per trap night (reproduced with permission from Biologisk Myggkontroll 2019).

Since the start of mosquito control in 2002, mosquito abundance has been significantly reduced and since 2013, the number of mosquitoes collected in the trap in central Österfärnebo has been kept below 500 individuals per trap-night. This shows the effectiveness of successful *Bti*-applications and the results are greatly appreciated by both the local people and the visitors.

3. PREVIOUSLY SUGGESTED ALTERNATIVE CONTROL METHODS

Swedish authorities, including the Swedish Environmental Protection Agency (Swedish EPA), realize the importance of floodwater mosquito control and the need for *Bti*-treatments, but nevertheless want to reduce the usage of VectoBac G® in favour of alternative methods. However, the enormous abundance of mosquitoes causing the nuisance makes this a quite complicated task, since a very dramatic reduction is needed to reach the less than 500 mosquitoes per trap-night required by the locals and visitors alike.

For example, in Österfärnebo village a 97.8% reduction was required to reduce the 23 000 mosquitoes per trap-night to the acceptable abundance of 500 mosquitoes per trap-night. With respect to the impressive flight range of blood-seeking *Ae. sticticus* females, this means in practical terms that at least 90% of the *Ae. sticticus* larvae in at least 95% of the shallow flooded areas within 5 km around the village need to be controlled (Schäfer and Lundström 2014).

The alternative methods specifically suggested by the Swedish authorities are 1) changing the hydrology of the River Dalälven, and 2) increasing the mowing and grazing in the wet meadows (Lundqvist et al. 2013).

3.1. Changing the Hydrology of the River Dalälven

Changing hydrology for floodwater mosquito control requires stabilizing the water level to avoid floods and such a hydrological regime can cause a very strong reduction in mosquito abundance if fully implemented. However, nature conservationists request increased flood magnitude and duration in May and June, which is also the major larval production period for *Ae. sticticus*. Clearly, a single hydrological strategy will not be able to achieve the requirements for both a stable water level and increased flooding. In addition, this topic involves a very large number of stakeholders and many laws and regulations that control the water flow and water levels in different parts of the river. Furthermore, implementing the whole process would be very costly and time-consuming. Thus, reducing flood-water mosquito nuisance by hydrological changes of the River Dalälven, without reducing biological diversity, is a highly complex task that could be considered almost impossible (Hedström-Ringvall et al. 2017).

3.2. Increased Mowing and Grazing in the Wet Meadows

The suggestion of using mowing and grazing as alternative mosquito control methods originates from the general opinion that mosquito nuisance was less severe more than a century ago, when mowing and grazing of the floodplain meadows were common.

One study could show fewer mosquito larvae in areas with mowing and grazing than in areas without these methods, but this study was restricted to one year (Östman et al. 2015). In an unpublished study comparing the numbers of eggs in areas with and without mowing, no difference was found, and thus no long-term effect of mowing or grazing on the abundance of floodwater mosquitoes can be expected (Östman 2013).

We obtained similar results when comparing our own larval surveillance data over 15 years from areas with and without mowing and grazing Thus, although mowing and grazing sometimes might result in lower abundances of floodwater mosquito larvae, these measures cannot serve as reliable mosquito control methods. The potential reduction of larval abundance is too low and too unpredictable.

The search for effective alternatives to control *Ae. sticticus* in the River Dalälven floodplains should identify methods that could fulfil three main criteria: 1) ability to obtain very strong population suppression, 2) being environmentally neutral, and 3) suitable for large-scale application in natural swamps, marshes and wet meadows.

Evidently there are few mosquito control methods, other than larviciding with *Bti*-based products, capable of inducing such a strong population reduction without adding substances that might represent a distinct risk to the environment. In addition, Sweden is a member of the European Union and should comply with the biocide directive (EU 2012).

4. NEW METHODS FOR MOSQUITO CONTROL

Several potential new mosquito population suppression methods have surfaced in the last decades including the Sterile Insect Technique (SIT), the Incompatible Insect Technique (IIT), the Release of Insects carrying Dominant Lethality (RIDL), and genetically modifications based on CRISPR-Cas9 technology (Huang et al. 2017). The RIDL and CRISPR-Cas9 technologies have potential for strong population suppression, but the genetic modifications forming the strategic base for the methods will probably induce very strong counter-reactions from the general public, the Swedish EPA and other environmental protection authorities.

The SIT is a species-specific and environmentally safe method for area-wide management of insect pests which relies on repeated release of a large number of sterile male insects (Knipling 1955, 1979, 1998; Krafsur 1998; Dyck et al. 2021). The population reduction effect is achieved after sterile males are released and mate with the wild females, which will then lay infertile eggs. If a surplus of sterile males is regularly released on an area-wide basis over a sufficient time period, and they successfully mate with the local females, ultimately this will result in suppression or local elimination of the target insect population. The necessary ratio of released sterile males to local fertile males depends on the biology of the target species, the initial wild population density, the risk of reinfestation from neighbouring areas, the competitiveness of the released sterile males, and the complementary control operations that can be performed (Dame et al. 2009). The SIT has and is being used in successful area-wide integrated pest management programmes (AW-IPM) against the New World screwworm fly *Cochliomya hominivorax* (Coquerel), the Mediterranean fruit fly *Ceratitis capitata* (Wiedemann) and other tephritid flies, tsetse flies, the codling moth *Cydia pomonella* (L.), the false codling moth *Thaumatotibia leucotreta* (Meyrick) and the pink bollworm *Pectinophora gossypiella* (Saunders) (Lindquist et al. 1992; Vreysen et al. 2007; Dyck et al. 2021; Boersma, this volume; Nelson, this volume; Staten and Walters, this volume).

SIT field trials in the 1970's and 1980's demonstrated that it could also work against mosquitoes (Patterson et al. 1970; Lofgren et al. 1974; Benedict and Robinson 2003; Dame et al. 2009). In the last decade, the Joint FAO/IAEA Programme has been the main driver for development of the mosquito SIT package (Lees et al. 2014; Bourtzis et al. 2016). The focus is on three mosquito vector species of major medical importance: the arbovirus vectors *Aedes aegypti* L. and *Aedes albopictus* (Skuse), and the malaria vector *Anopheles arabiensis* Patton.

Major technical improvements for the SIT against the two *Aedes* species have already resulted in successful SIT pilot-scale field studies, paving the way for the development of the SIT as a full-scale mosquito population suppression method (Lees et al. 2021). Successful field trials of the SIT for suppressing populations of *Aedes* mosquitoes are recorded for *Ae. albopictus* in Italy (Bellini et al. 2013), while a SIT/IIT combination was shown successful in suppressing a population of *Ae. albopictus* in China (Zhang et al. 2016; Zheng et al. 2019) and suppressing a population of *Ae. aegypti* in Thailand (Kittayapong et al. 2019).

The IIT relies on symbiotic bacteria of the genera *Wolbachia*, inherited in insects, and that can manipulate the reproductive system of their host insects (Kittayapong et al. 2002; Werren et al. 2008). The incompatibility of sperm from a *Wolbachia*-infected male that fertilizes the eggs of a non-infected female, or of a female that is infected with another *Wolbachia* strain, can be used for population suppression by the IIT. The technique was first developed in 1967 against the lymphatic filariasis vector *Culex pipiens fatigans* in Burma, and the IIT was shown capable of eliminating the local mosquito vector population (Laven 1967). More recent positive results have been obtained in field experiments with the IIT against *Aedes polynesiensis* Marks 1954, *Culex pipiens quinquefasciatus* Say 1823, *Ae. albopictus* and *Ae. aegypti* (Atyame et al. 2011, 2015; Moretti and Calvitti 2012; O´Connor et al. 2012; Ritchie et al. 2015; Mains et al. 2016; Strugarek et al. 2019).

The control action of both the SIT and the IIT relies on providing a surplus of sexually active males that upon mating with the local females cause infertility of their eggs. These eggs cannot hatch to larvae, thus precluding development of new mosquito generations and over time the local population declines and perhaps, if isolated, is even locally eliminated. With the SIT, this is achieved by the release of sexually active mosquito males that have been sterilized by radiation. With the IIT, infertile eggs are the consequence of incompatibility between released sexually active *Wolbachia*-transfected males mating with local females that either are uninfected by *Wolbachia* or are infected with a different *Wolbachia*-strain.

We consider the SIT and the IIT as interesting to evaluate as part of an integrated approach for area-wide population suppression of our target species, the floodwater mosquito *Ae. sticticus*. Being environmentally neutral, both the SIT and the IIT could, after population pre-treatment with VectoBac G, potentially meet the criteria of inducing a high level of population suppression, although they have not been tested against a floodwater mosquito species.

5. PREREQUISITES FOR A STERILE MALE STRATEGY TO CONTROL *Aedes sticticus*

Applying the SIT or the IIT for *Ae. sticticus* control requires a laboratory colony of the species, a facility for mass-rearing of mosquitoes, the sterilisation of males, a transport strategy, a dispersal system, monitoring systems for several life stages, and a method capable of reducing the population of this superabundant species before commencing the sterile male release.

Differences in ecology and behaviour will demand a partially different SIT strategy for *Ae. sticticus* than for more commonly considered species *Ae. aegypti* or

Ae. albopictus (Lees et al. 2021). The latter two have continuous reproduction during a major part of the year and larval habitats are small, cryptic and widely dispersed. SIT-based control of these species requires that sterile males are released at least once a week over many months over mosquito habitat in domestic and rural areas. In contrast, *Ae. sticticus* larval sites are well-defined temporary flooded areas with synchronised batches of larvae during a flood event. Thus, SIT-based control requires very focused release of males in conjunction with flood events. The synchronised emergence of *Ae. sticticus* in relation to floods indicate that the release of sterile males during this emergence period could be sufficient to induce a high percentage of egg infertility in local females. However, it is probably a safer strategy to continue with weekly releases of sterile males for an additional 3-4 weeks after each emergence.

Several supporting factors necessary for successful SIT or IIT application are already well established for *Ae. sticticus*, while other factors need to be dealt with. As shown on previous pages, an efficient method for large-scale larval suppression is available that can significantly reduce population size before sterile male release. One factor in favour of implementing the SIT or the IIT against *Ae. sticticus* is that mating occurs in or near well-defined larval breeding areas with specific relation to flood events. Another factor in favour of the SIT or the IIT is the availability of existing methods to measure gender, larvae and egg abundance. The following pages provide details on the major factors that need to be addressed to develop and test the SIT or the IIT against *Ae. sticticus* in Sweden.

5.1. Egg Storage and Hatching

The eggs of the floodwater mosquito *Ae. sticticus* range from 0.610 to 0.645 mm in length and from 0.180 to 0.215 mm in width (Gjullin et al. 1950). The eggs are extremely hardy and remain viable for several years (Gjullin et al. 1950; Trpis and Horsfall 1967), allowing for a long shelf-life and stockpiling of eggs during industrial mass-production year-round. Eggs could be stored at 4°C for long time periods.

The eggs will not hatch in clean tap water but hatch readily in a willow-leaf infusion or when amino acids are added to the water (Gjullin et al. 1950). A reduction in dissolved oxygen is the main hatching stimulus for the eggs (Gjullin et al. 1950) with increased eclosion when the hatching media is a nutrient rich broth (Trpis and Horsfall 1967). Hatching of eggs can occur at 8°C, but the hatching is more efficient and better synchronised at higher temperatures with optimum of about 21°C (Trpis and Horsfall 1967). Eggs of *Ae. sticticus* may have to be exposed to a period of winter before hatching (Horsfall and Trpis 1967).

5.2. Larval Rearing

The development of *Ae. sticticus* larvae depends on temperature, diet, larval density and water depth (Trpis and Horsfall 1969). Water temperatures of 8°C to 32°C were tested, and 21°C was considered the optimum rearing temperature with maximum percentage maturing in the shortest time interval. At 25°C larval development was accelerated by 1-2 days, but mortality increased.

The larvae were fed liver yeast suspended in water, and for a pan with 30 larvae in 1700 ml of water the optimal yield was achieved when the larvae were provided 110 mg of dry yeast equivalent per pan per day. Feeding every second day required doubled amount of food and feeding every third days resulted in increased mortality. The density of larvae, reared in 1700 ml of water at 25°C, influenced the developmental time. Pupation began and was completed on day 6 in pans with 30 larvae, while pupation occurred days 7 to 10 in pans with 60 larvae and on days 8 to 13 in pans with 90 larvae. Water depth was also important, especially at higher temperature both development and survival were best in very shallow water.

5.3. Adult Rearing and Mating

The rearing of adult *Ae. sticticus* may require relatively large cages of 1.0 x 2.0 x 2.5 m (5 m^3) to maintain a normal mating behaviour of the laboratory reared males. Mating of *Ae. sticticus* occurs in damp and shady areas among trees and bushes, but they are not forming any obvious swarms. The actual triggers of mating activity are unknown, and this is of course a potential obstacle when trying to establish a laboratory colony. Experience from colonization of other mosquito species showed that a combination of natural light cycle and a sufficiently sizeable cage triggered mating (Kuhn 2002; Lundström et al. 1990). Small cage size can induce a problem if mating couples split when not in the air, but this is not a problem with *Ae. sticticus*.

Photo documentation of a mating *Ae. sticticus* couple in the field show that they continue the sexual activity even after landing on the rubber boot of the observer (Fig. 6).

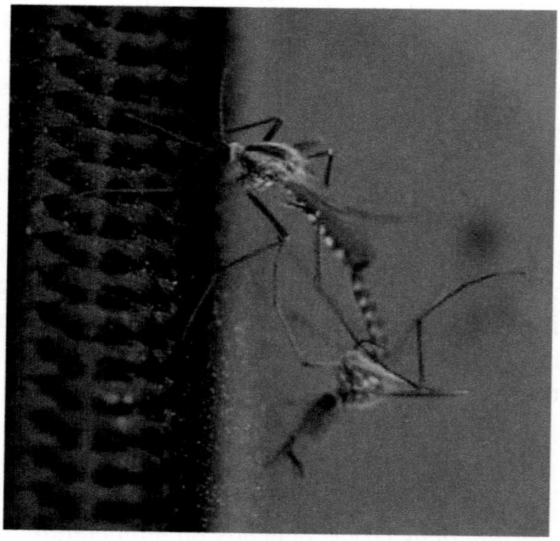

Figure 6. A mating couple of the floodwater mosquito Aedes sticticus, *that initiated mating in the air and continued after they landed on the photographer's rubber boot, in the Valmbäcken alder swamp in July 2015 (credit J. O. Lundström).*

Our strategy for colonization of *Ae. sticticus* will initially focus on evaluating the mating in relatively large cages with simulated natural dusk and dawn periods. If not successful, forced copulation could be used for a few generations. Experience from colonization of the floodwater mosquito *Ae. vexans* could provide additional suggestions (Kuhn 2002). The colonization of *Ae. vexans* was based on mosquitoes released in a walk-in cage of 5 m^3 and simulated dusk and dawn periods. Once the colony was established, *Ae. vexans* adapted to mating in smaller cages with a 1.4 m^3 volume. However, such changes in behaviour could be a disadvantage for the laboratory reared males in the competition for mating with wild females. We are also aware that the close ecological similarities between these two floodwater *Aedes* species of the northern hemisphere is no guarantee that colonization of *Ae. sticticus* will be successful.

The adult mosquitoes require a food regime with constant access to 10% sugar solution (males and females) while the females in addition need to be provided blood approximately once a week for egg production. Our practical experience is that Swedish *Ae. sticticus* readily feed on bovine blood heated to 38°C and provided from a membrane feeder. The initial unsuccessful trials to colonize *Ae. sticticus* have shown that there is no need for a specific egg substrate, since the females readily deposit the eggs on moist paper. However, many details concerning larval and adult rearing and mating will have to be optimized before efficient rearing and mass-production of high-quality males will be possible.

5.4. Sex Separation

The male pupae of *Ae. sticticus* are smaller than the female pupae, allowing for mechanical size-based sex separation in the pupal stage. The Fay-Morlan separator, a mechanical sex separation method (Fay and Morlan 1959; Sharma et al. 1972; Focks 1980), is the standard method for sex separation of *Ae. aegypti* and *Ae. albopictus* pupae. The sieves method to separate male and female pupae is also commonly used and both methods are potential options for *Ae. sticticus* sex separation.

The eventual contamination with some female mosquitoes among the sterile mosquito males, is a serious problem. The blood-seeking females could cause nuisance, which may cause public aversion that severely reduces the perceived effect of the control strategy. In addition, females mixed with the released sterile males could also divert some mating away from the target native females, thereby reducing the effect of the SIT intervention. Therefore, an efficient and secure method of separating males from females in the mass-production process is imperative to the success of the strategy.

5.5. Sterilisation by Ionizing Radiation

Sterilisation of male insects for the SIT can be done by ionizing radiation or by chemical treatment (Bakri et al. 2021). Sterilisation by ionizing radiation that randomly destroys fractions of DNA in the male gonads was the first tested method. Sterilisation by X-ray or gamma radiation from a ^{60}Co radiation chamber is nowadays the standard method of sterilizing male insects for SIT application. This is an extremely reliable method, suitable for industrial-scale insect sterilisation.

The use of gamma radiation for sterilisation of mosquitoes was first tested against *Ae. albopictus* in Italy (Bellini et al. 2013) and has later been more generally applied as the mosquito sterilisation method. However, the use of ^{60}Co requires special security measures, adequate regulation in the country, radiation protection protocols, and the initial investment is high. More recently, X-ray machines suitable for mass-sterilisation of male insects have been developed and are becoming available (Yamada et al. 2014; Bakri et al. 2021). The X-ray equipment is cheaper, requires no regulation in the country, and requires less security than the gamma radiation equipment. Also, X-ray is commonly used for medical purposes making it probably the least controversial method for sterilizing male *Ae. sticticus*.

5.6. Wolbachia *and* Aedes sticticus

The *Wolbachia* bacterial symbiont can be used to induce sterility through mating incompatibility (Bourtzis 2008; Werren et al. 2008; Rasić et al. 2014). Recently, the United States Environmental Protection Agency (US-EPA) has approved the release of *Wolbachia pipientis* transfected male *Ae. albopictus* (*w*Pip strain; ZAP males) for population suppression in the District of Columbia, and in 20 states of the USA (US-EPA 2017; Waltz 2017).

The use of *Wolbachia* for IIT implementation is dependent on knowledge about the eventual occurrence of natural infection in the target populations, as it requires that the wild female is either free of any *Wolbachia* bacteria symbionts or carries another strain of the bacteria than the infected and released males.

A preliminary study of the occurrence of *Wolbachia* was carried out with *Ae. sticticus* samples collected during the regular annual mosquito surveillance programme in the River Dalälven floodplains by Biologisk Myggkontroll (Schäfer et al. 2018). *Wolbachia*-specific PCR screening, as reported in Kittayapong et al. (2000), was carried out in 279 mosquitoes of 17 mosquito species for naturally occurring infections with this bacterial symbiont (Table 1). A total of 7 out of 17 species (41.2%) contained the *Wolbachia* symbiont. However, the PCR results indicated that the 20 *Ae. sticticus* individuals screened were free of the *Wolbachia* symbiont, indicating that the species is probably free from *Wolbachia* infection.

Based on this information, there is potential for using *Wolbachia*-transfected *Ae. sticticus* males for population suppression. This would require the establishment of a *Wolbachia*-transfected *Ae. sticticus* strain that could be produced in large numbers for male-only release. The use of a local mosquito strain as well as a local *Wolbachia* strain might also make it easier to receive the necessary permits from the authorities.

5.7. Transport

The production of sterile male *Ae. sticticus* could either be done in Sweden, or in another country within reach for timely delivery to the suggested pilot study area in the River Dalälven floodplain. Any decision on the location of such a production unit will require an evaluation of the costs and reliability for production and delivery, as well as the logistics for delivery, in relation to the reaction time from low-level maintenance production to full-scale production for release. Also, it has to be guaranteed that the long-distance shipment is not detrimental to the quality of the insects. A key factor is the availability of a provider with the knowledge and drive to perform the sterile male production.

Table 1. The occurrence of the bacterial symbiont Wolbachia *in mosquito species collected in the wetlands of the River Dalälven floodplains, central Sweden*

Species	No. Tested	No. *Wolbachia* positive specimen	Percent positive
Aedes annulipes (Meigen)	32	0	0
Aedes cantans (Meigen)	15	0	0
Aedes cinereus (Meigen)	30	5	16.67
Aedes communis (De Geer)	20	0	0
Aedes diantaeus (Howard, Dyar & Knab)	19	0	0
Aedes intrudens (Dyar)	18	0	0
Aedes punctor (Kirby)	20	0	0
***Aedes sticticus* (Meigen)**	20	0	0
Aedes vexans (Meigen)	20	0	0
Culiseta alaskensis Ludlow	13	0	0
Culiseta bergrothi Edwards	15	8	53.33
Culiseta morsitans (Theobald)	5	4	80.00
Culiseta ochroptera (Peus)	2	0	0
Culex pipiens L./ *torrentium* Martini	4	4	100.00
Coquillettidia richiardii Ficalbi	10	10	100.00
Anopheles maculipennis sl (Meigen)	22	9	40.91
Anopheles claviger (Meigen)	14	6	42.86

A recent SIT pilot study in Heidelberg, Germany, relied on sterile male *Ae. albopictus* produced in Italy and the transport time from the production unit to field release was 24 h by car (R. Bellini, personal communication). More recently, this sterile male transport is done by DHL delivery by air, shortening transport time and increasing reliability. Transport by air allows rapid long-distance transport between mosquito factory and the field release area.

Insects are poikilotherms and thus have about the same body temperature as their surrounding environment. This is reflected in slower activity at low temperature and increasing activity with rising temperature within certain temperature limits. Thus, it is possible to chill mosquitoes and thereby make them less active and less vulnerable to physical damage during transport. Since the chilling reduces all life processes, it allows for packing of very large number of insects in a small volume as long as they are in a chilled state.

The technique of chilling mosquitoes for long-distance transport has been tested for *Anopheles arabiensis*, *Ae. aegypti* and *Ae. albopictus* and the conclusion is that there is large variability in responses (Culbert et al. 2017, 2019). Apparently, the reaction to chilling needs to be established for each species and may even have to be evaluated for the specific population (Culbert et al. 2019). Although the northern floodwater mosquito *Ae. sticticus* is already used to an environment where immobilization by chilling is inevitable, there is an obvious need to test the species for optimal transportation temperature before practical use.

5.8. Male Mating Quality

Since mating success is central for the SIT and the IIT, there is a need to evaluate male mating quality on a regular basis. This is normally done using walk-in field cages to sufficiently frequently test for mating competition and success. This argues in the direction of creating large and efficient production units, supplemented with capacity for adequate male quality evaluation. Such facilities need capacity for timely delivery of quality sterile males for release.

5.9. Male Dispersal

Ground release of *Ae. sticticus* males in large and inaccessible natural wetland areas is not possible within a limited timeframe. Aerial release is the main alternative as it is rapid and allows high precision; it is also the least disturbing for vegetation and animals alike. The release should preferable be done by either helicopter or drone (unmanned aerial vehicle, see Benavente et al., this volume), but probably not by airplane, as the male mosquitoes are fragile and may become damaged if speed of the dispersing aircraft is too high.

The dispersal of *Ae. sticticus* males, in conjunction with mating, is important information when deciding on the male release strategy. Our preliminary observations provide evidence of *Ae. sticticus* mating in the same general area as the larval habitat. The synchronised hatching of *Ae. sticticus* eggs during a flood, the likewise synchronised development of larvae to pupal stage, and the emergence of males about one day before females provide opportunities for males to encounter emerging females without searching over extensive areas. Furthermore, the temporary flooded areas are very humid and thereby favourable environment for mosquitoes that are sensitive to desiccation.

Mating of *Ae. sticticus* was only observed in the shade under deciduous trees and bushes. The present level of biology knowledge indicates that the mating in this species occurs in the shaded terrestrial parts of the temporary flooded wet meadows and swamps. This information indicates that it is possible to develop a remote assessment method for locating actual mating areas. Release of sterile *Ae. sticticus* males could be concentrated in the defined mating environment and thus optimize impact in relation to time and costs.

5.10. Monitoring of Ae. sticticus *Males and Females*

The ability of released *Ae. sticticus* males to survive and disperse in the target areas, and the actual abundance of sterile males relative to native fertile males, are crucial to follow before, during and after the release. Discrimination of released sterile males from native males of the same species will require a marking system for released males, for example fluorescent powder or dye (Pal 1947; Verhulst et al. 2013). However, marking with fluorescent dye is not always recommended for mosquitoes as a coloured mosquito might have negative impact on the human population acceptance of the technique. Johnson et al. (2017) evaluated a new internal marking technique for mosquitoes using rhodamine B, showing that the marking remained after sugar-feeding and was visible for lifetime in *Ae. aegypti.* However, the authors recommend that small-scale mark-release-recapture experiments be performed to obtain more accurate estimates of male survival and mark persistence prior to adoption as an operational assessment (FAO/IAEA 2019).

Mosquito males are more difficult to collect than the females, and there is no trapping system available for specific sampling of *Ae. sticticus* males. As no obvious way of attracting and sampling males is available, we decided to test a more general strategy using the MosVac, a portable battery-operated, aspirator (Go Green Co., Ltd., Bangkok, Thailand).

The test was done in the Valmbäcken alder swamp, at the edge of the frequently flooded lake Färnebofjärden and in conjunction with a flood event, to make sure that males would occur in the study area. The test was carried out in June 2015, on the day we expected emergence of adult *Ae. sticticus* to commence. Only males were caught during the first sampling event, while on the following consecutive sampling days a mixture of males and females were collected (Fig. 7).

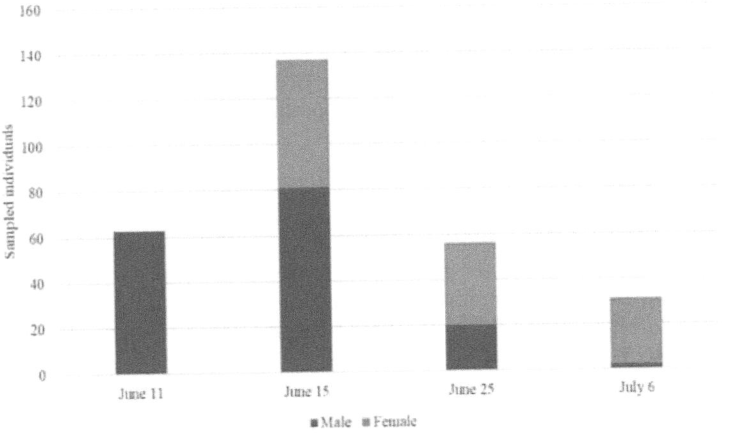

Figure 7. Abundance of male and female Aedes sticticus *measured by vacuuming grassy and bushy areas of the Valmbäcken alder swamp, River Dalälven floodplains, with a MosVac aspirator. The first sampling was done June 11, 2015, coinciding with the first day (D1) of adult emergence after the flood in May. Sampling was repeated June 15 (D5), June 25 (D15) and July 6 (D26).*

Only a few male *Ae. sticticus* were collected 26 days after initiation of emergence, when the male to female ratio was heavily distorted towards female dominance in the sample.

These preliminary data show that the MosVac aspirator was able to sample sufficient numbers of males for evaluating a pilot SIT study against *Ae. sticticus*. It also indicated that male abundance was reduced more rapidly than female abundance, with almost no males left approximately four weeks after adult emergence. Our intention is to carry out more detailed evaluations of male *Ae. sticticus* sampling strategy using MosVac aspirators, and other potential male sampling methods, to develop a standardized protocol for evaluation of sterile male releases.

Monitoring of female mosquito abundance is routinely done by Biologisk Myggkontroll (2019) in about 40 trap sites spread over the whole floodplain of the River Dalälven. Sampling is performed in all trap sites for one night every second week from spring to fall using CDC-traps baited with carbon dioxide as an attractant. This sampling will provide information on the relative abundance of the blood-seeking *Ae. sticticus* females that cause the nuisance.

5.11. Measuring the Abundance of Ae. sticticus *Larvae During Floods*

The *Ae. sticticus* larval abundance is routinely measured by dipping with a white plastic dipper on a long shaft. A large amount of data on larval abundance before and after each mosquito control operation in the River Dalälven floodplains are available since 2002. Such background data are very useful when evaluating the effect of the SIT on an *Ae. sticticus* population, because a subsequent reduction of larval abundance is expected if sufficient numbers of native females have mated with sterile males, resulting in infertile eggs.

Furthermore, as the egg bank of *Ae. sticticus* remains viable for many years (Gjullin et al. 1950), it will be useful for a SIT- or IIT-based intervention to continue measuring the larval abundance as a proxy for the abundance of fertile eggs. Declining abundance of larvae over the years will show a real population reduction. This will make it possible to observe if the actual population is reduced or even locally eliminated.

5.12. Potential IPM Strategy: Combination of Other Tools

The high population density of *Ae. sticticus* in the River Dalälven floodplains, without efficient control, is of a magnitude that would make it extremely costly and almost impossible to solely rely on SIT- or IIT-based control. However, the area-wide use of aerial dispersal of VectoBac G against the larvae is highly efficient and already induces about a 95%-99% reduction in blood-seeking female abundance (Schäfer and Lundström 2014). Such pre-treatment of the pilot area, reducing the target species population to a fraction, will make it possible to decrease the required number of sterile males to be released substantially, thereby boosting both the economics of the releases and their population suppression effect.

5.13. A Suggested Pilot Study of SIT or IIT Application Against Aedes sticticus

Before implementing new area-wide population suppression methods against *Ae. sticticus*, there is an obvious need for a pilot study of the techniques to be integrated against the target population in its natural setting. The results of the pilot study will provide guidance for evaluating efficacy and will also provide guidance on whether and how to proceed when expanding into AW-IPM programmes using the SIT or the IIT.

Two flood-prone and extremely productive areas for *Ae. sticticus* have been selected as suggested pilot study areas (Fig. 8). Former lakes Hallsjön and Karbosjön are located close to the village of Huddunge and have been subject to *Bti*-based mosquito control since 2005. All known floodwater mosquito breeding sites within flight distance are included in the routine *Bti*-treatments, thus the study areas are protected from massive reinvasion of *Ae. sticticus* and would function as isolated populations.

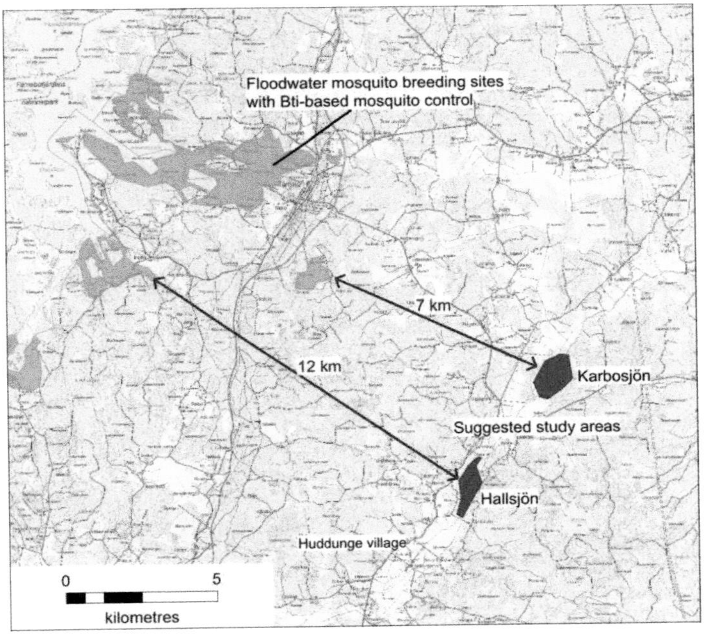

Figure 8. The two suggested areas, Hallsjön and Karbosjön for a pilot study of the SIT against the floodwater mosquito Aedes sticticus *are located northeast of Huddunge village and are isolated from breeding sites close to the River Dalälven. All breeding sites are included in routine* Bti-*based mosquito control which protects the study areas from massive reintroduction of* Ae. sticticus *females.*

The enormous mosquito nuisance problems around Lake Hallsjön were the motive to initiate mosquito control using VectoBac G spread from helicopter, as previously described. A first survey of mosquito abundance in the Hallsjön area was carried out in 2004, and in 2005 mosquito control was commenced. From 2005 onwards, the abundance of female *Ae. sticticus* is regularly monitored with CDC-trapping once every second week from May until September each year (Fig. 9).

The relative abundance of blood-seeking *Ae. sticticus* females before initiation of treatments was about 16 000 per CDC-trap and night. After several years of treatment with VectoBac G, the maximum number of female *Ae. sticticus* collected any time during summer is 11-44 per CDC-trap and night, representing a 99.97% reduction. This proves the excellent population suppression resulting from professional VectoBac G larviciding and confirms the almost total elimination of *Ae. sticticus* larvae as observed within 24 h after each treatment.

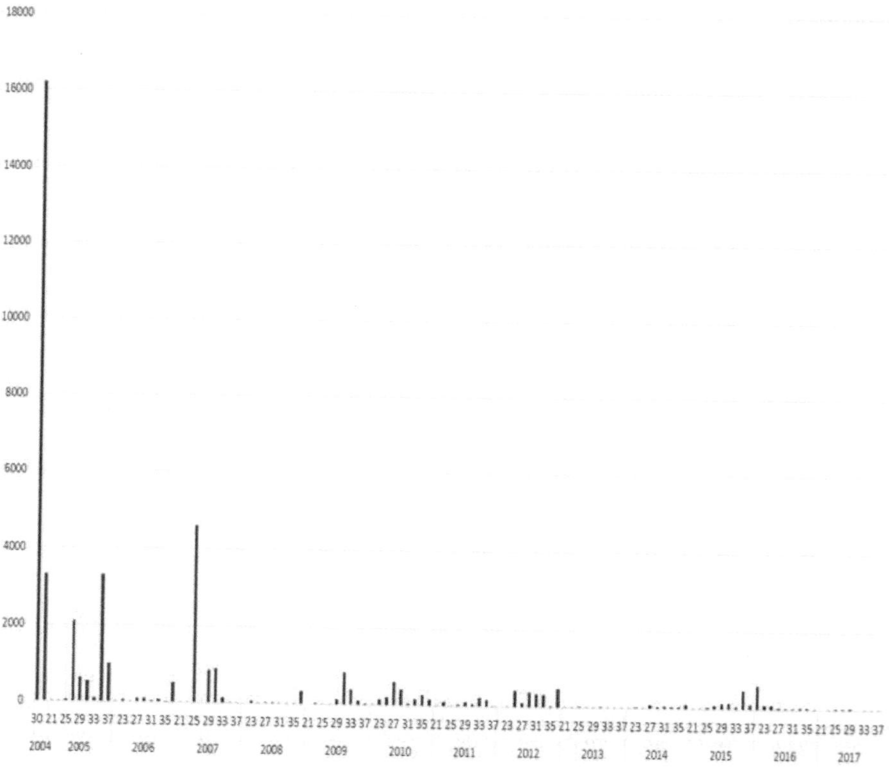

Figure 9. The abundance of blood-seeking Aedes sticticus *females at Hallsjön, central Sweden, for the years 2004 to 2017 as measured by fortnightly sampling with CDC miniature light traps baited with carbon dioxide. Mosquito control using VectoBac G commenced in 2005, inducing a very strong reduction in abundance, except for week 27 in 2007 when a mistake allowed massive mosquito emergence (data not shown).*

The size of the remaining population of surviving *Ae. sticticus* males and females after treatment is difficult to measure since there remain so few individuals and these are dispersed. However, the CDC-traps are positioned in areas with trees and bushes attracting the females produced in a much larger surrounding area of wet meadow estimated at 1-2 ha. If we use 44 females as a 10% fraction of actual numbers produced in 2 ha, the production could be supposed to be 220 females per productive ha. As males and females emerge in approximately the same abundance, this would provide us 220 males per productive ha. The total area of the two pilot areas is about 150 ha, and during a large flood it is estimated that about 75% of the area is producing *Ae. sticticus*. Thus, only 25 000 local fertile males remain in the study areas after a VectoBac G treatment. Based on the estimated abundance of local males, there will be a need for releasing approximately 250 000 sterile or incompatible males to obtain a sterile to wild male ratio of 10:1. As can be understood, these are very approximate estimates, although they probably catch the approximate general tendency.

The success of a SIT or IIT trial is crucially dependent on the release of a sufficient number of good quality sterile males to compete with the local fertile males for mating with the local females. Therefore, it might be useful to try other methods for population size estimates before the initial release of sterile or incompatible *Ae. sticticus* males in the two suggested pilot study areas.

6. PERSPECTIVES

The sterile male technique, either the SIT or the IIT, has potential to serve as an alternative solution to sole reliance on *Bti*-treatments against *Aedes sticticus,* but for an evaluation of the real potential, the suggested pilot test needs to be performed. In case of positive and encouraging results, new challenges arise.

Setting up the SIT or the IIT over the whole of the River Dalälven floodplains will require an integrated strategy with *Bti*-treatments for many years, thus the desired reduction of the control agent will not be achieved for quite some time. Sterile male release will have to start in defined subsets of the floodplains, for example in the easternmost lake system and then move westwards. There is also a risk that *Bti*-treatments will have to increase in the beginning since there are untreated areas in the current control programme. Complete area-wide coverage of breeding sites will be needed to ensure low mosquito population size and low risk for reinfestation. Thus, for approximately 10-15 years there will be a need for both large-scale *Bti*-treatments and sterile male releases. As a result of this intensive work, the use of the suppression agent may phase out completely, although it might be wise to keep the possibility of *Bti*-treatments as a backup. Release of sterile males will have to continue at a maintenance level since re-establishment of *Ae. sticticus* might occur.

In conclusion, integrating the sterile male technique into the management of a floodwater mosquito like *Ae. sticticus* means intensive work effort over many years, but the goal of an environmentally neutral mosquito control, eventually without using any suppression agent, is considered achievable. Nevertheless, this will require political decisions ensuring stakeholder commitment and the economic basis for such a project.

7. ACKNOWLEDGEMENTS

We would like to acknowledge the valuable comments and suggestions from two anonymous external reviewers, and the English language support provided by Christian Blue.

8. REFERENCES

Atyame, C. M, N. Pasteur, E. Dumas, P. Tortosa, M. Tantely, N. Pocquet, S. Licciardi, A. Bheecarry, B. Zumbo, M. Weill, and O. Duron. 2011. Cytoplasmic incompatability as a means to control *Culex pipiens quinquefasciatus* mosquito in the islands of the southwestern Indian Ocean. PLoS Neglected Tropical Diseases 5: e1440.

Atyame, C. M., J. Cattel, C. Lebon, O. Flores, J.-S. Dehecq, M. Weil, L. C. Gouagna, and P. Tortosa. 2015. Wolbachia-based population control strategy targetting *Culex quinquefasciatus* mosquitoes proves efficient under semi-field conditions. PLoS One 10(3): e01119288.

Bakri, A., K. Mehta, and D. R. Lance. 2021. Sterilizing insects with ionizing radiation, pp. 355–398. *In* V. A. Dyck, J. Hendrichs, and A. S. Robinson (eds.), Sterile Insect Technique – Principles and practice in Area-Wide Integrated Pest Management. Second Edition. CRC Press, Boca Raton, Florida, USA.

Becker, N. 1989. Life strategies of mosquitoes as an adaptation to their habitats. Bulletin of the Society for Vector Ecology 14: 6–25.

Becker, N. and H. W. Ludwig. 1981. Untersuchungen zur Faunistik und Ökologie der Stechmücken (Culicinae) und ihrer Pathogene im Oberrheingebiet. Mitteilungen der Deutschen Gesellschaft für allgemeine und angewandte Entomologie 2: 186–194.

Becker, N., D. Petric, M. Zgomba, C. Boase, M. Madon, C. Dahl, and A. Kaiser. 2010. Mosquitoes and their control, Second Edition. Springer Verlag, Berlin/Heidelberg, Germany. 577 pp.

Bellini, R., A. Medici, A. Puggioli, F. Balestrino, and M. Carrieri. 2013. Pilot field trials with *Aedes albopictus* irradiated sterile males in Italian urban areas. Journal of Medical Entomology 50: 317–325.

Benedict, M., and A. Robinson. 2003. The first release of transgenic mosquitoes: An argument for the Sterile Insect Technique. Trends in Parasitology 19: 349–355.

Biologisk Myggkontroll. 2019. Bekämpar översvämningsmyggor vid Nedre Dalälven.

Bourtzis, K. 2008. Wolbachia-based technologies for insect pest population control. Advances in Experimental Medicine and Biology 627: 104–113.

Bourtzis, K., R. S. Lees, J. Hendrichs, and M. J. B. Vreysen. 2016. More than one rabbit out of the hat: Radiation, transgenic and symbiont-based approaches for sustainable management of mosquito and tsetse fly populations. Acta Tropica 157: 115–130.

Brust, R. A. 1980. Dispersal behavior of adult *Aedes sticticus* and *Aedes vexans* (Diptera: Culicidae) in Manitoba. Canadian Entomologist 112: 31–42.

Caquet, T., M. Roucaute, P. Le Goff, and L. Lagadic. 2011. Effects of repeated field applications of two formulations of *Bacillus thuringiensis* var. israelensis on non-target saltmarsh invertebrates in Atlantic coastal wetlands. Ecotoxicology and Environmental Safety 74: 1122–1130.

Culbert, N. J., J. R. L. Gilles, and J. Bouyer. 2019. Investigating the impact of chilling temperature on male *Aedes aegypti* and *Aedes albopictus* survival. PLoS One 14: e0221822.

Culbert, N. J., R. S. Lees, M. J. B.Vreysen, A. C. Darby, and J. R. L. Gilles. 2017. Optimised conditions for handling and transport of male *Anopheles arabiensis*: Effects of low temperature, compaction, and ventilation on male quality. Entomologia Experimentalis et Applicata 164: 276–283.

Dame, D. A., C. F. Curtis, M. Q. Benedict, A. S. Robinson, and B. G. J. Knols. 2009. Historical applications of induced sterilisation in field populations of mosquitoes. Malaria Journal 8 (Suppl. 2): S2.

Dyck, V. A., J. Hendrichs, and A. S. Robinson (eds.). 2021. Sterile Insect Technique – Principles and practice in Area-Wide Integrated Pest Management. Second Edition. CRC Press, Boca Raton, Florida, USA. 1200 pp.

(EU) European Union. 2012. Regulation (EU) No 528/2012 of the European Parliament and of the Council of 22 May 2012 concerning the making available on the market and use of biocidal products. ISSN 1977. 2012. 677: 2985.

(FAO/IAEA). Food and Agriculture Organization of the United Nations/International Atomic Energy Agency. 2019. Guidelines for Mark-Release-Recapture procedures of *Aedes* mosquitoes. J. Bouyer, F. Balestrino, N. Culbert, M. Gómez, D. Petric, H. Yamada, R. Argilés, and R. Bellini (eds.). Vienna, Austria. 22 pp.

Fay, R., and H. Morlan. 1959. A mechanical device for separating the developmental stages, sexes and species of mosquitoes. Mosquito News 19:144–147.

Focks, D. A. 1980. An improved separator for the developmental stages, sexes, and species of mosquitoes (Diptera: Culicidae). Journal of Medical Entomology 17: 567–568.

Gjullin, C. M., W. W. Yates, and H. H. Stage. 1950. Studies on *Aedes vexans* (Meig.) and *Aedes sticticus* (Meig.), flood-water mosquitoes, in the lower Columbia River Valley. Annals of the Entomological Society of America 43: 262–275.

Hedström-Ringvall, A., C. Kjörk, K. Pettersson, M. Engström, D. Wisaeus, N. Hjerdt, J. Berglund, and P.-E. Sandberg. 2017. Ekologiskt anpassad årsreglering av Dalälven. 2017 Länsstyrelsen Dalarna Rapport 09: 1–170.

Horsfall, W. R., and M. Trpis. 1967. Eggs of floodwater mosquitoes. X. Conditioning and hatching of winterized eggs of *Aedes sticticus* (Diptera: Culicidae). Annals of the Entomological Society of America 60: 1021–1025.

Horsfall, W. R., H. W. J. Fowler, L. J. Moretti, and J. R. Larsen. 1973. Bionomics and embryology of the inland floodwater mosquito *Aedes vexans*. University of Illinois Press, Champaign, USA. 211 pp.

Huang, Y.-J. S., S. Higgs, and D. L. Vanlandingham. 2017. Review: Biological control strategies for mosquito vectors of arboviruses. Insects 8: 21.

Jaenson, T. G. T. 1986. Massförekomst av *Aedes rossicus* och andra stickmyggor vid Dalälven hösten 1985. Entomologisk Tidskrift 107: 51–52.

Johnson, B. J., S. N. Mitchell, C. J. Paton, J. Stevenson, K. M. Staunton, N. Snoad, N. Beebe, B. J. White, and S. A. Ritchie. 2017. Use of rhodamine B to mark the body and seminal fluid of male *Aedes aegypti* for mark-release-recapture experiments and estimated efficacy of sterile male releases. PLoS Neglected Tropical Diseases 11: e0005902.

Kittayapong, P., K. J. Baisley, V. Baimai, and S. L. O'Neill. 2000. The distribution and diversity of *Wolbachia* infections in Southeast Asian mosquitoes. Journal of Medical Entomology 37: 340–345.

Kittayapong, P., P. Mongkalangoon, V. Baimai, and S. L. O'Neill. 2002. Host age effects and expression of cytoplasmic incompatibility in field populations of the Wolbachia-superinfected *Aedes albopictus*. Heredity 88: 270–274.

Kittayapong, P., S. Ninphanomchai, W, Limohpasmanee, C. Chansang, U. Chansang, and P. Mongkalangoon. 2019. Combined Sterile Insect Technique and Incompatible Insect Technique: The first proof-of-concept to suppress *Aedes aegypti* vector populations in semi-rural settings in Thailand. PLoS Neglected Tropical Diseases 13(10): e0007771.

Knipling, E. 1955. Possibilities of insect control or eradication through use of sexually sterile males. Journal of Economic Entomology 48: 459–462.

Knipling, E. 1979. The basic principles of insect population suppression and management. USDA Agriculture Handbook No 512. Washington, DC, USA. 659 pp.

Knipling, E. 1998. Role of parasitoid augmentation and Sterile Insect Technique for area-wide management of agricultural insect pests. Journal of Agricultural Entomology 15: 273–301.

Krafsur, E. 1998. Sterile Insect Technique for suppressing and eradicating insect populations: 55 years and counting. Journal of Agricultural Entomology 15: 303–317.

Kuhn, R. 2002. Colonisation of the floodwater mosquito *Aedes vexans* (Meigen) (Diptera: Culicidae). European Mosquito Bulletin 12: 7–16.

Lagadic, L., M. Roucaute, and T. Caquet. 2013. *Bti* sprays do not adversely affect non-target aquatic invertebrates in French Atlantic coastal wetlands. Journal of Applied Ecology 51: 102–113.

Lagadic, L., R. B. Schäfer, M. Roucaute, E, Szös, S. Chouin, J. de Maupeou, C. Duchet, E. Franquet, B. le Hunsec, C. Bertrand, S. Fayolle, B. Frances, Y. Rozier, R. Foussadier, J.-B. Santoni, and C. Lagneau. 2016. No association between the use of *Bti* for mosquito control and the dynamics of non-target aquatic invertebrates in French coastal and continental wetlands. Science of the Total Environment 553: 486–494.

Laven, H. 1967. Eradication of *Culex pipiens fatigans* through cytoplasmic incompatibility. Nature 216: 383–384.

Lees, R. S., D. O. Carvalho, and J. Bouyer. 2021. Potential impact of integrating the Sterile Insect Technique into the fight against disease-transmitting mosquitoes, pp. 1081–1118. *In* V. A. Dyck, J. Hendrichs, and A. S. Robinson (eds.), Sterile Insect Technique – Principles and practice in Area-Wide Integrated Pest Management. Second Edition. CRC Press, Boca Raton, Florida, USA.

Lees, R. S., B. Knols, R. Bellini, M. Q. Benedict, A. Bheecarry, H. C. Bossin, D. E. Chadee, J. Charlwood, R. K. Dabire, L. Djogbenou, A. Egyir-Yawson, R. Gato, L. C. Gougna, M. M. Hassan, S. A. Khan, L. L. Koekermoer, G. Lemperiere, N. C. Manoukis, R. Mozuraitis, R. J. Pitts, F. Simard, and J. R. L. Gilles. 2014. Review: Improving our knowledge of male biology in relation to genetic control programmes. Acta Tropica 132S: S2–S11.

Lindquist, D. A., M. Abusowa, and M. J. Hall. 1992. The New World screwworm fly in Libya: A review of its introduction and eradication. Medical and Veterinary Entomology 6: 2–8.

Lofgren, C. S., D. A. Dame, S. G. Breeland, D. E. Weidhaas, G. M Jeffery, R. Kaiser, H. R. Ford, M. D. Boston, and K. F. Baldwin. 1974. Release of chemosterilized males for the control of *Anopheles albimanus* in El Salvador. III. Field methods and population control. American Journal of Tropical Medicine and Hygiene 23: 288–297.

Lundqvist, A.-C., M. Widemo, and I. Lindquist. 2013. Förslag till hur myggproblemet vid Nedre Dalälven kan hanteras på lång sikt. Redovisning av ett regleringsbrevsuppdrag. Länsstyrelsen Gävleborg Rapport 500-8033-13.

Lundström, J. O., B. Niklasson, and B. D. Francy. 1990. Swedish *Culex torrentium* and *Cx. pipiens* (Diptera: Culicidae) as experimental vectors of Ockelbo virus. Journal of Medical Entomology 27: 561–563.

Lundström, J. O., J. Chirico, A. Folke, and C. Dahl. 1996. Vertical distribution of adult mosquitoes (Diptera: Culicidae) in southern and central Sweden. Journal of Vector Ecology 21: 159–166.

Lundström, J. O., Y. Brodin, M. L. Schäfer, T. Z. P. Vinnersten, and O. Ostman. 2010a. High species richness of Chironomidae (Diptera) in temporary flooded wetlands associated with high species turn-over rates. Bulletin of Entomological Research 100: 433–444.

Lundström, J. O., M. L. Schäfer, E. Petersson, T. Z. P. Vinnersten, J. Landin, and Y. Brodin. 2010b. Production of wetland Chironomidae (Diptera) and the effects of using *Bacillus thuringiensis israelensis* for mosquito control. Bulletin of Entomological Research 100: 117–125.

Mains, J. W., C. L. Brelsfoard, R. I. Rose, and S. L. Dobson. 2016. Female adult *Aedes albopictus* suppression by *Wolbachia*-infected male mosquitoes. Nature Scientific Reports 6: 33846.

Merdic, E., and T. Lovakovic. 2001. Population dynamic of *Aedes vexans* and *Ochlerotatus sticticus* in flooded areas of the River Drava in Osijek, Croatia. Journal of the American Mosquito Control Association 17: 275–280.

Minar, J., I. Gelbic, and J. Olejnicek. 2001. The effect of floods on the development of mosquito populations in the middle and lower river Morava Regions. Acta Universitatis Carolinae Biologica 45: 139–146.

Moretti, R., and M. Calvitti. 2012. Male mating performance and cytoplasmic incompatibility in a wPip Wolbachia trans-infected line of *Aedes albopictus* (*Stegomyia albopicta*). Medical and Veterinary Entomology 27: 377–386.

O´Connor, L., C. Plichart, A. Cheong Sang, C. L. Brelsfoard, H. C. Bossin, and S. L. Dobson. 2012. Open release of male mosquitoes infected with a *Wolbachia* biopesticide: Field performance and infection containment. PLoS Neglected Tropical Diseases 6: e1797.

Östman, Ö. 2013. Hävdens betydelse för mängden översvämningsmyggor – del 2. Länsstyrelsen Gävleborg Rapport 24: 1–13.

Östman, Ö., Å. Wengström, U. Gradin, J. Wissman, M. Schäfer, and J. O. Lundström. 2015. Lower abundance of flood water mosquito larvae in managed wet meadows in the lower Dalälven floodplains, Sweden. Wetlands Ecology and Management 23: 257–267.

Patterson, R. S., D. E. Weidhaas, H. R. Ford, and C. S. Lofgren. 1970. Suppression and elimination of an island population of *Culex pipiens quinquefasciatus* with sterile males. Science 168: 1368–1369.

Pal, R. 1947. Marking mosquitoes with fluorescent compounds and watching them by ultra-violet light. Nature 160: 298–299.

Rasić, G., N. M. Endersby, C. Williams, and A. A. Hoffmann. 2014. Using *Wolbachia*-based release for suppression of *Aedes* mosquitoes: Insights from genetic data and population simulations. Ecological Application 24: 1226–1234.

Ritchie, S. A., M. Townsend, C. J. Paton, A. G. Callahan, and A. A. Hoffmann. 2015. Application of *w*MelPop *Wolbachia* strain to crash local populations of *Aedes aegypti*. PLoS Neglected Tropical Diseases 9: e0003930.

Schäfer, M. L., and J. O. Lundström. 2001. Comparison of mosquito (Diptera: Culicidae) fauna characteristics of forested wetlands in Sweden. Annals of the Entomological Society of America 94: 576–582.

Schäfer, M. L., and J. O. Lundström. 2014. Efficiency of *Bti*-based floodwater mosquito control in Sweden – Four examples. Journal of the European Mosquito Control Association 32: 1–8.

Schäfer, M. L., J. O. Lundström, and E. Petersson. 2008. Comparison of mosquito (Diptera: Culicidae) populations by wetlands type and year in the lower River Dalälven region, Central Sweden. Journal of Vector Ecology 33: 150–157.

Schäfer, M. L., Wahlqvist P., and J. O. Lundström. 2018. The Nedre Dalälven River landscape in central Sweden - A hot-spot for mosquito (Diptera: Culicidae) diversity. Journal of the European Mosquito Control Association 36: 17–22.

Sharma, V. P., R. S. Patterson, and H. R. Ford. 1972. A device for the rapid separation of male and female pupae. Bulletin of the WHO 47: 429–432.

Strugarek, M., H. Bossin, and Y. Dumont. 2019. On the use of the Sterile Insect Technique or the incompatible insect technique to reduce or eliminate mosquito populations. Applied Mathematical Modelling 68: 443–470.

Sudarić Bogojević, M., E. Merdić, and T. Bogdanović. 2011. The flight distances of floodwater mosquitoes (*Aedes vexans*, *Ochlerotatus sticticus* and *Ochlerotatus caspius*) in Osijek, Eastern Croatia. Biologia (Bratislava) 66: 678–683.

Trpis, M., and W. R. Horsfall. 1967. Eggs of floodwater mosquitoes (Diptera: Culicidae). XI. Effect of medium on hatching of *Aedes sticticus*. Annals of the Entomological Society of America 60: 1150–1152.

Trpis, M., and W. R. Horsfall. 1969. Development of *Aedes sticticus* (Meigen) in relation to temperature, diet, density and depth. Annales Zoologici Fennici 6:156–160.

Trpis, M., and J. A. Shemanchuk. 1970. Effect of constant temperature on the larval development of *Aedes vexans* (Diptera: Culicidae). Canadian Entomologist 102: 1048–1051.

(US-EPA) US Environmental Protection Agency. 2017. Final registration decision on the new active ingredient *Wolbachia pipientis* ZAP (*w*Pip) strain in *Aedes albopictus*. PC Code: 069035. EPA Reg. Number: 89668-4. Document EPA-HQ-OPP-2016-0205-0034. Washington, DC, USA.

Verhulst, N. O., J. A. C. M. Loonen, and W. Takken. 2013. Advances in methods for colour marking of mosquitoes. Parasites & Vectors 6: 200.

Vreysen, M. J. B., J. Gerado-Abaya, and J. P. Cayol. 2007. Lessons from Area-Wide Integrated Pest Management (AW-IPM) programmes with an SIT component: An FAO/IAEA perspective, pp 723–744. *In* M. J. B. Vreysen, A. S. Robinson, and J. Hendrichs (eds.), Area-wide control of insect pests: From research to field implementation. Springer, Dordrecht, The Netherlands.

Waltz, E. 2017. US government approves 'killer' mosquitoes to fight disease. US Environmental Protection Agency will allow release of insects in 20 states and Washington DC. 06 November 2017. Nature News.

Werren, J. H., L. Baldo, and M. E. Clark. 2008. *Wolbachia*: Master manipulators of invertebrate biology. Nature Reviews Microbiology 6: 741–751.

Wilson, G. R., and W. R. Horsfall. 1970. Eggs of floodwater mosquitoes. XII. Installment hatching of *Aedes vexans*. Annals of the Entomological Society of America 64: 1644–1647.

Yamada, H., A. G. Parker, C. F. Oliva, F. Balestrino, and J. R. L. Gilles. 2014. X-ray-induced sterility in *Aedes albopictus* (Diptera: Culicidae) and male longevity following irradiation. Journal of Medical Entomology 51: 811–816.

Zhang, D., R. S. Lees, Z. Xi, K. Bourtzis, and J. R. L. Gilles. 2016. Combining the Sterile Insect Technique with the Incompatible Insect Technique: III – Robust mating competitiveness of irradiated triple *Wolbachia*-infected *Aedes albopictus* males under semi-field conditions. PLoS One 11(3): e0151864.

Zheng, X., D. Zhang, Y. Li, C. Yang, Y. Wu, X. Liang, Y. Liang, X. Pan, L. Hu, Q. Sun, X. Wang, Y. Wei, J. Zhu, W. Qian, Z. Yan, A. G. Parker, J. R. L. Gilles, K. Bourtzis, J. Bouyer, M. Tang, B. Zheng, J. Yu, J. Liu, J. Zhuang, Z. Hu, M. Zhang, J. T. Gong, X. Y. Hong, Z. Zhang, L. Lin, Q. Liu, Z. Hu, Z. Wu, L. A. Baton, A. A. Hoffmann, and Z. Xi. 2019. Incompatible and sterile insect techniques combined eliminate mosquitoes. Nature 572: 56–61.

SECTION 3

CLIMATE CHANGE, GLOBAL TRADE AND INVASIVE SPECIES

BUFFALO FLIES (*Haematobia exigua*) EXPANDING THEIR RANGE IN AUSTRALIA FACILITATED BY CLIMATE CHANGE: THE OPPORTUNITY FOR AREA-WIDE CONTROLS

P. J. JAMES[1], M. MADHAV[1] AND G. BROWN[2]

[1]*University of Queensland, St Lucia, Queensland, Australia; p.james1@uq.edu.au*
[2]*Department of Agriculture and Fisheries Queensland, Queensland, Australia*

SUMMARY

Buffalo flies *Haematobia exigua* de Meijere were introduced to Australia in 1838 and have become major cattle pests in Australia's northern cattle industries. They have been steadily expanding their range southward and their spread is likely to be further facilitated by climate change. Control programmes consisting of compulsory chemical treatments and regulated cattle movements have proven unsuccessful in preventing the spread of buffalo flies and, without area-wide intervention, they are likely to become major cattle pests in Australia's southern beef and dairy industries. Buffalo flies do not have a pupal overwintering strategy but survive winter in localised foci of slowly cycling low level populations of flies. Populations increase and spread to infest surrounding areas when weather becomes favourable in summer. This suggests the potential for an area-wide control approach, targeting overwintering foci of the flies. A project has been initiated to transinfect buffalo flies with *Wolbachia,* determine the effects of *Wolbachia* infection in the flies, and assess the feasibility of control by *Wolbachia*-based approaches directly targeting overwintering foci of the flies.

Key Words: Bos indicus, Bos taurus, Bubalus bubalis, Chrysomya bezziana, Haematobia irritans, Lucilia cuprina, Stephanofilaria, Wolbachia, Muscidae, biological control, cattle, horn flies, invasive species

1. BUFFALO FLIES AND HORN FLIES

Buffalo flies *Haematobia exigua* de Meijere and horn flies *Haematobia irritans* (L.) are obligate parasites, living most of their lives on cattle, and leaving only to oviposit when cattle defecate. Both the male and the female subsist completely on blood, using their sharp mouthparts to pierce the animal's skin. If uncontrolled, infestations may reach several thousand flies per animal, each feeding up to 40 times daily, irritating cattle and causing production loss and welfare impacts.

J. Hendrichs, R. Pereira and M. J. B. Vreysen (eds.), Area-Wide Integrated Pest Management: Development and Field Application, pp. 463–482. CRC Press, Boca Raton, Florida, USA.

Horn flies have been estimated to cost the North American and Brazilian cattle industries close to USD 1000 million (Cupp et al. 1998) and USD 2540 million (Grisi et al. 2014) per annum respectively, whereas buffalo flies are estimated to cost the Australian beef cattle industry AUSD 99 million annually with losses presently confined mainly to the northern part of the country (Lane et al. 2015).

Buffalo fly feeding can lead to the development of lesions that are of significant welfare concern (Jonsson and Matchoss 1998), reduce hide value and make cattle less acceptable for the market (Guglielmone et al. 1999; Lane et al. 2015). These lesions can range in nature from dry and alopecic or scab encrusted, to severe open areas of ulceration (Johnson 1989). They are found most commonly beneath the eyes of cattle, but can also be prevalent on the neck, dewlap, belly and flanks (Sutherst et al. 2006) and their development and persistence has been associated with a currently unnamed species of filarial nematode (*Stephanofilaria* sp.), transmitted by buffalo flies (Johnson 1989; Shaw and Sutherland 2006). Although similar lesions are found associated with horn fly feeding, they are mainly abdominal in distribution and generally not nearly as severe as those associated with buffalo flies (Hibler 1966; Silva et al. 2010).

Skin lesions are most widespread in northern areas of Australia, where buffalo flies are present throughout the year, with up to 95% of cattle affected (Johnson et al. 1986). It is expected that the prevalence of lesions will increase in more southern parts of the fly range as global warming extends the length of the buffalo fly season and the intensity of fly attack. In a survey of Queensland dairy producers, buffalo flies were considered to be a greater problem for production than cattle ticks, and when asked what aspect of infestation concerned them most, 42% of producers noted the welfare effects (Jonsson and Matchoss 1998). In addition, the lesions present a potential focus for strikes by Old World screwworm *Chrysomya bezziana* (Villeneuve) flies, which are endemic in a number of Australia's nearest northern neighbouring countries and which are considered a major biosecurity risk for northern Australia (AHA 2017).

2. TAXONOMIC STATUS

Buffalo flies and horn flies are very closely related and have variously been considered as sub-species (Zumpt 1973; Pont 1973) and separate species (Skidmore 1985). The larval stages of the two species are extremely similar and are probably morphologically indistinguishable (Pont 1973). Morphological differentiation of the adults is also difficult, and the main distinguishing feature in the flies is the presence of 4 to 6 long curled hairs on the hind tarsi of male *H. exigua*, which are not found in *H. irritans* (Mackerras 1933; Iwasa and Ishiguro 2010). Kano et al. (1972) suggested a number of other morphological distinguishing features, but Iwasa and Ishiguro (2010) indicated that these varied with latitudinal gradient within each species. Snyder (1965), who studied many specimens from Micronesia, stated that even the bristling on the hind tarsi of *H. exigua* is variable (Zumpt 1973).

Urech et al. (2005), who measured buffalo flies cultured on cattle in Australia and horn flies from a laboratory colony in Florida, indicated that there were distinct differences in the cuticular hydrocarbons of the two groups of flies and suggested that these differences may support their status as separate species.

Iwasa and Ishiguro (2010) examined the mtDNA in the COI to COII genes of horn flies collected from two sites in Japan and buffalo flies from sites in Taiwan and Viet Nam and found sequence divergence of 1.8% - 1.9% between the two species. They concluded that the relative genetic divergence observed between and within the two species may indicate an intermediate status in species development. From a more recent study of molecular differentiation of the two species, using the mtDNA COI, cytochrome B (Cytb), NADH dehydrogenase subunit 5 (ND5) genes, and the nuclear and 18S and 28S ribosomal RNA regions, Low et al. (2014, 2017) concluded that the two species are genetically distinct and that the COI and Cytb genes were the most informative for distinguishing the two species. Regardless of whether or not buffalo and horn flies can be considered separate species, all indications are that they are extremely closely related.

3. INVASION AND DISTRIBUTION

Both buffalo and horn flies have proven to be extremely invasive species. Either *H. irritans* or *H. exigua* is now present in most major cattle production areas of the world, with the exception of sub-Saharan Africa, where the species *Haematobia thirouxi potans* (Bezzi) and *Haematobia minuta* (Bezzi) occupy this niche (Zumpt 1973) (Fig. 1).

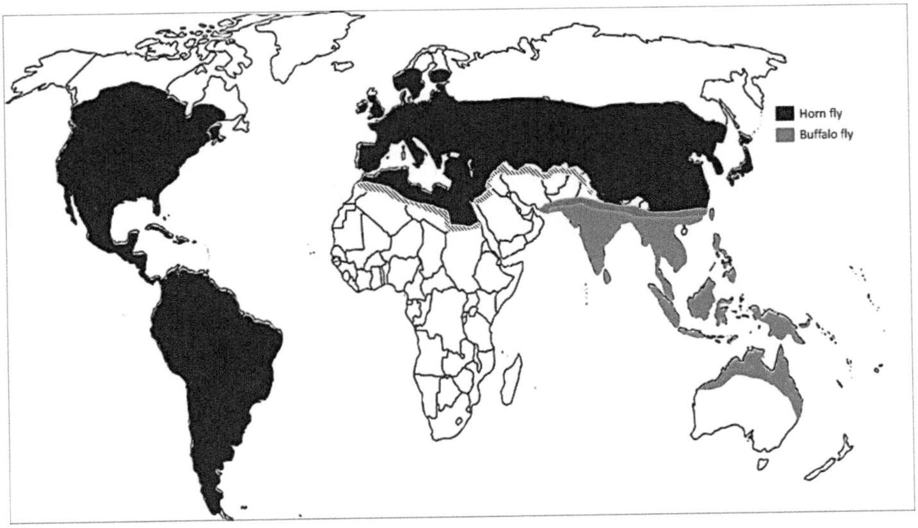

Figure 1. World distribution of horn fly H. irritans *and buffalo fly* H. exigua.

Horn flies were introduced to the east-coast of North America from Europe on imported cattle in 1885-86 (Butler and Okine 1999). They spread rapidly to reach California by 1893 and by 1900 had been reported from most of the USA, Canada and Puerto Rico. They were reported in South America from cattle in Colombia, Ecuador and Venezuela by 1937 (Mancebo et al. 2001), and were first reported in Brazil's northern-most State, Roraima in 1956.

By 1980, horn flies had spread south of the Amazon into Goiás state (Mancebo et al. 2001), by 1991 had reached the south of Brazil (Mancebo et al. 2001), and by 1993 had spread through Uruguay and all of the major cattle production areas in Argentina (22°S to 44°S) (Anziani et al. 1993; Guglielmone 1999).

On the western side of the continent horn flies were found in Bolivia before 1955 (Munro 1960), in Peru by at the latest 1973 (Zumpt 1973), and in Chile in 1967 (Gonzalez 1967), although it appears that they did not become a significant pest in Chile until 1993 (Campano and Avalos 1994). These records suggest that the southerly spread of horn flies in South America may have occurred independently on both the west and east sides of the continent.

3.1. Spread of Buffalo Flies in Australia

Buffalo flies have been similarly invasive in Australia, although their spread has occurred more slowly than for horn flies in the Americas, and has been limited at its southern extent by the inability of buffalo flies to undergo a winter pupal dormant phase, as occurs in horn flies (Ferrar 1969; Cook and Spain 1982). Buffalo flies entered mainland Australia near Darwin (12.5° S, 130.8° E) in 1838, probably on water buffalos (*Bubalus bubalis* L. 1758) introduced from Timor in 1825 (Tillyard 1931). Early spread occurred very slowly and coincided closely with the spread of buffalos (Hill 1917), which appear to be the preferred native host of the flies in Asia (Iwasa and Ishigura 2010) (Fig. 2).

It wasn't until 1928 that buffalo flies reached the Queensland border, approximately 1300 km southeast of their original point of introduction (Seddon 1967), subsequently spreading across the dry stretch of land south of the Gulf of Carpentaria to eastern Queensland during a series of wet years in 1939-41. From there, they spread rapidly to the east coast of Cape York in northern Australia and southwards along the eastern coast until they appeared to reach a southerly limit just north of Bundaberg (24.8°S latitude) by 1946. Here the spread paused, and no further southerly spread was observed for the next 30 years (Fig. 2).

Following a series of mild winters and wet years from 1973 onwards, changes to buffalo flies and tick regulatory programmes, possibly aided by changes in the chemicals used for cattle tick *Rhipicephalus australis* Fuller treatment, southerly range expansion recommenced and buffalo flies reached the Brisbane Valley and Nambour in 1977, the Tweed Valley in New South Wales in April 1978, and Bonville, south of Coffs harbour (30.4°S) in 1982 (Williams et al. 1985). Since then, the flies have continued their southerly spread with infestations seen as far south as Dubbo, Narromine and Maitland (32.7°S) in 2011 (Fig. 2). This represents an increase in their southerly range of approximately 1000 km in the last 40 years.

Following their first detection in New South Wales in 1978, the flies now survive the winter in many eastern parts of the state and have become a significant endemic cattle pest in these areas.

The impact of buffalo flies and the area affected in Australia varies significantly with season and weather conditions. During warm wet summers the distribution of the flies increases significantly in northern and north-eastern areas, and they may spread to affect cattle in an area potentially more than two times larger than the permanently infested range (Fig. 2).

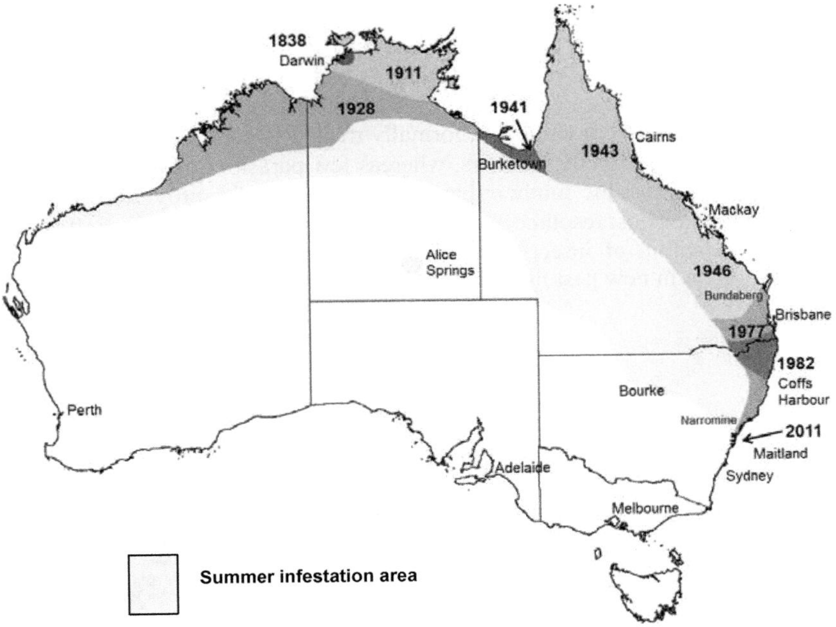

Figure 2: Spread over time of the buffalo fly H. exigua *in Australia.*

3.2. Effects of Climate Change

All indications to date suggest that climate change effects in Australia will facilitate the continued spread of buffalo flies into new areas and will increase the economic and welfare impacts in the southern parts of their current range (CSIRO/BOM 2016). Rising temperatures will enable more rapid *H. exigua* population growth, an increased number of generations each year, greater fly activity in many areas and longer seasons of cattle challenge. In addition, predicted rises in minimum temperatures and a reduction in the frequency of frosts will favour survival in marginal areas and further southerly extension of the flies' range. A possible increase in the summer incidence of rainfall in some areas of Australia may also favour the flies' breeding.

The results of CLIMEX modelling (R. Dobson personal communication 2015) suggest greater impacts from buffalo flies in the southern parts of their current range, including the potential for persisting fly populations to establish through most of the moist coastal belt of New South Wales and in foci as far south as South Australia and southern Western Australia (Fig. 3). In addition, increased weather variability and extreme rainfall events predicted under climate change may assist the spread of flies across inhospitable areas to new foci suitable for winter fly survival. Once established in these areas, new overwintering foci would provide a source for more extensive incursions during warm wet periods, similar to that seen in northern Australia.

The CLIMEX modelling does not account for factors such as a changing resource base, microclimate effects or changes in pest biology. In southern areas, the cattle industry is based largely on *Bos taurus* L. breeds that are more susceptible to buffalo flies than the *Bos indicus* L. cattle that predominate in northern areas (Frisch et al. 2000). In addition, northern cattle are normally treated to control cattle ticks, which can also impact on buffalo fly numbers, whereas few parasite treatments are applied to southern cattle. Thus, the southern beef and dairy industries provide a susceptible and largely untreated host resource extremely favourable for invasion by buffalo flies. Furthermore, adaption of insects at the edge of their range can be an important contributing factor in new pest invasions (Hill et al. 2011).

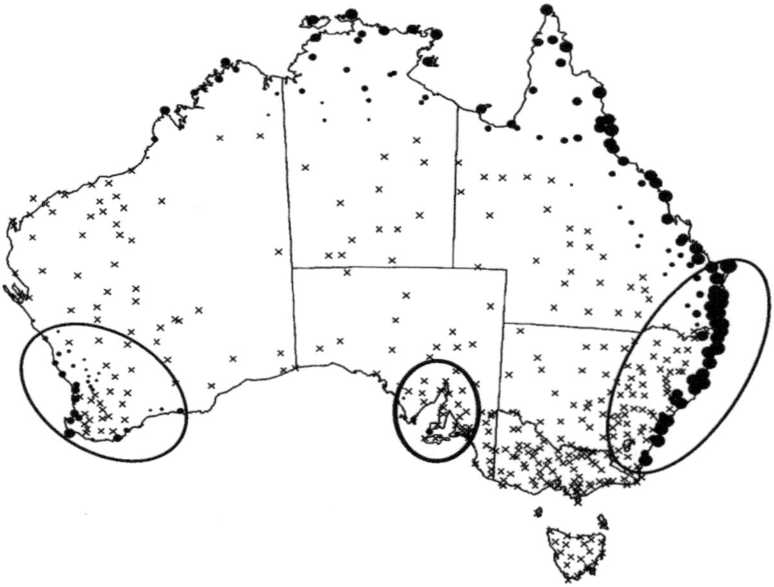

Figure 3. CLIMEX predictions of areas suitable for the establishment of persisting buffalo fly H. exigua *populations under predicted climate change. Size of solid circles indicates degree of favourability of areas for buffalo fly persistence; crosses indicate weather station sites not suitable for buffalo flies' persistence; the large open ellipses indicate areas of most significant range expansion (credit Rob Dobson).*

The degree to which genetic adaptation of buffalo flies to cooler temperatures has contributed to their southerly spread is uncertain. However, Iwasa and Ishigura (2010) note that the flies in their native range appear to prefer buffalo to cattle as hosts, and a period of adaptation to cattle and cooler Australian conditions may have contributed to their spread. Development of pupal overwintering capacity as they move south, a possibility given their close genetic relatedness with horn flies, is a concerning prospect and could see the species develop a temperate distribution in Australia, similar to the distribution seen for horn flies in the northern hemisphere and South America.

4. POTENTIAL FOR THE USE OF AREA-WIDE APPROACHES AGAINST BUFFALO FLIES IN AUSTRALIA

Currently, control of buffalo flies in Australia depends largely on chemical treatments, although techniques such as buffalo fly traps (Sutherst and Tozer 1995) and selection of more tolerant *Bos indicus* breeds (Frisch et al. 2000) are also used. In addition, dung beetles may assist the regulation of buffalo fly populations under some circumstances (Doube 1986). Treatments are applied almost exclusively on a herd-by-herd basis. However, modelling studies indicate the likely inefficiency of herd-by-herd approaches showing that the effects of invasion of pests from untreated areas can be devastating in compromising the effectiveness of control measures (Knipling 1972a). Application of control techniques on an area-wide basis, targeting the entire population rather than just individual properties or herds, can be much more efficient than more intensive programmes applied on a herd-by-herd basis. Area-wide approaches are expected to be particularly advantageous when pests are mobile and can readily auto-disseminate and therefore may not be easily controlled by property-based or herd-based programmes (Hendrichs et al. 2007), such as is the case with buffalo flies.

4.1. Chemical-based Programmes

Area-wide control programmes have historically been based mainly on the application of chemical insecticides by methods such as aerial spraying or intensive ground spraying, or in the case of diseases of livestock, by individual animal or herd treatments with quarantine controls and movement restrictions (Graham and Hourigan 1977).

At various stages in the spread of buffalo flies in Australia, regulatory programmes, supported by legislation, and which included movement controls and compulsory spraying of relocated cattle with insecticides, were used in an attempt to prevent their southerly incursion (Parliament of Queensland 1965). However, these programmes were not effective in stemming the southward spread of buffalo flies (Anonymous 1934; Roberts 1946; Eastaway 1974) and all were eventually abandoned in 1978 (Williams et al. 1985). However, buffalo flies remain a legislatively specified notifiable disease in some southern states of Australia, where the flies are not currently present (DAWR 2017).

4.2. Autocidal and Biologically-based Approaches

Programmes which require widespread application of insecticides are increasingly unacceptable on a community basis and can be compromised by the development of resistance or resurgence of pests from cattle that are not treated or where treatments are poorly applied. More biologically-based, species-specific and environment-friendly techniques which operate by disrupting biological processes of pests, generally find wider community acceptance and are often more effective than insecticide applications (Bourtzis et al. 2016). In addition, because of the ability of released insects to disperse into all areas occupied by the target field population and to actively search out and mate with target insects, biologically-based methods are often more effective against pests that can disperse autonomously, or which survive in cryptic habitats that are hard to reach with chemical sprays.

The most well-known of these approaches is the Sterile Insect Technique (SIT) in which insects of the target species are mass-reared and sterilised using low level ionizing radiation, followed by inundative releases of the sterilised insects (usually males) on an area-wide basis over the entire area of the target population (Vreysen and Robinson 2011; Dyck et al. 2021). The sterile males mate with field females, which consequently produce infertile eggs and through sequential releases the target population is suppressed, or under certain favourable conditions, eradicated.

Some of the most significant successes with the SIT have involved insect pests of livestock, including the eradication of New World screwworm *Cochliomyia hominivorax* (Coquerel) from North and Central America (Wyss 2000), eradication of an incursion of this pest in Libya (Lindquist et al. 1992), and removal of the tsetse fly *Glossina austeni* (Newstead) from the Island of Unguja in the Zanzibar archipelago (Vreysen et al. 2000, 2014). Localised eradication or suppression using the SIT has also been achieved on a number of occasions with other tsetse species in Africa (Vreysen et al. 2013). Successful eradication using this approach can be extremely cost efficient. For example, in the New World screwworm programme in the Americas it has been calculated that the direct benefits achieved each year from the programme are equal to or greater than the total cost of the sterile male release programme over the fifty years of its operation (Vreysen and Robinson 2011).

Horn flies are one of the species suggested by Knipling (1972b) as likely candidates for control by the SIT. Knipling considered that the close association of the flies with cattle and its consequent accessibility to control meant that fly populations could be readily reduced by insecticide treatment of cattle, then the remaining population eliminated using the SIT. In early trials with horn flies, cattle were sprayed with topical insecticides to reduce fly numbers. However, the subsequent sterile insect releases were compromised because the released flies were more susceptible to the insecticides used than were the field flies (Eschle et al. 1973, 1977). This was overcome by using methoprene, an insect growth regulator administered in drinking water, which targeted the larval stages of horn flies, and had no effect on the released adult flies. Trials on the isolated Kalaupapa peninsula of Molokai in Hawaii subsequently confirmed that a semi-isolated population could be effectively eradicated using this method, even in the very horn fly-favourable environment of Hawaii (Eschle et al. 1977). Unfortunately, the area was later reinfested by the introduction of infested cattle into the area.

Although the SIT is by far the most widely known and successful genetic technique used against livestock pests to date, SIT application is not always feasible and a range of other genetically-based techniques have also been tested, or are under contemplation. For example, in continental Australia the extensive areas of livestock production and the wide distribution of associated pest species, together with few natural geographic boundaries, made the use of the SIT impractical or at least of dubious cost-benefit for use against many livestock pests, as in the case of sheep blowfly *Lucilia cuprina* (Wiedemann). As a consequence, a range of other genetically-based techniques such as the use of compound chromosome and sex-linked translocation strains (field female killing systems), which were predicted to be more effective at lower release ratios, were developed and tested (Foster et al. 1985, 1988, 1991). Field testing showed promise for these other approaches, but for a number of reasons discussed by Scott (2014), they were never implemented for widespread use.

More recently, transgenic sexing strains of *L. cuprina* have been developed that carry a tetracycline-repressible female lethal genetic system that could form the basis for mass-production of only males of *L. cuprina,* and potentially other fly species, for use in genetic control programmes (Scott 2014). A range of other techniques such as the Release of Insects carrying Dominant Lethal (RIDL) genes, RNAi and homing endonuclease genes (HEG) are now also being considered for use with mosquitoes, tsetse flies and other species (reviewed by McGraw and O'Neill 2013; Bourtzis et al. 2016) and have proceeded to field testing in some instances (Harris et al. 2011). With increasing access to sequenced insect pest genomes (International *Glossina* Genome Initiative 2014; Anstead et al. 2015) and rapid advances in molecular technology, most notably the availability of new gene editing techniques such as CRISPR-Cas9, many new, purpose-designed approaches for control will likely emerge.

5. *Wolbachia* AND AREA-WIDE CONTROL OF BUFFALO FLIES

Other technologies of much current interest for use in area-wide control programmes are symbiont-based approaches (Bourtzis 2008; Bourtzis et al. 2016; McGraw and O'Neill 2013; Wilke and Marelli 2015), in particular the use of *Wolbachia*. *Wolbachia* are maternally transmitted intracellular bacteria in the family Alphabacteria, estimated to infect 40% of terrestrial arthropod species (Zug and Hammerstein 2012). *Wolbachia* are capable of spreading through insect populations by manipulating host reproductive processes and have many and varied other effects that present potential for use in buffalo fly control programmes (Hoffmann et al. 2015). These can be considered in three main groups:

1. Cytoplasmic incompatibility, which can be harnessed for population suppression, population replacement or potentially population elimination
2. Fitness effects induced by *Wolbachia* infection, and
3. Transmission blocking of secondary pathogens.

These strategies are considered below for their potential to reduce the impacts of buffalo flies or interrupt their spread into uninfected areas.

5.1. Cytoplasmic Incompatibility and Incompatible Insect Technique (IIT)

Wolbachia infection can interfere with insect reproduction in several ways, including through the induction of cytoplasmic incompatibility whereby matings between infected males and non-infected females or between males and females infected with incompatible *Wolbachia* strains (bidirectional incompatibility), produce infertile eggs. This approach when used as an insect suppression or eradication strategy has been termed the Incompatible Insect Technique (IIT) (Zabalou et al. 2009). The IIT method is similar in approach to SIT, with *Wolbachia*-infected males used as *de facto* sterile males. Since *Wolbachia* is not paternally transmitted, as long as similarly infected females are not also released, the *Wolbachia* strain present in the released males does not establish in the target population in the field. Thus, serial release of only the infected males can lead to population suppression or eradication.

The effectiveness of using *Wolbachia*-induced cytoplasmic incompatibility in this way was demonstrated as early as the 1960s when release of *Wolbachia*-infected male *Culex quinquefasciatus* Say mosquitoes, vectors of human filariasis, led to local eradication of this species from areas in Myanmar (Laven 1967). Since then, studies towards the use of IIT have been conducted with a range of mosquito species, including *Aedes polynesiensis* Marks (Brelsfoard et al. 2009; O'Connor et al. 2012), *Aedes albopictus* (Skuse) (Calvitti et al. 2010), *Anopheles stephensi* Liston (Bian et al. 2013) and *Culex pipiens pallens* (Coquillett) (Chen et al. 2013), as well as the veterinary pests *Glossina morsitans* Westwood (Alam et al. 2011; Bourtzis et al. 2016) and *Stomoxys calcitrans* (L.) (Kusmintarsih 2009).

Use of an IIT approach could be applicable for eradication of confined foci of overwintering populations of buffalo flies to prevent or retard southerly spread or to slow rates of re-colonisation of favourable northern areas in summer. The IIT method could also be used to eradicate buffalo flies that become established in relatively isolated areas as a result of climate change, such as those predicted in South Australia and south-western Western Australia (Fig. 2).

Ideally only male *Wolbachia*-infected buffalo flies would be released, but to date no method for accurate mass-sexing of horn or buffalo flies has been reported. In the case of the Hawaii SIT trials with horn flies, irradiated flies of both sexes were released (Eschle et al. 1977). Although this is usually undesirable, because it increases competition with field females for mates and can temporarily increase fly pressure on cattle, it did not compromise success in the case of the Hawaiian trial and may not be a consideration if used against low-level populations present in overwintering foci of buffalo flies.

Reduction of male mating competitiveness from the effects of irradiation is one of the difficulties sometimes experienced in SIT programmes (Zhang et al. 2015). As female flies are often sterilised at levels of radiation below that which causes reduction of competitiveness in males, this has led to the suggestion of the complementary simultaneous use of the SIT and the IIT, with *Wolbachia* used to induce functional sterility in the males and low-level irradiation used to sterilise the females thereby also assuring that the *Wolbachia* strain present in the released males does not establish in the target pest population (Brelsfoard et al. 2009; Zhang et al. 2015; Bourtzis et al. 2016). In the absence of a practical sexing method, a similar approach could be considered for buffalo flies.

Alternatively, the development of a self-sexing strain in stable flies *S. calcitrans* (Seawright et al. 1986), which are in the same subfamily as buffalo flies, the determination of a near infrared (NIR)-based method for sexing tsetse fly pupae (Dowell et al. 2005; Moran and Parker 2016), and the rapid advances with molecular techniques currently being made in other species (Scott 2014), suggest significant potential for the future development of a sexing method for buffalo flies.

Notwithstanding the potential added difficulties for artificial rearing, the use of a strain of *Wolbachia* that also confers a fitness disadvantage or inability to overwinter in infected flies, such as *w*MelPop (see below), is a further possibility to guard against the effects of inadvertent female release in a *Wolbachia*-based IIT programme.

5.2. *Using* Wolbachia-*Induced Fitness Effects to Collapse Overwintering Populations of Buffalo Flies*

Different strains of *Wolbachia* can induce a range of different effects on the fitness of infected hosts (Hoffmann et al. 2015). Some of these effects include reduced life span (McMeniman et al. 2009), mortality of eggs (McMeniman and O'Neill 2010), slowed larval development (Ross et al. 2014), and reduced overall fitness (Yeap et al. 2011, 2014; Ross et al. 2015). Infection with *Wolbachia* has also been shown to interfere with blood-feeding efficiency in mosquitoes (Moreira et al. 2009; Turley et al. 2009), and to affect locomotor activity in parasitoid wasps, *Drosophila* species, and some mosquitoes (Fleury et al. 2000; Peng et al. 2008; Evans et al. 2009). Similar effects in buffalo flies could also have deleterious effects on survival and mating efficiency, as well as the persistence of their populations, particularly during winter.

The most profound deleterious effects described have been from the 'popcorn' (*w*MelPop) strain of *Wolbachia*, initially isolated from laboratory populations of *Drosophila melanogaster* Meigen (Min and Benzer 1997). The *w*MelPop strain replicates in host cells, causing cellular damage, characteristic morphological changes in infected tissues, and a range of physiological effects. These effects reduce life span by approximately one-half in *D. melanogaster* and transinfected mosquitoes (Min and Benzer 1997; McMeniman et al. 2009). Reductions of life span of this magnitude, and other fitness characters, can have profound effects on the population dynamics of a species, particularly during unfavourable times of the year (Rasic et al. 2014). However, the effects of *Wolbachia* are highly strain-, host- and environment-dependent, and less profound effects on fitness have also been observed in other *Wolbachia*-host associations (Hoffmann et al. 2015).

Modelling conducted by Rasic et al. (2014) demonstrated potential for using fitness reductions induced by *Wolbachia* to suppress or eliminate *Aedes aegypti* L. populations, particularly in locally or seasonally variable environments. Their results suggested that the effects of *w*MelPop were not sufficient to reduce persistence of mosquito populations in the very favourable climates of north Queensland, but they were likely to cause local extinctions in the more mosquito-marginal environments of central Queensland. These predictions were supported by semi-field cage studies, which showed that reductions in the survival of desiccation-resistant eggs resulting from *w*MelPop infection, eliminated populations of *Ae. aegypti* during extended dry periods (Ritchie et al. 2015).

Wolbachia could also be used to drive co-inherited deleterious 'payload genes' in the genome of infected insects into the target pest population (Curtis and Sinkins 1998; Hoffmann and Turelli 2013; Champer et al. 2016). These genes could confer reduced fitness or conditionally lethal effects such as cold temperature sensitivity or insecticide susceptibility. Conversely the use of linked traits that confer a fitness advantage in certain circumstances might be used to facilitate the spread of *Wolbachia* strains into a population. For example, insecticide resistance that confers a competitive advantage under a spraying regime could be used to assist the spread of a *w*MelPop-infected strain that confers a pathogen blocking capability or seasonal lethality (Hoffmann and Turelli 2013).

As more pest insect genomes are characterised, along with the rapid advancement in molecular transformation technologies, it is expected that possibilities for this approach will grow rapidly. Using *Wolbachia* as the driving mechanism is expected to have greater public acceptance and less potential for unanticipated effects than transgenic gene drives (Champer et al. 2016). An attractive alternative approach is the direct transformation of *Wolbachia* genomes with genes to be driven into a pest population. Until recently, successful genetic transformation of *Wolbachia* had proved elusive, but the recent reporting of a phage-mediated system for the genetic modification of *Wolbachia* (Bordenstein and Bordenstein 2017) offers exciting possibilities in this area.

5.3. Stephanofilaria *Blocking*

Buffalo fly-associated lesions are of significant welfare and economic concern, with estimates of over 95% of cattle affected in northern areas of Australia (Johnson 1989). Although the exact etiology of buffalo fly-associated lesions is unclear, an unnamed species of filarial nematode (*Stephanofilaria* sp.), transmitted by buffalo flies and found in the lesions, is thought to play a role (Johnson et al. 1986; Johnson 1989). Surveys of buffalo flies collected from near Townsville in the 1980s found a 2.9% (range 0% - 9.3%) prevalence of *Stephanofilaria* in female flies (Johnson 1989), whereas a more recent study in 2004 measured infection rates between 29% and 57% in flies collected from four sites near Rockhampton (Shaw and Sutherland 2006).

Wolbachia infection has been demonstrated to reduce vectorial capacity of various species of mosquitoes for a range of pathogens, including filarial nematodes. Inhibition of development of filarial nematodes was seen with both *w*MelPop in *Ae. aegypti* (Kambris et al. 2009) and *w*AlbB in *Ae. polynesiensis* (Andrews et al. 2012) and resulted in a reduction in the prevalence of infective third stage nematodes in the mosquitoes. The mechanism of pathogen blocking is not completely understood but may be due to competition for host resources or modulation of host immune response, in particular reduction in levels of reactive oxygen species (Andrews et al. 2012). The *w*MelPop strain of *Wobachia* also reduces the efficiency of disease transmission by shortening the life span of vectors and reducing the likelihood that a pathogen will be able to complete its required extrinsic incubation period before host mortality.

The shortest incubation period seen for *Stephanofilaria* sp. in buffalo flies was 7 days (Johnson 1989), suggesting that the life-shortening effects of *w*MelPop

Wolbachia could also significantly affect the transmission dynamics of this filarial nematode species.

Lesions associated with horn fly-transmitted *Stephanofilaria stilesi* in North America appear to be less extensive and severe than buffalo fly-associated lesions in Australia (Hibler 1966). As horn flies are infected with *Wolbachia*, but buffalo flies are not, it is tempting to hypothesise that this difference may be associated with differences in the efficiency of *Stephanofilaria* transmission, although many other factors could also be involved. Disruption of the spread of *Stephanofilaria* or reduction in the severity of lesions by the introduction of a transmission-blocking *Wolbachia* strain into buffalo flies, would be a significant outcome for the Australian cattle industries from both economic and welfare perspectives.

6. BUFFALO FLY OVERWINTERING, A SUSCEPTIBLE STAGE FOR AREA-WIDE APPROACHES?

Horn flies have the ability to overwinter in the pupal phase, as pharate adults, whereas buffalo flies do not (Ferrar et al. 1969; Cook and Spain 1982), which is a major difference between the two species (Showler et al. 2014). In the northerly part of their range in North America, adult horn flies begin to disappear from cattle in autumn and do not reappear until the next spring. Overwintering dormancy allows horn flies to emerge and rapidly re-establish throughout the previous season's range when conditions become suitable in spring or summer. There is, however, significant plasticity in this response and at warmer latitudes horn fly populations continue cycling throughout the year (Showler et al. 2014).

In more marginal areas, horn flies may survive winter both as adults, with reduced activity, and in the pupal stage, with various levels of dormancy. Mendes and Linhares (1999) working in a warm winter climate in Brazil (21°30'S), verified diapause in 9.1% of winter pupae, even though horn flies were present on cattle year-round. These authors note that this dual overwintering mechanism could present difficulties for the design of cost-efficient eradication programmes for horn flies. The plasticity in overwintering response has most likely been a key factor allowing horn flies to disperse and become established in a wide range of environments.

In contrast to horn flies, buffalo flies die out through much of their summer range in winter (Fig. 2). Their range at the southern and continental edges in Australia is limited by cooler temperatures and low moisture levels in dung during winter (Cook and Spain 1982). Low temperatures either prevent development completely, or they slow the development of the larval stages to a degree that they can't be completed before moisture content in dung falls to lethal levels. The occurrence of frosts can also have a devastating effect on the survival of the soil stages, i.e. larvae and pupae (Cook and Spain 1982).

Williams et al. (1985) found that buffalo flies overwintered at the edge of their winter range as slowly cycling, low level fly populations in local areas of moderate microclimates. Most of these overwintering foci were in hilly, heavily timbered areas that were well-watered from either creeks, dams or swamps, and less exposed to low minimum temperatures or frosts than the low-lying surrounding areas. Nearly all of the overwintering sites identified were within 40 km of the coast, where temperatures were likely moderated by coastal influences. Re-colonisation of summer-suitable areas and southern range extension relied on overwintering of buffalo flies in these foci. When conditions became favourable each year, the flies built-up in numbers and either dispersed from these areas autonomously or were transported by cattle movements to reinfest their summer range (Fig. 2).

These localised overwintering foci provide a potential target for the application of *Wolbachia*-based approaches. The use of *Wolbachia* in either an IIT approach, to compromise *Stephanofilaria* transmission or to introduce a deleterious fitness factor, is likely to be most efficiently achieved at times of low fly populations, such as during overwintering, when suitable release ratios will be most readily achieved. Indeed, SIT and IIT approaches are often initiated when target populations are low, or involve population reduction by insecticide treatments prior to the release of infected flies.

Persistence of buffalo fly populations in overwintering foci is precarious and it is the soil stages that are most subject to adverse effects from low temperatures and dryness. Adult flies living with the warmth and blood provided by their cattle hosts are less affected by adverse winter conditions. Therefore, it is likely that released adult flies will be less exposed to the effects of winter conditions than the soil stages and able to persist for sufficient time to mate with overwintering adult flies and either interrupt reproduction or spread *Wolbachia* infection.

7. TOWARDS A *Wolbachia*-BASED APPROACH TO CONTROLLING BUFFALO FLIES

The effects of *Wolbachia* are most profound in new host associations (McGraw et al. 2002) and *Wolbachia*-based approaches to control require either the transinfection of *Wolbachia* into uninfected host populations, or transinfection of already infected populations with different strains of *Wolbachia* (O'Connor et al. 2012). Transinfection has most often been achieved by embryonic microinjection, but adult microinjection has also been successful in some instances (Hughes and Rasgon 2014). The success rate of microinjection is generally low, with subsequent loss of infection in newly injected hosts common, particularly in more distantly related host species. This is thought to be due to inability of the injected *Wolbachia* to adapt quickly enough to the new host environment. However, the probability of success can be increased by prior adaptation of *Wolbachia* in target host cell lines (McMeniman et al. 2008, 2009)

Although *Wolbachia* has not been found in buffalo flies, it is found widely in horn flies (Jeyaprakash and Hoy 2000; Floate et al. 2006; Zhang et al. 2009). The very close relatedness of buffalo and horn flies suggests it likely that the former will be a competent host for *Wolbachia* and that the likelihood of successful transinfection with suitable strains of *Wolbachia* is high.

We have successfully established cell lines for both horn and buffalo flies and have achieved persisting infections of *w*AlbB, *w*Mel, and *w*MelPop in cell lines for both species, also suggesting good potential for the successful transinfection of buffalo flies with *Wolbachia.*

We are undertaking a programme of microinjection towards the stable transinfection of buffalo flies with these three *Wolbachia* strains. We have also recently developed laboratory rearing methods for buffalo flies and have established a stable persisting laboratory colony (James et al. 2013). These methods will facilitate maintenance of transinfected strains and studies to determine the effects induced in these flies by infection with *w*AbB, *w*Mel and *w*MelPop, with a view to develop *Wolbachia*-based strategies for reducing buffalo fly spread and impacts.

8. CONCLUSION

Without intervention, buffalo flies are likely to become major cattle pests in Australia's southern beef and dairy industries, and also increase their impacts in northern herds. Their further southerly invasion is likely to be facilitated by the effects of climate change, together with the availability of a large, susceptible and mostly unprotected *Bos taurus* cattle population in the southern areas of Australia.

Previous regulatory procedures, based on spraying and cattle movement controls, have failed to prevent the southward spread of buffalo flies. Using an integrated area-wide approach incorporating use of a biological agent, such as *Wolbachia*, and focusing on the pest population rather than cattle, avoids potential disadvantages associated to widespread chemical use. In addition, *Wolbachia* are vertically transmitted from female flies to their eggs and restricted to living exclusively within host cells, thus minimising the potential for non-target effects. The use of *Wolbachia* has had good community acceptance in Australia to date (Kolopack et al. 2015) and importantly, a legislative framework for the release of *Wolbachia*-transinfected strains already exists in Australia (De Barro et al. 2011).

The design of optimal strategies will rely on a knowledge of the biological effects of candidate strains of *Wolbachia* in buffalo flies. A number of critical steps towards this end have been completed, including the establishment of an *in vitro* colony of buffalo flies as well as *Haematobia* cell lines transinfected with the *w*Alb, *w*Mel and *w*MelPop strains of *Wolbachia*.

We are currently undertaking embryonic and adult microinjection with these strains towards the establishment of transinfected buffalo fly lines. Successful completion of this step will allow characterisation of the effects of *Wolbachia* in buffalo flies towards the design of potential *Wolbachia*-based control strategies and an initial assessment of the likely feasibility of using a *Wolbachia*-based area-wide approach to reduce buffalo fly impacts in endemic areas and interrupt the southerly encroachment of buffalo flies.

9. ACKNOWLEDGEMENTS

We thank the Eliminate Dengue Programme for provision of *Wolbachia* strains; also Professors Elizabeth McGraw and Sassan Asgari for ongoing interest and advice, Professors Tim Kurrtii and Uli Munderloh for help in the establishment of the *Haematobia* cell lines, and Dr. Rob Dobson for permission to use the results of the buffalo fly CLIMEX modelling. We also gratefully acknowledge Meat and Livestock Australia for the provision of funding towards the research programme reported here.

10. REFERENCES

Alam, U., J. Medlock, C. Brelsfoard, R. Pais, C. Lohs, S. Balmand, J. Carnogursky, A. Heddi, P. Takac, A. Galvani, and S. Aksoy. 2011. *Wolbachia* symbiont infections induce strong cytoplasmic incompatibility in the tsetse fly *Glossina morsitans*. PLoS Pathogens 7(12): e1002415.

Andrews, E. S., P. R. Crain, Y. Q. Fu, D. K. Howe, and S. L. Dobson. 2012. Reactive oxygen species production and *Brugia pahangi* survivorship in *Aedes polynesiensis* with artificial *Wolbachia* infection types. PLoS Pathogens 8(12): e1003075.

(AHA) Animal Health Australia. 2017. Screwworm fly surveillance and preparedness program. Animal Health Australia, Canberra Australia.

Anonymous. 1934. Buffalo fly control in north-west Queensland. Queensland Agricultural Journal 42: 71–82.

Anstead, C. A., P. K. Korhonen, N. D. Young, R. S. Hall, A. R. Jex., S.C Murali, D. S. T Hughes, S. F. Lee, T. Perry, A. J. Stroehlein, B. R. E. Ansell, B. Breugelmans, A. Hofmann, J. Qu, S. Dugan, S. L. Lee, H. Chao, H. Dinh, Y. Han, H. V. Doddapaneni, K. C. Worley, D. M., Muzny, P. Ioannidis, R. M. Waterhouse, E. M. Zdobnov, P. J. James, N. H. Bagnall, A. C. Kotze, R. A. Gibbs, S. Richards, P. Batterham, and R.B. Gasser. 2015. *Lucilia cuprina* genome unlocks parasitic fly biology to underpin future interventions. Nature Communications 6: 7344.

Anziani, O. S., A. A. Guglielmone, A. R. Signorini, C. Aufranc, and A. J. Mangold. 1993. *Haematobia irritans* in Argentina. Veterinary Record 132: 588–588.

Bian, G. W., G. L. Zhou, P. Lu, and Z. Y. Xi. 2013. Replacing a native *Wolbachi*a with a novel strain results in an increase in endosymbiont load and resistance to dengue virus in a mosquito vector. PLoS Neglected Tropical Diseases 7: e2250.

Bordenstein, S. R., and S. R. Bordenstein. 2017. Phage-mediated manipulation of *Wolbachia*. International patent WO2017181043A1. Application filed by Vanderbilt University, USA.

Bourtzis, K. 2008. *Wolbachia*-based technologies for insect pest population control. *In* S. Aksoy (ed.), Transgenesis and the management of vector-borne Disease. Advances in Experimental Medicine and Biology 627: 104–113.

Bourtzis, K., R. S. Lees, J. Hendrichs, and M. J. B. Vreysen. 2016. More than one rabbit out of the hat: Radiation, transgenic and symbiont-based approaches for sustainable management of mosquito and tsetse fly populations. Acta Tropica 157: 115–130.

Brelsfoard, C. L., W. St Clair, and S. L. Dobson. 2009. Integration of irradiation with cytoplasmic incompatibility to facilitate a lymphatic filariasis vector elimination approach. Paras. & Vectors 2: 38.

Butler, J. F., and J. S. Okine. 1999. The horn fly, *Haematobia irritans* (L.): Review of programs on natural history and control, pp. 625–646. *In* J. F. Burger (ed.), Contributions to the knowledge of Diptera: A collection of articles on Diptera commemorating the life and work of Graham B. Fairchild. Associated Publishers, Gainesville, Florida, USA.

Calvitti, M., R. Moretti, E. Lampazzi, R. Bellini, and S. L. Dobson. 2010. Characterization of a new *Aedes albopictus* (Diptera: Culicidae)-*Wolbachia pipientis* (Rickettsiales: Rickettsiaceae) symbiotic association generated by artificial transfer of the *w*Pip strain from *Culex pipiens* (Diptera: Culicidae). Journal of Medical Entomology 47: 179–187.

Campano, S., and P. Avalos. 1994. Presence of *Haematobia irritans* (Diptera: Muscidae) in cattle from Chile. Parasitología al Día 18: 59–61.

Champer, J., A. Buchman, and O. S. Akbari. 2016. Cheating evolution: Engineering gene drives to manipulate the fate of wild populations. Nature Reviews Genetics 17: 146–159.

Chen, L., C. Zhu, and D. Zhang. 2013. Naturally occurring incompatibilities between different *Culex pipiens pallens* populations as the basis of potential mosquito control measures. PLoS Neglected Tropical Diseases 7: e2030.

Cook, I. M., and A. V. Spain. 1982. The effects of temperature and moisture on survival of the immature stages of the buffalo fly, *Haematobia irritans exigua* Demeijere (Diptera, Muscidae). Australian Journal of Zoology 30: 923–930.

Cupp, E. W., M. S. Cupp, J. M. C. Ribeiro, and S. E. Kunz. 1998. Blood-feeding strategy of *Haematobia irritans* (Diptera: Muscidae). Journal of Medical Entomology 35: 591–595.

Curtis, C. F., and S. P. Sinkins. 1998. *Wolbachia* as a possible means of driving genes into populations. Parasitology 116: S111–S115.

(CSIRO/BOM) Commonwealth Scientific and Industrial Research Organisation / Bureau of Meteorology. 2016. Australian climate change science program: Australia's changing climate. Department of the Environment and Energy. Australian Government. 10 pp.

(DAWR) Department of Agriculture and Water Resources Australia. 2017. State and territory notifiable animal diseases list. Department of Agriculture and Water Resources, Canberra, Australia.

De Barro, P. J., B. Murphy, C. C. Jansen, and J. Murray. 2011. The proposed release of the yellow fever mosquito, *Aedes aegypti,* containing a naturally occurring strain of *Wolbachia pipientis*, a question of regulatory responsibility. Journal of Consumer Protection and Food Safety 6: 33–40.

Doube, B. M. 1986. Biological control of the buffalo fly in Australia: The potential for southern Africa dung fauna. Miscellaneous Publications of the Entomological Society of America 61: 16–34.

Dowell, F. E., A. G. Parker, M. Q. Benedict, A. S. Robinson, A. B. Broce, and R. A. Wirtz. 2005. Sex separation of tsetse fly pupae using near-infrared spectroscopy. Bulletin of Entomological Research 95: 249–257.

Dyck, V. A., J. Hendrichs, A. S. Robinson (eds.). 2021. Sterile Insect Technique – Principles and practice in Area-Wide Integrated Pest Management. Second Edition. CRC Press, Boca Raton, Florida, USA. 1200 pp.

Eastaway, B. 1974. The buffalo fly in Queensland. Queensland Agricultural Journal 100: 221–224.

Eschle, J. L., J. A. Miller, and C. D. Schmidt. 1977. Insect growth-regulator and sterile males for suppression of horn flies. Nature 265: 325–326.

Eschle, J. L., S. E. Kunz, C. D. Schmidt, B. F. Hogan, and R. O. Drummond. 1973. Suppression of a population of horn flies with the sterile male technique. Environmental Entomology 2: 976–980.

Evans, O., E. P. Caragata, C. J. McMeniman, M. Woolfit, D. C. Green, C. R. Williams, C. E. Franklin, S. L. O'Neill, and E. A. McGraw. 2009. Increased locomotor activity and metabolism of *Aedes aegypti* infected with a life-shortening strain of *Wolbachia pipientis*. Journal of Experimental Biology 212: 1436–1441.

Ferrar, P. 1969. Colonisation of an island by buffalo fly, *Haematobia exigua*. Australian Veterinary Journal 45: 290–292.

Fleury, F., F. Vavre, N. Ris, P. Fouillet, and M. Bouletreau. 2000. Physiological cost induced by the maternally-transmitted endosymbiont *Wolbachia* in the *Drosophila* parasitoid *Leptopilina heterotoma*. Parasitology 121: 493–500.

Floate, K. D., G. K. Kyei-Poku, and P. C. Coghlin. 2006. Overview and relevance of *Wolbachia* bacteria in biocontrol research. Biocontrol Science and Technology 16: 767–788.

Foster, G. G., W. G. Vogt, and T. L. Woodburn. 1985. Genetic analysis of field trials of sex-linked translocation strains for genetic control of the Australian sheep blowfly *Lucilia cuprina* (Wiedemann). Australian Journal of Biological Sciences 38: 275–293.

Foster, G. G., G. L. Weller, and G. M. Clarke. 1991. Male crossing over and genetic sexing systems in the Australian sheep blowfly *Lucilia cuprina*. Heredity 67: 365–371.

Foster, G. G., W. G. Vogt, T. L. Woodburn, and P. H. Smith. 1988. Computer simulation of genetic control - Comparison of sterile males and field female killing systems. Theoretical and Applied Genetics 76: 870–879.

Frisch, J. E., C. J. O'Neill, and M. J. Kelly. 2000. Using genetics to control cattle parasites - The Rockhampton experience. International Journal for Parasitology 30: 253–264.

Gonzalez, R. H. 1967. *Haematobia irritans* in Chile. Revista Chilena de Entomología 6: 142.

Graham, O. H., and J. L. Hourrigan. 1977. Eradication programs for the arthropod parasites of livestock. Journal of Medical Entomology 13: 629–658.

Grisi, L., R. C. Leite, J. R. D. Martins, A. T. M. de Barros, R. Andreotti., D. Cancado, A. A. P. de Leon, J. B. Pereira, and H. S. Villela. 2014. Reassessment of the potential economic impact of cattle parasites in Brazil. Revista Brasileira de Parasitologia Veterinaria 23: 150–156.

Guglielmone, A. A., E. Gimeno, J. Idiart, W. P. Fisher, M. M. Volpogni, O. Quaino, O. S. Anziani, S. G. Flores, and O. Warnke. 1999. Skin lesions and cattle hide damage from *Haematobia irritans* infestations. Medical and Veterinary Entomology 13: 324–329.

Harris, A. F., D. Nimmo, A. R. McKemey, N. Kelly, S. Scaife, C. A. Donnelly, C. Beech, W. D. Petrie, and L. Alphey. 2011. Field performance of engineered male mosquitoes. Nature Biotechnology 29: 1034–1093.

Hendrichs, J., P. Kenmore, A. S. Robinson, and M. J. B. Vreysen. 2007. Area-Wide Integrated Pest Management (AW-IPM): Principles, practice and prospects, pp. 3-33. *In* M. J. B. Vreysen, A. S. Robinson, J. Hendrichs (eds.), Area-wide control of insect pests, from research to field implementation. Springer, Dordrecht, The Netherlands.

Hibler, C. P. 1966. Development of *Stephanofilaria stilesi* in horn fly. Journal of Parasitology 52: 890–898.

Hill, G. F. 1917. Some notes on the bionomics of the buffalo-fly (*Lyperosia exigua*, Meij.). Proceedings of the Linnean Society of New South Wales 41: 763–768.

Hill, J. K., H. M. Griffiths, and C. D. Thomas. 2011. Climate change and evolutionary adaptations at species' range margins. Annual Review of Entomology 56: 143–159.

Hoffmann, A. A., and M. Turelli. 2013. Facilitating *Wolbachia* introductions into mosquito populations through insecticide-resistance selection. Proc. Royal Society B-Biological Sciences 280: 2013.0371.

Hoffmann, A. A., P. A. Ross, and G. Rasic. 2015. *Wolbachia* strains for disease control: Ecological and evolutionary considerations. Evolutionary Applications 8: 751–768.

Hughes, G. L., and J. L. Rasgon. 2014. Transinfection: A method to investigate *Wolbachia*-host interactions and control arthropod-borne disease. Insect Molecular Biology 23: 141–151.

International *Glossina* Genome Initiative. 2014. Genome sequence of the tsetse fly (*Glossina morsitans*): Vector of African trypanosomiasis. Science 344: 380–386.

Iwasa, M., and N. Ishiguro. 2010. Genetic and morphological differences of *Haematobia irritans* and *H. exigua*, and molecular phylogeny of Japanese Stomoxyini flies (Diptera, Muscidae). Medical Entomology and Zoology 61: 335–344.

James, P. J. 2013. In vitro culture of buffalo fly. Final Report. Project B NBP 0488. Meat and Livestock Australia, North Sydney, Australia.

Jeyaprakash, A., and M. A. Hoy. 2000. Long PCR improves *Wolbachia* DNA amplification: WSP sequences found in 76% of sixty-three arthropod species. Insect Molecular Biology 9: 393–405.

Johnson, S. J. 1989. Studies on stephanofilariasis in Queensland. PhD thesis. James Cook University. Brisbane, Queensland, Australia. 191 pp.

Johnson, S. J., R. J. Arthur, and R. K. Shepherd. 1986. The distribution and prevalence of stephanofilariasis in cattle in Queensland. Australian Veterinary Journal 63: 121–124.

Jonsson, N. N., and A. L. Matchoss. 1998. Attitudes and practices of Queensland dairy farmers to the control of the cattle tick, *Boophilus microplus*. Australian Veterinary Journal 76: 746–751.

Kambris, Z., P. E. Cook, H. K. Phuc, and S. P. Sinkins. 2009. Immune activation by life-shortening *Wolbachia* and reduced filarial competence in mosquitoes. Science 326: 134–136.

Kano, R., S. Shinonaga, and T. Hasegawa. 1972. On the specific name of *Haematobia* (Diptera, Muscidae) from Japan. Japan Journal of Sanitary Zoology 23: 49–56.

Knipling, E. F. 1972a. Entomology and the management of man's environment. Journal of the Australian Entomological Society 11: 153–167.

Knipling, E. F. 1972b. Integrated control of livestock insect pests, pp. 379–397. *In* M. A. Khan, and W. O. Haufe (eds.), Toxicology, biodegradation and efficacy of livestock pesticides. Swets & Zeitlinger, Amsterdam, The Netherlands.

Kolopack, P. A., J. A. Parsons, and J. V. Lavery. 2015. What makes community engagement effective? Lessons from the Eliminate Dengue Program in Queensland Australia. PLoS Neglected Tropical Diseases 9: e0003713.

Kusmintarsih, E. S. 2009. Horizontal transfer of the "popcorn effect" strain of *Wolbachia* from *Drosophila melanogaster* to *Stomoxys calcitrans*. Microbiology Indonesia 3: 121–125.

Lane, J., T. Jubb, R. Shepherd, J. Webb-Ware, and G. Fordyce. 2015. Priority list of endemic diseases for the red meat industries. Final Report Project B.AHE.0010. Meat and Livestock Australia, North Sydney, New South Wales, Australia.

Laven, H. 1967. Eradication of *Culex pipiens fatigans* through cytoplasmic incompatibility. Nature 216: 383–384.

Lindquist, D. A., M. Abusowa, and M. J. R. Hall. 1992. The New World screwworm fly in Libya - A review of its introduction and eradication. Medical and Veterinary Entomology 6: 2–8.

Low, V. L., T. K. Tan, P. E. Lim, L. N. Domingues, S. T. Tay., Y. A. L Lim, T. G. Goh, C. Panchadcharam, P. Bathmanaban, and M. Sofian-Azirun. 2014. Use of COI, CytB and ND5 genes for intra- and inter-specific differentiation of *Haematobia irritans* and *Haematobia exigua*. Veterinary Parasitology 204: 439–442.

Low, V. L., T. K. Tan, B. K. Prakash, W. Y. Vinnie-Siow, S. T. Tay, R. Masmeatathip, U. K. Hadi, Y. A. L. Lim, C. D. Chen, Y. Norma-Rashid, and M. Sofian-Azirun. 2017. Contrasting evolutionary patterns between two haplogroups of *Haematobia exigua* (Diptera: Muscidae) from the mainland and islands of Southeast Asia. Scientific Reports 7: 5871.

Mackerras, I. M. 1933. The taxonomy of *Lyperosia exigua* De Mieijere (Diptera, Muscidae). The Annals and Magazine of Natural History 11: 58–64.

Mancebo, O. A., C. M. Monzon, and G. M. Bulman. 2001. *Haematobia irritans*: Una actualización de diez años de su introducción en Argentina. Veterinaria (Argentina) 43: 34–46.

McGraw, E. A., and S. L. O'Neill. 2013. Beyond insecticides: New thinking on an ancient problem. Nature Reviews Microbiology 11: 181–193.

McGraw, E. A., D. J. Merritt, J. N. Droller, and S. L. O'Neill. 2002. *Wolbachia* density and virulence attenuation after transfer into a novel host. Proceedings of the National Academy of Sciences of the United States of America 99: 2918–2923.

McMeniman, C. J., and S. L. O'Neill. 2010. A virulent *Wolbachia* infection decreases the viability of the dengue vector *Aedes aegypti* during periods of embryonic quiescence. PLoS Neglected Tropical Diseases 4: e748.

McMeniman, C. J., A. M. Lane, A. W. C. Fong, D. A. Voronin, I. Iturbe-Ormaetxe, R., Yamada, E. A., McGraw, and S. L. O'Neill. 2008. Host adaptation of a *Wolbachia* strain after long-term serial passage in mosquito cell lines. Applied and Environmental Microbiology 74: 6963–6969.

McMeniman, C. J., R. V. Lane, B. N. Cass, A. W. C. Fong, M. Sidhu, Y. F. Wang, and S. L. O'Neill. 2009. Stable introduction of a life-shortening *Wolbachia* infection into the mosquito *Aedes aegypti*. Science 323: 141–144.

Mendes, J., and A. X. Linhares. 1999. Diapause, pupation sites and parasitism of the horn fly, *Haematobia irritans*, in south-eastern Brazil. Medical and Veterinary Entomology 13: 185–190.

Min, K. T., and S. Benzer. 1997. *Wolbachia*, normally a symbiont of *Drosophila*, can be virulent, causing degeneration and early death. Proceedings of the National Academy of Sciences of the United States of America 94: 10792–10796.

Moran, Z. R., and A. G. Parker. 2016. Near infrared imaging as a method of studying tsetse fly (Diptera: Glossinidae) pupal development. Journal of Insect Science 16: 72.

Moreira, L. A., E. Saig, A. P. Turley, J. M. C. Ribeiro, S. L. O'Neill, and E. A. McGraw. 2009. Human probing behavior of *Aedes aegypti* when infected with a life-shortening strain of *Wolbachia*. PLoS Neglected Tropical Diseases 3: e568.

Munro, J. A. 1960. A special survey of Bolivian insects. United States Department of Agriculture Cooperative Economic Insect Report 10 (45): 1064–1072.

O'Connor, L., C. Plichart, A. C. Sang, C. L. Brelsfoard, H. C. Bossin, and S. L. Dobson. 2012. Open release of male mosquitoes infected with a *Wolbachia* biopesticide: Field performance and infection containment. PLoS Neglected Tropical Diseases 6: e1797.

Parliament of Queensland. 1965. Buffalo fly control acts, 1941 to 1965. Office of the Queensland Parliamentary Counsel. Brisbane, Australia.

Peng, Y., J. E. Nielsen, J. P. Cunningham, and E. A. McGraw. 2008. *Wolbachia* infection alters olfactory-cued locomotion in *Drosophila spp*. Applied and Environ. Microbiology 74: 3943–3948.

Pont, A. C. 1973. Studies on the Australian Muscidae (Diptera). A revision of the subfamilies Muscinae and Stomoxyinae. Australian Journal of Zoology (Suppl. Series) 21: 129–296.

Rasic, G., N. M. Endersby, C. Williams, and A. A. Hoffmann. 2014. Using *Wolbachia*-based release for suppression of *Aedes* mosquitoes: Insights from genetic data and population simulations. Ecological Applications 24: 1226–1234.

Ritchie, S. A., M. Townsend, C. J. Paton, A. G. Callahan, and A. A. Hoffmann. 2015. Application of *w*MelPop *Wolbachia* strain to crash local populations of *Aedes aegypti*. PLoS Neglected Tropical Diseases 9: e0003930.

Roberts, F. H. S. 1946. The buffalo fly. Queensland Agricultural Journal 63: 112–116.

Ross, P. A., N. M. Endersby, and A. A. Hoffmann. 2015. Substantial fitness costs for *Wolbachia* infection on the starvation resistance of *Aedes aegypti* larvae. American Journal of Tropical Medicine and Hygiene 93: 216–216.

Ross, P. A., N. M. Endersby, H. L. Yeap, and A. A. Hoffmann. 2014. Larval competition extends developmental time and decreases adult size of *w*MelPop *Wolbachia*-infected *Aedes aegypti*. American Journal of Tropical Medicine and Hygiene 91: 198–205.

Scott, M. J. 2014. Development and evaluation of male-only strains of the Australian sheep blowfly, *Lucilia cuprina*. BMC Genetics 15 (Suppl. 2): S3.

Seawright, J. A., B. K. Birky and B. J. Smittle. 1986. Use of a genetic technique for separating the sexes of the stable fly (Diptera, Muscidae). Journal of Economic Entomology 79: 1413–1417.

Seddon, H. R. 1967. Diseases of domestic animals in Australia. Part 2: Arthropod infestations (flies, lice and fleas). Department of Health, Commonwealth of Australia, Canberra, Australia. 152 pp.

Shaw, S. A., and I. A. Sutherland. 2006. The prevalence of *Stephanofilaria sp.* in buffalo fly, *Haematobia irritans exigua*, in central Queensland. Australian Journal of Entomology 45: 198–201.

Showler, A. T., W. L. A. Osbrink, and K. H. Lohmeyer. 2014. Horn fly, *Haematobia irritans irritans* (L.), overwintering. International Journal of Insect Science 6: 43–47.

Silva, L. A. F., R. E. Rabelo, M. I. de Moura, M. C. S. Fioravanti, L. M. F. Borges, and C. R. de Oliveira Lima. 2010. Epidemiological aspects and treatment of parasitic lesions similar to Stephanofilariasis disease in nursing cows. Semina: Ciencias Agrarias 31: 689–698.

Skidmore, P. 1985. The biology of the Muscidae of the world. Kluwer Academic Publishers, Dordrecht, The Netherlands. 550 pp.

Snyder, F. M. 1965. Diptera. Muscidae. Insects of Micronesia 13: 191–327.

Sutherst, R. W., and R. S. Tozer. 1995. Control of buffalo fly (*Haematobia irritans exigua* De Meijere) on dairy and beef cattle using traps. Australian Journal of Agricultural Research 46: 269–284.

Sutherst, R. W., A. S. Bourne, G. F. Maywald, and G. W. Seifert. 2006. Prevalence, severity, and heritability of *Stephanofilaria* lesions on cattle in central and southern Queensland, Australia. Australian Journal of Agricultural Research 57: 743–750.

Tillyard, R. J. 1931. The buffalo fly in Australia. Journal of the Council for Scientific and Industrial Research, Australia 4: 234–243.

Turley, A. P., L. A. Moreira, S. L. O'Neill, and E. A. McGraw. 2009. *Wolbachia* infection reduces blood-feeding success in the dengue fever mosquito, *Aedes aegypti.* PLoS Negl. Tropical Dis. 3: e516.

Urech, R., G. W. Brown, C. J. Moore, and P. E. Green. 2005. Cuticular hydrocarbons of buffalo fly, *Haematobia exigua*, and chemotaxonomic differentiation from horn fly, *H. irritans*. Journal of Chemical Ecology 31: 2451–2461.

Vreysen, M. J. B., and A. S. Robinson. 2011. Ionizing radiation and area-wide management of insect pests to promote sustainable agriculture: A review. Agronomy for Sust. Development 31: 233–250.

Vreysen, M. J. B., M. T. Seck, B. Sall, and J. Bouyer. 2013. Tsetse flies: Their biology and control using area-wide integrated pest management approaches. Journal of Invertebrate Pathology 112: S15–S25.

Vreysen, M. J., K. M. Saleh, M. Y. Ali, A. M. Abdulla, Z. R. Zhu, K. G. Juma, V. A. Dyck, A. R. Msangi, P. A. Mkonyi, and U. Feldmann. 2000. *Glossina austeni* (Diptera: Glossinidae) eradicated on the island of Unguja, Zanzibar, using the Sterile Insect Technique. Journal of Economic Entomology 93: 123–135.

Vreysen, M. J. B., K. Saleh, F. Mramba, A. Parker, U. Feldmann, V. A. Dyck, A. Msangi, and J. Bouyer. 2014. Sterile insects to enhance agricultural development: The case of sustainable tsetse eradication on Unguja Island, Zanzibar, using an Area-Wide Integrated Pest Management approach. PLoS Neglected Tropical Diseases 8: e2857.

Wilke, B. A. B., and M. T. Marrelli. 2015. Paratransgenesis: A promising new strategy for mosquito vector control. Parasites & Vectors 8: 342.

Williams, J. D., R. W. Sutherst, G. F. Maywald, and C. T. Petherbridge. 1985. The southward spread of buffalo fly (*Haematobia irritans exigua*) in eastern Australia and its survival through a severe winter. Australian Veterinary Journal 62: 367–369.

Wyss, J. H. 2000. Screwworm eradication in the Americas. Annals of the New York Academy of Sciences 916: 186–193.

Yeap, H. L., P. Mee, T. Walker, A. R. Weeks, S. L. O'Neill, P. Johnson, S. A. Ritchie, K. M. Richardson, C. Doig, N. M. Endersby, and A. A. Hoffmann. 2011. Dynamics of the "popcorn" *Wolbachia* infection in outbred *Aedes aegypti* informs prospects for mosquito vector control. Genetics 187: 583–595.

Yeap, H. L., J. K. Axford, J. Popovici, N. M. Endersby, I. Iturbe-Ormaetxe, S. A. Ritchie, and A. A. Hoffmann. 2014. Assessing quality of life-shortening *Wolbachia*-infected *Aedes aegypti* mosquitoes in the field based on capture rates and morphometric assessments. Parasites & Vectors 7(1): 58.

Zabalou, S., A. Apostolaki, I. Livadaras, G. Franz, A. S. Robinson, C. Savakis, and K. Bourtzis. 2009. Incompatible Insect Technique: Incompatible males from a *Ceratitis capitata* genetic sexing strain. Entomologia Experimentalis et Applicata 132: 232–240.

Zhang, B., E. McGraw, K. Floate, D., P. J. James, W. K. Jorgensen, and J. T. Rothwell. 2009. *Wolbachia* infection in Australasian and North American populations of *Haematobia irritans* (Diptera: Muscidae). Veterinary Parasitology 162: 350–353.

Zhang, D., R. S. Lees, Z. Y. Xi, J. R. L. Gilles, and K. Bourtzis. 2015. Combining the Sterile Insect Technique with *Wolbachia*-based approaches: II - A safer approach to *Aedes albopictus* population suppression programmes, designed to minimize the consequences of inadvertent female release. PLoS One 10: e0135194.

Zug, R., and P. Hammerstein. 2012. Still a host of hosts for *Wolbachia*: Analysis of recent data suggests that 40% of terrestrial arthropod species are infected. PLoS One 7: e38544.

Zumpt, F. 1973. The Stomoxyine biting flies of the world. Diptera: Muscidae. Taxonomy, biology, economic importance and control measures. Gustav Fischer Verlag, Stuttgart, Germany. 175 pp.

GIS-BASED MODELLING OF MEDITERRANEAN FRUIT FLY POPULATIONS IN GUATEMALA AS A SUPPORT FOR DECISION-MAKING ON PEST MANAGEMENT: EFFECTS OF ENSO, CLIMATE CHANGE, AND ECOLOGICAL FACTORS

E. LIRA[1] AND D. MIDGARDEN[2]

[1]USDA-APHIS-IS, Mediterranean Fruit Fly Programme, Guatemala City, Guatemala; estuardo.lira@aphis.usda.gov
[2]USDA-APHIS-IS, Santo Domingo, Dominican Republic; david.d.midgarden@aphis.usda.gov

SUMMARY

The regional Mediterranean Fruit Fly Programme (Moscamed) in Belize, Guatemala and southern Mexico has applied geographic information systems (GIS) in the analysis of Mediterranean fruit fly populations since 2004. GIS allow integration of trapping data, control activities and environmental information; when combined with expert knowledge/interpretation (entomologist, ecologist and technical managers), they allow spatio-temporal analysis to determine geographic and temporal patterns, and their relationships with ecological factors and control activities. Ecological factors impacting the distribution of Mediterranean fruit fly (or medfly) populations also allow projecting pest demographics under climate change. Most of the prediction models of climate change indicate that the temperature will increase in the coming years. Temperature is a key ecological factor for insects in general, and medfly is no exception. Auclair et al. (2008) used the climate-host-insect interaction to develop predictive tools related with El Niño Southern Oscillation (ENSO) conditions, under the hypothesis that increasing temperatures will also increase medfly populations. A combination of GIS, statistical analysis, and climate change predictions indicate that hot El Niño years increase the reproductive rate of the pest, whereas cold La Niña years will have the opposite effect. With the medfly prediction model, early warnings can be provided to high-level decision makers and programme managers to act in an effective and timely-manner, including shifting in programme strategies and assigning larger budgetary resources to the programme when expecting difficult years.

Key Words: Geographic information systems, spatial-temporal analyses, temperature, El Niño Southern Oscillation (ENSO), La Niña years, Tephritidae, *Ceratitis capitata,* Belize, Guatemala, Mexico, population behaviour, population distribution, prediction models

J. Hendrichs, R. Pereira and M. J. B. Vreysen (eds.), Area-Wide Integrated Pest Management: Development and Field Application, pp. 483–504. CRC Press, Boca Raton, Florida, USA.

1. INTRODUCTION

The Mediterranean Fruit Fly Programme (Moscamed), managed jointly by the governments of Guatemala, Mexico and the USA, has been operating since 1977 to contain the Mediterranean fruit fly (*Ceratitis capitata* Wied.) (or medfly) in Guatemala and to protect the areas free of this pest in Guatemala, Mexico, Belize and the USA (Gutiérrez Samperio 1976; Enkerlin et al. 2015, 2017). Moscamed conducts two main activities:

1. Surveillance, through pest monitoring in infested areas, as well as detection and delimitation of the pest in areas of low pest prevalence and pest free areas using a geo-referenced trapping system located in Guatemala, Mexico and Belize; and
2. Control, through area-wide integrated pest management (AW-IPM) for population suppression and eradication, using a combination of environment-friendly techniques including the Sterile Insect Technique (SIT), aerial and ground sprays of an organically-approved insecticidal bait (spinosad), bait stations, and quarantine checkpoints to monitor and reduce movement of infested fruit into medfly free areas.

In 2004, Moscamed implemented a geographic information system (GIS) to manage the information related with detection of the pest, sterile fly releases, and the other activities involved in the AW-IPM activities. Since that time, information about ecological factors, such as hosts, temperature and rainfall, has been incorporated into the GIS. The GIS dataset has enabled new perspectives of the ecology and behaviour of the medfly populations but understanding the population ecology of medfly in Guatemala remains a key programme challenge. The relationship between coffee, *Coffea arabica* L. (as the main host) and fly captures was explained by Midgarden and Lira (2008). In addition, Auclair et al. (2008) found relationships between El Niño and medfly outbreak years by combining trapping and weather information.

One of the main factors that regulate the medfly populations is temperature. As with most insects, the medfly generational time is determined by degree-day accumulation. About 328°C degree-days are needed to complete one life cycle from egg to adult (Grout and Stoltz 2007). The amount of time needed for this accumulation varies with temperature, and therefore also with altitude. Degree-days are the accumulation of heat units above a "base temperature" (the minimum needed for development) and below a thermal maximum (above which development is also halted) over a 24-hour period (Pedigo 1996). Below a minimum temperature threshold, no development takes place, but above it, heat units drive development. In the case of the medfly, USDA (2003) indicated that its lower threshold is ~12°C and its upper threshold is ~28°C.

Temperature is not only important for medfly, but the increase of temperature is also one of the indicators of climate change (IPCC 2018). Climate can be defined as the long-term statistics of the meteorological elements in one particular area (WMO 1992), thus climate change is a difference in the long-term statistics of a given area between two different periods. Rahmstorf et al. (2012) observed that global mean temperature has been increasing due the climate change at 0.16 °C per decade. In areas below the upper threshold above which the medfly development is limited, these increases will have an impact on medfly population growth by shortening the time required for its life cycle, allowing the fly to complete more generations in the same time period, resulting in a higher rate of population increase.

Auclair et al. (2008) used an analysis of the climate-host-insect interaction to develop predictive tools related with El Niño conditions, under the hypothesis that an increase in temperatures will also cause increases in medfly population growth. These models have been used in Moscamed as an early warning system of pest population's growth. More recently we integrated other ecological factors, including soil types, to generate maps of the potential distribution of medfly, and how this distribution will be affected if temperatures continue to increase.

This chapter describes how Moscamed uses GIS to integrate the medfly ecology with observed patterns of populations, and how this information can be incorporated into prediction models that consider climate-host-medfly interaction in support of the pest management decision-making process.

2. MEDITERRANEAN FRUIT FLY PROGRAMME

Medfly was reported first time in Guatemala in 1975. In 1977, the governments of Guatemala, Mexico and the USA established Moscamed, a joint programme with the objective of protecting and promoting the fruit production in all three countries by containing the medfly in Guatemala (Enkerlin et al. 2015, 2017). Currently Moscamed operates in the state of Chiapas in Mexico, Guatemala and Belize to protect the medfly free areas in these countries and in the USA. The geographic area where Moscamed operates is shown in Fig. 1.

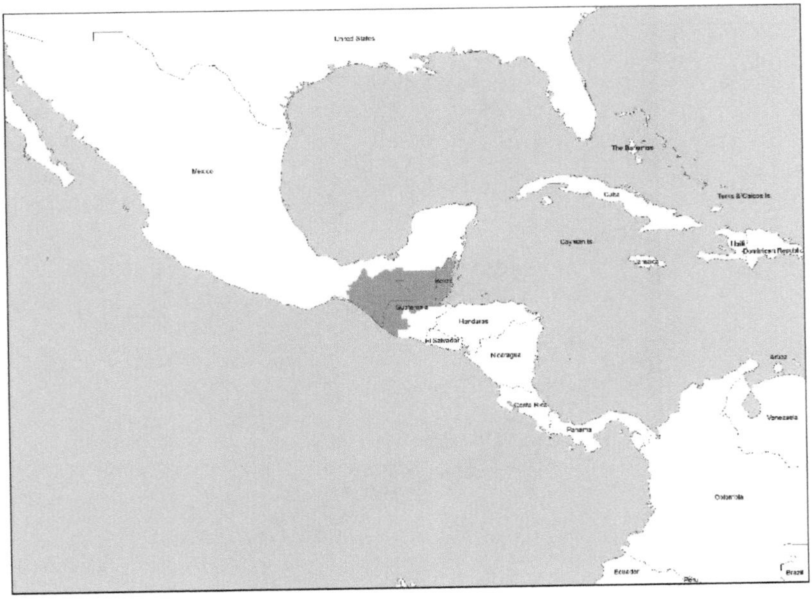

Figure 1. Area where the regional Moscamed programme operates in Belize, Guatemala and the state of Chiapas in Mexico to contain the invasive medfly, which is already established in Central and South America (credit Moscamed).

In order to detect the pest, Moscamed currently maintains a trapping network of 23 256 traps and conducts fruit sampling in strategic places. Coffee is the main and most abundant medfly host in this area. Moscamed, based on the reports of the coffee national institutions in Guatemala and Mexico, estimates that coffee covers an area of 5194 km^2. Therefore, a majority of the traps are installed in coffee production areas. The trap locations and the coffee areas are presented in Fig. 2. In 2017, Moscamed covered an area of 171 102 km^2, as can be seen in Fig. 3. Of that area, 87% (149 110 km^2) is considered as *Pest Free Area*, and most of the efforts and resources are oriented to maintain the pest free area status. A further 6% (9454 km^2) is considered as *Low Prevalence Area* and 7% (12 538 km^2) as *Suppression Area*.

The main control activity of the programme is the SIT, which is applied on an area-wide basis for prevention, eradication, and containment, depending on the presence of the pest (Hendrichs et al. 2021). The sterile fly densities (males/hectare) are determined for areas called "release blocks" based on the Rendón Method (Rendón 2008), with the aim of releasing higher densities in areas with higher pest population levels. Blocks are visited once or twice per week and evaluated on a weekly based to make adjustments in density, shape or location when needed. Weekly, Moscamed produces 1.4 billion sterile pupae and releases them in an area of around 5000 km^2 in Mexico and Guatemala. The SIT is combined with other control methods such as ground bait sprays, bait stations, and aerial bait sprays where populations are too high for only SIT releases. The distribution of sterile male release blocks and densities during one particular week in 2017 is presented in Fig. 4.

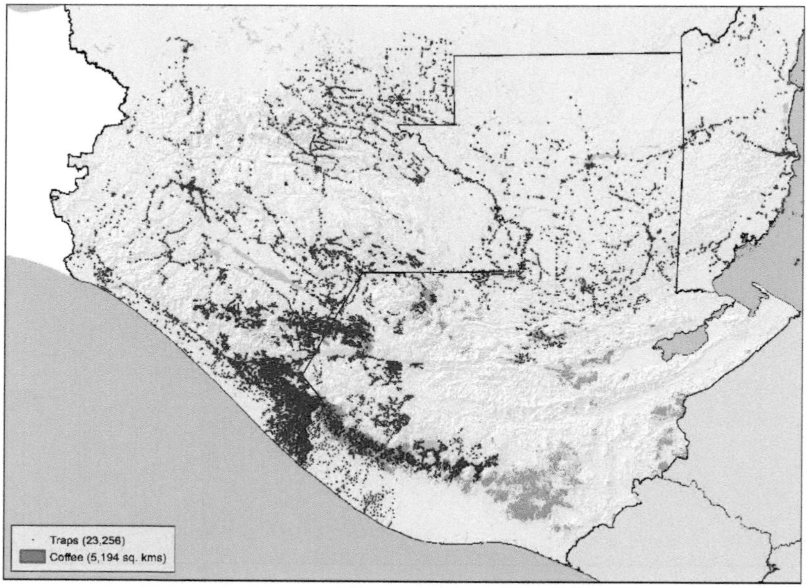

Figure 2. Moscamed trapping network in 2017, overlapping the main medfly host in Belize, Guatemala and the state of Chiapas in Mexico (credit Moscamed).

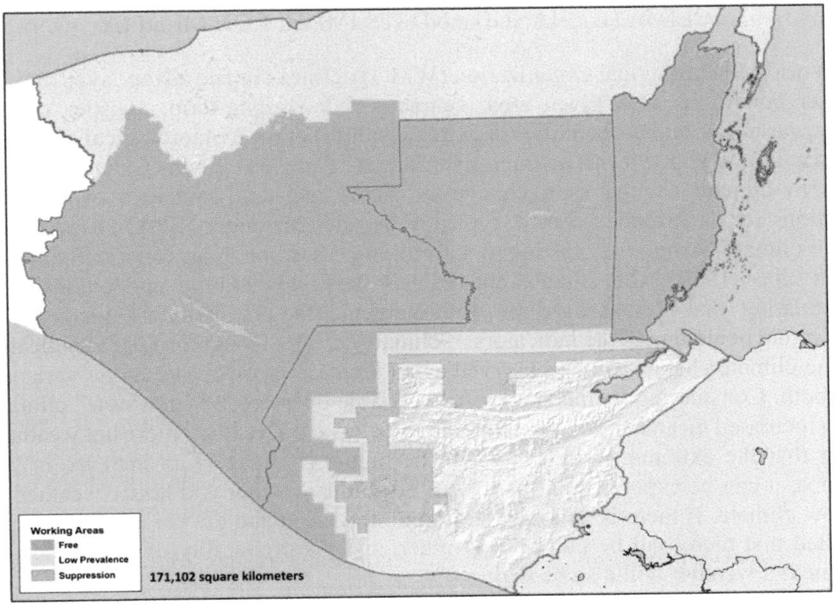

Figure 3. Moscamed working areas in 2017 in Belize, Guatemala and the state of Chiapas in Mexico (credit Moscamed).

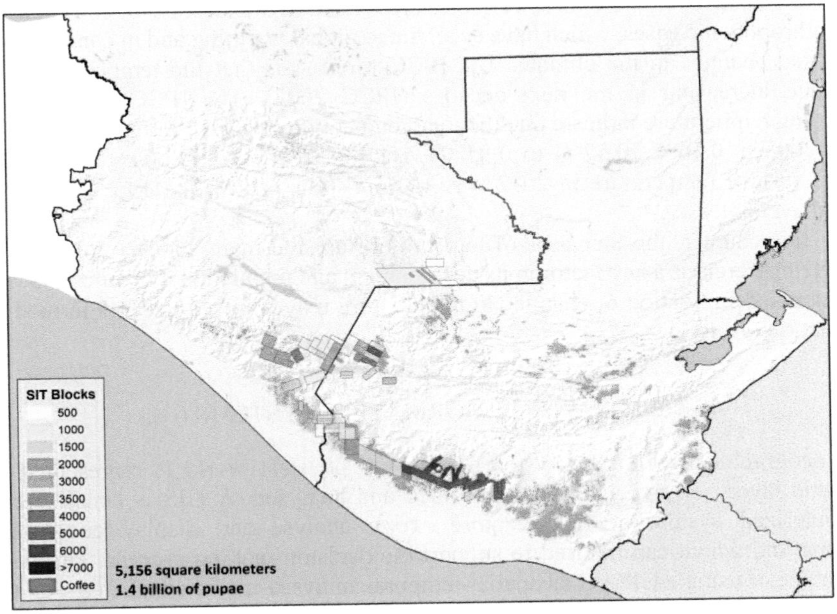

Figure 4. Sterile medfly release blocks in Guatemala and the state of Chiapas in Mexico. Numbers indicate fly release densities per hectare and per week (credit Moscamed).

3. CLIMATE CHANGE AND ITS IMPACT ON MEDFLY

The World Meteorological Organization (WMO) defines climate as the "synthesis of weather conditions in a given area, characterized by long-term statistics (mean values, variances, and probabilities of extreme values) of the meteorological elements in that area" (WMO 1992). The Intergovernmental Panel on Climate Change (IPCC) refers to climate change as a change in those long-term statistics of weather conditions for an extended period, typically decades or longer (IPCC 2018). Even though climate change may be due to natural processes, the main concern for IPCC is that since 1950′s the climate change has been accelerating and evidence is accumulating that it is caused by anthropogenic factors, with the increase of temperature being one of the indicators of climate change. Houghton (2015) indicates that the climate change can be observed as an increase in the mean, in the variance or in both. Considering temperature as the variable of interest, if the "new" climate has an increased mean temperature, this suggests less cold weather, more hot weather, and/or that the extreme hot weather will be higher. If there is an increase in the variance, it can be expected that there will be colder weather and hotter weather in the new climate. If there is an increase in both, the mean and the variance, then it is expected that there will be more hot weather, and the probability of occurrence of extreme hot weather will also be higher.

The IPCC considers that the three main lines of evidence of the climate change are: a) land and ocean surface temperature anomaly, b) sea level change, and c) greenhouse gas concentrations in the atmosphere (IPCC 2018). These three lines have shown an increase, which has become more evident after 1950. According to the IPCC, the changes in these three variables are related to the increase of emissions of the anthropogenic gases, which have accelerated global warming and in consequence catalysed changes in the climate. The IPCC forecast is that the temperature will continue increasing in the next decades (IPCC 2007). The IPCC observations regarding temperature indicate that the total temperature increase from 1850-1899 to 2001-2005 is 0.76°C [0.57°C to 0.95°C]. Hansen et al. (2013) observed that the global surface temperature in 2012 was +0.56°C (1°F) above the 1951-1980 base period average.

In this chapter, the increases of temperature are the main concern for medfly, since temperature is a key factor in its development and population dynamics. As will be discussed in Section 6, changes in temperature may trigger increases in medfly population levels.

4. GEOGRAPHIC INFORMATION SYSTEMS (GIS)

In a geographic information system (GIS), the physical world is represented as thematic layers, so that it can be described and analysed. A GIS is considered a computerized system used to acquire, store, analyse and display geographic information, which can be used to support the decision-making process. The main advantage of using a GIS is that spatial-temporal analyses can be conducted, and the results presented in an "easy-to-read" format such as maps, which are graphic and simplified representations of the reality.

With Moscamed in Guatemala establishing its GIS in 2004 (Lira 2010), the trapping network information generated by its operations is converted into geographic data and stored with other geographic layers such as land use, temperature, rainfall, altitude, and soil types in a digital format. Using different GIS operations, modelling and analyses of the medfly populations are conducted and different scenarios are generated. With the adequate cartographic techniques, i.e. generalization and symbolization, maps are provided to decision-makers to support their pest management decisions. According to Huisman and de By (2009) symbolization is the process to choose the visual design employed to communicate information on a map in an efficient manner by combining the visual variables of colour, intensity, size, orientation, transparency, and fill. The same authors indicate that generalization is the process of producing a graphic representation of a smaller scale from a larger original scale.

Midgarden et al. (2014) described that, for tephritid fruit fly programmes, GIS serve as a bridge between the trap samples and the spatial analysis methods. These methods enable: 1) improvement in the way to report and summarize the collected information in a more meaningful way; 2) identification of unrecognized patterns of population growth and spread, and 3) development of improved integrated pest management strategies. In the case of Moscamed, the use of GIS and improved understanding of medfly ecology allowed to change the containment and eradication strategies of the medfly in Guatemala. Starting in December of 2007, the Gradual Advance Plan (GAP) was implemented (McGovern et al. 2008). The GAP consists of pushing back the leading edge of the infestation by 10 to 20 km per year, with the subsequent movement of the low suppression and suppression areas into the adjacent infested areas in a strategy known as the "rolling carpet" approach (Hendrichs et al. 2021). The GAP allowed expanding the medfly-free area 150 km into Guatemala in less than four years, despite severe budget reductions (Enkerlin et al. 2017).

5. POPULATION BEHAVIOUR OF MEDITERRANEAN FRUIT FLY IN GUATEMALA

Over the year, in south-western Guatemala, the growth of medfly populations in infested areas has a logistic trend, with an S-shaped curve. Fig. 5 describes this behaviour in four phases. The population growth begins in November of each year, reaching maximum growth rate in January, and reaching the maximum population size in February. After that the growth rate decreases, and then the fly population gradually declines, and finally reaches a minimum in October/November. Because of the detection system used (mainly based on adult traps), this behaviour is measured as captures of adult flies, and it seems to occur independent of the availability of maturing coffee berries (Fig. 6). However, as explained by Midgarden and Lira (2008), the reason for this time-displacement is that the populations seen as adults were laid as eggs before, when coffee berries were available.

Depending on the altitude and the host availability, these 4 phases can occur earlier or later in the year, especially phases 2 and 3. But the general pattern is repeated every year. It can be said that the "medfly-year" runs from November to October.

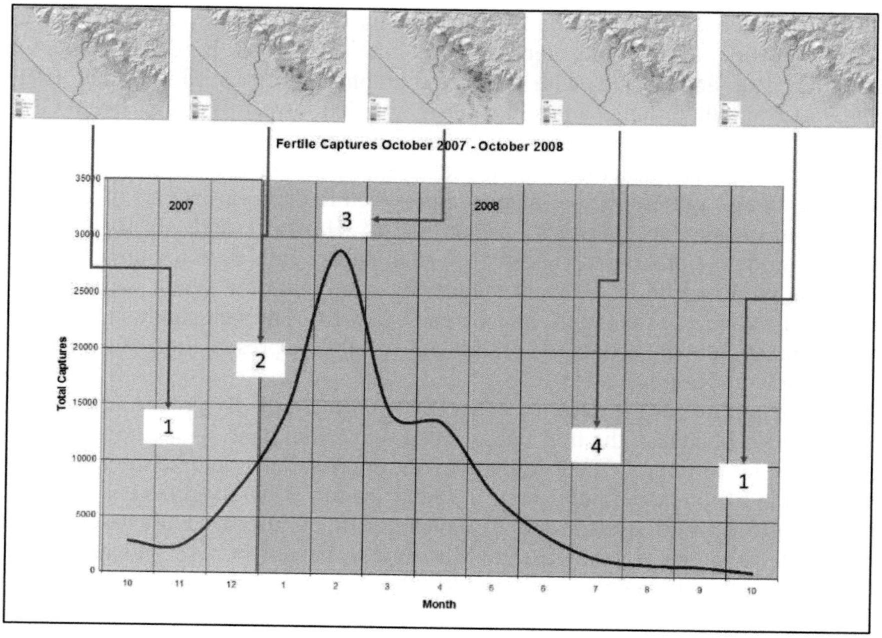

Figure 5: Four phases of medfly population behaviour in south-western Guatemala. 1. Beginning of population growth; 2. Maximum growth rate; 3. Maximum population size; 4. Decreasing rates and population declines, reaching a minimum size (credit Moscamed).

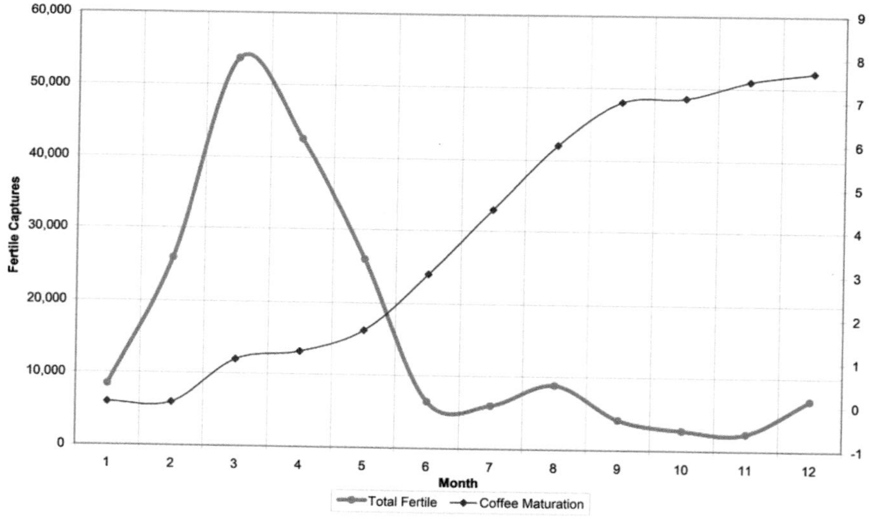

Figure 6. Relationship between captures of fertile medflies and coffee maturation throughout the year 2007 in the south-western region of Guatemala between 600 and 1500 m above sea level (the authors, based on Moscamed and ANACAFE 2008 data).

6. MEDITERRANEAN FRUIT FLY ECOLOGY: EFFECTS OF HOST, TEMPERATURE AND SOILS ON MEDFLY POPULATIONS

The population behaviour of medfly in Guatemala, described in the previous Section, can be explained by different ecological factors. We analysed three factors affecting medfly: host, temperature, and soils. The contribution of each of these ecological factors to medfly populations is discussed in this Section.

6.1. Main Host

Coffee is the main host of medfly in southern Mexico and Guatemala (Gutiérrez Samperio 1976). Midgarden and Lira (2008) explained how the coffee phenology (including events such as flowering and fructification) interacts with medfly biology. According to these authors, the adult fly population outbreaks may appear in one location after the coffee harvest, while the growing larval population was actually present at another location and at an earlier time (during the coffee berry fructification period).

This "shift" of the pest in time and space is related with the altitude gradient in which the main coffee production areas in Guatemala are located, with elevations varying from ~400 to ~2100 meters above the sea level. This altitudinal gradient drives a gradient in the time of maturation and harvesting of coffee; in consequence adult flies can infest the coffee berries at lower elevations in July, and gradually move up to higher elevations following the maturing phenology of coffee final harvest in December or later.

Midgarden and Lira (2008) observed that, due to the pupation time after coffee harvest, the emergence of the highest population of adult flies will occur in March-April of the next year. At that moment, coffee berries are scarce, likely resulting in extensive dispersal of mature adults to search for other available hosts including mandarin (*Citrus reticulata* L.), orange (*C. sinensis* L.), peach, (*Prunus persica* L.) and pear (*Pyrus communis* L.) at middle to high altitudes and guavas (*Psidium guajava* L.), caimito (*Chrysophilum caimito* L.) and tropical almond (*Terminalia catappa* L.) at lower altitudes (Enkerlin et al. 2016). In summary, Midgarden and Lira (2008) concluded that

> *"flies are captured in detection traps in March through April and can be seen as part of an ecological "shell game": the fly population outbreaks appear in one location in April (non-infested or host-poor areas west of the leading edge of the pest population), while the growing population was actually present at another location months earlier (e.g. December in untreated coffee areas to the East)".*

Fig. 6 summarizes the coffee-medfly relationship explained by Midgarden and Lira (2008) in relation to the months of the year. In the left Y axis, the number of fertile or wild fly captures per month is presented, and in the right Y axis the coffee maturation level is shown, in a scale of 0 to 8, being 0 no berries at all, and 8 total maturation. After December it is expected that most of the harvest occurs, "cleaning-up" mature coffee berries from the field.

6.2. Temperature

Regarding temperature, Ricalde et al. (2012) indicated that an insects' development depends on thermal requirements, with each insect species having an optimal temperature range for development, limited by lower and upper thresholds (base temperature (*Tb*) and upper limit (*Ts*)) plus a required thermal accumulation for developmental transition to complete a life cycle. The thermal accumulation between *Ts* and *Tb* in one day (24 hours) is measured in "degree-days". These are calculated as follows (example): if *Tb* of an insect is 10 °C, and temperature remains constant at 15 °C for 24 hours, 5 degree-days will be accumulated.

Ricalde et al. (2012) found that the base temperatures for medfly were between 8.47°C and 9.60°C and the degree-days required to complete the life cycle varies from 328 to 350, depending on the location. This is in accordance with Grout and Stoltz (2007), who found that for *C. capitata* be able to complete an egg-to-egg cycle (hatching from the egg, larvae growth, transformation into a pupae, emergence as an adult, reaching sexual maturity, copulation and laying of viable eggs), the thermal constants are: 337.8 degree-days, minimum development threshold of 9.6°C, maximum development threshold of 33.0°C, and optimum development threshold of 28.5°C. According to USDA (2003), the parameters for medfly are: a) ~ 328 degree-days for completing a life cycle, b) ~ 12°C as minimum threshold, and c) ~ 28°C as maximum. These estimates vary among them, but they can be used as reference to estimate the length of medfly life cycle.

Using the Grout and Stoltz (2007) thermal constants as reference, it is possible to estimate the length of the life cycle. If it is assumed that the daily temperature is constant at 28.5 °C, 17.87 days will be required to complete a life cycle, since each day 18.90 degree-days will be accumulated (28.5 °C minus 9.6 °C) to reach the needed 337.8 degree-days. If the temperature is constant at 20 °C, the number of days to complete a life cycle will be 32.48, since every day 10.40 degreed days will be accumulated (20 °C minus 9.6 °C). This dependency of insect development on temperature drives the population's behaviour: temperature speeds up or slows down the life cycles, and in consequence the resulting number of flies in a fixed period. If it is considered that the "medfly-year" runs from November to October (as indicated in Section 5) and the parameters proposed by USDA (2003) are applied to the average daily temperature of one weather station in one site of Moscamed's suppression area, it can be estimated that the number of life cycles for the "medfly-year" starting November 1st of 2012 to October 30th of 2013 is 7.74 (Fig. 7).

Based on these life cycles, and considering the other ecological factors as constant (host availability and soils) and an estimated population increase rate of 6x (Rendon 2008), from one wild female fly on day one of the medfly-year, after 365 days a medfly population of 1 021 780 flies can be expected (Fig. 8). It is important to stress that the quantitative estimate of the number of flies from one female does not relate directly to real population patterns (Fig. 5) because it considers only potential maximum, and not limitations from factors like host availability or predation.

To measure the impact of increasing temperatures on medfly populations, and assuming an increase of 1°C of temperature, the same estimation was made. That estimation indicates that with such a temperature increase, the number of life cycles in 365 days will be 8.86 (Fig. 9). Even though that it is only 1.11 more life cycles, under the same assumptions (host availability, soils, and population increase rate), with that increase of temperature, a population of 7 463 038 flies (more than 7 times higher) is to be expected after 365 days (Fig. 10). These estimations reflect the drastic effect of temperature on medfly populations.

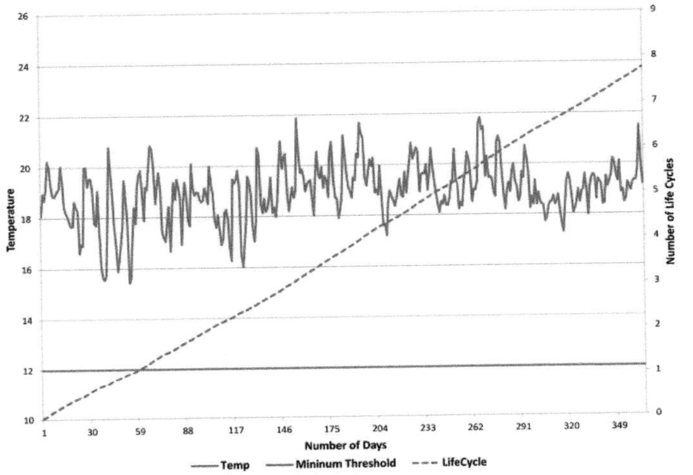

Figure 7. Average daily temperature and number of medfly life cycles expected from November 1st of 2012 to October 30th of 2013 in a coffee farm in the suppression area in Guatemala at 1600 meters above sea level (the authors, based on weather information of ANACAFE 2014).

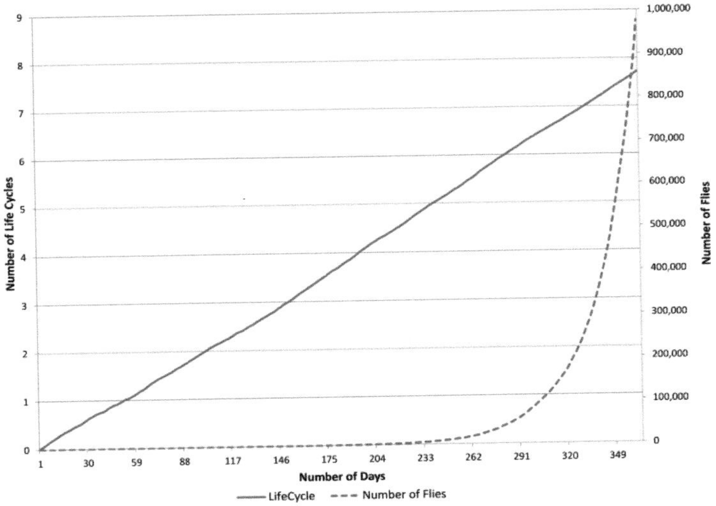

Figure 8. Number of life cycles and number of medflies (offspring from one wild female fly on day one) expected from November 1st of 2012 to October 31th of 2013 in a coffee farm in the suppression area in Guatemala at 1600 meters above sea level (the authors, based on weather information of ANACAFE 2014).

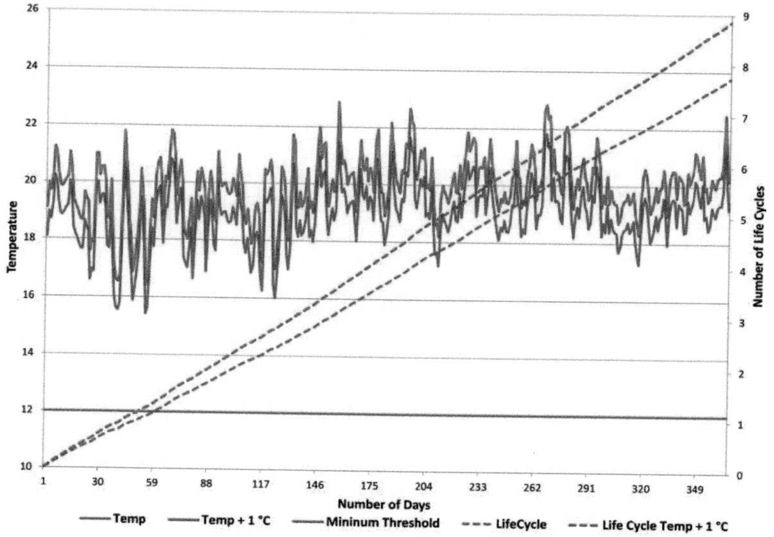

Figure 9. Number of medfly life cycles expected in a year in a coffee farm in the suppression area in Guatemala at 1600 meters above sea level in relation to an average daily temperature increase of 1°C (the authors, based on weather information of ANACAFE 2014).

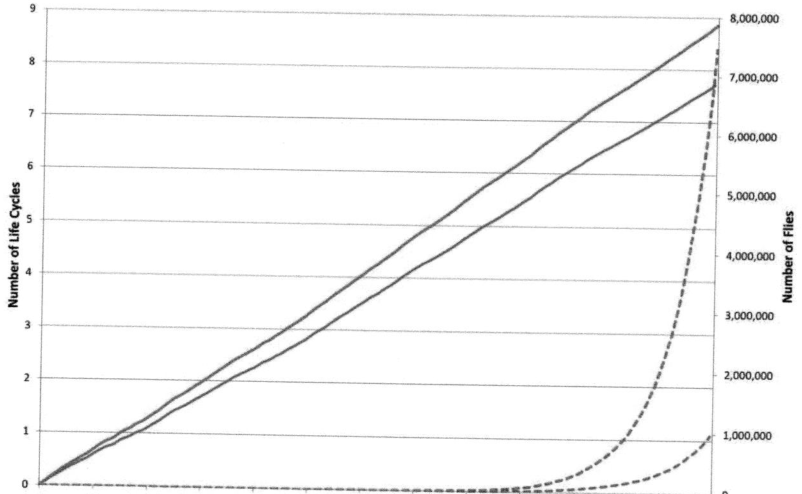

Figure 10. Number of medfly life cycles and number of medflies (offspring from one wild female fly on day one) expected in a year with an average daily temperature increase of 1°C (the authors, based on weather information of ANACAFE 2014).

6.3. *Soil Texture*

The last factor of the three factors considered here is soil. Under similar conditions of host availability and temperature, differences in the size of medfly populations have been found. Those differences might be explained by factors such as soil types. Larval and, mainly, pupal stages of the medfly occur in the soil, so soil conditions will affect medfly pupae survival. Eskafi and Fernandez (1999) found that pupal survival was negatively correlated with the soil bulk density, but positively with percentage of soil porous space and percentage of water saturation.

To estimate the relationship between soil texture and the presence of medfly 17 014 traps in Guatemala were used. For each trap, the maximum number of flies captured from 2004 to 2016 was obtained and the traps were overlaid with a map of soil textures (Simmons et al. 1959; MAGA 2000). The textures were classified from 1 to 10, according to the content of sand. In this classification, 1 included very clayey soils (almost no sand and high-bulk density) and 10 included very sandy soils (almost only sand and low high-bulk density). The result of overlaying the traps with the classified soil textures was that each trap had a texture class and the maximum number of flies captured. The average of the maximum number of flies captured per texture class classified by sand content was calculated and plotted indicating a positive relationship between fly captures and the content of sand (Fig. 11).

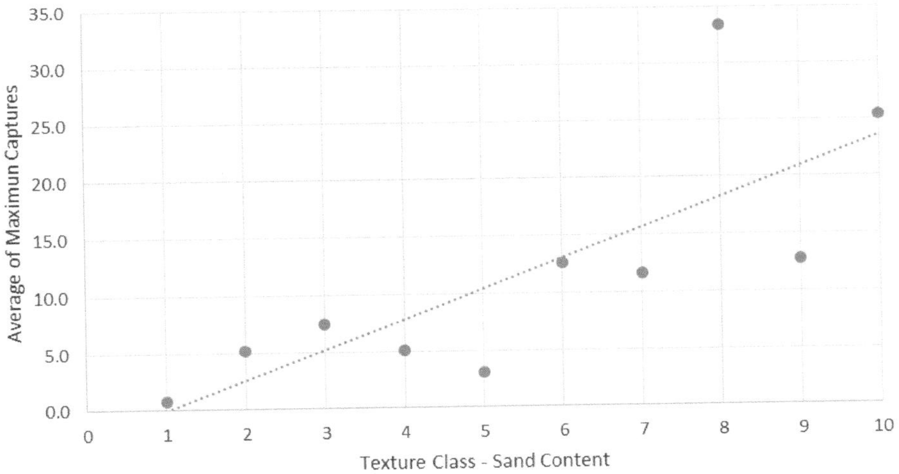

Figure 11. Relationship between medfly captures and soil texture classes (1 almost no sand and 10 very sandy soils) (the authors, based on data of Moscamed and the soil information of MAGA 2001).

From the three factors analysed, we can infer that the maximum potential for presence of medfly occurs in coffee production areas with yearly average temperatures of around 20°C, and sandy-loam soils; being the coffee availability for oviposition the main factor, either a) promoting population growth or b) restricting and decreasing population growth.

7. MEDITERRANEAN FRUIT FLY CAPTURES FROM 2004 TO 2016

Even though the described population curve is observed every year, the difference from one medfly-year to another is that the maximum population size might be higher or lower. Fig. 12 presents a sequence of the yearly average fly per trap per day (FTD) of the wild populations from 2004 to 2016.

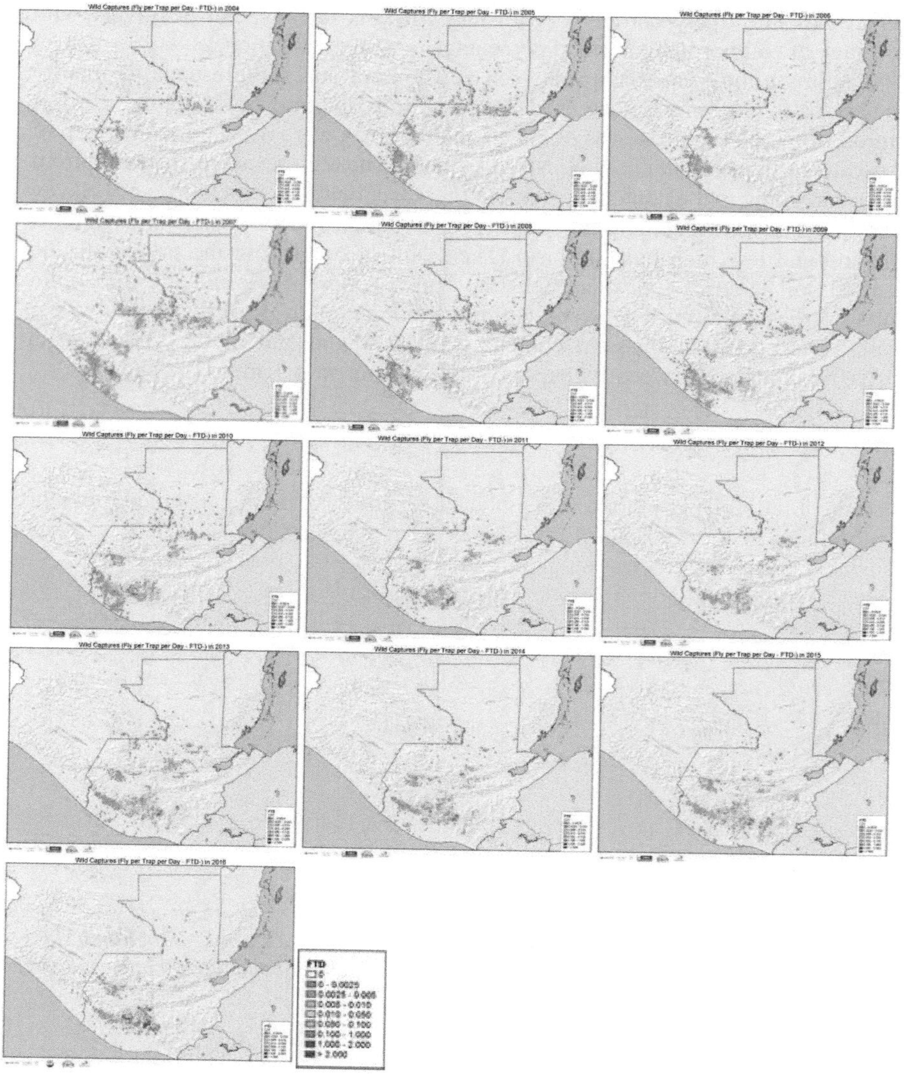

Figure 12. Average Fly per Trap per Day (FTD) numbers per year, from 2004 to 2016 in Guatemala and southern Mexico (credit Moscamed).

During this 13-year period it has been observed that there have been "good" and "bad" years regarding the number of flies captured. A "good-medfly-year" occurs when the maximum population size is low, as measured by relatively few captures in the infested areas, and as a consequence few or no finds in the neighbouring low prevalence and free areas. In the sequence shown in Fig. 12, 2004, 2006, 2012 are considered as "good" years. In contrast, a "bad-medfly-year" occurs when the maximum population size is higher than normal, and the number of captures is very high in the infested areas, spreading into the low prevalence and free areas. In the period of 2004 to 2016, years 2007 and 2016 are considered "bad" years.

Even though most of the control activities (SIT application, ground and aerial sprays, quarantine, and mechanical control) were conducted in a similar way between 2004 and 2016, a "jump" from one good year to a bad year was observed periodically, with no apparent reason. For example, between 2006 and 2007 a huge population increase occurred (Fig. 13). There are several hypotheses that have been advanced to explain this pattern.

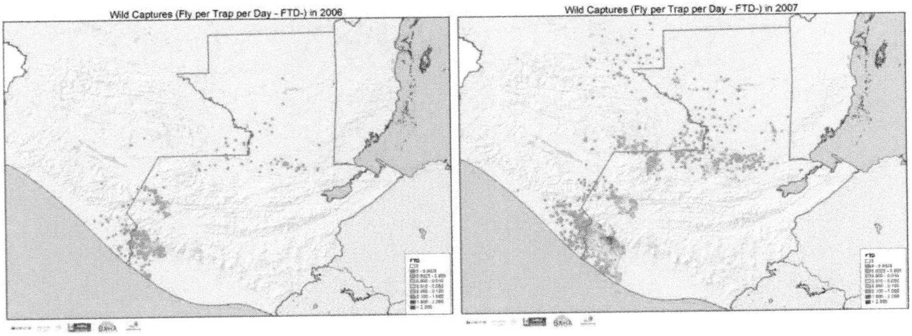

Figure 13. Comparison of wild medfly captures in the years 2006 and 2007, showing a drastic increase in the number of captures in 2007 (credit Moscamed 2017).

Auclair et al. (2008), suggested that the periodic changes are related to the "El Niño" pattern (El Niño Southern Oscillation or ENSO) (Wang et al. 2017). They analysed the trapping information and related it with temperature and rainfall, finding a correlation between the weather conditions in the "lead-in" year (6 months previous to the peak of the captures) and medfly population dynamics. The "bad" years were classified as "outbreak years" or "medfly storm years", and the "good" years as normal years. These relationships indicated that dry and hot "lead-in" years will produce a "medfly-storm" or an "outbreak year", while wet and cool "lead-in" years will produce lower than average trap captures or normal years. Auclair et al. (2008) concluded that before a medfly-storm year:

> *"rainfall was less and temperature was greater on average during the key months of population growth during the lead in years compared to average".*

These conditions, in the area in which Moscamed operates, are generated by the El Niño/La Niña cycle, with dry/hot years occurring with the El Niño phenomenon. In summary, an "El Niño Lead-in Year" will drive to a "medfly-storm year". With this pattern detected, Auclair et al. (2008), generated the "El Niño Forecast Plume", which basically suggests that before a year of interest that is expected to be a "medfly-storm" year, the signal of El Niño will increase.

This model, generated in 2008, was executed again in 2015 (Allan Auclair, personal communication). An updated version of the model, together with the ENSO data for 2015 led to a prediction that 2016 was going to be an outbreak year. The prediction by Auclair was borne out in 2016, which was indeed a fly-storm year. The maximum number of captures in 2016 was much above the normal captures and much higher compared with 2014 and 2015 (Fig. 14).

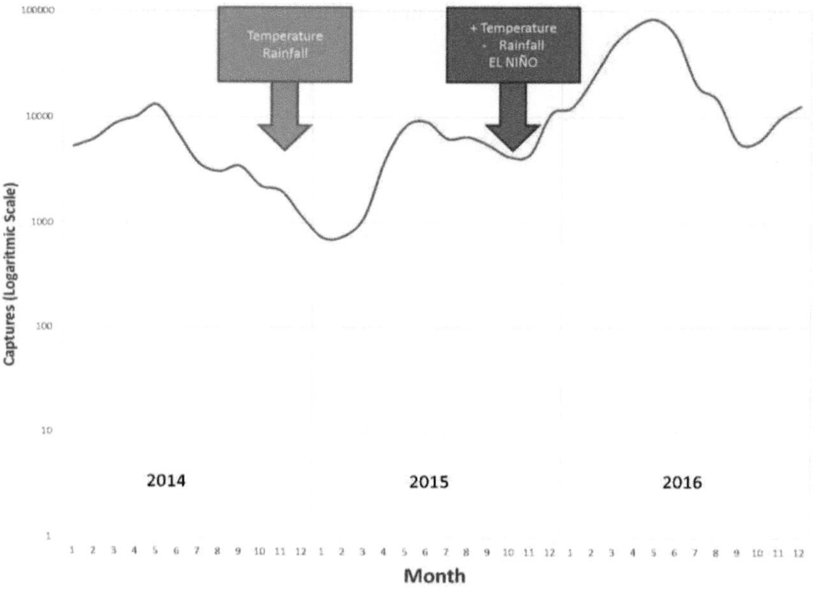

Figure 14. Number of fertile medfly captures per month from 2014 to 2016 in the region (the authors, based on the information of Moscamed).

By overlaying the medfly captures with the ENSO Anomalies from 2015 to 2017 and the Forecast Plume, it can be observed that what Auclair predicted in 2015 was correct; as shown in Fig. 15.

8. DISTRIBUTION AND PREDICTION MODELS

In this Section we generate medfly distribution models based on the information generated by Moscamed, and analysis of population demographics due to the ecological variables mentioned above (host, temperature and soil texture).

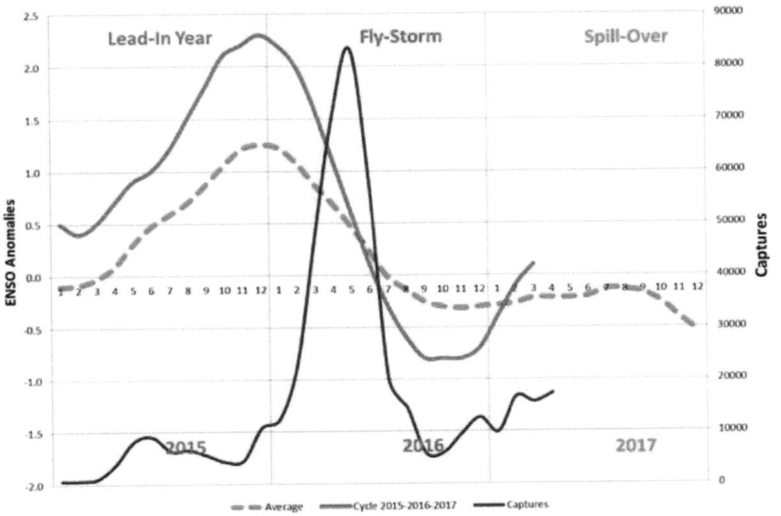

Figure 15. El Niño signal (left y-axis), outbreak years and medfly capture from 2015 to 2017 (right y-axis). The green dotted line is the "El Niño Forecast Plume" for an outbreak year generated by Auclair et al. (2008). The red solid line is the El Niño signal as reported by NOAA (2017). The brown continuous line is the total number of captures for medfly in the region from 2015 to 2017 (the authors, based on information of Moscamed, Auclair et al. 2008 and NOAA 2017).

Trapping information (a dataset of 17 014 traps serviced from 2004 to 2015) was overlaid with the ecological factors in order to estimate the distribution of the medfly in areas with no traps in Guatemala, using the maximum entropy software (Maxent) (Phillips et al. 2006). Maxent is a software widely used for modelling species distribution. It uses machine learning methods to statistically estimate the relationships within species presence locations (response variable) and a set of environmental predictors (explanatory variables). The response variable selected was the maximum number of flies captures in one trap in one week. Moscamed uses Jackson (trimedlure attractant) and Phase IV (open-bottom baited with a dry food-based synthetic attractant) traps for monitoring the wild populations.

According to Midgarden et al. (2004) there is no significant difference in the total number of wild flies captured for these two types of traps. So, both trap types were included in the analysis. This variable represents the maximum level of infestation in one site. The explanatory variables selected were:

1. Distance to coffee – as a measurement of the main host, generated using the coffee production areas from the land use map of Guatemala (GIMBOT 2014).
2. Temperature – variable related with the life cycle, generated from the INSIVUMEH weather stations in the digital database of MAGA (2001).
3. Soil texture – classified for clay to sandy, as a measurement of the effect of this condition on the larval/pupal stage, obtained from the soil maps of Guatemala (MAGA 2001).

With the GIS, the information was prepared to be able to use it in Maxent. The results of the modelling are presented in Fig. 16. The output is the logistic probability of presence of medfly, ranging between values from 0 to 1.

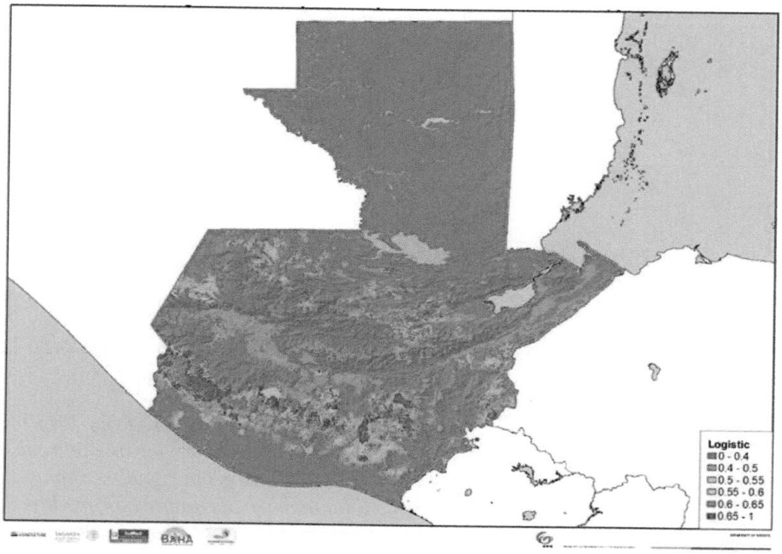

Figure 16. Logistic probability of medfly occurrence in Guatemala using the maximum entropy (Maxent) method (the authors, based on information of Moscamed).

The IPCC observations regarding temperature indicate that the total temperature increase from 1850–1899 to 2001–2005 is 0.76°C [0.57°C to 0.95°C] (IPCC 2007). Hansen et al. (2013) observed that the global surface temperature in 2012 was +0.56°C (1°F) warmer than the 1951-1980 base period average. Rahmstorf et al. (2012) observed that global mean temperature has been increasing due to climate change at a rate of 0.16°C/decade.

With these observations of consistent temperature increases and the wide range of medfly tolerances to temperatures (minimum and maximum), it is expected that fly populations will increase each year under normal conditions. To estimate the effect of increasing temperature due the climate change, the temperature in Maxent was modified by adding 1°C. The results of that estimation are shown in Fig. 17.

According to the modelling conducted, and the prediction of the increase of temperature, it seems that in some areas the probability of occurrence of medfly will increase at higher altitudes, mainly in the temperate areas, but in lower altitudes (subtropical areas) this probability will decrease. This prediction might be explained by the fact that medfly has lower and higher temperature thresholds. As indicated before, it is expected that the maximum potential for presence of medfly will occur in areas with yearly average temperatures of around 20°C.

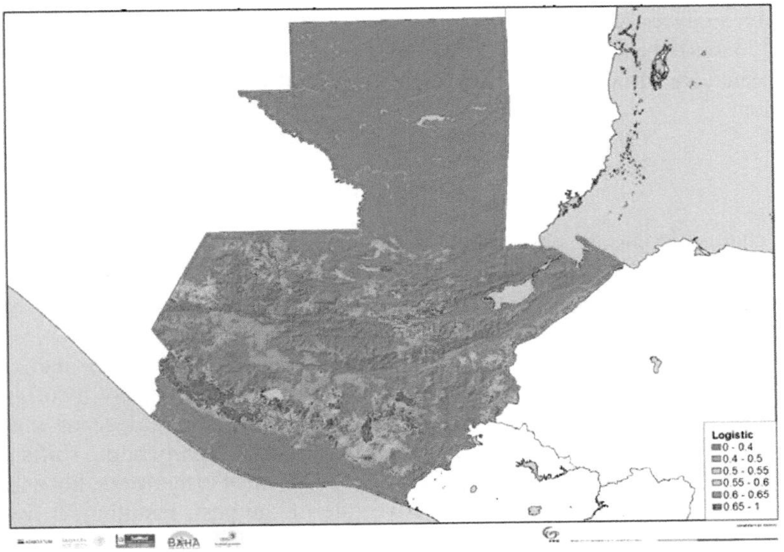

Figure 17. Logistic probability of medfly occurrence in Guatemala using the maximum entropy (Maxent) method and considering an increase in temperature of 1 °C (the authors, based on information of Moscamed).

In the subtropical areas, the temperature is higher than this temperature, in consequence the increase in temperature might decrease the probability of medfly occurrence. In contrast, in the temperate areas, the temperature is below the optimal temperature of 20 °C, and the expected increase in temperature might also increase the probability of medfly occurrence, since the temperature will be closer to the optimal. Fig. 18 shows the expanded area of south-western Guatemala, where this contrast can be appreciated.

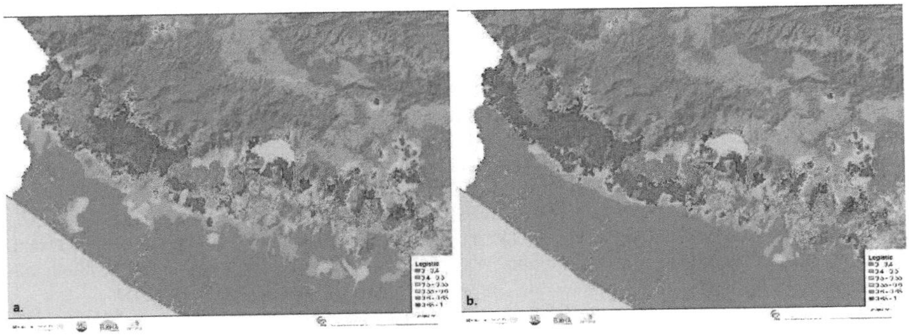

Figure 18. "Zoom-in" to the south-western area of Guatemala to see the differences between the modelling with current temperature (a.) and with an increase of temperature by 1°C (b.) (the authors, based on information of Moscamed).

Further analyses should include other levels of temperature increases and the effect of increased variance of temperature. Future analysis should also consider other explanatory variables, as well as using other modelling techniques for species distribution.

9. CONCLUSIONS

Climate change including temperature increase appears inevitable. Nevertheless, understanding how this will affect the population ecology of a pest will provide programme managers with key information for decisions on insect pest management that can minimize the negative effects of these changes.

Knowledge about the effects of climate, in particular temperature and other ecological factors, such as host phenology and the population trends of medfly, has allowed us to develop a predictive model that can be applied as a decision-making tool in support of effective medfly programme management. Temperature shifts from climate change have a direct impact on medfly populations. Furthermore, hot and dry El Niño years will increase the reproductive rate of the pest, resulting in overall population increase, whereas cold La Niña years will have the opposite effect, resulting in population reduction. Well-coordinated AW-IPM activities based on information analysis is crucial to avoid the increases of medfly populations. With the prediction models generated, early warnings can be provided to high-level decision-makers and programme managers to act in an effective and timely-manner, including shifting programme strategies and assigning larger budgetary resources to the programme when expecting difficult years.

10. ACKNOWLEDGEMENTS

The authors would like to thank Nick C. Manoukis and Walther Enkerlin for many suggestions that significantly improved this manuscript.

11. REFERENCES CITED

ANACAFE (National Association of Coffee). 2008. Personal communication with Oscar García, Technician of ANACAFE, Guatemala.

ANACAFE (National Association of Coffee). 2014. ANACAFE Weather Stations System.

Auclair, A., C. Chen, W. Macheel, P. Rendón, E. Lira, D. Midgarden, R. Magery, D. Enfield, and A. Anyamba. 2008. Using climate indices to predict Medfly outbreaks in Guatemala and Mexico. Poster presented at the 7th meeting of the Working Group on Fruit Flies of the Western Hemisphere, November 2008, Mazatlán, México.

Enkerlin, W., J. M. Gutiérrez-Ruelas, A. V. Cortes, E. C. Roldan, D. Midgarden, E. Lira, J. L. Z. López, J. Hendrichs, P. Liedo, and F. J. T. Arriaga. 2015. Area freedom in Mexico from Mediterranean fruit fly (Diptera: Tephritidae): A review of over 30 years of a successful containment program using an integrated Area-Wide SIT approach. Florida Entomologist 98: 665–681.

Enkerlin, W., A. Villaseñor, S. Flores, D. Midgarden, E. Lira, P. Rendon, J. Hurley, E. Salazar, W. Méndez, R. Castañeda, E. Cotoc, J. L. Zavala, H. Celedonio, and J. M. Gutiérrez Ruelas. 2016. Descriptive analysis of the factors affecting population fluctuation of the Mediterranean fruit fly (*Ceratitis capitata*, Wied.) in coffee areas located in Guatemala and its implications in IPM Strategies, pp. 46–63. *In* B. Sabater-Muñoz, T. Vera, R. Pereira, and W. Orankanok (eds.), Proceedings of the 9th International Symposium on Fruit Flies of Economic Importance, Bangkok, Thailand.

Enkerlin, W. R., J. M. Gutiérrez Ruelas, R. Pantaleon, C. Soto Litera, A. Villaseñor Cortés, J. L. Zavala López, D. Orozco Dávila, P. Montoya Gerardo, L. Silva Villarreal, E. Cotoc Roldán, F. Hernández López, A. Arenas Castillo, D. Castellanos Dominguez, A. Valle Mora, P. Rendón Arana, C. Cáceres Barrios, D. Midgarden, C. Villatoro, E. Lira Prera, O. Zelaya Estradé, R. Castañeda Aldana, J. López Culajay, F. Ramírez y Ramírez, P. Liedo Fernández, G. Ortíz Moreno, J. Reyes Flores, and J. Hendrichs. 2017. The Moscamed regional programme: Review of a success story of Area-Wide Sterile Insect Technique application. Entomologia Experimentalis et Applicata 164: 188–203.

Eskafi, F., and M. Fernandez. 1999. Larval–pupal mortality of Mediterranean fruit fly (Diptera: Tephritidae) from interaction of soil, moisture, and temperature. Environmental Entomology 19: 1666–1670.

Grout, T., and K. Stoltz. 2007. Developmental rates at constant temperatures of three economically important *Ceratitis* spp. (Diptera: Tephritidae) from Southern Africa. Environmental Entomology 36: 1310–1317.

GIMBOT (Grupo Interinstitucional de Monitoreo de Bosques y Uso de la Tierra). 2014. Mapa de bosques y uso de la tierra 2012 y mapa de cambios en uso de la tierra 2001 - 2010 para estimación de emisiones de gases de efecto invernadero. Documento informativo. 16 pp. Ministerio de Ambiente y Recursos Naturales. Gobierno de la República de Guatemala, Guatemala City, Guatemala.

Gutiérrez Samperio, J. 1976. La Mosca del Mediterraneo *Certatitis capitata* (Wiedeman) y los factores ecológicos que favorecerían su establecimiento y propagación en México. Dirección General de Sanidad Vegetal. SAGARPA. Talleres Gráficos de la Nación. México, D. F., México. pp. 223.

Hansen, J., M. Sato, and R. Ruedy. 2013. Global temperature update through 2012. Columbia University, New York, NY, USA.

Hendrichs, J., M. J. B. Vreysen, W. R. Enkerlin, and J. P. Cayol. 2021. Strategic options in the use of the Sterile Insect Technique, pp. 841–884. In V. A. Dyck, J. Hendrichs, and A. S. Robinson (eds.), Sterile Insect Technique – Principles and practice in Area-Wide Integrated Pest Management. Second Edition. CRC Press, Boca Raton, Florida, USA.

Houghton, J. T. 2015. Global warming: The complete briefing. Cambridge University Press. 5th Edition. Cambridge, UK. 454 pp.

Huisman, O., and R. de By. 2009. Principles of Geographic Information Systems: An introductory textbook. International Institute for Geo-Information Science and Earth Observation. Enschede, The Netherlands. 540 pp.

IPCC (Intergovernmental Panel on Climate Change). 2007. Summary for policymakers, pp. 1–17. *In* S. Solomon, D. Qin, M. Manning, Z. Chen, M. Marquis, K. B. Averyt, M. Tignor and H. L. Miller (eds.), Climate change 2007: The physical science basis. Contribution of working group I to the Fourth Assessment Report of the Intergovernmental Panel on Climate Change. Cambridge University Press, Cambridge, United Kingdom and New York, USA.

IPCC (Intergovernmental Panel on Climate Change). 2018.

Lira, E. 2010. Uso de sistemas de información geográfica, pp. 159–165. *In* P. Montoya, J. Toledo, and E. Hernandez (eds.), Moscas de la fruta: Fundamentos y procedimientos para su manejo. S y G Editores. México, D. F., México.

MAGA (Ministerio de Agricultura, Ganadería y Alimentación de Guatemala). 2000. First approximation to the taxonomic classification of the soils of the Republic of Guatemala, scale 1:250,000. Technical memory. Financial support of: Plan de acción forestal para Guatemala (PAFG); Instituto Nacional de Bosques (INAB). Guatemala, Guatemala.

MAGA (Ministerio de Agricultura, Ganadería y Alimentación). 2001. Base de datos espacial digital de Guatemala. Digital information generated by Banco Interamericano de Desarrollo (BID) – 1147/OC-GU, Unidad de Políticas e Información Estratégica (UPIE-MAGA), Guatemala, and Centro Agronómico Tropical de Investigación y Enseñanza (CATIE), Costa Rica.

McGovern, T., E. Lira, and P. Rendón. 2008. Medfly program operational update: Targeted suppression of population centers of *Ceratitis capitata* Wied. through the use of GIS/GPS technology, pp. 9–10. *In* P. Montoya, F. Fleischer, and S. Flores (eds.), Proceedings of the 7th meeting of the Working Group on Fruit Flies of the Western Hemisphere. Mazatlán, México.

Midgarden, D., and E. Lira. 2008. Ecological relationship of Medfly and coffee in Guatemala and Mexico, pp. 241–247. *In* R. Sugayama, R. Zucchi, S. Ovruski, and J. Sivinski (eds.), Fruit Flies of economic importance: From basic to applied knowledge. Proceedings of the 7th International Symposium on Fruit Flies of Economic Importance. 10-15 September 2006. Salvador, Bahia, Brazil.

Midgarden, D., E. Lira, and M. Silver. 2014. Spatial analysis of tephritid fruit fly traps, pp. 277–320. *In* T. Shelly, N. Epsky, E. Jang, J. Reyes-Flores, and R. Vargas (eds.), Trapping and the detection, control, and regulation of tephritid fruit flies: Lures, area-wide programs, and trade implications. Springer Science+Business Media, Dordrecht, The Netherlands.

Midgarden, D., O. Ovalle, N. Epsky, H. Puche, P. Kendra, P. Rendon, and R. Heath. 2004. Capture of Mediterranean fruit flies (Diptera: Tephritidae) in dry traps baited with a food-based attractant and Jackson traps baited with trimedlure during sterile male release in Guatemala. Journal of Economic Entomology 97: 2137–2143.

NOAA (U.S. National Oceanic and Atmospheric Administration). 2017. Historical El Niño / La Niña episodes (1950-present).

Pedigo, L. 1996. Entomology and pest management. Second Edition. Prentice Hall. New Jersey, USA. 679 pp.

Rahmstorf, S., G. Foster, and A. Cazenave. 2012. Comparing climate projections to observations up to 2011. Environmental Research Letters 7: 044035.

Phillips, S., R. Anderson, and R. Schapire. 2006. Maximum entropy modeling of species geographic distribution. Ecological Modelling 190: 231–259.

Rendón, P. 2008. Induction of sterility in the field. Proceedings of the 7th meeting of the Working Group on Fruit Flies of the Western Hemisphere, November 2-7, 2008, Mazatlán, México.

Ricalde, M. P., D. E. Nava, A. E. Loeck, and M. G. Donatti. 2012. Temperature-dependent development and survival of Brazilian populations of the Mediterranean fruit fly, *Ceratitis capitata*, from tropical, subtropical and temperate regions. Journal of Insect Science 12(1): 33.

Simmons, C., J. Tarano, and J. Pinto. 1959. Clasificación de reconocimiento de los suelos de la República de Guatemala. Instituto Agropecuario Nacional. Servicio Cooperativo Inter-Americano de Agricultura, Ministerio de Agricultura. Guatemala City, Guatemala. 1000 pp.

USDA (United States Department of Agriculture). 2003. Mediterranean fruit fly action plan. Animal and Plant Health Inspection Service (APHIS). Plant Protection and Quarantine.

Wang, C., C. Deser, J.-Y. Yu, P. DiNezio, and A. Clement. 2017. El Niño and Southern Oscillation (ENSO): A review, pp. 85-106. *In* P. W. Glynn, D. Manzello, and I. Enochs (eds.), Coral reefs of the eastern tropical Pacific. Coral reefs of the world 8. Springer Science+Business Media, Dordrecht, The Netherlands.

WMO (World Meteorological Organization). 1992. International meteorological vocabulary. WMO-No.182. Secretariat of the World Meteorological Organization, Geneva, Switzerland.

TRENDS IN ARTHROPOD ERADICATION PROGRAMMES FROM THE GLOBAL ERADICATION DATABASE, GERDA

D. M. SUCKLING[1,2,3], L. D. STRINGER[1,2,3] AND J. M. KEAN[3,4]

[1]The New Zealand Institute for Plant & Food Research Limited, PB 4704, Christchurch 8140, New Zealand; Max.Suckling@plantandfood.co.nz
[2]School of Biological Sciences, University of Auckland, Tāmaki Campus, Building 733 Auckland, New Zealand
[3]Better Border Biosecurity, New Zealand
[4]AgResearch, Hamilton, New Zealand

SUMMARY

The Global Eradication and Response Database (GERDA, http://b3.net.nz/gerda) documents representative incursion responses and eradication attempts against tephritid fruit flies of economic importance, Lepidoptera, tsetse flies, screwworm flies, mosquitoes, ants, beetles and other particular taxa of invasive arthropods since 1869. It includes cases where governments were quickly resigned to the inability to eradicate, as well as cases where a positive outcome was sought in a declared eradication programme. The distribution of pests is expanding well beyond what has been recorded in GERDA, but this information contains useful trends. The rate of eradication attempts continued to rise during the 20th and into the 21st century. In the case of Lepidoptera other than gypsy moth, 75% of programmes were started in the last 20 years. This is evidence for the rapid geographic range expansion under globalisation. It also indicates how active risk analysis and improved technology are increasingly enabling governments to attempt eradication to avoid projected substantial long-term costs of pest establishment. More than 80% of eradication programmes have been successful for arthropods in the database. For certain groups such as tephritid fruit flies of economic importance, the success rate is even higher, due to the experience gained from previous similar programmes, as well as the progress in the development of lures and suppression tools. A steady increase in the number of eradication programmes globally suggests that current exclusion measures for constraining the spread of invasive species are not adequate. Cost-benefit analysis based on prior pest behaviour indicates that additional mitigation against certain taxa are warranted (if possible). It is likely that all these reasons have led to this increase in the number of eradication programmes over time as a consequence of increases in travel and trade volumes from an expanding number of countries, a desire to maintain or reduce pest pressure on exotic and native commodities, and the development of new tools to increase the technical feasibility of eradication attempts. It is notable that arthropod eradication programmes still rely significantly on insecticides, but their importance is steadily decreasing when compared to the application of other tools.

J. Hendrichs, R. Pereira and M. J. B. Vreysen (eds.), Area-Wide Integrated Pest Management: Development and Field Application, pp. 505–518. CRC Press, Boca Raton, Florida, USA.

Key Words: GERDA, eradication attempts, incursion responses, globalisation, invasive species, arthropods, alien, non-indigenous, risk analysis

1. INTRODUCTION

Invasive insect pest species are spreading as a consequence of increasing global trade and continue to emerge as a threat to food production and ecosystem health (Liebhold et al. 2016). This includes insect pests that need to be controlled to avoid significant losses in cropping systems in all regions of the world (Vreysen et al. 2016). Failure to manage these species would have serious consequences for food production worldwide (Vreysen et al. 2007). Some pests have already become ubiquitous global pests, but many are still undergoing geographic range expansion (EPPO 2018). There are sometimes arguments over what constitutes the current range of some species, such as the debate over whether tephritid fruit flies have established below detectable levels in California (McInnis et al. 2017; Carey et al. 2017; Shelly et al. 2017). There is also evidence of a problem of "fake news" in at least one eradication programme in California (Lindeman 2013).

Government-led incursion response programmes can have either eradication or sometimes just delimitation and containment as the goal. Governments often conduct a risk analysis to assess whether the establishment of the unwanted organism is likely to exceed an economic, environmental or social impact threshold and require an attempt to eradicate (Tobin et al. 2014). As part of a project identifying factors affecting outcomes from arthropod eradication efforts (Tobin et al. 2014; Liebhold et al. 2016), a global eradication database called "GERDA" (Kean et al. 2017) has been collating official incursion response programmes that can range from doing nothing through attempting eradication.

GERDA input is based on volunteerism, and registration for a login to access the data is free. GERDA contains information on 1139 incursion responses, of which 1037 led to eradication attempts (Kean et al. 2017). While this is not a complete list of all global incursion response data, data are continuously being verified and entered into the online database. More than 430 registered users of GERDA are listed, from 43 countries. The base data are available along with references so that entries can be checked.

The GERDA data have been contributed by many individuals, who share the vision of everyone having access to information that will facilitate better decision-making with respect to incursion responses. The data, scope and definitions used in the database are available (e.g. Box 1, Kean et al. 2017) and have been used to review global trends (Tobin et al. 2014), as well as details for particular taxa. For example, 28 lepidopteran species were the target of 144 known government-led incursion responses, with effort spread across 12 moth families, dominated by the Lymantriinae and Tortricidae (Suckling et al. 2017). Likewise, Suckling et al. (2016) reviewed the eradication of fruit flies of economic importance covering more than 200 programmes across 16 species.

In this paper, we reinvestigated the database for trends in 811 arthropod eradications in 94 countries, with an additional 63 programmes added since Tobin et al. (2014).

2. RESULTS

2.1. Summary of Known Arthropod Eradications

To August 4, 2017, the database reported 1093 incursion responses including 972 eradication programmes in 105 countries, targeting 309 taxa, of which 166 were arthropods. A total of 768 arthropod eradications have been recorded. Of the 634 arthropod programmes for which the outcome is known, 514 (81%) were successful and 120 (19%) failed. The number of arthropod eradication programmes initiated climbed rapidly through the second half of the 20th century (Fig. 1).

Box 1. GERDA: Frequently Asked Questions
(Kean et al. 2017 (GERDA, http://b3.net.nz/gerda/faq.php)

The word "eradicate" originates from the Latin "to uproot" (*eradicatus*). In ecology, eradication is the intentional local extinction, or extirpation, of a particular taxon. This involves the killing or complete removal of every individual of a population of the target taxon from a defined area, i.e. achieving population size zero.

The target taxon is most often a population of a species but may sometimes be a subspecies or more than one closely related species. Eradication programmes almost always target populations of pestiferous invasive species in part of their invaded range.

The target area for an eradication programme may vary greatly in size. Often eradications are carried out in geographically-isolated areas, such as islands, to minimise the risk of reintroduction. Sometimes, however, eradications may target a particular part of a species' range because of the environmental, economic, or political benefits of removal, even if the species is likely to reinvade.

One of the biggest challenges of any eradication programme is demonstrating success. The International Plant Protection Convention (IPPC) specifies standards for plant pest eradications in its International Standards for Phytosanitary Measures No. 9 (FAO 2016a), but these are largely descriptive. Current international practice specifies that, provided that adequate surveillance activity has been carried out, eradication of most plant pests can be declared once there have been no detections for at least three times the normal generation time of the target taxon (FAO 2016b; Kean et al. 2017).

Of the 768 arthropod eradication programmes recorded so far in GERDA, the method of pest detection is known for 42% of the entries. A range of detection methods led to the start of these programmes (Fig. 2), of which more than half were based on specific traps and lures (e.g. pheromones). In contrast, insect traps without specific lures (e.g. a light trap) have not been the primary tool for detection of a pest that has subsequently been subjected to an eradication programme.

2.2. Clustering of Arthropod Eradications

Certain orders of insects were more frequently targeted for eradication, especially Diptera (Tephritidae and Culicidae in particular), followed by Coleoptera and Lepidoptera (Table 1). This is likely due to the potential significant impact that species from these groups have on primary agricultural production and human health, as well as the availability of eradication tools (Suckling 2015).

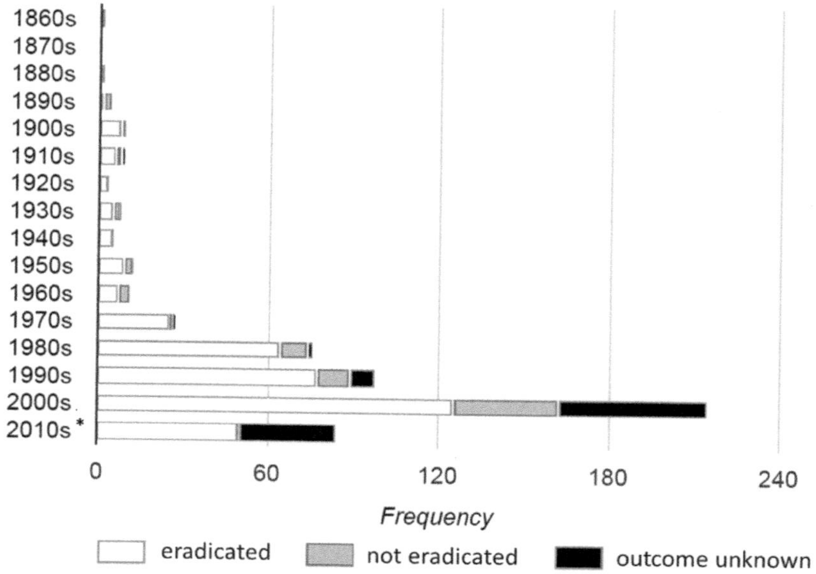

*Figure 1. Arthropod eradication programmes initiated each decade, reported in the Global Eradication and Response Database, GERDA (http://b3.net.nz/gerda) (accessed August 4, 2017). *Data for this decade are from January 2010 until August 4, 2017.*

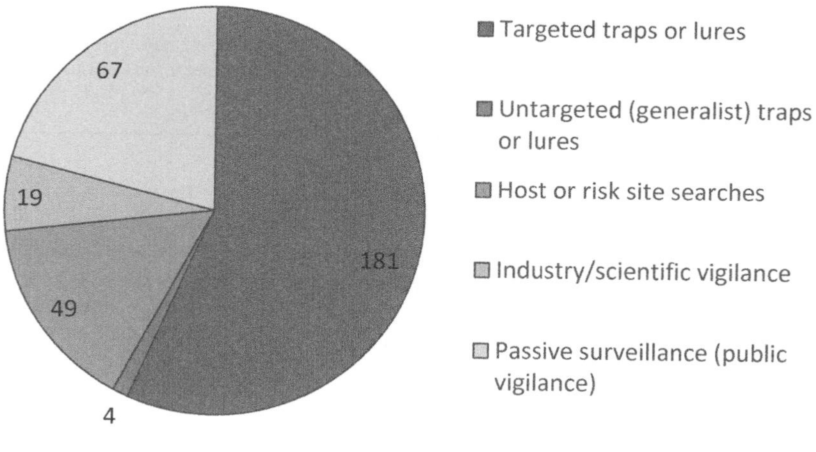

Figure 2. The number of times various detection methods led to the initiation of arthropod eradication programmes stated in the Global Eradication and Response Database (updated from Kean et al. 2017).

GERDA shows evidence of the same clusters forming around similar arthropod taxa as can be found in a database on attractants for pest management (Suckling 2015; El-Sayed 2018). GERDA illustrates the complexity of the different life histories that must be sufficiently understood before engaging in eradication attempts.

Once detection trapping systems are developed, it is frequently possible to use knowledge from previous programmes against similar threats to gain efficiencies. Some methods are applicable over a range of different taxa, whereas in other cases, methods available for closely related species can be easily adapted.

A number of major taxa showed a high degree of re-occurrence as targets of eradication programmes (Table 1; Fig. 3).

Table 1. Number of eradication programmes so far recorded in the Global Eradication and Response Database (GERDA) and success rate by insect order (August 4, 2017)

Order	Count	% Likely or Confirmed Eradication
Diptera	331	77.6
Coleoptera	139	42.4
Lepidoptera	135	75.5
Hymenoptera	70	62.8
Hemiptera	34	50.0
Isoptera	18	38.9
Thysanoptera	11	64.0
Other	33	72.0
Total	768	

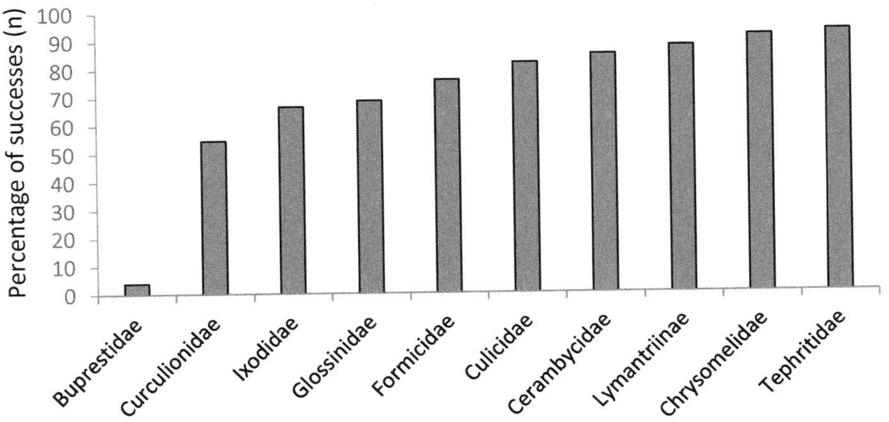

Figure 3. Percentage of successful arthropod eradication programmes by major target taxon (from GERDA, updated from Kean et al. 2017).

The eradication success rate was relatively high for most major target taxa. However, success rates were lower for weevils and buprestid beetles, notably the emerald ash borer, *Agrilus planipennis* Fairmaire (Buprestidae), which has proven difficult to eradicate with the current tools available. About 60% of tick (Ixodidae) and tsetse fly (Glossinidae) eradication programmes were successful, followed by ants (Formicidae) and mosquitoes (Culicidae), longhorn beetles (Cerambycidae), tussock moths (Erebidae), leaf beetles (Chrysomelidae) and fruit flies (Tephritidae) (Fig. 3). The few thrips were largely in glasshouse situations.

We acknowledge that the GERDA data probably contain some reporting bias, but contend that the results are nevertheless indicative of a trend whereby some insect taxa are more easily eradicated than others, due to their biology, invasion dynamics or the tools available to detect and manage them.

The clustering of certain targets at different taxonomic levels, such as the Tephritidae within the Diptera, rather than representatives of all families within the Diptera, indicates the patchiness of pests in certain taxonomic and economic clusters. Even though flies have a broad range of hosts, the vast majority are not pests, apart from those with fruit, animals or humans as hosts, which are well represented in the eradication data as targets on multiple occasions.

2.3. Increase in the Number of Arthropod Eradication Programmes

The annual number of eradication programmes across all invasive arthropods has increased steeply in recent decades (Fig. 1), but this has been accompanied by the development of more specific, environmentally-friendly and cost-effective tools. In fact, the sub-family Lymantriinae demonstrate how, with the development of new tools, patterns of eradications attempts can change. The 74 *Lymantria dispar* L. entries in GERDA are USA-dominated due to the "Slow the Spread Programme" for *L. dispar* in the USA (Sharov et al. 2002). For the 61 programmes against this pest for which the control tools are known, initial eradication and management attempts in the late 19th and early 20th centuries consisted of picking of egg masses by hand as one of the few options available (Myers et al. 2000). This was followed by the spraying of persistent insecticides, which showed a steady increase until the 1960s, when efforts were made to develop alternative control methods, such as mass-trapping approaches (Fig. 4).

The synthesis of the sex pheromone for *L. dispar* enabled the delimitation and monitoring of populations through the trapping of male moths attracted to traps (Bierl et al. 1970). Then in the 1980s, the biopesticide, *Bacillus thuringiensis kurstaki,* was extensively trialled in the USA for use against Lepidoptera (USFS 1994). Also, at this time, mating disruption became more readily available (Cardé, this volume), after being tested and reported from the late 1970s (Schwalbe et al. 1979). In addition, a number of Sterile Insect Technique (SIT) field trials were successfully conducted against gypsy moth in the 1970s and 1980's, but it was concluded that the method was not cost-effective (Simmons et al. 2021).

In the 1970's and 1980's there was a rapid increase in the number of *L. dispar* eradications attempted (Fig. 4). The biopesticide is now the only tool recorded as used for eradication for the past 20 years, probably because it is cheap to apply, and a large amount of the moth's range is over forested areas, so few people are affected by application, thus not opposed to its use. The effectiveness of the biopesticide and the acceptability of its use, is probably the reason why there has been a shift from multiple tools used in earlier years, to a single tool used in recent years for the eradication of *L. dispar* (Fig. 4).

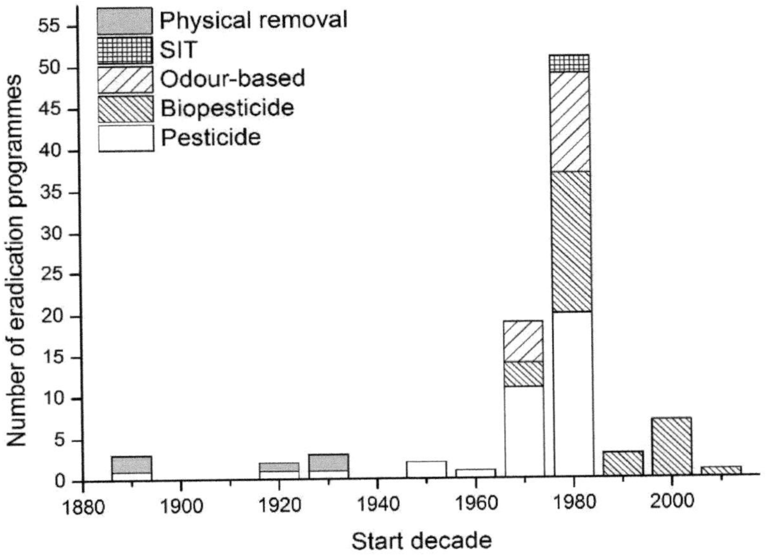

Figure 4. The type of suppression tools used for 61 Lymantria dispar *eradication programmes over time. Physical removal includes both removal of host and hand removal of the pest; odour-based includes mass-trapping, lure and kill, and mating disruption (from GERDA, updated from Kean et al. 2017).*

However, urban-based eradications can face public opposition to aerial use of the biopesticide, so mass-trapping, mating disruption used with reduced rates (Tcheslavskaia et al. 2005), or sterile insect release can present more favourable options (Gamble et al. 2010). In fact, mating disruption is widely used in the "Slow the Spread Programme" (Tcheslavskaia et al. 2005; Liebhold et al. this volume), although details of this suppression programme on the leading edge of gypsy moth infestation are not reported in GERDA due to its focus on eradication programmes.

It is not possible to identify a single cause of the increase in the number of arthropod eradication and response programmes, but it is likely due to a combination of several factors, including the ability to monitor and detect introduced pest populations with their sex pheromone, from the 1970s.

With the development of better or more publicly-accessible pest control technologies throughout time, a corresponding change in the type of management methods used is expected, and this appears to be the case (Fig. 5).

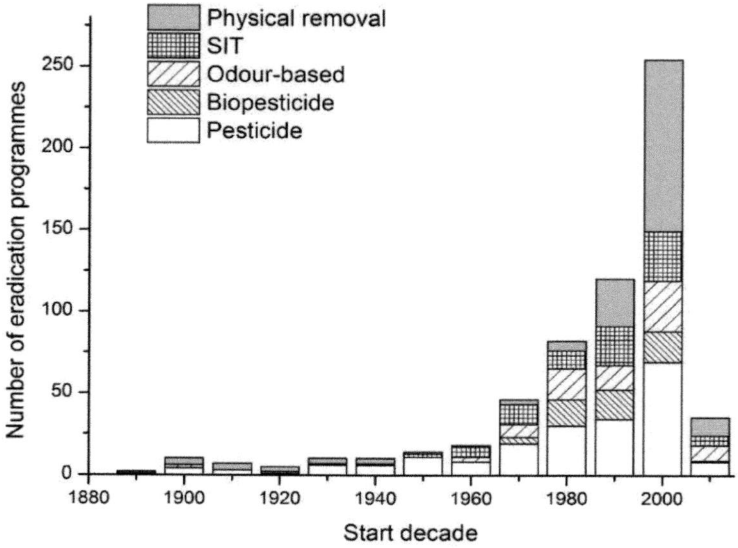

Figure 5. The number of programmes using different tools to eradicate populations of arthropods. Note some programmes used multiple tools. Physical removal includes both removal of host and hand removal of the pest; odour-based includes mass-trapping, lure and kill, and mating disruption (from GERDA, updated from Kean et al. 2017).

Insecticides still feature heavily in arthropod eradications, but they are steadily decreasing percentage-wise (for example, see 1930-1950s in Fig. 5 where they represented most of the tools used in the past versus today). Failed eradications suggest that pest management costs and area of insecticide treated crops is likely rising over time (ca. 20% of programmes). In the case of newly established pests, the new applications of insecticides can disrupt existing integrated pest management (IPM) programmes using biological control (Cameron et al. 2009) and force the return or initial establishment of broad-spectrum insecticide programmes (Berry et al. 2009; Vereijssen et al. 2015). This pattern has been repeated many times and one of the impacts of such invasive species is to remove the future opportunity for agricultural production using organic means, where demand and opportunities do exist in Western countries.

Unexpectedly, removal by hand and host removal was the primary tool in the early 2000s. This may be due to new pests being targeted, for which there was little information on alternatives, including tree-killing beetles, such as the emerald ash borer *A. planipennis* (Fig. 5).

With the development of odour-based lures, control options for greener technologies have increased primarily for Lepidoptera, but this approach has also been used for Coleoptera and Diptera species. The SIT is a technology without non-target impacts, but the need to maintain colonies of the target pest species and security requirements for radiation sources has likely led to the limited number of lepidopteran, coleopteran and dipteran species for which SIT has been used (Klassen et al. 2021). Nevertheless, this number is gradually increasing and recent studies found that it might be feasible to tackle other pest taxa with SIT integration in the future, such as the brown marmorated stink bug, *Halyomorpha halys* (Hemiptera: Pentatomidae) (Welsh et al. 2017), or *Drosophila suzukii* (Matsumura) (Diptera: Drosophilidae) (Lanouette et al. 2017).

New sources of radiation that can be switched off, such as X-ray, remove the need to secure or replenish decaying radioactive sources, an expensive exercise (Mastrangelo et al. 2010; Mehta and Parker 2011). This technology is also increasingly being used for post-harvest disinfestation (Follett and Weinert 2012), but there still remain issues of reliability, which are of serious concern for the SIT component of these programmes.

The cost of achieving eradication is positively related to the size of the infested area (Fig. 6), with successful outcomes more likely when the detected population is still small and at low density, but this is not necessarily guaranteed. However, the availability of detection tools (Fig. 2) probably allows for the earlier detection of the pest, better delimitation, and thus a quicker eradication (Tobin et al. 2014).

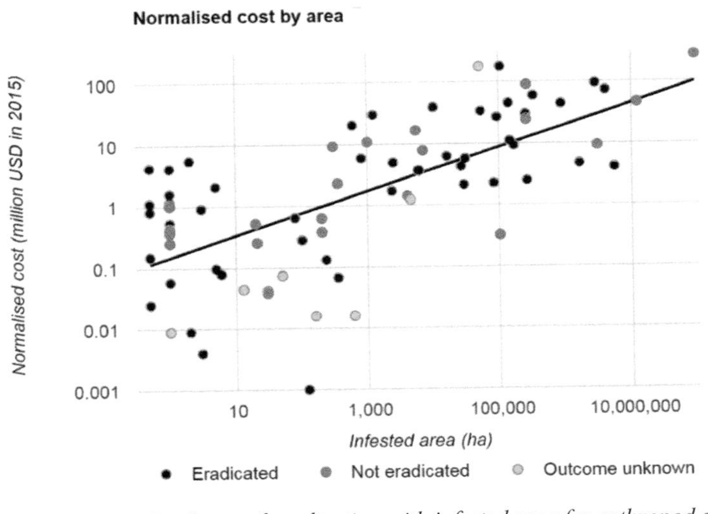

Figure 6. Rise in normalised cost of eradication with infested area for arthropod eradication programmes where the outcome is known or unknown (USD in 2015) (from GERDA, updated from Kean et al. 2017).

3. DISCUSSION

The rapid rise in non-native arthropod incursions and eradications is a concerning trend, because it has been clear for more than 10 years that passive vectoring from international travel and trade volumes, including non-commercial postal shipments, has risen to the stage where current biosecurity processes in most jurisdictions are inadequate to prevent or even reduce incursion frequencies (Liebhold et al. 2006). This has potential implications for food production in the cases of some of the worst invaders (Vreysen et al. 2007). In part, access to nutritious food supply becomes more challenging wherever horticultural crops become scarce due to fruit fly or other insect attacks and the diet shifts towards cereals which lack some of the 20 or so micronutrients required (Broadley and White 2010). The expansion of oriental fruit fly *Bactrocera dorsalis* (Hendel) (Diptera: Tephritidae) into Africa and islands of the Pacific region, and onto additional host plants (Grové et al. 2017) illustrates the point.

On the positive side, there has been a rapid rise in government-led eradication programmes, along with a rise in countries undertaking eradication efforts (to at least 90), which is encouraging in view that in the past there was often only resignation and no attempt to contain and eliminate an incursion. GERDA does not document all of the cases where governments were quickly resigned to the inability to eradicate, but it represents cases where a positive outcome was sought in a declared eradication programme. These undeclared cases add to the increasing burden of pests discussed here, and it should be noted that the distribution of some pests have expanded well beyond what has been recorded in GERDA, due to its focus on eradication, not pest management.

A number of trends affecting the probability of success are clear in the data. The type of invasive organism makes a difference to the outcome, which is directly linked to the type of surveillance that is possible. Tobin et al. (2014) reported that the availability of lures was an enormous (>22-fold) factor increasing the likelihood of early detection followed by eradication. The type of organisms that warrant eradication are clustered in groups, which frequently turn out to be targets of pest management in other jurisdictions (Suckling 2015). This implies recognized pest status rather than new offender status, although both cases exist. Novel and unwanted tussock moths (Lepidoptera: Erebidae: Lymantriinae) were detected in New Zealand, which led to several large-scale aerial urban biopesticide-based eradication programmes, but the particular species involved in the incursions were not anticipated (Brockerhoff et al. 2010).

The ability to conduct delimitation of an incursion is essential to the eradication success, so perhaps it is unsurprising that there is investment in developing and using insect attractants in many countries, although their commercial availability often lags behind the scientific reports (Baker et al. 2016). Across all arthropods, more than 80% of eradication programmes have been successful, although for certain groups, such as fruit flies of economic importance, the success rate is even higher (~90%). All sectors have been affected by invasive arthropods, although some have had a lot more experience than others, particularly sectors related to food production in horticulture or urban gardens, and human and animal health.

One of the species with a high colonisation rate as well as a record of eradication success is *B. dorsalis,* a tephritid fruit fly for which effective lures are available. In addition, the *Lymantria* species were well represented in this group, and they too have been eradicated many times. Attractants used to lure these two species, as well as other damaging pests, to traps have been identified as well as a number of control strategies. The advanced state of detection and control tools for *B. dorsalis* and *Lymantria* species is probably because of the large damage they cause and the rate at which they spread in the new range, which has prioritised research in this area. This has probably led to the unexpected result that a high dispersal rate is not a factor limiting success rate. As Tobin et al. (2014) highlighted, the existence of a detection tool, and thus likely a detection network programme, has had a bigger impact on success rate. Generally, the availability of multiple tools that can be used for eradications can lead to faster successful outcomes, but there are examples, such as a single or a few applications of *B. thuringiensis kurstaki* that resulted in successful eradication of *L. dispar* populations.

Tobin et al. (2014) reported 672 eradication programmes against arthropods (to 2010), and we have been able to compile additional cases, although of course current entries are best checked directly on the GERDA database, where users can find simple summary tools (Kean et al. 2017). Tobin et al. (2014) discussed several biases of data compiled in this database, e.g. more recent data are easier to locate and compile, and that successful programmes are reported more often due to the reluctance to publicize failures.

We contend, however, that the evolving database GERDA is robust to indicate the trends and drivers of eradication success and failure. For Lepidoptera other than gypsy moth, 75% of programmes were started in the last 20 years, suggesting rapid geographic range expansion under globalisation (Suckling et al. 2017). This entails that regular reviews of the trends in new data will be warranted, and/or the need of regulatory agencies to input their eradication data and use GERDA in order to remain updated. Operational biosecurity agencies involved in arthropod eradication are already invited to contribute their data to GERDA, to strengthen the evidence for conclusions and policy over time.

A fast increase in the number of eradication programmes globally suggests that current exclusion measures are not adequate to manage an increasing risk to global food security, due to an unrelenting supply of invasive species (Seebens et al. 2017). An increase in the biodiversity of invasive pests is emerging, increasing the challenge considerably. However, pest incursions have motivated the development of newer technologies that will make incursion responses and eradication success a more likely outcome in the future.

The role of public support cannot be over-stated, and as Lindeman (2013) states:

> "*the California Department of Food and Agriculture (CDFA) lost the battle over aerial spraying against the invasive light brown apple moth (LBAM,* Epiphyas postvittana *(Walker)) largely because of a report and other supporting grey literature documents that expressed highly disputable facts, evidence, and conclusions.*"

Vigilance over the facts is clearly needed by the scientific community.

4. ACKNOWLEDGMENTS

We thank the Better Border Biosecurity collaboration (www.b3nz.org), other sponsors including the Joint FAO/IAEA Division of Nuclear Techniques in Food and Agriculture, and others listed online who have contributed data to GERDA, as well as two anonymous reviewers.

5. REFERENCES

Baker, T. C., J. J. Zhu, and J. G. Millar. 2016. Delivering on the promise of pheromones. Journal of Chemical Ecology 42: 553–556.

Berry, N. A., M. K. Walker, and R. C. Butler. 2009. Laboratory studies to determine the efficacy of selected insecticides on tomato/potato psyllid. New Zealand Plant Protection 62: 145–151.

Bierl, B. A., M. Beroza, and C. W. Collier. 1970. Potent sex attractant of the gypsy moth: Its isolation, identification, and synthesis. Science 170: 87–89.

Broadley, M. R., and P. J. White. 2010. Eats roots and leaves. Can edible horticultural crops address dietary calcium, magnesium and potassium deficiencies? Proceedings of the Nutrition Society 69: 601–612.

Brockerhoff, E. B., A. Liebhold, B. Richardson, and D. M. Suckling. 2010. Eradication of invasive forest insects: Concept, methods, costs and benefits. New Zealand Journal of Forestry Science 40: S117–S135.

Cameron, P. J., G. P. Walker, A. J. Hodson, A. J. Kale, and T. J. B. Herman. 2009. Trends in IPM and insecticide use in processing tomatoes in New Zealand. Crop Protection 28: 421–427.

Carey, J. R., N. Papadopoulos, and R. Plant. 2017. The 30-year debate on a multi-billion-dollar threat: tephritid fruit fly establishment in California. American Entomologist 63: 100–113.

El-Sayed, A. M. 2018. The Pherobase: Database of insect pheromones and semiochemicals. HortResearch, Lincoln, New Zealand.

(EPPO) European Plant Protection Organization. 2018. Global database. Paris, France.

(FAO) Food and Agriculture Organization of the United Nations. 2016a. Guidelines for pest eradication programmes. International Standard for Phytosanitary Measures (ISPM) No. 9, International Plant Protection Convention (IPPC). FAO, Rome, Italy.

(FAO) Food and Agriculture Organization of the United Nations. 2016b. Fruit fly pest free areas. International Standard for Phytosanitary Measures (ISPM) No. 26, International Plant Protection Convention (IPPC). FAO, Rome, Italy.

Follett, P. A., and E. D. Weinert. 2012. Phytosanitary irradiation of fresh tropical commodities in Hawaii: Generic treatments, commercial adoption, and current issues. Radiation Physics and Chemistry 81: 1064–1067.

Gamble, J. C., T. Payne, and B. Small. 2010. Interviews with New Zealand community stakeholders regarding acceptability of current or potential pest eradication technologies. New Zealand Journal of Crop and Horticultural Science 38: 57–68.

Grové, T., K. De Jager, and M. S. De Beer. 2017. Indigenous hosts of economically important fruit fly species (Diptera: Tephritidae) in South Africa. Journal of Applied Entomology 141: 817–824.

Kean, J. M., D. M. Suckling, N. J. Sullivan, P. C. Tobin, L. D. Stringer, D. C. Lee, G. R. Smith, R. Flores Vargas, J. Fletcher, F. Macbeth, D. G. McCullough, and D. A. Herms. 2017. Global eradication and response database (GERDA).

Klassen, W., C. F. Curtis, and J. Hendrichs. 2021. History of the Sterile Insect Technique, pp. 1–44. *In* V. A. Dyck, J. Hendrichs, and A. S. Robinson (eds.), Sterile Insect Technique – Principles and practice in Area-Wide Integrated Pest Management. Second Edition. CRC Press, Boca Raton, Florida, USA.

Lanouette, G., J. Brodeur, F. Fournier, V. Martel, M. J. B. Vreysen, C. Cáceres, and A. Firlej. 2017. The Sterile Insect Technique for the management of the spotted wing drosophila, *Drosophila suzukii*: Establishing the optimum irradiation dose. PLoS One 12:e0180821.

Liebhold, A. M., T. T. Work, D. G. McCullough, and J. F. Cavey. 2006. Airline baggage as a pathway for alien insect species entering the United States. American Entomologist 52: 48–54.

Liebhold, A. M., L. Berec, E. G. Brockeroff, R. S. Epanchin-Niell, A. Hastings, D. A. Herms, J. M. Kean, D. G. McCullough, D. M. Suckling, P. C. Tobin, and T. Yamanaka. 2016. Eradication of invading insect populations: From concepts to applications. Annual Review of Entomology 61: 335-352.

Lindeman, N. 2013. Subjectivized knowledge and grassroots advocacy: An analysis of an environmental controversy in Northern California. Journal of Business and Technical Communication 27: 62–90.

Mastrangelo, T., A. G. Parker, A. Jessup, R. Pereira, D. Orozco-Dávila, A. Islam, T. Dammalage, and J. M. M. Walder. 2010. A new generation of X ray irradiators for insect sterilisation. Journal of Economic Entomology 103: 85–94.

Mehta, K., and A. G. Parker. 2011. Characterization and dosimetry of a practical X-ray alternative to self-shielded gamma irradiators. Radiation Physics and Chemistry 80: 107–113.

McInnis, D. O., J. Hendrichs, T. Shelly, N. Barr, K. Hoffman, R. Rodriguez, D. R. Lance, K. Bloem, D. M. Suckling, W. Enkerlin, P. Gomes, and K. H. Tan. 2017. Can polyphagous invasive tephritid pest populations escape detection for years under favorable climatic and host conditions? American Entomologist 63: 89–99.

Myers, J. H., D. Simberloff, A. M. Kuris, and J. R. Carey. 2000. Eradication revisited: Dealing with exotic species. Trends in Ecology & Evolution 15: 316–320.

Schwalbe, C. P., E. C. Paszek, R. E. Webb, B. A. Bierl-Leonhardt, J. R. Plimmer, C. W. McComb, and C. W. Dull. 1979. Field evaluation of controlled release formulations of disparlure for gypsy moth mating disruption. Journal of Economic Entomology 72: 322–326.

Seebens, H., T. M. Blackburn, E. E. Dyer, P. Genovesi, P. E. Hulme, J. M. Jeschke, S. Pagad, P. Pysek, M. Winter, M. Arianoutsou, S. Bacher, B. Blasius, G. Brundu, C. Capinha, L. Celesti-Grapow, W. Dawson, S. Dullinger, N. Fuentes, H. Jager, J. Kartesz, M. Kenis, H. Kreft, I. Kuhn, B. Lenzner, A. Liebhold, A. Mosena, D. Moser, M. Nishino, D. Pearman, J. Pergl, W. Rabitsch, J. Rojas-Sandoval, A. Roques, S. Rorke, S. Rossinelli, H. E. Roy, R. Scalera, S. Schindler, K. Stajerova, B. Tokarska-Guzik, M. van Kleunen, K. Walker, P. Weigelt, T. Yamanaka, and F. Essl. 2017. No saturation in the accumulation of alien species worldwide. Nature Communications 8: 14435.

Sharov, A. A., D. Leonard, A. M. Liebhold, E. A. Roberts, and W. Dickerson. 2002. "Slow the Spread": A national program to contain the gypsy moth. Journal of Forestry 100: 30–35.

Shelly, T. E., D. R. Lance, K. H. Tan, D. M. Suckling, K. Bloem, W. Enkerlin, K. Hoffman, K. Barr, R. Rodríguez, P. J. Gomes, and J. Hendrichs. 2017. To repeat: Can polyphagous invasive tephritid pest populations remain undetected for years under favorable climatic and host conditions? American Entomologist 63: 224–231.

Simmons, G. S., Bloem, K. A., S. Bloem, J. E. Carpenter, and D. M. Suckling. 2021. Impact of moth suppresion/eradication programmes using the Sterile Insect Techgnique or inherited sterility, pp. 1007–1050. *In* V. A. Dyck, J. Hendrichs, and A. S. Robinson (eds.). Sterile Insect Technique – Principles and practice in Area-Wide Integrated Pest Management. Second Edition. CRC Press, Boca Raton, Florida, USA.

Suckling, D. M. 2015. Can we replace toxicants, achieve biosecurity, and generate market position with semiochemicals? Frontiers in Ecology and Evolution 3: 1–7.

Suckling, D. M., J. M. Kean, L. D. Stringer, C. Cáceres-Barrios, J. Hendrichs, J. Reyes-Flores, and B. C. Dominiak. 2016. Eradication of tephritid fruit fly pest populations: Outcomes and prospects. Pest Management Science 72: 456–465.

Suckling, D. M., D. E. Conlong, J. E. Carpenter, K. A. Bloem, P. Rendón, and M. J. B. Vreysen. 2017. Global range expansion of pest Lepidoptera requires socially acceptable solutions. Biological Invasions 17: 1107–1119.

Tcheslavskaia, K. S., K. W. Thorpe, C. C. Brewster, A. A. Sharov, D. S. Leonard, R. C. Reardon, V. C. Mastro, P. Sellers, and E. A. Roberts. 2005. Optimization of pheromone dosage for gypsy moth mating disruption. Entomologia Experimentalis et Applicata 115: 355–361.

Tobin, P. C., J. M. Kean, D. M. Suckling, D. G. McCullough, D. A. Herms, and L. D. Stringer. 2014. Determinants of successful arthropod eradication programs. Biological Invasions 16: 410–414.

(USFS) United States Forest Service. 1994. *Bacillus thuringiensis* for managing gypsy moth: A review. R. Reardon, N. Dubois, and W. McLane (eds). United States Department of Agriculture, Forest Service, National Center of Forest Health Management. West Virginia, USA. pp. 33.

Vereijssen, J., A. M. Barnes, N. A. Berry, G. M. Drayton, J. D. Fletcher, J. M. E. Jacobs, N. Jorgensen, M. C. Nielsen, A. R. Pitman, I. A. W. Scott, G. R. Smith, N. M. Taylor, D. A. J. Teulon, S. E. Thompson, and M. K. Walker. 2015. The rise and rise of *Bactericera cockerelli* in potato crops in Canterbury. New Zealand Plant Protection 68: 85–90.

Vreysen, M. J. B., A. S. Robinson, and J. Hendrichs (eds.). 2007. Area-wide control of insect pests: From research to field implementation. Springer: Dordrecht, The Netherlands. 789 pp.

Vreysen, M. J. B., W. Klassen, and J. E. Carpenter. 2016. Overview of technological advances toward greater efficiency and efficacy in sterile insect-inherited sterility programs against moth pests. Florida Entomologist 99: 1–12.

Welsh, T. J., L. D. Stringer, R. Caldwell, J. E. Carpenter, and D. M. Suckling. 2017. Irradiation biology of male brown marmorated stink bugs: Is there scope for the Sterile Insect Technique? International Journal of Radiation Biology 93: 1357–1363.

SUCCESSFUL AREA-WIDE ERADICATION OF THE INVADING MEDITERRANEAN FRUIT FLY IN THE DOMINICAN REPUBLIC

J. L. ZAVALA-LÓPEZ[1], G. MARTE-DIAZ[2] AND F. MARTÍNEZ-PUJOLS[2]

[1]*FAO/IAEA, Technical Cooperation Expert; jlzavalalopez@gmail.com*
[2]*Ministry of Agriculture, Programa Moscamed–RD, Ministry of Agriculture, Dominican Republic*

SUMMARY

The presence of the Mediterranean fruit fly *Ceratitis capitata* (Wiedemann) (Tephritidae) in the Dominican Republic was officially reported in March 2015. Subsequent delimitation found that the pest had already spread to 2053 km^2 in the eastern part of the country, constituting a major outbreak. Trading partners imposed an immediate ban on most exports of fruit and vegetables listed as hosts of the pest, resulting in a loss of over USD 40 million over the remaining nine months of 2015. The outbreak was centred on Punta Cana, one of the busiest tourist destinations in the Caribbean. The agricultural production sites affected by the ban were more than 200 km away from the outbreak. The Dominican Government established the Moscamed Programme (Moscamed-RD) through its Ministry of Agriculture as an emergency response. This programme received the financial and operational support to carry out all required surveillance and control activities. The Food and Agriculture Organization of the United Nations (FAO), the International Atomic Energy Agency (IAEA), and the United States Department of Agriculture-Animal and Plant Health Inspection Service (USDA-APHIS) cooperated to assist the country in establishing a national monitoring network to determine the geographic extent of the outbreak and to initiate an eradication campaign with support from regional organizations such as the Organismo Internacional Regional de Sanidad Agropecuaria (OIRSA) and the Interamerican Institute for Cooperación on Agriculture (IICA). The regional Guatemala-México-USA Moscamed Programme played a major role in assisting through technology transfer, which included the application of the Sterile Insect Technique (SIT) and other integrated pest management components. An international Technical Advisory Committee (TAC), chaired by FAO/IAEA, provided technical oversight beginning in September 2015. The last fly was detected in January 2017 and official eradication was announced in July 2017 after six generations had passed with no detections of the pest. The Dominican Republic is now on the list of countries that have successfully eradicated the Mediterranean fruit fly and has substantially strengthened its fruit fly surveillance system and emergency response capacity.

J. Hendrichs, R. Pereira and M. J. B. Vreysen (eds.), Area-Wide Integrated Pest Management: Development and Field Application, pp. 519–537. CRC Press, Boca Raton, Florida, USA.

Key Words: *Ceratitis capitata*, medfly, Tephritidae, Sterile Insect Technique, SIT, IPM, fruit exports, Caribbean, invasive pest

1. INTRODUCTION

Agriculture contributes substantially to the Dominican Republic's GDP and is the primary employer of the labour force, as well as among the main sources of foreign currency. Fruit and vegetable production and exports make up a significant portion of these benefits, including the production of avocados, bell peppers, mangoes, and tomatoes. Exotic pests and diseases present a risk to agricultural production and exports, and international phytosanitary standards recommend continuous vigilance to prevent negative impacts on this sector of the economy. The Dominican Republic experienced the repercussions of the presence of an invasive pest for which it was largely unprepared.

The incursion of Mediterranean fruit fly *Ceratitis capitata* (Wiedemann) into Dominican Republic was of high importance, not only for the country, but for all other countries of the Caribbean Basin that are free of this major pest. Its presence was suspected by the Dominican Republic's Ministry of Agriculture in October 2014. After accurate identification was confirmed, the detection was officially reported in March 2015. Exports of listed Mediterranean fruit fly host fruit and vegetables were banned immediately by trading partners as the lack of an operational national detection system caused uncertainty about the extent and the distribution of the pest in the country. This reduction of exports resulted in a loss of more than USD 40 million over the remaining nine months of 2015, putting some 30 000 jobs at risk (Gil 2016).

1.1. Characteristics of the Outbreak

The outbreak was located in the Punta Cana region in the eastern Dominican Republic (Fig. 1), one of the most visited tourist destinations in the Caribbean, and therefore the pest was suspected to have been brought by tourists. Delimitation trapping confirmed high densities of the pest in the coastal areas of Punta Cana and adjacent Bávaro, with sporadic detections in several contiguous provinces within the surrounding area of 2053 km^2.

Fortunately, agricultural production of Mediterranean fruit fly hosts for export was non-existent in the affected area, with the major fruit and vegetable production sites affected by the ban located more than 200 km away from the outbreak. An additional characteristic of the outbreak was that certain known hosts, which are typically moved through commerce, such as mangos, citrus, guavas, cherries (acerola), and other host fruits common in backyards throughout the region were not attacked. Rather, larval finds were limited to three species of wild or ornamental fruits of no agricultural importance (see Section 2.2.2.).

1.2. Establishment of the Moscamed Programme in the Dominican Republic

The Government of the Dominican Republic, through its Ministry of Agriculture, responded to this emergency with the establishment of the Moscamed Programme in the Dominican Republic (Moscamed-RD), providing the required financial and operational support to perform all recommended delimitation and eradication activities.

The initial challenges were, among others: social and economic effects of the export ban; pressure of the media and private sector; answers demanded by stakeholders; questions by some on the need for eradication; mobilizing for financial and human resources; and streamlining external support, as assistance was being offered by a number of entities.

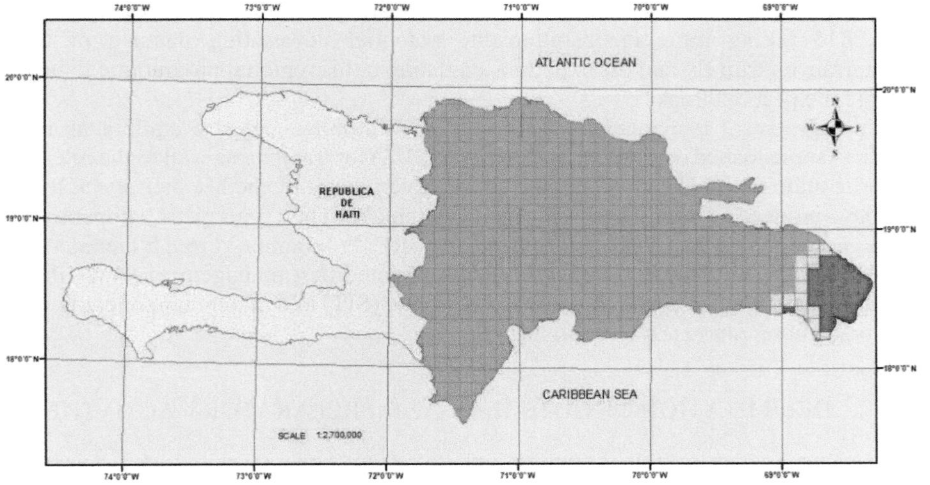

Figure 1. Map of the island of Hispaniola showing the Mediterranean fruit fly-infested area in 2015 in the eastern part of the Dominican Republic, covering parts of the provinces of La Altagracia, La Romana, San Pedro de Macorís, El Seibo and Hato Mayor (red= infested area; yellow= buffer area; green= free area).

The Food and Agriculture Organization of the United Nations (FAO), the International Atomic Energy Agency (IAEA), and the United States Department of Agriculture (USDA), through its Animal and Plant Health Inspection Service (APHIS), collaborated to assist the Ministry of Agriculture in establishing a national monitoring network to delimit the distribution of the outbreak and to initiate an eradication campaign. First steps were begun by USDA-APHIS, followed by a series of technical assistance and capacity building missions by the Guatemala-Mexico-USA Moscamed Programme (Enkerlin et al. 2015, 2017).

In view of the potential devastating damage of the Mediterranean fruit fly to the Dominican Republic and neighbouring countries of the Caribbean Basin, an international coordination meeting took place in Santo Domingo in May 2015 with the participation of FAO, IAEA, regional organizations such as the Instituto Interamericano de Cooperación para la Agricultura (IICA) and the Organismo Internacional Regional de Sanidad Agropecuaria (OIRSA), as well as USDA-APHIS. The objective was to coordinate technical and financial support, as well as the supply of some critical equipment and resources. In September the technical assistance was formalized under FAO/IAEA, which established and coordinated a Technical Advisory Committee (TAC) composed of international experts.

Authorities of the Ministry of Agriculture of the Dominican Republic, with encouragement of the USDA-APHIS and the FAO/IAEA, agreed to collaborate under a Cooperative Agreement with the Moscamed Regional Programme (Guatemala-Mexico-USA). The Letter of Understanding validating the agreement was signed in July 2015 taking into consideration the potential devastating damage of the Mediterranean fruit fly and the expertise available in this regional programme to help manage the pest outbreak.

This agreement facilitated not only continued training, but also equipment and supplies (some loaned or donated by USDA-APHIS) for trapping as well as the release of sterile male flies, also supplied on a cost recovery basis by the Mediterranean fruit fly mass-production facility in El Pino, Guatemala. This synergistic cooperation played a major role in assisting the Moscamed-RD Programme through technology transfer of all components of an area-wide integrated pest management (AW-IPM) approach that included the Sterile Insect Technique (SIT) as a major component in the final eradication phase (Dyck et al. 2021).

2. DELIMITATION OF OUTBREAK AND PREPARATORY ACTIVITIES

The eradication process followed during the 2015-2017 campaign, which is broadly summarized in Fig. 2, included these phases and actions:

2.1. Preparatory Pre-eradication Phase

Immediately after reporting the detection of the Mediterranean fruit fly in the eastern part of the Dominican Republic, trade restrictions were imposed on the export of Dominican fruit and vegetable host material. This was mainly due to the absence of a solid operational trapping network and resulting uncertainty about the geographic distribution of the pest.

Through the above-mentioned Cooperative Agreement, technology transfer and capacity building efforts were strengthened, and training of Moscamed-RD personnel continued in subjects such as detection, identification (taxonomic as well as sterile vs. wild), pest suppression and eradication, public relations, quarantine and other activities related to the implementation of AW-IPM programmes.

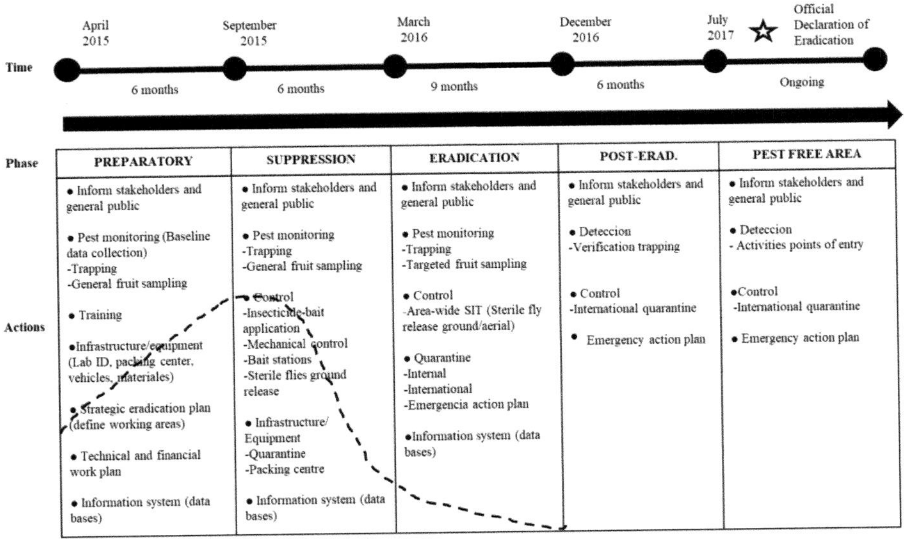

Figure 2. Phases and actions of the eradication process followed during the Mediterranean fruit fly eradication campaign 2015-2017 in the Dominican Republic (dotted line is a theoretical representation of population density) (source Walther Enkerlin, FAO/IAEA Insect Pest Control Section).

The detection system was gradually enhanced from the original limited and occasional trapping to an effective national surveillance system. Trap types used included Jackson traps baited with the male-attractant trimedlure, the female-biased Phase IV traps baited with the synthetic food lure Biolure, and Multilure traps baited with the more generic liquid protein baits such torula yeast and Ceratrap.

The trapping network during the preparatory or pre-eradication phase rapidly expanded from 1006 traps, mainly at points of entry (every two weeks with inspection levels of only about 25% of traps), to 14 589 traps country-wide (9936 male-specific Jackson traps and 4653 female biased Phase IV traps) that remained in place during 2015 (Table 1 and Fig. 3).

The country-wide trapping reached inspection levels of over 95% (weekly in the buffer and infested areas, or every two weeks in the Mediterranean fruit fly-free areas and points of entry).

Trapping and fruit sampling were significantly increased in the eastern region, aimed at accurately determining the distribution and potential spread of the infestation, locating any remnants of the infestation that may have been missed, as well as enabling sound decision-making and planning of suppression and eradication activities. The majority of traps in the country, 64% (or 9936 traps) were placed in the eastern region, where the initial detection occurred, consisting of 4687 Jackson Traps and 5249 food-based traps (Fig. 4).

Table 1. Maximum number of traps in the national Mediterranean fruit fly trapping network established in 2015 in the Dominican Republic

Traps/ Attractant	La Alta-gracia	La Romana	El Seibo	Hato Mayor	San Pedro de Macorís	Rest of Domi-nican Republic	Total
Jackson/ Trimedlure	2797	363	605	476	446	5249	9936
Phase IV/ Biolure	3015	430	577	372	259	0	4653
Total	5812	793	1182	848	705	5249	14 589

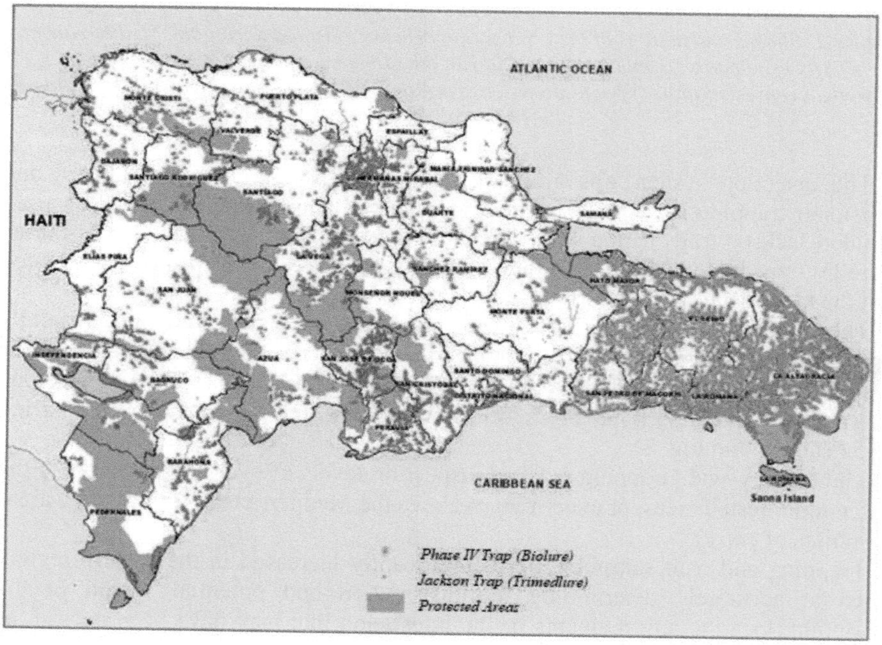

Figure 3. National Mediterranean fruit fly trapping network established in 2015 in the Dominican Republic (yellow triangles= Jackson traps, green circles= Phase IV/Biolure traps).

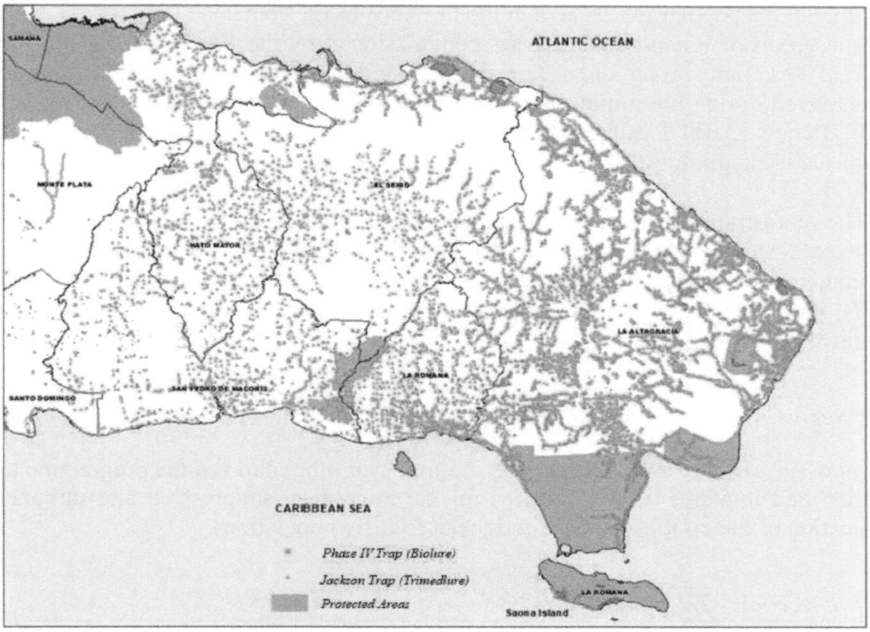

Figure 4. Trapping network in the eastern region of the Dominican Republic (yellow triangles= Jackson traps, green circles= Phase IV/Biolure traps).

During this preparatory phase, fruit sampling activities followed a general approach, systematically collecting a wide range of soft-skinned fruit species that could potentially be susceptible to Mediterranean fruit fly infestation. A total of 10 589 fruit samples were collected and dissected. Once the host range had been assessed for the Dominican Republic, a targeted stratified sampling protocol was implemented as explained in Section 2.2.2. (FAO/IAEA 2017a, and FAO/IAEA 2018).

The results of the surveillance system indicated that the infestation was concentrated in and around the coastal touristic areas of Punta Cana and Bávaro, with sporadic wide-spread detections throughout the eastern provinces. Most adult (1572) and immature stages of the fly (1189 larvae in 225 infested samples) were found in the 8 weeks after initial activities were begun in March 2015. The pest was found not only in the Punta Cana and Bávaro area, but later also in other areas in the Province of La Altagracia, and subsequently also in the nearby provinces of El Seibo, San Pedro

Macorís, and Hato Mayor (see Fig. 7). A reproducing and recurring population was also found in a popular tourist area in the Province of La Romana.

The group of international experts commissioned by the FAO/IAEA under the external TAC first met on-site in January 2016 and reviewed activities and results so far achieved during the initial months of the programme. The TAC confirmed that eradication was still feasible and recommended that an area-wide programme be established, integrating the SIT, as the core eradication activity, with other control methods.

The cooperation with several stakeholders, in particular experts from the Guatemala-México-USA Moscamed Programme and the FAO/IAEA, as well as the recommendations of the TAC (September 2015, January 2016, October 2016 and July 2017), served to guide the implementation of all activities and provide continuous technical back-stopping.

2.2. *Suppression and Eradication Phase*

The new surveillance system (trapping and fruit sampling) allowed the programme to develop and implement the strategies for the immediate suppression and ultimate eradication of the established Mediterranean fruit fly populations.

2.2.1. Detection - Trapping Networks

The goal during this suppression phase was a trap density of 2 traps per km^2 at a 1:1 ratio (Jackson trap to Phase IV/Multilure trap) in areas with host presence, as well as achieving high trap servicing levels (Fig. 5). Once the SIT was initiated, the trap ratio was adjusted in release areas to a 1:9 ratio to focus on wild female detection and to minimize sterile male recapture. Trap service intervals were changed to once every two weeks in the Mediterranean fruit fly-free areas and remained at once per week in the infested and buffer areas of the eastern region (Fig. 5).

For each Mediterranean fruit fly find, a high-density delimitation trapping was installed in a 9 km^2 area around the find for three life cycles, as indicated by international trapping protocols (FAO 2016).

Once the infestation on the island was well delimited and aerial sterile fly releases initiated in 2016, the total number of traps in the Mediterranean fruit fly-free areas of the eastern region was gradually reduced to allow concentrating more of the available resources on areas with suppression/eradication activities.

Overall, there was no real trapping network in the first quarter of 2015, and from 5 April 2015 to 14 January 2017, 4174 adults (3938 males and 236 females) were caught in 594 traps out of a total of 14 589 traps deployed country-wide. Adult detections were higher during the second and third quarter of the year, both in 2015 and 2016 (Fig. 6).

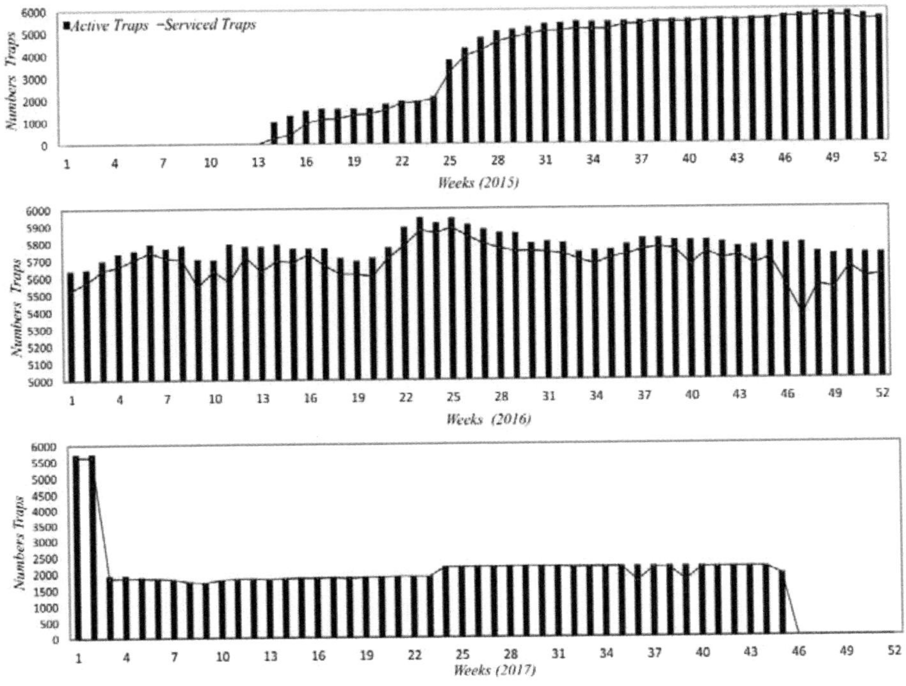

Figure 5. Numbers of installed traps (solid bars) and servicing levels of these traps (line) in the eastern region, including La Altagracia Province, during the 2015-2017 eradication campaign.

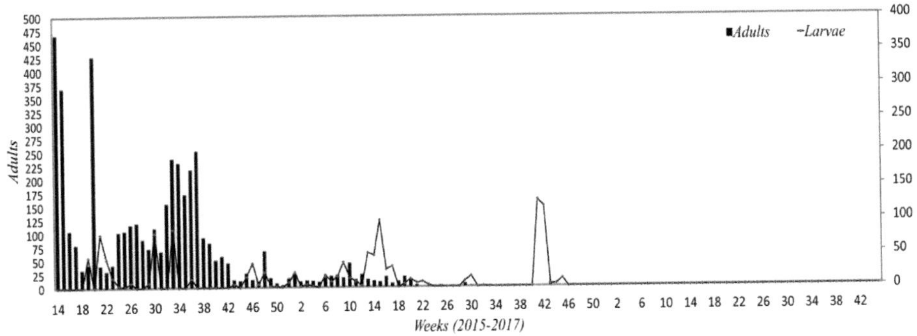

Figure 6. Numbers of detected wild adult flies (black bars) and larvae (line) of Ceratitis capitata *per week during the 2015-2017 eradication campaign in the eastern region of the Dominican Republic.*

The trapping network also provided valuable information on the spatial distribution of the pest, clearly showing that the population was concentrated on the eastern part of the country, mainly within the La Altagracia Province, with the highest numbers present in Punta Cana where the epicentre of the outbreak was located (Fig. 7). A second, incipient outbreak was also found in the Province of La Romana.

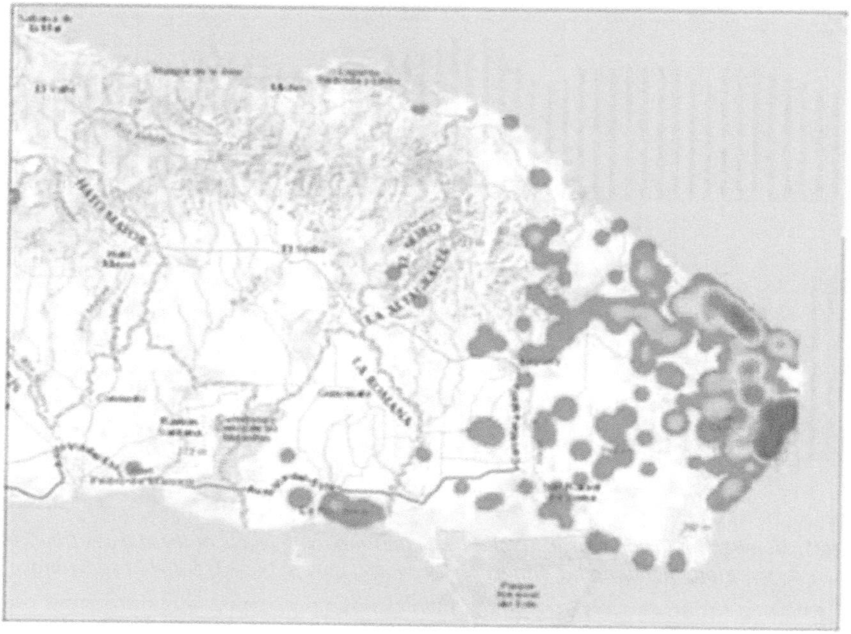

Figure 7. Locations where all Mediterranean fruit flies were captured in the Dominican Republic between 2015 and 2017 in the Provinces of La Altagracia, La Romana, El Seibo, Pedro de Macorís, and Hato Mayor. Colours represent the absolute numbers of wild fly detections per location (green dot= 1-2 flies; yellow= 3-5 flies; orange = 10 to 25 flies; red = >150 flies).

2.2.2. Detection - Fruit Sampling

As was done for the trapping, fruit sampling was adjusted during this phase to mainly target or direct sampling to the confirmed hosts. The general fruit sampling data indicated 95 infested samples of tropical almond *Terminalia catappa* L. and 19 infested samples of yellow caya *Sideroxylon foetidissimum* Jacq., locally known as "yellow caya" which therefore were the major *C. capitata* hosts, though larvae were also found in three samples of another wild host *Simarouba berteroana* Krug & Urb, locally known as "aceitunas = olives" or "black caya", with 3 infested samples. Based on this information, fruit sampling efforts in the infested area were mainly targeted to these three host species to increase the probability of detection.

Overall, 1189 larvae were detected in 10 589 fruit samples with a total mass of 34 789 kg of fruits. Consistent with the trapping results, the majority of larvae were detected during the second and third quarter of the year, both in 2015 and 2016 (Fig. 8). This figure shows that the peak infestation occurred during week 21 of the year 2015 with an average of 7.4 larvae per sample, although the highest number of larvae was obtained in week 41 of 2016 in a large localized infestation in the Bávaro area. The last larva was detected in week 46 of 2016.

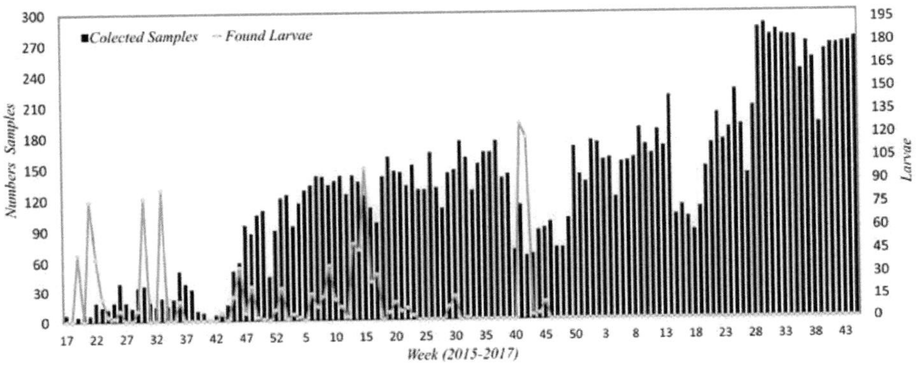

Figure 8. Numbers of fruit samples collected (black bars) and Mediterranean fruit fly larvae detected (brown line) during 2015-2017.

2.2.3. Mechanical/Cultural Control

Mechanical/cultural control consisted of the collection and disposal of *C. capitata* host fruit (on the ground and in the tree), as well as the elimination or severe pruning of host trees, mainly tropical almond, yellow caya and black caya, in the infested areas. A total of 1200 tons of fruit were collected and destroyed, mainly from tropical almond and yellow caya (Fig. 9).

2.2.4. Bait Spray Application

Bait sprays were mainly applied by ground to an area of one square kilometre surrounding hot spots (where repeated detections were made) in 2016 and 2017, although limited aerial bait spraying was also carried out in hot spot areas in 2015 (Fig. 10).

In addition, three scenarios for the aerial application of bait spray were proposed in preparation for the first quarter of 2016 in case of an increase in detections during the March-May period:

a) application of the bait spray on 32 241 ha, which covered all accumulated outbreaks,

b) only on 4342 ha, which covered active outbreaks, and

c) only on 1883 ha, which covered outbreaks from the last 4 weeks.

Due to reasons beyond the control of the programme, such as inadequate supply of GF-120, arrival of a large tourist population for Easter, and the socio-political situation in the area, it was not possible to carry out the aerial bait sprays as planned to expedite the eradication process.

Figure 9. Mechanical /cultural control activities consisting of the destruction of host fruit and the pruning or elimination of wild hosts in the infested area.

Figure 10. Aerial and ground insecticide-bait spray activities and placement of bait stations.

2.2.5. Bait Stations

Bait stations were used as part of the AW-IPM approach to support the ground sprays within the one square kilometre core area of the delimitation trapping area and to cover some locations outside of the core area (Piñero et al. 2014). They were used as a complement when the infested area could not be sprayed, where there was a lack of host trees to be treated, in the surrounding areas in cases of dense vegetation that was difficult to penetrate, and also when ground sprays were ineffective because of the heavy rains.

In total 28 176 stations baited with Ceratrap, 21 133 stations baited with GF-120, and 1513 prototype bait stations developed in Guatemala were installed in areas neighbouring outbreaks (Fig. 10). They were also used as a preventive measure in areas where larvae had been detected in fruit.

2.2.6. Quarantine and Exclusion Activities

A network of quarantine road stations was placed strategically on the main highways and exit points from the La Altagracia Province to prevent the movement of the pest through infested fruit to Mediterranean fruit fly-free areas (Fig. 11).

Apart from the internal quarantine stations, inspection at international points of entry was upgraded due to the large number of tourists (ca. 5 million per year) visiting the country. X-ray machines were installed at seaports and airports, with particular attention to the Punta Cana and La Romana airports. Careful supervision of exclusion activities at these points of entry continues to be crucial to prevent new fruit fly incursions into the country in consignments or in passenger luggage.

2.2.7. Sterile Insect Technique

Sterile male fly releases began in October 2015 after the Mediterranean fruit fly infestation was delimited and the populations in hot spots suppressed. For the first six months, the flies were emerged in paper bags and released by ground, beginning with 1 million pupae per week, increasing gradually to 15 million per week. Aerial release of sterile flies was initiated in March 2016, using the chilled adult release system following an area-wide approach in release blocks (FAO/IAEA 2017b). An existing Ministry of Agriculture building in Higuey (one-hour drive to the airport) was adapted to host the fly emergence and release facility. A cold room was installed adjacent in the facility and an average of 72 million good quality sterile male flies were emerged, chilled, packed and released each week (average of 82.1% of emergence, 91.7% flight ability and 87.3 absolute fliers).

The sterile flies were distributed by air over eight release blocks or polygons, covering a total of 42 000 ha in the provinces of La Altagracia and La Romana (Barclay et al. 2016; FAO/IAEA/USDA 2019). The total number of sterile males released throughout the campaign was 4062 million. USDA chilled release machines (single-box) were loaned to the programme from the APHIS Aircraft and Equipment Operations facility in Edinburg, Texas. Each machine was installed in a Beechcraft King Air 90 and loaded with a single 1 m tall release box with a maximum capacity of 14 million sterile medflies per flight.

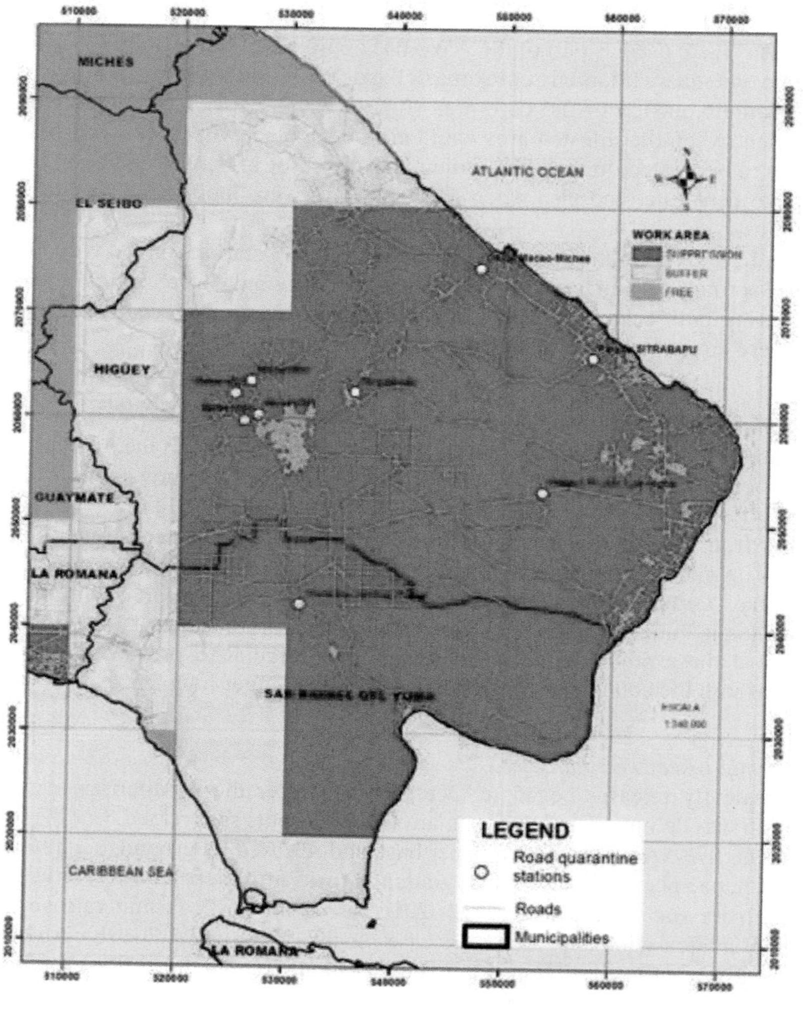

Figure 11. Distribution of road quarantine stations in the eastern part of Dominican Republic during the Mediterranean fruit fly eradication phase.

The distribution of the recaptured sterile flies (% traps with at least one capture) averaged 60%, which is below the recommended level of 85%. Release blocks located along the coastline were affected by strong dominant winds from the east, likely causing sterile fly drift (Fig. 12). Therefore, 15 million additional sterile male flies were released weekly by ground, on average, along the coast of Bávaro, Punta Cana and La Romana to achieve effective sterile to wild fly ratios in the main outbreak areas. Blocks showing low sterile fly distribution were further reinforced through ground releases specifically focussed on detection and outbreak sites.

Figure 12. Blocks of aerial and ground release of sterile flies and example of average sterile fly recapture in a week (red dot= 0-1 sterile fly; orange= 2-5; yellow= 6-10 sterile flies; light green= 11-25 sterile flies; green= >26 sterile flies).

The last fertile adult Mediterranean fruit fly was detected the second week of January 2017 after less than two years of beginning intensive control measures against the pest (Programa Moscamed-RD 2017). In May 2017, sterile releases were suspended once three fly generations had passed since the last wild fly catch, which was equivalent to at least 12 weeks or ca. 3 fly generations of zero catches after the last detection.

2.3. Post-Eradication Phase

Following another risk mapping analysis at the end of 2016, a further re-arrangement of the trapping network was carried out. The total number of traps in service was reduced to a total country-wide of 4630 in early 2017, of which 2835 were deployed in the eastern region and 1795 in the rest of the country (Fig. 5 and Table 2). In addition, Phase IV traps were replaced by Multilure and McPhail traps baited with torula yeast in view of their better performance under the conditions of the Dominican Republic.

In addition, verification trapping was conducted during the post-eradication phase after sterile male releases were terminated. Verification trapping implies that traps were placed at a higher density (5/km^2) in areas where infestations had previously been confirmed. More sensitive traps such as C&C (Cook & Cunningham) and yellow panel traps (Programa Regional Moscamed 2012) were included in the verification trapping (FAO/IAEA 2018). This was implemented in May-June 2017 as a final confirmation to support official declaration of eradication in July 2017, and then continued through October 2017 at the request of USDA-APHIS.

Table 2. Number of traps deployed in the relevant provinces in 2017 after the risk analysis

Traps (Attractant)	La Altagracia	La Romana	El Seibo	Hato Mayor	San Pedro de Macorís	Rest of Domi-nican Republic	Total
Jackson (Trimedlure)	734	204	75	56	123	1730	2922
Mc Phail (Torula yeast)	656	0	0	0	0	65	721
C&C	0	22	0	0	0	0	22
Yellow Panel	0	11	0	0	0	0	11
Multilure (Ceratrap)	489	201	79	72	113	0	954
Total	1879	438	154	128	236	175	4630

As a result of the successful implementation of the programme, the export ban for horticultural products in most western and central areas of the country was lifted in early 2016, only 9 months after intensive surveillance and suppression activities were begun. USDA-APHIS lifted the export ban for 23 provinces in January 7, 2016 and later for another 2 provinces in August 10, 2016.

The benefits of the programme in confining the invading pest to the eastern part of the country, which allowed opening some export markets, and then achieving eradication in early 2017 were immediate, with exports nearly recovering to pre-outbreak levels in 2016, and even significantly increasing in 2017 (Fig. 13).

Now that Mediterranean fruit fly has been eradicated, a reliable surveillance network is being maintained to detect future *C. capitata* and other fruit fly populations early, and trained personnel and supplies are in place to provide a rapid response to any future detection or outbreak.

International quarantines and trapping at ports of entry, suitable host areas, tourist sites, markets and those locations where pest presence was recurrent during the outbreak are also being strengthened to protect the Mediterranean fruit fly-free status and prevent further introductions.

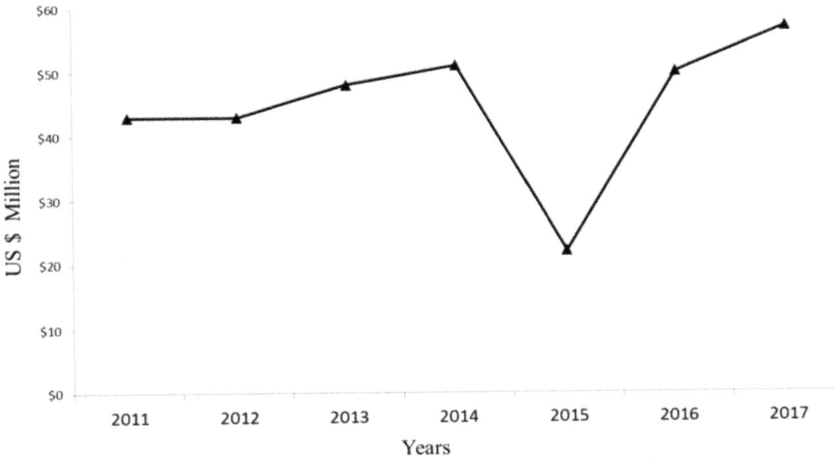

Figure 13. Exports of horticultural products from the Dominican Republic to the USA between 2011 and 2017, including the export ban in March 2015 because of the Mediterranean fruit fly invasion.

3. CONCLUSIONS

The last adult Mediterranean fruit fly was detected in the Dominican Republic in the second week of January 2017. Eradication of the pest from the Dominican Republic using an IPM approach including area-wide SIT application was confirmed in April 2017 after a period of at least three full life cycles with zero captures. Nevertheless, the official declaration of eradication took place in July 2017 after six generations of zero catches and an additional verification trapping network established in high risk areas, including previous detection sites. These additional detection efforts confirmed the absence of the pest.

Most importantly, the country has strengthened its quarantine procedures and developed the capacity for early detection and emergency response for invasive fruit fly pest incursions, as well as for area-wide application of the SIT. This valuable experience can now be shared with Haiti and other countries throughout the Caribbean region to strengthen their quarantine and surveillance systems for invasive fruit flies and other pests, and to prevent similar situations, which can result in serious economic and social losses for the whole region.

The Dominican Republic is now on the list of countries that have successfully eradicated the Mediterranean fruit fly, along with Chile, Mexico, and the USA, and others that have established Mediterranean fruit fly-free areas including Argentina, Australia, Guatemala and Peru on the American continent.

In view of the experience, the Dominican Republic has established a National Fruit Fly Programme with an assigned annual budget to maintain the gained expertise, manage native *Anastrepha* fruit flies, and maintain the surveillance and response capacities for invasive fruit flies and other pests.

4. ACKNOWLEDGEMENTS

The authors wish to acknowledge the highly qualified technical staff of the Moscamed Programme of the Governments of Guatemala, Mexico and the USA that played a major role in the transfer and area-wide implementation of the Sterile Insect Technique in the Dominican Republic.

Also, our appreciation of the valuable contributions of Walther Enkerlin, Jorge Hendrichs, Don McInnis, David Midgarden, Pedro Rendón, Jesus Reyes, and Ricardo Rodriguez, members of the Technical Advisory Committee that provided technical and strategic guidance throughout the eradication process.

Important in-kind contributions of national and international organizations including FAO, IAEA, IICA, OIRSA and USDA are also acknowledged. Our thanks to Jorge Melvin Arias from the Programa Moscamed–RD for the compilation and analysis of data. The valuable assistance improving an earlier version of the manuscript by David Midgarden, and in editing of figures and graphs by Alberto Gonzalez Lara from the Moscamed Programme in Mexico, is also acknowledged.

5. REFERENCES

Barclay, H. J., R. Steacy, W. Enkerlin, and P. den Driessche. 2016. Modelling diffusive movement of sterile insects released along aerial flight lines. International Journal of Pest Management 62: 228–244.

Dyck, V. A., J. Hendrichs, and A. S. Robinson (eds.). 2021. Sterile Insect Technique – Principles and practice in Area-Wide Integrated Pest Management. Second Edition. CRC Press, Boca Raton, Florida, USA. 1200 pp.

Enkerlin, W. R., J. M. Gutiérrez-Ruelas, A. Villaseñor-Cortés, E. Cotoc-Roldán, D. Midgarden, E. Lira, J. L. Zavala-López, J. Hendrichs, P. Liedo, and F. J. Trujillo-Arriaga. 2015. Area freedom in Mexico from Mediterranean fruit fly (Diptera: Tephritidae): A review of over 30 years of successful containment program using an integrated area-wide SIT approach. Florida Entomologist 98: 665–681.

Enkerlin, W. R., J. M. Gutiérrez Ruelas, R. Pantaleón, C. Soto-Litera, A. Villaseñor-Cortés, J.L. Zavala-López, D. Orozco-Dávila, P. Montoya-Gerardo, L. Silva-Villarreal, E. Cotoc-Roldán, F. Hernández-López, A. Arenas-Castillo, D. Castellanos-Domínguez, A. Valle-Mora, P. Rendón-Arana, C. Cáceres-Barrios, D. Midgarden, C. Villatoro-Villatoro, E. Lira-Prera, O. Zelaya-Estradé, R. Castañeda-Aldana, J. López-Culajay, P. Liedo-Fernández, G. Ortíz-Moreno, J. Reyes-Flores, and J. Hendrichs. 2017. The Moscamed regional programme: A success story of area-wide Sterile Insect Technique application. Entomologia Experimentalis et Applicata 164: 188–203.

(FAO) Food and Agriculture Organization of the United Nations. 2016. Requirements for the establishment and maintenance of pest free areas for tephritid fruit flies. International Standard for Phytosanitary Measures (ISPM) 26. International Plant Protection Convention, Rome, Italy.

(FAO/IAEA/USDA) Food and Agriculture Organization/International Atomic Energy Agency/United States Department of Agriculture. 2019. Product quality control for sterile mass-reared and released tephritid fruit flies. Version 7.0. International Atomic Energy Agency, Vienna, Austria. 148 pp.

(FAO/IAEA) Food and Agriculture Organization/International Atomic Energy Agency. 2017a. Fruit sampling guidelines for area-wide fruit fly programmes. W. R. Enkerlin, J. Reyes-Flores, and G. Ortiz (eds.). Food and Agriculture Organization of the United Nations. Vienna, Austria. 45 pp.

(FAO/IAEA) Food and Agriculture Organization/International Atomic Energy Agency. 2017b. Guideline for packing, shipping, holding and release of sterile flies in area-wide fruit fly control programmes. J. L. Zavala-López, and W. R. Enkerlin (eds.). Food and Agriculture Organization of the United Nations. Rome, Italy. 155 pp.

(FAO/IAEA) Food and Agriculture Organization/International Atomic Energy Agency. 2018. Trapping guidelines for area-wide fruit fly programmes. Second edition. W. R. Enkerlin, and J. Reyes-Flores (eds.). Rome, Italy. 65 pp.

Gil, L. 2016. Dominican Republic uses nuclear technology to win war against fruit flies. 25 November 2016. Office of Public Information and Communication. IAEA, Vienna, Austria.

Piñero, J. C., W. Enkerlin, and N. D. Epsky. 2014. Recent developments and applications of bait stations for Integrated Pest Management of tephritid fruit flies, pp. 457-492. *In* T. Shelly, N. Epsky, E. B. Jang, J. Reyes-Flores, and R. Vargas (eds.), Trapping and the detection, control, and regulation of tephritid fruit flies. Springer Science & Business Media, Dordrecht, The Netherlands.

Programa Moscamed-RD. 2017. Informes semanales 2015-2017 del Programa Moscamed-RD al Ministerio de Agricultura. Santo Domingo, República Dominicana.

Programa Regional Moscamed. 2012. Manual del sistema de detección por trampeo de la mosca del Mediterráneo (*Ceratitis capitata*, Wied). SAGARPA-MAGA-USDA, February 2012. Ciudad de Guatemala, Guatemala. 24 pp.

THE ERADICATION OF THE INVASIVE RED PALM WEEVIL IN THE CANARY ISLANDS

M. FAJARDO[1], X. RODRÍGUEZ[2], C. D. HERNÁNDEZ[2], L. BARROSO[2], M. MORALES[2], A. GONZÁLEZ[3] AND R. MARTÍN[3]

[1]*IPM Independent Consultant, Project Manager; fajardo_innfforma@yahoo.es*
[2]*Gestion del Medio Rural de Canarias, Santa Cruz de Tenerife, Spain*
[3]*Gobierno de Canarias, Santa Cruz de Tenerife, Spain*

SUMMARY

After the first detection in 2005 of the Red Palm Weevil (RPW), *Rhynchophorus ferrugineus,* in the Canary Islands, the Government of the archipelago established and implemented the RPW regional eradication programme. The area-wide application of different control measures in a coordinated and integrated way for 10 years has resulted in the eradication of this invasive pest in the archipelago. The last pest focus, located on the Island of Fuerteventura, was declared eradicated in May 2016. In this paper, the different control measures that were applied, as well as the way they were executed, are discussed. Special attention is given to the factors considered key to success. It is concluded that, with the knowledge and techniques available, the eradication of RPW is possible under favourable political and financial circumstances. The biggest threats to the success of this programme originated in human factors, rather than in intrinsic characters of the insect or the techniques used.

Key Words: Rhynchophorus ferrugineus, Phoenix spp., date palms, eradication programme, Gran Canaria, Fuerteventura, Tenerife, geographic information system, Spain

1. INTRODUCTION

The Canary Islands date palm, *Phoenix canariensis* hort. ex Chabaud is endemic to the Canary Islands, where it can be found naturally in valleys and ravines and as an ornamental tree in public and private gardens and parks. It is one of the most emblematic elements of biodiversity in the Canary Islands landscape.

In the first decade of the 21st century, real estate boomed in the Canary Islands and this led to a drastic increase in the import of adult palm trees, especially the date palm, *Phoenix dactylifera* L. This is how the red palm weevil (*Rhynchophorus ferrugineus* Olivier) (RPW) entered the Canary Islands, posing a serious threat to the conservation of *P. canariensis*.

J. Hendrichs, R. Pereira and M. J. B. Vreysen (eds.), Area-Wide Integrated Pest Management: Development and Field Application, pp. 539–550. CRC Press, Boca Raton, Florida, USA.

The RPW was first detected on the Island of Fuerteventura in September 2005 (Martín et al. 2013). This introduction most likely originated from the import of date palms from Egypt for ornamental purposes. Subsequently, inspections begun in the areas where *P. dactylifera* imports had taken place in the previous 6 years. In this way 11 new infested areas were found, 7 in Gran Canaria and 4 in Fuerteventura. Inside these 11 areas different phytosanitary measures were implemented, including surveillance (palm tree inspection and maintenance of a RPW trapping network) and removal and destruction of infested palm trees. However, the programme as outlined below was more than a sum of these activities.

The RPW regional eradication programme was initiated in September 2006. It was implemented by the Canary Islands Government public company 'Gestión del Medio Rural de Canarias', and co-funded by the Spanish Ministry of Agriculture.

2. THE PROGRAMME

2.1. Centralised Coordination and Organigramme

Especially in projects that involve separate and different geographic areas (e.g. different islands), each one with their responsible administration, there is always a tendency of projects to be implemented in a different way according to local ideas. Therefore, a centralised coordination unit, as well as a programme structure that was transparent, proved to be vital to reach the objectives of the project.

The organogram of the programme is shown in Fig. 1. The entire team consisted up to 35 people and each team on each of the three affected islands with RPW infestations (Gran Canaria, Fuerteventura and Tenerife, where the only RPW focus was detected in 2007) was headed by an island team leader. Efficient programme management proved to be the most difficult challenge and that was already obvious during the initial phases of the implementation of the project. Different aspects of project management and implementation resulted more challenging than the technical-scientific knowledge of the pest. These included establishing an efficient team, keeping track of the project objectives, efficient communication, effective coordination between institutions, and strict adherence to established protocols.

2.2. Legislative Measures

Since the detection of the RPW in Europe, all Governments, including the Canary Islands Government and the Island Councils, made legislative efforts, within the scope of their responsibilities, to arm themselves with legal instruments to control RPW. During de development of the eradication programme, the basic framework for the adopted measures was derived from:

APA/94/2006, 26 January, amending the Order of 12 March 1987 to prohibit the importation of plants of palm species (Palmae) of more than 5 cm of base diameter into the Autonomous Community of the Canary Islands (BOE No. 24 of 1/28/2006) (APA 2006).

Commission Decision 2007/365/EC of 25 May 2007 adopting emergency measures to prevent the introduction into and the spread within the European

Community of *Rhynchophorus ferrugineus* (Olivier) and its subsequent amendments. (OJ L139 / 24 of 31/05/2007) (OJ L266 / 14 of 07/10/2008) (17/08/2010 DOCE L) (European Commission 2007).

Decree of 29 October 2007 declaring the existence of pests produced by the harmful agents *Rhynchophorus ferrugineus* (Olivier) and *Diocalandra frumenti* (Fabricius) and establishing the phytosanitary measures for their control and eradication (Boletín Oficial de Canarias no. 222, dated 6.11.2007) from the Council of Agriculture, Livestock, Fisheries and Food of the Canary Government (BOC 2007).

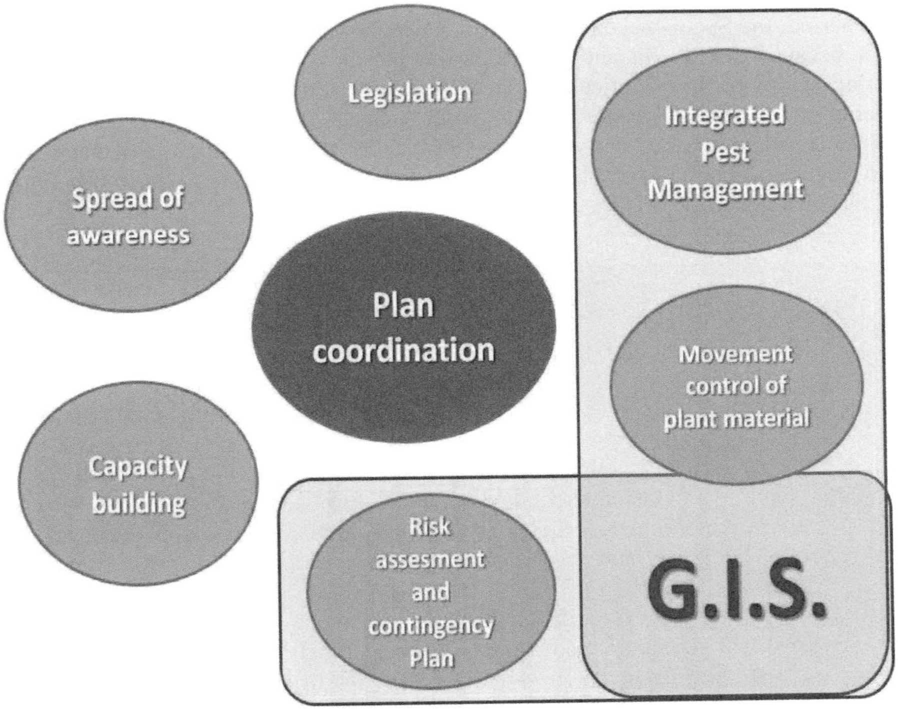

Figure 1. Red Palm Weevil Programme's organigramme.

The measures included in the regulations, at all levels, can be divided into two major groups (Gobierno de Canarias 2019a):

1. Measures that aimed at reducing man-assisted insect dispersal, including the prohibition of import of plants for planting, quarantine measures and regulations for the movement of plant material, and imposing related obligatory measures for nurseries and farmers, and

2. Measures that aimed at reducing the establishment of the pest and its natural spread after detection in a certain area.

The measures of the first group include the control of movement of plants for planting, promotion of stakeholder awareness, and stimulation of increased cooperation between institutions. Although these measures were the most difficult to enforce, they were crucial for the success of the programme. To ensure compliance with these regulatory measures, it was necessary to include staff with legal experience on phytosanitary regulations in the multidisciplinary teams of the eradication programme.

The global economic crisis resulting from the great recession of 2008, which affected Spain particularly hard, proved indirectly to be a bonus that made the implementation of the RPW quarantine and eradication measures easier. In the pre-crisis period, the Spanish economy was increasingly biased towards the construction sector because of a credit and real estate bubble (Jimeno and Santos 2014). The bursting of this bubble drastically reduced the number of requested permits for the construction of new real estate in Spain from close to a million per year before the crisis to less than 200 000 per year following the crisis. As a result, the demand for importing, transplanting and moving of palms from nurseries to new real estate sites was significantly reduced (Fig. 2).

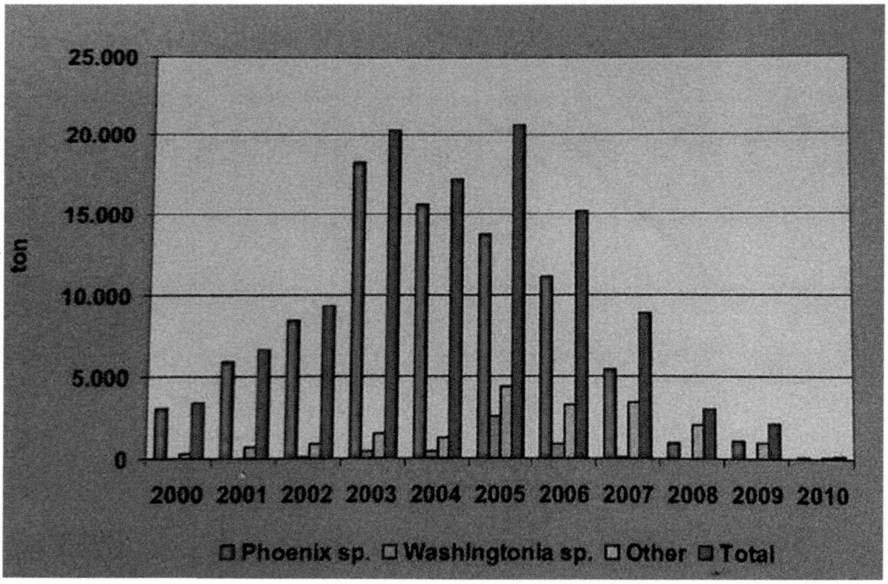

Figure 2. Tons of imported palm trees into Spain (source: Jose Maria Cobos, Spanish Ministry of Agriculture).

2.3. *Awareness Campaigns*

To involve as many people as possible in the eradication campaign, several information and awareness campaigns accompanied the programme to provide training and information to all stakeholders and citizens. These stakeholders included engineers, technicians and staff of the different public administrations, as well as gardeners in hotels, gardening companies, plant nurseries and the general public. The main goal was to achieve optimal support from the public in reporting and alerting the programme management board immediately when observing a palm tree with suspicious symptoms. This reporting system of the general public was crucial to implement quick follow-up actions. The following communication tools were used:

- A specific web page (Gobierno de Canarias 2019b)
- TV advertising campaigns
- Information on DVDs
- Conferences and special workshops held on each one of the islands
- Brochures.

One of the crucial factors in this communication campaign was complete transparency, and to openly disclose all available information. In this way, all stakeholders became "co-owners" of the programme and felt confident and involved.

2.4. *Specific Training*

The Order of October 2007 stipulated that anybody working on palm trees should be accredited as "*Specialist on labour on palm trees*". The objective of this accreditation was to ensure that any person or company carrying out any type of work on palm trees was conversant with all methods and protocols described in the legislation and had received the necessary training to ensure proper palm tree management. To achieve this goal, several courses were organized to train-the-trainers, who obtained thereafter the necessary accreditation. The programme encouraged all official administrations and relevant academic institutions to include courses on palm tree management in their curricula. The RPW programme edited and produced the "*Manual of good practices in palm trees*", which was used as textbook in these courses.

2.5. *Movement of Plant Material*

Import or movement of infested plants for planting is the main route of introduction and spread of the pest over long distances. Therefore, the programme deemed it crucial to restrict and control the movement of plants. In addition, to manage the movement of palm trees from nurseries, the programme required all nurseries to be registered. Nurseries were inspected periodically with emphasis on:

- The registration of the entry and exit of palm trees
- Visual inspections of possible RPW damage and symptoms
- Application of the mandatory chemical treatments.

A "*Phytosanitary Accreditation*" was created, consisting of a 6-month valid movement authorization for palms susceptible to RPW, except *P. canariensis, P. dactylifera* and *Washingtonia* spp., for which authorizations were requested for every single movement. It was mandatory to have an authorization of the Canary Islands Plant Protection Organization to perform transplantations of *Phoenix* spp. and *Washingtonia* spp. These could only be carried out by accredited companies. All requests for transplantation of palm trees located within a range of 5 km from infested plants were denied.

2.6. Integrated Pest Management Programme

The following activities were included in the integrated pest management programme: visual inspection of trees, chemical treatments, destruction of RPW-infested palms, monitoring/mass-trapping of RPW, and cultural measures such as pruning, which was only authorized for public safety reasons and should include the application of an oil painting or pruning mastic on the pruning scars. All of these are common measures implemented in control programmes world-wide (Abraham et al. 1998; Faleiro 2006; Dembilio and Jaques 2015).

2.6.1. Delimitation

Upon detection of a RPW infestation in a palm tree, or a group of palm trees, a new pest focus was declared, which was defined by two main areas:

Intensive surveillance area: The area with a 1 km radius from the outermost affected palm trees in the focus. All palm trees within this area were registered into batch, and only palms trees that were positively diagnosed with RPW were registered individually.

Guided surveillance area: The area within a 3 km radius from the border of the intensive surveillance area.

2.6.2. Inspection of Palm Trees

After testing all possible tree inspection methods available, intensive visual inspection was found to be the most effective detection method. The method consisted of a thorough observation of the stipe and all the bases of the crown's fronds. This type of inspection was performed by specialized personnel.

2.6.2.1. Inspection Inside Intensive Surveillance Area. In addition to regular inspections (about 3 inspection/palm/year) of all palm trees within the intensive surveillance area, visual inspections were carried out around each trap where RPW adults were caught, as well as around newly detected infested palm trees.

2.6.2.2. "Guided Inspections". Specialised technicians, responsible for the guided surveillance, carried out the visual inspections within the guided surveillance areas. These inspections allowed marking the location of affected palm trees outside the intensive surveillance areas that prevented the dispersal of the pest from infested areas.

2.6.2.3. Alert System. The programme established an alert system in which any citizen could report observations of palm trees with apparent symptoms of the RPW. Through this system, five new RPW outbreaks were detected very early, which made it very easy to bring these outbreaks under control. The success of the alert system measure was facilitated by the public outreach and awareness campaigns (see above in point 2.3.).

2.6.3. Removal of Infested Palm Trees

All palm trees suspected of being infested with RPW were removed. In most cases, the removal took place within 24 hours after detection. In those cases where this was not possible, palms were treated and enmeshed until removal.

The removal process followed a strict disposal protocol to avoid dispersal of adult weevils during the process. The tree stump was guarded and inspected for several days after the removal and a trap was deployed next to it for at least a week.

2.6.4. Chemical Treatments

Chemical treatments aimed to control the immature stages of RPW found in the most superficial part of the palms. Insecticides such as chlorpyrifos 48%, imidacloprid 20% and thiamethoxam 25% were sprayed on the tree at very low pressure, using about 15 litres of the mixture per palm.

Throughout the programme, chemical treatments were routinely applied (about 2 treatments/year) to all palm trees inside the intensive surveillance area, but also to palm trees around each newly detected affected palm, as well as around traps when adult specimens were caught.

2.6.5. Trapping Networks

Food and pheromone baited traps were deployed to maintain a trapping network following different strategies and objectives:

- Mass-trapping
- Adult weevil attraction to the centre of each pest foci
- Population monitoring
- Detection of new pest foci.

The traps were baited with 700 mg of *R. ferrugineus* attractant (4-methyl-5 nonanol 90% and 4-methyl-5-nonanon 10%, both purity>95%) and either ethyl acetate (kairomone) or fresh palm tissue. They were checked for weevils and serviced once a week. At an average temperature of 28ºC the attractant is released at a rate of 11mg/day making the trap effective for a period of 6 to 8 weeks (product label information). The self-made four window (4 cm diameter) 10 litre bucket traps with no opening on the lid were placed at more than 15 meters from any palm tree, and if possible, were buried half in the ground. At the onset of the programme, white traps were used, but starting in 2011 these were painted black (Ávalos and Soto 2010).

Different strategies were followed to manage the trapping networks. As recommended by Oehlschlager (1994), the programme started using a grid of 1 trap/ha in pest foci and surrounding guided surveillance areas. This was later replaced by 'dynamic micro-networks of traps', where traps were deployed at a density of 4 traps/ha in the polygons of the affected palm trees. Following this approach, no traps were deployed in areas around pest foci and their surrounding guided surveillance areas, where the presence of the pest was not proven. The objective was not to attract the weevils away from affected areas by placing traps into areas where they had not yet been observed. These networks were 'dynamic' and continuously adjusted and adapted in size based on (a) new detections of affected palms, (b) increased catches in certain areas, (c) the absence of newly infested palm trees, and (d) the absence of weevil catches. On islands with known RPW foci, traps were also placed around the areas where the infested palm trees had been disposed of.

In islands that had remained free of RPW, traps were placed in areas where imported date palm trees had been planted in the last 5 years, e.g. golf courses, hotels, newly constructed real estate projects, nurseries, etc. Using this approach, the 2007 outbreak in the Island of Tenerife was detected early.

2.7. *Contingency Planning*

As soon as a new focus was detected, a contingency plan was developed and implemented. The purpose of this contingency plan was to determine the origin of the outbreak, as well as to determine the location and to remove all infested palm trees. All human resources of the programme were dedicated to the new focus until the situation was brought under control.

3. A TOOL: GEOGRAPHIC INFORMATION SYSTEM (GIS)

A geographic information system (GIS) is a system designed to capture, store, manipulate, analyse, manage, and present spatial or geographic data (Foote and Lynch 1995). The eradication programme included a programming team (ITs), responsible for the development of the used GIS applications.

The GIS was the main tool supporting the decision-making process for three of the main activities of the programme (Fig. 1), i.e. the IPM programme, the control of movement of all plant material, and contingency plans.

The GIS was considered an essential tool for the planning and effective coordination of the eradication pest programme that allowed:

- Data and spatial analysis for optimal decision-making
- Efficient planning and use of resources
- Assessment of the programme (results, achievement of objectives) and workers from readily available quality information
- Improvement of the programme's internal and external communications.

The GIS consisted of four important elements: mobile applications, a database, a web application and a web viewer.

3.1. *Database*

The main objective of the database was to store and centralise all relevant information:

- Elements of the programme, e.g. pest foci, groups of palm trees, individual palm trees, traps, nurseries
- Activities of the programme, e.g. removal of infested palm trees, inspections, chemical treatments
- Results of the programme, e.g. trap catches, inspection data etc.
- Resources of the programme, e.g. workers, type of chemical products, type of traps, pheromones.

All this information was conveniently organized and related. All other software applications developed interacted with the database, either to introduce new values (e.g. field-collected data with the mobile application) or for the processing of information (web viewer, web application) to generate reports, customized maps, etc.

3.2. *Mobile Application*

An application for mobile devices was developed to facilitate data collection in the field. It was designed to avoid mistakes when entering data resulting in great efficiency, accuracy and high data quality.

Usually, at the end of each week, each island team leader summarized the collected data using the internet. These data were stored on a web server and automatically imported into the central database of the project.

3.3. *Web Application*

A web application was developed to use the database in a more friendly and efficient way. This application allowed:

- Data entry
- Data editing
- Performing queries
- Generation of tables
- Generating graphics and reports.

3.4. *Web Viewer*

The web viewer allowed observing and analysing all the spatial information collected by the field teams. As a result, it was possible to show on a map:

- Stored data, such as lots, affected palm trees, traps, farmers and nurseries
- Customized queries, e.g. palm trees removed by date ranges, palm lots in a range of 100 meter around a trap with catches, traps categorised by the number of catches or by a date range
- New layers, e.g. areas occupied by infested palm trees and traps.

4. MANAGEMENT OF HUMAN RESOURCES

The eradication programme as described above offered a framework to reach the eradication objectives. Nevertheless, for the correct implementation of all measures, it was essential to have an efficient management team. Probably the biggest challenge of any programme direction is to establish and manage this team.

The team was composed of members whose attitude towards work and internal training was considered exemplary. Efficiency in the programme was maintained as each team member was aware of the relevance of his/her role in the implementation of the programme and its ultimate success. This entailed that the objectives and procedures of each task had to be clearly defined.

To achieve the programme's objectives, great attention was given to continuous training, improved motivation of the group members and to always create and maintain a positive team spirit. At all times it was emphasised that the group members were the protagonists of the obtained results. A team member could always make suggestions and the proposals were always evaluated and sometimes incorporated into the procedures.

5. RESULTS

On three islands (Gran Canaria, Fuerteventura and Tenerife), sixteen RPW foci were detected and eradicated (Fig. 3). More than 70 000 palm trees were registered, 706 081 visual inspections were made, and 209 547 chemical treatments were carried out. A total of 681 RPW adults were caught in traps (Fig. 4) and 660 palm trees removed. In May 2016, 11 years after the pest was first detected, and after three years without finding affected palm trees or catching RPW in traps, the Canary Islands were declared free of the pest.

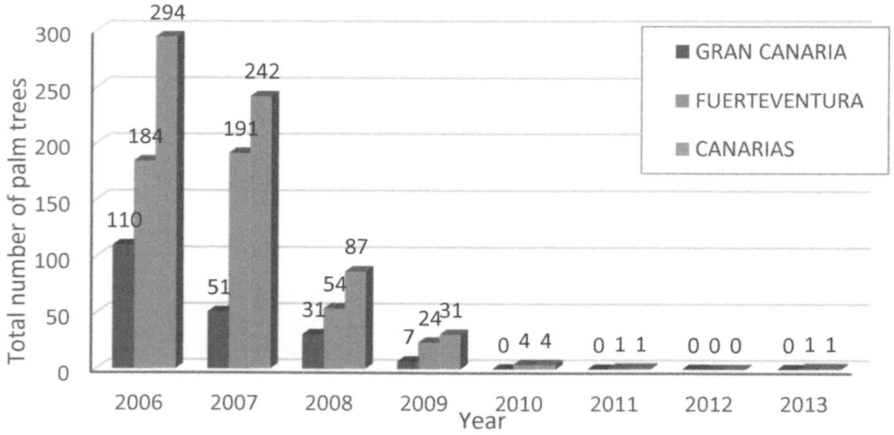

Figure 3. Affected palm trees between 2006 and 2013 on the islands of Gran Canaria and Fuerteventura, as well as total numbers for the Canaries.

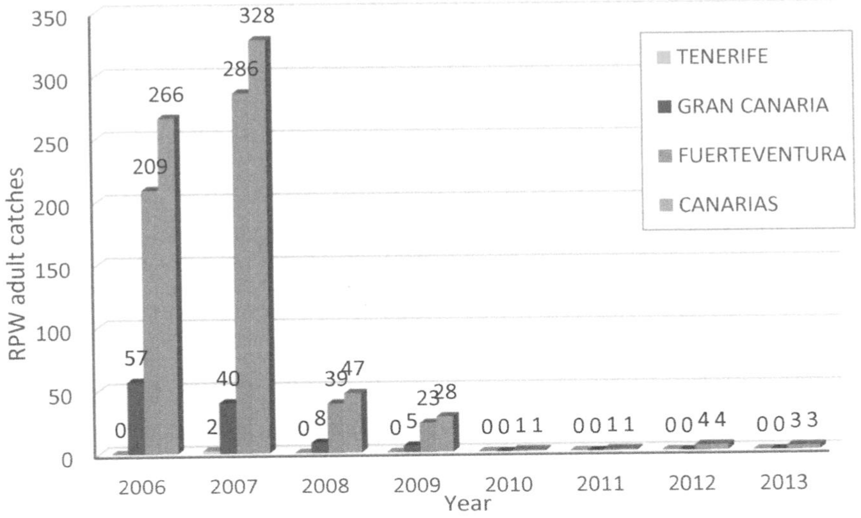

Figure 4. Red Palm Weevil adult catches in traps between 2006 and 2013 in Tenerife, Gran Canaria, and Fuerteventura, as well as total numbers for the Canaries.

The Canary Islands example shows that the presently available knowledge and control tactics can successfully eradicate the RPW. The main issue is not the lack of technical know-how, but the establishment of an efficient organization and its management to reach the objectives. The Canary Islands are now facing a new challenge and that is to maintain the motivation and support to prevent and detect early any new introduction of the RPW.

Taking into account the results and positive experience gained in the eradication programme of RPW in the Canary Islands, the minimum requirements for a successful eradication programme are as follows:

- Applying a programme in areas of recent introduction or where the RPW has been kept under control
- Identifying areas isolated from affected areas by a buffer zone with no susceptible palms or that are at least 10 km away from the nearest RPW focus, where no programme activities have to be applied
- Including adequate legislative measures and their enforcement aiming to avoid new introductions and the movement of plant material
- Correct design and integrated implementation of all programme´s activities and components
- Adequate budget available according to the number of pest foci and other requirements
- Centralised coordination (e.g. communication, decision-making) of area-wide programme activities
- Adequate human resources available and their management (e.g. training, attitude, motivation, constancy procedures)

- Effective use of GIS in support of data management and decision-making
- Public education and engagement, and involvement of all stakeholders
- Cooperation and coordination with public (Provinces, Autonomous Communities and affected Municipalities) and private entities (e.g. nurseries, gardening companies, hotels, farmers).

6. REFERENCES

Abraham, V. A., M. Al-Shuaibi, J. R. Faleiro, R. A. Abouzubairah, and P. S. P. V. Vidyasagar. 1998. An integrated approach for the management of red palm weevil *Rhynchophorus ferrugineus* Oliv. a key pest of date palm in the Middle East. Journal of Agricultural and Marine Sciences 3: 77–83.

APA (Ministerio de Agricultura, Pesca y Alimentación). 2006. Orden APA/94/2006, de 26 de enero, por la que se modifica la Orden de 12 de marzo de 1987, por la que se establecen para las islas Canarias, las normas fitosanitarias relativas a la importación, exportación y tránsito de vegetales y productos vegetales, para prohibir la importación de vegetales de especies de palmeras (Palmae) en la Comunidad Autónoma de Canarias. Documento Boletín Oficial del Estado-A-2006-1330.

Ávalos, J. A., and A. Soto. 2010. Mejora de la eficacia del trampeo de adultos de *Rhynchophorus ferrugineus* (Olivier) (Coleoptera: Dryophthoridae) mediante atracción cromática. Phytoma España: La Revista Profesional de Sanidad Vegetal 223: 38–42.

BOC (Boletín Oficial de Canarias) 2007. 833 - Orden 1833, de 29 de octubre de 2007, por la que se declara la existencia de las plagas producidas por los agentes nocivos *Rhynchophorus ferrugineus* (Olivier) y *Diocalandra frumenti* (Fabricius) y se establecen las medidas fitosanitarias para su erradicación y control. Boletín Oficial de Canarias número 222, martes 6 de noviembre de 2007.

Dembilio, O., and J. A. Jaques. 2015. Biology and management of red palm weevil, pp. 13-36. *In* W. Wakil, J. R. Faleiro, and T. A. Miller (eds.), Sustainable pest management in date palm: Current status and emerging challenges. Springer International Publishing, Switzerland.

European Commission. 2007. Commission Decision of 25 May 2007 on emergency measures to prevent the introduction into and the spread within the Community of *Rhynchophorus ferrugineus* (Olivier) (notified under document number C (2007) 2161) (2007/365/EC) (OJ L 139, 31.5.2007, p. 24). Amended by Commission Decision 2008/776/EC of 6 October 2008, and Commission Decision 2010/467/EU of 17 August 2010.

Faleiro, J. R. 2006. A review of the issues and management of the red palm weevil *Rhynchophorus ferrugineus* (Coleoptera: Rhynchophoridae) in coconut and date palm during the last one hundred years. International Journal of Tropical Insect Science 26: 135–154.

Foote, K. E., and M. Lynch. 1995. Geographic Information Systems as an Integrating Technology: Context, concepts, and definitions. Geographer's Craft Project, Department of Geography, The University of Colorado at Boulder, Colorado, USA.

Gobierno de Canarias. 2019a. El picudo rojo *Rhynchophorus ferrugineus* Olivier. Dossier informativo. 127 pp. Consejería de Agricultura, Ganadería, Pesca y Aguas. Consejería de Medio Ambiente y Ordenación Territorial.

Gobierno de Canarias. 2019b. Picudo rojo en Canarias. Consejería de Agricultura, Ganadería, Pesca y Aguas.

Jimeno, J. F., and T. Santos. 2014. The crisis of the Spanish economy. SERIEs 5: 125–141.

Martín, R., A. González, L. Barroso, M. Morales, C. D. Hernández, X. Rodríguez, and M. Fajardo. 2013. Plan de surveillance, de lutte et d'éradication du Charançon Rouge dans les Îles Canaries (Espagne). In Proceedings, Palm Pest Mediterranean Conference, Nice, France, 16-18 Janvier 2013, Alfortville, France. Association Française de Protection des Plantes (AFPP), France.

Oehlschlager, A. C. 1994. Use of pheromone baited traps in control of red palm weevil in the Kingdom of Saudi Arabia. Consultancy Report, Ministry of Agriculture, Riyadh, Saudi Arabia.

AREA-WIDE MANAGEMENT OF INVADING GYPSY MOTH (*Lymantria dispar*) POPULATIONS IN THE USA

A. M. LIEBHOLD[1], D. LEONARD[2], J. L. MARRA[3] AND S. E. PFISTER[4]

[1]*USDA Forest Service Northern Research Station, Morgantown, West Virginia 26505 USA; aliebhold@fs.fed.us*
[2]*USDA Forest Service, Forest Health Protection, Asheville, North Carolina 28802, USA*
[3]*Washington State Department of Agriculture, Olympia, Washington 98504, USA*
[4]*USDA/APHIS/PPQ/CPHST, Buzzards Bay, Massachusetts 02542, USA*

SUMMARY

The European strain of the gypsy moth, *Lymantria dispar* L. (Lepidoptera: Erebidae) was accidentally introduced to North America over 100 years ago and despite its explosive population growth there, the species still only occupies less than 1/3 of its potential range. While this slow rate of spread can be attributed in part to the limited dispersal capacities of this strain, its constrained distribution mainly reflects the success of efforts to limit range expansion of this species. Currently, two major area-wide programmes are operated to limit the spread of the gypsy moth in the USA, in addition to a third programme that suppresses gypsy moth outbreaks in the infested areas. The detection / eradication programme is led by the United States Department of Agriculture - Animal and Plant Health Inspection Service (USDA-APHIS) in cooperation with state governments and utilizes networks of pheromone traps to detect newly invaded populations of the gypsy moth in the uninfested portions of the USA. Over the last decades, hundreds of isolated populations have been detected and eradicated. Most eradication treatments in the USA are conducted using aerial sprays of *Bacillus thuringiensis*. The USDA Forest Service also operates another area-wide programme entitled "Slow the Spread" (STS) in cooperation with state agencies that operates at the edge of the generally infested area and aims to slow the gypsy moth's spread. This programme also uses grids of pheromone traps to locate isolated populations, which are then treated. The STS programme has adopted several major innovations that make it one of the most advanced area-wide programmes for managing invading species. Among these innovations, the STS programme adopts a complex geographic information system (GIS)-based decision algorithm for processing trap data, identifying treatment areas and evaluating programme efficacy. Also, the STS programme is unique in that it largely has adopted mating disruption to eradicate or suppress isolated populations ahead of the invading front.

Key Words: Lepidoptera: Erebidae, eradication, barrier zone, biological invasion, decision support

J. Hendrichs, R. Pereira and M. J. B. Vreysen (eds.), Area-Wide Integrated Pest Management: Development and Field Application, pp. 551–560. CRC Press, Boca Raton, Florida, USA.

1. INTRODUCTION

While many invasive species are rare in their native ranges, this is not the case for the gypsy moth *Lymantria dispar* L. (Lepidoptera: Erebidae). Across much of its native range, which spans most of temperate Asia, Europe and North Africa, this species episodically reaches outbreak levels, causing massive defoliation of host trees (Giese and Schneider 1979; Johnson et al. 2005). Similarly, across much of the region that the gypsy moth has invaded in North America it has caused considerable damage, i.e. >15 million ha have been defoliated in the USA during the last 30 years alone (USDA/USFS 2017). Forest defoliation caused by the gypsy moth can have severe impacts that include effects on aesthetics, particularly in forested residential areas, as well as triggering tree mortality and growth loss, ultimately leading to shifts in regional forest composition (Morin and Liebhold 2016).

The history of gypsy moth invasion in North America began in 1868-1869 when Étienne Léopold Trouvelot accidentally released the insect in the backyard of his house in Medford, Massachusetts (Liebhold et al. 1989; McManus 2007). At the time, Trouvelot was a commercial artist but had an amateur interest in entomology and was rearing a large assortment of insects in his garden. Though Trouvelot notified local authorities about the escaped insects, no action was taken until about 1880 when the first outbreak started in his neighbourhood, alarming residents. At that time, the state of Massachusetts embarked on a large eradication campaign, but the effort was abandoned in 1900 as a result of the lack of effective surveillance and control tools. Nevertheless, this programme was apparently the first attempted insect eradication in the world. Even though the gypsy moth has been in North America for almost 150 years, it still only occupies less than 1/3 of its potential range (Fig. 1) (Morin et al. 2004). One of the causes of this exceptionally slow invasion spread is that in the European strain of the gypsy moth, from which populations were introduced, females are incapable of flight and most spread is driven by accidental movement of life stages by humans (Liebhold et al. 1992). The other reason why spread has been so slow is that efforts to limit its spread have been successful. Even though the initial eradication campaign in Massachusetts was a failure, there have been numerous government-led barrier zones and other programmes aimed to contain this insect; these efforts have evolved over the last century, but the programmes currently in place represent state-of-the-art area-wide management and serve as excellent models for potential application to other insect invasions.

Currently there are three different large programmes that target gypsy moth populations in the USA and these programmes vary both in their objectives and in their geographic scope (Fig. 1). First, United States Department of Agriculture (USDA) Forest Service, Forest Health Protection works with various state agencies to manage the Gypsy Moth Cooperative Suppression Programme. This programme operates within the gypsy moth generally infested area to suppress outbreak populations of the gypsy moth; because decisions about treatments are made individually on a stand-by-stand basis, this cannot be considered a true area-wide management programme.

The second programme, the gypsy moth detection/eradication programme is led by USDA Animal and Plant Health Inspection Service (APHIS) Plant Protection and Quarantine (PPQ) works with state agencies and aims to exclude gypsy moth invasion in regions of the USA where the gypsy moth is not currently established. The third programme, the gypsy moth "Slow the Spread" (STS) programme is carried out by the USDA Forest Service in cooperation with state agencies and operates in the transition area between the infested and uninfested portions of the USA. The objective of this programme is to slow the gypsy moth's spread into the uninfested region. Both, the detection/eradication and the STS programmes are examples of area-wide management. These programmes represent the culmination of an evolution of technology and strategy and thus serve as model programmes for other area-wide efforts. Here we describe both programmes, including both strategic and methodological details.

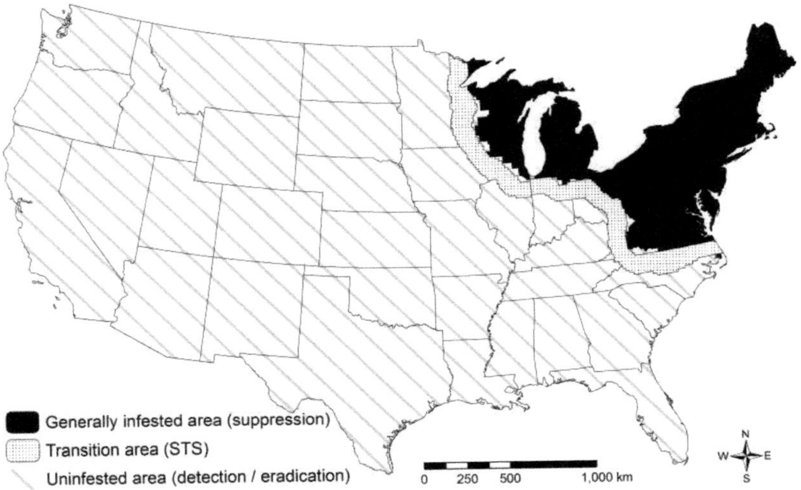

Figure 1. Map showing the spatial extent of invading gypsy moth populations (2017) in the USA and locations of generally infested area, the transition area (STS = "Slow the Spread" programme) and the uninfested area.

2. DETECTION / ERADICATION

Gypsy moth egg masses are often laid in cryptic locations and this behaviour leads to egg masses becoming associated with objects, such as cars, lawn furniture and firewood, that are transported during household moves. In the USA, intra-continental household moves are common, and this unfortunately results in gypsy moth egg masses being transported from outbreak areas in the generally infested area into uninfested states. If no action were taken, many of these translocated egg masses would found new populations that would grow and potentially damage forests in these regions. Fortunately, newly founded populations can be efficiently detected using pheromone-baited traps. Female gypsy moths produce a sex attractant, (+) cis-7,8-epoxy-2-methyloctadecane ("disparlure"), that was identified in the early 1970's (Bierl et al. 1970) and can be synthesized relatively inexpensively.

The general strategy of the detection/eradication programme is to detect and eradicate newly founded populations (Fig. 2). The strategy consists of the following steps: (1) regularly deploy an extensive network of traps to detect newly founded populations (Year 1); (2) deploy a dense grid of traps where moths were detected to confirm the persistence of the population and delimit its spatial extent (Year 2); (3) suppress the population below an extinction threshold (Year 3), and (4) deploy a dense grid of traps to confirm eradication or identify areas requiring additional treatment.

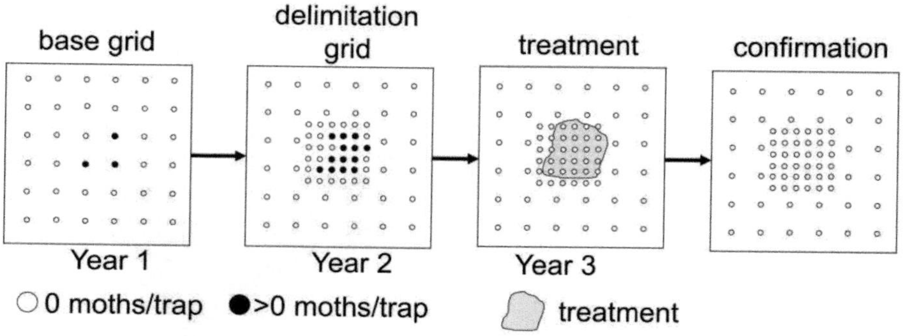

Figure 2. General strategy used in gypsy moth detection/eradication programmes.

Current protocols vary slightly among states, but in most states detection trapping for gypsy moth is conducted once a year in view that this pest has only one generation per year (USDA/APHIS/PPQ 2019). The density of trap deployment varies among land uses with different predicted risks of introduction. For example, affluent residential areas are considered to be high risk and traps are deployed at densities of ~0.4 trap/km^2 every other year, but unpopulated forests are considered low risk and traps are deployed at 0.1 trap/km^2 in such areas once every 4 years.

The vast majority of traps in base detection trapping capture no males, while capture of one or more males usually triggers delimitation trapping in the next phase (Year 2). This trapping serves both the purpose of delimiting the spatial extent of the invading population but also confirms the persistence of populations. In most cases, capture of one or more males in one year does not result in captures at the same location in the next year; low density populations are prone to extinction due to Allee effects or stochastic dynamics (Liebhold and Bascompte 2003).

The standard trap density in delimitation surveys is ~6 traps/km^2; delimitation traps are deployed in the area surrounding positive trap capture locations, extending to the nearest negative trap capture locations from the previous generation (USDA/APHIS/PPQ 2019). In addition to delimitation trapping, most states will visually examine a ~1000 m^2 vicinity around positive trap captures, searching for egg masses or other life stages. These searches serve two purposes: 1) the presence of immature life stages confirms the existence of a reproducing population, and 2) discovery of immature life stages usually is indicative of the "core" population that then becomes highest priority for treatment.

Once a population has been detected, found to persist and is delimited, then it is ready for treatment (Year 3). Because low-density gypsy moth populations are strongly affected by Allee effects (mostly arising from mate-location failure [Tobin et al. 2009]), treatments need not kill 100% of individuals; eradication can be achieved by reducing populations below an Allee threshold and residual populations can be anticipated to decline toward extinction (Liebhold and Bascompte 2003). Most gypsy moth eradication treatments use aerial sprays of *Bacillus thuringiensis kurstaki* (*Btk*). Usually at least 2 applications are made in a single year in order to maximize mortality; additional applications are made when there is uncertainty about the timing of susceptible early instars. In very small populations, ground treatments of *Btk* may be substituted for aerial treatment. Most treatments are applied over relatively small areas (i.e. < 100 ha) in urban areas so the non-target impacts (e.g. mortality of native Lepidoptera) are minimal. Mating disruption treatments are also quite effective against low-density gypsy moth populations (Thorpe et al. 2006), but they are infrequently used for eradication because mating disruption treatments shut-down trap captures that are used to confirm eradication success. Both types of treatments tend to be effective at eliminating populations in a single year.

Starting in the year of treatment, a grid of traps is deployed to detect residual populations post-treatment. This grid is comparable in density to delimitation grids and can be used to identify areas requiring additional treatments. Small populations are usually successfully eradicated by treatments in a single year, but in large populations it is not unusual to treat additional areas in subsequent years. Eradication success is generally declared after two or more years of no captures (Years 4-5).

With increasing trade between Asian countries and North America over the last decades, there has been an increasing flow of gypsy moth egg masses accidentally transported on ships, containers, bulk steel and cars imported from Asia. Unlike European populations, females from most Asian gypsy moth (AGM) populations are capable of at least some flight (Keena et al. 2008) and females are sometimes attracted in large numbers to brightly-illuminated seaports. Increased flight capabilities in AGM strains is a primary reason why invasions by such populations are anticipated to be more difficult to eradicate and contain. Given this risk, additional traps are deployed in areas adjacent to maritime terminals (both Pacific and Atlantic) and other high-risk locations. All trapped males (including those from the AGM high risk areas and ordinary detection survey traps) are returned to the USDA/APHIS/PPQ Laboratory in Buzzards Bay, Massachusetts where they are subjected to molecular analysis to determine their most likely origin.

Two genetic markers are used to assess the genotype of gypsy moth specimens: the nuclear marker FS1 (Garner and Slavicek 1996) and a mitochondrial marker (Bogdanowicz et al. 1993). Female flight is a trait that is considered to make a population more difficult to contain or eradicate. Consequently, policies for responding to AGM captures are more aggressive in that instead of waiting a year to delimit populations following initial capture (Fig. 2), eradication treatments are generally applied in the same year, directly following initial detection.

Similar practices for detection and eradication of both Asian and European strains of the gypsy moth are implemented in Canada.

3. SLOW THE SPREAD (STS)

As is the case with the spread of many other invading species, gypsy moth range expansion is not a continuous process. Instead, isolated populations sporadically develop ahead of the advancing population front. These isolated populations continuously expand and eventually coalesce with each other and the continuously infested population front (Sharov and Liebhold 1998). This pattern arises from a phenomenon, "stratified diffusion", in which dispersal is comprised of two different types of movement. In the case of the gypsy moth, spatially continuous short-range dispersal results from wind-borne movement of first instars and long-range stochastic dispersal occurs when humans accidentally move egg masses, larvae or pupae (e.g. movement of infested firewood).

In designing the STS programme, it was considered impractical to completely stop gypsy moth spread, but instead the objective shifted to slowing the rate of spread by detecting and suppressing new isolated populations ahead of the advancing front. This is accomplished by deploying a grid of pheromone traps along the transition area in order to detect isolated populations (Fig. 1). These populations are subsequently delimited and treated, much like the strategy used in the detection/eradication programme in the uninfested area (Fig. 2). However, unlike eradication programmes, the objective of treatments is not necessarily extinction of the isolated population, but instead the programme aims to suppress its growth. Population models demonstrate that suppressing these isolated populations ahead of the population front can have a substantial impact on reducing spread even though relatively small areas are treated (Sharov and Liebhold 1998).

The STS programme was initiated as a pilot programme in the states of North Carolina, Virginia, West Virginia and Michigan in 1992 and in 2000 was expanded to the entire population front running from the Atlantic Ocean to the Canadian border (Fig. 1) (favourable gypsy moth host type stops just north of the Canadian border with Minnesota).

The majority of the funding (USD 8 - 10 million / year) for STS comes from the USDA Forest Service, which grants most funds to the STS Foundation, which in turn grants funds to individual state governments to carry out trapping and treatments (Leonard 2007). The STS Foundation is a non-profit organization managed by a representative from each of the 11 states participating in STS. Structuring management of the programme around the Foundation increases partner state ownership of and accountability for the programme, promotes programme management based on biological rather than jurisdictional boundaries and facilitates uniform implementation of protocols, and decision-making.

Expenditure of these funds was justified by a benefit-cost analysis, which demonstrated that the economic benefit of postponing the initiation of gypsy moth impacts and management (e.g. the USDA Forest Service Gypsy Moth Cooperative Suppression Programme described above) expenses vastly exceeds the cost of the STS programme (Leuschner et al. 1996). Roughly half the STS budget is spent on trapping and half on treatments.

Trapping is conducted in a ca. 100 km band (coincident with the transition area shown in Fig. 1) termed the "action area", in which traps are deployed in a rectangular grid with 2-3 km spacing between traps (Roberts and Ziegler 2007). In a 70 km band located just inside the generally infested area and adjacent to the action area, trapping also takes place in an "evaluation area" in which traps are deployed in a grid with 5-8 km between traps; trapping data from the evaluation area play a crucial role in measuring reduction in spread rates as well as in locating boundaries for the action area. When base trapping in the action area indicates the location of an isolated population, a delimitation grid is deployed in the next year; delimitation trapping is conducted with traps placed on a grid with 250 to 500 m between traps. Much like detection/eradication, a delimitation grid is also deployed following treatment in order to evaluate treatment success.

While most treatments in gypsy moth detection/eradication programmes are conducted using *Btk*, most treatments in STS are conducted using mating disruption (Thorpe et al. 2006). This reflects, in part, the objective of minimizing the overall environmental impact of the STS programme considering that treatments are applied over a relatively large area (Table 1). But also, mating disruption treatments have lower overall costs. Historical data indicate that when applied against low-density populations in the STS programme, mating disruption treatments are equally as effective as *Btk* applications (Sharov et al. 2002). Occasionally, moderate density gypsy moth populations (> ~100 moths / trap) are detected in the STS action area and these populations are usually treated with *Btk* because of lower efficacy of mating disruption in such higher density populations.

Table 1. Traps deployed, populations treated and treatment areas in the gypsy moth Detection / Eradication and STS programmes 2010-2016

Year	Detection and Eradication			Slow the Spread (STS)		
	Traps deployed	Populations treated	Area treated (ha)	Traps deployed	Populations treated	Area treated (ha)
2010	102 795	2	525	89 950	231	216 125
2011	107 646	3	2340	83 800	221	208 750
2012	108 060	2	833	53 900	149	213 414
2013	144 925	1	421	47 850	130	164 441
2014	112 153	0	755	60 000	138	169 425
2015	123 938	2	674	60 000	182	205 561
2016	134 151	3	1484	65 000	176	179 084

The STS programme represents a highly innovative area-wide integrated pest management programme in many ways. One of the early innovations of the programme was its adoption of GIS technology so that all trap data are geo-referenced. But perhaps the most innovative aspect of the programme is its implementation of a highly standardized "decision algorithm" that is applied throughout the programme (Tobin et al. 2004; Tobin and Sharov 2007). The decision algorithm consists of computer code that processes data, mostly in the form of survey trap data, to make decisions on action (trapping and treatment) and to generate output used by STS managers to evaluate the effectiveness of the programme. The decision algorithm was developed, in part, from population models that simulated gypsy moth spread and decision-making to optimally reduce spread (Sharov and Liebhold 1998), but the decision algorithm has been continually fine-tuned to increase efficacy and reduce costs without losing efficiency.

The decision algorithm is applied every year to process trapping data from the field. It performs various quality control analyses in order to flag potential data quality problems. The most basic task it performs is the application of several different algorithms to locate potential isolated populations from the base trapping grid. Once these areas are located, the decision algorithm then highlights areas where delimiting trapping should be performed or where treatments are needed.

The other major feature of the decision algorithm is evaluation of programme efficacy. This starts with individual evaluations that are made for each treatment block. But the decision algorithm also measures spread rates along the entire action area in each year. While STS programme managers are constantly monitoring the efficacy of treatments, the ultimate success of the programme is based upon reduction of invasion spread. During the decades prior to the programme's initiation, gypsy moth range expansion averaged ~21 km/year (Liebhold et al. 1992), but since national implementation of the STS programme in 2000 spread has averaged about 4 km/year, which is an 80% reduction and exceeds the programme objective of 50% reduction in spread.

One of the key features of the decision algorithm is the flow of data and visualization of decision algorithm output (Fig. 3).

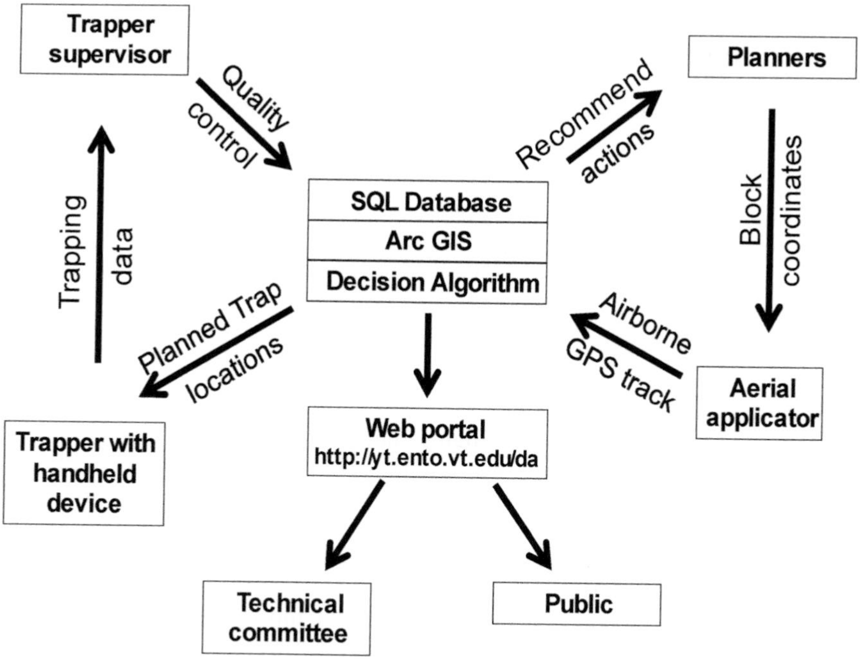

Figure 3. Flow of data and information in the STS programme. "SQL" refers to the structured query language.

Starting in 1996, the programme adopted the use of handheld GPS units by trappers to guide trap placement and record trap captures, but as technology advanced the GPS units have been replaced by handheld tablets. All of the decision output from the decision algorithm is reviewed each year by state and federal STS planners. For a variety of reasons, planners may decide to over-ride or modify the decision recommended by the decision algorithm. The performance of the programme is closely monitored by STS managers and the STS Technical Committee, and this information is used to adjust the decision algorithm as needed.

Finally, all trap and treatment locations, trap counts, and analyses of programme performance are freely available for viewing on a web portal (STS 2019). This web access makes the programme completely transparent so that any interested members of the public or government agencies can view programme activities.

4. CONCLUSION

The gypsy moth is one of the most damaging invasive pests in North America and consequently substantial effort has historically been expended to contain its spread. Over the course of these historical programmes, considerable knowledge has been gained both on understanding how this organism spreads, but also about how to maximize the efficacy of containment efforts. The fact that gypsy moth currently only occupies about 1/3 of its potential range in the USA, despite having become established here for nearly 150 years, reflects the success that has been achieved from these programmes.

A large fraction of the success in limiting the gypsy moth's spread in the USA can be attributed to certain specific technological developments which include the development of an inexpensive yet highly sensitive pheromone trap for this insect, the application of GIS technology and the perfection of various treatment technologies for eradication or suppression of high- and low-density populations.

With increasing globalization, the problem of biological invasions continues to grow as thousands of species are moved around the globe, often causing catastrophic impacts. Given this trend, there is increasing need for effective strategies not only for eradication of newly established populations, but also to contain populations when eradication is not feasible. Technological developments that have facilitated the highly successful area-wide programmes for limiting the spread of the gypsy moth serve as models for management of other invading organisms. Many components of these programmes can be readily applied in other containment systems and should help limit the impacts of invading organisms in the future.

5. REFERENCES

Bierl, B. A., M. Beroza, and C. W. Collier. 1970. Potent sex attractant of the gypsy moth: Its isolation, identification, and synthesis. Science 170(3953): 87–89.

Bogdanowicz, S. M., W. E. Wallner, J. Bell, T. M. Odell, and R. G. Harrison 1993. Asian gypsy moths (Lepidoptera: Lymantriidae) in North America: Evidence from molecular data. Annals of the Entomological Society of America 86: 710–715.

Garner, K. J., and J. M. Slavicek. 1996. Identification and characterization of a RAPD-PCR marker for distinguishing Asian and North American gypsy moths. Insect Molecular Biology 5: 81–91.

Giese, R. L., and M. L. Schneider. 1979. Cartographic comparisons of Eurasian gypsy moth distribution (*Lymantria dispar* L.; Lepidoptera: Lymantriidae). Entomological News 90: 1–16.

Johnson, D. M., A. M Liebhold, O. N. Bjørnstad, and M. L. McManus. 2005. Circumpolar variation in periodicity and synchrony among gypsy moth populations. Journal of Animal Ecology 74: 882–892.

Keena, M. A., M. J. Côté, P. S. Grinberg, and W. E. Wallner. 2008. World distribution of female flight and genetic variation in *Lymantria dispar* (Lepidoptera: Lymantriidae). Environmental Entomology 37: 636–649.

Leonard, D. S. 2007. Chapter 7. Project organization, planning, and operations. pp. 91–98. *In* P. C. Tobin, and L. M. Blackburn (eds.), Slow the Spread: A national program to manage the gypsy moth. USDA Forest Service Northern Research Station General Technical Report NRS–6. Morgantown, West Virginia, USA.

Leuschner, W. A., J. A. Young, S. A. Waldon, and F. W. Ravlin. 1996. Potential benefits of slowing the gypsy moth's spread. Southern Journal of Applied Forestry 20: 65–73.

Liebhold, A. M. and J. Bascompte. 2003. The Allee effect, stochastic dynamics and the eradication of alien species. Ecology Letters 6: 133–140.

Liebhold, A., V. Mastro, and P. W. Schaefer. 1989. Learning from the legacy of Leopold Trouvelot. Bulletin of the Entomological Society of America 35: 20–22.

Liebhold, A. M., J. Halverson, and G. Elmes. 1992. Quantitative analysis of the invasion of gypsy moth in North America. Journal of Biogeography 19: 513–520.

McManus, M. L. 2007. Chapter 1. In the beginning: Gypsy moth in the United States, pp. 3–13. *In* P.C. Tobin, and L. M. Blackburn (eds.), Slow the Spread: A national program to manage the gypsy moth. USDA Forest Service Northern Research Station General Technical Report NRS–6. Morgantown, West Virginia, USA.

Morin, R. S., and A. M. Liebhold. 2016. Invasive forest defoliator contributes to the impending downward trend of oak dominance in eastern North America. Forestry 89: 284–289.

Morin, R. S., A. M. Liebhold, E. R. Luzader, A. J. Lister, K. W. Gottschalk, and D. B. Twardus. 2004. Mapping host-species abundance of three major exotic forest pests. USDA Forest Service Northeastern Research Station Research Paper NE–726. Morgantown, West Virginia, USA.

Roberts, E. A., and A. H. Ziegler. 2007. Chapter 3. Gypsy moth population monitoring and data collection, pp. 33–46. *In* P.C. Tobin, and L. M. Blackburn (eds.), Slow the Spread: A national program to manage the gypsy moth. USDA Forest Service Northern Research Station General Technical Report NRS–6. Morgantown, West Virginia, USA.

Sharov, A. A., and A. M. Liebhold. 1998. Model of slowing the spread of gypsy moth (Lepidoptera: Lymantriidae) with a barrier zone. Ecological Applications 8: 1170–1179.

Sharov, A. A., D. S. Leonard, A. M. Liebhold, and N. S. Clemens. 2002. Evaluation of preventive treatments in low-density gypsy moth populations. Journal of Economic Entomology 95: 1205–1215.

(STS) Slow the Spread. 2019. Slow the Spread Foundation, Inc.

Thorpe, K., R. Reardon, K. Tcheslavskaia, D. Leonard, and V. Mastro. 2006. A review of the use of mating disruption to manage gypsy moth, *Lymantria dispar* (L.). Forest Health Technology Enterprise Team 2006–13. USDA Forest Service, Washington, DC, USA.

Tobin, P. C., and A. A. Sharov. 2007. Chapter 4. The decision algorithm: Selection of and recommendation for potential problem areas, pp. 47–6. *In* P.C. Tobin, and L. M. Blackburn (eds.), Slow the Spread: A national program to manage the gypsy moth. USDA Forest Service Northern Research Station General Technical Report NRS-6. Morgantown, West Virginia, USA.

Tobin, P. C., A. A. Sharov, A. M. Liebhold, D. S. Leonard, E. A. Roberts, and M. R. Learn. 2004. Management of the gypsy moth through a decision algorithm under the STS project. American Entomologist 50: 200–209.

Tobin, P. C., C. Robinet, D. M. Johnson, S. L. Whitmire, O. N. Bjørnstad, and A. M. Liebhold. 2009. The role of Allee effects in gypsy moth (*Lymantria dispar* (L.)) invasions. Population Ecology 51: 373–384.

(USDA/APHIS/PPQ) United States Department of Agriculture/Animal and Plant Health Inspection Service/Plant Protection and Quarantine. 2019. Gypsy moth program manual. Second edition.

(USDA/USFS) United States Department of Agriculture/United States Forest Service. 2017. Gypsy moth digest. Northeastern Area State and Private Forestry.

SUCCESSFUL AREA-WIDE PROGRAMME THAT ERADICATED OUTBREAKS OF THE INVASIVE CACTUS MOTH IN MEXICO

A. BELLO-RIVERA[1], R. PEREIRA[2], W. ENKERLIN[2], S. BLOEM[3], K. BLOEM[4], S. D. HIGHT[5], J. E. CARPENTER[6], H. G. ZIMMERMANN[7], H. M. SANCHEZ-ANGUIANO[8], R. ZETINA-RODRIGUEZ[9] AND F. J. TRUJILLO-ARRIAGA[1]

[1]Dirección General de Sanidad Vegetal, SENASICA-SAGARPA, Ciudad de México, C.P. 04530, México; arturo.bello@senasica.gob.mx
[2]Insect Pest Control Section, Joint FAO/IAEA Division of Nuclear Techniques in Food and Agriculture, Vienna, Austria
[3]NAPPO, 1730 Varsity Drive Suite 145, Raleigh, NC 27606, USA
[4]USDA-APHIS-PPQ, 1730 Varsity Drive Suite 400, Raleigh, NC 27606, USA
[5]USDA-ARS-CMAVE, Centre for Biological Control, FAMU, Tallahassee, FL 32308, USA
[6]Retired, Tifton, Georgia, USA (former USDA-ARS)
[7]Retired, Pretoria, South Africa (former ARC-LNR)
[8]OIRSA, Ciudad de México, C. P. 11590, México
[9]Representación Regional 6 SENASICA, San Francisco de Campeche, Campeche

SUMMARY

The spectacular success of the cactus moth, *Cactoblastis cactorum* (Berg), in controlling invasive *Opuntia* species has been cited often in biological control literature. This insect is highly damaging to plants of the Cactaceae family and has been regarded as one of the most successful classical weed biological control agents. In Australia, the cactus moth effectively controlled approximately 25 million hectares of non-native *Opuntia* species that had invaded cattle grazing lands. In Mexico, however, the Cactaceae are native, where they have their greatest diversity, and have major ecological and socio-economic importance. In the 1950s,

J. Hendrichs, R. Pereira and M. J. B. Vreysen (eds.), Area-Wide Integrated Pest Management: Development and Field Application, pp. 561–580. CRC Press, Boca Raton, Florida, USA.

the cactus moth was introduced into several Caribbean islands for management of *Opuntia* spp. on grazing lands. The moth arrived accidentally in Florida by 1989, and concerns were raised that this highly mobile moth could spread into the south-western USA and reach Mexico. Therefore, a cactus moth surveillance programme was established in Mexico in 2002 to detect and prevent the introduction and establishment of this invasive pest. Through this detection trapping network, two localized but large outbreaks of *C. cactorum* were detected off the coast of the Peninsula of Yucatán, Mexico; on Isla Mujeres, 10 August 2006, and Isla Contoy, 4 May 2007, both in the municipality of Isla Mujeres in the state of Quintana Roo. An eradication programme was immediately implemented by the National Service for Health, Safety and Food Quality (SENASICA) of the Ministry of Agriculture, Livestock, Rural Development, Fisheries and Food (SAGARPA) of Mexico, in close collaboration with other national and international organizations. An area-wide integrated pest management (AW-IPM) approach was implemented that included regulatory actions, outreach activities, surveillance, intensive sanitation and the release of sterile moths. Eradication was achieved in 2008 and officially declared in 2009. A surveillance network is currently maintained for detecting early pest incursions, allowing for a rapid response to any future incursion of the pest. This successful programme has resulted in the protection of the high value commercial *Opuntia* cacti industry in Mexico, as well as native *Opuntia* species in natural arid and semi-arid ecosystems where they are an essential element in maintaining biodiversity and soil conservation.

Key Words: Opuntia, prickly pear cactus, *Cactoblastis cactorum,* Sterile Insect Technique, sterile moths, biological control, AW-IPM, Yucatán, Quintana Roo

1. INTRODUCTION

In general, biological control is the use of one organism to reduce the population density of another organism. Biological control has been used by humans for about two millennia and has become widely used in pest management programmes since the end of the nineteenth century (DeBach 1964; van Lenteren and Godfray 2005). Classical biological control involves the introduction of a host-specific, non-native natural enemy adapted to a non-native organism that became a pest in its new homeland. Classical weed biological control implies the importation of natural enemies to control a non-native weed species.

The South American cactus moth, *Cactoblastis cactorum* (Berg) (Lepidoptera: Pyralidae), is a well-known example demonstrating the great success that can be achieved using plant-feeding insects as classical biological control agents of invasive plants (Dodd 1940; Pettey 1948, Zimmermann et al. 2004). The removal of highly invasive species of prickly pear cacti (*Opuntia* spp.) from millions of hectares (ha) of Australian farmland, rangeland, and natural habitat was a great early success in biological control of weeds. The dramatic "before and after" pictures of devastated dense cactus vegetation after releasing *C. cactorum* (Dodd 1940) are familiar examples of the stunning impact of successful classical biological weed control (DeBach et al. 1976).

Several programmes resulted in successes similar to the programme in Australia, including South Africa and Hawaii (Zimmermann et al. 2000, 2004). However, in 1956 the decision was made to release *C. cactorum* on the Caribbean island of Nevis, part of the Leeward Islands group of the West Indies, where *Opuntia* species occurred that were native. Control of *Opuntia* on Nevis Island was deemed important to help manage a complex of native prickly pears dominated by *Opuntia triacantha* (Willdenow) and introduced species that were considered a serious weed pest in over-grazed rangeland where they outcompeted grasses and caused injury to livestock and animal handlers (Simmonds and Bennett 1966). Three species of natural enemies, including *C. cactorum*, were shipped from South Africa and released on Nevis Island in early 1957. *C. cactorum* was apparently the only natural enemy that became established, spreading rapidly and causing the collapse of prickly pear populations on the island. This biological control programme was considered "outstandingly successful" (Simmonds and Bennett 1966).

Based upon these successes, *C. cactorum* was introduced on the islands of Montserrat and Antigua in 1960, where it also became established and caused substantial reduction of native prickly pear populations (Simmonds and Bennett 1966). Thereafter, *C. cactorum* continued spreading, either naturally or with intentional or unintentional human involvement, through many regions of the Caribbean, including Puerto Rico, the U.S. Virgin Islands, the Dominican Republic, the Bahamas, and Cuba, where it attacked native *Opuntia* spp. (Zimmermann et al. 2000).

In 1989, *C. cactorum* was discovered on native *Opuntia* spp. in southern Florida, USA (Habeck and Bennett 1990; Dickel 1991) and received considerable attention in view of its potential negative ecological and economic impact in the southern and western USA (Simberloff et al. 1996). How the moth arrived in Florida is unclear. The moth may have arrived through natural dispersal by flight, possibly storm-aided, since the moth was established in Cuba, only 128 km from the Florida Keys (Zimmermann et al. 2000). Perhaps the most compelling explanation for the expansion of *C. cactorum* was proposed by Pemberton (1995), who suggested that the moth may have been unintentionally introduced through commerce of ornamental cactus. During the 1980s, 300,000 *Opuntia* plants destined for nursery sales were shipped from the Dominican Republic to Miami every year. From 1981-1986, *C. cactorum* was intercepted 13 times at Miami ports, including larvae found in stems of *Opuntia* plants from the Dominican Republic (Pemberton 1995).

Molecular genetic analysis of *C. cactorum* specimens from Florida also supports the hypothesis that multiple introductions into Florida occurred from a location outside the insects' native range in South America (Simonsen et al. 2008; Marsico et al. 2011).

Thus *C. cactorum* had become an invasive species with a high biotic potential. Although the moth is not a typical long-distance flyer, with a maximum recorded dispersal distance of 24 km (S. D. Hight unpublished data), it has a high reproductive capacity with egg sticks usually consisting of 70-90 eggs, sometimes up to 120 eggs, and a female that produces three or four egg sticks during her lifetime (Zimmermann et al. 2004). "*A worm that turned*" is the title of a popular article that describes the threat of the same "miracle" insect when its host plants were no longer considered weeds, but native cacti of great economic, ecological and aesthetic value (Stiling 2000).

Over a period of 20 years, this invasive moth has been spreading from southern Florida as far north as central coastal South Carolina and as far west as coastal Louisiana (Hight et al. 2002; Hight and Carpenter 2009). In 2017, the moth was found in southern North Carolina, approximately 160 km from the last known location of *C. cactorum* in South Carolina (Jarred Driscoll, North Carolina Department of Agriculture, personal communication).

In view of this threat, already in 2002, Mexico initiated a surveillance campaign for *C. cactorum* under the *National Preventive Campaign against Cactus Moth*, with the purpose of early detection of any incursion. This surveillance network detected two outbreaks of *C. cactorum* in the state of Quintana Roo, Yucatán Peninsula, i.e. on 10 August 2006 on Isla Mujeres and on 4 May 2007 on Isla Contoy, both within the municipality of Isla Mujeres (Hight and Carpenter 2009) (Fig. 1).

Also because of concerns of the growing threat to native and cultivated *Opuntia* species with the continued spread of the cactus moth, the USA and Mexico developed a *Bi-National Cactus Moth Programme* in 2006 that was implemented with funding from both countries. Operations supported by the Bi-National Cactus Moth Programme were initially directed at suppressing populations and containing the leading edge of the infestation in the USA along the Gulf Coast of Florida, Alabama, and Mississippi (Bloem et al. 2007; Carpenter et al. 2008). When the two outbreaks occurred in Quintana Roo in 2006, the Bi-National Programme also supported eradication efforts in Mexico.

In this paper we describe the development of a surveillance and eradication programme that was implemented by the Government of Mexico in close collaboration with the USA and with other national and international organizations.

2. THE IMPORTANCE OF *OPUNTIA* CACTI IN MEXICO

The genus *Opuntia* is one of the most widely used plants in Mexico and Central America. Due to the high protein and fibre content in the stems, or pads, and the amount of water in their tissue (88-91%, Pimienta-Barrios 1990). *Opuntia* cacti have an extremely wide range of uses, from human and animal food to cosmetics and adhesives (Barbera 1995). In Mexico, traditional uses of *Opuntia* vary widely although there are two main products that account for the economic importance of *Opuntia* products: food and fodder (Pimienta-Barrios 1990; Barbera 1995).

Fodder is mainly for cattle and goats in all parts of Mexico, but its use for forage has been documented in many other parts of the world including the USA, northern and southern Africa, and several South American countries (Felker 1995). In Brazil, for example, close to 300 000 ha has been planted with *Opuntia* cacti to produce fodder for livestock (Barbera 1995). *Opuntia* spp. have been cultivated worldwide because of the value of the plants as ornamental and agricultural commodities, and their ability to adapt to various climatic conditions, particularly to semi-arid and arid areas (Hanselka and Paschal 1989).

Opuntia spp., which have been important cultural and agricultural plants in Mesoamerica since the pre-Hispanic era (Zimmermann and Pérez-Sandi-Cuen 2006), annually generate over 300 000 tons of fruit and vegetables in Mexico, and are cultivated on more than 70 000 ha (Pimienta-Barrios 1990; Flores-Valdez et al. 1995; Soberón et al. 2001). As food, *Opuntia* can be consumed as a vegetable (by dicing young pads – "nopalitos") or as a fruit (tuna or cactus pear). The fruits are produced in 15 out of 32 Mexican states employing close to 20 000 people, whereas vegetables are produced in 14 states and employ close to 8 000 people. In addition, most rural people consume prickly pear from local wild *Opuntia* populations and family-owned plantations maintained at various sizes. The average income generated by *Opuntia* products over the period 1990-1998 is approximately 50 million USD per year, with vegetable usage constituting more than half of the value (USD 27 million), followed by cactus pears (USD 20 million), and finally fodder (USD 1 million). In addition, the export market of *Opuntia* products is valued at USD 50 million per year. Exports are mainly to the U.S., Canada, Europe, and Japan (Soberón et al. 2001). If *C. cactorum* were to establish in Mexico, ca. 30 000 producers of cactus fruit and vegetable would be affected, as well as the nopal processing industries.

Mexico is recognized as the origin of the genus *Opuntia* (Esparza et al. 2004) and has the highest number of cultivated species and varieties of prickly pear in the world (19 cultivated types in total) (Flores-Valdez and Gallegos 1993). Mexico also has one of the highest diversity of species of the genus *Opuntia* (in addition to Cactaceae biodiversity of ca. 560 species) that cover an area of close to 3 000 000 ha (1.5% of Mexican territory). The actual number of *Opuntia* species varies in the literature, partially because of frequent hybridization between species and the lack of a standardized taxonomic classification. Bravo-Hollis (1978) recognized 104 species of *Opuntia* in Mexico, 56 of which are in the subgenus *Platyopuntia* (prickly pears), 38 of which are endemic. From an ecological point of view, loss of acreage covered by *Opuntia* species would accelerate soil loss in arid and semi-arid areas, and the loss of ecological niches to a variety of organisms (Ojeda 2004).

3. FIRST DETECTION IN MEXICO

The two *C. cactorum* outbreak sites (Isla Mujeres and Isla Contoy) were located in the state of Quintana Roo, one of the 32 states of Mexico. This state is part of the Yucatán Peninsula in the south-eastern part of the country, bordering to the north with the state of Yucatán and the Gulf of Mexico, to the east with the Caribbean Sea, to the south with Belize, and to the west with the state of Campeche. Its capital is Chetumal and its most populous city is Cancún.

First outbreak site: Isla Mujeres is a small island located in the Caribbean Sea, 13 km off the coast of the Yucatán Peninsula (Fig. 1). The island is 7 km long and on average 1 km wide (for a total of 455 ha), and it is part of one of the eleven municipalities in the state of Quintana Roo. The municipality of Isla Mujeres, which includes a part of continental territory, is located 13 km from the city of Cancún, the main tourist attraction in the region. Isla Mujeres has a permanent settlement of approximately 16 000 inhabitants whose primary livelihood is to support the tourist industry. The island is accessed from Cancún by vehicle and passenger ferryboats. Delimiting trapping revealed that the size of the outbreak area was 45 ha infested with moth populations following the spatial distribution of its *Opuntia* hosts scattered over much of the island but concentrated on the southern end.

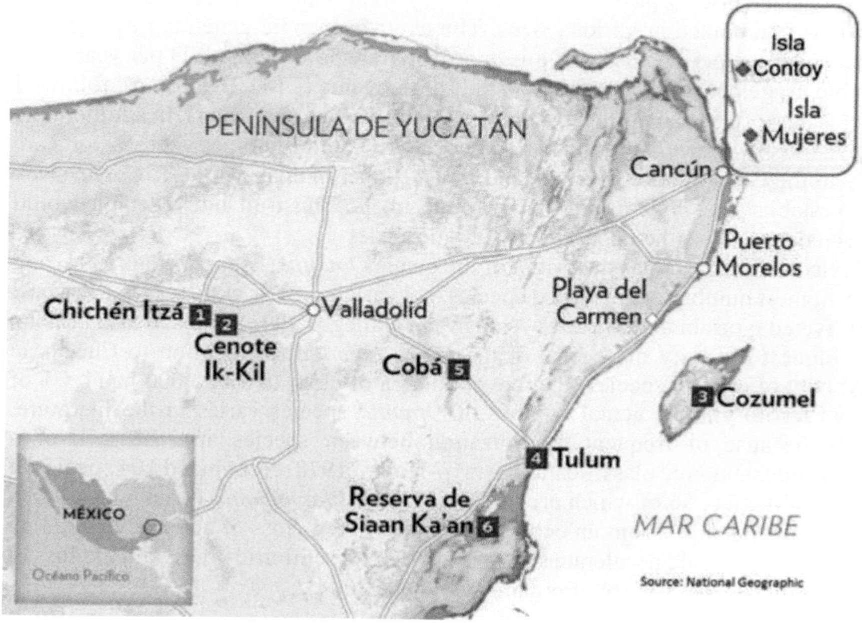

Figure 1. Location of Isla Mujeres and Isla Contoy near Cancún, Quintana Roo, in the Caribbean Sea, where populations of Cactoblastis cactorum *were detected, but were later eradicated.*

Second outbreak site: Isla Contoy (Fig. 1) is an even smaller island, also belonging to the municipality of Isla Mujeres and located 30 km north of Isla Mujeres. The island is 8.75 km long, and on average 0.5 km wide, and has a surface area of 317 ha. In 1961, the Mexican government declared the island a protected area and a bird sanctuary, and in 1998, it was declared a national park. The island is known as "The Island of the Birds" due to the large numbers of frigate birds and other oceanic avian species that use Isla Contoy as a nesting location.

To the south of the island is the Ixlaché reef that is part of the second largest barrier reef in the world. The National Park Service maintains a small visitor's centre and housing for visiting researchers on the island. Access by tourists to the island is controlled and monitored by the National Park Service. Delimiting trapping revealed that the size of the outbreak area was 3 ha with the moth population somewhat limited to dense stands of *Opuntia* host plants in the central part of the island.

4. FORMATION OF OPERATIONAL PROGRAMME

A number of national and international institutions and organizations joined efforts in the fight against these cactus moth outbreaks. These included: The Food and Agriculture Organization of the United Nations (FAO), the International Atomic Energy Agency (IAEA), the United States Department of Agriculture (USDA), the North American Plant Protection Organization (NAPPO), and the Mexican Government through the Ministry of Agriculture, Livestock, Rural Development, Fisheries and Food (SAGARPA), the National Service for Health, Safety and Food Quality (SENASICA) and the National Forestry Commission (CONAFOR). Technical and economic support was provided to research and development (R&D) and to the implementation of an area-wide eradication programme.

5. RESEARCH AND DEVELOPMENT

The necessary tools for implementation of the eradication programme were developed through effective research and development. These included: determination of the moth's mating behaviours (Hight et al. 2003), design and evaluation of trapping systems (Bloem et al. 2005a), pheromone identification and synthesis (Heath et al. 2006; Cibrián-Tovar 2009; Cibrián-Tovar et al. 2017), population ecology studies including dispersal (Hight et al. 2002; Bloem et al. 2005a; Sarvary et al. 2008), studies on biology of the pest (McLean et al. 2006), implementation of trials to assess susceptibility to insecticides (Bloem et al. 2005b), development of an artificial diet and mass-rearing tools (Marti and Carpenter 2008; Carpenter and Hight 2012), assessment of the feasibility of biological control (Paraiso et al. 2012) and the Sterile Insect Technique (SIT) (Carpenter et al. 2001; Hight et al. 2005), and assessment of the potential economic, social, and environmental impacts of *C. cactorum* both in the USA and in Mexico (Zimmermann et al. 2004; Simonson et al. 2005; Sánchez et al. 2007).

One of the essential tools that was developed and validated was the trapping system based on a standard (unpainted) wing trap (Bloem et al. 2005a), placed at a height of 2.0 m above ground, and baited with a rubber septum impregnated with a synthetic female sex pheromone (Heath et al. 2006). Support from the Joint FAO/IAEA Division was essential for transferring the tools for implementing an AW-IPM approach integrating the SIT (Dyck et al. 2021).

6. PROGRAMME OPERATIONS

The Mexican Government provided the necessary financial resources for programme operations aimed at eradicating the outbreaks on Islas Mujeres and Contoy. The financial resources to support programme implementation are summarized in Table 1. It is important to point out that when the pest was first detected in Mexico, resources were immediately allocated through the declaration of a National Emergency, which is an instrument of Mexican legislation that allows for timely response to this type of phytosanitary emergency.

To raise awareness of the potential impact of the presence of the cactus moth, the following sectors were alerted and informed: Political and administrative government offices and divisions, academic research and technical institutions, industry stakeholders, commercial agricultural enterprises, natural area managers, growers, farmers and ranchers, non-government agencies, amusement parks, hotels in tourist areas, and the public in general. It was also essential to carry out the necessary feasibility assessments and to support the preparation of public outreach materials such as booklets, videos, and calendars.

Table 1. Total financial resources (USD) made available from 2006 through 2010 for the implementation of cactus moth eradication programme activities by the Mexican Government through the National Service for Health, Safety and Food Quality (SENASICA) and the National Forestry Commission (CONAFOR)

Year	Contribution from SENASICA	Contribution from CONAFOR	Total
2006	192 926	181 818	374 744
2007	1 090 909	0	1 090 909
2008	772 727	34 386	807 114
2009	318 182	0	318 182
2010	240 000	0	240 000
Total	2 614 744	216 205	2 830 949

The cactus moth eradication programme in Mexico followed a similar AW-IPM approach and operational tactics as had been used to suppress and contain the pest in the USA along the Gulf of Mexico. These included: regulatory actions, raising public awareness, extensive surveys, sanitation of infested host plants through mechanical removal as well as removal of egg sticks, and finally the release of sterile moths produced at the USDA insectary in Tifton, Georgia and shipped to Mexico (Table 2). These actions eradicated *C. cactorum* populations in Mexico (Bloem et al. 2007; Hernández et al. 2007; Carpenter et al. 2008; NAPPO 2009; Hight and Carpenter 2016).

Table 2. Actions implemented to eradicate the cactus moth from Isla Mujeres and Isla Contoy

Actions/Activities	Isla Mujeres	Isla Contoy	Rest of the Quintana Roo	Yucatán and Campeche	Rest of the Country
Surveillance					
Trap deployment and weekly trap checks	X	X	X	X	X
Identification of adult captures with differentiation between wild and sterile adults	X	X	X		
Identification of sentinel sites and weekly checking of sentinel plants for egg-sticks oviposited by wild females	X	X			
Review of permanent observation points			X	X	X
Identification of suspect detections in cultivated and wild areas	X	X	X	X	X
Pest Control					
Host plant removal (*Opuntia dillenii* Ker Gawl, *Nopalea cochenillifera (L.)* Mill.)	X				
Eggstick removal	X	X			
Application of insecticides	X	X			
Application of herbicides	X				
Sterile moth releases	X	X			
Host census	X	X	X		
Exchange programme with homeowners of ornamental *Opuntia* plants or voluntary surrender by the inhabitants	X				
Regulation of host mobilization					
Prohibition of the move-ment of *Opuntia* plants or their parts off or within the island	X	X			
Review of transport in the sea ports	X	X			
Disclosure and training					
Training workshops for staff of programme	X	X	X	X	X
Development and dispersal of informative material	X	X	X	X	X
Radio and television spots to educate the public of the programme and dangers of *C. cactorum* outbreak	X	X	X	X	X
Meetings with municipal presidencies and local authorities	X	X	X	X	
Meetings with schools, hoteliers and managers of natural parks	X	X	X		
Supervision and evaluation visits with the following groups:					
SAGARPA State Delegation	X	X	X	X	X
Plant Health General Directorate	X	X	X	X	X
Experts from FAO/IAEA	X	X	X	X	
Experts from USDA	X	X	X		
Technical Group Meetings	X	X	X	X	X
Technical reports					
Data collection	X	X	X	X	X
Weekly reports to the Plant Health General Directorate	X	X	X	X	X
Reports to the local authorities	X	X	X		

The SENASICA-SAGARPA national surveillance system allowed early detection of pest incursions, followed immediately by the implementation of delimitation and eradication actions to eliminate the two outbreaks of this invasive species. This prevented further dispersal and introduction of the pest to other uninfested areas. The actions that were implemented to eradicate the cactus moth from Isla Mujeres and Isla Contoy are summarized in Table 2.

7. PROGRAMME RESULTS AND ERADICATION

7.1. Surveillance and Control on Isla Mujeres

Initially 66 traps were deployed on Isla Mujeres, primarily in and around the infested 45 ha; this was later expanded to a total of 115 traps. Trapping and host plant sampling and removal soon revealed high infestation levels throughout the island with thousands of egg sticks and larvae collected and destroyed and adult males trapped (Fig. 2). The effect of the rapid and aggressive response from SENASICA personnel was visible almost immediately, and the density of the outbreak population of *C. cactorum* was quickly reduced. A total of 4126 egg sticks were collected from plants during the eradication effort and 321 adult males were trapped.

Figure 2. Larvae and egg sticks of Cactoblastis cactorum *collected on Isla Mujeres during the initial stages of the eradication programme.*

Once trapping efforts revealed a dispersed infestation over much of the island, removal of above-ground host plant material was conducted to suppress additional *C. cactorum* population increase. As part of the mechanical control, 240 tons of potential and infested host plants were removed throughout the island. Only 27 sentinel host plants remained, which were plants that were easy to access and could be checked daily for the presence of egg sticks or damage caused by the larvae.

Two censuses for *Opuntia* spp. host plants were carried out in the backyards of homes throughout the island in March and June 2008. A total of 3050 sites with susceptible host plants were identified in private homes, public offices, hotels, and vacant lots. To facilitate the removal of host plants, a protocol was implemented to replace each host plant with a non-host ornamental plant, mainly species of palms (1140 specimens) and magueys (*Agave* spp.) (74 specimens). CONAFOR kindly provided the replacement plants for this part of the programme.

The population trend of *C. cactorum* on Isla Mujeres is indicated by the weekly total number of egg sticks collected (Fig. 3) and weekly total number of wild males caught (Fig. 4). A drastic reduction in the cactus moth population resulted from the intensive mechanical control and adult mass-trapping. This prepared the ground for the integration of the SIT on an area-wide basis and assured a competitive sterile to wild insect ratio. The targeted initial overflooding ratio was 10 sterile to 1 wild moths as identified in trap captures. In preparation for sterile moth releases, extensive quality control, flight ability, and dispersion tests had been conducted, and shipping techniques evaluated in the USA.

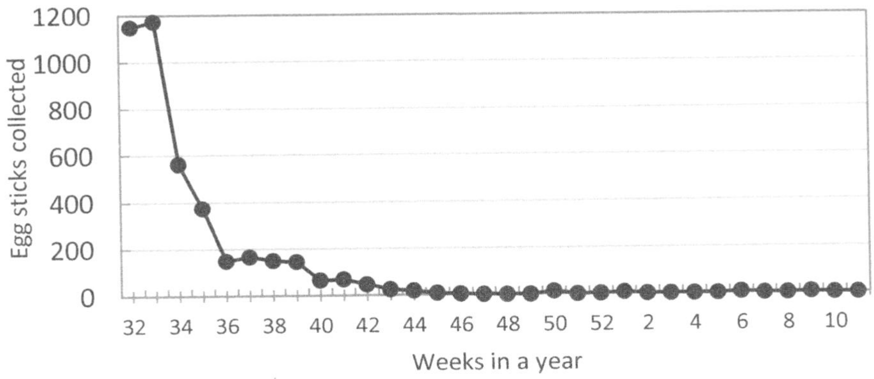

Figure 3. Total C. cactorum *egg sticks collected from* Opuntia spp. *plants each week from August 2006 to March 2007 on Isla Mujeres. Collections were concentrated in an area of 45 ha in the southern end of the island's 455 ha.*

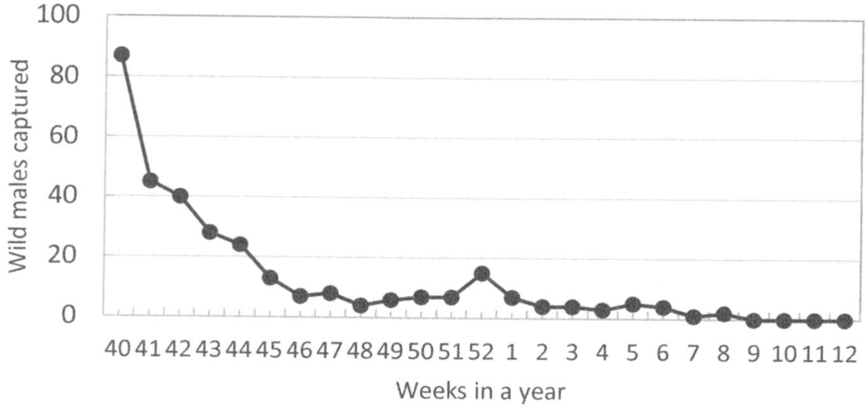

Figure 4. Total wild C. cactorum *male moths caught weekly in pheromone-baited traps from October 2006 to March 2007 on Isla Mujeres. Captures were concentrated in an area of 45 ha in the southern end of the island's 455 ha.*

The first trial of sterile adult moth releases (n = 1398) was carried out on Isla Mujeres on 1 November 2007; it primarily targeted to the area where the last egg sticks were collected and the last male *C. cactorum* were captured. Release time frame for packages of shipped sterile moths was within 48 hr; from insect packaging, delivery to a USA airport, receipt in Cancún, to release on Isla Contoy.

Given the fact that the last wild individuals were detected during the second week of March 2007 (Fig. 3), it is possible that during the time the sterile moths were released, eradication may have already been achieved. Nevertheless, application of the SIT gave assurance that no remnant of the invasive population of the cactus moth remained.

7.2. Surveillance and Control on Isla Contoy

After detection of the cactus moth on 4 May 2007, programme activities were immediately intensified on Isla Contoy. This however, required the processing of access permits, as the island is a protected natural reserve and the cactus vegetation could not be mechanically removed as it was on Isla Mujeres.

A total of 44 traps were deployed and the trap catches indicated that the outbreak was limited to the central part of the island. All larvae found (n = 1028) as well as all damaged plant parts (n = 122) were removed. Some wild cactus plants exhibited damage caused by another cactophagus moth, presumably of the genus *Melitara,* which made it difficult to differentiate the damage from that of the cactus moth.

A total of 46 *C. cactorum* egg sticks were collected on Isla Contoy (Fig. 5) and a total of 41 wild male adult cactus moths were caught in the traps (Fig. 6).

The first trial release of sterile adult moths (n=1281) was carried out on Isla Contoy on November 7, 2007. After this release, 18 sterile adults were caught in the traps (confirmed in the laboratory), suggesting that the dispersal behaviour and competitiveness of the sterile insects was adequate.

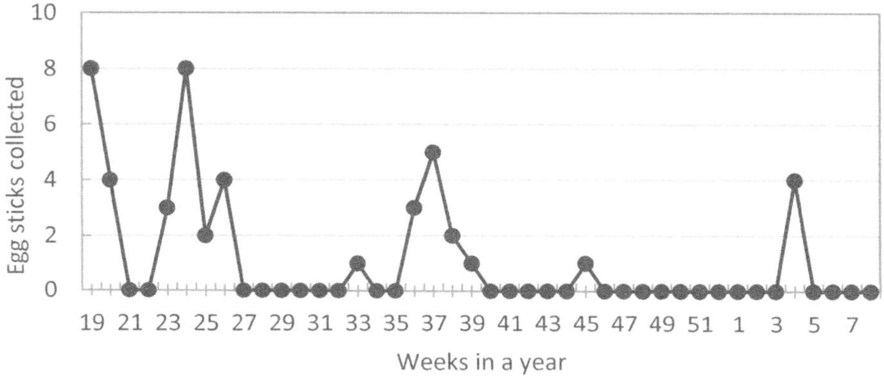

Figure 5. Total C. cactorum *egg sticks collected weekly from* Opuntia *spp. host plants from May 2007 to February 2008 on Isla Contoy. Collections were concentrated in an area of 3 ha of the island's 317 ha.*

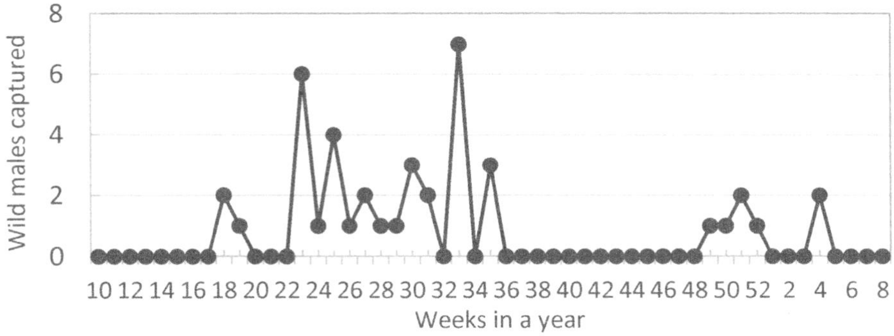

Figure 6. Total wild C. cactorum *male moths caught weekly in pheromone-baited traps from March 2007 to February 2008 on Isla Contoy. Captures were concentrated in an area of 3 ha of the island's 317 ha.*

The last wild moth was captured in January 24, 2008. After conducting various field release tests, the programme carried out continued weekly releases of sterile moths, considering the biological conditions of the pest and the availability of sterile insects, from March 1, 2008 until July 2, 2008. For both islands a total of 21 398 sterile moths were released, of which 75% were males.

Weekly releases of sterile moths during 4 consecutive months assured complete eradication of possible remnant *C. cactorum* populations. The sterile moths, reared and sterilized at the USDA-Agricultural Research Service (ARS) laboratory in Tifton, Georgia, were packaged in Petri dishes and shipped by commercial air cargo in insulated cardboard wrapped Styrofoam shipping boxes (KoolTempTM GTS-89 Shipping System, Cold Chain Technologies, Holliston, Massachusetts). Hight and Carpenter (2016) provide additional packaging details. Refrigerant Kool Guard II cooling bricks were included in the containers to maintain a targeted temperature around the sterile insects of 1–4°C.

7.3. Additional Programme Activities

Between January 6 and February 3, 2007, 6 wild adult cactus moths were caught in traps deployed in the continental area of Cancún, on the Quintana Roo mainland. In response, activities were immediately intensified on the mainland adjacent to Isla Mujeres and Isla Contoy. Three thousand additional traps were deployed and distributed among houses, hotels and shops in the coastal area covering a total area of 22.2 km^2. Wild host plants were examined in the vicinity of Nichupte Lagoon for possible detection of egg sticks or larvae. No further positive moth captures were reported after February 3, 2007, and no egg sticks were found on mainland Quintana Roo.

In the state of Quintana Roo, a trapping network was maintained with a total of 262 traps and 729 Permanent Observation Points (POPs) with *Opuntia* host plants present. In the neighbouring state of Yucatán (from El Cuyo to Celestum), 130 traps were deployed and 122 POPs were monitored; and in Campeche state (Rio Verde, Real de Salinas and Jaina) 79 traps were deployed and 50 POPs monitored. Traps and POPs were checked weekly and fortnightly, respectively.

Training and raising awareness activities where implemented in support of the programme. They consisted of 3 training courses, 500 radio spots, 700 posters and 4000 flyers distributed on Isla Contoy, 25 000 flyers and banners in English and Spanish, 107 videos in English and Spanish distributed in the area of the Yucatán Peninsula, 6 news broadcasts on local television stations and one on national television.

Since the beginning of the eradication programme (August 2006), 6 experts from the Mexico-USA Bi-National Cooperative Programme and from the FAO/IAEA Joint Division made multiple visits to the programme and provided technical advice.

8. PROGRAMME OUTCOME

In 2009, after a period equivalent to three estimated biological life cycles of the pest without any further detection, the two cactus moth outbreaks were considered eradicated from the state of Quintana Roo, maintaining Mexico free of this pest.

The following Official Agreements were published in the Official Gazette of the Mexican Federation:

- On 26 March 2009. "*Agreement by which the outbreak of cactus moth (*Cactoblastis cactorum *Berg) was declared eradicated from Isla Mujeres, Municipality of Isla Mujeres, State of Quintana Roo*" (SAGARPA 2009a).
- On 12 October 2009. "*Agreement by which the outbreak of cactus moth (*Cactoblastis cactorum *Berg) declared eradicated from Isla Contoy, Municipality of Isla Mujeres, State of Quintana Roo*" (SAGARPA 2009b).

Based on the above and in accordance with the International Standard for Phytosanitary Measures No.8 "Determination of pest status in an area" (FAO 2017), the cactus moth was considered eradicated after more than three generations without detections, and continued surveillance since then has confirmed its continuous absence from Mexico.

The National Preventive Campaign against Cactus Moth began in 2002 and continued until 2009, with eradication actions carried out under the coordination of National Plant Protection Directorate.

In 2010, the Phytosanitary Epidemiological Surveillance Programme was created and all surveillance actions against the cactus moth became part of this Programme, under the coordination of the National Phytosanitary Reference Centre. Activities under this Programme have continued and were expanded throughout the country since its creation, for timely detection of any cactus moth incursions. This preventive activity is conducted through exploratory actions, sentinel plots and trapping routes in the areas where *Opuntia* is commercially grown for vegetable, forage and prickly pear production.

From 2010-2017, the federal investment in surveillance activities amounts to more than USD 4.15 million (SAGARPA-SENASICA 2017) (Fig. 7).

To date there are 84 105 field observations/records on the genus *Opuntia* entered into the National Plant Protection Directorate database (SIRVEF 2018). These observations are distributed as follows: 2841 observations at sentinel plots, 65 103 observations for trapping routes (trapping with pheromone), and 11 497 observations on surveillance routes (sentinel plants where the technicians look for larval damage and egg sticks of the cactus moth). The observations/records are established in 842 sentinel plots, 131 surveillance routes, and a national trapping network of 1660 traps, installed in potential strategic risk areas for the establishment of the pest (SAGARPA-SENASICA 2016).

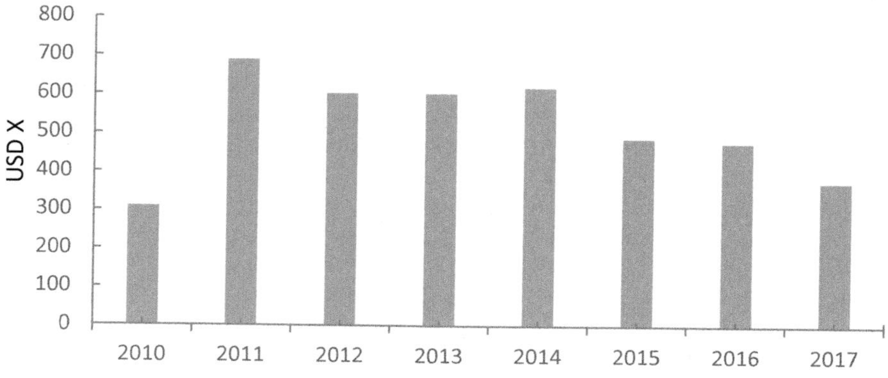

Figure 7. Yearly assigned budget for Cactoblastis cactorum *preventive surveillance programme in Mexico from 2010 up to 2017.*

As a result of these actions under the Phytosanitary Epidemiological Surveillance Programme, to date, no more specimens of the cactus moth have been detected anywhere in the country.

9. FUTURE PERSPECTIVE AND ONGOING SURVEILLANCE

It is almost inevitable that this insect will continue advancing along the USA Gulf Coast, where active containment activities are no longer ongoing. This could result in the pest eventually spreading into Texas and reaching areas adjacent to northeast Mexico, thus posing a constant threat of introduction and invasion of Mexican territory.

Pathway risk analysis has shown that the most likely routes of invasion to Mexico are from Florida including the Keys, along the Gulf Coast states to Texas, and from there to northern Mexico (Simonson et al. 2005). Likewise, Cuba is the closest island to the Caribbean coast of Mexico, and arrival of the cactus moth through this route is very likely due to its proximity with the Yucatán Peninsula, a region with large and wide-spread populations of *Opuntia* cacti (Sánchez et al. 2007).

Due to this permanent pest risk, it is essential to continue the preventive activities through the Phytosanitary Epidemiological Surveillance Programme. Emphasis should be placed on states that are more exposed to cactus moth entries and where *Opuntia* cacti are a valuable commodity.

10. CONCLUSIONS

The cactus moth programme in Mexico is an example of the successful eradication of outbreaks of an invasive pest species integrating the SIT (Hendrichs et al. 2021). The AW-IPM approach was able to combine technical, social, economic, human and political resources to achieve the objective.

Nevertheless, the threat prevails and the Mexican government, under the ongoing preventive Phytosanitary Epidemiological Surveillance Programme, is prepared for the early detection and timely response to a new cactus moth incursion.

11. ACKNOWLEDGMENTS

We would like to express our thanks to Jose Manuel Montiel Castelan for providing data for this paper, and Roberto Jose Gomez Pauza, Selene Bello Rivera, and two anonymous reviewers for their valuable comments to an earlier draft. Appreciation is also extended to the committed men and women who worked tirelessly on the Eradication Programme that removed the invasive *C. cactorum* from Mexico.

12. REFERENCES

Barbera, G. 1995. History, economic and agro-ecological importance, pp. 1–11. *In* G. Barbera, P. Inglese, and E. Pimienta-Barrios (eds.), Agro-ecology, cultivation and uses of cactus pear. FAO Plant Production and Protection Paper 132. Rome, Italy.

Bloem, S., S. D. Hight, J. E. Carpenter, and K. A. Bloem. 2005a. Development of the most effective trap to monitor the presence of the cactus moth (Lepidoptera: Pyralidae). Florida Entomologist 88: 300–306.

Bloem, S., R. F. Mizell, K. A. Bloem, S. D. Hight, and J. E. Carpenter. 2005b. Laboratory evaluation of insecticides for control of the invasive *Cactoblastis cactorum* (Lepidoptera: Pyralidae). Florida Entomologist 88: 395–400.

Bloem, K. A., S. Bloem, J. E. Carpenter, S. D. Hight, J. Floyd, and H. Zimmermann. 2007. Don't let cacto blast us: Development of a bi-national plan to stop the spread of the cactus moth *Cactoblastis cactorum* in North America, pp. 337–344 *In* M. J. B. Vreysen, A. S. Robinson, and J. Hendrichs (eds.), Area-wide control of insect pests. From research to field implementation. Springer, Dordrecht, The Netherlands.

Bravo-Hollis, H. H. 1978. Las cactáceas de México. Vol. 1. Universidad Nacional Autónoma de México, Mexico City, Mexico. 743 pp.

Carpenter, J. E., and S. D. Hight. 2012. Rearing the oligophagous *Cactoblastis cactorum* (Lepidoptera: Pyralidae) on meridic diets without host plant material. Florida Entomologist 95: 1132–1141.

Carpenter, J. E., K. A. Bloem, and S. Bloem. 2001. Applications of F_1 sterility for research and management of *Cactoblastis cactorum* (Lepidoptera: Pyralidae). Florida Entomologist 84: 531–539.

Carpenter, J. E., S. D. Hight, and A. Bello. 2008. Eradication and containment of *Cactoblastis cactorum* in Mexico and the United States. Abstract 1286. 23rd International Congress of Entomology Durban, South Africa, 6–12 July 2008.

Cibrián-Tovar, J. 2009. Complementing the identification of the sexual pheromone of the cactus moth, *Cactoblastis cactorum* Berg. Second report (February–May 2010), IAEA-USDA, Tifton, Georgia, USA. 21 pp.

Cibrián-Tovar, J., J. E. Carpenter, S. D. Hight, T. Potter, G. Logarzo, and J. C. Velázquez-González. 2017. A reinvestigation of *Cactoblastis cactorum* (Lepidoptera: Pyralidae) sex pheromone for improved attractiveness and greater specificity, pp. 119–131. *In* V. Shields (ed.), Biological control of pest and vector insects. IntechOpen, Open Access Books.

DeBach, P. 1964. Biological control of insect pests and weeds. Chapman and Hall, London, UK. 844 pp.

DeBach, P., C. B. Huffaker, and A. W. MacPhee. 1976. Evaluation of the impact of natural enemies, pp. 255–285. *In* C. B. Huffaker and P. S. Messenger (eds.), Theory and practice of biological control. Academic Press, New York, USA. 788 pp.

Dodd, A.P. 1940. The Biological campaign against prickly pear. Commonwealth Prickly Pear Board, Brisbane, Australia. 177 pp.

Dickel, T. S. 1991. *Cactoblastis cactorum* in Florida (Lepidoptera: Pyralidae: Phycitinae). Tropical Lepidoptera 2: 117–118.

Dyck, V. A., J. Hendrichs, and A. S. Robinson (eds.). 2021. Sterile Insect Technique – Principles and practice in Area-Wide Integrated Pest Management, Second Edition. CRC Press, Boca Raton, Florida, USA. 1200 pp.

Esparza, G., R. D. Valdez, and S. J. Méndez. 2004. El nopal: tópicos de actualidad. Universidad Autónoma de Chapingo, Mexico. 303 pp.

(FAO) Food and Agriculture Organization of the United Nations. 2017. Determination of pest status in an area. International Standard for Phytosanitary Measures (ISPM) No. 8. International Plant Protection Convention (IPPC), Rome, Italy.

Flores-Valdez, C. A., and C. Gallegos. 1993. Situación y perspectivas de la producción de tuna en la región centro norte de México. Universidad Autónoma de Chapingo-CIESTAAM. Chapingo, México.

Flores-Valdez, C. A., T. M. Esquivel De Luna, and P. P. Ramírez Moreno. 1995. Mercado mundial de la tuna. Universidad Autónoma de Chapingo-CIESTAAM. Chapingo, Mexico.

Felker, P. 1995. Forage and fodder production and utilization. Agro-ecology, cultivation and uses of cactus pear, pp. 144–154. *In* G. Barbera, P. Inglese and E. Pimienta-Barrios (eds.), FAO Plant Production and Protection Paper 132, Rome, Italy.

Habeck, D. H., and F. D. Bennett. 1990. *Cactoblastis cactorum* Berg (Lepidoptera: Pyralidae) a phycitine new to Florida. Entomology Circular 333, Division of Plant Industry, Florida Department of Agriculture and Consumer Services, Tallahassee, Florida, USA. 4 pp.

Hanselka, C. W., and J. C. Paschal. 1989. Developing prickly pear as a forage, fruit, and vegetable resource. Proceedings of Conference 14 July 1989. Kingsville, Texas. Texas Agric. Ext. Serv. College Station. Texas, USA.

Heath, R. R., P. E. A. Teal., N. D. Epsky, B. D. Dueben, S. D. Hight, S. Bloem, J. E. Carpenter, T. J. Weissling, P. E. Kendra, J. Cibrián-Tovar, and K. A. Bloem. 2006. Pheromone-based attractant for males of *Cactoblastis cactorum* (Lepidoptera: Pyralidae). Environmental Entomology 35: 1469–1476.

Hendrichs, J., W. R. Enkerlin, and R. Pereira. 2021. Invasive insect pests: Challenges and the role of the Sterile Insect Technique in their prevention, containment, and eradication, pp. 885–922. *In* V. A. Dyck, J. Hendrichs, and A. S. Robinson (eds.), Sterile Insect Technique – Principles and practice in Area-Wide Integrated Pest Management, Second Edition. CRC Press, Boca Raton, Florida, USA.

Hernández, J., H. M. Sánchez, A. Bello, and G. González. 2007. Preventive programme against the cactus moth *Cactoblastis cactorum* in Mexico, pp. 345-350. *In* M. J. B. Vreysen, A. S. Robinson, and J. Hendrichs (eds.), Area-wide control of insect pests. From research to field implementation. Springer, Dordrecht, The Netherlands.

Hight, S. D., and J. E. Carpenter. 2009. Flight phenology of male *Cactoblastis cactorum* (Lepidoptera: Pyralidae) at different latitudes in the southeastern United States. Florida Entomologist 92: 208–216.

Hight, S. D., and J. E. Carpenter. 2016. Performance improvement through quality evaluations of sterile cactus moths, *Cactoblastis cactorum* (Lepidoptera: Pyralidae), mass-reared at two insectaries. Florida Entomologist 99: 206–214.

Hight, S. D., J. E. Carpenter, K. A. Bloem, S. Bloem, R. W. Pemberton, and P. Stiling. 2002. Expanding geographical range of *Cactoblastis cactorum* (Lepidoptera: Pyralidae) in North America. Florida Entomologist 85: 527–529.

Hight, S. D., S. Bloem, K. A. Bloem, and J. E. Carpenter. 2003. *Cactoblastis cactorum* (Lepidoptera: Pyralidae): Observations of courtship and mating behaviors at two locations on the Gulf Coast of Florida. Florida Entomologist 86: 400–408.

Hight, S. D., J. E. Carpenter, S. Bloem, and K. A. Bloem. 2005. Developing a sterile insect release program for *Cactoblastis cactorum* (Berg) (Lepidoptera: Pyralidae): Effective overflooding ratios and release-recapture field studies. Environmental Entomology 34: 850–856.

Marti, O. G., and J. E. Carpenter. 2008. Rearing *Cactoblastis cactorum* (Lepidoptera: Pyralidae) on a factitious meridic diet at different temperatures and larval densities. Florida Entomologist 91: 679–685.

Marsico, T. D., L. E. Wallace, G. N. Ervin, C. P. Brooks, J. E. McClure, and M. E. Welch. 2011. Geographic patterns of genetic diversity from the native range of *Cactoblastis cactorum* (Berg) support the documented history of invasion and multiple introductions for invasive populations. Biological Invasions 13: 857–868.

McLean, S. C., K. A. Bloem, S. Bloem, S. D. Hight, and J. E. Carpenter. 2006. Effect of temperature and length of exposure time on percent egg hatch of *Cactoblastis cactorum* (Lepidoptera: Pyralidae). Florida Entomologist 89(3): 340–347.

(NAPPO) North American Plant Protection Organization. 2009. Detection and eradication of a cactus moth (*Cactoblastis cactorum* Berg) outbreak in Isla Contoy, municipality of Isla Mujeres, Quintana Roo, Mexico. Official Pest Reports. North American Plant Protection Organization. Ottawa, Canada.

Ojeda, A. A. 2004. Ficha de *Cactoblastis cactorum*. Subsecretaria de Gestión para la Protección Ambiental. Dirección General de Gestión Forestal y de Suelos. Dirección de Salud Forestal y Conservación de Recursos Genéticos. Subsecretaria de Medio Ambiente y Recursos Naturales. México.

Paraiso, O., S. D. Hight, M. T. K. Kairo, S. Bloem, J. E. Carpenter, and S. Reitz. 2012. Laboratory biological parameters of *Trichogramma fuentesi* (Hymenoptera: Trichogrammatidae), an egg parasitoid of *Cactoblastis cactorum* (Lepidoptera: Pyralidae). Florida Entomologist 95: 1–7.

Pemberton, R. W. 1995. *Cactoblastis cactorum* (Lepidoptera: Pyralidae) in the United States. An immigrant biological control agent or an introduction of the nursery industry? American Entomologist 41: 230–232.

Pettey, F. W. 1948. The biological control of prickly pear in South Africa. Science Bulletin, Department of Agriculture of the Union of South Africa 271: 1–163.

Pimienta-Barrios, E. 1990. El nopal tunero. Universidad de Guadalajara, Guadalajara, Jalisco, México. 246 pp.

(SAGARPA) Secretaría de Agricultura, Ganadería, Desarrollo Rural, Pesca y Alimentación. 2009a. Acuerdo mediante el cual se declara erradicado el brote de palomilla del nopal (*Cactoblastis cactorum* Berg.) en Isla Mujeres, Municipio de Isla Mujeres, Estado de Quintana Roo. Diario Oficial de la Federación, 26 de marzo de 2009. Ciudad de México, México.

(SAGARPA) Secretaría de Agricultura, Ganadería, Desarrollo Rural, Pesca y Alimentación. 2009b. Acuerdo mediante el cual se declara erradicado el brote de palomilla del nopal (*Cactoblastis cactorum* Berg.) en Isla Contoy, Municipio de Isla Mujeres, Estado de Quintana Roo. Diario Oficial de la Federación, 12 de octubre de 2009. Ciudad de México, México.

(SAGARPA-SENASICA) Secretaría de Agricultura, Ganadería, Desarrollo Rural, Pesca y Alimentación - Servicio Nacional de Sanidad, Inocuidad y Calidad Agroalimentaria. 2016. Palomilla del nopal *Cactoblastis cactorum* (Berg, 1885) (Lepidoptera: Pyralidae): Aviso público del riesgo y la situación actual. Ciudad de México, México. 7 pp.

(SAGARPA-SENASICA) Secretaría de Agricultura, Ganadería, Desarrollo Rural, Pesca y Alimentación - Servicio Nacional de Sanidad, Inocuidad y Calidad Agroalimentaria. 2017. Programas de trabajo de vigilancia epidemiológica fitosanitaria para los estados de Campeche, Coahuila, Chihuahua, Distrito Federal, México, Nuevo León, Quintana Roo, San Luis Potosí, Tabasco, Tamaulipas, Veracruz, Yucatán y Zacatecas. SENASICA-SAGARPA-PVEF. Ciudad de México, México.

Sánchez, A. H., J. Cibrián, C. J., Osorio, and C. Aldama. 2007. Impacto económico y social en caso de introducción y establecimiento de la palomilla del nopal (*Cactoblastis cactorum*) en México. Informe financiado por el Organismo Internacional de Energía Atómica (OIEA), (TC MEX/50/29) y presentado a la Dirección General de Sanidad Vegetal. Ciudad de México, México. 43 pp.

Sarvary, M. A., K. A. Bloem, S. Bloem, J. E. Carpenter, S. D. Hight, and S. Dorn. 2008. Diel flight pattern and flight performance of *Cactoblastis cactorum* (Lepidoptera: Pyralidae) measured on a flight mill: Influence of age, gender, mating status, and body size. Journal of Economic Entomology 101:314–324.

Simmonds, F. J., and F. D. Bennett. 1966. Biological control of *Opuntia* spp. by *Cactoblastis cactorum* in the Leeward Islands (West Indies). Entomophaga 11: 183–189.

Simonsen, T. J., R. L. Brown, and F. A. H. Sperling. 2008. Tracing an invasion: Phylogeography of *Cactoblastis cactorum* (Lepidoptera: Pyralidae) in the United States based on mitochondrial DNA. Annals of the Entomological Society of America 101: 899–905.

Simonson, E. S., J. T. Stohlgren, L. Tyler, P. W. Gregg, R. Muir, and J. G. Lynn. 2005. Preliminary assessment of the potential impacts and risks of the invasive cactus moth, *Cactoblastis cactorum* Berg, in the U.S. and Mexico. Final Report to the International Atomic Energy Agency, April 25, 2005. IAEA, Vienna, Austria.

(SIRVEF) Sistema Integral de Referencia para la Vigilancia Epidemiológica Fitosanitaria. 2018. SAGARPA-SENASICA, Mexico.

Soberón, J., J. Golubov, and J. Sarukhan. 2001. The importance of *Opuntia* in Mexico and routes of invasion and impact of *Cactoblastis cactorum* (Lepidoptera: Pyralidae). Florida Entomologist 84: 486-492.

Stiling, P. 2000. A worm that turned. Natural History 109(5): 40–43.

van Lenteren, J. C., and H. C. J. Godfray. 2005. European science in the Enlightenment and the discovery of the insect parasitoid life cycle in The Netherlands and Great Britain. Biological Control 32:12–24.

Zimmermann, H. G., and S. M. Pérez-Sandi-Cuen. 2006. The consequences of introducing the cactus moth *Cactoblastis cactorum* to the Caribbean and beyond. Transcontinental Reproductiones Fotomecánicas S.A. de C.V. Mexico City, Mexico. 65 pp.

Zimmermann, H. G., V. C. Moran, and J. H. Hoffmann. 2000. The renowned cactus moth, *Cactoblastis cactorum*: Its natural history and threat to native *Opuntia* floras in Mexico and the United States of America. Diversity and Distributions 6: 259–269.

Zimmermann H. G., S. Bloem, and H. Klein. 2004. Biology, history, threat, surveillance and control of the cactus moth, *Cactoblastis cactorum*. IAEA/FAO-BSC/CM. Vienna, Austria. 40 pp.

AREA-WIDE ERADICATION OF THE INVASIVE EUROPEAN GRAPEVINE MOTH *Lobesia botrana* IN CALIFORNIA, USA

G. S. SIMMONS[1], L. VARELA[2], M. DAUGHERTY[3], M. COOPER[4], D. LANCE[5], V. MASTRO[5], R. T. CARDE[3], A. LUCCHI[6], C. IORIATTI[7], B. BAGNOLI[8], R. STEINHAUER[9], R. BROADWAY[10], B. STONE SMITH[10], K. HOFFMAN[11], G. CLARK[12], D. WHITMER[13] AND R. JOHNSON[14]

[1]*USDA-APHIS-PPQ, CPHST California Station and Otis Lab, 1636 E. Alisal Road, Salinas, California 93905, USA; Gregory.S.Simmons@aphis.usda.gov*
[2]*University of California Cooperative Extension, Santa Rosa, California 95403, USA*
[3]*Department of Entomology, Riverside, California 92521, USA*
[4]*University of California Cooperative Extension, Napa, California 94559, USA*
[5]*Retired, USDA-APHIS-PPQ, CPHST 1398 W. Truck Rd, Buzzards Bay, Massachusetts 02542, USA*
[6]*Department of Agriculture, Food & Environment, University of Pisa, 56124 Pisa, Italy*
[7]*Center For Technology Transfer, FEM-IASMA, 38010 San Michele all'Adige, Trento, Italy*
[8]*Department for Innovation in Biological, Agro-Food and Forest Systems, University of Tuscia, via San Camillo de Lellis, 01100 Viterbo, Italy*
[9]*Wineland Consulting, St. Helena, California 94574, USA*
[10]*USDA-APHIS-PPQ-FO, Sacramento, California 95814, USA*
[11]*CDFA, Street 2800 Gateway Oaks Drive, Sacramento, California 95833, USA*
[12]*Napa County Department of Agriculture, Napa, California 94559, USA*
[13]*Retired, Napa County Department of Agriculture, Napa, California 94559, USA*
[14]*USDA-APHIS-PPQ, 4700 River Rd, Riverdale, Maryland 20737, USA*

J. Hendrichs, R. Pereira and M. J. B. Vreysen (eds.), Area-Wide Integrated Pest Management: Development and Field Application, pp. 581–596. CRC Press, Boca Raton, Florida, USA.

SUMMARY

In the fall of 2009, the first confirmed North American detection of the European grapevine moth (EGVM) *Lobesia botrana* (Denis & Schiffermüller) occurred in Napa County, California, USA. Based on its status as a significant grape pest in other parts of the world, the establishment of EGVM in California presented significant production and export issues for grapes, as well as for other fresh market agricultural commodities. Over the following seven years, an intensive California state-wide survey and area-wide eradication campaign was undertaken in partnership with agricultural officials at local, state and federal levels, university scientists and the wine, table grape and raisin industries. These efforts resulted in a dramatic decline in moth captures in pheromone traps from over 100 000 moths in 2010, to one in 2014, and none in 2015. In August of 2016, eradication was declared for all previously infested areas in California. The decision to pursue the eradication effort was based on the limited host range and geographic area of the EGVM infestation, the availability of effective tools for monitoring and control, and the strong support of the affected grape production and export industries. The eradication campaign employed coordinated logistical, regulatory, and technical efforts that included: 1) state-wide-monitoring using a network of pheromone-baited traps and in field monitoring; these findings were recorded in a geographic information system that was used to regularly communicate survey results to programme officials; 2) an area-wide application of mating disruption dispensers to infested vineyards, including use in urban environments within infested zones; 3) implementation by coordinators of area-wide insecticide treatments with application timing determined by degree-day modelling for each infested region; 4) a robust regulatory programme that initiated and maintained a quarantine of infested areas that regulated movement of fruit, farming equipment and winery processing waste; 5) an extensive outreach programme to grape growers, wineries, pest control specialists and the public; 6) formation of a technical working group that provided recommendations to the operational programme. An extensive methods development effort supported the programme. This included developing enhanced detection methods for vineyards under mating disruption, testing efficacy and residual control of insecticides, testing mating disruption formulations, evaluating the impacts of winery processing methods on EGVM mortality, developing methods to determine the timing of the development of successive EGVM generations (or biofix) under California conditions to improve degree-day models, developing EGVM rearing methods, testing the quality of pheromone lures and trap monitoring; and a spatial analysis of trapping data to determine programme effectiveness and to analyse invasion pathways.

Key Words: Lepidoptera, pheromone, surveillance, detection, mating disruption, invasive species, grape pests, degree-day models

1. INTRODUCTION

The European grapevine moth (EGVM), *Lobesia botrana* (Denis & Schiffermüller), is a tortricid moth that has historically been a pest of the Mediterranean regions of Europe, North Africa, and Asia. Recently it has been introduced into the Americas region with first detections in Chile in 2008, California, USA in 2009 and Argentina in 2010 (Ioriatti et al. 2011, 2012; Taret et al., this volume).

Grapevine flowers and berries are favoured hosts for the EGVM. Other hosts include olive flowers, blueberries and plums. *Daphne gnidium* L., an evergreen shrub from the Mediterranean region, is hypothesized to be the ancestral host (Thiéry and Moreau 2005). Although reported on these other plants, they appear to be used opportunistically only when principal EGVM hosts are in the same environment, though there are some areas in Italy where EGVM populations can sustain themselves exclusively on *D. gnidium* in the absence of grapevine (Lucchi and Santini 2011). The EGVM has multiple generations a year, starting in the spring from overwintering pupae with 3-4 generations observed in the Mediterranean regions and 3 generations documented in California, with the 3rd generation (and possible a 2nd generation) going into winter diapause as pupae (Ioriatti et al. 2012; Cooper et al. 2014). Successive generations target developing stages of grapes, with the first feeding on flower clusters, the second on green berries, and the third inside the bunches after veraison, the change of colour of the grape berries reflecting the onset of ripening. Webbing within the clusters may be apparent, along with excrement and shrivelled berries. Feeding on berries causes direct losses and leads to fungal infections that can cause extensive rot leading to total loss of clusters (Ioriatti et al. 2012).

After overwintering, the first generation starts in spring as eggs and are laid singly on flowers. Larvae hatch and form a feeding nest by webbing together groups of flowers. Larvae from later generations feed on green, ripening, and ripe grapes. Their feeding reduces yield and also affects quality of table grapes or wine grapes, with damage causing bunch rot and mould. Bunch rot causes bad flavours in wine, making heavily infested grapes unusable (Fermaud and Le Menn 1992).

In the fall of 2009, the first confirmed North American detection of the EGVM was made in Napa County, California (Gilligan et al. 2011). Based on EGVM's status as a significant grape pest in other parts of the world, its establishment in California presented significant production and export issues for grapes, as well as for other fresh market agricultural commodities. In response to this EGVM invasion, an extensive California state-wide survey, regulatory programme and area-wide eradication campaign was undertaken in partnership among local, state and federal agricultural officials, university scientists, and the wine, table grape and raisin industries (Cooper el al. 2014).

Here we describe the emergency response programme and results for the detection, regulatory action, initiation of an area-wide programme in 2010 and its coordination, communication and outreach, along with the methods development and research that was initiated to support the programme that led to successful EGVM eradication declared in August 2016.

2. FIRST DETECTION AND FORMATION OF AN OPERATIONAL PROGRAMME

On October 7, 2009, the United States Department of Agriculture (USDA) confirmed the presence of the EGVM in Napa County, California for the first time in North America (Gilligan et al. 2011). This area was in the heart of the Napa Valley grape production area close to Napa River, and growers reported extensive damage as well as near 100% losses caused by direct infestation or spoilage due to cluster rots from several vineyards near the site of the first detection (APHIS 2010). Growers had already reported problems in this area in 2008, with many clusters of fruit being rejected, but the damage was thought to be caused by another tortricid species (APHIS 2010). Larvae collected from the same region in September 2008 were not identified as EGVM at the time, but these were later confirmed also to be EGVM (Gilligan et al. 2011).

While the official recognition did not occur until 2009, given the extensive damage in that year, the damage and identification of larvae from 2008, and its widespread extent revealed by detection trapping in 2010, it is likely that the first EGVM arrived in Napa County at least a year or more before 2008, building up over time until extensive vineyard damage was observed.

In 2010, state and federal officials began an emergency response programme to delimit the extent of the infestation and to establish an agricultural quarantine. An extensive pheromone trap monitoring programme was established deploying traps at densities between 6 to 39 traps per km^2 throughout vineyards state-wide. Trapping density was dependent on whether trapping occurred within a delimitation area or was part of the detection trapping network (CDFA 2013). More than 60 000, traps were deployed with >10 000 in Napa and Sonoma counties alone (Fig. 1).

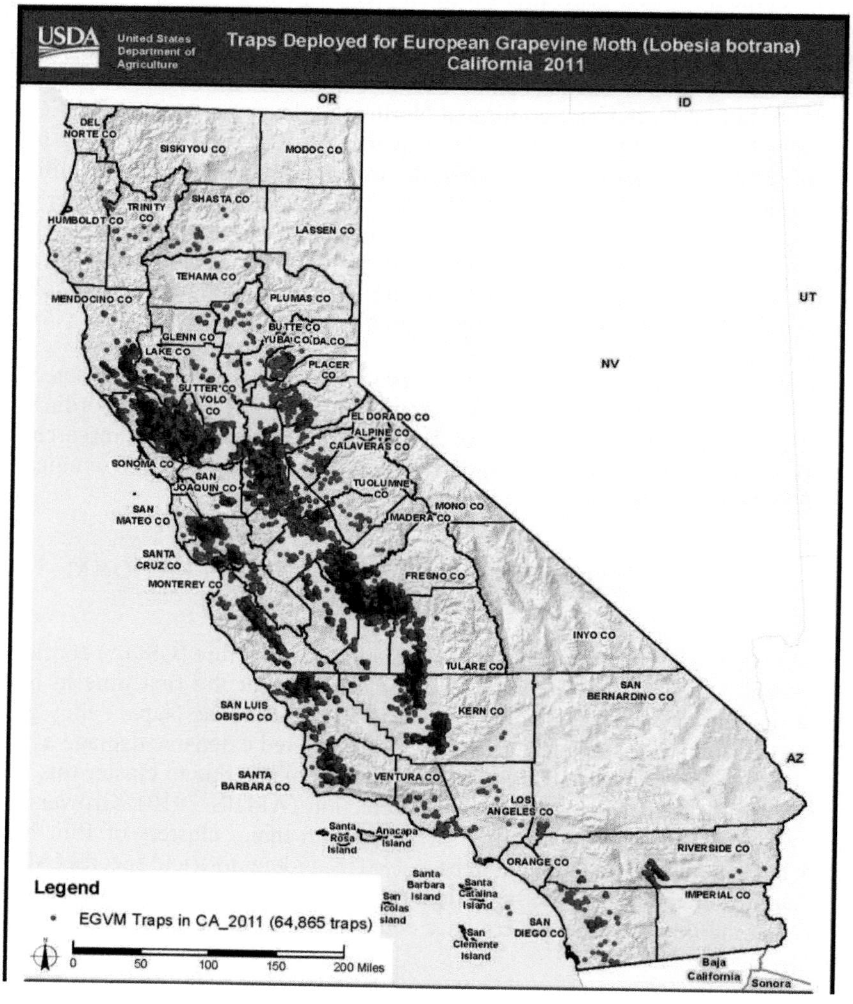

Figure 1. European grapevine moth (EGVM) trap distribution in California counties at the height of the detection effort in the eradication programme (source R. Broadway, USDA-APHIS-PPQ).

Urban regions in the quarantine area or near grape production areas were also monitored at the rate of 10 traps per km^2, and detection trapping was conducted in urban areas outside of quarantine areas at a rate of 2 traps per km^2 (CDFA 2013; Cooper et al. 2014).

The USDA-Animal and Plant Health Inspection Service (APHIS) - Plant Protection and Quarantine (PPQ) established a technical working group (EGVM-TWG) to make recommendations in support of the emergency response programme. The group was composed of national and international experts in grape pest entomology, area-wide control programmes, viticulture practices and Lepidoptera biology and control. Their primary role was to provide guidance to the programme on the operation of the emergency response and whether eradication of EGVM was feasible.

In 2010, over 100 000 male EGVMs were caught in pheromone traps. While the majority of these captures were located in Napa County, there were significant populations in adjacent Sonoma County and few smaller isolated populations elsewhere, which were attributed to movement of grapes to wineries and, in one case, recycled wooden vineyard trellis posts (Lance et al. 2016) (Figs. 2 and 3).

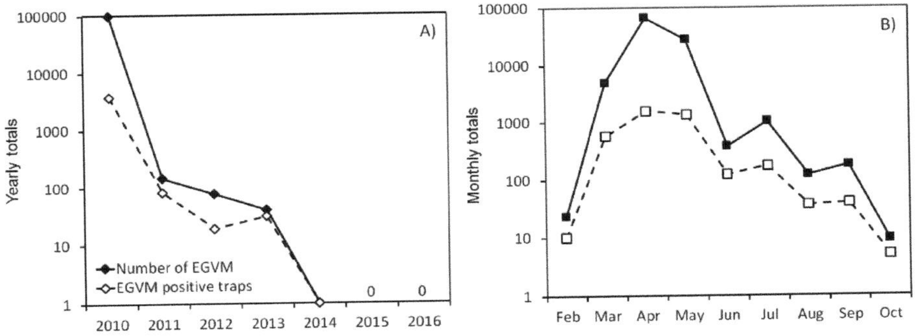

Figure 2. A) Yearly total, 2010 to 2016, and B) 2010 monthly total number of European grapevine moth (EGVM) males caught (solid lines) and total number of traps with at least one EGVM male caught (dashed lines). Y-axes are on a log10 scale.

In response to these detections, a cooperative eradication programme was initiated in 2010 with participation from growers, the wine industry, federal, state, county and University of California authorities and scientists (see Cooper et al. 2014; Lance et al. 2016).

3. EMERGENCY RESPONSE AND ERADICATION PROGRAMME

With technical input from the TWG, the EGVM programme initiated a comprehensive regulatory and area-wide eradication effort. The programme consisted of:

1. A state-wide detection network of pheromone traps and vineyard inspections
2. A centralised system to record and map data

3. Regulations for the movement of plant material, farming and winery equipment, harvested fruit, winery waste, nursery plants, and harvesting bins from and within quarantine areas
4. Area-wide coordinated treatments of mating disruption and insecticide sprays
5. A residential grapevine inspection and treatment programme
6. An extensive outreach and communication effort; and
7. A programme of research and methods development to support the needs of the programme.

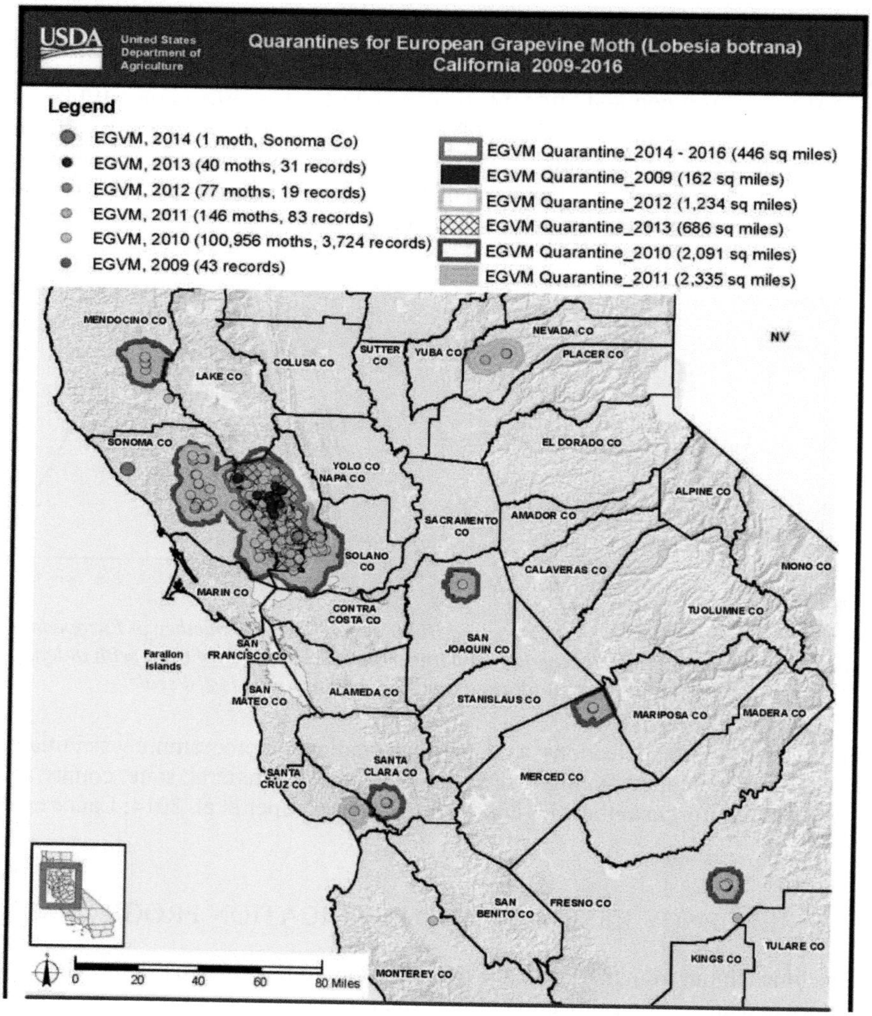

Figure 3. European grapevine moth (EGVM) detections and quarantine areas established in California counties during the eradication programme 2009-2016 (source R. Broadway, USDA-APHIS-PPQ).

4. RESEARCH AND METHODS DEVELOPMENT

The EGVM-TWG, composed of national and international experts, made recommendations to the operational programme on a range of technical issues concerning regulatory action, treatments and detection. While technical information for programme decision-making was available from TWG expertise or from the literature, the TWG also made recommendations on the required research and methods development activities needed to support a programme specific to California, because EGVM was a new pest in North America

Methods development became even more important once a goal of eradication had been established, as this would be one of the first efforts to eradicate populations of this species. A key early decision was to establish a colony of EGVM in the USDA-APHIS-PPQ quarantine facility at the APHIS laboratory in Cape Cod, Massachusetts. Later a second colony was established in a containment laboratory at the University of California at Davis, California. These two colonies supported the important research goals by providing material for research on treatment methods and other mitigation methods.

Other research work included developing enhanced detection methods for vineyards under mating disruption, testing efficacy and residual control of insecticides, testing mating disruption formulations and pheromone lures, evaluating EGVM infestation in other host plants besides grape, assessing the impacts of winery processing methods on EGVM mortality, validation of degree-day models for California conditions, developing EGVM rearing methods, and a spatial analysis of trapping data to determine programme effectiveness and to analyse invasion pathways (Lucchi et al. 2012; Van Steenwyk et al. 2013; Varela et al. 2013; Cooper et al. 2014; Daugherty et al. 2015).

Since EGVM represented a new invasive pest in California, a critical activity was the testing and validation of EGVM degree-day models from Europe for use in California (see EGVM models described in CABI 2019). This information was used to determine the start of the first spring generation, or biofix, to set the timetable for placement of traps, to determine treatment schedules and for decision-making for other programme operations requiring an accurate assessment of the EGVM life cycle in California.

Especially since EGVM was not yet widespread, an accurate model was needed to conduct programme operations throughout the state. The use of degree-day models became increasingly important as the programme progressed and EGVM populations were reduced, leaving fewer population cues available to make treatment decisions (Varela et al. 2011; Cooper et al. 2014).

5. DETECTION AND SURVEILANCE PROGRAMME

Traps were installed across the state in all grape growing areas (Fig. 1). Trapping levels varied across the state depending on the infestation levels, availability of host plants, as well as programme resources. In the first quarantine areas in Napa, Sonoma and Solano counties, traps were deployed in commercial vineyards at a density of 39 traps per km^2 and in residential areas at 10 traps per km^2 within a 5-km radius from a detection (CDFA 2013). Beyond 5 km from a quarantine, traps were deployed at 10 traps per km^2 in commercial vineyards and 2 traps per km^2 in residential areas (CDFA 2013).

For all commercial vineyards outside of quarantine areas, traps were deployed at 10 traps per km^2 (CDFA 2013). An additional protocol for detection trapping was to deploy traps at grape processing facilities in unregulated counties that received grapes from quarantine areas. These areas, determined to be "high hazard", were trapped within 0.8 km of the facility at 10 traps per km^2. There were 22 wineries in nine unregulated counties that fit these criteria (CDFA 2013).

In the first year of the programme, there was some testing of the EGVM pheromone blend, loading and emission rates. The resulting data supported the programme decision to use the single-component pheromone blend of (*Z, E*)-7-9-dodecadienyl acetate loaded at 1 mg on rubber septa. The septa lures were produced by a USDA laboratory the first year and afterward under commercial contract (CDFA 2013; Cooper et al. 2014).

Although different delta trap size and styles could be used for EGVM, a red paper delta trap was selected because, in part, the state had a large surplus supply of these traps available from another programme and because the red colour would limit the number of bees trapped as by-catch. These traps have all interior surfaces coated with biotac glue.

Traps were hung at canopy height, approximately one meter above the ground on vineyard wires or on vines at the end of rows along major vineyard roadways, permitting easy access for trappers. Traps in residential areas were placed on grapes if available or on secondary hosts such as *Prunus spp*. or olive. Vineyard traps were put into the field before bud break based on predictions by degree-day models, but in practice for northern California, the programme worked to have all traps deployed at the start of the growing season—near the end of February or early March. Traps were kept in the field until the end of September (CDFA 2013).

Traps were inspected biweekly and brought into the laboratory for identification of moths at local offices and suspected *L. botrana* finds were submitted to the California Department of Food and Agriculture (CDFA) State Diagnostics Laboratory for confirmation. Finds in a new area triggered an immediate establishment of a quarantine area, initiation of suppression treatments, and a new cycle of delimitation trapping in that area. Trap data were recorded with GPS coordinates for purposes of visualization with mapping programmes. These maps were provided to programme personnel to make operational decisions.

Included in the programme was a pre-season assessment of contracted lure quality, analysis of lure emission rates to determine trap service intervals, ongoing training to programme personnel regarding trap placement, and training to identify *L. botrana* and other moths in traps. Dead moths from a laboratory colony were used to seed traps in the field for quality checks on training and trap-checking frequency.

At the peak of the programme in 2011, there were 11 counties with detections for a quarantine area totalling 604 763 ha. Within this area, there were 325 000 ha of vineyards with > 60 000 traps deployed over all the vineyards in California (Fig. 2). Trapping has continued post-eradication at levels similar to those used throughout the programme in areas outside of quarantine (APHIS 2016).

6. REGULATORY PROGRAMME QUARANTINE AND DELIMITATION

State and federal quarantine areas were established around every EGVM detection consisting of an 8-kilometer radius. The standard of detection to establish a quarantine was defined as two moths found within 5 km, or a single larva or egg (CDFA 2013; Cooper et al. 2014). With the detection of an additional life stage, delimitation trapping would begin with placement of 39 traps per km^2 in the central 2.6 km^2 (1 square mile) and at 10 traps per km^2 in the 10-km^2 area surrounding the detection. An EGVM detection in the area surrounding the detection would trigger expansion of the quarantine to include the new find area (CDFA 2013).

To allow the movement of fruit from quarantine areas at harvest, requirements included checking compliance with previous control treatments, an inspection of fruit before harvest, covering truckloads with tarps and restricting the movement and processing of fruit to within an existing EGVM quarantine area. To further mitigate the risk of moving infested fruit, truckloads of grapes waiting at a winery needed to be processed within a specified short time period after arrival.

Early in the infestation it became clear that there was an association between the locations of outbreaks and where wine grapes were processed. Though it was unknown if standard grape processing techniques and handling of waste products would mitigate the risk of spreading EGVM, the programme implemented controls on the movement and processing of grapes for making wine, while concurrently evaluating EGVM mortality during wine-making.

Specifically, a series of experiments were conducted to determine if grape crushing, pressing and the fermentation process would kill EGVM larvae or pupae (see Varela et al. 2013; Smith et al. 2013; Cooper et al. 2014). This work showed that significant numbers of larvae could survive grape processing on harvesting and transport equipment, on green wastes from destemming, crushing and pressing, and on winery equipment (Smith et al. 2013).

This work led to requirements for winery-waste and green-waste management and treatment by composting at a regulated composting operation, on-site deep burial, or other destruction methods such as burning or heat treatment. Alternatively, if the grapes, clusters or other green wastes for white wine processing (that were not fermented) were pressed at 2.0 bars, the resulting winery waste could be returned to the original vineyard. Grapes in must (crushed grapes in juice) and pomace (a waste product after fermentation of red varieties) were not regulated as research showed that the process of fermentation would kill all EGVM life stages (Smith et al. 2013; Cooper et al. 2014).

Under conditions determined by the TWG, previously quarantined areas became eligible for deregulation after several conditions were met including: mating disruption used for a full year following the detection; insecticide treatments applied for the first and second generations for two years following the year of detection; a visual survey conducted in vineyards treated with mating disruption; mating disruption not used in the last two generations before deregulation; and trap density increased to 39 traps per km^2 in the years after mating disruption was removed. If no EGVM life stages were detected for six full generations after the last capture in the area, it could be removed from regulation (Cooper et al. 2014; APHIS 2015, 2016).

7. TREATMENTS

Commercial vineyards, residential plantings of grapes and other hosts were treated within 500 m of a EGVM detection. Treatments consisting of timed insecticide applications, mating disruption and fruit stripping were made on a coordinated area-wide basis. Treatments were continued for two full growing seasons following the year of detection (Cooper et al. 2014). University of California extension personnel using degree-day and crop stage reporting recommended when treatments should be applied.

The cooperative programme employed grower liaisons to help with outreach, to coordinate treatments in each county and to work with growers. These individuals, licensed pest control advisers, worked closely with all affected growers and operational programme personnel. As this was a voluntary programme, the grower liaison's work was crucial to ensure high levels of participation and were a key to successful eradication (Cooper et al. 2014).

7.1. Mating Disruption

In the California programme, mating disruption played a principal role in the management of EGVM in commercial vineyards. Plastic hollow-tube dispensers loaded with EGVM pheromone ((*E,Z*)-7,9-dodecadien-1-yl acetate (ISOMATE EGVM^{tm}) (Lance et al. 2016) were set out in all grape-bearing areas within a 500 m radius of any detection at the rate of 494 dispensers per hectare (Cooper et al. 2014; Lance et al. 2016). The goal was to deploy the dispensers before bud break, which was predicted by degree-day models. In practice, this occurred in February in the Napa County region.

Treatments with mating disruption were applied for at least two full flights following a moth detection. When moths were trapped in an area during the first flight of a season, mating disruption treatments were applied at that time. If moths were trapped in the second flight, mating disruption applications were made early in the following spring. A single pheromone application was sufficient for the season as field testing pheromone emission rates determined that these dispensers remained viable for the entire season under Napa conditions.

At the peak of ECVM suppression in 2012, mating disruption was used on 9340 ha in the core of the infested region of Napa and Sonoma counties (Cooper et al. 2014). Mating disruption was also used as a component of the residential treatments, with a peak of over 3000 properties treated in 2011 (Cooper et al. 2014).

Because the same pheromone was used for monitoring, widespread use of mating disruption caused monitoring programme traps to be less effective within treated areas. This was the primary reason the TWG recommended using mating disruption for a relatively short period of time as well as removing it for a period prior to deregulation to determine that an area was free of the EGVM. It also highlighted the need to test alternative attractants for their potential under conditions of mating disruption.

The use of mating disruption in combination with insecticides was recommended because the use of multi-tactic independent control measures in area-wide control programmes can be more effective by increasing the likelihood of Allee effects (Yamanaka and Liebhold 2009) and by helping to cover for possible gaps in treatment and for control of moths coming from undetected EGVM populations from nearby areas (Cardé, this volume; Liebhold et al. 2016).

The recent eradication of pink bollworm from the south-western USA and northern Mexico (Staten and Walters, this volume) is a convincing demonstration of this integrated approach and was considered successful using three or more control tactics on an area-wide basis (Tabashnik et al. 2010; Evenden 2016; Lance et al. 2016).

7.2. *Insecticide Treatments*

The programme specified coordinated treatments with insecticides made during the first and second generations for at least two complete growing seasons following the year of the detection. For the two counties at the heart of the infestation (Napa and Sonoma) a grower liaison was contracted to coordinate treatments in commercial areas. Degree-day models were used to target eggs and young larvae at the start of the first and second generations (Varela et al. 2011; Cooper et al. 2014). In practice, this resulted in a three-week treatment window for each flight. By the third generation, grape bunch closure can limit the effectiveness of insecticides so treatments to target the third generation were not recommended.

Materials used included conventional foliar insecticides: the growth regulator methoxyfenozide, and diamide chlorantraniliprole; other materials used included abamectin as well as *Bacillus thuringiensis* (*Bt*) and spinosyns for organic production vineyards (Daugherty et al. 2015). Treatments applied to commercial areas, while voluntary, had participation rates as high as 80%, with a peak of 12 306 ha treated in 2012 in Napa and Sonoma counties (Cooper el al. 2014). To meet eradication programme recommendations these treatments continued for several years during periods when there were no significant detections and when growers did not suffer any losses or direct impacts of EGVM infestation. The fact that participation rates were high during this period, and application costs were paid by individual growers, is a testament to the effectiveness of the coordination and the outreach provided to growers about programme needs. Personnel of CDFA coordinated and applied treatments in residential areas. CDFA officials were supported locally by the offices of the county Agricultural Commissioners, particularly as related to outreach at public meetings and consultation to gain permission from homeowners. These treatments also had a high rate of participation. They included application of *Bt*, fruit stripping and some uses of mating disruption (CDFA 2010a; Cooper et al. 2014).

8. OUTREACH AND COMMUNICATION

There was an extensive programme of outreach to grape growers, industry professionals, wineries and grape processing facilities, and the public at large. The outreach programme had several objectives. These included providing accurate technical information about the pest, helping to encourage participation with the programme and coordination of the area-wide programme treatments, and gaining public support for the eradication programme activities. As the EGVM was a new pest to California, communicating accurate pest biology and control information to growers, pest control advisers and grower liaisons, industry representatives and programme officials was a critical need and provided the linkage between the research effort and the operation of the eradication campaign (Cooper et al. 2014).

Information was provided through grower meetings and field days, public meetings, an email newsletter, University of California websites, communications with the grower liaisons, and university extension personnel, as well as local, state and federal government media campaigns using social media, blogs, mailings and local advertising (Fig. 4) (CDFA 2010b; Cooper et al. 2014; APHIS 2017; CDFA 2017; Napa County 2017; University of California 2017).

Figure 4. The European grapevine moth (EGVM) eradication outreach as a postcard from Napa County Agricultural Commissioner's office (used with permission of Napa County).

The outreach campaign was deemed critical to achieve programme success (Cooper et al. 2014; Daugherty et al. 2015). Indeed, in comparison to another recent unsuccessful eradication campaign, public support and engagement was considered essential for programme success (Zalom et al. 2013; Lance et al. 2016).

9. PROGRAMME RESULTS AND ERADICATION DECLARATION

The combined efforts of the programme resulted in a dramatic decline in moth captures from over 100 000 moths in 2010, to one moth in 2014, and no moth captures or larval finds by 2015 (Figs. 2 and 3). Using a step-wise process, the programme proceeded with deregulating large contiguous blocks once they met free-from-EGVM standards. This was a conservative approach and meant that some areas that had been free from EGVM for longer than required by the programme deregulation standards were kept under regulated status until larger associated geographic areas met the free-from-EGVM standard. With this approach, deregulation would not occur in a patchwork fashion (APHIS 2015).

In August of 2016, eradication was declared from all previously infested areas in California (CDFA 2016). At that time, the USA was declared free from this pest. At the end of a full 2017 trapping season, there have been no EGVM detections in the USA for over three years and all of the previously EGVM infested California grape production areas have been free from EGVM between 3-5 years depending on their location.

10. POST ERADICATION PHASE AND ONGOING VIGILANCE.

While the eradication campaign against EGVM was accomplished in a relatively short period, and grape producers in California can be confident that eradication has been achieved, the EGVM programme and industry officials drew up plans to conduct a post-eradication campaign for a period of at least three years after the eradication declaration (APHIS 2016).

Like other successful USA eradication campaigns (APHIS 2009; Cardé, this volume), adding a period of extra vigilance after an area is eradicated is considered a sensible safeguard. Early detection of any new EGVM incursions will be smaller and far cheaper to contain and eliminate then if the detection is made later at a time when the detection network has been reduced and fewer traps are deployed.

The pathway by which EGVM entered North America is still unknown, and the opportunities and conditions regarding international trade and possible entry pathways may remain the same (Cooper et al. 2014). The EGVM is a pest on the move and has been expanding its range. It is now present in parts of South America and is causing significant problems, which means there are additional possible invasion sources beyond European and Mediterranean countries (Ioriatti et al. 2012).

Besides the need to continue a post-eradication phase, there is a need for additional research and programme activity to enhance safeguarding of the USA grape industry. This work includes development of alternatives to pheromone detection methods that work under mating disruption treatment, such as kairomones or alternate pheromone blends. Work is underway to analyse detection data from the area-wide control programme using geospatial modelling techniques to evaluate landscape patterns of the invasion and to model trap-grid detection efficiency to help design lower-cost, long-term future detection strategies, in the post-eradication phase.

A second need is to conduct an economic analysis on the costs and benefits of continued monitoring for EGVM after the post-eradication phase has ended. This should include an analysis of possible harm of other potentially damaging economic grape pests not yet present in California and other North American grape production areas that may use similar pathways used by the EGVM. These include the European grape berry moth *Eupoecilia ambiguella* (Hübner), the grape tortrix *Argyrotaenia ljungiana* (Thunberg), the grape berry moth *Paralobesia viteana* (Clemens), the honeydew moth *Cryptoblabes gnidiella* (Millière), and others.

Lastly, there should be ongoing outreach to growers, field workers, trappers, and pest control advisers so they can continue to recognize the EGVM and for ongoing training by personnel involved in EGVM and other grape pest surveys.

11. ACKNOWLEDGEMENTS

Members of the EGVM Technical Working Group of APHIS-PPQ are Bruno Bagnoli, Department for Innovation in Biological, Agro-Food and Forest Systems, University of Tuscia, via San Camillo de Lellis, 01100 Viterbo, Italy; Ring T. Cardé, University of California, Riverside, California; Monica Cooper, University of California Cooperative Extension, Napa, California; Claudio Ioriatti, Istituto Agrario San Michele all'Adige (IASMA), Italy; David Lance (retired), USDA-APHIS-PPQ, Buzzards Bay, Massachusetts; Andrea Lucchi, Dipartimento di Scienze Agrarie, Alimentari ed Agroambientali, Pisa, Italy; Vic Mastro (retired), USDA-APHIS-PPQ, Buzzards Bay, Massachusetts; Gonçal Barrios, Departament d'Agricultura, Tarragona, Spain; Luis Sazo, University of Chile, Santiago de Chile, Chile; Gregory Simmons, USDA-APHIS-PPQ, Salinas, California; Robert Steinhauer, Wineland Consulting, LLC, St. Helena, California; and Lucia Varela, University of California Cooperative Extension, Santa Rosa, California. We thank Thomas Greene, USDA-APHIS-PPQ for literature research.

12. REFERENCES

(APHIS) Animal and Plant Health Inspection Service. 2009. Minimum standards for pink bollworm post-eradication. USDA. Accessed October 30, 2017.

(APHIS) Animal and Plant Health Inspection Service. 2010. Final report of the international technical working group for the European Grape Vine Moth (EGVM) in California, February 10, 2010. USDA.

(APHIS) Animal and Plant Health Inspection Service. 2015. Report of the international technical working group for the European grapevine moth program. USDA.

(APHIS) Animal and Plant Health Inspection Service. 2016. European grapevine moth post-eradication response guidelines. USDA.

(APHIS) Animal and Plant Health Inspection Service. 2017. European grapevine moth, USDA-APHIS Hungry Pest Website, Accessed October 22, 2017.

(CABI) Centre for Agriculture and Bioscience International. 2019. Datasheet *Lobesia botrana* (European grapevine moth). Invasive Species Compendium.

(CDFA) California Department of Food and Agriculture. 2010a. News Release #10-050: Fruit removal treatment for European grapevine moth scheduled for Solano county.

(CDFA) California Department of Food and Agriculture. 2010b. European grapevine moth (*Lobesia botrana*).

(CDFA) California Department of Food and Agriculture. 2013. European grapevine moth (EGVM) detection trapping guidelines fiscal year 2012-13 and fiscal year 2013–14.

(CDFA) California Department of Food and Agriculture. 2016. European grapevine moth eradicated from California. Agricultural officials confirm eradication of grape pest, lift quarantine restrictions.

(CDFA) California Department of Food and Agriculture. 2017. European grapevine moth (EGVM). Accessed October 22, 2017.

Cooper, M., L. G. Varela, R. Smith, D. Whitmer, G. S. Simmons, A. Lucchi, R. Broadway, and R. Steinhauer. 2014. Managing newly established pests: Growers, scientists and regulators collaborate on European grapevine moth program. California Agriculture 68: 125–133.

Daugherty, M., M. L. Cooper, G. S. Simmons, R. J. Smith, and L. G. Varela. 2015. Progress made on control of European grapevine moth: Present status and next steps. CAPCA Adviser Magazine, December, 18(6): 54–56.

Evenden, M. 2016. Mating disruption of moth pests in integrated pest management. A mechanist approach, pp. 365–393. *In* J. D. Allison, and R. T. Cardé (eds.), Pheromone communication in moths: Evolution, behavior and application. University of California Press, Berkeley, California, USA.

Fermaud, M., and R. Le Menn. 1992. Transmission of *Botrytis cinerea* to grapes by grape berry moth larvae. Phytopathology 82: 1393–1398.

Gilligan, T. M., M. E. Epstein, S. C. Passoa, J. A. Powell, and J. W. Brown. 2011. Discovery of *Lobesia botrana* [(Denis & Schiffermüller)] in California: An invasive species new to North America (Lepidoptera: Tortricidae). Proceedings of the Entomological Society of Washington 113: 14–30.

Ioriatti, C., G. Anfora, M. Tasin, A. De Cristofaro, P. Witzgall, and A. Lucchi. 2011. Chemical ecology and management of *Lobesia botrana* (Lepidoptera: Tortricidae). Journal of Economic Entomology 104: 1125–1137.

Ioriatti, C., A. Lucchi, and L. G. Varela. 2012. Grape berry moths in western European vineyards and their recent movement into the New World, pp. 339–359. *In* N. J. Bostanian, C. Vincent, and R. Isaacs (eds.), Arthropod management in vineyards: Pests, approaches, and future directions. Springer, Dordrecht, The Netherlands.

Lance, D. R., D. S. Leonard, V. C. Mastro, and M. L. Walters. 2016. Mating disruption as a suppression tactic in programs targeting regulated lepidopteran pests in US. Journal of Chemical Ecology 42: 590–605.

Liebhold, A. M., L. Berec, E. G. Brockerhoff, R. S. Epanchin-Neill, A. Hastings, D. A. Herms, J. M. Kean, D. G. McCullough, D. M. Suckling, P. C. Tobin, and T. Yamanaka. 2016. Eradication of invading insect populations: From concepts to applications. Annual Review of Entomology 61: 335–352.

Lucchi, A., and L. Santini. 2011. Life history of *Lobesia botrana* on *Daphne gnidium* in a Natural Park of Tuscany. IOBC/WPRS Bulletin 67: 197–202.

Lucchi, A., B. Bagnoli, M. L. Cooper, C. Ioriatti, and L. G. Varela. 2012. The successful use of sex pheromones to monitor and disrupt mating of *Lobesia botrana* in California. *In* Abstracts IOBC/WPRS Working Group meeting on "Pheromones and other semio-chemicals in integrated production", October 1-5, 2012 in Bursa, Turkey.

Napa County. 2017. European grapevine moth. County of Napa, California, USA. Accessed October 22, 2017.

Smith, R. J., M. L. Cooper, and L. G Varela. 2013. Determining the threat of dispersal of *Lobesia botrana* larvae in infested grapes processed for wine making. Practical Winery and Vineyard: 73.

Tabashnik, B. E., M. S. Sisterson, P, C. Ellsworth, T. J. Dennehy, L. Antilla, L. Liesner, M. Whitlow, R. T. Staten, J. A. Fabrick, and G. C. Unnithan. 2010. Suppressing resistance to *Bt* cotton with sterile insect releases. Nature Biotechnology 28: 1304–1307.

Thiéry, D., and J. Moreau. 2005. Relative performance of European grapevine moth (*Lobesia botrana*) on grapes and other hosts. Oecologia 143: 548–557.

University of California. 2017. UC IPM / Invasive and Exotic Pests / European grapevine moth. Accessed October 22, 2017.

Varela, L. G., M. L Cooper, W. J. Bentley, and R. J. Smith. 2011. Degree-day accumulations used to time insecticide treatments to control 1st generation European grapevine moth. Sonoma County, California, USA. University of California Cooperative Extension Publication.

Varela, L., A. Lucchi, B. Bagnoli, G. Nicolini, and C. Ioriatti. 2013. Impacts of standard wine-making process on the survival of *Lobesia botrana* larvae (Lepidoptera: Tortricidae) in infested grape clusters. Journal of Economic Entomology 106: 2349–2353.

Yamanaka, T., and A. M. Liebhold. 2009. Mate-location failure, the Allee effect, and the establishment of invading populations. Population Ecology 51: 337–340.

Zalom, F., J. Grieshop, M. Lelea, K. Jennifer, J. K. Sedell. 2013. Community perceptions of emergency responses to invasive species in California: Case studies of the light brown apple moth and the European grapevine moth. Cooperative Agreement #10-8100-1531-CA. USDA-APHIS & University of California, Davis, California, USA. Accessed October 30, 2017.

AREA-WIDE MANAGEMENT OF *Lobesia botrana* IN MENDOZA, ARGENTINA

G. A. A. TARET, G. AZIN AND M. VANIN

Instituto de Sanidad y Calidad Agropecuaria de Mendoza (ISCAMEN)
Mendoza, Argentina; iscamen@iscamen.com.ar

SUMMARY

The invasive European grapevine moth *Lobesia botrana* (Denis & Schiffermüller) was first detected in 2010 in the Province of Mendoza, Argentina, after it became established in neighbouring Chile in 2008. Foreseeing the threat to the major wine and table grape industry of the Province of Mendoza, the Institute for Agricultural Sanitation and Quality of Mendoza Province (ISCAMEN), established the "*Lobesia botrana* Control and Eradication Programme" that has been following an area-wide integrated pest management (AW-IPM) strategy. The mainstay of this programme is widespread mating disruption against the adult moth population with judicious application of selective biological and chemical insecticides targeting the immature stages of the first two generations of the pest. We describe the spread of the pest in the Province of Mendoza since its introduction in 2010, the control tools currently being used, and the results achieved so far in suppressing the pest in the four main oases of Mendoza Province, as well as the future prospects for control. Before *L. botrana* detection, the absence of a major, direct pest of grapes meant that limited interventions were required in the management strategies used in the vineyards of the Province of Mendoza. Therefore, the challenge was to maintain a high level of natural biological control while introducing a pest management programme for *L. botrana* (i.e. mating disruption and conventional insecticides) that would not impact the complex of beneficial insects well established in the region. The situation required a great effort involving training, knowledge and technology transfer from public organizations to the private sector (growers and industry), and communication with environmental organizations. After two intensive control seasons with area-wide suppression of the pest as the main objective, population's levels of *L. botrana* have decreased dramatically. It required a substantial financial investment and technical effort to cover vineyards over a total area of 160 000 ha. Decision making for suppression activities was supported by the use of mapping software (GIS) to visualize programme results every week in terms of moth population levels. Although the support from the industries and growers has been instrumental to the success of this area-wide programme, the challenge will be to maintain this support to further suppress the pest and, eventually in the future, to eradicate *L. botrana* from Mendoza Province.

Key Words: European grapevine moth, Tortricidae, invasive, integrated pest management, mating disruption, flowable pheromone, aerial sprays, immature stages, insect growth regulators, *Bacillus thuringiensis kurstaki*, *Streptomyces avermitilis*, grapes, wine industry

J. Hendrichs, R. Pereira and M. J. B. Vreysen (eds.), Area-Wide Integrated Pest Management: Development and Field Application, pp. 597–614. CRC Press, Boca Raton, Florida, USA.

1. INTRODUCTION

The European grapevine moth *Lobesia botrana* (Denis & Schiffermüller) is a key pest of grapes (*Vitis vinifera* L.), that greatly impacts wine and table grape industries where it becomes established (Coscollá Ramón 1981, 1998; Zangheri et al. 1992; Armendáriz et al. 2007). It has been introduced to new regions as a result of trade and travel and is a serious threat for all the vine-growing areas that are presently unaffected (Ioriatti et al. 2012; CABI 2019). In 2010, this invasive moth was first detected in the Province of Mendoza, Argentina (SENASA 2015), close to the Chilean border. It became first established in Chile in 2008 (González 2008; SAG 2010) and spread rapidly to all grape growing regions of that country.

The favourable agroecological profile of Mendoza Province, and the absence of significant grape pests, allowed the development of a major wine and table grape industry, largely free of insecticide use. Production only involved occasional fungicide applications due to the favourable dry desert climate. Establishment of *L. botrana* would create significant changes to the existing pest management programme. Consequently, the Instituto de Sanidad y Calidad Agropecuaria de Mendoza (ISCAMEN) established on the 3rd of March 2010 the "*Lobesia botrana* Control and Eradication Programme" to be implemented with the participation of producers and government organizations. It proposed an area-wide integrated pest management (AW-IPM) strategy based on the integration of mating disruption applied by ground and air against the adult moth population, and the application of selective biological and chemical insecticides directed at the immature stages of the first two generations of the pest.

Here we describe the introduction and spread of the pest in Mendoza Province, the control strategy and tools used with a focus on results achieved in 2017-2018, and prospects for control in future seasons.

2. CHARACTERISTICS OF THE PEST AND DAMAGE

The European grapevine moth is a multivoltine pest, with the number of generations a function of temperature and photoperiod (Thiéry 2008; CABI 2019). In Mendoza this pest has four generations per season, each with five larval instars. It overwinters in the pupal stage under the bark in vineyards. Adults emerge gradually in the spring and have crepuscular habits (Milonas et al. 2001; Gallardo et al. 2009). Females from the first generation lay their eggs in the flower corollas (Torres-Vila et al. 1997; Maher 2002), however females of the later generations oviposit their eggs directly on the developing grapes, with the larvae causing direct damage and the loss of commercial value due to rejections by wineries. Each female can lay approximately 100 eggs in its life. While there is little data about fecundity, mortality, and population growth rates of the European grapevine moth outside its native range, it is clear from results presented below and population growth models that the potential for population growth is staggering (Gabel and Mocko 1986; Briere and Pracros 1998; Gutierrez et al. 2012; Gilioli et at. 2016).

The damage caused by *L. botrana* larvae to floral buttons, flowers and fruit is illustrated in Fig. 1. Feeding of the pest not only affects yields, but also reduces quality. Larval feeding triggers infections by the fungus *Botrytis cinerea* Persoon and other organisms that are the gateway to diseases such as bunch rot in the grapes. Furthermore, the presence of fungal and faecal residues of larvae cause undesirable aromas and flavours in wines that can lead to technical problems in wine making.

Figure 1. Damage caused by Lobesia botrana *larvae to floral buttons, flowers and fruit, and the resulting bunch rot in grapes (credit ISCAMEN).*

3. SPREAD AND DISTRIBUTION OF EUROPEAN GRAPEVINE MOTH IN THE PROVINCE OF MENDOZA

Since the initial detection of *L. botrana* in 2010, ISCAMEN and the Servicio Nacional de Sanidad y Calidad Agroalimentaria (SENASA) have carried out monitoring activities to obtain critical incidence data necessary for the implementation of AW-IPM activities against the insect. Over 4000 pheromone-baited delta traps (E/Z 7,9 dodecadienyl acetate) are deployed in the main four oases of Mendoza Province. The traps are inspected weekly, except during the winter months when they are inspected every two weeks. ISCAMEN collects and disseminates these trapping data in the form of pest population distribution maps to support growers' pest management practices. The rapid spread of the pest in the four oasis regions until 2016 is presented in Figs. 2 a, b c.

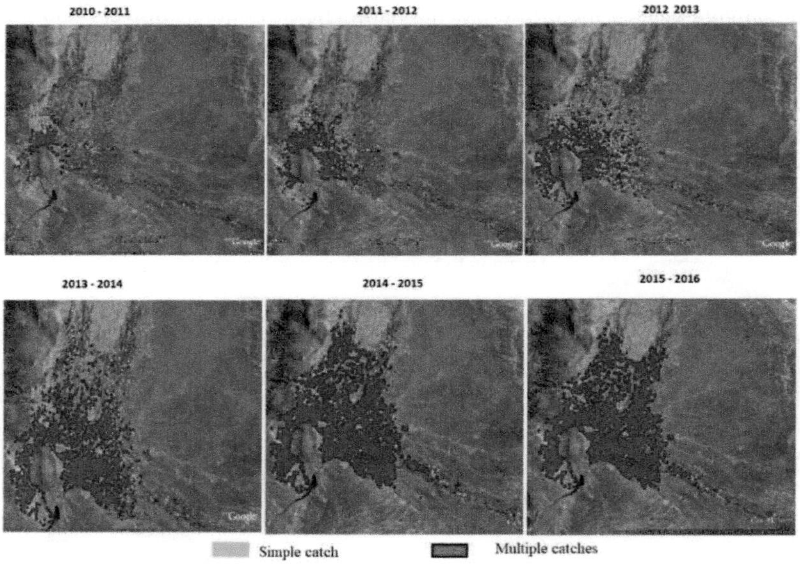

Figure 2a. The distribution of Lobesia botrana *between 2010 and 2016 based on moth catches in the Northern and Eastern Oases of Mendoza Province, Argentina (credit ISCAMEN).*

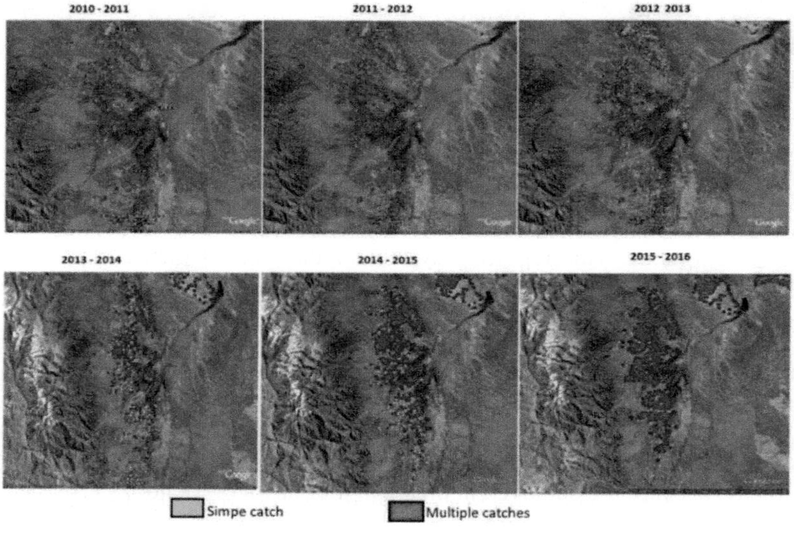

Figure 2b. The distribution of Lobesia botrana *between 2010 and 2016 based on moth catches in the Central Oasis of Mendoza Province, Argentina (credit ISCAMEN).*

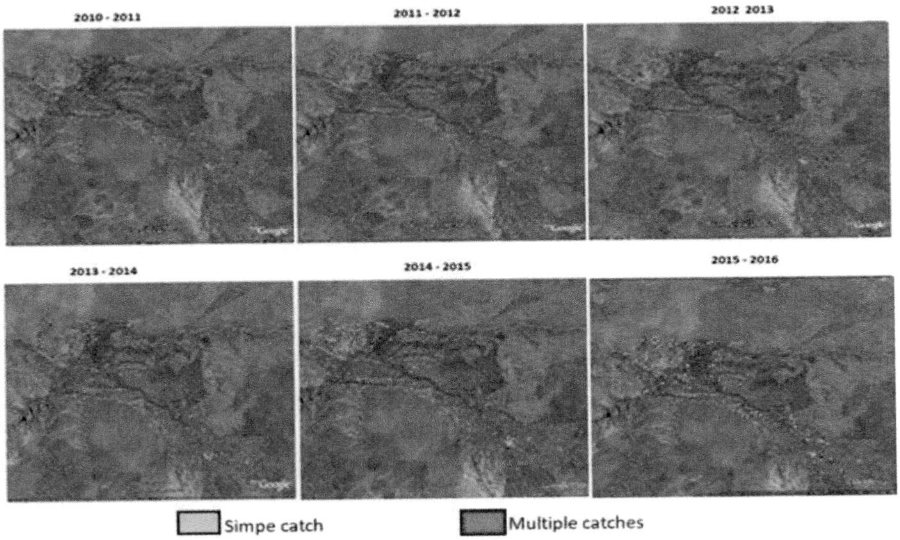

Figure 2c. The distribution of Lobesia botrana *between 2010 and 2016 based on moth catches in the Southern Oasis of Mendoza Province, Argentina (credit ISCAMEN).*

The results of this monitoring programme indicated a rapid spread and increase of the population of the European grapevine moth, which currently infests 150 000 ha of the total 160 000 ha of vineyards in the Province of Mendoza.

The most affected areas are the Northern and Eastern Oases (Fig. 2a), and to a lesser extent the Central Oasis (Fig. 2b). During the last seasons, the harvested grapes in these oases have shown variable damage. This was related to factors such as the degree of susceptibility of the different varieties, and different crop and pest management systems, among others.

The damage on floral clusters and grapes have caused yield losses ranging from 10 to 70% (based on information from growers). In some situations, the damage can be 100% at harvest depending on the environmental conditions (mainly rain and humidity) and the cultivars (Moreau et al. 2006; Xuéreb and Thiéry 2006). Compared to the other areas, the Southern Oasis still has low *L. botrana* population levels (Fig. 2c).

4. CONTROL STRATEGY FOR EUROPEAN GRAPEVINE MOTH IN MENDOZA: THE 2017-2018 SEASON

Since the European grapevine moth was detected in Argentina, the official state and federal organizations promulgated *Law 27.227*, that formalized the establishment of a Technical Committee composed of representatives from SENASA, the Instituto Nacional de Technología Agropecuaria (INTA) and the phytosanitary agencies of the wine production provinces (Mendoza, San Juan, La Rioja, Catamarca, Tucumán, Salta, Jujuy, Río Negro, Neuquén, La Pampa, and southern Buenos Aires).

The provinces where the pest is present (Mendoza, San Juan and Salta) coordinate control actions, while the provinces free of *L. botrana* focus on detection and quarantine activities (controlling the movement of trucks with grapes or containers used during harvesting, or harvest machines to avoid risk of dispersal).

The same law also established a Technical Advisory Committee (TAC) with technical members of the above-mentioned organizations, along with members of the Asociación de Cooperativas Vitivinícolas Argentinas (ACOVI), the Corporación Vitivinícola Argentina (COVIAR), the Federación de Cooperativas Vitivinícolas Argentinas (FECOVITA), the Instituto Nacional de Semillas (INASE), the Instituto Nacional de Vitivinicultura (INV), Argentina's wineries, association of table grape growers and others.

Using the results obtained in the previous seasons and the current pest status, the TAC developed a management strategy for each oasis of the province for the 2017-2018 season. This was authorized by SENASA after thorough evaluation and verification that these were compliant with the requirements of the national programme. The TAC also established protocols for monitoring, assessment of fruit damage, movement of fruit inside and outside the affected production and regulated areas, and sanitary controls within the quarantine areas.

Growers needed to register with the programme to receive assistance and the necessary control items and products. A total of 7903 growers were registered covering 11 400 properties over a total area of 132 674 ha of vineyards (Table 1). These numbers, out of the total 160 000 ha of vineyards, epitomize the severity of the *L. botrana* problem for the grape and wine industries.

Public and private entities were involved in providing training to this large number of growers that included workshops given to growers, technicians, professionals and winery owners. The training was essential to ensure the correct use of the control tools. A total of USD 20 million was invested to enable implementation of the management programme in the affected provinces.

Table 1. Number of vineyards, surface area and number of growers participating in the Lobesia botrana *management programme in the oases of Mendoza*

Oasis	Number of Vineyards	Area (ha)	Number of Growers
Northern and Eastern	7689	89 392*	4503
Central	1356	25 601	1002
Southern	2972	17 681	2398

* *Of which approximately 25 000 ha are abandoned or semi-abandoned, requiring special attention by SENASA*

5. CONTROL TOOLS USED

5.1. Mating Disruption

The decision to use mating disruption for supressing *L. botrana* in the Province of Mendoza was made because it is species-specific and there is an extremely low risk of developing resistance to the pheromone blend (although see Mochizuki et al. 2002). It does not produce insecticide residues, does not affect the beneficial fauna, and thereby limits the appearance of secondary pests. It is also compatible with other control methods and is accepted for use in organic orchards (Pérez Marín et al. 1995).

Mating disruption works by releasing synthetic versions of the sex-pheromones that female *L. botrana* use to attract con-specific males (Cardé, this volume). Pheromone dispensers release pheromone plumes that confuse the males. Though the exact mechanism(s) of mating disruption is debated, be it habituation to the pheromone, attenuation of the sensory organs, false trail following, or trail masking, it is clear that this technique reduces mating frequency and/or delays mating in such a way as to reduce overall fecundity, resulting in decreased damage to the grapes.

The semipermeable polymer of the pheromone dispensers (RAK from BASF or ISONET from Shin-Etsu) releases the pheromone mix at a constant rate, depending on the temperature and humidity, for a period of 6 months. This provides protection of the crop for 180 to 200 days, which covers the entire grape season. Pheromone dispensers must be deployed before the onset of adult activity in the season (September is the first flight of *L. botrana* in Mendoza, see Fig. 3) and they need to be evenly distributed in the vineyards at a density (typically 350 dispensers per ha).

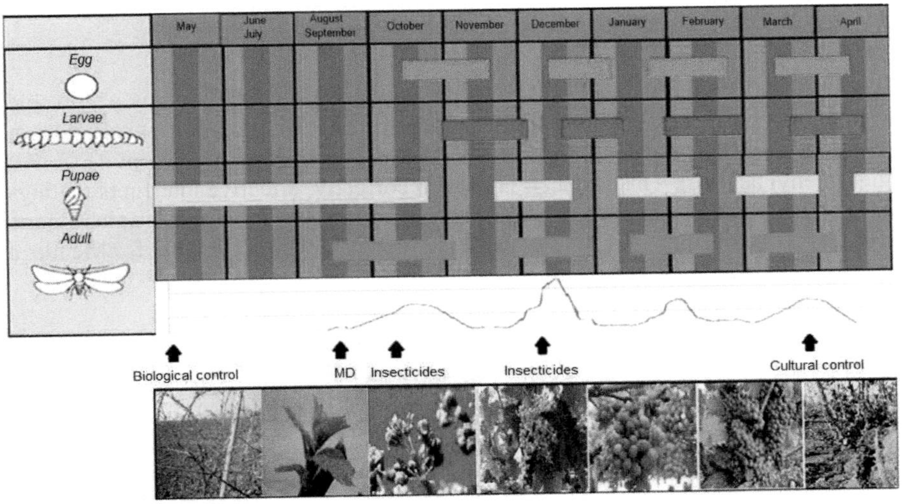

Figure 3. Scheme of a growing season illustrating the best time for applying the different control methods in relation to the biological life cycle of Lobesia botrana *and the development of the crop in Mendoza Province (MD= mating disruption) (credit ISCAMEN).*

Depending on the pest pressure, however, mating disruption may need to be supplemented with insecticide applications to control the immature stages of the insect. Mating disruption is a density-dependent management tool, being less effective when pest populations are high, or when there are untreated areas with migration of moths into nearby mating disruption-treated areas. Mating disruption is a fundamental control tactic for the European grapevine moth in all countries where it is present (Del Tío et al. 2001; Louis et al. 2002; Vassiliou 2009; Ioriatti et al. 2011). It was also one of the main components for achieving the eradication of the outbreak of this pest in northern California, USA (Mastro et al. 2010a, b, c; Varela et al. 2010; Lance et al. 2011; Lucchi et al. 2014; Simmons et al., this volume).

The Ministry of Agribusiness of the Nation procured 22 million mating disruption dispensers at a total cost of USD 9 344 500, which was enough to cover 62 857 ha. This amount was supplemented with USD 400 000 from the Province of Mendoza, which was enough to procure an additional 1 094 390 dispensers covering an area of approximately 3 126 ha. In the 2017-2018 growing season, mating disruption was therefore used on 65 983 ha of vineyards to manage populations of the European grapevine moth.

In the 2018-2019 season, the state again invested USD 7.2 million to cover 60 000 ha (120 USD/ha) with pheromone dispensers. In some areas, mating disruption was used as a single control tactic, whereas in others, it was combined with the application of specific insecticides for immature stages depending on the population levels. When populations remained below a seasonal average of 10 moths per day, only mating disruption was used. When more than 10 moths were caught, immature stages were suppressed with insecticide applications.

5.2. Aerial Applications of Pheromones

An alternative to hand-applied, polymer mating disruption dispensers is a flowable formulation that can be rapidly applied aerially. The product is a suspension of encapsulated particles containing 19.2% w/v of the active ingredient (E/Z 7,9 dodecadienyl acetate). This product, if applied correctly, effective interrupts matings for 30-60 days from the moment of its aerial application over the crop. The pheromone is applied at the beginning of the second flight, in mid to late November for Mendoza Province (Fig. 4).

Figure 4. Drops with pheromones on the crop foliage (credit ISCAMEN).

Flowable mating disruption does not last as long as polymer dispensers, however, aerial application allows for rapid, area-wide coverage, in response to building populations. This approach is also ideal to control moth populations in abandoned, semi-abandoned farms or mismanaged vineyards. These are the main refuges for this pest, where it proliferates and then emigrates to the treated orchards. No phytotoxicity symptoms were observed during its application (Fig. 4). The trapping data of the areas under treatment confirmed its effectiveness (Fig 5).

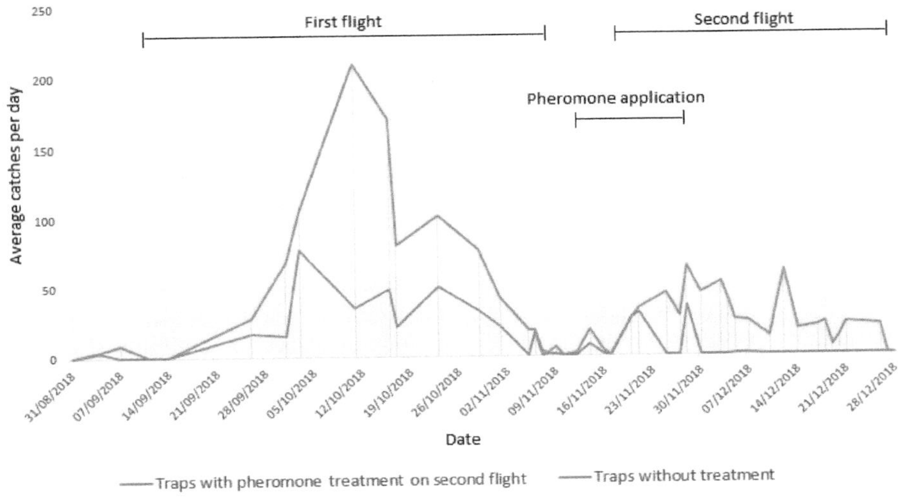

Figure 5. Effect of aerial pheromone sprays on the average catches of Lobesia botrana *per trap in areas with and without mating disruption treatment (credit ISCAMEN).*

5.3. Insecticides for the Control of Immature Stages

The invasion of European grapevine moth required the use of insecticides to control larval stages of the pest even though these historically have not been used in vineyards of the Province of Mendoza. Chemical control may be a necessary suppression tactic of agricultural pests in the framework of IPM, complementing biologically-based methods, such as biological control, mating disruption, together with mechanical and cultural controls.

In view of the characteristics of the host crop (grapes), and taking into consideration the importance of protecting the beneficial fauna in the vineyards, the proposed phytosanitary products to suppress immature *L. botrana* had to conform to the following characteristics (and hence, being compatible with an AW-IPM approach): 1) low environmental impact, 2) specificity for the control of the target pest; 3) no effect on pollinators; and 4) no risks to human health.

Aerial and ground applications of the selective insecticide Coragen® (chlorantraniliprole or rynaxpyr, an anthranilic diamide insecticide) were used over 138 000 ha of vineyards to control immature stages of the first generation of the pest.

In addition, the biological product Dipel DF® (*Bacillus thuringiensis kurstaki*) was applied by air over 68 000 ha of vineyards to control the immature stages of the first or second generation of the pest.

Finally, the selective insecticide Proclaim Forte® (emamectin benzoate, an avermectin insecticide, derived from a metabolite of the bacterium *Streptomyces avermitilis* Kim and Goodfellow) was applied from the ground over 14 500 ha of vineyards to control immature stages of the second generation. Both Coragen and Dipel DF were already registered with SENASA for aerial applications on different crops.

In the 2018/2019 growing season, the products Delegate (active ingredient spinetoram) and Intrepid SC (active ingredient methoxyfenozide) were also used over large areas as part of the control strategy (Sáenz de Cabezón et al. 2005). Growers used Delegate for ground spraying, whereas Intrepid SC was mainly used for aerial application to target the first generation of immature stages.

5.4. Aerial and Ground Application of Insecticides

5.4.1. Advantages of Aerial Application of Chemical and Biological Products

To obtain effective management of *L. botrana*, it is essential to apply the phytosanitary products during the various periods of insect susceptibility (Fig. 3). However, the growers of the affected areas are limited to ground spraying as they have no access to appropriate levels of mechanization. Since the detection of the pest in 2010, and depending on the available financial resources, SENASA assisted the growers in managing populations of *L. botrana* with insecticides of low environmental impact. Furthermore, it was essential to apply aerial sprays to the abandoned vineyards.

The aerial application of insecticides has several advantages, such as 1) to allow an area-wide strategy, covering commercial as well as abandoned or semi-abandoned orchards, 2) the control can be applied rapidly to target the most vulnerable stage of the pest, 3) the number of ground applications is significantly reduced, 4) the treatments with aircraft are traceable / auditable, as the aerial sprays are monitored and continuously registered with appropriate GPS technology, 5) the spray mixtures are prepared by a small number of highly trained professionals, rather than a large number of growers who infrequently mixed insecticides in the past, reducing the likelihood of errors in the application rate, 6) the aerial spraying is cost-effective and the application volumes and the quantity of active ingredient applied are significantly reduced per unit of surface, 7) the use of GPS technology allows the aircraft to interrupt the application over specific areas such as rural houses, and, 8) satellite imagery allows determining the criteria and application blocks at a geographic scale, as well as proper coordination of operations.

Prior to the aerial treatments, the planned activities are discussed and agreed upon with each community. The aerial operations require ground support equipment, and the phytosanitary organizations take samples to determine areas that need or don't need treatments. The data are communicated to the general public to ensure transparency. Provincial phytosanitary organizations assess population levels of the pest for each zone and the treatment products required for the aerial spraying.

A summary of the control strategy used during the 2018-2019 season, integrating the different control systems, are shown in Fig. 6. The products Coragen and Intrepid, are not certified for use in organic vineyards complying with NOP (National Organic Programme) norms established for exports to the USA. These products were not used when treating conventional orchards adjacent to NOP vineyards to avoid the possible drift of these insecticides. Mating disruption has been the focus in the control of the pest in the vineyards with organic certifications or NOP standards.

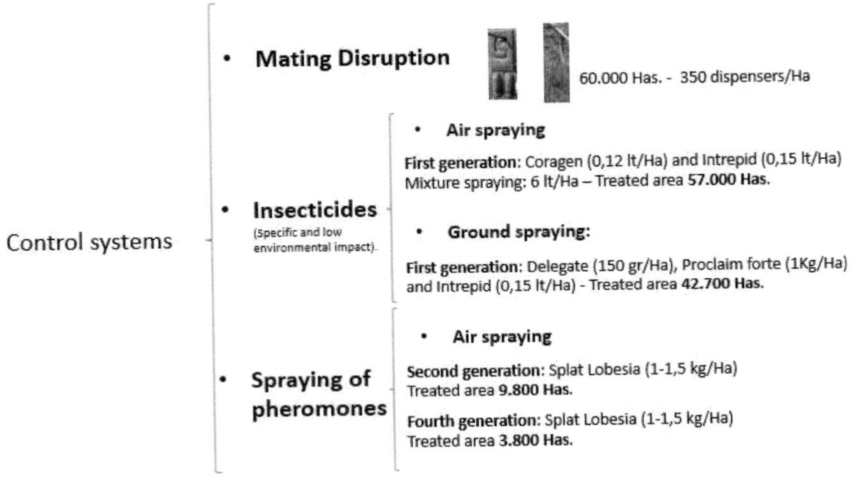

Figure 6. Strategy used for the control of Lobesia botrana *for the 2018/2019 season, in working areas based upon data from the monitoring programme.*

5.4.2. Ground Spraying of Chemicals by the Growers

Ground spraying by growers is used exceptionally and only in peri-urban areas or close to buildings where aerial applications are prohibited. For timing of interventions see Fig. 3:

- First alert: The appearance of eggs of the first generation of the pest when the floral clusters of the crop are being formed (length of 5-7 cm) signals the opportune moment to carry out the first treatment against immatures. This occurs in mid-October, depending on temperatures and the status of varieties in each area. Coragen was the product selected for controlling the first generation of the pest in ca. 23 000 ha and 42 700 ha in the 2017/2018 and 2018/2019 campaigns, respectively.

- Second alert: The timing of chemical treatments against the immatures stages of this second generation is zone-dependent and is influenced by the population dynamics of the pest, field surveys, and thermo-cumulative models. The starting date for treatment of the second generation of the pest has been around the middle of December. Proclaim Forte was the choice for the control of this generation of the pest in an area of ca. 18 500 ha during the 2017/2018 campaign.

5.5. *Cultural Control*

The main objective of cultural control is to prevent the insect completing its life cycle by removing infested grape clusters. The critical role of growers doing this post-harvest control is essential to managing this pest. Also, growers have to avoid the movement and transport of pruning wastes, as they can contain dormant stages of *L. botrana*. The implementation of these measures is supervised by the local phytosanitary organizations.

6. RESULTS OF CONTROL ACTIVITIES

6.1. *Impact of Aerial Spraying against Immatures*

The impact of aerial sprays on the pest population is shown in Fig. 7. It displays average weekly moth catches with 60 traps during the 2016/2017 season, when aerial spraying was not carried out, and during the 2017/2018 season when aerial spraying against immatures took place. The data clearly indicate a reduction in the population density of generations 2, 3 and 4. This is evidence of the benefit of an area-wide approach using aerial spraying combined with preventive control earlier in the season for this multivoltine species.

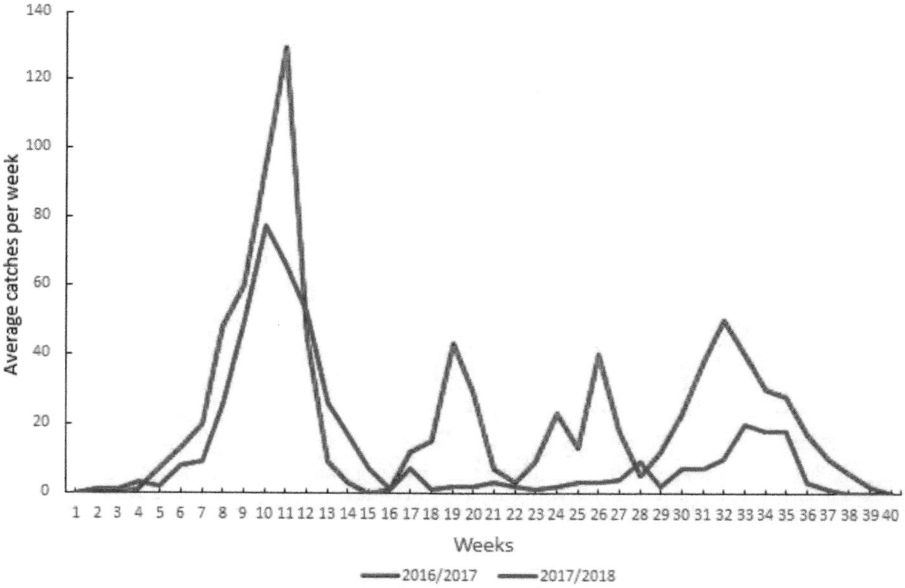

Figure 7. Average weekly catches of Lobesia botrana *with 60 traps during the 2016/2017 season (without aerial spraying) and during the 2017/2018 season (with aerial spraying against immatures in the first and second generations) in the Province of Mendoza.*

6.2. Management of Lobesia botrana *since the 2016/2017 Season*

6.2.1. Northern and Eastern Oases

At the end of the 2016-2017 season the pest was heterogeneously distributed throughout the Northern and Eastern Oases over a total area of approximately 50 000 ha (Fig. 8). Our experiences of using mating disruption as a control tactic, and those of other programmes, indicate the need for continuous treatment of the areas to achieve sustained reductions in the pest population (Cardé, this volume).

In the Northern and Eastern Oasis, mating disruption in the 2017/2018 and 2018/2019 seasons was applied over a total of almost 40 000 ha using fixed pheromone dispensers and in the 2018/2019 season over 14 000 ha with one aerial application of the pheromone. These areas were also treated with the aerial application of chemical/biological insecticides to control immatures resulting from the first flight of the insect. In the remaining 48 000 ha of the Northern-Eastern Oasis, chemical/biological products were sprayed by air and from the ground for the control of the first and second flight during the 2017-2018 and 2018-2019 seasons.

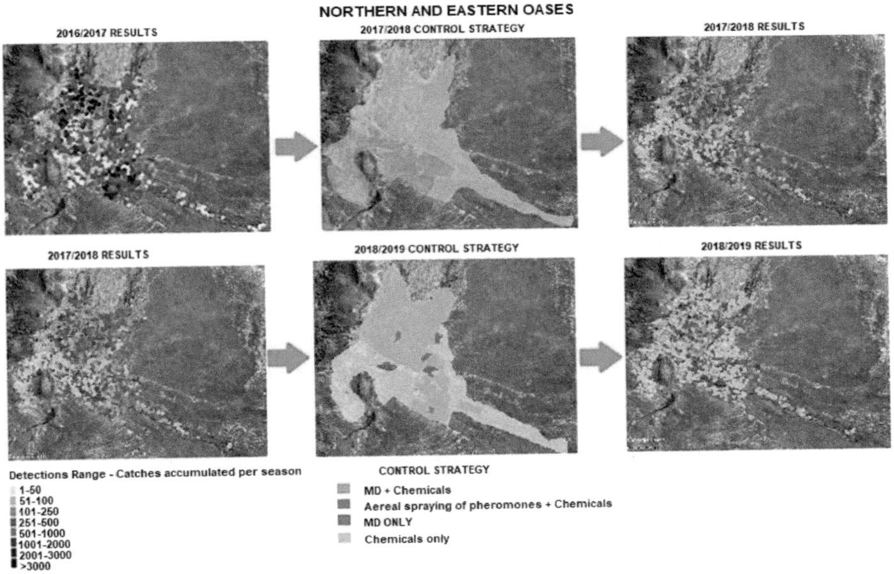

Figure 8. Distribution and levels of Lobesia botrana *populations in the Northern and Eastern Oases and control measures used in the 2016/2017, 2017/2018 and 2018/2019 seasons (credit ISCAMEN).*

6.2.2. Central Oasis

Based on the monitoring data of the 2016/2017 season, the Central Oasis showed a uniform distribution of the pest (Fig. 9). As in the northern and eastern Oases, the strategy was defined in accordance to the distribution and population density of the pest.

The Central Oasis has approximately 26 000 ha of vineyards and *L. botrana* is present throughout the entire area. During the 2017-2018 and 2018-2019 seasons different control tactics such as mating disruption and chemical control of the immature stages of the first generation were combined in areas with the highest population pressure. In the rest of the oasis, only chemical control was applied against the immature stages during the first and second flights of the pest.

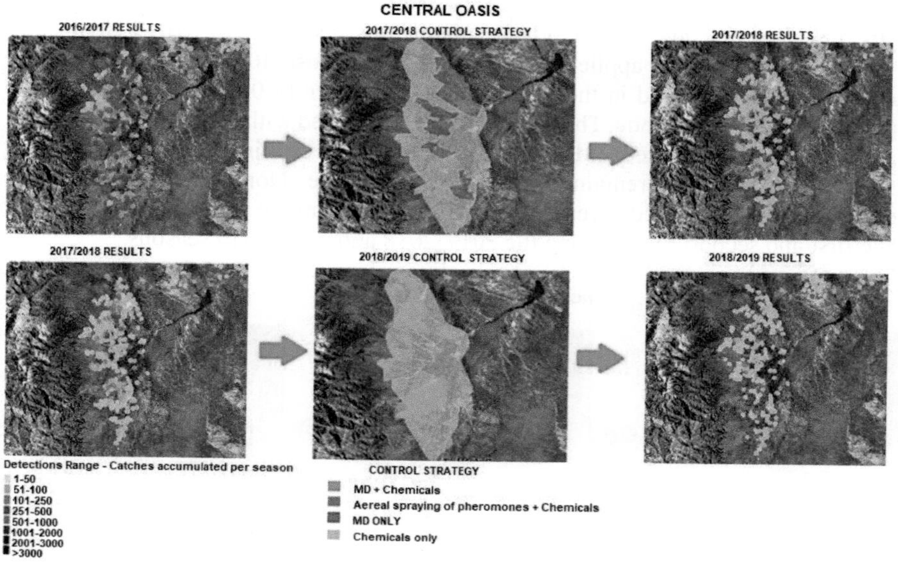

Figure 9. Distribution and levels of Lobesia botrana *populations in the Central Oasis and control measures used in the 2016/2017, 2017/2018 and 2018/2019 seasons (credit ISCAMEN).*

6.2.3. Southern Oasis

Population densities of the pest in the Southern Oasis 2016/2017 season were significantly lower than those observed in the Central and Northern-Eastern Oases (Fig. 10).

In the Southern Oasis the pest has been detected so far in 10 000 ha out of a total of 16 000 ha of vineyards. During the 2017/2018 season mating disruption was used over the entire area. During the 2018/2019 season, the densities of the pest insect were very low, and therefore it was decided to use only mating disruption in all infested areas, and to continuously monitor in the remaining zones.

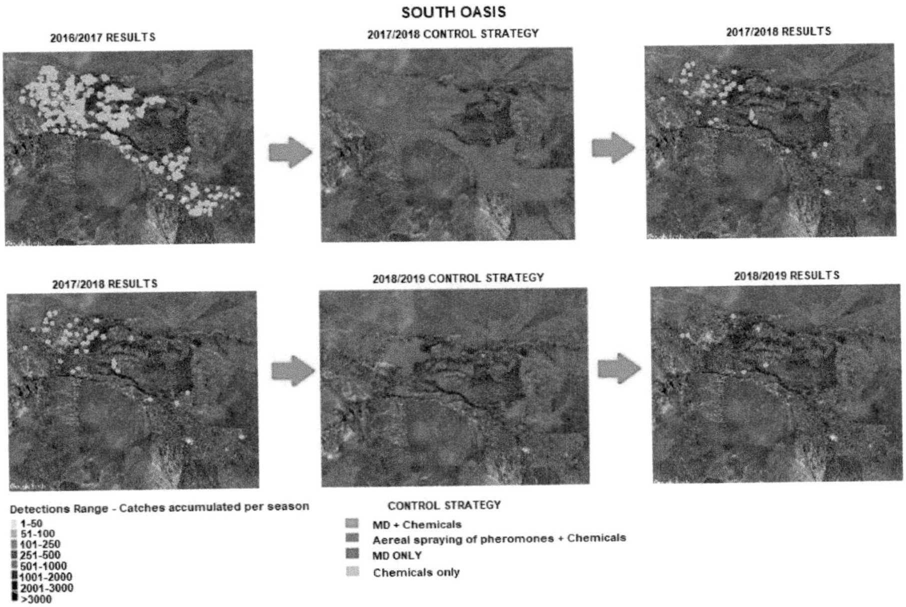

Figure 10. Distribution and levels of Lobesia botrana *populations in the Southern Oasis and control measures used in the 2016/2017, 2017/2018 and 2018/2019 seasons (credit ISCAMEN).*

6.3. Additional Activities during the 2017-2018 and 2018-2019 seasons

In the 2017/2018 and 2018/2019 seasons, mating disruption dispensers were applied as outlined above (see also Section 6.4.). However, there were several exceptions:

1. Table grapes growers who were registered in the framework of the IPM programme were also provided with mating disruption dispensers to minimize the risks of the pest dispersing to wine grape growing areas.
2. Mating disruption dispensers were provided to growers of table or organic wine grapes whose vineyards were smaller than 5 ha to ensure global coverage of mating disruption.
3. Certified organic growers (NOP certification) or growers who were transitioning from traditional to organic production were provided with mating disruption dispensers for their local deployment.

6.4. Comparison of Trap Catches in the Last Seasons

The comparison of *L. botrana* trap catches in the last nine seasons is presented in Fig. 11. According to the results obtained from the same 4200 traps kept from the first season to the end of the 2018/2019 season, an overall reduction of the target pest populations was reached. Nevertheless, there are still some foci where the actions must be intensified.

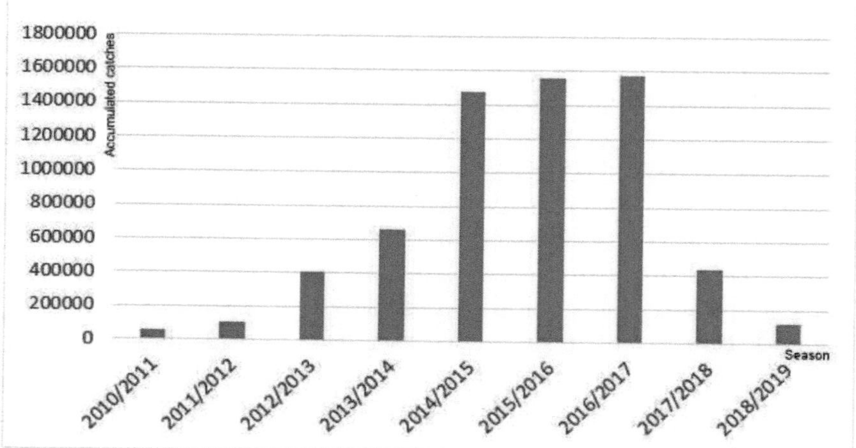

Figure 11. Comparison of total Lobesia botrana *captures per season in the Province of Mendoza from the 2010/2011 the 2018/2019 season.*

As in other insects with similar population dynamics, the control of the first generation of the insect will determine the level of pest population built-up later in the season, and therefore determines grape damage. Although these trap data are partially affected by the pheromones and other control measures applied, the reduction in the number of trap catches is an indicator of the effectiveness of the area-wide mating disruption. Some areas of the Southern Oasis begin to show discontinuity in wild moth catches even in the absence of any other kind of treatment, an indication of the potential towards achieving the final objective of eradication.

7. DEVELOPMENT OF BIOLOGICAL CONTROL AND STERILE INSECT TECHNIQUE AS ALTERNATIVE CONTROL SYSTEMS

During the last two years (starting in 2018), Argentina through ISCAMEN, has also been assessing the potential of integrating other control methods that are friendly to the environment. These are biological control and the Sterile Insect Technique (SIT) that are under development.

Biological control, through the mass-rearing and augmentative release of *Trichogramma* sp. is being developed at ISCAMEN's Pilot Experimental Unit. The efficiency of the microhymenopteran egg parasitoid was assessed on eggs of *Grapholita molesta* (Busck) eggs; however, eggs of *L. botrana* have also been parasitized.

On the other hand, research on the SIT has been initiated by developing the mass-rearing for *L. botrana.* Wild material was collected from the field and a colony established. The eggs collected from oviposition cages were evaluated in terms of egg hatch. A specific larval diet was used and evaluated in terms of the percentage recovery from eggs to adults. Further studies include the development of a system to separate mature larvae/pupae from the larval diet, determination of the irradiation dose, and methods for sterile moth packaging and release systems.

During the 2019/2020 season, sterile moths, produced at an ISCAMEN pilot facility, are being released in a 27-ha commercial vineyard. Recapture data are being collected and analysed to determine the optimal relationship between sterile males and wild *L. botrana* moths for increased SIT application.

Eventually the integration of the SIT, augmentative biological control and mating disruption could be validated as an insecticide-free IPM programme for the suppression of *L. botrana*.

8. REFERENCES

Armendáriz, I., G. Campillo, A. Pérez-Sanz, C. Capilla, J. S. Juárez, and L. Miranda. 2007. La polilla del racimo (*Lobesia botrana*) en la D.O. Arribes, años 2004 a 2006. Boletín de Sanidad Vegetal Plagas 33: 477–489.

Briere, J.-F., and P. Pracros. 1998. Comparison of temperature-dependent growth models with the development of *Lobesia botrana* (Lepidoptera: Tortricidae). Environmental Entomology 27: 94–101.

CABI. 2019. *Lobesia botrana* (European grapevine moth). Invasive Species Compendium.

Coscollá Ramón, R. 1981. Algunas consideraciones sobre la dinámica poblacional de *Lobesia botrana* Den. & Schiff. en las comarcas vitícolas valencianas. Boletín Servicios Plagas 7: 169–184.

Coscollá Ramón, R. 1998. Polillas del racimo (*Lobesia botrana* Den. & Schiff.). pp. 29–42. *In* Los parásitos de la vid, estrategias de protección razonada. Madrid, Spain.

Del Tío, R., J. L. Martínez, R. Ocete, and M. E. Ocete. 2001. Study of the relationship between sex pheromone trap catches of *Lobesia botrana* (Den. & Schiff.) (Lep., Tortricidae) and the accumulation of degree-days in Sherry vineyards (SW of Spain). Journal of Applied Entomology 125: 9–14.

Gabel, B., and V. Mocko. 1986. A functional simulation of European vine moth *Lobesia botrana* Den. & Schiff. (Lep., Torticidae) population development. Journal of Applied Entomology 101: 121–127.

Gallardo, A., R. Ocete, M. A. López, L. Maistrello, F. Ortega, A. Semedo, and F. J. Soria. 2009. Forecasting the flight activity of *Lobesia botrana* (Denis & Schiffermüller) (Lepidoptera, Torticidae) in Southwestern Spain. Journal of Applied Entomology 133: 626–632.

Gilioli, G., S. Pasquali, and E. Marchesini. 2016. A modelling framework for pest population dynamics and management: An application to the grape berry moth. Ecological Monitoring 320: 348-357.

González, R. H. 2008. Biología, desarrollo, caracterización de daños y manejo fitosanitario de la polilla europea de la vid, *Lobesia botrana* (D & S) (Lep., Tortricidae). Universidad de Chile, Dirección de Extensión, Santiago, Chile. 25 pp.

Gutierrez, A. P., L. Ponti, M. L. Cooper, G. Gilioli, J. Baumgärtner, and C. Duso. 2012. Prospective analysis of the invasive potential of the European grapevine moth *Lobesia botrana* (Den. & Schiff.) in California. Agricultural and Forest Entomology 14: 225–238.

Ioriatti, C., A. Lucchi, and L. G. Varela. 2012. Grape berry moths in Western European vineyards and their recent movement into the New World, pp. 339–359. *In* N. J. Bostanian, C. Vincent, and R. Isaacs (eds.), Arthropod Management in Vineyards: Pests, Approaches, and Future Directions. Springer Science + Business Media B.V.

Ioriatti, C., G. Anfora, M. Tasin, A. De Cristofaro, P. Witzgall, and A. Lucchi. 2011. Chemical ecology and management of *Lobesia botrana* (Lepidoptera: Tortricidae). Journal of Economic Entomology 104: 1125–1137.

Lance, D., R. T. Cardé, M. Cooper, A. Lucchi, L. Sazo, G. Simmons, R., Steinhauer, and L. G. Varela. 2011. Fourth report of the USDA APHIS International Technical Working Group for the European grapevine moth (EGVM) in California, USA. December 13-15, 2011. 9 pp.

Louis, F., A. Schmidt-Tiedemann, and K. J. Schirra. 2002. Control of *Sparganothis pilleriana* Schiff. and *Lobesia botrana* (Den. & Schiff.) in German vineyards using sex pheromone-mediated mating disruption. Bulletin IOBC/WPRS 25: 1–9.

Lucchi, A., B. Bagnoli, M. Cooper, C. Ioriatti, and L. Varela. 2014. The successful use of sex pheromones to monitor and disrupt mating of *Lobesia botrana* in California, pp. 45–48. *In* Pheromones and other semiochemicals. Bursa, Turkey, 2012. IOBC-WPRS Bulletin 99.

Maher, N. 2002. Sélection du site de ponte chez *Lobesia botrana* (Lepidoptera: Tortricidae): Influence de l'information chimique non-volatile présente sur les fruits de plantes hôtes. Thése N° 968, Université de Bordeaux 2. pp. 204.

Mastro, V., B. Bagnoli, R. T. Cardé, G. Wegner-Kiss, C. Ioriatti, U. Koch, D. Lance, A. Lucchi, G. Barrios, L. Sazo, R. Sforza, R. Steinhauer, and L. G. Varela. 2010a. First report of the USDA APHIS International Technical Working Group for the European grapevine moth (EGVM) in California, USA. February 10, 2010. 10 pp.

Mastro, V., R. T. Cardé, D. Lance, A. Lucchi, L. Sazo, R. Steinhauer, and L. G. Varela. 2010b. Second report of the USDA APHIS International Technical Working Group for the European grapevine moth in California, USA. May 14, 2010. 6 pp.

Mastro, V., R. T. Cardé, C. Ioriatti, D. Lance, A. Lucchi, L. Sazo, G. Simmons, R. Steinhauer, and L. G. Varela. 2010c. Third report of the International Technical Working Group for the European grapevine moth in California, USA. December 7, 2010. 7 pp.

Milonas, P. G., M. Savopoulou-Soultani, and D. G. Stavridis. 2001. Day-degree models for predicting the generation time and flight activity of local populations of *Lobesia botrana* (Den. & Schiff.) (Lep., Tortricidae) in Greece. Journal of Applied Entomology 125: 515–518.

Mochizuki, F., T. Fukumoto, H. Noguchi, H. Sugie, T. Morimoto, and K. Ohtani. 2002. Resistance to a mating disruptant composed of (Z)-11-tetradecenyl acetate in the smaller tea tortrix, *Adoxophyes honmai* (Yasuda) (Lepidoptera: Tortricidae). Applied Entomology and Zoology 37 (2): 299–304.

Moreau, J., B. Benrey, and D. Thiéry. 2006. Grape variety affects larval performance and also female reproductive performance of the European grapevine moth *Lobesia botrana* (Lepidoptera: Tortricidae). Bulletin Entomological Research 96: 205–212.

Pérez Marín, J. L., C. Ortega Sáenz, E. Palacios Ruiz, and C. Gil-Albarellos Marcos. 1995. Un nuevo método de control de la polilla del racimo de la vid: la confusión sexual. Boletín de Sanidad Vegetal - Plagas 21: 627–640.

(SAG) Servicio Agrícola y Ganadero. 2010. Modifica resolución N° 2.109, de 2008, que declara control obligatorio de la polilla del racimo de la vid (*Lobesia botrana).* Ministerio de Agricultura, Santiago, Chile.

Sáenz de Cabezón, F., V. Maron, F. Zalom, and I. Pérez-Moreno. 2005. Effects of methoxyfenozide on *Lobesia botrana* (Den. & Schiff.) (Lepidoptera: Torticidae) egg, larval and adult stages. Pest Management Science 61: 1133–1137.

(SENASA) Servicio Nacional de Sanidad y Calidad Agroalimentaria. 2015. El SENASA declaró la emergencia fitosanitaria por la detección de la plaga *Lobesia botrana*. Fecha de consulta: Septiembre de 2015.

Thiéry, D. 2008. Les Tordeuses nuisibles à la vigne, pp. 214-246. *In* Les ravageurs de la vigne. Second Edition. Féret, Bordeaux, France.

Torres-Vila, L. M., J. Stockel, R. Roehrich, and M. C. Rodríguez-Molina. 1997. The relation between dispersal and survival of *Lobesia botrana* larvae and their density in vine inflorescences. Entomologia Experimentalis et Applicata 84: 109–114.

Varela, L. G., R. J. Smith, M. L. Cooper, and R. W. Hoenisch. 2010. European grapevine moth, *Lobesia botrana*, in Napa Valley vineyards. Practical Winery & Vineyard. 30: 1–5.

Vassiliou, V. A. 2009. Control of *Lobesia botrana* (Lepidoptera: Tortricidae) in vineyards in Cyprus using the mating disruption technique. Crop Protection 28: 145–150.

Xuéreb, A, and D. Thiéry. 2006. Does natural larval parasitism of *Lobesia botrana* (Lepidoptera: Torticidae) vary between years, generation, density of the host and vine cultrivar? Bulletin of Entomological Research 96: 105–110.

Zangheri, S., G. Briolini, P. Cravedi, C. Duso, F. Molinari, and E. Pasqualini. 1992. *Lobesia botrana* (Denis & Schiffermüller), pp. 85–88. *In* Lepidotteri dei fruttiferi e della vite. Milan, Italy.

SECTION 4

REGULATORY AND SOCIO-ECONOMIC ISSUES

AREA-WIDE MANAGEMENT OF RICE PLANTHOPPER PESTS IN ASIA THROUGH INTEGRATION OF ECOLOGICAL ENGINEERING AND COMMUNICATION STRATEGIES

K. L. HEONG[1,6], Z. R. ZHU[1], Z. X. LU[2], M. ESCALADA[3], H. V. CHIEN[4], L. Q. CUONG[5] AND J. CHENG[1]

[1]*Institute of Insect Sciences, Zhejiang University, Zijingang, Hangzhou, China; klheong@yahoo.com*
[2]*Institute of Plant Protection and Microbiology, Zhejiang Academy of Agricultural Sciences (ZAAS), Hangzhou, China*
[3]*Visayas State University, Baybay, Leyte, Philippines*
[4]*Plant Protection Department, Mekong University, Vinh Long, Viet Nam*
[5]*Southern Regional Plant Protection Center, Long Dinh, Tien Giang Viet Nam*
[6]*Formerly at the International Rice Research Institute (IRRI), Los Baños, Philippines*

SUMMARY

Most rice insect pests are exogenous immigrants from either long distances or neighbouring areas. For their management to be economical and sustainable, an area-wide perspective is imperative. Key pests, like the planthoppers and stem borers, are highly dependent on rice for survival and reproduction. They multiply and move from one rice crop to another, sometimes carrying virus diseases such as ragged stunt, grassy stunt and rice stripe from source areas. The planthoppers are r strategists, unable to overwinter in northern China, Japan and Korea, and are known to "migrate" or are displaced by wind from southern China to temperate regions of China, Japan and the Korean peninsula. With adequate faunal biodiversity and biological control ecosystem services in a rice crop, immigrant pests have low chances of survival and growth capacities, and often remain a minor pest. However, when the local ecosystem services are compromised, often by unnecessary insecticide use or extreme weather conditions, such as droughts or floods, the immigrants show high survival and growth rates. Since, 2008 the Rice Bowl of Thailand suffered brown planthopper (BPH) (*Nilaparvata lugens* Stål) outbreaks for 14 consecutive rice seasons that caused

J. Hendrichs, R. Pereira and M. J. B. Vreysen (eds.), Area-Wide Integrated Pest Management: Development and Field Application, pp. 617–631. CRC Press, Boca Raton, Florida, USA.

losses of more than USD 200 million. Farmers were routinely applying insecticides as prophylactics and the BPH consequently "escaped" its natural control and populations increased 100 000-fold. Ecological engineering approaches involve practices that will build and restore biodiversity and ecosystem services, and reduce insecticide-induced threats to ecosystem services. An area-wide increase in floral biodiversity in the crop landscape provides shelter, nectar, alternate hosts and pollen (abbreviated as SNAP by Professor Wratten) to conserve the natural enemy fauna. Pioneered in Jin Hua, China with sesame plants grown on the rice bunds, ecological engineering is now practiced in China, the Philippines, Thailand, and Viet Nam, using several flower species. A multi-country, multi-year field trial conducted by scientists of the International Rice Research Institute (IRRI) in collaboration with researchers from Australia, China, Thailand and Viet Nam showed that the growing of flowers on rice bunds as an ecological engineering practice increased profits (by 7.5%), yields (by 5%), biological control (by 45%) and added aesthetic values to the rural landscape. At the same time the ecological engineering practice decreased insecticide use (by 70%), pest densities (by 30%) and farmers' chemical input costs (by 70%). Farmers are adopters and implementers of ecological engineering practices, and to reach and motivate the millions of farmers in Viet Nam, two TV serials developed using entertainment-education principles were launched to promote the establishment of flower strips and to reduce insecticide applications. The TV serials helped farmers to "see" and appreciate the role of parasitoids by linking these (termed locally as "small bees") to widely-known bees. Farmers that viewed the serials decreased their insecticide use by 24%, had 3.3% higher yields, increased their appreciation of parasitoids and gained positive attitudes towards the establishment of flower strips. To achieve sustainable area-wide pest management, ecological engineering practices have to be coupled with rational pesticide management through better pesticide policies, regulations and implementation, accurate pest diagnostics and timely professional advice to farmers. Aside from its proven impacts on pest control and more profitable farming, increasing biodiversity and ecosystem services in rice fields can also contribute towards climate change adaptation and a more resilient environment.

Key Words: Nilaparvata lugens, brown planthopper, rice, rice insect pests, ecosystem services, migration, mass-media, entertainment-education, habitat manipulation, biological control, natural enemies, parasitoids, predators, China, Philippines, South Korea, Thailand, Viet Nam

1. INTRODUCTION

Rice is the staple food for more than 3000 million people, grown on 159 million hectares (ha) in most Asian countries (IRRI 2013). It is often believed that insect pests are major constraints as there are hundreds of herbivore species that can potentially attack various parts of the rice plant (Heinrichs 1994). However, in reality only a few species are key pests that can cause economically-significant yield loss and among these, most species are either monophagous or oligophagous, and are highly dependent on rice for survival (Way and Heong 1994).

Besides herbivores, there are hundreds of predator, parasitoid and detritivore species that are beneficial to the rice ecosystem (Heong et al. 1991). The most destructive pests are the planthoppers, which are *r* strategists capable of multiplying rapidly. They damage crops directly by sucking, causing a symptom known as "hopperburn" and by transmitting virus diseases (Heong and Hardy 2009). Unable to overwinter and strictly monophagous, the planthoppers migrate or are displaced by winds over long distances from maturing fields to new rice crops, often from northern regions of Viet Nam and southern China to central and northern China, Japan and the Korean peninsula (Watanabe et al. 2009).

The biodiversity of predators and parasitoids present in rice ecosystems is usually adequate to limit the growth of these migratory pests. However, when the faunal biodiversity that provides the biological control ecosystem services are compromised and the immigrant populations are abnormally high, they can reach outbreak proportions.

Since 2008 the Rice Bowl of Thailand suffered brown planthopper (BPH) (*Nilaparvata lugens* Stål) outbreaks during 14 consecutive rice seasons, causing losses of more than USD 200 million (Heong et al. 2015a). Thailand's rice farmers had been routinely applying insecticides as prophylactics, primarily cypermethrin and abamectin, thereby killing natural enemies and as a result the BPH "escaped" its natural control and populations increased unchecked by more than 100 000-fold in just 2 months. At the same time, similar BPH outbreaks were also reported in Java, Indonesia caused by similar insecticide misuse (Fox 2014).

1.1. Biological Control Ecosystem Services

Ecosystem services are benefits that people obtain from ecosystems (MEA 2005). Biological control ecosystem services are among the regulatory services derived from predation and parasitism activities, delivered by a diverse and abundant complex of predator and parasitoid species present in the agroecosystem.

Tropical rice ecosystems are richly endowed with these predatory and parasitoid species that play significant roles in keeping the number of pest species low (Way and Heong 1994). As a consequence, the pest species rarely cause economically-significant damages, and even in intensified rice crops, insecticides are not regularly needed (FAO 2011).

1.2. Are There Productivity Gains from Farmers' Insecticide Usage?

Researchers have been questioning whether farmers' insecticide use, as promoted in the 1970s and 1980s under the Green Revolution, have had any productivity gains in rice (Heong et al. 2015c). Insecticides packaged together with fertilizer have been applied as prophylactics based on calendar schedules, and this practice has remained entrenched in most of Asian countries (Heong and Escalada 1997; Escalada et al. 2009).

Typically rice farmers would conduct overhead sprays in the early crop stages, using locally-made spray equipment with poor spray delivery. Most of these sprays do not impact the target pests and often roll off the rice plants into the water. One study found that more than 80% of rice farmers' sprays were found to be misuses of insecticides that resulted in no yield gain (Heong et al. 1995).

A replicated field plot experiment conducted by economists to assess cost-benefits from different pest management strategies found that fields with zero insecticide sprays offered the best option, while net losses increased with the number of insecticide sprays (Table 1) (Pingali et al. 1997).

Table 1. Comparison of insecticide spray strategies of farmers in 2 sites in the Philippines (after Pingali et al. 1997)

Sites	Management strategies	Number of sprays	Percent net benefits over "no spray"
Laguna	Complete protection	6	-11.7%
	Farmer's strategy	2	- 3.6%
	Integrated Pest Management	1	- 5.0%
	No spray	0	- 0.0%
Nueva Ecija	Complete protection	6	- 4.7%
	Farmer's strategy	2	- 3.1%
	Integrated Pest Management	1	- 3.5%
	No spray	0	- 0.0%

Moreover, yield-insecticide application relationships from 8 farm survey data-sets showed that average yields of farms, where no insecticides were used, did not differ from those with 3 or more sprays (Heong et al. 2015b).

In farmer participatory experiments, where farmers divided their fields into 2 plots, i.e. one that would not receive any sprays in the first 40 days after sowing and the other with normal practices, yields of the plots with reduced insecticide sprays were slightly higher (Huan et al. 2005). The prophylactic insecticide sprays, especially in the early crop stages, tended to destroy the biological control ecosystem services (Heong 2009) and increased subsequently crop vulnerability to planthopper outbreaks by about ten-fold (Heong et al. 2015b).

1.3. Ecological Engineering in Rice

Pioneered in China (Lu et al. 2015), ecological engineering in rice entails the promotion of insecticide reductions and the establishment of (nectar-rich) flowering plants on the bunds and field margins, with the ultimate goal to restore biodiversity and ecosystem services (Fig. 1).

The flora on the bunds provides shelter, nectar, alternate hosts and pollen (abbreviated SNAP) to conserve the natural enemy fauna and associated biological control services (Gurr et al. 2012). For instance, parasitoids of planthoppers live on alternative hosts on the bunds (see review by Gurr et al. 2010), crickets that are predators of pest eggs breed in bund habitats with the grass *Bracharia mutica* (Forssk.) Stapf and forage in rice fields at night (Kraker et al. 1999), and spiders also use such habitats for shelter and breeding. Coupled with withholding insecticide sprays in the early crop stages, biological control services are further enhanced through ecological engineering.

A multi-country and multi-season replicated field experiment in China, Thailand and Viet Nam showed that rice fields with flower strips as an ecological engineering practice required less insecticides (by 70%), had increased yields (by 5%) and profits (by 7.5%). In addition, the fields had increased biological control (by 45%), and lower pest abundances (by 30%) (Gurr et al. 2016).

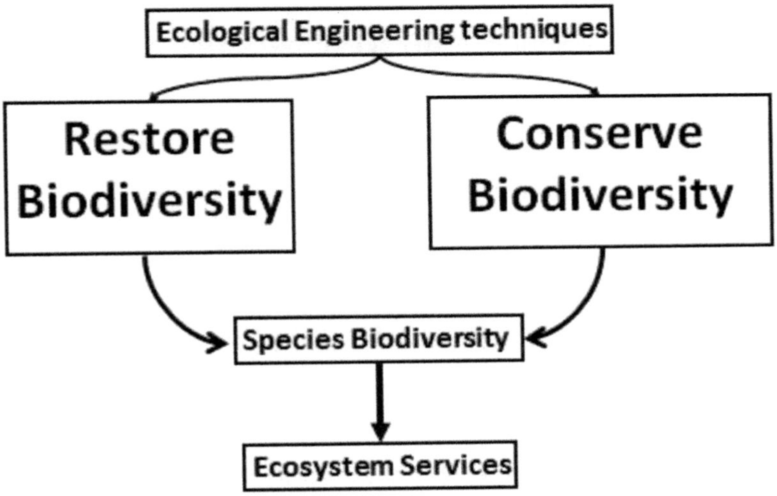

Figure 1. Ecological engineering techniques both restore and conserve biodiversity and ecosystem services (modified from Heong et al. 2014).

Another technique is to grow a trap-plant like vetiver grass (*Vetiveria zizanioides* (L.) Nash) on bunds before crop establishment. The grass will attract striped rice stemborer (*Chilo suppressalis* (Walker) females to lay eggs on the leaves of vertiver grass, but the larvae will not survive on them (Lu et al. 2017; Zheng et al. 2017).

In rice fields in Indonesia, Ives and Settle (1997) also recorded increases in natural enemy populations in asynchronous planting, where early-arriving generalist predators decimate incipient infestations of pests and suppress their populations.

2. AREA-WIDE IMPLEMENTATIONS OF ECOLOGICAL ENGINEERING PRACTICES

Most rice pests and especially the planthoppers are highly mobile, primarily monophagous, and the adults invade new rice crops from other fields. In order to manage such pests sustainably, control strategies need area-wide implementation. In Southeast Asia the size of most farms is small (less than 2 ha) and so a typical rice growing area, like the Mekong Delta, would easily be managed by about a million decision-makers.

To successfully implement ecological engineering practices in such large areas, there is a need to understand farmer decisions in order to help design communication strategies that can reach and motivate all of these decision-makers.

2.1. Understanding Farmer Decisions

In a series of social psychology studies to understand rice farmers' pest management decision-making, a psychometric model was developed (Escalada et al. 2006; Escalada and Heong 2012). In making resource management decisions, farmers are always faced with uncertainty, limited time and knowledge, and like most people they use the "satisficing" strategy rather than making decisions that would be maximizing outcomes (Simon 1982).

Farmers generally use "heuristics" (or rules of thumb) under conditions of limited time, knowledge and computational capacities. However, heuristics that farmers have developed through experience and guesswork about possible outcomes might have inherent faults and biases. Research to understand farmers' heuristics and their reasoning can help in developing communication strategies to frame alternative heuristics that can improve their decisions.

2.2. Designing Communication Strategies

Two mass-media based communication strategies, found to be effective in promoting pest management practices, were used to reach farmers and motivate their adoption of ecological engineering practices: multi-media campaigns and entertainment-education series on TV.

2.2.1. Multi-media Campaigns

Mass-media campaigns to motivate rice farmers to change practices have been highly successful in Viet Nam. The campaign to reduce early-season spraying has helped in easing farmers' loss aversion attitudes and consequently in lowering insecticide sprays by 53% (Heong et al. 1998; Escalada et al. 1999).

In a follow-up campaign, advocating farmers to reduce insecticide sprays, seed and fertilizer rates (locally named "Three Reductions"), farmers lowered their seed and fertilizer use by 10% and 7% respectively and their insecticide sprays by 33% (Huan et al. 2008).

In each of the campaigns a multi-stakeholder participatory planning and review process involving research, extension, mass-media, universities and local governments was used to develop materials and plan further media strategies (Heong et al. 2010). Prototype posters and leaflets for the ecological engineering campaigns were developed using the same process and were pre-tested before they were mass-produced and distributed. Fig. 2 shows examples of the posters used in Viet Nam and China to promote ecological engineering practices.

Figure 2. Posters developed for distribution to promote ecological engineering techniques in Tien Giang Province, Viet Nam (left) and San Men, China (right) (credits K. L. Heong and Z. R. Zhu).

In a post-campaign survey, conducted 8 months after the campaign launch in Viet Nam, significant changes in farmers' key beliefs related to ecological engineering were recorded (Table 2).

2.2.2. Entertainment-education on TV

Entertainment-education is the process of designing and implementing a programme to both entertain and educate so as to increase target audience members' knowledge, create favourable attitudes, shift norms and change behaviour (Singhal and Rogers 1999). A radio soap opera using this process to motivate change in farmers' pest management resulted in a reduction in insecticide sprays by 60% among the listeners (Heong et al. 2008).

Table 2. Changes in key beliefs related to ecological engineering before and after the mass-media campaign launch in Tien Giang Province, Viet Nam (after Heong et al. 2014)

Key belief items in Ecological Engineering	Percent farmers believing to be "always true"		Percent change
	Pre	Post	
Flowers on bunds attract bees and parasitoids	45.9	56.3	+22.7
Flowers on bunds are homes for spiders and natural enemies	31.1	43.9	+41.2
Flowers on bunds help us reduce insecticides	39.8	50.3	+26.4
Flowers on bunds help reduce BPH outbreaks	27.4	41.1	+50.0

To upscale the adoption of ecological engineering practices, a TV serial of 40 episodes, broadcast on Vinh Long TV twice a week was launched (Heong et al. 2014). Of the 593 farmers interviewed after the broadcast, about 41% had not watched any of the TV serial episodes (non-viewers). Of those who had watched the TV serial, most farmers (60%) watched five or fewer episodes and only 7% had watched more than fifteen episodes. Farmers who had viewed the TV serial sprayed significantly less insecticides (19% less), used less nitrogen fertilizer (6% less), and used lower seed rates (12% less). The yields of viewers were marginally higher than those of non-viewers (difference of 0.2 t/ha). Farmers who had watched the TV serial could recall what they learned from the series.

Table 3 shows the most common lessons farmers cited. There were significantly more farmers believing in statements that favour ecological engineering among the viewers than the non-viewers (Table 4).

3. CONSTRAINTS TO AREA-WIDE IMPLEMENTATION OF ECOLOGICAL ENGINEERING

While numerous research studies have demonstrated various successful methods and approaches, including ecological engineering, for reducing insecticide use in rice production, most of these have not been sustainable or adopted at an area-wide scale. Insecticide use in rice has continued to escalate despite the research showing that there are few or no productivity gains (Heong et al. 2015c) and that insecticides induce outbreaks of secondary pests like the planthoppers, that are more damaging (Bottrell and Schoenly 2012; Heong and Schoenly 1998).

The well-funded Farmer Field School (FFS) approach of the Food and Agriculture Organization of the United Nations (FAO), that trained millions of rice farmers in integrated pest management (IPM) (Matteson 2000), had indeed some successes. Graduates of FFS benefited from a statistically significant gain in knowledge of better pest management. However, such knowledge did not diffuse to other members of their villages (Feder et al. 2004). When the programme ended in Indonesia by the late 1990s the gains were rapidly lost (Thorburn 2013), insecticide use once again escalated, and insecticide imports increased by more than 5000% (Heong et al. 2015b).

Table 3. TV serial viewers' recalls of what they had learned from it (Heong et al. 2014)

What farmers learned from the TV serial	Percent farmers recalling*
Nectar flowers can attract natural enemies to help pest control	29.9
Nectar flowers can help reduce insecticide use	14.8
Flowers on the bunds can help protect the environment	11.3
Flowers help the beautify the rural landscape	10.1
Applying "3 reductions" methods	13.3
If insecticides are to be used, apply them correctly	11.9
Techniques in flower growing	9.6

* *Multiple responses*

Table 4. Key beliefs related to ecological engineering practices and comparison between viewers and non-viewers of the TV serial in percent of farmers who said that the statements were "always true" (Heong et al. 2014)

Belief statements	Percent farmers who believed the statement to be "always true"		χ^2
	Viewers	Non-viewers	
Flowers on bunds can attract bees and parasitoids to protect rice	32.2	21.1	25.7**
Flowers on bunds are homes for spiders and other predators	35.9	21.5	30.6**
Flowers on bunds help farmers reduce insecticide use	37.6	21.1	24.8**
Flowers on bunds can help reduce planthopper pest outbreaks	30.8	19.8	13.2*
Flowers on bunds make rice landscapes beautiful	68.3	55.4	19.6**

*χ^2 Chi Squared: * significant at 0.05%; ** significant at 0.01%*

Besides developing innovative pest management methods, the major constraints to adoption that need to be addressed are the weak regulations regarding insecticide marketing and the shortage of extension and field staff with knowledge and pest diagnostic capacities.

3.1. Pesticide Marketing in Asia

Using the "driving forces, pressures, states, impacts, responses" (DPSIR) analytical framework, Spangenberg et al. (2015) highlighted that the weak regulation of pesticide marketing is the root cause for the unsustainable implementation of ecologically-based practices. In Asia, the use of insecticides in rice is not driven by pest pressures but attributed mainly to the aggressive marketing strategies of insecticide companies, similar to those used to promote fast-moving consumer goods (FMCGs) (Heong et al. 2015b). Pesticides are readily available in the thousands of small retail shops across the countryside, where unlicensed shop owners simultaneously sell household items, provide credit and act as pest control advisers. Insecticides being sold as FMCGs are not consistent with the principles of IPM or rational insecticide use.

Such practices are rare in countries where pesticide use is well regulated, like Australia and Europe, and such practices are in complete violation of the FAO-WHO International Code of Conduct on Pesticide Management (FAO/WHO 2014), formally endorsed by most Member States of the United Nations and the pesticide industry.

Furthermore, insecticide marketing is routinely driven by attractive product packaging, deceptive brand names, such as "Admire", "Macho", "Fastac", "Venom" and "Warhawk", and sales incentives, whereas IPM requires knowledge-based choices and rational decision-making. Calendar-based applications are favoured through insecticide marketing, which appeal to farmers as their use is not based on an in-depth knowledge of the pest or the ecosystem, whereas IPM requires a sound understanding of the local environment, the agricultural ecosystems, the biological cycles of the pests, the relations of natural enemies and ecosystems, and their services in pest control. IPM promotes the need for a scientific rationale based on technical knowledge of ecosystems to guide insecticide use, while insecticide marketing strategies focus on mass-scale consumer use to maximize profits.

Pesticide sale agents often enjoy handsome incentives, such as cash, household equipment, holiday trips and even sponsored trips to Mecca to perform the Haj, based on sales volume. As a result, in most Asian countries, pesticide sale agents far outnumber government officials trying to promote (more sustainable) pest management alternatives. For instance, in Thailand there are about 200 government extension officers, while the pesticide industry employs more than 35 000 sales agents.

Some extension agents might also earn extra cash from chemical companies by promoting the use of their insecticides. For instance, agricultural extension agents in China generate most of their salaries and office operating costs through pesticide sales (Hamburger 2002). In Viet Nam, extension staff often earn extra money by selling inputs to farmers and thus tend to bias the information they provide to farmers (McCann 2005).

3.2. Acute Shortage of Pest Diagnostic Capacities and Professional Support to Farmers

The shortage of skills among plant protection technician is another major constraint for the smooth implementation and further up-scaling of ecological engineering tactics. In most Asian countries, government budget cuts and a decreased emphasis on agriculture in college education has led to drastic reductions in the pest and disease diagnostic capacities of younger extension staff. Many assigned to plant protection duties are often incapable of recognizing pests and diseases or their symptoms, and thus are unable to advice farmers. Wrong information and advice can lead to increased unnecessary insecticide spraying. They also have poor knowledge of insecticides, their modes of action, application details and ecological methods.

More attention needs to be given on the development of interactive tools for extension workers who want to learn and diagnose pest, disease, and other problems that occur in rice, and how to manage them, such as the Rice Doctor (2019) and several smart phone apps now available.

4. CONCLUDING REMARKS

Most key insect pests of rice are monophagous or oligophagous, and thus highly dependent on rice for survival. They are *r* strategists breeding in rice fields and migrating to invade new fields every season. The level of infestation in newly-established crops depends on the abundance of the immigrant populations and the generation survival rates.

In most cases rice farmers' insecticide applications have low and often negative productivity gains (Pingali et al. 1997; Heong et al. 2015c). Insecticide use, especially in the early crop stages, often used prophylactically or targeting leaf feeders, is unnecessary, has detrimental effects on ecosystem services, and promotes abnormal population growth and outbreaks of planthoppers (Heong 2009). The resulting high pest populations can readily invade neighbouring fields, where no insecticides are used, thus overwhelming resident natural enemy populations and inflicting huge damage and yield losses. In order to sustainably manage pest problems in rice production systems, area-wide perspectives and strategies are required.

The ecological engineering approach can be effectively employed to build or restore ecosystem services in rice production systems. This approach involves three ecological strategies to improve pest suppression (Gurr et al. 2012). The first is to minimize mortality of beneficial arthropods by reducing insecticide use, especially in the early crop stages. The second is to support the build-up of predators in the early crop stages by providing alternative food sources such as non-pest prey items, like chironomids. Avoiding insecticide use in the first 40 days after sowing (Way and Heong 1994) will also enhance this second strategy. The third is to enhance hymenopteran parasitoids through habitat manipulation, like growing nectar-rich flowering plants on the bunds. Bund flowers provide nectar for food, thus benefiting a huge diversity of parasitoids and other natural enemies (Gurr et al. 2010). Egg parasitoids of planthoppers, for instance, increase their attack capacities when they feed on sesame flowers (Zhu et al. 2013).

There are several ways to encourage farmers to adopt ecological engineering and to conserve biodiversity and ecosystem services through insecticide reduction. In light of the shortage of extension staff with pest diagnostic skills and the need for an area-wide adoption of pest management practices, communication strategies, such as multi-media campaigns and entertainment-education can be usefully employed to motivate farmers (Escalada and Heong 2012). Such strategic use of mass-media is relatively cheap, is able to reach tens of thousands of farmers at limited cost, in a short time and with frequent repetitions, and the resulting changes in behaviour can thus become sustainable. In addition, for area-wide pest management to be sustainable, pesticide management policies, especially those related to marketing, need to be in place, properly implemented and duly enforced.

As long as pesticide marketing tactics are not in compliance with the FAO-WHO Code of Conduct, and pesticides continue to be sold as FMCGs with numerous trade names in unlicensed retail outlets, it will be exceptionally challenging for non-pesticide management options, like ecological engineering or IPM, to be adopted sustainably.

Besides revising and enforcing pesticide marketing regulations, the enactment of environment-friendly laws will create the platform for ecological engineering techniques. In South Korea, for instance, the Environmentally Friendly Agriculture Promotion Act (EFA) was enacted and implemented in 1998 (Kim and Lim 2015). Under this Act, there were shifts in subsidies from chemical inputs to environment-friendly alternatives, like growing other crops or plants. As a result, there was a transformation of the rice production landscape in South Korea with farmers growing sesame, soybeans and flowering plants around their rice crops. In addition to benefiting from the subsidy shifts, farmers were also able to market products from other crops, especially sesame and soybeans. The government also established the EFA Department with staff to implement the Act and provide assistance, such as seeds of other crops.

5. ACKNOWLEDGEMENTS

The authors would like to thank the International Rice Research Institute (IRRI) where most of the work was carried out, and the Asian Development Bank (ADB) RETA 6489, and the Swiss Development Agency (SDC) for providing the funds for the research and networking activities, as well as the two anonymous reviewers for providing us with valuable suggestions.

6. REFERENCES

Bottrell, D. G., and K. G. Schoenly. 2012. Resurrecting the ghost of green revolutions past: The brown planthopper as a recurring threat to high-yielding rice production in tropical Asia. Journal of Asia-Pacific Entomology 15: 122–140.

Escalada, M. M. and K. L. Heong. 2012. Using farmer surveys and sociological tools to facilitate adoption of biodiversity-based pest management strategies, pp. 199–213. *In* G. M. Gurr, S. D. Wratten, W. E. Snyder, and D. M. Y. Read (eds.), Biodiversity and insect pests: Key issues for sustainable management. John Wiley & Sons, Ltd., UK.

Escalada, M. M., K. L. Heong, N. H. Huan, and V. Mai. 1999. Communications and behavior change in rice farmers' pest management: The case of using mass media in Vietnam. Journal of Applied Communications 83: 7–26.

Escalada, M. M., K. L. Heong, V. Sengsoulivong, and J. C. Schiller. 2006. Determinants of insecticide use decisions of lowland rice farmers in Laos, pp. 283–290. *In* J. C. Schiller, M. B. Champhengxay, B. Linguist, and S. Appa Rao (eds.), Rice in Laos. International Rice Research Institute, Los Baños Philippines.

Escalada, M. M., K. L. Heong, N. H. Huan, and H. V. Chien. 2009. Changes in rice farmers' pest management beliefs and practices in Vietnam: An analytical review of survey data from 1992 to 2007, pp. 447–456. *In* K. L. Heong, and B. Hardy (eds.), Planthoppers: New threats to the sustainability of intensive rice production systems in Asia. International Rice Research Institute, Los Baños, Philippines.

(FAO) Food and Agriculture Organization of the United Nations. 2011. Save and grow: A policymaker's guide to the sustainable intensification of smallholder crop production. FAO, Rome, Italy. 102 pp.

(FAO/WHO) Food and Agriculture Organization of the United Nations/World Health Organization. 2014. The international code of conduct on pesticide management. FAO, Rome, Italy. 37 pp.

Feder, G., R. Murgai, and J. B. Quizon. 2004. The acquisition and diffusion of knowledge: The case of pest management training in Farmer Field Schools, Indonesia. Journal of Agricultural Economics 55: 221–243.

Fox, J. J. 2014. Fast-breeding insect devastates Java's rice. Asian Scientist, March 2014 issue.

Gurr, G. M., K. L. Heong, J. A. Cheng, and J. L. Catindig. 2012. Ecological engineering against insect pests in Asian irrigated rice, pp. 214–229. *In* G. M. Gurr, S. D. Wratten, W. E. Snyder, and D. M. Y. Read (eds.), Biodiversity and insect pests: Key issues for sustainable management. John Wiley & Sons, Ltd., UK.

Gurr, G. M., J. Liu, D. M. Y. Read, J. L. A. Catindig, J. A. Cheng, L. P. Lan, and K. L. Heong. 2010. Parasitoids of Asian rice planthopper (Hemiptera: Delphacidae) pests and prospects for enhancing biological control. Annals of Applied Biology 158: 149–176.

Gurr, G. M., Z. X. Lu, X. S. Zheng, H. X. Xu, P. Y. Zhu, G. H. Chen, X. M. Yao, J. A. Cheng, Z. R. Zhu, J. L. Catindig, S. Villareal, H. V. Chien, L. Q. Cuong, C. Channoo, N. Chengwattana, P. H. Lan, L. H. Hai, J. Chaiwong, H. I. Nicol, D. J. Perovic, S. D. Wratten, and K. L. Heong. 2016. Multi-country evidence that crop diversification promotes ecological intensification of agriculture. Nature Plants 22 (2): 16014.

Hamburger, J. 2002. Pesticides in the People's Republic of China: A growing threat to food safety, public health and the environment. China Environment Series 5: 29–44.

Heinrichs, E. A. (ed.) 1994. Biology and management of rice pests. Wiley Eastern Ltd., New Delhi, India. 779 pp.

Heong, K. L. 2009. Are planthopper problems due to breakdown in ecosystem services? pp. 221–232. *In* K. L. Heong, and B. Hardy (eds.), Planthoppers: New threats to the sustainability of intensive rice production systems in Asia. International Rice Research Institute, Los Baños, Philippines.

Heong, K. L., and M. M. Escalada (eds). 1997. Pest management of rice farmers in Asia. International Rice Research Institute, Los Baños, Philippines. 245 pp.

Heong, K. L., and B. Hardy (eds.). 2009. Planthoppers: New threats to the sustainability of intensive rice production systems in Asia. International Rice Research Institute, Los Baños, Philippines. 460 pp.

Heong, K. L., and K. G. Schoenly. 1998. Impact of insecticides on herbivore-natural enemy communities in tropical rice ecosystems, pp. 381–403. *In* P. T. Haskell, and P. McEwen (eds.), Ecotoxicology: Pesticides and beneficial organisms. Chapman and Hall, London, U.K.

Heong, K. L., G. B. Aquino, and A. T. Barrion. 1991. Arthropod community structures of rice ecosystems in the Philippines. Bulletin of Entomological Research 81: 407–416.

Heong, K. L., M. M. Escalada, and A. A. Lazaro. 1995. Misuse of pesticides among rice farmers in Leyte, Philippines, pp. 97–108. *In* P. L. Pingali and P. A. Roger (eds.), Impact of pesticides on farmers' health and the rice environment. Kluwer Press, San Francisco, USA.

Heong, K. L., M. M. Escalada, N. H. Huan, and V. Mai. 1998. Use of communication media in changing rice farmers' pest management in South Vietnam. Crop Protection 17: 413–425.

Heong, K. L., M. M. Escalada, N. H. Huan, V. H. Ky Ba, L. V. Thiet, and H. V. Chien. 2008. Entertainment-education and rice pest management: A radio soap opera in Vietnam. Crop Protection 27: 1392–1397.

Heong, K. L., M. M. Escalada, N. H. Huan, H. V. Chien, and P, V. Quynh. 2010. Scaling out communication to rural farmers – Lessons from the "Three Reductions, Three Gains" campaign in Vietnam, pp. 207–220. *In* F. Palis, G. Singleton, and M. Casimero (eds.), Research to impact: Case studies for natural resources management of irrigated rice in Asia. International Rice Research Institute, Los Baños, Philippines.

Heong, K. L., M. M. Escalada, H. V. Chien, and L. Q. Cuong. 2014. Restoration of rice landscape biodiversity by farmers in Vietnam through education and motivation using media. *In* G. Mainguy (ed.), Special issue on large scale restoration of ecosystems. S.A.P.I.E.N.S (online) 7: (2) 29–35.

Heong, K. L., Cheng, J. A. and M. M. Escalada (eds.). 2015a. Rice planthoppers: Ecology, management, socio economics and policy. Zhejiang University Press, Hangzhou and Springer Science+Business Media Dordrecht. 231 pp.

Heong, K. L, L. Wong, and J. H. Delos Reyes. 2015b. Addressing planthopper threats to Asian rice farming and food security: Fixing insecticide misuse, pp. 69–80. *In* K. L. Heong, J. A. Cheng, and M. M. Escalada (eds.), Rice planthoppers: Ecology, management, socio economics and policy. Zhejiang University Press, Hangzhou and Springer Science+Business Media, Dordrecht, The Netherlands.

Heong, K. L., M. M. Escalada, H. V. Chien, and J. H. Delos Reyes. 2015c. Are there productivity gains from insecticide applications in rice production? pp. 179–190. *In* K. L. Heong, J. A. Cheng, and M. M. Escalada (eds.), Rice planthoppers: Ecology, management, socio economics and policy. Zhejiang University Press, Hangzhou and Springer Science+Business Media Dordrecht, The Netherlands.

Huan, N. H., L. V. Thiet, H. V. Chien, and K. L. Heong. 2005. Farmers' evaluation of reducing pesticides, fertilizers and seed rates in rice farming through participatory research in the Mekong Delta, Vietnam. Crop Protection 24: 457–464.

Huan, N. H., H. V. Chien, P. V. Quynh, P. S Tan, P. V. Du, M. M. Escalada, and K. L. Heong. 2008. Motivating rice farmers in the Mekong Delta to modify pest management and related practices through mass media. International Journal of Pest Management 54: 339–346.

(IRRI) International Rice Research Institute. 2013. Rice Almanac, 4th edition. International Rice Research Institute, Los Baños, Philippines. 283 pp.

Ives, A. R., and W. H. Settle. 1997. Metapopulation dynamics and pest control in agricultural systems. American Naturalist 149: 220–246.

Kraker, J. De, A. Van Huis, J. C. Van Lenteren, K. L. Heong, and R. Rabbinge. 1999. Egg mortality of rice leaffolders *Cnaphalocrocis medinalis* and *Marasmia patnalis* in irrigated rice fields. BioControl 44: 449–471.

Kim, C. G., and S. S. Lim. 2015. An evaluation of the environmentally friendly direct payment program in Korea. Journal of International Economic Studies 29: 3–22.

Lu, Z. X., P. Y. Zhu, G. M. Gurr, X. S. Zheng, G. H. Chen, and K. L. Heong. 2015. Rice pest management by ecological engineering: A pioneering attempt in China, pp. 161-178. *In* K. L. Heong, J. A. Cheng, and M. M. Escalada (eds.), Rice planthoppers: Ecology, management, socio economics and policy. Zhejiang University Press, Hangzhou and Springer Science+Business Media, Dordrecht, The Netherlands.

Lu, Y. H., K. Liu, X. S. Zheng, and Z. X. Lu. 2017. Electrophysiological responses of the rice striped stem borer *Chilo suppressalis* to volatiles of the trap plant vetiver grass (*Vetiveria zizanioides* L.). Journal of Integrative Agriculture 16(11): 2525–2533.

Matteson, P. C. 2000. Insect pest management in tropical Asian irrigated rice. Annual Review of Entomology 45: 549–574.

(MEA) Millennium Ecosystem Assessment. 2005. Ecosystems and human well-being: Synthesis. Island Press, Washington, DC, USA. 155 pp.

McCann, L. 2005. Transaction costs of agri-environmental policies in Viet Nam. Society and Natural Resources: An International Journal 18: 759–766.

Pingali, P. L., M. H. Hossain, and R. Gerpacio. 1997. Asian rice bowls: The returning crisis? International Rice Research Institute, Los Baños, Philippines and CABI International, Wallingford, UK. 341 pp.

Rice Doctor 2019. Interactive tool for extension workers, students, researchers and other users who want to learn and diagnose pest, disease, and other problems that can occur in rice; and how to manage them.

Simon, H. A. 1982. Models of bounded rationality: Empirically grounded economic reason. Volume 3. MIT Press, Cambridge, Massachusetts, USA. 336 pp.

Singhal, A., and E. M. Rogers. 1999. Entertainment–education: A communication strategy for social change. Lawrence Erlbaum Associates Publishers, Mahwah, New Jersey, USA. 280 pp.

Spangenberg, J. H., J. M. Douguet, J. Settele, and K. L. Heong. 2015. Escaping the lock-in of continuous insecticide spraying in rice: Developing an integrated ecological and socio-political DPSIR analysis. Ecological Modelling 295: 188–195.

Thorburn, C. 2013. Empire strikes back: The making and unmaking of Indonesia's national Integrated Pest Management program. Agroecology and Sustainable Food Systems 38: 3–24.

Watanabe, T., M. Matsumura, and A. Otuka. 2009. Recent occurrences of long-distance migratory planthoppers and factors causing outbreaks in Japan, pp. 179–190. *In* Heong, K. L. and B. Hardy (eds.), Planthoppers: New threats to the sustainability of intensive rice production systems in Asia. International Rice Research Institute, Los Baños, Philippines.

Way, M. J., and K. L. Heong. 1994. The role of biodiversity in the dynamics and management of insect pests of tropical irrigated rice - A review. Bulletin of Entomological Research 84: 567–587.

Zheng, X., Y. Lu, P. Zhu, F. C. Zhang, J. Tian, H. Xu, G. Chen, C. Nansen, and Z. Lu. 2017. Use of banker plant system for sustainable management of the most important insect pest in rice fields in China. Scientific Reports 7: 45581.

Zhu, P. Y., G. M. Gurr, Z. X. Lu, K. L. Heong, G. H. Chen, X. S. Zheng, H. X. Xu, and Y. J. Yang. 2013. Laboratory screening supports the selection of sesame (*Sesamum indicum* L.) to enhance *Anagrus* spp. parasitoids (Hymenoptera: Mymaridae) of rice planthoppers. Biological Control 64: 83–89.

BRIEF OVERVIEW OF THE WORLD HEALTH ORGANIZATION "VECTOR CONTROL GLOBAL RESPONSE 2017-2030" AND "VECTOR CONTROL ADVISORY GROUP" ACTIVITIES

R. VELAYUDHAN

Coordinator Vector Ecology and Management, Department of Control of Neglected Tropical Diseases, World Health Organization, 20 Avenue Appia, CH-1211 Geneva 27, Switzerland; VelayudhanR@who.int

SUMMARY

More than 80% of the world's population is at risk of vector-borne disease and many of these diseases are concentrated in the poorest communities in tropical and subtropical regions; they cause unacceptable mortality and morbidity and impede economic growth. A new and comprehensive approach to preventing diseases and responding to outbreaks is clearly needed — one that engages multiple sectors and communities and is based on the best available evidence base. The 70th World Health Assembly adopted the strategic approach outlined in a new WHO Global Vector Control Response (GVCR) for 2017–2030. The response aims to reduce the burden and threat of vector-borne diseases through effective, locally adapted, and sustainable vector control. It is an approach for tackling multiple vectors and diseases that requires action across many sectors beyond health, including environment, urban planning, education, etc. The recent outbreaks of *Aedes*-borne diseases call for concerted action to deal with increased urbanization, erratic water supply, increased movement of people and commodities, altered land use patterns, and other environmental changes, including climatic changes that extend the distribution of vectors and pathogen transmission to more temperate climes. The GVCR provides a unique opportunity to work together and address the increasing burden of vector-borne diseases. This new approach is supported by the Vector Control Advisory Group (VCAG) that was established by WHO in 2013 in order to carry out independent evaluation of the public health value of innovative new tools, technologies and approaches for vector control and to enable WHO to provide evidence-based advice to Member States on whether their deployment and strategy is the most appropriate for their specific circumstances.

Key Words: Vector-borne diseases, public health, emerging diseases, inter-sectoral collaboration, mosquitoes, strategic framework, integrated vector management (IVM), health education, *Aedes*, dengue, malaria, chikungunya, yellow fever, Zika, WHO

J. Hendrichs, R. Pereira and M. J. B. Vreysen (eds.), Area-Wide Integrated Pest Management: Development and Field Application, pp. 633–644. CRC Press, Boca Raton, Florida, USA.

1. BACKGROUND

Cognizant of recent major outbreaks of diseases such as dengue, malaria, chikungunya, yellow fever and Zika, as well as other emerging and persistently important diseases, Member States at the 139th executive board meeting of the World Health Organization (WHO) requested the director-general in May 2016 to devise an appropriate response.

Development of the Global Vector Control Response 2017-2030 (GVCR) (WHO 2017) commenced immediately through a fast-tracked broad consultative process that was co-led by three departments within the Communicable Diseases cluster: the Global Malaria Programme (GMP), the Department of Control of Neglected Tropical Diseases (NTD), and the Special Programme for Research and Training in Tropical Diseases (TDR).

The strategic approach was welcomed one year later by Member States at the 70th World Health Assembly, and a dedicated resolution was adopted (WHA 2017). The GVCR (WHO 2017) and associated resolution (WHA 2017) set out a new strategy to strengthen vector control worldwide through increased capacity, improved surveillance, better coordination and integrated action across diseases and sectors. It aims to reduce the burden and threat of vector-borne diseases through effective, locally adapted and sustainable vector control, and to prevent epidemics by 2030 in line with the United Nations Sustainable Development Goal 3 (Good Health and Well-being).

2. RATIONALE

- Major vector-borne diseases account for an estimated 17% of the global burden of all infectious diseases, and disproportionately affect human populations in poor countries.
- These diseases impede economic development through direct medical costs and indirect costs such as lost productivity and tourism.
- Social, demographic and environmental factors have caused increases in many vector-borne diseases in recent years, with major outbreaks of dengue, malaria, chikungunya, yellow fever and Zika since 2014.
- Most vector-borne diseases are preventable by various means, including vector control, if well implemented. Strong political commitment and improved investments have already led to major reductions in malaria, onchocerciasis and Chagas.
- The full impact of vector control against other diseases has yet to be achieved, but it is possible through re-alignment of programmes to optimise delivery of interventions that are tailored to a local context.
- This requires improved public health entomology capacity, a well-defined national research agenda, better coordination within and between sectors, strengthened monitoring systems, community involvement, and availability and use of more interventions with proven public health value.

3. NEED FOR A GLOBAL VECTOR CONTROL RESPONSE

Never has the need for a comprehensive approach to vector control to counter the impact of vector-borne diseases been more urgent. The unprecedented global spread of dengue and chikungunya viruses, as well as outbreaks of Zika and yellow fever in 2015-2016, clearly highlight the challenges faced by numerous countries. Transmission and risk of vector-borne diseases are rapidly changing due to unplanned urbanization, increased movement of people and goods, environmental changes and biological challenges, such as vectors populations becoming resistant to insecticides and evolving strains of pathogens. In particular, rapid, unplanned urbanization in tropical and sub-tropical areas renders large human populations at elevated risk of emergence and expansion of arboviral diseases spread by mosquitoes. Many countries are still unprepared to address these looming challenges (the current global distribution of seven major vector-borne diseases is shown in Fig. 1).

The strong influence of social and environmental factors on vector-borne pathogen transmission underscores the critical importance of flexible vector control delivery and monitoring, as well as evaluation systems that support locally tailored approaches. Re-alignment of national programmes to optimize implementation of interventions against multiple vectors and diseases will maximize the impact of available resources. Health systems must be prepared to detect and respond quickly and effectively to changes. This rapid response requires not only the availability of effective, evidence-based control interventions, but also well-trained staff who can build sustainable systems for their area-wide delivery. To achieve these goals, reforms to vector control programmatic structures are urgently needed.

Vector-borne diseases are everyone's problem, not just the health sector. Achievement of the United Nations Sustainable Development Goal 3 to ensure good health and well-being will rely on effective vector control, as will initiatives for clean water and sanitation (Goal 6), sustainable cities and communities (Goal 11) and climate action (Goal 13). Multiple approaches that are implemented by different sectors will be required for control and elimination of vector-borne disease, such as those promoting healthy environments (Pruss-Ustun et al. 2016). Engaging local authorities and communities as part of broad-based inter-sectoral collaboration will be key to improved vector control delivery, through tailoring of interventions to specific scenarios as informed by local entomological, epidemiological and ecological data. Building sustainable control programmes that are resilient in the face of technical, operational, climatic and financial challenges will require the engagement and collaboration of local communities.

Recent advances to modernize and develop new vector control and surveillance tools means that there has never been a better time to reinvigorate vector control. To be effective, strong political commitment and long-term investment are needed. The GVCR response seeks neither to replace or override existing disease-specific strategies nor to shift the focus from other essential interventions, such as vaccines against yellow fever, Japanese encephalitis and tick-borne encephalitis, mass-administration of medicines against lymphatic filariasis and human onchocerciasis, and artemisinin-based combination therapy against malaria. Rather, it aims to add to these efforts and help countries mount coherent and coordinated efforts to address the increasing burden and threat of vector-borne diseases.

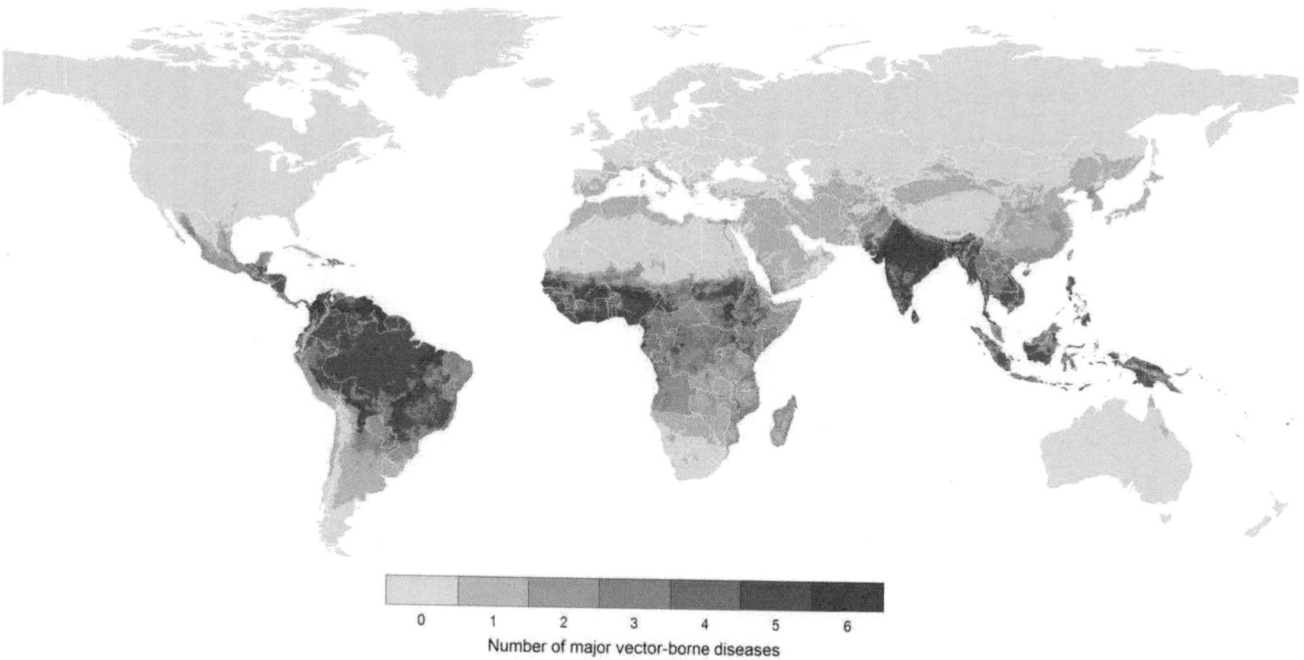

Figure 1. Combined global distribution of seven major vector-borne diseases for which integration of vector control programmes may be beneficial: malaria, lymphatic filariasis, leishmaniasis, dengue, Japanese encephalitis, yellow fever and Chagas disease transmission in 2015. Colours indicate the number of vector-borne diseases that pose a risk at each 5 × 5 km grid cell (from Golding et al. 2015).

The GVCR provides strategic guidance to countries and development partners to urgently strengthen vector control as a fundamental approach to preventing diseases and responding to outbreaks. To achieve this, re-alignment of vector control programmes is required, supported by increased technical capacity, strengthened monitoring and surveillance systems, and improved infrastructure. Ultimately, this will support implementation of a comprehensive approach to vector control that will enable the achievement of disease-specific national and global goals and contribute towards achievement of the United Nations Sustainable Development Goals.

4. THE GLOBAL VECTOR CONTROL RESPONSE AT A GLANCE

4.1. Vision

A world free of human suffering from vector-borne diseases.

4.2. Aim

To reduce the burden and threat of vector-borne diseases through effective locally-adapted and sustainable vector control.

4.3. Goals for 2020-2030

The goals, including 2020-2030 milestones and targets, are presented in Table 1.

Table 1. Goals for 2020-2030 on the global vector control response

Milestones	Targets		
	2020	2025	2030
Reduce mortality due to vector-borne diseases globally compared to 2016	At least 30%	At least 50%	At least 75%
Reduce case incidence due to vector-borne diseases globally compared to 2016	At least 25%	At least 40%	At least 60%
Prevent epidemics of vector-borne diseases*		In all countries without transmission in 2016	In all countries

* *Rapid detection of outbreaks and curtailment before spread beyond country or region*

4.4. Priority Activities for 2017-2022

For the period 2017-2022 the priority activities are as follows (they will be revised and updated for the subsequent period of 2023–2030):

1. National and regional vector control strategic plans developed and adapted to align with the global vector control response.
2. An assessment of national vector control needs carried out or updated, and a resource mobilization plan developed (including for outbreak response).
3. National entomology and cross-sectoral workforce appraised and enhanced to meet identified requirements for vector control, including for epidemic response and pre-emptive response.
4. Relevant staff from Ministries of Health and/or their supporting institutions trained in public health entomology and sustainable career pathway is established.
5. National and regional institutional networks to share data and support training and education in public health entomology, and technical support established and functioning.
6. National agendas for basic and applied research on entomology and vector control established and/or progress reviewed.
7. National inter-ministerial task forces and steering committee for multi-sectoral engagement in vector control established and functioning.
8. National vector surveillance systems strengthened and integrated with health information systems to guide vector control. Cross border exchange of information is encouraged.
9. National plans for effective community engagement and mobilization in vector control developed.
10. National targets for protection of at-risk populations with appropriate vector control aligned across vector-borne diseases.

5. BEYOND INTEGRATED VECTOR MANAGEMENT

Integrated vector management (IVM) is a rational decision-making process for the optimal use of resources for vector control, as presented in a WHO global strategic framework released in 2004, a WHO position statement issued in 2008, and other supporting documents (WHO 2004, 2008, 2016).

While this approach seeks to improve the efficacy, cost-effectiveness, ecological soundness and sustainability of disease-vector control, uptake has been poor, due to insufficient political buy-in for reorientation of programmes to support a harmonized approach to vector control across diseases. This has largely been due to limited human capacity to advocate, plan and implement, as well as fragmented global and national architecture to support a multi-disease approach.

Given the recent alarming increase in numerous vector-borne diseases and the serious threat posed to economic development, the GVCR response aims to reposition vector control as a key approach to prevent, manage, or eliminate vector-borne diseases. It builds on the basic concept of IVM with renewed focus on improved human capacity at national and sub-national levels, as well as strengthened infrastructure and systems, particularly in areas vulnerable to vector-borne disease upsurges (e.g. sustainable development, access to potable water, waste management, home construction, community design, water supply and solid waste disposal).

For sustainable impact in vector control, increased inter-sectorial and inter-disciplinary action is essential, linking efforts in environmental management, health education, and reorienting relevant government programmes around proactive strategies that will improve living conditions and control new and emerging threats. Critical attention is given in the GVCR to current opportunities available for leverage, as well as challenges that need to be addressed in order to enable effective and sustainable vector control with interventions that will work best for the specific/unique circumstances of each local situation.

6. OPPORTUNITIES FOR GLOBAL VECTOR CONTROL RESPONSE

Many opportunities exist to enhance the impact of vector control. The overall benefit of enhancing capacity and capability is a critical opportunity that is essential for a successful GVCR, will have a positive impact across diseases, and is essential for sustainability:

1. *Development.* Environmentally sustainable and resilient development in urban centres (Habitat III 2016; UNECE 2014-2015) that reduces poverty and improves living standards will reduce transmission of vector-borne pathogens.

2. *Recognition.* Existing global and regional strategies against vector-borne diseases demonstrate their importance in the global health agenda and in other sectors, and represent high-level commitment for their reduction, elimination and, for some, eradication.

3. *Expansion.* Recent successes in vector control, such as against vectors of malaria, onchocerciasis and lymphatic filariasis, have led to major reductions in vector-borne diseases. Further impacts could be achieved through sustained and expanded use of proven vector control interventions.

4. *Optimization.* Re-aligning national programmes to optimize implementation of vector control against multiple vectors and diseases, across geographic areas and human populations, will leverage available resources to maximum impact of IVM (Fig. 2).

5. *Collaboration.* Building on existing collaborations across ministries, sectors, partners and networks to share data and expertise will improve timely access to information and resources for the most effective vector-borne disease control. Regional and cross-border collaboration further helps in tracking outbreaks and responding in a timely manner.

6. *Adaptation.* The strong influence of social, demographic and environmental factors on vector-borne pathogen transmission underscores the critical importance of flexible vector control delivery, monitoring, and evaluation systems that support locally-tailored approaches that can be adapted to specific opportunities or challenges.

7. *Innovation.* Development of novel tools, technologies and approaches such as new insecticides and formulations, vector traps and baits, use of *Wolbachia* spp. and genetic modification for population reduction or replacement, other forms of vector sterilisation, larviciding via auto-dissemination, endectocides, spatial repellents and

vapor active insecticides, and housing improvements to exclude vectors and reduce favourable harbourages have the potential to further reduce disease burden.

8. *Technology.* Advances that support evidence-based vector control such as information communication technologies that support real-time data capture or social media, or risk stratification and predictive geo-informatics tools such as geographic information systems, remote sensing, and climatic models can be leveraged to further optimize planning and implementation.

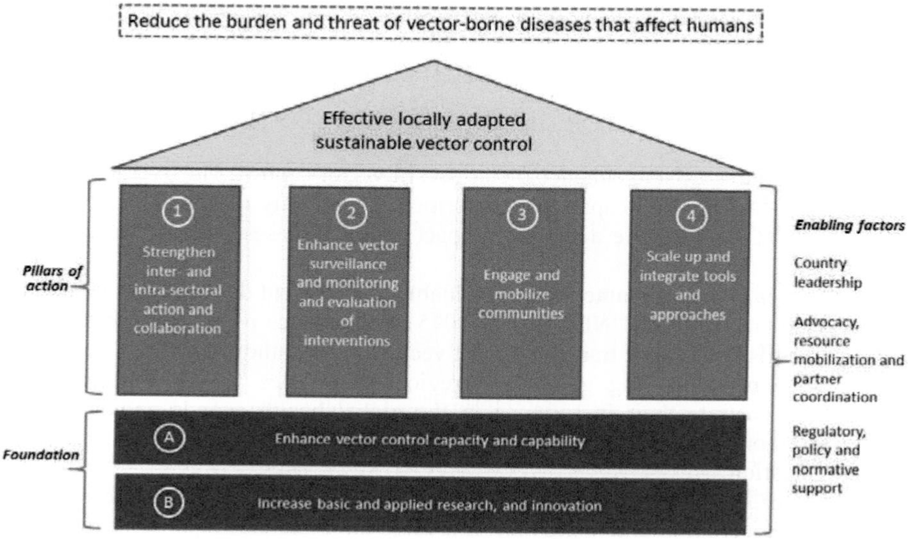

Figure 2. Response framework from the Global Vector Control Response (WHO 2017).

7. CHALLENGES OF GLOBAL VECTOR CONTROL RESPONSE

While there are many opportunities to enhance the impact of vector control, there are also multiple interconnected challenges that impede progress against vector-borne diseases and that need to be overcome. Threats to effective and impactful vector control can be grouped as systemic, structural, informational, environmental, human movement, and political and financial.

1. *Systemic.* Capacity for vector surveillance and control is insufficient in most countries at risk from vector-borne diseases. With a few notable exceptions, vector-borne disease prevention programmes at national and subnational levels have limited public health entomology capacity and poor infrastructure. Career structures for technical specialists and technicians within the health system are absent or weak. Establishing basic capacity provides enormous impact across multiple vector-borne diseases and sectors.

2. *Structural.* Many countries that are endemic for more than one major vector-borne disease have disease-specific programmes and strategies that do not optimally

leverage synergies and sometimes compete for resources. Well-funded programmes – such as for malaria in some countries of sub-Saharan Africa – are often expected to respond to outbreaks of other vector-borne diseases without adequate capacity and additional resources at the expense of routine, core activities.

3. *Informational.* The evidence base to support effective vector control is limited for most vector-borne diseases due to lack of research support. Vector surveillance remains weak in many countries despite insecticide resistance and changes in vector behaviour threatening the efficacy of current interventions. Available information is not used to its full potential to guide control interventions.

4. *Environmental.* Changes in vector habitats such as those due to rapid urbanization or alterations in land use, water management and farming practices are often unpredictable, uncontrollable, and complex. Climatic changes that extend the distribution of vectors and pathogen transmission to more temperate climes are also of concern. With two thirds of the global population expected to live in urban settings by 2050 (United Nations 2014), large populations in tropical and subtropical zones will be at increasing risk of *Aedes*-borne diseases.

5. *Movement of humans and goods.* Increased global human population movement due to changing travel patterns, migration for employment, or displacement resulting from humanitarian crises, as well as increased global trade are likely to accelerate the introduction of invasive species or pathogens into new receptive areas and expose non-immune populations to novel infections and disease.

6. *Political and financial.* Substantial financial support has been provided to scale up of the use of insecticide-treated bed nets and indoor residual spraying against malaria vectors since 2000. However, there has been insufficient focus on scale-up and effective delivery of vector control for other vector-borne diseases, especially in the absence of either large epidemics or high mortality rates.

8. THE GLOBAL VECTOR CONTROL RESPONSE IN SUMMARY

The GVCR aims to reduce the burden and threat of vector-borne diseases through effective locally-adapted, sustainable, and inter-sectorial vector control. Success will depend on the ability of countries to overcome current limitations and to strengthen the capacity, financial resources and inter-sectorial collaboration of their vector surveillance and control programmes. The WHO will support the development of regional action plans and country activities based on GVCR strategies and priorities identified in the Country Cooperation Strategy (CCS) that guides WHO work in each country to support its national health strategy.

The key areas of activity that intend to radically change the control of vector-borne diseases are:

- Aligning action across sectors, since vector control is more than the work done by the Ministry of Health (MOH). The MOH has to coordinate work with other relevant ministries and with city planners to reduce or build out habitat of urban vectors (*Aedes* mosquitoes, rodents, etc.).
- Engaging and mobilizing communities to become active partners in the GVCR, which will protect themselves and build resilience against future disease outbreaks (including the impact of climate change).

- Enhancing surveillance to trigger early responses to increased risk in disease or vector populations, and to identify when and why interventions are not working as expected.
- Encouraging pre-emptive programmes that use surveillance data to reduce or eliminate the vectors wherever feasible, and
- Scaling-up vector control tools (including the integration of new tools and innovative approaches) and using them in combination to maximize impact on disease, while minimizing impact on the environment.

Specifically, the new integrated approach calls for national programmes to be realigned so that public health workers can focus on the complete spectrum of relevant vectors and thereby control all the diseases they cause. Recognizing that efforts must be adapted to local needs and sustained, the success of the response will depend on the ability of countries to strengthen their vector control programmes and enhance their inter-sectorial collaborations, including with sectors that are not currently part of "traditional" vector control. National strategic plans need to be revised and country-specific targets defined till 2030. All the relevant GVCR documents and related links can be downloaded from WHO (2017).

9. VECTOR CONTROL ADVISORY GROUP

In 2013 the WHO established a Vector Control Advisory Group (VCAG) to assist WHO in assessing innovative approaches, new tools, and technologies for vector control, and to serve as an advisory body to WHO on new approaches of vector control for malaria and other vector-borne diseases. During the very early stages of innovation, scientists and developers of such new products or approaches can bring new ideas or new intervention concepts for feedback to the VCAG.

If the product developers wish, VCAG can provide advice on the type and depth of evidence that will likely be used for the assessment, providing an opportunity for product developers to align with VCAG on overall evidence requirements before the launch of resource-intensive activities such as large-scale epidemiological trials (randomized control trials with epidemiological end-point). The advice will be provided in individual discussions between the product developers and the group at the VCAG meeting. It may cover, for instance, the needs concerning evidence of epidemiological and vector control outcomes, epidemiological mode of action, economic feasibility or user acceptability. To support its deliberations, VCAG may consider the initial results of tests and studies carried out by the product developers.

The VCAG is jointly managed by the Global Malaria Programme (GMP), the Department of Control of Neglected Tropical Diseases (NTD), and the WHO Prequalification Team for the assessment of vector control products (VCAG 2019).

In summary the VCAG objectives are:

- To assess the public health value of new vector control tools, technologies and approaches submitted to WHO for evaluation.
- To provide guidance to product developers on data requirements and study designs to generate the evidence required for a VCAG assessment.

- To provide guidance to WHO and its policy advisory groups, the MPAC (Malaria Policy Advisory Committee) and the STAG (Strategic and Technical Advisory Group for neglected tropical diseases), on the public health value of new tools, technologies and approaches.

Once a relevant body of evidence has been presented to VCAG, which contains an indication of the epidemiological outcome of the new form of vector control, VCAG will review all available scientific evidence (which may also include available sources other than the data presented by the product developers). Based on this review, VCAG evaluates the public health value of the new intervention by answering questions such as: "Is this new intervention efficacious for some defined public health purpose (in disease prevention through vector control) and in some defined circumstances, and will it be useful to and feasible for its intended users?" The answer might in some instances requires additional evidence.

The new approaches or products ongoing assessment or that are in planning and are currently under consideration by VGAC are shown in Fig. 3.

Products under VCAG Review — World Health Organization

New Product - Variation	Disease Target	Step 3
ITN against IR Vector (extend ITN)	Malaria	ongoing
Treated walls against IR vector (extend IRS)	Malaria	N.A.
Peri-focal residual spraying (extend IRS)	Dengue and Aedes-borne arboviral diseases, Leishmaniasis	planning
Insecticide-treated curtain (extend ITN)	Malaria	N.A.
New Product Class – (chemical)	**Generic Exemplar**	
Attract and kill baits	Malaria	ongoing
Spatial repellents	Malaria, Dengue and Aedes-borne arboviral diseases	ongoing
ITM for specific risk groups	Malaria	N.A.
Vector traps	Dengue and Aedes-borne arboviral diseases	ongoing
Lethal house lures	Malaria	ongoing
Systemic insecticide	Leishmaniasis	planning
New Product Class – (biological)	**Generic Exemplar**	
Microbial control in adult vectors	Dengue and Aedes-borne arboviral diseases	ongoing
Pop. reduction through genetic manipulation	Dengue and Aedes-borne arboviral diseases (self limiting)	planning
Population reduction	Malaria (gene-drive)	N.A.
Pop. alteration of malaria vector mosquitoes	Malaria (gene-drive)	N.A.
SIT & incompatible insect technique (IIT)	Dengue and Aedes-borne arboviral diseases	planning

Figure 3. New products or innovative approaches under consideration by the Vector Control Advisory Group (VCAG) of WHO. Further details on the VCAG can be found under VGAG (2019). IIT = Incompatible Insect Technique using Wolbachia*; IR = insecticide resistant; IRS = indoor residual spraying; ITM = insecticide-treated material; ITN = insecticide-treated nets; N.A. =not applicable at present since it is yet to complete preliminary studies; SIT = Sterile Insect Technique. Step 3 refers to Phase 3 studies for assessing impact on the disease/infection or both.*

As soon as the VCAG decides that the answer to this question is "yes", and that proof-of-principle has indeed been established for the new form of vector control, responsibility within WHO for further review will pass to the advisory bodies MPAC and STAG for the WHO technical department(s) (e.g. Departments of Control of Neglected Tropical Diseases and Global Malaria Programme) responsible for the particular vector-borne disease(s) against which the new intervention is considered likely to be useful.

Hence, after validating the value of the new form of vector control, VCAG will present its results to WHO for consideration and submission to MPAC and/or STAG in their respective meetings, expressing its opinion on the usefulness of the new intervention.

10. REFERENCES

Golding, N., A. L. Wilson, C. L. Moyes, J. Cano, D. M. Pigott, R. Velayudhan, S. J. Brooker, D. L. Smith, S. I. Hay, and S. W. Lindsay. 2015. Integrating vector control across diseases. BMC Medicine 13: 249.

Habitat III. 2016. Habitat III new urban agenda. United Nations Conference on Housing and Sustainable Urban Development. Final draft: 10 September 2016.

Pruss-Ustun, A., J. Wolf, C. Corvalan, R. Bos, and M. Neira. 2016. Preventing disease through health environments: A global assessment of the burden of disease from environmental risks. World Health Organization. Geneva, Switzerland.

UNECE (United Nations Economic Commission for Europe). 2014-2015. United smart cities: Smart urban solutions for transition and developing countries. United Nations Economic Commission for Europe, United Nations Economic and Social Council. Geneva, Switzerland.

United Nations. 2014. World urbanization prospects: The 2014 revision, highlights (ST/ESA/SER.A/352). United Nations, Department of Economic and Social Affairs, Population Division. New York, USA.

(VCAG) Vector Control Advisory Group. 2019. Global Malaria Programme, the Department of Control of Neglected Tropical Diseases, and the WHO Prequalification Team for Vector Control Products. World Health Organization. Geneva, Switzerland.

(WHA) World Health Assembly. 2017. Global vector control response: An integrated approach for the control of vector-borne diseases. Agenda item 14.2 31 May 2017. Seventieth World Health Assembly, Resolution WHA70.16.

(WHO) World Health Organization. 2004. Global strategic framework for Integrated Vector Management. World Health Organization. Geneva, Switzerland.

(WHO) World Health Organization. 2008. WHO position statement on Integrated Vector Management. World Health Organization. Geneva, Switzerland.

(WHO) World Health Organization. 2016. A toolkit for integrated vector management in sub-Saharan Africa. World Health Organization. Geneva, Switzerland.

(WHO) World Health Organization. 2017. Global vector control response 2017–2030. World Health Organization. Geneva, Switzerland. 53 pp.

NEW MOLECULAR GENETIC TECHNIQUES: REGULATORY AND SOCIETAL CONSIDERATIONS

K. M. NIELSEN

Oslo Metropolitan University, Oslo, Norway; kaare.nielsen@hioa.no

SUMMARY

A rapidly expanding toolbox of techniques available for genome editing provides the basis for a new continuum in types of modifications that can be introduced into a genome and blur the bimodal GMO vs. non-GMO (genetically modified organism) divide. Site-directed nucleases (SDN) are now used to modify existing nucleotides within genomes instead of adding recombined DNA as transgenes. Moreover, new gene drive approaches are in development based on SDNs. A number of potential drive applications have been reported, but uncertainties in trait stability and limitations in knowledge of the affected system at various temporal and spatial levels slow down their current uses. Adoption of new genome targeted technology takes place in a social context. The context will vary between countries and cultures, expressed in values, ethics, politics and priorities — that are translated into different regulatory approaches. Some developed products using new genome editing techniques clearly fall under internationally negotiated regulations of GMOs. However, other product outcomes of editing techniques challenge our current understanding and definition of GMOs. There is an urgent need for further research, for building international consensus and harmonizing regulatory approaches to facilitate categorization, predictability, transparency, trust and trade.

Key Words: Transgenesis, genetic engineering techniques, genetically modified organisms, GMOs, living modified organisms, site-directed nucleases, CRISPR, gene drive, insect, regulation, genome editing, modern biotechnology, recombinant DNA, stacked transgenes

1. INTRODUCTION

1.1. Current Genetic Techniques

Established engineering techniques and transgenesis (see Table 1) have provided the basis for the range of *genetically modified organisms* (GMOs) available for commercial use today. GMOs or *living modified organisms* (LMOs) are any living organism that possesses a novel combination of genetic material obtained through the use of modern biotechnology (CBD 2000; FAO/IPPC 2017). The majority of these organisms are varieties of commodity crop plants grown in large-scale agriculture.

J. Hendrichs, R. Pereira and M. J. B. Vreysen (eds.), Area-Wide Integrated Pest Management: Development and Field Application, pp. 645–654. CRC Press, Boca Raton, Florida, USA.

The globally most commonly used transgenes encode insecticidal properties and herbicide-tolerance. These traits have been obtained through molecular recombination in the laboratory of DNA from several unrelated species. Example of GM-insects reaching field-scale applications are genetically modified male mosquitoes (with a dominant lethal gene) that have been released in various parts of the world (Alphey 2014; Evans et al. 2019) and diamondback moth in limited field releases in the USA (USDA 2017).

The GMO technologies that have reached commercial/field stage applications are based on the introduction of additional DNA to obtain the desired traits. The transgene denotes the inserted recombined DNA, usually consisting of a protein-encoding sequence recombined with several regulatory sequences that ensure expression of a new trait in the engineered organism. Current transgene-enabled technologies and products are typically protected through intellectual property rights (IPR) and sold under contractual agreements. Confidentiality claims limit independent peer review and transparency (Nielsen 2013). Similar constraints can be expected for organisms produced by the new techniques.

Established transgene delivery techniques, such as microinjection, particle gun and *Agrobacterium*-virulence, result in the random integration of the transgene in the genome of the modified host organisms. Newer commercial versions of plant GMOs are also typically "stacked", i.e. harbour several new-trait encoding transgenes after conventional crossing of plants containing single transgenes/insertion events.

The biological uncertainty caused by the technological opportunity to introduce recombined DNA from several unrelated species into an organism, the unpredictable random insertion of such DNA, the variable levels of containment of the resulting GMO, and the discourse around their cultural and ethical dimensions resulted in early developments of regulatory frameworks on a global scale.

International definitions and regulatory frameworks of GMOs include the Convention on Biological Diversity´s Cartagena Protocol on Biosafety (CBD 2000) and the World Health Organization/Food and Agriculture Organization of the United Nations (WHO/FAO) Codex Alimentarius (WHO/FAO 2009). At the national level, GMOs undergo volunteer or mandatory regulation and labelling, depending on the regulatory system in place in the respective country in which they are produced or imported.

1.2. New Molecular Genetic Techniques

The rapid development of new genetic techniques provides opportunities for editing existing DNA within genomes at specific locations. This is enabled through for instance the use of *site-directed nucleases* (SDN). Hence, genetic modifications at particular sites within a genome can be produced without necessarily adding exogenous genes from other organisms or species. The realization of this technology-enabled opportunity is currently limited by biological knowledge of relevant sites and nucleotide alterations that will produce organisms with desired phenotypes.

In addition to the opportunities for limited editing within genomes, SDN can also be used to add recombinant DNA such as transgenes into genomes, but now with site-specificity. This is in contrast to the random insertions expected from traditional gene-delivery techniques. The range of organisms that can be engineered with the new techniques remains somewhat limited to those amenable to traditional genetic modifications due to the need to engineer or produce germline cells with heritable traits.

Besides SDN, other techniques have been developed to engineer genomes at specific sites (Häcker and Schetelig, this volume). When combined with broader conceptual approaches, they may be part of a larger strategy to genetically alter, brand and market a particular organism (Table 1).

Table 1. Various genome-modification techniques, processes and conceptual approaches[a]

Base editing	Changing single nucleotides at targeted sites in the genome with site-directed nucleases (SDN)
Gene drive	Mechanism ensuring increase in a defined genotype in a population from one generation to the next
Genome editing	Introducing nucleotide substitutions, insertions, or deletions at targeted sites with SDN
Grafting	Process of transplanting rootstock with GM material or vice versa
Intragenesis and cisgenesis	Production and content of organisms with all-native DNA (intragenesis) and in addition conserved structure of transferred gene (cisgenesis)
Oligo-directed mutagenesis (ODM)	Transformation of a genome with short oligonucleotides to obtain site-specific base changes
Paratransgenesis[b]	The presence of e.g. GM-microorganisms in the gut of non-GM insects
Recombined or recombinant DNA	DNA molecules formed by laboratory methods of genetic recombination to bring together genetic material from multiple sources, creating sequences that would not otherwise be found in the genome
Reverse breeding	Breeding strategy where intermediate organisms are GM but not the final organism
RNA interference	Process where RNA molecules interfere with specific gene expression of the target organism by neutralizing targeted mRNA molecules
Synthetic organism[c]	Denotes organisms produced through assembled DNA fragments (synthetically produced) or with extensively modified or transplanted genomes
Transgenesis	Describes the process of inserting recombined and most often unrelated DNA into a genome, usually at random sites.

[a] *See e.g. Lusser et al. 2011; EFSA 2012a, 2012b; ACRE 2013; AGES 2013; Eckerstorfer et al. 2014*
[b] *Regulated also as microbial pesticides (see OECD 2004; Aguilera et al. 2011)*
[c] *Concept relates mostly to microorganisms and industrial production systems*

Here I consider two approaches of particular interest for genetic alteration of insect genomes and populations. Firstly, the use of genome editing by SDN for the modification of individual insect genomes, and secondly, the use of gene drive mechanisms to make deliberate changes in the genetic composition of targeted insect populations.

2. GENOME EDITING

2.1. *Site-directed Nucleases*

Enzymes that degrade or generate breaks within a genome are called nucleases. A high number of nucleases are known, of which some have site-specificity. The best known are bacterial restriction enzymes whose discovery initiated the genetic engineering era. Additional types of SDN have more recently been described and further developed for genome engineering purposes; these include the CRISPR-Cas9 system, transcription activator-like effector nucleases (TALEN), zinc finger nucleases (ZFN), and meganucleases. They all offer the opportunity to target genetic alterations to a particular site in the genome. However, they differ in how they recognize the target site, and how and how fast target site-specificity can be changed.

Target site-specificity is due to particular protein motifs (TALEN, ZFN, meganucleases) or RNA sequences (CRISPR-Cas9). The CRISPR-Cas9 system is the most used due to the ease of targeting specific genomic locations by changing the nucleotide composition of the single guide RNA sequence (sgRNA of approx. 20 bases) (Baker 2014; Hsu et al. 2014). Modifications made in the *Cas* protein has also improved site-specificity, as well as offering the choice between introducing double strand breaks at the target site, versus single-strand breaks and substitutions (Komor et al. 2016; Slaymaker et al. 2016). Due to the broad range of genetic outcomes produced (Hilscher et al. 2017) various names for the technique have been proposed including base, gene or genome editing, and gene targeting.

Nevertheless, since the same SDN technique also can be used to produce larger insertions that clearly fall under current GMO regulations, the use of SDN is not as such indicative of a particular genetic outcome or regulatory pathway. Depending on the extent to which DNA is added in the genome editing process, three classes of outcomes have been described (EFSA 2012b):

- If no new DNA is added in combination with the SDN, spontaneous repair of the double strand breaks in the genome can result in minor nucleotide changes, of which the desired one can be selected for.
- If homologous DNA templates are added with the nucleases, homology-directed repair may result in specific minor nucleotide changes being introduced at the targeted site.
- If larger DNA templates, with or without homologous flanking DNA, are provided with the nucleases, larger DNA fragments encoding several new traits can be inserted.

The nuclease function itself in the engineering process can be delivered as proteins, or encoded by temporal DNA templates, or as part of the inserted DNA.

The DNA repair of the strand breaks produced by the introduced SDN is necessary to obtain the desired genetic alterations. Several natural DNA repair mechanisms are known that deal with spontaneous errors occurring in the genome, such as strand breaks, including non-homologous end joining (NHEJ) of DNA strands and homologous recombination (HR) mediated mechanisms.

The much-desired site-specificity of nucleases is limited by the occurrence of unique target sites (nucleotide compositions) in the genome and the extent the particular nuclease used has off-target activity and unintentionally cleaves at other genomic sites with similar nucleotide compositions. The efficiency of DNA repair mechanisms may also vary between species. Whereas nucleases with higher site-specificity can be engineered, the biology and genetics of the targeted organisms are also important, including knowledge of relevant genetic variation and DNA repair mechanisms. In practical terms, there are also challenges to efficient gene delivery and regeneration systems, as well as chromosome copy number (ploidy), for stable alterations and inheritance.

2.2. *SDN in a Regulatory Context*

Products resulting from genetic techniques, such as SDN, challenge the suitability of the long-held bimodal division of organisms into GMO or non-GMO. A new continuum of product categories now arises because SDN-based engineering is used both in the absence of and in the presence of added DNA sequences. The possible outcomes therefore represent a continuum of engineered organisms with minor changes of existing DNA (e.g. removal of, changes in, or addition of a few nucleotides) to cases where extensive recombined DNA has been added into the genome. Thus, new organisms produced by SDN techniques may span from closely mimicking natural processes occurring through spontaneous mutations to those resembling GMOs produced through recombinant DNA and transgene-based genetic modifications (Podevin et al. 2012). Hence, in some cases the engineered organisms may pose similar biological uncertainties as traditional GMOs when considering the effects of the novel trait(s). In other cases, they are unlikely to be detectable without prior knowledge of the specific nucleotide change introduced. The European Food Safety Authority (EFSA 2012b) has therefore proposed three categories for organism/products produced by SDN techniques (Table 2).

Table 2. Categories for organisms produced by site-directed nucleases (SDN)

SDN-1:	Contains site-specific random mutations or short deletions
SDN-2:	Contains specific minor nucleotide changes produced by homologous recombination with an introduced DNA fragment with sequence similarity to the target site
SDN-3:	Contains an introduced (often exogenous/recombinant) DNA fragment that was integrated via non-homologous end joining or homologous recombination

While SDN-3 clearly falls within the current regulatory framework of GMOs as a recombinant DNA fragment is added to the genome in the process, the regulatory context of SDN-1 and SDN-2 based products remain to be clearly determined and harmonized across countries. Moreover, there is a clear need to develop a uniform taxonomy to ensure uncontested definitions, uniform communication, transparency, and regulatory clarity.

Some major reports on genome editing, SDN and related techniques include European Food Safety Authority (EFSA) guidance (EFSA 2012a, 2012b; Eckerstorfer et al. 2014; Nutfield 2016).

Site-specific engineering undoubtedly offers increased precision and control of the engineering process. Thus, a reduction in the biological uncertainty arising from the engineering process itself is expected. Some process related uncertainty may nevertheless remain; for instance, from the variable consistency of the outcomes of the suite of related SDN techniques applied across laboratories, and on different genomes, species, regeneration systems, and traits. Thus, some of the uncertainty arising from the engineering process, that formed part of the rationale for the current regulation of GMOs, is not a priori absent from all products produced by SDN.

Issues resembling those considered in the assessments of GMOs that may materialize in some applications of the new techniques are:

1. Incorporation of recombined DNA from multiple/unrelated species
2. Co-integration of vector sequences used for technology delivery
3. Off-target cleavage causing random breaks and genome rearrangements
4. Modified/insertion sites containing new open reading frames
5. Variable expression and stability of intended traits
6. Available history of safe use of traits (e.g. in donor versus new recipient), and
7. Effect of new trait on host biology.

The relevance of the above issues must be assessed on a case-by-case basis. The broad utility of SDN-based engineering approaches has clearly contributed to the broader discourse on process versus product-triggered regulation (Sprink et al. 2016). The *process-based approach* is triggered in a regulatory context by the techniques used, and the *product-based approach* is triggered by the novelty of the product.

In conclusion, SDN-based products may or may not resemble traditional GMOs and a globally harmonized approach has not yet been negotiated. A key issue is to reach a shared understanding of a process-based versus a product-based approach to risk and regulation. Risk assessment and regulation of SDN-based products that are not considered covered by current regulation of GMOs is currently done case-by-case at the national level.

3. GENE DISPERSAL AT THE POPULATION LEVEL

3.1. Self-sustaining Mechanisms for Gene Dispersal in Wild Populations

Gene drive is a mechanism that increases a specific genotype from one generation to the next—by ensuring non-Mendelian inheritance of chromosomes during sexual reproduction. Gene drives are naturally occurring mechanisms that have been adapted for purposes of genetic engineering of wild populations. Noteworthy, the genetic engineering approach is deliberately developed to be *self-sustaining*, e.g. to ensure the spread of defined genetic material through a natural population in the environment—in sharp contrast to most other *self-limiting* genetic engineering approaches today that seek to document containment of the engineered trait in cultivated or domesticated species.

New gene drive systems rely on the opportunities presented by SDN (CRISPR-Cas9) for the intentional spread of genes in wild populations (Esvelt et al. 2014 and references within). Gene drive systems today mostly remain at the idea/developmental stage in which the potential usages of the technology in the field are being explored. A key limitation for the broad uptake of gene drive systems include the extent the traits are sufficiently stable once present in wild populations (Callaway 2017).

It is noted that population control techniques such as the use of sterility (mainly induced by radiation) are not currently relying on gene drives. The Sterile Insect Technique (SIT) acts at the population scale, based on the systematic area-wide release of sterile insects that outcompete (in numbers) the fertile insects of the target pest population. The result being that the offspring is non-viable and the population declines. The SIT has been successfully employed for many decades and relies on radiation-based sterilisation (Dyck et al. 2021).

More recently, inducible male sterility has been achieved by recombinant DNA and traditional genetic engineering that results in mosquitoes and agricultural pests regulated as GMOs (Alphey 2014; WHO 2014; USDA 2017).

3.2. Gene Drive Systems in a Regulatory Context

The comparative approach that has proven useful in framing the assessment of self-limiting GMOs may be of little utility to assess intentional spread of self-sustaining DNA in wild populations by gene drive systems. The environmental impact assessment of gene drive technology requires robust knowledge of the affected population´s structure, size, behaviour, migration patterns, reproduction and generation time. Moreover, the assessment also rests on a robust knowledge of the population´s interactions with other organisms at various trophic levels in the receiving environment (see for instance USDA 2017). Limited availability of data on these aspects often reduces the opportunities for data-driven risk assessments (see van Lenteren et al. 2006).

It is currently not clear who is responsible for generating data that would allow a more in-depth understanding of wild populations targeted for gene drive approaches (e.g. agricultural or human health pests). It is also not fully clear how biological data collected at one location can be considered relevant for other areas. This will have an impact on the validity of risk assessment conclusions, that under situations of limited data, may partly rely on extrapolation between species, populations and environments. Trait stability is also a potential issue and resistance to an introduced gene drive mechanism in populations may develop (Callaway 2017).

Finally, gene drives must be considered in a tempo-spatial context. The spatial component also includes knowledge of a drive's dissemination pattern over time, within and between populations, and the potential for cross-border migration of gene drive-modified organisms.

Broad reviews of concepts and applications prior to release of GMOs with a gene drive system, as well as public discussions of technology that has the potential to affect the global commons, have been called for (Oye et al. 2014). An extensive report by the National Academy of Sciences of the USA (NAS 2016) concluded that there is insufficient evidence available to support the release of gene drive-modified organisms into the environment at this time. However, the report also acknowledges that gene drives have significant potential for basic and applied research, including implementing highly controlled field trials (NAS 2016).

In conclusion, current gene drive proposals are limited by uncertainties in the knowledge of the affected system and the target populations. Most new proposals rely on introducing self-limiting SDN-based transgenes and will fall under current GMO regulations. Many gene drive-modified organisms are yet at the exploratory stage and initial field releases will be regulated according to national GMO legislation.

4. TECHNOLOGICAL DEVELOPMENTS IN CONTEXT

Genome editing by SDN have been called a transformative and disruptive technology. The degree and speed of adoption of new technologies nevertheless takes place in a social context (Ishii and Araki 2017). That social context will vary between countries and include differences in values, ethics, politics and priorities. Differences in these variables translate into differences in regulation. Understanding drivers of technological innovation and broadly accepted frames for benefit-cost analyses is also of importance and may vary with culture and market opportunities.

The social context also includes the role of expert and expert cultures, which defines valid concerns for risk assessment and how scientific uncertainty and knowledge gaps are addressed and communicated at various levels (Nielsen and Myhr 2007).

Furthermore, standards and principles for approaches to risk and uncertainty are negotiated at both the national and international levels. Currently, such approaches are usually limited to defined biological risks, and internationally harmonized approaches to the broader issues encompassed in benefit-cost analyses are not yet available.

The broader issues include the ethical dimensions of importance for a broader societal acceptance. For instance, a clear understanding of how risk and benefits are or can be distributed among various stakeholders is needed (Nutfield 2016; ECNH 2016). In this context, there are both calls for more proportionate treatment of risks in a process versus product-based perspective, as well as calls for the need of balancing the exposure of risks with the beneficiaries.

In all cases, different types of uncertainty and how they are handled represent an integral component of the process of technology introductions. Procedures for transparency in the treatment and communication of risks identified in assessment have recently been proposed (EFSA 2017). Improved approaches to handling and communicating uncertainty may substantially improve transparency of processes and help further public trust in procedures.

The way forward for the technological opportunities presented here will be found in the landscape of biological, regulatory and political uncertainty as shaped by further scientific advances, values, and culture. The level of public trust created through the framing of risk, engagement and transparency seems essential to optimize the non-linear trajectory between technological opportunity and adoption.

5. REFERENCES

(ACRE) Advisory Committee on the Releases to the Environment. 2013. ACRE advice: New techniques used in plant breeding. UK. 44 pp.

Alphey, L. 2014. Genetic control of mosquitoes. Annual Review of Entomology 59: 205–224.

(AGES) Austrian Agency for Health and Food Safety. 2013. New plant breeding techniques. RNA-dependent methylation, reverse breeding. Federal Ministry of Health. Vienna, Austria. 72 pp.

Aguilera, J., A. Gomes, and K. M. Nielsen. 2011. Genetically modified microbial endosymbionts as insect pest controllers: Risk assessment through the European legislations. Journal of Applied Entomology 135: 494–502.

Baker, M. 2014. Gene editing at CRISPR speed. Nature Biotechnology 32: 309–312.

Callaway, E. 2017. Gene drives meet the resistance. Evolution could weaken technique's potential in the wild. Nature 542: 15.

(CBD) Convention on Biological Diversity. 2000. Cartagena Protocol on Biosafety to the Convention on Biological Diversity: Text and annexes. Secretariat of the Convention on Biological Diversity. Montreal, Canada. 31 pp.

Dyck, V. A., J. Hendrichs, and A. S. Robinson (eds.). 2021. Sterile Insect Technique – Principles and practice in Area-Wide Integrated Pest Management. Second Edition. CRC Press, Boca Raton, Florida, USA. 1200 pp.

Eckerstorfer, M., M. Miklau, and H. Gaugitsch. 2014. New plant breeding techniques and risks associated with their application. Report 0477, Umweltbundesamt GmbH. Vienna, Austria. 92 pp.

(ECNH) Federal Ethics Committee on Non-Human Biotechnology. 2016. New plant breeding techniques – Ethical considerations. Bern, Switzerland. 32 pp.

(EFSA) European Food Safety Authority. 2012a. Panel on Genetically Modified Organisms (GMO: Scientific opinion addressing the safety assessment of plants developed through cisgenesis and intragenesis. EFSA Journal 10: 2561.

(EFSA) European Food Safety Authority. 2012b. Panel on Genetically Modified Organisms (GMO): Scientific opinion addressing the safety assessment of plants developed using Zinc Finger Nuclease 3 and other Site-Directed Nucleases with similar function. EFSA Journal 10: 2943.

(EFSA) European Food Safety Authority. 2017. Scientific committee guidance on uncertainty in EFSA scientific assessment, scientific opinion. Draft version, European Food Safety Authority, Parma, Italy, 211 pp.

Esvelt, K., A. L. Smidler, F. Catteruccia, and G. M. Church. 2014. Emerging technology: Concerning RNA-guided gene drives for the alteration of wild populations. eLife 3: e03401.

Evans, B. R., P. Kotsakiozi, A. L. Costa-da-Silva, R. S. Ioshino, L. Garziera, M. C. Pedrosa, A. Malavasi, J. F. Virginio, M. L. Capurro, and J. R. Powell. 2019. Transgenic *Aedes aegypti* mosquitoes transfer genes into a natural population. Scientific Reports 9: 13047.

(FAO/IPPC) Food and Agriculture Organization of the United Nations/International Plant Protection Convention. 2017. Glossary of phytosanitary terms. International Standard for Phytosanitary Measures (ISPM) No. 5. IPPC, Rome, Italy.

Hilscher, J., H. Bürstmayer, and E. Stoger. 2017. Targeted modification of plant genomes for precision crop breeding. Biotechnology Journal 12: 1600173.

Hsu, P. D., E. S. Lander, and F. Zhang. 2014. Development and applications of CRISPR-Cas9 for genome engineering. Cell 157: 1262–1278.

Ishii, T., and M. Araki. 2017. A future scenario of the global regulatory landscape regarding genome-edited crops. GM Crops & Food 8: 44–56.

Komor, A. C., Y. B. Kim, M. S. Packer, J. A. Zuris, and D. R. Liu. 2016. Programmable editing of a target base in genomic DNA without double stranded DNA cleavage. Nature 533: 420–424.

Lusser, M., C. Parisi, D. Plan, and E. Rodriquez-Cerezo. 2011. New plant breeding techniques. State-of-the-art and prospects for commercial development. Joint Research Center, European Commission EUR 24760 EN. Publications Office of the European Union, Luxembourg.

NAS (National Academy of Sciences). 2016. Gene drives on the horizon: Advancing science, navigating uncertainty, and aligning research with public values. The National Academies Press. Washington, DC, USA. 230 pp.

Nielsen, K. M. 2013. Biosafety data as confidential business information. PLoS Biology 11: e1001499.

Nielsen, K. M., and A. Myhr. 2007. Understanding the uncertainties arising from technological interventions in complex biological systems: The case of GMOs, pp. 108-122. *In* T. Traavik, and L. L. Ching (eds.), Biosafety first: Holistic approaches to risk and uncertainty in genetic engineering and Genetically Modified Organisms. Tapir Academic Press, Trondheim, Norway.

(Nutfield) Nutfield Council on Bioethics. 2016. Genome editing: An ethical review. Nuffield Council on Bioethics, London, UK. 136 pp.

(OECD) Organisation for Economic Cooperation and Development. 2004. Guidance for information requirements for regulation of invertebrates as biological control agents. OECD Series on Pesticides 21, JT00156791. OECD, Paris. 22 pp.

Oye, K. A., K. Esvelt, E. Appleton, F. Catteruccia, G. Church, T. Kuiken, S. B-Y. Lightfoot, J. McNamara, A. Smidler, and J. P. Collins. 2014. Regulating gene drives. Science 345: 626–628.

Podevin, N., Y. Devos, H. V. Davies, and K. M. Nielsen. 2012. Transgenic or not? No simple answer! EMBO Reports 13: 1057–1061.

Slaymaker, I. M., L. Gao, B. Zetsche, D. A. Scott, W. X. Yan, and F. Zhang. 2016. Rationally engineered Cas9 nucleases with improved specificity. Science 351: 84–88.

Sprink, T., D. Eriksson, J. Schiemann, and F. Hartung. 2016. Regulatory hurdles for genome editing: Process- vs. product-based approaches in different regulatory contexts. Plant Cell Reporter 35: 1493–1506.

(USDA) United States Department of Agriculture. 2017. Proposal to permit the field release of genetically engineered diamondback moth in New York. Environmental assessment April 2017. USDA-APHIS, Riverdale, Maryland, USA.

van Lenteren, J. C., J. Bale. F. Bigler. H. M. T. Hokkanen, and A. J. M. Loomans. 2006. Assessing risks of releasing exotic biological control agents of arthropod pests. Annual Reviews of Entomology 51: 609–634.

(WHO) World Health Organization. 2014. Guidance framework for testing of genetically modified mosquitoes. WHO, Geneva, Switzerland.

(WHO/FAO) World Health Organization/Food and Agriculture Organization of the United Nations. 2009. Foods derived from modern biotechnology. Second edition. FAO, Codex Alimentarius Commission, Joint FAO/WHO Food Standards Programme. Rome, Italy. 76 pp.

WILL THE "NAGOYA PROTOCOL ON ACCESS AND BENEFIT SHARING" PUT AN END TO BIOLOGICAL CONTROL?

J. C. VAN LENTEREN

Laboratory of Entomology, Wageningen University, P.O. Box 6, 6700AA, Wageningen, The Netherlands; joop.vanlenteren@wur.nl

SUMMARY

Biological control is one of the most environmentally safe and economically profitable pest management methods. Beneficial organisms used in biocontrol can be of native or exotic origin. As invasive species are being accidentally introduced at an ever-increasing rate, deliberate introductions of non-native biocontrol agents are often needed for the area-wide management of these invasive pests. However, recent regulations have delayed or prevented prospecting for new, non-native natural enemies. A first phase of regulation started in the 1980s and concerned the development of risk analyses for non-native species. At this time, as commercial biocontrol became popular and the number of species of biocontrol agents on the market quickly increased, many thought that risk analyses were needed to prevent non-experts importing and commercializing insufficiently studied organisms. However, implementation of (environmental) risk assessments for biocontrol agents has resulted in a slowdown in the use of new non-native natural enemies, and in higher project costs caused by the need to prepare elaborate application dossiers. These regulations were mainly aimed at preventing potential negative effects of releasing non-native biocontrol agents and, thus, in increasing confidence in this pest management method. The second phase of regulations started more recently and deals with the question "Who owns biological control agents?" At the Convention on Biological Diversity (CBD) in Rio de Janeiro (Brazil) in 1993, one of the three objectives formulated was "the fair and equitable sharing of the benefits arising out of the utilization of genetic resources". Biocontrol agents are such genetic resources. The Nagoya Protocol, a supplementary agreement to the CBD, provides a framework for the effective implementation of the fair and equitable sharing of benefits (i.e. the Access and Benefit Sharing (ABS) regulations) arising out of the utilization of genetic resources. Signatories of the Protocol are required to develop a legal framework to ensure access to genetic resources, benefit-sharing and compliance. Recent applications of CBD principles have already created barriers to collection and export of natural enemies for biocontrol research in several countries. If the Nagoya Protocol is widely applied, it may seriously interfere with searching for and application of biocontrol agents against invasive pests. Therefore, the International Organization for Biological Control (IOBC) first of all made an appeal to those involved in developing the legal framework for ABS, to design regulations that support the biocontrol sector by facilitating the exchange of biocontrol agents, including clear guidelines. Secondly, the IOBC also strongly recommended that biocontrol agents should be considered as a special case under the CBD, by creating a non-financial ABS regime, mainly because classical biocontrol is a non-for-profit activity, and both developing and developed countries benefit from the use of the same biocontrol agents. Thirdly, as prospecting for new non-native natural enemies has currently been suspended if not terminated

J. Hendrichs, R. Pereira and M. J. B. Vreysen (eds.), Area-Wide Integrated Pest Management: Development and Field Application, pp. 655–667. CRC Press, Boca Raton, Florida, USA.

in many countries due to CBD and ABS procedures, the IOBC prepared a best practices guide to assist the biocontrol community to demonstrate due diligence in complying with ABS requirements. The best practices guide includes a draft ABS Agreement for collection and study of biocontrol agents that can be used for scientific research and non-commercial release into nature by countries having signed the Nagoya Protocol. If many countries decide to implement the IOBC proposal for an agreement for collection and study of natural enemies, biocontrol might face a bright future.

Key Words: Biocontrol, pest management, beneficial organisms, invasive species, non-native natural enemies, risk assessment, Convention on Biological Diversity, International Organization for Biological Control, best practices guide for biological control

1. INTRODUCTION

Biological control (hereafter "biocontrol") – the use of an organism to reduce the population density of another organism – is one of the most environmentally safe and economically profitable pest management methods (Barratt et al. 2018). In biocontrol, parasitoids, predators, pathogens, herbivores and antagonists are used to reduce populations of pests, diseases and weeds (van Lenteren et al. 2018). These beneficial organisms can be native, but are also often non-native, particularly when the pest (defined by FAO (2017) as including diseases, weeds and animal pests) is of non-native origin.

Invasive species are being introduced accidentally around the world at an increasing rate, caused by increasing travel, trade, and tourism, have led and will continue to result in the introduction of pests of foreign origin (Bacon et al. 2012; Seebens et al. 2017). In contrast, deliberate introductions of non-native biocontrol agents have resulted in permanent control of many pests, while having caused remarkably few problems (Cock et al. 2010). It is nowadays often required in biocontrol to perform a risk assessment for new agents prior to obtaining approval for introduction and release (van Lenteren et al. 2006). Until a few years ago, prospecting for new, non-native natural enemies after accidental introductions of non-native pests was possible, and usually occurred with the consent of the country where prospecting took place (Cock et al. 2010).

The Access and Benefit Sharing (ABS) requirement of the Nagoya Protocol (SCBD 2011) has brought about an almost complete stop to natural enemy exploration programmes, whereas introduction of non-native pests is continuing, resulting in eradication projects with a frequent input of chemical pesticides causing negative effects on biodiversity, the environment and human health (Suckling et al., this volume). Although biocontrol researchers recognize the importance of a proper ABS procedure, the current state of affairs is highly bureaucratic and does not acknowledge the mutual benefits of biocontrol projects for countries providing and receiving beneficial organisms, projects that have been carried out for more than 100 years (Cock et al. 2010).

Biocontrol does not fall under *bioprospecting* ("the search for plant and animal species from which medicinal drugs and other commercially valuable compounds can be obtained") or *biopiracy* ("bioprospecting without permission of the country that owns the genetic resources and which exploits plant and animal species by claiming patents to restrict their general use"). Bioprospecting and biopiracy are often concerned with products that can be protected with intellectual property rights in order to generate monetary profits for companies (e.g. pharmaceuticals), which is not the case with beneficial organisms because they cannot be patented. Up until now, many biologists, including taxonomists and biocontrol practitioners, are still unaware of the implications of the Nagoya Protocol under the Convention on Biological Diversity (CBD) for their field of research.

In this chapter, I will first summarize the achievements of biological control, then describe the first phase of regulations related to biocontrol, next explain the Nagoya Protocol and, finally, discuss the consequences of the Nagoya Protocol for the practice of biocontrol.

2. ACHIEVEMENTS OF BIOLOGICAL CONTROL

Biocontrol has been in use for at least 2000 years, but modern use started at the end of the 19th century (DeBach 1964; van Lenteren and Godfray 2005). Four different types of biological control are usually distinguished: natural, conservation, classical, and augmentative biocontrol (Cock et al. 2010):

- *Natural biocontrol* is an ecosystem service (Millennium Ecosystem Assessment 2005) whereby pest organisms are reduced by naturally occurring beneficial organisms. It occurs in all of the world's ecosystems without any human intervention and is the greatest contribution of biocontrol to agriculture when expressed in economic terms (Waage and Greathead 1988; Cock et al. 2012).
- *Conservation biocontrol* consists of human activities protecting and stimulating the performance of naturally occurring natural enemies. This form of biological control is currently receiving a lot of attention for pest control (Barratt et al. 2018; Heong et al. this volume).
- In *classical biocontrol* (CBC), natural enemies are collected in the area of origin of the pest and then released in areas where the pest invaded, and when successfully established, results in area-wide and permanent pest control providing large economic benefits (Cock et al. 2010, 2016a; Barratt et al. 2018).
- In *augmentative biocontrol* (ABC), native or non-native natural enemies are mass-reared for repeated release in large numbers to obtain immediate control of pests, usually in seasonal crops (van Lenteren et al. 2018).

For CBC, recent reviews provide detailed information about successes. Cock et al. (2010, 2016b) present achievements of CBC of insect pests with insect natural enemies. One of the many striking examples of how successful CBC can be is that of a scale pest, *Icerya purchasi* Maskell (Hemiptera: Monophlebidae). This citrus pest was accidentally introduced in the 1880s in California (USA). Chemical control was impossible and citrus production was expected to have to be terminated. However, entomologists identified the country of origin of the pest (Australia), collected its natural enemies, shipped them to California for release in citrus orchards. Within a few years the natural enemies, including the Vedalia beetle *Rodolia cardinalis* (Mulsant) (Coleoptera: Coccinellidae), spread over the citrus area and saved the citrus industry in California and later in many other scale-infested citrus areas.

The *Rodolia* beetle has already controlled the *Icerya* scale pest on citrus for more than 100 years in more than 50 countries without causing any negative side effect (Cock et al. 2010). Cock et al. (2010, 2016b) provide many other examples of successful CBC of arthropod pests, while Winston et al. (2014) and Shaw and Hatcher (2017) summarize numerous examples of CBC of weeds.

By the end of 2010, 6158 introductions, using 2384 different species of natural enemies against 588 pest species had been made in 148 countries, of which 2007 (32.6 %) led to establishment, and 620 (10.1 %) resulted in satisfactory control of 172 (29.3 %) different pest species. The 10% success rate can be considered as very high, particularly when compared with chemical control where 1 out of 20 000 – 1 000 000 candidate compounds may kill pest insects. Not only are the success rates in CBC impressive, but also the benefit-cost ratios are striking and in the order of 20:1–1000:1 (Cock et al. 2010; Barratt et al. 2018). These high benefit-cost ratios can be explained by the fact that once a good natural enemy has been found, pest control is permanent, unless it is disrupted by use of chemical pesticides that kill the biocontrol agent.

Cock et al. (2016b) mentioned that the number of CBC introductions has decreased each decade during the past 40 years. An important factor in the decrease during the past two decades is the development and implementation of the Nagoya Protocol (SCBD 2011).

Van Lenteren et al. (2018) recently summarized successes of ABC, which has been applied for more than 120 years in several cropping systems and is now estimated to be used on 30 million ha worldwide. Today, ABC is applied in many areas of agriculture, such as fruit and vegetable crops, cereals, maize, cotton, sugarcane, soybeans, grapes and many greenhouse crops. Since the 1970s, ABC has moved from a cottage industry to professional research and production facilities, as a result of which many efficient agents have been identified, quality control protocols, mass production, shipment and release methods matured, and adequate guidance for farmers and extension agents has been developed (van Lenteren 2003; Ravensberg 2011).

About 350 species of invertebrate natural enemies, as well as 209 microbial strains from 94 different species are currently commercially produced (van Lenteren et al. 2018). Recent successes of ABC with arthropod natural enemies include the virtually complete replacement of chemical insecticides by predators (mites and hemipterans) to control thrips and whiteflies on sweet peppers in greenhouses in Spain (Calvo et al. 2012), and the use of hemipteran predators to control new invasive pests like the South American tomato moth *Tuta absoluta* (Meyrick) (Lepidoptera: Gelechiidae) (Urbaneja et al. 2012). Another recent ABC success deals with the use of microbial control agents. The invasion of the cotton bollworm *Helicoverpa armigera* (Hübner) (Lepidoptera: Noctuidae) into Brazil in 2012 caused tremendous damage to corn, cotton, and soybeans, because insecticides were not effective due to pest resistance, or were simply not available. Emergency approvals of the entomopathogenic bacterium *Bacillus thuringiensis* Berliner (Bacilliaceae) and baculovirus products provided farmers with the only effective control method at the time (J. R. P. Parra, personal communication).

From 1900 to 1959 only exotic biocontrol agents were used in ABC in Europe. After 1960, the use of native biocontrol agents in ABC increased. Due to implementation of various regulations, and particularly as a result of the application of the Nagoya Protocol, prospecting for non-native natural enemies has practically come to a standstill and a very clear shift in use of native natural enemies has taken place since 2000 (Fig. 1).

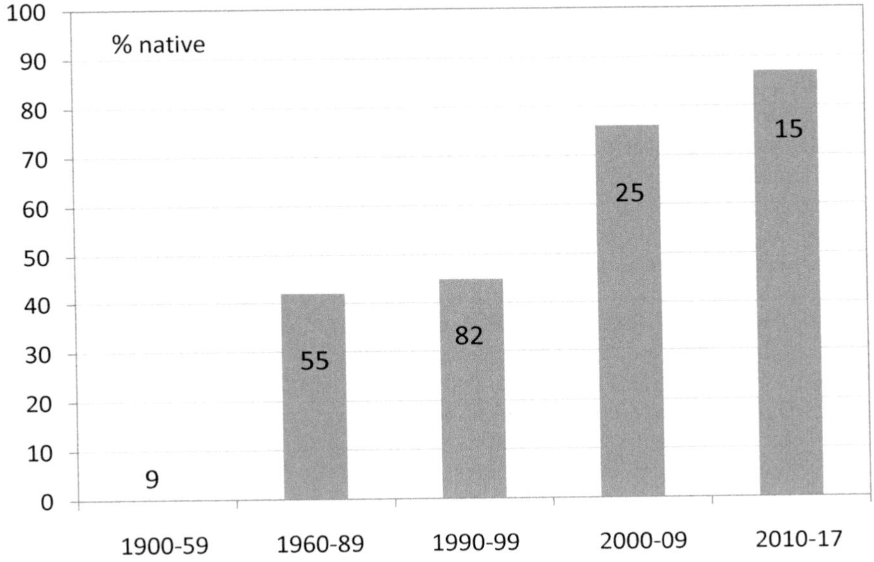

Figure 1. Percentage of new native natural enemies introduced on the European market for augmentative biocontrol through time (in columns; zero between 1900-1959). Total introductions per period are given in numbers (based on tables and supplementary material in van Lenteren et al. 2018).

3. FIRST PHASE OF REGULATION CONCERNING IMPORT AND RELEASE OF NATURAL ENEMIES

During the first century of "modern" biocontrol (1880 – 1980) few regulations existed concerning risk associated to insect biocontrol agents. However, most researchers in this field of ecology were well aware of the risk of importing certain types of natural enemies, particularly those that showed polyphagous predation behaviour and were, thus, often not only eating pests, but also other organisms (Boller et al. 2006).

In weed biocontrol this resulted in the so-called "centrifugal evaluation method" of potential biocontrol agents, whereby first species taxonomically closely related to the pest are tested and, if one or more of these are eaten by the herbivore, the next circle of related species is tested, and so on (Wapshere 1974).

Thorough risk analyses for natural enemies of arthropod pests were developed much later than for weed biocontrol, and pre-release environmental risk analyses for these natural enemies have only been applied since the 1980s. The reason might be that very few problems had been reported concerning negative effects of releases of non-native invertebrates for biological control (i.e. Follett and Duan 2000; Lynch et al. 2001). However, when ABC became popular in the 1980s and the number of species of biocontrol agents on the market quickly increased (see Fig. 2 in van Lenteren 2012), the need for pre-release environmental risk assessments for new natural enemies was realized, partly because non-experts also started to collect, import and release new biocontrol agents (Bigler et al. 2006).

In 2006, 20 countries had already implemented regulations for the import and release of biocontrol agents and many other countries followed. Initially, the Food and Agriculture Organization of the United Nations (FAO) together with the Centre for Agriculture and Bioscience International (CABI) and the International Organization for Biological Control (IOBC 2019) designed a code of conduct for import and release of biocontrol agents, but this non-compulsory guideline did not contain methods for risk assessment (FAO 1996, 2005). Next, IOBC took the initiative to develop standard methods that could be applied to produce data for risk assessment, as well as developing risk assessment methods including practical guidance on how to measure and evaluate effects leading to conclusions about risks and benefits of biocontrol agents under consideration, which are described in Bigler et al. (2006), van Lenteren et al. (2006), and many other papers.

The danger of current risk evaluations is that they only concentrate on possible negative effects caused by biocontrol agents, while not paying attention at the same time to:

1. The socio-economic effects of the damage brought about by the pest
2. The negative health and environmental effects produced by chemical pesticides, and
3. The many potential benefits from biocontrol (van Lenteren and Loomans 2006; Heimpel and Cock 2018).

Environmental risk assessments are now being used by a growing number of countries. Most risk assessments are characterized by the following general elements:

1. Characterization of natural enemy, i.e. information about the taxonomic status of the agent and its biology.
2. Human health risks, i.e. information about human health risks that is often much easier to obtain than for chemical pesticides, particularly for invertebrate natural enemies.
3. Environmental risks, i.e. collection of information on the environmental risks is usually the most time consuming aspect of the risk assessment and consists, amongst other things, of information on potential for establishment in the country of introduction, the prey/host range of the natural enemy (including tests of unrelated, beneficial, and rare and culturally valued species), dispersal capacity, and potential direct and indirect non-target effects that might be caused in the country of introduction. As collection of information for this element is generally very costly, a stepwise procedure for evaluation has been proposed, in order to be able to quickly eliminate obviously risky species (see Fig. 3 in van Lenteren et al. 2006).

4. Efficacy, i.e. a very different approach is proposed to obtain information about efficacy as compared with the one usually followed for chemical pesticides. Biocontrol agents are almost without exception used in integrated pest management (IPM) programmes, and therefore the efficacy does not necessarily have to be in the order of 95-100%. Any significant reduction in pest numbers by the biocontrol agents contributes to the overall pest management, and, thus, it suffices to show that an efficacious natural enemy is capable of significantly reducing pest populations.

The routine implementation of environmental risk assessments for biocontrol agents has resulted in a slowdown of newly marketed non-native ABC biocontrol agents (Figs. 1 and 2) and introductions for CBC (Fig. 3). The preparation of elaborate application dossiers has also resulted in higher developmental costs, but it did not bring prospecting for new non-native species to an end.

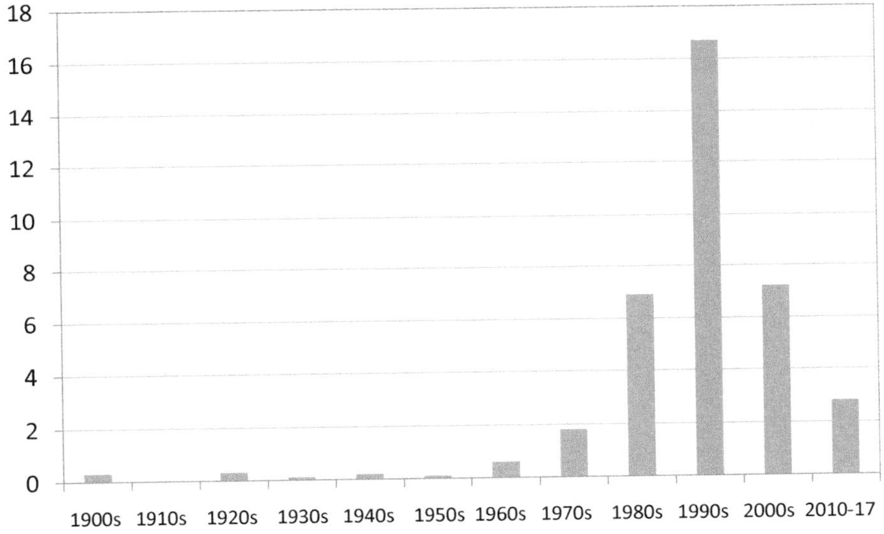

Figure 2. Average number of newly marketed natural enemy species per year within a decade for augmentative biocontrol (based on tables and supplementary material in van Lenteren et al. 2018).

The main problem in some parts of the world is a lack of harmonization of regulations. In Europe, for example, some countries apply no regulations at all, while in other countries they vary from rather easy to extremely complicated, and in our experience, the less expertise a country has in the field of biocontrol, the longer the evaluation of new biocontrol agents takes due to over-regulation.

Still, many biocontrol specialists consider it fair that risks of releasing non-native organisms need to be evaluated prior to import, that careful evaluation of risk and benefits will maintain or strengthen the excellent reputation of this sustainable and environment-friendly form of pest management and, that it in the end will result in even more confidence in biocontrol.

The dramatic decline in newly marketed natural enemies since 2000 is not the result of a decrease in the sale of biocontrol agents or of the area on which the agents are applied. On the contrary, the ABC market showed an annual increase of sales of 10% until 2005 and more than 15% per year since 2005 (Dunham 2015).

Fig. 3 shows the average number of natural enemy introductions per year within a certain decade for CBC. Cock et al. 2016b described an introduction as a unique combination of a biocontrol agent, target country and first year of introduction. Introduction of a certain species of natural enemy into five different countries is thus considered as five introductions. Since the start of CBC in the 1880s, a steady increase in number of introductions can be observed, with two periods showing lower numbers of introductions due to WWI and II.

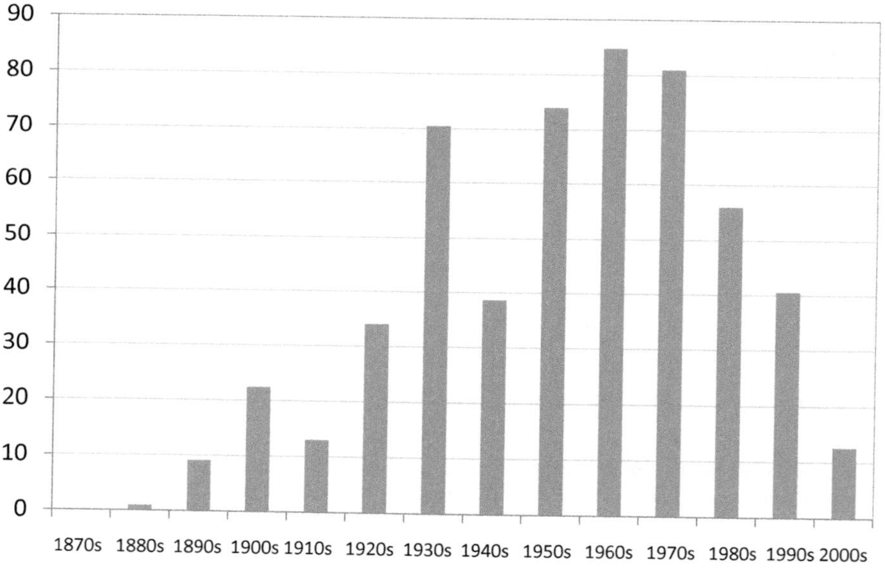

Figure 3. Average number of natural enemy introductions per year within a decade for classical biocontrol (based on Figure 1 in Cock et al. 2016b and data provided by M. Cock).

Then, starting in the 1980s the number of introductions decreases. According to Cock et al. 2016b this was caused by a growing risk-averse culture in many countries, which resulted in the need for more in-depth studies for non-target and environmental impact, and more recently, by the application of principles resulting from the Convention on Biological Diversity, which are discussed in the next Section.

4. SECOND PHASE OF REGULATIONS CONCERNING ACCESS AND BENEFIT SHARING

The first phase of regulations was aimed at improving biocontrol, preventing potential negative effects, and increasing confidence. The second phase of regulations has nothing to do with the science of biocontrol, but deals with the question "Who owns biological control agents?" This Section is based on recent papers written by Cock et al. (2010), Mason et al. (2018), as well as discussions among members of IOBC Global Commission on Access and Benefit Sharing (ABS 2019).

At the Rio CBD (CBD 1993) three objectives were formulated: conservation of biological diversity, sustainable use of its components, and fair and equitable sharing of the benefits arising from the utilization of genetic resources. The CBD is an international framework convention, and its provisions are binding for its contracting parties. However, the CBD cannot prescribe how decisions are to be implemented by the parties since different countries have different legal structures. Countries have sovereign rights over their biological resources, and agreements governing the access to these resources and the sharing of the benefits arising from them should be established and agreed between the parties involved.

In 2002, the Conference of Parties (COP) to the CBD adopted the 'Bonn Guidelines on Access to Genetic Resources and Fair and Equitable Sharing of Benefits Arising out of their Utilization - Decision VI/24' (SCBD 2002). Genetic resources are defined as genetic material, i.e. material containing functional units of heredity that is of actual or potential value (CBD Art. 2), so this includes all biocontrol agents taken from one country (provider) to another (recipient) (Cock et al. 2010). The value of the genetic resources need not be commercial (i.e. monetary) but may be scientific or academic in nature. As the CBD definition also includes the potential value of such resources, in effect all genetic material falls under the provisions of the ABS system (Cock et al. 2010).

The Nagoya Protocol came into force in October 2014 and is a supplementary agreement to the CBD that provides a framework for the effective implementation of the fair and equitable sharing of benefits arising from the utilization of genetic resources (SCBD 2011). Signatories and countries acceding to the Nagoya Protocol are required to develop a legal framework to ensure access to genetic resources, benefit-sharing and compliance. The biocontrol community needs to comply with ABS regulations arising under the Nagoya Protocol (Mason et al. 2018). The ABS is a potentially serious threat to the use of biocontrol for two important reasons:

1) the necessity of agreements governing access to genetic resources, and
2) sharing of benefits arising from their use between the parties involved.

Recent applications of CBD principles have already created barriers to collect and export natural enemies for biocontrol research and application in several countries. If the Nagoya Protocol is widely applied, it may seriously interfere with searching for and application of biocontrol agents against invasive pests. Therefore, the IOBC Global Commission on Biological Control and Access and Benefit Sharing first of all made an appeal for ABS regulations that support the biocontrol sector by facilitating the exchange of biocontrol agents, including clear guidelines (ABS 2019). These guidelines should also include fast track procedures for finding and applying biocontrol agents in case of humanitarian or emergency situations, such as after unintentional export of an invasive pest to a new area. The IOBC Commission also strongly recommended that biocontrol agents should be considered as a special case with respect to an ABS regime under the CBD (Cock et al. 2010) by creating a non-financial ABS regime for classical biocontrol, as countries providing biocontrol agents are also users of the technology, in view that:

1) biocontrol is widely used in both developing and developed countries often with the same biocontrol agents,

2) biocontrol agents cannot be patented,

3) collected information about biocontrol agents is shared publicly,

4) indigenous/traditional knowledge has not been relevant, and

5) the social benefits of biocontrol (such as increase of environmental and public health, reduction in pesticide use and costs of crop production), are a public good (Cock et al. 2010).

As prospecting for new non-native natural enemies has currently been suspended, if not terminated due to CBD and ABS procedures (or lack of procedures) in many countries, the IOBC Commission prepared a best practices guide to assist the biocontrol community to demonstrate due diligence in complying with ABS requirements, which should include the following components outlined in Mason et al. (2018): (1) collaborations to facilitate information exchange about what biocontrol agents are available and where they may be obtained, (2) knowledge sharing through freely available databases that document successes (and failures), (3) cooperative research to develop capacity in source countries, and (4) transfer of production technology to provide opportunities for small-scale economic activity.

Within the IOBC best practice guide for exchange of biocontrol genetic resources, the section concerning gaining access to biocontrol agents is of particular importance, and the IOBC Commission designed a draft ABS agreement for collection and study of biocontrol agents that can be used for scientific research and non-commercial release into nature by countries having signed the Nagoya Protocol. The IOBC Commission also proposed conditions under which biocontrol agents can be provided or should not be provided for countries where ABS regulations are not restrictive or do not exist. Details of the draft IOBC ABS agreement and the conditions can be found in Appendix 1 and 2 in Mason et al. (2018).

The purpose of the agreement is to set out the conditions for the use of naturally occurring biocontrol agent genetic resources, any associated traditional knowledge (TK), and the sharing of resulting benefits between the parties concerned in accordance with the CBD. The agreement is designed to promote non-commercial activities, such as research in taxonomy, ecology, and genetics, and to foster conservation and the environmentally sound and sustainable use of biocontrol agents. The objective of the agreement is to provide a sound basis for cooperation, transparency, communication and trust between the parties, taking into account the concerns of both providers and users of biocontrol agents. The agreement is based on prior informed consent and mutually agreed terms issued beforehand by the provider to the user for the access to the biocontrol agents, and specifies the terms for access to biocontrol agents, their utilization, their possible transfer to third parties and for sharing the benefits resulting from their utilization.

The IOBC will use this document to negotiate the design of regulations that support the biocontrol sector by facilitating the exchange of biocontrol agents with CBD, and encourages those involved in the practice of biocontrol to follow the IOBC best practice guide for exchange of biocontrol genetic resources as published in Mason et al. (2018).

5. THE FUTURE OF BIOLOGICAL CONTROL IN LIGHT OF THESE REGULATIONS

As explained above, the first phase of regulations was mainly aimed at preventing potential negative effects of release of non-native biocontrol agents and, thus, in increasing confidence in this pest management method. These first phase regulations have led to slower development of and higher costs to implement programmes that use new biocontrol agents, as well as a change in biocontrol approaches by first evaluating native natural enemies when a new pest arrives.

The second phase of regulation relates to implementation of the Nagoya Protocol and is having a much more drastic effect on the science of biocontrol. Although some countries have declined to place restrictions on access to their genetic resources for non-commercial research, including biological control, others have enacted legislation. However, many countries have yet to enact legislation and set up regulatory bodies, so that procedures are not in place to make decisions.

Prospecting for new natural enemies has been greatly reduced, biocontrol researchers risk imprisonment in some countries when collecting species, and the bureaucratic procedures to establish an ABS agreement are so unclear and time-consuming that none has been realized to date. This is a very unwelcome development for both developing and developed countries because a clean, safe and sustainable form of pest management may be replaced by environmentally polluting and health threatening chemical pesticides.

Moreover, the current state of affairs concerning ABS are creating serious risks when accidentally exported pests, disease and weeds invade new regions in the world, because their naturally occurring biocontrol agents in their country of origin can no longer be studied, imported and released in recently invaded areas.

Hopefully the IOBC proposal for an agreement for collection and study of biocontrol agents will be embraced and implemented by biocontrol scientists in many countries as well as by CBD, and soon result in renewed activities in both classical and augmentative biocontrol based on exchange of non-native biocontrol agents.

6. ACKNOWLEDGEMENTS

Barbara Barratt (AgResearch Invermay, Mosgiel, New Zealand), Jacques Brodeur (Institut de Recherche en Biologie Végétale, Université de Montréal, Montréal, Canada), Matthew Cock (CABI, Egham, UK), and Peter Mason (Agriculture and Agri-Food Canada, Ottawa, Canada) are thanked for reading and improving this chapter.

7. REFERENCES

(ABS) Commission on Biological Control and Access and Benefit Sharing. 2019. International Organization for Biological Control (IOBC).

Bacon, S. J., S. Bacher, and A. Aebi. 2012. Gaps in border controls are related to quarantine alien insect invasions in Europe. PLoS One 7(10): e47689.

Barratt, B. I. P., C. V. C. Moran, F. Bigler, and J. C. van Lenteren. 2018. The status of biological control and recommendations for improving uptake for the future. BioControl **63: 155–167.**

Bigler, F., D. Babendreier, and U. Kuhlmann. 2006. Environmental impact of invertebrates for biological control of arthropods: Methods and risk assessment. CABI Publishing, Wallingford, Oxon, UK. 288 pp.

Boller, E. F., J. C. van Lenteren, and V. Delucchi. 2006. International Organization for Biological Control of Noxious Animals and Plants (IOBC): History of the first 50 years (1956 – 2006). Ponsen & Looijen. Wageningen, The Netherlands.

Calvo, F. J., K. Bolckmans, and J. E. Belda. 2012. Biological control-based IPM in sweet pepper greenhouses using *Amblyseius swirskii* (Acari: Phytoseiidae). Biocontrol Science and Technology 22: 1398–1416.

Cock, M. J. W., J. C. van Lenteren, J. Brodeur, B. I. P. Barratt, F. Bigler, K. Bolckmans, F. L. Cônsoli, F. Haas, P. G. Mason, and J. R. P. Parra. 2010. Do new access and benefit sharing procedures under the Convention on Biological Diversity threaten the future of biological control? Biocontrol 55: 199–218.

Cock, M. J. W., J. C. Biesmeijer, R. J. C. Cannon, P. J. Gerard, D. Gillespie, J. J. Jiménez, P. M. Lavelle, and S. K. Raina. 2012. The positive contribution of invertebrates to sustainable agriculture and food security. CAB Reviews: Perspectives in Agriculture, Veterinary Science, Nutrition and Natural Resources 7(43): 1–27.

Cock, M. J. W., R. K. Day, H. L. Hinz, K. M. Pollard, S. E. Thomas, F. E. Williams, A. B. R. Witt, and R. H. Shaw. 2016a. The impacts of some classical biological control successes. CAB Reviews: Perspectives in Agriculture, Veterinary Science, Nutrition and Natural Resources 10(42): 1–58.

Cock, M. J. W., S. T. Murphy, M. T. K. Kairo, E. Thompson, R. J. Murphy, and A. W. Francis. 2016b. Trends in the classical biological control of insect pests by insects: An update of the BIOCAT database. BioControl 61: 349–363.

(CBD) Convention on Biological Diversity. 1993. Convention on Biological Diversity (with annexes). Concluded at Rio de Janeiro on 5 June 1992. United Nations - Treaty Series 1760(30619): 142–382.

DeBach, P. 1964. Biological control of insect pests and weeds. Chapman and Hall, London, UK. 844 pp.

Dunham, W. C. 2015. Evolution and future of biocontrol. Paper presented at the 10th Annual Biocontrol Industry Meeting (ABIM), October 20th, Basel, Switzerland.

(FAO) Food and Agriculture Organization of the United Nations. 1996. Code of conduct on the import and release of biological control agents. International Standards for Phytosanitary Measures No. 3. International Plant Protection Convention. FAO, Rome, Italy.

(FAO) Food and Agriculture Organization of the United Nations. 2005. Guidelines for the export, shipment, import and release of biological control agents and other beneficial organisms. International Standards for Phytosanitary Measures No. 3. International Plant Protection Convention. FAO, Rome, Italy.

(FAO) Food and Agriculture Organization of the United Nations. 2017. Glossary of phytosanitary terms. International Standards for Phytosanitary Measures No. 5. International Plant Protection Convention. FAO, Rome, Italy.

Follett, P. A., and J. J. Duan (eds.). 2000. Non-target effects of biological control. Springer Science+Business Media, LLC, New York, NY, USA. 316 pp.

Heimpel, G., and M. J. W. Cock. 2018. Shifting paradigms in the history of classical biological control. BioControl 63: 27–37.

(IOBC) International Organization for Biological Control / Organisation Internationale de Lutte Biologique (OILB). 2019. http://www.iobc-global.org

Lynch, L. D., H. M. T. Hokkanen, D. Babendreier, F. Bigler, G. Burgio, Z. H. Gao, S. Kuske, A. J. M. Loomans, I. Menzler-Hokkanen, M. B. Thomas, M. G. Tommasini, J. K. Waage, J. C. van Lenteren, and Q. Q. Zeng. 2001. Indirect effects in the biological control of arthropods with arthropods, pp. 99–125. *In* E. Wajnberg, J. C. Scott, and P. C. Quimby (eds.), Evaluating indirect ecological effects of biological control. CABI Publishing, Wallingford, Oxon, UK.

Mason, P. G., M. J. W. Cock, B. I. P. Barratt, J. Klapwijk, J. C. van Lenteren, J. Brodeur, K. A. Hoelmer, and G. E. Heimpel. 2018. Best practices for the use and exchange of invertebrate biological control genetic resources relevant for food and agriculture. BioControl 63: 149–154.

Millennium Ecosystem Assessment. 2005. Ecosystems and human well-being: Synthesis. World Resources Institute. Island Press, Washington, DC, USA. 137 pp.

Ravensberg, W. J. 2011. A roadmap to the successful development and commercialization of microbial pest control products for control of arthropods. Springer, Dordrecht, The Netherlands. 383 pp.

(SCBD) Secretariat of the Convention on Biological Diversity. 2002. Bonn guidelines on access to genetic resources and fair and equitable sharing of the benefits arising out of their utilization. Convention on Biological Diversity, Montreal, Canada.

(SCBD) Secretariat of the Convention on Biological Diversity. 2011. Nagoya protocol on access to genetic resources and the fair and equitable sharing of benefits arising from their utilization to the Convention on Biological Diversity: Text and annex. Convention on Biological Diversity, Montreal, Canada.

Seebens, H., T. M. Blackburn, E. E. Dyer, P. Genovesi, P. E., Hulme, J. M. Jeschke, S. Pagad, P. Pysek, M. Winter, M. Arianoutsou, S. Bacher, B. Blasius, G. Brundu, C. Capinha, L. Celesti-Grapov, W. Dawson, S. Dullinger, N. Fuentes, H. Jaeger, J. Kartesz, M. Kenis, H. Kreft, L. Kuehn, B. Lenzner, A. Liebhold, A. Mosena, D. Moser, M. Nishino, D. Pearman, J. Pergl, W. Rabitsch, J. Rojas-Sandoval, A. Roques, S. Rorke, S. Rossinelli, H. E. Roy, R. Scalera, S. Schindler, K. Stajerova, B. Tokarska-Guzik, M. van Kleunen, K. Walker, P. Weigelt, T. Yamanaka, and F. Essl. 2017. No saturation in the accumulation of alien species worldwide. Nature Communications 8 (14435): 1–9.

Shaw, R. H., and P. E. Hatcher. 2017. Weed biological control, pp. 215–243. *In* Hatcher, P. E. and R. J. Froud-Williams (eds.), Weed research: Expanding horizons. Wiley, UK.

Urbaneja, A., J. Gonzalez-Cabrera, J. Arno, and R. Gabarra. 2012. Prospects for the biological control of *Tuta absoluta* in tomatoes of the Mediterranean basin. Pest Management Science 68: 1215–1222.

van Lenteren, J. C. (ed.). 2003. Quality control and production of biological control agents: Theory and testing procedures. CABI Publishing, Wallingford, Oxon, UK.

van Lenteren, J. C. 2012. The state of commercial augmentative biological control: Plenty of natural enemies, but a frustrating lack of uptake. BioControl 57: 71–84.

van Lenteren, J. C., and H. C. J. Godfray. 2005. European science in the enlightenment and the discovery of the insect parasitoid life cycle in The Netherlands and Great Britain. Biological Control 32: 12–24.

van Lenteren, J. C., and A. J. M. Loomans. 2006. Environmental risk assessment: Methods for comprehensive evaluation and quick scan, pp. 254–272. *In* F. Bigler, D. Babendreier, U. Kuhlmann (eds.), Environmental impact of invertebrates for biological control of arthropods: Methods and risk assessment. CABI Publishing, Wallingford, Oxon, UK.

van Lenteren, J. C., J. Bale, F. Bigler, H. M. T. Hokkanen, and A. J. M. Loomans. 2006. Assessing risks of releasing exotic biological control agents of arthropod pests. Annual Review of Entomology 51: 609–634.

van Lenteren, J. C., K. Bolckmans, J. Köhl, W. Ravensberg, and A. Urbaneja. 2018. Biological control using invertebrates and microorganisms: Plenty of new opportunities. BioControl 63: 39–59.

Waage, J. K., and D. J. Greathead. 1988. Biological control: Challenges and opportunities. Philosophical Transactions Royal Society London Series B 318: 111–128.

Wapshere, A. J. 1974. A strategy for evaluating the safety of organisms for biological weed control. Annals of Applied Biology 77: 201–211.

Winston, R. L., M. Schwarzländer, H. L. Hinz, M. D. Day, M. J. W. Cock, and M. H. Julien (eds.). 2014. Biological control of weeds: A world catalogue of agents and their target weeds, 5th edition. USDA Forest Service, Forest Health Technology Enterprise Team, Morgantown, West Virginia, USA. 838 pp.

BARRIERS AND FACILITATORS OF AREA-WIDE MANAGEMENT INCLUDING STERILE INSECT TECHNIQUE APPLICATION: THE EXAMPLE OF QUEENSLAND FRUIT FLY

A. MANKAD[1], B. LOECHEL[1] AND P. F. MEASHAM[2]

[1]CSIRO Land & Water, GPO Box 2583, Brisbane QLD 4001, Australia; aditi.mankad@csiro.au; barton.loechel@csiro.au
[2]Hort Innovation, 527 Gregory Terrace, Fortitude Valley QLD 4006, Australia; penny.measham@horticulture.com.au

SUMMARY

The area-wide management (AWM) of highly mobile insect pests such as tephritid fruit flies requires an integrated understanding of technical, social and institutional processes that drive a coordinated approach within a defined area. Furthermore, the success of an AWM programme is dependent upon the coordinated efforts of key stakeholders within the designated area (e.g. growers, community members). Yet, public views regarding AWM may not reflect those views held by scientists or stakeholders. Public considerations for acceptance are likely varied and multidimensional. A series of qualitative (phases 1-2) and quantitative (phase 3) studies examined stakeholder and community attitudes towards AWM to manage Queensland fruit fly *Bactrocera tryoni* (Froggatt) (Tephritidae) and the novel use of the Sterile Insect Technique (SIT) as a possible component of AWM. Research was conducted over three regions of varying pest prevalence, ranging from zero to endemic; participants included growers, extension officers, industry and government representatives, and members of the general public. Participants in this research were asked to consider potential barriers and facilitators to the widespread uptake of AWM integrating the SIT, including any relevant institutional-level factors. Combined data revealed potential social barriers to AWM and SIT uptake. Most notably, there were perceptions of low efficacy in successfully coordinating key social groups for the purposes of an AW-IPM approach, and a concern for the possibility of 'free-riders' within an area-wide system. On the other hand, innovation complexity and observability of outcomes were important factors contributing to acceptance of AWM. Importantly, all participants were influenced by the attitudes and behaviours of important others. Participants also identified key facilitators that could assist in the uptake of AWM using the SIT. These facilitators include the importance of trustworthy information sources and harnessing the persuasive influence of community champions and central packing houses on commercial growers. Overall, there was high stated acceptance for the SIT, both on-farm and in towns, as long as SIT application was found to be economically feasible at individual farm or household level and the community was adequately consulted.

J. Hendrichs, R. Pereira and M. J. B. Vreysen (eds.), Area-Wide Integrated Pest Management: Development and Field Application, pp. 669–692. CRC Press, Boca Raton, Florida, USA.

Key Words: Bactrocera tryoni, Tephritidae, social science, biosecurity, integrated pest management, public attitudes, behavioural science, stakeholders, coordination, social barriers, technology uptake, Australia

1. INTRODUCTION

The area-wide management (AWM) of mobile insect pests such as tephritid fruit flies requires a multidimensional integrated understanding of technical, social and institutional processes that drive a coordinated approach within a defined area. Such management of agricultural pests is dependent upon the success and efficacy of the technical treatments associated with control, as well as the social environment within which the management takes place. Indeed, the central tenet of AWM is coordination amongst all relevant stakeholders. This inherently implies an element of social interaction to achieve effective pest management over a region. In the AWM context, people with varying interests must come together for a common goal. Thus, there must be an understanding of the social and institutional mechanisms in place to enable the implementation of coordinated activities.

This chapter focuses on a series of qualitative (phase 1-2) and quantitative (phase 3) studies examining stakeholder and community attitudes towards AWM to control Queensland fruit fly, *Bactrocera tryoni* (Froggatt), for the eventual integration of the Sterile Insect Technique (SIT) as a component of AWM. Through phases 1 and 2 of this social science research, contextual barriers and facilitators of AWM – as perceived by farmers and members of the general public – were identified. Predictions could also be made with respect to underlying attitudinal drivers of acceptance of AWM and the integration of the SIT, and intentions to participate in an AWM approach, at the completion of phase 3. While AWM can take place without the integration of the SIT, the focus of this project was AWM for eventual SIT implementation. Therefore, the results are discussed in this context.

1.1. *Queensland Fruit Fly: A Significant Horticultural Pest*

Prior to European settlement of Australia, the Queensland fruit fly was considered endemic to the tropical and subtropical rainforests of north-eastern Australia (Meats 1981; Reynolds and Orchard 2015). It has subsequently spread from this habitat and is now considered endemic in most of east-coast mainland Australia, including temperate areas, except where under regulatory control (Dominiak and Daniels 2012). In Queensland, *B. tryoni* occurs in high numbers year-round with 3-4 generations per year in the southern areas (Meats 1981). Meats (1981) believed that populations in temperate regions were transient, due to populations immigrating in each season but not establishing year-round. However, there is now clear evidence that the fly is permanently established in temperate eastern Australia (Dominiak and Daniels 2012; Reynolds and Orchard 2015; Agriculture Victoria 2017; Dominiak and Mapson 2017).

Many fruit industries in Australia are looking to increase exports in the future, which come with a concomitant need to be able to meet protocol requirements of trading partners, especially in crop monitoring and compliance. Achieving pest-free status and increased market access is important, but equally important is maintaining current trade through effective control strategies. With a very large host range, more than three-quarters of Australian fruit and vegetable exports are susceptible to Queensland fruit fly (Clarke et al. 2014). Not surprisingly, therefore, this pest causes significant difficulties for producing clean fruit and developing market access. On average, the annual value of fruit fly susceptible production in Australia is over AUD 5000 million (between 2006 and 2009) (Abdalla et al. 2012). Without fruit fly control, production losses due to *B. tryoni* have been estimated at 80-100% (Sutherst et al. 2000) and even with available management efforts, estimates of production losses in endemic areas range from 0.5 to 3% annually depending on crop type (Abdalla et al. 2012). Queensland fruit fly is also considered a significant barrier to market access for Australian horticultural products (Ekman 2015). Pest-free status provides producers within such declared regions a significant advantage; Tasmania, for example, exported nearly 50% of its total cherry yield to China, compared to less than 1% from mainland Australia (CGA 2015). In order to export to either regulated domestic or international markets, producers in *B. tryoni* endemic areas must comply with specific requirements; protocols generally dictate a Probit 9 end-point phytosanitary treatment, such as cold storage or fumigation with methyl bromide. These treatments assume growers implement a level of pre-harvest and post-harvest control, both incurring costs (Lloyd et al. 2010).

With the loss of regulatory fruit fly area-freedom in some parts of temperate eastern Australia (Agriculture Victoria 2017; Dominiak and Mapson 2017), and the regulatory loss of some chemical options for Queensland fruit fly (Dominiak and Ekman 2013; Florec et al. 2013), new control options for *B. tryoni* are needed. As AWM is an internationally recognised approach for mobile pests (Vreysen et al. 2007a), this seems a sensible approach to achieving effective control of Queensland fruit fly, while underpinning market access requirements. The application of an AWM approach within any chosen region or area requires the integration of control tactics (Chandler et al. 1999; Suckling et al. 2014), as the use of a single management tool is deemed insufficient to suppress effectively mobile pests (Vargas et al. 2010). AWM should also be seen as a long-term undertaking, with long-term solutions (Hendrichs et al. 2007; Vander Meer et al. 2007; Vreysen et al. 2007a; Yu and Leung 2011; Ogaugwu 2014). *B. tryoni* control and surveillance tools currently available in Australia and being used in-field include:

- Monitoring - assessing pest population trends through male pheromone traps
- Bait spraying - a protein (usually yeast-based) and toxicant, targeting female flies that are attracted to the food bait
- Male annihilation technique (MAT) - high density placement of a male attractant lure (cue-lure) and toxicant, targeting male flies (to be effective, MAT needs to "attract and kill" most males in a population)
- Orchard sanitation - the systematic removal of fallen and infested fruit
- Cover sprays - for some crops where registration allows, when pest pressure is high.

The general consensus is that bait-spraying and MAT are complementary measures, while sanitation is an essential key component of many fruit fly pest management approaches (Dominiak et al. 2015; Stringer et al. 2017). The release of sterile male flies embedded into a well-managed AWM programme that integrates the above control measures can be highly successful as shown in other parts of the world (Fisher et al. 1985; Fisher 1996; Enkerlin et al. 2005; Pereira et al. 2013), but has yet to be implemented on a large-scale in Australia.

1.2. Methodological Approaches

The methodological approaches chosen in this series of studies were complementary, where each phase built upon the previous phase:

Phase 1 comprised one-on-one interviews with fruit growers and industry representatives. Participants were asked about their attitudes towards Queensland fruit fly as a personal threat, as well as a threat to the fruit-growing industry.

Phase 2 was a series of focus groups with a wider range of "stakeholder" groups: fruit growers and packers, government and industry representatives, extension officers, and the general public. The content of the focus groups was an extension of the interview phase and provided more in-depth examination of the issues raised during interviews. Participants explored possible motivations for participating in AWM activities, as well as offering potential scenarios for rules, regulations and AWM and SIT funding. General public perspectives were also incorporated at phase 2.

Finally, phase 3 involved a broad-scale telephone survey of fruit growers and members of the general public. The purpose of the survey was to examine key factors, identified through the qualitative work, on a larger and more measurable scale. This was to determine the relative importance of different social and institutional factors on acceptance and intention around AWM and SIT implementation. However, the issues discussed in this chapter are predominantly derived from phases 1 and 2 of our research survey programme, which are rich in contextual data. It is also important to note that the scale of our social research was an examination of manageable farmer-farmer or neighbour-neighbour coordination options. Given the extensive length and breadth of Australian horticultural landscapes, our phase 1 results indicated that it was not feasible to consider interpersonal social factors of AWM collaboration beyond that scale.

This process of multi-method (quantitative and qualitative) data collection and analysis allowed an exploratory approach to identifying relevant social issues, without imposing any researcher bias on possible barriers or enablers that might be considered theoretically important. Therefore, a strength of this research is that the social factors covered within this chapter are participant-driven and derived from farmer and public perspectives. The associated theoretical underpinnings were chosen based on the emergent data, rather than preselecting a theoretical framework to analyse the social landscape.

1.3. Regional Descriptions

There were three study regions across south-eastern Australia, each of which is a dominant horticultural production area (see Fig. 1); however, the regions vary with respect to their 'fruit fly status'.

Region A is now recognised as an endemic area for Queensland fruit fly. Although some parts of region A have pockets of land where levels of Queensland fruit fly are undetectable, the region as a whole is not considered pest-free. Region B is a regulated region with traditionally low Queensland fruit fly prevalence, now with a suspended pest-free status. This is due to many recent outbreaks and signs of early invasion in many parts of the region.

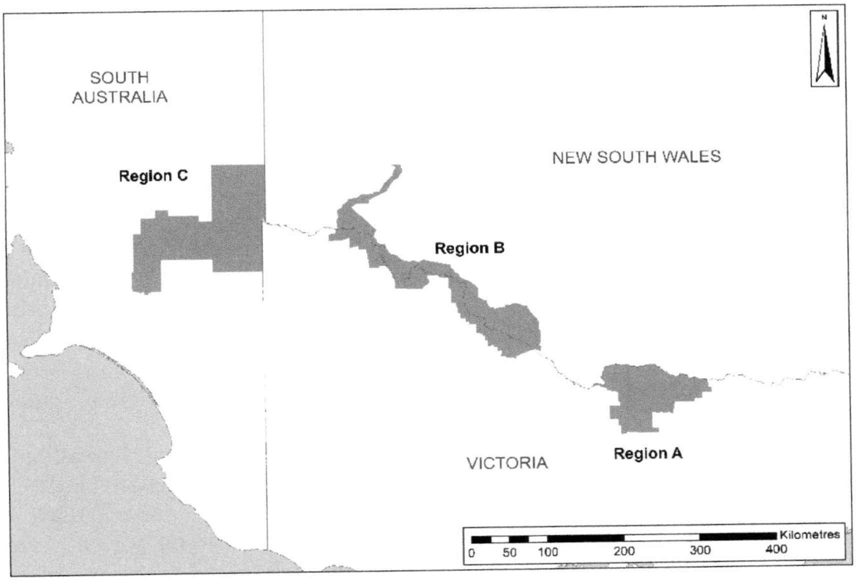

Figure 1. Map of the three study regions targeted for social science research across south-eastern Australia: Regions A and B, both spanning New South Wales and Victoria; and Region C, on the South Australian border (source Andy Hulthen, CSIRO).

Finally, region C is a recognised and regulated Queensland fruit fly free area in which any incursions are immediately eradicated by the relevant government authorities (as per International Standard for Phytosanitary Measures (ISPM) Number 26 (FAO 2016)). Until only a few years ago all three regions were under regulatory area freedom (Dominiak and Mapson 2017).

Potential institutional issues related to on-the-ground coordination for AWM in these three regions include regulatory boundaries such as local government areas, state borders, and other nationally recognised boundaries. Flies do not recognise or respect these geographic and administrative delineations that can complicate efforts to coordinate and fund AWM initiatives. Rules and regulations within jurisdictions, or a lack of legal authority (Hendrichs et al. 2007), can also frustrate AWM efforts. Differing organizational goals and rationales, such as those for private firms, government agencies and industry bodies, can further complicate coordinated action. Thus, getting the institutional arrangements 'right' appears a crucial element to effective AWM.

2. INSTITUTIONAL BARRIERS AND FACILITATORS FOR AWM

AWM can be viewed as a pest management innovation that is social, organizational and logistical in nature, as much as it is technological. While it incorporates biophysical technologies as part of its integrated pest management approach, it fundamentally requires cooperation between parties across geographic space. These parties can be diverse, ranging from commercial growing enterprises, to hobby-farmers, local councils, government agencies, and the general public. Contribution and coordination of these diverse groups can also occur at different levels (e.g. local, state, and national industry bodies).

AWM, theoretically, can also be implemented at different scales from activities organised at the local community level to regional and even national implementation programmes (Kruger 2017a, 2017b). Queensland fruit fly is reported to travel only a relatively short distance (compared to other mobile horticultural pests in Australia) in the presence of hosts and other resources (Dominiak 2012). This suggests grower-to-grower AWM for pest suppression would be effective at relatively small scales when growers are in close proximity and any adjoining non-commercial land was included in the programme (Barclay et al. 2011). Larger scale AWM initiatives for broader control may include the integration of multiple grower groups. This may include towns and settlements and their peri-urban fringes, and other host areas such as hobby-farms and nature reserves (Kruger 2016; Vreysen et al. 2007a). This may require government coordination.

Psychosocial factors that influence participation and cooperation, such as attitude, motivation and trust, are clearly important for sustained collective action. However, there are also a range of institutional factors are also important for coordination (Mankad 2016; Mankad et al. 2017). Institutional factors refer to the various forms and levels of rules and rule-sets that guide human behaviour and the arrangements between participants in a social system (Hodgson 2006). Social systems can be viewed as being comprised of structures, including formal institutions (as the effect or manifestation of rules) and participants or agents (individuals and organizations), whose behaviour is influenced by these institutions (Giddens 1984). Thus, institutions are referred to as the 'rules of the game' and counter-posed to agents as the 'players of the game' (North 1990). Rules are, of course, developed and enacted by people. Therefore, there is a dynamic relationship between institutions and actors (or social structures and agents) where rules influence the behaviour of actors, who at the same time are individually or collectively seeking to reshape the rules to meet their own particular goals (Giddens 1984; Hodgson 2007).

Institutions comprise a broad range of rule types, from formal to informal, such as policies, laws, regulations, protocols, guidelines, procedures, standards, conventions and norms. These establish particular ways of behaving and organising. Institutional factors enable a broad range of programmes and initiatives, such as raising awareness and education campaigns; research, development and extension programmes and facilities; pest monitoring networks, exclusion zones and diagnostic facilities.

This research identified a broad range of institutional factors that could potentially facilitate or constrain participation in AWM for Queensland fruit fly control. Typically, most of the constraints or barriers emerging from the research covered in this chapter relate to a deficit of those factors cited as important for the facilitation of AWM. Notable institutional facilitators of AWM that were cited included:

- Formal engagement of stakeholders and consistent communication by industry and government in raising awareness and knowledge building about Queensland fruit fly and AWM
- Integration of efforts across stakeholder groups (growers, government, industry and the general public)
- Cooperation within and between AWM groups and at higher levels of regional organization
- Leadership and governance aspects important to galvanising, guiding and sustaining these various levels of collective action
- Rules, regulation, legislation and enforcement required to achieve greater cooperation and compliance for broad pest suppression across multiple host areas and to raise funds to support AWM
- The potential of greater access to premium (typically export) markets and the role that market access 'protocols' play in this.

The following Sections provide some examples from three key areas that were identified as particularly important.

2.1. Stakeholder Cooperation and Integration

Cooperation between parties and integration of efforts is of crucial importance for AWM, whether at a local level (between neighbouring growers and other property managers), at a broader community level (between grower groups and a nearby town council/local government authority to suppress urban pest populations), or a whole-region level (involving multiple stakeholders representing industry, government and the general public to guide a regional approach). Cooperation at a local level is required mainly to ensure as many growers/property managers are controlling the fly as possible. Due to the often-close proximity of properties to one another, as a mobile pest, Queensland fruit fly can readily reinfest a 'managed' farm from an adjoining 'unmanaged' property. Coordinated timing of control operations is essential for area-wide suppression of a pest population, especially with regard to SIT application. Optimal conditions for SIT integration typically include an existing high level of sanitation and suppression, or low-level populations (Stringer et al. 2017), and avoiding insecticide spraying at the time of SIT releases to prevent destroying the sterile flies as they enter the environment.

Participants in the qualitative studies indicated a strong awareness of the need for cooperation across growers and other property managers. They often expressed concern that a small number of uncooperative growers (e.g. apathetic or negative), those unable to cooperate (e.g. elderly or absentee owners), or those for whom the benefits may be limited (e.g. host produce is relatively unaffected by fruit fly), could undermine the success of their own efforts. Concern was also expressed by growers regarding cooperation by urban communities. Rural towns and their backyards were generally perceived as the greatest host risk of population build-up in endemic areas, providing a permanent supply of the pest to infect surrounding commercial growing areas (that were generally perceived as having lower levels of infestation).

Cooperation by the local government authority or 'council' was perceived as important to the success of an AWM initiative. Councils were considered the main route to organising pest management activities in towns. This was confirmed by participant responses from town residents, and members of councils themselves. Coordinated activities generally involved gaining the cooperation of residents (property owners and/or rental tenants) in agreeing to deploying traps and/or the removal of host fruit/trees in their garden. General public participants in this research further indicated a high willingness to cooperate with their local council in undertaking Queensland fruit fly control practices on their property.

Stakeholder cooperation at a broader level was seen as important to guide and support a strategic approach to AWM at the regional level. This included the cooperation and integration of efforts of different fruit industries (citrus, stone fruit, grape, etc.), different industry segments (growers, fruit packing houses, industry associations, crop pest consultants and agricultural chemical suppliers), and different levels of government (federal, state and local). Cooperation was seen as essential to ensure a better integrated approach that provided a planned and strategic direction and consistency of messaging across the various stakeholder groups. This would ensure a common language and knowledge base, and thereby minimise confusion about what to do. A united approach at this higher stakeholder group level was also seen as important to gain more widespread action on the ground, at the grower level and amongst the general public, to minimise the incidence of untreated areas and 'social loafing', also referred to as 'free-riding'.

2.2. *Education, Engagement, Communication*

The importance of education, engagement and communication comprised a prominent theme in the qualitative findings of this study. Interviews and focus group participants strongly indicated that building high regional awareness of the threat of Queensland fruit fly and a strong understanding about its control, including through AWM and SIT integration, would be crucial to their success. For this to occur, education, communication and engagement activities were seen as paramount, especially using methods and forms of media likely to appeal to end-users.

A broad variety of media and formats were suggested, including: information seminars that included both technical specialists and growers experienced in dealing with the pest; support/extension personnel who could visit the farmer/ property owner with advice; various forms of media (TV, radio, websites, local newspapers, industry newsletters and magazines, council information mail-outs, etc.); as well as more detailed or scientific fact-sheets and best-practice guidelines. A primary concern, as indicated above, was that the messages conveyed through these various sources were consistent and thereby reinforced one another.

The quantitative survey sought to test the importance of knowledge and knowledge support (in the form of access to a technical or coordination support officer) with a much larger sample of growers and general public. Interestingly it found only a moderate level of influence on respondents' decisions to become involved in AWM (see also Section 3.4 in this chapter). Other factors were found to play a stronger role in decision-making, such as: perceptions of the level of threat posed by Queensland fruit fly, fairness in the implementation of AWM, costs and benefits of AWM, and social norms around what was the morally responsible thing to do (these psychosocial factors are described in detail below). Nevertheless, overall, it appears that the building of awareness and knowledge around Queensland fruit fly and its control through AWM, including the SIT, was considered a highly important component of any broader strategy to deal with it.

2.3. *Rules, Regulation and Enforcement*

A range of issues were identified in the qualitative studies relating to rules and regulations. Most issues concerned new or enhanced regulation considered necessary to improve success in implementing AWM to control Queensland fruit fly, while some related to current laws or regulations seen as likely to impede success. The main types of issues for which additional rules and regulations were seen as required have included:

- Ways to increase participation in AWM (e.g. mandating participation)
- Control of the pest in areas that were unmanaged or poorly controlled (e.g. removal of fruit-trees from abandoned orchards and backyards)
- Mandating specific control practices (e.g. orchard sanitation, backyard traps)
- Implementation of funding mechanisms to support AWM and SIT implementation (e.g. compulsory levies, accreditation schemes)
- Changes to market access protocols that would enable AWM and/or the SIT to be included.

A range of expectations were expressed around activities to limit the introduction of flies from elsewhere, even in areas where it had been declared endemic. In those regions where Queensland fruit fly had been declared endemic and the Fruit Fly Exclusion Zone (FFEZ) recently dismantled, there were calls for its reintroduction. In region C, a regulated Queensland fruit fly-free area, calls were often made for the state government to ensure greater vigilance in inspections and/or fund additional routes where checkpoints should be implemented.

Two main types of currently existing rules and regulations were considered an impediment to AWM. Privacy laws preventing release of data mapping the distribution and hot spots of Queensland fruit fly in a region were viewed as impeding timely control responses by growers. Also considered an impediment were laws preventing growers or members of the public removing host vegetation on public lands, such as fruit trees on roadsides adjoining commercial properties.

A number of participants contended that currently existing laws and penalties needed to be better communicated or more rigorously monitored and enforced. In contrast, at a broader level some study participants contended that reliance on rules and regulations would be counter-productive and a more positive approach relying on communication, engagement and incentives would be more helpful. Participants also reported they would value pest density maps and location of hot spots if these were made available in a timely manner. They were seen of benefit for individual farm pest management and also to socially "pressure" others with pest problems to participate in management activities.

Maintaining a Queensland fruit fly monitoring network of requisite density and ensuring timely reporting could, however, be expensive, raising questions of who should pay. Further, this "name and shame" approach was also considered potentially counter-productive by putting negligent growers off-side, further reducing collaborative behaviour.

3. SOCIAL-PSYCHOLOGICAL BARRIERS AND FACILITATORS FOR AWM AND THE SIT

An individual-level analysis of social psychological factors contributing to potential uptake of AWM practices and acceptance of the SIT was also conducted. This analysis provided important insight into how people living and working under institutional arrangements, discussed in Section 2, felt about AWM and the inclusion of the SIT in it. Furthermore, it was important to gain some baseline understanding of stakeholders' motivation to undertake AWM, possible factors affecting adherence to an AWM programme, and existing habitual on-farm behaviours for fruit fly management that would be affected by AWM. Overall, we found that being a farmer or a non-farmer was not an important driver of intentions to implement AWM or indeed accept the idea of AWM. Rather, it was the individual-level social psychological drivers that distinguished between different levels of acceptance of AWM and SIT integration, and intentions to implement AWM. Fig. 2 represents the varied individual factors that emerged from our research. Many of the highlighted social issues presented in this Section can be viewed as both barriers and facilitators to uptake of AWM, depending on the framing and messaging. A selection of dominant issues presented in Fig. 2 will be discussed here, and they will be described generally in terms of their influence on uptake of AWM and SIT.

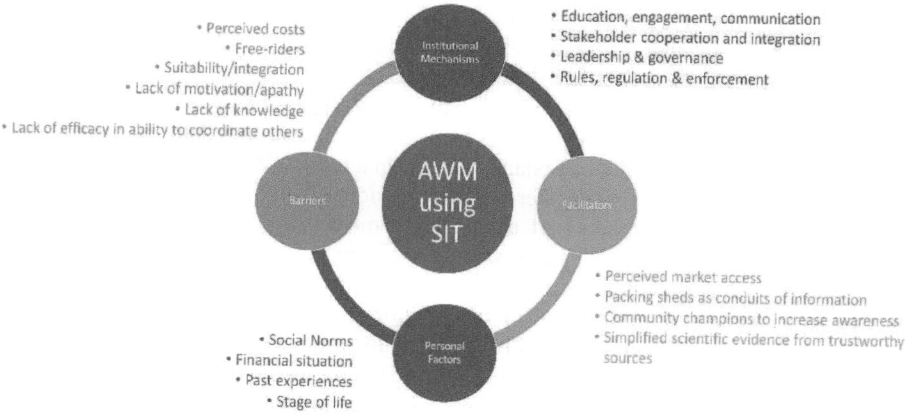

Figure 2. Key social factors influencing uptake of area-wide management (AWM) integrating the Sterile Insect Technique (SIT), framed as social barriers, facilitators, institutional mechanisms and personal factors.

3.1. Perceived Costs

The primary consideration for farmers taking part in our studies was the perceived cost of AWM and the SIT. Specifically, people were interested in how much it would financially cost them to change current practices to conform to recommended AWM behaviours (e.g. orchard sanitation and hygiene, baiting and trapping). But, more so than AWM, people were keen to know more about the financial viability of SIT implementation and whether this type of technology could be used at the individual grower level in a cost-effective way. There was also a consistent view amongst participants that if relative costs of the SIT were perceived as prohibitive, then the largest barrier to widespread involvement in AWM would be this perception that the SIT was unaffordable.

Importantly, perceived costs of engaging in AWM did not only include monetary costs, but also personal costs such as time and effort. This was particularly evident for perceived costs associated with coordination efforts with neighbours, as part of AWM. Some participants believed that trying to coordinate growers across differing landscapes and production types would be difficult, if not impossible. The effort associated with organization, as well as the personal costs of potentially acting for the benefit of others, were thought to be difficult to overcome. This perception of response costs was found to be consistent across the three different regions, which did not significantly differ in their perceptions of anticipated costs associated with involvement in AWM.

Past psychological research on perceived costs has shown that those who acknowledge both costs and benefits to change achieved better outcomes than those acknowledging only benefits or costs, or few costs and benefits (McEvoy and Nathan 2007). Perceived costs can also be interpreted from the perspective of 'sunk costs', when one has already invested time and effort in developing on-farm habits, and financially invested in certain practices. A perception of costs within this framework could reflect how well a potential behaviour change (e.g. AWM) meets one's goals and aspirations, without too much of a conflict (Gifford 2011). Our results suggest that if AWM activities are perceived as requiring very little behavioural adjustment on an individual level, and/or the end goals meet individual needs, then the perception of costs will be low. A lowered perception of cost is related to a higher intention to engage in novel behaviours (Rogers 1983).

To cover the costs of an AWM with SIT programme, participants in our focus groups suggested a range of mechanisms:

- Funding by government (federal, state and/or local – an approach that tended to be favoured by growers)
- Various form of levies or hypothecated taxes (revenue from a specific tax ear-marked for this purpose) on fruit production or fruit properties, urban households, or fruit sold to consumers, or
- A private user-pays model where fruit producers pay the entire cost.

Interestingly, while general public participants commonly suggested industry / growers and various levels of government should pay, they also appeared quite willing to pay something themselves, typically nominating local council taxes or levies imposed on retail fruit or urban households (see Nelson et al., this volume).

3.2. Compatibility of AWM with Existing On-farm Practices

According to Rogers' (2003) diffusion of innovation theory, perceptions of compatibility with respect to an innovation will affect its adoption. The innovative practice tested in the present study, using diffusion of innovation principles as predictors, was AWM. Compatibility refers to how consistent the innovation is with the values, experiences and needs of the potential adopters. In the farming context, this also refers to how compatible AWM activities are with the existing on-farm practices.

Perceived compatibility of AWM with existing farming practices emerged as a driver of AWM uptake through our qualitative study. Through stakeholder interviews, we found that participants were primarily concerned about the coordination aspect of AWM. Specifically, the concern was that engaging in coordinated pest management efforts would be difficult because of different harvesting times and spraying times for different production enterprises within a region. Some participants perceived a significant risk if one group of farmers were spraying to reduce numbers in preparation for SIT, when another group was at a different stage in their harvesting cycle and could not do the same. The perception of incompatibility was primarily focused on collaboratively preparing a farming region for the introduction of the SIT. There were queries as to whether SIT application would or could be compatible for a whole region, or whether it was more suited to smaller scale releases, such as individual farm scale. Another compatibility issue raised was the logistical strain of setting up a systematic monitoring grid, as well as removing unwanted and fallen fruit on a commercial farm. The latter activity was perceived to be an immense task in itself and potentially incompatible with existing resource allocations.

Compatibility was also subsequently identified as an important driver for the prediction of public acceptance for AWM. Our quantitative research suggested that a key factor likely to influence community adoption of AWM practices was the expectation that AWM activities should complement rather than impede or complicate their current pest management behaviours. For farmers, this could mean capitalising on existing peer communication networks and sources of information to ensure coordination aspects of AWM are built into established social systems. It could also include more aggressive social cues, such as building up peer pressure at the community level to encourage farmers to remove infested and fallen fruit or sharing of monitoring data. For the general public, this may mean ensuring the messaging around AWM at the household level demonstrates an understanding of existing backyard habits, to mobilize the public so that hosts or fruit are effectively eliminated by all in a timely way along roads and in backyards. It could also mean acknowledging that the community is comprised of individuals with varying levels of interest and buy-in for AWM. Thus, compatibility would rest on AWM tasks that are presented as flexible, adaptable for those with different levels of intentions and willingness to participate, and new behaviours that are not perceived as an additional burden to household activities.

3.3. *Free-riders and Fairness*

In a collaborative framework such as AWM, where the group benefits from shared action, there was an inherent concern for free-riders within the system. Certainly, much of the impetus driving social research in the context of an AWM scenario is a fear that a recalcitrant few could cause others to favour inaction in the name of fairness. The free-rider problem, at its core, is an economic problem describing those who benefit from goods, services or resources without contributing to payment (Baumol 1952). In the present study, we conceptualise the free-rider from a social psychological perspective. As such, 'free-riding' in this context comprises aspects of 'social loafing' (a psychological term), fairness, and conditional participation. Therefore, one is perceived to be free-riding if one is choosing to willingly disengage from a shared system but will still retain the resultant benefits of the shared system.

In psychological terms, the social loafing phenomenon describes a situation when an individual contributes less effort in a group setting, compared to when they work alone (Jackson and Harkins 1985; Karau and Williams 1993). In the context of AWM, that could be envisaged as a cluster of neighbouring growers working together to manage fruit flies. However, one grower in the middle of the region refuses to participate, whilst the others agree to implement coordinated activities. Consequently, the group coordinates and manages their pest management activities, while the lone grower in the middle refuses to implement any new practices because of different reasons. In such a scenario, the overall suppression of a pest will not be as strong compared to if all growers were participating in AWM. However, it is likely that the group will still work towards area-wide suppression even without the lone grower's involvement in the coordinated efforts of his/her neighbours. Thus, the lone grower is still likely to gain benefits from the collective efforts of the other growers, without exerting the same level of effort. This type of individualistic behaviour in the context of AWM can not only affect the success of managing a pest and coordination of efforts, but it can also be a deterrent for others to get involved. The importance of fairness in shared approaches such as AWM is paramount to its success.

The idea of perceived fairness emerged from our qualitative work as a key requirement for encouraging stakeholders to get involved in AWM. In particular, participants felt that growers who were harvesting non-dominant host crops, which would be less susceptible to Queensland fruit fly, would be more likely to question the fairness of AWM. These indirect beneficiaries of an AWM approach would likely need convincing of how AWM of the fruit fly was fair to people like themselves, who would directly benefit less from area-wide suppression of this pest. It was clear that, from a social justice perspective, AWM would need to appear advantageous to all involved, not just those growers who were growing Queensland fruit fly host plants. There was also a clear concern that growers of non-dominant host plants would be more likely to 'free-ride' the system, because of a perception that they may have less to gain.

While the qualitative research highlighted free-riding as a concern, our quantitative analysis helped to highlight the importance of fairness. Perceptions of fairness emerged as the single most dominant predictor of public acceptance for AWM. It was clear that acceptance for AWM amongst growers and members of the broader community were contingent upon how fair they believed AWM to be for everyone. In practical terms, this meant that no matter how simple or effective AWM might be, the strength of uptake may ultimately depend on all parties feeling as though AWM was a balanced approach.

3.4. Knowledge

Subjective or perceived knowledge (i.e. one's belief about personal knowledge on the subject matter), rather than actual knowledge, has been identified as a driver of protective action in the context of fear appeals (i.e. threats) in the psychological science literature (e.g. Martin et al. 2007; Nabi et al. 2008). Results from our present research found that the role of subjective knowledge could be examined more accurately by breaking it down into two: a) perceived knowledge of Queensland fruit fly, and b) knowledge of ways to manage this pest. Compared to members of the general public, growers had significantly greater knowledge of both Queensland fruit fly and technical methods of controlling Queensland fruit fly. However, subjective knowledge of Queensland fruit fly was moderate overall, and technical knowledge of control methods was low.

Qualitative results indicated that growers believed that they were already carrying out AWM. However, it was clear that growers had an inconsistent understanding of what AWM was, particularly the key role of coordination in AWM. Most growers interviewed believed that AWM was the management of one's whole property area, rather than a coordinated effort across potentially multiple properties within a given area. Predictive analyses further demonstrated that technical knowledge of Queensland fruit fly control was a direct predictor of intentions to implement AWM. Thus, those who felt they knew more about existing ways to control Queensland fruit fly were more likely to engage with AWM. Interestingly, however, subjective knowledge of Queensland fruit fly alone was not a predictor of intention to implement AWM.

An overall lack of knowledge of Queensland fruit fly amongst participants in our research may help to explain why perceived complexity of AWM was the second-most dominant predictor of acceptance for AWM (after fairness). This suggests that while people were happy to support the use of AWM as a way to manage fruit flies, the topic was an area in which they held limited knowledge. Thus, people indicated that they would respond more positively to uptake of AWM if the information was simple and personal involvement was not too demanding.

Another motivational construct affected by knowledge and awareness is that of *apathy*. Stakeholders such as industry representatives, agronomists, and farmers, all felt that a significant problem in galvanising people within a region – whether farmers or the general public – was a perceived lack of understanding of the relevance of Queensland fruit fly control. Stakeholders felt there was a general lack of knowledge about how fruit fly contributed to horticultural and financial losses and a lack of market access.

3.5. *Efficacy and Competence*

Self-efficacy, and the related concept of competence, have been found to be strong psychological drivers for behavioural action (Bandura 1977; Deci and Ryan 1995). The psychology literature describes self-efficacy as a belief in one's ability to accomplish a task. Self-efficacy can develop through observing modelled behaviour, feedback from others, and past experience or mastery of related behaviours. An important distinction to make, which will ultimately assist in the development of ways to promote self-efficacy on the ground, is that self-efficacy is not the same as self-confidence. The latter refers to a broader sense of personal esteem and general likelihood of succeeding; in comparison, self-efficacy is more specifically tied to one's ability to perform a task. Self-efficacy is therefore more closely linked with a perception of competence, however, the concepts remain distinct in psychological theories (e.g. Rodgers et al. 2014). Ryan and Deci (2000) have described perception of competence as a fundamental driver of human motivation. Perceptions of competence reflect a self-awareness of individual capabilities and one's ability to control a situation. Past psychological research has demonstrated that engagement in protective behaviours (e.g. AWM of a pest) is significantly linked with how efficacious – or, competent – one feels with respect to the required behaviours.

In the present examination of public intentions to implement AWM, we measured self-efficacy towards carrying out behaviours recommended as part of an AWM approach. People were asked how easily they could carry out tasks such as removing unwanted fruit from trees and the ground, hanging out fruit fly traps, and coordinating with neighbours. We found that perceptions of self-efficacy related to these AWM activities were moderately high, but that self-efficacy was the strongest driver of intention to participate when taking into account other motivational factors such as knowledge of Queensland fruit fly and its management, severity and vulnerability of the problem, perception of response costs, and perceived effectiveness of AWM. People were most likely to engage in AWM if they felt they could do so successfully and competently.

This finding, on the ground, translates to a developmental need amongst potential participants of AWM. That is, fulfilling a desire for people to feel competent in engaging with AWM tasks through training and by providing opportunities for experiential learning and mastery. It is generally understood that within a community or regional group, there are people with diverse physical capabilities and interests, particularly related to a niche topic such as Queensland fruit fly, or even the broader notion of pests in the garden. What is less understood is, fundamentally, that the need to feel competent or efficacious when carrying out tasks is a driver for all human behaviour. Therefore, encouraging people to participate in a potentially unfamiliar activity will require addressing this universal need to feel capable.

3.6. *Community Champions*

A strong finding in our social research was that farmers and members of the general public were putting their trust in 'important others'. Important others are peers, family and friends whose opinions the individual trusts, or whose personal approval is important. Important others can also be members of the community – whether that be the farming community or influential members of the public. They are seen as local leaders, championing local causes and effecting change. Our results highlighted that these key individuals within a community were often seen as thought leaders, who kept the community's best interests at the forefront.

Amongst farmers, extension personnel such as agronomists were consistently identified as trusted sources of information. They were thought to be influential in positively changing on-farm practices and in providing practical support on the ground. In addition, the role of packing houses was also found to be valuable in conveying important information. Packing houses involved in packing and/or marketing fruit must also comply with export protocols. For domestic and international trade, packing houses undertake in-line inspections and are subject to audits. Therefore, the level of influence packing houses have on growers is substantial (Kruger 2016). Packing houses are familiar with compliance measures of good agricultural practices through international quality assurance schemes, such as Global G.A.P. or Freshcare, or the new Harmonised Australian Retailer Produce Scheme (HARPS 2017). Thus, packing houses could directly influence change in grower behaviours through their fruit acceptance requirements and standards.

For the general public, a community champion was represented as their local government council rather than a specific individual or type of individual. In Australia, local councils often lead community-level initiatives for the direct benefit of the community. It was clear from our qualitative research that the general public expected the local council to lead Queensland fruit fly area-wide control efforts in their town and provide guidance to residents on what activities to do. Our results, therefore, provided a good example of how local groups could also champion community causes.

Importantly, it was evident that community champions were a valuable conduit for information that might otherwise not get transmitted or distributed to some sections of society. For example, members of the target population, difficult to reach using traditional or mainstream communication strategies, may be more receptive to personal interactions for gaining information. The functionality of community champions in the context of AWM participation and SIT adoption was seen as important. Given their high level of trust within the community, it was possible that community champions could use their local influence to emphasise shared goals in the community, such as working towards a mutually beneficial outcome (e.g. economic growth or a cleaner environment). They could also serve as influential "messengers" for information or scientific evidence, and act as informal project managers for AWM. This could enable some sort of community-led 'enforcement' of AWM to ensure that actions to prevent/control fruit fly were being done on the ground.

Community champions can also provide a vital social support role for those experiencing difficulty in carrying out AWM or those who are unsure of the impacts of Queensland fruit fly, AWM or SIT adoption on their livelihoods or lifestyles. Typically, those championing causes at the scale of AWM are not passive participants in the social process (Norris et al. 2008; Conrad and Hilchey 2011). Rather, they have carried out their own investigations of available information and have a level of personal investment or buy-in for the ideals represented by the initiative. As a result, community champions are likely aware of local concerns and can support the change process through context-specific knowledge they have gained.

3.7. *Evidence*

The theme emerging from our data highlights the importance of contextual evidence of success using AWM and the SIT for Queensland fruit fly control. Both growers and members of the general public indicated that for wide-scale cooperation and buy-in such as that required for AWM with an SIT component, they would need strong scientific evidence. This indicates the public values not only evidence of the science, but also an expectation that the public has access to evidence in a 'digestible' format (Blackstock et al. 2010; Garforth et al. 2013). This could take the form of public events conveying scientific results or creating science communication materials that highlight the science behind programmes such as AWM and the use of the SIT.

Experiential evidence was also an important factor for the acceptance of AWM as a pest control option, particularly for those living outside endemic areas. Growers stated they would be more likely to accept and participate in AWM if they could observe, first-hand, the presence of Queensland fruit fly, the damage it could do, the economic evidence that it was a problem, and that AWM was effective in reducing fly numbers. Similarly, industry stakeholder indicated that they would react more quickly on encouraging participation in AWM and the SIT use if they had evidence of what it means for their grower base, on-ground actions, and market access.

The SIT was also a technology that participants were unfamiliar with and while many growers and the general public believed it was an interesting idea in theory, they were clear that trusted and rigorous scientific evidence would be needed. Concerns remained around the notion that the SIT was safe, effective and worth the investment. It also appeared important for growers to see evidence that sterilisation works and that it is not a technique that is essentially still in trial stages. There was a fear that if the science is not correct, then growers may be used as unwilling guinea pigs on which to release untested sterile flies that may or may not be effective in controlling the wild fly population.

Notably, while scientific and experiential evidence was cited as being important, the source of such evidence would likely also be a significant factor (Twyman et al. 2008; Hernández-Jover et al. 2012). Therefore, any evidence communicated to the public would need to originate from trusted sources in order for evidence to serve as a facilitator for AWM and SIT adoption.

4. THE ROLE OF THE SIT AS PART OF QUEENSLAND FRUIT FLY AWM

Whilst the SIT has had a role in Queensland fruit fly outbreak responses for many years in south-eastern Australia (Reynolds et al. 2010), the SIT as a supplementary fruit fly management tool is a relatively novel concept for growers. Facts about the use of the SIT in managing Queensland fruit fly (e.g. increased efficacy under low pest prevalence conditions) are not well understood. However, the release of sterile male flies embedded into an AWM programme has proven to be highly successful overseas for other flies (Enkerlin et al. 2005; Pereira et al. 2013), as well as in Western Australia (Fisher et al. 1985; Fisher 1996).

The introduction of the SIT into AWM programmes for fruit fly suppression, containment or eradication is recommended for a number of reasons (Suckling et al. 2016). Permalloo et al. (2005) and Dyck et al. (2021) both consider the SIT as a sustainable pest management practice, and Hendrichs et al. (2007) consider it an effective environment-friendly tool with the ability to eradicate pests when used within an AWM programme. The SIT is considered pest-specific and compatible with existing tools (Vargas et al. 2008; Dyck et al. 2021; Enkerlin 2021).

McInnis et al. (2005) and Vargas et al. (2008) agree that male-only releases of the SIT is more effective than a bisexual line when using sterile flies; both for the mating ratio that is imposed, but also for the reduced risk of fruit damage by ovipositing female flies. McInnis et al. (2005) and Meats et al. (2003) suggest the sterile to wild fly ratio in trap catches can be effective in assessing the SIT, and McInnis et al. (2005) further posit that when sterile flies released are competitive, a sterile to wild fly ratio at 10:1 (in traps) shows effectiveness. This highlights the importance of a trapping network to monitor the distribution and abundance of both wild and sterile flies and the success of SIT programmes.

4.1. *Institutional Considerations*

The scope, scale and method of release of sterile flies will affect sterile fly quality and have implications for the management and governance arrangements for a SIT programme (Vreysen et al. 2007b). At the time of this study, details of these for the three study regions were limited, and a range of possibilities were canvassed in qualitative forums. Suggestions included:

- Small-scale on-farm release by individual growers coordinating at some level with other growers in an AWM approach
- AWM grower groups working together using farm vehicles or aircraft across a larger growing area
- Urban only releases to suppress pest populations in towns, or
- Broad-scale coverage of both commercial and urban areas, typically using aircraft.

Each approach is likely to require a different management arrangement and governance consideration.

Bottom-up approaches that rely on individual growers or AWM grower groups acting independently from a central authority, for example purchasing the sterile flies from a local supplier – much as they currently do for other Queensland fruit fly control techniques (bait spray, traps, chemical sprays, nets) – would, in principle, involve limited supervision. However, much experience with AWM with SIT programmes has shown they tend to be management intensive, requiring professionals dedicated to SIT implementation, and some form of central coordinating body (Vreysen et al. 2007b), such as a regional Queensland fruit fly control coordinating committee. Tasks include gathering information to monitor SIT releases in the region and compare sterile and wild fly populations to assess programme effectiveness.

Broader-scale AWM and releases managed by a central authority, such as a regional coordinating committee, will require an even higher level of management. Vreysen et al. (2007b) pose seven questions that need to be addressed by managers of any AWM with SIT scheme:

1. Is the programme management structure that includes full-time professionals sufficiently flexible and independent from the regional or federal government bureaucracy?
2. Is there sufficient commitment by key stakeholders so that there is a high degree of continuity in implementation of all essential programme components?
3. Are funding mechanisms established and available resources, manpower and institutional capacity sufficient in magnitude and quality to assure effective operations and sustainability of the programme?
4. Does the programme have the support of all stakeholders and firm commitments from those who must bear costs or conduct relevant operations?
5. Are essential high-quality data being collected and properly analysed in a timely manner to enable the programme management to provide feedback essential for corrective action by all key programme personnel?
6. Is the public awareness and public education programme of sufficient quality to help shape attitudes and behaviours in support of programme success?
7. Is the programme benefiting from timely and independent reviews?

At a broader level, SIT integration raised questions among participants in this study about the need for modification of market access protocols when including the SIT as a control strategy. This was perceived as a long and arduous process, as each market access protocol is specific to each of the many different countries of destination, and often requires unique specifications for each type of fruit and vegetable. Any protocol changes must be centrally managed through the federal department of agriculture in consultation with state and territory governments, whose own legislation may need modifying to accommodate changes. Furthermore, the federal department of agriculture is typically negotiating many other pest and biosecurity priorities with each country. For these reasons it was felt by some that the SIT may be of limited value in opening up new market access, at least in the short term.

Participants in our qualitative studies also raised questions on the implication of some aspects of SIT application that may present impediments to their acceptance by protocol markets. For example, increased fly counts in traps of monitoring grids, due to the trapping of sterile flies, would require willingness of protocol markets to accept assurances about distinctions with wild fly counts and count veracity. However, this perception may prove to be an unfounded fear, as many major horticultural exporting countries use the SIT extensively and presumably have found ways of addressing the issue (e.g. Enkerlin et al. 2015; Enkerlin 2021).

4.2. *Social Acceptance*

The evidence from our programme of social research has indicated a high level of social acceptance for the notion of adopting the SIT as part of AWM. Growers consistently perceived the SIT as another potential tool in their toolbox for suppressing or eradicating Queensland fruit fly. There was evidence that some growers were cautious in their assessment of the SIT, moderating personal expectations of the success of SIT application. These individuals cited that while the SIT sounded promising, their history with fruit fly showed them that there was no one answer to successfully controlling the pest. They correctly felt that the SIT could only be useful in conjunction with an AWM approach or other stringent control activities. Most growers felt that, other than cost, there would be no real barriers to the adoption of the SIT and pilot trials should begin. Associated with this was a sense of frustration amongst some growers, who believed that authorities 'holding on' to the technology, without implementing it in the field, left growers waiting. Thus, it seems that there was a complex association with the idea of SIT application amongst growers. On the one hand, growers were not keen to be trial subjects for a technology that according to them wasn't ready (and lacked the scientific evidence for the efficacy that they demanded). Yet, on the other hand, they were keen to see SIT implemented quickly, although it was generally not understood by them that an effective AWM system needs first to be in place.

An interesting result that emerged from our stakeholder interviews was a perception that the general public would be very accepting of the SIT. This was primarily driven by a belief that townspeople would not be unduly affected by the release of sterile flies, nor would they be involved in organising SIT releases. Therefore, there was nothing that townspeople could object to. The qualitative data were subsequently supported by our quantitative results, which indicated a high level of general acceptance for SIT integration amongst both growers and the general public. While general acceptance for SIT application was high overall, there were many questions about the sterile fly itself and, in particular, its sterility. However, these questions seemed to be driven by curiosity rather than a desire to question the legitimacy of the sterilisation technique.

Finally, when stakeholders and growers were asked about implementing a SIT programme, many respondents felt as though there was no real need for consultation because it was a simple concept to explain. Growers in particular felt that the SIT would be readily accepted amongst their peers. This is because participants believed that growers were so desperate for a long-term solution to Queensland fruit fly that anything was better than nothing at this point. The general public, however, did indicate that while general perceptions of the SIT were positive, they had many questions. Therefore, it was clear that general public acceptance for SIT application was conditional upon having the appropriate level of knowledge to make an informed decision.

5. CONCLUSIONS

This social science research into acceptance of AWM with a SIT component was undertaken to measure drivers and barriers to uptake of both a management style in the form of AWM, as well as a novel technology such as the SIT for Queensland fruit fly control. Most growers held an inconsistent understanding of AWM and were not aware of the significant coordination of effort required for true AWM. In fact, most growers felt that because they were managing pests on their own farms, they were effectively performing AWM. Overall, growers were willing to be involved in AWM as long as they felt competent to do so. Growers perceived a severe threat, and involvement in AWM was fair to everyone involved. Feelings of over-contributing with minimal benefits and a lack of cooperation, along with perceptions of financial costs, were seen to be potential barriers of uptake. Therefore, the crucial 'coordination' aspect of AWM was considered a significant barrier to overcome, depending on the scale of coordination required.

Aside from the involvement of growers, AWM in the regions examined within this study would also likely involve the participation of townspeople. Indeed, the survey results showed the general public were accepting AWM, as long as they were guided in the right direction and area-wide activities were simple. This perception of high general public interest was related to a belief that AWM in towns would require little investment at the individual level, such as time and cost. For example, simple backyard fruit fly traps could be subsidised by local government councils such that individual property owners would only contribute a nominal amount. In reality, this perception of a low level of engagement from householders may not be realistic and involvement in AWM may require greater effort than anticipated.

Funding the costs of implementing AWM with the SIT, whether applied only in urban areas or additionally in surrounding commercial fruit growing areas, will also require serious thought. Funding into the future will be needed, in particular for education efforts, technical and coordination personnel, subsidisation of monitoring grids and traps in urban areas and, in the case of SIT, for the supply and application of sterile flies. Fortunately, our research results indicated participants were willing to entertain a range of potential funding mechanisms, including levies on urban households.

Finally, the take-home message from our social research was that growers and the general public alike still need considerable education to have a realistic understanding of what AWM and SIT implementation entails. Fortunately, both participant groups were influenced by the behaviours of important others (i.e. social norms). If certain attitudes and behaviours were modelled for them, then growers and non-growers were likely to be more receptive to accepting AWM and SIT use, and also more likely to participate in the pest management approach. These findings were consistent across regions and social contexts, suggesting that the influence of social factors has the potential to contribute to an underlying understanding of drivers and barriers for AWM and SIT uptake.

6. ACKNOWLEDGMENTS

The Adaptive Area Wide Management of Queensland fruit fly using the SIT project is being delivered by Hort Innovation in partnership with CSIRO and is supported by funding from the Australian Government Department of Agriculture & Water Resources as part of its Rural R&D for Profit programme. Further partners include Agriculture Victoria, NSW DPI, PIRSA, QUT, SARDI, Wine Australia and BioFly. We also acknowledge the valuable contribution of reviewers in improving the quality of this manuscript.

7. REFERENCES

Abdalla, A., N. Millist, B. Buetre, and B. Bowen. 2012. Benefit-cost analysis of the National Fruit Fly Strategy Action Plan. ABARES report to client prepared for Plant Health Australia, December 2012. Canberra, Australia.

Agriculture Victoria. 2017. Queensland fruit fly: Declared outbreak zones. Retrieved May 22, 2018.

Bandura, A. 1977. Self-efficacy: Toward a unifying theory of behavioural change. Psychological Review 84: 191–215.

Barclay, H. J., R. Matlock, S. Gilchrist, D. M. Suckling, J. Reyes, W. R. Enkerlin, and M. J. B. Vreysen. 2011. A conceptual model for assessing the minimum size area for an Area-Wide Integrated Pest Management program. International Journal of Agronomy: 409328.

Baumol, W. J. 1952. The transactions demand for cash: An inventory theoretic approach. Oxford Journals 66: 545–556.

Blackstock, K. L., J. Ingram, R. Burton, K. M. Brown, and B. Slee. 2010. Understanding and influencing behaviour change by farmers to improve water quality. Science of the Total Environment 408: 5631–5638.

Chandler, L. D., M. M. Ellsbury, and W. D. Woodson. 1999. Area-wide management zones for insects. PPI Site-Specific Management Guidelines SSMG-19. South Dakota State University, Brookings, South Dakota, USA.

(CGA) Cherry Growers Association. 2015. Cherry export manual and biosecurity management program, Cherry Growers, Australia.

Clarke, A. R., K. James, J. Luck, M. Robinson, P. Taylor, and D. Barbour. 2014. The National Fruit Fly Research, Development & Extension Plan. Plant Biosecurity Cooperative Research Centre. Department of Industry and Science, Canberra, Australia.

Conrad, C. C., and K. G. Hilchey. 2011. A review of citizen science and community-based environmental monitoring: Issues and opportunities. Environmental Monitoring and Assessment 176: 273–291.

Deci, E. L., and R. M. Ryan. 1995. Human autonomy: The basis for true self-esteem, pp. 31–49. *In* M. Kernis (ed.), Efficacy, agency, and self-esteem. Plenum, New York, USA.

Dominiak, B.C., and D. Daniels. 2012. Review of the past and present distribution of Mediterranean fruit fly (*Ceratitis capitata* Wiedemann) and Queensland fruit fly (*Bactrocera tryoni* Froggatt) in Australia. Australian Journal of Entomology 51: 104–115.

Dominiak, B. C., and J. H. Ekman. 2013. The rise and demise of control options for fruit fly in Australia. Crop Protection 51: 57–67.

Dominiak, B. C., and R. Mapson. 2017. Revised distribution of *Bactrocera tryoni* in eastern Australia and effect on possible incursions of Mediterranean fruit fly: Development of Australia's eastern trading block. Journal of Economic Entomology 110: 2459–2465.

Dominiak, B. C., B. Wiseman, C. Anderson, B. Walsh, M. McMahon, and R. Duthie. 2015. Definition of and management for areas of low pest prevalence for Queensland fruit fly *Bactrocera tryoni* Froggatt. Crop Protection 72: 41–46.

Dyck, V. A., J. Hendrichs, and A. S. Robinson (eds.). 2021. Sterile Insect Technique – Principles and practice in Area-Wide Integrated Pest Management. Second Edition. CRC Press, Boca Raton, Florida, USA. 1200 pp.

Enkerlin, W. R. 2021. Impact of fruit fly control programmes using the Sterile Insect Technique, pp. 979–1006. *In* V. A. Dyck, J. Hendrichs, and A. S. Robinson (eds.), Sterile Insect Technique – Principles and practice in Area-Wide Integrated Pest Management. Second Edition. CRC Press, Boca Raton, Florida, USA.

Enkerlin, W., J. M. Gutiérrez-Ruelas, A. V. Cortes, E. C. Roldan, D. Midgarden, E. Lira, J. L. Z. López, J. Hendrichs, P. Liedo, and F. J. T. Arriaga. 2015. Area freedom in Mexico from Mediterranean fruit fly (Diptera: Tephritidae): A review of over 30 years of a successful containment program using an integrated area-wide SIT approach. Florida Entomologist 98: 665–681.

Ekman, J. 2015. Pests, diseases and disorders of sweet corn: A field identification guide. Applied Horticultural Research. Horticulture Innovation Australia Limited. Eveleigh, New South Wales, Australia.

Fisher, K. T., A. R. Hill, and A. N. Sproul. 1985. Eradication of *Ceratitis capitata* (Wiedemann) (Diptera: Tephritidae) in Carnarvon, Western Australia. Australian Journal of Entomology 24(3): 207–208.

Fisher, K. T. 1996. Queensland fruit fly (*Bactrocera tryoni*): Eradication from Western Australia, pp. 535-541. *In* B. A. McPheron, and G. J. Steck (eds.), Fruit fly pests, a world assessment of their biology and management. St. Lucie Press, Delray Beach, Florida, USA.

Florec, V., R. J. Sadler, B. White, and B. C. Dominiak. 2013. Choosing the battles: The economics of area wide pest management for Queensland fruit fly. Food Policy 38: 203–213.

(FAO) Food and Agriculture Organization of the United Nations. 2016. Establishment of pest free areas for fruit flies (Tephritidae). International Standard for Phytosanitary Measures (ISPM) No. 26. International Plant Protection Convention. Rome, Italy.

Garforth, C. J., A. P. Bailey, and R. B. Tranter. 2013. Farmers' attitudes to disease risk management in England: A comparative analysis of sheep and pig farmers. Preventive Veterinary Medicine 110: 456–466.

Giddens, A. 1984. The constitution of society: Outline of the theory of structuration. Polity Press. Cambridge, UK.

Gifford, R. 2011. The dragons of inaction: Psychological barriers that limit climate change mitigation and adaptation. American Psychology 66: 290–302.

(HARPS) Harmonised Australian Retailer Produce Scheme. 2017. Horticulture Innovation Australia.

Hendrichs, J., P. Kenmore, A. S. Robinson, and M. J. B. Vreysen. 2007. Area-Wide Integrated Pest Management (AW-IPM): Principals, practice and prospects, pp. 3-33. *In* M. J. B. Vreysen, A. S. Robinson, and J. Hendrichs (eds.), Area-wide control of insect pests: From research to field implementation. Springer, Dordrecht, The Netherlands.

Hernández-Jover, M., M. Taylor, P. Holyoake, and N. Dhand. 2012. Pig producers' perceptions of the influenza pandemic H1N1/09 outbreak and its effect on their biosecurity practices in Australia. Preventive Veterinary Medicine 106: 284–294.

Hodgson, G. M. 2006. What are institutions? Journal of Economic Issues 40: 1–25.

Hodgson, G. M. 2007. Institutions and individuals: Interaction and evolution. Organization Studies 28: 95–116.

Jackson, J. M., and S. G. Harkins. 1985. Equity in effort: An explanation of the social loafing effect. Journal of Personality and Social Psychology 49: 1199–1206.

Karau, S. J., and K. D. Williams. 1993. Social loafing: A meta-analytic review and theoretical integration. Interpersonal Relations and Group Processes 65: 681–706.

Kruger, H. 2016. Designing local institutions for cooperative pest management to underpin market access: The case of industry-driven fruit fly area-wide management. International Journal of the Commons 10: 176–199.

Kruger, H. 2017a. Creating an enabling environment for industry-driven pest suppression: The case of suppressing Queensland fruit fly through area-wide management. Agricultural Systems 156: 139–148.

Kruger, H. 2017b. Helping local industries help themselves in a multi-level biosecurity world – Dealing with the impact of horticultural pests in the trade arena. NJAS - Wageningen Journal of Life Sciences 83: 1–11.

Lloyd, A. C., E. L. Hamacek, R. A. Kopittke, T. Peek, P. M. Wyatt, C. J. Neale, M. Eelkema, and H. Gu. 2010. Area-wide management of fruit flies (Diptera: Tephritidae) in the Central Burnett district of Queensland, Australia. Crop Protection 29: 462–469.

Mankad, A. 2016. Psychological influences on biosecurity control and farmer decision-making. A review. Agronomy for Sustainable Development 36: 40.

Mankad, A., B. Loechel, and P. F. Measham. 2017. Psychosocial barriers and facilitators for area-wide management of fruit fly in southeastern Australia. Agronomy for Sustainable Development 37: 67.

Martin, I. M., H. Bender, and C. Raish. 2007. What motivates individuals to protect themselves from risks: The case of wildland fires. Risk Analysis 27: 887–900.

McEvoy, P. M., and P. Nathan. 2007. Perceived costs and benefits of behavioral change: Reconsidering the value of ambivalence for psychotherapy outcomes. Journal of Clinical Psychology 63: 1217–1229.

McInnis, D., R. F. L. Mau, S. Tam, R. Lim, J. Komatsu, L. Leblanc, D. Muromoto, R. Kurashima, and C. Albrecht. 2005. Development and field release of genetic sexing strains of the melon fly, *Bactrocera cucurbitae*, and Oriental fruit fly, *B. dorsalis*, in Hawaii. Extended synopses of the FAO/IAEA International Conference on Area-Wide Control of Insect Pests: Integrating the sterile insect and related nuclear and other techniques, Vienna, Austria.

Meats, A. 1981. The bioclimatic potential of the Queensland fruit fly, *Dacus tryoni*, in Australia. Proceedings of the Ecological Society of Australia 11: 151–161.

Meats, A. W., R. Duthie, A. D. Clift, and B. C. Dominiak. 2003. Trials on variants of the Sterile Insect Technique (SIT) for suppression of populations of the Queensland fruit fly in small towns neighbouring a quarantine (exclusion) zone. Australian Journal of Experimental Agriculture 43: 389–395.

Nabi, R. L., D. Roskos-Ewoldsen, and F. D. Carpentier. 2008. Subjective knowledge and fear appeal effectiveness: Implications for message design. Health Communications 23: 191–201.

Norris, F. H., S. P. Stevens, B. Pfefferbaum, K. F. Wyche, and R. L. Pfefferbaum. 2008. Community resilience as a metaphor, theory, set of capacities, and strategy for disaster readiness. American Journal of Community Psychology 41: 127–150.

North, D. C. 1990. Institutions, institutional change and economic performance. Cambridge University Press, Cambridge, UK.

Ogaugwu, C. E. 2014. Towards area-wide control of *Bactrocera invadens*: Prospects of the Sterile Insect Technique and molecular entomology. Pest Management Science 70: 524–527.

Pereira, R., B. Yuval, P. Liedo, P. E. A. Teal, T. E. Shelly, D. O. McInnis, and J. Hendrichs. 2013. Improving sterile male performance in support of programmes integrating the Sterile Insect Technique against fruit flies. Journal of Applied Entomology 137: 178–190.

Permalloo, S., S. I. Seewooruthun, P. Sookar, M. Alleck, and B. Gungah. 2005. Area-wide fruit fly control in Mauritius. *In* Extended synopses of the FAO/IAEA International Conference on Area-Wide Control of Insect Pests: Integrating the sterile insect and related nuclear and other techniques, Vienna, Austria.

Reynolds, O. L., B. C. Dominiak, and B. A. Orchard. 2010. Pupal release of the Queensland fruit fly, *Bactrocera tryoni* (Froggatt) (Diptera: Tephritidae), in the Sterile Insect Technique: Seasonal variation in eclosion and flight. Australian Journal of Entomology 49: 150–159.

Reynolds, O. L., and B. A. Orchard. 2015. Roving and stationary release of adult sterile Queensland fruit fly, *Bactrocera tryoni* (Froggatt) (Diptera, Tephritidae). Crop Protection 76: 24–32.

Rodgers, W. M., D. Markland, A. M. Selzler, T. C. Murray, and P. M. Wilson. 2014. Distinguishing perceived competence and self-efficacy: An example from exercise. Research Quarterly for Exercise and Sport 85: 527–539.

Rogers, E. M. 2003. Diffusion of Innovations Theory. New York Free Press 5th ed. New York, USA.

Rogers, R. W. 1983. Cognitive and physiological processes in fear appeals and attitude change: A revised theory of protection motivation, pp. 153–176. *In* J. Cacioppo, and R. Petty (eds.), Social psychophysiology. Guilford Press, New York, USA.

Ryan, R. M., and E. L. Deci. 2000. Self-determination theory and the facilitation of intrinsic motivation, social development, and well-being. American Psychology 55: 68–78.

Stringer, L. D., J. M. Kean, J. R. Beggs, and D. M. Suckling. 2017. Management and eradication options for Queensland fruit fly. Population Ecology 59: 259–273.

Suckling, D. M., L. D. Stringer, A. E. A. Stephens, B. Woods, D. G. Williams, G. Baker, and A. M. El-Sayed. 2014. From Integrated Pest Management to integrated pest eradication: Technologies and future needs. Pest Management Science 70: 179–189.

Suckling, D. M., J. M. Kean, L. D. Stringer, C. Caceres-Barrios, J. Hendrichs, J. Reyes-Flores, and B. C. Dominiak. 2016. Eradication of tephritid fruit fly pest populations: Outcomes and prospects. Pest Management Science 72: 456–465.

Sutherst, R. W., B. S. Collyer, and T. Yonow. 2000. The vulnerability of Australian horticulture to the Queensland fruit fly, *Bactrocera (Dacus) tryoni,* under climate change. Australian Journal of Agricultural Research 51: 467–480.

Twyman, M., N. Harvey, and C. Harries. 2008. Trust in motives, trust in competence: Separate factors determining the effectiveness of risk communication. Judgment and Decision Making 3: 111–120.

Vander Meer, R. K., R. M. Pereira, S. D. Porter, S. M. Valles, and D. H. Oi. 2007. Area-wide suppression of invasive fire ant *Solenopsis* spp. populations, pp. 487–496. *In* M. J. B. Vreysen, A. S. Robinson, and J. Hendrichs (eds.), Area-wide control of insect pests: From research to field implementation. Springer, Dordrecht, The Netherlands.

Vargas, R. I., R. F. L. Mau, E. B. Jang, R. Faust, and L. Wong. 2008. The Hawaii fruit fly area-wide pest management programme. Publications from USDA-Agricultural Research Service/University of Nebraska Lincoln Faculty Paper 656.

Vargas, R., M. Detto, D. D. Baldocchi, and M. F. Allen. 2010. Multiscale analysis of temporal variability of soil CO2 production as influenced by weather and vegetation. Global Change Biol. 16: 1589–1605.

Vreysen, M. J. B., A. S. Robinson, and J. Hendrichs (eds.). 2007a. Area-wide control of insect pests: From research to field implementation. Springer, Dordrecht, The Netherlands.

Vreysen, M. J. B., J. Gerardo-Abaya, and J. P. Cayol. 2007b. Lessons from Area-Wide Integrated Pest Management (AW-IPM) programmes with an SIT component: An FAO/IAEA perspective, pp. 723–744. *In* M. J. B. Vreysen, A. S. Robinson, and J. Hendrichs (eds.), Area-wide control of insect pests: From research to field implementation. Springer, Dordrecht, The Netherlands.

Yu, R., and P. Leung. 2011. Estimating the economic benefits of area-wide pest management: An extended framework with transport cost. Annals of Regional Science 46: 455–468.

INDUSTRY-DRIVEN AREA-WIDE MANAGEMENT OF QUEENSLAND FRUIT FLY IN QUEENSLAND AND NEW SOUTH WALES, AUSTRALIA: CAN IT WORK?

H. KRUGER

School of Sociology, Research School of Social Sciences, Australian National University, Canberra ACT 2601, Australia; heleen.kruger@agriculture.gov.au

SUMMARY

Queensland fruit fly, *Bactrocera tryoni* (Froggatt) (Tephritidae), is one of Australia's most problematic and costly horticultural pests. As key insecticides traditionally used to manage the pest have recently been restricted, area-wide management (AWM) of Queensland fruit fly is becoming a key recommended practice. The increased push for AWM coincides with several state governments reducing direct on-ground support for pest management. It is increasingly up to local industries to take the reins of implementing AWM programmes. This study explored the social and institutional aspects of industry-driven AWM to understand how these programmes can best be supported. The findings are based on AWM case studies in Queensland and New South Wales in Australia, as well as interviews with people who operate in Australia's broader fruit fly management innovation system. The findings reported here complement the prevailing techno-centric emphasis relating to Queensland fruit fly management. They are summarised in five principles: (i) the local social profile influences the prospects of successful AWM; (ii) AWM needs to be based on adaptive co-management; (iii) local industries need help to help themselves; (iv) AWM programmes in Queensland need strong two-way connectivity with the broader Queensland fruit fly management innovation system; and (v) industry-driven AWM programmes need institutional adjustment to share public roles and responsibilities. These principles are discussed, as well as their policy implications. The study concludes that industry-driven AWM is only possible in certain circumstances.

Key Words: Bactrocera tryoni, Tephritidae, community support, enabling environment, institutional design, fruit flies, social factors, fruit industry, horticulture, stakeholder involvement, adaptive co-management, international trade

J. Hendrichs, R. Pereira and M. J. B. Vreysen (eds.), Area-Wide Integrated Pest Management: Development and Field Application, pp. 693–708. CRC Press, Boca Raton, Florida, USA.

1. INTRODUCTION

In various countries there is an increasing push for local agricultural industries to be less dependent on direct government support and take on more responsibility for pest management (Higgins et al. 2016). One such pest in Australia is the Queensland fruit fly (*Bactrocera tryoni* (Froggatt)), which is established throughout parts of eastern Australia (Clarke et al. 2011). Queensland fruit fly is a particularly problematic pest as most fruit and vegetables are susceptible to infestation to varying degrees. Queensland fruit fly numbers can quickly soar under favourable conditions and the pest has the ability to cause considerable damage to crops (Dominiak and Ekman 2013). The pest can therefore have a huge economic impact, especially as several of Australia's horticulture international trading partners place costly requirements or restrictions on produce from Queensland fruit fly-infested areas (PHA 2008). In addition, the application of two key chemical insecticides that were traditionally used to manage the pest, fenthion and dimethoate, have been restricted (Dominiak and Ekman 2013).

Area-wide management (AWM) is seen as a key alternative Queensland fruit fly management strategy (NFFC 2016; PHA 2008). It promises a reduced need for insecticides and is acknowledged in international trade regulations as an acceptable phytosanitary approach for fruit fly (FAO 2016). However, many regions face reduced direct on-ground government support for Queensland fruit fly management resulting from cuts in pest monitoring and treatment activities, and extension services. Local industries now need to drive collaboration between various stakeholder groups and risk contributors (such as landholders with Queensland fruit fly hosts on their land) in order to initiate and maintain AWM programmes.

Responsibility for the management of Queensland fruit fly is addressed across a federated system involving delegated people in various organizations and roles. These organizations include the Australian Government, which oversees international trade matters that are affected by Queensland fruit fly. State/territory governments are responsible for providing support for Queensland fruit fly suppression and domestic market access impacted by Queensland fruit fly. Various public and private organizations conduct Queensland fruit fly-related research. Several recent initiatives were introduced to address the Queensland fruit fly problem, including the National Fruit Fly Council, the National Fruit Fly Research, Development and Extension Plan (PBCRC 2015), and the establishment of a Sterile Insect Technique consortium (*SITplus*). As these initiatives occurred after the initial field work of this research, their impact on the results presented here was limited.

This paper contains a summary of the key findings of a PhD project that investigated the social and institutional aspects of industry-driven Queensland fruit fly AWM in Australian horticulture industries, with special focus on three case studies in the states of Queensland and New South Wales. Traditionally, pest management has predominantly been approached as a technical issue in need of technical solutions (Schut et al. 2014). This chapter complements this techno-centric thinking.

2. METHODS

The study involved mixed methods research and included qualitative interviews and focus groups (facilitated small group discussions), and a quantitative grower survey. In phase 1 (2013-2014) an assessment was made of how the success of industry-driven AWM can be bolstered at the local on-ground level. It comprised three case studies (Table 1), together with a review of scholarly literature about socio-ecological systems (e.g. Ostrom 2005; Armitage et al. 2008) and community-based natural resource management (such as Berkes 2010; Klerkx et al. 2012; Curtis et al. 2014).

The online grower survey, involving the same three case study areas, was carried out between phases 1 and 2 towards the end of 2015. For a more detailed summary of the phase 1 methods see Kruger (2016a, 2016b).

In phase 2 (2015-2016), ways were identified to create an enabling environment for industry-driven AWM, i.e. the conditions needed that will support local industries to take the lead in AWM programmes. This was done both in terms of Queensland fruit fly suppression and market access. It involved 33 interviews with people operating in the broader fruit fly management innovation system (Table 2).

Table 2. An overview of the people interviewed in phase 2 about how to create an enabling environment for industry-driven fruit fly AWM

Organizational background	Number of interviewees[c]
State government[a]	
- Queensland fruit fly researcher	7
- Policy	6
- Industry support	4
- Operational management	1
Australian Government - Policy	3
University	2
Industry body	7
Local industry	2
Local government[b]	1
Consultant	3

[a] *New South Wales and Queensland only*
[b] *Five others were interviewed during phase 1 (see Table 1)*
[c] *Some interviews involved more than one interviewee*

These findings, together with the grower survey results, were analysed by applying agriculture innovation systems thinking (Klerkx et al. 2012; Schut et al. 2014). Such thinking sees innovation as a co-evolutionary process involving not only technical, but also social and institutional change that results in on-ground progress (Klerkx et al. 2012). For a more detailed summary of the phase 2 methods see Kruger (2017). For a detailed explanation of the full PhD study's methods see Kruger (2018).

Table 1. Overview of the three AWM (area-wide management) case studies investigated, including for each case study information about the areas they cover, the number of growers, kinds of crops produced, local support for AWM, indicators of success, factors that enable or hinder AWM, and the data collected from these regions (adjusted from Kruger 2016a)

	Central Burnett (Queensland)	**Riverina** (New South Wales)	**Young-Harden** (New South Wales)
Local Government Areas [a]	Along the Burnett River within the North Burnett Region Local Government Area	Local Government Areas of Carrathool, Griffith, Leeton, Murrumbidgee, Narrandera	Local Government Areas of Young and Harden
No. of growers	40 (2016 - mainly citrus) [b]	Approx. 420 [b] citrus; 372 [b] grape; 55 [b] prune, and many other growers of horticultural crops (2014)	Approx. 35 [b] cherry, 20 [b] grape (2015)
Crops produced and areas involved	2,266 ha [c] citrus (mainly mandarins); 370 ha table grapes (2010)	8,133 ha [c] citrus; 7,200 ha vegetables; 15,000 ha grapes; 1,365 ha walnuts; 885 ha prunes; 300 ha cherries; and smaller plantings of other hosts (approx. 30,000 ha of horticulture)	Estimated 650-700 ha [b] cherries; and up to 1,000 ha [b] of wine grapes
Support for AWM	▪ All growers implement rigorous Queensland fruit fly baiting ▪ Growers voluntary fund town treatments, but contributions are dwindling	▪ The local citrus industry pushes for broad-scale Queensland fruit fly management to support market access ▪ Other horticultural industries show little interest as Queensland fruit fly does not affect them economically	▪ Cherry industry pushes for AWM to support market access ▪ Grape growers are less engaged as unaffected economically by fruit fly
Indicators of AWM success	▪ Successful Queensland fruit fly suppression in season ▪ Queensland fruit fly management practices led to protocol ICA-28, increasing domestic market access	▪ Too early to tell	▪ Too early to tell
Factors enabling AWM	▪ Trust relationships between growers and consultants ▪ Small industry ▪ Growers have similar on-farm objectives ▪ Small, horticulture-dependent towns	▪ Cold winters ▪ Some packing houses insist that their grower suppliers provide proof of Queensland fruit fly management through spray records and receipts of fruit fly management inputs	▪ Cold winters ▪ Strong local government support ▪ Small industry ▪ Medium-sized towns
Factors hindering AWM	▪ Lack of enforcement mechanism to sustain grower contributions for town treatments ▪ Queensland fruit fly is endemic ▪ Disappointment with inaccessibility to overseas markets without cold treatment contributes to dwindling grower contributions for town treatments	▪ Many horticulture growers with multiple on-farm objectives ▪ Lack of representative leadership ▪ Many part-time, low-input growers ▪ Large/medium-sized towns dependent on multiple industries ▪ No local power to enforce cooperation ▪ Lack of communication channels to all horticulture growers	▪ Little local power to enforce cooperation ▪ Limited local communication and influence across all horticulture growers
Qualitative data	▪ Thirteen interviews [d] ▪ One focus group ▪ October 2013	▪ Twenty interviews [d] ▪ One focus group ▪ March 2014	▪ Nine interviews [d] ▪ One focus group ▪ September 2013
Grower survey response rate	70% (28/40) [e]	51% (50/98) [e]	63% (20/32) [e]

[a] *Local Government Areas are political entities that form the third tier of government in Australia*
[b] *Numbers provided by local industries*
[c] *Personal communication with Nathan Hancock, Citrus Australia Ltd (12 Feb 2016)*
[d] *Included local programme coordinators, growers, crop pest consultants and representatives of AWM programme management groups, packing houses, local councils and industry associations*
[e] *Number of respondents / total accessible grower population*

3. FINDINGS AND DISCUSSION

Fig. 1 represents the consolidation of the key findings of the PhD research, and shows that AWM programmes can be conceptualised as comprising three key components that stretch across various levels of activity:

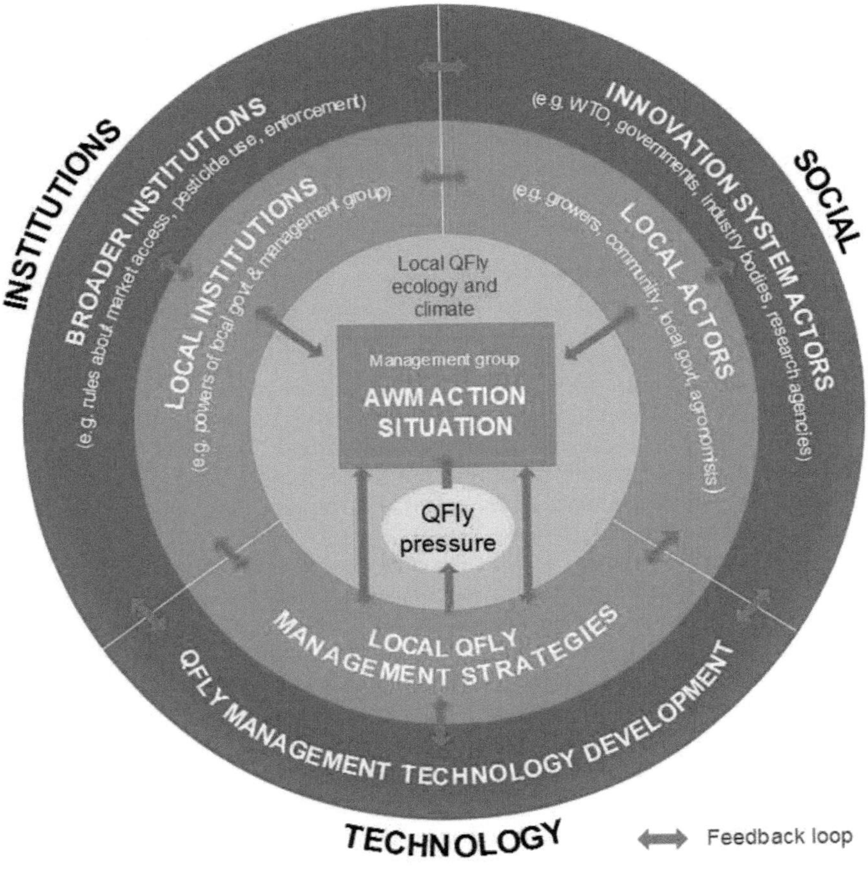

Figure 1. Conceptualised fruit fly management innovation system that will enable industry-driven AWM.

- *Social* – people and groups at local, state, national and international levels whose actions, interactions and decisions (or lack thereof) can either facilitate or hinder AWM.
- *Institutions* – formal (laws, regulations, standards) and informal (norms and accepted behaviour) that apply at local, state, national and international levels that influence the design and implementation of AWM.

- *Technology* – Queensland fruit fly-related technologies and information that are available at the local level and flow predominantly from the broader fruit fly management innovation system that stretches across state, national and international levels.

Following agriculture innovation systems thinking, each AWM programme is therefore embedded in a broader innovation system at higher levels comprising social, institutional and technological components that influence the feasibility of industry-driven AWM programmes.

The two-way arrows in Fig. 1 emphasise the importance of multi-directional information flow across levels and between the three key components. This is necessary to ensure that the different components are responsive to each other.

The key findings can be summarised in the following five principles that are also reflected in Fig. 1. A detailed explanation of the study's findings is contained in Kruger (2018).

3.1. Principle 1. The Local Social Profile Influences the Prospects of Successful AWM

With the international drive for harmonised phytosanitary measures, the impact due to the uniqueness of different local regional communities on their ability to achieve AWM can be easily underestimated. This research found four social variables that determine the degree of difficulty to achieve AWM, i.e. the transaction cost to develop local institutions (formal and informal rules) that support AWM that are widely supported by risk contributors (Kruger 2016b).

First variable: High heterogeneity of the contributors to Queensland fruit fly risk complicates setting local rules, such as identifying who needs to implement fruit fly management, what is required from them and how to win their commitment. Different on-farm objectives, pest impacts and market requirements for Queensland fruit fly management make it harder to identify a common AWM objective. For example, in the Riverina area in New South Wales, a small proportion of citrus growers is interested in exporting to premium Queensland fruit fly-sensitive markets. However, a large proportion is comprised of part-time growers supplying the domestic juice market. As their horticultural operations are not their primary income source, it lessens their incentive to participate in AWM, whereas exporters would like to see fruit fly management implemented to a high standard. Moreover, a heterogeneous mix of grower sectors means that the local AWM management group needs to establish trust and communication channels with more diverse stakeholders. This can challenge the management group's legitimacy and credibility across the region.

When stakeholder investment is not proportionate to the distribution of eventual benefits, they might view any or some rules as unfair, which discourages cooperation (Ostrom 2005). Identifying who will gain most, and therefore who ought to contribute most, is not always initially evident. For example, AWM is seen as a good candidate to be included in systems approaches for trade, e.g. applying at least two independent risk management measures that cumulatively achieve acceptable levels of pest risk (Dominiak and Ekman 2013; FAO 2018). However, in practice many markets remain wary of accepting systems approaches and achieving AWM may not necessarily render post-harvest treatments, such as cold sterilisation, obsolete, as is the case in Central Burnett in Queensland.

Second variable: High levels of social capital (i.e. personal trust-based amicable relationships) between key participants may facilitate the establishment of AWM, including established trust and communication networks, champions and leadership, such as in Central Burnett (see Box 1). However, it is important that social capital is not located in just some participating groups, as cliques may hinder progress (Pretty 2003).

Third variable: Existing social mechanisms that provide opportunities for monitoring on-farm Queensland fruit fly management, e.g. to keep a check on potential 'free-riders', can facilitate collective action and lower transaction costs. For example, in Central Burnett crop pest consultants making routine field visits enable low-cost monitoring of whether growers are managing Queensland fruit fly. In Riverina, some packing houses insist that their grower suppliers provide proof of Queensland fruit fly management. However, this could be thwarted in times of produce shortage when supply chains ease their grower requirements in order to secure supply.

Fourth variable: The ratio between the number of growers that are keen to pursue AWM and risk contributors who have little incentive to manage Queensland fruit fly (such as several town residents and part-time growers) affects the cost and effort needed to establish AWM. In Central Burnett growers were able to fund town treatments as towns are small relative to the production area. In the Riverina, with the large urban centre of Griffith, this would be less feasible, rendering AWM supporters reliant on awareness-raising to urge town residents to manage Queensland fruit fly on their land.

Policy Implications: The varying profiles of the three case studies illustrate that finding 'one-size-fits-all' local institutions ideal for all AWM programmes is unlikely. The rules relating to how the programme is run, what it involves and what it sets out to achieve are best negotiated locally and tailored to local circumstances. The AWM action situation (Fig. 1) needs to allow for Enticott's (2008) 'spaces of negotiation' to find ways to achieve wide stakeholder support.

3.2. Principle 2. AWM Needs to be Based on Adaptive Co-management

The formal requirements for fruit fly AWM (e.g. International Standards for Phytosanitary Measures (ISPMs) and the Queensland fruit fly Code of Practice (Jessup et al. 2007)), could suggest that AWM constitutes standardised 'spaces of prescription' (Enticott 2008), where AWM relies on introducing standard Queensland fruit fly management measures. Local industries' attempts to achieve AWM often involve social, institutional and technical complexity and uncertainty.

Aspects of AWM that involve complexity and uncertainty include its collaborative nature, various on-farm objectives and that not all stakeholders and risk contributors may have incentives to continually manage Queensland fruit fly, as was especially the case in the Riverina. Every local community is different. External expertise should not be privileged at the expense of local knowledge and learning processes rooted in local socio-economic, cultural and political contexts (Kruger 2016a).

Designing AWM programmes requires a good understanding of the regional Queensland fruit fly situation, including Queensland fruit fly behaviour in and amongst crops, and other local hosts, and this might vary across years with different weather conditions (Clarke et al. 2011).

In the system shown in Fig. 1, an AWM management group needs to continually monitor Queensland fruit fly pressure and adjust local Queensland fruit fly management strategies accordingly. Moreover, achieving market access is complex and uncertain. This was illustrated by the situation in Central Burnett where, despite a successful AWM programme, anticipated access to lucrative markets sensitive to Queensland fruit fly did not occur.

Global experience with a wide range of natural resource management situations has shown that complexity and uncertainty are best surmounted through adaptive co-management, i.e. a flexible process of 'learning by doing' that draws on different knowledge systems (Armitage et al. 2008). Agricultural innovation systems literature refers to the need for co-producing integrated knowledge between different stakeholder groups to provide a holistic understanding of how to best improve plant protection systems (e.g. Schut et al. 2014). For Queensland fruit fly this includes knowledge on the biology and behaviour of the pest, and its host distribution and phenology. It also requires knowledge on the effectiveness of cultural and other phytosanitary measures and their integration, trade requirements and politics, institutions, risk contributors' motivations and attitudes, and community engagement. Taking these factors into consideration will maximise the chances of designing a successful AWM programme tailored to local conditions.

In the case studies analysed, co-production of knowledge typically occurred within the local management group. Here, participants engaged in social learning, learning from the activities implemented and each other, to continually refine Queensland fruit fly-related management activities. From an adaptive co-management perspective, outcomes need to be closely monitored when changes are introduced to ensure that the system does not loose functionality. Social learning requires sound communication processes that are well-facilitated, both horizontally between local players and vertically across levels as illustrated in Fig. 1. Adaptive co-management was evident in the successful Central Burnett AWM programme (see Box 1).

Adaptive co-management does not offer a 'quick fix', and the investment (transaction cost) to engage with different players and learning processes can be considerable, especially in the early stages of an AWM programme. However, it does offer several important benefits over time (Kruger 2016a):

- Local knowledge of Queensland fruit fly behaviour in the local environment increases
- Management activities are suited to the local region and continually improve
- Local adaptive capacity strengthens, as lessons are learnt about what works and does not work in the local region and under different circumstances
- A common narrative develops between key stakeholders, which is fundamental to developing the much-needed shared vision for AWM
- A more sustainable, fit-for-context and locally-owned initiative is developed.

Box 1. Adaptive Co-management of Queensland fruit fly in Central Burnett

The successful Central Burnett AWM programme represents many aspects of adaptive co-management. Local crop consultants, state government researchers and citrus growers worked closely together on various regional projects from the 1990s, resulting in trust-based amicable relationships. Jointly they decided on research priorities and Queensland fruit fly-related activities for the region, and discussed findings, which subsequently shaped in-field activities. Research was carried out in the region, including in growers' orchards. Growers participated in some research activities, such as when the male annihilation technique was introduced. When AWM was launched in 2003, the local government assisted with community engagement to minimise Queensland fruit fly pressure originating from towns. Other local horticultural industries were engaged in the AWM effort. State government researchers were in close contact with their colleagues responsible for negotiating domestic trade protocols. This assisted in fine-tuning a domestic market access protocol to be accepted by some jurisdictions and be practical for growers to implement. Most growers employed a local crop consultant, with whom they regularly communicated, implying that they had the ear of key decision-makers, as all crop consultants were management group members. When changes were introduced, such as less intense town treatments, the results were closely monitored to allow for rapid response in case of unfavourable results.

Policy Implications: Adaptive co-management requires a mind-shift from focusing exclusively on implementing Queensland fruit fly management measures – such as when the now-restricted chemicals were still available – to implementing measures with the intention of continually learning and adjusting. This is illustrated in Fig. 1 by the two-way arrows between the AWM action situation, the local Queensland fruit fly management strategies and Queensland fruit fly pressure.

Adaptive co-management implies that grower groups need to actively build networks with others, including other growers (or their representatives), experts in Queensland fruit fly behaviour and market access, and community representatives, such as local government. Several of these may involve representatives of organizations that are active in the broader fruit fly management innovation system (Fig. 1).

Several aspects of market access represent 'top-down' elements where growers have little control, such as some trade protocol requirements that are negotiated bilaterally at the federal level. While it is important for those designing AWM programmes to take the technical requirements for market access into consideration as early as possible, it is best for AWM programmes to first focus on achieving Queensland fruit fly suppression. Unmet trade requirements relating to Queensland fruit fly can then be seen as 'bolt-on' components. Government and industry bodies can facilitate access to relevant experts. Other ways in which local industries can be supported are discussed below.

3.3. Principle 3. Local Industries Need Help to Help Themselves

An 'ideal type' of knowledge and capabilities that local industries need to access in order to achieve and maintain AWM, was developed during this research. It demonstrates the importance of dedicated staff to oversee and coordinate an AWM programme. In summary, these capabilities are:

- *Programme administration and management* – Effective programme implementation, including planning, implementation and monitoring; securing funding, sound financial management, understanding broader institutional requirements, organising and facilitating meetings, and record-keeping.
- *Stakeholder interaction* – Including achieving a shared local vision, maximising uptake across risk contributors, networking, advocating the programme to key stakeholders, conflict management and supporting growers with recommended practice implementation.
- *Understanding Queensland fruit fly behaviour and management* – Including general Queensland fruit fly biology and behaviour within the target region, on-ground management options, and consistently implementing regional Queensland fruit fly management strategies.
- *Understanding market access requirements:*
  - Phytosanitary measure options – including effectiveness and limitations
  - Formal market access standards – e.g. relevant ISPMs and Queensland fruit fly Code of Practice; and concepts such as Probit 9 levels, appropriate level of protection and risk management
  - Informal aspects and requirements – e.g. market expectations and politics
  - Market access application and approval process, including data gathering, and
  - Consistent implementation, e.g. monitoring and corrective actions, where needed.

Phase 1 respondents talked about the difficulty of establishing the needed networks to access the knowledge and capabilities required. Some phase 2 interviewees believed that stakeholders at higher levels may easily assume that growers have certain levels of knowledge or capabilities, but in reality, this varies. Interviewees across phases 1 and 2 spoke about issues that added cost and effort to achieve AWM that could be made easier. For example, a high level of government staff turnover contributes to growers struggling to maintain relationships built on a mutual understanding of their local Queensland fruit fly situation, and possible ways forward. Moving forward sometimes depends on joint decision-making between stakeholders from different agencies and 'getting them in one room' is difficult for industry.

Policy Implications: Given the considerable decline in public extension support in Australia, training could be offered to those who are likely to fill at least part of this gap, such as private crop consultants, key growers and other interested local people. This can strengthen local knowledge and capacity on issues such as trade; Queensland fruit fly biology, behaviour and management; and community engagement. However, training without strengthened intermediation between local level actors and other Queensland fruit fly management innovation system actors will do little to encourage 'upward' information flow (see *Principle 4*). Training can also be a tool to support collaborations across levels by more quickly facilitating in-depth conversations between growers and other stakeholders. Feedback from interviewees also stressed the importance of effective stakeholder coordination between states/territories and between government and industry bodies; minimising staff changes and fostering a client-oriented ethos in government departments. Innovative policy-making can contribute to overcoming local reliance on voluntary approaches (see *Principle 5*).

3.4. *Principle 4. AWM Programmes Need Strong Two-way Connectivity with the Broader Queensland fruit fly Management Innovation System*

This research found that local industries easily become disconnected from the broader system. Phase 1 interviewees spoke about the difficulty of establishing networks and finding information and guidance to instigate AWM. Some phase 2 interviewees spoke about inaccessible research findings, as much research is not publicly available, and many growers will not read scientific articles. The disconnect is intensified by the loss of public extension services. The 'grower voice' was considered as under-represented in national Queensland fruit fly management dialogues. Higher level governance bodies may underestimate local complexities and overestimate the knowledge and abilities of local stakeholders. While peak industry bodies make much contribution to filling this gap, not all growers may see their peak industry body as representing their concerns. Limited resources prevent industry bodies from being involved in all AWM attempts to develop an in-depth understanding of the local issues.

Innovation studies show that successful innovation that results in positive on-ground change tends to result from a co-evolutionary process involving concurrent technological, social, organizational, and institutional change. As such, growers tend to be partners and entrepreneurs in the innovation system. The prevailing innovation approaches that focus primarily on technology production are increasingly criticized for not achieving intended outcomes often due to a lack of adoption (Klerkx et al. 2012). Instead, a well-functioning innovation system, in this case one that promotes industry-driven Queensland fruit fly AWM, needs to deliver the required institutional, social, and technological change that will maximise the chances for it to flourish (Hekkert et al. 2007). This requires strong feedback loops between local AWM attempts and the broader Queensland fruit fly management innovation system (see Fig. 1).

Policy Implications: Australia's National Fruit Fly Council has made considerable effort to make information on Queensland fruit fly management more accessible since the research was conducted, including online. However, establishing knowledge brokering to enable feedback loops across levels (Fig. 1) can support co-producing integrated knowledge (Kruger 2017). This requires fulfilling key functions to support information flow and collaboration between stakeholder groups (Klerkx et al. 2012), which are easily overlooked as they are often invisible and hard to measure (Klerkx and Leeuwis 2009; Meyer 2010). These key functions are:

- Demand articulation – including assisting local industries with finding a shared vision to identify their technology, knowledge, funding, and policy needs (Klerkx and Leeuwis 2009; Meyer 2010).
- Network establishment – including local horizontal networks and vertical networks with policy-makers and researchers (Klerkx et al. 2012; Meyer 2010).
- Information translation – to connect 'external' information with the local context and growers' existing knowledge in summarized form and language that growers find useful (Klerkx et al. 2012); and local issues are 'translated' to other innovation system players to inform their decision-making (Klerkx and Leeuwis 2009; Meyer 2010).

- Innovation process management – working towards better arrangements in the multi-actor network, including facilitating cooperation and learning (Klerkx and Leeuwis 2009).

As in *Principle 2*, collaborations between heterogeneous stakeholders to learn from each other are needed to produce integrated knowledge. Such collaborations represent innovation platforms. Interconnected innovation platforms across levels can be created, e.g. by connecting local AWM groups with multi-stakeholder groups at state and national levels. It enables representation of a stronger 'grower voice' at higher level deliberations and decision-making. The design of such intervention is best negotiated amongst key stakeholders to meet their needs and expectations, ensure maximum buy-in, and to enable a good fit with existing structures. Queensland fruit fly *management coordinators* at regional, state and national levels can facilitate interconnected innovation platforms and can potentially be co-funded between government and industry.

3.5. Principle 5. Industry-driven AWM Programmes Need Institutional Adjustment to Share Public Responsibilities and Roles

A major challenge for industry-driven AWM is dealing with Queensland fruit fly pressure from host plants in town backyards, and on peri-urban and public land. Therefore, public support is vital for AWM success (Dyck et al. 2021). All case studies were reliant on voluntary approaches to address Queensland fruit fly pressure from towns, such as the awareness-raising activities in Young-Harden and Riverina. Currently, legislative power rests predominantly with state governments, but there may be reluctance to introduce enforceable measures that favour industry over other rural groups (NSW 2014). Many local governments have very limited powers to put in place enforceable rules and in the case of study areas they did not have powers to enter backyards without resident permission.

The case studies revealed barriers to behaviour change other than a lack of awareness, e.g. community apathy or recommended Queensland fruit fly management practices not making economic sense. For example, routine sanitation in orchards to remove fallen fruit is another essential AWM component that many growers are reluctant to fund. Other challenges included absentee landholders and derelict orchards.

At least 89 per cent of growers surveyed across all case studies agreed that Queensland fruit fly infestation in towns increases on-farm Queensland fruit fly pressure. However, only 42 per cent of respondents in the Riverina and 40 per cent in Young-Harden had a strong or some belief that regular educational activities would ensure that town residents would adequately manage Queensland fruit fly on their properties. This suggests a limited potential for growers voluntarily contributing to a reliance on awareness-raising activities in towns.

Another major concern is maintaining commitment and AWM programme funding over the longer term. Establishing an income stream often depends on voluntary contributions from growers. However, Central Burnett demonstrates that this is thwarted by 'free-riding' when some growers refuse to contribute while still benefiting from reduced Queensland fruit fly pressure from towns. This causes others to also ‘opt out’ of voluntary contributions. For example, 59 per cent of Central Burnet survey respondents said that they would contribute to town treatments only if others contributed too.

Studies about resource governance, involving resource-users taking the lead in setting the rules around resource usage, do not exclude complementary state intervention to back-up the collective action driven by resource users. In various contexts it is seen as important to sustain the trust among resource-users that others will also cooperate, and that a lack of cooperation will not jeopardise individual efforts (Ostrom 2005).

Policy Implications: A recommendation from this work is to apply ‘smart regulation’, i.e. using complementary policy instruments and behavioural interventions to assist in overcoming the weaknesses of individual instruments, while still capitalising on their strengths. For example, by combining approaches that draw on people’s intrinsic motivation to ‘do the right thing’ with legal instruments that can be enforced and that provide legitimacy to local industry AWM efforts.

Several policy instruments were explored as part of the PhD research project that were drawn from the case studies, as well as Australia’s Landcare Programme for natural resource management, and effective overseas AWM programmes. These instruments could be used in combination with others to ensure that the overall approach is locally tailored:

- *Community education and awareness-raising* – if well implemented, this approach can influence people who have an intrinsic motivation to manage the pest, such as households valuing their backyard produce. The investment needed to sustain effective awareness campaigns are easily under-estimated. However, such campaigns will do little to overcome behavioural change barriers beyond a lack of awareness (Curtis et al. 2014).
- *Broad-scale state regulation enforcing Queensland fruit fly management on all land* – while this can be applied consistently across risk contributors, it is very costly to monitor and enforce. There are also moral challenges, such as when landholders are physically or financially unable to manage Queensland fruit fly. Magistrate courts easily misunderstand the level of Queensland fruit fly control required and have rejected cases brought for prosecution (personal communication, state government interviewee, 24 September 2013).
- *Co-opting local stakeholders to support better implementation of government powers* – including authority to enter private property and/or prosecute non-compliant landowners. This lessens the monitoring and enforcement burden on state governments. It is applied in the US states of Oregon and Washington, for example under the 2011 Washington Code (Washington State Legislature 2011). However, some authorities may be reluctant to favour industry needs over those of other community groups.

- *Devolved power to enable industry, in partnership with local communities, to devise rules appropriate for the local context* – including possibly allowing enforcement as a 'back-up' mechanism. This aligns most closely with the underlying principles contained in much of the community engagement literature that values community involvement in the decision-making of issues affecting them. However, results elsewhere have been mixed in natural resource management (Berkes 2010). Potential adversarial effects include conflict and 'power grabs' by some groups (Berkes 2010). Dealing with uninformed people trying to influence the programme is difficult (Dyck et al. 2021). Such approaches tend to require considerable investment and skill.
- *Legislated cost-recovery structures* – A legislated income-stream can facilitate industry and other appropriate local stakeholders to implement on-going and reliable pest treatments in Queensland fruit fly risk areas in combination with community awareness activities. It can come from mandatory contribution from growers, the state and possibly town residents, such as in the successful OKSIR programme for codling moth control in British Columbia, Canada (Dyck et al. 2021; Nelson et al., this volume). However, such schemes could encounter resistance from those expected to contribute unless the contributions are adjusted to be proportionate to the expected benefits.

4. CONCLUSION

The research reported here explored whether industry-driven fruit fly area-wide management is feasible, with special focus on Queensland fruit fly AWM programmes in New South Wales and Queensland, Australia. This manuscript contains a summary of a social science PhD project that involved an investigation of three case studies of AWM programmes (or attempts thereof), a grower survey and interviews with people operating in the broader Queensland Fruit Fly management innovation system.

It found that the feasibility of industry-driven AWM depends on social and institutional factors at the local level and within the broader Queensland fruit fly management innovation system. Advantageous local factors include a favourable social profile, such as growers with relatively homogeneous on-farm goals and high levels of social capital, as well as the application of adaptive co-management. AWM programmes need to be adjusted to the local context, with market access requirements seen as 'bolt-on' components.

An enabling environment for industry-driven AWM requires a broader innovation system that is responsive to the needs of local industries. This requires strong two-way information flow between local programmes and other players in the innovation system, such as policy-makers and technology developers, which can be supported through knowledge-brokers and vertically interconnected innovation platforms. 'Smart regulation' can assist local industries to overcome the limitations of voluntary approaches (such as depending on awareness-raising and education alone), by influencing people through a combination of policy instruments tailored to local conditions.

5. REFERENCES

Armitage, D. R., R. Plummer, F. Berkes, R. I. Arthur, A. T. Charles, I. J. Davidson-Hunt, A. P. Diduck, N. C. Doubleday, D. S. Johnson, M. Marschke, P. McConney, E. W. Pinkerton, and E. K. Wollenberg. 2008. Adaptive co-management for social-ecological complexity. Frontiers in Ecology and the Environment 7: 95–102.

Berkes, F. 2010. Devolution of environment and resources governance: Trends and future. Environmental Conservation 37: 489–500.

Clarke, A. R., K. S. Powell, C. W. Weldon, and P. W. Taylor. 2011. The ecology of *Bactrocera tryoni* (Diptera: Tephritidae): What do we know to assist pest management? Annals of Applied Biology 158: 26–54.

Curtis, A., H. Ross, G. R. Marshall, C. Baldwin, J. Cavaye, C. Freeman, A. Carr, and G. J. Syme. 2014. The great experiment with devolved natural resource management governance: Lessons from community engagement in Australia and New Zealand since the 1980s. Australasian Journal of Environmental Management 21: 175–199.

Dominiak, B. C., and J. H. Ekman. 2013. The rise and demise of control options for fruit fly in Australia. Crop Protection 51: 57–67.

Dyck, V. A., J. Reyes Flores, M. J. B. Vreysen, E. E. Regidor Fernández, B. N. Barnes, M. Loosjes, P. Gómez Riera, T. Teruya, and D. Lindquist. 2021. Management of Area-Wide Integrated Pest Management programmes that integrate the Sterile Insect Technique, pp. 781–814. *In* V. A. Dyck, J. Hendrichs, and A. S. Robinson (eds.), Sterile Insect Technique – Principles and practice in Area-Wide Integrated Pest Management. Second Edition. CRC Press, Boca Raton, Florida, USA.

Enticott, G. 2008. The spaces of biosecurity: Prescribing and negotiating solutions to bovine tuberculosis. Environment and Planning A 40: 1568–1582.

(FAO) Food and Agriculture Organization of the United Nations. 2016. Establishment of pest free areas for fruit flies (Tephritidae). International Standard for Phytosanitary Measures (ISPM) No. 26. International Plant Protection Convention. FAO, Rome, Italy.

(FAO) Food and Agriculture Organization of the United Nations. 2018. Systems approach for pest risk management of fruit flies (Tephritidae). International Standard for Phytosanitary Measures (ISPM) No. 35. International Plant Protection Convention. FAO, Rome, Italy.

Hekkert, M. P., R. A. Suurs, S. O. Negro, S. Kuhlmann, and R. E. Smits. 2007. Functions of innovation systems: A new approach for analysing technological change. Technological Forecasting and Social Change 74: 413–432.

Higgins, V., M. Bryant, M. Hernández-Jover, C. McShane, L. and Rast. 2016. Harmonising devolved responsibility for biosecurity governance: The challenge of competing institutional logics. Environment and Planning A 48(6): 1131–1151.

Jessup, A. J., B. Dominiak, B. Woods, C. P. F. de Lima, A. Tomkins, and C. J. Smallridge. 2007. Area-wide management of fruit flies in Australia, pp. 685–697. *In* M. J. B. Vreysen, A. S. Robinson, and J. Hendrichs (eds.), Area-wide control of insect pests. From research to field implementation. Springer, Dordrecht, The Netherlands.

Klerkx, L., and C. Leeuwis. 2009. Establishment and embedding of innovation brokers at different innovation system levels: Insights from the Dutch agricultural sector. Technological Forecasting and Social Change 76: 849–860.

Klerkx, L., M. Schut, C. Leeuwis, and C. Kilelu. 2012. Advances in knowledge brokering in the agricultural sector: Towards innovation system facilitation. Institute of Development Studies Bulletin 43: 53–60.

Kruger, H. 2016a. Adaptive co-management for collaborative commercial pest management: The case of industry-driven fruit fly area-wide management. International Journal of Pest Management 62: 336–347.

Kruger, H. 2016b. Designing local institutions for cooperative pest management to underpin market access: The case of industry-driven fruit fly area-wide management. International Journal of the Commons 10: 176–199.

Kruger, H. 2017. Creating an enabling environment for industry-driven pest suppression: The case of suppressing Queensland fruit fly through area-wide management. Agricultural Systems (Issue C) 156: 139–148.

Kruger, H. 2018. The social and institutional aspects of industry-driven fruit fly area-wide management in Australian horticultural industries, Australian National University and Charles Darwin University, PhD thesis, 293 pp.

Meyer, M. 2010. The rise of the knowledge broker. Science Communication 32: 118–127.

(NFFC) National Fruit Fly Council. 2016. Area wide management. Plant Health Australia. Canberra, Australia.

(NSW) New South Wales Trade and Investment. 2014. Management of fruit flies in New South Wales. *In* Plant Biosecurity & Product Integrity. NSW Government, Sydney, Australia. 5 pp.

Ostrom, E. 2005. Understanding institutional diversity. Princeton Paperbacks. Princeton University Press, Princeton, New Jersey, USA. 376 pp.

(PBCRC) Plant Biosecurity Cooperative Research Centre. 2015. National Fruit Fly Research, Development and Extension Plan. Canberra, Australia. 56 pp.

(PHA) Plant Health Australia. 2008. Draft national fruit fly strategy. A Primary Industry Health Committee Commissioned Project. Plant Health Australia, Canberra, Australia. 64 pp.

Pretty, J. 2003. Social capital and the collective management of resources. Science 302(5652): 1912–1914.

Schut, M., J. Rodenburg, L. Klerkx, A. van Ast, and L. Bastiaans. 2014. Systems approaches to innovation in crop protection. A systematic literature review. Crop Protection 56: 98–108.

Washington State Legislature. 2011. Horticultural Pest and Disease Board. *In* Washington State Legislature, Title 15 Chapter 15.9. Olympia, Washington, USA.

A SUCCESSFUL COMMUNITY-BASED PILOT PROGRAMME TO CONTROL INSECT VECTORS OF CHAGAS DISEASE IN RURAL GUATEMALA

P. M. PENNINGTON[1], E. PELLECER RIVERA[1], S. M. DE URIOSTE-STONE[2], T. AGUILAR[1] AND J. G. JUÁREZ[1]

[1]*Center for Health Studies, Universidad del Valle de Guatemala, Guatemala; pamelap@uvg.edu.gt*
[2]*School of Forest Resources, University of Maine, Orono, Maine, USA*

SUMMARY

The adoption of novel integrated vector management (IVM) strategies requires proof-of-concept demonstrations. To implement a community-based intervention, for the control of vectors of Chagas disease in Guatemala, we engaged all relevant stakeholder groups. Based on this and previous experiences of the authors on engaged research and community-based interventions, several key factors can help facilitate effective integration of stakeholders in support of area-wide integrated vector management (AW-IVM) programmes. First and foremost, the diversity of stakeholders needs to be engaged early-on in the participatory action research and implementation processes, to provide ownership and contribute ideas on how to design and implement an intervention. Another important component, situational analysis regarding current pest control policies, practices and relevant stakeholders, is generated through interviews with key informants, at both national and local levels (governmental and non-governmental organizations); it can facilitate the joint identification of strengths, weaknesses, opportunities and threats regarding current pest control strategies and proposing solutions through an AW-IVM approach. In addition, successful AW-IVM can result from identifying locally relevant strategies to implement the proof-of-concept demonstrative project. Further, it is critical to maintain constant communication with the local and national leaders, involving them throughout the implementation and evaluation processes. Flexibility should also be built into the project to allow for community-driven changes in the strategy, through a cyclical joint reflective process. Periodic feedback of project development needs to be scheduled with key stakeholders to maintain rapport. Finally, the results of the evaluation should be shared and discussed with stakeholders to ensure long-term sustainability of the programme, intervention, or project. Here we present the citizen engagement procedures used to integrate community members, health officials, and non-governmental organization staff for Chagas disease control in a region of Guatemala. We demonstrate how these methods can be applied to support AW-IVM programmes, so that communities and authorities are actively involved in the development and implementation of a jointly agreed intervention. In 2012, we developed the IVM intervention in an area of Guatemala with persistent *Triatoma dimidiata* (Latreille) infestation that is associated with the presence of infected rodents (rats and mice), that act as reservoirs of the *Trypanosoma cruzi* Chagas parasites inside the households. Nine control communities received only the Ministry of Health insecticide application against the vector and nine intervention communities participated in the

J. Hendrichs, R. Pereira and M. J. B. Vreysen (eds.), Area-Wide Integrated Pest Management: Development and Field Application, pp. 709–727. CRC Press, Boca Raton, Florida, USA.

AW-IVM intervention. The intervention included a programme for rodent control by the community members, together with education about the risk factors for vector infestation, and insecticide application by the Ministry of Health. Entomological evaluations in 2014 and 2015 showed that vector infestation remained significantly lower in both intervention and control communities. In 2015, we found that there was a higher acceptance of vector surveillance activities in the intervention communities compared to control communities, suggesting that participatory activities increase programme sustainability. Finally, we found that there was a significant increase over time in the number of households with infected vectors in the control group, whereas there was no significant increase in the communities that participated in the programme. Thus, an AW-IVM programme including simultaneous rodent and vector control could reduce the risk of Chagas infection in communities with persistent vector infestation.

Key Words: Central America, Ministry of Health, Jutiapa, community-driven changes, citizen engagement, Reduviidae, *Trypanosoma cruzi*, *Rhodnius prolixus*, *Triatoma dimidiata*, *Triatoma infestans*, stakeholders, participatory action research, vector surveillance, area-wide integrated vector management (AW-IVM), peridomestic environments, insecticide application, rodent control

1. BACKGROUND ON CHAGAS DISEASE

1.1. Chagas Disease in Latin America

Chagas disease is widespread in the Americas, affecting 6-7 million people (WHO 2017). It is considered one of the most neglected tropical diseases with serious public health implications, causing the loss of more than 600 000 disability-adjusted life years in Latin America (Mathers et al. 2007). The causing agent, *Trypanosoma cruzi* Chagas, is transmitted primarily by a few species of blood-feeding triatomine insects of the Reduviidae family (Dias et al. 2002). Strategies to control the vector species associated with domestic environments have been successful in several Latin American regions.

There are three area-wide regional initiatives in the Americas for Chagas disease vector control: the Southern Cone, the Andean Pact, and the Central American Initiative (Dias et al. 2002). All three initiatives aim to reduce the incidence of Chagas disease through vertically coordinated multi-country vector control programmes, blood supply screening, and health education (Dias et al. 2002). The two area-wide South American initiatives targeted *Triatoma infestans* Klug for vector control, as it is the main vector for the transmission of the disease in these regions (Massad 2008). The control activities for *T. infestans* resulted in the interruption of vector-borne transmission in Brazil, Chile and Uruguay (Dias 2007). Also, it reduced the incidence of infection in children and young adults in its member countries (Moncayo and Silveira 2009).

The Central American Initiative focused on the coordinated use of indoor residual spraying (IRS) to eliminate *Rhodnius prolixus* (Stähl) and reduce *Triatoma dimidiata* (Latreille) domestic populations (PAHO 2011). This area-wide programme included an attack and a surveillance phase that was coordinated by the Ministers of Health of Central America and several cooperation agencies, including local universities and the Japanese International Cooperation Agency (JICA 2014).

The success of the IRS interventions resulted in the interruption of Chagas transmission by *R. prolixus* in Guatemala, Honduras and Nicaragua, and the elimination of the vector in Costa Rica, El Salvador and Mexico (Hashimoto and Schofield 2012). These initiatives were successful and led to a significant decrease in the incidence and prevalence of Chagas disease. However, as with many area-wide pest/vector control programmes (Vreysen et al. 2007), remaining foci with persistent infestations hinder regional success. To succeed in these areas will require approaches that integrate novel ecological, biological and social factors.

Since early 2000s, housing improvement was proposed for sustainable control in regions with persistent *T. dimidiata* infestation (Lucero et al. 2013). In 2009, the World Health Organization (WHO) promoted the development of novel interventions for the control of Chagas and dengue in Latin America (Sommerfeld and Kroeger 2015). The interventions included multi-sectoral and -disciplinary ecosystem management strategies. Our study was part of an initiative by several countries to develop novel approaches for the control of *T. dimidiata* and *T. infestans* (Gürtler and Yadon 2015). We aimed to develop a community-based strategy for sustainable control of an area with persistent *T. dimidiata* in south-eastern Guatemala (De Urioste-Stone et al. 2015).

1.2. Vector Control Programme in Jutiapa, Guatemala

Guatemala started a major vector control programme in 2000 as part of the National Strategic Plan for Chagas Control (Hashimoto and Schofield 2012). This included several rounds of IRS application in the endemic area that covered over 45 000 km^2 in the initial and second programme phase (2000-2005) (Hashimoto et al. 2012). In 2009, Guatemala was certified to have interrupted transmission of the disease by *R. prolixus*, and *T. dimidiata* infestation was reduced nine-fold in the endemic region (Hashimoto and Schofield 2012; Manne et al. 2012). Vector control activities during the 2000-2010 period were estimated to have reduced the number of seropositive school-age children from 5.3% (1998) to 1.3% (2005-2006) (Hashimoto et al. 2012). Thus, successful Chagas disease control was achieved in most of the endemic area.

The effectiveness of IRS was evaluated by Hashimoto et al. (2006) across the department of Jutiapa, located in the south-eastern region of Guatemala. In areas where the baseline infestation rates were originally 20%, one spraying cycle reduced infestation to a mean of 1.4% within 3-21 months (Fig. 1). However, the infestation levels increased to an average of 8.1% during a second screening 20-45 months after spraying. In villages with an initial 40% infestation rate, the first spraying cycle reduced it to an average of 12.2%. Given that the control programme aims to reduce infestation to below 5%, a second spraying cycle was carried out in these villages. This effort reduced infestation to 4.8% in 40 of 52 villages 3-10 months after spraying. However, 12 of the 52 villages showed higher than 5% infestation after two spraying cycles, necessitating a third cycle that reduced the infestation from 10.9% to 4.1% 3-5 months after spraying.

With uneven results of prevention and control efforts, emphasis shifted to exploring long-term sustainability of surveillance and control interventions (Schofield et al. 2006). Despite a growing recognition of the role of social, cultural, economic and political conditions as risk factors linked to the disease, research and control efforts focusing on these factors have remained scarce (Ventura-García et al. 2013). Vulnerable groups such as indigenous populations and groups living in poverty continue to be at a high risk for disease transmission due to cultural, social, political, and health system barriers (Dell'Arciprete et al. 2014).

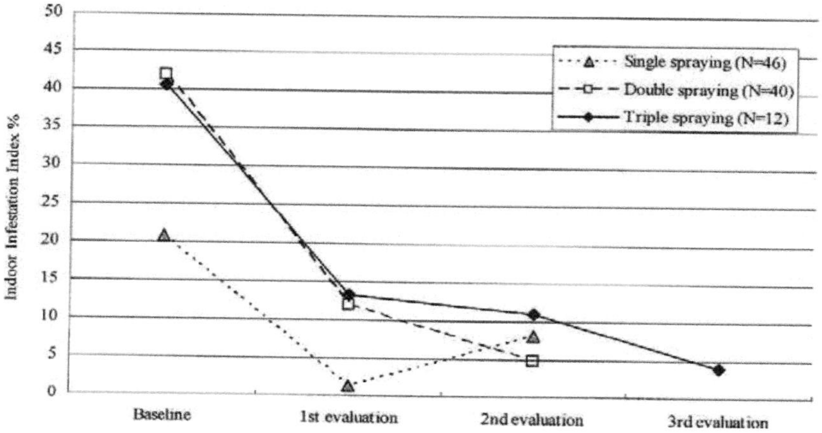

Figure 1. Changes of indoor infestation index for the single, double, and triple spraying areas (from Hashimoto et al. 2006).

2. DEVELOPMENT OF PROCEDURES AND SURVEYS

2.1. *Setting the Foundations*

2.1.1. *Area of Interest and Initial Field Visit*

The municipality of Comapa in the department of Jutiapa has been recognized as an endemic focus for Chagas disease, with persistent *T. dimidiata* infestation (Hashimoto et al. 2006) and high prevalence of *T. cruzi* infection (Rizzo et al. 2003). After multiple cycles of IRS and house improvement interventions (Bustamante et al. 2014), transmission continues at low levels in school-age children (Juárez et al 2018). A multidisciplinary team was formed that was composed of social and biological scientists with the aim to develop a project for improved Chagas disease control in this region. Our aim was to create a process that allowed stakeholders to contribute ideas, and implement an intervention that considered local conditions, resources, and concerns.

2.1.2. PRECEDE-PROCEED Framework

The PRECEDE-PROCEED model (PRECEDE: *Predisposing, Reinforcing, and Enabling Causes in Educational Diagnosis and Evaluation*; PROCEED: *Policy, Regulatory and Organizational Constructs in Educational and Environmental Development*) was used as a framework to guide the identification of risk factors through an analysis of the situation, and to develop and implement the intervention programme considering the risk conditions. This framework has been widely used for planning, implementing and evaluating health promotion programmes (Edberg 2007). It uses a multidisciplinary approach that includes disciplines such as sociology, psychology, epidemiology, business, and education (Blank 2006).

The steps required by the process before the intervention – PRECEDE – are based on a situational analysis: social, environmental, entomological, epidemiological, psychological, educational, and institutional assessments. A key component is the identification of risk factors (predisposing and reinforcing), as well as an informed development of the intervention proposal, with input from participants during group meetings. The intervention is based on the PRECEDE findings and implemented through the PROCEED steps in the formative and final evaluation of the intervention. As suggested by Edberg (2007), for the evaluation component, attention needs to be placed on assessing 1) the process of implementation, 2) the impact of the intervention (i.e. changes in knowledge, changes in practices and policies, changes in awareness), and 3) a limited number of outcomes due to the short timeframe of the project.

2.1.3. Defining the Study Design

In the PRECEDE stage of the project, a mixed methodology design was used to gain in-depth understanding of social, economic and environmental factors associated with persistent triatomine infestation (Bustamante et al. 2014). The approach allowed for a situational analysis of the community context, as well as for generalization of results and the credibility of the conclusions due to triangulation across research methods (Mertens 2014; Patton 2002).

Each method was selected based on its usefulness to the intervention, framed within the PRECEDE component of the framework. The methodology consisted of five stages: a) building rapport and gaining entry, b) mapping of households and sampling design, c) conducting baseline entomological and household surveys on knowledge, attitudes and practices regarding Chagas disease and possible risk factors, d) facilitating group meetings, and e) analysing documented evidence regarding local Chagas disease control activities and the socio-economic context.

Quantitative and qualitative data analyses allowed jointly developing a situational analysis report and an intervention proposal through a participatory process of individual and group learning and reflection (Bustamante et al. 2014; De Urioste-Stone et al. 2015). The PROCEED component of the study included a pre-test and post-test control group study design (De Urioste-Stone et al. 2015). Once again, a mixed methodology was used to generate both quantitative and qualitative measures during the intervention (Fig. 2).

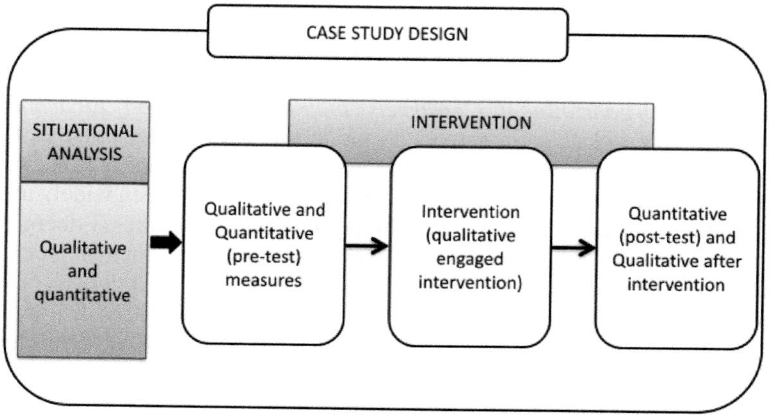

Figure 2. Case study diagram embedded mixed methods and design of the qualitative-engaged intervention (modified from Creswell and Plano Clark 2007).

One-year after completion of the intervention, an interim evaluation was performed in 2014 with household surveys and semi-structured interviews, and two-years after completion, in 2015, a final entomological survey was performed by the Ministry of Health (MoH) vector control programme. Given that change in behaviour requires time, this stage included pre- and post-surveys as indicators of change in knowledge and attitudes related to identified risk factors and to the disease. This allowed the study to determine the impact of participatory activities on these indicators. Small behavioural changes were measured qualitatively via participatory activities such as giving each participant a calendar to keep track of their activities (e.g. household cleaning and rodent trapping) throughout the month. Entomological indicators at baseline and follow-up were used as a proxy of reduced disease transmission.

2.1.4. Quality Assurance in Case Study Research

Trustworthiness strategies (Mertens 2014; Patton 2015) must be applied for quality assurance. Triangulation in case study research is key to enhance credibility (Creswell 1998). We used triangulation across stakeholders (Erlandson et al. 1993; Flick 1998; Mertens 2014; Patton 2015), and by generating information through different research techniques (Erlandson et al. 1993; Flick 1998; Patton 2002; De Urioste-Stone et al. 2015). We validated, through collective and iterative dialogue with the participants, our understanding and interpretation of the main concepts and ideas generated during participatory activities (Creswell 1998; Erlandson et al. 1993; Flick 1998; Mertens 2014). This was done immediately after generating data to enhance credibility.

2.1.5. Ethical Considerations of a Multidisciplinary Approach

A variety of informed consent forms were used; confidentiality was assured by creating coding systems with IDs (questionnaires, interviews, and group meetings), and careful and secure data management. Written consent forms were read and signed by the participants; in case the person could not sign the form, a fingerprint was requested, as well as a signature of a witness. For group activities, a written consent form was used; the consent form was read and signed by a representative of the group and a member of the research team (and included a list of the participants). Participants were requested to sign a consent form for photographs and videos during participatory group meetings. Potential benefits of the study were described and agreed upon prior to data collection.

The consent protocols for the cognitive study were collectively developed with the communities during the ethnographic phase, taking as model the consent forms used for interviews and signing a letter of conformity with the communities according to their own terms. We also were flexible in changing procedures and techniques to respond to contextual particularities in the communities of study. For example, the informed consent originally proposed was a verbal consent, but the communities requested to sign them during the first pilot study, so change to a written consent form was approved by the ethics committee.

After facilitation of each meeting, the research team went through a reflective process and preliminary analysis of results before planning the next group session. Hence, formative data analysis informed data collection.

2.2. Gaining Entry and Building Rapport

Gaining entry and building rapport are essential when conducting qualitative research (Ely et al. 1991). We considered this element of the research process essential for carrying out any type of intervention, and to ensure active collaboration of participants. Early on, a meeting was called by the leader of the Community Development Councils (COCODEs) for the entire community to share the study objectives, methods and expected participation at the community and individual level. For participatory activities, the entire community was invited.

For surveys and intervention phases, communities were randomly selected, and a set number of households was randomly selected to be included in the activities. The selection process was explained at these meetings to prevent any misunderstanding.

The following activities were carried out to enhance the success in gaining access and building trust with stakeholders in the study area:

- Periodic communication with gatekeepers from organizations and communities.
- Meetings in Comapa to present the intervention strategies and results from each stage of the project to local leaders, communities, participants and other stakeholders. In every meeting there was a space for dialogue to obtain feedback about concerns and ideas.
- Rapport with the national authorities was also enhanced, and the interim and final results were presented annually at the National Chagas Vector Control Programme evaluation meetings.

- Presentation of results from each stage of the study to communities and stakeholders.
- Flexibility to change procedures and techniques to respond to the situations, getting Institutional Review Board approval for protocol modifications as the project progressed.
- Respect and value for the time provided by co-participants according to their cultural norms and conception of time.
- Recognition, adaptation and respect for organizational and ethnic cultures and ways.
- Collaborative definition of the location, dates, and times to conduct meetings, interviews and workshops.
- Ongoing reflection on the processes of gaining entry and building trust.

Community meetings were conducted to share general results of the Knowledge, Attitude and Practices (KAP) questionnaires and entomological surveys after the PRECEDE and PROCEED stages. Interim and final results were presented annually at the National Chagas Vector Control Programme evaluation meetings.

2.2.1. Engaging National and Local Health Authorities

Before preparing the proposal, the idea was first presented to the head of the National Chagas Vector Control Programme. After approval at the national level, the idea was presented to the Jutiapa Health Area epidemiologist. During this meeting, a visit to the field site provided context and shaped the proposal for local relevance. This visit was critical to gain support from the local health authorities, who provided valuable input for the final proposal. Brainstorming sessions looked at potential stakeholders to approach, and potential collaboration activities and resources. Meetings were also conducted with leaders from the Municipality of Comapa. The main objective of these meetings was to gain stakeholder permission to conduct the research, to explain its different components and to discuss and gain feedback on how to implement them. Local authorities provided us with baseline data from the communities, maps, and an up-dated list of community leaders.

2.2.2. Engagement of Community Leaders

In 2002, the COCODEs were created in Guatemala to serve as the local organization that identifies and brings together community leaders. The aim of the COCODEs is to serve as a channel to facilitate the participation of the population, to plan and implement development efforts using a democratic approach (Congreso de la República de Guatemala 2002). COCODEs have usually facilitated introduction of projects (or any external initiative) to their communities, serving as the first contact to approach the community and gain access and project approval.

For our project, a list of communities and their leaders was generated at the municipality level. A meeting was organized with 74 members of the COCODEs from Comapa, where the leaders were informed about the duration and objectives of the project. The methodology was explained in detail, as well as the expected participation of the community members. Lists of problems related to Chagas and organizations working in the area were generated and the overall scope of knowledge and experiences with Chagas at the community level was detailed. The meeting with the COCODEs leaders was essential to gaining entry and ensure proper communication about the project.

However, the COCODEs are also an institution that can influence decision-making, which may lead to politicization of initiatives. New projects with no former experience in the area should pay attention to these political dynamics, which might be very challenging. In our case, some communities had one COCODE appointed by the municipality and another appointed by the community members. At the end two meetings were done for each community. We found throughout the study that identifying leaders who were recognized by the community allowed us to gain entry and build a relationship of trust with each community, without politicizing the project. In the absence of organizations such as COCODE, one option is to determine which other institutions collaborate with local leaders and which organizations (governmental and non-governmental) are working in the area. It is very likely that relevant structures are already in place, i.e. a network of leaders, youth teams, volunteers or women groups, that could be invited to collaborate, rather than starting interactions from scratch.

2.2.3. Community Engagement

Based on our experience, securing collaboration from the communities is essential to avoid a paternalistic approach. It was important to acknowledge that we did not have all the answers and that we could not propose all potential solutions using solely ideas generated in the office or by conducting experiments in the laboratory. We recognized the need to listen to the local community members, and value their opinions to make the intervention successful and sustainable.

After evaluating the eco-bio-social baseline information, we conducted participatory activities at three different communities, inviting all members of each community to each meeting through the local leader. The communities were selected based on different social organization characteristics and vector infestation levels. In these meetings, we actively listened to the ideas and concerns of local participants regarding their role in risk factor mitigation.

Based on this approach, we proposed potential strategies that were discussed with all stakeholders and could be modified later if required. Actively listening to ideas, experiences and respecting the knowledge of local populations was key to engaging and empowering participants. Through this engagement, the project moved from being an endeavour from a traditional research team, to becoming a collective undertaking with the active participation of local populations – this is essential for the activities to continue after the project has come to an end.

2.3. Stakeholders Analysis

2.3.1. Stakeholder Context

We developed a list of stakeholders at the national, regional and local levels that stipulated their roles in Chagas vector control. The different roles and areas of emphasis for stakeholders are shown in Fig. 3.

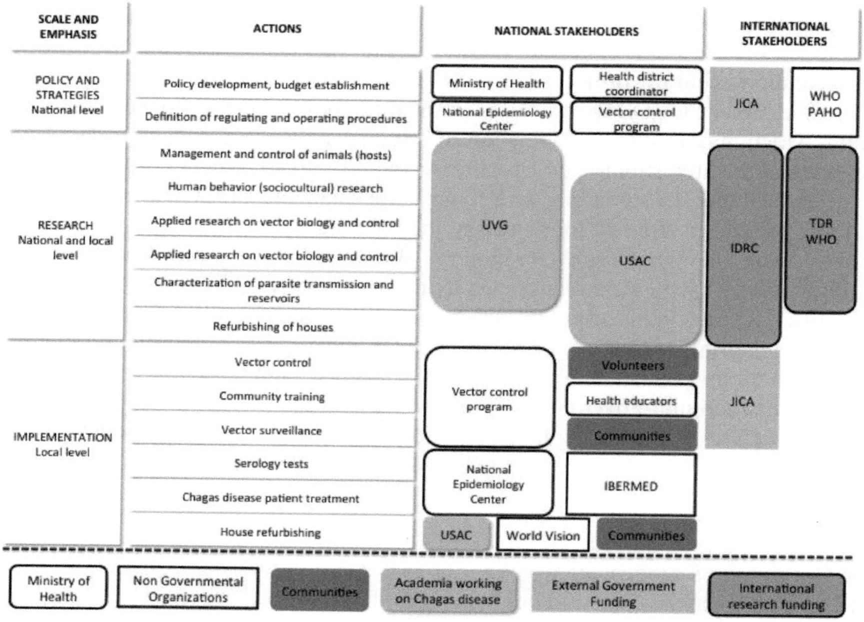

Figure 3. Stakeholder mapping of the National System for Chagas Control at the local and national levels (with permission from De Urioste-Stone et al. 2015). (Universidad del Valle de Guatemala: UVG; Universidad de San Carlos de Guatemala: USAC; International Development Research Centre: IDRC; Japan International Cooperation Agency (JICA); Tropical Disease Research-WHO: TDR-WHO; Pan American Health Organization: PAHO; Médicos con Iberoamérica: IBERMED).

The relationships we observed between the different stakeholders are portrayed in Fig. 4. This type of analysis is useful to understand and identify the persons who can positively contribute to the project and those who can potentially be obstructive. During the study, we continuously discussed newly developing power relationships between the different stakeholders to predict potential conflicts, but also opportunities for collaboration and to leverage resources.

We strived to understand the different roles and interactions among stakeholders to build the relationships for a participatory process. We acknowledged and considered knowledge about Chagas disease and its control, including interests, positions, alliances and relevance in Chagas disease control of those involved (Schmeer 1999).

2.3.2. Training of Personnel/Collaborators

Several training sessions were undertaken with local field staff of the MoH. The first efforts focused on standard operating procedures, use of maps and global positioning system (GPS) equipment. Field staff received training on how to apply the household/KAP questionnaire, specifically on how to approach interviewees, how to ask each of the questions, how to facilitate consent request and other ethical concerns.

Training sessions on the biosafety and handling of vector specimens were undertaken by all members of the vector control MoH team. These activities were crucial to generate understanding and empowerment related to the proposed activities for the intervention; all staff and collaborators need to fully understand the procedures to generate high quality data. This approach also allowed for collaboration across disciplines and forming a strong interdisciplinary team.

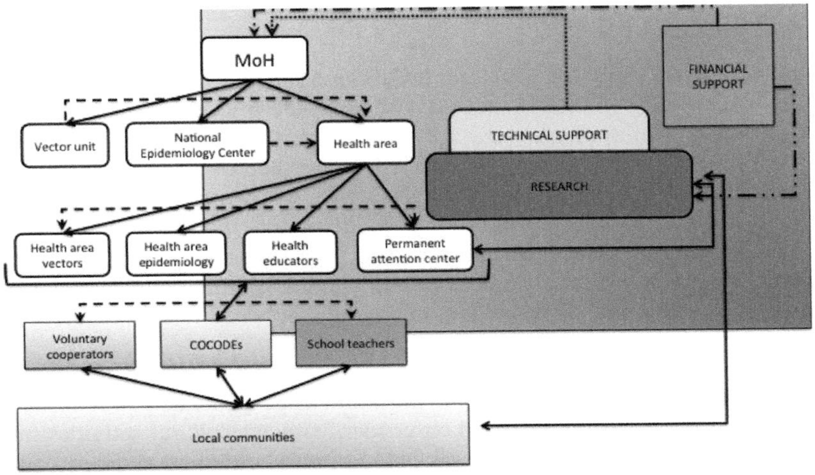

Figure 4. Mapping of stakeholder relationships. Arrows denote direction of relationships between stakeholders: unidirectional denoting supervision relationships, bidirectional being collaborative interactions. The dotted lines indicate secondary linkages through cooperation relationships, the solid lines indicate primary linkages through collaborative relationships.

2.3.3. Pre-testing of Survey Procedures and Questionnaire

We validated all survey procedures, questionnaires and the consent forms before starting data collection, in two communities from the same municipality that were not part of our study. These communities were selected from the sample of communities found over 850 meters above sea level that were not selected for the survey. Two validation rounds were conducted in two different communities with four households randomly selected in each. We obtained feedback to make the required modifications.

Each household was visited to assess the household/KAP and entomological forms, with the goal of reducing measurement error (Dillman et al. 2008; Krosnick et al. 2014). The reactions of interviewed community members to wording and ordering of questions was observed and analysed, and participants were asked to suggest better phrasing and more appropriate wording when questions were not clear. As a result, the accuracy of the questions was improved, several questions were added, and the order of questions revised. The order of the procedures was also modified to improve the interviewing process by performing the KAP survey instrument before the entomological survey.

The instruments were modified, and a third round took place to pilot test the effectiveness of the selected protocols, with four more households randomly selected. After the feedback from the pilot test was included, the household/KAP questionnaires were further reviewed and revised by staff of the MoH.

3. THE INTERVENTION

3.1. Developing the Intervention

Before the intervention, three communities of the Municipality of Comapa, selected based on their vector infestation levels (two with the highest infestation and one without persistent infestation), were invited to share information regarding animal management in relation to Chagas disease risk factors and to identify their problems and potential solutions (Bustamante et al. 2014). All community members were invited to these activities. We used a participatory model to gain in-depth understanding regarding local practices related to risk factors. Through a series of community reflection exercises, it became evident that community members wanted to better understand the disease, and to identify actions to change the current conditions.

An anthropological study was undertaken to better understand the economic production practices, further adding to the knowledge generated previously. We observed very distinct activities based on gender, with agricultural activities mainly carried out by men, whereas women carried out household chores, raised children, cared for peridomestic animals (e.g. chickens), and traded goods with other community members.

The first evaluation of the region showed that, with respect to Chagas disease, chicken management practices and the presence of dogs and rodents posed an important risk for the household (Bustamante et al. 2014). The findings indicated the importance of developing sustainable animal and environmental management practices that would modify behaviours of community members that posed risk factors in relation to triatomine infestation, as part of a gender-oriented education programme.

3.2. The Intervention Framework

The intervention framework included a close collaboration with the communities to generate an integral animal and environmental management programme. It also aimed to implement relevant components of another educational programme in the area called "Clean House, Clean Patio" developed by the MoH and JICA (De Urioste-Stone et al. 2015). The general objective was to provide a strategy to improve Chagas disease prevention in different ecological and social settings. This was achieved by better understanding the ecological, vector-biological and social ("eco-bio-social") determinants of peridomestic animal management in relation to vector infestation. As a result, we developed and evaluated a community-centred intervention to reduce habitats for rodents and chickens inside the household, both of which were found to be important blood sources and risk factors for triatomines.

Eighteen communities with baseline infestation levels above 15% were selected from the 30 communities surveyed in the situational analysis (De Urioste-Stone et al. 2015). Nine communities were randomly assigned to the intervention (AW-IVM) and the other nine communities to the control (only the MoH insecticide application) groups (Fig. 5).

3.3. Participation and Community Engagement during Intervention

We used some elements of Participatory Action Research (PAR) to guide changes in perception and behaviour at multiple levels, through "direct involvement, intervention or insertion in processes of social action" (Fals-Borda 2001). The model facilitated active listening to the concerns and ideas of participants, and it promoted a reflexive process about the implications of the interventions amongst researchers and participants. We relied on an iterative process of reflection and action (Freire 1970).

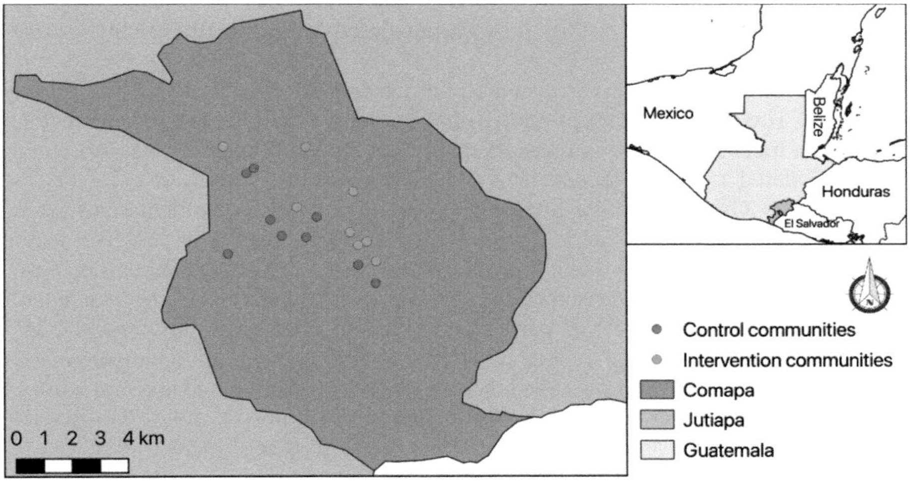

Figure 5. Distribution of communities assigned to intervention and control groups for the intervention phase, Comapa, Jutiapa, Guatemala, 2012.

We - participants, researchers and collaborating institutions (non-governmental organizations, MoH) - engaged in a cycle of sharing knowledge, following up changes in practices, and reflecting about the research process during the intervention (De Urioste et al. 2015). We believe that through this knowledge-reflection-action iterative method, we facilitated a process best described by Kemmis and McTaggart (2003) as:

> *"...systematic inquiry, process; participation and collaboration; bridging local and scientific knowledge; empowerment; and action".*

Participatory techniques (FAO 1990; Chambers 2002) were used as a vehicle to stimulate dialogue and reflection among participants about Chagas disease. Educational material about Chagas and topics related to chicken, dog, rodent/grain management were discussed and analysed with participants. Seven monthly participatory meetings were held within each of the intervention communities:

- *Meeting 1:* A full description of the intervention methodology and timeline was presented to the selected members of the intervention communities. Consent of all the participants was obtained. Methods were discussed and feedback of the participants was incorporated into the research tools to be developed in the following meetings. An informative brochure that contained results obtained in the baseline was developed and shared with the participants.
- *Meeting 2:* A community level Strengths, Weaknesses, Opportunities, Threats (SWOT) analysis was generated, to have a better understanding of the situation and to identify specific issues to be addressed in the following meetings. The analysis included topics such as (1) current practices to manage and control the triatomines, (2) current knowledge about *T. dimidiata* behaviour and presence in the houses, and (3) other institutions working directly or indirectly with the disease and the vector.
- *Meeting 3:* Group narrative related to knowledge and experience with Chagas disease. The research team presented information on the disease and its vector, forms of transmission, symptoms, effects and treatment. A calendar was presented and validated as a personal matrix to document household activities.
- *Meeting 4:* Group narrative related to knowledge regarding rodents and grain storage as risk factors for vector presence in the house. The research team presented information on the biology and ecology of rodents, the danger they pose in relation to Chagas disease, and proposals for control strategies. The calendars were delivered to the meeting participants for use as personal matrices to record the activities and practices related to rodent and grain storage, to be implemented during that month. During this meeting, rodent traps were delivered, and a practical session demonstrated the use of the traps, protocols to kill the rodents easily and ethically, and to manage and bury the carcasses.
- *Meeting 5:* Group evaluation with respect to the dynamics of the PAR process, i.e. assessment of the knowledge gained on the vector, the disease and the rodents, the activities proposed for rodent control, and about the use of the matrix (a calendar) to document rodent and grain storage practices. Discussion of needed changes in design and new personal commitments acquired to continue advancing.
- *Meeting 6:* Group discussion on the use of the matrix for rodent control activities and any related changes made. Sharing of knowledge and experiences related to chickens, vegetation and waste management as risk factors for vector presence. Presentation by the research team on the importance of chicken management, and proposal of a strategy for integral waste management, through compost production, a chicken coop and food production. Validation and discussion about the matrices to have a record of the activities and practices related to environmental management.

- *Meeting 7:* Group discussion regarding the use of the matrix for grain storage, and rodent, chicken, dog and environmental management practices and any changes made, and all steps of the PAR process. Closing presentation with findings, achievements and identification of possible knowledge gaps. Opinion survey among participants and local leaders regarding the PAR process.

4. RESULTS AND CONCLUDING REMARKS

Comparison of KAP survey results with Student's t test, in a pre-test in 2012 and post-test in 2014, showed that the intervention with participatory activities produced a significant change in protective practices against risk factors for persistent *T. dimidiata* infestation, including rodent control using mechanical traps and environmental management, as well as chicken management (Student's t test, p<0.001) (De Urioste-Stone 2015). An odds ratio comparison of entomological indices in 2012 and 2014 showed higher early instar reinfestations in the communities that received no treatment (control), compared to the intervention communities (OR 8.3, 95% CI 2.4-28.4). In addition, there was a significant reduction in rodent infestations in the intervention group over time (OR 1.9, 95% CI 1.09-3.45). The overall infestation levels were maintained below 10% in both intervention and control groups during the first evaluation in 2014 (Fig. 6).

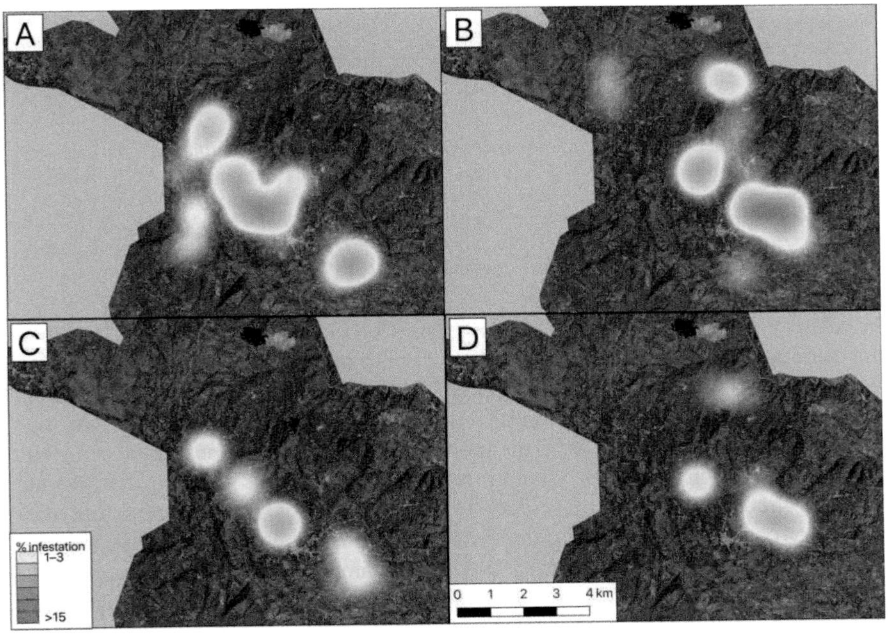

Figure 6. Reduction of Triatoma dimidiata *infestation in the Municipality of Comapa, Department of Jutiapa, Guatemala in the control and intervention groups pre- (2012) and post-intervention (2014). A) Control group 2012. B) Intervention group 2012. C) Control group 2014. D) Intervention group 2014.*

In 2015, an entomological survey was performed as part of the MoH surveillance activities. The survey was performed simultaneously with a serological survey of household inhabitants (Juarez et al 2018). During this 2015 survey, we observed that communities where Participatory Action Research was conducted had a higher participant retention, when comparing treatments (Pearson Chi^2= 3.298, p= 0.046, one-tailed test). A trend was also observed, at a 90% CI, for lower drop-out rates for the intervention communities with an Odds Ratio of 1.67 (90% CI= 1.05, 2.65) (Table 1). This suggests that the participatory process increases long-term community acceptance of MoH surveillance activities.

Table 1. Community recruitment and continued participation between control and intervention treatments in numbers (%) for the municipality of Comapa, Department of Jutiapa, Guatemala

	2012	2014	2014	2015	2015
Treatment	Recruitment	Drop-out	Remained	Drop-out	Remained
Control	215 (49.9)	15 (3.5)	200 (46.4)	37 (8.6)	178 (41.3)*
Intervention	216 (50.1)	23 (5.3)	193 (44.8)	24 (5.6)	192 (44.5)*
Total	431 (100)	38 (8.8)	393 (91.2)	61 (14.2)	370 (85.8)

* *Significant difference, $p<0.05$. We were unable to observe any statistical difference for domestic infestation levels between the control (21%; 8.5% and 15%) and intervention (20%; 8.3% and 12%) by year (2012-2014-2015). We did observe a trend that may suggest that intervention practices may prevent long-term reinfestation of the houses (n=192) that were originally infested (OR 2.5, 95% CI 0.93-6.99). On the other hand, the control group (n=178) showed a higher probability of finding infestation if the household was previously infested (OR 3.16, 95% CI 1.1-9.00)*

We also evaluated the effects of rodent infestation on vector infection by *T. cruzi*, for the triatomines collected in 2012 and 2015. Triatomine infections were confirmed using PCR for the parasite *T. cruzi* (Maddren 2018). We found that the proportion of infested houses with infected triatomines significantly increased over time in the control group (Table 2). However, in the intervention group the proportion of infested houses with infected triatomines did not increase over time. This suggests that rodent control may reduce the risk of infection in the households. It appears that the participatory process gave the household inhabitants the tools to reduce risk factors for infection. Through this community-based programme, we learned that complex health problems such as vector-borne zoonotic diseases require multidisciplinary and community-based approaches to develop innovative solutions that target ecological, socio-economic, cultural, institutional, and biophysical factors of risk.

To develop relevant solutions, an in-depth understanding of the dynamics of communities and the role/interactions among stakeholders are first needed to be able to understand the context that frames the public health issue. Once biophysical and social science data are collected and triangulated, and the problems are identified based on this data integration, solutions should include the input from those that will be involved and affected by the intervention.

Table 2. Percentage of sampled houses in the Municipality of Comapa with Triatoma dimidiata *infected with* Trypanosoma cruzi

Treatment	Year	Infested Houses	Percent of houses with *T. cruzi* infected *T. dimidiata* (95% CI)
Control	2012	30	65.1 (50.3-79.8)*
	2015	18	90.4 (78.2-100.0)*
Intervention	2012	17	68.1 (44.2-91.9)
	2015	20	77.2 (59.9-94.4)

* *Significant difference,* $Chi^2 = 5.8$, $p<0.05$

A participatory-reflexive process can help improve the effectiveness of strategies aimed at achieving changes in human social and behavioural contexts. These activities empower the community to engage in practices that are truly relevant and feasible to them. Introduction of AW-IVM innovations should engage citizens throughout the process to ensure public awareness and involvement in the projects.

In our case, given that these vectors can colonize peridomestic environments, it will be necessary to implement additional strategies in the future to reduce the availability of these habitats and prevent future domestic infestations. In addition, all members of the communities should be engaged in the process, to ensure that peridomestic environments do not become a source of infestation for neighbouring households and communities.

5. REFERENCES

Blank, M. E. 2006. El modelo precede/proceed: Un organizador avanzado para la reconceptualización del proceso de enseñanza-aprendizaje en educación y promoción de la salud. Salus Online 10: 18–25.

Bustamante, D. M., S. M. De Urioste-Stone, J. G. Juárez, and P. M. Pennington. 2014. Ecological, social and biological risk factors for continued *Trypanosoma cruzi* transmission by *Triatoma dimidiata* in Guatemala. PLoS One 9(8): e104599.

Chambers, R. 2002. Participatory workshops: A sourcebook of 21 sets of ideas and activities. Earthscan, London, UK. 236 pp.

Congreso de la República de Guatemala. 2002. Ley de los Consejos de Desarrollo Urbano y Rural. Decreto Numero 11-2002 del Congreso. Ciudad de Guatemala, Guatemala.

Creswell, J. W. 1998. Qualitative inquiry and research design, choosing among five traditions. SAGE Publications, Thousand Oaks, California, USA.

Creswell, J. W., and V. L. Plano Clark. 2007. Designing and conducting mixed methods research. SAGE Publications, Thousand Oaks, California, USA.

Dell'Arciprete, A., J. Braunstein, C. Touris, G. Dinardi, I. Llovet, and S. Sosa-Estani. 2014. Cultural barriers to effective communication between indigenous communities and health care providers in Northern Argentina: An anthropological contribution to Chagas disease prevention and control. International Journal for Equity in Health 13: 6.

De Urioste-Stone, S. M., P. M. Pennington, E. Pellecer, T. M. Aguilar, G. Samayoa, H. D. Perdomo, and J. G. Juárez. 2015. Development of a community-based intervention for the control of Chagas disease based on peridomestic animal management: An eco-bio-social perspective. Transactions of the Royal Society of Tropical Medicine and Hygiene 109: 159–167.

Dias, J. C. P. 2007. Southern Cone Initiative for the elimination of domestic populations of *Triatoma infestans* and the interruption of transfusion Chagas disease: Historical aspects, present situation, and perspectives. Memórias do Instituto Oswaldo Cruz 102 (Suppl. 1): 11-18. Epub August 31, 2007.

Dias, J. C. P., A. C. Silveira, and C. J. Schofield. 2002. The impact of Chagas disease control in Latin America: A review. Memórias do Instituto Oswaldo Cruz 97: 603–612.

Dillman, D., J. D. Smyth, and L. M. Christian. 2008. Internet, mail, and mixed-mode surveys: The tailored design method. John Wiley & Sons, Inc., Hoboken, New Jersey, USA.

Edberg, M. 2007. Essentials of health behavior: Social and behavioral theory in public health. Jones & Bartlett Publishers, Sudbury, Massachusetts, USA.

Ely, M., M. Anzul. T. Friedman, D. Garner, and A. McCormack Steinmetz. 1991. Doing qualitative research: Circles within circles. The Falmer Press, London, UK.

Erlandson, D. A., E. L. Harris, B. L. Skipper, and S. D. Allen. 1993. During naturalistic inquiry: A guide to methods. SAGE Publications, Newbury Park, California, USA.

(FAO) Food and Agriculture Organization of the United Nations. 1990. The Community's toolbox: The idea, methods and tools for participatory assessment, monitoring and evaluation in community forestry. FAO Regional Wood Energy Development Programme in Asia, Bangkok, Thailand.

Fals-Borda, O. 2001. Participatory (action) research in social theory: Origins and challenges, pp. 27–37. *In* P. Reason and H. Bradbury (eds.), Handbook of action research: Participative inquiry and practice. SAGE Publications, Thousand Oaks, California, USA.

Flick, U. 1998. An introduction to qualitative research. London: SAGE Publications, Thousand Oaks, California, USA.

Freire, P. 1970. Pedagogy of the oppressed. Herder and Herder. New York, New York, USA. 186 pp.

Gürtler, R. E., and Z. E. Yadon. 2015. Eco-bio-social research on community-based approaches for Chagas disease vector control in Latin America. Transactions of the Royal Society of Tropical Medicine and Hygiene 109: 91–98.

Hashimoto, K., and C. J. Schofield. 2012. Elimination of *Rhodnius prolixus* in Central America. Parasites & Vectors 5: 45.

Hashimoto, K., C. Cordon-Rosales, R. Trampe, and M. Kawabata. 2006. Impact of single and multiple residual sprayings of pyrethroid insecticides against *Triatoma dimidiata* (Reduviiade; Triatominae), the principal vector of Chagas disease in Jutiapa, Guatemala. The American Journal of Tropical Medicine and Hygiene 75: 226–230.

Hashimoto, K., H. Alvarez, J. Nakagawa, J. Juarez, C. Monroy, C. Cordon-Rosales, and E. Gil. 2012. Vector control intervention towards interruption of transmission of Chagas disease by *Rhodnius prolixus*, main vector in Guatemala. Memórias do Instituto Oswaldo Cruz 107: 877–887.

(JICA) Japan International Cooperation Agency. 2014. Summary best practices from the Chagas disease control in Guatemala, El Salvador, Honduras and Nicaragua. 33 pp.

Juárez, J. G., P. M. Pennington, J. P. Bryan, R. E. Klein, C. B. Beard, E. Berganza, N. Rizzo, and C. Cordon-Rosales. 2018. A decade of vector control activities: Progress and limitations of Chagas disease prevention in a region of Guatemala with persistent *Triatoma dimidiata* infestation. PLoS Neglected Tropical Diseases 12(11): e0006896.

Kemmis, S. and R. McTaggart. 2003. Participatory action research, pp. 336–396. *In* N. K. Denzin and Y. S. Lincoln (eds.), Strategies of qualitative inquiry. SAGE Publications, Thousand Oaks, California, USA.

Krosnick, J. A., P. J. Lavrakas, and N. Kim. 2014. Survey research, pp. 404–442. *In* H. T. Reis and C. M. Judd (eds.), Handbook of research methods in social and personality psychology. Second Edition. Cambridge University Press, Cambridge, UK.

Lucero, D. E., L. A. Morrissey, D. M. Rizzo, A. Rodas, R. Garnica, L. Stevens, D. M. Bustamante, and M.C. Monroy. 2013. Ecohealth interventions limit triatomine reinfestation following insecticide spraying in La Brea, Guatemala. The American Journal of Tropical Medicine and Hygiene 88: 630–637.

Maddren, R. 2018. Characterisation of domestic *T. cruzi* transmission in South-Eastern Guatemala: Genetic analysis of the major vector species *Triatoma dimidiate*. Master thesis, London School of Hygiene and Tropical Medicine, London, UK.

Manne, J., J. Nakagawa, Y. Yamagata, A. Goehler, J. S. Brownstein, and M. C. Castro. 2012. Triatomine infestation in Guatemala: Spatial assessment after two rounds of vector control. The American Journal of Tropical Medicine and Hygiene 86: 446–454.

Massad, E. 2008. The elimination of Chagas' disease from Brazil. Epidemiology and Infection 136: 1153–1164.

Mathers, C. D., M. Ezzati, and A. D. Lopez. 2007. Measuring the burden of neglected tropical diseases: The global burden of disease framework. PLoS Neglected Tropical Diseases 1(2): e114.

Mertens, D. M. 2014. Research and evaluation in education and psychology: Integrating diversity with quantitative, qualitative, and mixed methods (4th ed.). SAGE Publications, Thousand Oaks, California, USA.

Moncayo, Á., and A. C. Silveira. 2009. Current epidemiological trends for Chagas disease in Latin America and future challenges in epidemiology, surveillance and health policy. Memórias do Instituto Oswaldo Cruz 104 (Suppl. 1): 17–30.

(PAHO) Pan American Health Organization. 2011. Iniciativa de los países de América Central para la interrupción de la transmisión vectorial y transfusional de la enfermedad de Chagas (IPCA): Historia de 12 años de una iniciativa subregional 1998–2010. Representación de la Organización Panamericana/Mundial de la Salud en Honduras. Tegucigalpa, Honduras. 89 pp.

Patton, M. Q. 2002. Qualitative research & evaluative methods (3rd ed.). SAGE Publications, Thousand Oaks, California, USA.

Patton, M. Q. 2015. Qualitative research & evaluative methods: Integrating theory and practice (4th ed.). SAGE Publications, Thousand Oaks, California, USA.

Rizzo, N. R., B. A. Arana, A. Diaz, C. Cordon-Rosales, R. E. Klein, and M. R. Powell. 2003. Seroprevalence of *Trypanosoma cruzi* infection among school-age children in the endemic area of Guatemala. American Journal of Tropical Medicine and Hygiene 68: 678–682.

Schmeer, K. 1999. Guidelines for conducting a stakeholder analysis. Partnerships for Health Reform, Abt Associates Inc. Bethesda, Maryland, USA.

Schofield, C. J., J. Jannin, and R. Salvatella. 2006. The future of Chagas disease control. Trends in Parasitology 22: 583–588.

Sommerfeld, J. and A. Kroeger. 2015. Innovative community-based vector control interventions for improved dengue and Chagas disease prevention in Latin America: Introduction to the special issue. Transactions of The Royal Society of Tropical Medicine and Hygiene 109 (2): 85–88.

Ventura-García, L., M. Roura, C. Pell, E. Posada, J. Gascón, E. Aldasoro, J. Muñoz, and R. Pool. 2013. Socio-cultural aspects of Chagas disease: A systematic review of qualitative research. PLoS Neglected Tropical Diseases 7(9): 1–8.

Vreysen, M. J. B., A. S. Robinson, and J. Hendrichs (eds.). 2007. Area-wide control of insect pests: From research to field implementation. Springer Dordrecht, The Netherlands. 789 pp.

(WHO) World Health Organization. 2017. Fact sheet: Chagas disease (American trypanosomiasis). Washington, DC, USA.

CITIZEN SCIENCE AND ASIAN TIGER MOSQUITO: A PILOT STUDY ON PROCIDA ISLAND, A POSSIBLE MEDITERRANEAN SITE FOR MOSQUITO INTEGRATED VECTOR MANAGEMENT TRIALS

V. PETRELLA[1], G. SACCONE[1], G. LANGELLA[2,3], B. CAPUTO[4], M. MANICA[4,5], F. FILIPPONI[4], A. DELLA TORRE[4] AND M. SALVEMINI[1]

[1]*Department of Biology, University of Naples Federico II, Italy; marco.salvemini@unina.it*
[2]*National Research Council, Institute for Agricultural and Forest Systems in the Mediterranean (ISAFoM), Italy*
[3]*Department of Agricultural Sciences, University of Naples Federico II, Italy*
[4]*Department of Public Health and Infectious Diseases, University of Rome La Sapienza, Italy*
[5]*Department of Biodiversity and Molecular Ecology, Edmund Mach Foundation, San Michele all'Adige, Italy*

SUMMARY

During the past twenty years, the number of research projects involving people not trained as scientists, the so-called citizen science, has increased consistently, including mosquito monitoring and control projects. The involvement of citizens in mosquito monitoring programmes not only helps scientists during the data collection phase, but also raises public awareness on mosquito-transmitted diseases and educates citizens about virtuous behaviours that can help in reducing mosquito populations and their spread. The Asian tiger mosquito *Aedes albopictus* (Skuse) is an invasive species that became established in Europe starting in 1979, with Italy representing currently one of the most infested countries. Procida, a small Mediterranean island in the Naples gulf (Campania region, southern Italy) has unique and very interesting features facilitating the field testing of mosquito integrated vector management (IVM) approaches and control methods, including the Sterile Insect Technique (SIT). With the help of the local municipal administration, the Procida citizens are actively involved as volunteers in monitoring the seasonal and spatial distribution of the Asian tiger mosquito. The collected baseline data will be useful to implement a future island-wide integrated suppression trial of *Ae. albopictus,* including the release of sterile males, to be carried out in collaboration with the local municipal administration and with the technical support of the Joint FAO/IAEA Division in Vienna.

J. Hendrichs, R. Pereira and M. J. B. Vreysen (eds.), Area-Wide Integrated Pest Management: Development and Field Application, pp. 729–745. CRC Press, Boca Raton, Florida, USA.

Key Words: Baseline data, mosquito surveillance, *Aedes albopictus, Culex pipiens, Culex laticinctus,* IVM, SIT, Sterile Insect Technique, Italy

1. INTRODUCTION

According to the World Health Organization (WHO), mosquitoes are considered the deadliest animals on Earth, causing more than one million human deaths every year and representing a risk due to the diseases they transmit, and to which more than one third of the human population is exposed (WHO 2019).

In this scenario, during the last four decades, invasive mosquito species have played a significant role because of their confirmed or potential capabilities to vector an increasing number of diseases to humans and animals. Introductions of invasive mosquito species have increased world-wide as a result of the globalisation of trade and travel, climate change, and the capacity of the mosquitoes to adapt from their native areas to temperate regions, such as the European continent.

Among all the invasive mosquito species listed to date, *Aedes* species are of major concern, with at least five described species having become established in parts of Europe (Medlock et al. 2015). This includes the Asian tiger mosquito *Aedes albopictus* (Skuse 1894) (Diptera: Culicidae) that is considered a major threat to public health in Europe.

1.1. Aedes albopictus *Invades Europe*

Ae. albopictus originated from Southeast Asia and has spread world-wide during the last 40 years. In Europe, it was firstly recorded in Albania in 1979 and is present today in 24 European countries (ECDC 2018a). Species distribution models, combining eco-environmental and terrestrial cover variables, and future climatic scenarios, predict further spread of this species in Europe and other countries in the world (Fischer et al. 2011; Caminade et al. 2012; Cunze et al. 2016).

In Italy, the mosquito arrived in 1990 in Genova (Liguria region) (Sabatini et al. 1990) and has quickly spread over the whole Italian territory, in particular in the north-eastern area (Friuli-Venezia-Giulia region, and large parts of the Lombardia, Veneto and Emilia Romagna regions) and central and southern coastal areas, including major islands (Albieri et al. 2010; Marini et al. 2010; Valerio et al. 2010). *Ae. albopictus* has a very aggressive day-time human-biting behaviour (Valerio et al. 2010; Manica et al. 2016) and it is a competent vector for more than 20 arboviruses, including the dengue and chikungunya viruses, in addition to filarial nematodes of veterinary and zoonotic significance (Cancrini et al. 2003; Pietrobelli 2008; Bonizzoni et al. 2012).

The first European outbreak of chikungunya fever occurred in Italy in 2007 and was ascribed to the presence of the local *Ae. albopictus* populations. This event drastically increased awareness of the risk of new or re-emerging mosquito-borne diseases in Europe (Gasperi et al. 2012). In addition, the Zika virus (ZIKV) outbreak in South and North America starting in 2015, and the confirmation that *Ae. albopictus* is a competent vector of the disease (Grard et al. 2014; Di Luca et al. 2016; Heitmann et al. 2017), have further emphasised the importance to carefully monitor and sustainably manage this species also in European countries. During the summer of 2017, about 250 cases of chikungunya fever were reported in the urban and costal area of Lazio (Venturi et al. 2017).

In the Mediterranean region, this species is mainly active during the summer, and evidence has been collected that under specific climate conditions the populations show a bimodal distribution with peaks in July and September (Manica et al. 2016). Considering the heavy nuisance caused by the *Ae. albopictus* female day-time biting behaviour, the presence of this species is considered also a serious socio-economic threat for regions with a tourism-based economy (Roiz et al. 2008).

To face the increasing risk of the spread of vector-borne diseases, several European Union (EU) countries started mosquito monitoring, surveillance and control programmes and, in 2005, the European Centre for Disease Prevention and Control (ECDC) was established, i.e. an EU agency aimed at strengthening Europe's defences against infectious diseases, including vector-borne diseases (Zeller et al. 2013; ECDC 2018b).

Mosquito monitoring programmes usually are based on the use of special traps (ovitraps to collect mosquito eggs, gravitraps to collect gravid female mosquitoes, and adult traps to collect adult mosquitoes of both sexes) to determine the occurrence and the spatial-temporal distribution of the species. However, the management of an area-wide trap network, covering wide territories or a whole country, requires great financial resources as well as a significant labour force.

1.2. Involvement of Civil Society in Citizen Science

Recently, the involvement of civil society in research projects, also known as citizen science, has become increasingly popular (Dickinson et al. 2012; Bonney et al. 2014). To support mosquito monitoring performed by experts through conventional trapping (the so called "active" monitoring), community-based surveillance activities have been launched in some EU countries. These citizen science projects are based on the public participation through active monitoring, such as the "Mosquito Atlas" in Germany (Mückenatlas 2019) and the "Mosquito Recording Scheme" (MRS 2019) in the UK, or through smartphone-based mosquito data collection applications (the so called "passive" monitoring), such as the "Mosquito Radar" (Muggenradar 2019) in the Netherlands, the "Mosquito

Alert" (2019) (hunting the tiger) in Spain, the "iMoustique®" in France, the "MosquitoWEB" (2019) in Portugal, and the ZanzaMapp (2019) in Italy.

The results of some of these projects, recently reviewed by Kampen and colleagues, demonstrated that public mosquito surveillance, despite some limitations mainly represented by the inexperience of volunteers, can usefully supplement surveillance programmes by:

1. Substantially reducing the field work costs
2. Collecting data in such a quantity that conventional research groups would not be able to generate by themselves
3. Raising awareness and improving knowledge amongst citizens on invasive species and associated public health problems
4. Detecting the arrival and the spread of *Ae. albopictus* and other invasive mosquito species populations in various European areas (Kampen et al. 2015).

Hence, such citizen science projects can help public agencies with the monitoring and control efforts of invasive mosquito species (Jordan et al. 2017; Palmer at al. 2017).

Mosquito management activities are frequently ineffective because some mosquito species, as is the case of *Ae. albopictus*, breed in human-made water containers, mostly located within private-access properties and areas, making the required action within the target area (egg or adult monitoring, sanitization and control of larval breeding sites, etc.) very complex, if not impossible, to be achieved. A mosquito community-based monitoring network could facilitate the monitoring on such private properties, and also help in the successive implementation of area-wide mosquito population suppression programmes that can include eco-friendly approaches such as the Sterile Insect Technique (SIT) or the Incompatible Insect Technique (IIT).

The SIT, which is based on the mass-rearing and release of sterile male-only insects that induce sterility in the local population, has been successfully applied against the New World screwworm fly and several fruit flies, tsetse flies, and lepidopteran species (Dyck et al. 2021).

The IIT is an alternative population suppression strategy, based on cytoplasmic incompatibility (CI), widespread in many diplodiploid species. With IIT the sterility in the target population is achieved through the release and mating of males infected with a different *Wolbachia* strain, which results in embryonic lethality (Saridaki and Bourtzis 2010; Lees et al. 2015).

The absence of effective vaccines against mosquito-borne diseases and the problem of growing insecticide resistance in mosquito populations (Sokhna et al. 2013; Vontas et al. 2012) have made the SIT, the IIT, and related approaches potentially promising components of area-wide integrated vector management (AW-IVM) programmes for some key mosquito species (Lees et al. 2015; Bourtzis et al. 2016).

Ae. albopictus is a suitable candidate species for SIT and/or IIT application because of its relative ease of mass-rearing, its sexual dimorphism that facilitates sex separation, and its low biological dispersal potential (Bellini et al. 2007, 2013; Albieri et al. 2010; Marini et al. 2010; Balestrino et al. 2014; Gilles et al. 2014).

In this paper, we present a community-based mosquito monitoring approach that we are developing on Procida, a Mediterranean island in southern Italy. We are collecting, with citizen involvement, baseline data and setting-up the optimal social and technical working environment for future *Ae. albopictus* population suppression experiments by the SIT and/or the IIT (Bourtzis et al. 2016).

2. PROCIDA ISLAND

Procida is a small island of the Phlegraean archipelago, situated in the Naples gulf, about three km from both the mainland and Ischia Island. It is a flat volcanic island (average 27 m above sea level) with a 16 km-long jagged coastline which forms four capes and with a total surface area of only 4.1 km^2, including the uninhabited tiny satellite island of Vivara (0.4 km^2). Except for Vivara, which is a natural reserve, Procida's territory is quite urbanized and accessible.

Most of Procida's private properties include a garden with ornamental flowers, vegetable cultivations and/or orchards with citrus plants and family-type farming of chickens and rabbits. Despite its small surface, Procida has a very high and urban-like population density with 10.477 inhabitants (2459 inhabitants/km^2 - ISTAT 28/02/2017). This human population density approximately doubles during the summer months, because of tourism, which is the current main local economic activity.

According to the perception of residents, *Ae. albopictus* arrived on the island around the year 2000, most probably introduced by tourists and/or maritime transport of goods. Thanks to very favourable host and climatic conditions, with an average annual temperature of 16.2°C and an average annual precipitation of 797 mm (http://bit.ly/ecdata_procida; accessed: 07th May 2018), and very abundant availability of water containers in private gardens, *Ae. albopictus* spread quickly over the entire island, reaching high population densities in some areas and becoming a serious nuisance in the last years.

Procida has unique and very interesting features for field testing of mosquito IVM including the SIT or the IIT: a very small size, a completely urbanized and accessible territory, a high human population density and year-round presence of *Ae. albopictus*. The island has obtained a world-famous reputation, due to several novels and films that were set there, which could help provide wide media coverage in the case of very positive population suppression results, that could facilitate fundraising for future larger population control tests.

Furthermore, many Procida citizens are aware of the SIT approach and of its advantages and effectiveness in insect pest control programmes. In fact, during the 1970's and 1980's, Procida island was chosen as an experimental area to study the field performance of sterilized male Mediterranean fruit fly *Ceratitis capitata* from a genetic sexing strain (Robinson 2002) in a cooperative programme between the Italian National Committee for Research and Development of Nuclear Energy (ENEA) and the International Atomic Energy Agency (IAEA) (Cirio 1975; Cirio et al. 1987). About 20 million sterile Mediterranean fruit fly males were released on the island from April to July 1986 and the population suppression obtained, linked with the protection of citrus fruits, was positively perceived by the residents for several years thereafter.

3. FIRST RECORD OF *Aedes albopictus* ON PROCIDA ISLAND

Official data about Asian tiger mosquito presence in the Phlegraean islands (islands in the Gulf of Naples and the Campania region of southern Italy) are available only for the satellite island of Vivara, where *Ae. albopictus* was detected for the first time in 2002 (D'Antonio and Zeccolella 2007).

In September 2015, an entomological field survey was undertaken in five private properties and five tourist facilities to confirm the presence of *Ae. albopictus* on Procida island and to obtain preliminary data about its distribution (Fig. 1A).

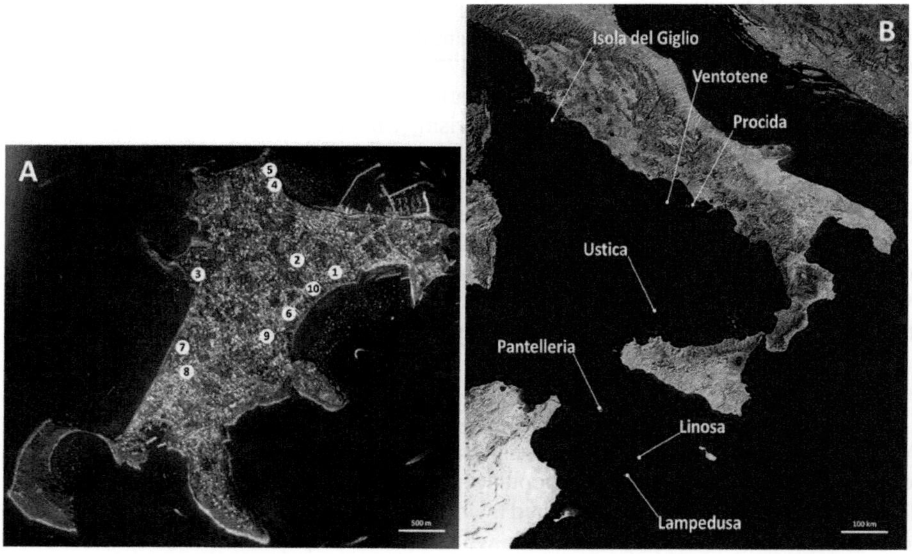

Figure 1. A) Aedes albopictus *field sampling sites on Procida island in 1-3 September 2015. Locations 1-10 are listed in Table 1. B) Italian minor islands were the presence of* Ae. albopictus *has been officially reported.*

Modified CDC light traps (CDC-LT) were used for the field sampling (Reisen et al. 2000; Li et al. 2016), without standard light source and baited with dry ice. CDC-LTs were placed in shaded locations in courtyards of private houses with rich vegetation for three days, in the period 1-3 September 2015 (26.3°C average temperature, 75% average humidity, 7.6 km/h average wind speed). Traps were activated each day from 8.00 h till 20.00 h. Deployment and inspection of the CDC-LTs were hampered by the frequent bites of *Ae. albopictus* to the trap operators.

A total of 240 adult mosquitoes were collected in eight out of 10 CDC-LTs. Adult *Aedes* mosquito specimens were identified using morphological characteristics (Schaffner et al. 2001) and a summary of the number and sex of *Ae. albopictus* trapped at each site is given in Table 1. A total of 216 *Ae. albopictus* adults were sampled (169 females and 47 males) with the remaining 24 mosquitoes identified as males and females of *Culex pipiens* L. and *Culex laticinctus* Edwards (Di Marco and Severini, unpublished results).

Table 1. Aedes albopictus *sampled with CDC-LTs in 1-3 September 2015 on Procida island*

Trap No.	Site Name	Coord. N.	Coord. E.	*Aedes albopictus* males	*Aedes albopictus* females	Other Mosquitoes*
1	Edificio Scolastico	40°45' 39.1"	14°01' 27.8"	3	68	22
2	Madonna della Libera	40°45' 43.1"	14°01' 13.1"	2	7	0
3	Camping_ Punta Serra	40°45' 38.6"	14°00' 36.3"	0	0	0
4	Tirreno Residence	40°46' 05.1"	14°01' 03.7"	1	15	1
5	Via Faro	40°46' 09.0"	14°01' 03.0"	1	2	0
6	Olmo	40°45' 27.0"	14°01' 10.3"	37	48	0
7	Chiaio-lella	40°45' 17.5"	14°00' 30.7"	0	3	1
8	Hotel Riviera	40°45' 10.9"	14°00' 32.3"	2	19	0
9	Camping Vivara	40°45' 21.2"	14°01' 02.2"	0	0	0
10	Via dei Bagni	40°45' 34.3"	14°01' 19.0"	1	7	0
			Total	47	169	24

* Culex pipiens *L. and* Culex laticinctus

Trap site number 6, named "Olmo", represents an interesting hot spot of *Ae. albopictus* density as well as a "prototype" of the typical family-type garden on Procida island. This site is comprised of a 0.09 ha garden with cultivation of vegetables and farming of chickens and rabbits. In this garden three big water containers were identified that were utilized for the irrigation of the vegetables. The containers contained hundreds of mosquito larvae of various developmental stages and pupae.

Larvae and pupae were collected, transported alive to the laboratory at Department of Biology, University of Naples Federico II, and reared until adulthood resulting in 57 males and 118 females of *Ae. albopictus*.

Our *Ae. albopictus* record represents the first official report of this species on Procida island and these data are added to the recent record of the occurrence of the Asian tiger mosquito on other six Italian minor islands: Isola del Giglio (Toscana), Ventotene (Lazio), and Ustica, Lampedusa, Linosa and Pantelleria (Sicilia) (Fig. 1B) (Romi et al. 2016; Toma et al. 2017).

Other Mediterranean islands with reported presence of *Ae. albopictus* mosquitoes are the Maltese islands (Gatt et al. 2009) and the Balearic Islands of Mallorca, Ibiza and Minorca (Spain) (Miquel et al. 2013; Barceló et al. 2015; Bengoa et al. 2016).

4. THE PUBLIC SURVEY ON PROCIDA ISLAND OF THE MOSQUITO PROBLEM

During the entomological survey on Procida Island in September 2015, a public survey was likewise conducted to evaluate the perception by Procida inhabitants of the Asian tiger mosquito problem and their interest to support and to participate in area-wide programmes to control this insect. The public survey was a crucial step to start informing citizens about our project and to create the first positive relationship with local people interested and sensitive to the mosquito problem.

We interviewed, using a paper-based questionnaire, 200 randomly selected people (about 2% of the total island population; see Table 2). We obtained a very high participation rate with 200 out of 213 people who accepted to participate in our survey (94%). A list of the questions of the survey and their responses is reported in Table 3.

According to 70% of respondents, the abundance and hence the problem of mosquitoes in Procida has increased in the past 10 years. Most inhabitants spend on average more than one hour a day in green areas or outdoors, and about 50% of respondents were forced to limit the time spent outdoor activities because of the mosquitoes. Seventy-seven percent of respondents were aware of the nature of mosquitoes as human disease vectors and about a quarter of respondents know people who needed medical assistance because of a mosquito bite. Eighty five percent of respondents attempted to limit mosquito bites using electric diffusers, mosquito nets or chemical repellents.

Table 2. Sample classification from the survey of Procida inhabitants on the Asian tiger mosquito problem

Survey sample characteristics (N = 200)				
Gender	Male	Female		
	103	97		
Residence	Local resident	Tourist		
	171	29		
Age	18-39 years	40-64 years	64+ years	
	85	90	25	
Occupation	Unemployed	Employed	Retired	Student
	40	120	26	14

By contrast, very few respondents make active efforts to curtail the proliferation of these insects through the reduction of larval breeding sites (only 3% of respondents use larvicide products to dissolve in water and only 11% remove water containers from their houses or gardens).

Eighty eight percent of respondents were in favour of a mosquito control programme on Procida, although only 44% of people surveyed agreed to the installation of monitoring traps on their private properties. A third of respondents agreed to contribute economically to the project, and 25% of respondents would be interested to commit themselves as volunteers to the realization of the project (contributing from one to 24 hours per week).

5. THE PROCIDA COMMUNITY-BASED APPROACH

Considering the specific features of Procida island and the positive response from locally interviewed inhabitants and tourists, in 2016 we decided to start an *Ae. albopictus* systematic monitoring project, involving the local municipal administration and citizens, with the aim to collect baseline data to fully characterize the Procida site for a future SIT-based area-wide suppression programme.

We developed a multi-step approach plan, to progressively increase the citizen and administrator participation in the project, which includes three main phases:

- Phase 1: Monitoring over one year using ovitraps to define the temporal dynamic of the *Ae. albopictus* population on the island.
- Phase 2: Higher-density ovitrap monitoring to capture the spatial distribution of the mosquito on the island, possibly identifying hot spots.
- Phase 3: Estimation of local mosquito population density by mark-release-recapture (MRR) experiments.

Table 3. Responses to survey questions on the Asian tiger mosquito problem by Procida inhabitants

Survey question	Responses (percentages)			
1. Over the past decade, the mosquito problem in Procida *has*:	increased	decreased	the same	don't know
	70.5	2.0	15.5	12.0
2. Is there a garden or green area in your estate?	Yes	No		
	77.5	22.5		
3. In your estate the presence of mosquitoes is:	abundant	medium	scarce	absent
	35.0	44.5	17.0	3.5
4. How much time do you spend on average every day in a green area/garden?	< 1 hour	> 1 hour	no time	
	18.0	71.5	10.5	
	Yes	**No**		
5. During summer, are you forced to give up outdoor activities because of the mosquitoes?	50.5	49.5		
6. Did It happen that you or any of your relatives and/or friends had to go to the doctor for a mosquito bite?	75.0	25.0		
7. Do you know that the Asian tiger mosquito can transmit viral diseases to humans?	76.5	23.5		
8. Do you use protective measures against mosquitoes?	84.5	15.5		
8a. Do you use electric diffusers?	57.5	42.5		
8b. Do you remove standing water?	10.5	89.5		
8c. Do you use mosquito nets?	53.5	46.5		
8d. Do you use larvicides?	3.0	97.0		
8e. Do you use insect repellents?	44.5	55.5		
9. Would you welcome a regional/municipal mosquito control programme?	88.0	12.0		
10. Would you agree to the installation in your property of traps for the capture and monitoring of mosquitoes?	43.5	56.5		
11. Would you agree to contribute personally to the financing of a mosquito control project?	33.0	67.0		
12. Are you interested in participating, as a volunteer, to a mosquito monitoring and control programme in Procida?	25.0	75.0		

Following a numbered list of planned steps to be carried out, the main results obtained to date are:

1. *Contact with local administration:* Consultations with the Procida major and municipal counsellors to explain the various aspects of the project and to request logistic support from the Procida municipal administration.
2. *Information campaign:* Implemented on the island by the distribution of pamphlets to inform Procida citizen about the project and to invite them to participate.
3. *Active mosquito monitoring by citizens – 1:* A public assembly was organized in collaboration with the Procida municipal administration to select volunteers for the first project phase. Twelve persons, including the major and two municipal counsellors were involved as volunteers in the ovitrap monitoring programme (April 2016-May 2017). These volunteers were trained over a one-week period after which they managed 16 out of 26 ovitraps autonomously over 13 months, reporting weekly mosquito collections (Fig. 2). During the 13 months of monitoring, a total of 44 245 *Ae. albopictus* eggs were collected that were subsequently transported to the Department of Biology of the University of Naples Federico II for counting and species identification.
4. *Media coverage:* In collaboration with the press office of the Procida municipal administration press releases about the project were issued. (http://bit.ly/press_release1; http://bit.ly/press_release2; http://bit.ly/press_release3; http://bit.ly/press_release4; http://bit.ly/press_release5).
5. *Official agreement between institutions:* An official memorandum of understanding (MoU) was signed between the Procida municipal administration and the Department of Biology of the University of Naples Federico II for the implementation of the project on the island.
6. *Crowdfunding campaign:* A crowdfunding campaign was launched to collect funds and to further diffuse the project between Procida inhabitants (http://bit.ly/crowdfunding_procida).
7. *Active mosquito monitoring by citizens - 2:* A second public assembly was organized in collaboration with the Procida municipal administration to select volunteers for project phase 2 from July 2016 to September 2016. We successfully involved 79 families to allow deployment of 101 ovitraps on their private properties all over the island. A total of 40 811 eggs were collected during two weeks in July and two weeks in September 2016.
8. *Passive mosquito monitoring by citizens:* We utilized the mobile application ZanzaMapp (2019) developed at the University of Rome La Sapienza, to involve citizen as well as tourists in the passive monitoring of mosquitoes. An entomological survey was performed on the island at the same time to validate the data obtained by the citizens and tourists using the ZanzaMapp app (September 2016).

Figure 2. Volunteers and researchers during Aedes albopictus *monitoring phase 1 on Procida island.*

9. *Active mosquito monitoring by citizens - 3:* A third public assembly was organized with citizens in collaboration with the Procida municipal administration to select volunteers for project phase 3. We successfully involved 12 families to allow placement in their gardens of ovitraps, CDC-light traps, Biogents BG-Sentinel traps, sticky traps and human landing catches stations to set up optimal conditions to perform a mark-release-recapture test in an area of the island selected as a potential SIT testing site (September 2017).
10. *School involvement:* Procida primary and secondary school students have been involved in a collaborative didactic project aimed at the diffusion of knowledge about Asian tiger mosquito. Students were asked to actively participate in a sanitation campaign to reduce *Ae. albopictus* larval breeding sites in their home properties (October 2017-June 2018).
11. *Active mosquito monitoring by citizens - 4:* Ten volunteers were selected for the management of a permanent mosquito monitoring network on the island using 20 ovitrap and 20 low-cost weather stations (April 2018-May 2019). Volunteers collected eggs weekly and eggs were counted using stereo-microscopes. Data resulting from counting were uploaded by students in an *ad-hoc* online database. The collected eggs were sent monthly to the University of Naples Federico II to validate the counts.

6. CONCLUSIONS

Mosquitoes are considered among the deadliest animals on earth and invasive mosquito species, such as the Asian tiger mosquito, represent a special concern in view of their ability to adapt to new habitats world-wide and to transmit several serious diseases to humans. Managing the public health threat represented by mosquitoes is not only a matter of vector control but also of influencing or modifying public behaviour.

The involvement of civil society in research projects, the so-called citizen science, represents an interesting opportunity in this context to mobilize support, with the possible reductions in working costs, as well as increasing capillarity in the target area. At the same time, it could increase public awareness about the mosquito problem and about virtuous control efforts that could limit mosquito spread (Fig. 2).

Procida Island has a unique combination of key features that makes it an ideal "open-space laboratory" to study the effectiveness of surveillance and suppression methods, including the SIT and similar tools as part of an AW-IVM approach against *Ae. albopictus*. The project on Procida Island began as a pilot study in 2015, with a limited group of volunteers and resources. We successfully involved the Procida administration and about 200 local inhabitants.

Through the action of a dedicated team of full-time professionals and the participation of an increased numbers of volunteers, we started the collection of baseline data about the spatial and temporal population dynamics of *Ae. albopictus*.

We aim in the near future to utilize these collected data, the know-how obtained, and the network of established interactions with citizen and the local municipal administration, to extend the approach to the whole territory of the island. Our long-term and most ambitious objective is to obtain a consistent suppression of the population of this invasive species on Procida to help the local economy, which is mainly based on tourism. Furthermore, we aim to promote the application of a SIT-based mosquito integrated management approach, empowered by active community engagement and participation, on other islands of Campania Region (Ischia and Capri), and eventually on the mainland.

7. ACKNOWLEDGEMENTS

This work was supported by a special grant by the Rector of University of Naples Federico II, Prof. Gaetano Manfredi.

We are deeply grateful for the invaluable support received from the Procida Major Raimondo Ambrosino and the municipal counsellors Rossella Lauro and Titta Lubrano.

We are very grateful to the Procida volunteers Costantino D'Antonio, Davide Zeccolella, Luigi "Corecane" D'Orio, Cesare Buoninconti, Amedeo Schiano, Michele Meglio, Alberto Salvemini, Max Noviello, Anna and Antonio Amalfitano, Emanuela Coppola, Biagio and Isa Coppola, Angela and Pasquale Lubrano, Claudia Riccio and Antonietta Pagano, who shared their time to support this project.

We thank Nella Scotto for the assistance during the survey carried out on the island.

We greatly thank Prof. Luciano Gaudio for his support and the following students of the University of Naples Federico II involved in this project: Luca Iesu, Brunella Bozzi, Angela Meccariello, Rita Colonna and Antonio Marino. We thank Marco Di Luca and Francesco Severini, Istituto Superiore di Sanità, Rome, Italy, for the help in the identification of *Culex* mosquito species.

We thank all the Procida citizens and tourist accommodations (Hotel Riviera, Hotel La Torre, Hotel Savoia, Camping Punta Serra, Camping Vivara) that granted us access to their private properties for the mosquito monitoring activities. We thank the Comitato di Gestione Isola di Vivara for the authorization to access the reserve.

8. REFERENCES

Albieri, A., M. Carrieri, P. Angelini, F. Baldacchini, C. Venturelli, S. M. Zeo, and R. Bellini, R. 2010. Quantitative monitoring of *Aedes albopictus* in Emilia-Romagna, Northern Italy: Cluster investigation and geostatistical analysis. Bulletin of Insectology 63: 209–216.

Balestrino, F., A. Puggioli, R. Bellini, D. Petric, and J. R. L. Gilles. 2014. Mass production cage for *Aedes albopictus* (Diptera: Culicidae). Journal of Medical Entomology 51: 155–163.

Barceló, C., M. Bengoa, M. Monerris, R. Molina, S. Delacour-Estrella, J. Lucientes, and M. A. Miranda. 2015. First record of *Aedes albopictus* (Skuse, 1894) (Diptera; Culicidae) from Minorca (Balearic Islands, Spain). Journal of the European Mosquito Control Association 33: 1–4.

Bellini, R., M. Calvitti, A. Medici, M. Carrieri, G. Celli, and S. Maini. 2007. Use of the Sterile Insect Technique against *Aedes albopictus* in Italy: First results of a pilot trial, pp. 505–515. *In* M. J. B. Vreysen, A. S. Robinson, and J. Hendrichs (eds.), Area-wide control of insect pests: From research to field implementation. Springer, Dordrecht, The Netherlands.

Bellini, R., A. Medici, A. Puggioli, F. Balestrino, and M. Carrieri. 2013. Pilot field trials with *Aedes albopictus* irradiated sterile males in Italian urban areas. Journal of Medical Entomology 50: 317–325.

Bengoa, M., S. Delacour-Estrella, C. Barceló, C. Paredes-Esquivel, M. Leza, J. Lucientes, R. Molina, and M. A. Miranda. 2016. First record of *Aedes albopictus* (Skuse, 1894) (Diptera; Culicidae) from Minorca (Balearic Islands, Spain). Journal of the European Mosquito Control Association 34: 5–9.

Bonizzoni, M., W. A. Dunn, C. L. Campbell, K. E. Olson, O. Marinotti, and A. A. James. 2012. Complex modulation of the *Aedes aegypti* transcriptome in response to dengue virus infection. PLoS One 7(11): e50512.

Bonney, R., J. L. Shirk, T. B. Phillips, A. Wiggins, H. L. Ballard, A. J. Miller-Rushing, and J. K. Parrish. 2014. Next steps for citizen science. Science (80) 343: 1436–1437.

Bourtzis, K., R. S. Lees, J. Hendrichs, and M. J. B. Vreysen. 2016. More than one rabbit out of the hat: Radiation, transgenic and symbiont-based approaches for sustainable management of mosquito and tsetse fly populations. Acta Tropica 157: 115–130.

Caminade, C., J. M. Medlock, E. Ducheyne, K. M. McIntyre, S. Leach, M. Baylis, and A. P. Morse. 2012. Suitability of European climate for the Asian tiger mosquito *Aedes albopictus*: Recent trends and future scenarios. Journal of The Royal Society Interface 9: 2708–2717.

Cancrini, G., A. Frangipane Di Regalbono, I. Ricci, C. Tessarin, S. Gabrielli, and M. Pietrobelli. 2003. *Aedes albopictus* is a natural vector of *Dirofilaria immitis* in Italy. Veterinary Parasitology 118: 195–202.

Cirio, U. 1975. The Procida medfly pilot experiment: Status of the medfly control after two years of sterile-insect releases, pp. 39-49. *In* Panel and Research Co-ordination Meeting on the Sterile-Male Technique for Control of Fruit Flies, 12 Nov 1973. FAO/IAEA, Vienna, Austria.

Cirio, U., M. Capparella, and A. P. Economopoulus. 1987. Control of medfly (*Ceratitis capitata* Wied.) by releasing a mass-reared genetic sexing strain, pp. 515–522. *In* A. P. Economopoulos (ed.), Fruit flies. Elsevier, Amsterdam, The Netherlands.

Cunze, S., J. Kochmann, L. K. Koch, and S. Klimpel. 2016. *Aedes albopictus* and its environmental limits in Europe. PLoS One 11(9): e0162116.

D'Antonio, C., and D. Zeccolella. 2007. "Attuali conoscenze della fauna terrestre della Riserva Naturale di Stato Isola de Vivara" *In* "Vivara. Viaggio alla scoperta della fauna terrestre di una piccola isola del Mediterraneo". Casa Editrice Autorinediti, Napoli, Italia.

Di Luca, M., F. Severini, L. Toma, D. Boccolini, R. Romi, M. E. Remoli, M. Sabbatucci, C. Rizzo, G. Venturi, G. Rezza, and C. Fortuna. 2016. Experimental studies of susceptibility of Italian *Aedes albopictus* to Zika virus. Eurosurveillance 21(18): 30223.

Dickinson, J. L., J. Shirk, D. Bonter, R. Bonney, R. L. Crain, J. Martin, T. Phillips, and K. Purcell. 2012. The current state of citizen science as a tool for ecological research and public engagement. Frontiers in Ecology and the Environment 10(6): 291–297.

Dyck, V. A., J. Hendrichs, and A. Robinson (eds.). 2021. Sterile Insect Technique – Principles and practice in Area-wide Integrated Pest Management. Second Edition. CRC Press, Boca Raton, Florida, USA. 1200 pp.

(ECDC) European Centre for Disease Prevention and Control. 2018a. *Aedes albopictus* - current known distribution in Europe, June 2018.

(ECDC) European Centre for Disease Prevention and Control. 2018b. Mosquito-borne diseases: An emerging threat.

Fischer, D., S. M. Thomas, F. Niemitz, B. Reineking, and C. Beierkuhnlein. 2011. Projection of climatic suitability for *Aedes albopictus* Skuse (Culicidae) in Europe under climate change conditions. Global and Planetary Change 78: 54–64.

Gasperi, G., R. Bellini, A. R. Malacrida, A. Crisanti, M. Dottori, and S. Aksoy. 2012. A new threat looming over the Mediterranean basin: Emergence of viral diseases transmitted by *Aedes albopictus* mosquitoes. PLoS Neglected Tropical Diseases 6(9): e1836.

Gatt, P., J. C. Deeming, and F. Schaffner. 2009. First record of *Aedes (Stegomyia) albopictus* (Skuse) (Diptera: Culicidae) in Malta. European Mosquito Bulletin 27: 56–64.

Gilles, J. R. L., M. F. Schetelig, F. Scolari, F. Marec, M. L. Capurro, G. Franz, and K. Bourtzis. 2014. Towards mosquito sterile insect technique programmes: Exploring genetic, molecular, mechanical and behavioural methods of sex separation in mosquitoes. Acta Tropica 132 (Suppl.): S178–187.

Grard, G., M. Caron, I. M. Mombo, D. Nkoghe, S. Mboui Ondo, D. Jiolle, D. Fontenille, C. Paupy, and E. M. Leroy. 2014. Zika virus in Gabon (Central Africa) - 2007: A new threat from *Aedes albopictus*? PLoS Neglected Tropical Diseases 8(2): e2681.

Heitmann, A., S. Jansen, R. Lühken, M. Leggewie, M. Badusche, B. Pluskota, N. Becker, O. Vapalahti, J. Schmidt-Chanasit, and E. Tannich. 2017. Experimental transmission of zika virus by mosquitoes from central Europe. Eurosurveillance 22(2): 30437.

Jordan, R. C., A. E. Sorensen, and S. Ladeau. 2017. Citizen science as a tool for mosquito control. Journal of the American Mosquito Control Association 33(3): 241–245.

Kampen, H., J. M. Medlock, A. Vaux, C. Koenraadt, A. van Vliet, F. Bartumeus, A. Oltra, C. A. Sousa, S. Chouin, and D. Werner. 2015. Approaches to passive mosquito surveillance in the EU. Parasites and Vectors 8(1): 9.

Lees, R. S., J. R. Gilles, J. Hendrichs, M. J. B. Vreysen, and K. Bourtzis. 2015. Back to the future: The Sterile Insect Technique against mosquito disease vectors. Current Opinion in Insect Science 10: 156–162.

Li, Y., X. Su, G. Zhou, H. Zhang, S. Puthiyakunnon, S. Shuai, S. Cai, J. Gu, X. Zhou, G. Yan, and X. G. Chen. 2016. Comparative evaluation of the efficiency of the BG-Sentinel trap, CDC light trap and Mosquito-oviposition trap for the surveillance of vector mosquitoes. Parasites and Vectors 9: 446.

Manica, M., F. Filipponi, A. D'Alessandro, A. Screti, M. Neteler, R. Rosà, A. Solimini, A. della Torre, and B. Caputo. 2016. Spatial and temporal hot spots of *Aedes albopictus* abundance inside and outside a South European metropolitan area. PLoS Neglected Tropical Diseases 10(6): e0004758.

Marini, F., B. Caputo, M. Pombi, G. Tarsitani, and A. Della Torre. 2010. Study of *Aedes albopictus* dispersal in Rome, Italy, using sticky traps in mark-release-recapture experiments. Medical and Veterinary Entomology 24: 361–368.

Medlock, J. M., K. M. Hansford, V. Versteirt, B. Cull, H. Kampen, D. Fontenille, G. Hendrickx, H. Zeller, W. Van Bortel, and F. Schaffner. 2015. An entomological review of invasive mosquitoes in Europe. Bulletin of Entomological Research 105: 637–663.

Miquel, M., R. Del Río, D. Borràs, C. Barceló, C. Paredes-Esquivel, J. Lucientes, and M. A. Miranda. 2013. First detection of *Aedes albopictus* (Diptera: Culicidae) in the Balearic Islands (Spain) and assessment of its establishment according to the ECDC guidelines. Journal of the European Mosquito Control Association 31: 8–11.

Mosquito Alert. 2019. Ciencia ciudadana para investigar y controlar mosquitos transmisores de enfermedades. Spain.

MosquitoWEB. 2019. Portugal. www.mosquitoweb.pt

(MRS) Mosquito Recording Scheme. 2019. Biological Records Centre. Public Health England.

Muggenradar. 2019. Meld mate van muggenoverlast. Nature Today, The Netherlands.

Mückenatlas. 2019. Deutschland kartiert die Stechmücken. Germany.

Palmer, J. R. B., A. Oltra, F. Collantes, J. A. Delgado, J. Lucientes, S. Delacour, M. Bengoa, R. Eritja, and F. Bartumeus. 2017. Citizen science provides a reliable and scalable tool to track disease-carrying mosquitoes. Nature Communications 2017 Oct 24, 8(1): 916.

Pietrobelli, M. 2008. Importance of *Aedes albopictus* in veterinary medicine. Parassitologia 50: 113–115.

Reisen, W. K., R. P. Meyer, R. F. Cummings, and O. Delgado. 2000. Effects of trap design and CO_2 presentation on the measurement of adult mosquito abundance using Centers for Disease Control-style miniature light traps. Journal of the American Mosquito Control Association 16: 13–18.

Robinson, A. S. 2002. Genetic sexing strains in medfly, *Ceratitis capitata*, Sterile Insect Technique programmes. Genetica 116: 5–13.

Roiz, D., R. Eritja, R. Molina, R. Melero-Alcibar, and J. Lucientes. 2008. Initial distribution assessment of *Aedes albopictus* (Diptera: Culicidae) in the Barcelona, Spain, area. Journal of Medical Entomology 45: 347–352.

Romi, R., S. D'Avola, D. Todaro, L. Toma, F. Severini, A. Stancanelli, F. Antoci, F. La Russa, D. Boccolini, S. Casano, S. D. Sotera, E. Carraffa, F. Schaffner, M. Di Luca, and A. Torina. 2016. *Aedes albopictus* (Skuse, 1894) in the southernmost limit of Europe: First record in Lampedusa, Linosa and Pantelleria islands and current distribution in Sicily, Italy, pp. 185. *In* Proceedings of the 29th SOIPA Congress, June 21–24, 2016, Bari, Italy.

Sabatini, A., V. Raineri, G. Trovato, and M. Coluzzi. 1990. *Aedes albopictus* in Italia e possibile diffusione della specie nell'area mediterranea. Parassitologia 32: 301–304.

Saridaki, A. and K. Bourtzis. 2010. *Wolbachia*: More than just a bug in insect genitals. Current Opinion in Microbiology 13: 67–72.

Schaffner, F., G. Angel, B. Geoffroy, J. P. Hervy, A. Rhaiem, and J. Brunhes. 2001. The mosquitoes of Europe/Les moustiques d'Europe. Programme d'identification et d'enseignement. IRD Éditions et EID Méditerranée (CD ROM). Montpellier, France.

Sokhna, C., M. O. Ndiath, and C. Rogier. 2013. The changes in mosquito vector behaviour and the emerging resistance to insecticides will challenge the decline of malaria. Clinical Microbiology and Infection 19(10): 902–907.

Toma, L., F. Toma, G. Pampiglione, M. Goffredo, F. Severini, and M. Di Luca. 2017. First record of *Aedes albopictus* (Skuse, 1894) (Diptera; Culicidae) from three islands in the Tyrrhenian Sea (Italy). Journal of the European Mosquito Control Association 35: 25–28.

Valerio, L., F. Marini, G. Bongiorno, L. Facchinelli, M. Pombi, B. Caputo, M. Maroli, and A. della Torre. 2010. Host-feeding patterns of *Aedes albopictus* (Diptera: Culicidae) in urban and rural contexts within Rome Province, Italy. Vector-Borne Zoonotic Diseases 10: 291–294.

Venturi, G., M. Di Luca, C. Fortuna, M. E. Remoli, F. Riccardo, F. Severini, L. Toma, M. Del Manso, E. Benedetti, M. G. Caporali, A. Amendola, C. Fiorentini, C. De Liberato, R. Giammattei, R. Romi, P. Pezzotti, G. Rezza, and C. Rizzo. 2017. Detection of a chikungunya outbreak in Central Italy, August to September 2017. Eurosurveillance 22(39): 11–14.

Vontas, J., E. Kioulos, N. Pavlidi, E. Morou, A. della Torre, and H. Ranson. 2012. Insecticide resistance in the major dengue vectors *Aedes albopictus* and *Aedes aegypti*. Pesticide Biochemistry and Physiology 104: 126–131.

(WHO) World Health Organization. 2019. Vector-borne diseases.

ZanzaMapp. 2019. Italy. http://www.zanzamapp.it/

Zeller, H., L. Marrama, B. Sudre, W. Van Bortel, and E. Warns-Petita. 2013. Mosquito-borne disease surveillance by the European Centre for Disease Prevention and Control (ECDC). Clinical Microbiology and Infection 19(8): 693–698.

COMMUNITY ENGAGEMENT FOR *Wolbachia*-BASED *Aedes aegypti* POPULATION SUPPRESSION FOR DENGUE CONTROL: THE SINGAPORE EXPERIENCE

C. LIEW, L. T. SOH, I. CHEN, X. LI, S. SIM AND L. C. NG

Environmental Health Institute, National Environment Agency, Singapore; NG_Lee_Ching@nea.gov.sg

SUMMARY

Despite a long-running and comprehensive national dengue control programme, Singapore continues to experience regular outbreaks of dengue. The Environmental Health Institute of the Singapore National Environment Agency (NEA) is thus evaluating a *Wolbachia*-based combined Incompatible and Sterile Insect Technique approach (IIT/SIT) as a dengue control strategy. This approach involves field releases of irradiated male *Wolbachia*-carrying *Aedes aegypti* (L.) mosquitoes, with the aim of further suppressing the urban *Aedes aegypti* mosquito population and reducing dengue transmission. Since the beginning of our project, we considered community education and engagement as a key factor for the success of the field studies. We have therefore conducted extensive groundwork to share and consult with, and engage a wide range of stakeholders, including residents at the study sites, the general public, the medical and scientific communities, and government agencies. In this chapter, we outline our strategy for engaging residents at the study sites and the general public (two primary stakeholder groups), focusing on the key principles around which we have built our approach.

Key Words: Mosquitoes, *Aedes albopictus, Wolbachia,* cytoplasmic incompatibility, Incompatible Insect Technique, Sterile Insect Technique, outreach, community education, public mobilisation, high-rise urban environment, public housing apartments

1. INTRODUCTION

Alongside rapid urbanisation and increased global travel, the worldwide incidence of dengue has risen dramatically in recent decades. Today, around 3900 million people in 128 countries are at risk of contracting the disease, with an estimated 390 million annual infections worldwide (Bhatt et al. 2013).

In Singapore, the four serotypes of dengue virus are transmitted between humans, mainly by *Aedes aegypti* (L.), the yellow fever mosquito, with *Aedes albopictus* (Skuse), the Asian tiger mosquito, as a secondary vector. Reflecting the global situation, dengue transmission in Singapore correlates with the presence of *Ae.*

J. Hendrichs, R. Pereira and M. J. B. Vreysen (eds.), Area-Wide Integrated Pest Management: Development and Field Application, pp. 747–761. CRC Press, Boca Raton, Florida, USA.

aegypti, whilst areas with *Ae. albopictus* alone are associated only with isolated dengue cases (Hapuarachchi et al. 2016; Ong et al. 2019).

Singapore's long-running and comprehensive dengue management programme, which has a strong focus on source reduction and environmental management, now maintains the *Aedes* House Index (the percentage of properties in which breeding sites are detected, Ong et al. 2019) at low levels of around 2%. Yet, despite effective suppression of the *Aedes* mosquito vector populations, Singapore continues to experience regular dengue outbreaks, with explosive ones—associated with switching of the predominant dengue virus serotype (Lee et al. 2010) —occurring in 2005, 2007, and 2013-2014.

The continued susceptibility of Singapore's population to dengue outbreaks may be attributed to multiple factors, including: a highly urbanised, high-density, and high-rise environment; reduced herd immunity after decades of low local dengue transmission; transmission outside homes, such as at schools and workplaces; and the presence of cryptic (or unusual) *Aedes* mosquito breeding sites that are difficult to detect (Dieng et al. 2012; Low et al. 2015), amongst others. These challenges, coupled with the absence of approved antivirals and an effective vaccine against dengue, highlight the urgent need for novel and sustainable vector and dengue control approaches.

One new approach, currently being tested in field trials in Asia, the Americas, Australia, and the Pacific islands, involves the release of *Ae. aegypti* mosquitoes carrying the *Wolbachia* bacterium (McMeniman et al. 2009). In Singapore, the Environmental Health Institute (EHI) of the National Environment Agency (NEA) is trialling since 2016 a *Wolbachia*-based combined Incompatible and Sterile Insect Technique (IIT/SIT) approach, involving releases of irradiated male *Wolbachia*-carrying *Ae. aegypti* (*Wolbachia-Aedes*) mosquitoes (Lees et al. 2015; Bourtzis et al. 2016; Dyck et al. 2021). As the eggs resulting from mating between released male *Wolbachia-Aedes* mosquitoes and female urban *Ae. aegypti* mosquitoes are non-viable, the initiative, termed 'Project *Wolbachia* – Singapore', ultimately aims to further suppress mosquito populations and hence the risk of dengue transmission (Liew and Ng 2019; NEA 2019a). Irradiation is used to sterilise the small percentage of female *Wolbachia-Aedes* mosquitoes remaining after sorting mass-reared pupae, thus preventing population replacement stemming from their release (Bourtzis et al. 2016; Lees et al. 2021).

Project *Wolbachia* – Singapore's IIT/SIT approach aligns well with Singapore's long-standing emphasis on mosquito population suppression, and is intended to complement traditional vector control measures, such as mosquito breeding habitat removal, space spraying with insecticides, and entomological surveillance. Existing Singapore legislation provides for NEA to produce *Wolbachia-Aedes* mosquitoes for research purposes, and the approval for releases of these male mosquitoes was obtained at the ministerial level.

Singapore has a high human population density of almost 8000 people per km^2, and more than 80% of the population lives in public housing apartments (SingStat 2019). This makes Project *Wolbachia* – Singapore the first trial in the world to use the technology in such a high-rise, high-density urban environment.

In addition to rigorous laboratory studies and risk assessment, community engagement has been an integral component of Project *Wolbachia* – Singapore since its inception. Since 2012 (well before field releases commenced in 2016), EHI has carried out extensive groundwork to share and to consult with, and engage stakeholders, including residents at the study sites, the general public, the medical and scientific communities, schools and tertiary institutions, and government agencies.

These efforts continue today, as Project *Wolbachia* – Singapore trials advance in stage and scope. As of December 2019, trial releases accompanied by community engagement activities have been conducted in two study sites of 163 and 121 high-rise apartment blocks respectively (Fig. 1), covering a total of 27 600 households. Here, we present an overview of our strategies for dialoguing with residents at the study sites, and the public, with an emphasis on the key community engagement principles that have guided our approach.

2. PROJECT *WOLBACHIA* – SINGAPORE COMMUNITY ENGAGEMENT PRINCIPLES

Community engagement is essential to the testing and future use of novel vector control technologies involving the release of modified mosquitoes. While there are differences in country contexts and technologies used, studies on the design, implementation and evaluation of such community engagement programmes emphasise common principles, such as inclusive public engagement and consultation, transparency, and tailoring the engagement to the local audience (Subramaniam et al. 2012; McNaughton and Duong 2014; Ernst et al. 2015; Kolopack et al. 2015).

To achieve effective and respectful outreach for Project *Wolbachia* – Singapore, we developed a framework for engaging residents and the public based on the principles laid out in this Section. This framework also builds upon existing professional collaborations in research and dengue management, as well as long-standing outreach efforts urging the general public to remove and destroy *Aedes* mosquito breeding habitats in their homes (NEA 2019b).

Whilst this chapter focuses on residents at the study sites and the general public, it should be noted that our community engagement efforts also extend to other stakeholder groups. For example, we conduct scientific seminars, lectures, and workshops to inform and consult the scientific and medical communities about *Wolbachia-Aedes* technology. We additionally consult with international experts to share knowledge and key findings, and also hold seminars and workshops to engage stakeholders in the government, including agencies with functions in public health and the environment. The feedback received from all stakeholder groups has been instrumental in shaping our outreach and deployment strategies.

Figure 1. Project Wolbachia *– Singapore study sites as of December 2019. Trial releases of male* Wolbachia*-carrying* Aedes aegypti *mosquitoes, accompanied by community engagement activities, have been conducted in 163 high-rise apartment blocks at Nee Soon East (top) and 121 at Tampines West (bottom), covering a total of 27 600 households.*

2.1. Transparent, Clear and Consistent

Project *Wolbachia* – Singapore involves the release of male *Wolbachia-Aedes* mosquitoes in close proximity to residential dwellings. We thus endeavoured to make our community engagement as accurate and clear as possible, so that residents and the public are well informed about the intervention taking place in their neighbourhoods and the possible effects that this may have on their households.

Importantly, our community engagement got off to an early start. We initiated outreach activities in 2012, in parallel with our laboratory studies and risk assessment of the *Wolbachia-Aedes* technology, well before the first male *Wolbachia-Aedes* mosquitoes were released in 2016. This allowed time for members of the public to familiarise themselves with *Aedes* mosquito biology and behaviour, such as the fact that male mosquitoes do not bite, and with *Wolbachia-Aedes* technology in general. We also had sufficient time to address potential issues raised by stakeholders; for example, we conducted laboratory studies to confirm that male *Wolbachia-Aedes* mosquitoes do not lose the *Wolbachia* bacterium as they age, a concern that was raised by an expert entomologist.

Given that public messaging in Singapore has for decades emphasised the importance of keeping mosquito populations low, the proposed strategy of releasing large numbers of male *Wolbachia-Aedes* mosquitoes may be seen by some as counterintuitive. Thus, we sought to ensure that Project *Wolbachia* – Singapore's purpose and goals—that the mosquito releases, together with existing control methods being applied, are compatible and intended to further suppress mosquito populations—were clearly communicated to the public.

To avoid conflicting messaging, consistency is also key. For example, residents at the release sites are encouraged to remain vigilant and continue practising standard mosquito control activities, such as turning over pails and flowerpots, and clearing roof gutters. Residents are also advised to kill adult mosquitoes as they normally would.

As Project *Wolbachia* – Singapore scaled up and progressed to more advanced stages (NEA 2018a, 2019c), we also delivered prompt updates to keep the public abreast of new developments. These developments included the releases of male *Wolbachia-Aedes* mosquitoes at higher floors, in addition to the releases at the ground floors; collaborations with private sector companies to incorporate technologies such as automated devices for mosquito production, sorting, and release (NEA 2018b); and the use of irradiation post-pupal sorting to sterilise any female *Wolbachia-Aedes* mosquitoes remaining from the rearing process among the males to be released (NEA 2018a).

We have endeavoured to provide members of the public with adequate information, so that they can develop informed opinions on the risks and benefits of Project *Wolbachia* – Singapore. At the same time, we also sought to present this information in a manner that is accessible to individuals without scientific training. Striking this balance is important so that key messages are not obscured by technical details and scientific jargon yet remain accurate.

2.2. Science-based and Educational

Misinformation is a common source of fear and doubt surrounding new technologies. Our community engagement thus aims to demystify the *Wolbachia-Aedes* technology, by equipping the public with a strong understanding of the scientific concepts behind it.

In our outreach materials and engagement sessions, we use accessible language and infographics to explain concepts such as the origin and natural occurrence of the *Wolbachia* bacterium, cytoplasmic incompatibility, and how the release of male *Wolbachia-Aedes* mosquitoes will lead to a reduction in *Ae. aegypti* mosquito populations. Project *Wolbachia* – Singapore scientists and technicians were also heavily involved in outreach and were often on-site during door-to-door house visits, dialogue/outreach sessions and roadshows, to answer any questions related to the technology.

Where possible, we also created hands-on experiences for the public to learn about *Wolbachia-Aedes* technology. For example, participants at our roadshows and dialogue/ outreach sessions were encouraged to place their hands into transparent Perspex boxes containing male *Wolbachia-Aedes* mosquitoes, so that they could experience for themselves that male mosquitoes do not bite. We also organise regular tours of NEA's *Wolbachia-Aedes* mosquito production facility (Fig. 2), where members of the public learn how mosquitoes are reared, sorted, and prepared for release.

Figure 2. Environmental Health Institute (EHI) researchers conducting a tour of the National Environment Agency's (NEA's) Wolbachia-Aedes *mosquito production facility for members of the public, during a Project* Wolbachia *– Singapore learning journey.*

As children and the youth community are often readily engaged and may also proactively help convey information to their families, we also hold sharing sessions at schools and tertiary institutions. Additionally, we have been engaging childcare centres (Fig. 3), and have been working with schools located within the release sites, to provide hands-on lessons related to Project *Wolbachia* – Singapore, including the release of male *Wolbachia-Aedes* pupae around school premises, and the identification of mosquitoes trapped by Gravitraps.

Figure 3. A Project Wolbachia *– Singapore show-and-tell session for young children at a childcare centre.*

Such citizen science initiatives have increased scientific understanding, and, by allowing participants to contribute directly, have helped build acceptance and a sense of ownership for the project.

2.3. Comprehensive and Inclusive

Given the need to reach out to people with diverse backgrounds, needs, and concerns, we set out to make our engagement programme as comprehensive and inclusive as possible. To reach a greater number of target groups, we employed multiple tiers of communication strategies, with varying levels of detail and engagement.

Informational brochures, distributed to all residents and stakeholders at study/ release sites (Fig. 4), are used to convey short, key messages explaining the *Wolbachia-Aedes* technology and addressing what residents can do to help keep mosquito populations low.

Figure 4. Informational brochures for distribution to residents and all stakeholders at the study/release sites.

These brochures use accessible language and infographics, and, given Singapore's multicultural landscape, are available in English, Mandarin, Malay and Tamil - the country's four official languages (Fig. 5).

More detailed information on the technology, releases, and trial results was made available through media releases (NEA 2018a, b, 2019c, d, e) and features in mainstream media (Co 2019; The Straits Times 2019), as well as via social media and other public communication channels (e.g. documentary features) (National Geographic Asia 2019). Members of the public interested in learning more can also visit NEA's *Wolbachia* website (NEA 2019a) for technical details, including scientific literature.

Besides disseminating information, we also conduct numerous face-to-face engagement activities, to engage residents more thoroughly and better understand their concerns. These activities include educational sessions in schools and pre-schools, door-to-door house visits, coffee chat sessions, apartment block and garden parties, roadshows in locations with heavy footfall (e.g. marketplaces, community centres, and shopping malls) (Fig. 6), and mosquito production facility tours.

Figure 5. A brochure for Phase 3 of Project Wolbachia *– Singapore, with information translated into Singapore's four official languages: English, Mandarin, Malay and Tamil.*

Figure 6. Outreach booths at public events/roadshows to showcase Project Wolbachia – *Singapore. Top: At a shopping mall. Bottom: At a community centre, with Singapore's Minister for the Environment and Water Resources, Mr. Masagos Zulkifli (far right).*

At the time of writing, we have held more than 100 public engagement sessions to raise awareness, answer queries, and solicit feedback about Project *Wolbachia* – Singapore. To further extend the reach of our message, we engaged members of the public who are interested to help out as NEA Dengue Prevention Volunteers (DPVs), and who can then help to convey accurate information on *Wolbachia-Aedes* technology to their own networks within the community.

Leveraging the increasing pervasiveness of digital media, we used websites, social media platforms, and videos (NEA 2019f, g, h) to complement our face-to-face outreach activities and increase exposure. As Project *Wolbachia* – Singapore scales up and the number of affected members of the public involved in the project increases, we foresee digital media playing an even larger role in our community engagement strategy.

In developing activities to engage different groups and individuals, we relied strongly on our own local knowledge and experience gained, as well as that of leaders and residents in the community. At the same time, over the course of our outreach, we also learned through experience which engagement methods are the most effective and adjusted our strategy accordingly. For example, after one dialogue session with poor attendance, we shifted our strategy towards bringing our roadshows to where people are, rather than getting residents to go to a specific dialogue.

2.4. Consultative and Responsive

To promote Project *Wolbachia* – Singapore, an important scientific initiative that may result in tangible public health benefits, we adopted a consultative approach early on to respect the concerns and opinions of the public, taking seriously any feedback received from residents, members of the public, experts, and other stakeholders. Valid concerns (such as safety of the technology, potential negative ecological impacts, niche replacement by other mosquitoes, and unintentional release of female mosquitoes) were channelled into our risk assessment track, a parallel effort that has resulted in the publication of a risk assessment of the technology (Ng et al. 2017; NEA 2019g).

Through house visits, roadshows, and dialogue sessions, we actively solicited feedback from residents and the public. We also consulted community or grassroots leaders, who have strong networks on the ground and a thorough understanding of residents' concerns. Following detailed engagement sessions to explain *Wolbachia-Aedes* technology and understand and address concerns, we were able to obtain leaders' support, approval and advocacy. The spectrum of views and concerns we received informed our release and engagement strategies later on.

We encouraged the community to learn more about the project and have also established mechanisms for the public to pose queries, voice concerns, and report incidents related to Project *Wolbachia* – Singapore. Feedback can be submitted via email, telephone hotline, an online reporting system, or verbally to our field officers. Our site managers also carry a dedicated mobile phone by which residents can reach them.

Our officers follow up on feedback received by contacting each case through house visits, telephone calls, or email, to learn more about and address their individual situations. During such follow-ups, we engage the respondents with information about Project *Wolbachia* – Singapore, as well as attempt to address the issues raised. For example, a few residents who reported increased irritation from male mosquitoes were advised to keep their windows and doors closed, or to kill the mosquitoes as they normally would. If residents report bites, field officers are typically deployed to inspect the area for mosquito breeding sites.

Whilst these mosquitoes resulting from local breeding sites are unrelated to the release of male *Wolbachia-Aedes* mosquitoes, such feedback from residents has enabled us to find and eliminate mosquito breeding sites, an outcome that is consistent with the aims of Project *Wolbachia* – Singapore.

By respectfully engaging dissenting views and responding promptly to public feedback, we hope to improve the ability of the public to make informed opinions, thereby enhancing acceptance of Project *Wolbachia* – Singapore.

3. IMPORTANCE OF SITUATIONAL AWARENESS

Like most field studies, trials of *Wolbachia-Aedes* technology may be affected by environmental factors such as the presence of mosquito breeding habitats, temperature and rainfall variations, and imported mosquito-borne infections. Proper situational awareness is therefore important for community engagement efforts to swiftly address concerns from the public when they arise.

For example, a coincidental rise in the populations of other mosquito species (such as *Culex quinquefasciatus* Say) in the community, and a resulting increase in bites experienced, could lead to public doubts concerning Project *Wolbachia* – Singapore. In such cases, data on the population trends of the other mosquito species, together with an understanding of their biting behaviour and the detection of their breeding locations on site, could provide the evidence needed to be able to dissociate these experiences from the project.

Dengue cases may also occur within study sites, complicating the messaging that Project *Wolbachia* – Singapore suppresses the dengue-transmitting mosquito population and hence potentially also the risk of disease transmission. A good analysis of the situation within and around each of these sites is thus critical, as most of these reported dengue cases may have resulted from infections outside the release sites — perhaps at school or at work— but were tagged to residential addresses within the sites.

Thus, our community engagement team is working closely with field officers to gather and respond to all information received on mosquito populations, mosquito sightings, and dengue cases within and around study sites, with the aim of assessing the cause of all issues raised by the community.

4. FUTURE DIRECTIONS AND CONCLUSION

Thanks to early and comprehensive community engagement, nationwide online and face-to-face street surveys conducted in 2016 (prior to the launch of Project *Wolbachia* – Singapore) found high levels of acceptance for *Wolbachia-Aedes* technology, with a high number of respondents indicating that they had no concerns with the release of male *Wolbachia-Aedes* mosquitoes in their neighbourhoods. As Project *Wolbachia* – Singapore scales up (NEA 2019c), an ongoing challenge will be to expand community engagement accordingly, whilst still maintaining its quality. Whilst face-to-face interactions will remain a mainstay of our engagement strategy, this is highly labour-intensive; in the future, engagement through digital and social media will be scaled up to reach larger segments of the population. To maintain quality and impact, it will be critical to continually evaluate our community education and engagement strategies, and to modify them when necessary (Jayawardene et al. 2011; Healy et al. 2014). To this end, Project *Wolbachia* – Singapore as a whole, including the community engagement component, is regularly reviewed by an external panel comprising local and international experts (NEA 2019c).

We will also continue to conduct population surveys to ascertain public awareness, perceptions and attitudes towards Project *Wolbachia* – Singapore, which will allow us to further fine-tune our outreach and deployment strategies. Prior to the release of male *Wolbachia-Aedes* mosquitoes, a nationwide online survey conducted in 2016 revealed that 94% of the population had no objections to the release of male *Wolbachia-Aedes* mosquitoes in their neighbourhoods, and a face-to-face street survey targeting older respondents (aged 40 years and above) revealed similar results – 89% had no objections to such releases in their neighbourhoods, and 31% had heard of the project through mainstream media. During the Phase 1 field study in 2016, a survey conducted showed that more than 70% of the households interviewed had heard of our project, and more than 90% had no concerns with the releases.

In summary, community engagement is increasingly recognised as a critical dimension of biomedical and global health research, as well as the social sciences, especially where novel technologies such as *Wolbachia*-based control methods are concerned. We believe that respectful, impactful community engagement is crucial for the success of *Wolbachia*-based technologies and is recognised by senior leadership as a key project priority.

5. ACKNOWLEDGEMENTS

We work closely with the Singapore NEA's Corporate Communications Department and the 3P (People, Public and Private) Network Division, and are grateful for their advice as well as communications and engagement support. We are also grateful to the Residents' Committees and Citizens' Consultative Committees at the release sites, for their support and for assistance with outreach activities. Special thanks to Tampines Town Council, Yishun Town Council, Tampines West Community Centre, and Nee Soon East Community Centre for help with publicity and outreach.

6. REFERENCES

Bhatt, S., P. W. Gething, O. J. Brady, J. P. Messina, A. W. Farlow, C. L. Moyes, J. M. Drake, J. S. Brownstein, A. G. Hoen, O. Sankoh, M. F. Myers, D. B. George, T. Jaenisch, G. R. Wint, C. P. Simmons, T. W. Scott, J. J. Farrar, and S. I. Hay. 2013. The global distribution and burden of dengue. Nature 496: 504–507.

Bourtzis, K., R. S. Lees, J. Hendrichs, and M. J. B. Vreysen. 2016. More than one rabbit out of the hat: radiation, transgenic and symbiont-based approaches for sustainable management of mosquito and tsetse fly populations. Acta Tropica 157: 115–130.

Co, C. 2019. New facility to produce up to 5 million *Wolbachia* mosquitoes weekly in fight against dengue. Channel News Asia, 2 December 2019.

Dyck, V. A., J. Hendrichs, and A. S. Robinson (eds.). 2021. Sterile Insect Technique – Principles and practice in Area-Wide Integrated Pest Management. Second Edition. CRC Press, Boca Raton, Florida, USA. 1200 pp.

Dieng, H., R. G. Saifur, A. H. Ahmad, M. C. Salmah, A. T. Aziz, T. Satho, F. Miake, Z. Jaal, S. Abubakar, and R. E. Morales. 2012. Unusual developing sites of dengue vectors and potential epidemiological implications. Asian Pacific Journal of Tropical Biomedicine 2: 228–232.

Ernst, K. C., S. Haenchen, K. Dickinson, M. S. Doyle, K. Walker, A. J. Monaghan, and M. H. Hayden. 2015. Awareness and support of release of genetically modified "sterile" mosquitoes, Key West, Florida, USA. Emerging Infectious Diseases 21(2): 320–324.

Hapuarachchi, H. C., C. Koo, J. Rajarethinam, C.-S. Chong, C. Lin, G. Yap, L. Liu, Y.-L. Lai, P. L. Ooi, J. Cutter, and L. C. Ng. 2016. Epidemic resurgence of dengue fever in Singapore in 2013-2014: A virological and entomological perspective. BMC Infectious Diseases 16: 300.

Healy, K., G. Hamilton, T. Crepeau, S. Healy, I. Unlu, A. Farajollahi, and D. M. Fonseca. 2014. Integrating the public in mosquito management: Active education by community peers can lead to significant reduction in peridomestic container mosquito habitats. PLoS One 9(9): e108504.

Jayawardene, W., D. K. Lohrmann, A. H. Youssefagha, and D. C. Nilwala. 2011. Prevention of dengue fever: An exploratory school-community intervention involving students empowered as change agents. The Journal of School Health 81: 566–573.

Kolopack, P. A., J. A. Parsons, and J. V. Lavery. 2015. What makes community engagement effective?: Lessons from the *Eliminate Dengue* program in Queensland Australia. PLoS Neglected Tropical Diseases 9(4): e0003713.

Lee, K.-S., Y.-L. Lai, S. Lo, T. Barkham, P. Aw, P.-L. Ooi, J. C. Tai, M. Hibberd, P. Johansson, S. P. Khoo, and L. C. Ng. 2010. Dengue virus surveillance for early warning, Singapore. Emerging Infectious Diseases 16: 847–849.

Lees, R. S., J. R. L. Gilles, J. Hendrichs, M. J. B. Vreysen, and K. Bourtzis. 2015. Back to the future: The sterile insect technique against mosquito disease vectors. Current Opinion in Insect Science 10: 156–162.

Lees, R. S., D. O. Carvalho, and J. Bouyer. 2021. Potential impact of integrating the Sterile Insect Technique into the fight against disease-transmitting mosquitoes, pp. 1081–1118. *In* V. A. Dyck, J. Hendrichs, and A. S. Robinson (eds.), Sterile Insect Technique – Principles and practice in Area-wide Integrated Pest Management. Second Edition. CRC Press, Boca Raton, Florida, USA.

Liew, C., and L. C. Ng. 2017. The buzz behind Project *Wolbachia* - Singapore. ENVISION Magazine 13: 82–83.

Low, S.-L., S. Lam, W.-Y. Wong, D. Teo, L. C. Ng, and L.-K. Tan. 2015. Dengue seroprevalence of healthy adults in Singapore: Serosurvey among blood donors, 2009. American Journal of Tropical Medicine and Hygiene 93: 40–45.

McMeniman, C. J., R. V. Lane, B. N. Cass, A. W. Fong, M. Sidhu, Y.-F. Wang, and S. L. O'Neill. 2009. Stable introduction of a life-shortening *Wolbachia* infection into the mosquito *Aedes aegypti*. Science 323: 141–144.

McNaughton, D., and T. T. H. Duong. 2014. Designing a community engagement framework for a new dengue control method: A case study from central Vietnam. PLoS Neglected Tropical Diseases 8(5): e2794.

National Geographic Asia. 2019. City of innovation: Singapore. Full Documentary.

(NEA) National Environment Agency, Singapore. 2018a. Objectives met for Phase 1 of *Wolbachia* field studies. NEA to commence Phase 2 *Wolbachia-Aedes* field study at three sites in April 2018.

(NEA) National Environment Agency, Singapore. 2018b. Project *Wolbachia* - Singapore welcomes new collaborator for ongoing Phase 2 field study.

(NEA) National Environment Agency, Singapore. 2019a. Project *Wolbachia* Singapore. *Wolbachia-Aedes* mosquito suppression strategy.

(NEA) National Environment Agency, Singapore. 2019b. Prevent *Aedes* mosquito breeding.

(NEA) National Environment Agency, Singapore. 2019c. Project *Wolbachia* - Singapore progresses to Phase 3 field study at two extended sites in February 2019.

(NEA) National Environment Agency, Singapore. 2019d. Successful study paves the way for doubling the size of male *Wolbachia-Aedes* mosquito release.

(NEA) National Environment Agency, Singapore. 2019e. New NEA facility to boost production of male *Wolbachia-Aedes aegypti* mosquitoes to benefit more residents.

(NEA) National Environment Agency, Singapore. 2019f. *Wolbachia* video series. Part I: Safety and efficacy studies; Part II: Behind the scenes; Part III: Engaging stakeholders and the community; Part IV: Small-scale field study; Part V: Project progresses to Phase 2 field study.

(NEA) National Environment Agency, Singapore. 2019g. *Wolbachia* in my neighbourhood.

(NEA) National Environment Agency, Singapore. 2019h. *Wolbachia* is SAFE.

Ng, L. C., C. Liew, R. Gutierrez, C. S. Chong, C. H. Tan, G. Yap, J. P. S. Wong, and I. M. Li. 2017. How safe is *Wolbachia* for *Aedes* control? A risk assessment for the use of male *Wolbachia*-carrying *Aedes aegypti* for suppression of the *Aedes aegypti* mosquito population. Epidemiological News Bulletin 43: 8–16.

Ong, J., X. Liu, J. Rajarethinam, G. Yap, D. Ho, and L. C. Ng. 2019. A novel entomological index, *Aedes aegypti* breeding percentage, reveals the geographical spread of the dengue vector in Singapore and serves as a spatial risk indicator for dengue. Parasites & Vectors 12: 17.

(SingStat) Singapore Department of Statistics. 2019. Households - Latest Data 2018.

Subramaniam, T. S. S., H. L. Lee, N. W. Ahmad, and S. Murad. 2012. Genetically modified mosquito: The Malaysian public engagement experience. Biotechnology Journal 7: 1323–1327.

The Straits Times. 2019. NEA to test releasing mosquitoes via drone as Project *Wolbachia* moves to Phase 4 of fight against dengue. 7 October 2019.

SECTION 5

NEW DEVELOPMENTS AND TOOLS FOR AREA-WIDE INTEGRATED PEST MANAGEMENT PROGRAMMES

TECHNICAL INNOVATIONS IN GLOBAL EARLY WARNING IN SUPPORT OF DESERT LOCUST AREA-WIDE MANAGEMENT

K. CRESSMAN

Plant Production and Protection Division, Food and Agriculture Organization of the United Nations (FAO), Rome, Italy; keith.cressman@fao.org

SUMMARY

Technical innovations can play an important role in the effective management of transboundary pests if they are well integrated with participation and collaboration by affected countries and coordinated by a centralised body. This is particularly relevant to those migratory pests that can easily and rapidly move across regions and continents to simultaneously threaten food security and livelihoods in numerous countries. Innovations to FAO's successful desert locust global monitoring and early warning system are highlighted and the lessons learned can be applied and adapted to other emerging transboundary threats such as the fall armyworm in Africa and the red palm weevil.

Key Words: Schistocerca gregaria, transboundary plant pests and diseases, remote sensing, surveillance, forecasting, early warning, geographic information systems, unmanned aerial vehicles

1. INTRODUCTION

Transboundary pests are migratory insects and disease vectors that easily move from one country to another and can rapidly traverse regions and travel great distances to threaten crop production throughout the world. While the desert locust *Schistocerca gregaria* Forskål is probably the most well-known and best studied of the migratory insect pests, there are other notable transboundary pests such as the migratory locust *Locusta migratoria* (L.), the Moroccan locust *Dociostaurus maroccanus* (Thunberg), and Italian locust *Calliptamus italicus* (L.), as well as the red palm weevil *Rhynchophorus ferrugineus* (Olivier) and the fall armyworm *Spodoptera frugiperda* (Smith) that are becoming an increasing threat to agriculture and livelihoods in Africa and Asia (Fig. 1).

J. Hendrichs, R. Pereira and M. J. B. Vreysen (eds.), Area-Wide Integrated Pest Management: Development and Field Application, pp. 765–778. CRC Press, Boca Raton, Florida, USA.

The desert locust is an ancient insect, dating from millions of years and coexisting with early man until the advent of cultivation, when it became what is considered today as the world's most dangerous migratory insect pest. The unique behaviour of the desert locust allows it to quickly take advantage of optimal environmental conditions by rapidly increasing in number and forming highly migratory swarms that can affect some 20 percent of the earth's land surface and livelihoods of millions of people in Africa, the Middle East and Asia.

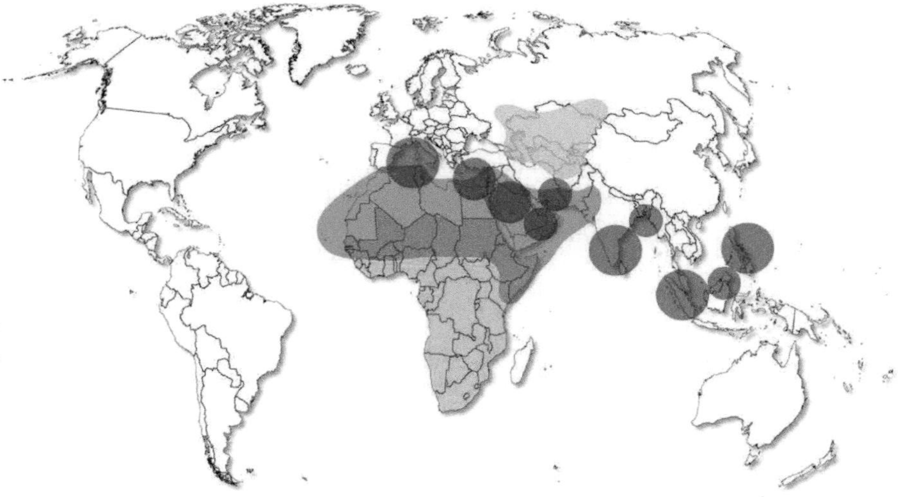

Figure 1. Major transboundary pests in Africa and Asia: Desert locust distribution area (red), fall armyworm (green), red palm weevil (blue), and migratory / Moroccan / Italian locusts in Central Asia (yellow) (source FAO DLIS).

It was not until the early twentieth century that efforts began in earnest to systematically monitor the pest across its vast desertic habitat stretching from West Africa to India, consisting of some of the world's driest and remotest areas. In 1943, the Desert Locust Information Service (DLIS) was established in the UK, which was the basis for the future global forecasting and early warning system. The DLIS was responsible for the systematic collection and mapping of desert locust infestations so that seasonal breeding areas could be identified, and a better understanding could be gained about the formation of swarms and their migratory patterns. In 1978, FAO assumed the centralised responsibility of the DLIS.

1.1. Desert Locust Biology and Behaviour

Low numbers of isolated desert locust are present somewhere within its vast habitat throughout the year. This area includes about 30 countries and covers some 16 million km^2 of desert in North Africa, the Arabian Peninsula and south-western Asia (Fig. 2).

The individualistic solitarious adults and hoppers (the nymphal stage) are well-camouflaged to blend in with their environment as a means of protection from predators. The adults are passive fliers at night, drifting up to 400 km downwind. Often the winds bring the adults into areas of recent rainfall, which wets the sandy soil sufficiently for females to lay eggs that hatch after about two weeks. The resulting hoppers shed their skins (moulting) on about a weekly basis six times before becoming an adult. The entire lifecycle lasts about three months, but may last up to a half year, as adults may remain immature for months in low temperatures or in the absence of regular rains. Desert locust do not have a dormant stage, they do not overwinter, and eggs cannot survive from one year to the next.

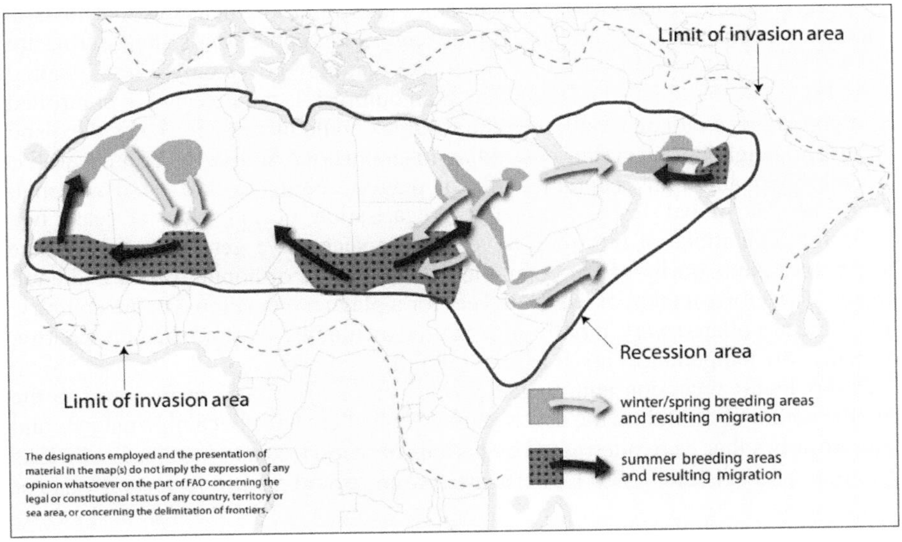

Figure 2. Desert locust recession and invasion areas, and their seasonal migration patterns (source FAO).

Locusts will quickly take advantage of exceptionally heavy and often short-lasting rainfall, whether localized or widespread, that cause ecological conditions to become favourable for breeding and survival. Once annual vegetation dries out, locusts tend to concentrate in those areas that still remain green, increase in density and form small groups that may fuse together and develop into hopper bands and adult swarms. This behavioural phase change is referred to as gregarisation. Swarms fly during the daylight hours, moving more than 200 km in a single day. This allows them to easily traverse Africa and to cross water bodies such as the Red Sea, the Mediterranean Sea and the Arabian Sea. They also can extend further north, south and east from their normal habitat and invade several dozen countries (Fig. 2). In October 1988, swarms migrated some 5,000 km across the Atlantic Ocean from West Africa to the Caribbean (Rosenberg and Burt 1999).

A single desert locust is able to consume its own weight, approximately 2 grams of food every day. A 1 km^2 sized swarm generally contains about 40 million adults (Pedgley 1981). The high number of locusts and their voracious appetite can pose a serious threat to agriculture and food security. To put this into perspective, a swarm the size of Bamako, Mali will eat the same amount of food in one day as half the entire population of Mali. Similarly, a swarm the size of Vienna will consume the same amount of food in one day as everyone in Austria and Switzerland, while a swarm the size of New York City will eat as much food in a single day as all the residents in the states of New York and Pennsylvania.

Desert locust plagues do not develop overnight. Instead, they evolve from a serious of events in which locust numbers steadily increase. It starts with a calm period of recession, when locusts are normally present at low densities in semi-arid or arid areas, causing no significant crop damage, and hopper bands and swarms are absent. This is followed by localized outbreaks that may cover only a few hundred square kilometres within a single country. If an outbreak is not detected or controlled and if good rains continue to fall, then the outbreak could increase in size and expand into several neighbouring countries, leading to an upsurge. Similarly, if an upsurge is not controlled and rains continue, then a plague could develop on a regional, continental, or global level. Normally, gregarisation occurs after at least two successful generations of breeding. Several more successive generations must take place before the majority of the populations consist only of hopper bands and adult swarms. Therefore, it takes at least one year for a plague to develop. On the contrary, a plague can collapse very rapidly in a matter of months due to effective control, unusually low temperatures and a lack of rainfall.

Desert locust recession and plague periods occur irregularly in response to the sporadic nature of rainfall in the desert. Since 1860, there have been nine plagues and major upsurges that were interrupted by periods of recessions and localised outbreaks (Symmons and Cressman 2001). These lasted from several months to several years or more (Fig. 3).

The last major upsurge or regional plague occurred in 2003-2005. In addition to an unusually cold winter, it took nearly USD 600 million and 13 million litres of insecticide sprayed by ground and aerial campaigns to bring the plague under control (Brader et al. 2006). In West Africa, more than 8 million people were affected, up to 100% of cereal crops were lost in some areas, and some 60% of household heads became indebted in Mauritania, while 90% of households in Burkina Faso received food aid.

1.2. Climate Change Impact

Changes in the climate will affect desert locust habitats, breeding, migration, and plague dynamics. It is well known that the current climate change is causing temperatures to increase. Warmer temperatures will extend the length of the summer, winter and spring breeding periods and allow desert locust eggs and hoppers to develop faster as long as this is associated with a continuation or increase of good rains. This is likely to be most pronounced during the winter and may allow an extra generation of breeding to take place (Cressman 2013).

Warmer temperatures could also potentially affect desert locust migration by allowing solitarious adults to fly longer during nights, especially during the colder portions of the year. Consequently, adults may arrive at a destination sooner or reach new areas further away. Warm temperatures could allow swarms to take off earlier in the morning, resulting in a longer period of flight during the day and a greater displacement distance. In other words, swarms could reach new places that have not been reachable up to now. Climate change could also allow swarms to fly higher than 1800 m, which is the general limit of flight due to temperature. If this is the case, then the Atlas Mountains in north-western Africa, the Hoggar Mountains of Algeria, the Jebel Akhdar Mountains in northern Oman, the mountains in the interior of Iran, and the mountain ranges along both sides of the Red Sea may no longer be natural barriers that impede migration. On the other hand, if warmer temperature regimes were to become extremely hot, for example above 50°C, then desert locust presence and survival could become limited in some areas of the Sahara and the Arabian Peninsula (Meynard et al. 2017).

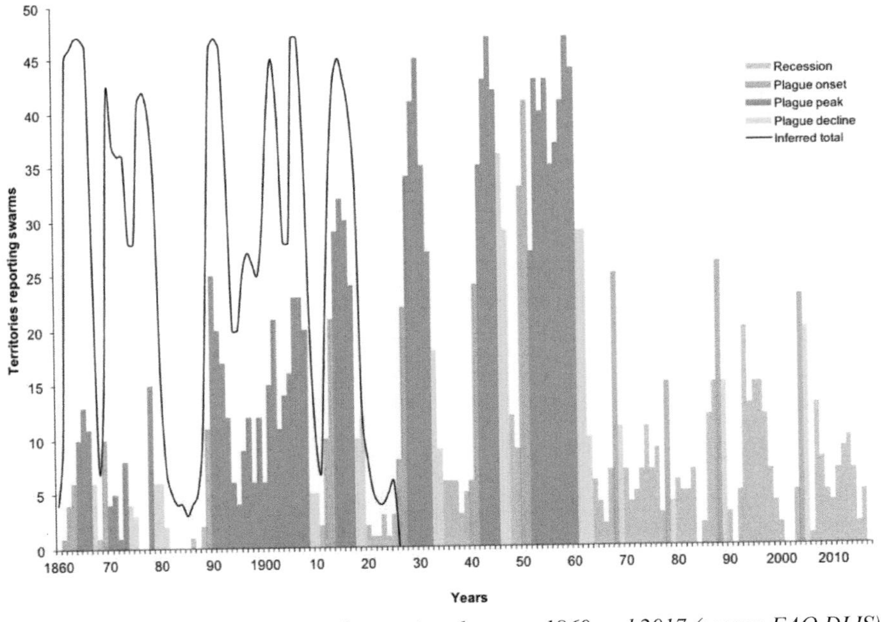

Figure 3. Desert locust plagues and recessions between 1860 and 2017 (source FAO DLIS).

The impact of climate change on the spatial variability of rainfall in desert locust breeding areas remains unclear. There is some evidence that rainfall could increase during the summer in seasonal breeding areas of the northern Sahel in West Africa and in the Yemen interior that could cause locust numbers to increase faster than normal. There appears to be more widespread agreement concerning an increase in extreme rainfall events.

It is worthy to note that unusually heavy rains have been responsible for locust plagues in the past, such as a cyclone in 1968 in Oman that caused a plague the following year and widespread heavy rains from Senegal to Morocco in October 2003 that led to a regional plague for the next two years.

The effects of wind are less certain, but any changes in wind speed, direction and circulation flows are likely to affect desert locust migration and could allow adults and swarms to reach new areas. Warmer temperatures in combination with shifting wind patterns and decreased rainfall could permit new migration routes into Southern Europe and Central Asia. In general, however, further research is required to better understand the impacts of climate change and variability on the desert locust population dynamics and migration.

2. MONITORING AND EARLY WARNING

Locust-affected countries and FAO have adopted a preventive control strategy for the area-wide management of desert locust in order to reduce the frequency, duration and intensity of plagues (Fig. 4). Successful preventive control requires effective early detection and warning, rapid response, good communications and contingency planning. The former consists of monitoring weather, ecological conditions and desert locust populations on a regular basis throughout the vast recession area that stretches from the Atlantic coast in West Africa to western India. This is accomplished through observations made by ground teams during survey and control operations that are recorded and sent to the national locust control centres and, from there, to FAO's centralised global system operated by the DLIS in Rome (Fig. 5).

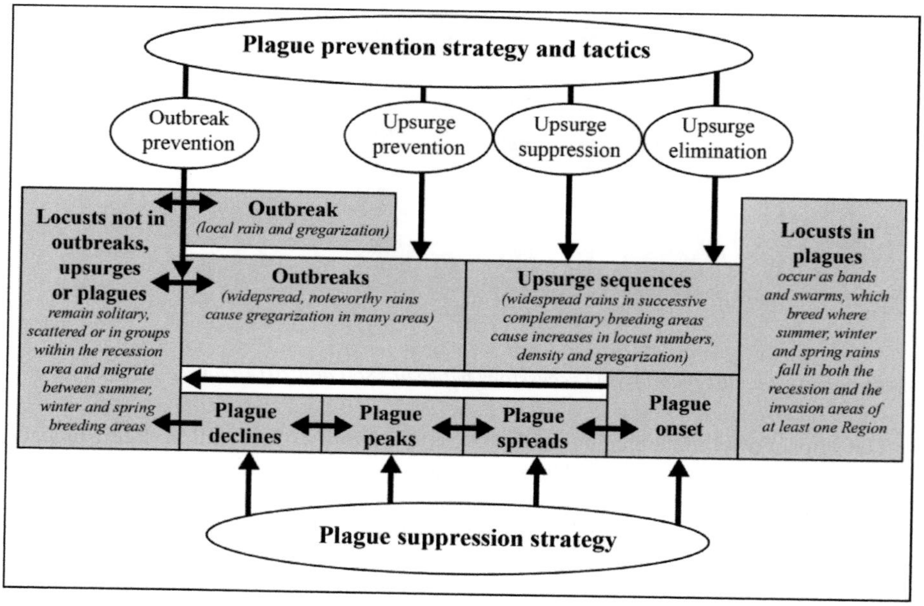

Figure 4. Desert locust preventive control strategy (source Magor et al. 2007).

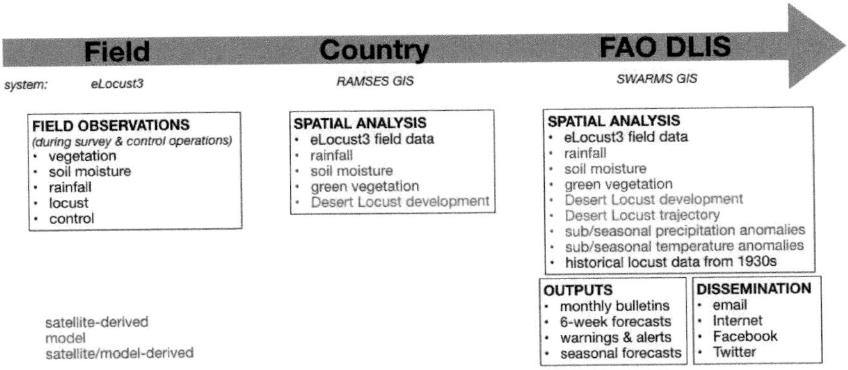

Figure 5. Information flow in the FAO global desert locust early warning system (source FAO DLIS).

Each locust-affected country is responsible for monitoring and controlling desert locust in its own territory. In most countries, a dedicated national locust centre within the Ministry of Agriculture and Plant Protection Department assumes these responsibilities. Specialized teams undertake ground surveys using four-wheeled-drive vehicles in the desert, looking for annual vegetation that may be green and harbouring desert locust hoppers or adults.

If important infestations are detected, then control operations are immediately implemented to prevent the locust population from increasing further and spreading. The obvious challenge is to find and treat these relatively small localities within a vast and remote area that is devoid of infrastructure and inhabitants. This is further complicated by increased conflict and insecurity in many places, preventing national locust teams from undertaking the necessary survey and control operations. The DLIS plays an important role to guarantee a coordinated action amongst countries during periods of simultaneous outbreaks and the presence of swarms in more than one country.

In the past, early warning was hampered by difficulties in accessing remote desert locust habitats and sending timely information. Although tremendous efforts were made to systematically map desert locust infestations and analyse changes in population dynamics, this was often hampered by information that usually took weeks or months to arrive from the field. As a result, it was impossible to provide advice for decisions to be made to allow responding on time.

An effective early warning system for any transboundary pest relies on the transmission of geo-referenced data in real- or near real-time from the field to a centralised collection point where data can be analysed rapidly in order to provide timely and meaningful advice for decision makers. Timeliness becomes even more essential when dealing with a transboundary pest such as the desert locust that has the ability to rapidly increase in number and form swarms that can quickly move from one area to another.

3. INNOVATIONS

The present state of the global FAO desert locust monitoring and early warning system is a direct function of continuous innovation. It has evolved alongside advances in communication, computer and spatial technologies during the past several decades.

3.1. Advances in Communications, Geographic Information Systems and Transport

At the beginning of the last major desert locust plague in 1987, telex was the primary communications means for countries to transmit data and reports from the field to the DLIS in Rome. Similarly, FAO used telex to send advice, warnings and monthly bulletins and forecasts. By the end of the plague in 1989, telex had been replaced by facsimile. Although this was an improvement because additional information formats such as tables and photos could be transmitted, it remained somewhat limited since not everyone had a fax machine. These limitations were overcome with the introduction of email in the mid-1990s and its widespread use from 2000 onwards. This was further expanded as a greater number of individuals began using personal computers and mobile phones.

Until the mid-1990s, field data were plotted and analysed manually using paper maps, transparencies and coloured pencils. This tedious and labour-intensive system was replaced by custom-made geographic information systems (GIS) to allow rapid mapping and detailed spatial data analysis. The *Schistocerca Warning and Management System* (SWARMS) in the DLIS was one of the first uses of GIS for operational monitoring rather than production of one-time static maps (Healey et al. 1996). In 2000, FAO developed a standard custom GIS, *Reconnaissance and Management System of the Environment of Schistocerca* (RAMSES), for locust-affected countries to manage and analyse their own survey and control results with remote sensing imagery and historical data. In 2014, RAMSES was redesigned as open-source software with a spatial database.

The remote locust habitats in the desert have become progressively easier to access in most countries due to improvements in transport and infrastructure. Four-wheeled-drive vehicles have replaced camels that were used by locust survey and control teams in the past. A proliferation of tracks and roads in the desert associated with increased development allow access to a greater number of places in a shorter amount of time. In this way, survey teams can monitor a larger area with the same amount of resources.

Even though countries could send and receive information faster and ground teams could reach desert locust habitats easier, the field teams were unable to transmit high quality data in real-time from the location of the survey or the control operations, often in the middle of nowhere, to national locust centres and the DLIS. FAO addressed this issue by developing a custom tool, *eLocust*, that is a rugged handheld tablet for recording field observations and sending them by satellite in real-time to the national locust centre. This was introduced in 2006 and it revolutionized desert locust early warning overnight. Suddenly, weather, ecological, locust and control data were available within a few minutes from anywhere in the desert between West Africa and India. National locust directors now could know at any given time the exact location of every field team and results of the survey and control operations.

An upgraded version in 2015 of the custom tool, *eLocust3*, allowed users in the field to enter more survey and control data, navigate to potential areas of green vegetation without the need for internet connectivity, and take photos and videos of the situation.

3.2. *Satellite-based Remote Sensing*

Despite such advances in communications and transport, the desert continues to be huge and vast. It remains unattainable for any single locust-affected country to have sufficient resources to scour each and every hectare in search of desert locust. Therefore, all efforts must be made to somehow delimit the large areas that need to be searched and prioritise them to those that have the greatest potential of containing important locust populations. Satellite-based remotely sensed imagery is routinely used to help guide field teams to such places. This is undertaken in a systematic manner by first determining those regions or areas where rains may have fallen by using satellite-based rainfall estimates. While model-derived estimates may be more accurate in terms of rainfall quantity, satellite-based estimates are a better spatial indicator of rainfall (Dinku et al. 2010).

Once a region of possible rainfall has been identified, then multi-temporal and multi-spectral image analysis that exploits the mid-infrared, near-infrared and red wavelengths is applied to daily observations from NASA's 250 m resolution Moderate-Resolution Imaging Spectroradiometer (MODIS, Aqua and Terra) to determine if there was a response of annual vegetation to the rainfall (Pekel et al. 2011). This is the vegetation required by desert locust for food and shelter. An automatic processing chain combines the daily imagery into a ten-day dynamic vegetation greenness map that shows the three-month greenness history of each 250 m pixel in order to monitor the development of green vegetation in those areas that received rainfall or runoff (Fig. 6). These maps are used by locust-affected countries to position and prioritize surveys and by FAO DLIS for analysis and forecasting. They can also be used by teams that are equipped with *eLocust3* to help navigate to green areas in the field.

The internet has become the *de facto* means of delivering imagery and other data to analysts and decision-makers. The DLIS is constantly seeking new ways to improve the timely distribution of remote sensing products. While some delays may be due to satellite reception, there are other delays that are attributed to processing the data into map products. The latter is being actively addressed by utilizing Google Earth Engine technology whereby the user can process the image online in less than a few minutes by taking advantage of parallel and cloud computer technologies. This on-demand system is not only faster but may be more sustainable in the future compared to traditional processing and delivery chains.

3.3. *Unmanned Aerial Vehicles for Surveillance and Focused Control*

In the past few years, unmanned aerial vehicles (UAVs or drones) have become increasingly available for public use, and the technology is rapidly improving and expanding to new fields of use. The use of fixed- and rotary-wing drones could potentially improve desert locust monitoring, early warning and rapid response control, while reducing the costs of survey and control operations. Drones could supplement current tools utilized for monitoring in order to help guide ground teams to green vegetation and locust infestations. For example, the latest satellite imagery would be analysed to identify regions or areas within a country where ecological conditions may be favourable for locusts, specifically, where recent rains have occurred and where green vegetation may be present.

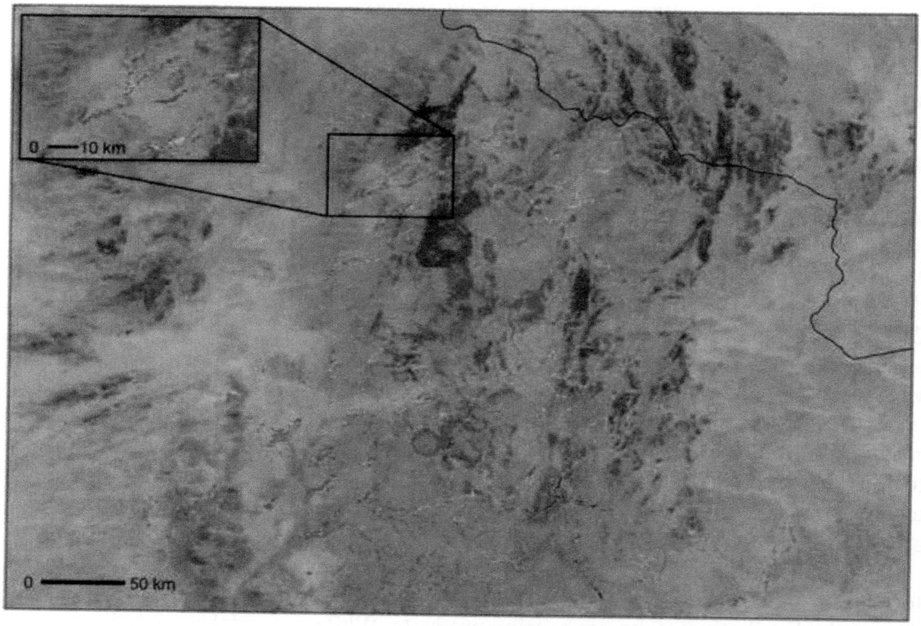

Figure 6. A 10-day dynamic NDVI map for 1–10 September 2017 indicating where annual vegetation in northern Mali has become green within the last 10 days (red), 20 days (orange) and 30 days (yellow) along dry river beds (wadis) in between the Adrar des Iforas hills. Green and darker colours indicate vegetation that has been green for several months, probably perennial vegetation such as trees, oases and forests. Each pixel is 250×250 m (0.0625 km^2) (source FAO DLIS).

This initial filtering will help to reduce the large and vast areas that need to be checked by ground teams. A team would then be deployed to this area to undertake surveys. The team would be equipped with a small, portable long-range fixed-wing drone. This drone should cover a radius of about 50-100 km. The team would programme the route itinerary of the drone and launch it. The drone would capture and process information along the route using optical and/or hyperspectral sensors to detect areas of green vegetation and moist soil, as well as the presence of any sizeable hopper or adult concentrations (groups, bands, and swarms). Once the drone returns to the survey team, the processed data would then be transferred to *eLocust3* and the team would use the results to go directly to the areas of interest or change direction if the results of the flight do not indicate the presence of favourable conditions for desert locust.

A ground team could also carry a small portable rotary drone with them during surveys to a specific location identified from above or an area that may contain vegetation or locusts. The drone would help the team get a better idea of the ecological conditions and the locust situation by taking low-level images of the area to identify the presence of green vegetation and locusts. If the team stops in an area with green vegetation or crops, the drone could look *in situ* for locusts and also determine the size of the potentially infested area. If the location is less precise, then the team could launch the rotary drone to look for any signs of green vegetation or favourable breeding areas within a 5 km radius. The drone could also be used to collect information from areas that are not accessible to the ground team due to topography or insecurity.

Lastly, a rotary drone could be used to undertake targeted control treatments of small infested areas and in areas that are difficult to access by ground teams (Benavente-Sánchez et al., this volume). This is highly desirable and advantageous because it makes control operations much safer and more precise. Field officers would avoid coming into contact with the chemicals as drones would be doing the spraying. Spot control would involve spraying only the specific locust infestation rather than treating the entire area, thus reducing pesticide usage and protecting the environment.

3.4. Forecasting the Time and Scale of Breeding and Early Warning Systems

The innovative tools mentioned up to this point are primarily utilized for directing national locust field teams in managing the current situation. Another set of technologies are exploited for locust prediction that involve the adaptation of cutting-edge methodologies to operational monitoring and forecasting.

Remote sensing imagery has traditionally been used for monitoring rainfall and vegetation. More recently, a new product has been developed to monitor ground soil moisture down to 10-15 cm below the surface. This moisture is critical for breeding to allow egg-laying and hatching, as well as the growth of annual vegetation necessary for locust feeding and shelter. Similar to other satellite-derived products, soil moisture maps are updated every ten days. This facilitates comparative analysis with rainfall vegetation imagery. The product will be another field tool to help delimit the large areas to be surveyed by guiding teams to specific places where breeding may be underway.

The DLIS utilizes seasonal predictions of monthly precipitation and temperature anomalies that are issued six months in advance and updated every month. These maps can help to forecast the timing and scale of locust breeding several months ahead, giving extra time for national locust programmes for planning and pre-positioning resources. Sub-seasonal predictions for up to four weeks in advance and updated twice a week can be used for short-term advice and warnings to assist in managing operations, especially during control campaigns. Both products are derived from several sophisticated weather models and may vary considerably from month to month. Hence, they can be less reliable at times and must be interpreted with caution.

At present, two models are used in desert locust early warning, an egg and hopper development model and a trajectory model. The former estimates the time required for egg and hopper development based on location. This is useful for forecasting the timing of breeding and planning field operations. The trajectory model is used to estimate the displacement of locust adults and swarms forward and backward in time. The model relies on meteorological data such as wind speed and direction, pressure and temperature provided by the European Centre for Medium-Range Weather Forecasts (ECMWF) every 12 hours for the desert locust recession area (Fig. 2.). This allows the forecaster to select the height or temperature level of flight and, by indicating the take-off or landing date and time, estimate where the adults or swarm came from or will go to.

4. SUCCESSES AND LESSONS LEARNED

The global desert locust early warning system operated by FAO is one example of successful area-wide pest management. Outbreaks, the precursors to upsurges and plagues, are now easier to detect and respond to on time due to improvements in monitoring and early warning (Fig. 4). This has resulted in a significant decrease in the duration, intensity and frequency of desert locust plagues. As mentioned previously, there were nine major plagues and ten major upsurges from 1860 to 2000, some of which lasted up to 14 years and affected 50 or more countries. Since 2000, there has been only one major upsurge and no plagues (Fig. 3).

This success can be attributed to a number of important factors. As the desert locust is such a well-known and old pest with a long history, countries are exceptionally aware of its importance and potential threat. This high visibility facilitates national interest at all levels and helps to engage the relevant stakeholders. Locust-affected countries understand the transboundary nature of the desert locust and, hence, willingly engage in joint monitoring and area-wide control activities as well as development projects to strengthen national capacities. Regardless of political relations, strong networks exist amongst countries that contribute to the regular and timely sharing of high-quality data and exchange of reliable information. Without this, an early warning system would not be possible.

The latest technologies are utilized and adapted in an innovative manner for the development of useful and well-focused tools that can be used by those involved in locust survey, control, reporting and forecasting. A bottom-up approach is used for these developments based on user requirements and feedback. The products, methodologies and tools are constantly updated to reflect changes in latest technologies and user needs. Substantial training and retraining are provided to each country using the train-the-trainers strategy, in which national master trainers are trained in the technology as well as in teaching and communications, so that they can provide essential training at all levels to national locust staff. Clear, concise, and imaginative training material and programmes are designed and updated to complement this process, supplemented by online videos for self-learning.

It is important to develop products and tools that are sustainable, can be maintained locally and used for as long as possible. Whenever possible, existing infrastructure, resources and expertise should be utilized and, if necessary, expanded rather than reinventing something that already exists. When developing an area-wide pest management programme, for example, that has a requirement for data collection in the field, it is far better to take advantage of the mobile phones that most farmers and extension agents already have, as well as the know-how to use them. This eliminates the need to develop, procure, distribute, and maintain a different device and to provide relevant training. Instead, it would only be necessary to develop an app for data collection and recording that works on the mobile phones. If data transmission is required, and there exists sufficient mobile telephone network coverage, then there is no need to rely on satellite communication, which is not only more expense but also requires additional hardware, maintenance, and training.

It is crucial to use technologies that are appropriate and relevant. For example, the Sterile Insect Technique (Dyck et al. 2021), requested by some countries, is not an appropriate control strategy for desert locust because of the size of its vast recession area and the fact that the released locusts would cause damage consuming crops. On the other hand, it has been extremely successful in contributing to the eradicating other pests such as the pink bollworm *Pectinophora gossypiella* (Saunders) in the south-western USA and northern Mexico (Staten and Walters, this volume), because it is not migratory, the released moths do not affect the cotton crop, and it was largely restricted to this host plant within this large geographic area.

Lastly, it is important to remain relevant by never ceasing to innovate in order to take advantage of the potential benefits and applications offered by the latest technologies. One example is the integration of drones and high-resolution satellite imagery into area-wide pest management programmes for monitoring and control. While this should not be a substitute for ground means, it can certainly supplement current efforts to monitor more effectively large areas and undertake safer and more precise control operations.

The innovative use of new technologies and tools in managing transboundary pests is unlikely to be entirely successful if it is not coupled with strong collaboration amongst beneficiary countries that is overseen by a centralised coordinating body. For example, 30 frontline countries affected by the desert locust share field data amongst themselves and with the DLIS. This allows DLIS to continual monitor the situation, forecast its developments, and provide timely and accurate early warning. A centralised coordinating structure also allows the implementation of standardized methodologies and training material in all countries as well as coordinated action between countries during outbreaks, upsurges, and plagues.

All of the lessons learned from the desert locust experience can be applied and adapted to other transboundary pests such as other locusts, as well as the red palm weevil and the fall armyworm, as a means to better manage these pests, protect crops, and enhance food security.

5. REFERENCES

Brader, L., H. Djibo, F. G. Faye, S. Ghaout, M. Lazar, P. N. Luzietoso, and M. A. Ould Babah. 2006. Towards a more effective response to desert locusts and their impacts on food security, livelihoods and poverty: Multilateral evaluation of the 2003-05 desert locust campaign. FAO, Rome, Italy.

Cressman, K. 2013. Climate change and locusts in the WANA region, pp. 131-143. *In* M. V. K Sivakumar, R. Lal, R. Selvaraju, and I. Hamdan (eds.), Climate change and food security in West Asia and North Africa. Springer, The Netherlands.

Dinku, T., P. Ceccato, K. Cressman, and S. J. Connor. 2010. Evaluating detection skills of satellite rainfall estimates over desert locust recession regions. Journal of Applied Meteorology and Climatology 49: 1322–1332.

Dyck, V. A., J. Hendrichs, and A. S. Robinson (eds.). 2021. Sterile Insect Technique – Principles and practice in Area-Wide Integrated Pest Management. Second Edition. CRC Press, Boca Raton, Florida, USA. 1200 pp.

Healey, R. G., S. G. Roberston, J. I. Magor, J. Pender, and K. Cressman. 1996. A GIS for desert locust forecasting and monitoring. International Journal of Geographical Information Systems 10: 117–136.

Magor, J. I., P. Ceccato, H. M. Dobson, J. Pender, and L. Ritchie. 2007. Preparedness to prevent desert locust plagues in the Central Region, an historical review. Desert Locust Technical Series, AGP/DL/TS/35. FAO, Rome, Italy. 184 pp.

Meynard, C. N., P. E. Gay, M. Lecoq, A. Foucart, C. Piou, and M. P. Chapuis. 2017. Climate-driven geographic distribution of the desert locust during recession periods: Subspecies' niche differentiation and relative risks under scenarios of climate change. Global Change Biology 23: 4739–4749.

Pedgley, D. 1981. Desert locust forecasting manual. Centre for Overseas Pest Research, Chatham, UK.

Pekel, J. F., P. Ceccato, C. Vancutsem, K. Cressman, E. Vanbogaert, and P. Defourny. 2011. Development and application of multi-temporal colorimetric transformation to monitor vegetation in the desert locust habitat. IEEE Journal of Selected Topics in Applied Earth Observations and Remote Sensing 4: 318–326.

Rosenberg, L. J., and P. J. A. Burt. 1999. Windborne displacements of desert locusts from Africa to the Caribbean and South America. Aerobiologia 15: 167–175.

Symmons, P., and K. Cressman. 2001. Desert locust guidelines. 1. Biology and behaviour. 2nd edition. FAO, Rome, Italy. 43 pp.

MATING DISRUPTION WITH PHEROMONES FOR CONTROL OF MOTH PESTS IN AREA-WIDE MANAGEMENT PROGRAMMES

R. T. CARDÉ

Department of Entomology, University of California, Riverside, California 92521, USA; carde@ucr.edu

SUMMARY

Mate finding in many insect groups is mediated by pheromones, particularly among moths where the male flies upwind along the pheromone plume to a calling female. The use of a formulated, synthetic copy of this message to disrupt this process dates to 1973 with the demonstration of protection of cotton from the pink bollworm *Pectinophora gossypiella* (Saunders). This method has been expanded to many moth pests, particularly tortricid moths that infest high-value pome fruits and grapes. Because most applications of mating disruptant are not directly lethal, an operational assumption is that efficacy is enhanced when the area under disruption is large enough to mitigate the effects of mated females immigrating into the protected area. Area-wide protocols thus should optimize efficacy of disruption and possibly permit effective control for some highly dispersive species such as heliothine moths that are unlikely to be controlled only by mating disruption in farms of even multiple hectare size. Successful area-wide programmes considered include those for the pink bollworm, codling moth *Cydia pomonella* (L.), oriental fruit moth *Grapholita molesta* (Busck), navel orangeworm *Amyelois transitella* (Walker), European grapevine moth *Lobesia botrana* (Denis & Schiffermüller), and gypsy moth *Lymantria dispar* (L.). Efficacious control, regardless of the magnitude of the crop area, typically requires an initially low population and therefore integration with other control measures.

Key Words: Lepidoptera, mate finding, pheromone plume, codling moth, pink bollworm, European grapevine moth, gypsy moth, oriental fruit moth, navel orangeworm, Sterile Insect Technique

1. INTRODUCTION

The French naturalist Jean-Henri Fabre, working with the giant peacock moth *Saturnia pyri* (Denis & Schiffermüller) in the 1870s, was one of the first to document that females release an alluring odour that draws in males from many metres downwind. In 1882 the New York state entomologist Joseph Albert Lintner, who was also amazed at the ability of a single female saturniid moth to attract many suitors, proposed that synthetic copies of these chemical messages might serve as a means of

J. Hendrichs, R. Pereira and M. J. B. Vreysen (eds.), Area-Wide Integrated Pest Management: Development and Field Application, pp. 779–794. CRC Press, Boca Raton, Florida, USA.

direct control of moth pests. The realisation of this method, however, had to await development of micro-analytical methods capable of identifying pheromones present in billionths of a gram per female (Roelofs 2016).

At present, female-produced pheromones have been chemically defined for several hundred moth species in many lineages (El-Sayed 2016; Löfstedt et al. 2016). Generally, these messages are comprised of blends of 2-3 components, although some moths apparently use a single component and a few species have blends as complex as 6 chemicals. Using these compounds to interrupt chemically triggered mate finding by males requires methods to gradually release these compounds into the crop to be protected. There are now many kinds of slow-release formulations that protect these usually labile chemicals until their release into the atmosphere.

Harry Shorey and his colleagues at the University of California, Riverside (Gaston et al. 1977) were the first to demonstrate in field trials that application of formulated synthetic pheromone could control a moth pest. In the 1970s, the pink bollworm, *Pectinophora gossypiella* (Saunders) (Gelechiidae), was a principal pest of cotton grown in southern California's Coachella Valley. Synthetic pheromone was formulated into open-ended, hollow plastic fibres and released at 10 µg per fibre per day; fibres deployed by hand at a density of approximately 1 per m^2 provided control by reduction in boll damage comparable to conventional practice in the insecticide-treated controls. This success provided an impetus for efforts to develop mating disruption for other key moth pests. Today this method has been commercialized for many of the most economically important moth pests (Evenden 2016) and we term this method of pest control 'mating disruption'. The successes and constraints of the mating disruption technique have been summarized by Cardé and Minks (1995), Cardé (2007), and Evenden (2016). Witzgall et al. (2010) reported that an estimated 770 000 ha yearly received mating disruption treatments for moths.

This review will explore the utility of this approach in large-scale applications where the goal of management is either to suppress a pest directly or in some cases, to eradicate a population over an entire region. To understand the prospects for successful disruption in area-wide programmes, it will be useful to consider how a broadcast application of a disruptant interferes with mate finding and how formulation type and a precise matching of the disruptant to the natural pheromone influences efficacy of disruption.

Most moth pheromones are straight-chain, C_{10} to C_{18} compounds with either an acetate, alcohol, or aldehyde moiety, and 1-3 positions of unsaturation (El-Sayed 2016; Löfstedt et al. 2016). These chemicals, and generally most other moth pheromones, are subject in the field to degradation by processes such as isomerization, oxidation and polymerization, and consequently one objective of all formulations is to provide protection against degradation while releasing the active ingredients into the atmosphere, ideally at a fairly constant rate until depletion. Most formulations use a plastic matrix (e.g. microcapsules, open-ended hollow fibres, sealed plastic tubes ('ropes'), PVC capsules, laminates, etc.) to retain the disruptant. Other formulations use dollops of a waxy emulsion into which the disruptant is mixed or the disruptant is released as an aerosol in timed puffs from widely spaced, sealed cans (Table 1).

Table 1. Examples of formulations used in mating disruption of moths (modified from Cardé 2007); the densities and methods of application, field longevities, and probable modes of action are general examples

Formulation*	Density ha^{-1}	Application method	Longevity	Mode of Action
Atomizer 'puffer'	<1 to several	hand-placed	season-long	sensory impairment; camouflage; competition
Sealed plastic tubes	hundreds	hand-placed	season-long	sensory impairment; camouflage; competition
Open-ended, hollow fibres, laminate 'flakes'	≈10 000	aerial	weeks	sensory impairment; camouflage; competition
Waxy dollops	100-10 000	hand-applied, aerial, speciali-sed equipment	weeks to season-long	sensory impairment; camouflage; competition
Microcapsules spray	millions	conventional	days to several weeks	sensory impairment; camouflage
'Attracticide' (e.g. fibres with insecticide in sticker)	≈1000	specialized equipment	weeks	direct toxicity; impairment of orientation; competition

* *When the formulation is not comprised of the full (attractive) pheromone blend or it contains an antagonist, it may not evoke competition (for an example see Section 3.3)*

In field applications these differing formulation types produce a large range of densities of disruptant sources and release rates and consequently in the atmospheric concentrations and spatial distributions they generate. Some formulations are point sources that match or exceed the attractiveness of a calling (pheromone-emitting) female. Other disruptant formulations are either intrinsically non-attractive because they are an incomplete copy of the pheromone blend (lacking components required for good attraction), or they contain synthetic by-products that are antagonistic and so either reduce or even eliminate attractiveness of the formulated product (Cardé 2007).

As formulations can be expected to be applied repeatedly to the same area and therefore themselves could be a source of pollution, a formulation's degradability over time should be a factor in its selection.

2. MECHANISMS OF MATING DISRUPTION

Disruptants can interfere with mate location in 3 principal ways:

1. *Competition*: Males may spend time and energy orienting to sources of formulation. The efficacy of this mechanism should be dependent on the ratio and comparative attractiveness of these sources to calling females (see Miller et al. 2006 for a theoretical consideration of this mechanism). A variant on this method adds insecticide to point sources of pheromone, an "attract and kill" strategy (Cork 2016).

2. *Sensory Impairment*: Exposure of males to disruptant causes either a diminution in responsiveness by raising the threshold for response, or by altering navigational ability, or exposure may simply eliminate responsiveness to the pheromone. Generally, such impairment can be due to adaptation of either sensory receptors or habituation, which is a central nervous system phenomenon, or both factors.

3. *Camouflage*: The pheromone plume from a calling female becomes imperceptible amongst the background of disruptant.

When the formulation mimics the natural pheromone, all of these mechanisms could contribute to efficacy and they could act additively or synergistically. There are other supplementary ('minor') mechanisms such as delay in mating that also could contribute to efficacy (see for examples: Cardé 2007 and Evenden 2016).

As will be documented, the susceptibility to disruption seems to vary with species, formulation type and application rate (Table 1), and whether the active ingredients match the full natural pheromone. Some of the behavioural traits that promote or interfere with successful mating disruption are listed in Tables 2 and 3.

Table 2. Male behavioural traits expected to confer higher or lower susceptibility to mating disruption matched with moths thought to possess these traits (see Cardé 2007 for further details on mechanisms)

Higher Susceptibility	Lower Susceptibility
Readily habituated [oriental fruit moth]	Difficulty to habituate [pink bollworm]
Slow to dishabituate [oriental fruit moth, codling moth]	Rapid dishabituation [pink bollworm]
Poor ability to navigate along plumes within a background of pheromone	Good ability to navigate along plumes within a background of pheromone
Rhythm of response imprecisely coordinated with female calling [pink bollworm]	Male and females mating rhythms well-coordinated
Males rely principally on pheromone for orientation and mating [pink bollworm]	Visual and tactile cues facilitate orientation and mating [oriental fruit moth]

Although formulation type, the match of its active ingredient to the natural pheromone, and its application rate all are quite important to disruption efficacy, ecological factors also are crucial: what is the pest's density at the start of application, does it need to be suppressed to a suitable level before application, and how migratory are mated and unmated females?

The first issue influences efficacy in two ways: the higher the density of moths, the closer they are to each other spatially and phenologically; at very low population densities, both factors should diminish the probability of females mating, essentially an Allee effect, in which a population at very low density could collapse because of a failure to find a mate (Liebhold and Bascompte 2003).

In most management protocols, it will be necessary to use additional control tactics such as cultural methods (e.g. sanitation) or insecticide application to reduce the population to a level amenable to mating disruption. For those moths that are notable adult nectar feeders (such as noctuids), a bait of adult feeding stimulants laced with insecticide provides another tactic to integrate with mating disruption (Gregg et al. 2018).

Moth movement is a second ecological factor to consider. If the species is highly migratory (such as many heliothine moths) and the area under management is near a source population, then the area under mating disruption may suffer crop damage due to the immigration of mated females. Possibly too, virgin females could leave an area under disruption, mate, and then reinvade the crop. The density and proximity of the outside population and the tendency of mated females to migrate should govern in part the programme's success. Migratory tendency and capability vary considerably among moth species and, even within a species, migration can be modulated by changes in crop suitability and season. Application of the principles of density and movement will be useful to interpreting several case studies.

Table 3. Female behavioural traits expected to confer higher or lower susceptibility to mating disruption matched with moths thought to possess these traits (see Cardé 2007 for further details on mechanisms)

Higher Susceptibility	Lower Susceptibility
Low pheromone emission rate [oriental fruit moth]	High pheromone emission rate
Calling from within plant canopy	Calling from top of canopy [pink bollworm]
Rhythm of response imprecisely -coordinated with female calling [pink bollworm]	Male and females mating rhythms well-coordinated
Low migratory tendency [North American gypsy moth]	Migratory [navel orangeworm]

Of particular relevance to the use of mating disruption in area-wide programmes will be assessing how size and isolation or distance of treated areas from outside infestation influences efficacy. Given the logistical difficulties and cost of evaluating whether area-wide programmes enhance management of a 'keystone' moth pest over conventional field-by-field or grower-by-grower practices, much of what we can conclude will be by inference rather than by referring to empirical tests.

The following case studies are examples of successful management using mating disruption, often combined with other techniques. Each example is considered briefly and specific documentation of the field evidence pertinent to area-wide use is cited. Mating disruption has been studied in all of the following pest species over many years.

3. CASE STUDIES

3.1. Pink Bollworm

The first field-scale use of mating disruption was against the pink bollworm, *Pectinophora gossypiella* (Gaston et al. 1977), and in 1978 it was the first pheromone to be registered by the USA Environmental Protection Agency (US-EPA) as an insecticide. The active ingredients are a 1:1 mix of the two components of the female's pheromone, (*Z,Z*)- and (*Z*,E)-7,11-hexadecadienyl acetates. There are now several formulation products in use (e.g. hollow fibres, sealed plastic 'ropes').

There is one notable field study of its use in area-wide management. In Arizona in the 1970s, control of this pest relied largely on frequent night-time aerial sprays of insecticide aimed at the adult moth. This approach was becoming untenable, because of increasing insecticide resistance and costs. The Parker Valley is isolated from other cotton-growing areas by surrounding desert (Fig. 1), thereby eliminating influx of mated females from outside the treated area and thus constituting an ideal location to evaluate an area-wide programme. Growers in the Parker Valley of Arizona, in collaboration with the Animal and Plant Health Inspection Service (APHIS) arm of the United States Department of Agriculture (USDA), devised a multi-year programme to attempt to directly manage pink bollworm with mating disruption and some use of insecticide in all cotton fields (detailed in Staten et al. 1996).

Figure 1. Aerial view of the Parker Valley agricultural area along the Colorado River in Arizona. The growing area was well isolated from other cotton production by surrounding desert (courtesy Google Maps).

Parker Valley's 11 250 ha of cotton comprised a significant size and, prior to this programme, intensive application of insecticide provided unacceptable levels of control (> 25% boll infestation in the year prior to the programme's start). Mating disruptant applications began in 1990 and the programme ended in 1993. Throughout, approximately one third of the cotton received a hand application of rope dispensers and the rest of the applications were by air and used fibres or fibres plus an insecticide overspray.

Over the project's four-year span, the use of insecticide dropped from half of the aerial applications to less than a tenth. There were approximately 650 fields in the programme and 45 of these were selected at random for assessing boll damage. From 2000 to 3000 cotton bolls were cracked weekly to check infestation. The highest recorded weekly larval count per boll was 21% in 1990, dropping to 7% in 1991, 2% in 1992 and zero in 1993, when over 20 000 bolls were examined season-long and not a single pink bollworm larva was found. Clearly this technology offered a new paradigm for management of a pest that was becoming very difficult to control with conventional insecticides. One remarkable feature of this demonstration was that it succeeded even though initial infestation levels were high and this species is considered rather migratory (Stern 1979). These sorts of programmes lack replication and direct experimental controls, but the substantial infestation levels across the Parker Valley in the years prior to the programme serve to verify efficacy. Despite the success of this programme, mating disruption for pink bollworm is now in limited use worldwide (≈ 50 000 ha annually according to Witzgall et al. (2010), or ≈3% of cotton worldwide), mainly due to the availability of *Bacillus thuringiensis (Bt)*-cotton, which has supplanted mating disruption and conventional insecticides for control of this and other moths.

Mating disruption also played a central role in the successful area-wide effort to eradicate pink bollworm from the south-western USA (Texas to California) and adjoining areas in northern Mexico. This audacious undertaking was reviewed by Lance et al. (2016). The programme began in 1994 and involved at various stages and regions the application of mating disruptant, release of sterile moths, some application of insecticides and, beginning 1997-1998, planting of *Bt*-cotton. Since 2008 no larvae have been detected anywhere in the entire cotton-growing regions of the south-western USA and northern Mexico (Staten and Walters, this volume). The success of this programme (which continues with monitoring using pheromone traps) is difficult to attribute to any single control technique and it is most probable that the combination of methods was needed to suppress reproduction. Its success was likely also guaranteed by the integration of these methods on an area-wide basis over all cotton growing areas in the region and the fact that this moth is a cotton specialist.

3.2. Codling Moth

Cydia pomonella (L.) (Tortricidae), infests many kinds of pome fruits and walnuts, and it is particularly troublesome in apple orchards. Most work has characterised its pheromone as a single component (*E,E*)-8,10, dodecadienol. Probably more field tests on mating disruption have been conducted on this species than any other moth. As of 2010, about 210 000 ha were treated yearly with mating disruptant for this pest worldwide (Witzgall et al. 2010). A general conclusion is that this species is difficult to disrupt in the sense that for success initial populations need to be low and isolated from the influx of mated females from untreated areas; in any case, growers will not tolerate more than a small percentage of fruit damage.

Witzgall et al. (2008) reviewed these studies and provided two general conclusions: 1) prior to treating with disruptant, initial populations must be low and therefore a remedial application of insecticide or some other method could be necessary to achieve the desired population level; and 2) the larger the area under pheromone management and the greater its isolation from outside sources of infestation, the greater the level of protection and economic benefit per hectare (McGhee et al. 2011). Given that apples typically are produced in orchards of varying size and proximity to other orchards and non-commercial hosts (a 'patchwork' pattern), coordination of a common strategy across many growers in a region is a logistical and sociological challenge (Brunner et al. 2002).

No definitive field studies have established the minimum plot size for maximum achievable efficacy (i.e. the point where no further reduction in mating can be expected) and certainly efficacy will vary somewhat with application rate and formulation type (e.g. puffers vs. hand-applied, point-source), how thoroughly the formulation disperses disruptant throughout the canopy, canopy structure, initial population density, climate, and topography (see Witzgall et al. 2008). The principal lessons to be drawn from numerous field studies with codling moth are that for mating disruption to succeed, initial populations must be very low (if need be requiring remedial treatment with an insecticide) and sufficiently isolated from sources of mated females.

3.3. Oriental Fruit Moth

Grapholita molesta (Busck) (Tortricidae) is a troublesome pest on peaches, pears, nectarines, plums and apples; its first generation also can inflict significant damage to shoot tips. Its pheromone is a 3-component blend: (*Z*)-8- and (*E*)-8-dodecenyl acetates (95:5) with 3-10% of (Z)-8-dodecenol added. This moth seems especially susceptible to mating disruption, with a wide variety of formulation types (microdispersibles to aerosol puffers) of the pheromone providing excellent fruit protection (Cardé 2007; Evenden 2016). As of 2010, about 50 000 ha were treated yearly with mating disruptant for this moth (Witzgall et al. 2010).

In Australia, Il'ichev et al. (2002) set out to establish that an area-wide approach would provide enhanced control over an orchard-by-orchard management. In the project's first year an area of over 800 ha including 18 orchards were treated with hand-applied dispensers at 1000 per ha; in the second year over 1000 ha comprising 40 orchards were treated with 500 dispensers per ha.

In the first year, growers decreased insecticide application by half and during the second year most growers did not apply any insecticides for this pest. Areas previously identified as 'hot spots' of infestation also were brought under control. Il'ichev and colleagues concluded that area-wide management with pheromone was highly effective for oriental fruit moth, but cautioned that infestations could linger at the edges of disruptant-treated blocks if the permeation of air-borne disruptant there was incomplete due to wind patterns or if neighbouring orchards harboured oriental fruit moths (Il'ichev et al. 2002).

3.4. *European Grapevine Moth*

The pheromone of *Lobesia botrana* (Denis & Schiffermüller) (Tortricidae) is usually considered to be a single compound, (*Z*,*E*)-7-9-dodecadienyl acetate. Historically this moth has been a grape pest in Southern Europe, North Africa, Anatolia and the Caucasus; it was recently introduced into South America (Taret et al., this volume). As of 2010, Witzgall et al. indicated that 100 000 ha in the European Union, Argentina and Chile were treated yearly with mating disruptant for this moth.

In 2009, it was discovered in northern California with a nexus in Napa County. Following the discovery of this pest, its distribution was mapped in 2010 with a network of traps throughout the grape growing regions of California. Over 100 000 males were trapped, mainly in Napa County, but with significant populations in adjoining Sonoma County, and isolated pockets elsewhere, these being attributed to movement of material such as stakes from the original area of infestation rather than natural dispersal. A multi-pronged eradication programme with support from growers and organizational direction from federal, state, county and extension agencies was initiated in 2011 (for detailed reviews see Lance et al. 2016, and Simmons et al., this volume).

In 2011 and 2012, mating disruptant was applied to ≈160 km^2 in Napa County, generally in the core-infested area, and to ≈16 km^2 in Sonoma County, with smaller application areas in 2013 and 2014. Since 2015, extensive pheromone trapping has not caught any males, and the moth has been declared eradicated. It is fair to note that conventional insecticides were used in many but not all vineyards. Therefore, the area-wide demise of this pest cannot be attributed solely to mating disruption, but it was viewed as a key component of successful eradication (Lance et al. 2016).

3.5. *Navel Orangeworm Moth*

Amyelois transitella (Walker) (Pyralidae) is a key pest of almonds, pistachios and walnuts in California. This moth can be controlled using mating disruption as part of an integrated pest management programme. Aerosol dispensers, deployed at one per 0.8 ha, emit (*Z*,*Z*)-11-13-hexadecadienal, one of the 4-components of its pheromone, but which alone is at best slightly attractive compared to the full blend. Mating disruptant generally provides control levels similar to insecticide-alone regimes and the combination of disruption and insecticide generally resulted in better pest management and even lower levels of damage (Higbee and Burks 2008).

Based on flight mill studies, virgin navel orangeworm females appear capable of migrating several km (Sappington and Burks 2014), although whether mated females would naturally migrate such distances when they are within a host orchard remains to be determined. Evidence that such moderate-distance migration is an important consideration stems from the field study of Higbee and Seigel (2009), who found that navel orangeworm damage in almond blocks was correlated with distance from pistachios (with traditonally higher populations): some spillover was detectable even with a 5-km separation. The migratory capacity of navel orangeworms demonstrated so far suggests that crop protection should be enhanced as the area under treament is scaled up.

In California, almond and pistachios orchards often extend in contiguous plantings over many km and therefore movement of mated females from untreated orchards to adjoining blocks treated with disruptant is probable. This pest is an ideal candidate for an area-wide management programme using mating disruption. Careful monitoring in a 4-year demonstration project in 1050 ha of almonds, showed that 3 insecticide sprays yielding 5-10% damage could be replaced by mating disruption, producing only 0.5% damage (Higbee 2010). Registration and use of its complete pheromone blend might provide higher levels of nut protection than the currently used aldehyde (Higbee et al. 2017). Another rationale for area-wide application is that several percent nut damage occured in 16-ha mating disruptant plots (Higbee et al. 2017); presumably, increasing greatly the area under disruption would enhace protection by limiting influx of mated females. Overall, in 2016 mating distuption was used in California on over 80 000 ha of almonds and pistachios (Higbee personal communication).

3.6. *Gypsy Moth*

The potentially expanding range of the gypsy moth, *Lymantria dispar* (L.) (Erebidae), in North America and introductions of its Asian form are monitored by a network of more than 200 000 pheromone traps (Lance et al. 2016). Traps are baited with the pheromone, (*7R,8S*)-cis-7,8-epoxy-2-methyloctadecane, called '(+)-disparlure', but, as optically pure disparlure is quite expensive, the (+)- enantiomer is only used in survey lures. Mating disruption uses the inexpensive racemate as a disruptant. The racemate is about one tenth as attractive because its (-)-enantiomer is an antagonist of attraction.

Aerially applied racemic disparlure is being used to retard the advance of gypsy moths to the south and west of low-level populations along the edge of its advancing USA range. Mating disruption is deployed in a 100-km-wide band from North Carolina to Minnesota in a programme called "Slow the Spread" (reviewed by Lance et al. 2016; Liebhold et al., this volume). In this zone, trap capture is low (between 0-1 per trap yearly throughout most of the band, rising to several or more per trap near the edge adjoining the established population). Prior studies have established efficacy of aerially applied formulations (e.g. Thorpe et al. 2006), verified by substantial reductions in capture in pheromone-baited traps in subsequent years and mating of tethered, sentinel females.

From 2000 to the present, an average of about 180 000 ha have been treated yearly (Lance et al. 2016). The migration biology of the North American gypsy moth differs from most moths in that the female is flightless and natural population spread occurs through wind-borne 'ballooning' of first instar larvae. (Anthropogenic transport, particularly of egg masses, remains an important regulatory and practical issue.) Thus, female movement does not influence the success of mating disruption in a given area. Larval movement (ballooning 1st instars), however, will influence the size of the area wherein mating disruption will be effective.

4. IS EFFICACY INFLUENCED BY THE SIZE OF THE AREA TO BE TREATED?

A reoccurring theme in all studies of mating disruption is the assumption that the larger the plot size, the more efficacious mating disruption, because large size mitigates the influx of mated females from outside the treated area. Experimental evaluation of formulations and active ingredients for efficacy can be accomplished, at least with some moths, with replicated small plots (ca. 25 by 15 m) (Roelofs and Novak 1981) and with small field cages (ca. 8 m^3) (Koch et al. 2009) by testing for disruption of attraction to pheromone lures and to females. It is generally accepted that proof of crop protection, however, needs to be assessed in replicated plots that are large enough to minimize or eliminate the immigration of mated females from untreated areas, whereas disruption of attraction to a lure or female in a trap are insufficient metrics. There have been several notable studies using replicated large plots [Brockerhoff et al. 2012 with the light brown apple moth, *Epiphyas postvittana* (Walker); Onufrieva et al. 2018 with the gypsy moth; and Higbee and Burks 2008, Higbee et al. 2017 with the navel orangeworm moth]. With such large experimental plots, some useful information on the infiltration of mated females may be found by monitoring infestation or disruption levels from a plot's edge to its centre.

As the distance that mated females migrate varies with species and also can be influenced by habitat, host availability and season, there is no set answer as to how large a plot needs to be to guard against such immigration from the periphery. What is clear is that efficacy should be enhanced with enlarged treatment areas and becomes optimal if all areas that harbour the population of the pest are treated. Crucial to understanding these interactions will be characterisation of the moth's migratory capacity and the conditions in the field that trigger this behaviour. There are not, however, replicated field experiments that have defined optimal plot size directly, and given the sizes required, these are unlikely to be undertaken.

5. INTEGRATION OF MATING DISRUPTION WITH STERILE INSECT RELEASE

The Sterile Insect Technique or SIT is another environment-friendly method for insect suppression and in some cases eradication (Dyck et al. 2021). The SIT efficacy is largely contingent on the ratio of sterile to native insects (the overflooding ratio) and the competitiveness of sterile insects. Because mating disruption seeks to diminish mating success, it seems counterintuitive to combine these two approaches, as application of disruptant would diminish the probability that sterile insects would mate with native insects. However, because both techniques reduce the number of fertile females, in combination they may enhance population suppression over either method used alone. Furthermore, where use of either mating disruption or SIT fails to provide sufficient control (e.g. because of gaps in coverage), then the other method could be employed. There are many examples of incomplete population suppression using mating disruption, even with prior application of conventional insecticides to suppress the population to a low level (e.g. Witzgall et al. 2008; Evenden 2016; Higbee et al. 2017).

The SIT generally aims to achieve a particular overflooding ratio (e.g. a 10 to 1 ratio of sterile insects to native ones) and one challenge is determining the density of the native population. One method for evaluating a SIT programme is to compare the ratio of internally-marked sterile to native males captured in pheromone traps (Vreysen 2021). A difficulty, however, in combining mating disruption and SIT is that application of disruptants makes the monitoring of population levels and the overflooding ratio with pheromone traps problematic. Of course, absence or very low numbers of trap catch in pheromone-baited traps often is used to verify that male orientation to females is being suppressed and therefore that mating disruption is protecting the crop.

For evaluation of population density in some species (e.g. pink bollworm and codling moth), males can be trapped in mating disruption areas by using a lure with high rate of pheromone emission, that is, 10-fold higher than would normally be used in population monitoring (Doane and Brooks 1981; Witzgall et al. 2008), and so these high-dose traps could be used for evaluation of the SIT component. In a few cases, there are kairomone lures that are effective in sampling adult moths in mating disruption plots (e.g. Knight 2010; Burks 2017).

The combination of these two methods has of course been limited to those few moth species for which the SIT package and mating disruption have been fully developed and tested for field efficacy. With pink bollworm, aerial release of sterile insects was long used to suppress this pest in the San Joaquin Valley of California, with daily releases of up to 18 million moths. As considered earlier in Section 3.1, mating disruption, SIT and other measures were combined in an ambitious and successful area-wide eradication programme in the south-western USA and northern Mexico (Lance et el. 2016; Staten and Walters, this volume). As both mating disruption, SIT and in the later years *Bt*-cotton were combined, it is impossible to parse out the precise contribution of each tactic and especially the extent to which either mating disruption or SIT contribute to suppression. It also is possible that sequential application of these techniques (SIT preceding mating disruption or vice versa) would yield higher levels of eventual suppression than simultaneous application.

The SIT is the mainstay in the management of the codling moth in the Okanagan Valley of British Columbia, Canada (Nelson et al., this volume). Over the past 20 years, this programme has produced a dramatic reduction in insecticide use in apples (some growers have not applied an insecticide for codling moth in 15 years). Fruit infestation is extremely low (<0.2%) in >90% of the orchards. Like successful programmes using mating disruption with codling moth (Section 3.2), a key factor is early intervention with insecticides to reduce populations to the low levels where SIT can provide a final reduction and continued control (Bloem et al. 2007).

An effort in the same region to combine SIT with mating disruption, Judd and Gardiner (2005) established in organic orchards that these two measures coupled with removal of overwintering larvae using tree bands (cultural control) in concert suppressed codling moths to non-detectable levels within several years. Judd and Gardiner (2005) proposed that mating disruption and SIT together was as effective as SIT and some initial insecticide.

The area-wide integration of mating disruption with SIT application was also effective south of the Canadian-USA border (Calkins et al. 2000). This finding as well as the successful eradication efforts with pink bollworm both point to a favourable interaction of mating disruption with SIT, but precisely how these two seemingly competitive approaches either complement or interfere with each other remains to be determined.

Whether combining mating disruption and SIT enhances crop protection over either method alone thus remains an outstanding question. Before widespread implementation of these combined tactics, it would be valuable for future programmes to compare in replicated plots the levels of crop protection provided by each technique alone and in combination. Perhaps modelling how these two processes interact would provide insight into the value of combining SIT and mating disruption.

It also might be feasible to develop through either conventional selection or genetic engineering strains for SIT application that have males that are 'resistant' to mating disruption (e.g. by reducing their tendency to habituate response to pheromone, possibly by altering their biogenic amines levels—see Linn and Roelofs 1986). Males from such a strain would be more apt than their wild counterparts to find females amidst a background of mating disruptant. How such males perform in a non-disruptant environment would be important to understand, but presumably such a trait would render modified moths less competitive than their wild counterparts, because this is a trait that would not be favoured under the constraints of traditional natural selection in the field. Presumably, as well, interbreeding with wild moths would not produce offspring.

In field-cage observations using laboratory strains of codling moth, a single encounter with a point source of pheromone can render a moth unresponsive to pheromone for 24 hours (Stelinski et al. 2006). In contrast, the strain of pink bollworm used in the SIT programme remains pheromone responsive ('resistant' to habituation) after a 24-hour exposure to pheromone (Cardé et al.1998). Although we do not know how released sterile pink bollworm males behave in disruptant plots, these observations suggest that they should search for females and therefore the field efficacy of mating disruptants in this strain should be at least partially contingent on a mechanism of competition between pheromone sources from deployed formulations and those from wild and also sterile females.

Cardé et al. (1998), interpreting the mechanisms of mating disruption in the pink bollworm, assumed that wild and sterile pink bollworm males were behaviourally equivalent. An open question, however, is whether continuous exposure to high pheromone levels during many generations under mass-rearing has altered its pheromone response characteristics.

An obvious criterion for evaluating sterile moths is how competitive they are in finding mates. If the SIT is to be integrated with mating disruption, then another criterion may be the ability of sterile males to find wild females amidst a background of disruptant and whether this is a trait that can be enhanced by selection.

6. CONCLUSIONS

The principle and its practical application that formulated pheromone can control many kinds of moth pests are well established. The susceptibility of a given moth pest to mating disruption, however, varies with characteristics of its communication system (e.g. rate of pheromone release from the female, male sensitivity to sensory interference) and formulation rate of release, its match to the natural pheromone, and type (e.g. widely-spaced aerosol dispensers, point sources mimicking females, non-attractive disruptants). The acceptance of this method for management of many species of moths signifies both its efficacy and cost-effectiveness.

Incorporation of mating disruption into area-wide integrated management programmes would seem to be a straightforward process, simply requiring scale-up of established methods, with the benefit that efficacy is expected to be enhanced, given that a large area of application (hundreds of ha) mitigates the influx of mated females from the periphery. One concern in widespread use of mating disruption is that the goal, suppression of mating, also suppresses capture in pheromone-baited traps, which can be a useful tool for population monitoring.

Replicated field demonstrations that the larger the disruptant-treated area, the greater the efficacy of mating disruption generally, however, will remain an elusive and perhaps intractable experimental goal. We will continue to rely on replicated, small-scale trials to establish that formulations are sufficiently efficacious to warrant area-wide application. Ultimately, the success of any area-wide approach depends not just on effective tools for insect control, but cost-effectiveness, environment-friendliness, social acceptance, and management protocols for implementation (Hendrichs et al. 2007).

7. ACKNOWLEDGEMENTS

I thank Charles Burk, Brad Higbee and Greg Simmons for providing valuable comments and suggestions.

8. REFERENCES

Bloem, S., A. McCluskey, R. Fugger, S. Arthur, S. Wood, and J. Carpenter. 2007. Suppression of the codling moth *Cydia pomonella* in British Columbia, Canada using an area-wide integrated approach with an SIT component, pp. 591–601. *In* M. J. B. Vreysen, A. S. Robinson, and J. Hendrichs (eds.), Area-wide control of insect pests: From research to field implementation. Springer, Dordrecht, The Netherlands.

Brockerhoff, E. G., D. M. Suckling, M. Kimberley, B. Richardson, G. Coker, S. Gous, J. L. Kerr, D. M. Cowan, D. R. Lance, T. Strand, and A. Zhang. 2012. Aerial application of pheromones for mating disruption of an invasive moth as a potential eradication tool. PLoS One 7(8)343767.

Brunner, J., S. Welter, C. Calkins, R. Hilton, E. Beers, J. Dunley, T. Unruh, A. Knight, R. Steenwyk, and P. Van. 2002. Mating disruption of codling moth: A perspective from the Western United States, IOBC/WPRS Bulletin 25(9): 11–20.

Burks, C. S. 2017. Combination phenyl propionate/pheromone traps for monitoring navel orangeworm (Lepidoptera: Pyralidae) in almonds in the vicinity of mating disruption. Journal of Entomology 110: 438–446.

Calkins, C. O., A. L. Knight, G. Richardson, and K. A. Bloem. 2000. Area-wide population suppression of codling moth, pp. 215-219. *In* K. H. Tan (ed.), Area-wide control of fruit flies and other insect pests. University Sains Malaysia. Pulau Penang, Malaysia.

Cardé, R. T. 2007. Using pheromones to disrupt mating of moth pests, pp. 122–169. *In* M. Kogan and P. Jepson (eds.), Perspectives in ecological theory and Integrated Pest Management, Cambridge University Press. Cambridge, UK.

Cardé, R. T., and A. K. Minks. 1995. Control of moth pests by mating disruption: Successes and constraints. Annual Review of Entomology 40: 559–585.

Cardé, R. T., R. T. Staten, and A. Mafra-Neto. 1998. Behaviour of pink bollworm males near high-dose, point sources of pheromone in field wind tunnels: Insights into mechanisms of mating disruption. Entomologia Expermentalis et Applicata 89: 35–46.

Cork, A. 2016. Pheromone as management tools. Mass-trapping and lure-and-kill, pp. 349-363. *In* J. D. Allison and R. T. Cardé (eds.), Pheromone communication in moths: Evolution, behavior and application. University of California Press, Berkeley, California, USA.

Doane, C. C., and T. W. Brooks. 1981. Research and development of pheromones for insect control with emphasis on the pink bollworm, pp. 285–303. *In* E. R. Mitchell (ed.), Management of insect pests with semiochemicals. Concepts and practice, pp. 229–242. Plenum, New York, NY, USA.

Dyck, V. A., J. Hendrichs, and A. S. Robinson (eds.). 2021. Sterile Insect Technique – Principles and practice in Area-Wide Integrated Pest Management. Second Edition. CRC Press, Boca Raton, Florida, USA. 1200 pp.

El-Sayed, A. M. 2016. The pherobase: Database of pheromones and semiochemicals.

Evenden, M. 2016. Mating disruption of moth pests in Integrated Pest Management. A mechanist approach, pp. 365-393. *In* J. D. Allison and R. T. Cardé (eds.), Pheromone communication in moths: Evolution, behavior and application. University of California Press, Berkeley, California, USA.

Gaston, L. K., R. S. Kaae, H. H. Shorey, and D. Sellers. 1977. Controlling pink bollworm by disrupting sex pheromone communication between adult moths. Science 196: 904–905.

Gregg, P. C., A. P. Del Socorro, and P. J. Landolt. 2018. Advances in attack-and kill for agricultural pests: beyond pheromones. Annual Review of Entomology 63: 465–470.

Hendrichs, J., P. Kenmore, A. S. Robinson, and M. J. B. Vreysen. 2007. Area-wide Integrated Pest Management (AW-IPM): Principles, practice and prospects. pp. 3–33. *In* M. J. B. Vreysen, A. S. Robinson, and J. Hendrichs (eds.), Area-wide control of insect pests: From research to field implementation. Springer, Dordrecht, The Netherlands.

Higbee, B. S. 2010. NOW Areawide Projects in Kern County – Progress Report on the Lost Hills and Santa Fe navel orangeworm areawide projects - July 2010, unpublished report.

Higbee, B. S., and C. S. Burks. 2008. Effects of mating disruption treatments on navel orangeworm (Lepidoptera: Pyralidae) sexual communication and damage in almonds and pistachios. Journal of Economic Entomology 101: 1633–1642.

Higbee, B. S., and J. P. Seigel. 2009. Navel orangeworm sanitation standards could reduce almond damage. California Agriculture 63: 24-28.

Higbee, B. S., C. S. Burks, and R. T. Cardé. 2017. Mating disruption of the navel orangeworm (Lepidoptera: Pyralidae) using widely spaced, aerosol dispensers: Is the pheromone blend the most efficacious disruptant? Journal of Economic Entomology 110: 22056–2061.

Il'ichev, A. L., L. Gut, D. G. Williams, M. S. Hossain, and P. H. Jerie. 2002. Area-wide approach for improved control of the Oriental fruit moth *Grapholita molesta* (Busck) (Lepidoptera: Tortricidae). General and Applied Entomology 31: 7–15.

Judd, G. J. R., and M. G. T. Gardiner. 2005. Toward eradication of codling moth in British Columba by complimentary actions of mating disruption, tree banding and Sterile Insect Technique: Five-year study in organic orchards. Crop Protection 24: 718–733.

Knight, A. 2010. Improved monitoring of female codling moth (Lepidoptera: Tortricidae) with pear ester plus acetic acid in sex pheromone-treated orchards. Environmental Entomology 39: 1283–1290.

Koch, U. T., E Doye, K. Schumann, and U. Andrick. 2009. CIRCE – an addition to the toolbox for assessment / improvement of mating disruption. Pheromones and other semiochemicals. IOBC/WPRS Bulletin 41: 17–24.

Lance, D. R., D. S. Leonard, V. C. Mastro, and M. L. Waters. 2016. Mating disruption as a suppression tactic in programs targeting regulated lepidopteran pests in US. Journal of Chemical Ecology 42: 590–605.

Liebhold, A. M. and J. Bascompte. 2003. The Allee effect, stochastic dynamics and eradication of alien species. Ecological Letters 6: 133–140.

Linn, C. E., and W. L. Roelofs. 1986. Modulatory effects of octopamine and serotonin on male sensitivity and periodicity of response to sex pheromone in the cabbage looper moth, *Trichoplusia ni*. Archives of Insect Biochemistry and Physiology 3: 161–171.

Löfstedt, C., N. Walberg, and J. G. Millar. 2016. Evolutionary patterns of evolutionary diversity in Lepidoptera, pp. 43–78. *In* J. D. Allison and R. T. Cardé (eds.), Pheromone communication in moths: Evolution, behavior and application. University of California Press, Berkeley, California, USA.

Miller, J. R., L. J. Gut, F. M. de Lame, and L. L. Stelinski. 2006. Differentiation of competitive vs. non-competitive mechanisms mediating disruption of moth sexual communication by point sources of sex pheromone (part I): Theory. Journal of Chemical Ecology 32: 2089–2014.

McGhee, P. S., D. L. Epstein, and L. J. Gut. 2011. Quantifying the benefits of areawide pheromone mating disruption programs that target codling moth (Lepidoptera: Tortricidae). American Entomologist 57(2): 94–100.

Onufrieva, K., S., A. D. Hickman, D. S. Leonard, and P. C. Tobin. 2018. Relationship between efficacy of mating disruption and gypsy moth density. International Journal of Pest Management 65: 44-52.

Roelofs, W. L. 2016. Reminisces of the early days, pp. 3–9. *In* J. D. Allison and R. T. Cardé (eds.), Pheromone communication in moths: Evolution, behavior and application. University of California Press, Berkeley, California, USA.

Roelofs, W. L., and M. A. Novak. 1981. Small-plot disruption for screening potential disruptants, pp. 229–242. *In* E. R. Mitchell (ed.), Management of insect pests with semiochemicals. Concepts and practice. Plenum, New York, NY, USA.

Sappington, T. W., and C. S. Burks 2014. Patterns of flight behavior and capacity of unmated navel orangeworm (Lepidoptera: Pyralidae) adults related to age, gender, and wing size. Environmental Entomology 43: 696–707.

Staten, R. T., O. El-Lissy, and L. Antilla. 1996. Successful area-wide program to control pink bollworm by mating disruption, pp. 383–396. *In* R. T. Cardé and A. K. Minks (eds.), Insect pheromone research. New directions. Chapman Hall, New York, NY, USA.

Stern, V. M. 1979. Long and short range dispersal of the pink bollworm, *Pectinophora gossypiella* over southern California. Environmental Entomology 8: 524–527.

Stelinski, L. L., L. J. Gut, and J. R. Miller. 2006. Orientational behaviors and EAG responses of male codling moth after exposure to synthetic sex pheromone from various dispensers. Journal of Chemical Ecology 32: 1527–1538.

Thorpe, K., R. Reardon, K. Tcheslavskaia, D. Leonard, and V. Mastro. 2006. A review of the use of mating disruption to manage gypsy moth, *Lymantria dispar* (L.). United States Department of Agriculture, Forest Service, Forest Health Technology Enterprise Team (FHTET), Morgantown, West Virginia, USA. 86 pp.

Vreysen, M. J. B. 2021. Monitoring sterile and wild insects in area-wide integrated pest management programmes, pp. 485–528. *In* V. A. Dyck, J. Hendrichs, and A. S. Robinson (eds.), Sterile Insect Technique – Principles and practice in Area-Wide Integrated Pest Management. Second Edition. CRC Press, Boca Raton, Florida, USA.

Witzgall, P., L. Stelinski, L. Gut, and D. Thompson. 2008. Codling moth management and chemical ecology. Annual Review of Entomology 53: 503–522.

Witzgall, P., P. Kirsch, and A. Cork. 2010. Sex pheromones and their impact on pest management. Journal of Chemical Ecology 36: 80–100.

CRISPR-BASED GENE DRIVES FOR COMBATTING MALARIA: NEED FOR AN EARLY STAGE TECHNOLOGY ASSESSMENT

W. LIEBERT

Institute of Safety and Risk Sciences (ISR),
University of Natural Resources and Life Sciences (BOKU), Vienna, Austria;
liebert@boku.ac.at

SUMMARY

The potential power of CRISPR-based gene drives makes it necessary to engage in science and technology assessment already in early stages of research and development. In order to argue for efforts to address this urgent need, gene drives to combat malaria-transmitting mosquitoes are discussed using the concept of prospective technology assessment. First, development risks are described, followed by considerations about anticipatable risks and irreversible consequences, as well as unforeseeable effects and uncertainties. Afterwards, fundamental problems in connection with the development of gene drives against malaria mosquitoes are raised. Opportunities for shaping technology are briefly discussed, before alternatives, in particular the World Health Organization's elimination strategy, are considered. Finally, several normative questions are put forward.

Key Words: Anticipation, gene drive, gene flow, malaria, malaria vector, mutagenic chain reaction, normative questions, prospective technology assessment, risk, selfish genetic element

1. THE MALARIA CHALLENGE

Malaria is caused by *Plasmodium* parasites which infected female mosquitoes can transmit to humans. Most of the malaria control strategies have been aimed at managing the vector of the malaria pathogen, i.e. the relevant mosquitoes of the genus *Anopheles*. However, not only the main malaria vector *Anopheles gambiae* Giles can transmit malaria, but in sum more than 30 *Anopheles* species (with highly varying regional distribution), i.e. roughly one percent of all known mosquito species can do so (NAS 2016).

J. Hendrichs, R. Pereira and M. J. B. Vreysen (eds.), Area-Wide Integrated Pest Management: Development and Field Application, pp. 795–808. CRC Press, Boca Raton, Florida, USA.

The *Plasmodium* cycle first takes place in an infected mosquito (in particular inside the salivary glands) and after transmission in the liver and blood of a human. Some of the pathogens reach a sexual stage so that the whole malaria-parasite cycle (mosquito–human–mosquito) can start again after a bite of another mosquito which gets infected (White et al. 2014). Malaria is still endemic in many regions of the world. The World Health Organization (WHO) estimates that currently about 200 million humans are infected annually. In 2015, death cases due to malaria were in the range of 438 000, primarily in Africa (WHO 2015a).

The fight against malaria is regarded as one of the grand global challenges and was appropriately included in the UN Agenda for Sustainable Development in 2015. Control of the mosquito vector is mainly hampered by the development of resistance against the commonly used insecticides. Therefore, novel biotechnological strategies are under development and are coming into operation (for an overview cf. e.g. Alphey 2014). One such novel approach is a new tool of genetic engineering, i.e. the mutagenic chain reaction, mostly called *gene drive,* which comprises a variety of different technologies.

2. GENE DRIVE AND PROSPECTIVE TECHNOLOGY ASSESSMENT

Sexual reproduction provides, in principle, a 50 % probability for the propagation of the parental genotypes to the progeny in accordance with Mendelian heredity rules. This evolutionary mechanism can be circumvented if specific parts of the genome can be transformed from heterozygote to homozygote, which has been observed in nature in some exceptional cases. Then, certain genotypes can be inherited with quite a high probability (significantly above 50% and up to 100%) even if the fitness of the offspring is adversely affected.

In 2003, the British researcher Austin Burt proposed to genetically engineer such gene drives and to use them against malaria-transmitting mosquitoes (Burt 2003). The key idea was "to drive novel genes or mutations into wild populations" (Bull 2015). Theoretically, a rapid and targeted manipulation of entire mosquito populations would be possible while circumventing Mendelian rules. First attempts were made in the laboratory (e.g. Windbichler et al. 2011), but a research boom started only in 2013 with the discovery of the novel gene scissor CRISPR-Cas9 (Clustered Regularly Interspaced Short Palindromic Repeats–CRISPR-associated 9). The first CRISPR-based manipulation of the fruit fly *Drosophila melanogaster* Meigen in the laboratory was named as a mutagenic chain reaction by the researchers involved (Gantz and Bier 2015). Meanwhile, gene drives have been technically realized in the laboratory in yeast (DiCarlo et al. 2015) and in the malaria-transmitting mosquitoes *Anopheles stephensi* Liston (Gantz et al. 2015) and *An. gambiae* (Hammond et al. 2016).

Considering the malaria case in particular, gene drive approaches can be divided into *suppression drives* and *modification/manipulation drives*. Suppression drives aim at dramatically reducing or eradicating malaria-transmitting insect populations, regionally or globally. Modification/manipulation drives strive to genetically modify or manipulate mosquitoes in a way that malaria infection of humans is reduced or disabled.

Gene drives could become extremely powerful tools for humans to make dramatic intentional or unintentional changes in populations and entire ecosystems. Therefore, it is necessary to engage in appropriate procedures of science and technology assessment in good time before such technologies are mature. An appropriate concept to guide such efforts is the approach of *prospective technology assessment* (ProTA) (Liebert and Schmidt 2010, 2015), which includes:

- analysis of scientific-technological development at an early stage, anticipating what might be relevant for science-based mid-term assessments and for (participatory) discourse inside and outside science
- assessment of intentions/visions, potentials, risks and unintended consequences, realistic potentials versus unrealistic visions and promises, uncertainties (and ignorance)
- characterization of the type of technology involved
- analysis/assessment of opportunities for shaping science and technology and of technical or socio-technical alternatives, so that desired potentials can be exploited
- reflection on normative issues, values and interests involved.

ProTA is more than an accompanying exercise of socio-economic research on societal or industrial acceptability, but it is partly impossible without analysing the scientific-technological core itself. ProTA differs also from outlining a development pathway for gene drive mosquitoes from research to use in the wild, including safe and efficient field-testing, regulations and post-implementation monitoring (James et al. 2018). In the following, above mentioned aspects of ProTA will be elaborated on. In doing so, sometimes more questions will be raised than answered.

3. INTENTIONS, POTENTIAL AND DEVELOPMENT RISKS

In general, research aiming to realize the potential of CRISPR-Cas9-based gene drives to reduce or eradicate malaria seems to be justified. The suffering of malaria victims worldwide, and in several African regions in particular, cannot be ignored by the international community. However, malaria is not only a naturally occurring phenomenon and a serious plague for humankind, but also an unpleasant result of human evolution that has provoked a long-lasting human and humanitarian struggle against it.

It is not only a question of scientific understanding and development of technical tools. Social, political and economic factors are also of great importance. Social organization and behaviour, access to modern healthcare, a functioning healthcare system in the regions concerned, preventive and curative measures and availability of suitable means are of utmost importance when trying to manage malaria. What we currently observe is an unjust societal and global divide with respect to the malaria burden, with prevention and medication affordable for the rich, but not for the poor. It is important to recognize that the remarkable social stratification of the malaria burden has almost nothing to do with new technological approaches to fight against malaria.

Nevertheless, new biotechnical tools might provide helpful contributions to better manage malaria. Production and release of sterilized or *Wolbachia*-infected mosquitoes are already being validated in pilot projects to locally manage mosquito populations (Bourtzis et al. 2016). At first glance, CRISPR-based gene drives manipulating the genetic code of mosquitoes, at least theoretically, seem to be most promising in terms of efficacy. The release of one single modified organism could, in the medium-term, result in the modification of all individuals of a specific mosquito population (Noble et al. 2017). However, it turned out that in all of the above-mentioned laboratory experiments the chain reactions are being reduced after a few generations. The engineered gene drives effectively became unstable (Gantz et al. 2015; Champer et al. 2016, 2017), which is not the case in naturally occurring selfish genes. Indeed, in the drive systems described so far, resistance alleles were detected after a few generations, highlighting a fundamental instability of engineered homing-based systems[1]. Thus, the potential of gene drives, which could be exploited in principle, has been made visible, but not more. Instead, development risks have emerged questioning whether the objective of a functional gene drive can be achieved. The observed slowdown of gene drives is partly due to the fact that, after a cleavage of targeted DNA due to CRISPR-Cas9, competing repair mechanisms are coming into play. With relevant probability non-homologous end joining (NHEJ) is sometimes faster and more prevalent than homology-directed repair (HDR), which only leads to the engineered homing endonuclease with self-replicating characteristics passing on the new genetic information from generation to generation. NHEJ can produce resistance against the drive and eventually stop the mutagenic chain reaction (Unckless et al. 2017; Champer et al. 2017).

It is not clear whether an evolutionary stable homing-based gene drive can be engineered (Bull 2015). Limited efficiency of gene drives could be regarded as something positive, in particular if one is afraid of a virtually unlimited efficacy of such engineered systems. However, it is possible to precisely predict the non-linear behaviour of gene drives in the wild? How exactly can mathematical modelling reproduce the actual complex dynamics? Can a persistent behaviour of a gene drive construct, which was originally predicted as eventually self-limiting, be excluded? It is being proposed to overcome resistance phenomena by combining several drives and targeting different DNA sites simultaneously. But that has not been demonstrated so far and it could turn out that complexity and instability would again be increased instead.

It is also questioned whether homologous recombination, which is essential for gene drives, is error-free (Guirouilh-Barbat et al. 2014). There are also doubts that gene drives in wild mosquito populations will be feasible.[2] Although CRISPR-Cas9 is much easier, cheaper, faster, and more precise in usage than older tools of targeted genetic engineering, e.g. ZINC finger or TALEN (NAS 2016, p. 24-31; Häcker and Schetelig, this volume), off-target effects due to an incomplete specificity of constructed guide-RNAs cannot be excluded, which are associated with off-target DNA cleavages. These can contribute to an evolutionary instability of gene drives. Furthermore, cross-fertilization has been reported among various mosquito species (Tripet et al. 2011). Thus, one insect species could take the role of transmitting diseases from another.

At the same time, it is also well known that *Anopheles* mosquitoes possess a large genetic diversity and evolutionary adaptability. In particular, that is the case for *An. gambiae*, the primary carrier of *Plasmodium* in sub-Saharan Africa (The *Anopheles gambiae* 1000 Genomes Consortium 2017). Hence, the genetic variability of targeted mosquito species is a hurdle for an efficient gene drive strategy and a single gene drive might not be effective so that the development and use of many gene drive constructs might be needed to cover the range of genetic variants of *Anopheles*. Hence, functioning of mutagenic chain reactions or gene drives in malaria-transmitting mosquito populations is questioned from various sides. Development risks are numerous, but scientists eager to take on the challenges could ultimately overcome the obstacles. Therefore, one should be careful in stating that gene drives are not feasible; rather, one should anticipate that they can become a reality in the near future, and therefore there is an urgent need to study the risks and consequences involved.

4. RISKS, CONSEQUENCES, AND UNCERTAINTIES

What risks would be involved using gene drives to suppress insect pests in the wild? Unleashing a highly potent mutagenic chain reaction in insects like malaria-transmitting mosquitoes could eventually spread across national borders. The population extinction programme could potentially work globally. This begs the following pertinent questions: who has the legitimate right to decide the initiation of such a mission and who will regulate it? Who will take part in the decision-making? What to do if approval is obtained in one country but not in the neighbouring country? Fundamentally, this would concern all living beings, and humankind as a whole should be asked. Could that be organized?

What, if modified mosquito genes mutate and further evolve, creating possibly unwanted variants? Could that be induced by off-target effects due to the fact that the specificity of guide RNAs targeting cleavage points in the DNA is not 100%? Polymorphism as a genetic variation inside a population could be induced causing unclear consequences (Araki et al. 2014). Recently, a scientific debate has started over the importance of off-target effects. The creation of unintended single-nucleotide variants – not only small insertions and deletions (indels) – would fuel concerns. That can only be detected by whole-genome sequencing after gene-editing with CRISPR-Cas9. Unfortunately, that is seldom done by the researchers so far.[3]

Is it possible that after implementing the genetically engineered population suppression or modification against *An. gambiae*, *Plasmodium falciparum* Welch and/or other malaria parasites find a way to change their currently favoured or most important host mosquito? The ecological niche left open could be filled again. We already know about 30 mosquito species that can transmit malaria, but what would be the next move? Would this require efforts to eradicate more and more mosquito populations or species by engineering and unleashing more and more gene drives in nature? Moreover, the reduction of transmitting rates could lead to increased selection pressure on the pathogen itself, which could in turn develop an increased virulence (David et al. 2013).

As cross-fertilization between different mosquito species has been observed (Fang 2010), an interspecific gene flow could affect other non-target mosquitoes or insects. This entails the potential of a gene drive-based eradication programme to jump over to other species (David et al. 2013), in particular if target sequences are equal. An effective elimination programme, originally targeting one species, could then result in significant consequences for ecosystems.

If targeted mosquito species play important roles in ecosystems or if other species are non-intentionally also affected, unwanted cascading effects in ecosystems are possible. Is it scientifically irrefutably clarified that malaria mosquitoes only have a damaging function in nature by infecting humans and other hosts? Do they have instead also important beneficial or indispensable roles in the food chain or in pollination processes? For mosquitoes in general, there are examples showing such vital purposes, like in the Camargue, in Nordic Arctic, or in aquatic systems (Fang 2010), but mainly we have to admit ignorance about side effects of eradicating malaria-transmitting *Anopheles* mosquitoes (David et al. 2013). However, a comparison in this respect with other methods of malaria vector control must not turn out negatively for gene drives.

Questions related to modification/manipulation drives indirectly aimed at the *Plasmodium* parasite or rendering the parasite harmless for humans are quite similar, even though the strategy would be different, i.e. not suppression or eradication. Off-target effects could also have negative consequences and unwanted genetically engineered variants could emerge and reproduce. Accidentally, also the pathogenicity of the parasite could be increased.

It is well known that parasites evolve quickly and in unforeseen ways in relation to their hosts (Wijayawardena et al. 2013. A genetically engineered intervention into this interplay could result in unexpected or even damaging consequences. Could a mutating parasite like *Plasmodium* escape the grasp of newly engineered characteristics of the host mosquito? Would that just cheat the gene drive or probably even facilitate malaria transmission pathways or render the infections′ impact on humans more likely or worse? Would that maybe provoke the engineering of another gene drive attempt and after the next one, another one, and so on?

Some scientists involved in gene drive research and development are warning that modified/manipulated drives are highly invasive in wild populations, even if their efficiency is low or resistance against the drive occurs (Noble et al. 2018). The question, among others, is what consequences will the spread of newly introduced alleles have? Is it possible to clarify this in advance? Is that also relevant for suppression drives, even if they are more or less limited in their potential to drive to fixation?

Thus, several risks and consequences of gene drives can be anticipated in all clarity. However, there are also plenty of uncertainties and unknowns due to the dynamic complexity of natural systems. Some important dynamical features are probably unknowable. A report of the US National Academy of Sciences (NAS 2016) has confirmed that many gaps exist in our knowledge on off-target effects of gene drives inside targeted organisms and on non-intended effects on other species and the environment.

5. FUNDAMENTAL PROBLEMS AND QUESTIONS RELATED TO GENE DRIVES

If it is possible to engineer gene drives working effectively in a natural environment, then any single release of one gene drive modified organism could have irreversible consequences. The characteristics of the modified organism or the ramifications of the induced mutagenic chain reaction could also turn out as "wrong" or become detrimental, what is probably unknown prior to its release. Therefore, there is no tolerable limited release, as long as unwanted impacts cannot be completely excluded. The dual-use potential of gene drives seems to be obvious. Once it is sufficiently clarified that gene drives can be reliably engineered, small competent groups could (with only a limited amount of money) covertly pursue a strategy to engineer and use gene drives to the harm of others. One example could be an attacker (state or non-state actor) who decides to manipulate an important organism that is beneficial for agriculture and that can only be found regionally or locally, with the aim to harm an adversary. Also, weapon-like effects of engineered gene drives are imaginable.[4] Studying such possibilities in detail is therefore necessary prior to major investments in the development of gene drive technology.

With mutagenic chain reactions becoming a reality, the already significant depth of intervention by humans into natural processes would be massively increased. It has been proposed to differentiate genetic methods of mosquito control into more or less *self-limiting* approaches, and in increasingly *self-sustaining* invasive tools, which can or should persist in nature (Alphey 2014). Gene drives, in principle, correspond to the latter category and could become extremely powerful technologies, but with a high risk as unintended changes could become fully irreversible.

The precautionary principle, which is a guideline at least in the European context, would require an in-depth study of the risks of this new technology prior to any development in the laboratory and even more so prior to any consideration about application in nature. A number of serious risks and probable hazardous events due to gene drives have been already identified (cf. e.g. Hayes et al. 2018, p. S143). Those must be scrutinized and "Scientists must remain mindful that great power entails equally great responsibility, and take precautions accordingly" (Min et al. 2018, p. S54).

Unlike genetically modified organisms (GMOs), animals manipulated by gene drives are engineered with selfish genetic elements deliberately designed to spread in the environment and to operate autonomously in nature. Could these constructed or affected organisms evolve in a way not anticipated or even not anticipatable by their designers? Must we realize that a new form of technology is emerging?

Jan Schmidt (2015, 2016) has coined the term *late-modern technology* ("nach-moderne Technik") to identify a remarkable paradigm-shift. This new type of technology is based on the concept of self-organization and is linked to instabilities which can be triggered in non-linear dynamical systems. In contrast to the classic-modern type of technology, which is related to concepts of stability, linearization, predictability and controllability of functions and outcomes, in late-modern technology evolutionary, self-organizational, and non-linear features are exploited, which could lead in principle to intrinsic limits with respect to predictability and controllability.

Organisms manipulated by an engineered gene drive are not only intentionally self-replicating but also capable of further evolutionary changes. This technology has to go through instabilities and has to trigger instabilities. It is a type of technology which acts nature-like in nature. However, this kind of built-in dynamic is provoking limitations with regard to the possibility of a stable construction of the techno-biological system and also with regard to subsequent monitoring and control of the technology. A remarkable difference to classical modern and rational technology concepts is emerging.

Engineering gene drives to fight malaria could turn out as a harbinger of much more. Could gene drives against other mosquitoes, or against other insects in general or rodents transmitting infectious diseases be next? Gene drives could also be engineered against any so-called pest animals. Various tephritid fruit flies or plague locusts (Acrididae) could be a target. Also, non-native invasive species (plants as well as animals) could be attacked, eventually eradicated. All that is already mentioned in the NAS (2016) report. The research, development and use of gene drives against malaria-transmitting mosquitoes, which can be argued for based on convincing humanitarian objectives, could be the door-opener for a new dimension of human campaigns in and probably against nature or its biodiversity.

What starts with the fight against malaria could end up with nature being totally in the hands of humans ("Natur unter Menschenhand"), nature under complete management of humans, as the biologist and influential science manager Hubert Markl propagated already 20 years ago (Markl 1998, p. 147). Gene drives will dramatically change the way humans can interact with nature. Humans will have a tool then, to deliberately steer evolution – with less time for the ecosystems to adapt to the induced turbulence, in contrast to changes due to current tools and naturally occurring mutations. Humans can decide what species they like or dislike, which one has the right to survive in a given form, etc.

One of the young shooting-stars on the scientific scene (emerging from the group around George B. Church), Kevin Esvelt, already named his new working group at the Massachusetts Institute of Technology "Sculpting Evolution". Despite the far-reaching visions, however, we are faced with huge unknowns of the complex, non-linear interactions in genetic transfer, living cells and organisms, populations, ecosystems, and sensitive global life connections.

It appears that an intrinsic logic might underlie gene drive approaches against malaria-transmitting mosquitoes, as well as the other above-mentioned objectives: it is conceivable that after a first gene drive attempt more and more gene drives have to be engineered and released to correct or improve what wasn´t achieved in the first step. As one gene drive will not suffice, the pressure for more will "naturally" be generated to bring about human control – which could turn out as being unachievable in the end. Do we have to expect a chain reaction of mutagenic chain reactions? A pathway towards life on earth totally in the hand of humans, prone to human errors and ignorance?

6. SHAPING OPPORTUNITIES FOR GENE DRIVES?

Is there a chance to shape gene drives in a way that serious risks associated with them could be eliminated? Engineering "reversal drives" has been proposed that can undo results of a drive if necessary. But how to deal with off-target effects of both the original and the reversal drives? One is stuck in a principle problem. Therefore, even for several scientists involved in gene drive research, engineering reversal drives is not a convincing concept because a potential fundamental irreversibility of gene drives is admitted, and a reversal drive could only be a second drive which again could be ill-targeted.

Recently, self-limiting CRISPR-based drives have been proposed (Noble et al. 2019). But, so-called *daisy-chain drives* are just theoretical models and do not reflect on the real complex dynamics in real life. Furthermore, this concept seems to be more a test-bed approach, where drives at first (hopefully) are locally and timely restricted, but later a full-fleshed global release is intended. Furthermore, my impression is that it is more likely that all attempts to improve gene drives by additional features, proposed so far, increase the complexity and non-linearity of the engineered constructs and thus it is more likely that they would increase concerns.

7. ALTERNATIVES TO GENE DRIVES

If one hesitates to believe in gene drives as the new silver bullet against malaria, one has to refer to the alternatives. Several ideas for technical alternatives are currently discussed, researched, validated and are partly in use (Alphey 2014; Bourtzis et al. 2016; Fasulo et al., this volume; Häcker and Schetelig, this volume). One example is the infection of *Aedes* or *Anopheles* mosquitoes with *Wolbachia* bacteria (Incompatible Insect Technique or IIT), which are maternally inherited and affect reproduction capabilities, leading to strategies to suppress or replace mosquito populations (Bourtzis et al. 2016). Another example of population suppression is the release of genetically modified mosquitoes passing on dominant lethal factors to their offspring (RIDL) (Alphey 2014; but see Evans et al. (2019).

Another approach, which has been in use for over 50 years on all continents against major agricultural pest insects, is the Sterile Insect Technique (SIT), where large numbers of the target insects are mass-reared, sterilized by radiation, and then released in order to negatively influence the reproduction of insect populations (Dyck et al. 2021).

Multiple releases of sufficient quantities of manipulated mosquitoes are necessary in all these cases (when population suppression rather than replacement is the objective), to obtain the wanted results. Such approaches can also be debated and must be carefully assessed. Obeying the precautionary principle is also mandatory for these technologies. But, in principle, one could say, that unpleasant risks and other ramifications of these alternative technologies might be less severe than in the case of gene drives. In the case of the SIT, there are decades of track record of successful large-scale application against many pest insects. As the released insects are sterile, they cannot become established, and thus there is no irreversibility.[5]

8. GLOBAL PROGRAMME TO ELIMINATE MALARIA

Of particular importance with regard to alternatives is the global programme to eliminate malaria. The United Nations and the WHO have declared the intent of reaching this goal by 2030 (WHO 2015b) using the following classical methods:

- vector control, in particular by distribution of long-lasting insecticidal bed nets; indoor residual spraying; mosquito screening, surveillance and monitoring; and education of citizens in endemic regions
- prevention, in particular preventive intermittent chemoprophylactic treatment of pregnant women and children under 5 years, especially in many African regions
- better access and use of diagnostic testing and appropriate medical treatment (e.g. artemisinin-based combination therapy).

In this century, important successes in the fight against malaria have already been recorded: a reduction of malaria incidence by 37% and of mortality rate by 60% (WHO 2015a). Therefore, the hopes are high that the malaria elimination strategy can be accomplished, if sufficient funding can be raised (much more than USD 2.5 billion annually, which was the global financing for malaria control in 2014, will be needed) and if a concerted effort and enduring engagement of all stakeholders can be achieved over the next coming years. In this struggle, unglamorous tasks such as improvement or set up a minimally functional health care system are critical, including durable and affordable access to diagnostics and pharmaceuticals, as well as educating and empowering communities so that they can reduce the risk themselves. A success with this strategy would also have further positive ramifications not just with respect to malaria.

It is clear that the global programme to eliminate malaria involves much more than just high technological means such as gene drives. The malaria challenge is not only a problem solvable with scientific-technological approaches, but also social, political and economic factors have to be addressed. Not only is vector control crucial, but in the end, control of *P. falciparum* and other parasites is of utmost importance. The parasite cycle has to be interrupted in a sustainable manner, which is much more an issue of health care, access to suitable simple measures, societal development, etc. than using sophisticated novel technologies. Without a somewhat stable (minimal) health care system in the endemic regions of concern, elimination of malaria is impossible. Also, the socio-economic conditions of the disease's origin, besides the natural-scientific causes for malaria, need to be scrutinized in order to find appropriate means to support the transition process towards malaria elimination. Population growth of mosquitoes has also to do with socio-cultural or techno-economic change, for example the rapid increase of plastic containers used for food distribution or scrap tyres provide some mosquitoes with ideal breeding grounds.

One should also remark that in several countries past (successful) elimination campaigns also had harmful side effects on the environment by massive use of DDT; furthermore, first resistances of mosquitoes against insecticides developed at that time. Improper use of malaria drugs, which mostly have to be taken in suitable combinations, has also led to resistance of the malaria parasite.

9. NORMATIVE QUESTIONS

Many issues and values involved in the necessary debate on gene drives are mentioned in the NAS (2016) study "Gene drives on the horizon". Many scientists state: "risk has to be balanced against benefit", which can lead to a purely utilitarian position. Clearly, weighing positive and negative consequences is a relevant part, but ethical reflection should heed all ramifications and also fundamental problematics.

Derived from Hans Jonas´ principle of *responsibility* (Jonas 1979), which aims at achieving a "conservative" preservation of our lifeworld, the precautionary principle focuses mainly on objective reasoning in respect to serious risks, notwithstanding that benefits might also be possible. On the other hand, an unfolding principle, which strives for "*alliance technology*" serving humankind and being concurrently in harmony with nature (Bloch 1959), requests a positive and socially just developmental progression of humankind by using new technology that is bound to its alliance with nature, in harmony with nature. Both principles which may be contradictory at a first glance can be used as normative orientation[6], already in the process of research and development, when striving for the eradication of malaria (probably using gene drives).

The perceived role of humans in nature is highly relevant: what position has humankind within nature, still being a central part of our common lifeworld which we share with other living beings? One position claims a human role as the manager of all life on earth, man as "master of nature" (Descartes), the other sees humankind embedded into nature and as partner of life on earth, or as Albert Schweitzer has formulated it in his principle *respect for life* (Schweitzer 1966): *I am life that wants to live in the middle of life that wants to live.*

How should members of the scientific community and of our societies behave in between these diametrically opposite positions? Schweitzer´s position seems to be incompatible with an approach where humankind feels entitled to steer evolution on earth. Who is entitled to change nature in a way that it could irreversibly affect all life on earth? A single researcher, the scientific community, a competent national agency, a nation?

In any case, the whole fabric of risks, uncertainties and ignorance, and the possible dramatic consequences of mutagenic chain reactions, will (and must) have a massive influence on ethical discourse debating responsible conduct in gene drive R&D directed towards practical use, and other technology-based malaria elimination strategies. In the end, it has to be assessed which procedure for malaria elimination seems to be promising, associated with low risk, and is globally, societally and ethically acceptable. I deem it as obvious that the WHO strategy cannot be replaced by anything else. But maybe additional new measures or tools could be helpful.

All potentials, risks, uncertainties, ramifications of gene drive R&D have to be made transparent within science and to the broader public as early as possible. When a gene drive technology against malaria-transmitting mosquitoes seems to be mature – and that could soon be the intuition of several researchers and funding organizations (like the Gates Foundation) involved – it would probably be too late to be withheld. Then it will be no longer possible to stop its use in a region with serious malaria burden.

The hope of promoters is that some advantages will predominate notwithstanding those arguments describing possible or anticipatable negative side effects or long-term consequences, serious uncertainties and ramifications that have been put forward. All the concerns will then be covered-up by promises and hopes.

In conclusion, using the example of combatting malaria, this contribution tried to substantiate why prospective technology assessment in the field of gene drive research and development is urgent and what issues should be assessed in order to provide input for decision making inside and outside science. The pertinent questions to be answered are going beyond tailored disciplinary research fields of scientific specialty. As scientists involved in gene drive research have put it:

> *"Determining whether, when, and how to develop gene drive interventions responsibly will be a defining challenge of our time"* (Min et al. 2018, p. S40).

10. NOTES

(1) More precisely: homing endonuclease genes (HEGs) or homing-based gene drives.
(2) The entomologist Flaminia Catteruccia is cited with: "...you can have the fanciest technology on earth, the perfect gene drive, but if your laboratory mosquitoes can´t mate with wild mosquitoes, then it´s not going to work at all" (Shaw 2016).
(3) Schaefer et al. (2017) had reported in Nature an unexpected high number of unintended single-nucleotide variants after genome editing with CRISPR-Cas9 in a laboratory population of mice. After several criticisms, the Nature editors have withdrawn the paper because it could not be shown beyond any doubt that the use of the gene scissor was the cause of these effects. However, they admitted that Schaefer et al. "did not examine only predicted sites", as in many other studies, "but looked at the entire genome" and that the "work of Schaefer et al. highlights limitations in the current literature that should be considered" (Nature Methods 15 (4): 229-230).
(4) At end of 2017, it has been revealed that the US Defense Advanced Research Projects Agency (DARPA) is investing 100 million USD in gene drive technology (http://genedrivefiles.synbiowatch.org/).
(5) Therefore, sterile insects are accepted as beneficial organisms by the International Plant Protection Convention (FAO 2005), to which 183 countries are signatories.
(6) More about these viewpoints can be found in Liebert and Schmidt (2015).

11. REFERENCES

Alphey, L. 2014. Genetic control of mosquitoes. Annual Review of Entomology 59: 2015–224.

Araki, M., K. Nojima, and T. Ishii. 2014. Caution required for handling genome editing technology. Trends in Biotechnology 52 (5): 234–237.

Bourtzis, K., R. S. Lees, J. Hendrichs, and M. J. B. Vreysen. 2016. More than one rabbit out of the hat: Radiation, transgenic and symbiont-based approaches for sustainable management of mosquito and tsetse fly populations. Acta Tropica 157: 115–130.

Bloch, E. 1959. Prinzip Hoffnung. Suhrkamp-Verlag, Frankfurt, Germany. English translation: The principle of hope. Cambridge, Massachusetts, USA, 1986.

Bull, J. 2015. Evolutionary decay and the prospects for long-term disease intervention using engineered insect vectors. Evolution, Medicine, and Public Health: 152–166.

Burt, A. 2003. Site-specific selfish genes as tools for the control and genetic engineering of natural populations. Proceedings of the Royal Society B: Biological Sciences 270: 921–928.

Champer, J., A. Buchman, and O. Akbari. 2016. Cheating evolution: Engineering gene drives to manipulate the fate of wild populations. Nature 17: 146-159.

Champer, J., R. Reeves, S. Y. Oh, C. Liu, J. Liu, A. G. Clark, and P. W. Messer. 2017. Novel CRISPR/Cas9 gene drive constructs reveal insights into mechanisms of resistance allele formation and drive efficiency in genetically diverse populations. PLoS Genetics 13 (7): e1006796.

David, A., J. Kaser, A. Morey, A. Roth, and D. Andow. 2013. Release of genetically engineered insects: A framework to identify potential ecological effects. Ecology and Evolution 3: 4000–4015.

DiCarlo, J., A. Chavez, S. Dietz, K. Esvelt, and G. Church. 2015. Safeguarding CRISPR-Cas9 gene drives in yeast. Nature Biotechnology 33: 1250–1255.

Dyck, V. A., J. Hendrichs, A. S. Robinson (eds.). 2021. Sterile Insect Technique – Principles and practice in Area-Wide Integrated Pest Management. Second Edition. CRC Press, Boca Raton, Florida, USA. 1200.

Evans, B. R., P. Kotsakiozi, A. L. Costa-da-Silva, R. S. Ioshino, L. Garziera, M. C. Pedrosa, A. Malavasi, J. F. Virginio, M. L. Capurro, and J. R. Powell. 2019. Transgenic *Aedes aegypti* mosquitoes transfer genes into a natural population. Scientific Reports 9: 13047.

Fang, J. 2010. A world without mosquitoes. Nature 466: 432–434.

(FAO) Food and Agriculture Organization of the United Nations. 2005. International Standard for Phytosanitary Measures (ISPM) No. 3, Guidelines for the export, shipment, import and release of biological control agents and other beneficial organisms. FAO, Rome, Italy.

Gantz, V., and E. Bier. 2015. The mutagenic chain reaction: A method for converting heterozygous to homozygous mutations. Science 348: 442–444.

Gantz, V., N. Jasinskiene, O. Tatarenkova, A. Fazekas, V. Macias, E. Bier, and A. James. 2015. Highly efficient Cas9-mediated gene drive for population modification of the malaria vector mosquito *Anopheles stephensi*. Proceedings of the National Academy of Sciences 112: 6736–6743.

Guirouilh-Barbat, J., S. Lambert, P. Bertrand, and B. Lopez. 2014. Is homologous recombination really an error-free process? Frontiers in Genetics 5: 175.

Hammond, A., R. Galizi, K. Kyrou, A. Simoni, C. Siniscalchi, D. Katsanos, M. Gribble, D. Baker, E. Marois, S. Russell, A. Burt, N. Windbichler, A. Crisanti, and T. Nolan. 2016. A CRISPR-Cas9 gene drive system targeting female reproduction in the malaria mosquito vector *Anopheles gambiae*. Nature Biotechnology 34: 78–83.

Hayes, K., G. Hosack, G. Dana, S. Foster, J. Ford, R. Thresher, A. Ickowicz, D. Peel, M. Tizard, P. De Barro, T. Strive, and J. Dambacher. 2018. Identifying and detecting potentially adverse ecological outcomes associated with the release of gene-drive modified organisms. Journal of Responsible Innovation 5(S1): S139–S158.

James, S., F. H. Collins, P. A. Welkhoff, C. Emerson, H. C. J. Godfray, M. Gottlieb, B. Greenwood, S. W. Lindsay, C. M. Mbogo, F. O. Okumu, H. Quemada, M. Savadogo, J. A. Singh, K. H. Tountas, and Y. T. Toure. 2018. Pathway to deployment of gene drive mosquitoes as a potential biocontrol tool for elimination of malaria in Sub-Saharan Africa: Recommendations of a scientific working group. American Journal of Tropical Medicine and Hygiene 98 (Supplement 6): 1–49.

Jonas, H. 1979. Das Prinzip Verantwortung. Suhrkamp-Verlag, Frankfurt, Germany. English translation: The imperative of responsibility: In Search for an ethics of the technological age. Chicago, USA, 1984.

Liebert, W., and J. Schmidt. 2010. Towards a prospective technology assessment: Challenges and requirements for technology assessment in the age of technoscience. Poiesis & Praxis 7: 99–116.

Liebert, W., and J. Schmidt. 2015. Demands and challenges of a prospective technology assessment, pp. 331-340. *In* C. Scherz, T. Michalek, L. Hennen, L. Hebáková, J. Hahn, and S. Seitz (eds.), The next horizon of technology assessment. Technology Centre ASCR, Prague, Czech Republic.

Markl, H. 1998. Wissenschaft gegen Zukunftsangst. Hanser-Verlag, München/Wien, Germany/Austria.

Min, J., A. Smidler, D. Najjar, and K. Esvelt. 2018. Harnessing gene drive. Journal of Responsible Innovation 5(S1): S40–S65.

(NAS) National Academies of Sciences. 2016. Gene drives on the horizon: Advancing science, navigating uncertainty, and aligning research with public values. The National Academies Press. Washington, DC, USA.

Noble, C., J. Olejarz, K. Esvelt, G. Church, and M. Nowak. 2017. Evolutionary dynamics of CRISPR gene drives. Science Advances 3(4): e1601964

Noble, C., B. Adlam, G. Church, K. Esvelt, and M. Nowak. 2018. Current CRISPR gene drive systems are likely to be highly invasive in wild populations. eLife 7: e33423

Noble, C., J. Min, J. Olejarz, J. Buchtahl, A. Chavez, A. Smidler, E. DeBenedictis, G. Church, M. Nowak, and K. Esvelt. 2019. Daisy-chain gene drives for the alteration of local populations. Proceedings of the National Academy of Sciences 116: 8275–8282.

Schaefer, K., W.-H. Wu, D. Colgan, S. Tsang, A. Bassuk, and V. Mahajan. 2017. Unexpected mutations after CRISPR-Cas9 editing in vivo. Nature Methods 14 (6): 547–548.

Schmidt, J. C. 2015. Synthetic biology as late-modern technology, pp. 1-30. *In* B. Giese, C. Pade, H. Wigger, and A. von Gleich (eds.), Synthetic biology, character and impact. Springer-Verlag, Cham/Heidelberg, Switzerland/Germany.

Schmidt, J. C. 2016. Philosophy of late-modern technology, pp. 13-29. *In* J. Boldt (ed.), Synthetic biology, metaphors, worldviews, ethics, and law. Springer-Verlag, Cham/Heidelberg, Switzerland/ Germany.

Schweitzer, A. 1966. Die Ehrfurcht vor dem Leben. Grundtexte aus fünf Jahrzehnten. Verlag C. H.Beck, Nördlingen, Germany. 167 pp.

Shaw, J. 2016. Editing an end to malaria? The promise, and possible perils, of a new genetic tool. Harvard Magazine, May-June 2016.

The *Anopheles gambiae* 1000 Genomes Consortium. 2017. Genetic diversity of the African malaria vector *Anopheles gambiae*. Nature 552: 96–100.

Tripet, F., P. Lounibos, D. Robbins, J. Moran, N. Nishimura, and E. Blosser. 2011. Competitive reduction by satyrization? Evidence for interspecific mating in nature and asymmetric reproductive competition between invasive mosquito vectors. American Journal of Tropical Medicine and Hygiene 85: 265–270.

Unckless, R., A. Clark, and P. Messer. 2017. Evolution of resistance against CRISPR-Cas9 gene drive. Genetics 205: 827–841.

White, N., S. Pukrittayakamee, T. Hien, A. Faiz, O. Mokuolu, and A. Dondorp. 2014. Malaria. The Lancet 383: 723–735.

(WHO) World Health Organization. 2015a. World malaria report 2015. WHO Press, Geneva, Switzerland.

(WHO) World Health Organization. 2015b. Global technical strategy for malaria 2016-2030. WHO Press, Geneva, Switzerland.

Wijayawardena, B., D. Minchella, and J. DeWoody. 2013. Hosts, parasites and horizontal gene transfer. Trends in Parasitology 29: 329–338.

Windbichler, N, M. Menichelli, P. Papathanos, S. Thyme, H. Li, U. Ulge, B. Hovde, D. Baker, R. Monnat, A. Burt, and A. Crisanti. 2011. A synthetic homing endonuclease-based gene drive system in the human malaria mosquito. Nature 473: 212–217.

GENOME EDITING AND ITS APPLICATIONS FOR INSECT PEST CONTROL: CURSE OR BLESSING?

I. HÄCKER[1,2] AND M. F. SCHETELIG[1,2]

[1]Justus-Liebig-University Giessen, Institute for Insect Biotechnology, Heinrich-Buff-Ring 26-32, 35392 Giessen, Germany; marc.schetelig@agrar.uni-giessen.de
[2]Fraunhofer Institute for Molecular Biology and Applied Ecology (IME), Project Group Bioresources, Winchesterstrasse 2, 35394 Giessen, Germany

SUMMARY

Gene and genome editing are described as cutting-edge research tools with the potential to tackle urgent global challenges in the management of agricultural pests and human disease vectors such as mosquitoes. The field is defined by the chances and challenges to interlink the disciplines of insect genomics, molecular biology, and pest control together with the need for clear risk assessment, policy development and public approval of the application of such novel technologies. The goal is to generate innovative and sustainable pest control solutions applied in the best interest for the environment and human society. Here, starting from available genome editing technologies, the current strategies and applications for insect pest control are discussed, including approaches to overcome the evolution of resistance alleles and other potential pitfalls to be expected from selective pressures resulting from gene drive applications. They are supplemented by views on regulatory, policy and ethical considerations that in our opinion will be necessary to define how the different tools can be used in the future in a safe and responsible way.

Key Words: CRISPR-Cas9, gene editing, gene drive, genome engineering, GMO, transgenic, insect genomics, engineered nucleases, meganucleases, zinc finger nucleases, TALEN, homing endonucleases, regulations, risk assessment, ethics

J. Hendrichs, R. Pereira and M. J. B. Vreysen (eds.), Area-Wide Integrated Pest Management: Development and Field Application, pp. 809–842. CRC Press, Boca Raton, Florida, USA.

1. INTRODUCTION

Europe faces serious problems related to insect pests in two key areas: health and agriculture. The European Centre for Disease Prevention and Control has warned European authorities of the increased risk of local transmission cycles and epidemics of mosquito-borne infectious diseases (ECDC 2018). Locally transmitted cases of malaria, for instance, have already been reported in Greece, while in other regions in the world the disease claims nearly half a million lives annually. Similar threats are emerging with viral diseases such as chikungunya and viral haemorrhagic fevers including Hantaan, dengue, and yellow fever. European authorities have responded to this new threat with recommendations for vector control measures, as well as increased financing for relevant research, demonstrating the interest and concern of the European Union (EU).

Additionally, endemic and invasive agricultural insect pests cause tremendous losses in agricultural yield and revenues. Worldwide, insects cause up to 40% loss of potential crop yields, and the European Environment Agency reported the presence of over 10 000 invasive pest species for Europe in 2017 (EEA 2017). Significant damage caused by these pests affect the economy and negatively impact biodiversity. Invasive species directly cost Europe billions of Euros per year. Additionally, heavy reliance on pesticides to control them generates significant unquantified costs. These are costs associated with environmental pollution, e.g. soil and water. Other costs arise from resistant pests and include those required to develop alternative new effective pesticides. Also, loss in biodiversity, reintroducing biodiversity, and the impact on public health from pesticide residues or direct exposure to pesticides add to the expenses.

For these reasons, innovative pest management methods and strategies are urgently needed. Numerous stakeholders have expressed this urgency to prevent total crop losses and combat diseases transferred by mosquito vectors, but, at the same time, demand clear information about the risks and benefits of such novel technologies.

Novel vector and pest control solutions based on genome editing tools have the potential to improve the lives of millions of people, both by offering adequate protection against insect-borne diseases and by preventing crop and livestock damages caused by invasive agricultural pests. By using these novel technologies, different approaches are now possible that were only fiction in the past. While they sound promising, their possible side effects have to be considered and detailed evaluations prepared concerning the applicability of such new systems. Moreover, societal questions need to be answered like 'Should we use some of the technologies at all?' In the end, it is a combination of science, technology, ethics, policy, and communication that will define the feasibility and applicability of one or the other technology.

In this chapter, we want to review the technological genome editing options that are available and have been developed in the field so far. We also discuss their 'curse or blessing', together with considerations for their safe and responsible application for insect pest control tactics and strategies that do not pose risks and are friendly to the environment. The views presented here are the personal opinion of the authors on this important topic, without the claim for completeness or being the only or best possible solution. More considerations on the evaluation of gene editing and gene drive technologies are presented in the chapters by Liebert (this volume) and Nielsen (this volume).

2. AVAILABLE GENOME EDITING TECHNOLOGIES

Genome editing technologies can be divided into three categories: homologous recombination, engineered nucleases, and the CRISPR/Cas system.

2.1. Homologous Recombination

The early genome editing trials were performed by *homologous recombination* (HR), based on observations from the yeast *Saccharomyces cerevisiae* Meyen ex E.C. Hansen, where HR occurs at a high frequency. A specific sequence can be inserted into the genome at a defined position by flanking the sequence with homology arms identical to the sequences left and right of the selected insertion position. The construct is then injected into cells and inserted into the genome by recombination of the homology arms with the corresponding genomic sequences. However, the recombination rate in most cell types is meagre. In higher plants, it is estimated to be in the range of 0.01-0.1% (Puchta 2002; Hanin and Paszkowski 2003; Reiss 2003). In mammalian cells such as mouse embryonic stem cells it can be 1% or higher, but it can as well be as low as one in more than a million events (Vasquez et al. 2001). Moreover, the integration often occurs at unspecific sites, leading to off-targeting with a frequency of one in 10^2 to 10^4 treated cells (Vasquez et al. 2001). Design of the homology arms (sequence and length) seems crucial for the off-target rate in human cells, as the genome contains a lot of repetitive elements such as LINEs and SINEs (Long and Short Interspersed Nuclear Elements), which comprise 36% of the human genome. The longer the arms, the higher the probability of including such repetitive elements, which can cause recombination at unspecific sites (Ishii et al. 2014).

2.2. Engineered Nucleases

A more efficient way to edit genomes was established with the use of *engineered nucleases*. These nucleases act like molecular scissors, inducing double strand breaks at specific genomic positions that are then repaired by one of the two cellular repair pathways: non-homologous end joining (NHEJ) or homology-directed repair (HDR).

Three different nuclease families have been engineered for genome editing: meganucleases, zinc finger nucleases (ZFNs), and transcription activator-like effector-based nucleases (TALEN).

Meganucleases are predominantly found in microbes, and it is almost impossible to find a natural meganuclease targeting the specified sequence. Therefore, scientists applied different strategies, including random mutagenesis and high throughput screening, fusion of different nucleases, as well as rational design, to modify the binding specificity of the enzymes to expand the rather limited choice of target sequences (Sussman et al. 2004; Arnould et al. 2006; Rosen et al. 2006). With a recognition sequence of 14-40 nucleotides, meganucleases have a high target site specificity.

The ZFNs and TALENs are fusion proteins consisting of a non-specific DNA cutting enzyme, the restriction endonuclease Fok1, which is linked to a zinc finger (ZF) or transcription activator-like effector (TALE) domain. These peptides recognize specific DNA sequences and thereby confer sequence specificity to the endonuclease. Like meganucleases, the choice of naturally occurring target sites of ZFs and TALEs is limited and can be extended by protein engineering to theoretically bind nearly any desired sequence.

Targeted genome editing including TALENs was named the 2011 method of the year by Nature Methods (Becker 2012). Protein engineering, however, is time consuming, cumbersome and costly. Therefore, while ZFNs and TALENs were successfully applied to modify different insect genomes (Bozas et al. 2009; Takasu et al. 2010; Liu et al. 2012; Aryan et al. 2013; Smidler et al. 2013; Takasu et al. 2013), a widespread use of these endonucleases was prevented by the need for a new engineered protein for each new genomic target site.

2.3. The CRISPR-Cas System

The discovery of *Clustered Regularly Interspaced Short Palindromic Repeats (CRISPR)* has tremendously advanced the field of genome editing. CRISPR sequences were first discovered in bacteria and archaea in the 1990s and were identified as a prokaryotic equivalent to the eukaryotic acquired immune system. Upon infection of a bacterial or archaeal cell with a pathogen, the cell incorporates a short sequence of the foreign DNA (e.g. virus or plasmid) into its genome. Such foreign DNA sequences are collected in clusters and separated by short repeat sequences. Small clusters of Cas (CRISPR associated) genes are located next to the repeat-spacer arrays (for comprehensive explanations and illustrations see Horvath and Barrangou 2010; Marraffini and Sontheimer 2010; Bhaya et al. 2011). Upon reinfection, the arrays are transcribed and processed by one family of Cas proteins into short CRISPR RNAs (crRNA), which are bound by another class of Cas proteins.

The crRNAs then guide these Cas proteins to the foreign DNA for degradation by endonucleolytic cleavage in a mechanism similar to RNA interference in higher eukaryotes (Marraffini and Sontheimer 2010). The target site specificity of the Cas endonuclease is determined by its bound crRNA. Therefore, Cas proteins can be programmed to target nearly any genomic site by adjusting the crRNA sequence. As the crRNA sequence adjustment is easy and cost-effective, CRISPR-Cas is an ideal and versatile tool for genome editing and has essentially abolished the use of ZFNs and TALENs.

The Cas proteins are categorized into two classes. The class I systems use multiple Cas proteins for the degradation of foreign nucleic acids, whereas the class II systems consist of one single protein. Therefore, the class II systems are more suitable for research and application purposes. The most used nuclease for CRISPR genome editing is the multifunctional class II protein *Cas9*. In addition to the crRNA, Cas9 requires a transactivating CRISPR RNA (tracrRNA) to function. To streamline its application in the laboratory, scientist fused the crRNA and tracrRNA into one single-guide RNA (sgRNA) (Jinek et al. 2012). CRISPR-Cas can be used to knock out existing genes making use of the cell's NHEJ repair pathway or to introduce new DNA sequences including whole genes and transgene constructs by adding a repair template containing the respective sequence information and homology regions to the Cas9 target site. The cell's HDR pathway then uses the repair template to repair the Cas-induced double-strand break.

Since its first application for genome editing in 2012/2013, more class II Cas proteins suitable for genome editing purposes have been identified (Zetsche et al. 2015; Abudayyeh et al. 2016; Yang et al. 2016), and the system is continuously being adjusted and optimized for different applications and purposes. These efforts include modifications to decrease the off-target rate of the CRISPR-Cas system (Fu et al. 2014; Kleinstiver et al. 2016; Nowak et al. 2016).

Shortly after its first application for genome editing, CRISPR-Cas was used in *Drosophila melanogaster* Meigen (Ren et al. 2013; Yu et al. 2013; Bassett et al. 2014). Within just four years, it was subsequently applied to several other insect species including the lepidopterans *Bombyx mori* L. (Wang et al. 2013; Ma et al. 2014), *Spodoptera litura* (F.) (Bi et al. 2016), and *Danaus plexippus* (L.) (Markert et al. 2016), the orthopteran *Gryllus bimaculatus* (De Geer) (Awata et al. 2015), and importantly also to several dipteran vector and pest species, namely *Aedes aegypti* (L.) (Kistler et al. 2015), *Anopheles gambiae* (Giles) (Hammond et al. 2016), *Ceratitis capitata* (Wiedemann) (Meccariello et al. 2017; Aumann et al. 2018) and *Drosophila suzukii* (Matsumura) (Li and Scott 2016; Kalajdzic and Schetelig 2017; Li and Handler 2017), as well as in the non-pest dipteran, *Musca domestica* L. (Heinze et al. 2017).

The CRISPR-Cas genome editing system also transformed gene drive research and development. Gene drives are genetic drive mechanisms that can be used to spread a genetic trait through a population by biasing its inheritance beyond the Mendelian inheritance rate of 50%. Gene drives were initially designed using naturally occurring selfish genetic elements like *Medea* (maternal-effect dominant embryonic arrest) (Chen et al. 2007; Buchman et al. 2018) or homing endonucleases

(HE) (Burt 2003; Windbichler et al. 2011). HE recognize and cut short DNA sequences. These sequences are only located on chromosomes different from the one on which the HE is located, and additionally on the homologous chromosome exactly at the location of the HE. In a heterozygous individual, after introduction of a double-strand brake by the HE on the homologous chromosome, the cell's repair mechanism uses the chromosome containing the HE to repair the brake, thereby copying the HE gene onto the homologous chromosome and converting the heterozygote into a homozygote. This process is called *homing*.

The CRISPR-Cas system can be programmed to act like an HE by targeting its genomic integration site on the homologous chromosome using a respective homing guide RNA. Identical to the HE, the induced cut is repaired using the allele containing the CRISPR construct as a template, thereby copying it onto the homologous chromosome. Thus, in contrast to genome editing via CRISPR, where only the CRISPR-induced molecular changes, but not the CRISPR components themselves are passed on to the next generation, in CRISPR gene drives also the genes coding for the CRISPR components are incorporated into the genome and passed on to the offspring. With the inherent programmable target site specificity of the CRISPR-Cas system, nearly any genomic position can be selected for the placement of the CRISPR homing construct. If homing occurs in the germline, then theoretically all the offspring will carry the CRISPR construct instead of only 50% like in normal Mendelian inheritance.

The potential use of these genome editing technologies, especially of CRISPR-Cas gene drives for insect pest control, is discussed below.

3. CURRENT STRATEGIES AND APPLICATIONS INVOLVING GENOME EDITING FOR INSECT PEST CONTROL

A promising and proven, sustainable and species-specific method to manage insect pest populations is the Sterile Insect Technique (SIT) (Dyck et al. 2021). It is based on the mass-production and release of sterilized males of the target species in the affected area. The sterile males mate with wild females in the field, which will not result in viable offspring, thereby reducing the wild population size of the pest in the next generation. By repeated releases, the population can be decreased to a manageable level. The SIT is most successfully applied in area-wide programmes against several pest species of agricultural importance, including the Mediterranean fruit fly, the Mexican fruit fly, *Anastrepha ludens* Loew, or the New World screwworm, *Cochliomyia hominivorax* (Coquerel).

The effectiveness of such programmes can in some cases be increased with the establishment of so-called sexing systems to eliminate females to allow male-only releases. Male-only releases are desirable for some agricultural pests and are a prerequisite for vector insects such as mosquitoes. In both cases, early elimination of females enables more cost-effective mass-rearing and release.

Most importantly, the efficiency of male-only releases is superior due to the lack of undesired mating between sterilized males and sterilized females (Franz et al. 2021).

In mosquito control programmes, the release of only male insects is essential as it precludes the risk of increasing the number of disease-transmitting individuals by releasing females. Therefore, mass-reared males and females must be separated at large-scale or females be eliminated at some point during the production process. This sexing of the insects (needed up to one billion male insects per week) is a significant bottleneck for the application of the SIT to new insect species. There is, for example, no effective sexing system to date for any of the vector mosquitoes.

For recent field trials with transgenic *Ae. aegypti* mosquitoes carrying a conditional lethal RIDL system (Release of Insects carrying a Dominant Lethal) that kills the offspring in the late larval stage (Thomas et al. 2000; Phuc et al. 2007), male and female pupae were separated mechanically by hand, resulting in the production of 0.5-1.5 million males per week and a female contamination of the released insects of less than 1% (Carvalho et al. 2015). This labour-intensive and time-consuming method, however, does not allow large-scale programmes beyond the field trial scale. Therefore, coordinated international research efforts are ongoing to establish sexing systems in different *Anopheles* and *Aedes* species (Gilles et al. 2014; Bourtzis and Tu 2018). Genome editing, combined with available classical genetics or transgenic technologies, can help to solve these and other issues related to insect pest control.

3.1. Unravelling Sex Determination Pathways in Insects

Genome editing is being used in basic research of insect pests, for example, to uncover gene functions and thereby better understand the target insect's biology and physiology. One major point of interest concerning sexing is the elucidation of the sex determination pathways in pest species to identify the responsible gene(s) for male/female development. CRISPR technology was used to knock out the candidate gene for the male-determining factor *Nix* in *Ae. aegypti* (Hall et al. 2015). Knockout of *Nix* resulted in feminized males, whereas ectopic expression resulted in masculinized females, identifying *Nix* as necessary and sufficient for determining maleness in the yellow fever mosquito.

Similarly, the Cas9-mediated knockout of the candidate M-factor gene *Mdmd* in *M. domestica* confirmed the key role of this gene for male gonadal and germline development in the house fly (Sharma et al. 2017). Once the sex determination pathways are understood for pest species of interest, the knockout or overexpression of the sex determination genes via CRISPR-Cas could be used to create female lethality, or for the conversion of females into phenotypic males to create strains for large-scale sexing (Meccariello et al. 2019).

3.2. Site-directed Mutagenesis in Pest Insects to Enable Population Control

For several economically important insect species, transgenic insect strains have been established to demonstrate the ability to generate male-only populations for control programmes. The first strains consisted of transgenic conditional female-lethal systems in agricultural and livestock pests like *C. capitata* (Fu et al. 2007; Ogaugwu et al. 2013), *Anastrepha suspensa* Loew (Schetelig and Handler 2012b), *A. ludens* (Schetelig et al. 2016), and *L. cuprina* (Yan et al. 2017). In these systems, under permissive conditions, a lethal gene is explicitly expressed in females to kill them at an early embryonic stage via apoptosis (Schetelig and Handler 2012b; Ogaugwu et al. 2013; Schetelig et al. 2016; Yan et al. 2017), or later in development by the ubiquitous accumulation of tTA (Fu et al. 2007), resulting in a 100% male cohort for release. Similar systems have been created expressing the lethal cassette in both sexes to produce genetic sterility (Gong et al. 2005; Schetelig et al. 2009; Schetelig and Handler 2012a). A release of these insects would result in biologically fertile matings, but the offspring would die between the early embryonic (Schetelig et al. 2009; Schetelig and Handler 2012a) and the late larval stage (Gong et al. 2005). All the transgenic systems described here, and many others, have commonly been introduced into the insect genomes via transposable elements, which integrate into the genome randomly. Integration at an unfavourable genomic site, however, can have adverse effects on insect fitness due to insertional mutagenesis, as well as on transgene expression levels due to nearby regulatory genomic elements. Genome editing technologies now allow inserting the transgene construct at specific, characterized genomic positions, thus avoiding harmful side effects of random integrations and improving the quality of transgenic strains for pest control.

Genome editing technologies furthermore offer the possibility of recreating a genetic trait from one species in another, which is not possible with any other technology with similar efficiency and ease. Mediterranean fruit fly SIT programmes, for example, rely on a conditional, temperature sensitive lethal (*tsl*) mutation, obtained by classical mutagenesis and breeding, to eliminate female embryos during mass-rearing via heat shock, resulting in the release of only sterile males (Franz et al. 2021). It has been tried to generate such *tsl* strains in other insect pests via classical mutagenesis (Ndo et al. 2018), or to link other selectable makers to one sex via radiation-induced translocations. Most successful have been pupal colour markers in the Mediterranean fruit fly and the Mexican fruit fly, as well as dieldrin resistance in *An. gambiae* (Curtis et al. 1976), *An. arabiensis* (Curtis 1978) and *Anopheles albimanus* Wiedemann (Seawright et al. 1978). In *An. arabiensis*, classical mutagenesis was used again years later to induce insecticide resistance in males as a tool for female elimination (Yamada et al. 2012). All these sexing tools have been developed either by classical mutagenesis or selection of a naturally occurring phenotype and linking it to one sex. The underlying mutations and mechanism of the corresponding phenotype in most cases are unknown, although it would be extremely valuable to know. Once such a mutation created by classical mutagenesis (e.g. the conditional *tsl* in the Mediterranean fruit fly) is identified, CRISPR-Cas genome editing could be used to precisely edit the genome of another pest species to create an identical mutation in the respective gene. The mutation created in this way would not be different from the mutations induced by the approved classical mutagenesis techniques, not involving any transgene or foreign DNA (but usually rather small deletions or single nucleotide changes).

Li and Handler (2017) recently used CRISPR-Cas to create point mutations into the transformer sex determination gene in the fruit pest *D. suzukii,* thereby recreating the temperature-sensitive *D. melanogaster transformer-2* mutations (*tra-2*^{ts1} and *tra-2*^{ts2}) in a proof-of-principle experiment. The CRISPR-Cas system was also used to create an X-shredder in *An. gambiae* mosquitoes to achieve female elimination. The Cas9 is expressed gonadally to target X-chromosomal ribosomal RNA sequences, thereby destroying the X chromosome in X gametes, resulting in predominantly Y gametes and therefore predominantly male offspring (Galizi et al. 2016). The Cas9-based X-shredder is based on the original idea from Galizi et al. (2014) using the endonuclease I-*Ppo*I to cut X-chromosomal ribosomal RNA sequences.

Besides the important topic of sexing, genome editing could be used to improve any other aspect of insect pest control programmes, for example, to enhance insect fitness to overcome deficits induced in the insects by the mass-rearing process such as low competitiveness and short life span. Alternatively, males could be modified to improve mate-seeking success, all of which could improve the efficacy of SIT control programmes.

3.3. Gene Drive Systems for Population Suppression or Replacement

Finally, genome editing technologies like CRISPR also open a new path to insect pest control via gene drives. Gene drives could be used for pest control in two ways, via insect population suppression or population replacement. Like other approaches based on the release of sterile insects, population suppression via gene drives results in population size reduction in the next generation. However, population suppression drives do not use reproductive or genetic sterility in the classical sense, as this would prevent the spread of the trait into the population, as all the offspring carrying the trait would die before reproduction. Therefore, the genetic trait conferring population size reduction is linked to a gene drive component. Upon a one-time release of a seed population, the gene drive component drives the trait into the population with each successive generation, abolishing the need of repeated mass-releases. A population suppression gene drive could, for example, be a genetic modification killing females and resulting in only male offspring. These males will carry the "sterility" construct and pass it on to their sons upon mating with wild type females, thereby decreasing the population size further with each generation as no females are produced, until the population collapses, at which point the gene drive construct would disappear.

Another approach could be a modification that reduces female fertility. Such a system has been developed in the vector mosquito *An. gambiae* by targeting different putative female fertility genes. The gene drive components were designed to home in the germline of both sexes to ensure that all female offspring is affected (Hammond et al. 2016). In a different project, female sterility was achieved by targeting a female-specific exon of the sex determination gene *doublesex* in *An. gambiae*. Disruption of this exon by Cas9 did not affect male development but resulted in females with an intersex phenotype that were completely sterile (Kyrou et al. 2018). Most gene drive research has been performed in *Anopheles* mosquitoes so far, where it might be applied first in a control programme. There is only one report in an agricultural pest, *D. suzukii*, showing the functionality of a synthetic *Medea* gene drive system that could be used to spread a cargo gene into the population, for example for population suppression purposes (Buchman et al. 2018).

Population replacement gene drives mostly make sense for vector insects. The idea is to replace the wild type population by insects that are refractory to the infection with the pathogen, thereby interrupting the disease transmission cycle. In the first CRISPR-based gene drive in mosquitoes, Gantz et al. (2015) developed a drive system in which the Cas9 homing construct is expressed in the male and female germline of *An. stephensi*. The construct further contains previously identified dual anti-pathogen effector genes conferring resistance to *Plasmodium* infection, which are expressed somatically (Isaacs et al. 2011, 2012).

Gene drive technology holds great promise to solve problems caused by harmful insects, and the initial enthusiasm expressed in view of the options for gene drive design that opened up with the discovery of CRISPR was huge. It triggered statements that there would be the first mosquito gene drives out in field trials within less than two years. This enthusiasm was dampened, however, when the technology hit a sudden roadblock, making it very clear that we have to understand much more about drive mechanisms and their potential pitfalls before releases could be considered.

Two predominant issues arose in laboratory experiments: first, loss of function due to the evolution of resistant alleles (Hammond et al. 2017; Marshall et al. 2017; Unckless et al. 2017; KaramiNejadRanjbar et al. 2018), which is mostly a problem for homing-based CRISPR-Cas gene drives (i.e. mutations evolving at the homing gRNA target site), and second, failure of the drive due to population genetic diversity (Drury et al. 2017; Buchman et al. 2018) (i.e. sequence variation at the gene drive target site throughout a population; this affects CRISPR-based as well as other gene drives). Evolution of resistance alleles increases with the selective pressure put onto the gene drive-carrying insect by the drive itself or a linked genetic construct and is inverse proportional to the conservation level of the target sequence (see Section 7.3.1. for approaches being followed to overcome the technological pitfalls identified). The CRISPR-Cas technology with its versatility will be instrumental to overcome the identified evolutionary pitfalls, and in developing new drive systems to study and improve this technology for safe applications in the future.

4. REGULATORY CONSIDERATIONS OF GENOME EDITING

Genome editing technologies did not play a significant role in the development of genetically modified organisms (GMO) for agriculture, the food industry or other applications before the discovery of CRISPR and its potential for large-scale and high throughput, affordable genome editing. Therefore, questions of how organisms resulting from genome editing should be regulated were not relevant. This changed with the introduction of CRISPR, triggering large-scale research not only in crop optimization but also in modifying farm animals to be leaner, to develop faster, have longer wool, or a differently coloured coat. It raised the urgent question of how such products should be evaluated and regulated for bringing them to the market, and how currently existing regulations apply to genome editing. Should genome-edited products be classified and treated as GMO or not?

In the European Union (EU), the deliberate release of GMO is regulated by the 2001 EU directive (2001/18/EC). The directive specifies the procedures required in the EU for the evaluation and authorization of GMO releases. While the directive covers all kinds of GMOs, it so far has been applied only for the regulation of genetically modified (GM) crops. This selective implementation has been described as the 'plant paradigm'. The 2001 directive states that an organism is characterized as GMO if its genetic material has been altered in a way that could not have occurred naturally by mating or recombination. Therefore, also organisms developed by non-transgenic methods can be classified as GMO. At the same time, however, conventional mutagenesis techniques like radiation or chemical mutagenesis, that were considered safe in 2001, are exempt from the GMO directive as long as they do not involve recombinant DNA (the mutagenesis exemption). Genomic modifications created by genome editing technologies can be changes on the level of whole genes (including the insertion of transgenic constructs) but can also be the introduction of small mutations (insertions or deletions of a few base pairs) all the way down to single nucleotide changes, as they could also occur by classical mutagenesis and breeding techniques. Therefore, from a rational and scientific point of view, such small mutations induced by genome editing technologies should be treated similarly to mutations induced by classical mutagenesis.

To clarify if and to which extent the rules of the 2001 directive would apply to the new genome editing technologies, the French Council of State sent an inquiry to the European Court of Justice (ECOJ) to interpret the directive for the new technologies. The ECOJ's answer was provided in the form of a complex 'Advocate General's opinion in Case C-528/16' statement, released in January 2018. It principally stated that changes induced by genome editing that could also have occurred by conventional mutagenesis should be regulated in the same way. However, scientifically unexpected, the court ruled in July 2018 against the Advocate General's advice and stated that all products resulting from genome editing are subject to the directive and are to be treated as GMOs.

This decision raises different questions and uncertainty for the application of the new technologies. Interestingly, however, from a scientific point of view, it also puts the regulations for the established and safe methods in question. In an open letter to the Federal Ministry of Education and Research in Germany, the German Life Sciences Association stated that the inquiry was answered by the judges according to the legal conditions, but was not based on scientific facts available from EU authorities (the Scientific Advice Mechanism (SAM) and the European Food Safety Authority (EFSA)) as well as from a large number of scientists worldwide.

This EU decision follows years of complicated communication attempts in the area of regulation of GM organisms and products, which is still not consistent between countries. Regulation of GMO in the EU considers and evaluates the *process*, not the *product*. Therefore, two products with identical traits developed by different technologies are regulated differently. The recent court ruling on genome-edited organisms is a perfect representative of this evaluation approach. Genome editing as a process would be regulated, not the resulting product, meaning that all genome-edited products would be regulated without exception, even if the same trait could have been obtained by approved technologies without regulatory requirements. Moreover, in the EU, only the potential risks of the GMO are considered, while the prospective benefits are not considered.

The USA essentially takes the opposite approach. There, only the product is evaluated, independent of the method used to create it (Global Legal Research Center 2014). Consequently, the USA does not regulate any products that could as well be the result of traditional mutagenesis or breeding techniques, summarized by the United States Department of Agriculture (USDA) in the following statement on agricultural products on its website:

> "*Under its biotechnology regulations, USDA does not currently regulate or have any plans to regulate plants that could otherwise have been developed through traditional breeding techniques as long as they are developed without the use of a plant pest as the donor or vector and they are not themselves plant pests*" (USDA 2018).

Thus, small deletions or single nucleotide substitutions are not regulated in the USA. Even the introduction of larger nucleotide sequences that could have also occurred by cross-breeding is not regulated. In a March 2018 press release, the USDA specifically stated that this includes changes made by genome editing technologies. This approach to product evaluation resulted in the recent clearing for commercialization of a mushroom edited by CRISPR-Cas (Waltz 2016). The clearance was given without the regulation for GMOs, as only a few base pairs in the polyphenol oxidase gene were removed to enhance the shelf-life. Another significant difference in the USA approach for risk assessment compared to the EU is that also potential benefits of the commercial GMO are considered.

The existence of globally diverse approaches to the evaluation and regulation of GMO not only causes problems in international trade, as products will be classified differently in different countries, but also pose an enormous challenge for the release of modified insects. With their big range of motion and dispersal, insects will not stop at borders. Therefore, especially for insects carrying a drive mechanism with the potential to spread through whole populations, a common international ground for evaluation and regulation has to be found.

5. THE IMPORTANCE OF INFORMED DECISION-MAKING ON GENOME EDITING TECHNOLOGIES

The long ongoing GM crop debate in the EU essentially led to a moratorium in the EU on GM crops from 1998 to 2010, which in 2003 triggered a case filing by the USA and other countries with the World Trade Organization (WTO) against the EU. The ban on GM crops combined with unfavourable and sometimes biased news coverage created fear and insecurity in the EU, resulting in strong opposition against GM technologies by the public. This stalled the corresponding research and caused the industry in the EU to step by step pull out of GM crop research and production, e.g. Bayer Crop Science and BASF closing their GM crop research in Germany between 2004 and 2011.

With the most recent ruling by the ECOJ that also classifies all genome-edited products as GMO, the required checks and controls needed to develop such products for the market would be too expensive for research institutes and small companies. Moreover, the classification of all genome-edited organisms as GMO will probably lead again to strong opposition and general rejection by the public, adding another hurdle for the marketing of genome-edited products. Therefore, in our opinion, decisions like the one by the ECOJ equal a moratorium on genome-edited products in the EU with far-reaching consequences. Again, it will incapacitate a whole biotechnology industry sector that will flourish in countries with less restrictive regulation. It will prevent innovation and the development of new technologies in the EU and will cause biotechnology companies to withdraw their respective departments from the EU market, which in turn will prevent the creation of or destroy jobs. Also, in the international biotechnology sector, there is concern that the EU will continue to lag behind countries such as the USA, where the regulations are more favourable for GMOs and the GM crop market. The ECOJ decision on genome-edited organisms will also influence trade markets, as occurred already during the EU GM moratorium, which experienced in the early 2000s negative impacts on the agricultural export revenue of countries such as the USA, Canada, and Argentina (Disdier and Fontagne 2010).

Regarding pest control programmes intending the area-wide use of genome-edited insects, this could be most problematic for agricultural pests that feed inside fruits/crops like the Mediterranean fruit fly. Marketing of such crops in the EU might be restricted by the control measures for GMO contamination in food, for which the EU tolerance levels are as low as 0.9%. For a polyphagous insect like the Mediterranean fruit fly, which feeds on many different fruits, multiple agricultural products in the treatment area might be affected by the EU import restrictions (Max-Planck-Gesellschaft 2017). Use of genome-edited insects for pest control will therefore require the establishment of a definition of insect contamination in food and a decision on tolerance levels. This general framework will be critical for exporting countries to decide if the trade-off between the possible import restrictions and concomitant loss of markets on one side, and the reduced crop production costs as well as more abundant harvests on the other side, are worth the use of genome-edited insects for pest control.

Besides the import restrictions, however, genome-edited insects might not be regulated by the 2001 EU directive on the release of GMO if the insects are 100% sterile. Sterile insects do not fit the definition of an organism and therefore aren't a GMO either. Thus, 100% sterilized mosquitoes released as part of a SIT programme seem not to be regulated in the EU (HCB 2017). The release of radiation-sterilized *Ae. albopictus* mosquitoes is carried out in Italy and Germany as part of SIT mosquito suppression trials. To our knowledge, in Germany, these releases did not require prior authorization and risk assessment. Thus, one could speculate that scientists might be allowed to use genome editing, for example, to develop sexing strains for a pest species, allowing to remove female insects during rearing and to release the corresponding males after (100%) sterilisation by irradiation.

6. ETHICAL ASPECTS OF GENOME EDITING AND GMO RELEASES

The CRSIPR/Cas system allows genome editing with an ease and effectiveness that appeared to be some way in the future just a few years ago. Now that "everything" seems feasible, the always present question of what is socially, environmentally and also ethically responsible to do is coming into the focus of discussions. As we are no experts on social sciences and bioethics, we want to just briefly touch on some questions that have been raised in different panels and newsgroups.

A major topic evolves around the moral aspect of genome editing. This involves the fundamental questions if we are allowed to edit an organism's genome at all, and in case of area-wide pest management programmes, if we have the right to eliminate invasive or native populations or even a species as a whole. Both questions have been raised before but became more prominent with the discovery of the possibilities that opened up by CRISPR-Cas. While these questions are important to discuss, it is also important to consider the impact that currently applied strategies and technologies have.

The classical mutagenesis and breeding techniques, for example, that have been broadly applied for decades and are widely accepted and mostly haven't been questioned, can randomly change the genome of an organism at multiple positions, and the induced changes on the molecular level are commonly not known. What can be achieved with CRISPR genome editing, depending on the modification, would not be different from what has been done for decades, but can be done now in a less random process.

Similarly, the use of insecticides has the potential to eliminate a pest population in the targeted area, with the side-effect of not only eliminating the target species but also affecting many other beneficial insects, besides the environmental impact. With the recent alarming news on the decline of overall insect numbers and biodiversity (Hallmann et al. 2017), the extensive use of insecticides has come once more into the focus of widespread criticism and public concern.

Another important point for ethical discussion deals with questions concerning constitutional rights, individual expectations, democratic decisions, and also questions about how we can balance the potential elimination of a species against the decreased burden for the human population (e.g. the decrease of infectious diseases, or reduced insecticide use and crop losses due to agricultural pests). The application of GM insects for pest control directly affects the people in the target area by releasing the insects into their air space. Constitutionally everyone has the personal right to decide on things affecting one's own life. Moreover, different groups involved in and affected by the release of genome-edited insects (scientists, companies, authorities, producers, the public) will be guided by a variety of (contradicting) motivations concerning the release. Different perceptions of what is a desirable future and fear of new technological developments due to lack of understanding further complicate the situation.

Thus, it will be essential to identify and involve all the different stakeholders at an early stage of a project. An open and honest discussion of the limits of a technology and of the scientific knowledge, as well as public education will be crucial to build trust in science, in involved organizations, and in the decision-making process. Open discussion forums, where representatives of different stakeholders meet, could help to promote dialogue and the mutual understanding of goals, motivations, expectations and concerns.

Finally, the involvement of the public should not be limited to information and education campaigns. It should also include the collection of concerns of the educated public for discussion and consideration in the decision-making process.

7. POTENTIAL RISKS, CHANCES, AND CHALLENGES OF GENOME EDITING IN INSECT PEST CONTROL

7.1. Are Genome Editing Technologies for Insect Pest Control a Curse?

In general, genome editing could be used to create insect strains for population suppression or replacement approaches. Strains for population suppression could contain a trait that for example allows sexing (female elimination during rearing), and the strain would then be used in a SIT approach (i.e. radiation sterilisation and subsequent repeated mass-releases of the sterile males). On the other hand, they could contain a gene drive construct in combination with a trait allowing population reduction, for example by targeting female fertility genes. Such strains would then be used in a one-time limited release and the trait for population reduction would be driven through the population until a critical population density threshold is reached where effective reproduction is impaired. At this point the population would collapse in the targeted region and the genetic trait would disappear from the environment.

The situation would be different for the release of population replacement gene drives, as these genetic elements are designed to remain in the environment. Such genome-edited insects typically would have the gene drive component in their genome combined with a genetic modification that for example makes a mosquito population refractory to pathogen infection. The goal of such a construct would be to drive the immunity against the pathogen through the mosquito population until all or the majority of the insects are immune, which would interrupt the pathogen transmission cycle between the human host and the mosquito.

7.1.1. Use of Genome Edited Insects for Population Suppression in SIT-like Approaches

From a scientific point of view, the risk of using genome-edited insects for population suppression in SIT-like approaches is comparable to strains developed by classical mutagenesis approaches. The genome-edited trait disappears from the environment with the death of each released generation, as these insects cannot reproduce.

For both genome modifying technologies, genome editing and classical mutagenesis, a random mutation could arise in the generated strains that inactivates the trait. In the example of a genomic modification that kills specifically females for sexing purposes, a revertant mutation could occur at the modification site that allows females to survive again. Studies in *D. melanogaster* showed that such revertant mutations could occur at a frequency of 10^{-7} or less (Chovnick et al. 1971; Handler 2016). Therefore, at a mass-rearing scale of up to 1-3 x 10^9 insects per week, a few revertant insects are expected. Depending on the numbers, this could result in an efficiency concern, the release of females, or the contamination of the wild population with a marker. Nevertheless, all of those insects can be sterilized before release and would therefore not interfere with the overall success of the programme.

Critics might articulate concerns about off-target effects caused by genome editing technologies (for a review on CRISPR off-target effects see Zhang et al. 2015), resulting in additional mutations at other positions in the genome than the intended one. However, classical mutagenesis using chemicals or radiation also causes multiple unknown mutations and chromosome breaks in the genome in addition to the one causing the selected trait. In the case of classical mutagenesis this fact is accepted and not considered as a risk or potential problem, and the respective organisms can be deployed without regulations.

7.1.2. Self-perpetuating Gene Drive Systems Used for Population Suppression

In the case of a self-perpetuating gene drive systems used for population suppression, the success of a programme depends, besides the stability of the population reduction trait, on the stability of the gene drive. Resistance development against the drive would abolish the spread of the population reduction trait in the population, going back to normal Mendelian inheritance. This would not have any harmful consequences except that the suppression approach will not work anymore, and the modified insects will decline until completely gone, unless the mutation not only inactivates the drive but also lends a selective advantage over unmodified insects. However, even then, the non-functional insect is very unlikely to pose a threat to the environment or human health. For further population reduction, a new drive system would have to be developed.

7.1.3. Gene Drive Systems Used for Population Replacement

In the case of population replacement drives, potential resistance development on several levels could impair the success of the approach. First, the mosquito could develop a resistance against the drive, stopping the spread of the immunity trait and resulting in its eventual loss. Second, the pathogen could develop resistance against the immunity trait. In this case, the trait would further spread through the population due to the gene drive component, but it would not have the immunizing effect anymore.

The selection of resistant alleles, in general, is to be expected in each approach that puts a selective or survival pressure on a population of an organism. Such pressure would be applied to the target insect in case of population suppression approaches, and on the pathogen in replacement approaches spreading immunity through a mosquito population. This resistance development mechanism, however, is in no way connected to the use of genome editing technologies. It is already happening globally with the observed pathogen resistance developing against antivirals, antibiotics, or anti-malarial drugs, with the increasing resistance of insects against insecticides, or with behavioural resilience, for example the change in biting behaviour of mosquitoes from night to day to avoid bed nets (Liu 2015; Thomsen et al. 2017). It will always be an arms race, and we have to be aware that every human activity applying survival pressure on an organism will select for an evading reaction.

7.1.4. Risk of Overestimating the Understanding of Gene Functions

Genome editing experiments in different mammals have shown that there is a risk of overestimating our understanding of gene functions. It is often limited to one or two single functions, when instead genes often are part of large regulatory networks or have different functions throughout development. Modification of their expression can lead to unforeseen and unwanted consequences besides the desired effect. A variety of genome editing studies have been performed to knock out the *Myostatin* (*MSTN*) gene in different mammals. *MSTN* controls muscle growth, and existing knockout mutants show higher muscle mass and leaner meat. However, while resulting in the desired higher muscle mass, the *MSTN* knockout also caused unwanted side effects like additional thoracic vertebra in piglets (Qian et al. 2015), or rabbits with enlarged tongues and severe health problems like high rates of stillbirth and early stage death (Guo et al. 2016).

These results show that despite an increasing number of sequenced genomes, we still don't know much about gene functions. Therefore, careful studies will be needed for genome editing approaches aiming at population replacement, which will involve editing of specific genes to produce a certain phenotype – envisioned beneficial – that should stably persist in the target population without side-effects. This is of less concern when genome editing is used for population suppression applications, for example in scenarios where:

- A conditional female-killing mutation like the Mediterranean fruit fly *tsl* is recreated in another species to establish a sexing system
- A sex determination gene is knocked out to produce single-sex offspring for population reduction purposes, or
- A transgene construct is introduced into a specific genomic position previously characterized not to be disadvantageous to the insect.

In these scenarios, the quality of sterilized males in terms of mating performance, reproduction, and life span is the main characteristic to evaluate the usability of any strain.

7.1.5. Other General Risks

There are other general risks associated with area-wide pest or vector management approaches. These are, again, not specific to the use of genome edited insects, but apply to any control approach. In case of vector-borne diseases, the local eradication of a disease maintained over several generations can pose a risk to the human population in case of the reintroduction of the disease. Since the pathogen would not have challenged the human immune system for several generations, a reintroduction of the disease into that population could result in severe outcomes of the infection.

For population suppression approaches targeting endogenous species, the consequences of local eradication for the ecosystem are mostly unknown, as the species' role in the ecosystem is often not well studied. Therefore, possible consequences on the food chain, on competing species, and the possible long-term consequences for the human population, can only be guessed.

One concern in mosquito elimination approaches, for example, is that the niche opened by the (local) elimination of the target species could be filled quickly by another vector species that might transmit the same or even other diseases.

7.2. Are Genome Editing Technologies for Insect Pest Control a Blessing?

Currently the most used strategy to fight agricultural pests and vector insects is the application of insecticides. While they can be very effective in achieving rapid suppression in local applications, insecticides have many disadvantages. A major concern is their lack of species-specificity and the concomitant negative impact on many non-target insect species, or potentially even representing a risk to the environment. Furthermore, increasing resistance to insecticides is being observed in a rapidly growing number of pest insects worldwide, requiring an increase in application doses or the combination of different insecticides to still have an effect. This, however, is a dead end as it will lead to even stronger impacts on the environment and ultimately to complete resistance and loss of function of existing substances. Excessive use of insecticides in combination with other factors has already caused a dramatic decrease in the overall insect mass in Europe (Hallmann et al. 2017) and has led to the drastic decline of beneficial insects like pollinators as described in newspaper articles for some areas of China. An additional limitation of insecticides is that they cannot be applied on an area-wide basis due to public opposition. As a result, remote pest breeding sites are not accessible for treatment, thereby representing a constant, untouchable reservoir for the resurgence of the pest population after or despite of insecticide treatment.

Area-wide control programmes based on the release of modified insects are a promising strategy for the sustainable and species-specific pest control without the adverse side effects of insecticides. 'Modified' in this context can mean anything from sterilized by radiation for pest control using the SIT approach, all the way up to genetically modified insects being, for example, refractory to pathogen infection and carrying a gene drive construct for population replacement. Any method based on the release of modified insects has the substantial advantage of being highly specific to the target species and therefore not negatively impacting other non-target organisms, even those closely related. They also do not have toxic or other adverse side effects for the environment. Methods based on the release of modified insects are therefore environmentally safe and sustainable. Moreover, they allow an area-wide treatment against the whole population of the insect pest, and the mobility of the released insects also allows reaching remote habitats or protected areas that are inaccessible or not open for other approaches like the spraying of insecticides.

Approaches based on the release of modified insects also allow reaching the next generation of pest insects, like the drought-resistant eggs of *Aedes* mosquitoes, that are not affected by insecticide spraying and thus lead to a resurgence of the population in the next rainy season. The presence of modified insects throughout the egg hatching period, in contrast, would intercept the newly emerging generation and thus prevent a resurgence of the wild population.

Currently, the application of the SIT against new pest species is limited by a few bottlenecks that can be overcome. Besides a suitable mass-rearing system, an efficient sexing system that can be applied at large-scale is essential. With genome editing technologies one can envision the generation of such systems. For example, once suitable mutations obtained by classical mutagenesis in one target species are uncovered, genome editing could be used to reproduce these mutations in homologous genes in other insect pests.

One prominent example would be the above-mentioned *tsl* mutation in the Mediterranean fruit fly that specifically kills female embryos upon heat shock, allowing 100% sexing of this pest species at a large scale, which has been the key for Mediterranean fruit fly SIT programmes since more than 25 years and could allow building sexing systems in other pest insects (Robinson 2002; Franz et al. 2021). This mutation is an excellent example of a classically generated mutation that, once identified, can pave the way for similar technologies to be built in other pest insects.

Genome editing technologies are moreover the key to any approach intending to spread, via the use of a gene drive, sterility or pathogen refractoriness in a population without continuous mass-releases of the modified insects. Such strategies can only be pursued with the use of genome editing. The use of a gene drive to spread a lethality or sterility trait into a population for suppression approaches would have the advantage that optimally a one-time release of a seed population would be sufficient to suppress a population compared to continuous mass-releases required for the SIT approach.

Genome editing technologies will therefore be able to support and advance environment-friendly and sustainable pest control methods on various levels, and will thus help protect beneficial insect species, the biological diversity, and healthy ecosystems.

7.3. Challenges and Possible Solutions for Genome Editing Strategies

Genome editing technologies face multiple challenges that have to be understood and evaluated before they will gain the status of being acceptable and safe for use in insect pest control.

7.3.1. Technological Challenges

One of the challenges is the development and selection of resistant alleles observed with different (mostly CRISPR homing) gene drive approaches (Hammond et al. 2017; Reed 2017; Unckless et al. 2017; KaramiNejadRanjbar et al. 2018). While the drives in general work with a satisfying efficiency in laboratory experiments, the evolution of resistance alleles that are not recognized anymore by the gRNA of the CRISPR-Cas system was generally observed with CRISPR homing drives within just a few generations (Champer et al. 2018), resulting in the inactivation of the drive. Much effort is therefore being invested into the development of new drive strategies and improvement of the existing systems. Approaches include the use of multiple gRNAs, the expression of the drive exclusively in the male germline, or the targeting of highly conserved genes (Kyrou et al. 2018).

The rationale of the first approach is that a mutation in one of the gRNA target sites will not be able to inactivate the drive as the other target sites are still functional. A simultaneous mutation of multiple target sites is much less likely (even though it is not impossible), thereby preventing or at least strongly postponing drive inactivation (Prowse et al. 2017). Activating the drive only in the male germline should limit resistance-allele formation post-fertilization, as the sperm should transmit only low amounts of the Cas endonuclease into the embryo, thereby preventing mutagenic events in the embryo. Targeting highly conserved genes like the sex determination gene *doublesex* for homing has been effective in building a functional gene drive system in the laboratory (Kyrou et al. 2018). It can prevent resistance allele formation as mutations in highly conserved sequences would likely be deleterious to the organism, and the resistance alleles would not persist (Esvelt et al. 2014; Champer et al. 2016; Noble et al. 2017; Champer et al. 2018; Nash et al. 2018).

Other proposed systems reduce resistance potential in so-called daisy-chain drives and toxin-antidote systems (Champer et al. 2016; Noble et al. 2019). Current studies and simulations seem to indicate, however, that resistance evolution will remain an issue that can't be completely suppressed, just slowed down, and might require the combination of different strategies to design successful drives (Callaway 2017; Champer et al. 2018). A drive will be successful if it can spread through the whole population before the first resistance allele formation occurs.

Besides the development of resistance to gene drives, any genome editing approach can be affected by the appearance of spontaneous inactivating or revertant mutations due to the natural mutation rate in the genomes of insects. Most of such inactivating mutations would be point mutations, deletions, or changes by recombination events or moving transposons. With a mutation frequency of 10^{-5} to 10^{-7} per base pair, depending on the mutation event (Chovnick et al. 1971; Tobari and Kojima 1972; Bender et al. 1983; Neel 1983; Woodruff et al. 1983; Handler 2016), the occurrence of multiple mutations is to be expected as mass-rearing is scaled up towards 10^9 insects per week.

This phenomenon is not specific to genome editing technologies, however. It is a concern for any genetic modification used in insect mass-rearing approaches, including transgenic technologies or modifications induced by classical mutagenesis. Thus, already now, safeguard technology is used for strain maintenance like the filter rearing system used for ongoing fruit fly SIT programmes, to prevent the occurrence and persistence of inactivating mutations in the release population (Caceres 2005; Franz et al. 2021). Furthermore, besides such filter rearing systems, the safe use of transgenic insects for large-scale programmes will require backup systems. In case of failure of one system due to a random mutation, the other will serve as a safeguard to either preserve the strain function or prevent the strain from surviving in the wild due to the inactivating mutation (Eckermann et al. 2014; Handler 2016).

7.3.2. Ecologic and Economic Challenges

The success of any area-wide control programme involving, but not limited to, genome-edited insects will also depend on understanding the pest insect's population ecology. Population suppression, as well as replacement, strategies will be influenced to a different extent by factors like the dynamics within the population, insect migration behaviour and range, the geographic situation, environmental factors influencing the target population, as well as density threshold levels for successful reproduction or effective disease transmission, but also the adaptive or evasive potential of the target species. For a successful insect pest control programme with regard to environmental and human safety, it will additionally be essential to understand the insect's role in the food chain and its interaction or competition with other species.

Critics of population elimination approaches claim that the target pest could be replaced by a competing pest, causing the same or other problems. However, except for some rare cases, such as a very strong and stable gene drive that in theory should be capable of eliminating a whole population, most strategies for population control will only lead to a reduction in population size. Even if potentially they can lead to elimination in large areas, they won't have the potential for global eradication of a species, and a resurgence of the targeted species is expected as soon as the treatment is stopped.

The commercial applicability of genome-edited insects also depends on the crop and the pest complex threatening this crop. If one major pest species threatens the crop, the use of a species-specific strategy involving genome-edited insects is a practical approach. However, in a situation where a crop is equally affected by two or more pest species, the specific suppression or elimination of one of them could lead to population increases of the remaining pest species. In this case, any species-specific insect release strategy would also need to simultaneously address the other pest(s), but such an approach may be less economically viable.

7.3.3. Regulatory Challenges

The release of genome-edited insects for area-wide pest control programmes will face significant challenges concerning existing diverging regulations for genome-edited organisms. As discussed above in Section 5, there is no uniform international approach to the evaluation, risk assessment and the release of GM and genome-edited organisms, with every country pursuing an individual approach to regulation. Consequently, one country might allow the release of a genome-edited insect after a thorough evaluation of the product for stability and safety, while a neighbouring country might generally prohibit the release of any genome-edited insect. This will pose a problem for releasing countries and their agricultural trade, as the insects will not stop at their country's borders, and they would not be able to guarantee the confinement of the genome-edited insects within their territory, even if the insects are released far from the border (Reeves and Phillipson 2017). While this is already a difficult situation for non-disseminating approaches such as GM crops and insects, it would practically require inter-country regulation of any strategy relying on gene drives which have the potential and the purpose to spread.

Besides the regulations concerning the release of genome-edited insects, some food safety regulations will pose a challenge for the use of these insects in agricultural pest control. This will be most relevant for insects that feed inside of the fruits/crops. Such crops could be contaminated with the genome-edited insects, at least for the time of the releases, which could restrict marketing of these products in other countries, leading to the possible (temporary) loss or shifting of markets (Max-Planck-Gesellschaft 2017). In this case, the producing country would have to decide if the application of the genome-edited insects is economically worthwhile. It will also influence the approval processes for new genome-edited insect strains by the governments and regulatory bodies. How regulatory authorities in different countries will classify genome-edited insect material in agricultural products will, therefore, have a strong impact on international trade.

7.3.4. Identification and Involvement of Stakeholders

Area-wide pest management approaches involving the release of genome-edited insects will affect many different stakeholders. Identifying these stakeholders and their interests, and involving them in the planning and decision-making process, will be crucial for the success of any genome-edited insect deployment; it will also strongly impact the future of pest management projects. We summarize some important points here and refer the interested reader to the more in depths analyses performed by Gould (2008) and Baltzegar et al. (2018) in relation to the deployment of genome-edited insects in agriculture and human health.

Ecological, economic, regulatory, and social contexts need to be understood in depth in a case-by-case assessment to be able to determine all the groups and subgroups that will benefit or be negatively affected by each intervention with genome-edited insects. Regarding the ecological impact of the genome-edited insect release, different groups might have to be involved in the process on this level, including for example nature conservation organizations, ecologists, entomologists, farmer associations, public health workers, as well as environmental activists or NGOs opposing or not the use of GM or genome-edited organisms.

A central question for the identification of stakeholders will be who pays for the programme and who profits from it. Moreover, are the ones paying also the ones profiting from the deployment? Considering the case of agricultural pest control, who will benefit from the genome-edited insect releases will strongly depend on the production systems, but also on the crop(s) and the pest complexes threatening these crops. It can be assumed that a conventional farmer will mostly profit from the release due to the lower pest burden, resulting in lower costs for insecticide treatments and higher quality harvests with better yields and access to more profitable markets.

In contrast, for the organic farmers, the situation is complex. While they will also profit from the reduced pest insect burden, they might not be allowed to sell their products as they may contain residues of genome-edited material (for example in the form of insect larvae feeding inside the product). In many countries, there is a zero tolerance for GM material in organic food. Depending on how genome-edited material in organic food will be classified and regulated by individual authorities, the use of genome-edited insects might have a negative impact on organic farmers as their product might not be marketable anymore.

Thus, in case of a financing model that involves the producers in the application area, all would pay but not all would benefit. On the other hand, due to the insects' dispersal capacities, neighbouring producers might also profit from the decreased insect burden without paying, for example, if their production areas are located in proximity to a local programme.

Potential sales problems due to food contamination with genome-edited material will not be restricted to local markets but also affect international agricultural trade, where the different tolerance levels for GM/genome-edited material in food will affect the accessibility of markets. The negative impact will therefore not stop at the level of the producers, but it will also affect all downstream links in the trade chain and therefore many more stakeholders (Max-Planck-Gesellschaft 2017). On the one hand, the use of genome-edited insects might close some markets for certain production types (like organic farming), at least for the period of the release. On the other hand, the use of genome-edited insects could also lead to the reopening of markets, if the pest control measure results in the elimination of a quarantine pest, whose presence in the product would have prevented the export or required expensive additional post-harvest measures such as phytosanitary treatments.

Finally, a strong group of stakeholders that needs to be involved is the public, as the insects will be released in their airspace. Involvement of the public in the past has been handled very diversely. The first releases of GM insects were carried out with at best unidirectional public information campaigns but without any possibility to influence the decisions. This caused protest and considerable mistrust against companies and organizations promoting releases of GMOs (Subbaraman 2011; Baltzegar et al. 2018). Such experiences, step by step, initiated an important reconsideration and change in behaviour towards the public by scientists and organizations. In 2015, the release of GM diamondback moths was halted in New York, USA by the responsible authorities due to public disapproval despite the regulatory approval for the release (Boor 2015; USDA/APHIS 2015; Baltzegar et al. 2018). Similarly, the release of genetically modified mosquitoes by Oxitec Ltd. has been opposed in the target area in the Florida Keys, USA causing Oxitec to finally withdraw the application with the US Environmental Protection Agency (US-EPA) (Klingener 2018).

In Brazil, a major public awareness and engagement campaign was successfully conducted before initiating the release of transgenic mosquitoes in an urban area, emphasizing participatory action and a community-based programme (Capurro et al. 2016). Reconsideration of public involvement goes as far as claims from scientists to make biotechnology research that will affect "everyone" completely transparent from the beginning to build trust (Esvelt 2016, 2017). Certainly, for the success of a programme and its safe, economically worthwhile, and socially and environmentally responsible application, representatives of all relevant stakeholder groups should be identified and involved early in the decision-making process.

8. RECOMMENDATIONS FOR THE SAFE APPLICATION OF GENETICALLY ENGINEERED INSECTS FOR PEST CONTROL

8.1. Evaluation of the Product, not the Technology

From a scientific point of view, the evaluation process of a genome-edited insect should focus on the product, not the technology that was used to create the product. A technology like genome editing can be used to create a variety of genetic modifications, as discussed above, from point mutations and small insertions or deletions, to the introduction of whole transgene constructs. These modifications can also be achieved with other technologies like classical mutagenesis or transgenic approaches, although with different mechanisms. Therefore, the primary evaluation criteria for the product should be the introduced genetic trait, its properties, the projected consequences and benefits, as well as the potential risks in case of field deployment. In a second step, the evaluation should also consider the molecular mechanisms of the technology used, to be able to investigate potential side effects that are characteristic for each technology, including classical mutagenesis and breeding approaches.

A thorough product evaluation in our opinion should include the following points (without claim for completeness):

- Introduced genetic trait and its phenotypic properties – does it fulfil the aspired task?
- Stability of the genetic modification over time, and potential reasons for the trait to fail
- Consequences of a failure of the trait
- Molecular mechanism of the technology and its potential for side effects
- Are off-target mutations present and do the identified changes to the genome have consequences for the trait and its stability, for the genomic stability of the product in the environment upon release, or for the biological quality of the insect?
- Projected benefits of the release of the modified insect
- Potential risks of the deployment of the modified insect into the environment
- Benefits and risks (if any) of current pest control strategies
- Weighting of the benefits and risks of the new strategy against those of the current control strategy and against no intervention
- Possible ecological consequences (e.g. species reduction or elimination).

8.2. Product Evaluation and Risk Assessment

All the above points should be considered in the risk assessment of the product. For the best possible decision concerning a positive impact on society as well as nature, however, a thorough scientific product evaluation and risk assessment will not be enough. It will be crucial to have a wholistic decision-making process that will identify and involve representatives of all possible stakeholders and openly discuss their motivation, expectations and concerns early in the process.

The involvement of as many perspectives and sources of knowledge as possible from the start of the decision-making process will allow a more comprehensive process, that will also have a stronger democratic legitimization (Hartley et al. 2016).

One important point will be the honest, transparent and open discussion of the limits and gaps of the technology and the scientific knowledge and the consequences of a release. This will help to build trust in the scientific process and counteract the hype that can surround new biotechnologies (Caulfield and Condit 2012; Hartley et al. 2016). The 2016 publication by Hartley, based on the discussions of an international and interdisciplinary workshop entitled "Responsible Risk? Achieving Good Governance of Agricultural Biotechnology" held in Norway in November 2015, identifies five points that should be implemented in the scientifically and socially responsible development and application of agricultural biotechnology products: (1) commitment to candour, (2) recognition of underlying values and assumptions, (3) involvement of a broad range of knowledge and actors, (4) consideration of a range of alternatives, and (5) preparedness to respond (Hartley et al. 2016). In the end it should be a common decision of all stakeholders, if the expected positive impact of the product on society and nature is worth taking the risk, even if small.

9. CONCLUSIONS

Genome editing has a high potential for improvement of diverse human life situations. It could be the solution to control many vector-borne diseases, or safely and species-specifically control devastating agricultural pests, to reduce the burden on the human population worldwide. Nevertheless, it has to be studied thoroughly to the point where the technology and its potential drawbacks and side effects are very well understood, current technological roadblocks have been overcome, and the products are evaluated carefully according to well-defined regulations.

Regulations for genome-edited insects and other genome-edited products should be in the best interest of society based on scientific data acquired, rather than being based on an opinion from groups that are not independent. Therefore, also the underlying interests of institutions funding genome editing research should be critically analysed before they are allowed to give "public" views at any stage in the decision-making process. False information policies lead to the stalling of academic as well as industrial research, and in the case of Germany killing important innovation and economic development that will instead move to more open and innovation-friendly countries.

Genome editing is still in its infancy, and the consequences of tampering with gene function are not yet well understood, even when editing genes that researchers thought to be well studied. The gene networks are far more complex than so far assumed and meddling with them at this stage can have unforeseen consequences. While this lack of full understanding urges progressing with caution, it poses at the same time a strong demand for more research to understand the mechanisms involved, learn about side effects, and re-engineer the technology accordingly to make genome editing the safest possible technology. What would be counterproductive is a substantial restriction or even shutdown of the research due to uninformed or even misinformed fears and false information, which prevents better understanding and improvement of gene editing technologies.

In no way this should be a charter to do anything that is possible, but to base decision-making of every new technology on transparent science, facts, and comparing them to existing technologies to allow decisions like "are they an improvement or not?" This train of thought has been used in plant breeding for a long time, where only seeds with proven improvements over the existing ones are allowed on the market.

10. REFERENCES

Abudayyeh, O. O., J. S. Gootenberg, S. Konermann, J. Joung, I. M. Slaymaker, D. B. Cox, S. Shmakov, K. S. Makarova, E. Semenova, L. Minakhin, K. Severinov, A. Regev, E. S. Lander, E. V. Koonin, and F. Zhang. 2016. C2c2 is a single-component programmable RNA-guided RNA-targeting CRISPR effector. Science 353(6299): aaf5573.

Arnould, S., P. Chames, C. Perez, E. Lacroix, A. Duclert, J. C. Epinat, F. Stricher, A. S. Petit, A. Patin, S. Guillier, S. Rolland, J. Prieto, F. J. Blanco, J. Bravo, G. Montoya, L. Serrano, P. Duchateau, and F. Paques. 2006. Engineering of large numbers of highly specific homing endonucleases that induce recombination on novel DNA targets. Journal of Molecular Biolology 355(3): 443–458.

Aryan, A., M. A. Anderson, K. M. Myles, and Z. N. Adelman. 2013. TALEN-based gene disruption in the dengue vector *Aedes aegypti*. PLoS One 8(3): e60082.

Aumann, R. A., M. F. Schetelig, and I. Häcker. 2018. Highly efficient genome editing by homology-directed repair using Cas9 protein in *Ceratitis capitata*. Insect Biochemistry and Molecular Biology 101: 85–93.

Awata, H., T. Watanabe, Y. Hamanaka, T. Mito, S. Noji, and M. Mizunami. 2015. Knockout crickets for the study of learning and memory: Dopamine receptor *Dop1* mediates aversive but not appetitive reinforcement in crickets. Scientific Reports 5: 15885.

Baltzegar, J., J. C. Barnes, J. E. Elsensohn, N. Gutzmann, M. S. Jones, S. King, and J. Sudweeks. 2018. Anticipating complexity in the deployment of gene drive insects in agriculture. Journal of Responsible Innovation 5: S81–S97.

Bassett, A. R., C. Tibbit, C. P. Ponting, and J. L. Liu. 2014. Highly efficient targeted mutagenesis of *Drosophila* with the CRISPR/Cas9 System. Cell Reports 6(6): 1178–1179.

Becker, M. 2012. Method of the Year 2011. Nature Methods 9(1): 1.

Bender, W., M. Akam, F. Karch, P. A. Beachy, M. Peifer, P. Spierer, E. B. Lewis, and D. S. Hogness. 1983. Molecular genetics of the bithorax complex in *Drosophila melanogaster*. Science 221(4605): 23–29.

Bhaya, D., M. Davison, and R. Barrangou. 2011. CRISPR-Cas systems in bacteria and archaea: Versatile small RNAs for adaptive defense and regulation. Annual Review of Genetics 45: 273–297.

Bi, H. L., J. Xu, A. J. Tan, and Y. P. Huang. 2016. CRISPR/Cas9-mediated targeted gene mutagenesis in *Spodoptera litura*. Insect Science 23(3): 469–477.

Boor, K. J. 2015. Cage trials for diamondback moth product test to begin. June 11 2015. Cornell University, New York, USA.

Bourtzis, K., and Z. J. Tu. 2018. Joint FAO/IAEA Coordinated Research Project on "Exploring genetic, molecular, mechanical and behavioural methods of sex separation in mosquitoes" - An introduction. Parasites & Vectors 11(Suppl 2): 653.

Bozas, A., K. J. Beumer, J. K. Trautman, and D. Carroll. 2009. Genetic analysis of zinc-finger nuclease-induced gene targeting in *Drosophila*. Genetics 182(3): 641–651.

Buchman, A., J. M. Marshall, D. Ostrovski, T. Yang, and O. S. Akbari. 2018. Synthetically engineered *Medea* gene drive system in the worldwide crop pest *Drosophila suzukii*. Proceedings National Academy of Sciences USA 115(18): 4725–4730.

Burt, A. 2003. Site-specific selfish genes as tools for the control and genetic engineering of natural populations. Proceedings of the Royal Society B: Biological Sciences 270(1518): 921–928.

Caceres, C. 2005. Filter rearing system for sterile insect technology. J. L. Capinera (ed.), Encyclopedia of entomology. Springer, Dordrecht, The Netherlands.

Callaway, E. 2017. Gene drives thwarted by emergence of resistant organisms. Nature 542(7639): 15.

Capurro, M. L., D. O. Carvalho, L. Garziera, M. C. Pedrosa, I. Damasceno, I. Lima, B. Duarte, J. F. Virginio, R. S. Lees, and A. Malavasi. 2016. Description of social aspects surrounding releases of transgenic mosquitoes in Brazil. International Journal of Recent Scientific Research 7(4): 10363–10369.

Carvalho, D. O., A. R. McKemey, L. Garziera, R. Lacroix, C. A. Donnelly, L. Alphey, A. Malavasi, and M. L. Capurro. 2015. Suppression of a field population of *Aedes aegypti* in Brazil by sustained release of transgenic male mosquitoes. PLoS Neglegted Tropical Diseases 9(7): e0003864.

Caulfield, T., and C. Condit. 2012. Science and the sources of hype. Public Health Genomics 15:209–217.

Champer, J., A. Buchman, and O. S. Akbari. 2016. Cheating evolution: Engineering gene drives to manipulate the fate of wild populations. Nature Reviews Genetics 17(3): 146–159.

Champer, J., J. Liu, S. Y. Oh, R. Reeves, A. Luthra, N. Oakes, A. G. Clark, and P. W. Messer. 2018. Reducing resistance allele formation in CRISPR gene drive. Proceedings National Academy of Sciences USA 115(21): 5522–5527.

Chen, C. H., H. Huang, C. M. Ward, J. T. Su, L. V. Schaeffer, M. Guo, and B. A. Hay. 2007. A synthetic maternal-effect selfish genetic element drives population replacement in *Drosophila*. Science 316(5824): 597–600.

Chovnick, A., G. H. Ballantyne, and D. G. Holm. 1971. Studies on gene conversion and its relationship to linked exchange in *Drosophila melanogaster*. Genetics 69(2): 179–209.

Curtis, C. F. 1978. Genetic sex separation in *Anopheles arabiensis* and the production of sterile hybrids. Bulletin World Health Organization 56(3): 453–454.

Curtis, C. F., J. Akiyama, and G. Davidson. 1976. Genetic sexing system in *Anopheles gambiae* species A. Mosquito News 36(4): 492–498.

Disdier, A. C., and L. Fontagne. 2010. Trade impact of European measures on GMOs condemned by the WTO panel. Review of World Economics 146(3): 495–514.

Drury, D. W., A. L. Dapper, D. J. Siniard, G. E. Zentner, and M. J. Wade. 2017. CRISPR/Cas9 gene drives in genetically variable and nonrandomly mating wild populations. Science Advances 3(5): e1601910.

Dyck, V. A., J. Hendrichs, and A. S. Robinson (eds.). 2021. Sterile Insect Technique – Principles and practice in Area-Wide Integrated Pest Management. Second Edition. CRC Press, Boca Raton, Florida, USA. 1200 pp.

(ECDC) European Centre for Disease Prevention and Control. 2018. Rapid risk assessment: Local transmission of dengue fever in France and Spain. 22 October 2018. Solna, Sweden.

Eckermann, K. N., S. Dippel, M. KaramiNejadRanjbar, H. M. Ahmed, I. M. Curril, and E. A. Wimmer. 2014. Perspective on the combined use of an independent transgenic sexing and a multifactorial reproductive sterility system to avoid resistance development against transgenic Sterile Insect Technique approaches. BMC Genetics 15 Suppl 2: S17.

(EEA) European Environment Agency. 2017. Invasive alien species in Europe. Last modified 26 Oct 2017. Copenhagen, Denmark.

Esvelt, K. M. 2016. Gene editing can drive science to openness. Nature 534(153).

Esvelt, K. M. 2017. Precaution: Open gene drive research. Science 355(6325): 589–590.

Esvelt, K. M., A. L. Smidler, F. Catteruccia, and G. M. Church. 2014. Concerning RNA-guided gene drives for the alteration of wild populations. Elife 3: e03401.

Franz, G., K. Bourtzis, and C. Cáceres. 2021. Practical and operational genetic sexing systems based on classical genetic approaches in fruit flies, an example for other species amenable to large-scale rearing for the Sterile Insect Technique, pp. 575–604. *In* V. A. Dyck, J. Hendrichs, and A. S. Robinson (eds.), Sterile Insect Technique – Principles and practice in Area-Wide Integrated Pest Management. Second Edition. CRC Press, Boca Raton, Florida, USA.

Fu, G., K. C. Condon, M. J. Epton, P. Gong, L. Jin, G. C. Condon, N. I. Morrison, T. H. Dafa'alla, and L. Alphey. 2007. Female-specific insect lethality engineered using alternative splicing. Nature Biotechnology 25(3): 353–357.

Fu, Y., J. D. Sander, D. Reyon, V. M. Cascio, and J. K. Joung. 2014. Improving CRISPR-Cas nuclease specificity using truncated guide RNAs. Nature Biotechnology 32(3): 279–284.

Galizi, R., L. A. Doyle, M. Menichelli, F. Bernardini, A. Deredec, A. Burt, B. L. Stoddard, N. Windbichler, and A. Crisanti. 2014. A synthetic sex ratio distortion system for the control of the human malaria mosquito. Nature Communications 5: 3977.

Galizi, R., A. Hammond, K. Kyrou, C. Taxiarchi, F. Bernardini, S. M. O'Loughlin, P. A. Papathanos, T. Nolan, N. Windbichler, and A. Crisanti. 2016. A CRISPR-Cas9 sex-ratio distortion system for genetic control. Scientific Reports 6: 31139.

Gantz, V. M., N. Jasinskiene, O. Tatarenkova, A. Fazekas, V. M. Macias, E. Bier, and A. A. James. 2015. Highly efficient Cas9-mediated gene drive for population modification of the malaria vector mosquito *Anopheles stephensi*. Proceedings National Academy of Sciences USA 112(49): E6736–6743.

Gilles, J. R., M. F. Schetelig, F. Scolari, F. Marec, M. L. Capurro, G. Franz, and K. Bourtzis. 2014. Towards mosquito Sterile Insect Technique programmes: Exploring genetic, molecular, mechanical and behavioural methods of sex separation in mosquitoes. Acta Tropica 132 (Suppl.): S178–187.

Global Legal Research Center. 2014. Restrictions on genetically modified organisms. The Law Library of Congress, Washington, DC, USA.

Gong, P., M. J. Epton, G. Fu, S. Scaife, A. Hiscox, K. C. Condon, G. C. Condon, N. I. Morrison, D. W. Kelly, T. Dafa'alla, P. G. Coleman, and L. Alphey. 2005. A dominant lethal genetic system for autocidal control of the Mediterranean fruit fly. Nature Biotechnology 23(4): 453–456.

Gould, F. 2008. Broadening the application of evolutionarily based genetic pest management. Evolution 62(2): 500–510.

Guo, R., Y. Wan, D. Xu, L. Cui, M. Deng, G. Zhang, R. Jia, W. Zhou, Z. Wang, K. Deng, M. Huang, F. Wang, and Y. Zhang. 2016. Generation and evaluation of *Myostatin* knock-out rabbits and goats using CRISPR/Cas9 system. Scientific Reports 6: 29855.

Hall, A. B., S. Basu, X. Jiang, Y. Qi, V. A. Timoshevskiy, J. K. Biedler, M. V. Sharakhova, R. Elahi, M. A. Anderson, X. G. Chen, I. V. Sharakhov, Z. N. Adelman, and Z. Tu. 2015. A male-determining factor in the mosquito *Aedes aegypti*. Science 348(6240): 1268–1270.

Hallmann, C. A., M. Sorg, E. Jongejans, H. Siepel, N. Hofland, H. Schwan, W. Stenmans, A. Muller, H. Sumser, T. Horren, D. Goulson, and H. de Kroon. 2017. More than 75 percent decline over 27 years in total flying insect biomass in protected areas. PLoS One 12(10): e0185809.

Hammond, A., R. Galizi, K. Kyrou, A. Simoni, C. Siniscalchi, D. Katsanos, M. Gribble, D. Baker, E. Marois, S. Russell, A. Burt, N. Windbichler, A. Crisanti, and T. Nolan. 2016. A CRISPR-Cas9 gene drive system targeting female reproduction in the malaria mosquito vector *Anopheles gambiae*. Nature Biotechnology 34(1): 78–83.

Hammond, A. M., K. Kyrou, M. Bruttini, A. North, R. Galizi, X. Karlsson, N. Kranjc, F. M. Carpi, R. D'Aurizio, A. Crisanti, and T. Nolan. 2017. The creation and selection of mutations resistant to a gene drive over multiple generations in the malaria mosquito. PLoS Genetics 13(10): e1007039.

Handler, A. M. 2016. Enhancing the stability and ecological safety of mass-reared transgenic strains for field release by redundant conditional lethality systems. Insect Science 23(2): 225–234.

Hanin, M., and J. Paszkowski. 2003. Plant genome modification by homologous recombination. Current Opinion Plant Biology 6(2): 157–162.

Hartley, S., F. Gillund, L.van Hove, and F. Wickson. 2016. Essential features of responsible governance of agricultural biotechnology. PLoS Biology 14(5): e1002453-e1002453.

(HCB) Haute Conseil de Biotechnologies, France (2017). Scientific Opinion of the High Council for Biotechnology concerning use of genetically modified mosquitoes for vector control in response to the referral of 12 October 2015 (Ref. HCB-2017.06.07). Paris, France. 142 pp.

Heinze, S. D., T. Kohlbrenner, D. Ippolito, A. Meccariello, A. Burger, C. Mosimann, G. Saccone, and D. Bopp. 2017. CRISPR-Cas9 targeted disruption of the yellow ortholog in the housefly identifies the brown body locus. Scientific Reports 7(1): 4582.

Horvath, P., and R. Barrangou. 2010. CRISPR/Cas, the immune system of bacteria and archaea. Science 327(5962): 167–170.

Isaacs, A. T., F. Li, N. Jasinskiene, X. Chen, X. Nirmala, O. Marinotti, J. M. Vinetz, and A. A. James. 2011. Engineered resistance to *Plasmodium falciparum* development in transgenic *Anopheles stephensi*. PLoS Pathogens 7(4): e1002017.

Isaacs, A. T., N. Jasinskiene, M. Tretiakov, I. Thiery, A. Zettor, C. Bourgouin, and A. A. James. 2012. Transgenic *Anopheles stephensi* coexpressing single-chain antibodies resist *Plasmodium falciparum* development. Proceedings National Academy of Sciences USA 109(28): E1922–1930.

Ishii, A., A. Kurosawa, S. Saito, and N. Adachi. 2014. Analysis of the role of homology arms in gene-targeting vectors in human cells. PLoS One 9(9): e108236.

Jinek, M., K. Chylinski, I. Fonfara, M. Hauer, J. A. Doudna, and E. Charpentier. 2012. A programmable dual-RNA-guided DNA endonuclease in adaptive bacterial immunity. Science 337(6096): 816–821.

Kalajdzic, P., and M. F. Schetelig. 2017. CRISPR/Cas-mediated gene editing using purified protein in *Drosophila suzukii*. Entomologia Experimentalis et Applicata 164(3): 350–362.

KaramiNejadRanjbar, M., K. N. Eckermann, H. M. M. Ahmed, C. H. Sanchez, S. Dippel, J. M. Marshall, and E. A. Wimmer. 2018. Consequences of resistance evolution in a Cas9-based sex conversion-suppression gene drive for insect pest management. Proceedings National Academy of Sciences USA 115(24): 6189–6194.

Kistler, K. E., L. B. Vosshall, and B. J. Matthews. 2015. Genome engineering with CRISPR-Cas9 in the mosquito *Aedes aegypti*. Cell Reports 11(1): 51–60.

Kleinstiver, B. P., V. Pattanayak, M. S. Prew, S. Q. Tsai, N. T. Nguyen, Z. Zheng, and J. K. Joung. 2016. High-fidelity CRISPR-Cas9 nucleases with no detectable genome-wide off-target effects. Nature 529(7587): 490–495.

Klingener, N. 2018. GMO mosquito application withdrawn — But another is on the way. November 28, 2018. WLRN, Miami, Florida, USA.

Kyrou, K., A. M. Hammond, R. Galizi, N. Kranjc, A. Burt, A. K. Beaghton, T. Nolan, and A. Crisanti. 2018. A CRISPR-Cas9 gene drive targeting doublesex causes complete population suppression in caged *Anopheles gambiae* mosquitoes. Nature Biotechnology 36(11): 1062–1066.

Li, F., and M. J. Scott. 2016. CRISPR/Cas9-mediated mutagenesis of the *white* and *Sex lethal* loci in the invasive pest, *Drosophila suzukii*. Biochemical and Biophysical Research Communications 469(4): 911–916.

Li, J., and A. M. Handler. 2017. Temperature-dependent sex-reversal by a transformer-2 gene-edited mutation in the spotted wing drosophila, *Drosophila suzukii*. Scientific Reports 7(1): 12363.

Liu, J., C. Li, Z. Yu, P. Huang, H. Wu, C. Wei, N. Zhu, Y. Shen, Y. Chen, B. Zhang, W. M. Deng, and R. Jiao. 2012. Efficient and specific modifications of the *Drosophila* genome by means of an easy TALEN strategy. Journal of Genetics & Genomics 39(5): 209–215.

Liu, N. 2015. Insecticide resistance in mosquitoes: Impact, mechanisms, and research directions. Annual Review of Entomology 60: 537–559.

Ma, S., J. Chang, X. Wang, Y. Liu, J. Zhang, W. Lu, J. Gao, R. Shi, P. Zhao, and Q. Xia. 2014. CRISPR/Cas9 mediated multiplex genome editing and heritable mutagenesis of BmKu70 in *Bombyx mori*. Scientific Reports 4: 4489.

Markert, M. J., Y. Zhang, M. S. Enuameh, S. M. Reppert, S. A. Wolfe, and C. Merlin. 2016. Genomic access to monarch migration using TALEN and CRISPR/Cas9-mediated targeted mutagenesis. G3: Genes, Genomes, Genetics (Bethesda) 6(4): 905–915.

Marraffini, L. A., and E. J. Sontheimer. 2010. CRISPR interference: RNA-directed adaptive immunity in bacteria and archaea. Nature Review of Genetics 11(3): 181–190.

Marshall, J. M., A. Buchman, C. H. Sanchez, and O. S. Akbari. 2017. Overcoming evolved resistance to population-suppressing homing-based gene drives. Scientific Reports 7(1): 3776.

Max-Planck-Gesellschaft. 2017. Genetically modified insects could disrupt international food trade: Genetically modified organisms for pest control could end up as contaminants in agricultural products throughout the globe. ScienceDaily, 1 February 2017.

Meccariello, A., S. M. Monti, A. Romanelli, R. Colonna, P. Primo, M. G. Inghilterra, G. Del Corsano, A. Ramaglia, G. Iazzetti, A. Chiarore, F. Patti, S. D. Heinze, M. Salvemini, H. Lindsay, E. Chiavacci, A. Burger, M. D. Robinson, C. Mosimann, D. Bopp, and G. Saccone. 2017. Highly efficient DNA-free gene disruption in the agricultural pest *Ceratitis capitata* by CRISPR-Cas9 ribonucleoprotein complexes. Scientific Reports 7(1): 10061.

Meccariello, A., M.Salvemini, P. Primo, B. Hall, P. Koskinioti, M. Dalíková, A. Gravina, M. A. Gucciardino, F. Forlenza, M. E. Gregoriou, D. Ippolito, S. M. Monti, V. Petrella, M. M. Perrotta, S. Schmeing, A. Ruggiero, F. Scolari, E. Giordano, K. T. Tsoumani, F. Marec, N. Windbichler, J. Nagaraju, K. P. Arunkumar, K. Bourtzis, K. D. Mathiopoulos, J. Ragoussis, L. Vitagliano, Z. Tu, P. A. Papathanos, M. D. Robinson, and G. Saccone. 2019. Maleness-on-the-Y (MoY) orchestrates male sex determination in major agricultural fruit fly pests. Science 365(6460): 1457–1460.

Nash, A., G. M. Urdaneta, A. K. Beaghton, A. Hoermann, P. A. Papathanos, G. K. Christophides, and N. Windbichler. 2018. Integral gene drives for population replacement. Biology Open: Bio.037762.

Ndo, C., Y. Poumachu, D. Metitsi, H. P. Awono-Ambene, T. Tchuinkam, J. L. R. Gilles, and K. Bourtzis. 2018. Isolation and characterization of a temperature-sensitive lethal strain of *Anopheles arabiensis* for SIT-based application. Parasites & Vectors 11 (Suppl 2): 659.

Neel, J. V. 1983. Frequency of spontaneous and induced point mutations in higher eukaryotes. Journal of Heredity 74(1): 2–15.

Noble, C., J. Olejarz, K. M. Esvelt, G. M. Church, and M. A. Nowak. 2017. Evolutionary dynamics of CRISPR gene drives. Science Advances 3(4): e1601964.

Noble, C., J. Min, J. Olejarz, J. Buchthal, A. Chavez, A. L. Smidler, E. A. DeBenedictis, G. M. Church, M. A. Nowak, and K.M. Esvelt. 2019. Daisy-chain gene drives for the alteration of local populations. Proceedings of the National Academy of Sciences USA 116(17): 8275–8282.

Nowak, C. M., S. Lawson, M. Zerez, and L. Bleris. 2016. Guide RNA engineering for versatile Cas9 functionality. Nucleic Acids Research 44(20): 9555–9564.

Ogaugwu, C. E., M. F. Schetelig, and E. A. Wimmer. 2013. Transgenic sexing system for *Ceratitis capitata* (Diptera: Tephritidae) based on female-specific embryonic lethality. Insect Biochemistry and Molecular Biology 43(1): 1–8.

Phuc, H. K., M. H. Andreasen, R. S. Burton, C. Vass, M. J. Epton, G. Pape, G. Fu, K. C. Condon, S. Scaife, C. A. Donnelly, P. G. Coleman, H. White-Cooper, and L. Alphey. 2007. Late-acting dominant lethal genetic systems and mosquito control. BMC Biology 5: 11.

Prowse, T. A. A., P. Cassey, J. V. Ross, C. Pfitzner, T. A. Wittmann, and P. Thomas. 2017. Dodging silver bullets: Good CRISPR gene-drive design is critical for eradicating exotic vertebrates. Proceedings of the Royal Society B: Biological Sciences 284(1860).

Puchta, H. 2002. Gene replacement by homologous recombination in plants. Plant Molecular Biology 48(1–2): 173–182.

Qian, L., M. Tang, J. Yang, Q. Wang, C. Cai, S. Jiang, H. Li, K. Jiang, P. Gao, D. Ma, Y. Chen, X. An, K. Li, and W. Cui. 2015. Targeted mutations in *Myostatin* by zinc-finger nucleases result in double-muscled phenotype in Meishan pigs. Scientific Reports 5: 14435.

Reed, F. A. 2017. CRISPR/Cas9 gene drive: Growing pains for a new technology. Genetics 205(3): 1037–1039.

Reeves, R., and M. Phillipson. 2017. Mass releases of genetically modified insects in area-wide pest control programs and their impact on organic farmers. Sustainability 9: 59.

Reiss, B. 2003. Homologous recombination and gene targeting in plant cells. International Review of Cytology 228: 85–139.

Ren, X., J. Sun, B. E. Housden, Y. Hu, C. Roesel, S. Lin, L. P. Liu, Z. Yang, D. Mao, L. Sun, Q. Wu, J. Y. Ji, J. Xi, S. E. Mohr, J. Xu, N. Perrimon, and J. Q. Ni. 2013. Optimized gene editing technology for *Drosophila melanogaster* using germ line-specific Cas9. Proceedings National Academy of Sciences USA 110(47): 19012–19017.

Robinson, A. S. 2002. Genetic sexing strains in medfly, *Ceratitis capitata*, Sterile Insect Technique programmes. Genetica 116(1): 5–13.

Rosen, L. E., H. A. Morrison, S. Masri, M. J. Brown, B. Springstubb, D. Sussman, B. L. Stoddard, and L. M. Seligman. 2006. Homing endonuclease I-CreI derivatives with novel DNA target specificities. Nucleic Acids Research 34(17): 4791–4800.

Schetelig, M. F., and A. M. Handler. 2012a. Strategy for enhanced transgenic strain development for embryonic conditional lethality in *Anastrepha suspensa*. Proceedings National Academy of Sciences USA 109(24): 9348–9353.

Schetelig, M. F., and A. M. Handler. 2012b. A transgenic embryonic sexing system for *Anastrepha suspensa* (Diptera: Tephritidae). Insect Biochemistry and Molecular Biology 42(10): 790–795.

Schetelig, M. F., C. Caceres, A. Zacharopoulou, G. Franz, and E. A. Wimmer. 2009. Conditional embryonic lethality to improve the Sterile Insect Technique in *Ceratitis capitata* (Diptera: Tephritidae). BMC Biology 7: 4.

Schetelig, M. F., A. Targovska, J. S. Meza, K. Bourtzis, and A. M. Handler. 2016. Tetracycline-suppressible female lethality and sterility in the Mexican fruit fly, *Anastrepha ludens*. Insect Molecular Biology 25(4): 500–508.

Seawright, J. A., P. E. Kaiser, D. A. Dame, and C. S. Lofgren. 1978. Genetic method for the preferential elimination of females of *Anopheles albimanus*. Science 200(4347): 1303–1304.

Sharma, A., S. D. Heinze, Y. Wu, T. Kohlbrenner, I. Morilla, C. Brunner, E. A. Wimmer, L. van de Zande, M. D. Robinson, L. W. Beukeboom, and D. Bopp. 2017. Male sex in houseflies is determined by *Mdmd*, a paralog of the generic splice factor gene *CWC22*. Science 356(6338): 642–645.

Smidler, A. L., O. Terenzi, J. Soichot, E. A. Levashina, and E. Marois. 2013. Targeted mutagenesis in the malaria mosquito using TALE nucleases. PLoS One 8(8): e74511.

Subbaraman, N. 2011. Science snipes at Oxitec transgenic-mosquito trial. Nature Biotechnology 29: 9–11.

Sussman, D., M. Chadsey, S. Fauce, A. Engel, A. Bruett, R. Monnat, Jr., B. L. Stoddard, and L. M. Seligman. 2004. Isolation and characterization of new homing endonuclease specificities at individual target site positions. Journal of Molecular Biology 342(1): 31–41.

Takasu, Y., I. Kobayashi, K. Beumer, K. Uchino, H. Sezutsu, S. Sajwan, D. Carroll, T. Tamura, and M. Zurovec. 2010. Targeted mutagenesis in the silkworm *Bombyx mori* using zinc finger nuclease mRNA injection. Insect Biochemistry and Molecular Biology 40(10): 759–765.

Takasu, Y., S. Sajwan, T. Daimon, M. Osanai-Futahashi, K. Uchino, H. Sezutsu, T. Tamura, and M. Zurovec. 2013. Efficient TALEN construction for *Bombyx mori* gene targeting. PLoS One 8(9): e73458.

Thomas, D. D., C. A. Donnelly, R. J. Wood, and L. S. Alphey. 2000. Insect population control using a dominant, repressible, lethal genetic system. Science 287(5462): 2474–2476.

Thomsen, E. K., G. Koimbu, J. Pulford, S. Jamea-Maiasa, Y. Ura, J. B. Keven, P. M. Siba, I. Mueller, M. W. Hetzel, and L. J. Reimer. 2017. Mosquito behavior change after distribution of bednets results in decreased protection against malaria exposure. Journal Infectious Diseases 215(5): 790–797.

Tobari, Y. N., and K. I. Kojima. 1972. A study of spontaneous mutation rates at ten loci detectable by starch gel electrophoresis in *Drosophila melanogaster*. Genetics 70(3): 397–403.

(USDA/APHIS) United States Department of Agriculture / Animal and Plant Health Inspection Service. 2015. National environmental policy act decision and finding of no significant impact. Field release of gentically engineered diamondback moths strains OX4319L-Pxy, OX4319N-Pxy, and OX4767A-Pxy.

USDA (United States Department of Agriculture). 2018. Details on USDA plant breeding innovations. Last modified: June 14, 2018. Animal and Plant Health Inspection Service, United States Department of Agriculture, Washington, DC, USA.

Unckless, R. L., A. G. Clark, and P. W. Messer. 2017. Evolution of resistance against CRISPR/Cas9 gene drive. Genetics 205(2): 827–841.

Vasquez, K. M., K. Marburger, Z. Intody, and J. H. Wilson. 2001. Manipulating the mammalian genome by homologous recombination. Proceedings National Academy of Sciences USA 98(15): 8403–8410.

Waltz, E. 2016. Gene-edited CRISPR mushroom escapes US regulation. Nature 532(7599): 293.

Wang, Y., Z. Li, J. Xu, B. Zeng, L. Ling, L. You, Y. Chen, Y. Huang, and A. Tan. 2013. The CRISPR/Cas system mediates efficient genome engineering in *Bombyx mori*. Cell Research 23(12): 1414–1416.

Windbichler, N., M. Menichelli, P. A. Papathanos, S. B. Thyme, H. Li, U. Y. Ulge, B. T. Hovde, D. Baker, R. J. Monnat, Jr., A. Burt, and A. Crisanti. 2011. A synthetic homing endonuclease-based gene drive system in the human malaria mosquito. Nature 473(7346): 212–215.

Woodruff, R., B. Slatko, and J. Thompson. 1983. Factors affecting mutation rates in natural populations, pp. 37–124. *In* M. Ashburner, H. L. Carson, and J. N. Thompson (eds.), The genetics and biology of *Drosophila* Vol. 3c. Academic Press, London, UK.

Yamada, H., M. Q. Benedict, C. A. Malcolm, C. F. Oliva, S. M. Soliban, and J. R. Gilles. 2012. Genetic sex separation of the malaria vector, *Anopheles arabiensis*, by exposing eggs to dieldrin. Malaria Journal 11: 208.

Yan, Y., R. J. Linger, and M. J. Scott. 2017. Building early-larval sexing systems for genetic control of the Australian sheep blow fly *Lucilia cuprina* using two constitutive promoters. Scientific Reports 7(1): 2538.

Yang, H., P. Gao, K. R. Rajashankar, and D. J. Patel. 2016. PAM-dependent target DNA recognition and cleavage by C2c1 CRISPR-Cas endonuclease. Cell 167(7): 1814–1828.e12.

Yu, Z., M. Ren, Z. Wang, B. Zhang, Y. S. Rong, R. Jiao, and G. Gao. 2013. Highly efficient genome modifications mediated by CRISPR/Cas9 in *Drosophila*. Genetics 195(1): 289–291.

Zetsche, B., J. S. Gootenberg, O. O. Abudayyeh, I. M. Slaymaker, K. S. Makarova, P. Essletzbichler, S. E. Volz, J. Joung, J. van der Oost, A. Regev, E. V. Koonin, and F. Zhang. 2015. Cpf1 is a single RNA-guided endonuclease of a class 2 CRISPR-Cas system. Cell 163(3): 759–771.

Zhang, X. H., L. Y. Tee, X. G. Wang, Q. S. Huang, and S. H. Yang. 2015. Off-target effects in CRISPR/Cas9-mediated genome engineering. Molecular Therapy Nucleic Acids 4: e264.

SYNTHETIC SEX RATIO DISTORTERS BASED ON CRISPR FOR THE CONTROL OF HARMFUL INSECT POPULATIONS

B. FASULO[1], A. MECCARIELLO[1], P. A. PAPATHANOS[2] AND N. WINDBICHLER[1]

[1]Imperial College London, Department of Life Sciences, Imperial College Road, London, SW7 2AZ, UK; nikolai.windbichler@imperial.ac.uk
[2]Department of Experimental Medicine, University of Perugia, Perugia, Italy

SUMMARY

Since the overall reproductive output of a population is typically determined by the fertility of its females, which are rate-limiting in gamete production, a successful way to genetically control a population should involve artificially biasing the sex ratio towards males. In male heterogametic species, this could be achieved by the expression of a transgene-encoded endonuclease during spermatogenesis that would target and "shred" the X chromosome at several loci. This would prevent the transmission of X chromosome bearing gametes to the progeny, generating only males. Recent developments in molecular and synthetic biology have provided genome editing tools with great potential to engineer the genome of different species. Given the targeting flexibility of CRISPR-based endonucleases, it may now be possible to test whether X chromosome shredding has the potential to become a universal strategy to genetically control a wide variety of insect pests, of both agricultural and public health relevance.

Key Words: Meiotic drive, sex ratio distorters, biased sex ratio, gene editing, genome editing, X-shredding, genetic control, spermatogenesis, male-determining genes, gene drive, selfish chromosomes

1. INTRODUCTION

In most sexually reproducing organisms, males and females are generated in approximately the same numbers. Fisher's principle states that the sex ratio is in equilibrium when an individual spends the same amount of energy to produce equal numbers of males and females. When the ratio is different from the equilibrium, the less predominant sex, or rather genes determining development towards this sex, will have an advantage that will last until the equilibrium is re-established (Fisher 1958). As the balanced sex ratio is approached, the advantage associated with producing the

J. Hendrichs, R. Pereira and M. J. B. Vreysen (eds.), Area-Wide Integrated Pest Management: Development and Field Application, pp. 843–855. CRC Press, Boca Raton, Florida, USA.

rarer sex wanes, and the equilibrium is re-established. Consequently, novel genetic traits that bias the sex ratio towards one sex gain a short-term advantage but are eventually counterbalanced by a neutralizing evolutionary force in the form of drive suppressors that evolve on the autosomes or on the Y chromosome. Sex ratios can therefore be portrayed as the dynamic outcome of an ongoing evolutionary "tug of war" (Argasinski 2013). Fisher's principle only applies in those cases where the sex-ratio is controlled by genes acting in the homogametic sex, or by autosomal genes acting in the heterogametic sex, in XX-XY and ZW-ZZ systems.

In a population of a sexually reproducing organism, a dramatic sex bias towards one of the two sexes usually decreases the population's overall fertility. Since the overall reproductive output of a population is typically determined by the fertility of its females, which are rate-limiting in gamete production, a successful way to genetically control the population could involve artificially biasing the sex ratio towards males (Hamilton 1967). Such a genetic control strategy could reduce the size of harmful animal populations, such as agricultural insect pests or disease-vector species, or even result in the suppression or collapse of the population before suppressor alleles can arise to re-establish a balanced sex ratio. In control programmes, a male-biased sex ratio would also be favourable because females are often responsible for the damage (e.g. they are often the vectors of human parasites or viruses (e.g. mosquitoes) or lay eggs in agricultural products (e.g. fruit flies).

Hamilton was among the first to suggest how genetic sex ratio distorters (SRDs) could be applied to eradicate mosquito populations, and that under certain conditions, non-Fisherian sex ratios could arise and yet be maintained. He considered a heterogametic species with males (XY) harbouring a mutant Y chromosome that can bias fertilization in its favour at the cost of the X chromosome. Males carrying such a Y mutation would only produce sons. As a result, the mutant Y chromosome would gain a selective advantage and spread within the population, rendering it increasingly male-biased at each generation. The decline in female numbers would result in a decrease in size and eventually the collapse of the population (Hamilton 1967). In this example of SRD, an invasive Y chromosome interferes with the production of X-bearing gametes during spermatogenesis and spreads through the population in a self-sustaining manner. Hamilton's thought experiment was inspired by observations of natural populations of mosquitoes, where Craig et al. (1960) reported that a SRD transmitted by males was responsible for a male bias in *Aedes aegypti* (L.), the yellow fever mosquito. Observations in other species since then have also identified an X chromosome bias and thus more female progeny in some populations of *Drosophila simulans* Sturtevant (Mercot et al. 1995). Importantly, it is often difficult to identify the existence of a SRD in a population because local suppressors evolved to counteract it – had they not the population would have likely disappeared.

For the purpose of this chapter, it is important to distinguish such driving (invasive) SRDs from traits that bias the sex ratio, but that are inherited in a non-driving standard Mendelian fashion (Fig. 1). Both types of SRDs, in turn, are distinct from female killing systems (FK) that also result in a biased sex ratio in progeny, but they do so at the expense of reproductive output because of post-zygotically lethality of female offspring.

Unlike the bias generated by female killing systems (FK), SRDs operate pre-zygotically during gametogenesis and thus do not result in an overall reduction of male fertility. The main advantage of driving SRDs, as far as genetic population control is concerned, is their invasiveness. In the absence of resistance against the drive, the driving Y chromosome will eliminate the X chromosome in sperm and eventually lead to population collapse due to the lack of females.

Nevertheless, non-invasive SRD traits also have significant potential for genetic control, although they are eventually lost in the absence of continuous releases. They are advantageous in particular when compared to other forms of inundative genetic control approaches such as the Sterile Insect Technique (SIT) (Dyck et al. 2021) or its transgenic cousin, the Release of Insects carrying a Dominant Lethal (RIDL technology) (Thomas et al. 2000) as they require smaller effective releases (Schliekelman et al. 2005).

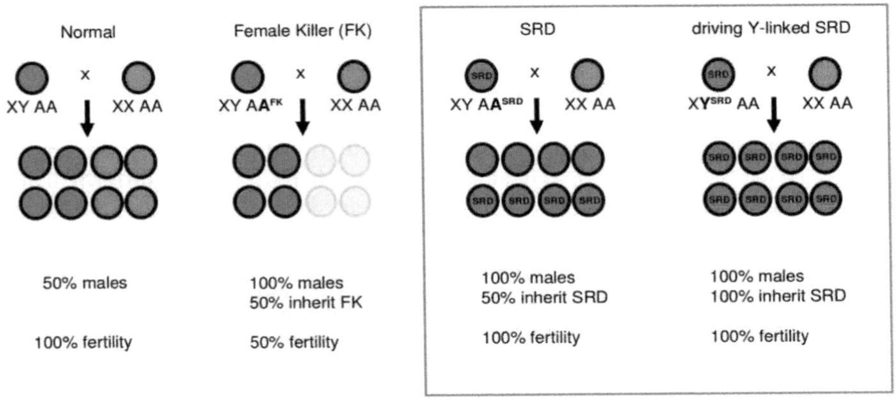

Figure 1. Male-biased sex ratio distorters (SRDs), unlike female killing systems (FK), bias the reproductive sex ratio towards males, while ideally maintaining full fertility.

SRD is a form of segregation distortion (SD) or meiotic drive, a term that also encompasses transmission of anomalies that are not strictly meiotic but that alter the normal process of meiosis, generating a gametic pool with one type of allele in excess (Zimmering et al. 1970). When SDs are physically linked to sex-determining loci or sex chromosomes, meiotic drive will result in an unbalanced sex ratio in the next generation. Reduced recombination between sex chromosomes favours the emergence of meiotic drive systems along them and indeed sex chromosome SDs are abundant in nature (Hammer 1991; Lyttle 1991).

SRDs occur mostly in association with male heterogamy, and usually, it is the X chromosome that drives against the Y chromosome, as a consequence of which males produce a strongly female-biased progeny. However, SRD systems in which the X drives against the Y chromosome, as in *Drosophila*, are not efficient in insect control programmes. Not only does population collapse proceed at a slower pace and is delayed by male polygamy, but it could result in short-term spikes in population size due to the female-biased progeny (Novitski 1947; Hamilton 1967).

Y-linked SRDs have been found to occur in nature in the culicine mosquitoes *Ae. aegypti* L. and *Culex pipiens* L., both of which harbour homomorphic sex chromosomes (Gilchrist and Haldane 1974; Fontaine et al. 2017). Males of these two species are heterozygous at the sex-determining locus (Mm) which is located in chromosome 1. The gene determining male sex was recently described in *Ae. aegypti* (Hall et al. 2015; Turner et al. 2018). The meiotic drive locus, M^D, is closely associated to M, acts in *trans,* distorting expression at a responder locus that is proximal to and indistinguishable from m, the locus that is homozygous in females. Subtle enhancers and suppressors of M^D strength have been discovered on all autosomes of *Ae. aegypti,* as have responder loci of varying sensitivity (Wood and Ouda 1987; Wood and Newton 1991; Cha et al. 2006).

Specific crosses involving the field-caught T37 strain of *Ae. aegypti*, result in a male-biased population of about 85% (Shin et al. 2012). Initial cage-population experiments to assess the suitability of the M^D locus for controlling natural populations of *Ae. aegypti,* showed females with developed resistance to M^D. Although the mechanism responsible for the resistance remains unknown, the level of distortion ultimately attained was insufficient to achieve effective population control (Hickey and Craig 1966; Robinson 1983).

Since natural resistance to M^D is common in the field, only native populations that are highly sensitive could be targeted. The efforts involved in the many attempts to apply M^D for insect control have highlighted the problems that could arise by using naturally occurring distorters for which resistance or rather counteracting alleles are already in existence. Also, their potential to be transferred to other target species is unclear as SRD and responder loci are expected to have co-evolved. Thus, efforts have intensified to develop entirely synthetic SRD strategies, which will be the focus for the remainder of this chapter.

2. SYNTHETIC SEX RATIO DISTORTERS BASED ON THE X CHROMOSOME SHREDDING MODEL

Synthetic distorters have the advantage of being unaffected by some, or all of the suppressor alleles that may exist to counteract naturally circulating distorter alleles. This is the case of synthetic SRD systems designed to circumvent the established sex determination pathway, operating independently of it. In the naturally occurring SRDs in *Cx. pipiens* and *Ae. aegypti* mosquitoes, cytological observations revealed that during the early stages of male meiosis, the X-equivalent chromosome harbouring the m locus are fragmented. This is accompanied by an increase in M-bearing gametes and a reduction in the number of females born in the next generation (Newton et al. 1976; Sweeny and Barr 1978).

The observations of X chromosome fragmentation suggested that a similar system for SRD could be artificially created through endonuclease-mediated cleavage of the X chromosome during male meiosis. In male heterogametic species, this could be achieved by the expression of a transgene-encoded endonuclease during spermatogenesis that would target and cut the X chromosome at several loci (Fig. 2). Consequently, only gametes with the Y chromosome would be produced or would be functional to achieve fertilization and only males would be generated (Burt 2003).

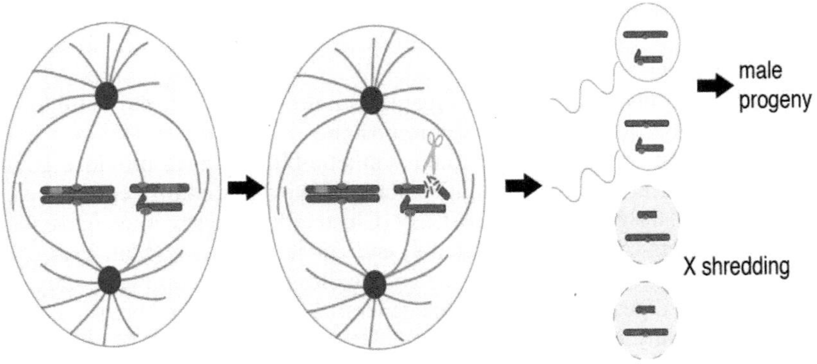

Figure 2. A multicopy target sequence (green bars) on the X chromosome is targeted by an endonuclease (scissors) during spermatogenesis. Shredding of the X chromosome favours the unaffected Y-bearing sperm and results in a male-biased progeny. Blue and red boxes are endonuclease and gRNA genes, respectively.

2.1. *I-PpoI as the First Synthetic Sex Ratio Distorter in the Malaria Mosquito*

A transgenic SRD trait was first developed and tested in the malaria mosquito *Anopheles gambiae* Giles, expressed from autosomal locations (Windbichler et al. 2007, 2008; Klein et al. 2012). It utilized the *Physarum polycephalum* Schwein I-PpoI endonuclease, which was driven by the *β2-tubulin* promoter that is specific to spermatogenesis. I-PpoI is an intron-encoded endonuclease mapping in the 28S rDNA locus that selectively cleaves ribosomal DNA sequences.

In *An. gambiae*, the ribosomal repeats are localised exclusively on the X chromosome making the targeting and cutting highly specific to the X chromosome. The expression of wild-type I-PpoI during spermatogenesis resulted in the cleavage of the X chromosome, but also in male sterility. It was found that during fertilisation, mature sperm carrying stable I-PpoI transfers endonuclease protein to the egg, determining the shredding of the maternal X chromosome and thus sterility.

To reduce the *in-vivo* half-life of I-PpoI and limit its activity to male meiosis, Galizi and colleagues (Galizi et al. 2014) generated a series of protein variants by site-directed mutagenesis. They modified amino acid residues involved in the zinc-binding core formation, protein packaging and protein dimerization. Next, they generated transgenic mosquitoes carrying autosomal insertions of each variant of the I-PpoI protein and analysed hatch rate and percentage of SRD in the progeny of hemizygous males crossed to wild-type females. W124A strains, with a mutation in the dimerization domain of I-PpoI, showed the highest hatch rate and SRD in the progeny corresponding to 97.4% males and a hatch rate close to the control. The SRD phenotype was stably inherited for four consecutive generations by transgenic sons. In five independent laboratory cage experiments with such males it was at least two orders of magnitude more efficient than sterile males. The cage studies also showed that releases of hemizygous distorter males at an overflooding ratio of 3x to wild-type males, was sufficient to eliminate the *An. gambiae* cage populations within six generations.

2.2. CRISPR-Cas9-Based X Chromosome Shredding to Induce Sex Ratio Distortion

While the I-PpoI system is a working model for an X chromosome shredder in *An. gambiae*, it is not transferable to other organisms unless they share the same location of the target rDNA sequences on the X chromosome.

Recent developments in molecular and synthetic biology have provided genome editing tools with great potential to engineer the genome of different species. The most promising is the RNA-guided CRISPR-Cas9 (Clustered Regularly Interspersed Short Palindromic Repeats-CRISPR-associated 9) endonuclease system (Mali et al. 2013). Here, a guide RNA (gRNA) "guides" the endonuclease to the complementary DNA sequence, which is digested by the enzyme generating double-strand breaks. Cas9 can be used to cleave any complementary target DNA harbouring a PAM (Protospacer Adjacent Motif) sequence, a motif consisting of the three-base-pair, NGG (any nucleobase followed by two guanine nucleobases).

In addition, other RNA-guided endonucleases have recently been discovered, such as Cpf1, smaller than Cas9 and A/T rich genomic-region specific, that may also prove to be as versatile as the Cas9 system (Zetsche et al. 2015). Compared to systems such as zinc-finger nuclease (ZFN), TALE nucleases (TALEN) and homing endonuclease genes (HEGs), RNA-guided systems do not require alteration of the protein to recognize the target sequences. Instead, gRNAs consisting of 18-20 RNA nucleotides are sufficient to lead the endonuclease to its target. Thus, endonucleases with novel specificities can be generated more easily and a larger number of genomic loci can be targeted and cleaved in parallel. Lastly, different gRNAs can be used to target multiple sites simultaneously, thus decreasing the odds of developing resistance alleles.

The CRISPR-Cas9 system, because of its specificity and flexibility, has been tailored to modify the genomes of different organisms, including yeast (DiCarlo et al. 2015), plants (Li et al. 2013), worms (Frokjaer-Jensen 2013), fruit flies (Gratz et al. 2013; Gantz and Bier 2015), the jewel wasp *Nasonia vitripennis* Walker (Li et al. 2017), mosquitoes (Basu et al. 2015; Gantz et al. 2015; Kistler et al. 2015; Hammond et al. 2016), zebrafish (Hwang et al. 2013), mice (Shen et al. 2013), monkeys (Niu et al. 2014), and human cells (Cong et al. 2013).

Recently, CRISPR-Cas9 was successfully used to act as a SRD system in *An. gambiae*. As a follow-up of the I-PpoI work, Galizi and colleagues, designed a transformation construct containing the *cas9* gene under the control of the *β-tubulin* gene promoter, and a gRNA targeting an X chromosome-linked rDNA sequence distinct from the original I-PpoI target (Galizi et al. 2016). This site was selected because it was conserved among closely related species of the *An. gambiae* complex, but crucially, was absent from more distantly related insects such as *Drosophila*. Similarly, to the observations with I-PpoI, the progeny of males from all tested transgenic lines displayed a high SRD with values ranging from 86.1% to 94.8% males. The fertility of transgenic males of almost all lines was similar to the wild-type, with hatching rates between 83.6% and 93.2%. Furthermore, although the *β-tubulin* promoter was used to regulate the expression of both Cas9 and I-PpoI (discussed above) endonucleases, the Cas9 protein was not paternally carried over into the fertilised embryo, thus increasing the hatching values and simplifying the generation of SRD.

Similar to I-PpoI, five consecutive generations of transgenic males stably inherited the SRD phenotype. As was observed for I-PpoI, however, the paternal X chromosome of rare survivor females exhibited both target sequence repeats that were intact indicating that not all of the targets were cleaved, as well as target sequence repeats that were found to be resistant to *in-vitro* re-cleavage as a possible consequence of DNA miss-repair. This is consistent with CRISPR-Cas9-mediated generation of resistant alleles by non-homologous end joining (NHEJ) or microhomology mediated end joining (MMEJ), repair mechanisms in DNA double-strand breaks that have been shown to adversely affect CRISPR-based gene drive systems (Hammond et al. 2016). For CRISPR-based SRDs, the likelihood of miss-repair leading to overall resistance to X chromosome shredding and consequently to the failure of population control is greatly reduced as a higher number of X chromosomal sites are targeted simultaneously.

2.3. Potential for Establishing CRISPR-Cas9-Based Sex Ratio Distorters in Other Species

Given the flexibility allowed by CRISPR-based endonucleases, it may now be possible to test whether X chromosome shredding has the potential to become a universal strategy to genetically control a wide variety of heterogametic insect pests, of both agricultural and public health relevance. However, the proof-of-principle in *An. gambiae* relies on the peculiar localisation of the mosquito's rDNA genes, which are confined to the X chromosome, and the availability of an endonuclease that targets these repeats, two characteristics unlikely to be shared by many other insect species.

For X chromosome shredding to become a more widely used approach, what is required is a method to identify motifs repeated exclusively on the X chromosome, and the ability to engineer endonucleases able to target and cleave these sequences in the male germline.

Recently, a bioinformatic pipeline called Redkmer (for Repeat Extraction and Detection based on kmers, all the possible sub-sequences of length k from a read obtained through sequencing) used for DNA sequence analysis and comparison, was developed for X chromosome target sequences shredding by CRISPR endonucleases (Papathanos and Windbichler 2018). It relies on long and short read sequencing technology and can identify highly abundant X-specific sequences. Because genome assemblies typically exclude highly repetitive sequences such as satellite DNA, it was designed to not rely on high-quality assembled genomes. Indeed, for many target insect species, high-quality genome assemblies are not available.

The Redkmer pipeline requires as input only whole genome sequencing (WGS) data based on long (e.g. PacBio) and short (e.g. Illumina) unassembled reads of the relevant insect. WGS for the long reads is performed only on male individuals, while short read data are generated from both, male and female individuals, independently. For the selection of highly abundant X chromosome kmers, the pipeline relies on two features: (1) WGS elements mapping on the X chromosome that occur on average twice as often in female compared to male data; (2) overall sequence abundance with X chromosome specificity.

Preliminary data in *An. gambiae* confirmed the potential of the bioinformatic pipeline, revealing X-specific and abundant kmers that overlapped with the known rDNA cluster. The now affordable costs of next-generation sequencing and the increase in computer power, combined with the flexibility of the CRISPR-Cas9 system, will ease the application of SRD to genetically control disease vectors and economically or ecologically damaging invasive species.

3. CONSIDERATIONS ON THE USE OF SYNTHETIC SEX RATIO DISTORTERS

SRD systems are subject to two essential limitations. First, SRDs will only work in sexually reproducing species. Second, like the SIT, maintaining SRD traits within a target population requires scheduled releases of new transgenic individuals that need to be continuously reared in considerable numbers rendering the approach costly. As mentioned, this could be eased by linking the SRD trait to the male Y chromosome. Thus, in this case, all male offspring will inherit the transgene that will spread invasively.

For scaling up and potential field testing of the non-driving SRD traits described above, transgene expression should be conditionally repressed. This would enable mass-rearing, as constant backcrossing of a constitutively expressed SRD trait at each generation is unlikely to be a practical method. Conditional expression systems have been successfully used in the field, in combination with dominant lethality, but should be adapted to work in conjunction with CRISPR-based constructs.

3.1. Difficulty in Turning Non-driving SRD into Driving-SRD Traits

All synthetic SRD systems described so far are non-driving autosomal distorters. Moving autosomal distorters to the Y chromosome is, however, far from the straightforward proposition it appeared to be initially. All active distorter transgenes examined by Galizi et al. (unpublished) had the construct inserted on an autosome. Lines with the transgene integrated on either the X or Y chromosome did not show a significant level of I-PpoI expression in the testes. This was explained by the repressive effect of meiotic sex chromosome inactivation (MSCI) that silences unpaired chromatin during meiotic stages of spermatogenesis. Inactivation of transgenes integrated on the X chromosome has been described before in mosquitoes (Magnusson et al. 2012; Papa et al. 2017). MSCI may turn out to be a significant obstacle in turning non-driving SRD into driving-SRD traits, as its biological function may be that of policing SD traits.

3.2. Advantages of the Standard Autosomal SRD Systems

One advantage of the standard autosomal SRD systems based on RNA-guided endonucleases is that they can be engineered to work, at least in principle, in most heterogametic species and therefore suppress the population of many target organisms.

Recently the precise deletion of chromosomes in cell lines of mammals (Adikusuma et al. 2017) has been demonstrated and indicates that this system could also work in higher organisms. From an ecological standpoint, a second advantage is the reduced risk of invasion compared to driving Y-linked SRDs or compared to other gene drive elements that are being developed for the purpose of population replacement (Gantz and Bier 2016; Hammond et al. 2016). Once released, the frequency of the SD alleles decreases instead of increasing in the population, a desirable feature to avoid permanent ecosystems alterations (Oye et al. 2014).

3.3. *Engineering Multiple gRNAs to Reduce Generation of Resistance Alleles*

In RNA-guided SRD systems there is the need to understand the likelihood at which resistant alleles will develop in the target sequences (Bull 2015). The tendency of natural selection to favour equal sex ratios exerts pressure on the NHEJ (non-homologous end joining) repair system that may generate resistance alleles via insertion and deletions in target sequences. In addition, natural sequence polymorphism between individuals of the same population could also prevent cutting. Therefore, to reduce the growth of resistance alleles, it is critical to engineer different gRNAs cutting simultaneously, including some that target conserved regions of essential genes. The degree to which X chromosome shredding systems can rely on the existence of numerous repetitive sequences of the same target on the X chromosome, may directly determine the likelihood of the rise of resistance alleles, although further research is required in this area.

3.4. *Environmental, Ecological, and Regulatory Challenges*

The release of transgenically modified organisms for population control is challenged by a series of environmental, ecological, and regulatory difficulties. Transgenic males should be able to mate and scout for females to the same extent as wild-type males (Lacroix et al. 2012). A different behaviour would drop the chances of spreading to remote regions reducing the success of population control. Different factors can decrease the competitiveness of males such as mass-rearing, inbreeding, transgene expression and its insertion site in the genome (Catteruccia et al. 2003; Reed et al. 2003; Baeshen et al. 2014). The use of the PhiC31 integration system that provides precisely mapped docking sites, has helped to reduce the position effect on gene expression (Amenya et al. 2010).

Genetic variation of wild-type individuals is another feature that can influence the success of the release operation. Polymorphisms in the gRNA target site can compromise the ability of the endonuclease to cleave the DNA. Different strains can have slightly dissimilar mating behaviours or live in distinctive ecological niches making their control even more difficult. These are all features that should be considered when applying genetic engineering to control vector-borne disease or agricultural pests.

Ultimately, the release of transgenically modified individuals causes environmental and safety challenges that should be addressed in each individual case. Ecological and molecular containment strategies should be considered when designing RNA-guided SRD systems. The chances of the SRD trait spreading to non-targeted species, and horizontal transfer of the transgenes must be safely reduced before releasing the modified individuals (David et al. 2013; Nielsen, this volume). Species-specific targeting sequences and promoters to regulate the endonuclease expression should also prevent lateral gene transfer beyond target populations (Oye et al. 2014).

Confined laboratories, with high containment levels and only small-field tests, should initially be used to determine the safety and specificity of the transgene modification. An open discussion with regulatory agencies, the scientific community and the public is fundamental to inform on the risks and benefits of using genetic-engineering technologies to control vector-borne diseases and alleviate the economic burden inflicted by agricultural pests (Oye et al. 2014).

4. OTHER APPROACHES TO ACHIEVE SYNTHETIC SEX RATIO DISTORTION

We have focussed our discussion on X chromosome shredding as a paradigm for developing synthetic SRDs. However, other approaches are also conceivable to achieve the same goal. Recently, the long-elusive male-determining genes of a number of important pest and vector species have been identified. These include the Y chromosome-linked male-determining genes in mosquitoes *An. gambiae* and *An. stephensi* Liston, the M-locus linked gene in *Ae. aegypti*, the mobile splicing factor in *M. domestica* L. and the MoY factor in *Ceratitis capitata* (Wiedemann) (Hall et al. 2015; Criscione et al. 2016; Krzywinska et al. 2016; Sharma et al. 2017; Meccariello et al. 2019).

Knowing the sex-determining genes in each of these species will allow designing synthetic SRDs. In addition, the use of a nuclease-based gene drive could ensure the transmission of the synthetic SRDs to the entire progeny and bring more rapidly the population to collapse (Kyrou et al. 2018; Häcker and Schetelig, this volume).

Finally, caution should be used when sex-determining genes play functional roles in other essential pathways, such as in dosage compensation. Their multiple functions may interfere with some strategies to manipulate sex ratios, for example through the lethality of transgenic male-determination signals on chromosomes other than the Y chromosome.

5. REFERENCES

Adikusuma, F., N. Williams, F. Grutzner, J. Hughes, and P. Thomas. 2017. Targeted deletion of an entire chromosome using CRISPR/Cas9. Molecular Therapy 25: 1736–1738.

Amenya, D. A., M. Bonizzoni, A. T. Isaacs, N. Jasinskiene, H. Chen, O. Marinotti, G. Yan, and A. A. James. 2010. Comparative fitness assessment of *Anopheles stephensi* transgenic lines receptive to site-specific integration. Insect Molecular Biology 19: 263–269.

Argasinski, K. 2013. The dynamics of sex ratio evolution: From the gene perspective to multilevel selection. PLoS One 8: e60405.

Baeshen, R., N. E. Ekechukwu, M. Toure, D. Paton, M. Coulibaly, S. F. Traore, and F. Tripet. 2014. Differential effects of inbreeding and selection on male reproductive phenotype associated with the colonization and laboratory maintenance of *Anopheles gambiae*. Malaria Journal 13: 19.

Basu, S., A. Aryan, J. M. Overcash, G. H. Samuel, M. A. Anderson, T. J. Dahlem, K. M. Myles, and Z. N. Adelman. 2015. Silencing of end-joining repair for efficient site-specific gene insertion after TALEN/CRISPR mutagenesis in *Aedes aegypti*. Proceedings of the National Academy of Sciences of the USA 112: 4038–4043.

Bull, J. J. 2015. Evolutionary decay and the prospects for long-term disease intervention using engineered insect vectors. Evolution Medicine and Public Health 2015: 152–166.

Burt, A. 2003. Site-specific selfish genes as tools for the control and genetic engineering of natural populations. Proceedings of the Royal Society B: Biological Sciences 270: 921–928.

Catteruccia, F., H. C. Godfray, and A. Crisanti. 2003. Impact of genetic manipulation on the fitness of *Anopheles stephensi* mosquitoes. Science 299: 1225–1227.

Cha, S. J., D. D. Chadee, and D. W. Severson. 2006. Population dynamics of an endogenous meiotic drive system in *Aedes aegypti* in Trinidad. American Journal of Tropical Medicine and Hygiene 75: 70–77.

Cong, L., F. A. Ran, D. Cox, S. Lin, R. Barretto, N. Habib, P. D. Hsu, X. Wu, W. Jiang, L. A. Marraffini, and F. Zhang. 2013. Multiplex genome engineering using CRISPR/Cas systems. Science 339: 819–823.

Craig, G. B., W. A. Hickey, and R. C. VandeHey. 1960. An inherited male-producing factor in *Aedes aegypti*. Science 132: 1887–1889.

Criscione, F., Y. Qi, and Z. Tu. 2016. GUY1 confers complete female lethality and is a strong candidate for a male-determining factor in *Anopheles stephensi*. eLife 5: e19281.

David, A. S., J. M. Kaser, A. C. Morey, A. M. Roth, and D. A. Andow. 2013. Release of genetically engineered insects: A framework to identify potential ecological effects. Ecology and Evolution 3: 4000–4015.

DiCarlo, J. E., A. Chavez, S. L. Dietz, K. M. Esvelt, and G. M. Church. 2015. Safeguarding CRISPR-Cas9 gene drives in yeast. Nature Biotechnology 33: 1250–1255.

Dyck, V. A., J. Hendrichs, and A. S. Robinson (eds.). 2021. Sterile Insect Technique – Principles and practice in Area-Wide Integrated Pest Management. Second Edition. CRC Press, Boca Raton, Florida, USA. 1200 pp.

Fisher, R. A. 1958. The genetical theory of natural selection. Dover, New York, NY, USA. 291 pp.

Fontaine, A., I. Filipovic, T. Fansiri, A. A. Hoffmann, C. Cheng, M. Kirkpatrick, G. Rasic, and L. Lambrechts. 2017. Extensive genetic differentiation between homomorphic sex chromosomes in the mosquito vector, *Aedes aegypti*. Genome Biology and Evolution 9: 2322–2335.

Frokjaer-Jensen, C. 2013. Exciting prospects for precise engineering of *Caenorhabditis elegans* genomes with CRISPR/Cas9. Genetics 195: 635–642.

Galizi, R., L. A. Doyle, M. Menichelli, F. Bernardini, A. Deredec, A. Burt, B. L. Stoddard, N. Windbichler, and A. Crisanti. 2014. A synthetic sex ratio distortion system for the control of the human malaria mosquito. Nature Communications 5: 3977.

Galizi, R., A. Hammond, K. Kyrou, C. Taxiarchi, F. Bernardini, S. M. O'Loughlin, P. A. Papathanos, T. Nolan, N. Windbichler, and A. Crisanti. 2016. A CRISPR-Cas9 sex-ratio distortion system for genetic control. Scientific Reports Nature 6: 31139.

Gantz, V. M., and E. Bier. 2015. Genome editing. The mutagenic chain reaction: A method for converting heterozygous to homozygous mutations. Science 348: 442–444.

Gantz, V. M., and E. Bier. 2016. The dawn of active genetics. Bioessays 38: 50–63.

Gantz, V. M., N. Jasinskiene, O. Tatarenkova, A. Fazekas, V. M. Macias, E. Bier, and A. A. James. 2015. Highly efficient Cas9-mediated gene drive for population modification of the malaria vector mosquito *Anopheles stephensi*. Proceedings of the National Academy of Sciences of the USA 112: E6736-6743.

Gilchrist, B. M., and J. B. S. Haldane. 1974. Sex linkage and sex determination in a mosquito *Culex molestus*. Hereditas 33: 175–190.

Gratz, S. J., A. Cummings, J. N. Nguyen, D. C. Hamm, L. K. Donohue, M. M. Harrison, J. Wildonger, and K. M. O'Connor-Giles. 2013. Genome engineering of *Drosophila* with the CRISPR RNA-guided Cas9 nuclease. Genetics 194: 1029–1035.

Hall, A. B., S. Basu, X. Jiang, Y. Qi, V. A. Timoshevskiy, J. K. Biedler, M. V. Sharakhova, R. Elahi, M. A. Anderson, X. G. Chen, I. V. Sharakhov, Z. N. Adelman, and Z. Tu. 2015. SEX DETERMINATION. A male-determining factor in the mosquito *Aedes aegypti*. Science 348: 1268–1270.

Hamilton, W. D. 1967. Extraordinary sex ratios. A sex-ratio theory for sex linkage and inbreeding has new implications in cytogenetics and entomology. Science 156: 477–488.

Hammer, M. F. 1991. Molecular and chromosomal studies on the origin of t haplotypes in mice. The American Naturalist 137: 359–365.

Hammond, A., R. Galizi, K. Kyrou, A. Simoni, C. Siniscalchi, D. Katsanos, M. Gribble, D. Baker, E. Marois, S. Russell, A. Burt, N. Windbichler, A. Crisanti, and T. Nolan. 2016. A CRISPR-Cas9 gene drive system targeting female reproduction in the malaria mosquito vector *Anopheles gambiae*. Nature Biotechnology 34: 78–83.

Hickey, W. A., and G. B. Craig. 1966. Genetic distortion of sex ratio in a mosquito, *Aedes aegypti*. Genetics 53: 1177–1196.

Hwang, W. Y., Y. Fu, D. Reyon, M. L. Maeder, S. Q. Tsai, J. D. Sander, R. T. Peterson, J. R. Yeh, and J. K. Joung. 2013. Efficient genome editing in zebrafish using a CRISPR-Cas system. Nature Biotechnology 31: 227–229.

Kistler, K. E., L. B. Vosshall, and B. J. Matthews. 2015. Genome engineering with CRISPR-Cas9 in the mosquito *Aedes aegypti*. Cell Reports 11: 51–60.

Klein, T. A., N. Windbichler, A. Deredec, A. Burt, and M. Q. Benedict. 2012. Infertility resulting from transgenic I-PpoI male *Anopheles gambiae* in large cage trials. Pathogens and Global Health 106: 20–31.

Krzywinska, E., N. J. Dennison, G. J. Lycett, and J. Krzywinski. 2016. A maleness gene in the malaria mosquito *Anopheles gambiae*. Science 353: 67–69.

Kyrou, K., A. M. Hammond, R. Galizi, N. Kranjc, A. Burt, A. K. Beaghton, T. Nolan, and A. Crisanti. 2018. A CRISPR-Cas9 gene drive targeting *doublesex* causes complete population suppression in caged *Anopheles gambiae* mosquitoes. Nature Biotechnology 36: 1062–1066.

Lacroix, R., A. R. McKemey, N. Raduan, L. Kwee Wee, W. Hong Ming, T. Guat Ney, A. A. S. Rahidah, S. Salman, S. Subramaniam, O. Nordin, A. T. N. Hanum, C. Angamuthu, S. Marlina Mansor, R. S. Lees, N. Naish, S. Scaife, P. Gray, G. Labbé, C. Beech, D. Nimmo, L. Alphey, S. S. Vasan, L. Han Lim, A. N. Wasi, and S. Murad. 2012. Open field release of genetically engineered sterile male *Aedes aegypti* in Malaysia. PLoS One 7(8): e42771.

Li, J. F., J. E. Norville, J. Aach, M. McCormack, D. Zhang, J. Bush, G. M. Church, and J. Sheen. 2013. Multiplex and homologous recombination-mediated genome editing in *Arabidopsis* and *Nicotiana benthamiana* using guide RNA and Cas9. Nature Biotechnology 31: 688–691.

Li, M., L. Y. C. Au, D. Douglah, A. Chong, B. J. White, P. M. Ferree, and O. S. Akbari. 2017. Generation of heritable germline mutations in the jewel wasp *Nasonia vitripennis* using CRISPR/Cas9. Scientific Reports Nature 7: 901.

Lyttle, T. W. 1991. Segregation distorters. Annual Review of Genetics 25: 511–557.

Magnusson, K., G. J. Lycett, A. M. Mendes, A. Lynd, P. A. Papathanos, A. Crisanti, and N. Windbichler. 2012. Demasculinization of the *Anopheles gambiae* X chromosome. BMC Evolutionary Biology 12: 69.

Mali, P., K. M. Esvelt, and G. M. Church. 2013. Cas9 as a versatile tool for engineering biology. Nature Methods 10: 957–963.

Meccariello, A., M. Salvemini, P. Primo, B. Hall, P. Koskinioti, M. Dalíková, A. Gravina, M. A. Gucciardino, F. Forlenza, M. E. Gregoriou, D. Ippolito, S. M. Monti, V. Petrella, M. M. Perrotta, S. Schmeing, A. Ruggiero, F. Scolari, E. Giordano, K. T. Tsoumani, F. Marec, N. Windbichler, J. Nagaraju, K. P. Arunkumar, K. Bourtzis, K. D. Mathiopoulos, J. Ragoussis, L. Vitagliano, Z. Tu, P. A. Papathanos, M. D. Robinson, and G. Saccone. 2019. Maleness-on-the-Y (MoY) orchestrates male sex determination in major agricultural fruit fly pests. Science 365 (6460): 1457–1460:

Mercot, H., B. Llorente, M. Jacques, A. Atlan, and C. Montchamp-Moreau. 1995. Variability within the Seychelles cytoplasmic incompatibility system in *Drosophila simulans*. Genetics 141: 1015–1023.

Newton, M. E., R. J. Wood, and D. I. Southern. 1976. A cytogenetic analysis of meiotic drive in the mosquito *Aedes aegypti*. Genetica 46: 297–318.

Niu, Y, B. Shen, Y. Cui, Y. Chen, J. Wang, L. Wang, Y. Kang, X. Zhao, W. Si, W. Li, A. P. Xiang, J. Zhou, X. Guo, Y. Bi, C. Si, B. Hu, G. Dong, H. Wang, Z. Zhou, T. Li, T. Tan, X. Pu, F. Wang, S. Ji, Q. Zhou, X. Huang, W. Ji, and J. Sha. 2014. Generation of gene-modified cynomolgus monkey via Cas9/RNA-mediated gene targeting in one-cell embryos. Cell 156: 836–843.

Novitski, E. 1947. Genetic analysis of anomalous sex ratio condition in *Drosophila affinis*. Genetics 32: 526–534.

Oye, K. A., K. Esvelt, E. Appleton, F. Catteruccia, G. Church, T. Kuiken, S. B. Lightfoot, J. McNamara, A. Smidler, and J. P. Collins. 2014. Biotechnology. Regulating gene drives. Science 345: 626–628.

Papa, F., N. Windbichler, R. M. Waterhouse, A. Cagnetti, R. D'Amato, T. Persampieri, M. K. N. Lawniczak, T. Nolan, and P. A. Papathanos. 2017. Rapid evolution of female-biased genes among four species of *Anopheles* malaria mosquitoes. Genome Research 27: 1536–1548.

Papathanos, P. A., and N. Windbichler. 2018. Redkmer: An assembly-free pipeline for the identification of abundant and specific X-chromosome target sequences for X-shredding by CRISPR endonucleases. CRISPR Journal 1: 88–98.

Reed, D. H., E. H. Lowe, D. A. Briscoe, and R. Frankham. 2003. Fitness and adaptation in a novel environment: Effect of inbreeding, prior environment, and lineage. Evolution 57: 1822–1828.

Robinson, A. S. 1983. Sex-ratio manipulation in relation to insect pest control. Annual Review Genetics 17: 191–214.

Schliekelman, P., S. Ellner, and F. Gould. 2005. Pest control by genetic manipulation of sex ratio. Journal of Economic Entomology 98: 18–34.

Sharma, A., S. D. Heinze, Y. Wu, T. Kohlbrenner, I. Morilla, C. Brunner, E. A. Wimmer, L. van de Zande, M. D. Robinson, L. W. Beukeboom, and D. Bopp. 2017. Male sex in houseflies is determined by *Mdmd*, a paralog of the generic splice factor gene *CWC22*. Science 356: 642–645.

Shen, B., J. Zhang, H. Wu, J. Wang, K. Ma, Z. Li, X. Zhang, P. Zhang, and X. Huang. 2013. Generation of gene-modified mice via Cas9/RNA-mediated gene targeting. Cell Research 23: 720–723.

Shin, D., A. Mori, and D. W. Severson. 2012. Genetic mapping a meiotic driver that causes sex ratio distortion in the mosquito *Aedes aegypti*. Journal of Heredity 103: 303–307.

Sweeny, T. L., and A. R. Barr. 1978. Sex ratio distortion caused by meiotic drive in a mosquito, *Culex pipiens* L. Genetics 88: 427–446.

Thomas, D. T., C. A. Donnelly, R. J. Wood, and L. S. Alphey. 2000. Insect population control using a dominant, repressible, lethal genetic system. Science 287: 2474–2476.

Turner, J., R. Krishna, A. E. van't Hof, E. R. Sutton, K. Matzen, and A. C. Darby. 2018. The sequence of a male-specific genome region containing the sex determination switch in *Aedes aegypti*. Parasites and Vectors 11: 549.

Windbichler, N., P. A. Papathanos, F. Catteruccia, H. Ranson, A. Burt, and A. Crisanti. 2007. Homing endonuclease mediated gene targeting in *Anopheles gambiae* cells and embryos. Nucleic Acids Research 35: 5922–5933.

Windbichler, N., P. A. Papathanos, and A. Crisanti. 2008. Targeting the X chromosome during spermatogenesis induces Y chromosome transmission ratio distortion and early dominant embryo lethality in *Anopheles gambiae*. PLoS Genetics 4: e1000291.

Wood, R. J., and N. A. Ouda. 1987. The genetic basis of resistance and sensitivity to the meiotic drive gene D in the mosquito *Aedes aegypti* L. Genetica 72: 69–79.

Wood, R. J., and M. E. Newton. 1991. Sex-ratio distortion caused by meiotic drive in mosquitoes. The American Naturalist 137: 379–391.

Zetsche, B, J. S. Gootenberg, O. O. Abudayyeh, I. M. Slaymaker, K. S. Makarova, P. Essletzbichler, S. E. Volz, J. Joung, J. van der Oost, A. Regev. E. V. Koonin, and F. Zhang. 2015. Cpf1 is a single RNA-guided endonuclease of a class 2 CRISPR-Cas system. Cell 163: 759–771.

Zimmering, S, L. Sandler, and B. Nicoletti. 1970. Mechanisms of meiotic drive. Annual Review Genetics 4: 409–436.

THE USE OF SPECIES DISTRIBUTION MODELLING AND LANDSCAPE GENETICS FOR TSETSE CONTROL

M. T. BAKHOUM[1], M. J. B. VREYSEN[1] AND J. BOUYER[1,2,3]

[1]Insect Pest Control Laboratory, Joint FAO/IAEA Division of Nuclear Techniques in Food and Agriculture, Vienna, Austria; J.Bouyer@iaea.org
[2]Unité Mixte de Recherche ASTRE 'Animal, Santé, Territoires, Risques et Ecosystèmes', Campus International de Baillarguet, Centre de Coopération Internationale en Recherche Agronomique pour le Développement (CIRAD), Montpellier, France
[3]Intertryp, IRD, CIRAD, Université de Montpellier, Montpellier, France

SUMMARY

Trypanosomosis is one of the major constraints to rural development in sub-Saharan Africa. Studies of the species distribution and landscape genetics allows for targeting tsetse-infested areas and optimizing the management of tsetse fly populations. In this chapter, a review is provided on 1) available spatial tools and landscape genetics, 2) the use of tsetse distribution models to rank populations according to their level of isolation to assist identifying populations that can be targeted for eradication, and 3) the use of tsetse distribution models to optimize tsetse control efforts within an area-wide integrated pest management context.

Key Words: Glossina, habitat suitability, African trypanosomosis, nagana, sleeping sickness, sub-Saharan Africa, area-wide integrated pest management, spatial modelling, Maxent, distribution models, friction models, population genetics

1. INTRODUCTION – THE TSETSE AND TRYPANOSOMOSIS PROBLEM

Tsetse flies (*Glossina* spp.) are widely distributed in sub-Saharan Africa and inhabit semi-arid, sub-humid and humid lowlands in 37 countries across the continent with a potential distribution range of some 8.7 million km^2 (Rogers and Robinson 2004). They transmit trypanosomes, the causative agents of sleeping sickness (human African trypanosomosis, HAT) and nagana (African animal trypanosomosis, AAT).

J. Hendrichs, R. Pereira and M. J. B. Vreysen (eds.), Area-Wide Integrated Pest Management: Development and Field Application, pp. 857–868. CRC Press, Boca Raton, Florida, USA.

Together, the animal and human diseases pose health threats and a great economic burden to vast regions of sub-Saharan Africa where they are endemic (Swallow 1999; Diall et al. 2017).

The number of HAT cases has substantially declined in the last 15 years, mainly through increased disease surveillance and treatment of affected patients (WHO 2013), and since 2012 in *Trypanosoma brucei gambiense* HAT foci, thanks to increased vector control (Courtin et al. 2015; Mahamat et al. 2017). Whereas in 2000 there were more than 25 000 newly reported cases, this had decreased to 7106 new cases by 2012. In 2018, less than 1000 new cases of HAT were reported to the World Health Organization (WHO 2019), which represents the lowest number of sleeping sickness cases ever recorded. In some countries with ongoing conflicts, there is obviously the likelihood of under-reporting of cases. There is, however, general consensus that the complete elimination of HAT as a public health problem can only be possible through the inclusion of an effective vector management component (Solano et al. 2013; Feldmann et al. 2021).

Contrary to the advances made with the management of HAT, the AAT continues to represent the greatest animal health constraint to improved livestock production in sub-Saharan Africa, causing economic losses amounting to USD 4.75 billion annually and putting approximately 50 million cattle at risk (i.e. milk and meat production) (Swallow 1999; Scoones 2016). The disease also prevents the integration of crop farming and livestock keeping, a crucial component for the development of sustainable agricultural systems (Alsan 2015).

Early death can result in chronically infected animals if AAT is not treated, and at least three million cattle and other domestic animals succumb to the disease each year (Hursey and Slingenbergh 1995). When not lethal, AAT brings livestock into a chronically debilitating condition and reduces fertility, weight gain, meat and milk production by at least 50%, as well as the work efficiency of oxen used to cultivate the land (Budd 1999; Swallow 1999; Shaw 2004). The disease has also indirect negative effects on the development of commercial domestic and livestock production, i.e. it discourages the use of more-productive exotic and cross-bred cattle, depresses the growth and distribution of livestock populations, reduces the potential opportunities for integration of livestock keeping and crop production through less draught power to cultivate land and to transport farm products to market, and less manure to fertilize (in an environment-friendly way) soils for better crop production.

In addition, the scarcity of domestic animals leads to a serious shortage of animal protein for human consumption and as people tend to avoid areas infested with tsetse flies, they affect human settlement (Shaw 2004). Consequently, tsetse flies and the AAT it transmits are considered as one of the root causes of hunger and poverty in about one third of the African continent (Feldmann et al. 2021). AAT therefore is an important limiting factor to reach Sustainable Development Goals 1 (No Poverty) and 2 (Zero Hunger) (UN 2019).

Understanding the distribution of tsetse flies in space and time is essential in selecting the most appropriate intervention strategies for the area-wide management of tsetse populations in different parts of sub-Saharan Africa. Despite substantial efforts for more than a century, deliberate efforts to reduce the vast tsetse belt have had very limited success (Vreysen et al. 2013).

In past decades, spraying of insecticides was effective in controlling tsetse fly populations in certain areas of the African continent, e.g. to stop the advancing front of tsetse flies in south-eastern Zimbabwe and south-western Mozambique between 1962 and 1974 (Robertson and Kluge 1968; Lovemore 1972, 1973, 1974; Robertson et al. 1972), and to eliminate tsetse flies in Zimbabwe (Jordan 1986), Nigeria (MacLennan and Kirby 1958), Botswana (Davies 1980), Zambia (Paynter and Brady 1992), the Zambezi region (formerly Caprivi Strip) of Namibia, and in Malawi (Vale 1999). However, although the spraying of non-residual pyrethroids as ultra-low volume formulations either using ground fogging techniques or the sequential aerosol technique (SAT) was until recently still used in Burkina Faso, Ethiopia, Ghana, Guinea and Zimbabwe (Bouyer and Vreysen 2018), the spraying of residual insecticides is no longer recommended anymore on environmental grounds.

Whereas in the past, suppression methods were often used alone and against only certain segments of the tsetse population, in the last decades it has become evident that more sustainable tsetse population management can be obtained when applying two compatible strategies, i.e. area-wide integrated pest management (AW-IPM) (Vreysen et al. 2007) and the phased conditional approach (PCA) (Bouyer and Vreysen 2018). The management of tsetse fly populations can be implemented using two basic approaches, i.e. on a localised field-by-field basis or on an area-wide basis (total population management) (AW-IPM) (Hendrichs et al. 2007; Klassen and Vreysen 2021). AW-IPM is an approach that consists of a coordinated effort against all sub-units of a target pest population in an ecosystem before the pest population has reached damaging proportions. Local field-by-field pest control is a reactive effort when the pest population reaches damaging levels and is carried out individually and independent of the action of neighbouring farmers. These two strategies have different objectives. Whereas population suppression (the reduction of the insect pest density below a threshold preventing damage or disease transmission) can be the objective in both cases, local population elimination (eradication would signify the elimination of all populations of a given insect species from the planet) is only possible using the second approach, with higher costs but proportional longer-term impacts (Vreysen 2006; Bouyer et al. 2013).

The AW-IPM approach minimizes the risk of reinvasion, as areas that are of no interest to the farmers are also targeted. This approach usually requires several years of planning and a specialised organization with dedicated staff to implement the control activities, in an adaptive management scheme. The AW-IPM can benefit from advanced technologies such as geographic information systems (GIS), population genetics (increasingly being used for designing and implementing tsetse control efforts; Bouyer et al. 2021), remote sensing and aerial dissemination techniques (Vreysen 2006; Dicko et al. 2014; Klassen and Vreysen 2021).

This chapter highlights the integration of species distribution modelling and landscape genetics to facilitate the management of tsetse populations. First, it presents a general overview of spatial tools and landscape genetics. Second, it presents how modelling the distribution of tsetse populations and ranking them according to their level of isolation can help to identify populations that can be targeted for eradication. Finally, it presents how tsetse distribution models can be used to optimize tsetse control efforts within an AW-IPM context.

2. GENERAL OVERVIEW OF SPATIAL TOOLS AND LANDSCAPE GENETICS

2.1. Spatial Modelling and Geographic Information Systems

Spatial tools have long been important to natural resource applications. The GIS, global positioning system (GPS), and remote sensing (RS) are spatial tools that have become more and more important for decision-making in the control of diseases, i.e. to locate important target sites, to predict population change based on climatic trends, to report potential anomalies, but also to analyse landscape patterns, disaster management, etc. Data can, in many instances, only be fully understood when they can be placed in a geographic context. Hence the benefits that can be derived from using GIS, which are computer-based tools that analyse, store, manipulate and visualize geographic information, usually in a map (Bouyer et al. 2021). GPS is a satellite navigation system used to determine the exact position of an object, whereas RS aims at providing access to a range of satellite-derived data products about the earth's surface using electromagnetic sensors.

These geospatial tools have made the design and the implementation of AW-IPM programmes and disease control much more effective and cost-efficient. The importance of geomatic tools to assist various stages of planning and application of the Sterile Insect Technique (SIT) as part of an AW-IPM approach is presented elsewhere (Bouyer et al. 2021) and includes the selection of project sites, planning of pre-intervention surveys and feasibility assessments before the start of the operational implementation campaign. Geomatic tools are also essential to monitor and analyse insect populations during area-wide control efforts to be able to implement adaptive management (Vreysen et al. 2013).

2.2. Landscape Genetics Approach

Landscape genetics, that associates tsetse population genetics with spatial tools, is an innovative and emerging approach that enables understanding how geographic and environmental features structure genetic variation at the population and individual scales (Feldmann and Ready 2014). Genetic markers with varying temporal or spatial resolution can be used to implement landscape genetics, depending on ecological questions (De Meeûs et al. 2007). This approach can not only be used for improving ecological knowledge, but also for explaining observed spatial genetic patterns as clines, isolation by distance, genetic boundaries to gene flow, metapopulations and random patterns (Manel et al. 2003), in order to manage properly the genetic diversity.

Various approaches have been used to quantify the spatial structure relying on landscape ecology. However, most involve the incorporation of the notion of landscape resistance or friction (i.e. the impediments to gene flow) caused by landscape features. The most common approach employed involves measuring the "cost" distance between populations sampled based on one or more alternative landscape resistance models (Manel et al. 2003).

Informing on the resistance of landscape to movement is essential to refine species distribution models. The geographic distribution of a given species can be seen as the intersection between biotic (B), abiotic (A) and movement (M) factors in a BAM diagram (Soberón and Peterson 2005). However, most species distribution models neglect the latter (Barve et al. 2011), often because they are based on presence/absence data and rarely include genetic data. It is however essential to account for movement, as a given landscape can be suitable for a given species, but not inhabited because it is out of reach for this species. Moreover, friction should not be mapped based on expert knowledge, as this is very subjective and therefore essentially unpredictable.

Tsetse studies have demonstrated that the environmental parameters driving landscape suitability are totally different from those driving landscape friction (Bouyer et al. 2015). This can be described as the "salamander paradox", i.e. if a forest salamander finds itself at 100 m from the edge of a forest, in an area where the ground is bare and very inhospitable to its survival, it will either die quickly or move as fast as possible to the forest. Conversely, if the salamander finds itself in the forest, it will be in a location with all suitable conditions and it will not disperse much. Therefore, the friction of the forest will be higher than that of the bare ground, which seems counter-intuitive (Peterman et al. 2014).

Insect pest populations can be structured at micro-geographic scales, which must be accounted for to optimize control. For example, the inclusion of population genetics data in control programmes against tsetse populations of the *palpalis* group in West Africa provided information on the level of genetic isolation of the target populations from the neighbouring ones, which allowed informed decisions for developing control strategies. However, for population genetic tools to provide accurate inferences, individuals must be sampled at the smallest scale possible and the molecular markers carefully selected (Solano et al. 2010a).

3. APPLICATION OF DISTRIBUTION MODELS AND LANDSCAPE GENETICS FOR TSETSE CONTROL

3.1. *Mapping Landscape to Identify Isolated Tsetse Populations*

Understanding how geographic and environmental features structure genetic variation of tsetse populations is essential for the development of intervention strategies of these cyclical vectors of HAT and AAT in sub-Saharan Africa. Evidence of restricted or absence of gene flow allows genetically isolated islands to be identified (Solano et al. 2009), or isolated ecological population islands (Solano et al. 2010b), from where the tsetse populations present could be eradicated without risk of reinvasion.

For example, two environment-friendly tsetse eradication campaigns achieved the creation of a sustainable tsetse-free zone, i.e. (1) on the Island of Unguja, Zanzibar where an AW-IPM strategy was used to sustainably remove an isolated population of *Glossina austeni* Newstead using the integration of insecticide-impregnated screens, insecticide pour-on on livestock and the SIT (Vreysen et al. 2000), and (2) in the Okavango Delta of Botswana, where an isolated population of *Glossina morsitans centralis* (Machado) was sustainably removed using the SAT in combination with traps and targets in the barrier zones (Vreysen et al. 2000; Kgori et al. 2006).

Recently, landscape genetics has established itself as an important area of research/investigation in the field of tsetse fly control (Bouyer et al. 2015; Bouyer and Lancelot 2018; Saarman et al. 2018). This has allowed the identification of potentially isolated tsetse populations, which offers the opportunity of:

1) selecting the most appropriate intervention strategies for stage 1 of the progressive control pathway (PCP) for AAT (Diall et al. 2017), a stepwise approach leading to their reduction, elimination and finally, vector eradication,

2) planning an integrated management approach (stage 2 of the PCP), and

3) the choice of suppression and elimination activities (stages 3 and 4 of the PCP).

Bouyer et al. (2015) developed a friction map between 37 populations of *Glossina palpalis gambiensis* (Robineau-Desvoidy) in different areas of West Africa by iterating linear regression models of genetic distance between the populations and environmental data as predictors and by determining least-cost dispersal paths. The effect of environmental factors on genetic distance was studied using a linear regression model to estimate the relationship between genetic distance and a set of environmental factors. The main variables influencing genetic distance were:

1) the geographic distance,

2) being located within the same river basin or not, and

3) three metrics of habitat fragmentation, namely the patch density, the surface of suitable area, and the maximum distance between the habitat patches (Bouyer et al. 2015).

A density-based clustering algorithm, applied to the Maxent open-source software output (Phillips et al. 2019), identified eight potentially isolated clusters of suitable habitats containing tsetse populations that were located at least 10 km away from the main tsetse belt (Fig. 1) (Bouyer et al. 2015). This is essential for selecting potential target areas that contain isolated tsetse fly populations that could potentially be eradicated in a sustainable way.

Moreover, the population with the highest predicted genetic distance from the main tsetse belt ($P = 0.003$) was located in the Niayes area of Senegal and is the target of an ongoing eradication campaign (Vreysen et al., this volume).

In line with Bouyer et al. (2015)'s approach for identifying isolated tsetse populations, Saarman et al. (2018) developed methods to create a connectivity surface to identify isolated habitat areas reflecting the genetic and ecological connectivity at a spatial scale of interest. By integrating genetic data from 38 samples, remotely sensed environmental data, and hundreds of field-survey observations from northern Uganda, the approach of Saarman et al. allowed the identification of isolated habitat of *Glossina fuscipes fuscipes* Newstead. To identify isolated habitats, the methodological framework (1) first identifies environmental parameters in correlation with genetic differentiation, (2) predicts spatial connectivity using field-survey observations and the most predictive important environmental parameter(s), and (3) overlays the connectivity surface onto a habitat suitability map (Saarman et al. 2018).

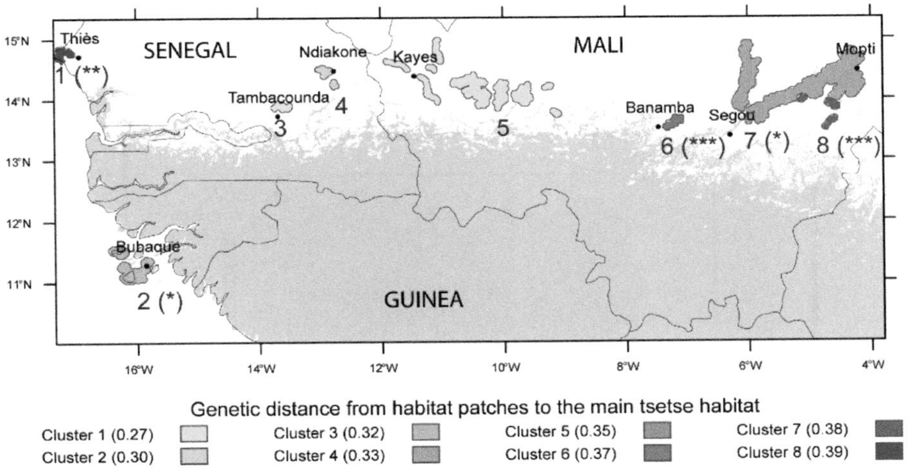

Figure 1. Identification of 8 isolated populations of Glossina palpalis gambiensis *in West Africa. The main tsetse belt predicted by Maxent*[1] *for a sensitivity of 0.90 is in grey and habitat patches are shown as filled, red shapes. Contours and shapes of isolated patches were defined as 5-km radius buffers around pixels of habitat patches. The genetic distance of these patches to the main tsetse belt (reddish scale) was predicted by the AICc-best regression model along least-cost paths. Star symbols after cluster numbers represent the p values for the friction between the patches and the general habitat: (***) $p = 10-3$, (**) $10-3 \leq p < 10-2$, (*) $10-2 \leq p < 5\ 10-2$ (modified from Bouyer et al. 2015).*

[1]The open-source Maxent software is based on the maximum-entropy approach for modelling species niches and distributions (Phillips et al. 2019).

The results from this approach indicate that net photosynthesis is the most powerful predictor of genetic differentiation for *G. f. fuscipes* in northern Uganda. Of the 40 distinct landscape patches of adequate size and distance (purple outlines in Fig. 2a), the resulting connectivity area identified a large, well-connected habitat area in north-western Uganda, as well as 24 plots that contained habitat that was for > 25% considered suitable for *G. f. fuscipes* according to the model (purple outlines in Fig. 2b). These 24 isolated plots were selected as possible candidates to locally create tsetse-free zones and / or testing of new control methods or approaches.

Landscape genetics may also be used to locate areas of high friction where barriers to tsetse dispersal such as insecticide targets or traps are more likely to isolate the target areas.

3.2. Tsetse Distribution Models to Optimize Vector Control

Tsetse distribution models are not only used to map the risk of AAT (Dicko et al. 2015) but are also very useful to optimize tsetse control operations. These models are very useful for selecting priority intervention areas and guiding the management of the vector control operations during all stages of the PCP for addressing AAT (Diall et al. 2017). For example, these models were applied in pilot studies of tsetse control targeting one riverine tsetse species, *G. palpalis gambiensis* in the Niayes area in

Senegal (Dicko et al. 2014), and two savannah species, *G. morsitans morsitans* Westwood and *G. pallidipes* Austin in the Masoka area, mid-Zambezi valley in Zimbabwe (Chikowere et al. 2017).

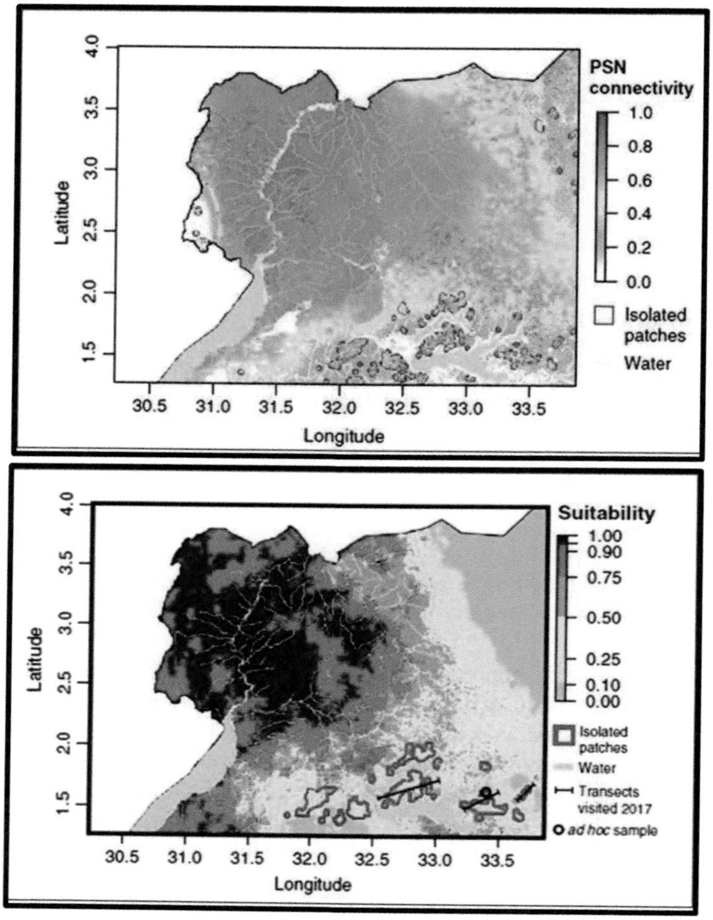

Figure 2a (top). Location of discrete isolated patches in purple and identification of 24 isolated patches of Glossina fuscipes fuscipes *in northern Uganda. A Maxent model was used to produce a connectivity surface, using the environmental variables significantly correlated with genetic differentiation from the previous step and field-survey presence data from 317 traps from northern Uganda.*

Figure 2b (bottom). Habitat suitable for Glossina fuscipes fuscipes *predicted by updating the habitat suitability map obtained with 317 presence data and 12 environmental variables relevant to tsetse ecology by Maxent model. Twenty-four isolated patches identified by the model (purple polygons), the three transects (black lines) used for the field survey, and the location of the tsetse sample from one of the isolated patches used to validate the method (modified from Saarman et al. 2018).*

Using a regularized logistic regression and Maxent, Dicko et al. (2014) compared the probability of presence of *G. palpalis gambiensis* and habitat suitability, respectively. The nature of predictions differed between regularized logistic regression (probability) and Maxent (index). The result provided a better understanding of the relationship between tsetse presence and various environmental parameters as measured by RS. Maxent predicted very well suitable areas considered the most important for an eradication objective, based on an expert-based landscape classification, as some suitable patches can be unoccupied at a certain time and colonized later (Peck 2012), but must nevertheless be included in the target area when applying an AW-IPM strategy (Cecilia et al. 2019).

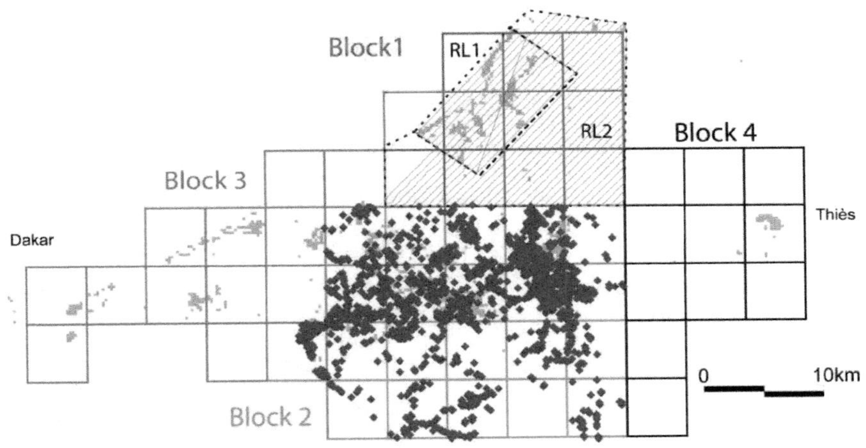

Figure 3. Optimization of the integrated control strategy using model predictions. Maxent model was used to predict the suitable habitats for Glossina palpalis gambiensis *in Senegal. In block 1, the suitable habitats allowed to delimitate two polygons for aerial releases (RL1 and RL2) where the number of sterile males released per km² was adapted to the availability of suitable habitat based on Maxent predictions. In block 1, the green and grey lines present the track flying records of aerial releases on 14th April 2014 in RL1 and 11th April 2014 in RL2 respectively. In block 2, 1,347 insecticide-impregnated traps were deployed from December 2012 to February 2013 in the predicted suitable sites (blue diamonds) to suppress the* Glossina palpalis gambiensis *populations (modified from Dicko et al. 2014).*

Maxent predictions were used throughout the eradication campaign in the Niayes area of Senegal to make the entire operation more efficient in terms of deployment of insecticide-treated targets, release density of sterile males, and the selection of sites to deploy the monitoring traps used for programme evaluation (Fig. 3). Thereby, Maxent predictions allowed optimizing efficiency and reducing the cost of the eradication campaign.

4. CONCLUSIONS

Species distribution modelling and landscape genetics are crucial for planning and optimizing tsetse fly control programmes, especially when sustainable eradication is the selected strategy. Potentially isolated clusters of tsetse fly habitats were identified based on species distribution models and ranked according to their predicted genetic distance to the main tsetse population to locate potential target populations for eradication.

Integrating distribution models and genetic studies in feasibility studies for control operations may also be useful to apply the PCP for controlling AAT, a stepwise approach leading to the suppression, elimination and eradication of vector populations and disease (Diall et al. 2017). Furthermore, distribution models can help improve the efficiency of control activities, leading to a reduction in costs. Finally, these distribution modelling and landscape genetics approaches should be integrated in the future, not only into tsetse control efforts, but might also help in the area-wide management of other pests as well.

5. REFERENCES

Alsan, M. 2015. The effect of the tsetse fly on African development. American Economic Review 105: 382–410.

Barve, N., V. Barve, A. Jiménez-Valverde, A. Lira-Noriega, S. P. Maher, A. T. Peterson, J. Soberón, and F. Villalobos. 2011. The crucial role of the accessible area in ecological niche modeling and species distribution modeling. Ecological Modelling 222: 1810–1819.

Bouyer, J., and R. Lancelot. 2018. Using genetic data to improve species distribution models. Infection, Genetics and Evolution 63: 292–294.

Bouyer, J., and M. J. B. Vreysen. 2018. Vectors: Tsetse flies, pp. 77. *In* J. A. W. Coetzer, G. R. Thomson, N. J. Maclachlan, and M.-L. Penrith (eds.), Infectious Diseases of Livestock. Anipedia, J. A. W Coetzer, and P. Oberem (Directors).

Bouyer, F., A. Dicko, M. T. Seck, B. Sall, M. Lo, M. J. B. Vreysen, J. Bouyer, E. Chia, and A. Wane. 2013. A cost-benefit analysis of tsetse eradication in the Niayes area of Senegal. *In* 32nd ISCTRC Conference, 8-12 September, Khartoum, Soudan.

Bouyer, J., A. Dicko, G. Cecchi, S. Ravel, L. Guerrini, P. Solano, M. J. B. Vreysen, T. De Meeûs, and R. Lancelot. 2015. Mapping landscape friction to locate isolated tsetse populations that are candidates for elimination. Proceedings of the National Academy of Sciences USA 112: 14575–14580.

Bouyer, J., J. S. H. Cox, L. Guerrini, R. Lancelot, A. H. Dicko, J. St. H. Cox, and M. J. B. Vreysen. 2021. Use of geographic information systems and spatial modelling in area-wide integrated pest management programmes that integrate the Sterile Insect Technique, pp. 703–730. *In* V. A. Dyck, J. Hendrichs, and A. S. Robinson (eds.), Sterile Insect Technique – Principles and practice in Area-Wide Integrated Pest Management. Second Edition. CRC Press, Boca Raton, Florida, USA.

Budd, L. 1999. DFID-funded tsetse and trypanosome research and development since 1980. Volume 2: An economic analysis. DFID Livestock Production Programme. Natural Resources International, Chatham Maritime, UK. 123 pp.

Cecilia, H., S. Arnoux, S. Picault, A. Dicko, M. T. Seck, B. Sall, M. Bassène, M. Vreysen, S. Pagabeleguem, A. Bancé, J. Bouyer, and P. Ezanno. 2019. Environmental heterogeneity drives tsetse fly population dynamics and control. bioRxiv: 493650 (recommended by Peer Community in Ecology).

Chikowere, G., A. Dicko, P. Chinwada, M. Zimba, W. Shereni, F. Roger, J. Bouyer, and L. Guerrini. 2017. A pilot study to delimit tsetse target populations in Zimbabwe. PLoS Neglected Tropical Diseases 11(5): e0005566.

Courtin, F., M. Camara, J. B. Rayaisse, M. Kagbadouno, E. Dama, O. Camara, I. S. Traoré, J. Rouamba, M. Peylhard, M. B. Somda, M. Leno, M. J. Lehane, S. J. Torr, P. Solano, V. Jamonneau, and B. Bucheton. 2015. Reducing human-tsetse contact significantly enhances the efficacy of sleeping sickness active screening campaigns: A promising result in the context of elimination. PLoS Neglected Tropical Diseases 9(8): e0003727.

Davies, J. E. 1980. The history of tsetse fly control in Botswana. Report to the Director of Veterinary Services, Botswana.

De Meeûs, T., K. D. McCoy, F. Prugnolle, C. Chevillon, P. Durand, S. Hurtrez-Boussès, and F. Renaud. 2007. Population genetics and molecular epidemiology or how to "débusquer la bête". Infection, Genetics and Evolution 7: 308–332.

Diall, O., G. Cecchi, G. Wanda, R. Argilés-Herrero, M. J. B. Vreysen, G. Cattoli, G. J. Viljoen, R. Mattioli, and J. Bouyer. 2017. Developing a progressive control pathway for African animal trypanosomosis. Trends in Parasitology 33: 499–509.

Dicko, A., R. Lancelot, M. T. Seck, L. Guerrini, B. Sall, M. Lo, M. J. B. Vreysen, T. Lefrançois, W. M. Fonta, S. L. Peck, and J. Bouyer. 2014. Using species distribution models to optimize vector control in the framework of the tsetse eradication campaign in Senegal. Proceedings of the National Academy of Sciences USA 111: 10149–10154.

Dicko, A. H., L. Percoma, A. Sow, Y. Adam, C. Mahama, I. Sidibé, G.-K. Dayo, S. Thévenon, W. Fonta, S. Sanfo, A. Djiteye, E. Salou, V. Djohan, G. Cecchi, and J. Bouyer. 2015. A spatio-temporal model of African animal trypanosomosis risk. PLoS Neglected Tropical Diseases 9 (7): e0003921.

Feldmann, U., and P. D. Ready. 2014. Applying GIS and population genetics for managing livestock insect pests: Case studies of tsetse and screwworm flies. Acta Tropica 138 (Supplement): S1–S5.

Feldmann, U., V. A. Dyck, R. C. Mattioli, J. Jannin, and M. J. B. Vreysen. 2021. Impact of tsetse fly eradication programmes using the sterile insect technique, pp. 1051–1080. *In* V. A. Dyck, J. Hendrichs, and A. S. Robinson (eds.), Sterile Insect Technique – Principles and practice in Area-Wide Integrated Pest Management. Second Edition. CRC Press, Boca Raton, Florida, USA.

Hendrichs, J., A. S. Robinson, P. Kenmore, and M. J. B. Vreysen. 2007. Area-wide integrated pest management (AW-IPM): Principles, practice and prospects, pp. 3–33. *In* M. J. B. Vreysen, A. S. Robinson and J. Hendrichs (eds.), Area-wide control of insect pests. From research to field implementation. Springer, Dordrecht, The Netherlands.

Hursey, B. S., and J. Slingenbergh. 1995. The tsetse fly and its effects on agriculture in sub-Saharian Africa. Revue Mondiale de Zootechnie 84: 67–73.

Jordan, A. M. 1986. Trypanosomiasis control and African rural development. Longman Group Ltd., London, UK. pp. 357.

Kgori, P. M., S. Modo, and S. J. Torr. 2006. The use of aerial spraying to eliminate tsetse from the Okavango Delta of Botswana. Acta Tropica 99: 184–199.

Klassen, W., and M. J. B. Vreysen. 2021. Area-wide integrated pest management and the Sterile Insect Technique, pp. 75–112. *In* V. A. Dyck, J. Hendrichs, and A. S. Robinson (eds.), Sterile Insect Technique – Principles and practice in Area-Wide Integrated Pest Management. Second Edition. CRC Press, Boca Raton, Florida, USA.

Lovemore, D. F. 1972. Annual report of the Branch of Tsetse and Trypanosomiasis Control, Department of Veterinary Services, Ministry of Agriculture, Rhodesia, for the year ended 30th September 1971. Government Printing and Stationery, Salisbury, Rhodesia. 65996-2-300.

Lovemore, D. F. 1973. Annual report of the Branch of Tsetse and Trypanosomiasis Control, Department of Veterinary Services, Ministry of Agriculture, Rhodesia, for the year ended 30th September 1972. Government Printing and Stationery, Salisbury, Rhodesia. 68789-6-300.

Lovemore, D. F. 1974. A note on tsetse and trypanosomiasis control in Rhodesia during the year ended 30th September 1974. Type-written internal report prepared for the Director of Veterinary Services. Salisbury, Rhodesia.

MacLennan, K. J. R., and W. W. Kirby. 1958. The eradication of *Glossina morsitans submorsitans* Newst. in part of a river flood plain in Northern Nigeria by chemical means. Bulletin of Entomological Research 49: 123–131.

Mahamat, M. H., M. Peka, J. B. Rayaisse, K. S. Rock, M. A. Toko, J. Darnas, G. M. Brahim, A. B. Alkatib, W. Yoni, I. Tirados, F. Courtin, C. Nersy, I. O. Alfaroukh, S. J. Torr, M. J. Lehane, and P. Solano. 2017. Adding tsetse control to medical activities allows a decrease in transmission of sleeping sickness in the Mandoul focus (Chad). PLoS Neglected Tropical Diseases 11(7): e0005792.

Manel, S., M. K. Schwartz, G. Luikart, and P. Taberlet. 2003. Landscape genetics: Combining landscape ecology and population genetics. Trends in Ecology & Evolution 18: 189–197.

Paynter, Q., and J. Brady. 1992. Flight behaviour of tsetse flies in thick bush (*Glossina pallidipes* (Diptera: Glossinidae)). Bulletin of Entomological Research 82: 513–516.

Peck, S. L. 2012. Networks of habitat patches in tsetse fly control: implications of metapopulation structure on assessing local extinction. Ecological Modeling 246: 99–102.

Peterman, W. E., G. M. Connette, R. D. Semlitsch, and L. S. Eggert. 2014. Ecological resistance surfaces predict fine-scale genetic differentiation in a terrestrial woodland salamander. Molecular Ecology 23: 2402–2413.

Phillips, S. J., M. Dudík, R. E. Schapire. 2019. Maxent software for modeling species niches and distributions (Version 3.4.1).

Robertson, A. G., and E. B. Kluge. 1968. The use of insecticide in arresting an advance of *Glossina morsitans* Westwood in the south-east lowveld of Rhodesia. Proceedings and Transactions of the Rhodesia Scientific Association 53: 17–33.

Robertson, A. G., E. B. Kluge, D. A. Kritzinger, and A. E. De Sousa. 1972. The use of residual insecticides in the reclamation of the Rhodesia– Mozambique border region between the Sabi/Save and Limpopo rivers from *Glossina morsitans* Westwood. Proceedings and Transactions of the Rhodesia Scientific Association 55: 34–62.

Rogers, D. J., and T. P. Robinson. 2004. Tsetse distribution, pp. 139–179. *In* I. Maudlin, P. H. Holmes, and M. A. Miles (eds.), The Trypanosomiases. CABI International Wallingford, Oxfordshire, UK.

Saarman, N., M. Burak, R. Opiro, C. Hyseni, R. Echodu, K. Dion, E. A. Opiyo, A. W. Dunn, G. Amatulli, S. Aksoy, and A. Caccone. 2018. A spatial genetics approach to inform vector control of tsetse flies (*Glossina fuscipes fuscipes*) in Northern Uganda. Ecology and Evolution 8: 5336–5354.

Scoones, I. 2016. Contested histories: Power and politics in trypanosomiasis control, pp. 78–93. *In* K. Bardosh (ed.), Science, politics and zoonotic disease in Africa, Routledge, New York, USA.

Shaw, A. P. M. 2004. Economics of African trypanosomosis, pp. 369–402. *In* I. Maudlin, P. H. Holmes, and M. A. Miles (eds.), The Trypanosomiases. CABI Publishing, Wallingford, Oxfordshire, UK.

Soberón, J., and A. T. Peterson. 2005. Interpretation of models of fundamental ecological niches and species' distributional areas. Biodiversity Informatics 2: 1–10.

Solano, P., S. Ravel, and T. De Meeûs. 2010a. How can tsetse population genetics contribute to African trypanosomiasis control? Trends in Parasitology 26: 255–263.

Solano, P., S. J. Torr, and M. J. Lehane. 2013. Is vector control needed to eliminate *gambiense* human African trypanosomiasis? Frontiers in Cellular and Infection Microbiology 3: 33.

Solano, P., S. Ravel, J. Bouyer, M. Camara, M. S. Kagbadouno, N. Dyer, L. Gardes, D. Herault, M. J. Donnelly, and T. De Meeûs. 2009. The population structure of *Glossina palpalis gambiensis* from island and continental locations in coastal Guinea. PLoS Neglected Tropical Diseases 3(3): e392.

Solano, P., D. Kaba, S. Ravel, N. A. Dyer, B. Sall, M. J. B. Vreysen, M. T. Seck, H. Darbyshir, L. Gardes, M. J. Donnelly, T. De Meeûs, and J. Bouyer. 2010b. Population genetics as a tool to select tsetse control strategies: Suppression or eradication of *Glossina palpalis gambiensis* in the Niayes of Senegal. PLoS Neglected Tropical Diseases 4(5): e692.

Swallow, B. M. 1999. Impacts of trypanosomiasis on African agriculture. PAAT Technical and Scientific Series 2. FAO, Rome, Italy. ISSN 1020-7163.

(UN) United Nations. 2019. Sustainable development goals. New York, NY, USA.

Vale, G. A. 1999. Joint tests of insecticides for use on cattle. Final Report. Regional Tsetse and Trypanosomiasis Control Programme, Malawi, Mozambique, Zambia and Zimbabwe. Harare, Zimbabwe.

Vreysen, M. J. B. 2006. Prospects for area-wide integrated control of tsetse flies (Diptera: Glossinidae) and trypanosomosis in sub-Saharan Africa. Revista Sociedad Entomológica Argentina 65 (1-2): 1-21.

Vreysen, M. J. B., A. S. Robinson and J. Hendrichs (eds.). 2007. Area-wide control of insect pests. From research to field implementation. Springer, Dordrecht, The Netherlands. 789 pp.

Vreysen, M. J. B., M. T. Seck, B. Sall, and J. Bouyer. 2013. Tsetse flies: Their biology and control using area-wide integrated pest management approaches. Journal of Invertebrate Pathology 112 (Supplement): S15–S25.

Vreysen, M. J. B., K. M. Saleh, M. Y. Ali, A. M. Abdulla, Z. R. Zhu, K. G. Juma, V. A. Dyck, A. R. Msangi, P. A. Mkonyi, and H. U. Feldmann. 2000. *Glossina austeni* (Diptera: Glossinidae) eradicated on the island of Unguja, Zanzibar, using the sterile insect technique. Journal of Economic Entomology 93: 123–135.

(WHO) World Health Organization & WHO Expert Committee on the Control and Surveillance of Human African Trypanosomiasis. 2013. Control and surveillance of human African trypanosomiasis: report of a WHO expert committee. WHO, Geneva, Switzerland. 237 pp.

(WHO) World Health Organization. 2019. Trypanosomiasis, human African (sleeping sickness). 11 October 2019.

AGENT-BASED SIMULATIONS TO DETERMINE MEDITERRANEAN FRUIT FLY DECLARATION OF ERADICATION FOLLOWING OUTBREAKS: CONCEPTS AND PRACTICAL EXAMPLES

N. C. MANOUKIS AND T. C. COLLIER

Tropical Crop and Commodity Protection Research Unit, Daniel K. Inouye United States Pacific Basin Agricultural Research Center, United States Department of Agriculture – Agricultural Research Service, Hilo, Hawaii 96720 USA; Nicholas.Manoukis@USDA.GOV

SUMMARY

Areas of the world that do not have established populations of the Mediterranean fruit fly *Ceratitis capitata* (Wiedemann) and other invasive pestivorous Tephritidae are sometimes subject to incursions due to increasing travel and trade. When these occur, control programmes are put in place often including quarantine and additional measures until eradication of the outbreak is declared. A critical practical question that arises is how long to maintain the eradication programme and associated area-wide measures after the last sampling of the invading Mediterranean fruit fly. Current practice is usually to maintain measures and increased monitoring until enough time has passed for three generations of flies without another fly catch; generation times are calculated via thermal unit accumulation ("Degree Day"). A recent alternative or complementary approach is to model the invading population using an Agent-Based Simulation (ABS). This chapter outlines the use of *MEDiterranean fruit Fly Outbreak and Eradication Simulation* (MED-FOES), an ABS implementation aimed at modelling invading Mediterranean fruit fly populations to determine effective duration of quarantine and other eradication measures following the last detection of an incursion. Basic concepts are described, together with a description of major functions and use of thousands of individual simulations to encompass the range of demographic possibilities. Finally, specific examples from Santiago, Chile and California, USA are offered to show how the ABS can provide useful information for programme managers setting eradication programme durations.

Key Words: Invasion biology, incursions, outbreaks, quarantine, eradication, Tephritidae, *Ceratitis capitata*, medfly, monitoring, detections, California, Chile, degree-days, model, modelling, individual-based simulations, multi-agent simulations, MED-FOES

J. Hendrichs, R. Pereira and M. J. B. Vreysen (eds.), Area-Wide Integrated Pest Management: Development and Field Application, pp. 869–888. CRC Press, Boca Raton, Florida, USA.

1. INTRODUCTION

Incursions of the polyphagous pest Mediterranean fruit fly *Ceratitis capitata* (Wiedemann) occur in urban and agricultural areas around the world. The Global Eradication and Response Database (GERDA), although not a complete compilation of all invasions and eradication efforts, currently lists 117 Mediterranean fruit fly incursions and eradication programmes in a wide variety of geographic areas, from the USA to Australia, Chile, Mexico, and New Zealand among others. Responding to each of these outbreaks costs an average of USD 12 million (normalized to 2012 USD) (Kean et al. 2012; Suckling et al. 2016). While this cost is small compared with the estimated damage from a Mediterranean fruit fly establishment in many of these areas (USD 1500 million in the US state of California alone [Siebert and Cooper 1995]), it is still important to optimize responses to incursions. This is because mounting and maintaining an effective response and quarantine is a significant burden and organizational challenge for state and private organizations.

Quarantine measures are often put into place following the detection of an incursion of an invasive pest (FAO 2016a). In the case of Mediterranean fruit fly, quarantine measures involve designating an area around the find where fresh fruits are restricted from leaving. If the area includes commercial host fruit production, then the quarantine measures can result in serious economic losses. Therefore, the period to maintain the quarantine and associated measures against the invading Mediterranean fruit fly population becomes a critical practical question.

Current practice in many parts of the world is based on ISPM 26 (FAO 2016b), such as the protocol used by the California Department of Food and Agriculture (CDFA), whereby the quarantine is suspended when three full generations of the fly have passed without another find; generation times are calculated via thermal unit accumulation (California Code of Regulations 2017). The thermal accumulation development models most often used simply posit that development from egg to adult requires accumulation of a specific amount of heat above a base threshold, measured in units of degree-days (Roltsch et al. 1999). Various calculation methods exist, but all approximate the integral of temperature over time for temperatures above a given base temperature. The simplicity of degree-days calculations is attractive, and thermal accumulation models are widely used with impressive accuracy for predicting developmental timing in many agricultural contexts. However, it may not be entirely appropriate for Mediterranean fruit fly eradication and quarantine duration determination. "Degree-days" is a development model, not a population or eradication model. Moreover, the requirement for three generations of degree-days to pass is difficult to justify theoretically.

An alternative approach to estimating eradication programme length or duration since the last detection is to use an Agent-Based Simulation (ABS; also called "Individual-Based" or "Multi-Agent") to simulate the arrival of *C. capitata* in a new area as an insect invasion (Manoukis and Hoffman 2014). In an ABS, individual flies are described as unique and autonomous, and where they usually interact with each other and their environment at a local level (Railsback and Grimm 2012). An important characteristic is that individual members of the simulation ("agents") are represented independently in computer memory via the simulation software, and these have behaviours (functions) and characteristics (parameters) that may make them unique. The behaviours and characteristics of the entire system are not explicitly coded by the programme rather they emerge from the interactions and behaviours of the system's constituent agents.

The ABS approach has not received a lot of attention for studies on insect invasions (Crespo-Pérez et al. 2011; although see Vinatier et al. 2011 for an example of an ABS of an agricultural pest), but it has been more widely used to address questions on physiological ecology, foraging networks, ant nest choice, and disease vector dynamics among others (Jackson et al. 2004; Pratt et al. 2005; Almeida et al. 2010; Radchuk et al. 2013).

This chapter is concerned with the concepts, implementation and use of an ABS entitled MED-FOES (*MED-Fly Outbreak and Eradication Simulation*; MED-FOES 2019), that was designed to simulate the invasion, programmatic response to an outbreak, and extirpation of a population of Mediterranean fruit fly in a pest free area (Barclay et al. 2021). The original goal of the ABS and its implementation is to provide an estimate of eradication programme duration that is roughly independent of the values determined by thermal unit accumulation.

In an area-wide context, which means addressing a total pest population within a defined area and not a localised field-by-field approach (Hendrichs et al. 2007), modelling eradication programme lengths is important for several reasons:

First, programmes against Tephritidae tend to encompass large areas, and many of the measures taken following a Mediterranean fruit fly incursion are likewise applied over these same vast areas, such as intense surveillance, restriction of fruit movement and in most cases SIT application. These approaches are not effective in an uncoordinated property-by-property setting.

Second, there are area-wide programme costs to be considered; these can grow to be large if excessively long programmes are implemented, in particular if quarantines and commercial production areas are involved.

Third, eradication efforts are often directly connected to area-wide programmes (Smith 1998; Myers et al. 2000). If eradication programme lengths against Mediterranean fruit fly are too short, there is potential for survival of remnants of the invading population, which would require a new eradication campaign or in a worst case scenario, result in the establishment of the pest species in a free area (Carey 1991; Papadopoulos et al. 2013; McInnis et al. 2017; Shelly et al. 2017).

MED-FOES is based on a modelling framework that simulates the process of Mediterranean fruit fly population extirpation/eradication (Manoukis and Hoffman 2014). A description of the important functions included in the simulation is given in Section 2, with minimal mathematical background. The use of thousands of simulations to understand the range of possible outcomes following an outbreak is the focus of Section 3. In Section 4, two examples of historical incursions are analysed to illustrate the use of MED-FOES for determining eradication programme length. Finally, Section 5, includes broad conclusions and some suggestions for future work. The actual installation and mechanics of using the software are not covered here; the reader is referred to the software manual for these details (MED-FOES 2019).

2. CONCEPTS AND IMPLEMENTATION

2.1. Simulation Description

Mediterranean fruit fly agents in the simulation can be created, and they develop, reproduce and die. The major challenge is to have the agents behave in a biologically realistic manner so that the results of thousands of executions of the simulation can be usefully related to flies in the real world.

Figure 1 gives a graphical representation of the developmental stages and states fly agents in MED-FOES can occupy as well as their connections. During each hour of the simulation, flies may move from one developmental stage to another, or not. They also may go from being alive to being non-reproductive (mated with a sterile male) or dead, when they are removed from the simulation and no longer considered.

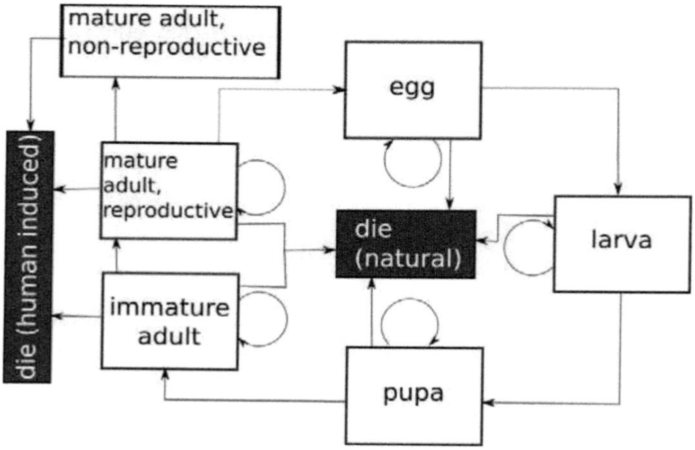

Figure 1. Graphical representation of the developmental stages and states of Mediterranean fruit fly agents in MED-FOES. Arrows indicate changes in stage or state per time step (adapted from Manoukis and Hoffman 2014).

2.2. Initialization

The simulation is initialized based on an estimated number of potential remnant adult female flies in the area of an incursion that has been subject to area-wide application of eradication measures. Given this number, the estimated number of female flies in all other stages is determined based on the expected stable age distribution for the Mediterranean fruit fly from the literature (Carey 1982; Vargas et al. 1997). These flies are then iterated through hourly time steps; for each time step, they may develop, reproduce, die of natural causes or die of human-induced causes.

The current version of MED-FOES includes a simple model of the initial stages of the simulation. There is a time between when a detection that exceed the threshold needed to trigger a response during which full counter-measures are not in place. In general, the number of flies detected until the point of declaration is used to estimate the initial population size range. During the short period between detection exceeding threshold and counter-measure implementation (2-5 days in many jurisdictions), the population of agents can grow, and no human-induced mortality is considered. At the time R (in days) counter-measures initiate and the population begins its demographic trajectory to eradication. This representation of the early stages is unrealistic, as in real incursions detections may occur asynchronously, trapping densities may be changed in response, and individual counter-measures are applied at different times. However, the simulations are generally not highly sensitive to the length of R or initial population size (Manoukis and Hoffman 2014). Future versions of MED-FOES are planned to have the ability to include more realistic initial stages and the ability to model time varying interventions.

2.3. Hourly Temperatures

The only time-dependent input to the ABS is an hourly air temperature time series. Ideally these data should be acquired from a long-term weather station in close proximity to the outbreak being simulated. Data from airports archived in the Integrated Surface Data repository of the National Oceanic and Atmospheric Administration (NOAA) are often appropriate and methods for processing them have been published (Collier and Manoukis 2017).

Care should be taken to ensure that the temperature data are accurate as the ABS (as well as simpler thermal accumulation models) can be sensitive to biases and errors in the data. For example, a sensor improperly shielded from the sun would tend to report slightly higher temperatures, and the cumulative nature of a thermal accumulation development model would amplify the effect over time. Similarly, spurious very high or very low values produced by sensor errors could lead to erroneous high mortality events in the ABS. Actual temperature data often contain gaps, errors, and a non-uniform sampling rate, and therefore require some 'cleaning' before they can be used.

A cleaning process typically starts with the detection and removal of erroneous outlier values. A large variety of methods exist for outlier removal, but a simple method which works well for temperature time series data is computing the running standard deviation (σ) over a window spanning several days, computing the running median (m) over a several hour window, and rejecting individual temperature values (t) for which the absolute value of $(t - m)/\sigma$ exceeds a predefined threshold.

The next step of cleaning is the identification and filling of gaps in the data. Because of the daily cycle of temperature data, it is appropriate to use different methods to fill small (less than a few hours) and large gaps (a few hours to several days). Small gaps may be filled by simple interpolation, which can also serve to simultaneously resample the data to a true hourly frequency. Large gaps, however, are more appropriately filled by interpolating across observations from the same time of day across days. Gaps larger than several days would likely require special treatment dependent on the particulars of the location and season. If actual hourly temperature data are not available, then an approximation may be inferred from daily minimum and maximum temperatures using established methods such as the one implemented in the "TemperatureEstimator.java programme" distributed with MED-FOES (Reicosky et al. 1989; Campbell and Norman 1997).

2.4. Development

The single factor that determines the probability of stage transition or ovarian maturation (development) for flies in the simulation is temperature. The simulation requires a base developmental temperature T_{min} and time-to-transition K for each of the following transitions: egg to larva, larva to pupa, pupa to adult and adult to mature adult.

Several reports in the literature give the mean time to transition (d) for these stages as affected by a range of constant fixed temperatures (T). The common practice is to regress the developmental rate ($= 1/d$) against fixed temperatures, which gives a clear linear relationship for temperatures between about 16 and 30 °C. The linear regression model would then be

$$\frac{1}{d} = a + bT$$

where a is the intercept and b the slope. The parameters required by the simulation are then calculated as follows:

$$T_{min} = -\frac{a}{b}$$

and

$$K = \frac{1}{b}$$

There are two methods available for calculating the probability of stage transition. The simplest is a "uniform" model, where the probability of transition at any given hour is determined only by the temperature during that hour. For each hour, a random number between zero and one is drawn from a uniform distribution for each insect that is not a mature adult. If this number is lower than the developmental rate for that temperature, then a stage transition occurs. The average time for a transition at a given temperature is thus equal to $1/d$.

Note that the "uniform" method described above does not take into account the time each insect has spent in a given stage. Therefore, it is possible (though unlikely for most parameter combinations and realistic temperatures) that an agent could go from being an egg to an adult in a few hours. A more realistic developmental model is the "thermal summation" approach (Fletcher 1989), where each degree above T_{min} for each hour counts towards a required threshold C for stage transition. Variation is included for each individual fly agent when it is created in the form of a variable γ, which is the standard deviation of the variation in development time as a proportion of the development time for each stage. Thus, when

$$C + \gamma < \sum_{t=0}^{i} T_i - T_{min}$$

from the time of insect creation (0) to the current time i, stage transition occurs. Note that the value of C is stage-specific, and γ constant across stages.

2.5. Mortality

There are two ways that death is implemented in the simulation. The simplest approach is to set a fixed stage-specific daily death rate, denoted as M_x. Each hour of the simulation, for each insect at stage x, a random double precision floating point number between 0 and 1 is drawn and if it is lower than $M_x^{1/24}$, then the insect dies. This method is completely temperature-independent and is a useful approximation when the effect of temperature on mortality is unknown.

In the case of the Mediterranean fruit fly reliable data are available on the effect of temperature on daily mortality rates. MED-FOES uses the stage-specific quadratic relationships from Gutierrez and Ponti (2011) for this relationship. When this mode it used, the run-specific parameters M_x are the death rate at the optimum temperature (20–25 °C) rather than the average mortality per unit time.

Additional mortality on adults is introduced at a given number of days after the simulation starts. The probability of human-induced death is fixed per simulation and the same for mature and immature adults; it represents human-induced mortality as a result of counter measures. Human induced mortality is currently limited to the adult stage, though some control measures, such as fruit stripping, may affect immature flies. After human intervention, it is optionally possible to simulate the effect of trapping separate from mortality induced by other countermeasures. This is discussed in the next Section.

2.6. Trapping

Spatially explicit consideration of the effect of trapping on agent mortality is possible as of MED-FOES version 0.6, using the approach of Manoukis et al. (2014) as implemented in the software "TrapGrid" (Manoukis et al. 2014). For parameters relevant to the Mediterranean fruit fly, see Manoukis et al. (2015). A brief description of the trapping model is given here.

TrapGrid is an implementation of a landscape-level, spatially explicit model of trap networks that incorporates variable attractiveness of traps and a movement model for dispersion. TrapGrid simulates susceptible insect capture by placing traps in a rectangular area. Each trap has a parameter indicating its attractiveness (λ). Using this value, the escape probability for a given insect at a given distance from the trap can be calculated. The calculation of escape probability can be conducted for many points in the trapping grid, yielding an instantaneous estimate of the escape probability. Note that probability of capture is simply 1- {probability of escape}.

We calculate the distance to a given trap as:

$$d = \sqrt{(x_t - x)^2 + (y_t - y)^2},$$

where (x_t, y_t) is the position of the trap and (x, y) is the position of the fly. For $d >= 0$, we use an exponential decay with a logistic ($H(d)$) to model the probability of being captured:

$$p = f(d, \lambda) = \lambda e^{-\lambda d} H(d)$$

where

$$H(d) = \frac{1}{1 + e^{-2\lambda d}}$$

These can be combined, producing the hyperbolic secant:

$$f(d,\lambda) = \frac{2e^{\lambda d}}{1 + e^{2\lambda d}}$$

Fig. 2 shows how the probability of capture changes with distance from the trap given λ. Each trap in a TrapGrid model represents the spatial relationship between distance from a lure-baited trap and probability of capture in the very near future. The parameter λ is the attractiveness of the trap, with smaller values representing a more attractive trap.

One important feature of the capture model used is that, for a given value of λ, $1/\lambda$ is the distance at which there is a 65% chance of capturing a susceptible insect. This allows easy comparison of trap attraction between species and lures. Movement in TrapGrid as used by MED-FOES is simple diffusion (Skellam 1951; Kareiva 1983), through which the probability of capture over time (p_t) is calculated.

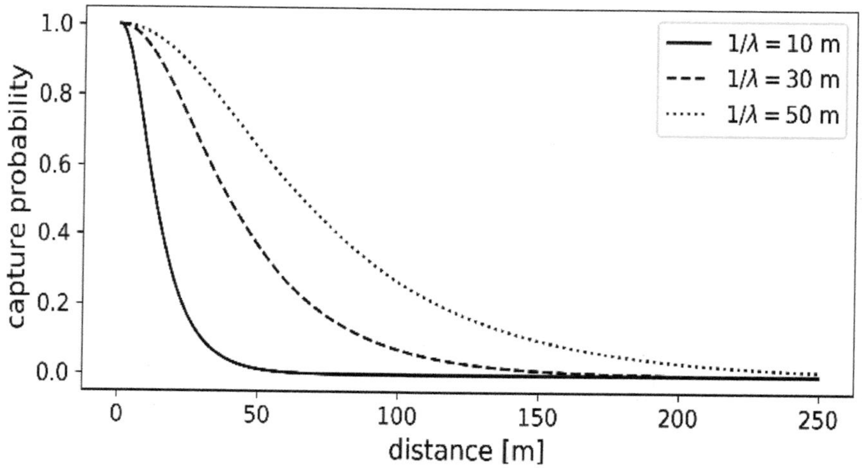

Figure 2. Capture probability for an individual insect versus distance from a trap, where attraction (1/λ) = 10 (solid), 30 (dashed) and 50 (dotted) meters. 1/λ represents the distance at which there is a 0.65 probability of capture (adapted from Manoukis et al. 2014).

In order to make an estimate of p_t, consider the net movement of a fly over the time period in question (usually a day), represented by the parameter D in diffusion models. Estimates of D from the literature are around 1 x 10^4 m^2/day (Corbett and Plant 1993). We can use the net movement per day to model the proportion of the arena space that an individual Mediterranean fruit fly might "experience" per day.

The diffusion in two dimensions is well studied, and has the following form:

$$\frac{dN(x,y,t)}{dt} = D\left[\frac{d^2N}{dx^2} + \frac{d^2N}{dy^2}\right]$$

where N is the population density, t is time, x, y are spatial coordinates and D is the diffusion coefficient. This process produces a bivariate normal distribution of density over time, the probability density function (pdf) of which is the basis of our model of the fly population and its spread over time:

$$g(x,y) = \frac{1}{2\pi\sigma^2} e^{-\frac{1}{2}\left(\frac{x^2}{\sigma^2}+\frac{xy^2}{\sigma^2}\right)}$$

Here, σ is the standard deviation, assuming that

$$\mu = 0,\ \sigma x = \sigma y = \sigma \text{ and that } cor(x, y) = 0.$$

Further description is beyond the scope of this document, but it is important to note that the diffusion parameter D in m^2 per day is the single factor that determines how quickly the simulated flies in TrapGrid will spread.

MED-FOES can run a set of TrapGrid simulations before creating agents. This set of spatial simulations is used to determine the average daily probability of death from trapping given a trapping network following a detection, and then this mortality is scaled to hourly time steps and applied.

2.7. Reproduction

Every mature reproductive Mediterranean fruit fly will oviposit eggs every 24 hours prior to intervention time t_S. In this sense, the simulation only includes females. The mean number of eggs and variance in reproductive output are set by the variables r and r_{var}. After human intervention, reproduction may be curtailed at a set daily rate, which is denoted r_{red}. This variable is included based on modelling of the Sterile Insect Technique (SIT) and its effects on host populations (Knipling 1979).

In the ABS, each mature adult fly that exists at t_S is subject to loss of reproductive ability each day with probability r_{red}. In the case studies below, flies emerging after t_S are assumed to not be able to reproduce. Mating is currently not required for reproduction, so the difficulty in finding a mate that comes with small population sizes is not modelled. In addition, the Allee effect, where there is reduced mean individual fitness in small populations, is also not considered.

3. ANALYSING AN OUTBREAK: MULTIPLE SIMULATIONS

Until this point, we have focused on a single simulation and how biological and population processes are represented. Individual simulations can vary from one to the next, even if the same parameters are used, due to stochasticity (random events) built into the model. However, in order to obtain useful and actionable information on a real-world incursion, a single set of parameters is insufficient because there is uncertainty on the values of critical parameters. Thus, it is necessary to execute a set of simulations, varying parameters between runs, in order to obtain the range of possible outcomes following an incursion. MED-FOES has this functionality built-in via a separate executable programme called "med-foes-p.jar" (MED-FOES 2019).

3.1. Exploring the Parameter Space

MED-FOES includes a large number of parameters that may be set over a range, including aspects of mortality, development, reproduction, and control measures. Even if a fixed number of discrete levels over all these ranges were selected, a full factorial experiment would be computationally prohibitive.

Latin Hypercube Sampling (LHS) is used to select parameter sets through the parameter hyperspace (Blower and Dowlatabadi 1994). The details of LHS are beyond the scope of this document, save that it ensures an even distribution of the combinations of parameters used for individual runs.

The number of simulations executed can be set by the user. A number in the thousands will usually provide sufficient resolution on the range of possible outcomes.

3.2. Summarizing Output

MED-FOES produces a summary output file that is named MED-FOESp_{timestamp}_summary.txt, where {timestamp} is the date and time the programme was run to avoid inadvertently overwriting previous output. This file contains the parameters used as well as results, including the mean, standard deviation, and 25, 50, 95, and 99% quantiles of the length of runs (individual simulations). Since runs nominally start close to the time of last detection and terminate upon eradication of the simulated population, these run length statistics summarize time to eradication across the various parameter sets tested.

Additionally, the mean and standard deviation of number of flies at the end of the simulations are reported and can be used to detect situations where some runs ended without reaching eradication due to an insufficient amount of input temperature data or exceeding the maximum number of flies allowed. Summary figures are produced that give a quick visual summary of the outcomes of the set of simulations (Fig. 3).

A MED-FOESp_{timestamp}_details.txt file is also produced. This is a tab-delimitated table reporting for each run the length of the run in simulated hours, the end condition (eradication, out of temperature data, or maximum flies exceeded), and total number of flies at the end. From this file, arbitrary statistics for the time to eradication can be produced.

Finally, for each individual run a summary and details file is produced under the "runs" directory. The summary file gives the specific parameters used for that particular simulation, in contrast to the MED-FOESp summary file, which gives the ranges sampled by the LHS procedure. It also gives summary results: number of hours simulated, number of flies at the end of the simulation, cumulative number of eggs, and cumulative number of deaths.

The details file is a comma-delimited file with a row for each day of the simulation containing: time, mean temperature, minimum temperature, maximum temperature, cumulative death, cumulative birth, the number of flies in each life stage (egg, larvae, pupae, adult), and total number of flies. The collection of runs details files can be used to produce a wide variety of detailed outputs, such as Figs. 4 and 5 shown in the next Section.

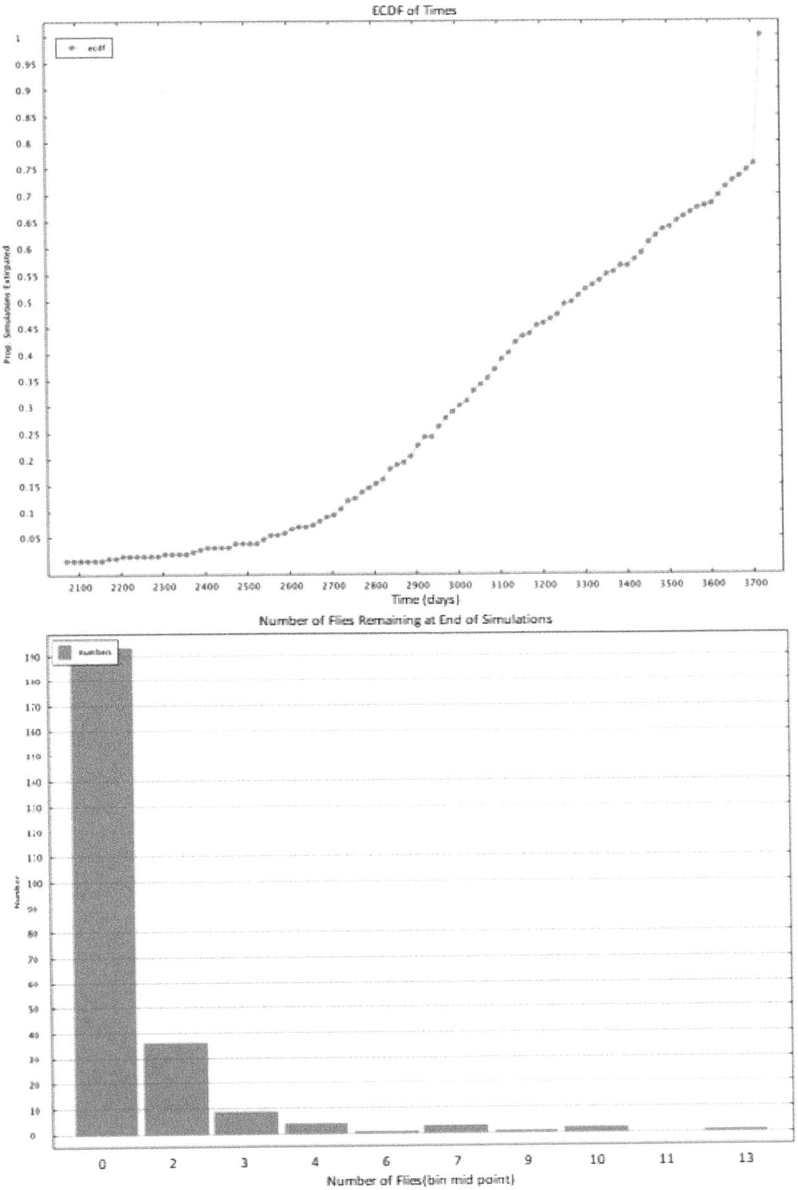

Figure 3. Example of graphical output from MED-FOES showing 250 runs simulating an outbreak of Mediterranean fruit fly in California. Upper panel: Empirical cumulative frequency distribution of the number of simulations showing extirpation over time. Bottom panel: Histogram of the number of surviving agents at the end of the 250 simulations. This simple summary graphic can be used in internal reports without the need of post-simulation data manipulation.

4. PRACTICAL EXAMPLES

4.1. Santiago, Chile

Chile is internationally recognized to be free of tephritids of economic importance, including species in economically important genera *Anastrepha, Bactrocera*, and *Ceratitis*. This is facilitated by Chile's geographic isolation, stemming from the presence of the Pacific Ocean to the West and the Andes mountains in the East, as well as its excellent pest exclusion infrastructure. Despite this isolation, occasionally Mediterranean fruit fly outbreaks are detected in the country, usually via the network of about 14 000 traps.

One such detection occurred in the neighbourhood of Independencia in Santiago, in 2011. The first fly was detected in a Jackson trap baited with trimedlure on 14 October, and the response including quarantine, enhanced trapping, bait spray (GF-120) applications, soil treatment under host trees, and fruit stripping was put in place by 18 October. Another fly was sampled on 18 October in a McPhail trap about 800 m from the first detection; thereafter, no further detections of adults were made. The response programme was concluded on 9 February 2012 after three generations (F_3) of degree-day development as calculated by the Chilean Agricultural and Livestock Service (SAG) using the method of Tassan et al. (1983). This occurred 115 days after the first find.

For the ABS analysis, conducted with version 0.6.2 of MED-FOES, critical parameters that had to be evaluated included: initial population size, reduction in fecundity over time, and hourly temperature data. Starting with the third of these, hourly temperature data were acquired from NOAA's online ISD-Lite dataset derived from the Integrated Surface Database for the weather station at the nearby (approx. 12 km from the outbreak neighbourhood) Comodoro Arturo Merino Benítez International Airport (SCEL). The air temperature data for one year starting on 18-10-2011 (day 0) were extracted and cleaned using the method described previously and saved to a comma separated format (csv) file for MED-FOES to use as input.

The SAG programme did not include SIT for the 2011 outbreak. This might mean that the modeller decides not to include the parameter on fertility reduction per day ("rred"). However, due to the intense fruit stripping (100% in a 400 m radius from each find) plus the soil drenches (not used in California currently), it was estimated that there was probably an effective reduction in the probability of reproduction over a wide range, 0.2-1.0.

Finally, the initial population size was estimated based on the surveillance network in Chile at the time of the outbreak. This consisted of one trimedlure trap per 25 ha and one protein trap per 100 ha, comparable to the California values of one trimedlure and one protein trap per 52 ha. This gives a rough detection sensitivity of 2-3 % of the adult population (Manoukis and Hoffman 2014). Assuming a stable age distribution (from Table 3 in Carey 1982), gives estimated numbers in the other life stages. Though these figures are rough, they are probably sufficient since the model is not very sensitive to initial numbers (Manoukis and Hoffman 2014).

The complete set of parameters used to generate the runs are given below:

T	SCEL_2011-10-18.csv	Hourly temperature data file
Ni	66,100	Initial population size (range)
Ad	29.8,49.7,15.6,1.8,3.1	Initial age distribution (from Carey 1982)
R	4	Days before intervention is implemented
S	0.05,0.15	Daily human-induced mortality (range)
rred	.2,1	Reduction of reproduction (range)
Sai	true	Sterility after intervention
TEL	9.6,12.5,27.27,33.80	Transition parameters, egg to larva (range)
TLP	5.0,10.8,94.50,186.78	Transition parameters, larva to pupa (range)
TPA	9.1,13.8,123.96,169.49	Transition parameters, pupae to adult (range)
TIM	7.9,9.9,58.20,105.71	Transition parameters, adult to sexually mature (range)
Me	0.0198,0.1211	Daily natural mortality of eggs (range)
Ml	0.0068,0.0946	Daily natural mortality of larvae (range)
Mp	0.0016,0.0465	Daily natural mortality of pupae (range)
Ma	0.0245,0.1340	Daily natural mortality of adults (range)
tdm	true	Use temperature dependent mortality
r	5,35	Eggs produced per reproduction event (range)
rvar	3.57	Variance in eggs produced per event
Dm	1	Development model; 0 = uniform, 1 = thermal summation
TuSD	0.05	Variation in thermal unit transition
Tmax	35	Maximum temperature for development
o	Run_2011-10-18	Output directory for results
nR	2500	Total number of simulations to run
nT	20	Number of threads to employ
Mx	500000	Maximum number of flies allowed
seed	4354885	Random number seed
q	true	Suppress progress output to terminal
pr	false	Produce only LHS parameters
plot	false	Generate summary plots

The parameters above when executed on a command line on a computer with a quad-core processor would be invoked roughly as follows (for more details please refer to the programme manual, distributed with MED-FOES):

```
java -jar med-foes-p.jar -T temps_SCEL_2011-10-18.csv -Ni 66,100 -R 4 -S 0.05,0.15 -rred 0.2,1 -Sai true -TEL 9.6,12.5,27.27,33.8 -TPA 9.1,13.8,123.96,169.49 -TIM 7.9,9.9,58.2,105.71 -TLP 5.0,10.8,94.5,186.78 -Me 0.0198,0.1211 -Ma 0.0245,0.134 -Ml 0.0068,0.0946 -Mp 0.0016,0.0465 -tdm true -r 5,35 -rvar 3.57 -Dm 1 -TuSD 0.05 -Tmax 35 -nR 2500 -nT 4 -Mx 500000 -seed 4354885 -q true -pr false -plot true -o Run_2011-10-18
```

The parameters and ranges above were used to execute 2500 runs. The number of living individual agents per day and mean trend from 100 of these simulations (reduced from 2500 for clarity) is shown in Fig. 4.

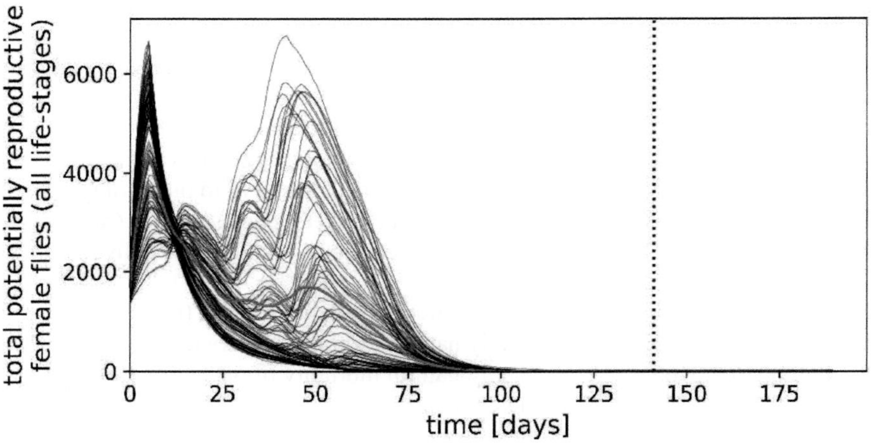

Figure 4. Total number of potentially reproductive female flies including all life stages for 100 simulations, last fly was detected 18-10-2011 (day 0), Santiago, Chile. Mean trend is shown by the red line. The dashed vertical line shows the point at which 95% of the simulations show eradication (141 days).

Fig. 4 shows some interesting characteristics, including parameter sets that seem to lead to increasing population sizes through a second and even third generation. This can be interpreted as a function of favourable temperatures for initial adult survival and reproduction, and for immatures to progress to adulthood from the initial population and from eggs laid before time *R*. Clearly a maximum population size of nearly 7000 individuals is a maximal (and unlikely) scenario, but it is indicated by a small number of the parameter sets simulated; the average time from declaration to eradication was 109 days.

In terms of indicated duration of the quarantine and other control measures, 95% of the simulations showed eradication after 141 days, 18% longer than the F_3 degree-day calculation that led to the actual 115-day control and quarantine period.

It is common for the ABS to predict longer durations than simple thermal accumulation models for outbreaks persisting through the summer in temperate areas. Since the ABS incorporates other factors which may mitigate the rapid generational turnover shown by thermal accumulation in hot weather, or the near halt in development seen in cold weather such as illustrated by the next example, it will often show less extreme seasonal swings than simple degree-day models (Collier and Manoukis 2017).

4.2. *Perris, California USA*

California has been the site of multiple *C. capitata* detections and outbreaks over the last four decades or so, leading some to propose that the species and other Tephritidae are established there (Carey 1991; Papadopoulos et al. 2013). However, this theory has not been accepted by most phytosanitary experts, internal and external California trading partners, or customers of horticultural commodities (McInnis et al. 2017; Shelly et al. 2017). As in Chile, California maintains a large (>90 000) trapping network for detecting Mediterranean fruit fly incursions. Additionally, since 1996 California has been conducting a preventive sterile male release programme over the high-risk Los Angeles Basin, where most of incursions of this pest are detected.

In December of 2014, Mediterranean fruit fly was detected in the city of Perris, that is located in Riverside County east of Los Angeles. The city is about 27 km outside the zone the preventive release programme covers. The initial detection was a find of two unmated adult females on 10 December 2014 in McPhail traps, which was followed by eight other finds in the same residential area over the next few weeks, including a find of larvae on 14 December 2014. The final find occurred on 29 December 2014. Mitochondrial genotyping indicated the AAAB mitotype consistent with a Central American source.

Eradication efforts started quickly, within one day, and included fruit removal, spinosad foliar bait spraying, and inundative releases of approximately 1.5 million sterile male flies every three or four days. The total quarantine area established was 215 km^2, with the sterile male releases targeting a 33.4 km^2 core area.

The weather station at March Air Reserve Bases (KRIV) is approximately 12 km from the find sites. This station has good data going back to the 1940s, available through NOAA's ISD archive, which allows not only modelling the 2014 outbreak, but also putting it in the context of how a similar outbreak would have progressed if it started on the same day of the year in previous years. Specifically, data from the ISD-lite data from 1950 through 2015 were cleaned as described earlier in this chapter and used to run MED-FOES simulations.

MED-FOES v0.6.2 was run with the same parameters as the Santiago model, except for the initial population (Ni) of 25 to 133 adult females, a delay before the start of SIT releases (R) of 1 day, and the input temperature data. By day 151 after the last fly was detected, 95% of the ABS simulations predicted eradication (Fig. 5). This is more than one month shorter than the quarantine and control period that was actually implemented of 189 days, or 187 days produced by recalculating the degree-day based three generation time (F_3) using the same KRIV temperature data used for the ABS.

In contrast to the ABS results for the 2011 Santiago outbreak, the 2014 Perris, simulations show no evidence of large population sizes after control measures were started. There is a very small increase in number of flies around day 80 for a few of the simulation parameter sets, but the overall character is rapid decline followed by a small population resisting final eradication for a relatively long time. Remembering that any eggs laid after time *R* (counter-measure start) are sterile, only agents that existed before that time could reproduce and account for population growth later in the simulations. Unfavourable environmental conditions for those individuals, leading to their rapid and early mortality, would cause the difference observed in overall numbers between this case and the Santiago case presented above.

The ABS results indicate eradication occurring significantly earlier than the simple thermal accumulation calculation (151 vs 187 days). This is a common finding for outbreaks covering cold periods in temperate climates, since cold temperatures in the degree-day model just slow development. In the ABS model, however, mortality occurs even with slow development and may even be increased due to particularly cold periods.

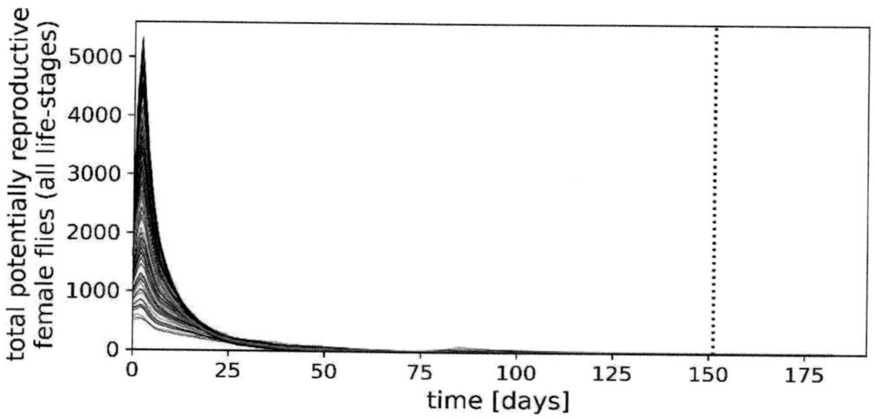

Figure 5. Total number of potentially reproductive female flies including all life stages for 100 simulations, last fly was detected 29-12-2014 (day 0), Perris, California. Mean trend is shown by the red line. The dashed vertical line shows the point at which 95% of the simulations show eradication (151 days).

In addition to simulating the particular outbreak, the ABS can be used to simulate what would happen if the same outbreak occurred at a different time. One application of this is to simulate the outbreak using temperate data from the same date in previous years for either historical context or to produce predictions for ongoing outbreaks. Fig. 6 shows both the time required for 95% of the ABS simulations to reach eradication and the degree-day thermal accumulation-based three generation time (F_3) computed using temperature values from 1950 to 2014. The values for the actual outbreak year of 2014 (filled markers) show that the degree-day calculation is exceptionally short while the ABS is almost the median value of previous years.

5. CONCLUSION

In the context of an area-wide programme aimed at achieving zero pest prevalence, determining the duration of eradication programmes following incursions by invasive tephritid fruit flies is as critical as it is difficult. Critical because failure to eradicate invading fruit flies will lead to increased costs incurred under follow-up programmes once population sizes increase again or are established in a different area, to say nothing of the costs of establishment in areas where they are not present (Siebert and Cooper 1995). In addition to these risks from control/quarantine periods that are too short, it is also important not to set overly long periods as these could be unnecessarily burdensome to producers and also lead to excessive programme costs and losses. The difficulty, however, lies in estimating the size of a possible remnant population of flies that contains so few detectable individuals (nominally adults responsive to the lure being used in the trapping array) that it is unlikely to catch any initially (Carey et al. 2017; McInnis et al. 2017).

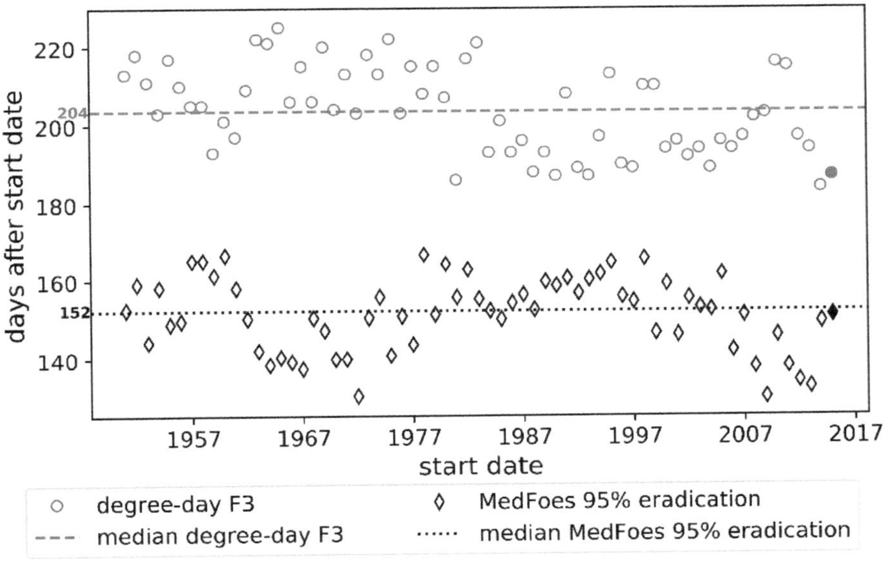

Figure 6. Times required for 95% of the ABS simulations to reach eradication in comparison to the three generation times (F_3) based on the degree-day thermal accumulation computed for hypothetical Perris, CA outbreaks staring on 29 December for the years 1950 through 2014. The actual outbreak in 2014 is shown by filled markers. Degree-day based values are circles, and ABS based values are diamonds. Dashed lines are medians.

Though the task is difficult, the established method of using simple thermal accumulation models to guide control activity durations and ending quarantines after the equivalent of three generations of degree-day development have elapsed without another find seems to work well in practice in many places around the world, including Chile and California as described here, though there are limits in more temperate areas (Collier and Manoukis 2017).

Agent-Based Simulations, as modelled in MED-FOES, can be a useful complement for programme managers and state entities as they provide a semi-independent estimate of the duration that control and quarantine activities should continue after the detection of the last fly find. The software is relatively easy to use and with increasing computational power in multi-core desktop computers it is feasible to run a large number of simulations in a relatively short time.

It is important to stress also the drawbacks of using ABS methods in general. These include the need for range estimates of many constituent parameters. In the case of Mediterranean fruit fly over 100 years of research on development, reproduction and mortality of this species is helpful, but some details remain poorly resolved, especially under natural conditions. A second general problem with individual-based methods is that simulating combinations of parameters to exhaustively test the range of possible outcomes can be computationally expensive. Here, again, current computer systems are powerful enough to mitigate this issue, but this problem can increase as models become more complex. A final point here is that the stochastic nature of the simulation and variable output necessitates statistical methods to handle results, and the mechanisms that drive outcomes may not be immediately clear.

One possible approach that integrates the current approach and the ABS would be to set the initial eradication programme duration following the last detection after an outbreak using the three-generation calculation on historical temperatures (already done in at least Florida and California USA, Gilbert et al. 2013), and then update with both degree-day calculations using measured hourly temperature data and with ABS simulations. The combination of both estimates will give improved insight to the outbreak dynamics, might suggest programmatic responses of effort modulation, and increase confidence in declarations of eradication.

MED-FOES instantiates a relatively simple model and therefore has several notable weaknesses. Control measures are simulated as constant average effects instead of time-varying, while eradication programmes typically incorporate discrete high-intensity efforts such as fruit removal, which will have different impacts depending on the age structure of the population. Additionally, quarantines are held for a period of time (at least one degree-day generation for *C. capitata* in California) after the end of suppression efforts during which intensive trapping continues. Finally, incorporating information on host plant availability, when known, would likely have significant effects. However, a major strength of the ABS approach is that all of these factors and more could be incorporated into the model in a straight-forward manner given the relevant input data.

Future developments for MED-FOES aimed at facilitating declaration of eradication after outbreaks include improved reporting incorporating analysis of historical temperature profiles. The analysis of temperatures in past years will be helpful for tracking anomalously long or short quarantine/control periods after the last detection, again increasing confidence in lengths. The flexibility and extensibility of the ABS framework also allows other applications of this model to address questions around the invasion biology of *C. capitata.*

One of the new applications of MED-FOES currently being explored is to estimate the potential growth rate of Mediterranean fruit fly populations in different areas and at different times of the year. This would yield results that could be compared to those from standard climatic suitability methods such as CLIMEX and Maxent (Webber et al. 2011). The specificity of MED-FOES to the biology of Mediterranean fruit fly could serve to refine the more general spatial and temporal analysis of patterns of detections and outbreaks when compared to the standard approaches based on environmental suitability modelling, especially when combined with incursion risk data.

6. ACKNOWLEDGEMENTS

Special thanks to Ricardo Rodriguez of SAG (Chile) for information on the Santiago outbreak and permission to share our analysis. Thanks also to Kevin Hoffman of CDFA for details on the Perris, California outbreak.

7. REFERENCES

Almeida, S. J. de, R. P. Martins Ferreira, Á. E. Eiras, R. P. Obermayr, and M. Geier. 2010. Multi-agent modeling and simulation of an *Aedes aegypti* mosquito population. Environmental Modelling and Software 25: 1490–1507.

Barclay, H. J., J. W. Hargrove, A. Meats, and A. Clift. 2021. Procedures for declaring pest free status, pp. 921–946. *In* V. Dyck, J. Hendrichs, and A. S. Robinson (eds.), Sterile Insect Technique – Principles and practice in Area-Wide Integrated Pest Management. Second Edition. CRC Press, Boca Raton, Florida, USA.

Blower, S. M., and H. Dowlatabadi. 1994. Sensitivity and uncertainty analysis of complex models of disease transmission: An HIV model, as an example. International Statistical Review 2: 229–243.

California Code of Regulations. 2017. Mediterranean fruit fly interior quarantine regulation. Title 3 Section 3406. Sacramento, California, USA.

Campbell, G. S., and J. M. Norman. 1997. An introduction to environmental biophysics, 2nd ed. Springer, New York, USA.

Carey, J. R. 1982. Demography and population dynamics of the Mediterranean fruit fly. Ecological Modeling 16: 125–150.

Carey, J. R. 1991. Establishment of the Mediterranean fruit fly in California. Science 253: 1369–1373.

Carey, J. R., N. Papadopoulos, and R. E. Plant. 2017. The 30-Year Debate on a multi-billion-dollar threat: Tephritid fruit fly establishment in California. American Entomologist 63: 100–113.

Collier, T., and N. Manoukis. 2017. Evaluation of predicted Medfly (*Ceratitis capitata*) quarantine length in the United States utilizing degree-day and agent-based models. F1000Research 6: 1863.

Corbett, A., and R. E. Plant. 1993. Role of movement in the response of natural enemies to agroecosystem diversification: A theoretical evaluation. Environmental Entomology 22: 519–531.

Crespo-Pérez, V., F. Rebaudo, J. F. Silvain, and O. Dangles. 2011. Modeling invasive species spread in complex landscapes: The case of potato moth in Ecuador. Landscape Ecology 26: 1447–1461.

Fletcher, B. S. 1989. Temperature-development rate relationships of the immature stages and adults of tephritid fruit flies, pp. 273–289. *In* A. S. Robinson, and G. Hooper (eds.), Fruit flies, their Biology, natural enemies and control. Elsevier, Amsterdam, The Netherlands.

Gilbert, A., R. Bingham, M. Nicolas, and R. Clark. 2013. Insect trapping guide, 13th edition. California Department of Food and Agriculture, Sacramento California, USA.

Gutierrez, A. P., and L. Ponti. 2011. Assessing the invasive potential of the Mediterranean fruit fly in California and Italy. Biological Invasions 13: 2661–2676.

(FAO) Food and Agriculture Organization of the United Nations. 2016a. Guidelines for pest eradication programmes. International Standard for Phytosanitary Measures (ISPM) 9. International Plant Protection Convention (IPPC). FAO, Rome, Italy.

(FAO) Food and Agriculture Organization of the United Nations. 2016b. Establishment of pest free areas for fruit flies (Tephritidae). International Standard for Phytosanitary Measures (ISPM) No. 26. International Plant Protection Convention. Rome, Italy.

Hendrichs, J., P. Kenmore. A. S. Robinson, and M. J. B. Vreysen. 2007. Area-wide integrated pest management (AW-IPM): Principles, practice and prospects, pp. 3–33. *In* M. J. B. Vreysen, A. S. Robinson, and J. Hendrichs (eds.), Area-wide control of insect pests: From research to field implementation. Springer, Dordrecht, The Netherlands.

Jackson, D. E., M. Holcombe, and F. L. Ratnieks. 2004. Trail geometry gives polarity to ant foraging networks. Nature 432: 907–909.

Kareiva, P. M. 1983. Local movement in herbivorous insects: Applying a passive diffusion model to mark-recapture field experiments. Oecologia 57: 322–327.

Kean, J. M., P. C. Tobin, D. C. Lee, G. R. Smith, L. D. Stringer, R. F. Vargas, D. G. McCullough, D. A. Herms, J. Fletcher, D. M. Suckling, and others. 2012. Global eradication and response database. GERDA.

Knipling, E. F. 1979. The basic principles of insect population suppression and management, Agriculture Handbook. U.S. Department of Agriculture, Washington DC, USA.

Manoukis, N. C., and K. Hoffman. 2014. An agent-based simulation of extirpation of *Ceratitis capitata* applied to invasions in California. Journal of Pest Science 87: 39–51.

Manoukis, N. C., B. Hall, and S. M. Geib. 2014. A computer model of insect traps in a landscape. Scientific Reports 4: 7015.

Manoukis, N. C., M. Siderhurst, and E. B. Jang. 2015. Field estimates of attraction of *Ceratitis capitata* to trimedlure and *Bactrocera dorsalis* to methyl eugenol in varying environments. Environmental Entomology 44: 695–703.

McInnis, D. O., J. Hendrichs, T. Shelly, N. Barr, K. Hoffman, R. Rodriguez, D. R. Lance, K. Bloem, D. M. Suckling, W. Enkerlin, P. Gomes, and K. H. Tan. 2017. Can polyphagous invasive tephritid pest populations escape detection for years under favorable climatic and host conditions? American Entomologist 63: 89–99.

MED-FOES. 2019. MED-FOES: MED-Fly Outbreak and Extirpation Simulation. An Agent-Based Simulation of Invasive Pest Insects.

Myers, J. H., D. Simberloff, A. M. Kuris, and J. R. Carey. 2000. Eradication revisited: Dealing with exotic species. Trends in Ecology and Evolution 15: 316–320.

Papadopoulos, N. T., R. E. Plant, and J. R. Carey. 2013. From trickle to flood: the large-scale, cryptic invasion of California by tropical fruit flies. Proceedings of the Royal Society B: Biological Sciences 280: 20131466.

Pratt, S. C., D. J. T. Sumpter, E. B. Mallon, and N. R. Franks. 2005. An agent-based model of collective nest choice by the ant *Temnothorax albipennis*. Animal Behaviour 70: 1023–1036.

Radchuk, V., K. Johst, J. Groeneveld, V. Grimm, and N. Schtickzelle. 2013. Behind the scenes of population viability modeling: Predicting butterfly metapopulation dynamics under climate change. Ecological Modeling 259: 62–73.

Railsback, S. F., and V. Grimm. 2012. Agent-based and individual-based modeling: A practical introduction. Princeton University Press, Princeton, New Jersey, USA.

Reicosky, D. C., L. J. Winkelman, J. M. Baker, and D. G. Baker. 1989. Accuracy of hourly air temperatures calculated from daily minima and maxima. Agricultural and Forest Meteorology 46: 193–209.

Roltsch, W. J., F. G. Zalom, A. J. Strawn, J. F. Strand, and M. J. Pitcairn. 1999. Evaluation of several degree-day estimation methods in California climates. International Journal Biometeorology 42: 169–176.

Shelly, T. E., D.R. Lance, K.H. Tan, D.M. Suckling, K. Bloem, W. Enkerlin, K. Hoffman, N. Barr, R. Rodríguez, P.J. Gomes, and J. Hendrichs. 2017. To repeat: Can polyphagous invasive tephritid pest populations remain undetected for years under favorable climatic and host conditions? American Entomologist 63: 224–231.

Siebert, J., and T. Cooper. 1995. If medfly infestation triggered a trade ban: Embargo on California produce would cause revenue, job loss. California Agriculture 49: 7–12.

Skellam, J. G. 1951. Random dispersal in theoretical populations. Biometrika 38: 196–218.

Smith, J. W. 1998. Boll weevil eradication: Area-wide pest management. Annals of the Entomological Society of America 91: 239–247.

Suckling, D. M., J. M. Kean, L. D. Stringer, C. Cáceres-Barrios, J. Hendrichs, J. Reyes-Flores, and B. Dominiak. 2016. Eradication of tephritid fruit fly pest populations: Outcomes and prospects. Pest Management Science 72(3): 546–465.

Tassan, R. L., K. S. Hagen, A. Cheng, T. K. Palmer, G. Feliciano, and T. L. Blough. 1983. Mediterranean fruit fly life cycle estimations for the California eradication program, pp. 564–570. *In* R. Cavalloro (ed.), Fruit flies of economic importance. Balkema, Rotterdam, The Netherlands.

Vargas, R. I., W. A. Walsh, D. Kanehisa, E. B. Jang, and J. W. Armstrong. 1997. Demography of four Hawaiian fruit flies (Diptera: Tephritidae) reared at five constant temperatures. Annals of the Entomological Society of America 90: 162–168.

Vinatier, F., P. Tixier, P. F. Duyck, and F. Lescourret. 2011. Factors and mechanisms explaining spatial heterogeneity: A review of methods for insect populations. Methods in Ecology and Evolution 2: 11–22.

Webber, B. L., C. J. Yates, D. C. Le Maitre, J. K. Scott, D. J. Kriticos, N. Ota, A. McNeill, J. J. Le Roux, and G. F. Midgley. 2011. Modelling horses for novel climate courses: Insights from projecting potential distributions of native and alien Australian acacias with correlative and mechanistic models. Diversity and Distribution 17: 978–1000.

REAL-TIME INSECT DETECTION AND MONITORING: BREAKING BARRIERS TO AREA-WIDE INTEGRATED MANAGEMENT OF INSECT PESTS

N. A. SCHELLHORN[1,2] AND L. K. JONES[1,2]

[1]*Commonwealth Scientific and Industrial Research Organisation (CSIRO), Brisbane, Qld, 4001 Australia*
[2]*RapidAIM Pty Ltd; nancy@rapidaim.io*

SUMMARY

Area-wide integrated pest management (AW-IPM) of insect pests relies on surveillance and communication to estimate wild population size, guide targeted control, and determine the effectiveness of any pest control action. However, knowing where and when pests arrive in real-time, communicating the information quickly, and delivering insect pest control in a coordinated manner are potential barriers to achieving area-wide management. Agricultural technology is creating opportunities to remove these barriers, which in turn will facilitate the adoption of AW-IPM. Technology advances in insect surveillance (detection and monitoring), data flow and information communication are being realized, and increasingly becoming commercially available. This technological change is largely being driven by macro-economic trends of increased cost of labour, international agricultural trade and shifting consumer demands, and a confluence of new hardware technologies that free computation from the desktop. As professionals and practitioners of pest management, there is an opportunity to shape technological solutions to remove barriers to AW-IPM, and to achieve sustainable pest management across commodities and pests. Yet, the success of the technological solution and its area-wide implementation will depend on the way that we think about the problem (innovation), and the solutions (engineering).

Key Words: Insect remote monitoring, automated surveillance, pest trapping, data flow, data visualisation, agricultural technology, innovation, fruit flies, Tephritidae

1. INTRODUCTION

Pests of global significance regularly cause economic loss due to their transboundary nature and because they are difficult to manage. They are often highly mobile, fail to recognize property boundaries, reproduce rapidly, and require labour-intensive actions to monitor their arrival and to control their populations. Coordinated, well-

J. Hendrichs, R. Pereira and M. J. B. Vreysen (eds.), Area-Wide Integrated Pest Management: Development and Field Application, pp. 889–902. CRC Press, Boca Raton, Florida, USA.

timed delivery of pest control options is often seen as a solution. This approach is referred to as area-wide integrated pest management (AW-IPM); essentially the control of total target pest populations within a delimited area (Klassen and Vreysen 2021). Successful AW-IPM of insect pests requires often highly coordinated effort, involves a regulatory framework, and integrates multiple pest control tactics (Hendrichs et al. 2007).

Evidence of success of the coordinated actions of many farmers was first shown theoretically by Levins (1969). He introduced the concept of metapopulations and distinguished between the dynamics of a single population and a set of local populations. Levin's theoretical model was motivated by, and applied to, a pest control situation over a large region, in which local populations would fluctuate in asynchrony and generations overlap. The output of the model showed that control measures should be applied synchronously throughout to achieve suppression. Many practical examples, successful and unsuccessful, of pest control applied at an area-wide basis have followed (Vreysen et al. 2007; Schellhorn et al. 2015). However, there are many potential barriers to efficient and sustainable AW-IPM such as the inability to know where and when pests arrive, communicating the information on time, and delivering pest control as a rapid response in a coordinated manner.

The AW-IPM relies on three key elements: a) accurate estimates of the pest population across a contiguous area of production and continuously over time, b) efficient communication of the population estimates to pest control managers, and c) dynamic coordination and delivery of the management action to suppress or eradicate the pests. Population estimates are generated by regular inspection of traps, plants, or sentinel animals at fixed or random locations (Southwood 1978). Weekly or fortnightly, the presence of pests or their absence is usually recorded on data sheets in the field, and later entered into an electronic database. Depending on the end user, the information may be communicated within 48 hours of weekly / fortnightly monitoring (e.g. crop agronomists to farm manager) or accumulated in a database for record management needed for historical reflection or a random audit (e.g. government biosecurity).

However, for each of these three elements there are several challenges. Manual inspection of traps is labour-intensive, tedious, and causes delays between insect detection and communication to pest managers. The sampling schedule of 7-21 days allows for pest population persistence and increase without intervention. Communicating insect data in a timely manner from across a contiguous area is unrealistic, unless there is an extensive workforce collecting the insect data and providing it quickly to the pest managers (Enkerlin et al. 2017). Once the end users have the data, delivering pest control in a coordinated manner, which can achieve area-wide suppression, is often logistically challenging, and costly. Up to now, many of these challenges have been mitigated by conducting each element with a centrally organized programme almost exclusively led by governments (Kean et al. 2019). This is certainly true in the case of those integrating the delivery of sterile insects (Dyck et al. 2021).

Pest population estimates, communication, and coordination are organized centrally with many stakeholder participants, which is seen as essential to the success of area-wide programmes, including those with a Sterile Insect Technique (SIT) component. Compared to the number of pests of global significance, there are relatively few examples of AW-IPM of endemic or established pests (Vreysen et al. 2007) because of these barriers, but that is changing. Barriers to more efficient:

a) pest detection, delimitation and monitoring,
b) data flow,
c) information communication, and
d) coordination of pest control,

are being removed through innovations in digital agriculture technology.

Increasingly the inputs and outputs in primary productions systems are being tracked, measured and analysed by automated or partially automated systems. These new technologies increase efficiencies, reduce labour costs and speed up capabilities for decision making. The pressure to continue developing these technologies is driven by global food and fibre demand, market access and traceability issues, and the approaching horizon of resource limitation (FAO 2017). The result of this is that more and more farms are connected and networked, and that data flow is moving away from the notebook and into digital information systems. This is critical in the case of early detection and rapid response to invasive species.

Venture capital investment in agricultural technology is on an exponential growth trajectory. For example, investment in early stage companies in Australia, Canada, Israel, and New Zealand has increased from USD 5.8 million in 2010 to USD 89.5 million in 2017. Growth areas include crop protection and input management, precision agriculture and imagery (Finistere Ventures 2018).

Technology is providing opportunities to achieve greater efficiencies for on-farm- and AW-IPM. Advances have been made in data flow, real-time insect surveillance (detection and monitoring), information communication, and mating disruption by using automated pheromone dispensers. Over that last decade, automation of pest surveillance and taxonomic identification, especially for tephritid fruit flies, has emerged in research institutions and the commercial market (Jiang et al. 2008; Liu et al. 2009; Faria et al. 2014; Philimis 2015; Doitsidis et al. 2017; Goldshtein et al. 2017; Potamitis et al. 2017a, 2017b; Shaked et al. 2018). However, in order for the implementation and adoption of area-wide automated insect surveillance to occur, two spheres will need to align: the way we think about the problem (innovation), and the way we solve the problem (engineering). How we bring these two spheres together will be critical to achieving efficient detection of exotic pest incursions, monitoring of endemic populations, and their control.

Here we focus on *why* we want to bring innovation and engineering together to achieve efficient and effective AW-IPM, *what* methodologies are currently on the market, and *how* current and future technological solutions will contribute to more efficient and effective pest suppression and eradication.

2. THE PROBLEM AND THE SOLUTION

2.1. Significant Barriers to Area-wide Suppression

> *"Innovation in all aspects of modern life is seen as a socio-economic cure for many of the troubles of modern societies"* (Ferguson et al. 2014).

One of the great troubles of modern society is achieving environmentally acceptable and economically sustainable pest management of food and fibre crops. This is especially true for managing highly mobile, invasive insect pests. The AW-IPM of economically important pests is viewed as a promising solution. Yet, to achieve area-wide pest suppression one needs to overcome significant potential barriers including:

a) support from many stakeholders, the community and public (including standardised approaches/methodologies across diverse stakeholder groups),

b) knowing where and when pests show up, and

c) the dynamics of pest populations in target landscapes that are heterogeneous in space and time.

Technological innovation offers solutions to overcome these challenges, making AW-IPM accessible for the management of numerous pests by communities and grower groups, and as well as increasing the efficiency and effectiveness of programmes that integrate the SIT. Logistical barriers can be reduced and eliminated to improve insect detection and monitoring, data flow, communication, and coordination of area-wide pest control.

2.2. The Innovative Solutions to Barriers of Efficient and Effective AW-IPM

The methodologies for in-field data collection of insect populations have barely changed since estimates began and these are well captured in classic references such as Southwood's 1978 "Ecological Methods"; visual counts, sweep nets, pitfall traps, destructive sampling of fruits, and lure-based insect traps are the standard. Generating population density estimates using these approaches is tedious, laborious, and restricted to snapshots in time and space. The data are collected in the field, then returned to the laboratory for manual entry into spreadsheets for later data checking and analysis of trends; a slow in-flow of data, which means a slow out-flow of information.

One of the most recent advances to speed up data flow is in-field electronic data entry. As of 2017 there have been approximately 30 downloadable mobile scouting apps for in-field data collection, a large proportion of which can be used to collect arthropod population data (Hopkins 2017). This technology is increasingly being used across the public and private sectors. Tablet or smartphone applications allow government officers and commercial pest control advisers to enter geo-referenced information at the location of data collection. By scanning a barcode on an insect trap, or geo-referencing a field scout's location, field staff can enter the insect count data on the device, which is then uploaded on a server in an easily readable format such as an Excel spreadsheet.

End users now have the option to use open source software to tailor data collection applications. A widely used example is the Open Data Kit (ODK) Project developed at The University of Washington (users range from individual researchers to Google and the World Health Organization). The project provides tools for a community of users to both create data collection apps and to contribute to the development of the software code. Blogs and forums facilitate an iterative process whereby ODK is continually updated by the feedback loop of users and developers actively engaging (ODK 2018). Fit-for-purpose packages are increasingly available as semi-commercial and commercial applications like those developed and used by different government agencies for fruit fly in-field data collection / communication (Table 1).

Table 1. Examples of end-user dashboards for in-field electronic pest data collection

Name	Location	Application	Target
CalTraps http://caltrap-info.com/faqs	Los Angeles County, USA originally, now covering most California counties	Pest detection data management applications	Tephritid fruit flies
Trapbase https://www.agriculture.vic.gov.au	Victoria, Australia	A Victorian Government online database for recording fruit fly trap locations, monitoring and mapping of fruit fly detections	Tephritid fruit flies
OpenScout www.dtn.com	Indiana, USA, DTN, formerly Spensa Technologies	Field scouting app; allows end user to select insects from a customizable list	Agronomic platform with scouting for insect pests and decision making
AgWorld Scout https://www.farms.com/agriculture-apps/crops//agworld-scout	Perth, Western Australia	In-field assessment to monitor crop health and pest pressure	Custom list of insects
Scoutpro www.scoutpro.org	Urbandale, Iowa, USA	Scouting platform with built-in ID keys	Developed for soybean, corn and wheat
Farm Dog https://farmdog.ag	Salinas, California, USA	Pest and disease management software	Custom list of insects
Koppert iPM Scout app www.koppertipm.com	Berkel en Rodenrijs, The Netherlands	Pest and disease scouting	Customised template
CropScout https://www.agric.wa.gov.au /apps/cropscout	Western Australian Government	Pest counts relative to spray thresholds	Canola aphid

In-field electronic pest data collection applications improve data flow and remove delays caused by manual data entry of a paper-based collection system, which then has to be entered into an electronic spreadsheet, collated, checked for errors, and presented in a meaningful visualization, which is rarely spatially explicit. In turn these new applications can improve communication among pest control managers and other stakeholders; pest location and density can be visualized spatially across an area allowing for a targeted response and review.

However, two significant barriers remain – the spatial and temporal resolution of the information is coarse due to the limitations of 'boots' on-the-ground, both in terms of the number of locations that can be sampled, and the frequency at which information can be collected. Data interpolation has been the primary means of getting around this issue. Yet, the benefit of the visualization of interpolated data does not translate into location-specific pest management. The scale of interpolated data is too coarse to achieve more targeted insecticide or biocide application. Automated insect detection and monitoring has the potential to address this problem effectively.

3. THINKING ABOUT THE PROBLEM

Thorough consideration of the barriers to pest detection and monitoring is needed in order to propose and develop the most useful automated solutions and process for adoption. An engineering solution alone, without the full context of an area-wide management programme, is unlikely to have the anticipated uptake and impact. As one example, trapping grids for pest detection are labour-intensive, therefore a simple solution would be to automate an insect trap to reduce the needs for field visits and labour. However, the result may be a reduced workforce with limited ability to respond rapidly when borders are breached, and an exotic pest invades.

3.1. Detecting Rare Events and Early Incursions

Following a detection of fruit flies in a fruit fly free zone, there is a requirement to move quickly and identify the area infested and the area that has remained free from fruit fly. This can be achieved by rapidly deploying traps at a higher density (10-fold increase) and checking the traps more frequently. Such an event often requires additional staff, and the process continues until the area is declared pest-free. Arguably, the problem is the low probability to detect a rare event or early incursion, the challenge to quickly delimit the infested area and deliver control of the infestation, and ultimately to provide sufficient evidence of pest absence.

Solving the problem of detecting rare events and early incursions for a rapid response can minimize the size and duration of the management response, the cost, and the amount of time that markets are closed (Suckling et al. 2016). As two small examples, approximately USD 720 000 / week was spent to tackle the Queensland fruit fly, *Bactrocera tryoni* (Froggatt) incursion into the pest-free state of northern Tasmania, Australia in autumn 2018 (Beavis 2018). In Miami-Dade County Florida, USA, the authorities spent USD 3.5 million in a few months eliminating an outbreak of *Bactrocera dorsalis* (Hendel), and an estimate of ~USD 25 million was incurred from cost and losses (Alvarez et al. 2016). An outbreak of *B. tryoni* in New Zealand in 2015 cost USD 9.72 million (BBC 2015).

These examples highlight that an innovative solution for biosecurity is a pest surveillance system that increases the probability of detection of rare events and early incursions; a solution that provides greater spatial and temporal resolution. This would allow a rapid response to contain the geographic extent of the outbreak, and to have confidence of area-freedom during and after control. Technological innovations that can quickly provide solutions to costly problems, such as these, are likely to have tremendous positive impact.

Insect detection solutions that are fast and easy to deploy, provide real-time data flow, and are cost-effective, will be even more significant for managing pests that are endemic and widespread. Pest control solutions across commodities mostly rely on the 'sample, spray and pray' approach, with sampling often being *ad hoc* (Zalucki et al. 2009). Searching for pests in orchards and crops requires training for correct identification, can be imprecise (how much searching is required to make an informed decision), is tedious and tiring (long periods of time on hands and knees looking at plants) and those who are hired to complete that task can be incentivized from sales of insecticide, not from not spraying. Advocates of IPM promote sustainable practices, but limited adoption of AW-IPM is the reality. The 'sample, spray and pray' principle is standard, and will remain so until practical solutions for insect monitoring and communication are provided. Technological innovation can play a role in breaking down the challenges of poor information about pest numbers, pest locations and effectiveness of insecticide sprays.

3.2. Trapping Guidelines

Beyond the technical challenges of automating insect surveillance, consideration must be given to the staffing needs and workplace culture when changes are made to long standing surveillance practices. For major insect pests, especially those that are barriers to trade such as fruit flies, insect trapping guidelines are well established, and harmonized based on the trapping objective, e.g. detecting, delimiting, monitoring, and the desired pest control outcome (FAO/IAEA 2018; ISPM 2018). These recent trapping guidelines are comprehensive and provide a level of detail rarely available for other insect pests.

One of the more challenging aspects of trapping guidelines is trapping density, its dynamic, and changes according to the survey objectives, and the pest species. Lure-based insect traps provide relative estimates expressed as numbers per unit effort and are dependent on many factors. In general, the probability of detecting an individual insect is based on the sampling effort (the number of traps deployed), the size of the target insect population, the activity of the insects, and the insect attractant and trapping device efficiency (Vreysen 2021; FAO/IAEA 2018).

3.3. Lure-based Trapping Methods

Lure-based trapping methods are used for many of the major global insect pests for monitoring endemic populations, and for detecting exotic pest incursions, such as major fruit fly species (e.g. oriental fruit fly, Queensland fruit fly, Mediterranean fruit fly (*Ceratitis capitata* (Wiedemann)), other invasive insects (e.g. Japanese beetle, *Popillia japonica* Newman; tropical gypsy moth, *Lymantria pelospila* Turner; Asian long-horned beetle, *Anoplophora glabripennis* (Motschulsky); Khapra beetle, *Trogoderma granarium* Everts); major endemic pests of field crops (e.g. noctuid armyworm, *Pseudaletia unipuncta* (Haworth); fall armyworm, *Spodoptera frugiperda* (Smith); corn earworm, *Helicoverpa zea* (Boddie)), and pome fruit pests (codling moth, *Cydia pomonella* (L.); oriental fruit moth, *Grapholita molesta* (Busck); oblique-banded leafroller, *Choristoneura rosaceana* (Harris)). These lure-based trapping methods also provide the cornerstone of many SIT programmes, both for delivery of sterile insects (determining release rates) and demonstrating efficacy of control of wild populations.

Lure-based traps estimate the relative density, unless the target's physiological response to the attractant is quantified (Taylor 2018). Relative estimates are expressed as numbers per unit effort and are advantageous over visual estimates by saving time (checking a point location is quicker than random searching of plants and animals), thus increasing efficiency of detection. Lure-based trapping, to generate tephritid population estimates, has limitations in that it is primarily males that are lured, and trapping efficacy can be influenced by various factors, including environmental differences, demographic factors, and the behaviour and physiological state of the individual (Taylor 2018).

Nevertheless, lure-based trapping has become highly specialized and efficient for tephritids (FAO/IAEA 2018). Trapping continues to provide a reliable and easy to use methodology for surveying pests, but also represents a costly and inefficient component of any insect survey programme, be it for government biosecurity or pest management. As an example, the trapping component of SIT-based AW-IPM programmes for Mediterranean fruit fly is estimated to represent 18-25% of programme costs, which is only partially due to the sorting of captured wild from sterile flies (Enkerlin et al. 1996; FAO/IAEA 2018). The trapping cost can be further reduced with automation, and with the technology of today it may even support SIT programmes by helping to prioritise workflow of staff.

4. SOLVING THE PROBLEM

Freeing computation from the desktop has been a key driver that has enabled the development of innovative solutions to automate insect monitoring and detection.

4.1. Automated Insect Monitoring and Communication Technology

Thus far, the majority of solutions include camera/s focused on dead insects caught in the bottom of traps or stuck to sticky cards (Table 2). As with manual traps, all of the automated solutions include off-the-shelf pheromones or attractants of the target insects. The camera takes pictures at fixed intervals (e.g. Trapview, Semios), and the software displays the images for the end user to confirm. For some systems, the confirmation process feeds back into the software as a component of the machine learning algorithm for improved automated diagnostics (e.g. Trapview).

Optical photosensors are being used to discriminate wing-beat frequency of insects entering a trap (e.g. Farmsense and AgroPestAlert). A company established by the authors of this paper use a type of capacitance sensing, similar to a behavioural fingerprint of insects, which detects and discriminates an insect as it enters the trap and delivers the information to mobile app in real-time (Table 2).

Table 2. Examples of automated insect monitoring and communication technology

Company Name	Location	Type	Target	Product Stage
DTN (formerly Spensa Tech)	Purdue, Indiana, USA	Sentinel; camera-fitted trap	Lepidopterap ests of pome fruit	Acquired by DTN April 2018; Commercial www.dtn.com
DTN (formerly Spensa Tech)	Purdue, Indiana, USA	Z-Trap; electromagnetic current inside delta trap	Lepidopterap ests of pome fruit	Acquired by DTN April 2018; Commercial www.dtn.com
TrapView	Hrusevje, Slovenia	Camera-fitted traps in various designs	Lepidoptera and Diptera	Commercial www.trapview.com
SnapTrap	Victoria, Australia	Camera-fitted traps	Fruit fly	Commercial www.snaptrap.com
SemiosBio Technologies Inc.	Vancouver, British Columbia, Canada	Variable rate mating disruption and automated pest camera trap	Lepidopterap ests of fruit and nuts; CA red scale	Commercial www.semios.com
iScout (Pessl Instruments)	Graz, Austria	Camera-fitted sticky trap	Non-specific	Commercial http://www.pesslinstru ments.com/
RapidAIM Pty Ltd.	Brisbane, Queensland, Australia	Novel capacitance-type sensor	Fruit fly for 1st product	Commercial www.rapidaim.io
AgroPestAlert	Tudela, Spain	Photosensor – wing-beat frequency	Non-specific	Prototype http://agropestalert.com /
Farmsence Inc.	Riverside, CA, USA	Optical – wing-beat frequency	Lepidoptera	Prototype www.farmsense.io

Electromagnetic current is used by Spensa Tech for monitoring lepidopteran orchard pests. The current from the 'Z-trap' surrounds and kills the insect, and the amount of current provides an indication of the insect size, as a surrogate for the insect's identity.

There are key considerations when adopting automated monitoring and communication technology, and these will differ depending on the objectives of the programme, and of the end users. However, general criteria include the reliability of detection, reliability of communicating the detection, efficacy of insect capture compared to an industry standard, efficiency compared to manual trapping, and the added value, for example, how real-time data flow allows for rapid response, which ultimately reduces costs of managing outbreaks, and minimizes disruption to trade.

Technology is rarely neutral, and the expectations of the benefits will need to align with the problem that is being solved. For example, if the expectation is that automation will replace humans, then it's unlikely that automation will ever be developed at a level that is cost-effective. However, if the expectation is that automation is providing better information for more informed pest management decisions, improving workflow, and reducing the harm from exotic pest incursions, then there are already technologies that can advance pest detection and monitoring (Table 2). Automated systems are already improving data flow into geographic information system (GIS) databases, in turn allowing real-time visualisation for managers and stakeholders of insect population hot spots and areas under control, and ultimately providing tighter feedback in detection, communication, control application and validation (Fig. 1).

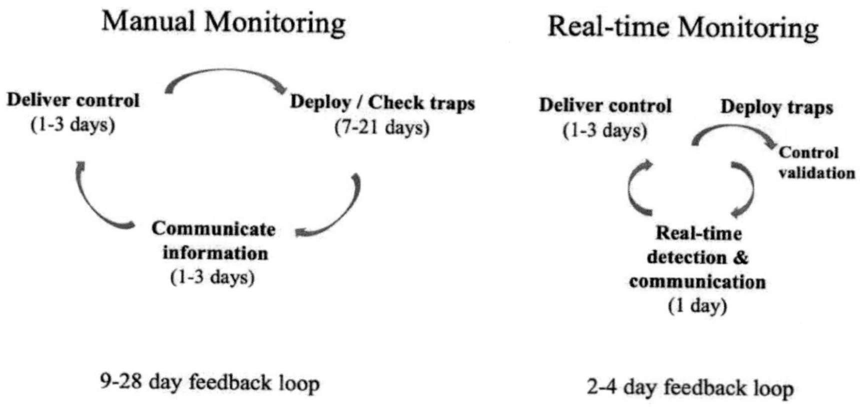

Figure 1. Feedback loop comparing manual and real-time monitoring.

4.2. Coordinating Data Flow, Control and Evaluation

For many decades, theory has demonstrated that for pest population control measures to be effective within AW-IPM, synchronicity at temporal and spatial scales is paramount (Levins 1969; Ives and Settle 1997). Surveillance methodology and strategy, if it is to be effective within AW-IPM, needs also to be synchronous across insect present and potential distribution areas. Some of the pitfalls of manual monitoring are the variations in methodology between stakeholders, as well as human error, and the large and complex chain that data progress through from field collection to management action.

Digitising and standardising data flow at all levels may greatly reduce the loss of data fidelity and allow for the level of dynamism that is required in large complex systems. Examples of where this model has been effectively rolled out at large scales are found within clinical data systems. The transition of medical data records from paper to personal computers to medical data collection apps has accelerated in the past 10 years (Zhang et al. 2017). So much so that clinician end user demand has driven the standardisation of digital health record-keeping at all levels of the data supply chain (FDA 2018).

Empirical examples have shown the benefit of synchronicity in pest control and include coordinated timing of insecticide application (Smith 1998; Lloyd et al. 2010), coordinated growing of trap crops (Sequeira 2001), area-wide release of sterile insects (Hendrichs et al. 1995), coordinated orchard hygiene (Lloyd et al. 2010), and sanitation methods to destroy a life stage, such as pupae in soil, to achieve area-wide reduction in pest populations (Duffield 2004; Lloyd et al. 2008). Some of these examples achieve coordination by default due to the narrow window of suitability of the action, e.g. as defined by the license for growing genetically modified cotton in Australia. However, area-wide data collection, data flow, and coordinating pest management of insecticide/biocide/biological application (sterile insects and natural enemies) has been logistically impossible without a central coordinating body and a regulatory framework. This has resulted in independent, asynchronous delivery of insecticide for marginal gain.

Fruit and vegetable production regions are often characterised by long growing seasons, diverse commodities, and polyphagous pests (such as tephritid fruit flies) that take advantage of these spatially and temporally heterogeneous landscapes that span the urban and rural habitats (Schwarzmueller et al. 2019). Often urban landscapes provide permanent low-density resources, such as a variety of backyard trees as potential pest hosts. Even though host plant density and diversity is much lower than in a commercial orchard setting, these urban environments can significantly contribute to high and persistent pest populations because populations are often uncontrolled and backyard hosts can fill the seasonal gaps by providing continuity of resources for pests (Schwarzmueller et al. 2019).

Beyond the biophysical challenges of AW-IPM in diverse landscapes are the psychosocial barriers for those who are directly (growers) and indirectly (regional towns people) affected by the pests. One of the greatest barriers to acceptance of communities participating in area-wide management of Queensland fruit fly was lack of social cooperation amongst growers such as insufficient care and responsibility about their role in the broader horticultural and social systems (Mankad et al. 2017). Increased transparency of the problems and issues for all actors, including better insights into knowing when pests show up, and where pests are located, may enable coordinated control and validation as to whether the AW-IPM actions are effectively suppressing the target population.

5. CONCLUSIONS

The inability to know where and when pests arrive, as well as their population dynamics in space and time, the inability to communicate this information in a timely manner and delivering coordinated pest control are barriers to AW-IPM, and in turn, potential barriers to efficient and sustainable pest management.

Increasingly technologies that free complex computation from desktop computing to in-field mobile devices are central to the development of in-field data collection, and automated insect detection and monitoring. Each of these will enable information communication among stakeholders, and the area-wide coordination of pest control.

Engineering solutions to overcome these barriers are many and varied. For government biosecurity, detecting rare events of pest incursions or escapees can minimize the size and duration of the management response, the cost, and the amount of time that markets are closed. For pest management of endemic species, automated insect detection solutions that are fast and easy to deploy, that provide real-time data flow, and that are cost-effective, can have a significant impact by improving estimates of pest numbers, pest locations and effectiveness of integrated control.

Even if it is early days, current in-field data collection and insect automation technology is already providing better information for more informed pest management decisions, improved workflow, and has the potential to reduce the harm from exotic pest incursions.

6. REFERENCES

Alvarez, S., E. A. Evans, and A. W. Hodges. 2016. Estimated costs and regional economic impacts of the Oriental fruit fly (*Bactrocera dorsalis*) outbreak in Miami-Dade County, Florida. University of Florida IFAS Extension. Fe988. Gainesville, Florida, USA.

(BBC) British Broadcasting Corporation. 2015. New Zealand: Officials declare victory over fruit fly.

Beavis, L. 2018. Tasmania's fruit fly fight hits $5.5 million bill, but Minister expects victory.

Doitsidis, L., N. Fouskitakis, N. K. Varikou, I. I. Rigakis, S. A. Chatzichristofis, A. K. Papfilippaki, and A. E. Birouraki. 2017. Remote monitoring of the *Bactorcera oleae* (Gmelin) (Diptera: Tephritidae) population using an automated McPhail trap. Computers and Electronics in Agriculture 137: 69–78.

Duffield, S. 2004. Evaluation of the risk of overwintering *Helicoverpa* spp. pupae under irrigated summer crops in southwestern Australia and the potential for area-wide management. Annals of Applied Biology 144: 17–26.

Dyck, V. A., J. Reyes Flores, M. J. B. Vreysen, E. E. Regidor Fernández, B. N. Barnes, M. Loosjes, P. Gómez Riera, T. Teruya, and D. Lindquist. 2021. Management of area-wide pest management programmes that integrate the Sterile Insect Technique, pp. 781–814. *In* V. A. Dyck, J. Hendrichs, and A. S. Robinson (eds.), Sterile Insect Technique – Principles and practice in Area-Wide Integrated Pest Management. Second Edition. CRC Press, Boca Raton, Florida, USA.

Enkerlin, W. R., L. Lopez, and H. Celedonio. 1996. Increased accuracy in discrimination between captured wild unmarked and released dye-marked adults in fruit fly (Diptera: Tephritidae) sterile released programs. Journal of Economic Entomology 89: 946–949.

Enkerlin, W. R., J. M. Gutiérrez Ruelas, R. Pantaleon, C. Soto Litera, A. Villaseñor Cortés, J. L. Zavala López, D. Orozco Dávila, P. Montoya Gerardo, L. Silva Villarreal, E. Cotoc Roldán, F. Hernández López, A. Arenas Castillo, D. Castellanos Dominguez, A. Valle Mora, P. Rendón Arana, C. Cáceres Barrios, D. Midgarden, C. Villatoro Villatoro, E. Lira Prera, O. Zelaya Estradé, R. Castañeda Aldana, J. López Culajay, F. Ramírez y Ramírez, P. Liedo Fernández, G. Ortíz Moreno, J. Reyes Flores, and J. Hendrichs. 2017. The Moscamed regional programme: Review of a success story of area-wide Sterile Insect Technique application. Entomological Experimentalis et Applicata 164: 188–203.

Faria, F. A., P. Perre, R. A. Zucchi, L. R. Jorge, T. M. Lewinsohn, A. Rocha, and R. da Silva Torres. 2014. Automatic identification of fruit flies (Diptera: Tephritidae). Journal of Visual Communication Image Research 25: 1516–1527.

(FAO) Food and Agriculture Organization of the United Nations. 2017. The future of food and agriculture - Trends and challenges. Rome, Italy.

(FAO/IAEA) Food and Agriculture Organization of the United Nations/International Atomic Energy Agency. 2018. Trapping guidelines for area-wide fruit fly programmes. W. R. Enkerlin, and J. Reyes-Flores, J. (eds.). Second Edition. Rome, Italy. 65 pp.

Ferguson, D. M., S. Purzer, M. W. Ohland, and K. Jablokow. 2014. The traditional engineer vs the innovative engineer. *In* 121st American Society of Engineering Education Annual Conference & Exposition Paper ID #8751. June 15-18, 2014. Indianapolis, Indiana, USA.

(FDA) Food and Drug Administration. 2018. Use of electronic health record data in clinical investigations: Guidance for industry. US Department of Health and Human Services.

Finistere Ventures. 2018. Early-stage Agtech report. Palo Alto, California, USA.

Goldshtein, E., Y. Cohen, A. Hetzroni, Y. Gazit, D. Timar, L. Rosenfeld, Y. Grinshpon, A. Hoffman, and A. Mizrach. 2017. Development of an automatic monitoring trap for Mediterranean fruit fly (*Ceratitis capitata*) to optimize control applications frequency. Computers and Electronics in Agriculture 139: 115–125.

Hendrichs, J., G. Franz, and P. Rendon. 1995. Increased effectiveness and applicability of the Sterile Insect Technique through male-only releases for control of Mediterranean fruit flies during fruiting season. Journal of Applied Entomology 119: 371–377.

Hendrichs, J., P. Kenmore, A. S. Robinson, and M. J. B. Vreysen. 2007. Area-Wide Integrated Pest Management (AW-IPM): Principles, practice and prospects, pp. 3–34. *In* M. J. B. Vreysen, A. S. Robinson, and J. Hendrichs (eds.), Area-wide control of pests: From research to field implementation. Springer Dordrecht, The Netherlands.

Hopkins, M. 2017. 17 field scouting apps for precision agriculture. Meister Media Worldwide.

(ISPM) International Standard for Phytosanitary Measures. 2018. Establishment of pest free areas for fruit flies (Tephritidae). ISPM 26, International Plant Protection Convention, FAO. Rome, Italy.

Ives, A. R., and W. H. Settle. 1997. Metapopulation dynamics and pest control in agricultural systems. American Naturalist 149: 220–246.

Jiang, J.-A., C.-L. Tseng, F.-M. Lu, E.-C. Yang, Z.-S. Wu, C.-P. Chen, S. Lin, L.-C. Lin, and C.-S. Liao. 2008. A GSM-based remote wireless automated monitoring system for field information: A case study for ecological monitoring of the Oriental fruit fly (*Bactrocera dorsalis* Hendel). Computers and Electronics in Agriculture 62: 243–259.

Kean, J. M., D. M. Suckling, N. J. Sullivan, P. C. Tobin, L. D. Stringer, G. R. Smith, B. Kimber, D. C. Lee, R. Flores Vargas, J. Fletcher, F. Macbeth, D. G. McCullough, D. A. Herms, et al. 2019. Global eradication and response database (GERDA).

Klassen, W., and M. J. B. Vreysen. 2021. Area-Wide Integrated Pest Management and the Sterile Insect Technique, pp. 75–112. *In* V.A. Dyck, J. Hendrichs, and A.S, Robinson (eds.), Sterile Insect Technique – Principles and practice in Area-Wide Integrated Pest Management. Second Edition. CRC Press, Boca Raton, Florida, USA.

Levins, R. 1969. Some demographic and genetic consequences of environmental heterogeneity for biological control. Bulletin of Entomological Society of America 15: 237–240.

Liu, Y., J. Zhang, M. Richards, B. Pham, P. Roe, and A. Clarke. 2009. Towards continuous surveillance of fruit flies using sensor networks and machine vision. 5th International Conference on Wireless Communications, Networking and Mobile Computing. September 2009, Beijing, China.

Lloyd, A. C., E. I. Hamacek, R. A. Kopittke, T. Peek, P. M. Wyatt, C. J. Neale, M. Eelkema, and H. Gu. 2010. Area-wide management of fruit flies (Diptera: Tephritidae) in the Central Burnett district of Queensland, Australia. Crop Protection 29: 462–469.

Lloyd, R. J., D. A. H. Murray, J. E. and Hopkinson. 2008. Abundance and mortality of overwintering pupae of *Helicoverpa armigera* (Hubner) (Lepidoptera: Noctuidae) on the Darling Downs, Queensland, Australia. Australia Journal of Entomology 47: 297–306.

Mankad, A., B. Loechel, and P. F. Measham. 2017. Psycosocial barriers and facilitators for area-wide management of fruit fly in southeastern Australia. Agronomy for Sustainable Development 37: 67(1–12).

(ODK) Open Data Kit Community. 2018. Open Data Kit. Washington, DC, USA.

Philimis, P. 2015. E-FLYWATCH Development of an innovative automated and wireless trap with warning and monitoring modules for integrated management of the Mediterranean (*Ceratitis capitata*) & olive (*Dacus oleae*) fruit flies. Final report project reference: 262362. CORDIS - EU Research Projects under Horizon 2020. European Commission.

Potamitis, I., P. Eliopoulos, and I. Rigakis. 2017a. Automated remote insect surveillance at a global scale and the internet of things. Robotics 6: 19.

Potamitis, I., I. Rigakis, and N.-A. Tatlas. 2017b. Automated surveillance of fruit flies. Sensors 17: 110.

Schellhorn, N. A., H. R. Parry, S. Macfadyen, Y. Wang, and M. P. Zalucki. 2015. Connecting scales: Achieving in-field pest control from areawide and landscape ecology studies. Insect Science 22: 35–51.

Schwarzmueller, F., N. A. Schellhorn, and H. R. Parry. 2019. Resource landscapes and movement strategy shape Queensland fruit fly population dynamics. Landscape Ecology 34: 2807–2822.

Sequeira, R. 2001. Inter-seasonal population dynamics and cultural management of *Helicoverpa* spp. in a central Queensland cropping system. Australian Journal of Experimental Agriculture 41: 49–247.

Shaked, B., A. Amore, C. Ioannou, F. Valdés, B. Alorda, S. Papanastasiou, E. Goldshtein, C. Shenderey, M. Leza, C. Pontikakos, D. Perdikis, T. Tsiligiridis, M. R. Tabilio, A. Sciarretta, C. Barceló, C. Athanassiou, M. A. Miranda, V. Alchanatis, N. Papadopoulos, and D. Nestel 2018. Electronic traps for detection and population monitoring of adult fruit flies (Diptera: Tephritidae). Journal of Applied Entomology 142: 43–51.

Smith, J. W. 1998. Boll weevil eradication: Area-wide pest management. Annals of the Entomological Society of America 91: 239–247.

Suckling, D. M., J. M. Kean, L. D. Stringer, C. Cáceres-Barrios, J. Hendrichs, J. Reyes-Flores, and B. C. Dominiak. 2016. Eradication of tephritid fruit fly pest populations: Outcomes and prospects. Pest Management Science 72: 456–465.

Southwood, T. R. E. 1978. Ecological methods: With particular reference to the study of insect populations. Chapman & Hall. London, UK. 548 pp.

Taylor, R. A. J. 2018. Spatial distribution, sampling efficiency and Taylor's power law. Ecological Entomology 43: 215–225.

Vreysen, M. J. B. 2021. Monitoring sterile and wild insects in area-wide Integrated Pest Management programmes, pp. 485–528. *In* V. A. Dyck, J. Hendrichs, and A. S. Robinson (eds.), Sterile Insect Technique – Principles and practice in Area-Wide Integrated Pest Management. Second Edition. CRC Press, Boca Raton, Florida, USA.

Vreysen, M. J. B., A. S. Robinson, and J. Hendrichs (eds.). 2007. Area-wide control of pests: From research to field implementation. Springer Dordrecht, The Netherlands. 789 pp.

Zhang, J, L. Sun, Y. Liu, H. Wang, N. Sun, and P. Zhang. 2017. Mobile device-based electronic data capture system used in a clinical randomised controlled trial: Advantages and challenges. Journal of Medical Internet Research 19 (3): e66.

Zalucki, M. P., D. Adamson, and M. J. Furlong. 2009. The future of IPM: Whither or wither? Australian Journal of Entomology 48: 85–96.

PROSPECTS FOR REMOTELY PILOTED AIRCRAFT SYSTEMS IN AREA-WIDE INTEGRATED PEST MANAGEMENT PROGRAMMES

D. BENAVENTE-SÁNCHEZ[1], J. MORENO-MOLINA[1] AND R. ARGILÉS-HERRERO[2]

[1]*Embention, Alicante, Spain; embention@embention.com*
[2]*Insect Pest Control Section, Joint FAO/IAEA Division of Nuclear Techniques in Food and Agriculture, Vienna, Austria; R.Argiles-Herrero@iaea.org*

SUMMARY

The scientific and technical advances achieved in recent years in the technology for *remotely piloted aircraft systems* (RPAS) and the approval of new regulatory frameworks in several countries have allowed the commercial expansion and use of these flying robots for different civil applications, including agriculture. The present review discusses the opportunities for the use of the RPAS in area-wide integrated pest management (AW-IPM) programmes within the current technical and legal constraints. These include targeted insecticide applications of hotspots in fruit fly and mosquito pest management programmes, aerial release of sterile males in mosquito and tsetse control programmes, and aerial release of parasitoids. The advantages of the RPAS technology - accuracy, increased safety and cost-efficiency for small and medium scale operations - are counterbalanced by its limitations at the technical level - reduced payload and flight duration - as well as at the regulatory level - mandatory special operational permits from regulatory agencies for operations beyond the visual line of sight.

Key Words: RPAS, unmanned aircraft, UAS, drones, regulation, sterile insects, parasitoids, pest control

1. INTRODUCTION

Area-wide integrated pest management (AW-IPM) programmes aim at controlling a given pest at a geographic scale, targeting the whole pest population (Hendrichs et al. 2007). By definition, there is an intrinsic spatial dimension in all area-wide programmes. The best way to operate at this geographic/spatial dimension is by combining geographic information systems (GIS) technology with aerial intervention tools (IAEA 2006). Until now, aircrafts have been widely used in AW-IPM implementation, and their deployment has major comparative advantages for those programmes operating over large areas of difficult topography and that lack road networks (Vreysen et al. 2007; Dyck et al. 2021). Although the technology of *remotely piloted aircraft systems* (RPAS) continues

J. Hendrichs, R. Pereira and M. J. B. Vreysen (eds.), Area-Wide Integrated Pest Management: Development and Field Application, pp. 903–916. CRC Press, Boca Raton, Florida, USA.

improving for many applications, it is already technically sufficiently mature to allow carrying out many of the activities of an AW-IPM operational programme by air. The term RPAS is the official International Civil Aviation Organization (ICAO) term for such aircraft, whereas terms such as *unmanned aircraft systems* (UAS) are being less used in view that regulations prescribe that all aircraft need to be piloted, even when the pilot is not on board (FAA 2017; CASA 2018). As the term *drone*, common in francophone countries, is increasingly identified with military applications, it is not used here.

The main current applications of RPAS in agriculture are air-borne scouting of field crops and of ranging livestock through remote sensing, as well as precision delivery systems and aerial spraying with low or ultra-low volumes in some crops. The advantages of the RPAS technology are mainly in the fields of safety, accuracy and cost efficiency. However, as it usually happens with technology involving a shift of paradigm in the way society deals with problems, the regulatory framework is not fully developed and has not evolved at the required speed. This has led some pioneers to develop applications with an *ad hoc* certificate of authorization (COA) or, in some cases, to operate at the fringes of the legal regulation. The future of RPAS will therefore depend as much on decisions made by regulators as it does on technological advances, and for example France's relatively permissive regulation has put it at the forefront of the agricultural use of unmanned aircrafts (The Economist 2017). In recent months, however, more appropriate regulatory frameworks have been adopted in different countries, clarifying the requirements for the operation of UAV and opening the field for commercial applications of RPAS, while at the same time, driving technology adaptation to the new regulations.

2. THE TECHNOLOGY

2.1. Elements of a Remotely Piloted Aircraft System (RPAS)

A RPAS is made of three main parts: the remote-control station, the datalink and the remotely piloted aircraft or (RPA) unmanned aerial vehicle (UAV) (Fig. 1). The control station contains all the elements that permit operators to manage the UAV flight. This includes the software that interacts with the aircraft via datalink, but also the computer (which can be just a tablet) and the joystick that controls the RPA in manual mode.

The datalink refers to the main elements that allow communications between the control station and the aircraft. The core element is the radio, which must use a frequency authorized by national authorities (for civil aviation purposes, usually 2.4 GHz or 900 MHz for remote control, along with 5.8 GHz for video and audio links) and can include helpful methods to avoid interferences, such as the "Frequency Hopping Spread Spectrum" (FHSS) technique. Communications can be digital or analogue: while the first provides higher quality, especially for video transmission, the range of the second one is higher for low-cost systems. The use of antennas and amplifiers should be considered too, especially when the payload and power capacity of the aircraft is too limited. For very long-range communications, it is highly recommended to use a tracking antenna: a directional antenna with embedded control actuators and installed encoders that automatically points to the aircraft with high precision, maximizing the communications range.

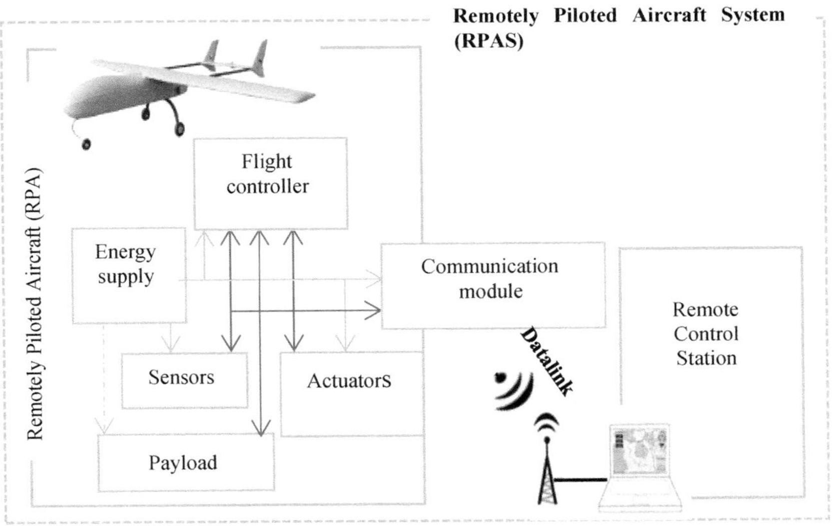

Figure 1. Main components of Remotely Piloted Aircraft Systems (RPAS) (modified after Wikipedia 2019).

Different options of unmanned vehicles are available to fulfil the required operation in the most efficient way: from fixed-wing aircrafts to multirotors, helicopters, blimps, parafoils or hybrid Vertical Take-Off and Landing (VTOL) platforms, among many others (Fig. 2).

There is also a wide variety of engines available for these vehicles, from electrical to fuel powered engines. Although not commonly used yet, remarkable progress has been achieved in the development of hydrogen fuel cell powered engines, which are expected to extend the flight endurance. The choice of the engine will depend on the autonomy requirements but also on the size and weight of the aircraft; sometimes altitude can be a limiting condition due to lower air density at high elevations above sea level, requiring for instance the use of electrical fuel injection (EFI) engines for higher efficiency. The payload capacity and flight duration are also key elements for the majority of operations and must be carefully balanced when selecting the RPA for any specific task.

Commonly related to the three mentioned elements of RPAS, autopilots are key in any unmanned operation, especially when the aircraft should perform autonomous missions without the constant control of operators. The most sophisticated autopilots have embed a suite of sensors and processors together with Line of Sight (LOS) and Beyond Radio Line of Sight (BRLOS) Machine to Machine (M2M) datalink radios, all with reduced size and weight to minimise the use of the payload capacity of the aircraft.

These autopilots usually permit the connection with external peripherals such as transponders and radars, increasing the capabilities of the platform during fully unmanned missions through the provision, for instance, of reliable and autonomous Detect & Avoid (DAA) tools. These refer to the technology that provides to unmanned vehicles the capability to detect obstacles on their route (such as other aircrafts) and to immediately find a way to prevent collision.

When they are aimed at professional purposes, autopilot systems must be developed according to international standards in order to be compliant with national regulations for the professional operation of RPAS.

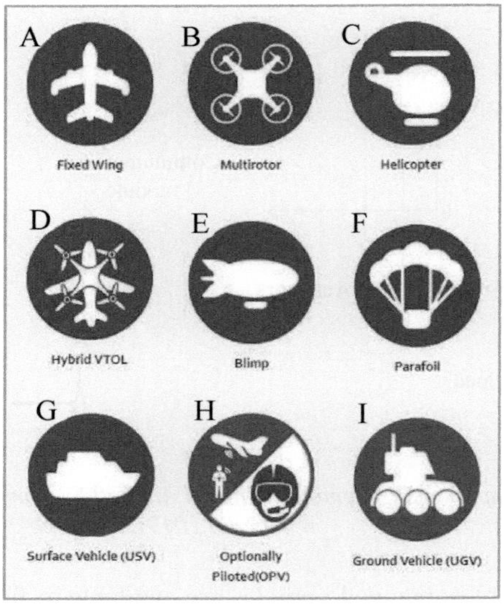

Figure 2. Types of unmanned vehicles: (A) Fixed-wing; (B) Multirotor; (C) Helicopter; (D) Hybrid Vertical Take-Off and Landing (VTol); (E) Blimp; (F) Parafoil; (G) Unmanned surface vehicle (USV); (H) Optionally piloted vehicle (OPV); and (I) Unmanned ground vehicle (UGV) (credit Embention).

2.2. Safety Considerations

In addition to the design and development of the RPAS according to aerial international standards, redundant systems greatly reduce the risk of failure. Redundancy usually is focused on the autopilot (normally with triple redundancy). Triple redundancy for autopilots refers to the inclusion of three autopilot units in the same flight controller. These three autopilots operate as if they were a single unit dealing with the control of the unmanned platform. A dissimilar arbiter (a piece of electronics included in the redundant autopilot) decides which of those three autopilots is in command attending to the efficiency and coherence of their orders. Therefore, in case any autopilot unit fails, there will be still two units capable to deal with the whole operation. For critical operations performed by large aircraft, such as originally manned helicopters that were adapted into unmanned platforms, redundancy is commonly extended to actuators, as long as the platform can tolerate the increase in weight caused by the redundant actuators and the budget of the project can deal with their price.

Autopilot systems have several configurable safety routines that can be triggered under specific situations. Some of these safety routines include landing if the battery/fuel is below a defined threshold of its capacity, go back to the take-off point if the datalink is lost, and reduce altitude if the wind speed is higher than a predefined value. Also, configurations aimed to end the flight in the safest possible way can be configured into the Flight Termination System (FTS): an arbiter microprocessor within the autopilot that activates the flight termination in case of failure of the main microprocessor or any of the motors (e.g. by releasing a parachute).

The most sophisticated autopilot systems also include sensors such as radars, lidars and/or transponders that permit the use of DAA technologies. An officially certified DAA system is currently mandatory to operate beyond visual line of sight in most of the countries, but due to the novelty and impact of this technology in the industry, currently the certification process of this kind of system consists of a series of negotiations with national authorities who normally request clear and irrefutable proof about the reliability of the system.

3. CURRENT APPLICATIONS OF THE RPAS TECHNOLOGY IN AGRICULTURE

Precision agriculture, construction and public safety are the most promising commercial and civil aviation markets for RPAS (The Economist 2017). RPAS are used in agriculture for low altitude remote sensing (LARS) of crop conditions with multi-spectral and hyperspectral cameras. High spatial resolution information on crop hydric stress (Berni et al. 2009), crop vegetation index (Primicerio et al. 2012), weed detection (Hardin et al. 2007) and yield mapping (Swain et al. 2010) can be obtained with the proper data acquisition and analysis.

When compared to satellite remote sensing, RPAS-based remote sensing has a higher spatial and temporal resolution, lower costs and is not affected by cloud coverage among other advantages (Nansen and Elliott 2016). However, the calibration of the sensors under field conditions and post-processing of the acquired information remain complex and require high level of expertise.

After analysing the data collected by RPAS, farmers can tailor their use of irrigation, pesticides, herbicides and fertilizers, applying by ground variable rate technologies and following the principles of precision-agriculture (Zhang and Kovacs 2012).

At the research level, remote sensing with RPAS has also been used to detect host plant responses to insect infestations, like the analysis of the variation in the vegetation index (Normalized Difference Vegetation Index or NDVI) in wheat fields caused by an attack of fall armyworm *Spodoptera frugiperda* J.E. Smith (Zhang et al. 2014).

RPAS are also used for low and ultra-low volume aerial spraying of insecticides and herbicides. In Japan, one of the world-leading manufacturers of agriculture RPAS (Yamaha Motor Company) estimates that about one million ha of rice, around 35% of the Japanese rice fields, are currently sprayed with their unmanned helicopters (Yamaha 2017).

In China, several commercial companies offer off-the-shelf commercial RPAS models for crop spraying with variable tank capacities and spraying methods, from thermal fogging to electrostatic spraying (DJI 2017; Joyance 2017). The mission planning and operation of these spraying RPAS is relatively simple thanks to dedicated remote controllers and software. Variable rates of pesticides, fertilizers and herbicides can be applied.

4. REGULATORY FRAMEWORK

4.1. Classification of Remotely Piloted Aircrafts and Operations

Although different legal frameworks exist in each country/region, most of the existing regulations classify the RPAs in relation to their weight (ICAO 2015; and other regulations listed in Table 1):

1. Micro-RPA if less than 2 kg of maximum take-off mass (MTOM)
2. Small RPA between 2 and 25 kg of MTOM and
3. Large RPA between 25 and 150 kg of MTOM.

They also classify them in relation to the type of operation:

1. Within the visual line of sight (VLOS) when the distance to the operator is less than 500 m with good visibility conditions and flight elevation below 120 m
2. Beyond visual line of sight (BVLOS) and flight elevation below 150m over the ground (also called Very Low Level or VLL operations)
3. Beyond visual line of sight (BVLOS) and flight elevation above 150 m over the ground; and
4. A fourth and intermediate class, which is the Extended Visual Line of Sight (EVLOS) for those operations where the main pilot has no visual contact with the RPA, but other pilots/observers, in communication with the main one, are capable to see it.

4.2. Different Regulations in Each Country

The regulatory framework for the operation of RPAs is not fully developed in all the countries and not fully harmonized across the countries and regions. However, some common ground among the regulations of different countries can be found:

1. Regarding the weight of the RPAs, those over 2 kg MTOM must be operated by certified operators; those over 25 kg MTOM must be registered and follow either a risk-based approach or obtain a type certificate.
2. Regarding the type of operation, RPAs should be operated within the VLOS. Operations BVLOS are only allowed under especial conditions (when the aircraft is under 2 kg, when operating in special areas of the airspace, or when using DAA technology or UAS Traffic Management (UTM)/U-Space systems (set of new services relying on a high level of digitalisation and automation of functions and specific procedures designed to support safe, efficient and secure access to airspace for large numbers of drones) (SESAR-JU 2019).
3. Operations over restricted areas (e.g. populated areas) must be performed by platforms with a certificate of airworthiness and also with permission from public authorities.
4. RPAS operators and RPAs above 250 g must be registered.
5. Most, if not all, RPAS operations must be insured.

Some examples of the relevant regulations for RPAS operators in different countries can be found in Table 1. Due to the early stage of the unmanned industry and regulations, waivers by the competent national authority against these regulations are possible.

Table 1. Regulatory framework for the use of remotely piloted aerial systems (RPAS) in different countries/regions

Country / Region	Organism	Document	Release Date
Australia	Civil Aviation Safety Authority (CASA 2018)	- General information - Regulation: Direction-operation of certain unmanned aircraft	2017-10
European Union	European Aviation Safety Agency (EASA 2017)	- General information - Opinion No 01/2018 Introduction of a regulatory framework for the operation of unmanned aircraft systems in the 'open' and 'specific' categories	2017-05
Japan	Civil Aviation Bureau (CAB 2016)	- General information	2015-10
Mexico	Dirección General de Aeronáutica Civil (DGAC 2017)	- General Information - Regulation: CO AV-23/10 R4 que establece requerimientos para operar un sistema de aeronave pilotada a distancia (RPAS)	2017-07
Spain	Agencia Estatal de Seguridad Aérea (AESA 2018)	- General information - Regulation: Utilización de RPAS	2017-12
USA	Federal Aviation Administration (FAA 2017)	- General information - JO 7200.23: Unmanned Aircraft Systems (UAS)	2016-10

4.3. Certificate of Airworthiness

The certificate of airworthiness is issued by the competent agency in charge of aviation safety in each country. Considering that it depends on national rules, this certificate may be mandatory if the maximum take-off mass (MTOM) of the remotely piloted aircraft is over 25 kg, but there are other situations where the certificate may be requested such as performing operations over populated areas or if the operation will take place BVLOS.

The request for a certificate needs to include every element of the RPAS, which includes a study about the risk of operating the system, the results of the tests made to the autopilot (hardware and software) and the structure of the aircraft and its resistance to harsh environments, such as extreme temperatures, humidity or dust conditions.

4.4. Pilots Qualifications

Regulations in most countries require RPAS pilots to be of legal age and their medical condition certified by an official aeronautical centre. In addition, the pilot must have the theoretical knowledge (an official license is required, such as an ultralight flight pilot license) and the practical knowledge on the specific RPAS (obtained from the manufacturer of the RPAS or by an organization delegated by the manufacturer) (ICAO 2015; and other regulations listed in Table 1).

5. ADVANTAGES AND OPPORTUNITIES FOR RPAS IN AW-IPM WITH THE CURRENT LEGAL REGULATIONS

One of the main advantages of the technology is in the field of safety, avoiding the risks for pilots inherent to any operation with manned aircrafts in agriculture. In case of accident, the severity of harm is minimised thanks to the absence of an onboard pilot and the lower kinetic energy of the aircraft.

Accuracy of the operations is very high (the same as in regular aircrafts), since the mission events are preconfigured based on geographic information and triggered automatically during the flight by the GPS information, avoiding human errors during operation. The parameters of the operation, such as the dose to be applied per surface unit during an aerial spraying operation or insect release, are controlled directly by the microprocessors that will adjust the actuators to obtain a constant value regardless of the variations of the flight speed (DJI 2017).

Cost-efficiency is also a major advantage of the RPAS technology thanks to the relatively low investment, maintenance and operational costs. However, cost-efficiency varies depending on the scale of the operation and detailed analysis should be conducted on a case by case basis.

Within the current legal framework, some of the activities that are part of the AW-IPM programmes for the control of different pest insects can be carried out using RPAS (Table 2). Some of the operations, such as the suppression of fruit fly hotspots, do not require a specific authorization from the civil aviation authority (CAA). Others, like those who require operation BVLOS, require a special operation permit by the relevant CAA based on a specific operations risk assessment (JARUS 2017).

5.1. Suppression of Fruit Fly Hotspots and Mosquito Larval Breeding Sites

One common activity in AW-IPM programmes against fruit flies is the suppression of hotspots by bait-spraying. Hotspots are detected and delimited by weekly trapping surveys and later normally treated by ground with bait sprays. For example, the Mediterranean fruit fly, *Ceratitis capitata* (Wiedemann) control programme in Valencia (Spain) uses a fleet of all-terrain vehicles (ATV) equipped with ground sprayers for this purpose (Argilés and Tejedo 2007) (Fig. 3). However, the low volume of insecticide required by bait sprays compared to cover sprays and the limited size of the area to be treated – the size of hotspots is usually some tens of ha – make this task technically suitable for aerial application by RPAS.

Currently, off-the-shelf RPAS for spraying have been developed by several commercial companies with specific software and remote controller for this task. From the regulation point of view, the operation in VLOS over commercial crops outside restricted areas and with a maximum take-off mass below 25 kg exempt this application in most countries of the requirement for a specific permit or certificate of airworthiness from the national civil aviation authorities. The area covered per day can range between 25 to 50 ha depending on the insecticide application specifications, which is several times higher than can be achieved by ground spraying.

Table 2. Summary of technical parameters for different RPAS applications

Activity	Weight of adults		Release density		Biomass of insects per surface		Area to be covered per flight		Swath widths	
	min	max	min	max	min	max	min	max	min	max
Release of sterile *Aedes* mosquitoes	0.9 mg	1.1 mg	1000 males/ha	2000 males/ha	90 g/km^2	220 g/km^2	40 ha	60 ha	75 m	100 m
Release of sterile tsetse flies	20 mg	30 mg	0.5 males/ha	1.0 males/ha	1 g/km^2	3 g/km^2	3000 ha	8000 ha	300 m	400 m
Fruit fly hotspot bait spraying	-	-	-	-	-	-	3 ha	6 ha	8 m	10 m

Activity	Flight distance (without ferry)		Speed during operation	Flight time (without ferry)		Payload	Type of aircraft	Type of operation	Specific authorization by relevant CAA
	min	max		min	max				
Release of sterile *Aedes* mosquitoes	5.3 km	6.0 km	10 m/s	9 min	10 min	300 g	rotationary; micro	VLOS; urban area	mandatory
Release of sterile tsetse flies	100 km	200 km	20 m/s	83 min	167 min	900 g	fixed-wing/ helicopter; medium	BVLOS; rural area	mandatory
Fruit fly hotspot bait spraying	3.8 km	6 km	8 m/s	8 min	13 min	10 kg	rotationary; medium	VLOS; rural area	exempted

Similarly, mosquito larvicides can be applied aerially to larval breeding areas, replacing the current ground and helicopter treatments, as demonstrated in the ongoing tests by L'Entente Interdépartementale pour la Démoustication du Littoral Méditerranéen in France (EID Méditerranée 2018).

Figure 3. (A) Delimiting the extent of hot spots with GIS-based field surveys of monitoring traps. (B) Green tracks showing the bait spray ground treatment done in different citrus orchards; only orchards with a ripening variety are treated. (C) ATV applying a bait spray treatment by ground. (D) Spraying RPAS with a 10 litre tank.

5.2. Release of Sterile Aedes Mosquitoes in SIT Programmes

The weight of adult *Aedes spp.* mosquitoes is low (ca. 1 mg). They also have a low dispersal behaviour, which compared to other pests, makes it possible to apply area-wide control programmes over relatively small urban areas. These two conditions make the aerial release of sterile mosquito males technically feasible using RPAS.

Flying over the urban or peri-urban scenario typical of an *Aedes* spp. mosquito control programmes, however, requires a mandatory permit and certificate of airworthiness for the specific operation by the local aerial safety agency. To facilitate obtaining these, the use of micro-RPAS, i.e. under 2 kg of maximum take-off mass, is advised.

An aerial platform of less than 1 kg, including the payload with the release machine and an insect cargo of around 75 000 sterile males, can suffice to release mosquitoes in VLOS over 50 ha with a flight duration of less than ten minutes per flight and a flight elevation under 100 m.

5.3. Release of Sterile Tsetse Flies in SIT Programmes

Tsetse fly SIT programmes are applied at a scale of very large areas, usually covering several hundred or thousand km^2 of rural landscape (Feldmann et al. 2021). The required release density of sterile males is only between 50 to 100 sterile males per km^2 and per week (equivalent to 1 to 3 g of insects per km^2 and per week, depending on the tsetse species). This is a very low release rate when compared to other pest species, namely fruit flies (Hendrichs et al. 2021). Release swaths of 300-400 m are common in view that tsetse flies disperse sufficiently between flight lines.

To conduct the aerial releases under these conditions, a remotely piloted aircraft with a flying endurance of at least 100 km and a payload of approximately 1 kg for the chilled insect release machine and the sterile insect cargo is needed (for an example see Fig. 4). Several fixed-wing and helicopter models currently available on the market with a MTOM below 25 kg meet these technical requirements. The release operation will necessarily be BVLOS, and thus will require specific authorization by the local aerial safety agency. In addition, the take-off and landing point will need to be as close as possible or within the release area to save ferry flying time, which may require a field insectary.

Figure 4. Unmanned aircraft developed for the Southern Tsetse Eradication Project in Ethiopia. The pods under the wings carry the sterile tsetse males in chilled conditions.

5.4. Release of Parasitoids in AW-IPM Programmes

Several AW-IPM programmes release parasitoids to control different pests, such as the aerial releases of chilled *Diachasmimorpha longicaudata* (Ashmead) parasitoids mass-produced by the Moscafrut programme in Mexico to control *Anastrepha ludens* (Loew) and *An. obliqua* (Macquart) fruit flies (Montoya et al. 2007); or ground releases in Costa Rica of *Spalangia endius* Walker to control the cattle pest *Stomoxys calcitrans* (L.) by Costa Rica's National Institute of Agricultural Technology, Innovation and Transfer (INTA) (Solorzano-Arroyo et al. 2017). The release of these tiny wasps can be carried out with RPAs under 25 kg, although to be more cost-effective, the operation needs to be BVLOS, and therefore requires the specific authorization by the local aerial safety agency based on a specific operation risk assessment (JARUS 2017).

In Mexico, aerial releases of *Tamarixia radiata* Waterston, the parasitoid of *Diaphorina citri* Kuwayama, the vector of Huanglongbing, are being conducted with an

RPAS by the company Mubarqui within the area-wide programme of the Servicio Nacional de Sanidad, Inocuidad y Calidad Agroalimentaria (SENASICA) that manages the citrus greening programme in Mexico (García-Ávila et al., this volume).

5.5. *Release of Sterile Fruit Flies in SIT Programmes*

The release of sterile fruit flies in SIT operational programmes is carried out with small aircrafts covering tens of thousands of ha per flight and with a cargo above 100 kg of biomass (FAO/IAEA 2017). RPAS doesn't seem suited for this job, since this would require large RPAs with a complex certification process.

Small pilot projects could possibly benefit of RPAS for aerial releases provided they apply a shift of strategy for the aerial releases, e.g. several short flights with a low load of sterile insects from a mobile base within the release area.

5.6. *Strategy and Logistics for the Releases of Sterile Insects*

As described, RPAS can potentially be used in some AW-IPM programmes to release sterile insects. To do so, some adaptations to the current release procedures are needed to compensate for the lower payload and endurance capacities of the RPA aircraft. Currently, large numbers of sterile insects are released in SIT operational programmes by manned aircrafts in a single flight covering large areas (for example, tens of millions of sterile insects over hundreds of km^2 in the case of fruit fly programmes).

If RPAS are to be used for the aerial release of sterile insects, multiple and shorter flights are needed for the same purpose. Currently, the manned aircrafts take-off from a runway located close to the sterile fly emergence and release facility. Release by RPAS will require that the sterile insects be transported by ground to the take-off point within the release area, or at least in the very close vicinity. This will require the use of a mobile take-off station from which the RPAS is operated and where sterile insects are kept under chilled conditions waiting for their turn to be released.

5.7. *Release Machines*

Apart from the mandatory requirements of using light materials, the release machines also require a chilling system capable of maintaining the temperature between 4°C and 8°C in the insect holding container, depending on the insect species, during the flight. Passive chilling systems like those based on the use of phase-change materials (PCM) can provide such conditions at a low payload cost. Relative humidity must be kept under the dew point to avoid condensation on the surface of the insect bodies. Some insects like mosquitoes or some species of parasitoid wasps have fragile legs and antennae and require higher holding temperatures and special equipment to avoid mechanical damage by compaction and vibrations during the flight and by friction while going through the release mechanism (Montoya et al. 2012).

Accurate releases with variable release rates can be continuous, along a flight path or discrete, focused on individual waypoints with predefined coordinates.

6. CONCLUSIONS

The RPAS technology is widely and increasingly used in agriculture for remote sensing and pesticide spraying of crops, mainly at the level of individual larger farms. However, RPAS are still not routinely used in operational AW-IPM programmes, such as fruit fly SIT programmes, because the scale of these programmes is usually too large for the current technical and regulatory limitations of RPAS.

Nevertheless, as discussed in this review, the legal regulations recently approved in several countries have opened the possibility to use the existing commercial RPAS technology for some of the field activities of AW-IPM programmes. This is the case of the chemical treatment of hotspots in fruit flies or mosquito control programmes, providing high application accuracy.

Some other applications, such as the release of sterile males of tsetse or mosquitoes, respectively over large or urban areas, require a specific risk assessment prior to the authorization by the relevant civil aviation authority in the country. In this regard, the development of the novel DAA technology and UTM/U-Space systems, and their official certification, will facilitate the authorization of applications and operations beyond the pilot VLOS, as commonly needed in area-wide programmes.

7. REFERENCES

(AESA) Agencia Estatal de Seguridad Aerea. 2018. Drones legal framework. Spanish Aviation Safety Agency. Ministerio de Fomento, Gobierno de España, Spain.

Argilés, R., and J. Tejedo. 2007. La lucha de la mosca de la fruta mediante la técnica del insecto estéril en la Comunitat Valenciana. Levante Agrícola 385: 157–162.

Berni, J. A. J., P. J. Zarco-Tejada, L. Suarez, V. Gonzalez-Dugo, and E. Fereres. 2009. Remote sensing of vegetation from UAV platforms using lightweight multispectral and thermal imaging sensors. International Archives Photogrammetry, Remote Sensing and Spatial Information Science: 38 (6): 6.

(CAB) Civil Aviation Bureau. 2016. Japan's safety rules on Unmanned Aircraft (UA)/Drone. Japan Ministry of Land, Infrastructure, Transport and Tourism, Tokyo, Japan.

(CASA) Civil Aviation Safety Authority. 2018. Flying drones/remotely piloted aircraft in Australia. Australian Government, Canberra, Australia.

(DGAC) Dirección General de Aeronáutica Civil. 2017. Circular Obligatoria CO AV-23/10 R4 que establece los requerimientos para operar un sistema de aeronave pilotada a distancia (RPAS). Secretaria de Comunicaciones y Transporte, México. 55 pp.

DJI (Da-Jiang Innovations Science and Technology Co., Ltd). 2017. Agras MG-1, DJI's first agricultural drone. DJI Company, Hangzhou, China.

Dyck, V. A., J. Hendrichs, and A. S. Robinson (eds.). 2021. Sterile Insect Technique – Principles and practice in Area-Wide Integrated Pest Management. Second Edition. CRC Press, Boca Raton, Florida, USA. 1200 pp.

(EASA) European Aviation Safety Agency. 2017. Notice of proposed amendment 2017-05 (A): Introduction of a regulatory framework for the operation of drones. European Union. 128 pp.

(EID Méditerranée) Entente Interdépartemental pour la Démoustication du Littoral Méditerranéen. 2018. Traitement drone bientôt opérationnel? 4 janvier 2018. Montpellier, France.

(FAA) Federal Aviation Administration. 2017. Unmanned aircraft systems (UAS). Regulations & policies. United States Department of Transportation, Washington, DC, USA.

(FAO/IAEA) Food and Agriculture Organization of the United Nations/International Atomic Energy Agency. 2017. Guideline for packing, shipping, holding and release of sterile flies in area-wide fruit fly control programmes. J. L. Zavala-López, and W. R. Enkerlin (eds.). Second edition, Rome, Italy. 140 pp.

Feldmann, U., V. A. Dyck, R. C. Mattioli, and J. Jannin. 2021. Potential impact of tsetse fly control involving the Sterile Insect Technique, pp. 1051–1088. *In* V. A. Dyck, J. Hendrichs, and A. S. Robinson (eds.), Sterile Insect Technique – Principles and practice in Area-Wide Integrated Pest Management. Second Edition. CRC Press, Boca Raton, Florida, USA.

Hardin, P. J., M. W. Jackson, V. J. Anderson, and R. Johnson. 2007. Detecting squarrose knapweed (*Centaurea virgata* Lam. Ssp. Squarrosa Gugl.) using a remotely piloted vehicle: A Utah case study. GIScience & Remote Sensing 44: 203–219.

Hendrichs, J., P. Kenmore, A. S. Robinson, and M. J. B. Vreysen. 2007. Area-Wide Integrated Pest Management (AW-IPM): Principles, practice and prospects, pp. 3–33. *In* M. J. B. Vreysen, A. S. Robinson, and J. Hendrichs (eds.), Area-wide control of insect pests: From research to field implementation. Springer, Dordrecht, The Netherlands.

Hendrichs, J., M. J. B. Vreysen, W. R. Enkerlin, and J. P. Cayol. 2021. Strategic options in using sterile insects for Area-Wide Integrated Pest Management, pp. 839–882. *In* V. A. Dyck, J. Hendrichs, and A.S. Robinson (eds.), Sterile Insect Technique – Principles and practice in Area-Wide Integrated Pest Management. Second Edition. CRC Press, Boca Raton, Florida, USA.

(IAEA) International Atomic Energy Agency. 2006. Designing and implementing a Geographical Information System: A guide for managers of area-wide pest management programmes. Joint FAO/IAEA Programme of Nuclear Techniques in Food and Agriculture. IAEA, Vienna, Austria.

(ICAO) International Civil Aviation Organization. 2015. Manual on Remotely Piloted Aircraft Systems (RPAS). Doc 10019 AN/507. Montréal, Quebec, Canada.

(JARUS) Joint Authorities for Rulemaking of Unmanned Systems. 2017. Guidelines on Specific Operations Risk Assessment. Document JAR-DEL-WG6-D.04. 48 pp.

Joyance. 2017. Spraying drones. Joyance Company, Shandong, China.

Montoya, P., J. Cancino, and L. Ruiz. 2012. Packing of fruit fly parasitoids for augmentative releases. Insects 3: 889–899.

Montoya, P., P. López, J. Cruz, F. López, C. Cadena, J. Cancino, and P. Liedo. 2017. Effect of *Diachasmimorpha longicaudata* releases on the native parasitoid guild attacking *Anastrepha* spp. larvae in disturbed zones of Chiapas, Mexico. BioControl 62: 581–593.

Montoya, P., J. Cancino J., M. Zenil, G. Santiago, and J. M. Gutierrez. 2007. The augmentative biological control component in the Mexican National Campaign against *Anastrepha* spp. fruit flies, pp. 661-670. *In* M. J. B. Vreysen, A. S. Robinson, and J. Hendrichs (eds.), Area-wide control of insect pests. Springer, Dordrecht, the Netherlands.

Nansen, C., and N. Elliott. 2016. Remote sensing and reflectance profiling in entomology. Annual Review of Entomology 61: 139–158.

Primicerio, J., S. F. D. Gennaro, E. Fiorillo, L. Genesio, E. Lugato, A. Matese, and F. P. Vaccari. 2012. A flexible unmanned aerial vehicle for precision agriculture. Precision Agriculture 13: 517–523.

(SESAR-JU) Single European Sky Air Traffic Management Research Joint Undertaking. 2019. U-space. European Commission for Transport. Bietlot, Belgium.

Swain, K. C., S. J. Thomson, and H. P. W. Jayasuriya. 2010. Adoption of an unmanned helicopter for low altitude remote sensing to estimate yield and total biomass of a rice crop. Transactions of the American Society of Agricultural and Biological Engineers 53: 21–27.

Solorzano-Arroyo, J. A., H. Mena, R. Romero, J. Trevino, J. Gilles, C. Geden, C. Taylor, H. Skovgaard. 2017. Biological control of livestock pest biting fly *Stomoxys calcitrans* at agriculture pineapple residues using the parasitoid *Spalangia endius* reared on irradiated Mediterranean fruit fly: Assessment of parasitism in field and laboratory in Costa Rica. Abstracts of the Third FAO/IAEA International Conference on Area-wide Management of Insect Pests. May 22–26, 2017, Vienna, Austria.

The Economist. 2017. The future of drones depends on regulation, not just technology. Engineers and regulators will have to work together to ensure safety as drones take to the sky. Technology Quarterly June 10, 2017. The Economist, London, UK.

Vreysen, M. J. B., A. S. Robinson, and J. Hendrichs (eds.). 2007. Area-wide control of insect pests: From research to field implementation. Springer, Dordrecht, The Netherlands. 789 pp.

Wikipedia. 2019. Unmanned aerial vehicle.

Yamaha. 2017. RMAX Type IG/Type II unmanned helicopter. Yamaha Motor Corporation, Iwata, Shizuoka, Japan.

Zhang, C., and J. M. Kovacs. 2012. The application of small unmanned aerial systems for precision agriculture: A review. Precision Agriculture 13: 693–712.

Zhang, C., D. Walters, and J. M. Kovacs. 2014. Applications of low altitude remote sensing in agriculture upon farmers' requests – A case study in northeastern Ontario, Canada. PLoS One 9(11): e112894.

Enterobacter: ONE BACTERIUM - MULTIPLE FUNCTIONS

P. M. STATHOPOULOU*, E. ASIMAKIS* AND G. TSIAMIS

Department of Environmental Engineering, University of Patras, 30100, Agrinio, Greece; gtsiamis@upatras.gr
**Both authors contributed equally to this work*

SUMMARY

Insects represent the most successful taxon of eukaryotic life, being able to colonize almost all environments. Microbial symbiomes associated with insects, impact important physiologies, and influence nutritional and immune system status, and ultimately, fitness. A variety of bacterial phyla are commonly present in insect guts, including Gammaproteobacteria, Alphaproteobacteria, Betaproteobacteria, Bacteroidetes, Firmicutes, Clostridia, Spirochetes, Verrucomicrobia, Actinobacteria, and others. Among them, the genus *Enterobacter* has been recognized as a dominant inhabitant of the gut for several important insect species, indicating an essential functional role for this taxon. Here, we review the known *Enterobacter* functional diversity among insects with respect to insect development, host exploitation, reproduction, and interactions with other organisms, in an attempt to provide an overview of the traits that have resulted in their evolutionary success. Many strains of *Enterobacter* species are not simply insect commensals but confer beneficial traits to their hosts that primarily fall into two categories: provision and degradation of nutrients and protection from pathogens.

Key Words: Microbial symbiomes, endosymbiotic bacteria, bacterial functional diversity, mutualisms, commensalisms, pathogenic relationships, nutrient degradation, nutrient provision, pathogen protection

1. INTRODUCTION

Microorganisms are well recognized as essential members of the biosphere. Over billions of years they have evolved into every conceivable niche on the planet. Microorganisms reshaped the oceans and atmosphere and gave rise to conditions conducive to the development of multicellular organisms (Gibbons and Gilbert 2015). Microbial diversity and distribution were mostly inaccessible and underestimated before the advent of molecular fingerprinting and high-throughput sequencing technologies, which now allow us to circumvent culture-based approaches (Fierer and Lennon 2011; Whitman et al. 1998).

J. Hendrichs, R. Pereira and M. J. B. Vreysen (eds.), Area-Wide Integrated Pest Management: Development and Field Application, pp. 917–945. CRC Press, Boca Raton, Florida, USA.

Bacteria and archaea estimates continue to rise within an increasing number of environments sequenced using advanced molecular techniques, with the number of existing microbial species predicted to reach millions (Brown et al. 2015; Williams et al. 2017; Zaremba-Niedzwiedzka et al. 2017).

The term symbiosis originates from the Greek word "symbioun" meaning "to live together"; it was defined by Anton de Bary in 1879 as "the living together of two dissimilar organisms, usually in intimate association, and usually to the benefit of at least one partner" (Bary 1879). Symbiotic bacteria are omnipresent in all types of ecosystems, having a significant impact on eukaryotic evolution and diversity (Martin and Schwab 2012; McFall-Ngai 2007; Ruby et al. 2004). Although considerable attention has been given to pathogenic bacteria, pathogens are a minute minority of animal symbionts.

Microbial symbioses are generally categorized as parasitism, commensalism, or mutualism, though some relationships may wander across these defined boundaries depending on evolutionary processes, changes in environmental conditions and/or health state of the host/symbionts (Webster 2014). Host development, defence, nutrient assimilation and disease, in humans and other animals, is influenced by microbial mutualisms, commensalisms and pathogenic relationships (Eren et al. 2015). The development of next-generation DNA sequencing platforms has allowed an in-depth understanding of the composition of the microbial populations inhabiting different hosts and symbiosis research has focused on associations that (i) have economic importance, (ii) have implications for human health, or (iii) offer ecologically fascinating insights (Webster 2014).

It is estimated that the majority of members of the largest class of invertebrates, i.e. Insecta, are involved in some type of symbiosis, with most of these relations being shared with bacteria. Microorganisms can colonize the insect exoskeleton, the gut and haemocoel, and are present within some insect cells. The microbiota may account for 1–10% of the insect biomass, implying that the insect, as well as any other higher organism, can be regarded as a multi-organismal entity (Douglas 2015). Bacterial symbionts are equally prevalent in mammals and insects; however, the bacterial diversity in insect digestive tracts is generally low and rarely exceeds a few tens of species (Colman et al. 2012). Several immunological, physiological and morphological hypotheses have been proposed to explain that fact (Broderick and Lemaitre 2012; Engel and Moran 2013). The lack of adaptive immune function in invertebrates might partly explain this low diversity (McFall-Ngai 2007).

2. INSECT SYMBIONTS

Insect-associated microorganisms cover an immense range of functions and are known to upgrade nutrient-poor diets; aid digestion of recalcitrant food components; protect from predators, parasites, and pathogens; contribute to inter- and intraspecific communication; affect efficiency of disease vectors; and govern mating and reproductive systems (Engel and Moran 2013; Gil and Latorre 2019).

2.1. *Obligate Mutualistic Symbionts*

Insects that depend exclusively on nutritionally restricted diets such as plant sap, vertebrate blood and woody material, commonly possess obligate mutualistic symbionts that are involved in the provision of essential nutrients or the degradation of food materials (Engel and Moran 2013; Engel et al. 2016; Latorre and Manzano-Marín 2017). At least 10% of all insect species depend on obligate nutritional symbioses where bacteria are required to synthesize nutrients that are absent in the diets of their hosts (Wernegreen 2004). Aphids provide an example of such obligate endosymbiosis since all aphids harbour endosymbiotic bacteria (i.e. microorganisms that live inside host cells or tissues) of the genus *Buchnera* in specialized host cells called 'bacteriocytes' (Manzano-Marín et al. 2016). Other obligatory endosymbionts include *Wigglesworthia* in tsetse flies (Akman et al. 2002), *Blochmannia* in carpenter ants (Williams and Wernegreen 2010), *Baumannia* in sharpshooters (Wu et al. 2006), *Carsonella* in psyllids (Nakabachi et al. 2006) and *Tremblaya* in mealybugs (Szabó et al. 2017). In a broader definition in the same category, we can include *Candidatus* Erwinia dacicola in the olive fruit fly *Bactrocera oleae* (Capuzzo et al. 2005; Ben-Yosef et al. 2014). These intracellular mutualists commonly have the following biological features: (a) they are located inside bacteriocytes, (b) are essential for fitness, (c) are transmitted maternally, and (d) display strict host-symbiont co-evolutionary patterns (Bourtzis and Miller 2008).

2.2. *"Facultative"* Symbiotic *Bacteria*

In addition to obligate symbionts, many insects harbour "facultative" symbiotic bacteria, which are not essential for either host survival or reproduction and are typically maintained with a patchy distribution in host populations. Some facultative symbionts, like *Wolbachia*, *Spiroplasma* and *Cardinium*, are famous as reproductive manipulators in insects, affecting host reproduction by inducing male-killing, feminization, parthenogenesis or cytoplasmic incompatibility (Zchori-Fein and Perlman 2004; Doudoumis et al. 2013, 2017; Mateos et al. 2018). It has been speculated that these reproductive manipulators are not only harmful agents but also accelerators of host speciation due to reproductive isolation between infected and uninfected hosts. At the same time, *Wolbachia* has been shown, under certain conditions, to protect their hosts from insect pathogenic viruses as well as to prevent the establishment and transmission of major human pathogens like dengue and chikungunya viruses and the malaria parasite *Plasmodium* (Schmidt et al. 2017; Tan et al. 2017; Ant et al. 2018).

2.3. *Gut Microbiota*

Research conducted mainly in the last ten years, has resulted in tremendous progress in the field of gut microbiota and their impact on host metabolism. In general, gut microorganisms provide several nutritional functions to their hosts, and in return, hosts provide symbiotic microorganisms with a stable, protected, and nutrient-rich environment (Kohl et al. 2014). All animals assemble and maintain a diverse but host-specific gut microbial community. Aside from the ubiquitous gut microflora in vertebrates, numerous invertebrates harbour endosymbiotic microorganisms inside their body cavity (Feldhaar 2011). It is believed that the gut microbiota can be considered as a bacterial organ, which is integrated into the biological system of the host (Sandeva et al. 2018). Gut microorganisms are also considered endosymbionts (Moya et al. 2008), although they are localized extracellularly within the gut lumen and their persistence within the gut could range from transient visitors to permanent inhabitants (Feldhaar 2011). Gut microorganisms produce a diverse metabolite repertoire from the anaerobic fermentation of undigested dietary components that reach the colon, and from endogenous compounds that are generated by the microorganisms themselves and their hosts (Rooks and Garrett 2016). Increased availability of technologies that profile microbial communities revealed that gut microorganisms are distinct from those of other characterized habitats in the biosphere, which indicates that strong selective forces differentiated gut-dwelling bacteria regardless of their lineage (Ley et al. 2008).

The basic structure of the digestive tract is similar across insects, although they possess a diversity of modifications associated with adaptations to specialized niches and feeding habits, and many of these specializations have evolved for housing gut microorganisms in specific gut compartments (Engel and Moran 2013; Pereira and Berry 2017). A variety of bacterial phyla are commonly present in insect guts, including Gammaproteobacteria, Alphaproteobacteria, Betaproteobacteria, Bacteroidetes, Firmicutes, including *Lactobacillus* and *Bacillus* species, Clostridia, Spirochetes, Verrucomicrobia, Actinobacteria, and others (Colman et al. 2012). The contribution of microorganisms, particularly gut microorganisms, to insect function is highly relevant from several perspectives, relevant to medicine, agriculture, and ecology (Douglas 2015).

2.4. Unveiling the Black Box of Symbiotic Associations

Characterization, exploitation and management of the insect-microbial symbiotic associations can contribute significantly to the control of agricultural pests and disease vectors. Recent advances in "omics", such as metagenomics, metatranscriptomics, and metaproteomics, have gradually unveiled the black box of symbiotic systems. However, capturing the properties of insect microbiota has been challenging due to the high variability in composition between individuals and closely related species (Pernice et al. 2014). Several biological and ecological factors such as age, genetics and environment have been proposed to explain insect microbiota composition. However, diet is one of the main factors driving variation in the composition of the gut microbiota in vertebrates and invertebrates. In insects, host diet composition has been shown to shape the microbiota composition regardless of taxonomy and geography of the specimens (Colman et al. 2012; Chandler et al. 2011; Yun et al. 2014).

Food itself can be a vector of commensals, and different diets will provide microbial inoculates of different community compositions. Also, one major difference between the human and insect microbiota is that the majority of the insect bacteria seem to be aerobic and therefore capable of digesting food outside the insect. This function may contribute to explaining why host diet is such a key driver of microbiota composition and explain why bacteria commonly associated with insects are very rare in diverse mammalian species and vice versa (Chandler et al. 2011; Pernice et al. 2014). Some insect-associated microorganisms are related to microbial taxa found in other animals, e.g. Enterobacteriaceae and other Gammaproteobacteria, Firmicutes and Bacteroidetes, but others are absent from any other animal and the external environment (e.g. many protists of the class Parabasalia are found exclusively within termites (Brune and Dietrich 2015)).

3. *ENTEROBACTER* A DOMINANT TAXON OF GUT MICROBIOTA WITH MULTIPLE ROLES

Members of the Enterobacteriaceae family are facultatively anaerobic, Gram-negative rods that are catalase-positive and oxidase-negative (Brenner et al. 1984). Currently, the family comprises 51 genera (Table 1) and 238 species. The number of species displays a wide divergence with a range of 1 to 22 per genus. Twenty-two genera contain only one species, while seven genera have more than ten species (Octavia and Lan 2014).

*Table 1. The genera (51) of the family Enterobacteriaceae**

Arsenophonus (1991), *Biostraticola (2008), Brenneria (1999),* ***Buchnera (1991)****, Budvicia (1985), Buttiauxella (1982), Cedecea (1981),* ***Citrobacter (1932),*** *Cosenzaea (2011), Cronobacter (2008), Dickeya (2005), Edwardsiella (1965),* ***Enterobacter (1960)****, Erwinia (1920),* ***Escherichia (1919),*** *Ewingella (1984), Gibbsiella (2011), Hafnia (1954),* ***Klebsiella (1885), Kluyvera (1981),*** *Leclercia (1987), Leminorella (1985), Lonsdalea (2012), Mangrovibacter (2010), Moellerella (1984),* ***Morganella (1943)****, Obesumbacterium (1963),* ***Pantoea (1989), Pectobacterium (1945),*** *Phaseolibacter (2013), Photorhabdus (1993), Plesiomonas (1962), Pragia (1988),* ***Proteus (1885), Providencia (1962),*** *Rahnella (1981),* ***Raoultella (2001),*** *Saccharobacter (1990),* ***Salmonella (1990),*** *Samsonia (2001),* ***Serratia (1823), Shigella (1919),*** *Shimwellia (2010),* ***Sodalis (1999),*** *Tatumella (1982), Thorsellia (2006), Trabulsiella (1992),* ***Wigglesworthia (1995), Xenorhabdus (1979),*** *Yersinia (1944), and Yokenella (1985).*

**The year the genus was proposed is listed in parentheses after the genus name; also, the genera of the family that have been characterized in insects are highlighted with bold text*

Numerous applications use members of Enterobacteriaceae, including biocontrol in agriculture, production of recombinant proteins and non-protein products, control of infectious diseases, anticancer agents, biowaste recycling, and bioremediation (Watanabe et al. 2000; Zhu et al. 2011; Zhang et al. 2012; Zhao et al. 2012).

The family Enterobacteriaceae is ubiquitous, and many species can exist as free-living in diverse ecological niches, both terrestrial and aquatic, and some are associated with animals, plants, or insects only (Octavia and Lan 2014). For convenience, the members are broadly categorized into three types: (1) those that can cause human infections or are primarily associated with human/animals and the environment, (2) those that are associated with plants or plant pathogens and the environment, and (3) those that are associated with insects or are endosymbionts.

The Enterobacteriaceae (Proteobacteria) are considered as one of the most dominant bacteria families associated with insects (Drew and Lloyd 1987; Behar et al. 2005; Jurkevitch 2011; Rizzi et al. 2013; Wang et al. 2014; Cambon et al. 2018). They gave rise to a variety of symbiotic forms, from the loosely associated commensals, often designated as secondary (S) symbionts, to obligate mutualists, called primary (P) symbionts (Husník et al. 2011). Many enterobacteria are also opportunistic pathogens.

The genus *Enterobacter* was first described by Hormaeche and Edwards (1960) and was named for the organisms' predominant natural habitat, the intestines of animals (from Greek *enteron*, meaning "intestine"). *Enterobacter* is a genus of common Gram-negative, facultatively anaerobic, rod-shaped, non-spore-forming bacteria of the family Enterobacteriaceae. The genus is polyphyletic based on the 16S rDNA sequence, with 14 lineages scattered across the 16S rDNA tree.

The taxonomy of *Enterobacter* has a complicated history, with several species transferred to and from this genus. The older species were assigned to the genus based on DNA-DNA hybridization values and phenotypic data; whereas the more recently described taxa rely on 16S rRNA gene- and *rpoB*-sequencing for genus allocation. It has been previously acknowledged that *Enterobacter* contains species which should be transferred to other genera. The polyphyletic nature of the genus makes it difficult to assign novel species to *Enterobacter* unless the strains cluster with the type species (*E. cloacae*) of the genus (Brady et al. 2013).

Members of the *Enterobacter* can be found in soil, water, sewage, vegetable and fruits, plants, terrestrial and aquatic environments. They can be isolated from the intestinal tracts of humans and other animals as commensals, but they are also significant human pathogens (Hoffmann and Roggenkamp 2003; Hoffmann et al. 2005; Davin-Regli and Pagès 2015).

Interestingly, *Enterobacter* spp. have been recognized as inhabitants of the gut of several insect species, suggesting that this genus may play various and vital roles (Vasanthakumar et al. 2006; Geiger et al. 2009; Jiang et al. 2012; Gujjar et al. 2017). In this respect, we reviewed the known *Enterobacter* diversity among insects in relation to the functional role in insect development, host exploitation, reproduction, and interactions with other organisms, in order to recognise the traits that have resulted in their evolutionary success. Many *Enterobacter* strains are not simply insect commensals but confer beneficial traits to their hosts that primarily fall into two categories: provision/degradation of nutrients and protection from pathogens.

3.1. Enterobacter *and Nutrient Bioavailability*

Many insects derive nutritional advantage from persistent associations with microorganisms that variously synthesize essential nutrients or digest and detoxify ingested food (Douglas 2009). These persistent relationships are symbioses, and the species of the genus *Enterobacter* often plays a leading role (Table 2). Endosymbiotic *Enterobacter* species add nutritional supplements to their host diet, most of the times in cooperation with other members of the insect gut microbiota, like *Klebsiella, Pectobacterium* or *Citrobacter*. In these interactions, the primary nutrient contributor may switch between the various bacterial species of these genera.

Table 2. Enterobacter *species provide nutritional supplements to their host diet*

Species	Strain Accession Number	Host		Function	Reference
E. agglomerans	Not available	*Coptotermes formosanus* Shiraki	Formosan subterranean termite	Nitrogen-fixing	Potrikus And Breznak 1977
E. sp.	Not available	*Coptotermes acinaciformis* (Frogatt), *Coptotermes lacteus* (Froggatt), *Cryptotennes primus* Hill, *Heterotermes ferox* (Froggatt), *Mastotennes darwiniensis* Froggatt, *Nasutitermes graveolus* (Hill), *Schedorhinotermes intermedius* (Brauer)	Termites	Nitrogen-fixing	Eutick et al. 1978
E. agglomerans	Not available	*Dendroctonus terebrans* (Olivier)	Bark beetles	Nitrogen-fixing	Bridges 1981
*E.*sp.		*D. frontalis* Zimmermann			
E. aerogenes		*Ips avulsus* (Eichhoff)			
E. cloacae	Not available	*Bactrocera tryoni (Froggatt)*	Queensland fruit fly	Nitrogen-fixing	Drew and Lloyd 1987; Murphy et al. 1994
E. sp.	Not available	*Anoplophora chinensis (Forster)*	Wood-boring beetle	Nitrogen-fixing	Rizzi et al. 2013
E. sp.	AB673459 AB673457	*Reticulitermes speratus (Kolbe),*	Termites	Recycling of uric acid nitrogen	Thong-On et al. 2012
	AB673458 AB673456	*Glyptotermes fuscus* (Oshima), *Cryptotermes domesticus* (Haviland)			
E. sp.	AY847157 AY847162	*Ceratitis capitata* (Wiedemann)	Mediterranean fruit fly	Nitrogen-fixing	Behar et al. 2005
E. hormaechei	KC759162	*Reticulitermes chinensis* (Snyder)	Termite	Lignin degradation	Zhou et al. 2017
E. sp.	JQ864378	*Reticulitermes chinensis* (Snyder)	Termite	Recycling of uric acid nitrogen	Fang et al. 2016
	JX291546				
E. aerogenes	Not available	*Mastotermes darwiniensis* Froggatt	Termite	Lignin degradation	Kuhnigk et al. 1994
E. cloacae	Not available	*Reticulitermes flavipes* (Kollar)			

E. aerogenes	Not available	*Mastotermes darwiniensis* Froggatt	Termites	Hemicellulose degradation	Schäfer et al. 1996
E. sakazaki	Not available	*Reticulitermes flavipes* (Kollar)			
E. aerogenes	Not available	*Coptotermes formosanus* Shiraki	Formosan subterra-nean termite	Cellulose/Hemi-cellulose degradation	Adams and Boopathy 2005
E. cloacae	Not available				
E. aerogenes	EU305608	*Coptotermes curvignathus* (Holmgren)	Termite	Lignocellulose degradation	Ramin et al. 2008
E. cloacae	EU305609				
E. cloacae	MF185378	*Rhynchophorus ferrugineus* (Olivier)	Red palm weevil	Cellulose degradation	Muhammad et al. 2017
E. sp.	JN167548	*Rhynchophorus ferrugineus* (Olivier)	Red palm weevil	Cellulose/Hemi-cellulose degradation	Butera et al. 2012
E. sp.	LN829595	Coleoptera	Wood-feeding beetle larvae	Lignocellulose degradation	Manfredi et al. 2015
E. sp.	HM771088	Cerambycidae, Elateridae,	Coleoptera larvae	Cellulose/Hemi-cellulose degradation	Rojas-Jiménez and Hernández 2015
	HM771065 HM771070 HM771068 HM771104 HM771106 HM771101 HM771061 HM771048 GU827537	Passalidae and Scarabaeidae			
E. sp.	HM235482 HM235485 HM235491 HM235497	*Sirex noctilio* Fabricius	Wood-feeding wasp	Cellulose degradation	Adams et al. 2011
E. asburiae, E. cloacae, E. sp.	KT957438-43	*Plutella xylostella* (Linnaeus)	Diamond-back moth	Cellulose/Hemi-cellulose degradation	Xia et al. 2017
E. sp.	Not available	*Diestrammena asynamora* (Adelung)	Greenhouse camel cricket	Cellulose/Hemi-cellulose degradation	Mathews et al. 2019
E. sp.	AB872360-384	*Holotrichia parallela* (Motschulsky)	Dark black chafer beetle	Cellulose/Hemi-cellulose degradation	Sheng et al. 2015
E. cloacae	Not available	*Coptotermes curvignathus* (Holmgren)	Termite	Cellulose degradation	Toczyłowska-Mamińska et al. 2015
E. cloacae	KC512248	*Plutella xylostella* (Linnaeus)	Diamond-back moth	Trehalose hydrolysis	Adhav et al. 2019

E. cloacae	Not available	*Thermobia domestica* (Packard)	Firebrat	Water and nutrients	Woodbury and Gries 2013a,b
E. sp.	LJAN0100 0071 locus_taq:A C520_4801	*Bactrocera oleae* (Rossi)	Olive fruit fly	Amino acids and vitamins	Estes et al. 2018a
E. sp.	KR232642	*Ceratitis capitata* (Wiedemann)	Mediterra-nean fruit fly	Probiotic	Augustinos et al. 2015
E. cloacae, E. asburiae	KY810513 KY810514	*Ceratitis capitata* (Wiedemann)	Mediterra-nean fruit fly	Probiotic	Hamden et al. 2013
E.agglome-rans	Not available	*Ceratitis capitata* (Wiedemann)	Mediterra-nean fruit fly	Probiotic	Niyazi et al. 2004
E. sp.	Not available	*Zeugodacus cucurbitae* (Coquillett)	Melon fly	Probiotic	Yao et al. 2017
E. sakazakii	DQ228421-22	*Stomoxys calcitrans* Linnaeus	Stable fly	Probiotic	Mramba et al. 2007
E. cloacae	Not available	*Plutella xylostella* (Linnaeus)	Diamond-back moth	Probiotic	Somerville et al. 2019

3.1.1. Nitrogen Fixation

A well-established contribution of *Enterobacter* species to nutrition is nitrogen fixation. Nitrogen fixation is a metabolic capability that is absent from the ancestral eukaryote and has been acquired by multiple eukaryotic groups through symbiosis. Many insects are known to live on low nitrogen diets, and microorganisms have been suggested to provide availability of these essential foodstuffs in various ways (Douglas 2009).

Potrikus and Breznak (1977) identified two *Enterobacter agglomerans* strains from the guts of Formosan subterranean termites (*Coptotermes formosanus*). Nitrogen fixation appears to play an essential role in termite biology by helping them overcome deficiencies related to their nitrogen-poor diet (wood). The process was found to be linked to the termite gut bacteria since antibiotic treatment eliminated the function (Breznak et al. 1973). Strains C-1 and C-2 were characterized with electron microscopy and numerous biochemical assays, including sugar fermentation tests. The nitrogen-fixing ability of the strains was verified with acetylene reduction tests, in both aerobic and anaerobic (100% N_2 or 100% Ar) growing conditions, using different sources of nitrogen (peptone, NH_4Cl, KNO_3). In media lacking peptone, NH_4Cl or KNO_3, nitrogen fixation by both strains occurred only under anaerobic conditions when 100% N_2 was provided. When peptone, NH_4Cl, or KNO_3 was provided in the media, fixation was taking place in both aerobic and anaerobic conditions (Potrikus and Breznak 1977).

Nitrogen fixation by *Enterobacter* species was also identified in the gut of termite species from Australia (in seven out of nine species tested) (Eutick et al. 1978). The *Enterobacter* strains could grow on nitrogen-free media under anaerobic conditions, but also under aerobic conditions in the presence of H_2SO_4 (Eutick et al. 1978).

Some years later, nitrogen-fixing *Enterobacter agglomerans* and *Enterobacter* spp. were isolated from various species of bark beetles, including *Dendroctonus terebrans*, *D. frontalis* and *Ips avulsus* (Bridges 1981). Even though large populations of nitrogen-fixing bacteria were found in the beetles, the *in-situ* assays in larvae did not reveal any acetylene reduction (Bridges 1981).

In the wood-boring beetle *Anoplophora chinensis*, enrichment studies of adult gut homogenates for nitrogen-fixing revealed the presence of four *Enterobacter* strains (Rizzi et al. 2013). The adult insects used in the enrichment cultures were isolated from *Acer saccharinum* L. and *Alnus* host trees (Rizzi et al. 2013).

Many microorganisms are valuable to the insect for their more comprehensive metabolic capabilities, including their capacity to utilize insect nitrogenous waste compounds (e.g. uric acid), synthesize 'high value' nitrogenous compounds (e.g. essential amino acids) and fix nitrogen (Douglas 2009). Uric acid is another substance utilized as a nitrogen source in insects feeding on nitrogen-poor diets composed of plant material (Potrikus and Breznak 1981). In such a case, four uricolytic *Enterobacter* strains were isolated from the guts of three wood-feeding termite species (*Reticulitermes speratus*, *Glyptotermes fuscus* and *Cryptotermes domesticus*) (Thong-On et al. 2012). Bacteria with uricolytic activity were grown anaerobically on plates containing various concentrations of uric acid. Isolated strains, RsN-1, GfU-1, Cd20b and Cd15a, showed >98% identity with *Enterobacter amnigenus*, *E. aerogenes*, *E. asburiae* and *E. cowanii*, respectively (Thong-On et al. 2012).

Enterobacter cloacae strains isolated from field-collected, and laboratory-reared *Bactrocera tryoni* (Froggatt) fruit flies showed dinitrogenase activity and were able to fix atmospheric nitrogen (Drew and Lloyd 1987; Murphy et al. 1994). Diazotrophic *Enterobacter* strains were also isolated from gut tissue of field-collected Mediterranean fruit flies *Ceratitis capitata* (Behar et al. 2005; Augustinos et al. 2021). However, the most potent nitrogen-fixing effect was produced by *Klebsiella* and *Citrobacter* species. Nitrogen fixation activity was estimated by a variety of methods, including acetylene reduction assays in live flies and bacteria grown on nitrogen-deficient media, amplification of the nitrogenase iron protein gene (*nifH*) from gut extracts and isolated colonies, as well as *in situ* expression and amplification of the *nifH* gene (Behar et al. 2005).

3.1.2. Degradation of Plant Cell Wall Components

Various *Enterobacter* strains are also able to provide nutrients to insects through the degradation of plant cell wall components (lignin, cellulose, hemicellulose) (König et al. 2006). Lignocellulolytic *Enterobacter* species have been identified in plant or wood-feeding insects, like beetles, termites or wasps.

A ligninolytic *Enterobacter* strain was isolated from abdomens of adult *Reticulitermes chinensis* termites that use the plant cell wall polysaccharide in their diet (Zhou et al. 2017). Strain PY12, classified as *E. hormaechei* based on its 16S rRNA sequence, produced a lignin-modifying peroxidase (LiP), a key component in the lignin degradation pathway (Zhou et al. 2017; Janusz et al. 2017). The decolourization of six different dyes determined Lip activity by spectrophotometry. Since the study focused only on one of the ligninolytic enzymes in the pathway, it was not entirely possible to elucidate how strain PY12 contributes to lignin degradation and the possible synergies it develops (Zhou et al. 2017).

Seven *Enterobacter* strains with the ability to degrade lignin and related aromatic compounds were identified in hindguts of various laboratory-reared and field-collected species of termites (Kuhnigk et al. 1994). Degradation was observed under aerobic conditions, while in the absence of oxygen, only slight modifications of the side group of aromatic compounds occurred (Kuhnigk et al. 1994). The strains were characterized as *E. aerogenes* (Km3 and KAn8), *E. cloacae* (Rt5a, Rt5b and Rt5c) and *Enterobacter* sp. (Rt3a and Rt3b). The first two were isolated from *Mastotermes darwiniensis* and the remaining five from *Reticulitermes flavipes* termites. Among the seven strains, Km3 and KAn8 exhibited the most significant degradation potential, with the ability to degrade nine out of 13 substrates tested (Kuhnigk et al. 1994). Apart from degrading lignin, strain Km3 also showed hemicellulose degrading activity (Schäfer et al. 1996). Hemicellulose was also degraded by *Enterobacter sakazakii* strain RA2 that was isolated from hindguts of wild *R. flavipes* termites from France (Schäfer et al. 1996). Interestingly, strain RA2 exhibited all (four out of four) enzyme activities that were tested and were related to polysaccharide degradation. On the other hand, strain Km3 lacked 1,4-β-xylanase activity (Schäfer et al. 1996). In a particular environment, though, bacteria lacking members of the metabolic cascade could act synergistically with other strains present to fully degrade substances.

The Formosan subterranean termite, *C. formosanus* contained *E. aerogenes* and *E. cloacae* strains that effectively utilized xylose as a carbon source (Adams and Boopathy 2005). Two strains isolated from *Coptotermes curvignathus* termites from Malaysia and identified as *E. aerogenes* and *E. cloacae* were able to degrade carboxymethylcellulose (CMC) and cellobiose (Ramin et al. 2008). Other lignocellulolytic strains identified in higher and lower termites belonged to the species *E. aerogenes*, *E. agglomerans* and *E. cloacae* (König et al. 2006; Mannesmann and Piechowski 1989).

3.1.3. Degradation and Biosynthesis of Other Nutrients

In the diamondback moth *Plutella xylostella*, Adhav et al. (2019) characterized, structurally and biochemically, a trehalase from *Enterobacter cloacae* which assists in the hydrolysis of trehalose, a significant energy source in insect metabolism. Firebrat nymphs (*Thermobia domestica*) acquire water and nutrients by consuming *Enterobacter cloacae*, facilitating their growth and survival in the absence of food or water for up to 22 days (Woodbury and Gries 2013a,b).

Genome sequencing of *Enterobacter* sp. OLF (Table 3), isolated from wild California olive fruit flies, revealed genes related to the biosynthesis of amino acids, vitamins and co-factors, degradation pathways, nitrogen metabolism, as well as the production of energy and precursor metabolites (Estes et al. 2018a, 2018b).

This characterization could mean that the strain could potentially supply its host, in cooperation with other symbionts, with amino acids, vitamins or other nutritional compounds missing from the olive fruit diet (Estes et al. 2018a, 2018b).

Table 3. Available genome sequences of Enterobacter *symbionts from insects*

Accession number	*Enterobacter* Strain	Host	Status	Submission Date
PRJNA420145	*E. cancerogenus* CR-Eb1	*Galleria mellonella* (Linnaeus)	complete	2018
PRJNA288712	*E. sp. OLF*	*Bactrocera oleae* (Rossi)	complete	2018
PRJNA340971	*E. Larv1_ips*	*Ips typographus* (Linnaeus)	draft	2019
PRJNA390046	*E.* sp. 10-1	Cerambycidae	draft	2017
PRJNA364290	*E. JKS000234*	Formicidae	draft	2017
PRJNA364289	*E. JKS000233*	Formicidae	draft	2017
PRJNA179500	*E. hormaechei YT2*	*Tenebrio molitor* (Linnaeus)	draft	2012
PRJNA180991	*E. hormaechei* YT3	*Zophobas morio* Fabricius	draft	2013
PRJNA180988	*E. cancerogenus* YZ1	*Tenebrio molitor* (Linnaeus)	draft	2013
PRJNA169065	*E. sp.* Ag1	*Anopheles gambiae* Giles	draft	2012

3.1.4. Probiotic Effects of Enterobacter

The most recent contribution of *Enterobacter* strains to insect fitness and nutrition is their emerging role as probiotics in diets of mass-reared *C. capitata* and *Zeugodacus cucurbitae* flies. Such properties can be particularly useful in Sterile Insect Technique (SIT) applications where irradiation treatment takes a heavy toll on insect competitiveness if the dose administered is high. In this regard, strains that were provided as probiotics in diets greatly improved various fitness parameters of the flies including pupal weight, longevity, adult size, flight ability, and adult emergence (Niyazi et al. 2004; Ben Ami et al. 2009; Hamden et al. 2013; Yuval et al. 2013; Augustinos et al. 2015, 2021; Kyritsis et al. 2017; Yao et al. 2017; Cáceres et al. 2019). For example, the provision of *Enterobacter* sp. strain AA26 resulted in increased production of pupae and adults, as well as reduced rearing duration in various developmental stages (from egg to pupa, pupal stage and from egg to adult), particularly for male *C. capitata* flies (Augustinos et al. 2015). However, it did not affect pupal weight, sex ratio, male mating competitiveness, flight ability or life span under food and water deprivation (Augustinos et al. 2015; Kyritsis et al. 2017).

Hamden et al. (2013) on the other hand observed increased pupal weight, male sexual performance and survival rates under food deprivation when they enriched the larval diet of mass-reared *C. capitata* with *Enterobacter* sp. and other beneficial bacteria (*Klebsiella pneumoniae* and *Citrobacter freundii*). In terms of mating competitiveness, irradiated males reared on probiotics achieved more matings and transferred larger quantities of sperm to females. Finally, in addition to pupal weight, probiotics also resulted in increased body size for adult males (head width, abdomen and thorax length) (Hamden et al. 2013). Niyazi et al. observed a diet-dependent probiotic effect on the mate-calling activity, mating success, life expectancy, and survival of mass-reared male *C. capitata* flies (Niyazi et al. 2004). The probiotic effect of *E. agglomerans* and *K. pneumoniae* on each of these parameters differed among the four diets that were tested (two standard adult diets and two enhanced experimental formulations), ranging from significant to non-existent (Niyazi et al. 2004). Differences in the probiotic effect were also observed between the laboratory trials and the field-cage assay (Niyazi et al. 2004).

Similarly, Yao et al. (2017) tested gut-associated *Enterobacter* spp. as probiotics in the larval diet of irradiated laboratory-reared *Z. cucurbitae* flies. Both live and autoclaved bacteria were tested to distinguish between insect-bacteria interactions and plain nutritional value of the probiotic effect. Application of the live bacteria increased female and male pupal weight, various morphological traits of adult flies, including head width and thorax length, as well as survival (Yao et al. 2017). In the case of survival rate, autoclaved bacteria resulted in a greater increase compared to live bacteria (Yao et al. 2017).

Mramba and colleagues studied the effects of *E. sakazakii* on stable fly *Stomoxys calcitrans* development (Mramba et al. 2007). They observed that sterile media did not support any fly development and that the effect was reversed with the addition of *E. sakazakii*. However, an improved effect for fly development was mainly observed in cooperation with other microbial partners in non-sterilized media. The combination of non-sterile media and *E. sakazakii* resulted in a slight increase in the survival of larvae to the pupal stage, in pupal weight and adult emergence and also displayed increased duration of the larval and pupal stages compared to the non-sterile media without the inoculum with *E. sakazakii* (Mramba et al. 2007).

A similar probiotic effect was observed in transgenic diamondback moth larvae (Somerville et al. 2019). The larvae that were grown on an aseptic diet (sterile diet and addition of streptomycin) and were inoculated with streptomycin-resistant *E. cloacae* JJBC exhibited increased pupal weight and production of progeny compared to an aseptic diet without the inoculum (Somerville et al. 2019).

In *Apis mellifera jemenitica* Ruttner, however, the *E. kobei* strain that was examined for its probiotic effect did not manage to reduce mortality of bee larvae infected with *Paenibacillus* spores (Al-Ghamdi et al. 2018).

3.2. Enterobacter *and Protective Functions*

Insects live in close coalition with microorganisms, which immensely influence their ecology and evolution. Microorganisms, such as bacteria, archaea, fungi, protozoa, viruses, can be associated with their insect host permanently or transiently, and such associations may be beneficial or harmful to the insects' fitness (Gurung et al. 2019). For instance, endosymbionts tend to be dependent on the hosts for obtaining nutrients, whereas they can provide fitness advantages in terms of nutritional components (see above), overcoming host defences, and protection from pathogens, parasites, or environmental stressors (Engel and Moran 2013; Mereghetti et al. 2017).

The insect microbiota are also able to provide protective functions to their hosts, thereby influencing the proliferation of important human or animal pathogens (Table 4), including *Plasmodium*, *Trypanosoma*, dengue, Zika, yellow fever or chikungunya viruses, inside the insect body with a variety of mechanisms, either indirectly, by causing innate responses of the insect immune system or directly, through competition for resources with the pathogen or the production of substances with anti-pathogenic effects (Dong et al. 2009; Moreira et al. 2009; Bian et al. 2010, 2013; Cirimotich et al. 2011b; Walker et al. 2011; Weiss and Aksoy 2011; van den Hurk et al. 2012; Zhang et al. 2013; Dennison et al. 2014; Frentiu et al. 2014; Aliota et al. 2016a, 2016b; Dutra et al. 2016; Tan et al. 2017; Kalappa et al. 2018).

Table 4. Enterobacter *species can provide protective functions to their hosts, aiding them to cope with abiotic and biotic stress*

Enterobacter Species	Strain Accession Number	Host	
Enterobacter sp.	JF690924	*Anopheles arabiensis* Patton	Mosquito
Enterobacter cloacae	Not available	*Anopheles stephensi* Liston	Mosquito
Enterobacter sp.	JQ680715	*Anopheles gambiae* Giles	Mosquito
Enterobacter cloacae, Enterobacter amnigenus	Not available	*Anopheles albimanus* Wiedemann	Mosquito
Enterobacter ludwigii	MF084966 MF084975	*Delia antiqua* (Meigen)	Onion fly
Enterobacter sp.	EU693561 EU693573	*Leptinotarsa decemlineata* Say	Colorado potato beetle
Enterobacter ludwigii, Enterobacter asburiae	KX398658 KX398648 KX398650 KX398654 KX398657	*Helicoverpa zea* (Boddie)	Corn earworm
Enterobacter agglomerans	Not available	*Rhagoletis pomonella* (Walsh)	Apple maggot fly
Enterobacter asburiae, Enterobacter cloacae, Enterobacter sp.	KT957438 KT957439 KT957440 KT957441 KT957442 KT957443	*Plutella xylostella* (Linnaeus)	Diamondback moth
Enterobacter sp.	KX117074	*Leptinotarsa juncta* Germar	False potato beetle
Enterobacter sp.	JX296530 KC977257	*Leptinotarsa decemlineata* Say	Colorado potato beetle
Enterobacter cloacae	KM878728 KM886372 KM878717	*Coptotermes formosanus* Shiraki	Formosan subterranean termite
Enterobacter cloacae	Not available	*Glyphodes duplicalis* Inoue et al.	Mulberry pyralid
Enterobacter amnigenus	Not available	*Anopheles dirus* Peyton & Harrison	Mosquito

These effects continuously draw attention as a means of controlling vector-based infectious diseases, like malaria or dengue virus, that are transmitted by *Anopheles* and *Aedes* mosquito vectors (Cirimotich et al. 2011a; Jayakrishnan et al. 2018; Saldaña et al. 2017).

3.2.1. Enterobacter *Anti-pathogenic Effects in Mosquitoes*

Among various bacterial strains that show anti-pathogenic effects, one of the best characterized is *Enterobacter* strain *Esp_Z*, which was isolated from the midgut of wild *Anopheles arabiensis* mosquitoes in Zambian populations. The *Enterobacter* strain inhibited *Plasmodium* development in the midgut by 98%, 99% and 99% before the ookinete, oocyst, and sporozoite stage respectively (Cirimotich et al. 2011c). This response proved to be independent of the mosquito immune system since anti-*Plasmodium* genes, that are usually triggered during *Plasmodium* infections, showed similar activity in mosquitoes infected with *Esp_Z* or with non-inhibitory bacteria.

Additionally, silencing of anti-*Plasmodium* genes with RNAi in mosquitoes infected with *Esp_Z* did not result in *Plasmodium* oocyst development (Cirimotich et al. 2011c). The response also proved to be independent of bacterial retention of mosquito factors that are essential for *Plasmodium* development like, xanthurenic acid, iron and substances involved in fatty acid metabolism. Moreover, the inhibition effect was dose-dependent, both *in vitro* and *in vivo* conditions, with 10^4 bacterial cells providing near-complete protection against parasite infection and coinciding with active bacteria replication (Cirimotich et al. 2011c).

In view of the above, the inhibition activity was therefore hypothesized to be the result of anti-pathogenic substances. Indeed, the *Enterobacter* strain was found to produce *Plasmodium*-killing (Luckhart et al. 1998; Kumar et al. 2003; Peterson et al. 2007; Molina-Cruz et al. 2008) reactive oxygen species (Cirimotich et al. 2011c). These molecules were not detected in non-inhibitory bacteria, and their effect was neutralized by antioxidants in *Esp_Z*-containing insects, such as vitamin C and glutathione (Cirimotich et al. 2011c).

On the other hand, Eappen et al. (2013) identified *Enterobacter cloacae* strains that inhibited *Plasmodium* development by activating a component of the immune system of *Anopheles stephensi* mosquitoes. More specifically, *E. cloacae* was able to induce the expression of a serine protease inhibitor (SRPN6) in the mosquito midgut. The effect was linked to bacteria that were acquired through feeding and were exposed to the luminal side of the midgut epithelium, whereas bacteria injected in the haemocoel and exposed to the basal surface of the midgut epithelium were incapable of SRPN6 induction (Eappen et al. 2013). Unlike strain *Esp_Z*, the *E. cloacae* strains did not interfere with *Plasmodium* ookinete formation, but with the differentiation of ookinetes into oocysts after they traversed the midgut epithelium. Silencing of the SRPN6 gene resulted in an increase in the number of *Plasmodium* oocysts and increased prevalence of infection, implying that additional immune components are likely to participate in the inhibition process by *E. cloacae* (Eappen et al. 2013).

In field-collected *Anopheles gambiae* mosquitoes from Cameroon, *Enterobacter* spp. bacteria isolated from their midgut reduced *Plasmodium falciparum* oocyst intensity and prevalence by 35% and 15%, respectively. However, the reduction in infection was lower when parasite exposure was higher (Tchioffo et al. 2013).

Enterobacter bacteria that were isolated from the midgut of field-collected *Anopheles albimanus* mosquitoes in Mexico suppressed *Plasmodium vivax* infection in the insectary and field-collected samples (Gonzalez-Ceron et al. 2003). After seven days of treatment, *E. cloacae* and *E. amnigenus* 2 reduced *Plasmodium* infection by 17 and 53 times compared to control mosquitoes, respectively. Additionally, *E. cloacae* infected mosquitoes showed 2.5 times lower *Plasmodium* oocyst density than the controls (Gonzalez-Ceron et al. 2003).

3.2.2. Enterobacter *Effects in Herbivorous Insects*

Two *Enterobacter ludwigii* strains (B424 and B539) that were part of the natural gut microbiota of the onion fly *Delia antiqua* collected from garlic fields in China, showed inhibitory effects against *Beauveria bassiana* (Balsamo), an entomopathogenic fungus that is used as a biocontrol agent in pest management applications (Valero-Jiménez et al. 2014; Zhou et al. 2019). The vital role of the bacterial symbionts of *D. antiqua* on the larvicidal potential of *B. bassiana*, was determined by first examining axenic (bacteria-free) and non-axenic larvae infected with the fungus (Zhou et al. 2019). In this case, treatment with the pathogen did not affect the survival of non-axenic larvae but indicated a significant reduction in the survival of axenic larvae. Furthermore, axenic larvae inoculated with microbiota were characterized by significantly higher survival rates than non-inoculated axenic larvae, when both were infected with the fungus (Zhou et al. 2019). Following the above tests, symbiotic bacteria were isolated from the surface and guts of the onion fly, and certain strains were tested for their anti-fungal effect. The *Enterobacter* strains strongly inhibited (ca. 70-99%) conidia germination and the hyphal growth of the entomopathogen fungus *B. bassiana*. Additionally, as expected from the previous tests, inoculation of axenic larvae with strain B424 significantly increased survival rate against the fungal infection (Zhou et al. 2019).

Similar observations were made with the Colorado potato beetle *Leptinotarsa decemlineata* fourth-instar larvae that were field-collected in Maryland and Virginia, USA (Blackburn et al. 2008). *Enterobacter* sp. strains 2B1C and 2B2D that were isolated from larval faecal fluid were found to inhibit two entomopathogens, *Photorhabdus temperata* Fischer-Le Saux et al. and the fungus *B. bassiana*. *P. temperata* is a bacterial symbiont of the entomopathogenic nematode *Heterorhabditis marelatus* Liu & Berry which can be used to suppress the Colorado potato beetle. However, in the presence of the *Enterobacter* strains the nematode fails to complete its reproductive cycle in the beetle, possibly due to the inhibition of *P. temperata* (Armer et al. 2004). The *in vitro* tests were based on comparing on PP3 agar plates (Bacto® Proteose Peptone) the growth of *P. temperata* and *B. bassiana* conidia in the presence of each *Enterobacter* strain. Both strains reduced the growth of *P. temperata* by approximately 33% while at the same time, strain 2B1C and 2B2D reduced fungal growth by almost 80% and 42.5% respectively (Blackburn et al. 2008).

Enterobacter strains have also been recognized as essential factors in shaping interactions between herbivorous insects and plants, by either activating or suppressing plant defences. In such a case of activation, laboratory-reared *Helicoverpa zea* larvae inoculated orally with an *Enterobacter ludwigii* strain were found to activate tomato plant *Solanum lycopersicum* defences (Wang et al. 2017). The strain was isolated from oral secretions, the regurgitant, of field-collected *H. zea* larvae in Rock Springs, Pennsylvania, USA. The *E. ludwigii* strain along with an *E. asburiae* strain were found to increase the activity of glucose oxidase (GOX), a molecular signal that induces plant defences in response to herbivory, in the labial glands of laboratory-reared larvae (Wang et al. 2017). Even though direct application of both strains on wounded tomato plants resulted in suppression of polyphenol oxidase (PPO) activity, a plant defence component which is regulated by jasmonic acid (JA), when tomato plants were damaged by *E. ludwigii*-inoculated larvae, they showed an increase in JA-regulated genes (proteinase inhibitor II (*Pin2*), cysteine proteinase inhibitor (*CysPI*) and polyphenol oxidase F (PPOF)), and a suppression in salicylic acid (SA) response (pathogenesis-related protein 1 (P4), *Pr1a* (P4)). Suppression of JA-mediated response proved to be dose-dependent, with high bacteria concentrations producing significant results when applied directly on the plant.

In the diamondback moth, two *Enterobacter* species, *E. asburiae* and *E. cloacae*, encoded all the genes related to the aerobic pathway for catechol degradation and could significantly degrade phenol *in vitro* (Xia et al. 2017). These compounds are important plant defence compounds of *Brassica* plants as they are toxic to insects (Xia et al. 2017).

In a contrary case of suppressing plant defences, in antibiotic-treated laboratory-reared false potato beetle (*Leptinotarsa juncta*) larvae, *Enterobacter* inoculation reduced JA-regulated gene expression (PPO) in tomato leaves (non-preferred host, *S. lycopersicum*), but did not have any effect on horsenettle (preferred host, *Solanum carolinense* Linnaeus) (Wang et al. 2016). Damage to leaves by untreated false potato beetle larvae or application of their oral secretions to wounded leaves resulted in reduced PPO activity in both plants compared to antibiotic-treated larvae (Wang et al. 2016). Additionally, direct application of *Enterobacter* on wounded leaves suppressed JA-regulated PPO and *CysPI* expression and increased SA-regulated *Pr1* expression in tomato, but not in horsenettle (Wang et al. 2016).

Similarly, Chung et al. (2013) observed that damage caused to tomato plants by antibiotic-treated Colorado potato beetle larvae increased JA-regulated gene expression (*CysPI* and *PPOF/B*) and a decrease in the SA defence response (*Pr1* (P4)) compared to untreated larvae. Additionally, lower *PPO* activity was observed when oral secretions of untreated larvae were applied on mechanically wounded plants. These observations suggested that bacteria in oral secretions of the Colorado potato beetle suppressed the JA-mediated defence mechanism. Indeed, an *Enterobacter* strain that was isolated from larval oral secretions suppressed JA-mediated antiherbivore defence response (PPO) in tomato plants when it was inoculated in antibiotic-treated Colorado potato beetle larvae (Chung et al. 2013).

4. BACTERIAL PARATRANSGENESIS WITH *ENTEROBACTER*

As *Enterobacter* strains form extremely stable relations with insects by constituting a prevalent portion of their gut microbiota, they can function as vectors for the introduction of functional genes that could be useful for multiple applications, including pest population control (Wilke and Marrelli 2015).

Such a system was examined in *Enterobacter cloacae* species isolated from the guts of Formosan subterranean termites in Hawaii (Husseneder and Grace 2005). The bacteria were transformed with a recombinant plasmid containing genes encoding ampicillin resistance and green fluorescent protein (GFP) and were fed to termite workers from three colonies (Husseneder and Grace 2005). The infection was established rapidly, with 90-100% of termite workers containing transformed bacteria 12 h after feeding, which persisted in the population for up to 11 weeks. Moreover, fluorescent microscopy revealed that the ingested bacteria expressed the GFP gene in the termite gut. Transformed bacteria were efficiently transferred through a colony, with infection rates reaching 80–100% after six days, even when the initial ratio between infected and uninfected termites was low. Recombinant bacteria were also transferred into the soil by infected termites but declined rapidly within four to five weeks (Husseneder and Grace 2005).

In another example, *Enterobacter cloacae* strain WBMH-3-CM^r was transformed with plasmids containing the ice nucleation gene *inaA* of *Erwinia ananas* IN10 and were ingested by larvae of mulberry pyralid *Glyphodes duplicalis* (Watanabe et al. 2000). Ice nucleation genes increase the supercooling points and reduce the tolerance of overwintering insects to cold, resulting in death by freezing. The supercooling points of mulberry pyralid larvae treated with the transgenic *E. cloacae* strain were higher than *E. ananas*-treated larvae, and their mortality rate reached 64.3% after 2 h exposure at -5°C and eventually increased to 95.2–100% after 6 and 18 h of exposure (Watanabe et al. 2000).

In *Anopheles dirus* larvae, *Enterobacter amnigenus* bacteria isolated from their gut tissue were transformed with the mosquito-larvicidal toxin of *Bacillus sphaericus* 2297 (Khampang et al. 2001). The fragment encoding the binary toxin was inserted into various plasmids, under different promoters in order to improve expression levels. *E. amnigenus* carrying a recombinant plasmid with the native *B. sphaericus* promoter exhibited the highest toxicity among the plasmids that were tested and proved to be 20 times more effective than *B. sphaericus* 2297 (Khampang et al. 2001).

5. INSECTICIDAL ACTIVITY OF *ENTEROBACTER*

Apart from their beneficial impact on host fitness as members of the insect gut microbiota, *Enterobacter* strains may also exhibit insecticidal activity even against their hosts, functioning as biocontrol agents in pest management applications (Table 5).

The shift in the behaviour of otherwise beneficial symbionts could be due to changes in the composition or location of the insect gut microbiota. Several examples of entomopathogenic *Enterobacter* strains have been documented. Among them, *E. cloacae* strain SL11 caused 30-73% mortality, depending on bacteria concentration, when fed to larvae of its host *Spodoptera litura* (Thakur et al. 2015). The strain also had a negative impact on essential fitness parameters of *S. litura* progeny, including development from larva to adult, growth rate, life span, morphology and reproduction. The pathogenic *E. cloacae* strain dominated the gut microbiota in infected insects, caused a reduction in the number of haemocytes and produced immune responses of phenoloxidase and lysozyme (Thakur et al. 2015).

Table 5. Enterobacter *species providing insecticidal activity*

Enterobacter Species	Strain Accession Number	Host		Function	Reference
Enterobacter cloacae	KP058541	*Spodoptera litura* (Fabricius)	Tobacco cutworm	Insecticidal effect	Thakur et al. 2015
Enterobacter hormaechei, Enterobacter sp.	GU906249 GU906251	*Rhynchites bacchus* (Linnaeus)	Leafroller weevil beetles	Insecticidal effect	Gokce et al. 2010
Enterobacter cloacae	Not available	*Bemisia argentifolii* Bellows & Perring	Silverleaf whitefly	Insecticidal effect	Davidson et al. 2000
Enterobacter cancerogenus	Not available	*Oberea linearis* (Linnaeus)	Hazelnut and walnut twig borer	Insecticidal effect	Bahar and Demirbağ 2007
Enterobacter sp.	KM979225	*Cimbex quadrimaculatus* (Müller)	Sawfly	Insecticidal effect	Cakici et al. 2015
Enterobacter sp.	JQ066777	*Spodoptera littoralis* (Boisduval)	Cotton leafworm	Insecticidal effect	Cakici et al. 2015
Enterobacter aerogenes	Not available	*Euproctis chrysorrhoea* (Linnaeus)	Brown-tail moth	Insecticidal effect	Yaman et al. 2000
Enterobacter aerogenes	AF306521	*Myrmeleon bore* (Tjeder)	Antlion	Insecticidal effect	Yoshida et al. 2001

Enterobacter strains with pathogenic effect against their host cause mortality in variable degree. In leafroller weevil beetles *Rhynchites bacchus* two *Enterobacter* strains (*E. hormaechei* Rb3 and *Enterobacter* sp. Rb5) caused 13% mortality (Gokce et al. 2010) in the silverleaf whitefly, *Bemisia argentifolii*, *E. cloacae* WFA73 produced 34% adult *B. argentifolii* mortality after 24 h and 75% after 48 h (Davidson et al. 2000). *E. cancerogenus* Ol11 caused 15% mortality in larvae of the beetle *Oberea linearis* ten days after treatment (Bahar and Demirbağ 2007).

The larvicidal effect of *Enterobacter* sp. CQ4 resulted in 58% mortality within ten days of third instar larvae of *Cimbex quadrimaculatus* (Çakici et al. 2015). In the cotton leafworm *Spodoptera littoralis*, *Enterobacter* sp. SL4 showed low larvicidal activity (17%) against third instar larvae, ten days after treatment (Çakici et al. 2014). Finally, an *Enterobacter aerogenes* isolate from the brown-tail moth *Euproctis chrysorrhoea* caused 45% mortality in third/fourth instar larvae (Yaman et al. 2000).

Yoshida et al. (2001) identified a mechanism of insecticidal activity based on a paralysing toxin that was produced by *Enterobacter aerogenes* bacteria contained in the saliva of *Myrmeleon bore* larvae. The 63 kDa protein assists larvae in capturing their prey and shows homology to the heat shock protein GroEL of *Escherichia coli*. The two proteins share similar sequences that contain only a few differences, in 11 residues and the carboxy terminus. A series of individual mutations revealed crucial residues for toxicity. Specific substitutions resulted in the loss of insecticidal activity of the *E. aerogenes* protein or the development of toxic effect by the otherwise harmless GroEL chaperone (Yoshida et al. 2001).

6. *ENTEROBACTER* GENOMICS

The number of available genome sequences from the *Enterobacter* taxon has not reached a threshold where comparative genomics can drive hypotheses and experiments. Recent progress in the genomics era resulted in one complete genome sequence of *Enterobacter cancerogenus* CR-Eb1 isolated from the larval gut of the greater wax moth *Galleria mellonella* (Chung et al. 2018). Also, there are several draft genome sequences of *Enterobacter* strains available from other insects (Table 3). In order to fully utilize the powerful analysis of comparative genomics and draft genomes from *Enterobacter* strains, more genomic and transcriptomic data are required from *Enterobacter* strains covering diverse functional roles. The availability of these genomes will be able to assist us in understanding their functional roles by defining their differences and similarities in gene content (Khamis et al. 2019).

The availability of *Enterobacter* sequenced bacterial genomes will also allow a more profound understanding of their organizational features that are related to fundamental cellular processes such as coordinated gene expression, chromosome replication and cell division. Genomic data will provide the means to characterize the fluidity of bacterial chromosomes, including genome rearrangements that imperil the selective features of chromosome order. Furthermore, a high-density microarray can be developed for the analysis of expression and genome content in a wide variety of *Enterobacter* strains, both sequenced and not sequenced.

Comparative genomics of more complete *Enterobacter* genomes will also allow studying the integration process from free-living to endosymbiont. Usually, symbiotic bacteria undergo drastic genetic, phenotypic, and biochemical changes, which can be detected by comparison with free-living relatives (Lo et al. 2016). Genome reduction, in some cases, is extremely strong, and it has generated the smallest bacterial genomes found to date; gene contents are so limited that their status as cellular entities is questionable (Latorre and Manzano-Marín 2017). It would be exciting to further improve our understanding of *Enterobacter* species' diversity in terms of their evolutionary history.

7. CONCLUSIONS

More than a hundred years of biological research has demonstrated the importance of microorganisms in the health and disease of higher organisms. Similarly, insects have symbiotic interactions that enable them to exploit unusually limited nutritional resources. In particular, recent findings suggest that symbiotic associations between insects and *Enterobacter* species may be beneficial to host fitness because of their various abilities to hydrolyse and ferment carbohydrates, catalyse nitrogen fixation, and produce vitamins and pheromones.

Also, the dominance of *Enterobacter* indicates an essential role in the protection of the insect host, or its nutritional resources, against parasitoids or predators, and also in terms of the interference in the transmission of malaria and other vector-borne diseases.

This review summarized our current knowledge of the relationship between *Enterobacter* and insect hosts. Characterization, exploitation and management of the insect-*Enterobacter* symbiotic associations can significantly contribute to and support integrated pest management applications for the control of agricultural pests and disease vectors.

To further decipher the *Enterobacter*-based symbiotic interactions we propose a systems biology approach in which, *in silico* predictions based on genomic analyses and phylogenetic information will be assessed by transcriptomic, proteomic, and metabolomic analyses. This information is considered essential for the success of downstream field applications.

8. REFERENCES

Adams, L., and R. Boopathy. 2005. Isolation and characterization of enteric bacteria from the hindgut of Formosan termite. Bioresource Technology 96: 1592–1598.

Adams, A. S., M. S. Jordan, S. M. Adams, G. Suen, L. A. Goodwin, K. W. Davenport, C. R. Currie, and K. F. Raffa. 2011. Cellulose-degrading bacteria associated with the invasive woodwasp *Sirex noctilio*. The International Society for Microbial Ecology (ISME) Journal 5: 1323–1331.

Adhav, A., S. Harne, A. Bhide, A. Giri, P. Gayathri, and R. Joshi. 2019. Mechanistic insights into enzymatic catalysis by trehalase from the insect gut endosymbiont *Enterobacter cloacae*. The Federation of European Biochemical Societies (FEBS) Journal 286: 1700–1716.

Akman, L., A. Yamashita, H. Watanabe, K. Oshima, T. Shiba, M. Hattori, and S. Aksoy. 2002. Genome sequence of the endocellular obligate symbiont of tsetse flies, *Wigglesworthia glossinidia*. Nature Genetics 32: 402–407.

Al-Ghamdi, A., K. Ali Khan, M. Javed Ansari, S. B. Almasaudi, and S. Al-Kahtani. 2018. Effect of gut bacterial isolates from *Apis mellifera jemenitica* on *Paenibacillus* larvae infected bee larvae. Saudi Journal of Biological Sciences 25: 383–387.

Aliota, M. T., S. A. Peinado, I. D. Velez, and J. E. Osorio. 2016a. The *w*Mel strain of *Wolbachia* reduces transmission of Zika virus by *Aedes aegypti*. Scientific Reports 6: 28792.

Aliota, M. T., E. C. Walker, A. Uribe Yepes, I. Dario Velez, B. M. Christensen, and J. E. Osorio. 2016b. The *w*Mel strain of *Wolbachia* reduces transmission of Chikungunya virus in *Aedes aegypti*. PLoS Neglected Tropical Diseases 10.

Ant, T. H., C. S. Herd, V. Geoghegan, A. A. Hoffmann, and S. P. Sinkins. 2018. The *Wolbachia* strain *w*Au provides highly efficient virus transmission blocking in *Aedes aegypti*. PLoS Pathogens 14: e1006815.

Armer, C. A., R. E. Berry, G. L. Reed, and S. J. Jepsen. 2004. Colorado potato beetle control by application of the entomopathogenic nematode H*eterorhabditis marelata* and potato plant alkaloid manipulation. Entomologia Experimentalis et Applicata 111: 47–58.

Augustinos, A. A., G. A. Kyritsis, N. T. Papadopoulos, A. M. M. Abd-Alla, C. Cáceres, and K. Bourtzis. 2015. Exploitation of the medfly gut microbiota for the enhancement of Sterile Insect Technique: Use of *Enterobacter* sp. in larval diet-based probiotic applications. PLoS One 10: e0136459.

Augustinos, A. A., G. A. Kyritsis, C. Cáceres, and K. Bourtzis. 2021. Insect symbiosis in support of the Sterile Insect Technique, pp. 605–630. *In* V. A. Dyck, J. Hendrichs, and A. S. Robinson (eds.), Sterile Insect Technique – Principles and practice in Area-Wide Integrated Pest Management. Second Edition. CRC Press, Boca Raton, Florida, USA.

Bahar, A., and Z. Demirbağ. 2007. Isolation of pathogenic bacteria from *Oberea linearis* (Coleptera: Cerambycidae). Biologia 62: 13–18.

Bary, A. de. 1879. Die Erscheinung der Symbiose: Vortrag. Karl J. Trübner Verlag. Strassburg, Germany.

Behar, A., B. Yuval, and E. Jurkevitch. 2005. Enterobacteria-mediated nitrogen fixation in natural populations of the fruit fly *Ceratitis capitata*. Molecular Ecology 14: 2637–2643.

Ben Ami, E., B. Yuval, and E. Jurkevitch. 2009. Manipulation of the microbiota of mass-reared Mediterranean fruit flies *Ceratitis capitata* (Diptera: Tephritidae) improves sterile male sexual performance. The International Society for Microbial Ecology (ISME) Journal 4: 28.

Ben-Yosef, M., Z. Pasternak, E. Jurkevitch, and B. Yuval. 2014. Symbiotic bacteria enable olive flies (*Bactrocera oleae*) to exploit intractable sources of nitrogen. Journal of Evolutionary Biology 27: 2695–2705.

Bian, G., Y. Xu, P. Lu, Y. Xie, and Z. Xi. 2010. The endosymbiotic bacterium *Wolbachia* induces resistance to dengue virus in *Aedes aegypti*. PLoS Pathogens 6: 1–10.

Bian, G., D. Joshi, Y. Dong, P. Lu, G. Zhou, X. Pan, Y. Xu, G. Dimipoulos, and Z. Xi. 2013. *Wolbachia* invades *Anopheles stephensi* populations and induces refractoriness to *Plasmodium* infection. Science 340: 748–751.

Blackburn, M. B., D. E. Gundersen-Rindal, D. C. Weber, P. A. W. Martin, and R. R. Farrar. 2008. Enteric bacteria of field-collected Colorado potato beetle larvae inhibit growth of the entomopathogens *Photorhabdus temperata* and *Beauveria bassiana*. Biological Control 46: 434–441.

Bourtzis, K., and T. A. Miller (eds.). 2008. Insect symbiosis. CRC Press, Boca Raton, Florida, USA. 324 pp.

Brady, C., I. Cleenwerck, S. Venter, T. Coutinho, and P. De Vos. 2013. Taxonomic evaluation of the genus *Enterobacter* based on Multilocus Sequence Analysis (MLSA): Proposal to reclassify *E. nimipressuralis* and *E. amnigenus* into *Lelliottia* gen. nov. as *Lelliottia nimipressuralis* comb. nov. and *Lelliottia amnigena* comb. nov., respectively, *E. gergoviae* and *E. pyrinus* into *Pluralibacter* gen. nov. as *Pluralibacter gergoviae* comb. nov. and *Pluralibacter pyrinus* comb. nov., respectively, *E. cowanii*, *E. radicincitans*, *E. oryzae* and *E. arachidis* into *Kosakonia* gen. nov. as *Kosakonia cowanii* comb. nov., *Kosakonia radicincitans* comb. nov., *Kosakonia oryzae* comb. nov. and *Kosakonia arachidis* comb. nov., respectively, and *E. turicensis*, *E. helveticus* and *E. pulveris* into *Cronobacter* as *Cronobacter zurichensis* nom. nov., *Cronobacter helveticus* comb. nov. and *Cronobacter pulveris* comb. nov., respectively, and emended description of the genera *Enterobacter* and *Cronobacter*. Systematic and Applied Microbiology 36: 309–319.

Brenner, D. J., G. R. Fanning, and J. K. Leete Knutson. 1984. Attempts to classify herbicola group-*Enterobacter agglomerans* strains by deoxyribonucleic acid hybridization and phenotypic tests. International Journal of Systematic Bacteriology 34: 45–55.

Breznak, J. A., W. J. Brill, J. W. Mertins, and H. C. Coppel. 1973. Nitrogen fixation in termites. Nature 244: 577.

Bridges, J. R. 1981. Nitrogen-fixing bacteria associated with bark beetles. Microbial Ecology 7: 131–137.

Broderick, N. A., and B. Lemaitre. 2012. Gut-associated microbes of *Drosophila melanogaster*. Gut Microbes 3: 307–321.

Brown, C. T., L. A. Hug, B. C. Thomas, I. Sharon, C. J. Castelle, A. Singh, M. J. Wilkins, K. C. Wrighton, K. H. Williams, and J. B. Banfield. 2015. Unusual biology across a group comprising more than 15% of domain Bacteria. Nature 523: 208–211.

Brune, A., and C. Dietrich. 2015. The gut microbiota of termites: Digesting the diversity in the light of ecology and evolution. Annual Review of Microbiology 69: 145–166.

Butera, G., C. Ferraro, S. Colazza, G. Alonzo, and P. Quatrini. 2012. The culturable bacterial community of frass produced by larvae of *Rhynchophorus ferrugineus* Olivier (Coleoptera: Curculionidae) in the Canary island date palm. Letters in Applied Microbiology 54: 530–536.

Cáceres, C., G. Tsiamis, B. Yuval, E. Jurkevitch, and K. Bourtzis. 2019. Joint FAO/IAEA coordinated research project on "Use of symbiotic bacteria to reduce mass-rearing costs and increase mating success in selected fruit pests in support of SIT application". BMC Microbiology 19: 284.

Çakici, F. O., A. Sevim, Z. Demirbağ, and İ. Demir. 2014. Investigating internal bacteria of *Spodoptera littoralis* (Boisd.) (Lepidoptera: Noctuidae) larvae and some *Bacillus* strains as biocontrol agents. Turkish Journal of Agriculture and Forestry 38: 99–110.

Çakici, F. O., I. Ozgen, H. Bolu, Z. Erbas, Z. Demirbağ, and I. Demir. 2015. Highly effective bacterial agents against *Cimbex quadrimaculatus* (Hymenoptera: Cimbicidae): Isolation of bacteria and their insecticidal activities. World Journal of Microbiol Biotechnology 31: 59–67.

Cambon, M., J.-C. Ogier, A. Lanois, J.-B. Ferdy, and S. Gaudriault. 2018. Changes in rearing conditions rapidly modify gut microbiota structure in *Tenebrio molitor* larvae. bioRxiv: 423178.

Capuzzo, C., G. Firrao, L. Mazzon, A. Squartini, and V. Girolami. 2005. '*Candidatus* Erwinia dacicola', a coevolved symbiotic bacterium of the olive fly *Bactrocera oleae* (Gmelin). International Journal of Systematic and Evolutionary Microbiology 55: 1641–1647.

Chandler, J. A., J. M. Lang, S. Bhatnagar, J. A. Eisen, and A. Kopp. 2011. Bacterial communities of diverse *Drosophila* species: Ecological context of a host–microbe model system. PLoS Genetics 7: e1002272.

Chung, J., H. Jeong, and C.-M. Ryu. 2018. Complete genome sequences of *Enterobacter cancerogenus* CR-Eb1 and Enterococcus sp. strain CR-Ec1, isolated from the larval gut of the greater wax moth, *Galleria mellonella*. Genome Announcements 6: e00044-18.

Chung, S. H., C. Rosa, E. D. Scully, M. Peiffer, J. F. Tooker, K. Hoover, D. S. Luthe, and G. W. Felton. 2013. Herbivore exploits orally secreted bacteria to suppress plant defenses. Proceedings of the National Academy of Sciences of the USA 110: 15728–15733.

Cirimotich, C. M., A. M. Clayton, and G. Dimopoulos. 2011a. Low- and high-tech approaches to control *Plasmodium* parasite transmission by *Anopheles* mosquitoes. Journal of Tropical Medicine 2011: 891342.

Cirimotich, C. M., J. L. Ramirez, and G. Dimopoulos. 2011b. Native microbiota shape insect vector competence for human pathogens. Cell Host Microbe 10: 307–310.

Cirimotich, C. M., Y. Dong, A. M. Clayton, S. L. Sandiford, J. A. Souza-Neto, M. Mulenga, and D. Dimopoulos. 2011c. Natural microbe-mediated refractoriness to *Plasmodium* infection in *Anopheles gambiae*. Science 332: 855–858.

Colman, D. R., E. C. Toolson, and C. D. Takacs-Vesbach. 2012. Do diet and taxonomy influence insect gut bacterial communities? Molecular Ecology 21: 5124–5137.

Davidson, E. W., R. C. Rosell, and D. L. Hendrix. 2000. Culturable bacteria associated with the whitefly, *Bemisia argentifolii* (Homoptera: Aleyrodidae). The Florida Entomologist 83: 159–171.

Davin-Regli, A., and J.-M. Pagès. 2015. *Enterobacter aerogenes* and *Enterobacter cloacae*; versatile bacterial pathogens confronting antibiotic treatment. Frontiers in Microbiology 6: 392.

Dennison, N. J., N. Jupatanakul, and G. Dimopoulos. 2014. The mosquito microbiota influences vector competence for human pathogens. Current Opinion in Insect Science 3: 6–13.

Dong, Y, F. Manfredini, and G. Dimopoulos. 2009. Implication of the mosquito midgut microbiota in the defense against malaria parasites. PLoS Pathogens 5: e1000423.

Douglas, A. E. 2009. The microbial dimension in insect nutritional ecology. Functional Ecology 23: 38–47.

Douglas, A. E. 2015. Multiorganismal insects: Diversity and function of resident microorganisms. Annual Review of Entomology 60: 17–34.

Doudoumis, V., A. M. M. Abd-Alla, G. Tsiamis, C. Brelsfoard, S. Aksoy, and K. Bourtzis. 2013. Tsetse-*Wolbachia* symbiosis: Comes of age and has great potential for pest and disease control. Journal of Invertebrate Pathology 112: S94-S103.

Doudoumis, V., F. Blow, A. Saridaki, A. Augustinos, N. A. Dyer, I. Goodhead, P. Solano, J.-B. Rayaisse, P. Takac, S. Mekonnen, A. G. Parker, A. M. M. Abd-Alla, A. Darby, K. Bourtzis, and G. Tsiamis. 2017. Challenging the *Wigglesworthia*, *Sodalis*, *Wolbachia* symbiosis dogma in tsetse flies: *Spiroplasma* is present in both laboratory and natural populations. Scientific Reports 7: 4699.

Drew, R. A. I., and A. C. Lloyd. 1987. Relationship of fruit flies (Diptera: Tephritidae) and their bacteria to host plants. Annals of the Entomological Society of America 80: 629–636.

Dutra, H. L. C, M. N. Rocha, F. B. S. Dias, S. B. Mansur, E. P. Caragata, and L. A. Moreira. 2016. *Wolbachia* blocks currently circulating zika virus isolates in Brazilian *Aedes aegypti* mosquitoes. Cell Host & Microbe 19: 771–774.

Eappen, A. G., R. C. Smith, and M. Jacobs-Lorena. 2013. *Enterobacter*-activated mosquito immune responses to *Plasmodium* involve activation of SRPN6 in *Anopheles stephensi*. PLoS One 8: e62937.

Engel, P., and N. A. Moran. 2013. The gut microbiota of insects – Diversity in structure and function. Federation of European Microbiological Societies (FEMS) Microbiology Reviews 37: 699–735.

Engel, P., W. K. Kwong, Q. McFrederick, K. E. Anderson, S. M. Barribeau, J. A. Chandler, R. Scott Cornman, J. Dainat, J. R. de Miranda, V. Doublet, O. Emery, J. D. Evans, L. Farinelli, M. L. Flenniken, F. Granberg, J. A. Grasis, L. Gauthier, J. Hayer, H. Koch, S. Kocher, V. G. Martinson, N. Moran, M. Munoz-Torres, I. Newton, R. J. Paxton, E. Powell, B. M. Sadd, P. Schnid-Hempel, S. J. Song, R. S. Schwarz, D. vanEngelsdorp, and B. Dainat. 2016. The bee microbiome: Impact on bee health and model for evolution and ecology of host-microbe interactions. mBio 7: e02164-15.

Eren, A. M., M. L. Sogin, H. G. Morrison, J. H. Vineis, J. C. Fisher, R. J. Newton, and S. L. McLellan. 2015. A single genus in the gut microbiome reflects host preference and specificity. The International Society for Microbial Ecology (ISME) Journal 9: 90–100.

Estes, A. M., D. J. Hearn, S. Agrawal, E. A. Pierson, and J. C. Dunning Hotopp. 2018a. Comparative genomics of the *Erwinia* and *Enterobacter* olive fly endosymbionts. Scientific Reports 8: 15936.

Estes, A. M., D. J. Hearn, S. Nadendla, E. A. Pierson, and J. C. Dunning Hotopp. 2018b. Draft genome sequence of *Enterobacter* sp. strain OLF, a colonizer of olive flies. Microbiology Resource Announcements 7: e01068-18.

Eutick, M. L., R. W. O'Brien, and M. Slaytor. 1978. Bacteria from the gut of Australian termites. Applied Environmental Microbiology 35: 823–828.

Fang, H., W. Chen, B. Wang, X. Li, S.-J. Liu, and H. Yang. 2016. Cultivation and characterization of symbiotic bacteria from the gut of *Reticulitermes chinensis*. Applied Environmental Biotechnology 1: 3–12.

Feldhaar, H. 2011. Bacterial symbionts as mediators of ecologically important traits of insect hosts. Ecological Entomology 36: 533–543.

Fierer, N., and J. T. Lennon. 2011. The generation and maintenance of diversity in microbial communities. American Journal of Botany 98: 439–448.

Frentiu, F. D., T. Zakir, T. Walker, J. Popovici, A. T. Pyke, A. van den Hurk, E. A. McGraw, and S. L. O'Neill. 2014. Limited dengue virus replication in field-collected *Aedes aegypti* mosquitoes infected with Wolbachia. PLoS Neglected Tropical Diseases 8: e2688.

Geiger, A., M.-L. Fardeau, P. Grebaut, G. Vatunga, T. Josénando, S. Herder, G. Cuny, P. Truc, and B. Ollivier. 2009. First isolation of *Enterobacter*, *Enterococcus*, and *Acinetobacter* spp. as inhabitants of the tsetse fly (*Glossina palpalis palpalis*) midgut. Infection, Genetics and Evolution 9: 1364–1370.

Gibbons, S. M., and J. A. Gilbert. 2015. Microbial diversity - exploration of natural ecosystems and microbiomes. Current Opinion in Genetics & Development 35: 66–72.

Gil, R., and A. Latorre. 2019. Unity makes strength: A review on mutualistic symbiosis in representative insect clades. Life (Basel) 9: e21.

Gokce, C., A. Sevim, Z. Demirbağ, and I. Demir. 2010. Isolation, characterization and pathogenicity of bacteria from *Rhynchites bacchus* (Coleoptera: Rhynchitidae). Biocontrol Science and Technology 20: 973–982.

Gonzalez-Ceron, L., F. Santillan, M. H. Rodriguez, D. Mendez, and J. E. Hernandez-Avila. 2003. Bacteria in midguts of field-collected *Anopheles albimanus* block *Plasmodium vivax* sporogonic development. Journal of Medical Entomology 40: 371–374.

Gujjar, N. R., S. Govindan, A. Verghese, S. Subramaniam, and R. More. 2017. Diversity of the cultivable gut bacterial communities associated with the fruit flies *Bactrocera dorsalis* and *Bactrocera cucurbitae* (Diptera: Tephritidae). Phytoparasitica 45: 453–460.

Gurung, K., B. Wertheim, and J. F. Salles. 2019. The microbiome of pest insects: It is not just bacteria. Entomologia Experimentalis et Applicata 167: 156–170.

Hamden, H., M. N. Guerfali, S. Fadhl, M. Saidi, and C. Chevrier. 2013. Fitness improvement of mass-reared sterile males of *Ceratitis capitata* (Vienna 8 strain) (Diptera: Tephritidae) after gut enrichment with probiotics. Journal of Economic Entomology 106: 641–647.

Hoffmann, H., and A. Roggenkamp. 2003. Population genetics of the nomenspecies *Enterobacter cloacae*. Applied Environmental Microbiology 69: 5306–5318.

Hoffmann, H., S. Stindl, A. Stumpf, A. Mehlen, D. Monget, J. Heesemann, K. H. Schleifer, and A. Roggenkamp. 2005. Description of *Enterobacter ludwigii* sp. nov., a novel *Enterobacter* species of clinical relevance. Systematic and Applied Microbiology 28: 206–212.

Hormaeche, E., and P. R. Edwards. 1960. A proposed genus *Enterobacter*. International Bulletin of Bacteriological Nomenclature and Taxonomy 10: 71–74.

Husník, F., T. Chrudimský, and V. Hypša. 2011. Multiple origins of endosymbiosis within the Enterobacteriaceae (γ-Proteobacteria): Convergence of complex phylogenetic approaches. BMC Biology 9: 87.

Husseneder, C., and J. K. Grace. 2005. Genetically engineered termite gut bacteria (*Enterobacter cloacae*) deliver and spread foreign genes in termite colonies. Applied Microbiology and Biotechnology 68: 360–367.

Janusz, G., A. Pawlik, J. Sulej, U. Świderska-Burek, A. Jarosz-Wilkołazka, and A. Paszczyński. 2017. Lignin degradation: Microorganisms, enzymes involved, genomes analysis and evolution. Federation of European Microbiological Societies (FEMS) Microbiology Reviews 41: 941–962.

Jayakrishnan, L., A. V. Sudhikumar, and E. M. Aneesh. 2018. Role of gut inhabitants on vectorial capacity of mosquitoes. Journal of Vector Borne Diseases 55: 69.

Jiang, J., C. Alvarez, P. Kukutla, W. Yu, and J. Xu. 2012. Draft genome sequences of *Enterobacter* sp. isolate Ag1 from the midgut of the malaria mosquito *Anopheles gambiae*. Journal of Bacteriology 194: 5481–5481.

Jurkevitch, E. 2011. Riding the Trojan horse: Combating pest insects with their own symbionts. Microbial Biotechnology 4: 620–627.

Kalappa, D. M., P. A. Subramani, S. K. Basavanna, S. K. Ghosh, V. Sundaramurthy, S. Uragayala, S. Tiwari, A. R. Anvikar, and N. Valecha. 2018. Influence of midgut microbiota in *Anopheles stephensi* on *Plasmodium berghei* infections. Malaria Journal 17: 385.

Khamis, F. M., P. O. Mireji, F. L. O. Ombura, A. R. Malacrida, E. O. Awuoche, M. Rono, S. A. Mohamed, C. M. Tanga, and S. Ekesi. 2019. Species-specific transcriptional profiles of the gut and gut microbiome of *Ceratitis quilicii* and *Ceratitis rosa* sensu stricto. Scientific Reports 9: 18355.

Khampang, P., P. Luxananil, S. Tanapongpipat, W. Chungjatupornchai, and S. Panyim. 2001. Recombinant *Enterobacter amnigenus* highly toxic to *Anopheles dirus* mosquito larvae. Current Microbiology 43: 448–451.

Kohl, K. D., J. Amaya, C. A. Passement, M. D. Dearing, and M. D. McCue. 2014. Unique and shared responses of the gut microbiota to prolonged fasting: A comparative study across five classes of vertebrate hosts. Federation of European Microbiological Societies (FEMS) Microbiology Ecology 90: 883–894.

König, H., J. Fröhlich, and H. Hertel. 2006. Diversity and lignocellulolytic activities of cultured microorganisms, pp. 271–301. *In* H. König, and A. Varma (eds.), Intestinal microorganisms of termites and other invertebrates. Springer, Berlin-Heidelberg, Germany.

Kuhnigk, T., E.-M. Borst, A. Ritter, P. Kämpfer, A. Graf, H. Hertel, and H. König. 1994. Degradation of lignin monomers by the hindgut flora of xylophagous termites. Systematic and Applied Microbiology 17: 76–85.

Kumar, S., G. K. Christophides, R. Cantera, B. Charles, Y. S. Han, S. Meister, G. Dimipoulos, F. C. Kafatos, and C. Barilla-Mury. 2003. The role of reactive oxygen species on *Plasmodium melanotic* encapsulation in *Anopheles gambiae*. Proceedings National Academy of Sciences of the USA 100: 14139–14144.

Kyritsis, G. A., A. A. Augustinos, C. Cáceres, and K. Bourtzis. 2017. Medfly gut microbiota and enhancement of the Sterile Insect Technique: Similarities and differences of *Klebsiella oxytoca* and *Enterobacter* sp. AA26 probiotics during the larval and adult stages of the VIENNA 8D53+genetic sexing strain. Frontiers in Microbiology 8: 2064.

Latorre, A., and A. Manzano-Marín. 2017. Dissecting genome reduction and trait loss in insect endosymbionts. Annals of the New York Academy of Sciences 1389: 52–75.

Lauzon, C. R., S. E. Potter, and R. J. Prokopy. 2003. Degradation and detoxification of the dihydrochalcone phloridzin by *Enterobacter agglomerans*, a bacterium associated with the apple pest, *Rhagoletis pomonella* (Walsh) (Diptera: Tephritidae). Environmental Entomology 32: 953–962.

Ley, R. E., C. A. Lozupone, M. Hamady, R. Knight, and J. I. Gordon. 2008. Worlds within worlds: Evolution of the vertebrate gut microbiota. Nature Reviews Microbiology 6: 776–788.

Lo, W.-S., Y.-Y. Huang, and C.-H. Kuo. 2016. Winding paths to simplicity: Genome evolution in facultative insect symbionts. Federation of European Microbiological Societies (FEMS) Microbiology Reviews 40: 855–874.

Luckhart, S., Y. Vodovotz, L. Cui, and R. Rosenberg. 1998. The mosquito *Anopheles stephensi* limits malaria parasite development with inducible synthesis of nitric oxide. Proceedings of the National Academy of Sciences of the USA 95: 5700–5705.

Manfredi, A. P., N. I. Perotti, and M. A. Martínez. 2015. Cellulose degrading bacteria isolated from industrial samples and the gut of native insects from Northwest of Argentina. Journal of Basic Microbiology 55: 1384–1393.

Mannesmann, R., and B. Piechowski. 1989. Verteilung von Gärkammerbakterien einiger Termitenarten. Materials and Organisms 24: 161–177.

Manzano-Marín, A., J.-C. Simon, and A. Latorre. 2016. Reinventing the wheel and making it round again: Evolutionary convergence in *Buchnera-Serratia* symbiotic consortia between the distantly related Lachninae aphids *Tuberolachnus salignus* and *Cinara cedri*. Genome Biology and Evolution 8: 1440–1458.

Martin, B., and E. Schwab. 2012. Current usage of symbiosis and associated terminology. International Journal of Biology 5: 32.

Mateos, M., H. Martinez, S. B. Lanzavecchia, C. Conte, K. Guillen, B. M. Moran-Aceves, J. Toledo, P. Liedo, E. D. Asimakis, V. Doudoumis, G. A. Kyritsis, N. T. Papadopoulos, A. A. Avgoustinos, D. F. Segura, G. Tsiamis, and K. Bourtzis. 2018. *Wolbachia pipientis* associated to tephritid fruit fly pests: From basic research to applications. bioRxiv: 358333.

Mathews, S. L., M. J. Epps, R. K. Blackburn, M. B. Goshe, A. M. Grunden, and R. R. Dunn. 2019. Public questions spur the discovery of new bacterial species associated with lignin bioconversion of industrial waste. Royal Society Open Science 6: 180748.

McFall-Ngai, M. 2007. Adaptive immunity: Care for the community. Nature 445: 153.

Mereghetti, V., B. Chouaia, and M. Montagna. 2017. New insights into the microbiota of moth pests. International Journal of Molecular Sciences 8: 2450.

Molina-Cruz, A., R. J. DeJong, B. Charles, L. Gupta, S. Kumar, G. Jaramillo-Gutierrez, and C. Barillas-Mury. 2008. Reactive oxygen species modulate *Anopheles gambiae* immunity against bacteria and *Plasmodium*. The Journal of Biological Chemistry 283: 3217–3223.

Moreira, L. A., I. Iturbe-Ormaetxe, J. A. Jeffery, G. Lu, A. T. Pyke, L. M. Hedges, B. C. Rocha, S. Hall-Mendelin, A. Day, M. Riegler, L. E. Hugo, K. N. Johnson, B. H. Kay, E. A. McGraw, A. F. van den Hurk, P. A, Ryan, and S. L. O'Neill. 2009. A *Wolbachia* symbiont in *Aedes aegypti* limits infection with dengue, Chikungunya, and *Plasmodium*. Cell 139: 1268–1278.

Moya, A., J. Peretó, R. Gil, and A. Latorre. 2008. Learning how to live together: Genomic insights into prokaryote–animal symbioses. Nature Reviews Genetics 9: 218–229.

Mramba, F., A, B. Broce, and L. Zurek. 2007. Vector competence of stable flies, *Stomoxys calcitrans* L. (Diptera: Muscidae), for *Enterobacter sakazakii*. Journal of Vector Ecology 32: 134–139.

Muhammad, A., Y. Fang, Y. Hou, and Z. Shi. 2017. The gut entomotype of red palm weevil *Rhynchophorus ferrugineus* Olivier (Coleoptera: Dryophthoridae) and their effect on host nutrition metabolism. Frontiers in Microbiology 8: 2291.

Murphy, K. M., D. S. Teakle, and I. C. MacRae. 1994. Kinetics of colonization of adult Queensland fruit flies (*Bactrocera tryoni*) by dinitrogen-fixing alimentary tract bacteria. Applied Environmental Microbiology 60: 2508–2517.

Nakabachi, A., A. Yamashita, H. Toh, H. Ishikawa, H. E. Dunbar, N. A. Moran, and M. Hattori. 2006. The 160-kilobase genome of the bacterial endosymbiont *Carsonella*. Science 314: 267.

Niyazi, N., C. R. Lauzon, and T. E. Shelly. 2004. Effect of probiotic adult diets on fitness components of sterile male Mediterranean fruit flies (Diptera: Tephritidae) under laboratory and field cage conditions. Journal of Economic Entomology 97: 1570–1580.

Octavia, S., and R. Lan. 2014. The family Enterobacteriaceae, pp. 225–286. *In* E. Rosenberg, E. F. DeLong, S. Lory, E. Stackebrandt, and F. Thompson (eds.), The Prokaryotes: Gammaproteobacteria. Springer, Berlin-Heidelberg, Germany.

Pereira, F. C., and D. Berry. 2017. Microbial nutrient niches in the gut. Environmental Microbiology 19: 1366–1378.

Pernice, M., S. J. Simpson, and F. Ponton. 2014. Towards an integrated understanding of gut microbiota using insects as model systems. Journal of Insect Physiology 69: 12–18.

Peterson, T. M. L., A. J. Gow, and S. Luckhart. 2007. Nitric oxide metabolites induced in *Anopheles stephensi* control malaria parasite infection. Free Radical Biology and Medicine 42: 132–142.

Potrikus, C. J., and J. A. Breznak. 1977. Nitrogen-fixing *Enterobacter agglomerans* isolated from guts of wood-eating termites. Applied Environmental Microbiology 33: 392–399.

Potrikus, C. J., and J. A. Breznak. 1981. Gut bacteria recycle uric acid nitrogen in termites: A strategy for nutrient conservation. Proceedings of the National Academy of Sciences of the USA 78: 4601–4605.

Ramin, M., A. R. Alimon, N. Abdullah, J. M. Panandam, and K. Sijam. 2008. Isolation and identification of three species of bacteria from the termite *Coptotermes curvignathus* (Holmgren) present in the vicinity of Universiti Putra Malaysia. Research Journal of Microbiology 3: 288–292.

Rizzi, A., E. Crotti, L. Borruso, C. Jucker, D. Lupi, M. Colombo, and D. Daffonchio. 2013. Characterization of the bacterial community associated with larvae and adults of *Anoplophora chinensis* collected in Italy by culture and culture-independent methods. BioMed Research International 2013: 420287.

Rojas-Jiménez, K., and M. Hernández. 2015. Isolation of fungi and bacteria associated with the guts of tropical wood-feeding Coleoptera and determination of their lignocellulolytic activities. International Journal of Microbiology 2015: 285018.

Rooks, M. G., and W. S. Garrett. 2016. Gut microbiota, metabolites and host immunity. Nature Reviews Immunology 16: 341–352.

Ruby, E., B. Henderson, and M. McFall-Ngai. 2004. We get by with a little help from our (little) friends. Science 303: 1305–1307.

Saldaña, M. A., S. Hegde, and G. L. Hughes. 2017. Microbial control of arthropod-borne disease. Memórias do Instituto Oswaldo Cruz 112: 81–93.

Sandeva, R., G. Sandeva, B. Chakarova, and A. Dimitrova. 2018. Gut microbiota - Our new important organ. Albanian Journal of Agricultural Sciences (Proceedings of ICOALS 2018): 175–181.

Schäfer, A., R. Konrad, T. Kuhnigk, P. Kämpfer, H. Hertel, and H. König. 1996. Hemicellulose-degrading bacteria and yeasts from the termite gut. Journal of Applied Bacteriology 80: 471–478.

Schmidt, T. L., N. H. Barton, G. Rašić, A. P. Turley, B. L. Montgomery, I. Iturbe-Ormaetxe, P. E. Cook, P. A. Ryan, S. A. Ritchie, A. A. Hoffmann, S. L. O'Neill, and M. Turelli. 2017. Local introduction and heterogeneous spatial spread of dengue-suppressing *Wolbachia* through an urban population of *Aedes aegypti*. PLoS Biology 15: 2001894.

Sheng, P., Y. Li, S. D. G. Marshall, and H. Zhang. 2015. High genetic diversity of microbial cellulase and hemicellulase genes in the hindgut of *Holotrichia parallela* larvae. International Journal of Molecular Science 16: 16545–16559.

Somerville, J., L. Zhou, and B. Raymond. 2019. Aseptic rearing and infection with gut bacteria improve the fitness of transgenic diamondback moth, *Plutella xylostella*. Insects 10: 89.

Szabó, G., F. Schulz, E. R. Toenshoff, J.-M. Volland, O. M. Finkel, S. Belkin, and M. Horn. 2017. Convergent patterns in the evolution of mealybug symbioses involving different intrabacterial symbionts. The International Society for Microbial Ecology (ISME) Journal 11: 715–726.

Tan, C. H., P. J. Wong, M. I. Li, H. Yang, L. C. Ng, and S. L. O'Neill. 2017. *w*Mel limits zika and chikungunya virus infection in a Singapore *Wolbachia*-introgressed *Ae. aegypti* strain, *w*Mel-Sg. PLoS Neglected Tropical Diseases 11: e. 0005496.

Tchioffo, M. T., A. Boissière, T. S. Churcher, L. Abate, G. Gimonneau, S. E. Nsango, P. H. Awono-Ambéné, R. Christen, A. Berry, and I. Morlais. 2013. Modulation of malaria infection in *Anopheles gambiae* mosquitoes exposed to natural midgut bacteria. PLoS One 8: e81663.

Thakur, A., P. Dhammi, H. S. Saini, and S. Kaur. 2015. Pathogenicity of bacteria isolated from gut of *Spodoptera litura* (Lepidoptera: Noctuidae) and fitness costs of insect associated with consumption of bacteria. Journal of Invertebrate Pathology 127: 38–46.

Thong-On, A., K. Suzuki, S. Noda, J. Inoue, S. Kajiwara, and M. Ohkuma. 2012. Isolation and characterization of anaerobic bacteria for symbiotic recycling of uric acid nitrogen in the gut of various termites. Microbes and Environments 27: 186–192.

Toczyłowska-Mamińska, R., K. Szymona, H. Madej, W. W. Zhen, A. Bala, W. Brutkowski, K. Krajewski, P. S. H'ng, and M. L. Maminski. 2015. Cellulolytic and electrogenic activity of *Enterobacter cloacae* in mediatorless microbial fuel cell. Applied Energy 160: 88–93.

Valero-Jiménez, C. A., A. J. Debets, J. A. van Kan, S. E. Schoustra, W. Takken, B. J. Zwaan, and C. J. M. Koenraadt. 2014. Natural variation in virulence of the entomopathogenic fungus *Beauveria bassiana* against malaria mosquitoes. Malaria Journal 13: 479.

van den Hurk, A. F., S. Hall-Mendelin, A. T. Pyke, F. D. Frentiu, K. McElroy, A. Day, S. Higgs, and S. L. O'Neill. 2012. Impact of *Wolbachia* on infection with chikungunya and yellow fever viruses in the mosquito vector *Aedes aegypti*. PLoS Neglected Tropical Diseases 6: e1892.

Vasanthakumar, A., I. Delalibera, J. Handelsman, K. D. Klepzig, P. D. Schloss, and K. F. Raffa. 2006. Characterization of gut-associated bacteria in larvae and adults of the southern pine beetle, *Dendroctonus frontalis* Zimmermann. Environmental Entomology 35: 1710–1717.

Walker, T., P. H. Johnson, L. A. Moreira, I. Iturbe-Ormaetxe, F. D. Frentiu, C. J. McMeniman, Y. S. Leong, Y. Dong, J. Axford, P. Kriesner, A. L. Lloyd, S. A. Ritchie, S. L. O'Neill, and A. A. Hoffmann. 2011. The *w*Mel *Wolbachia* strain blocks dengue and invades caged *Aedes aegypti* populations. Nature 476: 450–455.

Wang, H., L. Jin, T. Peng, H. Zhang, Q. Chen, and Y. Hua. 2014. Identification of cultivable bacteria in the intestinal tract of *Bactrocera dorsalis* from three different populations and determination of their attractive potential. Pest Management Science 70: 80–87.

Wang, J., S. H. Chung, M. Peiffer, C. Rosa, K. Hoover, R. Zeng, and G. W. Felton. 2016. Herbivore oral secreted bacteria trigger distinct defense responses in preferred and non-preferred host plants. Journal of Chemical Ecology 42: 463–474.

Wang, J., M. Peiffer, K. Hoover, C. Rosa, R. Zeng, and G. W. Felton. 2017. *Helicoverpa zea* gut-associated bacteria indirectly induce defenses in tomato by triggering a salivary elicitor(s). New Phytologist 214: 1294–1306.

Watanabe, K., K. Abe, and M. Sato. 2000. Biological control of an insect pest by gut-colonizing *Enterobacter cloacae* transformed with ice nucleation gene. Journal of Applied Microbiology 88: 90–97.

Webster, N. S. 2014. Cooperation, communication, and co-evolution: Grand challenges in microbial symbiosis research. Frontiers in Microbiology 5: 164.

Weiss, B., and S. Aksoy. 2011. Microbiome influences on insect host vector competence. Trends in Parasitology 27: 514–522.

Wernegreen, J. J. 2004. Endosymbiosis: Lessons in conflict resolution. PLoS Biology 2: e68.

Whitman, W. B., D, C. Coleman, and W. J. Wiebe. 1998. Prokaryotes: The unseen majority. Proceedings of the National Academy of Sciences of the USA 95: 6578–6583.

Wilke, A. B. B., and M. T. Marrelli. 2015. Paratransgenesis: A promising new strategy for mosquito vector control. Parasites & Vectors 8: 342.

Williams, L. E., and J. J. Wernegreen. 2010. Unprecedented loss of ammonia assimilation capability in a urease-encoding bacterial mutualist. BMC Genomics 11: 687.

Williams, T. A., G. J. Szöllősi, A. Spang, P. G. Foster, S. E. Heaps, B. Boussau, T. J. G. Ettema, and T. M. Embley. 2017. Integrative modeling of gene and genome evolution roots the archaeal tree of life. Proceedings of the National Academy of Sciences of the USA 114: E4602–4611.

Woodbury, N., and G. Gries. 2013a. How firebrats (Thysanura: Lepismatidae) detect and nutritionally benefit from their microbial symbionts *Enterobacter cloacae* and *Mycotypha microspora*. Environmental Entomology 42: 860–867.

Woodbury, N., and G. Gries. 2013b. Firebrats, *Thermobia domestica*, aggregate in response to the microbes *Enterobacter cloacae* and *Mycotypha microspora*. Entomologia Experimentalis et Applicata 147: 154–159.

Wu, D., S. C. Daugherty, S. E. Van Aken, G. H. Pai, K. L. Watkins, H. Khouri, L. J. Tallon, J. M. Zaborsky, H. E. Dunbar, P. L. Tran, N. A. Moran, and J. A. Eisen. 2006. Metabolic complementarity and genomics of the dual bacterial symbiosis of sharpshooters. PLoS Biology 4: e188.

Xia, X., G. M. Gurr, L. Vasseur, D. Zheng, H. Zhong, B. Qin, J. Lin, Y. Wang, F.-Q. Song, Y. Li, H. Lin, and M. You. 2017. Metagenomic sequencing of diamondback moth gut microbiome unveils key holobiont adaptations for herbivory. Frontiers in Microbiology 8: 663.

Yaman, M., Z. Demirbağ, and A. O. Beldüz. 2000. Isolation and insecticidal effects of some bacteria from *Euproctis chrysorrhoea* L. (Lepidoptera: Lymantriidae). Acta Microbiologica Polonica 49: 217–224.

Yao, M., H. Zhang, P. Cai, X. Gu, D. Wang, and Q. Ji. 2017. Enhanced fitness of a *Bactrocera cucurbitae* genetic sexing strain based on the addition of gut-isolated probiotics (*Enterobacter* spec.) to the larval diet. Entomologia Experimentalis et Applicata 162: 197–203.

Yoshida, N., K. Oeda, E. Watanabe, T. Mikami, Y. Fukita, K. Nishimura, K. Komai, and K. Matsuda. 2001. Protein function: Chaperonin turned insect toxin. Nature 411(6833): 44.

Yun, J.-H., S. W. Roh, T. W. Whon, M. J. Jung, M.-S. Kim, D.-S. Park, C. Yoon, Y.-D. Nam, Y.-J. Kim, J.-H. Choi, J.-Y. Kim, N.-R. Shin, S.-H. Kim, W.-J. Lee, and J.-W. Bae. 2014. Insect gut bacterial diversity determined by environmental habitat, diet, developmental stage, and phylogeny of host. Applied Environmental Microbiology 80: 5254–5264.

Yuval, B., E. Ben-Ami, A. Behar, M. Ben-Yosef, and E. Jurkevitch. 2013. The Mediterranean fruit fly and its bacteria – potential for improving Sterile Insect Technique operations. Journal of Applied Entomology 137: 39–42.

Zaremba-Niedzwiedzka, K., E. F. Caceres, J. H. Saw, D. Bäckström, L. Juzokaite, E. Vancaester, K. W. Seitz, K. Anantharaman, P. Stamawski, K. U. Kjeldsen, M. B. Sott, T. Nunoura, J. F. Banfield, A. Schramm, B. J. Baker, A. Spang, and T. J. Ettema. 2017. Asgard archaea illuminate the origin of eukaryotic cellular complexity. Nature 541: 353–358.

Zchori-Fein, E., and S. J. Perlman. 2004. Distribution of the bacterial symbiont *Cardinium* in arthropods. Molecular Ecology 13: 2009–2016.

Zhang, H., W. Mu, Z. Hou, X. Wu, W. Zhao, X. Zhang, H. Pan, and S. Zhang. 2012. Biodegradation of nicosulfuron by the bacterium *Serratia marcescens* N80. Journal of Environmental Science Health B. 47: 153–160.

Zhang, G., M. Hussain, S. L. O'Neill, and S. Asgari. 2013. *Wolbachia* uses a host microRNA to regulate transcripts of a methyltransferase, contributing to dengue virus inhibition in *Aedes aegypti*. Proceedings of the National Academy of Sciences of the USA 110: 10276–10281.

Zhao, C., N. Tang, Y. Wu, Y. Zhang, Z. Wu, W. Li, X. Qin, J. Zhao, and G. Zhang. 2012. First reported fatal *Morganella morganii* infections in chickens. Veterinary Microbiology 156: 452–455.

Zhou, H., W. Guo, B. Xu, Z. Teng, D. Tao, Y. Lou, and Y. Gao. 2017. Screening and identification of lignin-degrading bacteria in termite gut and the construction of LiP-expressing recombinant *Lactococcus lactis*. Microbial Pathogenesis 112: 63–69.

Zhou, F., X. Wu, L. Xu, S. Guo, G. Chen, and X. Zhang. 2019. Repressed *Beauveria bassiana* infections in *Delia antiqua* due to associated microbiota. Pest Management Science 75: 170–179.

Zhu, B., M.-M. Lou, G.-L. Xie, G.-F. Wang, Q. Zhou, F. Wang, Y. Fang, T. Su, B. Lu, and Y.-P. Duan. 2011. *Enterobacter mori* sp. nov., associated with bacterial wilt on *Morus alba* L. International Journal of Systematic and Evolutionary Microbiology 61: 2769–2774.

Author Index

A

B

C

D

E

F

G

H

Scientific Name Index

A

B

C

D

E

R

S

T

Subject Index

B

D

E

J

K

M

N

T

U

W